Instructor's Solutions Manual

Volume 2

to Accompany Swokowski's

Calculus: The Classic Edition

Jeffery A. Cole
Anoka-Ramsey Community College

Gary K. Rockswold
Minnesota State University, Mankato

Australia • Canada • Mexico • Singapore • Spain • United Kingdom • United States

COPYRIGHT© 1991 by Brooks/Cole
A division of Thomson Learning
The Thomson Learning logo is a trademark used herein under license.

For more information about this or any other Brooks/Cole products, contact:
BROOKS/COLE
511 Forest Lodge Road
Pacific Grove, CA 93950 USA
www.brookscole.com
1-800-423-0563 (Thomson Learning Academic Resource Center)

All rights reserved. No part of this work may be reproduced, transcribed, or used in any form or by any means—graphic, electronic, or mechanical, including photocopying, recording, taping, Web distribution, or information storage and/or retrieval systems—without the prior written permission of the publisher.

Printed in the United States of America.

5 4 3 2

ISBN 0-534-38274-6

PREFACE

This manual contains answers to all exercises in Chapters 9 through 19 of the text, *Calculus: The Classic Edition*, by Earl W. Swokowski. For most problems, a reasonably detailed solution is included. We have tried to correlate the length of the solutions with their difficulty. It is our hope that by merely browsing through the solutions, the instructor will save time in determining appropriate assignments for their particular class.

All figures are new for this edition. Most function values have been plotted using computer software, and we are very happy with the high precision provided by this method. We would appreciate any feedback concerning errors, solution correctness, solution style, or manual style. These and any other comments may be sent directly to us or in care of the publisher.

We would like to thank: Editor Dave Geggis, for entrusting us with this project and continued support; Earl Swokowski, for his assistance; Sally Lifland and Gail Magin of Lifland, et al., Bookmakers, for assembling the final manuscript; and George and Bryan Morris, for preparing the new figures. We dedicate this book to our wives, Joan and Wendy, and thank them for their support and understanding.

Jeffery A. Cole
Anoka-Ramsey Community College
11200 Mississippi Blvd. NW
Coon Rapids, MN 55433

Gary K. Rockswold
Minnesota State University, Mankato
P.O. Box 41
Mankato, MN 56002

Table of Contents

In the review sections, the solutions are abbreviated since more detailed solutions were given in sections. In easier groups of exercises, representative solutions are shown. When appropriate, only the answer is listed. When possible, we tried to make each piece of art with the same scale to show a realistic and consistent graph. This manual was done using EXP: *The Scientific Word Processor.*

The following notations are used in the manual.

Note: Notes to the instructor pertaining to hints on instruction or conventions to follow.

{ }	{ comments to the reader are in braces }
LHS, RHS	{ Left Hand Side, Right Hand Side – used for identities }
$\Rightarrow$	{ implies, next equation, logically follows }
$\Leftrightarrow$	{ if and only if, is equivalent to }
•	{ bullet, used to separate problem statement from solution or explanation }
★	{ used to identify the answer to the problem }
§	{ section references }
$\forall$	{ For all, i.e., $\forall x$ means "for all x". }
$\mathbb{R} - \{a\}$	{ The set of all real numbers except a. }
$\therefore$	{ therefore }

(continued on next page)

The following notations are defined in the manual, and also listed here for convenience.

DNE { Does Not Exist }

L, I, S { the original limit, integral, or series }

T, S { the result is obtained from using the trapezoidal rule or Simpson's rule }

$\stackrel{\text{A}}{=}$ { integration by parts has been applied— the parts are defined following the solution }

$\{\frac{\infty}{\infty}\}$, $\{\frac{0}{0}\}$ { L'Hôpital's rule is applied when this symbol appears }

C, D { converges or convergent, diverges or divergent }

AC, CC { absolutely convergent, conditionally convergent }

DERIV { see notes in §11.8 and §11.9 for this notation }

V, F, l { vertex, focus, and directrix of a parabola }

C, V, F, M { center, vertices, foci, and end points of the minor axis of an ellipse }

C, V, F, W { center, vertices, foci, and end points of the conjugate axis of a hyperbola }

D { discriminant value $(B^2 - 4AC)$ in §12.4 }

VT, HT { vertical tangent, horizontal tangent }

$\uparrow$, $\downarrow$ { increasing, decreasing }

CN { critical number(s) }

PI { point(s) of inflection }

CU, CD { concave up, concave down }

MAX, MIN { absolute maximum or minimum }

SP { saddle point }

$[\![n]\!]$ { equation number n }

$\mathbb{S}$ { surface area }

J { Jacobian }

$\mathbb{A}$ { the value of $\mathbf{F} \cdot < -f_x, -f_y, 1 >$ }

IF { integrating factor }

LMAX, LMIN { local maximum or minimum }

VA, HA, OA { vertical, horizontal, or oblique asymptote }

QI, QII, QIII, QIV { quadrants I, II, III, IV }

NTH, INT, BCT, LCT, RAT, ROT, AST { various series tests: nth-term, integral, basic comparison, limit comparison, ratio, root, the alternating series }

INSTRUCTOR'S SOLUTIONS MANUAL
VOLUME 2

Chapter 9: Techniques of Integration

Exercises 9.1

Note: The symbol $\stackrel{A}{=}$ indicates that integration by parts has taken place using substitution A. The substitutions for u, du, dv, and v are listed at the end of the problem.

1 $I \stackrel{A}{=} -xe^{-x} - \int -e^{-x}\,dx = -xe^{-x} - e^{-x} + C = -(x+1)e^{-x} + C$

A. $u = x,\ du = dx,\ dv = e^{-x}\,dx,\ v = -e^{-x}$

2 $I \stackrel{A}{=} -x\cos x - \int -\cos x\,dx = -x\cos x + \sin x + C$

A. $u = x,\ du = dx,\ dv = \sin x\,dx,\ v = -\cos x$

3 $I \stackrel{A}{=} \frac{1}{3}x^2e^{3x} - \frac{2}{3}\int xe^{3x}\,dx \stackrel{B}{=} \frac{1}{3}x^2e^{3x} - \frac{2}{3}\left[\frac{1}{3}xe^{3x} - \frac{1}{3}\int e^{3x}\,dx\right] =$

$\frac{1}{3}x^2e^{3x} - \frac{2}{9}xe^{3x} + \frac{2}{27}e^{3x} + C = \frac{1}{27}e^{3x}(9x^2 - 6x + 2) + C$

A. $u = x^2,\ du = 2x\,dx,\ dv = e^{3x}\,dx,\ v = \frac{1}{3}e^{3x}$

B. $u = x,\ du = dx,\ dv = e^{3x}\,dx,\ v = \frac{1}{3}e^{3x}$

4 $I \stackrel{A}{=} -\frac{1}{4}x^2\cos 4x + \frac{1}{2}\int x\cos 4x\,dx \stackrel{B}{=} -\frac{1}{4}x^2\cos 4x + \frac{1}{2}\left[\frac{1}{4}x\sin 4x - \frac{1}{4}\int \sin 4x\,dx\right] =$

$-\frac{1}{4}x^2\cos 4x + \frac{1}{8}x\sin 4x + \frac{1}{32}\cos 4x + C$

A. $u = x^2,\ du = 2x\,dx,\ dv = \sin 4x\,dx,\ v = -\frac{1}{4}\cos 4x$

B. $u = x,\ du = dx,\ dv = \cos 4x\,dx,\ v = \frac{1}{4}\sin 4x$

5 $I \stackrel{A}{=} \frac{1}{5}x\sin 5x - \frac{1}{5}\int \sin 5x\,dx = \frac{1}{5}x\sin 5x + \frac{1}{25}\cos 5x + C$

A. $u = x,\ du = dx,\ dv = \cos 5x\,dx,\ v = \frac{1}{5}\sin 5x$

6 $I \stackrel{A}{=} -\frac{1}{2}xe^{-2x} + \frac{1}{2}\int e^{-2x}\,dx = -\frac{1}{4}e^{-2x}(2x+1) + C$

A. $u = x,\ du = dx,\ dv = e^{-2x}\,dx,\ v = -\frac{1}{2}e^{-2x}$

7 $I \stackrel{A}{=} x\sec x - \int \sec x\,dx = x\sec x - \ln|\sec x + \tan x| + C$

A. $u = x,\ du = dx,\ dv = \sec x\tan x\,dx,\ v = \sec x$

8 $I = -\frac{1}{3}x\cot 3x + \frac{1}{3}\int \cot 3x\,dx = -\frac{1}{3}x\cot 3x + \frac{1}{9}\ln|\sin 3x| + C$

A. $u = x,\ du = dx,\ dv = \csc^2 3x\,dx,\ v = -\frac{1}{3}\cot 3x$

9 $I \stackrel{A}{=} x^2\sin x - 2\int x\sin x\,dx$. By Exercise 2, $I = x^2\sin x + 2x\cos x - 2\sin x + C$.

A. $u = x^2,\ du = 2x\,dx,\ dv = \cos x\,dx,\ v = \sin x$

10 $I \stackrel{A}{=} -x^3e^{-x} + 3\int x^2e^{-x}\,dx \stackrel{B}{=} -x^3e^{-x} + 3\left[-x^2e^{-x} + 2\int xe^{-x}\,dx\right]$

$= -x^3e^{-x} - 3x^2e^{-x} + 6\int xe^{-x}\,dx.$

By Exercise 1, $I = -e^{-x}(x^3 + 3x^2 + 6x + 6) + C$.

A. $u = x^3,\ du = 3x^2\,dx,\ dv = e^{-x}\,dx,\ v = -e^{-x}$

B. $u = x^2,\ du = 2x\,dx,\ dv = e^{-x}\,dx,\ v = -e^{-x}$

11 $I \overset{A}{=} x\tan^{-1}x - \int \frac{x}{1+x^2}\,dx = x\tan^{-1}x - \frac{1}{2}\ln(1+x^2) + C$

A. $u = \tan^{-1}x,\ du = \frac{1}{1+x^2}\,dx,\ dv = dx,\ v = x$

12 $I \overset{A}{=} x\sin^{-1}x - \int \frac{x}{\sqrt{1-x^2}}\,dx = x\sin^{-1}x + \sqrt{1-x^2} + C$

A. $u = \sin^{-1}x,\ du = \frac{1}{\sqrt{1-x^2}}\,dx,\ dv = dx,\ v = x$

13 $I \overset{A}{=} \frac{2}{3}x^{3/2}\ln x - \frac{2}{3}\int x^{1/2}\,dx = \frac{2}{9}x^{3/2}(3\ln x - 2) + C$

A. $u = \ln x,\ du = \frac{1}{x}\,dx,\ dv = x^{1/2}\,dx,\ v = \frac{2}{3}x^{3/2}$

14 $I \overset{A}{=} \frac{1}{3}x^3\ln x - \frac{1}{3}\int x^2\,dx = \frac{1}{9}x^3(3\ln x - 1) + C$

A. $u = \ln x,\ du = \frac{1}{x}\,dx,\ dv = x^2\,dx,\ v = \frac{1}{3}x^3$

15 $I \overset{A}{=} -x\cot x + \int \cot x\,dx = -x\cot x + \ln|\sin x| + C$

A. $u = x,\ du = dx,\ dv = \csc^2 x\,dx,\ v = -\cot x$

16 $I \overset{A}{=} \frac{1}{2}x^2\tan^{-1}x - \frac{1}{2}\int \frac{x^2}{1+x^2}\,dx = \frac{1}{2}x^2\tan^{-1}x - \frac{1}{2}\int\left(1 - \frac{1}{1+x^2}\right)dx =$

$\frac{1}{2}x^2\tan^{-1}x - \frac{1}{2}x + \frac{1}{2}\tan^{-1}x + C$

A. $u = \tan^{-1}x,\ du = \frac{1}{1+x^2}\,dx,\ dv = x\,dx,\ v = \frac{1}{2}x^2$

17 $I \overset{A}{=} -e^{-x}\sin x + \int e^{-x}\cos x\,dx \overset{B}{=} -e^{-x}\sin x + \left[-e^{-x}\cos x - \int e^{-x}\sin x\,dx\right] \Rightarrow$

$2I = -e^{-x}\sin x - e^{-x}\cos x \Rightarrow I = -\frac{1}{2}e^{-x}(\sin x + \cos x) + C$

A. $u = \sin x,\ du = \cos x\,dx,\ dv = e^{-x}\,dx,\ v = -e^{-x}$

B. $u = \cos x,\ du = -\sin x\,dx,\ dv = e^{-x}\,dx,\ v = -e^{-x}$

18 $I \overset{A}{=} \frac{1}{3}e^{3x}\cos 2x + \frac{2}{3}\int e^{3x}\sin 2x\,dx \overset{B}{=} \frac{1}{3}e^{3x}\cos 2x + \frac{2}{3}\left[\frac{1}{3}e^{3x}\sin 2x - \frac{2}{3}\int e^{3x}\cos 2x\,dx\right]$

$= \frac{1}{3}e^{3x}\cos 2x + \frac{2}{9}e^{3x}\sin 2x - \frac{4}{9}\int e^{3x}\cos 2x\,dx \Rightarrow \frac{13}{9}I = \frac{1}{3}e^{3x}\cos 2x + \frac{2}{9}e^{3x}\sin 2x$

$\Rightarrow I = \frac{1}{13}e^{3x}(3\cos 2x + 2\sin 2x) + C$

A. $u = \cos 2x,\ du = -2\sin 2x\,dx,\ dv = e^{3x}\,dx,\ v = \frac{1}{3}e^{3x}$

B. $u = \sin 2x,\ du = 2\cos 2x\,dx,\ dv = e^{3x}\,dx,\ v = \frac{1}{3}e^{3x}$

19 Let $y = \cos x$. Then $dy = -\sin x\,dx$.

By Example 3, $I = -\int \ln y\,dy = -y\ln y + y + C = y(1 - \ln y) + C$.

20 $I \overset{A}{=} \left[-\frac{1}{2}x^2e^{-x^2}\right]_0^1 + \int_0^1 xe^{-x^2}\,dx = \left[-\frac{1}{2}x^2e^{-x^2}\right]_0^1 + \left[-\frac{1}{2}e^{-x^2}\right]_0^1 =$

$-\frac{1}{2}(e^{-1} - 0) + (-\frac{1}{2})(e^{-1} - 1) = \frac{1}{2} - e^{-1} \approx 0.13$

A. $u = x^2,\ du = 2x\,dx,\ dv = xe^{-x^2}\,dx,\ v = -\frac{1}{2}e^{-x^2}$

[21] I $\stackrel{A}{=} -\csc x \cot x - \int \csc x \cot^2 x\,dx = -\csc x \cot x - \int \csc x(\csc^2 x - 1)\,dx$

$= -\csc x \cot x - \int \csc^3 x\,dx + \int \csc x\,dx \Rightarrow 2\mathrm{I} = -\csc x \cot x + \ln|\csc x - \cot x|$

$\Rightarrow \mathrm{I} = -\frac{1}{2}\csc x \cot x + \frac{1}{2}\ln|\csc x - \cot x| + C$

A. $u = \csc x,\ du = -\csc x \cot x\,dx,\ dv = \csc^2 x\,dx,\ v = -\cot x$

[22] I $\stackrel{A}{=} \tan x \sec^3 x - 3\int \sec^3 x \tan^2 x\,dx = \tan x \sec^3 x - 3\int \sec^3 x(\sec^2 x - 1)\,dx$

$= \tan x \sec^3 x - 3\int \sec^5 x\,dx + 3\int \sec^3 x\,dx.$

By Example 6, $4\mathrm{I} = \tan x \sec^3 x + \frac{3}{2}\tan x \sec x + \frac{3}{2}\ln|\sec x + \tan x|$.

Thus, $\mathrm{I} = \frac{1}{4}\tan x \sec^3 x + \frac{3}{8}(\tan x \sec x + \ln|\sec x + \tan x|) + C.$

A. $u = \sec^3 x,\ du = 3\sec^3 x \tan x\,dx,\ dv = \sec^2 x\,dx,\ v = \tan x$

[23] I $\stackrel{A}{=} \left[x^2\sqrt{x^2+1}\right]_0^1 - 2\int_0^1 x\sqrt{x^2+1}\,dx = \left[x^2\sqrt{x^2+1}\right]_0^1 - \left[\frac{2}{3}(x^2+1)^{3/2}\right]_0^1 =$

$(\sqrt{2} - 0) - \frac{2}{3}(2\sqrt{2} - 1) = \frac{1}{3}(2 - \sqrt{2}) \approx 0.20$

A. $u = x^2,\ du = 2x\,dx,\ dv = \dfrac{x}{\sqrt{x^2+1}}\,dx,\ v = \sqrt{x^2+1}$

[24] I $\stackrel{A}{=} x \sin\ln x - \int \cos\ln x\,dx \stackrel{B}{=} x \sin\ln x - \left[x\cos\ln x + \int \sin\ln x\,dx\right] \Rightarrow$

$2\mathrm{I} = x\sin\ln x - x\cos\ln x \Rightarrow \mathrm{I} = \frac{1}{2}x(\sin\ln x - \cos\ln x) + C$

A. $u = \sin\ln x,\ du = \dfrac{\cos\ln x}{x}\,dx,\ dv = dx,\ v = x$

B. $u = \cos\ln x,\ du = -\dfrac{\sin\ln x}{x}\,dx,\ dv = dx,\ v = x$

[25] I $\stackrel{A}{=} \left[-\frac{1}{2}x\cos 2x\right]_0^{\pi/2} + \frac{1}{2}\int_0^{\pi/2}\cos 2x\,dx = \left[-\frac{1}{2}x\cos 2x\right]_0^{\pi/2} + \left[\frac{1}{4}\sin 2x\right]_0^{\pi/2} =$

$-\frac{1}{2}(-\frac{\pi}{2} - 0) + \frac{1}{4}(0 - 0) = \frac{\pi}{4} \approx 0.79$

A. $u = x,\ du = dx,\ dv = \sin 2x\,dx,\ v = -\frac{1}{2}\cos 2x$

[26] I $\stackrel{A}{=} \frac{1}{5}x\tan 5x - \int \frac{1}{5}\tan 5x\,dx = \frac{1}{5}x\tan 5x + \frac{1}{25}\ln|\cos 5x| + C$

A. $u = x,\ du = dx,\ dv = \sec^2 5x\,dx,\ v = \frac{1}{5}\tan 5x$

[27] I $\stackrel{A}{=} \frac{1}{200}x(2x+3)^{100} - \frac{1}{200}\int (2x+3)^{100}\,dx$

$= \frac{1}{200}x(2x+3)^{100} - \frac{1}{200\cdot 101\cdot 2}(2x+3)^{101} + C$

$= \frac{1}{40{,}400}(2x+3)^{100}\left[202x - (2x+3)\right] + C = \frac{1}{40{,}400}(2x+3)^{100}(200x - 3) + C$

A. $u = x,\ du = dx,\ dv = (2x+3)^{99}\,dx,\ v = \frac{1}{200}(2x+3)^{100}$

[28] I $\stackrel{A}{=} -\frac{2}{3}x^3\sqrt{1-x^3} + 2\int x^2\sqrt{1-x^3}\,dx = -\frac{2}{3}x^3\sqrt{1-x^3} - \frac{4}{9}(1-x^3)^{3/2} + C$

$= -\frac{2}{9}\sqrt{1-x^3}\left[3x^3 + 2(1-x^3)\right] + C = -\frac{2}{9}\sqrt{1-x^3}\,(x^3+2) + C$

A. $u = x^3,\ du = 3x^2\,dx,\ dv = \dfrac{x^2}{\sqrt{1-x^3}}\,dx,\ v = -\frac{2}{3}\sqrt{1-x^3}$

[29] $I \overset{A}{=} -\frac{1}{5}e^{4x}\cos 5x + \frac{4}{5}\int e^{4x}\cos 5x\,dx$

$\overset{B}{=} -\frac{1}{5}e^{4x}\cos 5x + \frac{4}{5}\left[\frac{1}{5}e^{4x}\sin 5x - \frac{4}{5}\int e^{4x}\sin 5x\,dx\right]$

$= -\frac{1}{5}e^{4x}\cos 5x + \frac{4}{25}e^{4x}\sin 5x - \frac{16}{25}\int e^{4x}\sin 5x\,dx \Rightarrow$

$\frac{41}{25}I = -\frac{1}{5}e^{4x}\cos 5x + \frac{4}{25}e^{4x}\sin 5x \Rightarrow I = \frac{1}{41}e^{4x}(4\sin 5x - 5\cos 5x) + C$

A. $u = e^{4x},\ du = 4e^{4x}\,dx,\ dv = \sin 5x\,dx,\ v = -\frac{1}{5}\cos 5x$

B. $u = e^{4x},\ du = 4e^{4x}\,dx,\ dv = \cos 5x\,dx,\ v = \frac{1}{5}\sin 5x$

[30] $I \overset{A}{=} \frac{1}{2}x^2\sin(x^2) - \int x\sin(x^2)\,dx = \frac{1}{2}x^2\sin(x^2) + \frac{1}{2}\cos(x^2) + C$

A. $u = x^2,\ du = 2x\,dx,\ dv = x\cos(x^2)\,dx,\ v = \frac{1}{2}\sin(x^2)$

[31] $I \overset{A}{=} x(\ln x)^2 - 2\int \ln x\,dx$. By Example 3, $I = x(\ln x)^2 - 2x\ln x + 2x + C$.

A. $u = (\ln x)^2,\ du = \frac{2\ln x}{x}\,dx,\ dv = dx,\ v = x$

[32] $I \overset{A}{=} \frac{x2^x}{\ln 2} - \int \frac{2^x}{\ln 2}\,dx = \frac{x2^x}{\ln 2} - \frac{2^x}{(\ln 2)^2} + C = \frac{2^x}{\ln 2}\left(x - \frac{1}{\ln 2}\right) + C$

A. $u = x,\ du = dx,\ dv = 2^x\,dx,\ v = \frac{2^x}{\ln 2}$

[33] $I \overset{A}{=} x^3\cosh x - 3\int x^2\cosh x\,dx \overset{B}{=} x^3\cosh x - 3\left[x^2\sinh x - 2\int x\sinh x\,dx\right]$

$\overset{C}{=} x^3\cosh x - 3x^2\sinh x + 6(x\cosh x - \int\cosh x\,dx)$

$= x^3\cosh x - 3x^2\sinh x + 6x\cosh x - 6\sinh x + C$

A. $u = x^3,\ du = 3x^2\,dx,\ dv = \sinh x\,dx,\ v = \cosh x$

B. $u = x^2,\ du = 2x\,dx,\ dv = \cosh x\,dx,\ v = \sinh x$

C. $u = x,\ du = dx,\ dv = \sinh x\,dx,\ v = \cosh x$

[34] $I \overset{A}{=} \frac{1}{4}(x+4)\sinh 4x - \frac{1}{4}\int\sinh 4x\,dx = \frac{1}{4}(x+4)\sinh 4x - \frac{1}{16}\cosh 4x + C$

A. $u = x + 4,\ du = dx,\ dv = \cosh 4x\,dx,\ v = \frac{1}{4}\sinh 4x$

[35] $I = \int \sqrt{x}\cdot\frac{\cos\sqrt{x}}{\sqrt{x}}\,dx \overset{A}{=} 2\sqrt{x}\sin\sqrt{x} - \int\frac{\sin\sqrt{x}}{\sqrt{x}}\,dx = 2\sqrt{x}\sin\sqrt{x} + 2\cos\sqrt{x} + C$

A. $u = \sqrt{x},\ du = \frac{1}{2\sqrt{x}}\,dx,\ dv = \frac{\cos\sqrt{x}}{\sqrt{x}}\,dx,\ v = 2\sin\sqrt{x}$

[36] $I \overset{A}{=} x\tan^{-1}3x - \int\frac{3x}{1+9x^2}\,dx = x\tan^{-1}3x - \frac{1}{6}\ln(1+9x^2) + C$

A. $u = \tan^{-1}3x,\ du = \frac{3}{1+9x^2}\,dx,\ dv = dx,\ v = x$

[37] $I \overset{A}{=} x\cos^{-1}x + \int\frac{x}{\sqrt{1-x^2}}\,dx = x\cos^{-1}x - \sqrt{1-x^2} + C$

A. $u = \cos^{-1}x,\ du = -\frac{1}{\sqrt{1-x^2}}\,dx,\ dv = dx,\ v = x$

[38] $\text{I} \overset{\text{A}}{=} \frac{1}{11}(x+2)(x+1)^{11} - \frac{1}{11}\int (x+1)^{11}\,dx$

$= \frac{1}{11}(x+2)(x+1)^{11} - \frac{1}{11\cdot 12}(x+1)^{12} + C$

$= \frac{1}{132}(x+1)^{11}\big[12(x+2) - (x+1)\big] + C = \frac{1}{132}(x+1)^{11}(11x+23) + C$

A. $u = x+2,\ du = dx,\ dv = (x+1)^{10}\,dx,\ v = \frac{1}{11}(x+1)^{11}$

[39] $\text{I} \overset{\text{A}}{=} x^m e^x - m\int x^{m-1}e^x\,dx$ A. $u = x^m,\ du = mx^{m-1}\,dx,\ dv = e^x\,dx,\ v = e^x$

[40] $\text{I} \overset{\text{A}}{=} -x^m \cos x + m\int x^{m-1}\cos x\,dx$

A. $u = x^m,\ du = mx^{m-1}\,dx,\ dv = \sin x\,dx,\ v = -\cos x$

[41] $\text{I} \overset{\text{A}}{=} x(\ln x)^m - m\int (\ln x)^{m-1}\,dx$

A. $u = (\ln x)^m,\ du = \dfrac{m(\ln x)^{m-1}}{x}\,dx,\ dv = dx,\ v = x$

[42] $\text{I} \overset{\text{A}}{=} \sec^{m-2}x\tan x - (m-2)\int \sec^{m-2}x\tan^2 x\,dx$

$= \sec^{m-2}x\tan x - (m-2)\int \sec^{m-2}x(\sec^2 x - 1)\,dx$

$= \sec^{m-2}x\tan x - (m-2)\int \sec^m dx + (m-2)\int \sec^{m-2}x\,dx \Rightarrow$

$(m-1)\text{I} = \sec^{m-2}x\tan x + (m-2)\int \sec^{m-2}x\,dx \Rightarrow$

$$\text{I} = \frac{\sec^{m-2}x\tan x}{m-1} + \frac{m-2}{m-1}\int \sec^{m-2}x\,dx \text{ for } m \neq 1.$$

A. $u = \sec^{m-2}x,\ du = (m-2)\sec^{m-2}x\tan x\,dx,\ dv = \sec^2 x\,dx,\ v = \tan x$

[43] $\text{I} = x^5e^x - 5\int x^4e^x\,dx$

$= x^5e^x - 5\big[x^4e^x - 4\int x^3e^x\,dx\big]$

$= x^5e^x - 5x^4e^x + 20\big[x^3e^x - 3\int x^2e^x\,dx\big]$

$= x^5e^x - 5x^4e^x + 20x^3e^x - 60\big[x^2e^x - 2\int xe^x\,dx\big]$

$= x^5e^x - 5x^4e^x + 20x^3e^x - 60x^2e^x + 120\big[xe^x - \int e^x\,dx\big]$

$= e^x(x^5 - 5x^4 + 20x^3 - 60x^2 + 120x - 120) + C$

[44] $\text{I} = x(\ln x)^4 - 4\int (\ln x)^3\,dx$

$= x(\ln x)^4 - 4\big[x(\ln x)^3 - 3\int (\ln x)^2\,dx\big]$

$= x(\ln x)^4 - 4x(\ln x)^3 + 12\big[x(\ln x)^2 - 2\int \ln x\,dx\big]$

$= x(\ln x)^4 - 4x(\ln x)^3 + 12x(\ln x)^2 - 24(x\ln x - x) + C$ { by Example 3 }

$= x\big[(\ln x)^4 - 4(\ln x)^3 + 12(\ln x)^2 - 24\ln x + 24\big] + C$

[45] $A = \int_0^{\pi^2} \sin\sqrt{x}\,dx = \int_0^{\pi^2} \sqrt{x}\cdot\frac{\sin\sqrt{x}}{\sqrt{x}}\,dx \overset{\text{A}}{=} \Big[-2\sqrt{x}\cos\sqrt{x}\Big]_0^{\pi^2} + \int_0^{\pi^2}\frac{\cos\sqrt{x}}{\sqrt{x}}\,dx$

$= \Big[-2\sqrt{x}\cos\sqrt{x} + 2\sin\sqrt{x}\Big]_0^{\pi^2} = (2\pi + 0) - (0 - 0) = 2\pi$

A. $u = \sqrt{x},\ du = \dfrac{1}{2\sqrt{x}}\,dx,\ dv = \dfrac{\sin\sqrt{x}}{\sqrt{x}}\,dx,\ v = -2\cos\sqrt{x}$

[46] Using disks, $V = \pi \int_0^{\pi/2} (x\sqrt{\sin x})^2 \, dx = \pi \int_0^{\pi/2} x^2 \sin x \, dx$

$$\stackrel{\text{A}}{=} \pi \Big[-x^2 \cos x \Big]_0^{\pi/2} + 2\pi \int_0^{\pi/2} x \cos x \, dx$$

$$\stackrel{\text{B}}{=} \pi \Big[-x^2 \cos x + 2x \sin x \Big]_0^{\pi/2} - 2\pi \int_0^{\pi/2} \sin x \, dx$$

$$= \pi \Big[-x^2 \cos x + 2x \sin x + 2\cos x \Big]_0^{\pi/2} = \pi(\pi - 2) \approx 3.59.$$

A. $u = x^2$, $du = 2x\,dx$, $dv = \sin x\,dx$, $v = -\cos x$

B. $u = x$, $du = dx$, $dv = \cos x\,dx$, $v = \sin x$

[47] Using shells, $V = 2\pi \int_1^e x \ln x \, dx \stackrel{\text{A}}{=} 2\pi \Big[\frac{1}{2}x^2 \ln x \Big]_1^e - 2\pi \int_1^e \frac{1}{2}x \, dx =$

$$2\pi \Big[\tfrac{1}{2}x^2 \ln x - \tfrac{1}{4}x^2 \Big]_1^e = \tfrac{\pi}{2} \Big[x^2(2\ln x - 1) \Big]_1^e = \tfrac{\pi}{2}(e^2 + 1) \approx 13.18.$$

A. $u = \ln x$, $du = \frac{1}{x}\,dx$, $dv = x\,dx$, $v = \frac{1}{2}x^2$

[48] $W = \int_0^1 f(x)\,dx = \int_0^1 x^5 \sqrt{x^3 + 1}\,dx \stackrel{\text{A}}{=} \Big[\frac{2}{9}x^3(x^3 + 1)^{3/2} \Big]_0^1 - \frac{2}{3}\int_0^1 x^2(x^3 + 1)^{3/2}\,dx$

$$= \Big[\tfrac{2}{9}x^3(x^3 + 1)^{3/2} - \tfrac{4}{45}(x^3 + 1)^{5/2} \Big]_0^1 = \Big[\tfrac{2}{9}(2)^{3/2} - \tfrac{4}{45}(2)^{5/2} + \tfrac{4}{45} \Big] =$$

$$\Big[\tfrac{4}{9}\sqrt{2} - \tfrac{16}{45}\sqrt{2} + \tfrac{4}{45} \Big] = \tfrac{4}{45}(\sqrt{2} + 1) \approx 0.21$$

A. $u = x^3$, $du = 3x^2\,dx$, $dv = x^2\sqrt{x^3 + 1}\,dx$, $v = \frac{2}{9}(x^3 + 1)^{3/2}$

[49] Let $f(x) = e^x$, $g(x) = 0$, and $\rho = 1$ in (6.25). $m = \int_0^{\ln 3} e^x\,dx = 2.$

$$M_x = \tfrac{1}{2}\int_0^{\ln 3} (e^x)^2\,dx = 2. \quad M_y = \int_0^{\ln 3} xe^x\,dx \stackrel{\text{A}}{=} \Big[xe^x - e^x \Big]_0^{\ln 3} = 3\ln 3 - 2.$$

$$\bar{x} = \frac{M_y}{m} = \frac{3\ln 3 - 2}{2} \approx 0.65 \text{ and } \bar{y} = \frac{M_x}{m} = \frac{2}{2} = 1.$$

A. $u = x$, $du = dx$, $dv = e^x\,dx$, $v = e^x$

[50] $s(t) = \int v(t)\,dt = \int te^{-2t}\,dt \stackrel{\text{A}}{=} -\frac{1}{2}te^{-2t} + \frac{1}{2}\int e^{-2t}\,dt = -\frac{1}{2}te^{-2t} - \frac{1}{4}e^{-2t} + C.$

$$s(0) = 0 \Rightarrow -\tfrac{1}{4} + C = 0 \Rightarrow C = \tfrac{1}{4} \text{ and } s(t) = -\tfrac{1}{2}te^{-2t} - \tfrac{1}{4}e^{-2t} + \tfrac{1}{4}.$$

A. $u = t$, $du = dt$, $dv = e^{-2t}\,dt$, $v = -\frac{1}{2}e^{-2t}$

[51] Substituting $(v + C)$ for v in (9.1) yields $\int u\,dv = u(v + C) - \int (v + C)\,du =$

$$uv + uC - \int v\,du - Cu = uv - \int v\,du, \text{ which is (9.1).}$$

[52] Since f' is only positive or only negative, $g = f^{-1}$ exists on $[a, b]$.

Using *Figure 52a* and disks, $V = \pi \int_a^b [f(x)]^2\,dx \stackrel{\text{A}}{=} \pi \Big[x[f(x)]^2 \Big]_a^b - 2\pi \int_a^b xf(x)f'(x)\,dx$

$$= \pi bd^2 - \pi ac^2 - 2\pi \int_a^b xf(x)f'(x)\,dx. \text{ (cont.)}$$

Using *Figure 52b* and shells, V = volume of cylinder $+ 2\pi\int_c^d y\left[b - g(y)\right]dy =$

$\pi c^2(b - a) + 2\pi\left[\frac{1}{2}by^2\right]_c^d - 2\pi\int_c^d y\,g(y)\,dy$. Since $y = f(x)$, $g(y) = f^{-1}(y) = x$, and

$dy = f'(x)\,dx$. Thus, $V = \pi c^2 b - \pi c^2 a + \pi b d^2 - \pi b c^2 - 2\pi\int_a^b f(x)\cdot x\cdot f'(x)\,dx =$

$$\pi b d^2 - \pi a c^2 - 2\pi\int_a^b x f(x) f'(x)\,dx.$$

A. $u = \left[f(x)\right]^2$, $du = 2f(x)f'(x)\,dx$, $dv = dx$, $v = x$

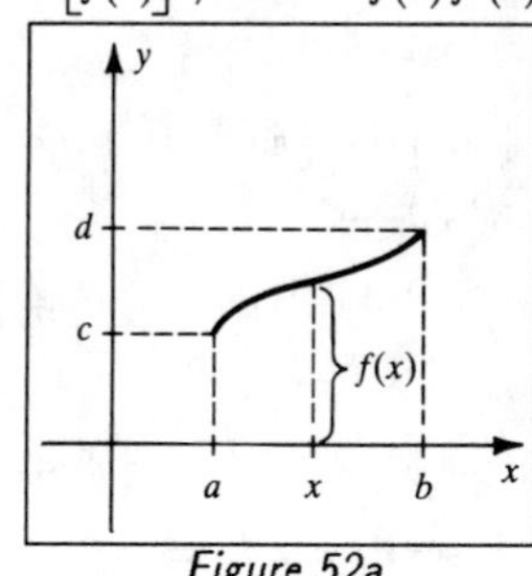

Figure 52a

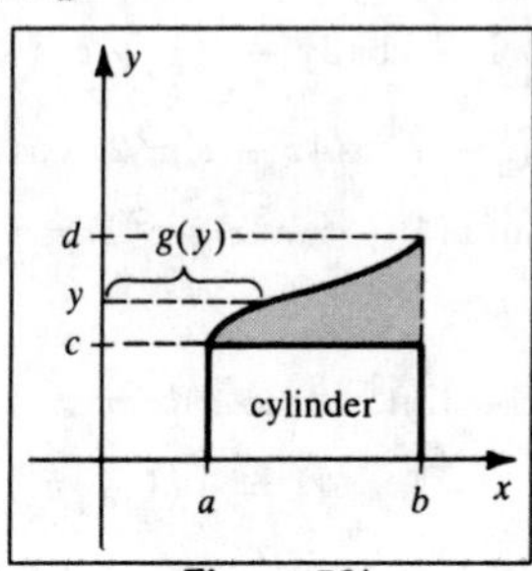

Figure 52b

53 $\int\frac{1}{x}\,dx = 1 + \int\frac{1}{x}\,dx \Rightarrow \ln|x| + C_1 = 1 + \ln|x| + C_2 \Rightarrow C_1 - C_2 = 1.$

Note that indefinite integrals represent a class of functions which differ by a constant.

54 Suppose F is an antiderivative of the indefinite integral $\int v\,du$. By (9.1), $uv - F$ represents an antiderivative of $\int u\,dv$. Thus, by the fundamental theorem of calculus,

$$\int_a^b u\,dv = \left[uv - F\right]_a^b = \left[uv\right]_a^b - \left[F(b) - F(a)\right] = \left[uv\right]_a^b - \int_a^b v\,du.$$

Exercises 9.2

1 $\mathrm{I} = \int(1 - \sin^2 x)\cos x\,dx$; $u = \sin x$, $du = \cos x\,dx \Rightarrow$

$$\mathrm{I} = \int(1 - u^2)\,du = u - \tfrac{1}{3}u^3 + C$$

2 $\mathrm{I} = \int\frac{1}{2}(1 - \cos 4x)\,dx = \frac{1}{2}x - \frac{1}{8}\sin 4x + C$

3 $\mathrm{I} = \int\frac{1}{4}(1 - \cos 2x)(1 + \cos 2x)\,dx = \int\frac{1}{4}(1 - \cos^2 2x)\,dx =$

$$\int\left[\tfrac{1}{4} - \tfrac{1}{8}(1 + \cos 4x)\right]dx = \int\left(\tfrac{1}{8} - \tfrac{1}{8}\cos 4x\right)dx = \tfrac{1}{8}x - \tfrac{1}{32}\sin 4x + C$$

4 $\mathrm{I} = \int(1 - \sin^2 x)^3\cos x\,dx$; $u = \sin x$, $du = \cos x\,dx \Rightarrow$

$$\mathrm{I} = \int(1 - u^2)^3\,du = \int(1 - 3u^2 + 3u^4 - u^6)\,du = u - u^3 + \tfrac{3}{5}u^5 - \tfrac{1}{7}u^7 + C$$

5 $\mathrm{I} = \int(1 - \cos^2 x)\cos^2 x\sin x\,dx$; $u = \cos x$, $-du = \sin x\,dx \Rightarrow$

$$\mathrm{I} = -\int(u^2 - u^4)\,du = -\tfrac{1}{3}u^3 + \tfrac{1}{5}u^5 + C$$

6 $\mathrm{I} = \int\sin^5 x(1 - \sin^2 x)\cos x\,dx$; $u = \sin x$, $du = \cos x\,dx \Rightarrow$

$$\mathrm{I} = \int(u^5 - u^7)\,du = \tfrac{1}{6}u^6 - \tfrac{1}{8}u^8 + C$$

[7] $\mathrm{I} = \int\left(\frac{1 - \cos 2x}{2}\right)^3 dx = \frac{1}{8}\int(1 - 3\cos 2x + 3\cos^2 2x - \cos^3 2x)\,dx$

$= \frac{1}{8}\int\left[1 - 3\cos 2x + \frac{3}{2}(1 + \cos 4x) - (1 - \sin^2 2x)\cos 2x\right]dx$

$= \frac{1}{8}\int(\frac{5}{2} - 4\cos 2x + \frac{3}{2}\cos 4x + \sin^2 2x\cos 2x)\,dx$

$= \frac{1}{8}(\frac{5}{2}x - 2\sin 2x + \frac{3}{8}\sin 4x + \frac{1}{6}\sin^3 2x) + C$

[8] $\mathrm{I} = \int\left(\frac{1 - \cos 2x}{2}\right)^2\left(\frac{1 + \cos 2x}{2}\right)dx = \frac{1}{8}\int(1 - \cos 2x - \cos^2 2x + \cos^3 2x)\,dx$

$= \frac{1}{8}\int\left[1 - \cos 2x - \frac{1}{2}(1 + \cos 4x) + (1 - \sin^2 2x)\cos 2x\right]dx$

$= \frac{1}{8}\int(\frac{1}{2} - \frac{1}{2}\cos 4x - \sin^2 2x\cos 2x)\,dx = \frac{1}{8}(\frac{1}{2}x - \frac{1}{8}\sin 4x - \frac{1}{6}\sin^3 2x) + C$

[9] $\mathrm{I} = \int\tan^3 x(1 + \tan^2 x)\sec^2 x\,dx;\ u = \tan x,\ du = \sec^2 x\,dx \Rightarrow$

$\mathrm{I} = \int(u^3 + u^5)\,du = \frac{1}{4}u^4 + \frac{1}{6}u^6 + C$

[10] $\mathrm{I} = \int(1 + \tan^2 x)^2\sec^2 x\,dx;\ u = \tan x,\ du = \sec^2 x\,dx \Rightarrow$

$\mathrm{I} = \int(1 + u^2)^2\,du = \int(1 + 2u^2 + u^4)\,du = u + \frac{2}{3}u^3 + \frac{1}{5}u^5 + C$

[11] $\mathrm{I} = \int(\sec^2 x - 1)\sec^2 x\sec x\tan x\,dx;\ u = \sec x,\ du = \sec x\tan x\,dx \Rightarrow$

$\mathrm{I} = \int(u^4 - u^2)\,du = \frac{1}{5}u^5 - \frac{1}{3}u^3 + C$

[12] $\mathrm{I} = \int(\sec^2 x - 1)^2\sec x\tan x\,dx;\ u = \sec x,\ du = \sec x\tan x\,dx \Rightarrow$

$\mathrm{I} = \int(u^2 - 1)^2\,du = \int(u^4 - 2u^2 + 1)\,du = \frac{1}{5}u^5 - \frac{2}{3}u^3 + u + C$

[13] $\mathrm{I} = \int\tan^4 x(\sec^2 x - 1)\,dx = \int\left[\tan^4 x\sec^2 x - \tan^2 x(\sec^2 x - 1)\right]dx$

$= \int(\tan^4 x\sec^2 x - \tan^2 x\sec^2 x + \sec^2 x - 1)\,dx$

$= \frac{1}{5}\tan^5 x - \frac{1}{3}\tan^3 x + \tan x - x + C$

[14] $\mathrm{I} = \int\cot^2 x(\csc^2 x - 1)\,dx = \int\left[\cot^2 x\csc^2 x - (\csc^2 x - 1)\right]dx =$

$-\frac{1}{3}\cot^3 x + \cot x + x + C$

[15] $\mathrm{I} = \int(\sin x)^{1/2}(1 - \sin^2 x)\cos x\,dx;\ u = \sin x,\ du = \cos x\,dx \Rightarrow$

$\mathrm{I} = \int(u^{1/2} - u^{5/2})\,du = \frac{2}{3}u^{3/2} - \frac{2}{7}u^{7/2} + C$

[16] $\mathrm{I} = \int(\sin x)^{-1/2}(1 - \sin^2 x)\cos x\,dx;\ u = \sin x,\ du = \cos x\,dx \Rightarrow$

$\int(u^{-1/2} - u^{3/2})\,du = 2u^{1/2} - \frac{2}{5}u^{5/2} + C$

[17] $\mathrm{I} = \int(\tan^2 x + 2 + \cot^2 x)\,dx = \int\left[(\tan^2 x + 1) + (1 + \cot^2 x)\right]dx =$

$\int(\sec^2 x + \csc^2 x)\,dx = \tan x - \cot x + C$

[18] $\mathrm{I} = \int(\csc^2 x - 1)\csc^2 x\csc x\cot x\,dx;\ u = \csc x,\ -du = \csc x\cot x\,dx \Rightarrow$

$\mathrm{I} = -\int(u^4 - u^2)\,du = \frac{1}{3}u^3 - \frac{1}{5}u^5 + C$

[19] $\mathrm{I} = \int_0^{\pi/4}(1 - \cos^2 x)\sin x\,dx = \left[-\cos x + \frac{1}{3}\cos^3 x\right]_0^{\pi/4} = \frac{2}{3} - \frac{5}{6\sqrt{2}} \approx 0.08$

[20] $\mathrm{I} = \int_0^1\left[\sec^2(\frac{\pi}{4}x) - 1\right]dx = \left[\frac{4}{\pi}\tan(\frac{\pi}{4}x) - x\right]_0^1 = \frac{4}{\pi} - 1 \approx 0.27$

Note: Exercises 21-24 use the trigonometric product-to-sum formulas.

[21] $\mathrm{I} = \frac{1}{2}\int(\cos 2x - \cos 8x)\,dx = \frac{1}{2}(\frac{1}{2}\sin 2x - \frac{1}{8}\sin 8x) + C$

[22] $\mathrm{I} = \frac{1}{2}\int_0^{\pi/4}\left[\cos 6x + \cos(-4x)\right]dx = \frac{1}{2}\left[\frac{1}{6}\sin 6x + \frac{1}{4}\sin 4x\right]_0^{\pi/4} = -\frac{1}{12}$

[23] $\mathrm{I} = \frac{1}{2}\int_0^{\pi/2}\left[\sin 5x + \sin x\right]dx = \frac{1}{2}\left[-\frac{1}{5}\cos 5x - \cos x\right]_0^{\pi/2} = \frac{1}{2}\left[0 - (-\frac{6}{5})\right] = \frac{3}{5}$

[24] $\mathrm{I} = \frac{1}{2}\int\left[\sin 7x + \sin x\right]dx = -\frac{1}{2}(\frac{1}{7}\cos 7x + \cos x) + C$

[25] $\mathrm{I} = \int\cot^4 x(1 + \cot^2 x)\csc^2 x\,dx$; $u = \cot x$, $-du = \csc^2 x\,dx \Rightarrow$

$$\mathrm{I} = -\int(u^4 + u^6)\,du = -\frac{1}{5}u^5 - \frac{1}{7}u^7 + C$$

[26] $\mathrm{I} = \int\left[1 + 2(\cos x)^{1/2} + \cos x\right]\sin x\,dx$; $u = \cos x$, $-du = \sin x\,dx \Rightarrow$

$$\mathrm{I} = -\int(1 + 2u^{1/2} + u)\,du = -u - \frac{4}{3}u^{3/2} - \frac{1}{2}u^2 + C$$

[27] $u = 2 - \sin x$, $-du = \cos x\,dx \Rightarrow \mathrm{I} = -\int\frac{1}{u}\,du = -\ln u + C,\ u > 0$

[28] $\mathrm{I} = \int(\sin^2 x - \cos^2 x)\,dx = \int -\cos 2x\,dx = -\frac{1}{2}\sin 2x + C$

[29] $u = 1 + \tan x$, $du = \sec^2 x\,dx \Rightarrow \mathrm{I} = \int u^{-2}\,du = -\frac{1}{u} + C$

[30] $\mathrm{I} = \int\tan^5 x\sec x\,dx = \int(\sec^2 x - 1)^2\sec x\tan x\,dx$; $u = \sec x$, $du = \sec x\tan x\,dx \Rightarrow$

$$\mathrm{I} = \int(u^2 - 1)^2\,du = \int(u^4 - 2u^2 + 1)\,du = \frac{1}{5}u^5 - \frac{2}{3}u^3 + u + C$$

[31] Using symmetry and disks, $V = 4\int_0^{\pi/2}\pi(\cos^2 x)^2\,dx = 4\pi\int_0^{\pi/2}\left[\frac{1}{2}(1 + \cos 2x)\right]^2 dx =$

$$\pi\int_0^{\pi/2}\left[1 + 2\cos 2x + \frac{1}{2}(1 + \cos 4x)\right]dx = \pi\left[\frac{3}{2}x + \sin 2x + \frac{1}{8}\sin 4x\right]_0^{\pi/2} =$$

$$\frac{3\pi^2}{4} \approx 7.40.$$

[32] Using disks, $V = \int_0^{\pi/4}\pi(\tan^2 x)^2\,dx = \pi\int_0^{\pi/4}\tan^2 x(\sec^2 x - 1)\,dx =$

$$\pi\int_0^{\pi/4}\left[\tan^2 x\sec^2 x - (\sec^2 x - 1)\right]dx = \pi\left[\frac{1}{3}\tan^3 x - \tan x + x\right]_0^{\pi/4} =$$

$$\pi(\frac{\pi}{4} - \frac{2}{3}) \approx 0.37.$$

[33] Since $v(t) = \cos^2\pi t \geq 0$, the distance traveled in *any* 5-second interval is given by

$$s(x + 5) - s(x) = \int_x^{x+5} v(t)\,dt = \int_x^{x+5}\cos^2\pi t\,dt = \int_x^{x+5}\frac{1}{2}(1 + \cos 2\pi t)\,dt =$$

$$\left[\frac{1}{2}t + \frac{1}{4\pi}\sin 2\pi t\right]_x^{x+5} = \left[\frac{1}{2}(x + 5) + \frac{1}{4\pi}\sin\left[2\pi(x + 5)\right]\right] - \left[\frac{1}{2}x + \frac{1}{4\pi}\sin(2\pi x)\right] =$$

$$\frac{5}{2} + \frac{1}{4\pi}\left[\sin(2\pi x + 10\pi) - \sin(2\pi x)\right] = \frac{5}{2},\text{ since }\sin(2\pi x + 10\pi) = \sin(2\pi x).$$

[34] $a(t) = \sin^2 t\cos t \Rightarrow v(t) = \int\sin^2 t\cos t\,dt = \frac{1}{3}\sin^3 t + C$. $v(0) = 10 \Rightarrow C = 10$.

$$s(t) = \int(\frac{1}{3}\sin^3 t + 10)\,dt = \frac{1}{3}\int(1 - \cos^2 t)\sin t\,dt + \int 10\,dt =$$

$$-\frac{1}{3}\cos t + \frac{1}{9}\cos^3 t + 10t + D.\ s(0) = 0 \Rightarrow -\frac{1}{3} + \frac{1}{9} + D = 0 \Rightarrow D = \frac{2}{9}.$$

[35] Using the trigonometric product-to-sum formulas, we have the following:

(a) $\text{I} = \int \sin mx \sin nx\, dx = \frac{1}{2}\int \left[\cos(m - n)x - \cos(m + n)x\right] dx.$

If $m \neq n$, then $\text{I} = \dfrac{\sin(m - n)x}{2(m - n)} - \dfrac{\sin(m + n)x}{2(m + n)} + C.$

If $m = n$, then $\text{I} = \dfrac{1}{2}\displaystyle\int (1 - \cos 2mx)\, dx = \dfrac{x}{2} - \dfrac{\sin 2mx}{4m} + C.$

(b) $\text{I}_1 = \int \sin mx \cos nx\, dx = \frac{1}{2}\int \left[\sin(m + n)x + \sin(m - n)x\right] dx.$

If $m \neq n$, then $\text{I}_1 = -\dfrac{\cos(m + n)x}{2(m + n)} - \dfrac{\cos(m - n)x}{2(m - n)} + C.$

If $m = n$, then $\text{I}_1 = \dfrac{1}{2}\displaystyle\int \sin 2mx\, dx = -\dfrac{\cos 2mx}{4m} + C.$

$\text{I}_2 = \int \cos mx \cos nx\, dx = \frac{1}{2}\int \left[\cos(m + n)x + \cos(m - n)x\right] dx.$

If $m \neq n$, then $\text{I}_2 = \dfrac{\sin(m + n)x}{2(m + n)} + \dfrac{\sin(m - n)x}{2(m - n)} + C.$

If $m = n$, then $\text{I}_2 = \dfrac{1}{2}\displaystyle\int (\cos 2mx + 1)\, dx = \dfrac{x}{2} + \dfrac{\sin 2mx}{4m} + C.$

[36] (a) If $m = n$, then $\text{I} = \displaystyle\int_{-\pi}^{\pi} \sin mx \sin nx\, dx = \left[\frac{x}{2} - \frac{\sin 2mx}{4m}\right]_{-\pi}^{\pi} = \pi.$

If $m \neq n$, then $\text{I} = \left[\dfrac{\sin(m - n)x}{2(m - n)} - \dfrac{\sin(m + n)x}{2(m + n)}\right]_{-\pi}^{\pi} = 0,$

since $(m - n)$ and $(m + n)$ are integers.

(b) Using Exercise 35(b), we have the following:

(i) If $m = n$, then $\text{I}_1 = \displaystyle\int_{-\pi}^{\pi} \sin mx \cos nx\, dx = \left[-\frac{\cos 2mx}{4m}\right]_{-\pi}^{\pi} = 0.$

If $m \neq n$, then $\text{I}_1 = \left[-\dfrac{\cos(m + n)x}{m + n} - \dfrac{\cos(m - n)x}{m - n}\right]_{-\pi}^{\pi} = 0.$

Note: Since $\sin mx$ is odd and $\cos nx$ is even,

we have an odd integrand on a symmetric interval. Hence, $\text{I} = 0$.

(ii) If $m = n$, then $\text{I}_2 = \displaystyle\int_{-\pi}^{\pi} \cos mx \cos nx\, dx = \left[\frac{x}{2} + \frac{\sin 2mx}{4m}\right]_{-\pi}^{\pi} = \pi.$

If $m \neq n$, then $\text{I}_2 = \left[\dfrac{\sin(m + n)x}{2(m + n)} + \dfrac{\sin(m - n)x}{2(m - n)}\right]_{-\pi}^{\pi} = 0.$

Exercises 9.3

Note: In (9.4), the resulting expressions are $a\cos\theta$, $a\sec\theta$, and $a\tan\theta$, respectively.

We use these results without mention.

[1] See *Figure 1*. $x = 2\sin\theta$, $dx = 2\cos\theta\, d\theta \Rightarrow \text{I} = \displaystyle\int \frac{1}{(2\sin\theta)(2\cos\theta)} 2\cos\theta\, d\theta =$

$$\frac{1}{2}\int \csc\theta\, d\theta = \frac{1}{2}\ln|\csc\theta - \cot\theta| + C = \frac{1}{2}\ln\left|\frac{2}{x} - \frac{\sqrt{4 - x^2}}{x}\right| + C$$

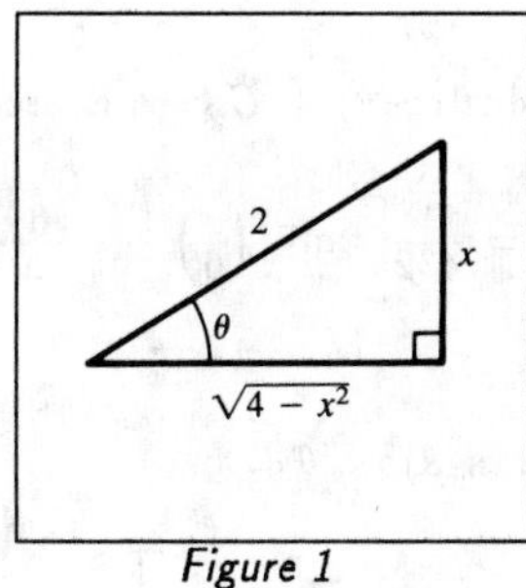

Figure 1

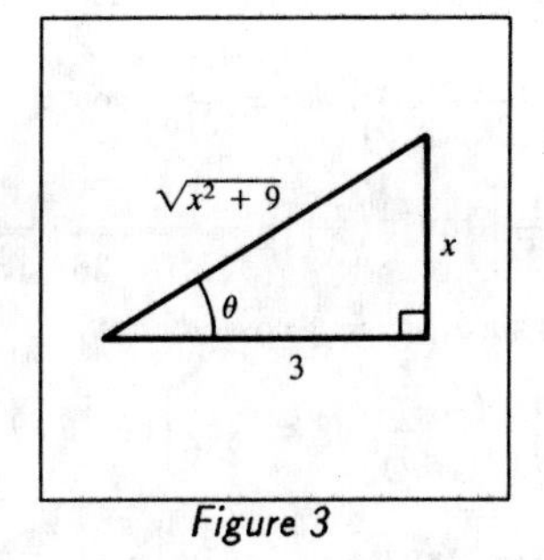

Figure 3

[2] See *Figure 1*. $x = 2\sin\theta,\ dx = 2\cos\theta\,d\theta \Rightarrow \mathrm{I} = \int \frac{2\cos\theta}{4\sin^2\theta}2\cos\theta\,d\theta = \int \cot^2\theta\,d\theta =$

$$\int(\csc^2\theta - 1)\,d\theta = -\cot\theta - \theta + C = -\frac{\sqrt{4-x^2}}{x} - \sin^{-1}\left(\frac{x}{2}\right) + C$$

[3] See *Figure 3*. $x = 3\tan\theta,\ dx = 3\sec^2\theta\,d\theta \Rightarrow \mathrm{I} = \int \frac{3\sec^2\theta}{(3\tan\theta)(3\sec\theta)}\,d\theta =$

$$\frac{1}{3}\int\csc\theta\,d\theta = \tfrac{1}{3}\ln\left|\csc\theta - \cot\theta\right| + C = \tfrac{1}{3}\ln\left|\frac{\sqrt{x^2+9}}{x} - \frac{3}{x}\right| + C$$

[4] See *Figure 3*. $x = 3\tan\theta,\ dx = 3\sec^2\theta\,d\theta \Rightarrow$

$$\mathrm{I} = \int\frac{3\sec^2\theta}{(9\tan^2\theta)(3\sec\theta)}\,d\theta = \frac{1}{9}\int\csc\theta\,\cot\theta\,d\theta = -\tfrac{1}{9}\csc\theta + C = -\frac{\sqrt{x^2+9}}{9x} + C$$

Note: The trigonometric functions for θ in Exercises 5–22 can be determined using a right triangle as in *Figure 1* and *Figure 3* above, and in Examples 1–4 in the text.

[5] $x = 5\sec\theta,\ dx = 5\sec\theta\tan\theta\,d\theta \Rightarrow$

$$\mathrm{I} = \int\frac{5\sec\theta\,\tan\theta}{(25\sec^2\theta)(5\tan\theta)}\,d\theta = \frac{1}{25}\int\cos\theta\,d\theta = \frac{1}{25}\sin\theta + C = \frac{\sqrt{x^2-25}}{25x} + C$$

[6] $x = 5\sec\theta,\ dx = 5\sec\theta\,\tan\theta\,d\theta \Rightarrow \mathrm{I} = \int\frac{5\sec\theta\,\tan\theta}{(125\sec^3\theta)(5\tan\theta)}\,d\theta = \frac{1}{125}\int\cos^2\theta\,d\theta =$

$\frac{1}{125}\int\frac{1}{2}(1 + \cos 2\theta)\,d\theta = \frac{1}{250}(\theta + \frac{1}{2}\sin 2\theta) + C = \frac{1}{250}(\theta + \sin\theta\,\cos\theta) + C =$

$$\frac{1}{250}\left[\sec^{-1}\left(\frac{x}{5}\right) + \frac{\sqrt{x^2-25}}{x}\cdot\frac{5}{x}\right] + C = \frac{1}{250}\left[\sec^{-1}\left(\frac{x}{5}\right) + \frac{5\sqrt{x^2-25}}{x^2}\right] + C$$

[7] $u = 4 - x^2,\ -\frac{1}{2}\,du = x\,dx \Rightarrow \mathrm{I} = -\frac{1}{2}\int u^{-1/2}\,du = -\sqrt{u} + C$

[8] $u = x^2 + 9,\ \frac{1}{2}\,du = x\,dx \Rightarrow \mathrm{I} = \frac{1}{2}\int\frac{1}{u}\,du = \frac{1}{2}\ln u + C,\ u > 0$

[9] $x = \sec\theta,\ dx = \sec\theta\tan\theta\,d\theta \Rightarrow$

$$\mathrm{I} = \int\frac{\sec\theta\,\tan\theta}{(\tan\theta)^3}\,d\theta = \int\cot\theta\,\csc\theta\,d\theta = -\csc\theta + C = -\frac{x}{\sqrt{x^2-1}} + C$$

[10] $2x = 5\sec\theta$ or $x = \frac{5}{2}\sec\theta,\ dx = \frac{5}{2}\sec\theta\,\tan\theta\,d\theta \Rightarrow \mathrm{I} = \int\frac{\frac{5}{2}\sec\theta\,\tan\theta}{5\tan\theta}\,d\theta =$

$$\frac{1}{2}\int\sec\theta\,d\theta = \tfrac{1}{2}\ln\left|\sec\theta + \tan\theta\right| + C = \tfrac{1}{2}\ln\left|\frac{2x}{5} + \frac{\sqrt{4x^2-25}}{5}\right| + C$$

[11] $x = 6\tan\theta,\ dx = 6\sec^2\theta\, d\theta \Rightarrow$

$$\text{I} = \int \frac{6\sec^2\theta}{(36\sec^2\theta)^2}\, d\theta = \frac{1}{216}\int \cos^2\theta\, d\theta = \frac{1}{432}(\theta + \sin\theta\cos\theta) + C \,\{\text{see Exercise 6}\}$$

$$= \frac{1}{432}\left[\tan^{-1}\left(\frac{x}{6}\right) + \frac{x}{\sqrt{x^2+36}}\cdot\frac{6}{\sqrt{x^2+36}}\right] + C = \frac{1}{432}\left[\tan^{-1}\left(\frac{x}{6}\right) + \frac{6x}{x^2+36}\right] + C$$

[12] $x = 4\sin\theta,\ dx = 4\cos\theta\, d\theta \Rightarrow$

$$\text{I} = \int \frac{4\cos\theta}{(4\cos\theta)^5}\, d\theta = \frac{1}{256}\int \sec^4\theta\, d\theta = \frac{1}{256}\int (1 + \tan^2\theta)\sec^2\theta\, d\theta$$

$$= \frac{1}{256}\left[\tan\theta + \frac{1}{3}\tan^3\theta\right] + C = \frac{1}{256}\left[\frac{x}{\sqrt{16-x^2}} + \frac{x^3}{3(16-x^2)^{3/2}}\right] + C$$

[13] $\displaystyle\int \frac{1}{\sqrt{9-x^2}}\, dx = \int \frac{1}{\sqrt{3^2-x^2}}\, dx = \sin^{-1}\left(\frac{x}{3}\right) + C$, by (8.9)(i). ***Note:*** In the first printing, the radicand was $9 - 4x^2$. The answer is then $\frac{1}{2}\sin^{-1}\left(\frac{2}{3}x\right) + C$.

[14] $\displaystyle\int \frac{1}{49+x^2}\, dx = \int \frac{1}{7^2+x^2}\, dx = \frac{1}{7}\tan^{-1}\left(\frac{x}{7}\right) + C$, by (8.9)(ii).

[15] $u = 16 - x^2,\ -\frac{1}{2}\, du = x\, dx \Rightarrow \text{I} = -\frac{1}{2}\int u^{-2}\, du = \dfrac{1}{2u} + C$

[16] $u = x^2 - 9,\ \frac{1}{2}\, du = x\, dx \Rightarrow \text{I} = \frac{1}{2}\int u^{1/2}\, du = \frac{1}{3}u^{3/2} + C$

[17] $3x = 7\tan\theta$ or $x = \frac{7}{3}\tan\theta,\ dx = \frac{7}{3}\sec^2\theta\, d\theta \Rightarrow \text{I} = \displaystyle\int \frac{(\frac{7}{3}\tan\theta)^3(\frac{7}{3}\sec^2\theta)}{7\sec\theta}\, d\theta$

$$= \frac{7^3}{3^4}\int \tan^2\theta\,\sec\theta\,\tan\theta\, d\theta = \frac{7^3}{3^4}\int (\sec^2\theta - 1)\sec\theta\,\tan\theta\, d\theta$$

$$= \frac{7^3}{3^4}\left[\frac{1}{3}\sec^3\theta - \sec\theta\right] + C = \frac{7^3}{3^4}\left[\frac{(9x^2+49)^{3/2}}{3(7)^3} - \frac{\sqrt{9x^2+49}}{7}\right] + C =$$

$$\frac{1}{243}(9x^2+49)^{3/2} - \frac{49}{81}\sqrt{9x^2+49} + C$$

[18] $5x = 4\tan\theta$ or $x = \frac{4}{5}\tan\theta,\ dx = \frac{4}{5}\sec^2\theta\, d\theta \Rightarrow \text{I} = \displaystyle\int \frac{\frac{4}{5}\sec^2\theta}{(\frac{4}{5}\tan\theta)(4\sec\theta)}\, d\theta =$

$$\frac{1}{4}\int \csc\theta\, d\theta = \frac{1}{4}\ln\left|\csc\theta - \cot\theta\right| + C = \frac{1}{4}\ln\left|\frac{\sqrt{25x^2+16}}{5x} - \frac{4}{5x}\right| + C$$

[19] $x = \sqrt{3}\sec\theta,\ dx = \sqrt{3}\sec\theta\tan\theta\, d\theta \Rightarrow$

$$\text{I} = \int \frac{\sqrt{3}\sec\theta\,\tan\theta}{(9\sec^4\theta)(\sqrt{3}\tan\theta)}\, d\theta = \frac{1}{9}\int \cos^3\theta\, d\theta = \frac{1}{9}\int (1 - \sin^2\theta)\cos\theta\, d\theta$$

$$= \frac{1}{9}(\sin\theta - \frac{1}{3}\sin^3\theta) + C = \frac{1}{9}\left[\frac{\sqrt{x^2-3}}{x} - \frac{(x^2-3)^{3/2}}{3x^3}\right] + C$$

$$= \frac{1}{9}\left[\frac{3x^2(x^2-3)^{1/2} - (x^2-3)^{3/2}}{3x^3}\right] + C = \frac{(3+2x^2)\sqrt{x^2-3}}{27x^3} + C$$

[20] $3x = \sin\theta$ or $x = \frac{1}{3}\sin\theta$, $dx = \frac{1}{3}\cos\theta\,d\theta \Rightarrow$

$$\text{I} = \int \frac{(\frac{1}{3}\sin\theta)^2(\frac{1}{3}\cos\theta)}{(\cos\theta)^3}\,d\theta = \frac{1}{27}\int \tan^2\theta\,d\theta = \frac{1}{27}\int(\sec^2\theta - 1)\,d\theta =$$

$$\frac{1}{27}(\tan\theta - \theta) + C = \frac{x}{9\sqrt{1-9x^2}} - \frac{1}{27}\sin^{-1}3x + C$$

[21] $\text{I} = \int \frac{16 + 8x^2 + x^4}{x^3}\,dx = \int(16x^{-3} + 8x^{-1} + x)\,dx = -\frac{8}{x^2} + 8\ln|x| + \frac{1}{2}x^2 + C$

[22] $x = \sin\theta$, $dx = \cos\theta\,d\theta \Rightarrow \text{I} = \int \frac{3\sin\theta - 5}{\cos\theta}\cos\theta\,d\theta = \int(3\sin\theta - 5)\,d\theta =$

$$-3\cos\theta - 5\theta + C = -3\sqrt{1-x^2} - 5\sin^{-1}x + C$$

[23] Using shells, $V = \int_0^5 2\pi x\frac{x}{\sqrt{x^2+25}}\,dx = 2\pi\int_0^5 \frac{x^2}{\sqrt{x^2+25}}\,dx$. $x = 5\tan\theta \Rightarrow$

$dx = 5\sec^2\theta\,d\theta$. $x = 0, 5 \Rightarrow \theta = 0, \frac{\pi}{4}$. Thus, $V = 2\pi\int_0^{\pi/4}\frac{(25\tan^2\theta)(5\sec^2\theta)}{5\sec\theta}\,d\theta$

$$= 50\pi\int_0^{\pi/4}\tan^2\theta\sec\theta\,d\theta = 50\pi\int_0^{\pi/4}(\sec^3\theta - \sec\theta)\,d\theta \text{ \{Example 6, §9.1\}}$$

$$= 50\pi\left[(\tfrac{1}{2}\sec\theta\tan\theta + \tfrac{1}{2}\ln|\sec\theta + \tan\theta|) - \ln|\sec\theta + \tan\theta|\right]_0^{\pi/4}$$

$$= 25\pi\left[\sec\theta\tan\theta - \ln|\sec\theta + \tan\theta|\right]_0^{\pi/4} = 25\pi\left[\sqrt{2} - \ln(\sqrt{2}+1)\right] \approx 41.85$$

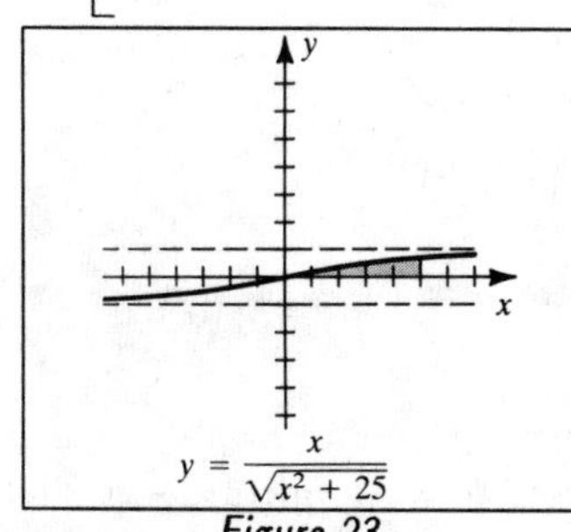

Figure 23

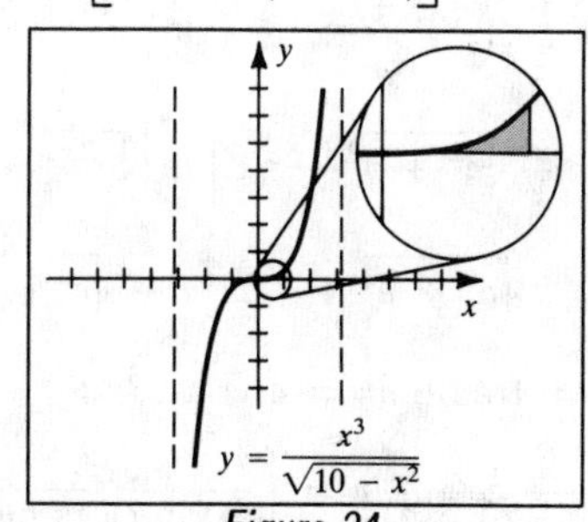

Figure 24

[24] $x = \sqrt{10}\sin\theta \Rightarrow dx = \sqrt{10}\cos\theta\,d\theta$. $x = 0, 1 \Rightarrow \theta = 0, \sin^{-1}(\frac{1}{\sqrt{10}})$ {call this α}.

Thus, $A = \int_0^1 \frac{x^3}{\sqrt{10-x^2}}\,dx = \int_0^{\alpha}\frac{(\sqrt{10}\sin\theta)^3(\sqrt{10}\cos\theta)}{\sqrt{10}\cos\theta}\,d\theta$

$$= 10\sqrt{10}\int_0^{\alpha}(1-\cos^2\theta)\sin\theta\,d\theta = 10\sqrt{10}\left[-\cos\theta + \tfrac{1}{3}\cos^3\theta\right]_0^{\alpha}$$

$$= 10\sqrt{10}\left[\left(-\frac{3}{\sqrt{10}} + \frac{1}{3}\left(\frac{3}{\sqrt{10}}\right)^3\right) - \left(-1 + \frac{1}{3}\right)\right] \left\{\sin\theta = \tfrac{1}{\sqrt{10}} \Rightarrow \cos\theta = \tfrac{3}{\sqrt{10}}\right\}$$

$$= 10\sqrt{10}\left(-\frac{21}{10\sqrt{10}} + \frac{2}{3}\right) = \frac{20}{3}\sqrt{10} - 21 \approx 0.08.$$

25 $x\,dy - \sqrt{x^2 - 16}\,dx \Rightarrow \frac{dy}{dx} = f'(x) = \frac{\sqrt{x^2 - 16}}{x} \Rightarrow f(x) = \int \frac{\sqrt{x^2 - 16}}{x}\,dx.$

$x = 4\sec\theta,\ dx = 4\sec\theta\tan\theta\,d\theta \Rightarrow$

$f(x) = \int \frac{(4\tan\theta)(4\sec\theta\,\tan\theta)}{4\sec\theta}\,d\theta = 4\int \tan^2\theta\,d\theta = 4\int (\sec^2\theta - 1)\,d\theta =$

$4(\tan\theta - \theta) + C = 4\left[\tfrac{1}{4}\sqrt{x^2 - 16} - \sec^{-1}\tfrac{x}{4}\right] + C.$

$f(4) = 0 \Rightarrow C = 0$, and $f(x) = y = \sqrt{x^2 - 16} - 4\sec^{-1}\frac{x}{4}.$

26 $\sqrt{1 - x^2}\,dy = x^3\,dx \Rightarrow \frac{dy}{dx} = f'(x) = \frac{x^3}{\sqrt{1 - x^2}} \Rightarrow f(x) = \int \frac{x^3}{\sqrt{1 - x^2}}\,dx.$

$x = \sin\theta,\ dx = \cos\theta\,d\theta \Rightarrow f(x) = \int \frac{(\sin\theta)^3(\cos\theta)}{\cos\theta}\,d\theta = \int (1 - \cos^2\theta)\sin\theta\,d\theta =$

$-\cos\theta + \tfrac{1}{3}\cos^3\theta + C = -\sqrt{1 - x^2} + \tfrac{1}{3}(1 - x^2)^{3/2} + C.\ f(0) = 0 \Rightarrow$

$-1 + \tfrac{1}{3} + C = 0 \Rightarrow C = \tfrac{2}{3}.$ Thus, $f(x) = y = -\sqrt{1 - x^2} + \tfrac{1}{3}(1 - x^2)^{3/2} + \tfrac{2}{3} =$

$\tfrac{1}{3}\{\sqrt{1 - x^2}\left[-3 + (1 - x^2)\right] + 2\} = \tfrac{1}{3}\left[2 - (x^2 + 2)\sqrt{1 - x^2}\right].$

27 $u = a\tan\theta,\ du = a\sec^2\theta\,d\theta \Rightarrow \int \sqrt{a^2 + u^2}\,du = \int (a\sec\theta)(a\sec^2\theta)\,d\theta =$

$a^2\int \sec^3\theta\,d\theta = \tfrac{1}{2}a^2(\sec\theta\,\tan\theta + \ln|\sec\theta + \tan\theta|) + C$ {see Example 6, §9.1}

$= \tfrac{1}{2}a^2\left(\frac{\sqrt{a^2 + u^2}}{a}\cdot\frac{u}{a} + \ln\left|\frac{\sqrt{a^2 + u^2}}{a} + \frac{u}{a}\right|\right) + C_1$

$= \frac{u^2}{2}\sqrt{a^2 + u^2} + \tfrac{1}{2}a^2\left[\ln\left|\sqrt{a^2 + u^2} + u\right| - \ln a\right] + C_1$

$= \frac{u}{2}\sqrt{a^2 + u^2} + \frac{a^2}{2}\ln\left|u + \sqrt{a^2 + u^2}\right| + C$, where $C = C_1 - \tfrac{1}{2}a^2\ln a$

28 $u = a\tan\theta,\ du = a\sec^2\theta\,d\theta \Rightarrow$

$\int \frac{1}{u\sqrt{a^2 + u^2}}\,du = \int \frac{a\sec^2\theta}{(a\tan\theta)(a\sec\theta)}\,d\theta = \frac{1}{a}\int \csc\theta\,d\theta = \frac{1}{a}\ln|\csc\theta - \cot\theta| + C =$

$\frac{1}{a}\ln\left|\frac{\sqrt{a^2 + u^2}}{u} - \frac{a}{u}\right| + C = -\frac{1}{a}\ln\left|\frac{u}{\sqrt{a^2 + u^2} - a}\right| + C =$

$-\frac{1}{a}\ln\left|\frac{u}{\sqrt{a^2 + u^2} - a}\cdot\frac{\sqrt{a^2 + u^2} + a}{\sqrt{a^2 + u^2} + a}\right| + C = -\frac{1}{a}\ln\left|\frac{\sqrt{a^2 + u^2} + a}{u}\right| + C$

29 $u = a\sin\theta,\ du = a\cos\theta\,d\theta \Rightarrow \int u^2\sqrt{a^2 - u^2}\,du = \int (a\sin\theta)^2(a\cos\theta)(a\cos\theta)\,d\theta =$

$a^4\int \sin^2\theta\,\cos^2\theta\,d\theta = a^4(\tfrac{1}{8}\theta - \tfrac{1}{32}\sin 4\theta) + C$ {see Exercise 3, §9.2} $=$

$a^4\left[\tfrac{1}{8}\theta - \tfrac{1}{32}(2\sin 2\theta\,\cos 2\theta)\right] + C = a^4\left[\tfrac{1}{8}\theta - \tfrac{1}{16}(2\sin\theta\,\cos\theta)(1 - 2\sin^2\theta)\right] + C$

$= \frac{a^4}{8}\sin^{-1}\frac{u}{a} - \frac{a^4}{8}\left(\frac{u}{a}\right)\left(\frac{\sqrt{a^2 - u^2}}{a}\right)\left(1 - \frac{2u^2}{a^2}\right) + C$ (cont.)

$= \frac{a^4}{8}\sin^{-1}\frac{u}{a} - \frac{a^2}{8}(u)\sqrt{a^2 - u^2}\left(\frac{a^2 - 2u^2}{a^2}\right) + C$

$= \frac{a^4}{8}\sin^{-1}\frac{u}{a} + \frac{u}{8}(2u^2 - a^2)\sqrt{a^2 - u^2} + C$

[30] $u = a\sin\theta,\ du = a\cos\theta\, d\theta \Rightarrow \int \frac{1}{u^2\sqrt{a^2 - u^2}}\, du = \int \frac{a\cos\theta}{(a\sin\theta)^2(a\cos\theta)}\, d\theta =$

$\frac{1}{a^2}\int \csc^2\theta\, d\theta = -\frac{1}{a^2}\cot\theta + C = -\frac{\sqrt{a^2 - u^2}}{a^2 u} + C$

[31] $u = a\sec\theta,\ du = a\sec\theta\tan\theta\, d\theta \Rightarrow \int \frac{\sqrt{u^2 - a^2}}{u}\, du = \int \frac{a\tan\theta}{a\sec\theta}\, a\sec\theta\tan\theta\, d\theta =$

$a\int \tan^2\theta\, d\theta = a\int(\sec^2\theta - 1)\, d\theta = a\tan\theta - a\theta + C = \sqrt{u^2 - a^2} - a\sec^{-1}\frac{u}{a} + C$

[32] $u = a\sec\theta,\ du = a\sec\theta\tan\theta\, d\theta \Rightarrow \int \frac{u^2}{\sqrt{u^2 - a^2}}\, du = \int \frac{(a\sec\theta)^2(a\sec\theta\tan\theta)}{a\tan\theta}\, d\theta =$

$a^2\int \sec^3\theta\, d\theta = \frac{1}{2}a^2(\sec\theta\tan\theta + \ln|\sec\theta + \tan\theta|) + C_1$ {see Example 6, §9.1} $=$

$\frac{1}{2}a^2\left(\frac{u}{a}\cdot\frac{\sqrt{u^2 - a^2}}{a} + \ln\left|\frac{u}{a} + \frac{\sqrt{u^2 - a^2}}{a}\right|\right) + C_1 =$

$\frac{u}{2}\sqrt{u^2 - a^2} + \frac{a^2}{2}\ln\left|u + \sqrt{u^2 - a^2}\right| + C$, where $C = C_1 - \frac{1}{2}a^2\ln a$.

Exercises 9.4

Note: In this section, K denotes the constant of integration.

[1] $\frac{5x - 12}{x(x - 4)} = \frac{A}{x} + \frac{B}{x - 4} \Rightarrow 5x - 12 = A(x - 4) + Bx.\ x = 0 \Rightarrow A = 3$

and $x = 4 \Rightarrow B = 2$. Thus, $I = \int\left[\frac{3}{x} + \frac{2}{x - 4}\right]dx = 3\ln|x| + 2\ln|x - 4| + K.$

[2] $\frac{x + 34}{(x - 6)(x + 2)} = \frac{A}{x - 6} + \frac{B}{x + 2} \Rightarrow x + 34 = A(x + 2) + B(x - 6).$

$x = 6 \Rightarrow A = 5$ and $x = -2 \Rightarrow B = -4.$

Thus, $I = \int\left[\frac{5}{x - 6} + \frac{-4}{x + 2}\right]dx = 5\ln|x - 6| - 4\ln|x + 2| + K.$

[3] $\frac{37 - 11x}{(x + 1)(x - 2)(x - 3)} = \frac{A}{x + 1} + \frac{B}{x - 2} + \frac{C}{x - 3} \Rightarrow$

$37 - 11x = A(x - 2)(x - 3) + B(x + 1)(x - 3) + C(x + 1)(x - 2).$

$x = -1 \Rightarrow A = 4,\ x = 2 \Rightarrow B = -5$, and $x = 3 \Rightarrow C = 1$. Thus,

$I = \int\left[\frac{4}{x + 1} + \frac{-5}{x - 2} + \frac{1}{x - 3}\right]dx = 4\ln|x + 1| - 5\ln|x - 2| + \ln|x - 3| + K.$

[4] $\frac{4x^2 + 54x + 134}{(x - 1)(x + 5)(x + 3)} = \frac{A}{x - 1} + \frac{B}{x + 5} + \frac{C}{x + 3} \Rightarrow$

$4x^2 + 54x + 134 = A(x + 5)(x + 3) + B(x - 1)(x + 3) + C(x - 1)(x + 5).$

$x = 1 \Rightarrow 192 = 24A \Rightarrow A = 8,\ x = -5 \Rightarrow -36 = 12B \Rightarrow$

$B = -3$, and $x = -3 \Rightarrow 8 = -8C \Rightarrow C = -1$. Thus,

$I = \int\left[\frac{8}{x - 1} + \frac{-3}{x + 5} + \frac{-1}{x + 3}\right]dx = 8\ln|x - 1| - 3\ln|x + 5| - \ln|x + 3| + K.$

5 $\dfrac{6x-11}{(x-1)^2} = \dfrac{A}{x-1} + \dfrac{B}{(x-1)^2} \Rightarrow 6x - 11 = A(x-1) + B.$

$x = 1 \Rightarrow B = -5.$ Equating x terms: $6 = A.$

$$\text{Thus, I} = \int\left[\frac{6}{x-1} + \frac{-5}{(x-1)^2}\right]dx = 6\ln|x-1| + \frac{5}{x-1} + K.$$

6 $\dfrac{-19x^2+50x-25}{x^2(3x-5)} = \dfrac{A}{x} + \dfrac{B}{x^2} + \dfrac{C}{3x-5} \Rightarrow -19x^2 + 50x - 25 =$

$Ax(3x-5) + B(3x-5) + Cx^2.$ $x = 0 \Rightarrow B = 5$ and $x = \frac{5}{3} \Rightarrow$

$\frac{50}{9} = C(\frac{25}{9}) \Rightarrow C = 2.$ Equating x^2 terms: $-19 = 3A + C \Rightarrow A = -7.$

$$\text{Thus, I} = \int\left[\frac{-7}{x} + \frac{5}{x^2} + \frac{2}{3x-5}\right]dx = -7\ln|x| - \frac{5}{x} + \frac{2}{3}\ln|3x-5| + K.$$

7 $\dfrac{x+16}{x^2+2x-8} = \dfrac{x+16}{(x-2)(x+4)} = \dfrac{A}{x-2} + \dfrac{B}{x+4} \Rightarrow$

$x + 16 = A(x+4) + B(x-2).$ $x = 2 \Rightarrow A = 3$ and $x = -4 \Rightarrow B = -2.$

$$\text{Thus, I} = \int\left[\frac{3}{x-2} + \frac{-2}{x+4}\right]dx = 3\ln|x-2| - 2\ln|x+4| + K.$$

8 $\dfrac{11x+2}{2x^2-5x-3} = \dfrac{11x+2}{(2x+1)(x-3)} = \dfrac{A}{2x+1} + \dfrac{B}{x-3} \Rightarrow$

$11x + 2 = A(x-3) + B(2x+1).$ $x = -\frac{1}{2} \Rightarrow A = 1$ and $x = 3 \Rightarrow B = 5.$

$$\text{Thus, I} = \int\left[\frac{1}{2x+1} + \frac{5}{x-3}\right]dx = \frac{1}{2}\ln|2x+1| + 5\ln|x-3| + K.$$

9 $\dfrac{5x^2-10x-8}{x^3-4x} = \dfrac{5x^2-10x-8}{x(x-2)(x+2)} = \dfrac{A}{x} + \dfrac{B}{x-2} + \dfrac{C}{x+2} \Rightarrow$

$5x^2 - 10x - 8 = A(x-2)(x+2) + Bx(x+2) + Cx(x-2).$

$x = 0 \Rightarrow A = 2,$ $x = 2 \Rightarrow B = -1,$ and $x = -2 \Rightarrow C = 4.$

$$\text{Thus, I} = \int\left[\frac{2}{x} + \frac{-1}{x-2} + \frac{4}{x+2}\right]dx = 2\ln|x| - \ln|x-2| + 4\ln|x+2| + K.$$

10 $\dfrac{4x^2-5x-15}{x^3-4x^2-5x} = \dfrac{4x^2-5x-15}{x(x-5)(x+1)} = \dfrac{A}{x} + \dfrac{B}{x-5} + \dfrac{C}{x+1} \Rightarrow$

$4x^2 - 5x - 15 = A(x-5)(x+1) + Bx(x+1) + Cx(x-5).$

$x = 0 \Rightarrow A = 3,$ $x = 5 \Rightarrow B = 2,$ and $x = -1 \Rightarrow C = -1.$

$$\text{Thus, I} = \int\left[\frac{3}{x} + \frac{2}{x-5} + \frac{-1}{x+1}\right]dx = 3\ln|x| + 2\ln|x-5| - \ln|x+1| + K.$$

11 $\dfrac{2x^2-25x-33}{(x+1)^2(x-5)} = \dfrac{A}{x+1} + \dfrac{B}{(x+1)^2} + \dfrac{C}{x-5} \Rightarrow$

$2x^2 - 25x - 33 = A(x+1)(x-5) + B(x-5) + C(x+1)^2.$

$x = -1 \Rightarrow B = 1$ and $x = 5 \Rightarrow -108 = 36C \Rightarrow C = -3.$

Equating x^2 terms: $2 = A + C \Rightarrow A = 5.$ Thus,

$$\text{I} = \int\left[\frac{5}{x+1} + \frac{1}{(x+1)^2} + \frac{-3}{x-5}\right]dx = 5\ln|x+1| - \frac{1}{x+1} - 3\ln|x-5| + K.$$

[12] $\dfrac{2x^2 - 12x + 4}{x^3 - 4x^2} = \dfrac{2x^2 - 12x + 4}{x^2(x-4)} = \dfrac{A}{x} + \dfrac{B}{x^2} + \dfrac{C}{x-4} \Rightarrow$

$2x^2 - 12x + 4 = Ax(x-4) + B(x-4) + Cx^2$. $x = 0 \Rightarrow B = -1$ and

$x = 4 \Rightarrow -12 = 16C \Rightarrow C = -\frac{3}{4}$. Equating x^2 terms: $2 = A + C \Rightarrow A = \frac{11}{4}$.

$$\text{Thus, I} = \int \left[\frac{\frac{11}{4}}{x} + \frac{-1}{x^2} + \frac{-\frac{3}{4}}{x-4}\right] dx = \frac{11}{4}\ln|x| + \frac{1}{x} - \frac{3}{4}\ln|x-4| + K.$$

[13] $\dfrac{9x^4 + 17x^3 + 3x^2 - 8x + 3}{x^4(x+3)} = \dfrac{A}{x} + \dfrac{B}{x^2} + \dfrac{C}{x^3} + \dfrac{D}{x^4} + \dfrac{E}{x+3} \Rightarrow$

$9x^4 + 17x^3 + 3x^2 - 8x + 3 =$

$$Ax^3(x+3) + Bx^2(x+3) + Cx(x+3) + D(x+3) + Ex^4.$$

$x = 0 \Rightarrow 3 = 3D \Rightarrow D = 1$ and $x = -3 \Rightarrow 324 = 81E \Rightarrow E = 4$.

By equating coefficients of like powers we have: x^4: $9 = A + E \Rightarrow A = 5$;

x^3: $17 = 3A + B \Rightarrow B = 2$; x^2: $3 = 3B + C \Rightarrow C = -3$.

$$\text{Thus, I} = \int \left[\frac{5}{x} + \frac{2}{x^2} + \frac{-3}{x^3} + \frac{1}{x^4} + \frac{4}{x+3}\right] dx =$$

$$5\ln|x| - \frac{2}{x} + \frac{3}{2x^2} - \frac{1}{3x^3} + 4\ln|x+3| + K.$$

[14] $\dfrac{5x^2 + 30x + 43}{(x+3)^3} = \dfrac{A}{x+3} + \dfrac{B}{(x+3)^2} + \dfrac{C}{(x+3)^3} \Rightarrow$

$5x^2 + 30x + 43 = A(x+3)^2 + B(x+3) + C$. $x = -3 \Rightarrow C = -2$.

Equating x^2 terms: $5 = A$, and x terms: $30 = 6A + B \Rightarrow B = 0$.

$$\text{Thus, I} = \int \left[\frac{5}{x+3} + \frac{-2}{(x+3)^3}\right] dx = 5\ln|x+3| + \frac{1}{(x+3)^2} + K.$$

[15] Since the degree of the numerator is greater than or equal to the degree of the denominator, we must first use long division to change the form of the integrand.

$\dfrac{x^3 + 6x^2 + 3x + 16}{x^3 + 4x} = 1 + \dfrac{6x^2 - x + 16}{x^3 + 4x}$; $\dfrac{6x^2 - x + 16}{x(x^2+4)} = \dfrac{A}{x} + \dfrac{Bx + C}{x^2 + 4} \Rightarrow$

$6x^2 - x + 16 = A(x^2 + 4) + (Bx + C)x$. $x = 0 \Rightarrow 16 = 4A \Rightarrow A = 4$.

Equating x terms: $-1 = C$, and x^2 terms: $6 = A + B \Rightarrow B = 2$.

$$\text{Thus, I} = \int \left[1 + \frac{4}{x} + \frac{2x - 1}{x^2 + 4}\right] dx = x + 4\ln|x| + \ln(x^2 + 4) - \frac{1}{2}\tan^{-1}\left(\frac{x}{2}\right) + K.$$

[16] $\dfrac{2x^2 + 7x}{x^2 + 6x + 9} = 2 + \dfrac{-5x - 18}{x^2 + 6x + 9}$; $\dfrac{-5x - 18}{(x+3)^2} = \dfrac{A}{x+3} + \dfrac{B}{(x+3)^2} \Rightarrow$

$-5x - 18 = A(x+3) + B$. $x = -3 \Rightarrow -3 = B$. Equating x terms: $-5 = A$.

$$\text{Thus, I} = \int \left[2 + \frac{-5}{x+3} + \frac{-3}{(x+3)^2}\right] dx = 2x - 5\ln|x+3| + \frac{3}{x+3} + K.$$

[17] Since $x^3 + 5x^2 + 4x + 20 = x^2(x + 5) + 4(x + 5) = (x^2 + 4)(x + 5)$, we have

$\dfrac{5x^2 + 11x + 17}{(x^2 + 4)(x + 5)} = \dfrac{Ax + B}{x^2 + 4} + \dfrac{C}{x + 5} \Rightarrow 5x^2 + 11x + 17 =$

$(Ax + B)(x + 5) + C(x^2 + 4)$. $x = -5 \Rightarrow 87 = 29C \Rightarrow C = 3$.

Equating x^2 terms: $5 = A + C \Rightarrow A = 2$, and x terms: $11 = B + 5A \Rightarrow B = 1$.

$$\text{Thus, I} = \int\left[\frac{2x + 1}{x^2 + 4} + \frac{3}{x + 5}\right]dx = \ln(x^2 + 4) + \tfrac{1}{2}\tan^{-1}\left(\tfrac{x}{2}\right) + 3\ln|x + 5| + K.$$

[18] $\dfrac{4x^3 - 3x^2 + 6x - 27}{x^2(x^2 + 9)} = \dfrac{A}{x} + \dfrac{B}{x^2} + \dfrac{Cx + D}{x^2 + 9} \Rightarrow 4x^3 - 3x^2 + 6x - 27 =$

$Ax(x^2 + 9) + B(x^2 + 9) + (Cx + D)x^2$. $x = 0 \Rightarrow -27 = 9B \Rightarrow B = -3$.

Equating coefficients we have: x^2: $-3 = B + D \Rightarrow D = 0$;

x: $6 = 9A \Rightarrow A = \frac{2}{3}$; x^3: $4 = A + C \Rightarrow C = \frac{10}{3}$.

$$\text{Thus, I} = \int\left[\frac{\frac{2}{3}}{x} + \frac{-3}{x^2} + \frac{\frac{10}{3}x}{x^2 + 9}\right]dx = \tfrac{2}{3}\ln|x| + \frac{3}{x} + \tfrac{5}{3}\ln(x^2 + 9) + K.$$

[19] $\dfrac{x^2 + 3x + 1}{(x^2 + 4)(x^2 + 1)} = \dfrac{Ax + B}{x^2 + 4} + \dfrac{Cx + D}{x^2 + 1} \Rightarrow$

$x^2 + 3x + 1 = (Ax + B)(x^2 + 1) + (Cx + D)(x^2 + 4)$.

Equating coefficients $\{x^0$ is used for constants$\}$ we have:

x^3: $0 = A + C$ and x: $3 = A + 4C \Rightarrow C = 1$ and $A = -1$;

x^2: $1 = B + D$ and x^0: $1 = B + 4D \Rightarrow D = 0$ and $B = 1$. Thus,

$$\text{I} = \int\left[\frac{-x + 1}{x^2 + 4} + \frac{x}{x^2 + 1}\right]dx = -\tfrac{1}{2}\ln(x^2 + 4) + \tfrac{1}{2}\tan^{-1}\left(\tfrac{x}{2}\right) + \tfrac{1}{2}\ln(x^2 + 1) + K.$$

[20] $u = x^2 + 1$, $2\,du = 4x\,dx \Rightarrow \text{I} = 2\int u^{-3}\,du = -\dfrac{1}{u^2} + K$.

[21] $\dfrac{2x^3 + 10x}{(x^2 + 1)^2} = \dfrac{Ax + B}{x^2 + 1} + \dfrac{Cx + D}{(x^2 + 1)^2} \Rightarrow$

$2x^3 + 10x = (Ax + B)(x^2 + 1) + Cx + (B + D)$. Equating coefficients we have:

x^3: $2 = A$; x^2: $0 = B$; x^0: $0 = B + D \Rightarrow D = 0$; x: $10 = A + C \Rightarrow C = 8$.

$$\text{Thus, I} = \int\left[\frac{2x}{x^2 + 1} + \frac{8x}{(x^2 + 1)^2}\right]dx = \ln(x^2 + 1) - \frac{4}{x^2 + 1} + K.$$

[22] $\dfrac{x^4 + 2x^2 + 4x + 1}{(x^2 + 1)^3} = \dfrac{Ax + B}{x^2 + 1} + \dfrac{Cx + D}{(x^2 + 1)^2} + \dfrac{Ex + F}{(x^2 + 1)^3} \Rightarrow$

$x^4 + 2x^2 + 4x + 1 = (Ax + B)(x^2 + 1)^2 + (Cx + D)(x^2 + 1) + (Ex + F)$.

Equating coefficients we have: x^5: $0 = A$; x^4: $1 = B$; x^3: $0 = 2A + C \Rightarrow C = 0$; x^2: $2 = 2B + D \Rightarrow D = 0$; x: $4 = A + E \Rightarrow E = 4$; x^0: $1 = B + D + F \Rightarrow F = 0$.

$$\text{Thus, I} = \int\left[\frac{1}{x^2 + 1} + \frac{4x}{(x^2 + 1)^3}\right]dx = \tan^{-1}x - \frac{1}{(x^2 + 1)^2} + K.$$

[23] $\dfrac{x^3 + 3x - 2}{x^2 - x} = x + 1 + \dfrac{4x - 2}{x^2 - x}$. Thus, $\text{I} = \int(x + 1)\,dx + 2\displaystyle\int\frac{2x - 1}{x^2 - x}\,dx =$

$$\tfrac{1}{2}x^2 + x + 2\ln|x^2 - x| + K = \tfrac{1}{2}x^2 + x + 2\ln|x| + 2\ln|x - 1| + K.$$

[24] $\dfrac{x^4+2x^2+3}{x^3-4x} = x + \dfrac{6x^2+3}{x^3-4x}$; $\dfrac{6x^2+3}{x^3-4x} = \dfrac{A}{x} + \dfrac{B}{x-2} + \dfrac{C}{x+2} \Rightarrow$

$6x^2+3 = A(x-2)(x+2) + Bx(x+2) + Cx(x-2)$. $x=0 \Rightarrow A = -\frac{3}{4}$,

$x = 2 \Rightarrow 27 = 8B \Rightarrow B = \frac{27}{8}$, and $x = -2 \Rightarrow 27 = 8C \Rightarrow C = \frac{27}{8}$.

Thus, I $= \displaystyle\int\left[x - \frac{\frac{3}{4}}{x} + \frac{\frac{27}{8}}{x-2} + \frac{\frac{27}{8}}{x+2}\right]dx =$

$$\tfrac{1}{2}x^2 - \tfrac{3}{4}\ln|x| + \tfrac{27}{8}\ln|x-2| + \tfrac{27}{8}\ln|x+2| + K.$$

[25] $\dfrac{x^6-x^3+1}{x^4+9x^2} = x^2 - 9 + \dfrac{-x^3+81x^2+1}{x^4+9x^2}$;

$\dfrac{-x^3+81x^2+1}{x^2(x^2+9)} = \dfrac{A}{x} + \dfrac{B}{x^2} + \dfrac{Cx+D}{x^2+9} \Rightarrow$

$-x^3+81x^2+1 = Ax(x^2+9) + B(x^2+9) + (Cx+D)x^2$. Equating coefficients we have: x^0: $1 = 9B \Rightarrow B = \frac{1}{9}$; x: $0 = 9A \Rightarrow A = 0$; x^2: $81 = B + D \Rightarrow D = \frac{728}{9}$;

x^3: $-1 = A + C \Rightarrow C = -1$. Thus, I $= \displaystyle\int\left[x^2 - 9 + \frac{\frac{1}{9}}{x^2} + \frac{-x+\frac{728}{9}}{x^2+9}\right]dx =$

$$\tfrac{1}{3}x^3 - 9x - \frac{1}{9x} - \tfrac{1}{2}\ln(x^2+9) + \tfrac{728}{27}\tan^{-1}(x/3) + K.$$

[26] $\dfrac{x^5}{(x^2+4)^2} = \dfrac{x^5}{x^4+8x^2+16} = x - \dfrac{8x^3+16x}{x^4+8x^2+16}$;

$\dfrac{8x^3+16x}{(x^2+4)^2} = \dfrac{Ax+B}{x^2+4} + \dfrac{Cx+D}{(x^2+4)^2} \Rightarrow$

$$8x^3+16x = (Ax+B)(x^2+4) + (Cx+D).$$

Equating coefficients we have:

x^3: $8 = A$; x^2: $0 = B$; x: $16 = 4A + C \Rightarrow C = -16$; x^0: $0 = 4B + D \Rightarrow D = 0$.

Thus, I $= \displaystyle\int\left[x - \frac{8x}{x^2+4} + \frac{16x}{(x^2+4)^2}\right]dx = \frac{1}{2}x^2 - 4\ln(x^2+4) - \frac{8}{x^2+4} + K.$

[27] $\dfrac{2x^3-5x^2+46x+98}{(x+4)^2(x-3)^2} = \dfrac{A}{x+4} + \dfrac{B}{(x+4)^2} + \dfrac{C}{x-3} + \dfrac{D}{(x-3)^2} \Rightarrow$

$2x^3 - 5x^2 + 46x + 98 =$

$$A(x+4)(x-3)^2 + B(x-3)^2 + C(x+4)^2(x-3) + D(x+4)^2.$$

$x = -4 \Rightarrow -294 = 49B \Rightarrow B = -6$ and $x = 3 \Rightarrow 245 = 49D \Rightarrow D = 5$.

Equating x^3 terms: $2 = A + C$, and x^0 terms: $98 = 36A + 9B - 48C + 16D \Rightarrow$

$72 = 36A - 48C \Rightarrow 6 = 3A - 4C$. Solving yields $A = 2$ and $C = 0$. Thus,

$$\text{I} = \int\left[\frac{2}{x+4} + \frac{-6}{(x+4)^2} + \frac{5}{(x-3)^2}\right]dx = 2\ln|x+4| + \frac{6}{x+4} - \frac{5}{x-3} + K.$$

28 $\dfrac{-2x^4 - 3x^3 - 3x^2 + 3x + 1}{x^2(x + 1)^3} = \dfrac{A}{x} + \dfrac{B}{x^2} + \dfrac{C}{x + 1} + \dfrac{D}{(x + 1)^2} + \dfrac{E}{(x + 1)^3} \Rightarrow$

$-2x^4 - 3x^3 - 3x^2 + 3x + 1 =$

$$Ax(x + 1)^3 + B(x + 1)^3 + Cx^2(x + 1)^2 + Dx^2(x + 1) + Ex^2.$$

$x = 0 \Rightarrow B = 1$ and $x = -1 \Rightarrow E = -4$.

Equating coefficients we have: x: $3 = A + 3B \Rightarrow A = 0$;

x^4: $-2 = A + C \Rightarrow C = -2$; x^2: $-3 = 3A + 3B + C + D + E \Rightarrow D = 0$.

Thus, $I = \displaystyle\int\left[\frac{1}{x^2} + \frac{-2}{x + 1} + \frac{-4}{(x + 1)^3}\right]dx = -\frac{1}{x} - 2\ln|x + 1| + \frac{2}{(x + 1)^2} + K.$

29 $\dfrac{4x^3 + 2x^2 - 5x - 18}{(x - 4)(x + 1)^3} = \dfrac{A}{x - 4} + \dfrac{B}{x + 1} + \dfrac{C}{(x + 1)^2} + \dfrac{D}{(x + 1)^3} \Rightarrow$

$4x^3 + 2x^2 - 5x - 18 =$

$$A(x + 1)^3 + B(x - 4)(x + 1)^2 + C(x - 4)(x + 1) + D(x - 4).$$

$x = -1 \Rightarrow -15 = -5D \Rightarrow D = 3$ and $x = 4 \Rightarrow 250 = 125A \Rightarrow A = 2$.

Equating coefficients we have: x^3: $4 = A + B \Rightarrow B = 2$;

x^0: $-18 = A - 4B - 4C - 4D \Rightarrow C = 0$. Thus, $I =$

$$\int\left[\frac{2}{x - 4} + \frac{2}{x + 1} + \frac{3}{(x + 1)^3}\right]dx = 2\ln|x - 4| + 2\ln|x + 1| - \frac{3}{2(x + 1)^2} + K.$$

30 $\dfrac{10x^2 + 9x + 1}{x(x + 1)(2x + 1)} = \dfrac{A}{x} + \dfrac{B}{x + 1} + \dfrac{C}{2x + 1} \Rightarrow$

$10x^2 + 9x + 1 = A(x + 1)(2x + 1) + Bx(2x + 1) + Cx(x + 1)$.

$x = 0 \Rightarrow A = 1$, $x = -1 \Rightarrow B = 2$, and $x = -\frac{1}{2} \Rightarrow -1 = -\frac{1}{4}C \Rightarrow C = 4$.

Thus, $I = \displaystyle\int\left[\frac{1}{x} + \frac{2}{x + 1} + \frac{4}{2x + 1}\right]dx = \ln|x| + 2\ln|x + 1| + 2\ln|2x + 1| + K.$

31 $\dfrac{x^3 + 3x^2 + 3x + 63}{(x - 3)^2(x + 3)^2} = \dfrac{A}{x - 3} + \dfrac{B}{(x - 3)^2} + \dfrac{C}{x + 3} + \dfrac{D}{(x + 3)^2} \Rightarrow$

$x^3 + 3x^2 + 3x + 63 =$

$$A(x - 3)(x + 3)^2 + B(x + 3)^2 + C(x - 3)^2(x + 3) + D(x - 3)^2.$$

$x = 3 \Rightarrow 126 = 36B \Rightarrow B = \frac{7}{2}$ and $x = -3 \Rightarrow 54 = 36D \Rightarrow D = \frac{3}{2}$.

Equating x^3 terms: $1 = A + C$, and x^0 terms: $63 = -27A + 9B + 27C + 9D \Rightarrow$

$7 = -3A + B + 3C + D \Rightarrow 3A - 3C = -2$. Solving yields $A = \frac{1}{6}$ and $C = \frac{5}{6}$.

Thus, $I = \displaystyle\int\left[\frac{\frac{1}{6}}{x - 3} + \frac{\frac{7}{2}}{(x - 3)^2} + \frac{\frac{5}{6}}{x + 3} + \frac{\frac{3}{2}}{(x + 3)^2}\right]dx =$

$$\frac{1}{6}\ln|x - 3| - \frac{7}{2(x - 3)} + \frac{5}{6}\ln|x + 3| - \frac{3}{2(x + 3)} + K.$$

32 $\dfrac{x^5 - x^4 - 2x^3 + 4x^2 - 15x + 5}{(x^2 + 1)^2(x^2 + 4)} = \dfrac{Ax + B}{x^2 + 1} + \dfrac{Cx + D}{(x^2 + 1)^2} + \dfrac{Ex + F}{x^2 + 4} \Rightarrow$

$x^5 - x^4 - 2x^3 + 4x^2 - 15x + 5 =$

$(Ax + B)(x^2 + 1)(x^2 + 4) + (Cx + D)(x^2 + 4) + (Ex + F)(x^2 + 1)^2 =$

$(A + E)x^5 + (B + F)x^4 + (5A + C + 2E)x^3 + (5B + D + 2F)x^2 +$

$(4A + 4C + E)x + (4B + 4D + F).$

Equating coefficients and solving yields $A = 0$, $B = 2$, $C = -4$, $D = 0$, $E = 1$, and

$F = -3$. Thus, $\mathrm{I} = \displaystyle\int\left[\frac{2}{x^2 + 1} + \frac{-4x}{(x^2 + 1)^2} + \frac{x - 3}{x^2 + 4}\right]dx =$

$2\tan^{-1}x + \dfrac{2}{x^2 + 1} + \frac{1}{2}\ln(x^2 + 4) - \frac{3}{2}\tan^{-1}\left(\frac{x}{2}\right) + K.$

33 $\dfrac{1}{a^2 - u^2} = \dfrac{A}{a + u} + \dfrac{B}{a - u} \Rightarrow 1 = A(a - u) + B(a + u).$

$u = -a \Rightarrow A = \frac{1}{2a}$ and $u = a \Rightarrow B = \frac{1}{2a}$. Thus, I =

$\dfrac{1}{2a}\displaystyle\int\left[\frac{1}{a + u} + \frac{1}{a - u}\right]du = \frac{1}{2a}(\ln|a + u| - \ln|a - u|) + K = \frac{1}{2a}\ln\left|\frac{a + u}{a - u}\right| + K.$

34 $\dfrac{1}{u(a + bu)} = \dfrac{A}{u} + \dfrac{B}{a + bu} \Rightarrow 1 = A(a + bu) + Bu.$

$u = 0 \Rightarrow A = \frac{1}{a}$ and $u = -\frac{a}{b} \Rightarrow B = -\frac{b}{a}$. Thus,

$\mathrm{I} = \dfrac{1}{a}\displaystyle\int\left[\frac{1}{u} + \frac{-b}{a + bu}\right]du = \frac{1}{a}(\ln|u| - \ln|a + bu|) + K = \frac{1}{a}\ln\left|\frac{u}{a + bu}\right| + K.$

35 $\dfrac{1}{u^2(a + bu)} = \dfrac{A}{u} + \dfrac{B}{u^2} + \dfrac{C}{a + bu} \Rightarrow 1 = Au(a + bu) + B(a + bu) + Cu^2.$

$u = 0 \Rightarrow B = \frac{1}{a}$ and $u = -\frac{a}{b} \Rightarrow C = \frac{b^2}{a^2}$. Equating u terms: $0 = Aa + Bb =$

$Aa + (\frac{1}{a})b \Rightarrow A = -\frac{b}{a^2}$. Thus, $\mathrm{I} = \displaystyle\int\left[\frac{-b/a^2}{u} + \frac{1/a}{u^2} + \frac{(b/a)^2}{a + bu}\right]du =$

$-\frac{b}{a^2}\ln|u| - \frac{1}{au} + \frac{b}{a^2}\ln|a + bu| + K = -\frac{1}{au} + \frac{b}{a^2}\ln\left|\frac{a + bu}{u}\right| + K.$

36 $\dfrac{1}{u(a + bu)^2} = \dfrac{A}{u} + \dfrac{B}{a + bu} + \dfrac{C}{(a + bu)^2} \Rightarrow$

$1 = A(a + bu)^2 + Bu(a + bu) + Cu$. $u = 0 \Rightarrow A = \frac{1}{a^2}$ and $u = -\frac{a}{b} \Rightarrow$

$C = -\frac{b}{a}$. Equating u^2 terms: $0 = Ab^2 + Bb \Rightarrow B = -\frac{b}{a^2}$.

Thus, $\mathrm{I} = \displaystyle\int\left[\frac{1/a^2}{u} + \frac{-b/a^2}{a + bu} + \frac{-b/a}{(a + bu)^2}\right]du =$

$\frac{1}{a^2}\ln|u| - \frac{1}{a^2}\ln|a + bu| + \dfrac{1}{a(a + bu)} + K = \dfrac{1}{a(a + bu)} - \frac{1}{a^2}\ln\left|\frac{a + bu}{u}\right| + K.$

[37] $f(x) = \dfrac{x}{x^2 - 2x - 3} = \dfrac{x}{(x-3)(x+1)} = \dfrac{A}{x-3} + \dfrac{B}{x+1} \Rightarrow$

$x = A(x+1) + B(x-3)$. $x = 3 \Rightarrow A = \frac{3}{4}$ and $x = -1 \Rightarrow B = \frac{1}{4}$.

Since $f(x) \le 0$ on $[0, 2]$, $A = -\displaystyle\int_0^2 f(x)\,dx = -\int_0^2 \left[\frac{\frac{3}{4}}{x-3} + \frac{\frac{1}{4}}{x+1}\right] dx =$

$$-\left[\tfrac{3}{4}\ln|x-3| + \tfrac{1}{4}\ln|x+1|\right]_0^2 = -\left[(\tfrac{1}{4}\ln 3) - (\tfrac{3}{4}\ln 3)\right] = \tfrac{1}{2}\ln 3 \approx 0.55.$$

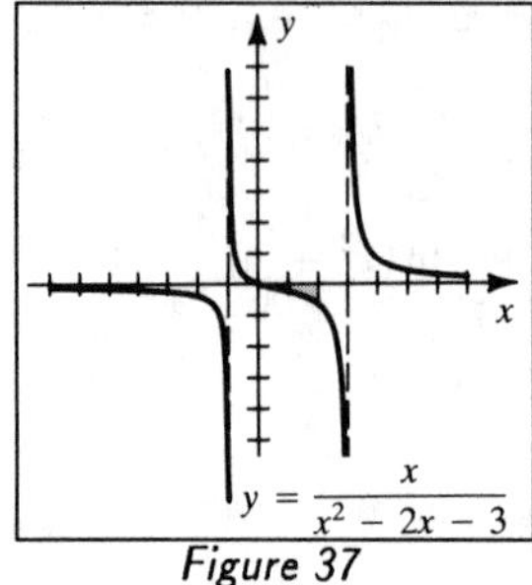

Figure 37

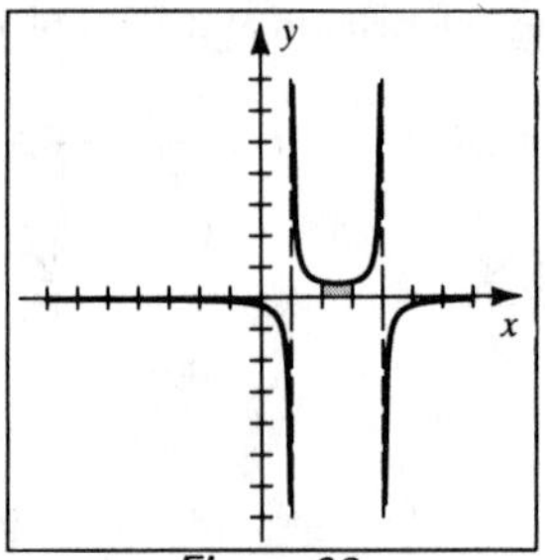

Figure 38

[38] Using shells, $V = 2\pi \displaystyle\int_2^3 \frac{x}{(x-1)(4-x)}\,dx$. $\dfrac{x}{(x-1)(4-x)} = \dfrac{A}{x-1} + \dfrac{B}{4-x} \Rightarrow$

$x = A(4-x) + B(x-1)$. $x = 1 \Rightarrow A = \frac{1}{3}$ and $x = 4 \Rightarrow B = \frac{4}{3}$.

Thus, $V = 2\pi \displaystyle\int_2^3 \left[\frac{\frac{1}{3}}{x-1} + \frac{\frac{4}{3}}{4-x}\right] dx = 2\pi\left[\frac{1}{3}\ln|x-1| - \frac{4}{3}\ln|4-x|\right]_2^3 =$

$$2\pi\left[(\tfrac{1}{3}\ln 2) - (-\tfrac{4}{3}\ln 2)\right] = \tfrac{10\pi}{3}\ln 2 \approx 7.26.$$

[39] Using disks, $V = \pi \displaystyle\int_2^3 \frac{1}{(x-1)^2(4-x)^2}\,dx$.

$\dfrac{1}{(x-1)^2(4-x)^2} = \dfrac{A}{x-1} + \dfrac{B}{(x-1)^2} + \dfrac{C}{4-x} + \dfrac{D}{(4-x)^2} \Rightarrow$

$1 = A(x-1)(4-x)^2 + B(4-x)^2 + C(x-1)^2(4-x) + D(x-1)^2$.

$x = 1 \Rightarrow B = \frac{1}{9}$ and $x = 4 \Rightarrow D = \frac{1}{9}$. Equating x^3 terms: $0 = A - C \Rightarrow A = C$,

and x^0 terms: $1 = -16A + 16B + 4C + D$. Solving yields $A = C = \frac{2}{27}$.

Thus, $V = \pi \displaystyle\int_2^3 \left[\frac{\frac{2}{27}}{x-1} + \frac{\frac{1}{9}}{(x-1)^2} + \frac{\frac{2}{27}}{4-x} + \frac{\frac{1}{9}}{(4-x)^2}\right] dx =$

$\pi\left[\dfrac{2}{27}\ln|x-1| - \dfrac{1}{9(x-1)} - \dfrac{2}{27}\ln|4-x| + \dfrac{1}{9(4-x)}\right]_2^3 =$

$$\pi\left[(\tfrac{2}{27}\ln 2 + \tfrac{1}{18}) - (-\tfrac{2}{27}\ln 2 - \tfrac{1}{18})\right] = \tfrac{\pi}{27}(4\ln 2 + 3) \approx 0.67.$$

40 We assume A, $f(t)$ and $B - f(t)$ are positive, and write $\dfrac{f'(t)}{f(t)\left[B - f(t)\right]} = A$.

$$\frac{f'(t)}{f(t)\left[B - f(t)\right]} = \frac{D}{f(t)} + \frac{E}{B - f(t)} \Rightarrow f'(t) = D\left[B - f(t)\right] + Ef(t).$$

$$f(t) = 0 \Rightarrow f'(t) = DB \Rightarrow D = \frac{f'(t)}{B}. \ f(t) = B \Rightarrow f'(t) = EB \Rightarrow E = \frac{f'(t)}{B}.$$

$$A = \frac{f'(t)}{f(t)\left[B - f(t)\right]} = \frac{f'(t)}{Bf(t)} + \frac{f'(t)}{B\left[B - f(t)\right]} \Rightarrow \frac{f'(t)}{f(t)} + \frac{f'(t)}{B - f(t)} = AB.$$

Integrating, $\ln f(t) - \ln\left[B - f(t)\right] = \ln\left[\dfrac{f(t)}{B - f(t)}\right] = ABt + k \Rightarrow$

$$\frac{f(t)}{B - f(t)} = Ke^{ABt}, \text{ where } K = e^k. \ f(0) = C \Rightarrow \frac{C}{B - C} = K.$$

Hence, $\dfrac{f(t)}{B - f(t)} = \dfrac{C}{B - C}\,e^{ABt} \Rightarrow \dfrac{B - f(t)}{f(t)} = \dfrac{(B - C)e^{-ABt}}{C} \Rightarrow$

$$\frac{B}{f(t)} = 1 + \frac{(B - C)e^{-ABt}}{C} = \frac{C + (B - C)e^{-ABt}}{C} \Rightarrow f(t) = \frac{BC}{C + (B - C)e^{-ABt}}.$$

41 $\mathrm{I} = \displaystyle\int\left(\frac{1}{ax^2 + bx}\cdot\frac{1/x^2}{1/x^2}\right)dx = \int\frac{1/x^2}{a + b/x}\,dx;$

$$u = a + \frac{b}{x},\ -\frac{1}{b}\,du = \frac{1}{x^2}\,dx \Rightarrow \mathrm{I} = -\frac{1}{b}\int\frac{1}{u}\,du = -\frac{1}{b}\ln|u| + K.$$

42 $\mathrm{I} = \displaystyle\int\left(\frac{1}{ax^n + bx}\cdot\frac{1/x^n}{1/x^n}\right)dx = \int\frac{1/x^n}{a + b/x^{n-1}}\,dx;\ u = a + \frac{b}{x^{n-1}},$

$$-\frac{1}{(n-1)b}\,du = \frac{1}{x^n}\,dx \Rightarrow \mathrm{I} = -\frac{1}{(n-1)b}\int\frac{1}{u}\,du = \frac{1}{b(1-n)}\ln|u| + K.$$

43 By (9.5), $\dfrac{f(x)}{g(x)} = \dfrac{A_1}{x - c_1} + \dfrac{A_2}{x - c_2} + \cdots + \dfrac{A_n}{x - c_n} \Rightarrow$

$(x - c_k)\dfrac{f(x)}{g(x)} = \dfrac{A_1(x - c_k)}{x - c_1} + \dfrac{A_2(x - c_k)}{x - c_2} + \cdots + A_k + \cdots + \dfrac{A_n(x - c_k)}{x - c_n}$ for

$k = 1, 2, \ldots, n.\ A_k = \displaystyle\lim_{x\to c_k}\left[\frac{(x - c_k)f(x)}{g(x)}\right] = \lim_{x\to c_k}\left[\frac{f(x)}{\dfrac{g(x) - g(c_k)}{x - c_k}}\right]$

(since $g(c_k) = 0$) $= \dfrac{f(x)}{g'(c_k)}$, for $k = 1, 2, \ldots, n$.

Note that $g'(c_k) = (c_k - c_1)(c_k - c_2)\cdots(c_k - c_{k-1})(c_k - c_{k+1})\cdots(c_k - c_n)$ and $g'(c_k) \neq 0$ since each c_k is unique.

44 $g(x) = x(x^4 - 5x^2 + 4) = x(x^2 - 1)(x^2 - 4) = x(x - 1)(x + 1)(x - 2)(x + 2).$

$g'(x) = 5x^4 - 15x^2 + 4$ and $f(x) = 2x^4 - x^3 - 3x^2 + 5x + 7 \Rightarrow$

$$\frac{f(0)}{g'(0)} = \frac{7}{4},\ \frac{f(1)}{g'(1)} = \frac{10}{-6},\ \frac{f(-1)}{g'(-1)} = \frac{2}{-6},\ \frac{f(2)}{g'(2)} = \frac{29}{24}, \text{ and } \frac{f(-2)}{g'(-2)} = \frac{25}{24}.$$

$$\text{Thus, } \frac{f(x)}{g(x)} = \frac{\frac{7}{4}}{x} + \frac{-\frac{5}{3}}{x - 1} + \frac{-\frac{1}{3}}{x + 1} + \frac{\frac{29}{24}}{x - 2} + \frac{\frac{25}{24}}{x + 2}.$$

Exercises 9.5

Note: { PF } indicates that the partial fractions method was used.

[1] $u = x + 1,\ du = dx \Rightarrow \int \frac{1}{(x + 1)^2 + 4}\,dx = \int \frac{1}{u^2 + 2^2}\,du = \frac{1}{2}\tan^{-1}\frac{u}{2} + C.$

[2] $u = x - 3,\ du = dx \Rightarrow \int \frac{1}{\sqrt{16 - (x - 3)^2}}\,dx = \int \frac{1}{\sqrt{4^2 - u^2}}\,du = \sin^{-1}\frac{u}{4} + C.$

[3] $x^2 - 4x + 8 = (x - 2)^2 + 4.\ u = x - 2,\ du = dx \Rightarrow$

$$\int \frac{1}{x^2 - 4x + 8}\,dx = \int \frac{1}{(x - 2)^2 + 4}\,dx = \int \frac{1}{u^2 + 2^2}\,du = \frac{1}{2}\tan^{-1}\frac{u}{2} + C.$$

[4] $x^2 - 2x + 2 = (x - 1)^2 + 1.\ u = x - 1,\ du = dx \Rightarrow$

$$\int \frac{1}{x^2 - 2x + 2}\,dx = \int \frac{1}{(x - 1)^2 + 1}\,dx = \int \frac{1}{u^2 + 1}\,du = \tan^{-1}u + C.$$

[5] $4x - x^2 = -(x^2 - 4x) = -(x^2 - 4x + 4) + 4 = 4 - (x - 2)^2.\ u = x - 2,$

$$du = dx \Rightarrow \int \frac{1}{\sqrt{4x - x^2}}\,dx = \int \frac{1}{\sqrt{4 - (x - 2)^2}}\,dx = \int \frac{1}{\sqrt{2^2 - u^2}}\,du = \sin^{-1}\frac{u}{2} + C.$$

[6] $7 + 6x - x^2 = 7 - (x^2 - 6x) = 7 - (x^2 - 6x + 9) + 9 = 16 - (x - 3)^2.$

$u = x - 3,\ du = dx \Rightarrow$

$$\int \frac{1}{\sqrt{7 + 6x - x^2}}\,dx = \int \frac{1}{\sqrt{16 - (x - 3)^2}}\,dx = \int \frac{1}{\sqrt{4^2 - u^2}}\,du = \sin^{-1}\frac{u}{4} + C.$$

[7] $9 - 8x - x^2 = 9 - (x^2 + 8x + 16) + 16 = 25 - (x + 4)^2.\ u = x + 4,\ du = dx,$

and $x = u - 4 \Rightarrow \mathrm{I} = \int \frac{2(u - 4) + 3}{\sqrt{25 - u^2}}\,du = \int \frac{2u}{\sqrt{25 - u^2}}\,du - 5\int \frac{1}{\sqrt{25 - u^2}}\,du =$

$$-2\sqrt{25 - u^2} - 5\sin^{-1}(\tfrac{1}{5}u) + C = -2\sqrt{9 - 8x - x^2} - 5\sin^{-1}\frac{x + 4}{5} + C.$$

[8] $9x^2 + 6x + 17 = (9x^2 + 6x + 1) + 16 = (3x + 1)^2 + 16.\ u = 3x + 1,\ \frac{1}{3}\,du = dx,$

and $x = \frac{1}{3}(u - 1) \Rightarrow \mathrm{I} = \frac{1}{3}\int \frac{\frac{1}{3}(u - 1) + 5}{u^2 + 16}\,du = \frac{1}{3}\int \frac{\frac{1}{3}u}{u^2 + 16}\,du + \frac{1}{3}\int \frac{\frac{14}{3}}{u^2 + 16}\,du =$

$\frac{1}{3}(\frac{1}{3})(\frac{1}{2})\ln(u^2 + 16) + \frac{1}{3}(\frac{14}{3})(\frac{1}{4})\tan^{-1}(\frac{1}{4}u) + C =$

$$\frac{1}{18}\ln(9x^2 + 6x + 17) + \frac{7}{18}\tan^{-1}\frac{3x + 1}{4} + C.$$

[9] $x^2 + 4x + 5 = (x + 2)^2 + 1.\ u = x + 2,\ du = dx \Rightarrow$

$\mathrm{I} = \int \frac{1}{(u^2 + 1)^2}\,du.$ Let $u = \tan\theta$ and $du = \sec^2\theta\,d\theta$. Thus, $\mathrm{I} = \int \frac{\sec^2\theta}{(\sec^2\theta)^2}\,d\theta =$

$\int \cos^2\theta\,d\theta = \frac{1}{2}\int (1 + \cos 2\theta)\,d\theta = \frac{1}{2}\theta + \frac{1}{4}\sin 2\theta + C = \frac{1}{2}\theta + \frac{1}{2}\sin\theta\cos\theta + C =$

$$\frac{1}{2}\left(\tan^{-1}u + \frac{u}{\sqrt{u^2 + 1}} \cdot \frac{1}{\sqrt{u^2 + 1}}\right) + C = \frac{1}{2}\left[\tan^{-1}(x + 2) + \frac{x + 2}{x^2 + 4x + 5}\right] + C.$$

[10] $x^2 - 6x + 34 = (x-3)^2 + 25$. $u = x - 3$, $du = dx \Rightarrow \mathrm{I} = \int \frac{1}{(u^2+25)^{3/2}}\,du$.

Let $u = 5\tan\theta$ and $du = 5\sec^2\theta\,d\theta$. Thus, $\mathrm{I} = \int \frac{5\sec^2\theta}{(5\sec\theta)^3}\,d\theta = \frac{1}{25}\int \cos\theta\,d\theta =$

$$\frac{1}{25}\sin\theta + C = \frac{1}{25}\left(\frac{u}{\sqrt{u^2+25}}\right) + C = \frac{x-3}{25\sqrt{x^2-6x+34}} + C.$$

[11] $x^2 + 6x + 13 = (x+3)^2 + 4$. $u = x + 3$, $du = dx \Rightarrow \mathrm{I} = \int \frac{1}{(u^2+4)^{3/2}}\,du$.

Let $u = 2\tan\theta$ and $du = 2\sec^2\theta\,d\theta$. Thus, $\mathrm{I} = \int \frac{2\sec^2\theta}{(2\sec\theta)^3}\,d\theta = \frac{1}{4}\int \cos\theta\,d\theta =$

$$\frac{1}{4}\sin\theta + C = \frac{1}{4}\left(\frac{u}{\sqrt{u^2+4}}\right) + C = \frac{x+3}{4\sqrt{x^2+6x+13}} + C.$$

[12] $x(6-x) = 9 - (x-3)^2$. $u = x - 3$, $du = dx \Rightarrow \mathrm{I} = \int\sqrt{9-u^2}\,du$. Let

$u = 3\sin\theta$, $du = 3\cos\theta\,d\theta$. Thus, $\mathrm{I} = \int(3\cos\theta)(3\cos\theta)\,d\theta = \frac{9}{2}\int(1+\cos 2\theta)\,d\theta =$

$\frac{9}{2}\theta + \frac{9}{4}\sin 2\theta + C = \frac{9}{2}\theta + \frac{9}{2}\sin\theta\cos\theta + C =$

$$\frac{9}{2}\sin^{-1}\frac{u}{3} + \frac{9}{2}\left(\frac{u}{3}\right)\left(\frac{\sqrt{9-u^2}}{3}\right) + C = \frac{9}{2}\sin^{-1}\frac{x-3}{3} + \frac{1}{2}(x-3)\sqrt{x(6-x)} + C.$$

[13] $2x^2 - 3x + 9 = 2(x^2 - \frac{3}{2}x + \frac{9}{16}) + 9 - \frac{9}{8} = 2\left[(x-\frac{3}{4})^2 + \frac{63}{16}\right]$.

$u = x - \frac{3}{4}$, $du = dx \Rightarrow$

$$\mathrm{I} = \frac{1}{2}\int\frac{1}{u^2+(\sqrt{63}/4)^2}\,du = \frac{1}{2}\left(\frac{4}{\sqrt{63}}\right)\tan^{-1}\left(\frac{4}{\sqrt{63}}u\right) + C = \frac{2}{3\sqrt{7}}\tan^{-1}\frac{4x-3}{3\sqrt{7}} + C.$$

[14] $x^2 + 2x + 5 = (x+1)^2 + 4$. $u = x + 1$, $du = dx \Rightarrow$

$$\mathrm{I} = \int\frac{2u-2}{(u^2+4)^2}\,du = \int\frac{2u}{(u^2+4)^2}\,du - \int\frac{2}{(u^2+4)^2}\,du = -\frac{1}{u^2+4} - \int\frac{2}{(u^2+4)^2}\,du.$$

For the last integral, let $u = 2\tan\theta$, $du = 2\sec^2\theta\,d\theta$. Hence, $\int\frac{2}{(u^2+4)^2}\,du =$

$$2\int\frac{2\sec^2\theta}{(4\sec^2\theta)^2}\,d\theta = \frac{1}{4}\int\cos^2\theta\,d\theta = \frac{1}{8}\int(1+\cos 2\theta)\,d\theta = \frac{1}{8}(\theta + \frac{1}{2}\sin 2\theta) + C =$$

$$\frac{1}{8}(\theta + \sin\theta\cos\theta) + C = \frac{1}{8}\left(\tan^{-1}\frac{u}{2} + \frac{u}{\sqrt{u^2+4}}\cdot\frac{2}{\sqrt{u^2+4}}\right) + C.$$

Thus, $\mathrm{I} = -\frac{1}{x^2+2x+5} - \frac{1}{8}\left[\tan^{-1}\frac{x+1}{2} + \frac{2(x+1)}{x^2+2x+5}\right] + C =$

$$-\frac{1}{8}\tan^{-1}\frac{x+1}{2} - \frac{x+5}{4(x^2+2x+5)} + C.$$

[15] $u = e^x$, $du = e^x\,dx \Rightarrow \mathrm{I} = \int\frac{e^x}{(e^x+2)(e^x+1)}\,dx = \int\frac{1}{(u+2)(u+1)}\,du =$

$$\int\left[\frac{1}{u+1} - \frac{1}{u+2}\right]du\ \{\mathrm{PF}\} = \ln\left|\frac{u+1}{u+2}\right| + C = \ln\left(\frac{e^x+1}{e^x+2}\right) + C.$$

[16] $x^2 + 10x = (x+5)^2 - 25$. $u = x + 5$, $du = dx \Rightarrow \mathrm{I} = \int \sqrt{u^2 - 5^2}\,du$.

Let $u = 5\sec\theta$ and $du = 5\sec\theta\tan\theta\,d\theta$. Thus, $\mathrm{I} = \int (5\tan\theta)(5\sec\theta\tan\theta)\,d\theta$

$= 25\int\left[\sec\theta\,(\sec^2\theta - 1)\right]d\theta = 25\int(\sec^3\theta - \sec\theta)\,d\theta$ {by Example 6, §9.1}

$= 25\left[\tfrac{1}{2}\sec\theta\tan\theta + \tfrac{1}{2}\ln|\sec\theta + \tan\theta| - \ln|\sec\theta + \tan\theta|\right] + C_1$

$= \tfrac{25}{2}\left[\sec\theta\tan\theta - \ln|\sec\theta + \tan\theta|\right] + C_1$

$$= \frac{25}{2}\left[\frac{u}{5}\cdot\frac{\sqrt{u^2-25}}{5} - \ln\left|\frac{u}{5} + \frac{\sqrt{u^2-25}}{5}\right|\right] + C_1$$

$= \tfrac{1}{2}(x+5)\sqrt{x^2+10x} - \tfrac{25}{2}\ln\left|x + 5 + \sqrt{x^2+10x}\right| + C$, where $C = C_1 + \tfrac{25}{2}\ln 5$.

[17] $\dfrac{(x^2 - 4x + 5) + 1}{x^2 - 4x + 5} = 1 + \dfrac{1}{x^2 - 4x + 5}$. $x^2 - 4x + 5 = (x-2)^2 + 1$.

$u = x - 2$, $du = dx \Rightarrow \mathrm{I} = \displaystyle\int_0^1\left[1 + \frac{1}{u^2+1}\right]du = \left[u + \tan^{-1}u\right]_0^1 = 1 + \tfrac{\pi}{4} \approx 1.79.$

[18] $x^2 + x + 1 = (x + \tfrac{1}{2})^2 + \tfrac{3}{4} \Rightarrow$

$$\mathrm{I} = \int_0^1 \frac{(x+\frac{1}{2}) - \frac{3}{2}}{(x+\frac{1}{2})^2 + \frac{3}{4}}\,dx = \int_0^1 \frac{x + \frac{1}{2}}{(x+\frac{1}{2})^2 + \frac{3}{4}}\,dx - \frac{3}{2}\int_0^1 \frac{1}{(x+\frac{1}{2})^2 + \frac{3}{4}}\,dx$$

$= \mathrm{I}_1 - \tfrac{3}{2}\mathrm{I}_2$. $\mathrm{I}_1 = \tfrac{1}{2}\left[\ln\left|(x+\tfrac{1}{2})^2 + \tfrac{3}{4}\right|\right]_0^1 = \left[\tfrac{1}{2}\ln|x^2 + x + 1|\right]_0^1 = \tfrac{1}{2}\ln 3$.

$u = x + \tfrac{1}{2}$, $du = dx \Rightarrow \mathrm{I}_2 = \displaystyle\int_{1/2}^{3/2}\frac{1}{u^2 + (\sqrt{3}/2)^2}\,du = \left[\frac{2}{\sqrt{3}}\tan^{-1}\frac{2u}{\sqrt{3}}\right]_{1/2}^{3/2} =$

$\tfrac{2}{\sqrt{3}}(\tan^{-1}\sqrt{3} - \tan^{-1}\tfrac{1}{\sqrt{3}}) = \tfrac{2}{\sqrt{3}}(\tfrac{\pi}{3} - \tfrac{\pi}{6}) = \tfrac{\pi}{3\sqrt{3}}$. Thus, $\mathrm{I} = \tfrac{1}{2}\ln 3 - \tfrac{\sqrt{3}\,\pi}{6} \approx -0.36$.

[19] $x^2 + 4x + 29 = (x+2)^2 + 25$. $u = x + 2$, $du = dx \Rightarrow A =$

$$\int_{-2}^{3}\frac{1}{(x+2)^2 + 25}\,dx = \int_0^5 \frac{1}{u^2 + 5^2}\,du = \left[\frac{1}{5}\tan^{-1}\frac{u}{5}\right]_0^5 = \frac{1}{5}\left(\frac{\pi}{4} - 0\right) = \frac{\pi}{20} \approx 0.16.$$

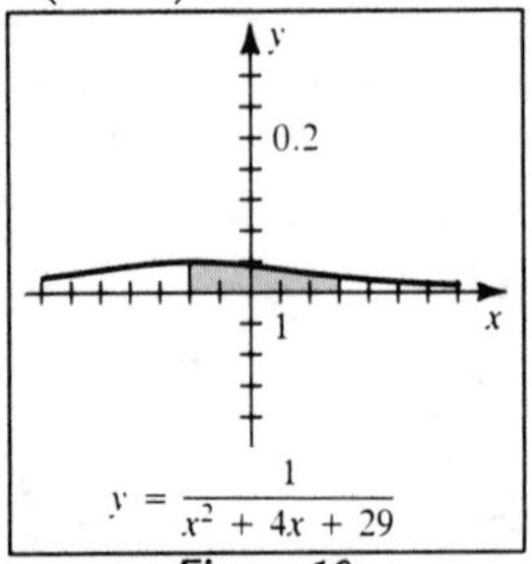

y, 0.2, 1, x

$y = \dfrac{1}{x^2 + 4x + 29}$

Figure 19

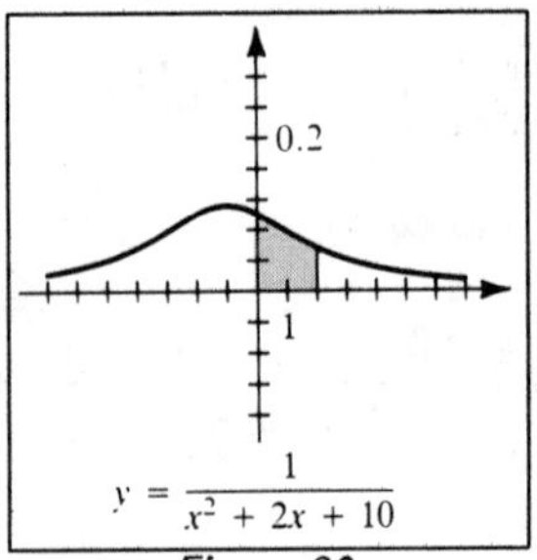

0.2, 1

$y = \dfrac{1}{x^2 + 2x + 10}$

Figure 20

[20] $V = \pi\displaystyle\int_0^2 \frac{1}{(x^2 + 2x + 10)^2}\,dx = \pi\int_0^2 \frac{1}{\left[(x+1)^2 + 9\right]^2}\,dx$. $u = x + 1$, $du = dx \Rightarrow$

$V = \pi\displaystyle\int_1^3 \frac{1}{(u^2 + 9)^2}\,du$. $u = 3\tan\theta$ and $du = 3\sec^2\theta\,d\theta$ {let $\alpha = \tan^{-1}\tfrac{1}{3}$} $\Rightarrow$

(cont.)

$$V = \pi\int_{\alpha}^{\pi/4} \frac{3\sec^2\theta}{(9\sec^2\theta)^2}\,d\theta = \frac{\pi}{27}\int_{\alpha}^{\pi/4} \cos^2\theta\,d\theta = \tfrac{\pi}{54}\Big[\theta + \sin\theta\cos\theta\Big]_{\alpha}^{\pi/4} =$$

$$\tfrac{\pi}{54}\Big[(\tfrac{\pi}{4} + \tfrac{1}{2}) - (\tan^{-1}\tfrac{1}{3} + \tfrac{1}{\sqrt{10}}\cdot\tfrac{3}{\sqrt{10}})\Big] = \tfrac{\pi}{54}(\tfrac{\pi}{4} + \tfrac{1}{5} - \tan^{-1}\tfrac{1}{3}) \approx 0.04.$$

Exercises 9.6

Note: {IP} indicates that integration by parts was used.

[1] $u = (x+9)^{1/3}$, $x = u^3 - 9$, and $dx = 3u^2\,du \Rightarrow$

$$\text{I} = \int (u^3 - 9)(u)(3u^2)\,du = \int (3u^6 - 27u^3)\,du = \tfrac{3}{7}u^7 - \tfrac{27}{4}u^4 + C.$$

[2] $u = (2x+1)^{1/2}$, $x = \frac{1}{2}(u^2 - 1)$, and $dx = u\,du \Rightarrow$

$$\text{I} = \int \tfrac{1}{4}(u^2 - 1)^2(u)(u)\,du = \tfrac{1}{4}\int (u^6 - 2u^4 + u^2)\,du = \tfrac{1}{28}u^7 - \tfrac{1}{10}u^5 + \tfrac{1}{12}u^3 + C.$$

[3] $u = (3x+2)^{1/5}$, $x = \frac{1}{3}(u^5 - 2)$, and $dx = \frac{5}{3}u^4\,du \Rightarrow$

$$\text{I} = \int \frac{\frac{1}{3}(u^5 - 2)(\frac{5}{3}u^4)}{u}\,du = \frac{5}{9}\int (u^8 - 2u^3)\,du = \frac{5}{81}u^9 - \frac{5}{18}u^4 + C.$$

[4] $u = (x+3)^{1/3}$, $x = u^3 - 3$, and $dx = 3u^2\,du \Rightarrow$

$$\text{I} = \int \frac{5(u^3 - 3)(3u^2)}{u^2}\,du = 15\int (u^3 - 3)\,du = \frac{15}{4}u^4 - 45u + C.$$

[5] $u = \sqrt{x} + 4$, $x = (u-4)^2$, and $dx = 2(u-4)\,du \Rightarrow$

$$\text{I} = 2\int_6^7 \frac{u-4}{u}\,du = 2\int_6^7 \Big(1 - \frac{4}{u}\Big)\,du = 2\Big[u - 4\ln|u|\Big]_6^7 = 2(1 + 4\ln\tfrac{6}{7}) \approx 0.767.$$

[6] $u = 4 + \sqrt{x}$, $x = (u-4)^2$, and $dx = 2(u-4)\,du \Rightarrow$

$$\text{I} = \int_4^9 \frac{2u - 8}{u^{1/2}}\,du = \int_4^9 (2u^{1/2} - 8u^{-1/2})\,du = \Big[\tfrac{4}{3}u^{3/2} - 16u^{1/2}\Big]_4^9 = \tfrac{28}{3}.$$

[7] $u = x^{1/6}$, $x = u^6$, and $dx = 6u^5\,du \Rightarrow$

$$\text{I} = \int \frac{(u^3)(6u^5)}{1+u^2}\,du = 6\int \frac{u^8}{1+u^2}\,du = 6\int\Big(u^6 - u^4 + u^2 - 1 + \frac{1}{1+u^2}\Big)\,du \text{ \{long division\}} = \tfrac{6}{7}u^7 - \tfrac{6}{5}u^5 + 2u^3 - 6u + 6\tan^{-1}u + C.$$

[8] $u = x^{1/12}$, $x = u^{12}$, and $dx = 12u^{11}\,du \Rightarrow \text{I} = \displaystyle\int \frac{12u^{11}}{u^3 + u^4}\,du = 12\int \frac{u^8}{1+u}\,du$

$$= 12\int\Big(u^7 - u^6 + u^5 - u^4 + u^3 - u^2 + u - 1 + \frac{1}{1+u}\Big)\,du$$

$$= 12(\tfrac{1}{8}u^8 - \tfrac{1}{7}u^7 + \tfrac{1}{6}u^6 - \tfrac{1}{5}u^5 + \tfrac{1}{4}u^4 - \tfrac{1}{3}u^3 + \tfrac{1}{2}u^2 - u + \ln|1+u|) + C$$

$$= \tfrac{3}{2}x^{2/3} - \tfrac{12}{7}x^{7/12} + 2x^{1/2} - \tfrac{12}{5}x^{5/12} + 3x^{1/3} - 4x^{1/4} + 6x^{1/6} - 12x^{1/12} + 12\ln(1 + x^{1/12}) + C.$$

[9] $u = \sqrt{x-2}$, $x = u^2 + 2$, and $dx = 2u\,du \Rightarrow$

$$\text{I} = \int \frac{2u}{(u^2+3)(u)}\,du = 2\int \frac{1}{u^2+3}\,du = \frac{2}{\sqrt{3}}\tan^{-1}\frac{u}{\sqrt{3}} + C.$$

10 $u = \sqrt{1 + 2x}$, $x = \frac{1}{2}(u^2 - 1)$, and $dx = u\,du \Rightarrow$

$$\text{I} = \int_1^3 \frac{\frac{1}{2}(u^2 - 1) + 3}{u}\,u\,du = \int_1^3 (u^2 + 2)\,du = \left[\tfrac{1}{3}u^3 + 2u\right]_1^3 = \tfrac{38}{3}.$$

11 $u = (x + 4)^{1/3}$, $x = u^3 - 4$, and $dx = 3u^2\,du \Rightarrow$

$$\text{I} = \int \frac{(u^3 - 3)(3u^2)}{u}\,du = 3\int (u^4 - 3u)\,du = \frac{3}{5}u^5 - \frac{9}{2}u^2 + C.$$

12 $u = x^{1/3}$, $x = u^3$, and $dx = 3u^2\,du \Rightarrow \text{I} = \displaystyle\int \frac{(u + 1)(3u^2)}{u - 1}\,du = 3\int \frac{u^3 + u^2}{u - 1}\,du =$

$$3\int \left(u^2 + 2u + 2 + \frac{2}{u - 1}\right) du = u^3 + 3u^2 + 6u + 6\ln|u - 1| + C.$$

13 $u = 1 + e^x$, $e^x = u - 1$, and $e^x\,dx = du \Rightarrow \text{I} = \int e^{2x}\sqrt{1 + e^x}\,(e^x)\,dx =$

$\int (u - 1)^2(u^{1/2})\,du = \int (u^{5/2} - 2u^{3/2} + u^{1/2})\,du = \frac{2}{7}u^{7/2} - \frac{4}{5}u^{5/2} + \frac{2}{3}u^{3/2} + C.$

14 $u = (1 + e^x)^{1/3}$, $e^x = u^3 - 1$, and $e^x\,dx = 3u^2\,du \Rightarrow$

$$\text{I} = \int \frac{e^x}{\sqrt[3]{1 + e^x}}\,e^x\,dx = \int \frac{(u^3 - 1)(3u^2)}{u}\,du = 3\int (u^4 - u)\,du = \frac{3}{5}u^5 - \frac{3}{2}u^2 + C.$$

15 $u = e^x + 4$, $e^x = u - 4$, and $e^x\,dx = du \Rightarrow \text{I} = \displaystyle\int \frac{u - 4}{u}\,du = u - 4\ln|u| + C_1 =$

$$e^x + 4 - 4\ln|e^x + 4| + C_1 = e^x - 4\ln(e^x + 4) + C, \text{ where } C = C_1 + 4.$$

16 $u = 1 + \sin x$, $\sin x = u - 1$, and $\cos x\,dx = du \Rightarrow$

$$\text{I} = \int \frac{2\sin x\cos x}{\sqrt{1 + \sin x}}\,dx = 2\int \frac{u - 1}{u^{1/2}}\,du = 2\int (u^{1/2} - u^{-1/2})\,du = \tfrac{4}{3}u^{3/2} - 4u^{1/2} + C.$$

17 $u = \sqrt{x + 4}$, $x = u^2 - 4$, and $dx = 2u\,du \Rightarrow$

$$\text{I} = \int \sin u\,(2u)\,du = 2\int u\sin u\,du = 2\sin u - 2u\cos u + C\ \{\text{IP}\}.$$

18 $u = \sqrt{x}$, $x = u^2$, and $dx = 2u\,du \Rightarrow$

$$\text{I} = \int ue^u\,(2u)\,du = 2\int u^2e^u\,du = 2u^2e^u - 4ue^u + 4e^u + C\ \{\text{IP}\}.$$

19 $u = x - 1$, $dx = du \Rightarrow$

$$\text{I} = \int_1^2 \frac{u + 1}{u^6}\,du = \int_1^2 (u^{-5} + u^{-6})\,du = \left[-\frac{1}{4u^4} - \frac{1}{5u^5}\right]_1^2 = \tfrac{137}{320}.$$

20 $u = 3x + 4$, $x = \frac{1}{3}(u - 4)$, and $dx = \frac{1}{3}\,du \Rightarrow \text{I} = \displaystyle\int \frac{\frac{1}{9}(u^2 - 8u + 16)}{u^{10}}\,\tfrac{1}{3}\,du =$

$$\frac{1}{27}\int (u^{-8} - 8u^{-9} + 16u^{-10})\,du = \frac{1}{27}\left(-\frac{1}{7u^7} + \frac{1}{u^8} - \frac{16}{9u^9}\right) + C.$$

21 $u = \cos x$, $-du = \sin x\,dx \Rightarrow \text{I} = -\displaystyle\int \frac{1}{u(u - 1)}\,du = -\int \left[\frac{-1}{u} + \frac{1}{u - 1}\right] du\ \{\text{PF}\} =$

$$\int \left(\frac{1}{u} + \frac{1}{1 - u}\right) du = \ln|u| - \ln|1 - u| + C = \ln|\cos x| - \ln(1 - \cos x) + C.$$

[22] $u = \sin x$, $du = \cos x\, dx \Rightarrow$

$$\mathrm{I} = \int \frac{1}{u^2 - u - 2}\, du = \int \frac{1}{(u-2)(u+1)}\, du = \int \left[\frac{\frac{1}{3}}{u-2} + \frac{-\frac{1}{3}}{u+1} \right] du \text{ \{PF\}} =$$

$$\tfrac{1}{3}\ln|u-2| - \tfrac{1}{3}\ln|u+1| + C = \tfrac{1}{3}\ln(2 - \sin x) - \tfrac{1}{3}\ln(\sin x + 1) + C.$$

[23] $u = e^x$, $du = e^x\, dx \Rightarrow$

$$\mathrm{I} = \int \frac{1}{u^2 - 1}\, du = \int \frac{1}{(u-1)(u+1)}\, du = \int \left[\frac{\frac{1}{2}}{u-1} + \frac{-\frac{1}{2}}{u+1} \right] du \text{ \{PF\}} =$$

$$\tfrac{1}{2}\ln|u-1| - \tfrac{1}{2}\ln|u+1| + C = \tfrac{1}{2}\ln|e^x - 1| - \tfrac{1}{2}\ln(e^x + 1) + C.$$

[24] $u = e^x$, $du = e^x\, dx \Rightarrow \mathrm{I} = \displaystyle\int \frac{1}{u + u^{-1}} \cdot \frac{du}{u} = \int \frac{1}{u^2 + 1}\, du = \tan^{-1} u + C.$

[25] $u = \sin x$, $du = \cos x \Rightarrow$

$$\mathrm{I} = \int \frac{2 \sin x \cos x}{\sin^2 x - 2\sin x - 8}\, dx = \int \frac{2u}{u^2 - 2u - 8}\, du = 2\int \left[\frac{\frac{2}{3}}{u-4} + \frac{\frac{1}{3}}{u+2} \right] du \text{ \{PF\}} =$$

$$\tfrac{4}{3}\ln|u-4| + \tfrac{2}{3}\ln|u+2| + C = \tfrac{4}{3}\ln(4 - \sin x) + \tfrac{2}{3}\ln(\sin x + 2) + C.$$

[26] $u = \cos x$, $-du = \sin x\, dx \Rightarrow \mathrm{I} = -\displaystyle\int \frac{1}{u(5+u)}\, du = -\int \left[\frac{\frac{1}{5}}{u} + \frac{-\frac{1}{5}}{5+u} \right] du \text{ \{PF\}} =$

$$-\tfrac{1}{5}\ln|u| + \tfrac{1}{5}\ln|5+u| + C = \tfrac{1}{5}\ln(5 + \cos x) - \tfrac{1}{5}\ln|\cos x| + C.$$

[27] $\mathrm{I} = \displaystyle\int \frac{1}{2 + \frac{2u}{1+u^2}} \cdot \frac{2}{1+u^2}\, du = \int \frac{1}{u^2 + u + 1}\, du$ {see I_2 in Exercise 18, §9.5} $=$

$$\frac{2}{\sqrt{3}}\tan^{-1}\frac{2(u + \frac{1}{2})}{\sqrt{3}} + C = \frac{2}{\sqrt{3}}\tan^{-1}\frac{2\tan(x/2) + 1}{\sqrt{3}} + C.$$

[28] $\mathrm{I} = \displaystyle\int \frac{1}{3 + 2 \cdot \frac{1-u^2}{1+u^2}} \cdot \frac{2}{1+u^2}\, du = 2\int \frac{1}{u^2 + 5}\, du =$

$$\frac{2}{\sqrt{5}}\tan^{-1}\frac{u}{\sqrt{5}} + C = \frac{2}{\sqrt{5}}\tan^{-1}\frac{\tan(x/2)}{\sqrt{5}} + C.$$

[29] $\mathrm{I} = \displaystyle\int \frac{1}{1 + \frac{2u}{1+u^2} + \frac{1-u^2}{1+u^2}} \cdot \frac{2}{1+u^2}\, du = \int \frac{1}{u+1}\, du =$

$$\ln|u+1| + C = \ln\left|\tan(x/2) + 1\right| + C.$$

[30] $\mathrm{I} = \displaystyle\int \frac{\cos x}{\sin x + \sin x \cos x}\, dx = \int \frac{(1-u^2)/(1+u^2)}{\frac{2u}{1+u^2} + \frac{2u}{1+u^2} \cdot \frac{1-u^2}{1+u^2}} \cdot \frac{2}{1+u^2}\, du$

$$= \int \frac{1-u^2}{u(1+u^2) + u(1-u^2)}\, du = \int \frac{1-u^2}{2u}\, du = \frac{1}{2}\int \left(\frac{1}{u} - u \right) du =$$

$$\tfrac{1}{2}\ln|u| - \tfrac{1}{4}u^2 + C = \tfrac{1}{2}\ln\left|\tan(x/2)\right| - \tfrac{1}{4}\tan^2(x/2) + C.$$

[31] $\mathrm{I} = \displaystyle\int \frac{1}{4\cos x - 3\sin x}\, dx = \int \frac{1}{4 \cdot \frac{1-u^2}{1+u^2} - 3 \cdot \frac{2u}{1+u^2}} \cdot \frac{2}{1+u^2}\, du$

$$= -\int \frac{1}{2u^2 + 3u - 2}\, du = \int \left[\frac{-\frac{2}{5}}{2u-1} + \frac{\frac{1}{5}}{u+2} \right] du \text{ \{PF\}} = -\tfrac{1}{5}\ln|2u - 1| +$$

$$\tfrac{1}{5}\ln|u+2| + C = -\tfrac{1}{5}\ln\left|2\tan(x/2) - 1\right| + \tfrac{1}{5}\ln\left|\tan(x/2) + 2\right| + C.$$

32 $\text{I} = \int \frac{1}{\frac{2u}{1+u^2} - \sqrt{3}\cdot\frac{1-u^2}{1+u^2}} \cdot \frac{2}{1+u^2}\,du = \int \frac{2}{2u - \sqrt{3}(1-u^2)}\,du$

$$= \frac{2}{\sqrt{3}}\int \frac{1}{u^2 + \frac{2}{\sqrt{3}}u - 1}\,du = \frac{2}{\sqrt{3}}\int \frac{1}{(u - \frac{1}{\sqrt{3}})(u + \sqrt{3})}\,du$$

$$= \frac{2}{\sqrt{3}}\int \left[\frac{\sqrt{3}/4}{u - \frac{1}{\sqrt{3}}} - \frac{\sqrt{3}/4}{u + \sqrt{3}}\right]du\ \{\text{PF}\} = \tfrac{1}{2}\ln\left|u - \tfrac{1}{\sqrt{3}}\right| - \tfrac{1}{2}\ln\left|u + \sqrt{3}\right| + C =$$

$$\tfrac{1}{2}\ln\left|\tan(x/2) - \tfrac{1}{\sqrt{3}}\right| - \tfrac{1}{2}\ln\left|\tan(x/2) + \sqrt{3}\right| + C.$$

33 $\text{I} = \int \frac{1}{\cos x}\,dx = \int \frac{1+u^2}{1-u^2}\cdot\frac{2}{1+u^2}\,du = \int \frac{2}{(1+u)(1-u)}\,du =$

$$\int\left[\frac{1}{1+u} + \frac{1}{1-u}\right]du\ \{\text{PF}\} = \ln|1+u| - \ln|1-u| + C =$$

$$\ln\left|\frac{1+u}{1-u}\right| + C = \ln\left|\frac{1+\tan\frac{1}{2}x}{1-\tan\frac{1}{2}x}\right| + C.$$

34 $\text{I} = \int \frac{1}{\sin x}\,dx = \int \frac{1+u^2}{2u}\cdot\frac{2}{1+u^2}\,du = \int \frac{1}{u}\,du = \ln|u| + C = \ln\left|\tan\tfrac{1}{2}x\right| + C.$

$$\left|\tan\tfrac{1}{2}x\right| = \left|\frac{\sin\frac{1}{2}x}{\cos\frac{1}{2}x}\right| = \sqrt{\frac{1-\cos x}{1+\cos x}} \Rightarrow \text{I} = \ln\left(\frac{1-\cos x}{1+\cos x}\right)^{1/2} + C =$$

$$\tfrac{1}{2}\ln\left(\frac{1-\cos x}{1+\cos x}\right) + C.$$

Exercises 9.7

1 $u = 3x$, $x = \frac{1}{3}u$, and $dx = \frac{1}{3}\,du \Rightarrow \text{I} = \int \frac{\sqrt{4+u^2}}{u}\,du.$

Using Formula 23 with $a = 2$, $\text{I} = \sqrt{4+u^2} - 2\ln\left|\frac{2+\sqrt{4+u^2}}{u}\right| + C.$

2 $u = \sqrt{3}x$, $x = \frac{1}{\sqrt{3}}u$, and $dx = \frac{1}{\sqrt{3}}\,du \Rightarrow \text{I} = \int \frac{1}{u\sqrt{2+u^2}}\,du.$

Using Formula 27 with $a = \sqrt{2}$, $\text{I} = -\frac{1}{\sqrt{2}}\ln\left|\frac{\sqrt{2+u^2}+\sqrt{2}}{u}\right| + C.$

3 Using Formula 37 with $a = 4$, $\text{I} = -\frac{x}{8}(2x^2 - 80)\sqrt{16 - x^2} + 96\sin^{-1}\frac{x}{4} + C.$

4 $u = 2x$, $x = \frac{1}{2}u$, and $dx = \frac{1}{2}\,du \Rightarrow \text{I} = \frac{1}{8}\int u^2\sqrt{u^2 - 16}\,du$. Using Formula 40 with

$a = 4$, $\quad \text{I} = \frac{1}{8}\left[\frac{u}{8}(2u^2 - 16)\sqrt{u^2 - 16} - 32\ln\left|u + \sqrt{u^2 - 16}\right|\right] + C$

$$= \tfrac{x}{4}(x^2 - 2)\sqrt{4x^2 - 16} - 4\ln\left|2x + \sqrt{4x^2 - 16}\right| + C.$$

5 Using Formula 54 with $a = 2$ and $b = -3$,

$$\text{I} = \tfrac{2}{135}(-9x - 4)(2 - 3x)^{3/2} + C = -\tfrac{2}{135}(9x + 4)(2 - 3x)^{3/2} + C.$$

6 Using Formula 60 with $a = 5$, $b = 2$, and $n = 2$,

$$\text{I} = \tfrac{2}{14}\left[x^2(5 + 2x)^{3/2} - 10\int x\sqrt{5 + 2x}\,dx\right].$$

Using Formula 54 for the last integral yields (cont.)

$$\text{I} = \tfrac{1}{7}x^2(5 + 2x)^{3/2} - \tfrac{10}{7}\Big[\tfrac{1}{30}(6x - 10)(5 + 2x)^{3/2}\Big] + C$$

$$= (5 + 2x)^{3/2}\Big[\tfrac{1}{7}x^2 - \tfrac{1}{21}(6x - 10)\Big] + C$$

$$= \tfrac{1}{21}(5 + 2x)^{3/2}(3x^2 - 6x + 10) + C.$$

[7] $u = 3x$, $\frac{1}{3}\,du = dx \Rightarrow \text{I} = \frac{1}{3}\int \sin^6 u\,du$. Using Formula 73 three times,

$$\text{I} = \tfrac{1}{3}\Big[-\tfrac{1}{6}\sin^5 u\cos u + \tfrac{5}{6}\int \sin^4 u\,du\Big]$$

$$= -\tfrac{1}{18}\sin^5 u\cos u + \tfrac{5}{18}\Big[-\tfrac{1}{4}\sin^3 u\cos u + \tfrac{3}{4}\int \sin^2 u\,du\Big]$$

$$= -\tfrac{1}{18}\sin^5 u\cos u - \tfrac{5}{72}\sin^3 u\cos u + \tfrac{5}{24}\Big[-\tfrac{1}{2}\sin u\cos u + \tfrac{1}{2}u\Big] + C$$

$$= -\tfrac{1}{18}\sin^5 u\cos u - \tfrac{5}{72}\sin^3 u\cos u - \tfrac{5}{48}\sin u\cos u + \tfrac{5}{48}u + C.$$

[8] $u = x^2$, $\frac{1}{2}\,du = x\,dx \Rightarrow \text{I} = \frac{1}{2}\int \cos^5 u\,du$. Using Formula 74 twice,

$$\text{I} = \tfrac{1}{2}\Big[\tfrac{1}{5}\cos^4 u\sin u + \tfrac{4}{5}\int \cos^3 u\,du\Big] = \tfrac{1}{10}\cos^4 u\sin u + \tfrac{2}{5}\Big[\tfrac{1}{3}\cos^2 u\sin u + \tfrac{2}{3}\int \cos u\,du\Big]$$

$$= \tfrac{1}{10}\cos^4 u\sin u + \tfrac{2}{15}\cos^2 u\sin u + \tfrac{4}{15}\sin u + C.$$

[9] Using Formula 78, $\text{I} = -\frac{1}{3}\cot x\csc^2 x + \frac{2}{3}\int \csc^2 x\,dx = -\frac{1}{3}\cot x\csc^2 x - \frac{2}{3}\cot x + C.$

[10] Using Formula 81 with $a = 5$ and $b = 3$, $\text{I} = -\dfrac{\cos 2x}{4} - \dfrac{\cos 8x}{16} + C.$

[11] Using Formula 90, $\text{I} = \dfrac{2x^2 - 1}{4}\sin^{-1}x + \dfrac{x\sqrt{1 - x^2}}{4} + C.$

[12] Using Formula 95 with $n = 2$, $\text{I} = \dfrac{1}{3}\left[x^3\tan^{-1}x - \displaystyle\int \frac{x^3}{1 + x^2}\,dx\right] =$

$$\tfrac{1}{3}x^3\tan^{-1}x - \frac{1}{3}\int\left[x - \frac{x}{1 + x^2}\right]dx = \tfrac{1}{3}x^3\tan^{-1}x - \tfrac{1}{6}x^2 + \tfrac{1}{6}\ln(1 + x^2) + C.$$

[13] Using Formula 98 with $a = -3$ and $b = 2$, $\text{I} = \frac{1}{13}e^{-3x}(-3\sin 2x - 2\cos 2x) + C.$

[14] Using Formula 101 with $n = 5$, $\text{I} = \frac{1}{36}x^6(6\ln x - 1) + C.$

[15] $u = 3x$, $x = \frac{1}{3}u$, and $dx = \frac{1}{3}\,du \Rightarrow \text{I} = \displaystyle\int \frac{\sqrt{\frac{5}{3}u - u^2}}{u}\,du$. Using Formula 115 with

$$a = \tfrac{5}{6},\ \text{I} = \sqrt{\tfrac{5}{3}u - u^2} + \frac{5}{6}\cos^{-1}\left(\frac{\frac{5}{6} - u}{\frac{5}{6}}\right) + C = \sqrt{5x - 9x^2} + \frac{5}{6}\cos^{-1}\frac{5 - 18x}{5} + C.$$

[16] $u = \sqrt{2}x$, $x = \frac{1}{\sqrt{2}}u$, and $dx = \frac{1}{\sqrt{2}}\,du \Rightarrow \text{I} = \displaystyle\int \frac{1}{u\sqrt{(3/\sqrt{2})u - u^2}}\,du.$

Using Formula 120 with $a = \dfrac{3}{2\sqrt{2}}$, $\text{I} = -\dfrac{\sqrt{(3/\sqrt{2})u - u^2}}{\frac{3}{2\sqrt{2}}u} + C = -\dfrac{2\sqrt{3x - 2x^2}}{3x} + C.$

[17] $u = \sqrt{5}x^2$, $\frac{1}{2\sqrt{5}}\,du = x\,dx \Rightarrow \text{I} = \dfrac{1}{2\sqrt{5}}\displaystyle\int \frac{1}{u^2 - 3}\,du.$

Negating both sides of Formula 19 and using this formula with $a = \sqrt{3}$ gives us

$$\text{I} = \frac{1}{2\sqrt{5}}\left[\frac{1}{2\sqrt{3}}\ln\left|\frac{u - \sqrt{3}}{u + \sqrt{3}}\right|\right] + C = \frac{1}{4\sqrt{15}}\ln\left|\frac{\sqrt{5}x^2 - \sqrt{3}}{\sqrt{5}x^2 + \sqrt{3}}\right| + C.$$

[18] $u = \sin x$, $du = \cos x\,dx \Rightarrow \mathrm{I} = \int \sqrt{u^2 - \frac{1}{4}}\,du$.

Using Formula 39 with $a = \frac{1}{2}$, $\mathrm{I} = \frac{u}{2}\sqrt{u^2 - \frac{1}{4}} - \frac{1}{8}\ln\left|u + \sqrt{u^2 - \frac{1}{4}}\right| + C$.

[19] $u = e^x$, $du = e^x\,dx \Rightarrow \mathrm{I} = \int e^x \cos^{-1} e^x\,(e^x)\,dx = \int u\cos^{-1}u\,du$.

Using Formula 91, $\mathrm{I} = \frac{1}{4}(2u^2 - 1)\cos^{-1}u - \frac{1}{4}(u\sqrt{1 - u^2}) + C$.

[20] Using Formula 86 (second form), $\mathrm{I} = \frac{1}{5}(\sin^3 x\cos^2 x) + \frac{2}{5}\int \sin^2 x\cos x\,dx =$

$\frac{1}{5}(\sin^3 x\cos^2 x) + \frac{2}{5}(\frac{1}{3}\sin^3 x) + C = \frac{1}{5}\sin^3 x\cos^2 x + \frac{2}{15}\sin^3 x + C$.

[21] Using Formula 60 three times, $\mathrm{I} = \frac{2}{9}\left[x^3(2 + x)^{3/2} - 6\int x^2\sqrt{2 + x}\,dx\right]$

$$= \tfrac{2}{9}x^3(2 + x)^{3/2} - \tfrac{4}{3}\left\{\tfrac{2}{7}\left[x^2(2 + x)^{3/2} - 4\int x\sqrt{2 + x}\,dx\right]\right\}$$
$$= \tfrac{2}{9}x^3(2 + x)^{3/2} - \tfrac{8}{21}x^2(2 + x)^{3/2} + \tfrac{32}{21}\left\{\tfrac{2}{5}\left[x(2 + x)^{3/2} - 2\int\sqrt{2 + x}\,dx\right]\right\}$$
$$= \tfrac{2}{9}x^3(2 + x)^{3/2} - \tfrac{8}{21}x^2(2 + x)^{3/2} + \tfrac{64}{105}x(2 + x)^{3/2} - \tfrac{256}{315}(2 + x)^{3/2} + C$$
$$= \tfrac{2}{315}(35x^3 - 60x^2 + 96x - 128)(2 + x)^{3/2} + C.$$

[22] Using Formula 61, $\mathrm{I} = 7\left[\dfrac{2x^3\sqrt{2 - x}}{-7} - \dfrac{12}{-7}\displaystyle\int\frac{x^2}{\sqrt{2 - x}}\,dx\right]$. Using Formula 56,

$$\mathrm{I} = -2x^3\sqrt{2 - x} + 12\left[\tfrac{2}{-15}(32 + 3x^2 + 8x)\sqrt{2 - x}\,\right] + C$$
$$= -2x^3\sqrt{2 - x} - \tfrac{8}{5}(32 + 3x^2 + 8x)\sqrt{2 - x} + C.$$

[23] $u = \sin x$, $du = \cos x\,dx \Rightarrow \mathrm{I} = \displaystyle\int\frac{2\sin x\cos x}{4 + 9\sin x}\,dx = 2\int\frac{u}{4 + 9u}\,du$.

Using Formula 47, $\mathrm{I} = \frac{2}{81}(4 + 9u - 4\ln|4 + 9u|) + C$.

[24] $u = \sec x$, $du = \sec x\tan x\,dx\left\{\frac{du}{u} = \tan x\,dx\right\} \Rightarrow$

$\mathrm{I} = \displaystyle\int\frac{1}{u\sqrt{4 + 3u}}\,du$. Using Formula 57, $\mathrm{I} = \dfrac{1}{2}\ln\left|\dfrac{\sqrt{4 + 3u} - 2}{\sqrt{4 + 3u} + 2}\right| + C$.

[25] Using Formula 58, $\mathrm{I} = 2\sqrt{9 + 2x} + 9\displaystyle\int\frac{1}{x\sqrt{9 + 2x}}\,dx$.

Using Formula 57, $\mathrm{I} = 2\sqrt{9 + 2x} + 3\ln\left|\dfrac{\sqrt{9 + 2x} - 3}{\sqrt{9 + 2x} + 3}\right| + C$.

[26] *Note*: $8x^3 - 3x^2 > 0 \Leftrightarrow x > \frac{3}{8} > 0$, so $\sqrt{8x^3 - 3x^2} = x\sqrt{-3 + 8x}$.

Using Formula 54 with $a = -3$ and $b = 8$, $\mathrm{I} = \int x\sqrt{-3 + 8x}\,dx =$

$\frac{2}{960}(24x + 6)(8x - 3)^{3/2} + C = \frac{1}{80}(4x + 1)(8x - 3)^{3/2} + C$.

[27] $u = x^{1/3}$, $x = u^3$, and $dx = 3u^2\,du \Rightarrow \mathrm{I} = \displaystyle\int\frac{3}{u(4 + u)}\,du$.

Using Formula 49 with $a = 4$ and $b = 1$, $\mathrm{I} = \dfrac{3}{4}\ln\left|\dfrac{u}{4 + u}\right| + C$.

[28] $u = x^{1/2}$, $x = u^2$, and $dx = 2u\,du \Rightarrow \text{I} = \int \frac{2u}{2u^3 + 5u^4}\,du = 2\int \frac{1}{u^2(2 + 5u)}\,du.$

Using Formula 50, $\text{I} = 2\left(-\frac{1}{2u} + \frac{5}{4}\ln\left|\frac{2 + 5u}{u}\right|\right) + C = -\frac{1}{u} + \frac{5}{2}\ln\left|\frac{2 + 5u}{u}\right| + C.$

[29] $u = \sec x$, $du = \sec x \tan x\,dx\ \left\{\frac{du}{u} = \tan x\,dx\right\} \Rightarrow$

$\text{I} = \int \frac{\sqrt{16 - u^2}}{u}\,du.$ Using Formula 32, $\text{I} = \sqrt{16 - u^2} - 4\ln\left|\frac{4 + \sqrt{16 - u^2}}{u}\right| + C.$

[30] $u = \csc x$, $du = -\csc x \cot x\,dx\ \left\{-\frac{du}{u} = \cot x\,dx\right\} \Rightarrow$

$\text{I} = -\int \frac{1}{u\sqrt{4 - u^2}}\,du.$ Using Formula 35, $\text{I} = \frac{1}{2}\ln\left|\frac{2 + \sqrt{4 - u^2}}{u}\right| + C.$

9.8 Review Exercises

[1] $\text{I} \overset{\text{A}}{=} \frac{1}{2}x^2\sin^{-1}x - \frac{1}{2}\int \frac{x^2}{\sqrt{1 - x^2}}\,dx.$ In the last integral, I_1, $x = \sin\theta$, $dx = \cos\theta\,d\theta \Rightarrow$

$\text{I}_1 = \int \frac{\sin^2\theta}{\sqrt{\cos^2\theta}}\cos\theta\,d\theta = \int \sin^2\theta\,d\theta = \frac{1}{2}\int(1 - \cos 2\theta)\,d\theta = \frac{1}{2}(\theta - \sin\theta\cos\theta) + C$

$= \frac{1}{2}(\sin^{-1}x - x\sqrt{1 - x^2}) + C.$ Thus, $\text{I} = \frac{1}{2}x^2\sin^{-1}x - \frac{1}{4}\sin^{-1}x + \frac{1}{4}x\sqrt{1 - x^2} + C.$

A. $u = \sin^{-1}x$, $du = \frac{1}{\sqrt{1 - x^2}}\,dx$, $dv = x\,dx$, $v = \frac{1}{2}x^2$

[2] $\text{I} \overset{\text{A}}{=} \frac{1}{3}\tan 3x \sec 3x - \int \tan^2 3x \sec 3x\,dx = \frac{1}{3}\tan 3x \sec 3x - \int(\sec^2 3x - 1)\sec 3x\,dx \Rightarrow$

$2\text{I} = \frac{1}{3}\tan 3x \sec 3x + \frac{1}{3}\ln|\sec 3x + \tan 3x| + C \Rightarrow$

$\text{I} = \frac{1}{6}\tan 3x \sec 3x + \frac{1}{6}\ln|\sec 3x + \tan 3x| + C.$

A. $u = \sec 3x$, $du = 3\sec 3x \tan 3x\,dx$, $dv = \sec^2 3x\,dx$, $v = \frac{1}{3}\tan 3x$

[3] $\text{I} \overset{\text{A}}{=} \Big[x\ln(1 + x)\Big]_0^1 - \int_0^1 \frac{x}{1 + x}\,dx = \Big[x\ln(1 + x)\Big]_0^1 - \int_0^1\left[1 - \frac{1}{1 + x}\right]dx =$

$\Big[x\ln(1 + x) - x + \ln(1 + x)\Big]_0^1 = 2\ln 2 - 1 \approx 0.39.$

A. $u = \ln(1 + x)$, $du = \frac{1}{1 + x}\,dx$, $dv = dx$, $v = x$

[4] $z = \sqrt{x}$, $z^2 = x$, and $2z\,dz = dx \Rightarrow$

$\text{I} = 2\int_0^1 ze^z\,dz \overset{\text{A}}{=} 2\Big[ze^z\Big]_0^1 - 2\int_0^1 e^z\,dz = 2\Big[ze^z - e^z\Big]_0^1 = 2.$

A. $u = z$, $du = dz$, $dv = e^z\,dz$, $v = e^z$

[5] $u = \sin 2x$, $\frac{1}{2}\,du = \cos 2x\,dx \Rightarrow \text{I} = \frac{1}{2}\int(1 - u^2)\,u^2\,du = \frac{1}{6}u^3 - \frac{1}{10}u^5 + C.$

[6] $\text{I} = \int\left[\frac{1}{2}(1 + \cos 2x)\right]^2 dx = \frac{1}{4}\int(1 + 2\cos 2x + \cos^2 2x)\,dx =$

$\frac{1}{4}\int\left[1 + 2\cos 2x + \frac{1}{2}(1 + \cos 4x)\right]dx = \frac{3}{8}x + \frac{1}{4}\sin 2x + \frac{1}{32}\sin 4x + C.$

[7] $\text{I} = \int \sec^4 x(\sec x \tan x)\,dx = \frac{1}{5}\sec^5 x + C.$

[8] $\mathrm{I} = \int \sec^5 x(\sec x \tan x)\, dx = \frac{1}{6}\sec^6 x + C.$

[9] $x = 5\tan\theta,\ dx = 5\sec^2\theta\, d\theta \Rightarrow$

$$\mathrm{I} = \int \frac{5\sec^2\theta}{(5\sec\theta)^3}\, d\theta = \frac{1}{25}\int \cos\theta\, d\theta = \frac{1}{25}\sin\theta + C = \frac{x}{25\sqrt{x^2 + 25}} + C.$$

[10] $x = 4\sin\theta,\ dx = 4\cos\theta\, d\theta \Rightarrow$

$$\mathrm{I} = \int \frac{4\cos\theta}{(4\sin\theta)^2(4\cos\theta)}\, d\theta = \frac{1}{16}\int \csc^2\theta\, d\theta = -\frac{1}{16}\cot\theta + C = -\frac{\sqrt{16 - x^2}}{16x} + C.$$

[11] $x = 2\sin\theta,\ dx = 2\cos\theta\, d\theta \Rightarrow$

$$\mathrm{I} = \int \frac{(2\cos\theta)(2\cos\theta)}{2\sin\theta}\, d\theta = 2\int \frac{1 - \sin^2\theta}{\sin\theta}\, d\theta = 2\int (\csc\theta - \sin\theta)\, d\theta$$

$$= 2\left[\ln|\csc\theta - \cot\theta| + \cos\theta\right] + C = 2\ln\left|\frac{2 - \sqrt{4 - x^2}}{x}\right| + \sqrt{4 - x^2} + C.$$

[12] $u = x^2 + 1,\ \frac{1}{2}\, du = x\, dx \Rightarrow \mathrm{I} = \frac{1}{2}\int u^{-2}\, du = -\frac{1}{2u} + C.$

[13] $\dfrac{x^3 + 1}{x(x - 1)^3} = \dfrac{A}{x} + \dfrac{B}{x - 1} + \dfrac{C}{(x - 1)^2} + \dfrac{D}{(x - 1)^3} \Rightarrow$

$x^3 + 1 = A(x - 1)^3 + Bx(x - 1)^2 + Cx(x - 1) + Dx.$

$x = 1 \Rightarrow D = 2$ and $x = 0 \Rightarrow A = -1$. Equating x^3 terms: $1 = A + B \Rightarrow$

$B = 2$, and x^2 terms: $0 = -3A - 2B + C \Rightarrow C = 1$. Thus,

$$\mathrm{I} = \int \left[\frac{-1}{x} + \frac{2}{x - 1} + \frac{1}{(x - 1)^2} + \frac{2}{(x - 1)^3}\right] dx$$

$$= -\ln|x| + 2\ln|x - 1| - \frac{1}{x - 1} - \frac{1}{(x - 1)^2} + K =$$

$$2\ln|x - 1| - \ln|x| - \frac{x}{(x - 1)^2} + K.$$

[14] $\dfrac{1}{x(1 + x^2)} = \dfrac{A}{x} + \dfrac{Bx + C}{1 + x^2} \Rightarrow 1 = A(1 + x^2) + (Bx + C)x.\ \ x = 0 \Rightarrow 1 = A.$

Equating x terms: $0 = C$, and x^2 terms: $0 = A + B \Rightarrow B = -1$.

Thus, $\mathrm{I} = \displaystyle\int \left[\frac{1}{x} + \frac{-x}{1 + x^2}\right] dx = \ln|x| - \frac{1}{2}\ln(1 + x^2) + C.$

[15] $\dfrac{x^3 - 20x^2 - 63x - 198}{(x - 3)(x + 3)(x^2 + 9)} = \dfrac{A}{x - 3} + \dfrac{B}{x + 3} + \dfrac{Cx + D}{x^2 + 9} \Rightarrow x^3 - 20x^2 - 63x - 198$

$= A(x + 3)(x^2 + 9) + B(x - 3)(x^2 + 9) + (Cx + D)(x - 3)(x + 3).$

$x = 3 \Rightarrow A = -5$ and $x = -3 \Rightarrow B = 2$. Equating x^3 terms:

$A + B + C \Rightarrow C = 4$, and x^2 terms: $-20 = 3A - 3B + D \Rightarrow D = 1$.

Thus, $\mathrm{I} = \displaystyle\int \left[\frac{-5}{x - 3} + \frac{2}{x + 3} + \frac{4x + 1}{x^2 + 9}\right] dx =$

$$-5\ln|x - 3| + 2\ln|x + 3| + 2\ln(x^2 + 9) + \tfrac{1}{3}\tan^{-1}(x/3) + K.$$

[16] $u = x + 2$, and $dx = du \Rightarrow$

$$\mathrm{I} = \int \frac{(u-2)-1}{u^5}\,du = \int \frac{u-3}{u^5}\,du = \int (u^{-4} - 3u^{-5})\,du = -\frac{1}{3u^3} + \frac{3}{4u^4} + C.$$

[17] $4 + 4x - x^2 = 4 - (x^2 - 4x + 4) + 4 = 8 - (x-2)^2$; $u = x - 2$, $du = dx \Rightarrow$

$$\mathrm{I} = \int \frac{u+2}{\sqrt{8-u^2}}\,du = \int \frac{u}{\sqrt{8-u^2}}\,du + \int \frac{2}{\sqrt{8-u^2}}\,du =$$

$$-(8-u^2)^{1/2} + 2\sin^{-1}\frac{u}{\sqrt{8}} + C = -\sqrt{4+4x-x^2} + 2\sin^{-1}\frac{x-2}{\sqrt{8}} + C.$$

[18] $x^2 + 6x + 13 = (x+3)^2 + 4$; $u = x + 3$, $du = dx \Rightarrow \mathrm{I} = \displaystyle\int \frac{u-3}{u^2+4}\,du =$

$$\tfrac{1}{2}\ln(u^2+4) - \tfrac{3}{2}\tan^{-1}\tfrac{u}{2} + C = \tfrac{1}{2}\ln(x^2+6x+13) - \tfrac{3}{2}\tan^{-1}\frac{x+3}{2} + C.$$

[19] $u^3 = x + 8$; $3u^2\,du = dx \Rightarrow \mathrm{I} = \displaystyle\int \frac{u(3u^2)}{u^3-8}\,du = 3\int\left[1 + \frac{8}{u^3-8}\right]du.$ $u^3 - 8 =$

$(u-2)(u^2+2u+4) = (u-2)\left[(u+1)^2+3\right]$. $z = u + 1$, $dz = du \Rightarrow$

$$\mathrm{I} = 3\int\left[1 + \frac{8}{(z-3)(z^2+3)}\right]dz = 3\int dz + 24\int\left[\frac{\frac{1}{12}}{z-3} - \frac{\frac{1}{12}z + \frac{1}{4}}{z^2+3}\right]dz \;\{\mathrm{PF}\}$$

$$= 3z + 2\ln|z-3| - \ln|z^2+3| - \frac{6}{\sqrt{3}}\tan^{-1}\frac{z}{\sqrt{3}} + C$$

$$= 3(u+1) + 2\ln|u-2| - \ln|u^2+2u+4| - \frac{6}{\sqrt{3}}\tan^{-1}\frac{u+1}{\sqrt{3}} + C$$

$$= 3(x+8)^{1/3} + \ln\left[(x+8)^{1/3} - 2\right]^2 - \ln\left|(x+8)^{2/3} + 2(x+8)^{1/3} + 4\right| - \frac{6}{\sqrt{3}}\tan^{-1}\frac{(x+8)^{1/3}+1}{\sqrt{3}} + C.$$

[20] $u = 2\cos x + 3$, $-\frac{1}{2}\,du = \sin x\,dx \Rightarrow \mathrm{I} = -\frac{1}{2}\int \frac{1}{u}\,du = -\frac{1}{2}\ln|u| + C.$

[21] $\mathrm{I} \overset{\mathsf{A}}{=} \frac{1}{2}e^{2x}\sin 3x - \frac{3}{2}\int e^{2x}\cos 3x\,dx \overset{\mathsf{B}}{=} \frac{1}{2}e^{2x}\sin 3x - \frac{3}{2}\left[\frac{1}{2}e^{2x}\cos 3x + \frac{3}{2}\mathrm{I}\right] \Rightarrow$

$$\tfrac{13}{4}\mathrm{I} = \tfrac{1}{2}e^{2x}\sin 3x - \tfrac{3}{4}e^{2x}\cos 3x \Rightarrow \mathrm{I} = \tfrac{1}{13}e^{2x}(2\sin 3x - 3\cos 3x) + C.$$

A. $u = \sin 3x$, $du = 3\cos 3x\,dx$, $dv = e^{2x}\,dx$, $v = \frac{1}{2}e^{2x}$

B. $u = \cos 3x$, $du = -3\sin 3x\,dx$, $dv = e^{2x}\,dx$, $v = \frac{1}{2}e^{2x}$

[22] $\mathrm{I} \overset{\mathsf{A}}{=} x\cos(\ln x) + \int \sin(\ln x)\,dx \overset{\mathsf{B}}{=} x\cos(\ln x) + x\sin(\ln x) - \int \cos(\ln x)\,dx \Rightarrow$

$$2\mathrm{I} = x\cos(\ln x) + x\sin(\ln x) \Rightarrow \mathrm{I} = \tfrac{1}{2}x\left[\cos(\ln x) + \sin(\ln x)\right] + C.$$

A. $u = \cos(\ln x)$, $du = -\dfrac{\sin(\ln x)}{x}\,dx$, $dv = dx$, $v = x$

B. $u = \sin(\ln x)$, $du = \dfrac{\cos(\ln x)}{x}\,dx$, $dv = dx$, $v = x$

[23] $\mathrm{I} = \int \sin^3 x(1 - \sin^2 x)\cos x\,dx = \frac{1}{4}\sin^4 x - \frac{1}{6}\sin^6 x + C.$

[24] $\mathrm{I} = \int (\csc^2 3x - 1)\,dx = -\frac{1}{3}\cot 3x - x + C.$

[25] $u = 4 - x^2$, $-\frac{1}{2}\,du = x\,dx \Rightarrow \mathrm{I} = -\frac{1}{2}\int u^{-1/2}\,du = -\sqrt{u} + C.$

26 $3x = 2\tan\theta \Rightarrow x = \frac{2}{3}\tan\theta;\ dx = \frac{2}{3}\sec^2\theta\, d\theta \Rightarrow \mathrm{I} = \int \frac{\frac{2}{3}\sec^2\theta}{(\frac{2}{3}\tan\theta)(2\sec\theta)}\, d\theta =$

$$\frac{1}{2}\int \csc\theta\, d\theta = \frac{1}{2}\ln|\csc\theta - \cot\theta| + C = \frac{1}{2}\ln\left|\frac{\sqrt{9x^2+4} - 2}{3x}\right| + C.$$

27 $\mathrm{I} = \int\left[x^2 - 2x + 3 + \frac{-6x^2+1}{x^3+2x^2}\right]dx = \int\left[x^2 - 2x + 3 + \frac{-\frac{1}{4}}{x} + \frac{\frac{1}{2}}{x^2} + \frac{-\frac{23}{4}}{x+2}\right]dx$

$$\{\mathrm{PF}\} = \tfrac{1}{3}x^3 - x^2 + 3x - \tfrac{1}{4}\ln|x| - \frac{1}{2x} - \tfrac{23}{4}\ln|x+2| + C.$$

28 $\mathrm{I} = \int\left[1 + \frac{3x^2 - 9x + 27}{(x-3)(x^2+9)}\right]dx = \int\left[1 + \frac{\frac{3}{2}}{x-3} + \frac{\frac{3}{2}x - \frac{9}{2}}{x^2+9}\right]dx\ \{\mathrm{PF}\} =$

$$x + \tfrac{3}{2}\ln|x-3| + \tfrac{3}{4}\ln(x^2+9) - \tfrac{3}{2}\tan^{-1}(x/3) + C.$$

29 $u = x^{1/2},\ u^2 = x$, and $2u\, du = dx \Rightarrow$

$$\mathrm{I} = \int \frac{2u}{u^3+u}\, du = 2\int \frac{1}{u^2+1}\, du = 2\tan^{-1}u + C.$$

30 $u = x + 5,\ du = dx \Rightarrow \mathrm{I} = \int \frac{2(u-5)+1}{u^{100}}\, du = \int \frac{2u - 9}{u^{100}}\, du =$

$$\int (2u^{-99} - 9u^{-100})\, du = -\frac{2}{98u^{98}} + \frac{9}{99u^{99}} + C = \frac{1}{u^{99}}\left(\frac{1}{11} - \frac{1}{49}u\right) + C.$$

31 $u = e^x,\ du = e^x\, dx \Rightarrow \mathrm{I} = \int \sec u\, du = \ln|\sec u + \tan u| + C.$

32 $u = x^2,\ \frac{1}{2}du = x\, dx \Rightarrow \frac{1}{2}\int \tan u\, du = -\frac{1}{2}\ln|\cos u| + C.$

33 $\mathrm{I} \overset{\mathsf{A}}{=} -\frac{1}{5}x^2\cos 5x + \frac{2}{5}\int x\cos 5x\, dx \overset{\mathsf{B}}{=} -\frac{1}{5}x^2\cos 5x + \frac{2}{5}\left[\frac{1}{5}x\sin 5x - \frac{1}{5}\int \sin 5x\, dx\right] =$

$-\frac{1}{5}x^2\cos 5x + \frac{2}{25}x\sin 5x + \frac{2}{125}\cos 5x + C =$

$$\tfrac{1}{125}\left[10x\sin 5x - (25x^2 - 2)\cos 5x\right] + C.$$

A. $u = x^2,\ du = 2x\, dx,\ dv = \sin 5x\, dx,\ v = -\frac{1}{5}\cos 5x$

B. $u = x,\ du = dx,\ dv = \cos 5x\, dx,\ v = \frac{1}{5}\sin 5x$

34 $\mathrm{I} = \int (2\sin x\cos x)\cos x\, dx = 2\int \cos^2 x\sin x\, dx = -\frac{2}{3}\cos^3 x + C.$

35 $\mathrm{I} = \int \sin x(1 - \cos^2 x)\cos^{1/2}x\, dx = \frac{2}{7}\cos^{7/2}x - \frac{2}{3}\cos^{3/2}x + C.$

36 $\mathrm{I} = \int \cos 3x\, dx = \frac{1}{3}\sin 3x + C.$

37 $u = e^x,\ du = e^x\, dx \Rightarrow \mathrm{I} = \int (1+u)^{1/2}\, du = \frac{2}{3}(1+u)^{3/2} + C.$

38 $u = 4x^2 + 25,\ \frac{1}{8}du = x\, dx \Rightarrow \mathrm{I} = \frac{1}{8}\int u^{-1/2}\, du = \frac{1}{4}u^{1/2} + C.$

39 $2x = 5\tan\theta$ or $x = \frac{5}{2}\tan\theta,\ dx = \frac{5}{2}\sec^2\theta\, d\theta \Rightarrow$

$$\mathrm{I} = \int \frac{(\frac{5}{2}\tan\theta)^2(\frac{5}{2}\sec^2\theta)}{5\sec\theta}\, d\theta = \frac{25}{8}\int \tan^2\theta\sec\theta\, d\theta = \frac{25}{8}\int (\sec^2\theta - 1)\sec\theta\, d\theta$$

$$= \tfrac{25}{8}\left[\tfrac{1}{2}\sec\theta\tan\theta + \tfrac{1}{2}\ln|\sec\theta + \tan\theta| - \ln|\sec\theta + \tan\theta|\right] + C_1\{\text{Exam. 6, §9.1}\}$$

$$= \frac{25}{16}\left[\frac{\sqrt{4x^2+25}}{5}\cdot\frac{2x}{5} - \ln\left|\frac{\sqrt{4x^2+25}}{5} + \frac{2x}{5}\right|\right] + C_1$$

$$= \tfrac{1}{16}\left[2x\sqrt{4x^2+25} - 25\ln(\sqrt{4x^2+25} + 2x)\right] + C.$$

40 $x^2 + 8x + 25 = (x+4)^2 + 9$; $u = x + 4$, $du = dx \Rightarrow$

$$\mathrm{I} = \int \frac{3(u-4)+2}{u^2+9}\,du = \int \frac{3u-10}{u^2+9}\,du = \frac{3}{2}\ln(u^2+9) - \frac{10}{3}\tan^{-1}\frac{u}{3} + C$$

$$= \frac{3}{2}\ln(x^2+8x+25) - \frac{10}{3}\tan^{-1}\frac{x+4}{3} + C.$$

41 $u = \tan x$, $du = \sec^2 x\,dx \Rightarrow \mathrm{I} = \int u^2\,du = \frac{1}{3}u^3 + C.$

42 $\mathrm{I} = \int \sin^2 x(1-\sin^2 x)^2 \cos x\,dx$

$$= \int (\sin^2 x - 2\sin^4 x + \sin^6 x)\cos x\,dx = \tfrac{1}{3}\sin^3 x - \tfrac{2}{5}\sin^5 x + \tfrac{1}{7}\sin^7 x + C.$$

43 $\mathrm{I} \overset{\text{A}}{=} -x\csc x + \int \csc x\,dx = -x\csc x + \ln|\csc x - \cot x| + C.$

A. $u = x$, $du = dx$, $dv = \cot x\csc x\,dx$, $v = -\csc x$

44 $\mathrm{I} = \int (1 + 2\csc 2x + \csc^2 2x)\,dx = x + \ln|\csc 2x - \cot 2x| - \frac{1}{2}\cot 2x + C.$

45 $u = 8 - x^3$, $-\frac{1}{3}\,du = x^2\,dx \Rightarrow \mathrm{I} = -\frac{1}{3}\int u^{1/3}\,du = -\frac{1}{4}u^{4/3} + C.$

46 $\mathrm{I} \overset{\text{A}}{=} \frac{1}{2}x^2(\ln x)^2 - \int x\ln x\,dx$

$$\overset{\text{B}}{=} \tfrac{1}{2}x^2(\ln x)^2 - \tfrac{1}{2}x^2\ln x + \tfrac{1}{2}\int x\,dx = \tfrac{1}{2}x^2(\ln x)^2 - \tfrac{1}{2}x^2\ln x + \tfrac{1}{4}x^2 + C.$$

A. $u = (\ln x)^2$, $du = \dfrac{2\ln x}{x}\,dx$, $dv = x\,dx$, $v = \frac{1}{2}x^2$

B. $u = \ln x$, $du = \dfrac{1}{x}\,dx$, $dv = x\,dx$, $v = \frac{1}{2}x^2$

47 $z = \sqrt{x}$, $z^2 = x$, and $2z\,dz = dx \Rightarrow \mathrm{I} = \int z\sin z(2z)\,dz = 2\int z^2\sin z\,dz.$

$$\mathrm{I} \overset{\text{A}}{=} 2\Big[-z^2\cos z + 2\int z\cos z\,dz\Big] \overset{\text{B}}{=} -2z^2\cos z + 4\Big[z\sin z - \int \sin z\,dz\Big] =$$

$$-2z^2\cos z + 4z\sin z + 4\cos z + C = -2x\cos\sqrt{x} + 4\sqrt{x}\sin\sqrt{x} + 4\cos\sqrt{x} + C.$$

A. $u = z^2$, $du = 2z\,dz$, $dv = \sin z\,dz$, $v = -\cos z$

B. $u = z$, $du = dz$, $dv = \cos z\,dz$, $v = \sin z$

48 $u = (5 - 3x)^{1/2}$, $x = \frac{1}{3}(5 - u^2)$, and $dx = -\frac{2}{3}u\,du \Rightarrow$

$$\mathrm{I} = \int \tfrac{1}{3}(5-u^2)(u)(-\tfrac{2}{3}u)\,du = -\tfrac{2}{9}\int (5u^2 - u^4)\,du = -\tfrac{10}{27}u^3 + \tfrac{2}{45}u^5 + C.$$

49 $u = e^x$, $du = e^x\,dx \Rightarrow \mathrm{I} = \displaystyle\int \frac{e^{2x}}{1+e^x}(e^x)\,dx = \int \frac{u^2}{1+u}\,du = \int \Big[u - 1 + \frac{1}{1+u}\Big]\,du$

$$= \tfrac{1}{2}u^2 - u + \ln|1+u| + C.$$

50 $u = e^{2x}$, $\frac{1}{2}\,du = e^{2x}\,dx \Rightarrow$

$$\mathrm{I} = \int \frac{1}{4 + (e^{2x})^2}(e^{2x})\,dx = \frac{1}{2}\int \frac{1}{4+u^2}\,du = \frac{1}{4}\tan^{-1}\frac{u}{2} + C.$$

51 $\mathrm{I} = \int (x^{3/2} - 4x^{1/2} + 3x^{-1/2})\,dx = \frac{2}{5}x^{5/2} - \frac{8}{3}x^{3/2} + 6x^{1/2} + C.$

52 $u = \sqrt{1 + \sin x}$, $u^2 = 1 + \sin x$, $2u\,du = \cos x\,dx$, and $\cos^2 x = 1 - \sin^2 x =$

$$1 - (u^2-1)^2 = 2u^2 - u^4 \Rightarrow \mathrm{I} = \int \frac{2u^2 - u^4}{u}(2u)\,du = \frac{4}{3}u^3 - \frac{2}{5}u^5 + C.$$

[53] $u = \sqrt{16 - x^2}$, $u^2 = 16 - x^2$, and $-u\,du = x\,dx \Rightarrow$

$$\mathrm{I} = \int \frac{x^2}{\sqrt{16 - x^2}}\,x\,dx = \int \frac{16 - u^2}{u}(-u)\,du = \tfrac{1}{3}u^3 - 16u + C.$$

[54] $u = 25 - 9x^2$, $-\frac{1}{18}\,du = x\,dx \Rightarrow \mathrm{I} = -\frac{1}{18}\int \frac{1}{u}\,du = -\frac{1}{18}\ln|u| + C.$

[55] $\mathrm{I} = \int \left[\dfrac{\frac{11}{2}}{x + 5} - \dfrac{\frac{15}{2}}{x + 7}\right] dx\ \{\mathrm{PF}\} = \frac{11}{2}\ln|x + 5| - \frac{15}{2}\ln|x + 7| + C.$

[56] $x^2 - 6x + 18 = (x - 3)^2 + 9$; $u = x - 3$, $du = dx \Rightarrow$

$$\mathrm{I} = 7\int \frac{1}{u^2 + 3^2}\,du = \frac{7}{3}\tan^{-1}\frac{u}{3} + C.$$

[57] $\mathrm{I} \overset{\mathsf{A}}{=} x\tan^{-1}5x - \displaystyle\int \frac{5x}{1 + 25x^2}\,dx = x\tan^{-1}5x - \frac{1}{10}\ln(1 + 25x^2) + C.$

A. $u = \tan^{-1}5x$, $du = \dfrac{5}{1 + 25x^2}\,dx$, $dv = dx$, $v = x$

[58] $\mathrm{I} = \int \left[\frac{1}{2}(1 - \cos 6x)\right]^2 dx = \frac{1}{4}\int (1 - 2\cos 6x + \cos^2 6x)\,dx =$

$$\tfrac{1}{4}\int \left[1 - 2\cos 6x + \tfrac{1}{2}(1 + \cos 12x)\right] dx = \tfrac{3}{8}x - \tfrac{1}{12}\sin 6x + \tfrac{1}{96}\sin 12x + C.$$

[59] $u = \tan x$, $du = \sec^2 x\,dx \Rightarrow \mathrm{I} = \int e^{\tan x}\sec^2 x\,dx = \int e^u\,du = e^u + C.$

[60] $u = 5x^2$, $\frac{1}{10}\,du = x\,dx \Rightarrow \mathrm{I} = \int \sin(5x^2)\,x\,dx = \frac{1}{10}\int \sin u\,du = -\frac{1}{10}\cos u + C.$

[61] $\sqrt{5}x = \sqrt{7}\tan\theta$ or $x = \sqrt{\frac{7}{5}}\tan\theta$, $dx = \sqrt{\frac{7}{5}}\sec^2\theta\,d\theta \Rightarrow \mathrm{I} = \displaystyle\int \frac{\sqrt{\frac{7}{5}}\sec^2\theta}{\sqrt{7}\sec\theta}\,d\theta = \frac{1}{\sqrt{5}}\int \sec\theta\,d\theta$

$$= \frac{1}{\sqrt{5}}\ln|\sec\theta + \tan\theta| + C_1 = \frac{1}{\sqrt{5}}\ln\left|\frac{\sqrt{7 + 5x^2}}{\sqrt{7}} + \frac{\sqrt{5}x}{\sqrt{7}}\right| + C_1 =$$

$$\frac{1}{\sqrt{5}}\ln\left|\sqrt{7 + 5x^2} + \sqrt{5}x\right| + C, \text{ where } C = C_1 - \frac{1}{\sqrt{5}}\ln\sqrt{7}.$$

[62] $\mathrm{I} = \displaystyle\int \frac{2x}{x^2 + 4}\,dx + \int \frac{3}{x^2 + 4}\,dx = \ln(x^2 + 4) + \frac{3}{2}\tan^{-1}\frac{x}{2} + C.$

[63] $\mathrm{I} = \int \cot^4 x(\csc^2 x - 1)\,dx = \int \left[\cot^4 x\csc^2 x - \cot^2 x(\csc^2 x - 1)\right] dx$

$= \int \left[\cot^4 x\csc^2 x - \cot^2 x\csc^2 x + \csc^2 x - 1\right] dx$

$= -\frac{1}{5}\cot^5 x + \frac{1}{3}\cot^3 x - \cot x - x + C.$

[64] $\mathrm{I} = \int \cot^4 x(\cot x\csc x)\,dx = -\frac{1}{5}\cot^5 x + C$

[65] $u = \sqrt{x^2 - 25}$, $u^2 + 25 = x^2$, and $u\,du = x\,dx \Rightarrow$

$$\mathrm{I} = \int (u^2 + 25)(u)\,u\,du = \tfrac{1}{5}u^5 + \tfrac{25}{3}u^3 + C.$$

[66] $u = \cos x$, $-du = \sin x\,dx \Rightarrow \mathrm{I} = -\int 10^u\,du = -\dfrac{10^u}{\ln 10} + C.$

[67] $\mathrm{I} = \frac{1}{3}x^3 - \frac{1}{4}\tanh 4x + C.$

[68] $\mathrm{I} \overset{\mathsf{A}}{=} x\sinh x - \int \sinh x\,dx = x\sinh x - \cosh x + C.$

A. $u = x$, $du = dx$, $dv = \cosh x\,dx$, $v = \sinh x$

[69] $\text{I} \overset{\text{A}}{=} -\frac{1}{4}x^2 e^{-4x} + \frac{1}{2}\int xe^{-4x}\,dx \overset{\text{B}}{=} -\frac{1}{4}x^2 e^{-4x} + \frac{1}{2}\left[-\frac{1}{4}xe^{-4x} + \frac{1}{4}\int e^{-4x}\,dx\right] =$

$$-\tfrac{1}{4}x^2 e^{-4x} - \tfrac{1}{8}xe^{-4x} - \tfrac{1}{32}e^{-4x} + C.$$

A. $u = x^2,\ du = 2x\,dx,\ dv = e^{-4x}\,dx,\ v = -\frac{1}{4}e^{-4x}$

B. $u = x,\ du = dx,\ dv = e^{-4x}\,dx,\ v = -\frac{1}{4}e^{-4x}$

[70] $u = \sqrt{x^3 + 1},\ x^3 = u^2 - 1$, and $\frac{2}{3}u\,du = x^2\,dx \Rightarrow \text{I} = \int x^3\sqrt{x^3 + 1}\,x^2\,dx =$

$$\int (u^2 - 1)(u)\left(\tfrac{2}{3}u\right)du = \tfrac{2}{3}\int (u^4 - u^2)\,du = \tfrac{2}{15}u^5 - \tfrac{2}{9}u^3 + C.$$

[71] $11 - 10x - x^2 = 11 - (x^2 + 10x + 25) + 25 = 36 - (x + 5)^2.$

$$u = x + 5,\ du = dx \Rightarrow \text{I} = \int \frac{3}{\sqrt{6^2 - u^2}}\,du = 3\sin^{-1}\frac{u}{6} + C.$$

[72] $\text{I} = \int (12x^{-1} + 7x^{-3})\,dx = 12\ln|x| - \dfrac{7}{2x^2} + C.$

[73] $\text{I} = \int \sin 7x\,dx = -\frac{1}{7}\cos 7x + C.$

[74] $\text{I} = \int e^1 e^{\ln 5x}\,dx = e\int 5x\,dx = \frac{5e}{2}x^2 + C.$

[75] $\text{I} = \displaystyle\int \left[\frac{-9}{x - 1} + \frac{18}{x - 2} + \frac{-5}{x - 3}\right] dx$ {PF}

$= -9\ln|x - 1| + 18\ln|x - 2| - 5\ln|x - 3| + C.$

[76] $x = 4\sin\theta,\ dx = 4\cos\theta\,d\theta \Rightarrow \text{I} = \displaystyle\int \frac{4\cos\theta}{(4\sin\theta)^4(4\cos\theta)}\,d\theta = \frac{1}{256}\int \csc^4\theta\,d\theta =$

$\frac{1}{256}\int (\cot^2\theta + 1)\csc^2\theta\,d\theta = \frac{1}{256}\left[-\frac{1}{3}\cot^3\theta - \cot\theta\right] + C =$

$$-\frac{1}{768}\cot^3\theta - \frac{1}{256}\cot\theta + C = -\frac{(16 - x^2)^{3/2}}{768x^3} - \frac{(16 - x^2)^{1/2}}{256x} + C.$$

[77] $\text{I} = \int x^3\cos x\,dx + \int \cos x\,dx = \text{I}_1 + \text{I}_2.\ \ \text{I}_1 \overset{\text{A}}{=} x^3\sin x - 3\int x^2\sin x\,dx \overset{\text{B}}{=}$

$x^3\sin x - 3\left[-x^2\cos x + 2\int x\cos x\,dx\right] = x^3\sin x + 3x^2\cos x - 6\int x\cos x\,dx \overset{\text{C}}{=}$

$x^3\sin x + 3x^2\cos x - 6\left[x\sin x - \int \sin x\,dx\right] =$

$x^3\sin x + 3x^2\cos x - 6x\sin x - 6\cos x + C$. Thus, $\text{I} = \text{I}_1 + \text{I}_2 =$

$$x^3\sin x + 3x^2\cos x - 6x\sin x - 6\cos x + \sin x + C.$$

A. $u = x^3,\ du = 3x^2\,dx,\ dv = \cos x\,dx,\ v = \sin x$

B. $u = x^2,\ du = 2x\,dx,\ dv = \sin x\,dx,\ v = -\cos x$

C. $u = x,\ du = dx,\ dv = \cos x\,dx,\ v = \sin x$

[78] $u = x - 3,\ du = dx \Rightarrow \text{I} = \int u^2(u + 4)\,du = \frac{1}{4}u^4 + \frac{4}{3}u^3 + C.$

[79] $2x = 3\sin\theta$ or $x = \frac{3}{2}\sin\theta,\ dx = \frac{3}{2}\cos\theta\,d\theta \Rightarrow$

$$\text{I} = \int \frac{3\cos\theta}{\left(\frac{3}{2}\sin\theta\right)^2}\left(\tfrac{3}{2}\cos\theta\right)d\theta = 2\int \cot^2\theta\,d\theta = 2\int (\csc^2\theta - 1)\,d\theta = -2\cot\theta - 2\theta + C$$

$$= -\frac{\sqrt{9 - 4x^2}}{x} - 2\sin^{-1}\frac{2x}{3} + C.$$

[80] $I = \int \left[\frac{4}{x+3} - \frac{4}{(x+3)^2} + \frac{1}{(x-3)^2} \right] dx \ \{\text{PF}\} =$

$$4\ln|x+3| + \frac{4}{x+3} - \frac{1}{x-3} + C.$$

[81] $I = \int (25 - 10\cot 3x + \cot^2 3x)\, dx$

$= \int (25 - 10\cot 3x + \csc^2 3x - 1)\, dx = 24x - \frac{10}{3}\ln|\sin 3x| - \frac{1}{3}\cot 3x + C.$

[82] $u = x^2 + 5,\ \frac{1}{2}\,du = x\,dx \Rightarrow I = \frac{1}{2}\int u^{3/4}\,du = \frac{2}{7}u^{7/4} + C.$

[83] $u = \sqrt[4]{x},\ u^4 = x$, and $4u^3\,du = dx \Rightarrow$

$$I = \int \frac{4u^3}{u^4(u^2+u)}\,du = 4\int \frac{1}{u^2(u+1)}\,du = 4\int \left[-\frac{1}{u} + \frac{1}{u^2} + \frac{1}{u+1} \right] du \ \{\text{PF}\} =$$

$$-4\ln|u| - \frac{4}{u} + 4\ln|u+1| + C = -\ln x - \frac{4}{\sqrt[4]{x}} + 4\ln(\sqrt[4]{x} + 1) + C.$$

[84] $I = \int x\sec^2 4x\,dx \overset{A}{=} \frac{1}{4}x\tan 4x - \frac{1}{4}\int \tan 4x\,dx = \frac{1}{4}x\tan 4x + \frac{1}{16}\ln|\cos 4x| + C.$

A. $u = x,\ du = dx,\ dv = \sec^2 4x\,dx,\ v = \frac{1}{4}\tan 4x$

[85] $u = 1 + \cos x,\ -du = \sin x\,dx \Rightarrow I = -\int u^{-1/2}\,du = -2\sqrt{u} + C.$

[86] $I = \int \left[\frac{3x}{x^2+4} + \frac{1}{x-2} \right] dx \ \{\text{PF}\} = \frac{3}{2}\ln(x^2+4) + \ln|x-2| + C.$

[87] $I \overset{A}{=} -\frac{x}{2(25+x^2)} + \frac{1}{2}\int \frac{1}{25+x^2}\,dx = -\frac{x}{2(25+x^2)} + \frac{1}{10}\tan^{-1}\frac{x}{5} + C.$

A. $u = x,\ du = dx,\ dv = \frac{x}{(25+x^2)^2}\,dx,\ v = -\frac{1}{2(25+x^2)}$

[88] $I = \int \sin^4 x(1 - \sin^2 x)\cos x\,dx = \frac{1}{5}\sin^5 x - \frac{1}{7}\sin^7 x + C.$

[89] $I = \int (\sec^2 x - 1)\tan x\sec x\,dx = \frac{1}{3}\sec^3 x - \sec x + C.$

[90] $u = 4 + 9x^2,\ \frac{1}{18}\,du = x\,dx \Rightarrow I = \frac{1}{18}\int u^{-1/2}\,du = \frac{1}{9}\sqrt{u} + C.$

[91] $I = \int \left[\frac{7}{x^2+5} + \frac{2x-3}{x^2+4} \right] dx \ \{\text{PF}\} = 7\int \frac{1}{x^2+5}\,dx - 3\int \frac{1}{x^2+4}\,dx + \int \frac{2x}{x^2+4}\,dx$

$= \frac{7}{\sqrt{5}}\tan^{-1}\frac{x}{\sqrt{5}} - \frac{3}{2}\tan^{-1}\frac{x}{2} + \ln(x^2+4) + C.$

[92] $u = 1 + \cos x,\ -du = \sin x\,dx \Rightarrow I = -\int u^{-3}\,du = \frac{1}{2u^2} + C.$

[93] $I = \int (x^3 - 4x + 4x^{-1})\,dx = \frac{1}{4}x^4 - 2x^2 + 4\ln|x| + C.$

[94] $I = \int (\csc^2 x - 1)\csc x\,dx = \int (\csc^3 x - \csc x)\,dx =$

$-\frac{1}{2}\csc x\cot x + \frac{1}{2}\ln|\csc x - \cot x| - \ln|\csc x - \cot x| + C$ {see Exercise 21, §9.1} $=$

$$-\frac{1}{2}\csc x\cot x - \frac{1}{2}\ln|\csc x - \cot x| + C.$$

[95] $I \overset{A}{=} \frac{2}{5}x^{5/2}\ln x - \frac{2}{5}\int x^{3/2}\,dx = \frac{2}{5}x^{5/2}\ln x - \frac{4}{25}x^{5/2} + C.$

A. $u = \ln x,\ du = \frac{1}{x}\,dx,\ dv = x^{3/2}\,dx,\ v = \frac{2}{5}x^{5/2}$

[96] $u = x^{1/3},\ u^3 = x$, and $3u^2\,du = dx \Rightarrow$

$$I = \int \frac{u^3}{u-1}(3u^2)\,du = 3\int \frac{u^5}{u-1}\,du = 3\int \left[u^4 + u^3 + u^2 + u + 1 + \frac{1}{u-1} \right] du$$

$$= \frac{3}{5}u^5 + \frac{3}{4}u^4 + u^3 + \frac{3}{2}u^2 + 3u + 3\ln|u-1| + C.$$

[97] $u = (2x+3)^{1/3}$, $x = \frac{1}{2}(u^3 - 3)$, and $dx = \frac{3}{2}u^2\,du \Rightarrow$

$$\mathrm{I} = \int \frac{\frac{1}{4}(u^3-3)^2}{u}\left(\tfrac{3}{2}u^2\right)du = \frac{3}{8}\int (u^7 - 6u^4 + 9u)\,du = \frac{3}{64}u^8 - \frac{9}{20}u^5 + \frac{27}{16}u^2 + C.$$

[98] $\mathrm{I} = \displaystyle\int\left(\tan x - \frac{\sin^2 x}{\cos x}\right)dx = \int\left(\tan x - \frac{1-\cos^2 x}{\cos x}\right)dx$

$= \int(\tan x - \sec x + \cos x)\,dx = \ln|\sec x| - \ln|\sec x + \tan x| + \sin x + C.$

[99] $z = x^2$, $\frac{1}{2}\,dz = x\,dx \Rightarrow \mathrm{I} = \frac{1}{2}\int ze^z\,dz \overset{\text{A}}{=} \frac{1}{2}ze^z - \frac{1}{2}\int e^z\,dz + C =$

$\frac{1}{2}ze^z - \frac{1}{2}e^z + C = \frac{1}{2}x^2 e^{(x^2)} - \frac{1}{2}e^{(x^2)} + C = \frac{1}{2}e^{(x^2)}(x^2 - 1) + C.$

A. $u = z$, $du = dz$, $dv = e^z\,dz$, $v = e^z$

[100] $u = x + 1$, $du = dx \Rightarrow \mathrm{I} = \int (u+1)^2 u^{10}\,du = \int (u^{12} + 2u^{11} + u^{10})\,du =$

$\frac{1}{13}u^{13} + \frac{1}{6}u^{12} + \frac{1}{11}u^{11} + C.$

Chapter 10: Indeterminate Forms and Improper Integrals

Exercises 10.1

Note: Let L denote the indicated limit, and DNE denote *does not exist.* The notation $\{\frac{0}{0}\}$ or $\{\frac{\infty}{\infty}\}$ indicates the form of the limit, and that L'Hôpital's rule was applied to obtain the next limit, as opposed to a simplification of the limit.

1. $L\ \{\frac{0}{0}\} = \lim_{x\to 0} \frac{\cos x}{2} = \frac{1}{2}$

2. $L\ \{\frac{0}{0}\} = \lim_{x\to 0} \frac{5}{\sec^2 x} = \frac{5}{1} = 5$

3. $L\ \{\frac{0}{0}\} = \lim_{x\to 5} \frac{\frac{1}{2}(x-1)^{-1/2}}{2x} = \frac{1/4}{10} = \frac{1}{40}$

4. $L\ \{\frac{0}{0}\} = \lim_{x\to 4} \frac{1}{\frac{1}{3}(x+4)^{-2/3}} = \frac{1}{1/12} = 12$

5. $L\ \{\frac{0}{0}\} = \lim_{x\to 2} \frac{4x-5}{10x-7} = \frac{3}{13}$

6. $L\ \{\frac{0}{0}\} = \lim_{x\to -3} \frac{2x+2}{4x+3} = \frac{-4}{-9} = \frac{4}{9}$

7. $L = \frac{1-3+2}{1-2-1} = \frac{0}{-2} = 0$

8. $L = \frac{4-10+6}{8-2-7} = \frac{0}{-1} = 0$

9. $L\ \{\frac{0}{0}\} = \lim_{x\to 0} \frac{\cos x - 1}{\sec^2 x - 1}\ \{\frac{0}{0}\} = \lim_{x\to 0} \frac{-\sin x}{2\sec^2 x \tan x} = -\frac{1}{2}\lim_{x\to 0} \cos^3 x = -\frac{1}{2}$

10. $L\ \{\frac{0}{0}\} = \lim_{x\to 0} \frac{\cos x}{1-\sec^2 x} = -\infty$, since $(\cos x) \to 1$ and $(1-\sec^2 x) \to 0^-$ as $x \to 0$.

11. $L\ \{\frac{0}{0}\} = \lim_{x\to 0} \frac{1-e^x}{2x}\ \{\frac{0}{0}\} = \lim_{x\to 0} \frac{-e^x}{2} = \frac{-1}{2} = -\frac{1}{2}$

12. $L\ \{\frac{0}{0}\} = \lim_{x\to 0^+} \frac{1-e^x}{3x^2}\ \{\frac{0}{0}\} = \lim_{x\to 0^+} \frac{-e^x}{6x} = -\infty$,

    since $(-e^x) \to -1$ and $(6x) \to 0^+$ as $x \to 0^+$.

13. $L\ \{\frac{0}{0}\} = \lim_{x\to 0} \frac{1-\cos x}{3x^2}\ \{\frac{0}{0}\} = \lim_{x\to 0} \frac{\sin x}{6x}\ \{\frac{0}{0}\} = \lim_{x\to 0} \frac{\cos x}{6} = \frac{1}{6}$

14. $L\ \{\frac{0}{0}\} = \lim_{x\to \pi/2} \frac{-\cos x}{-\sin x} = \frac{0}{-1} = 0$

15. $L = \infty$, since $(\cos^2 x) \to 0^+$ and $(1+\sin x) \to 2$ as $x \to \frac{\pi}{2}$.

16. $L = \infty$, since $(\cos x) \to 1$ and $(x) \to 0^+$ as $x \to 0^+$.

17. $L\ \{\frac{\infty}{\infty}\} = \lim_{x\to (\pi/2)^-} \frac{\sec x \tan x}{3\sec^2 x} = \frac{1}{3}\lim_{x\to (\pi/2)^-} \sin x = \frac{1}{3}$

18. $L\ \{\frac{\infty}{\infty}\} = \lim_{x\to 0^+} \frac{1/x}{-\csc^2 x} = \lim_{x\to 0^+} \left(-\sin x \cdot \frac{\sin x}{x}\right) = 0 \cdot 1 = 0$

    Note: No distinction is made between ∞ and $-\infty$ for use in L'Hôpital's rule.

19. $L\ \{\frac{\infty}{\infty}\} = \lim_{x\to \infty} \frac{2x}{1/x} = \lim_{x\to \infty} 2x^2 = \infty$

20. $L\ \{\frac{\infty}{\infty}\} = \lim_{x\to \infty} \frac{1/x}{2x} = \lim_{x\to \infty} \frac{1}{2x^2} = 0$

[21] $L\ \{\frac{\infty}{\infty}\} = \lim_{x\to 0^+} \frac{\cos x/\sin x}{2\cos 2x/\sin 2x} = \lim_{x\to 0^+} \frac{\cos x \sin 2x}{2\cos 2x \sin x} =$

$$\lim_{x\to 0^+} \frac{(\cos x)(2\sin x \cos x)}{2\cos 2x \sin x} = \lim_{x\to 0^+} \frac{\cos^2 x}{\cos 2x} = \frac{1}{1} = 1$$

[22] $L\ \{\frac{0}{0}\} = \lim_{x\to 0} \frac{2}{1/(1+x^2)} = \frac{2}{1} = 2$

[23] $L\ \{\frac{0}{0}\} = \lim_{x\to 0} \frac{e^x + e^{-x} - 2\cos x}{x\cos x + \sin x}\ \{\frac{0}{0}\} =$

$$\lim_{x\to 0} \frac{e^x - e^{-x} + 2\sin x}{-x\sin x + \cos x + \cos x} = \frac{0}{2} = 0$$

[24] $L\ \{\frac{0}{0}\} = \lim_{x\to 2} \frac{1/(x-1)}{1} = \frac{1}{1} = 1$

[25] $L = \infty$, since $(x\cos x + e^{-x}) \to 1$ and $(x^2) \to 0^+$ as $x \to 0$.

[26] $L = \infty$, since $(2e^x - 3x - e^{-x}) \to 1$ and $(x^2) \to 0^+$ as $x \to 0$.

[27] $L\ \{\frac{\infty}{\infty}\} = \lim_{x\to\infty} \frac{4x+3}{10x+1}\ \{\frac{\infty}{\infty}\} = \lim_{x\to\infty} \frac{4}{10} = \frac{2}{5}$

[28] $L\ \{\frac{\infty}{\infty}\} = \lim_{x\to\infty} \frac{3x^2+1}{9x^2}\ \{\frac{\infty}{\infty}\} = \lim_{x\to\infty} \frac{6x}{18x}\ \{\frac{\infty}{\infty}\} = \lim_{x\to\infty} \frac{6}{18} = \frac{1}{3}$

[29] $L\ \{\frac{\infty}{\infty}\} = \lim_{x\to\infty} \frac{1+\ln x}{1+1/x} = \infty$, since $(1+\ln x) \to \infty$ and $(1+1/x) \to 1$ as $x \to \infty$.

[30] $L\ \{\frac{\infty}{\infty}\} = \lim_{x\to\infty} \frac{3e^{3x}}{1/x} = \lim_{x\to\infty} 3xe^{3x} = \infty$

[31] Let n be an integer > 0. After n applications of L'Hôpital's rule, $L = \lim_{x\to\infty} \frac{n!}{e^x} = 0$.

[32] $L = \infty$. The fraction is the reciprocal of the fraction in Exercise 31.

[33] $L\ \{\frac{0}{0}\} = \lim_{x\to 2^+} \frac{1/(x-1)}{2(x-2)} = \lim_{x\to 2^+} \frac{1}{2(x-1)(x-2)} = \infty,$

since $\left[2(x-1)(x-2)\right] \to 0^+$ as $x \to 2^+$.

[34] $L\ \{\frac{0}{0}\} = \lim_{x\to 0} \frac{2\sin x\cos x - 2\sin x}{-2\cos x\sin x - x\cos x - \sin x} =$

$$\lim_{x\to 0} \frac{\sin 2x - 2\sin x}{-\sin 2x - x\cos x - \sin x}\ \{\frac{0}{0}\} = \lim_{x\to 0} \frac{2\cos 2x - 2\cos x}{-2\cos 2x - 2\cos x + x\sin x} = \frac{0}{-4} = 0$$

[35] $L\ \{\frac{0}{0}\} = \lim_{x\to 0} \frac{2/\sqrt{1-4x^2}}{1/\sqrt{1-x^2}} = \frac{2}{1} = 2$

[36] $L\ \{\frac{\infty}{\infty}\} = \lim_{x\to\infty} \frac{1/(x\ln x)}{1/x} = \lim_{x\to\infty} \frac{1}{\ln x} = 0$

[37] $\lim_{x\to 0} \frac{\tan x - \sin x}{x^3 \tan x} = \lim_{x\to 0} \frac{\sin x - \sin x\cos x}{x^3 \sin x} = \lim_{x\to 0} \frac{1-\cos x}{x^3}\ \{\frac{0}{0}\}$

$$= \lim_{x\to 0} \frac{\sin x}{3x^2}\ \{\frac{0}{0}\} = \lim_{x\to 0} \frac{\cos x}{6x}.$$

The last limit DNE since the left-hand and right-hand limits do not agree.

[38] $L = \frac{0}{-1} = 0$

[39] $L = \frac{3-0}{5-0} = \frac{3}{5}$

[40] $\text{L}\ \{\tfrac{0}{0}\} = \lim_{x\to 0} \dfrac{-e^x + e^{-x}}{2\cos x \sin x} = \lim_{x\to 0} \dfrac{e^{-x} - e^x}{\sin 2x}\ \{\tfrac{0}{0}\} =$

$$\lim_{x\to 0} \frac{-e^{-x} - e^x}{2\cos 2x} = \frac{-2}{2} = -1$$

[41] $\text{L}\ \{\tfrac{0}{0}\} = \lim_{x\to 1} \dfrac{4x^3 - 3x^2 - 6x + 5}{4x^3 - 15x^2 + 18x - 7}\ \{\tfrac{0}{0}\} =$

$$\lim_{x\to 1} \frac{12x^2 - 6x - 6}{12x^2 - 30x + 18}\ \{\tfrac{0}{0}\} = \lim_{x\to 1} \frac{24x - 6}{24x - 30} = \frac{18}{-6} = -3$$

[42] $\text{L}\ \{\tfrac{0}{0}\} = \lim_{x\to 1} \dfrac{4x^3 + 3x^2 - 6x - 1}{4x^3 - 15x^2 + 18x - 7}\ \{\tfrac{0}{0}\} = \lim_{x\to 1} \dfrac{12x^2 + 6x - 6}{12x^2 - 30x + 18} =$

$$\lim_{x\to 1} \frac{12x^2 + 6x - 6}{6(x-1)(2x-3)}.$$

The last limit DNE since the left-hand and right-hand limits do not agree.

[43] $\text{L}\ \{\tfrac{0}{0}\} = \lim_{x\to 0} \dfrac{1 - \dfrac{1}{1+x^2}}{x\cos x + \sin x} = \lim_{x\to 0} \dfrac{x^2}{(1+x^2)(x\cos x + \sin x)}\ \{\tfrac{0}{0}\} =$

$$\lim_{x\to 0} \frac{2x}{2x(x\cos x + \sin x) + (1+x^2)(-x\sin x + \cos x + \cos x)} = \frac{0}{2} = 0$$

[44] $\text{L} = \dfrac{0}{1+0} = 0$

[45] $\text{L}\ \{\tfrac{\infty}{\infty}\} = \lim_{x\to\infty} \dfrac{\frac{3}{2}x^{1/2} + 5}{1 + \ln x}\ \{\tfrac{\infty}{\infty}\} = \lim_{x\to\infty} \dfrac{3/(4\sqrt{x})}{1/x} = \frac{3}{4}\lim_{x\to\infty} \sqrt{x} = \infty$

[46] $\text{L}\ \{\tfrac{0}{0}\} = \lim_{x\to 0} \dfrac{x(1-x^2)^{-1/2} + \sin^{-1} x}{1 - \cos x}\ \{\tfrac{0}{0}\} =$

$$\lim_{x\to 0} \frac{(1-x^2)^{-1/2} - \frac{1}{2}x(1-x^2)^{-3/2}(-2x) + (1-x^2)^{-1/2}}{\sin x} = \lim_{x\to 0} \frac{2 - x^2}{(1-x^2)^{3/2}\sin x}.$$

The last limit DNE since the left-hand and right-hand limits do not agree.

[47] $\text{L} = \infty$, since $\lim_{x\to\infty} \tan^{-1} x = \frac{\pi}{2}$.

[48] $\dfrac{\tan x}{\cot 2x} = \tan x \tan 2x = \dfrac{\sin x \sin 2x}{\cos x \cos 2x} = \dfrac{\sin x(2\sin x \cos x)}{\cos x \cos 2x} = \dfrac{2\sin^2 x}{\cos 2x}.$

Thus, as $x \to \frac{\pi}{2}^-$, $\text{L} = \dfrac{2}{-1} = -2$.

[49] $\text{L}\ \{\tfrac{\infty}{\infty}\} = \lim_{x\to\infty} \dfrac{6e^{3x} + 1/x}{3e^{3x} + 2x}\ \{\tfrac{\infty}{\infty}\} = \lim_{x\to\infty} \dfrac{18e^{3x} - 1/x^2}{9e^{3x} + 2}\ \{\tfrac{\infty}{\infty}\} =$

$$\lim_{x\to\infty} \frac{54e^{3x} + 2/x^3}{27e^{3x}} = \lim_{x\to\infty} \left(2 + \frac{2}{27x^3e^{3x}}\right) = 2$$

[50] Let $u = 1/x$.

Then, $\lim_{x\to 0^+} \dfrac{e^{-1/x}}{x} = \lim_{u\to\infty} \dfrac{e^{-u}}{1/u} = \lim_{u\to\infty} \frac{u}{e^u}\ \{\tfrac{\infty}{\infty}\} = \lim_{u\to\infty} \frac{1}{e^u} = 0.$

[51] $\text{L} = \lim_{x\to\infty} \left(1 - \frac{\cos x}{x}\right) = 1 - 0 = 1$. *Note*: If we apply L'Hôpital's rule we find that

$\text{L} = \lim_{x\to\infty} (1 + \sin x)$ and DNE. For L'Hôpital's rule to apply,

this limit must exist or equal $\pm\infty$. Thus, L'Hôpital's rule does not apply.

[52] $L\ \{\frac{\infty}{\infty}\} = \lim_{x\to\infty} \frac{1 + \sinh x}{2x}\ \{\frac{\infty}{\infty}\} = \lim_{x\to\infty} \frac{\cosh x}{2} = \infty$

[53] Let $f(x) = \frac{\ln(\tan x + \cos x)}{\sqrt{\ln(x^2 + 1)}}$. $f(10^{-1}) \approx 0.9129$, $f(10^{-2}) \approx 0.9901$,

$f(10^{-3}) \approx 0.9990$, $f(10^{-4}) \approx 0.9999$. We predict that $\lim_{x\to 0^+} f(x) \approx 1$.

[54] Let $f(x) = \frac{\tan^2(\sin^{-1} x)}{1 - \cos[\ln(1 + x)]}$.

$f(10^{-1}) \approx 2.2256$, $f(10^{-2}) \approx 2.0202$, $f(10^{-3}) \approx 2.0020$, $f(10^{-4}) \approx 2.0002$.

$f(-10^{-1}) \approx 1.8215$, $f(-10^{-2}) \approx 1.9802$, $f(-10^{-3}) \approx 1.9980$, $f(-10^{-4}) \approx 1.9998$.

We predict that $\lim_{x\to 0} f(x) \approx 2$.

[55] $$\lim_{k\to 0^+} v(t) = \lim_{k\to 0^+} \frac{mg(1 - e^{-kt/m})}{k}\ \{\tfrac{0}{0}\} = \lim_{k\to 0^+} \frac{mg(\frac{t}{m}e^{-kt/m})}{1} = mg(\tfrac{t}{m}) = gt.$$

[56] $$\lim_{k\to 0^+} s(t) = \lim_{k\to 0^+} \frac{m \ln \cosh (\frac{gt^2}{m}k)^{1/2}}{k}\ \{\tfrac{0}{0}\}\ \{\text{let } c = gt^2\}$$

$$= \lim_{k\to 0^+} \frac{m \tanh (\frac{c}{m}k)^{1/2} \cdot \frac{1}{2}(\frac{c}{m}k)^{-1/2} \cdot \frac{c}{m}}{1} = \lim_{k\to 0^+} \frac{(\frac{1}{2}c) \tanh (\frac{c}{m}k)^{1/2}}{(\frac{c}{m}k)^{1/2}}\ \{\tfrac{0}{0}\}$$

$$= \lim_{k\to 0^+} \frac{(\frac{1}{2}c) \operatorname{sech}^2 (\frac{c}{m}k)^{1/2} \cdot \frac{1}{2}(\frac{c}{m}k)^{-1/2} \cdot \frac{c}{m}}{\frac{1}{2}(\frac{c}{m}k)^{-1/2} \cdot \frac{c}{m}}$$

$$= \lim_{k\to 0^+} \left[(\tfrac{1}{2}c) \operatorname{sech}^2 (\tfrac{c}{m}k)^{1/2}\right] = \tfrac{1}{2}c = \tfrac{1}{2}gt^2$$

[57] $$\lim_{\omega\to\omega_0} s = \lim_{\omega\to\omega_0} \frac{A\omega^2(\cos\omega t - \cos\omega_0 t)}{\omega_0^2 - \omega^2}\ \{\tfrac{0}{0}\}$$

$$= \lim_{\omega\to\omega_0} \frac{A\omega^2(-t\sin\omega t) + 2A\omega(\cos\omega t - \cos\omega_0 t)}{-2\omega}$$

$$= \lim_{\omega\to\omega_0} \frac{A\omega(-t\sin\omega t) + 2A(\cos\omega t - \cos\omega_0 t)}{-2} = \tfrac{1}{2}A\omega_0 t \sin\omega_0 t,$$

which can be made arbitrarily large for large values of t.

That is, the resulting oscillations increase in magnitude.

[58] $\lim_{t\to\infty} \frac{K}{1 + ce^{-rt}} = \frac{K}{1 + 0} = K$;

the population will attain the carrying capacity over a long period of time.

$$\lim_{K\to\infty} y(t) = \lim_{K\to\infty} \frac{K}{1 + \frac{K - y(0)}{y(0)} e^{-rt}} = \lim_{K\to\infty} \frac{K y(0) e^{rt}}{y(0)e^{rt} + [K - y(0)]}\ \{\tfrac{\infty}{\infty}\} =$$

$$\frac{y(0)e^{rt}}{1} = y(0)e^{rt};$$

the population will grow exponentially if the carrying capacity is unbounded.

59 (a) $\text{L}\ \{\tfrac{0}{0}\} = \lim\limits_{x\to 0} \dfrac{Si'(x)}{1} = \lim\limits_{x\to 0} \dfrac{(\sin x)/x}{1} = \lim\limits_{x\to 0} \dfrac{\sin x}{x} = 1$

(b) $\text{L}\ \{\tfrac{0}{0}\} = \lim\limits_{x\to 0} \dfrac{\frac{\sin x}{x} - 1}{3x^2}\ \{\tfrac{0}{0}\} = \lim\limits_{x\to 0} \dfrac{(x\cos x - \sin x)/x^2}{6x} =$

$\frac{1}{6} \lim\limits_{x\to 0} \dfrac{x\cos x - \sin x}{x^3}\ \{\tfrac{0}{0}\} = \frac{1}{6} \lim\limits_{x\to 0} \dfrac{-x\sin x}{3x^2} = -\frac{1}{18} \lim\limits_{x\to 0} \dfrac{\sin x}{x} = -\frac{1}{18}(1) = -\frac{1}{18}$

60 (a) $\text{L}\ \{\tfrac{0}{0}\} = \lim\limits_{x\to 0} \dfrac{C'(x)}{1} = \lim\limits_{x\to 0} \cos x^2 = 1$

(b) $\text{L}\ \{\tfrac{0}{0}\} = \lim\limits_{x\to 0} \dfrac{C'(x) - 1}{5x^4} = \lim\limits_{x\to 0} \dfrac{\cos x^2 - 1}{5x^4}\ \{\tfrac{0}{0}\} = \lim\limits_{x\to 0} \dfrac{-2x\sin x^2}{20x^3} =$

$-\frac{1}{10} \lim\limits_{x\to 0} \dfrac{\sin x^2}{x^2} = -\frac{1}{10}(1) = -\frac{1}{10}$

61 $C(x) = \displaystyle\int_0^x \cos(u^2)\, du$. Let $f(x) = \cos(x^2)$.

Then, $C(\frac{1}{4}) \approx$

$\text{S} = \frac{1/4 - 0}{3(4)}\left[f(0) + 4f(\frac{1}{16}) + 2f(\frac{1}{8}) + 4f(\frac{3}{16}) + f(\frac{1}{4})\right]$

$\approx \frac{1}{48}(11.9953) \approx 0.2499$. Similarly, $C(\frac{1}{2}) \approx 0.4969$,

$C(\frac{3}{4}) \approx 0.7266$, and $C(1) \approx 0.9045$.

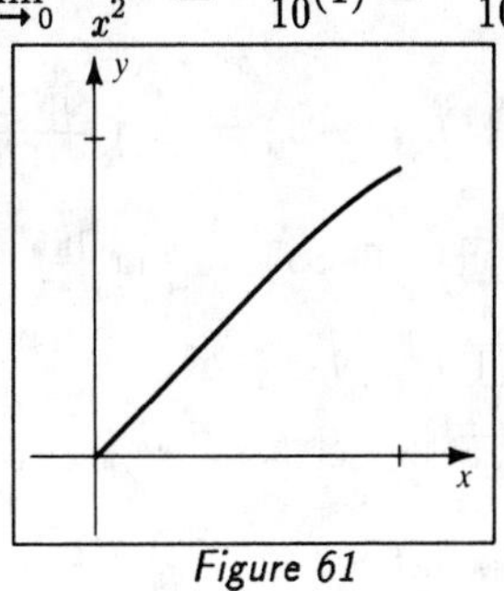

Figure 61

62 $V = \pi \displaystyle\int_0^1 [C(x)]^2\, dx$. Let $f(x) = [C(x)]^2$.

$V \approx \text{S} = (\pi)\frac{1-0}{3(4)}\left[f(0) + 4f(\frac{1}{4}) + 2f(\frac{1}{2}) + 4f(\frac{3}{4}) + f(1)\right]$

$\approx \frac{\pi}{12}\left[(0)^2 + 4(0.2499)^2 + 2(0.4969)^2 + 4(0.7266)^2 + (0.9045)^2\right] \approx 0.9617$

63 $\lim\limits_{n\to -1} \displaystyle\int_1^x t^n\, dt = \lim\limits_{n\to -1} \left[\frac{t^{n+1}}{n+1}\right]_1^x = \lim\limits_{n\to -1} \frac{x^{n+1} - 1}{n+1}\ \{\tfrac{0}{0}\} =$

$\lim\limits_{n\to -1} \dfrac{x^{n+1}\ln x}{1} = \ln x = \displaystyle\int_1^x t^{-1}\, dt$

64 $\lim\limits_{x\to\infty} \dfrac{f(x)}{g(x)} = \lim\limits_{x\to\infty} \dfrac{\int_0^x e^{(t^2)}\, dt}{e^{(x^2)}}\ \{\tfrac{\infty}{\infty}\} = \lim\limits_{x\to\infty} \dfrac{e^{(x^2)}}{2xe^{(x^2)}} = \lim\limits_{x\to\infty} \dfrac{1}{2x} = 0$

Exercises 10.2

Note: As in §10.1, the notation $\{\tfrac{0}{0}\}$ or $\{\tfrac{\infty}{\infty}\}$ indicates the form of the limit and that L'Hôpital's rule was applied to obtain the next limit. We have also included the notations $\{0\cdot\infty\}$, $\{0^0\}$, $\{\infty^0\}$, $\{1^\infty\}$, and $\{\infty - \infty\}$ to indicate the form of the limit.

1 $\text{L}\ \{0\cdot\infty\} = \lim\limits_{x\to 0^+} \dfrac{\ln x}{1/x}\ \{\tfrac{\infty}{\infty}\} = \lim\limits_{x\to 0^+} \dfrac{1/x}{-1/x^2} = \lim\limits_{x\to 0^+} (-x) = 0$

2 $\text{L}\ \{\infty\cdot 0\} = \lim\limits_{x\to(\pi/2)^-} \dfrac{\ln\sin x}{\cot x}\ \{\tfrac{0}{0}\} = \lim\limits_{x\to(\pi/2)^-} \dfrac{\cos x/\sin x}{-\csc^2 x} =$

$\lim\limits_{x\to(\pi/2)^-} (-\cos x\sin x) = 0$

[3] $L\ \{\infty\cdot 0\} = \lim_{x\to\infty}\frac{x^2-1}{e^{(x^2)}}\ \{\frac{\infty}{\infty}\} = \lim_{x\to\infty}\frac{2x}{2xe^{(x^2)}} = \lim_{x\to\infty}\frac{1}{e^{(x^2)}} = 0$

[4] $L\ \{\infty\cdot 0\} = \lim_{x\to\infty}\frac{e^{1/x}-1}{1/x}\ \{\frac{0}{0}\} = \lim_{x\to\infty}\frac{(-1/x^2)e^{1/x}}{(-1/x^2)} = \lim_{x\to\infty}e^{1/x} = e^0 = 1$

[5] $L = 1\cdot 0 = 0$

[6] $L = \infty$, since $\lim_{x\to-\infty}\tan^{-1}x = -\frac{\pi}{2}$.

[7] $L\ \{0\cdot\infty\} = \lim_{x\to 0^+}\frac{\ln\sin x}{\csc x}\ \{\frac{\infty}{\infty}\} = \lim_{x\to 0^+}\frac{\cos x/\sin x}{-\csc x\cot x} = \lim_{x\to 0^+}(-\sin x) = 0$

[8] $L\ \{\infty\cdot 0\} = \lim_{x\to\infty}\frac{\frac{\pi}{2}-\tan^{-1}x}{1/x}\ \{\frac{0}{0}\} = \lim_{x\to\infty}\frac{-1/(1+x^2)}{-1/x^2} = \lim_{x\to\infty}\frac{x^2}{1+x^2}\ \{\frac{\infty}{\infty}\} =$

$\lim_{x\to\infty}\frac{2x}{2x} = \lim_{x\to\infty}1 = 1$

[9] $L\ \{\infty\cdot 0\} = \lim_{x\to\infty}\frac{\sin(1/x)}{1/x}\ \{\frac{0}{0}\} = \lim_{x\to\infty}\frac{(-1/x^2)\cos(1/x)}{-1/x^2} = \lim_{x\to\infty}\cos(1/x) = 1$

[10] $L\ \{0\cdot\infty\} = \lim_{x\to\infty}\frac{\ln x}{e^x}\ \{\frac{\infty}{\infty}\} = \lim_{x\to\infty}\frac{1/x}{e^x} = \lim_{x\to\infty}\frac{1}{xe^x} = 0$

[11] $L = 0\cdot 1 = 0$

[12] $L = (\cos 0)^{0+1} = (1)^1 = 1$

[13] $\lim_{x\to\infty}(1+\frac{1}{x})^{5x}$ is of the form 1^∞.

Let $y = (1+\frac{1}{x})^{5x}$ and hence $\ln y = 5x\ln(1+\frac{1}{x})$. $\lim_{x\to\infty}\ln y\ \{\infty\cdot 0\} =$

$\lim_{x\to\infty}\frac{\ln(1+\frac{1}{x})}{1/(5x)}\ \{\frac{0}{0}\} = \lim_{x\to\infty}\frac{(-1/x^2)/(1+\frac{1}{x})}{-1/(5x^2)} = \lim_{x\to\infty}\frac{5}{1+\frac{1}{x}} = 5.$

Thus, $\lim_{x\to\infty}\ln y = \ln\lim_{x\to\infty}y = 5 \Rightarrow \lim_{x\to\infty}y = e^5$.

[14] $\lim_{x\to 0^+}(e^x+3x)^{1/x}$ is of the form 1^∞. $y = (e^x+3x)^{1/x} \Rightarrow \ln y = \frac{1}{x}\ln(e^x+3x)$.

$\lim_{x\to 0^+}\ln y\ \{\infty\cdot 0\} = \lim_{x\to 0^+}\frac{\ln(e^x+3x)}{x}\ \{\frac{0}{0}\} = \lim_{x\to 0^+}\frac{e^x+3}{e^x+3x} = 4$. Thus, $L = e^4$.

[15] $\lim_{x\to 0^+}(e^x-1)^x$ is of the form 0^0. $y = (e^x-1)^x \Rightarrow \ln y = x\ln(e^x-1)$.

$\lim_{x\to 0^+}\ln y\ \{0\cdot\infty\} = \lim_{x\to 0^+}\frac{\ln(e^x-1)}{1/x}\ \{\frac{\infty}{\infty}\} = \lim_{x\to 0^+}\frac{e^x/(e^x-1)}{-1/x^2} =$

$\lim_{x\to 0^+}\frac{-x^2e^x}{e^x-1}\ \{\frac{0}{0}\} = \lim_{x\to 0^+}\frac{-x^2e^x-2xe^x}{e^x} = \lim_{x\to 0^+}(-x^2-2x) = 0.$

Thus, $L = e^0 = 1$.

[16] $y = x^x \Rightarrow \ln y = x\ln x$. $\lim_{x\to 0^+}\ln y\ \{0\cdot\infty\} = \lim_{x\to 0^+}\frac{\ln x}{1/x}\ \{\frac{\infty}{\infty}\} = \lim_{x\to 0^+}\frac{1/x}{-1/x^2} =$

$\lim_{x\to 0^+}(-x) = 0$. Thus, $L = e^0 = 1$.

[17] $\lim_{x\to\infty}x^{1/x}$ is of the form ∞^0. $y = x^{1/x} \Rightarrow \ln y = \frac{1}{x}\ln x$.

$\lim_{x\to\infty}\ln y\ \{0\cdot\infty\} = \lim_{x\to\infty}\frac{\ln x}{x}\ \{\frac{\infty}{\infty}\} = \lim_{x\to\infty}\frac{1/x}{1} = 0$. Thus, $L = e^0 = 1$.

[18] $\lim_{x\to(\pi/2)^-} (\tan x)^{\cos x}$ is of the form ∞^0. $y = (\tan x)^{\cos x} \Rightarrow \ln y = \cos x \ln \tan x$.

$$\lim_{x\to(\pi/2)^-} \ln y \{0\cdot\infty\} = \lim_{x\to(\pi/2)^-} \frac{\ln \tan x}{\sec x} \{\tfrac{\infty}{\infty}\} = \lim_{x\to(\pi/2)^-} \frac{\sec^2 x/\tan x}{\sec x \tan x} =$$

$$\lim_{x\to(\pi/2)^-} \frac{\cos x}{\sin^2 x} = \tfrac{0}{1} = 0. \text{ Thus, } L = e^0 = 1.$$

[19] $\lim_{x\to(\pi/2)^-} (\tan x)^x = \infty$, since it is of the form $\infty^{\pi/2}$,

which is not an indeterminate form.

[20] $L = 0^2 = 0$

[21] $\lim_{x\to 0^+} (2x + 1)^{\cot x}$ is of the form 1^∞. $y = (2x + 1)^{\cot x} \Rightarrow \ln y = \cot x \ln(2x + 1)$.

$$\lim_{x\to 0^+} \ln y \{\infty\cdot 0\} = \lim_{x\to 0^+} \frac{\ln(2x + 1)}{\tan x} \{\tfrac{0}{0}\} = \lim_{x\to 0^+} \frac{2/(2x + 1)}{\sec^2 x} = \tfrac{2}{1} = 2.$$

Thus, $L = e^2$.

[22] $\lim_{x\to 0^+} (1 + 3x)^{\csc x}$ is of the form 1^∞. $y = (1 + 3x)^{\csc x} \Rightarrow \ln y = \csc x \ln(1 + 3x)$.

$$\lim_{x\to 0^+} \ln y \{\infty\cdot 0\} = \lim_{x\to 0^+} \frac{\ln(1 + 3x)}{\sin x} \{\tfrac{0}{0}\} = \lim_{x\to 0^+} \frac{3/(1 + 3x)}{\cos x} = \tfrac{3}{1} = 3.$$

Thus, $L = e^3$.

[23] $$L\{\infty - \infty\} = \lim_{x\to\infty} \frac{2x^2}{x^2 - 1} \{\tfrac{\infty}{\infty}\} = \lim_{x\to\infty} \frac{4x}{2x} = \lim_{x\to\infty} 2 = 2$$

[24] $$L\{\infty - \infty\} = \lim_{x\to 1^+} \frac{\ln x - x + 1}{(x - 1)\ln x} \{\tfrac{0}{0}\} = \lim_{x\to 1^+} \frac{\frac{1}{x} - 1}{\frac{x - 1}{x} + \ln x} =$$

$$\lim_{x\to 1^+} \frac{1 - x}{x - 1 + x\ln x} \{\tfrac{0}{0}\} = \lim_{x\to 1^+} \frac{-1}{1 + 1 + \ln x} = -\tfrac{1}{2}$$

[25] $$L\{\infty - \infty\} = \lim_{x\to 0^-} \frac{\sin x - x}{x \sin x} \{\tfrac{0}{0}\} = \lim_{x\to 0^-} \frac{\cos x - 1}{x\cos x + \sin x} \{\tfrac{0}{0}\} =$$

$$\lim_{x\to 0^-} \frac{-\sin x}{-x\sin x + 2\cos x} = \tfrac{0}{2} = 0$$

[26] $$L\{\infty - \infty\} = \lim_{x\to(\pi/2)^-} \frac{1 - \sin x}{\cos x} \{\tfrac{0}{0}\} = \lim_{x\to(\pi/2)^-} \frac{-\cos x}{-\sin x} = \frac{0}{-1} = 0$$

[27] $\lim_{x\to 1^-} (1 - x)^{\ln x}$ is of the form 0^0. $y = (1 - x)^{\ln x} \Rightarrow \ln y = \ln x \ln(1 - x)$.

$$\lim_{x\to 1^-} \ln y \{0\cdot\infty\} = \lim_{x\to 1^-} \frac{\ln(1 - x)}{1/\ln x} \{\tfrac{\infty}{\infty}\} = \lim_{x\to 1^-} \frac{-1/(1 - x)}{-1/[x(\ln x)^2]} =$$

$$\lim_{x\to 1^-} \frac{x(\ln x)^2}{1 - x} \{\tfrac{0}{0}\} = \lim_{x\to 1^-} \frac{2\ln x + (\ln x)^2}{-1} = 0. \text{ Thus, } L = e^0 = 1.$$

[28] $\lim_{x\to\infty} (1 + e^x)^{e^{-x}}$ is of the form ∞^0. $y = (1 + e^x)^{e^{-x}} \Rightarrow \ln y = e^{-x} \ln(1 + e^x)$.

$$\lim_{x\to\infty} \ln y \{0\cdot\infty\} = \lim_{x\to\infty} \frac{\ln(1 + e^x)}{e^x} \{\tfrac{\infty}{\infty}\} = \lim_{x\to\infty} \frac{e^x/(1 + e^x)}{e^x} = \lim_{x\to\infty} \frac{1}{1 + e^x} =$$

0. Thus, $L = e^0 = 1$.

[29] $L = 1 - \infty = -\infty$.

[30] $$L\{\infty - \infty\} = \lim_{x\to 0} \left[\cot^2 x - (1 + \cot^2 x)\right] = \lim_{x\to 0} (-1) = -1$$

[31] L $\{\infty\cdot 0\} = \lim_{x\to 0^+} \frac{\tan^{-1}x}{\tan 2x}\ \{\frac{0}{0}\} = \lim_{x\to 0^+} \frac{1/(1+x^2)}{2\sec^2 2x} = \frac{1}{2}$

[32] L $\{\infty\cdot 0\} = \lim_{x\to\infty} \frac{x^3}{2^x}\ \{\frac{\infty}{\infty}\} = \lim_{x\to\infty} \frac{3x^2}{(\ln 2)\,2^x}\ \{\frac{\infty}{\infty}\} = \lim_{x\to\infty} \frac{6x}{(\ln 2)^2\, 2^x}\ \{\frac{\infty}{\infty}\} =$

$$\lim_{x\to\infty} \frac{6}{(\ln 2)^3\, 2^x} = 0$$

[33] Since $\lim_{x\to 0} \cot^2 x = \infty$ and $\lim_{x\to 0} e^{-x} = 1$, L $= \infty$.

[34] Since $\lim_{x\to\infty} \sqrt{x^2+4} = \infty$ and $\lim_{x\to\infty} \tan^{-1}x = \frac{\pi}{2}$, L $= \infty$.

[35] L has the form 1^∞. $y = (1+\cos x)^{\tan x} \Rightarrow \ln y = \tan x \ln(1+\cos x)$.

$$\lim_{x\to(\pi/2)^-} \ln y\ \{\infty\cdot 0\} = \lim_{x\to(\pi/2)^-} \frac{\ln(1+\cos x)}{\cot x}\ \{\tfrac{0}{0}\} =$$

$$\lim_{x\to(\pi/2)^-} \frac{-\sin x/(1+\cos x)}{-\csc^2 x} = \frac{-1/1}{-1} = 1. \text{ Thus, L} = e^1.$$

[36] L has the form 1^∞. $y = (1+ax)^{b/x} \Rightarrow \ln y = \frac{b}{x}\ln(1+ax)$.

$$\lim_{x\to 0^+} \ln y\ \{\infty\cdot 0\} = \lim_{x\to 0^+} \frac{b\ln(1+ax)}{x}\ \{\tfrac{0}{0}\} = \lim_{x\to 0^+} \frac{ba}{1+ax} = ba. \text{ Thus, L} = e^{ab}.$$

[37] L $\{-\infty+\infty\} = \lim_{x\to -3^-} \left[\frac{x}{(x-1)(x+3)} - \frac{4(x-1)}{(x-1)(x+3)}\right] =$

$$\lim_{x\to -3^-} \frac{-3x+4}{(x-1)(x+3)}.$$

The last limit equals ∞ since it increases without bound as $x \to -3^-$.

[38] L $\{\infty-\infty\} = \lim_{x\to\infty} \left[\left(\sqrt{x^4+5x^2+3} - x^2\right)\cdot \frac{\sqrt{x^4+5x^2+3}+x^2}{\sqrt{x^4+5x^2+3}+x^2}\right] =$

$$\lim_{x\to\infty} \left(\frac{5x^2+3}{\sqrt{x^4+5x^2+3}+x^2}\right) = \lim_{x\to\infty} \left(\frac{5x^2/x^2+3/x^2}{\sqrt{x^4/x^4+5x^2/x^4+3/x^4}+x^2/x^2}\right)$$

$$= \frac{5+0}{\sqrt{1+0+0}+1} = \frac{5}{2}$$

[39] L has the form 1^∞. $y = (x+\cos 2x)^{\csc 3x} \Rightarrow \ln y = \csc 3x \ln(x+\cos 2x)$.

$$\lim_{x\to 0^+} \ln y\ \{\infty\cdot 0\} = \lim_{x\to 0^+} \frac{\ln(x+\cos 2x)}{\sin 3x}\ \{\tfrac{0}{0}\} = \lim_{x\to 0^+} \frac{1-2\sin 2x}{x+\cos 2x}\cdot\frac{1}{3\cos 3x} =$$

$$\tfrac{1}{1}\cdot\tfrac{1}{3} = \tfrac{1}{3}. \text{ Thus, L} = e^{1/3}.$$

[40] L $\{\infty\cdot 0\} = \lim_{x\to(\pi/2)^-} \frac{\cos 3x}{\cos x}\ \{\frac{0}{0}\} = \lim_{x\to(\pi/2)^-} \frac{-3\sin 3x}{-\sin x} = \frac{3}{-1} = -3$

[41] L $\{\infty-\infty\} = \lim_{x\to\infty} \sinh x\left(1 - \frac{x}{\sinh x}\right)$ {factor out the fastest growing function}

$$= \lim_{x\to\infty} \sinh x \left\{\text{since } \lim_{x\to\infty} \frac{x}{\sinh x}\ \{\tfrac{\infty}{\infty}\} = \lim_{x\to\infty} \frac{1}{\cosh x} = 0\right\} = \infty$$

[42] L $\{\infty-\infty\} = \lim_{x\to\infty} \ln\frac{4x+3}{3x+4} = \ln \lim_{x\to\infty} \frac{4x+3}{3x+4} = \ln\frac{4}{3}$

43 From the figure, $\lim_{x\to 0} f(x) \approx 1$.

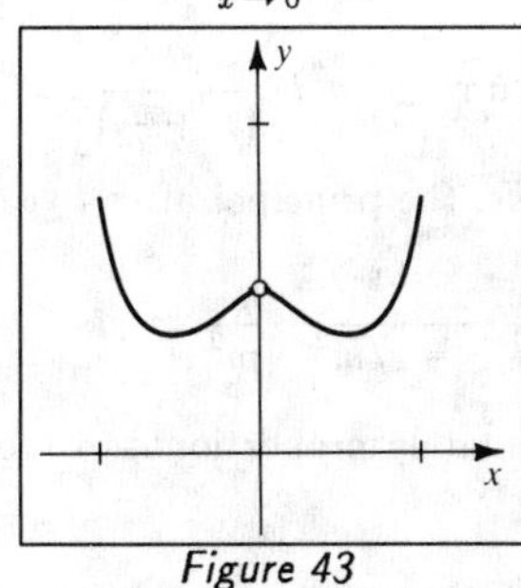

Figure 43

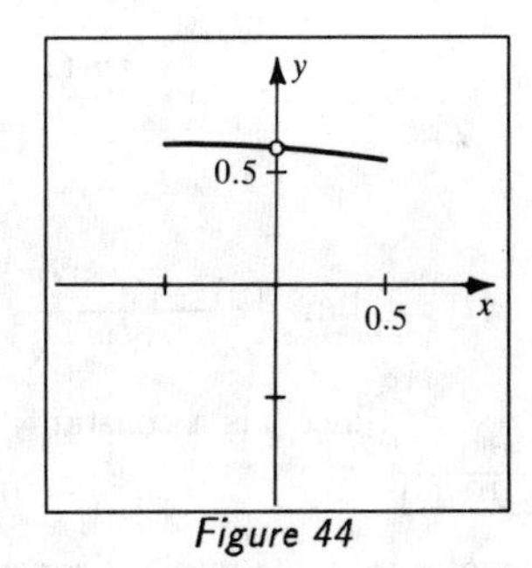

Figure 44

44 From the figure, $\lim_{x\to 0} f(x) \approx 0.61$.

45 (a) $f(x) = x^{1/x} = e^{(1/x)\ln x} \Rightarrow f'(x) = e^{(1/x)\ln x}\left(\frac{1}{x^2} - \frac{1}{x^2}\ln x\right) =$

$\frac{1}{x^2}e^{(1/x)\ln x}(1 - \ln x)$. $f'(x) = 0 \Rightarrow x = e$. $f'(x) > 0$ on $(0, e)$ and

$f'(x) < 0$ on $(e, \infty) \Rightarrow f(e) = e^{1/e} \approx 1.44$ is a *LMAX*. $y = x^{1/x} \Rightarrow$

$\ln y = \frac{1}{x}\ln x$ and $\lim_{x\to 0^+} \frac{\ln x}{x} = -\infty$. Hence, $\lim_{x\to 0^+} x^{1/x} = \lim_{x\to -\infty} e^x = 0$.

(b) From Exercise 17, $\lim_{x\to\infty} x^{1/x} = 1$. Thus, $y = 1$ is a horizontal asymptote.

(c) *Note*: There are *PI* at $x \approx 0.58$ and $x \approx 4.37$.

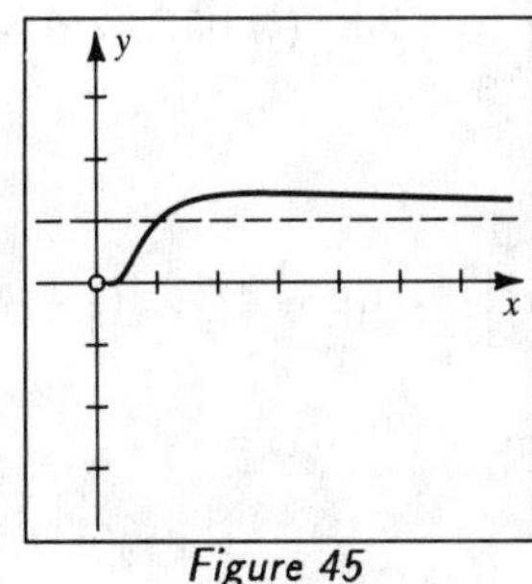

Figure 45

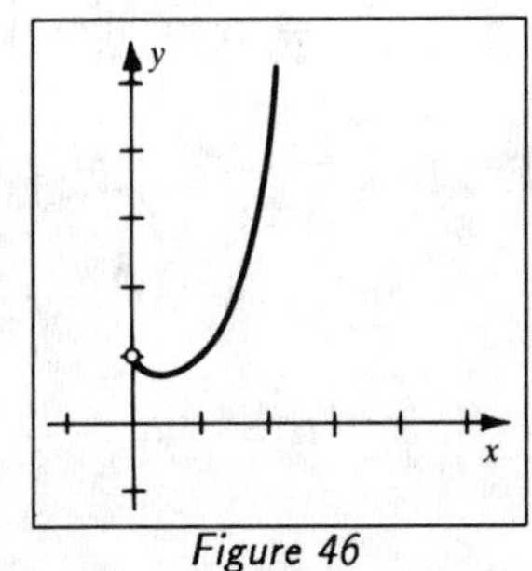

Figure 46

46 (a) $f(x) = x^x = e^{x\ln x} \Rightarrow f'(x) = e^{x\ln x}(1 + \ln x)$. $f'(x) = 0$ at $x = e^{-1} \approx 0.37$.

$f'(x) < 0$ on $(0, e^{-1})$ and $f'(x) > 0$ on $(e^{-1}, \infty) \Rightarrow$

$f(e^{-1}) = e^{-1/e} \approx 0.69$ is a *LMIN*. From Exercise 16, $\lim_{x\to 0^+} x^x = 1$.

(b) $\lim_{x\to\infty} x^x = \infty$. There are no horizontal asymptotes.

47 $y = \left(\frac{a^{1/x} + b^{1/x}}{2}\right)^x \Rightarrow \ln y = x\ln\left[\frac{1}{2}(a^{1/x} + b^{1/x})\right]$. $\lim_{x\to\infty} \ln y\ \{\infty\cdot 0\} =$

$$\lim_{x\to\infty} \frac{\ln\left[\frac{1}{2}(a^{1/x} + b^{1/x})\right]}{1/x}\ \{\tfrac{0}{0}\} = \lim_{x\to\infty}\left[\frac{-x^2}{1}\cdot\frac{2}{a^{1/x} + b^{1/x}}\cdot\frac{a^{1/x}\ln a + b^{1/x}\ln b}{-2x^2}\right]$$

$= \frac{1}{2}(\ln a + \ln b) = \frac{1}{2}\ln(ab) = \ln(ab)^{1/2} = \ln\sqrt{ab}$. Thus, $L = e^{\ln\sqrt{ab}} = \sqrt{ab}$.

48 $y = (1 + \frac{r}{m})^{mt} \Rightarrow \ln y = mt \ln(1 + \frac{r}{m})$.

$$\lim_{m\to\infty} \ln y \,\{\infty\cdot 0\} = \lim_{m\to\infty} \frac{t\ln(1 + \frac{r}{m})}{1/m} \,\{\tfrac{0}{0}\} = \lim_{m\to\infty}\left[-m^2 t\cdot\frac{-r/m^2}{(1 + \frac{r}{m})}\right] = rt.$$

Thus, the principal after t years is Pe^{rt}.

49 $$\lim_{m\to\infty} v(t) = \lim_{m\to\infty} \frac{g}{k}\cdot\frac{1 - e^{-kt/m}}{1/m} \,\{\tfrac{0}{0}\} = \lim_{m\to\infty}\frac{g}{k}\cdot\frac{-e^{-kt/m}}{-1/m^2}\cdot\frac{kt}{m^2} = gt.$$

Since g is a constant, $v(t)$ is approximately proportional to t for large m.

Exercises 10.3

Note: C denotes that the integral converges, D denotes that it diverges.

1 $$\mathrm{I} = \lim_{t\to\infty}\int_1^t x^{-4/3}\,dx = \lim_{t\to\infty}\left[-3x^{-1/3}\right]_1^t = (-3)\lim_{t\to\infty}\left[\frac{1}{\sqrt[3]{t}} - 1\right] = -3(0 - 1) = 3;\ \mathrm{C}$$

2 $$\mathrm{I} = \lim_{t\to-\infty}\int_t^0 (x - 1)^{-3}\,dx = \lim_{t\to-\infty}\left[-\frac{1}{2(x - 1)^2}\right]_t^0 = \left(-\frac{1}{2}\right)\lim_{t\to-\infty}\left[1 - \frac{1}{(t - 1)^2}\right] = -\tfrac{1}{2}(1 - 0) = -\tfrac{1}{2};\ \mathrm{C}$$

3 $$\mathrm{I} = \lim_{t\to\infty}\int_1^t x^{-3/4}\,dx = \lim_{t\to\infty}\left[4x^{1/4}\right]_1^t = 4\lim_{t\to\infty}(t^{1/4} - 1) = \infty;\ \mathrm{D}$$

4 $$\mathrm{I} = \lim_{t\to\infty}\int_0^t \frac{x}{1 + x^2}\,dx = \lim_{t\to\infty}\left[\tfrac{1}{2}\ln(1 + x^2)\right]_0^t = \tfrac{1}{2}\lim_{t\to\infty}\left[\ln(1 + t^2) - 0\right] = \infty;\ \mathrm{D}$$

5 $$\mathrm{I} = \lim_{t\to-\infty}\int_t^2 \frac{1}{5 - 2x}\,dx = \lim_{t\to-\infty}\left[-\tfrac{1}{2}\ln|5 - 2x|\right]_t^2 = \lim_{t\to-\infty}(\tfrac{1}{2}\ln|5 - 2t|) = \infty;\ \mathrm{D}$$

6 $$\mathrm{I} = \lim_{t\to-\infty}\int_t^0 \frac{x}{x^4 + 9}\,dx + \lim_{s\to\infty}\int_0^s \frac{x}{x^4 + 9}\,dx =$$
$$\lim_{t\to-\infty}\left[\tfrac{1}{6}\tan^{-1}\tfrac{1}{3}x^2\right]_t^0 + \lim_{s\to\infty}\left[\tfrac{1}{6}\tan^{-1}\tfrac{1}{3}x^2\right]_0^s = \tfrac{1}{6}(0 - \tfrac{\pi}{2}) + \tfrac{1}{6}(\tfrac{\pi}{2} - 0) = 0;\ \mathrm{C}$$

7 $$\mathrm{I} = \lim_{t\to\infty}\int_0^t e^{-2x}\,dx = \lim_{t\to\infty}\left[-\tfrac{1}{2}e^{-2x}\right]_0^t = -\frac{1}{2}\lim_{t\to\infty}(e^{-2t} - 1) = -\tfrac{1}{2}(0 - 1) = \tfrac{1}{2};\ \mathrm{C}$$

8 $$\mathrm{I} = \lim_{t\to-\infty}\int_t^0 e^x\,dx = \lim_{t\to-\infty}\left[e^x\right]_t^0 = \lim_{t\to\infty}(1 - e^t) = 1 - 0 = 1;\ \mathrm{C}$$

9 $$\mathrm{I} = \lim_{t\to-\infty}\int_t^{-1} x^{-3}\,dx = \lim_{t\to-\infty}\left[-\frac{1}{2x^2}\right]_t^{-1} = -\frac{1}{2}\lim_{t\to-\infty}\left[1 - (-\tfrac{1}{t^2})\right] = -\tfrac{1}{2}(1 - 0) = -\tfrac{1}{2};\ \mathrm{C}$$

10 $$\mathrm{I} = \lim_{t\to\infty}\int_0^t (x + 1)^{-1/3}\,dx = \lim_{t\to\infty}\left[\tfrac{3}{2}(x + 1)^{2/3}\right]_0^t = \tfrac{3}{2}\lim_{t\to\infty}\left[(t + 1)^{2/3} - 1\right] = \infty;\ \mathrm{D}$$

11 $$\mathrm{I} = \lim_{t\to-\infty}\int_t^0 (x - 8)^{-2/3}\,dx = \lim_{t\to-\infty}\left[3(x - 8)^{1/3}\right]_t^0 = 3\lim_{t\to-\infty}\left[-2 - (t - 8)^{1/3}\right] = \infty;\ \mathrm{D}$$

12 $$\mathrm{I} = \lim_{t\to\infty}\int_1^t \frac{x}{(1 + x^2)^2}\,dx = \lim_{t\to\infty}\left[-\frac{1}{2(1 + x^2)}\right]_1^t = -\tfrac{1}{2}(0 - \tfrac{1}{2}) = \tfrac{1}{4};\ \mathrm{C}$$

[13] $\mathrm{I} = \lim_{t\to\infty} \int_0^t \frac{\cos x}{1 + \sin^2 x}\,dx = \lim_{t\to\infty} \Big[\tan^{-1}(\sin x)\Big]_0^t = \lim_{t\to\infty} \Big[\tan^{-1}(\sin t) - 0\Big].$

This limit DNE since $\lim_{t\to\infty} \sin t$ DNE. D

[14] $\mathrm{I} = \lim_{t\to-\infty} \int_t^2 \frac{1}{x^2 + 4}\,dx = \lim_{t\to-\infty} \Big[\tfrac{1}{2}\tan^{-1}\tfrac{1}{2}x\Big]_t^2 = \frac{1}{2}\Big[\frac{\pi}{4} - \left(-\frac{\pi}{2}\right)\Big] = \frac{3\pi}{8}$; C

[15] $\mathrm{I} = \lim_{t\to-\infty} \int_t^0 xe^{-x^2}\,dx + \lim_{s\to\infty} \int_0^s xe^{-x^2}\,dx =$

$\lim_{t\to-\infty} \Big[-\tfrac{1}{2}e^{-x^2}\Big]_t^0 + \lim_{s\to\infty} \Big[-\tfrac{1}{2}e^{-x^2}\Big]_0^s = -\tfrac{1}{2}(1 - 0) - \tfrac{1}{2}(0 - 1) = 0$; C

[16] $\mathrm{I} = \lim_{t\to\infty} 2\int_0^t \cos^2 x\,dx$ {even integrand} $= 2\lim_{t\to\infty} \Big[\tfrac{1}{2}(x + \tfrac{1}{2}\sin 2x)\Big]_0^t = \infty$; D

[17] $\mathrm{I} = \lim_{t\to\infty} \int_1^t \frac{\ln x}{x}\,dx = \lim_{t\to\infty} \Big[\tfrac{1}{2}(\ln x)^2\Big]_1^t = \infty$; D

[18] $\mathrm{I} = \lim_{t\to\infty} \int_3^t \frac{1}{x^2 - 1}\,dx = \lim_{t\to\infty} \int_3^t \left[\frac{\frac{1}{2}}{x - 1} - \frac{\frac{1}{2}}{x + 1}\right]$ {PF} $dx =$

$\lim_{t\to\infty} \Big[\frac{1}{2}\ln|x - 1| - \frac{1}{2}\ln|x + 1|\Big]_3^t = \frac{1}{2}\lim_{t\to\infty} \left[\ln\left|\frac{x - 1}{x + 1}\right|\,\right]_3^t = \frac{1}{2}(0 - \ln\tfrac{1}{2}) = \frac{1}{2}\ln 2$; C

[19] $\mathrm{I} = \lim_{t\to\infty} \int_0^t \cos x\,dx = \lim_{t\to\infty} \Big[\sin x\Big]_0^t = \lim_{t\to\infty} \sin t.$

This limit DNE since the sine function oscillates between $+1$ and -1. D

[20] $\mathrm{I} = \lim_{t\to-\infty} \int_t^{\pi/2} \sin 2x\,dx = \lim_{t\to-\infty} \Big[-\tfrac{1}{2}\cos 2x\Big]_t^{\pi/2} = -\frac{1}{2}\lim_{t\to-\infty} \Big[1 - \cos 2t\Big] =$

DNE; D

[21] $\mathrm{I} = 2\lim_{t\to\infty} \int_0^t \operatorname{sech} x\,dx$ {even integrand}

$= 2\lim_{t\to\infty} \Big[\tan^{-1}(\sinh x)\Big]_0^t$ {Formula 107 or 8.3.42} $= 2(\frac{\pi}{2} - 0) = \pi$; C

[22] $\mathrm{I} = \int_0^t xe^{-x}\,dx = \lim_{t\to\infty} \Big[-xe^{-x} - e^{-x}\Big]_0^t$ {IP} $= \left(\lim_{t\to\infty} \frac{-t}{e^t} - 0\right) - (0 - 1) = 1$; C

Note: $\lim_{x\to\infty} \frac{x^n}{e^x} = 0$ {10.1.31} is used often, without mention, from this point on.

[23] $\mathrm{I} = \lim_{t\to-\infty} \int_t^0 \left[\frac{1}{x - 2} - \frac{1}{x - 1}\right] dx$ {PF} $= \lim_{t\to-\infty} \left[\ln\left|\frac{x - 2}{x - 1}\right|\,\right]_t^0 =$

$\ln|2| - \ln|1| = \ln 2$; C

[24] $\mathrm{I} = \lim_{t\to\infty} \int_4^t \left[\frac{3}{x - 3} - \frac{2}{x + 4}\right] dx$ {PF} $= \lim_{t\to\infty} \left[\ln\left|\frac{(x - 3)^3}{(x + 4)^2}\right|\,\right]_4^t = \infty$; D

[25] $0 \le \frac{1}{1 + x^4} \le \frac{1}{x^4}$ on $[1, \infty)$ and $\int_1^\infty \frac{1}{x^4}\,dx = \lim_{t\to\infty} \left[-\frac{1}{3x^3}\right]_1^t =$

$-\frac{1}{3}(0 - 1) = \frac{1}{3} \Rightarrow$ I *converges* by (i).

[26] $0 \le \frac{1}{\sqrt[3]{x^2}} \le \frac{1}{\sqrt[3]{x^2 - 1}}$ on $[2, \infty)$ and $\int_2^\infty x^{-2/3}\,dx = \lim_{t\to\infty} \Big[3x^{1/3}\Big]_2^t = \infty \Rightarrow$

I *diverges* by (ii).

[27] $0 \le \frac{1}{x} \le \frac{1}{\ln x}$ on $[2, \infty)$ and $\int_2^\infty \frac{1}{x}\,dx = \lim_{t\to\infty}\Big[\ln|x|\Big]_2^t = \infty \Rightarrow$ I *diverges* by (ii).

[28] $0 \le e^{-x^2} \le e^{-x}$ on $[1, \infty)$ and $\int_1^\infty e^{-x}\,dx = \lim_{t\to\infty}\Big[-e^{-x}\Big]_1^t = 0 + e^{-1} = \frac{1}{e} \Rightarrow$

I *converges* by (i).

[29] (a) $A = \int_1^\infty \frac{1}{x}\,dx = \lim_{t\to\infty}\Big[\ln|x|\Big]_1^t = \infty$. Not possible.

(b) Using disks, $V = \pi\int_1^\infty \left(\frac{1}{x}\right)^2 dx = \pi\lim_{t\to\infty}\Big[-\frac{1}{x}\Big]_1^t = -\pi(0 - 1) = \pi$.

[30] (a) $A = \int_1^\infty \frac{1}{\sqrt{x}}\,dx = \lim_{t\to\infty}\Big[2\sqrt{x}\Big]_1^t = \infty$. Not possible.

(b) Using disks, $V = \pi\int_1^\infty \left(\frac{1}{\sqrt{x}}\right)^2 dx = \pi\lim_{t\to\infty}\Big[\ln|x|\Big]_1^t = \infty$. Not possible.

[31] (a) $A = \int_4^\infty x^{-3/2}\,dx = \lim_{t\to\infty}\Big[-2x^{-1/2}\Big]_4^t = -2(0 - \frac{1}{2}) = 1$.

(b) Using disks, $V = \pi\int_4^\infty (x^{-3/2})^2\,dx = \pi\lim_{t\to\infty}\Big[-\frac{1}{2}x^{-2}\Big]_4^t = -\frac{\pi}{2}(0 - \frac{1}{16}) = \frac{\pi}{32}$.

[32] (a) $A = \int_8^\infty x^{-2/3}\,dx = \lim_{t\to\infty}\Big[3x^{1/3}\Big]_8^t = \infty$. Not possible.

(b) Using disks, $V = \pi\int_8^\infty (x^{-2/3})^2\,dx = \pi\lim_{t\to\infty}\Big[-3x^{-1/3}\Big]_8^t = -3\pi(0 - \frac{1}{2}) = \frac{3\pi}{2}$.

[33] Using shells, $V = \int_0^\infty 2\pi\, x e^{-x^2}\,dx = \lim_{t\to\infty}\int_0^t 2\pi\, x e^{-x^2}\,dx = \lim_{t\to\infty}\Big[-\pi e^{-x^2}\Big]_0^t = \pi$.

[34] $S = \lim_{t\to\infty}\int_0^t 2\pi\, e^{-x}\sqrt{1 + e^{-2x}}\,dx$. $u = e^{-x}$, $-du = e^{-x}\,dx \Rightarrow$

$S = -2\pi\int_1^0 \sqrt{1 + u^2}\,du = 2\pi\Big[\frac{u}{2}\sqrt{1 + u^2} + \frac{1}{2}\ln\left|u + \sqrt{1 + u^2}\right|\Big]_0^1$ {Formula 21} $=$

$\pi\Big[\sqrt{2} + \ln(1 + \sqrt{2})\Big] \approx 7.21$.

[35] (a) See Exercise 29(b).

(b) Using shells, $V = \int_1^\infty 2\pi\, x\frac{1}{x}\,dx = \lim_{t\to\infty}\int_1^t 2\pi\,dx = \lim_{t\to\infty}\Big[2\pi x\Big]_1^t = \infty$. No.

(c) $S = \int_1^\infty 2\pi\frac{1}{x}\sqrt{1 + (-\frac{1}{x^2})^2}\,dx = \int_1^\infty 2\pi\frac{1}{x}\sqrt{1 + \frac{1}{x^4}}\,dx$;

$0 \le \frac{2\pi}{x} \le \frac{2\pi}{x}\sqrt{1 + \frac{1}{x^4}}$ and $\int_1^\infty \frac{2\pi}{x}\,dx = \lim_{t\to\infty}\Big[2\pi\ln|x|\Big]_1^t = \infty \Rightarrow S$ DNE.

[36] (a) $\int_0^\infty R(t)\,dt$ represents the total amount of fuel (in grams) that can be burned.

(b) Never, since $R(t) > 0$ for all t. *Note:* $\lim_{T\to\infty}\int_0^T mke^{-kt}\,dt = m \Rightarrow$

the fuel is not used in any *finite* amount of time.

[37] If $F(x) = \frac{k}{x^2}$, then $W = \int_1^\infty \frac{k}{x^2}\,dx = \lim_{t\to\infty}\Big[-\frac{k}{x}\Big]_1^t = -k(0 - 1) = k$ J.

[38] $W = \int_a^{\infty}\left[\frac{-kq}{(x-\frac{1}{2}d)^2} + \frac{kq}{(x+\frac{1}{2}d)^2}\right]dx = \lim_{t\to\infty}\left[\frac{kq}{x-\frac{1}{2}d} - \frac{kq}{x+\frac{1}{2}d}\right]_a^t =$

$$-\frac{kq}{a-\frac{1}{2}d} + \frac{kq}{a+\frac{1}{2}d} = \frac{-4kqd}{4a^2-d^2}.$$ *Note:* $f(x) < 0$ on $[a, \infty) \Rightarrow W < 0$.

[39] (a) $R(t) = e^{-kt} \Rightarrow R'(t) = -ke^{-kt}$. Average time $= \int_0^{\infty}(-t)\,R'(t)\,dt =$

$$\int_0^{\infty} kte^{-kt}\,dt = \lim_{s\to\infty}\left[-te^{-kt} - \tfrac{1}{k}e^{-kt}\right]_0^s \{\text{IP}\} = \lim_{s\to\infty}\left[\tfrac{-s}{e^{ks}} - \tfrac{1}{ke^{ks}}\right] + \tfrac{1}{k} = \tfrac{1}{k}.$$

(b) If it is possible, then $R(t) = \frac{1}{t+1} \Rightarrow R'(t) = -\frac{1}{(t+1)^2}$. Average time $=$

$$\int_0^{\infty}\frac{t}{(t+1)^2}\,dt = \int_0^{\infty}\left[\frac{1}{t+1} - \frac{1}{(t+1)^2}\right]dt\ \{\text{PF}\} = \lim_{s\to\infty}\left[\ln|t+1| + \frac{1}{t+1}\right]_0^s$$

$= \infty$. This is a contradiction since it indicates that the average repair time is unbounded. Thus, it is not possible.

[40] (a) $f(t) = 12{,}000$ and $T = 20 \Rightarrow A = \int_{20}^{\infty} 12{,}000e^{-0.08t}\,dt = \lim_{s\to\infty}\left[\frac{12{,}000}{-0.08}e^{-0.08t}\right]_{20}^s$

$$= 150{,}000e^{-1.6} \approx \$30{,}284.48.$$

(b) $f(t) = 12{,}000e^{0.04t}$ and $T = 20 \Rightarrow A = \int_{20}^{\infty}(12{,}000e^{0.04t})e^{-0.08t}\,dt =$

$$\int_{20}^{\infty} 12{,}000e^{-0.04t}\,dt = \lim_{s\to\infty}\left[\frac{12{,}000}{-0.04}e^{-0.04t}\right]_{20}^s = 300{,}000e^{-0.8} \approx \$134{,}798.69.$$

[41] (a) $\text{I} \overset{\text{A}}{=} \lim_{t\to\infty}\left[-\frac{x}{2a}e^{-ax^2}\right]_0^t + \frac{1}{2a}\int_0^{\infty} e^{-ax^2}\,dx = 0 + \frac{1}{2a}\int_0^{\infty} e^{-ax^2}\,dx.$

$$u = \sqrt{a}\,x,\ \frac{1}{\sqrt{a}}\,du = dx \Rightarrow \text{I} = \frac{1}{2a^{3/2}}\int_0^{\infty} e^{-u^2}\,du.$$

A. $u = x$, $du = dx$, $dv = xe^{-ax^2}\,dx$, $v = -\frac{1}{2a}e^{-ax^2}$

(b) $\int_0^{\infty} cv^2e^{-mv^2/(2kT)}\,dv = 1 \Rightarrow c\cdot\frac{1}{2(\frac{m}{2kT})^{3/2}}\cdot\frac{\sqrt{\pi}}{2} = 1 \Rightarrow c = \frac{4}{\sqrt{\pi}}\cdot(\frac{m}{2kT})^{3/2}$

[42] $F_C\left[e^{-ax}\right] = \int_0^{\infty} e^{-ax}\cos sx\,dx = \lim_{t\to\infty}\left[\frac{e^{-ax}(-a\cos sx + s\sin sx)}{a^2+s^2}\right]_0^t \{\text{IP}\} = \frac{a}{a^2+s^2}$

[43] $L[1] = \int_0^{\infty} e^{-sx}(1)\,dx = \lim_{t\to\infty}\left[-\frac{1}{s}e^{-sx}\right]_0^t = -\frac{1}{s}(0-1) = \frac{1}{s},\ s > 0$

[44] $L[x] = \int_0^{\infty} e^{-sx}(x)\,dx\ \{\text{IP}\} = \lim_{t\to\infty}\left[-\frac{xe^{-sx}}{s} - \frac{e^{-sx}}{s^2}\right]_0^t = -\frac{1}{s^2}(0-1) = \frac{1}{s^2},\ s > 0$

[45] $L[\cos x] = \int_0^{\infty} e^{-sx}(\cos x)\,dx\ \{\text{IP}\} = \lim_{t\to\infty}\left[\frac{e^{-sx}(-s\cos x + \sin x)}{s^2+1}\right]_0^t =$

$$\frac{1}{s^2+1}\left[0 - (-s)\right] = \frac{s}{s^2+1},\ s > 0$$

46 $L[\sin x] = \int_0^\infty e^{-sx}(\sin x)\,dx\ \{\text{IP}\} = \lim_{t\to\infty}\left[\frac{e^{-sx}(-s\sin x - \cos x)}{s^2+1}\right]_0^t =$

$$\frac{1}{s^2+1}\big[0-(-1)\big] = \frac{1}{s^2+1},\ s>0$$

47 $L[e^{ax}] = \int_0^\infty e^{-sx}(e^{ax})\,dx = \int_0^\infty e^{(a-s)x}\,dx = \lim_{t\to\infty}\left[\frac{e^{(a-s)x}}{a-s}\right]_0^t =$

$$\frac{1}{a-s}(0-1)\ \{\text{if } s>a\} = \frac{1}{s-a}$$

48 $L[\sin ax] = \int_0^\infty e^{-sx}(\sin ax)\,dx\ \{\text{IP}\} = \lim_{t\to\infty}\left[\frac{e^{-sx}(-s\sin ax - a\cos ax)}{s^2+a^2}\right]_0^t =$

$$\frac{1}{s^2+a^2}\big[0-(-a)\big] = \frac{a}{s^2+a^2},\ s>0$$

49 (a) $\Gamma(1) = \int_0^\infty e^{-x}\,dx = \lim_{t\to\infty}\left[-e^{-x}\right]_0^t = 1.$

$\Gamma(2) = \int_0^\infty xe^{-x}\,dx\ \{\text{IP}\} = \lim_{t\to\infty}\left[-xe^{-x} - e^{-x}\right]_0^t = 1.$

$\Gamma(3) = \int_0^\infty x^2e^{-x}\,dx\ \{\text{IP}\} = \lim_{t\to\infty}\left[-x^2e^{-x} - 2xe^{-x} - 2e^{-x}\right]_0^t = 2.$

(b) $\Gamma(n+1) = \int_0^\infty x^n e^{-x}\,dx \overset{\text{A}}{=} \lim_{t\to\infty}\left[-x^n e^{-x}\right]_0^t + \int_0^\infty nx^{n-1}e^{-x}\,dx = 0 + n\Gamma(n)$

A. $u = x^n,\ du = nx^{n-1}\,dx,\ dv = e^{-x}\,dx,\ v = -e^{-x}$ $\qquad = n\Gamma(n).$

(c) $\Gamma(1) = 1 = 1!$ Assume $\Gamma(n+1) = n!$ for some $n \geq 1$. Then $\Gamma(n+2) =$

$(n+1)\Gamma(n+1) = (n+1)\,n!$ {by the induction hypothesis} $= (n+1)!$.

50 Let $u = ax$ so that $\text{I} = \int_0^\infty cx^k e^{-ax}\,dx$ has the form of the integral given in Exercise 49. $\frac{1}{a}\,du = dx \Rightarrow \text{I} = c\int_0^\infty (\frac{u}{a})^k e^{-u}\left(\frac{1}{a}\right)du = \frac{c}{a^{k+1}}\int_0^\infty u^k e^{-u}\,du = \frac{c}{a^{k+1}}\Gamma(k+1).$

$$\text{I} = 1 \Rightarrow c = \frac{a^{k+1}}{\Gamma(k+1)}.$$

51 $u = \frac{1}{x} \Rightarrow x = \frac{1}{u}$ and $dx = -\frac{1}{u^2}\,du;\ x = 2 \Rightarrow u = \frac{1}{2}$ and $x\to\infty \Rightarrow u\to 0^+$.

$\text{I} = \int_2^\infty \frac{1}{\sqrt{x^4+x}}\,dx = \int_{1/2}^0 \frac{1}{\sqrt{(1/u^4)+(1/u)}}\left(-\frac{1}{u^2}\right)du = \int_0^{1/2}\frac{1}{\sqrt{1+u^3}}\,du.$

Let $f(u) = \frac{1}{\sqrt{1+u^3}}$. $\text{I} \approx \text{S} = \frac{0.5-0}{3(4)}\left[f(0) + 4f(\frac{1}{8}) + 2f(\frac{1}{4}) + 4f(\frac{3}{8}) + f(\frac{1}{2})\right] \approx$

0.4926.

[52] $u = \frac{1}{x} \Rightarrow x = \frac{1}{u}$ and $dx = -\frac{1}{u^2}\,du$; $x = -10 \Rightarrow u = -\frac{1}{10}$ and $x \to -\infty \Rightarrow$

$u \to 0^-$. $\mathrm{I} = \int_{-\infty}^{-10} \frac{\sqrt{|x|}}{x^3 + 1}\,dx = \int_0^{-0.1} \frac{\sqrt{1/|u|}}{(1/u^3) + 1}\left(\frac{-1}{u^2}\right) du =$

$\int_{-0.1}^{0} \frac{u\sqrt{1/|u|}}{1 + u^3}\,du = \int_{-0.1}^{0} \frac{\sqrt{|u|}}{1 + u^3}\,du$. Let $f(u) = \frac{\sqrt{|u|}}{1 + u^3}$.

$\mathrm{I} \approx \mathrm{S} = \frac{0-(-0.1)}{3(4)}\left[f(-0.1) + 4f(-0.075) + 2f(-0.05) + 4f(-0.025) + f(0)\right]$

$\approx \frac{1}{120}(2.4922) \approx 0.0208.$

Exercises 10.4

[1] $\mathrm{I} = \lim_{t \to 0^+} \int_t^8 x^{-1/3}\,dx = \lim_{t \to 0^+} \left[\frac{3}{2}x^{2/3}\right]_t^8 = \frac{3}{2}(4 - 0) = 6$; C

[2] $\mathrm{I} = \lim_{t \to 0^+} \int_t^9 x^{-1/2}\,dx = \lim_{t \to 0^+} \left[2x^{1/2}\right]_t^9 = 2(3 - 0) = 6$; C

[3] $\mathrm{I}_1 = \lim_{t \to 0^-} \int_{-3}^{t} x^{-2}\,dx = \lim_{t \to 0^-} \left[-\frac{1}{x}\right]_{-3}^{t} = \infty$; D

Note: It is only necessary that one integral diverges for I to diverge.

In future solutions, we will compute one integral, and if it converges, continue until we can draw a conclusion.

[4] $\mathrm{I} = \lim_{t \to -2^+} \int_t^{-1} (x + 2)^{-5/4}\,dx = \lim_{t \to -2^+} \left[-\frac{4}{(x + 2)^{1/4}}\right]_t^{-1} = \infty$; D

[5] $\mathrm{I} = \lim_{t \to (\pi/2)^-} \int_0^t \sec^2 x\,dx = \lim_{t \to (\pi/2)^-} \left[\tan x\right]_0^t = \infty$; D

[6] $\mathrm{I} = \lim_{t \to 0^+} \int_t^1 \frac{e^{\sqrt{x}}}{\sqrt{x}}\,dx = \lim_{t \to 0^+} \left[2e^{\sqrt{x}}\right]_t^1 = 2(e - 1)$; C

[7] $\mathrm{I} = \lim_{t \to 4^-} \int_0^t (4 - x)^{-3/2}\,dx = \lim_{t \to 4^-} \left[\frac{2}{\sqrt{4 - x}}\right]_0^t = \infty$; D

[8] $\mathrm{I} = \lim_{t \to -1^+} -\int_t^0 (x + 1)^{-1/3}\,dx = \lim_{t \to -1^+} \left[-\frac{3}{2}(x + 1)^{2/3}\right]_t^0 = -\frac{3}{2}(1 - 0) = -\frac{3}{2}$; C

[9] $\mathrm{I} = \lim_{t \to 4^-} \int_0^t (4 - x)^{-2/3}\,dx = \lim_{t \to 4^-} \left[-3(4 - x)^{1/3}\right]_0^t = -3(0 - \sqrt[3]{4}) = 3\sqrt[3]{4}$; C

[10] $\mathrm{I} = \lim_{t \to 1^+} \int_t^2 \frac{x}{x^2 - 1}\,dx = \lim_{t \to 1^+} \left[\frac{1}{2}\ln\left|x^2 - 1\right|\right]_t^2 = \infty$; D

[11] $\mathrm{I}_1 = \lim_{t \to -1^-} \int_{-2}^{t} (x + 1)^{-3}\,dx = \lim_{t \to -1^-} \left[\frac{-2}{(x + 1)^2}\right]_{-2}^{t} = -\infty$; D

[12] $\mathrm{I}_1 = \lim_{t \to 0^-} \int_{-1}^{t} x^{-4/3}\,dx = \lim_{t \to 0^-} \left[\frac{-3}{\sqrt[3]{x}}\right]_{-1}^{t} = \infty$; D

[13] $\mathrm{I} = \lim_{t\to -2^+} \int_t^0 \frac{1}{\sqrt{4 - x^2}}\,dx = \lim_{t\to -2^+} \left[\sin^{-1}\frac{x}{2}\right]_t^0 = 0 - (-\frac{\pi}{2}) = \frac{\pi}{2}$; C

[14] $\mathrm{I} = \lim_{t\to -2^+} \int_t^0 \frac{x}{\sqrt{4 - x^2}}\,dx = \lim_{t\to -2^+} \left[-\sqrt{4 - x^2}\right]_t^0 = -2 - 0 = -2$; C

[15] $\mathrm{I}_1 = \lim_{t\to 0^-} \int_{-1}^t \frac{1}{x}\,dx = \lim_{t\to 0^-} \left[\ln|x|\right]_{-1}^t = -\infty$; D

[16] $\mathrm{I}_1 = \int_0^2 \frac{1}{(x-2)(x+1)}\,dx = \lim_{t\to 2^-} \int_0^t \left[\frac{\frac{1}{3}}{x-2} - \frac{\frac{1}{3}}{x+1}\right] dx$ {PF} $=$

$$\lim_{t\to 2^-} \left[\frac{1}{3}\ln\left|\frac{x-2}{x+1}\right|\right]_0^t = -\infty; \text{ D}$$

[17] $\mathrm{I} = \lim_{t\to 0^+} \int_t^1 x\ln x\,dx = \lim_{t\to 0^+} \left[\frac{1}{2}x^2\ln x - \frac{1}{4}x^2\right]_t^1$ {IP}

$= \frac{1}{4}\lim_{t\to 0^+} \left[x^2(2\ln x - 1)\right]_t^1 = -\frac{1}{4} - \frac{1}{4}\lim_{t\to 0^+} \left[t^2(2\ln t - 1)\right]$ $\{0\cdot\infty\}$

$= -\frac{1}{4} - \frac{1}{4}\lim_{t\to 0^+} \frac{2\ln t - 1}{1/t^2}$ $\{\frac{\infty}{\infty}\}$ $= -\frac{1}{4} - \frac{1}{4}\lim_{t\to 0^+} \frac{2/t}{-2/t^3} = -\frac{1}{4} - \frac{1}{4}\lim_{t\to 0^+}(-t^2)$

$= -\frac{1}{4} - 0 = -\frac{1}{4}$; C

[18] $\mathrm{I} = \lim_{t\to (\pi/2)^-} \int_0^t \tan^2 x\,dx = \lim_{t\to (\pi/2)^-} \int_0^t (\sec^2 x - 1)\,dx = \lim_{t\to (\pi/2)^-} \left[\tan x - x\right]_0^t =$

∞; D

[19] $\lim_{t\to (\pi/2)^-} \int_0^t \tan x\,dx = \lim_{t\to (\pi/2)^-} \left[\ln|\sec x|\right]_0^t = \infty$; D

[20] Simplifying the integrand, $\frac{1}{1 - \cos x}\cdot\frac{1 + \cos x}{1 + \cos x} = \frac{1 + \cos x}{\sin^2 x} = \frac{1}{\sin^2 x} + \frac{\cos x}{\sin^2 x}$.

Thus, $\mathrm{I} = \lim_{t\to 0^+} \int_t^{\pi/2} (\csc^2 x + \cot x\csc x)\,dx = \lim_{t\to 0^+} \left[-\cot x - \csc x\right]_t^{\pi/2} =$

$-1 + \lim_{t\to 0^+} \left[\cot t + \csc t\right] = \infty$. D

[21] $\mathrm{I} = \lim_{t\to 4^-} \int_2^t \left[\frac{\frac{1}{3}}{x-1} + \frac{\frac{2}{3}}{x-4}\right] dx$ {PF} $= \lim_{t\to 4^-} \left[\frac{1}{3}\ln|x-1| + \frac{2}{3}\ln|x-4|\right]_2^t =$

$-\infty$; D

[22] Since $\frac{1}{e} < 1 < e$, $\mathrm{I}_1 = \lim_{t\to 1^-} \int_{1/e}^t \frac{1}{x(\ln x)^2}\,dx = \lim_{t\to 1^-} \left[-\frac{1}{\ln x}\right]_{1/e}^t = \infty$; D

[23] $\mathrm{I}_1 = \lim_{t\to 0^-} \int_{-1}^t \frac{1}{x^2}\cos\frac{1}{x}\,dx = \lim_{t\to 0^-} \left[-\sin\frac{1}{x}\right]_{-1}^t$. The limit DNE. D

[24] $\mathrm{I}_1 = \lim_{t\to (\pi/2)^-} \int_0^t \sec x\,dx = \lim_{t\to (\pi/2)^-} \left[\ln|\sec x + \tan x|\right]_0^t = \infty$; D

[25] $\mathrm{I}_1 = \lim_{t\to (\pi/2)^-} \int_0^t \frac{\cos x}{\sqrt{1 - \sin x}}\,dx = \lim_{t\to (\pi/2)^-} \left[-2\sqrt{1 - \sin x}\right]_0^t = 0 - (-2) = 2$.

$\mathrm{I}_2 = \lim_{t\to (\pi/2)^+} \int_t^\pi \frac{\cos x}{\sqrt{1 - \sin x}}\,dx = \lim_{t\to (\pi/2)^+} \left[-2\sqrt{1 - \sin x}\right]_t^\pi = -2 - 0 = -2$.

$\mathrm{I} = \mathrm{I}_1 + \mathrm{I}_2 = 2 + (-2) = 0$. C

[26] $I_1 = \lim_{t\to 1^-} \int_0^t \frac{x}{\sqrt[3]{x-1}}\,dx = \lim_{t\to 1^-} \left[\frac{3}{5}(x-1)^{5/3} + \frac{3}{2}(x-1)^{2/3}\right]_0^t =$

$\left[0 - (-\frac{3}{5} + \frac{3}{2})\right] = -0.9.$

$I_2 = \lim_{t\to 1^+} \int_t^9 \frac{x}{\sqrt[3]{x-1}}\,dx = \lim_{t\to 1^+} \left[\frac{3}{5}(x-1)^{5/3} + \frac{3}{2}(x-1)^{2/3}\right]_t^9 =$

$\left[\frac{3}{5}(32) + \frac{3}{2}(4) - 0\right] = 25.2.$

$I = I_1 + I_2 = (-0.9) + 25.2 = 24.3.$ C

Note: To integrate, let $u = (x-1)^{1/3}$, $x = u^3 + 1$, and $dx = 3u^2\,du$.

[27] There are discontinuities at $x = 1$ and $x = 3$.

$I_1 = \int_0^1 \frac{1}{(x-1)(x-3)}\,dx = \lim_{t\to 1^-} \int_0^t \left[\frac{\frac{1}{2}}{x-3} - \frac{\frac{1}{2}}{x-1}\right] dx$ {PF} $=$

$\lim_{t\to 1^-} \left[\frac{1}{2}\ln|x-3| - \frac{1}{2}\ln|x-1|\right]_0^t = \infty;$ D

[28] There are discontinuities at $x = \pm 1$. Choose $x = 0$ as a value between -1 and 1.

$I_1 = \int_{-1}^0 \frac{x}{\sqrt[3]{x^2-1}}\,dx = \lim_{t\to -1^+} \left[\frac{3}{4}(x^2-1)^{2/3}\right]_t^0 = \frac{3}{4}(1-0) = \frac{3}{4}.$

$I_2 = \int_0^1 \frac{x}{\sqrt[3]{x^2-1}}\,dx = \lim_{t\to 1^-} \left[\frac{3}{4}(x^2-1)^{2/3}\right]_0^t = \frac{3}{4}(0-1) = -\frac{3}{4}.$

$I_3 = \int_1^3 \frac{x}{\sqrt[3]{x^2-1}}\,dx = \lim_{t\to 1^+} \left[\frac{3}{4}(x^2-1)^{2/3}\right]_t^3 = \frac{3}{4}(4-0) = 3.$

$I = I_1 + I_2 + I_3 = \frac{3}{4} + (-\frac{3}{4}) + 3 = 3.$ C

[29] $I = \int_0^4 \frac{1}{(x-4)^2}\,dx + \int_4^5 \frac{1}{(x-4)^2}\,dx + \int_5^\infty \frac{1}{(x-4)^2}\,dx.$

$I_1 = \int_0^4 \frac{1}{(x-4)^2}\,dx = \lim_{t\to 4^-} \left[-\frac{1}{x-4}\right]_0^t = \infty;$ D

[30] $I = \int_{-\infty}^{-3} \frac{1}{x+2}\,dx + \int_{-3}^{-2} \frac{1}{x+2}\,dx + \int_{-2}^0 \frac{1}{x+2}\,dx.$

$I_1 = \int_{-\infty}^{-3} \frac{1}{x+2}\,dx = \lim_{t\to -\infty} \left[\ln|x+2|\right]_t^{-3} = -\infty;$ D

[31] $0 \le \frac{\sin x}{\sqrt{x}} \le \frac{1}{\sqrt{x}}$ on $(0, \pi]$ and $\int_0^\pi \frac{1}{\sqrt{x}}\,dx = \lim_{t\to 0^+} \left[2\sqrt{x}\right]_t^\pi = 2\sqrt{\pi} \Rightarrow$

I *converges* by (i).

[32] $0 \le \frac{1}{x^3} \le \frac{\sec x}{x^3}$ on $(0, \frac{\pi}{4}]$ and $\int_0^{\pi/4} \frac{1}{x^3}\,dx = \lim_{t\to 0^+} \left[-\frac{1}{2x^2}\right]_t^{\pi/4} = \infty \Rightarrow$

I *diverges* by (ii).

[33] $0 \le \frac{1}{(x-2)^2} \le \frac{\cosh x}{(x-2)^2}$ on $[0, 2)$ and $\int_0^2 \frac{1}{(x-2)^2}\,dx = \lim_{t\to 2^-} \left[-\frac{1}{x-2}\right]_0^t = \infty \Rightarrow$

I *diverges* by (ii).

[34] $0 \le \frac{e^{-x}}{x^{2/3}} \le \frac{1}{x^{2/3}}$ on $(0, 1]$ and $\int_0^1 x^{-2/3}\,dx = \lim_{t\to 0^+} \left[3x^{1/3}\right]_t^1 = 3(1-0) = 3 \Rightarrow$

I *converges* by (i).

[35] (i) If $n \geq 0$, $\mathrm{I} = \int_0^1 x^n\,dx = \left[\frac{x^{n+1}}{n+1}\right]_0^1 = \frac{1}{n+1}$; C

(ii) If $-1 < n < 0$, then $n + 1 > 0$ and

$$\mathrm{I} = \lim_{t\to 0^+} \int_t^1 x^n\,dx = \lim_{t\to 0^+} \left[\frac{x^{n+1}}{n+1}\right]_t^1 = \frac{1}{n+1}; \text{ C}$$

(iii) If $n = -1$, $\mathrm{I} = \lim_{t\to 0^+} \left[\ln|x|\right]_t^1 = \infty$; D

(iv) If $n < -1$, then $n + 1 < 0$ and $\mathrm{I} = \lim_{t\to 0^+} \left[\frac{x^{n+1}}{n+1}\right]_t^1 = \infty$; D

Thus, I converges iff $n > -1$.

[36] (i) If $n \neq -1$, $\mathrm{I} = \int_0^1 x^n \ln x\,dx = \lim_{t\to 0^+} \left[\frac{x^{n+1}}{n+1}\ln x - \frac{x^{n+1}}{(n+1)^2}\right]_t^1$ {IP}.

If $n + 1 > 0$, then I converges. If $n + 1 < 0$, then I diverges.

(ii) If $n = -1$, $\mathrm{I} = \lim_{t\to 0^+} \left[\frac{1}{2}(\ln x)^2\right]_t^1 = -\infty$; D. Thus, I converges iff $n > -1$.

[37] (a) $A = \int_0^1 \frac{1}{\sqrt{x}}\,dx = \lim_{t\to 0^+} \left[2\sqrt{x}\right]_t^1 = 2(1 - 0) = 2.$

(b) Using disks, $V = \pi\int_0^1 \left(\frac{1}{\sqrt{x}}\right)^2 dx = \pi \lim_{t\to 0^+} \left[\ln|x|\right]_t^1 = \infty$. Not possible.

[38] (a) $A = \int_0^1 \frac{1}{\sqrt[3]{x}}\,dx = \lim_{t\to 0^+} \left[\frac{3}{2}x^{2/3}\right]_t^1 = \frac{3}{2}(1 - 0) = \frac{3}{2}.$

(b) Using disks, $V = \pi\int_0^1 (x^{-1/3})^2\,dx = \pi \lim_{t\to 0^+} \left[3x^{1/3}\right]_t^1 = 3\pi(1 - 0) = 3\pi.$

[39] (a) $A = \int_{-4}^4 \frac{1}{x+4}\,dx = \lim_{t\to -4^+} \left[\ln|x+4|\right]_t^4 = \infty$. Not possible.

(b) Using disks, $V = \pi\int_{-4}^4 \left(\frac{1}{x+4}\right)^2 dx = \pi \lim_{t\to -4^+} \left[-\frac{1}{x+4}\right]_t^4 = \infty$. Not possible.

[40] (a) $A = \int_1^2 \frac{1}{x-1}\,dx = \lim_{t\to 1^+} \left[\ln|x-1|\right]_t^2 = \infty$. Not possible.

(b) Using disks, $V = \pi\int_1^2 \left(\frac{1}{x-1}\right)^2 dx = \pi \lim_{t\to 1^+} \left[-\frac{1}{x-1}\right]_t^2 = \infty$. Not possible.

[41] $u = \sqrt{x}$, $u^2 = x$, and $2u\,du = dx \Rightarrow \int_0^1 \frac{\cos x}{\sqrt{x}}\,dx = \int_0^1 \frac{\cos(u^2)}{u}2u\,du = 2\int_0^1 \cos(u^2)\,du.$

Let $f(x) = \cos(x^2)$. $\mathrm{T} = (2)\frac{1-0}{2(4)}\{f(0) + 2\left[f(\frac{1}{4}) + f(\frac{1}{2}) + f(\frac{3}{4})\right] + f(1)\} \approx 1.7915.$

[42] Let $f(x) = (\sin x)/x$.

Since $\lim_{x\to 0} f(x) = 1$, f has a removeable discontinuity at $x = 0$. Let $f(0) = 1$.

$$\mathrm{S} = \tfrac{1-0}{3(4)}\left[f(0) + 4f(\tfrac{1}{4}) + 2f(\tfrac{1}{2}) + 4f(\tfrac{3}{4}) + f(1)\right] \approx \tfrac{1}{12}(11.3530) \approx 0.9461.$$

43 (a) $mv\frac{dv}{dy} + ky = 0 \Rightarrow mv\,dv = -ky\,dy \Rightarrow \int mv\,dv = -\int ky\,dy \Rightarrow$

$\frac{1}{2}mv^2 = -\frac{1}{2}ky^2 + C_1 \Rightarrow v^2 = -\frac{k}{m}y^2 + C_2$. $v(\pm c) = 0 \Rightarrow v^2(\pm c) = 0 \Rightarrow$

$-\frac{k}{m}c^2 + C_2 = 0 \Rightarrow C_2 = \frac{k}{m}c^2$. Thus, $v^2 = \frac{k}{m}(c^2 - y^2)$.

(b) From Example 5, $T = 2\int_{-c}^{c} \frac{1}{v(y)}\,dy = 2\int_{-c}^{c} \sqrt{\frac{m}{k}} \cdot \frac{1}{\sqrt{c^2 - y^2}}\,dy$

$= 2\lim_{t\to c^-} 2\sqrt{\frac{m}{k}}\int_0^t \frac{1}{\sqrt{c^2 - y^2}}\,dy$ {even integrand}

$= 4\sqrt{\frac{m}{k}}\lim_{t\to c^-}\left[\sin^{-1}\frac{y}{c}\right]_0^t = 4\sqrt{\frac{m}{k}}\left(\frac{\pi}{2} - 0\right) = 2\pi\sqrt{\frac{m}{k}}$.

44 (a) $v\frac{dv}{d\theta} + \frac{g}{L}\sin\theta = 0 \Rightarrow v\,dv = -\frac{g}{L}\sin\theta\,d\theta \Rightarrow \int v\,dv = \int -\frac{g}{L}\sin\theta\,d\theta \Rightarrow$

$\frac{1}{2}v^2 = \frac{g}{L}\cos\theta + C_1 \Rightarrow v^2 = \frac{2g}{L}\cos\theta + C_2$. $v(\pm\theta_0) = 0 \Rightarrow v^2(\pm\theta_0) = 0 \Rightarrow$

$\frac{2g}{L}\cos\theta_0 + C_2 = 0 \Rightarrow C_2 = -\frac{2g}{L}\cos\theta_0$. Thus, $v^2 = \frac{2g}{L}(\cos\theta - \cos\theta_0)$.

(b) Suppose the mass m turns through an angle $\Delta\theta$.

Period T = twice the time t from $-\theta_0$ to θ_0 = 4 times the time t from 0 to θ_0.

$T = 4t = 4(\frac{d}{v})$ {since $d = rt$ or $d = vt$}

$= 4 \cdot \frac{L\,\Delta\theta}{\sqrt{2g/L}\sqrt{\cos\theta - \cos\theta_0}}$ {arc length = radius · radian measure}

$= 2\sqrt{\frac{2L}{g}}\frac{\Delta\theta}{\sqrt{\cos\theta - \cos\theta_0}}$. Summing the times from $\theta = 0$ to $\theta = \theta_0$ yields

$$T = 2\sqrt{\frac{2L}{g}}\int_0^{\theta_0} \frac{1}{\sqrt{\cos\theta - \cos\theta_0}}\,d\theta.$$

45 (a) From the formula for T, $0 \le y \le y_0$.

$y = 0 \Rightarrow y_0 e^{-kt} = 0 \Rightarrow e^{-kt} = 0$ and t is undefined.

(b) $y = y_0 e^{-kt} \Rightarrow \frac{y}{y_0} = e^{-kt} \Rightarrow t = -\frac{1}{k}\ln\frac{y}{y_0} \Rightarrow T = \frac{1}{y_0}\int_0^{y_0} -\frac{1}{k}\ln\frac{y}{y_0}\,dy$.

$z = \frac{y}{y_0}$ and $dz = \frac{1}{y_0}\,dy \Rightarrow T = -\frac{1}{k}\int_0^1 \ln z\,dz = -\frac{1}{k}\lim_{t\to 0^+}\left[z\ln z - z\right]_t^1 =$

$\frac{1}{k} + \frac{1}{k}\lim_{t\to 0^+}\frac{\ln t}{1/t}\ \{\frac{\infty}{\infty}\} = \frac{1}{k} + 0 = \frac{1}{k}$. Time τ occurs when

$\frac{y}{y_0} = \frac{1}{2}$ or $t\,\{= \tau\} = -\frac{1}{k}\ln\frac{1}{2} = \frac{\ln 2}{k}$. Thus, $T = \frac{1}{k} = \frac{1}{\ln 2/\tau} = \frac{\tau}{\ln 2}$.

46 (a) From Exercise 45, $T = \frac{1}{k} = \frac{1}{0.2} = 5$ yr.

(b) $N = \frac{N_0}{1 + kN_0 t} \Rightarrow 1 + kN_0 t = \frac{N_0}{N} \Rightarrow t = \frac{1}{kN_0}\left(\frac{N_0}{N} - 1\right)$.

$T = \frac{1}{N_0}\int_0^{N_0} \frac{1}{kN_0}\left(\frac{N_0}{N} - 1\right)dN = \frac{1}{kN_0^2}\lim_{t\to 0^+}\left[N_0\ln N - N\right]_t^{N_0}$,

which diverges for all k. No, it is not possible.

10.5 Review Exercises

1. $\text{L} = \dfrac{\ln 2}{1+1} = \frac{1}{2}\ln 2$

2. $\text{L}\,\{\frac{0}{0}\} = \lim\limits_{x\to 0} \dfrac{2\cos 2x - 2\sec^2 2x}{2x} = \lim\limits_{x\to 0} \dfrac{\cos 2x - \sec^2 2x}{x}\,\{\frac{0}{0}\} =$

$$\lim_{x\to 0} \frac{-2\sin 2x - 4\sec^2 2x\tan 2x}{1} = \frac{0}{1} = 0$$

3. $\text{L}\,\{\frac{\infty}{\infty}\} = \lim\limits_{x\to\infty} \dfrac{2x+2}{1/(x+1)} = \lim\limits_{x\to\infty} (2x+2)(x+1) = \infty$

4. $\text{L}\,\{\frac{0}{0}\} = \lim\limits_{x\to 0} \dfrac{1/(1+x^2)}{1/(1-x^2)^{1/2}} = \frac{1}{1} = 1$

5. $\text{L}\,\{\frac{0}{0}\} = \lim\limits_{x\to 0} \dfrac{2e^{2x} + 2e^{-2x} - 4}{3x^2}\,\{\frac{0}{0}\} = \lim\limits_{x\to 0} \dfrac{4e^{2x} - 4e^{-2x}}{6x}\,\{\frac{0}{0}\} =$

$$\lim_{x\to 0} \frac{8e^{2x} + 8e^{-2x}}{6} = \frac{16}{6} = \frac{8}{3}$$

6. $\text{L} = \lim\limits_{x\to(\pi/2)^-} \tan x\cos x = \lim\limits_{x\to(\pi/2)^-} \sin x = 1$

7. L = 0 by Exercise 31 in §10.1.

8. $\text{L}\,\{0\cdot\infty\} = \lim\limits_{x\to(\pi/2)^-} \dfrac{\ln\cos x}{\sec x}\,\{\frac{\infty}{\infty}\} = \lim\limits_{x\to(\pi/2)^-} \dfrac{-\tan x}{\sec x\tan x} = \lim\limits_{x\to(\pi/2)^-} -\cos x = 0$

9. $\lim\limits_{x\to\infty} (1 - 2e^{1/x})x = (1-2)(\infty) = -\infty$

10. $\text{L}\,\{0\cdot\infty\} = \lim\limits_{x\to 0^+} \dfrac{\tan^{-1}x}{\sin x}\,\{\frac{0}{0}\} = \lim\limits_{x\to 0^+} \dfrac{1/(1+x^2)}{\cos x} = \frac{1}{1} = 1$

11. L has the form 1^∞. $y = (1+8x^2)^{1/x^2} \Rightarrow \ln y = \dfrac{1}{x^2}\ln(1+8x^2)$. $\lim\limits_{x\to 0}\ln y\,\{\infty\cdot 0\}$

$$= \lim_{x\to 0} \frac{\ln(1+8x^2)}{x^2}\,\{\tfrac{0}{0}\} = \lim_{x\to 0} \frac{16x/(1+8x^2)}{2x} = \lim_{x\to 0} \frac{8}{1+8x^2} = 8. \text{ Thus, L} = e^8.$$

12. L has the form 0^0. $y = (\ln x)^{x-1} \Rightarrow \ln y = (x-1)\ln\ln x$.

$$\lim_{x\to 1^+} \ln y\,\{0\cdot\infty\} = \lim_{x\to 1^+} \frac{\ln\ln x}{1/(x-1)}\,\{\tfrac{\infty}{\infty}\} = \lim_{x\to 1^+} \frac{1/(x\ln x)}{-1/(x-1)^2} =$$

$$\lim_{x\to 1^+} \frac{-(x-1)^2}{x\ln x}\,\{\tfrac{0}{0}\} = \lim_{x\to 1^+} \frac{-2(x-1)}{\ln x + 1} = \frac{0}{1} = 0. \text{ Thus, L} = e^0 = 1.$$

13. L has the form ∞^0. $y = (e^x+1)^{1/x} \Rightarrow \ln y = \frac{1}{x}\ln(e^x+1)$. $\lim\limits_{x\to\infty} \ln y\,\{0\cdot\infty\}$

$$= \lim_{x\to\infty} \frac{\ln(e^x+1)}{x}\,\{\tfrac{\infty}{\infty}\} = \lim_{x\to\infty} \frac{e^x}{e^x+1}\,\{\tfrac{\infty}{\infty}\} = \lim_{x\to\infty} \frac{e^x}{e^x} = 1. \text{ Thus, L} = e^1.$$

14. $\text{L}\,\{\infty - \infty\} = \lim\limits_{x\to 0^+} \left(\dfrac{\cos x}{\sin x} - \dfrac{1}{x}\right) = \lim\limits_{x\to 0^+} \dfrac{x\cos x - \sin x}{x\sin x}\,\{\frac{0}{0}\}$

$$= \lim_{x\to 0^+} \frac{-x\sin x + \cos x - \cos x}{x\cos x + \sin x} = \lim_{x\to 0^+} \frac{-x\sin x}{x\cos x + \sin x}\,\{\tfrac{0}{0}\}$$

$$= \lim_{x\to 0^+} \frac{-x\cos x - \sin x}{-x\sin x + \cos x + \cos x} = \frac{0}{2} = 0$$

15. $\text{L} = \lim\limits_{x\to\infty} \dfrac{\sqrt{x^2/x^2 + 1/x^2}}{x/x} = \dfrac{\sqrt{1+0}}{1} = 1$

[16] $\text{L}\ \{\frac{\infty}{\infty}\} = \lim_{x \to \infty} \dfrac{3^x(\ln 3) + 2}{3x^2}\ \{\frac{\infty}{\infty}\} = \lim_{x \to \infty} \dfrac{3^x(\ln 3)^2}{6x}\ \{\frac{\infty}{\infty}\} = \lim_{x \to \infty} \dfrac{3^x(\ln 3)^3}{6} = \infty$

[17] $\text{I} = \lim_{t \to \infty} \int_4^t x^{-1/2}\,dx = \lim_{t \to \infty} \Big[2\sqrt{x}\Big]_4^t = \infty;\ \text{D}$

[18] $\text{I} = \lim_{t \to \infty} \int_4^t x^{-3/2}\,dx = \lim_{t \to \infty} \Big[-\frac{2}{\sqrt{x}}\Big]_4^t = -2(0 - \frac{1}{2}) = 1;\ \text{C}$

[19] $\text{I} = \int_{-\infty}^{-3} \frac{1}{x+2}\,dx + \int_{-3}^{-2} \frac{1}{x+2}\,dx + \int_{-2}^{0} \frac{1}{x+2}\,dx = \text{I}_1 + \text{I}_2 + \text{I}_3.$

$$\text{I}_3 = \lim_{t \to -2^+} \int_t^0 \frac{1}{x+2}\,dx = \lim_{t \to -2^+} \Big[\ln|x+2|\Big]_t^0 = \infty;\ \text{D}$$

[20] $\text{I} = \lim_{t \to \infty} \int_0^t \sin x\,dx = \lim_{t \to \infty} \Big[-\cos x\Big]_0^t$. This limit DNE. D

[21] $\text{I}_1 = \lim_{t \to 0^-} \int_{-8}^t x^{-1/3}\,dx = \lim_{t \to 0^-} \Big[\frac{3}{2}x^{2/3}\Big]_{-8}^t = \frac{3}{2}(0 - 4) = -6.$

$\text{I}_2 = \lim_{t \to 0^+} \int_t^1 x^{-1/3}\,dx = \lim_{t \to 0^+} \Big[\frac{3}{2}x^{2/3}\Big]_t^1 = \frac{3}{2}(1 - 0) = \frac{3}{2}.$

$$\text{I} = \text{I}_1 + \text{I}_2 = -6 + \tfrac{3}{2} = -\tfrac{9}{2}.\ \text{C}$$

[22] $\text{I} = \lim_{t \to -4^+} \int_t^0 \frac{1}{x+4}\,dx = \lim_{t \to -4^+} \Big[\ln|x+4|\Big]_t^0 = \infty;\ \text{D}$

[23] $\text{I}_1 = \lim_{t \to 1^-} \int_0^t \frac{x}{(x^2-1)^2}\,dx = \lim_{t \to 1^-} \Big[-\frac{1}{2(x^2-1)}\Big]_0^t = \infty;\ \text{D}$

[24] $\text{I} = \lim_{t \to 1^+} \int_t^2 \frac{1}{x\sqrt{x^2-1}}\,dx = \lim_{t \to 1^+} \Big[\sec^{-1}x\Big]_t^2 = \frac{\pi}{3} - 0 = \frac{\pi}{3};\ \text{C}$

[25] $\int_{-\infty}^{\infty} \frac{1}{e^x + e^{-x}}\,dx = 2\int_0^{\infty} \frac{1}{e^x + e^{-x}} \cdot \frac{e^x}{e^x}\,dx = 2\lim_{t \to \infty} \int_0^t \frac{e^x}{e^{2x}+1}\,dx =$

$$2\lim_{t \to \infty} \Big[\tan^{-1} e^x\Big]_0^t = 2(\tfrac{\pi}{2} - \tfrac{\pi}{4}) = \tfrac{\pi}{2};\ \text{C}$$

[26] $\text{I} = \lim_{t \to -\infty} \int_t^0 xe^x\,dx\ \{\text{IP}\} = \lim_{t \to -\infty} \Big[xe^x - e^x\Big]_t^0 = -1 - 0 = -1;\ \text{C}$

[27] $\text{I} = \lim_{t \to 0^+} \int_t^1 \frac{\ln x}{x}\,dx = \lim_{t \to 0^+} \Big[\frac{1}{2}(\ln x)^2\Big]_t^1 = -\infty;\ \text{D}$

[28] $\text{I} = \lim_{t \to 0^+} \int_t^{\pi/2} \csc x\,dx = \lim_{t \to 0^+} \Big[\ln|\csc x - \cot x|\Big]_t^{\pi/2} =$

$$0 - \lim_{t \to 0^+} \Big[\ln\Big|\frac{1-\cos t}{\sin t}\Big|\Big]\ \{\tfrac{0}{0}\} = -\lim_{t \to 0^+} \Big[\ln\Big|\frac{\sin t}{\cos t}\Big|\Big] = \infty;\ \text{D}$$

29 $u = \frac{1}{x}$, $x = \frac{1}{u}$, and $dx = -\frac{1}{u^2}\,du$. $x = 1 \Rightarrow u = 1$ and $x \to \infty \Rightarrow u = 0^+$.

$$\mathrm{I} = \int_1^\infty e^{-x^2}\,dx = \lim_{t\to 0^+}\int_1^t e^{-1/u^2}\left(-\frac{1}{u^2}\right)du = \lim_{t\to 0^+}\int_t^1 \frac{e^{-1/u^2}}{u^2}\,du.$$

Let $f(u) = \frac{e^{-1/u^2}}{u^2}$. Since $\lim\limits_{u\to 0^+}\frac{e^{-1/u^2}}{u^2} = \lim\limits_{v\to\infty} v^2 e^{-v^2} = 0$

(where $v = \frac{1}{u}$ and L'Hôpital's rule was used), define $f(0) = 0$.

$$\mathrm{S} = \tfrac{1-0}{3(4)}\left[f(0) + 4f(\tfrac{1}{4}) + 2f(\tfrac{1}{2}) + 4f(\tfrac{3}{4}) + f(1)\right] \approx 0.1430.$$

30 $u = \frac{1}{x}$, $x = \frac{1}{u}$, and $dx = -\frac{1}{u^2}\,du$. $x = 1 \Rightarrow u = 1$ and $x \to \infty \Rightarrow u = 0^+$.

$$\mathrm{I} = \int_1^\infty e^{-x}\sin\sqrt{x}\,dx = \lim_{t\to 0^+}\int_1^t e^{-1/u}\sin\sqrt{\tfrac{1}{u}}\left(-\frac{1}{u^2}\right)du =$$

$$\lim_{t\to 0^+}\int_t^1 e^{-1/u}\sin\sqrt{\tfrac{1}{u}}\left(\frac{1}{u^2}\right)du.$$ Let $f(u) = e^{-1/u}\sin\sqrt{\frac{1}{u}}\left(\frac{1}{u^2}\right)$.

Since $\left|\lim\limits_{u\to 0^+} e^{-1/u}\sin\sqrt{\frac{1}{u}}\left(\frac{1}{u^2}\right)\right| = \left|\lim\limits_{v\to\infty}\frac{v^2\sin\sqrt{v}}{e^v}\right| \le \lim\limits_{v\to\infty}\frac{v^2}{e^v} = 0$

(where $v = \frac{1}{u}$ and L'Hôpital's rule was used), define $f(0) = 0$.

$$\mathrm{S} = \tfrac{1-0}{3(4)}\left[f(0) + 4f(\tfrac{1}{4}) + 2f(\tfrac{1}{2}) + 4f(\tfrac{3}{4}) + f(1)\right] \approx \tfrac{1}{12}(4.1594) \approx 0.3466.$$

31 Since $(\sin t)^{2/3} \ge 0$ and generates an infinite area between its graph and the x-axis as

$x \to \infty$, $\lim\limits_{x\to\infty} f(x) = \int_1^x (\sin t)^{2/3}\,dt = \infty$, $\lim\limits_{x\to\infty}\frac{f(x)}{g(x)}\ \{\tfrac{\infty}{\infty}\} = \lim\limits_{x\to\infty}\frac{f'(x)}{g'(x)} =$

$\lim\limits_{x\to\infty}\frac{(\sin x)^{2/3}}{2x} = 0$. Note that $\left|(\sin x)^{2/3}\right| \le 1$ and $2x \to \infty$.

32 $$\lim_{x\to\infty} e^{(x^2)}\left[1 - \operatorname{erf}(x)\right] = \lim_{x\to\infty}\frac{1 - \operatorname{erf}(x)}{e^{-x^2}}\ \{\tfrac{0}{0}\} = \lim_{x\to\infty}\frac{-\frac{2}{\sqrt{\pi}}e^{-x^2}}{-2xe^{-x^2}} =$$

$$\lim_{x\to\infty}\frac{1}{\sqrt{\pi}\,x} = 0$$

Chapter 11: Infinite Series

Exercises 11.1

Note: In Exercises 1–16, the first four terms are found by substituting 1, 2, 3, and 4 for n in the nth term.

[1] $\frac{1}{5}, \frac{2}{8} = \frac{1}{4}, \frac{3}{11}, \frac{4}{14} = \frac{2}{7}$; $\lim_{n\to\infty}\left(\frac{n}{3n+2}\cdot\frac{1/n}{1/n}\right) = \lim_{n\to\infty}\frac{1}{3+2/n} = \frac{1}{3}$

[2] $\frac{1}{6}, \frac{7}{11}, \frac{13}{16}, \frac{19}{21}$; $\lim_{n\to\infty}\left(\frac{6n-5}{5n+1}\cdot\frac{1/n}{1/n}\right) = \lim_{n\to\infty}\frac{6-5/n}{5+1/n} = \frac{6}{5}$

[3] $\frac{3}{5}, -\frac{9}{11}, -\frac{29}{21}, -\frac{57}{35}$; $\lim_{n\to\infty}\frac{7-4n^2}{3+2n^2} = \lim_{n\to\infty}\frac{7/n^2-4}{3/n^2+2} = \frac{-4}{2} = -2$

[4] $4, \frac{4}{-6} = -\frac{2}{3}, -\frac{4}{13}, \frac{4}{-20} = -\frac{1}{5}$; $\lim_{n\to\infty}\frac{4}{8-7n} = \lim_{n\to\infty}\frac{4/n}{8/n-7} = \frac{0}{-7} = 0$

[5] $-5, -5, -5, -5$; $\lim_{n\to\infty}(-5) = -5$

[6] $\sqrt{2}, \sqrt{2}, \sqrt{2}, \sqrt{2}$; $\lim_{n\to\infty}\sqrt{2} = \sqrt{2}$

[7] $\frac{4}{2} = 2, \frac{21}{9} = \frac{7}{3}, \frac{50}{28} = \frac{25}{14}, \frac{91}{65} = \frac{7}{5}$; $\lim_{n\to\infty}\frac{(2n-1)(3n+1)}{n^3+1} = \lim_{n\to\infty}\frac{6n^2-n-1}{n^3+1} = 0$

[8] $9, 17, 25, 33$; $\lim_{n\to\infty}(8n+1) = \infty$; DNE

[9] $\frac{2}{\sqrt{10}}, \frac{2}{\sqrt{13}}, \frac{2}{\sqrt{18}}, \frac{2}{\sqrt{25}} = \frac{2}{5}$; $\lim_{n\to\infty}\frac{2}{\sqrt{n^2+9}} = 0$

[10] $\frac{100}{5} = 20, \frac{200}{2\sqrt{2}+4} = \frac{100}{\sqrt{2}+2}, \frac{300}{3\sqrt{3}+4}, \frac{400}{8+4} = \frac{100}{3}$;

$$\lim_{n\to\infty}\frac{100n}{n^{3/2}+4} = \lim_{n\to\infty}\frac{100}{n^{1/2}+4/n} = 0$$

[11] $\frac{3}{10}, -\frac{6}{17}, \frac{9}{26}, -\frac{12}{37}$;

$$\lim_{n\to\infty}\left|(-1)^{n+1}\frac{3n}{n^2+4n+5}\right| = 0 \Rightarrow \lim_{n\to\infty}(-1)^{n+1}\frac{3n}{n^2+4n+5} = 0, \text{ by (11.8).}$$

[12] $\frac{1}{2}, -\frac{\sqrt{2}}{3}, \frac{\sqrt{3}}{4}, -\frac{2}{5}$; $\lim_{n\to\infty}\left|(-1)^{n+1}\frac{\sqrt{n}}{n+1}\right| = 0 \Rightarrow \lim_{n\to\infty}(-1)^{n+1}\frac{\sqrt{n}}{n+1} = 0$, by (11.8).

[13] $1.1, 1.01, 1.001, 1.0001$; $\lim_{n\to\infty}\left[1+(0.1)^n\right] = 1 + 0 = 1$

[14] $\frac{1}{2}, \frac{3}{4}, \frac{7}{8}, \frac{15}{16}$; $\lim_{n\to\infty}\left(1-\frac{1}{2^n}\right) = 1 - 0 = 1$

[15] $2, 0, 2, 0$; $\lim_{n\to\infty}\left[1+(-1)^{n+1}\right]$ DNE since it does not converge to a real number L.

[16] $2, \frac{3}{\sqrt{2}}, \frac{4}{\sqrt{3}}, \frac{5}{2}$; $\lim_{n\to\infty}\frac{n+1}{\sqrt{n}} = \infty$; DNE

Note: Let C denote *converges* and D denote *diverges*.

Also, let L denote the limit of the sequence, if it exists.

[17] L $= 6\lim_{n\to\infty}\left(-\frac{5}{6}\right)^n = 6\cdot 0 = 0$ by (11.6)(i); C

[18] L $= 8 - 0 = 8$; C

[19] L $= \frac{\pi}{2}$; C

[20] L $= 0$; C, since $\tan^{-1} n \to \frac{\pi}{2}$ as $n \to \infty$

[21] L $= -\infty$; D

[22] L $= \infty$; D, since $(1.0001)^n \to \infty$ as $n \to \infty$.

[23] Using (11.5)(i), $\lim_{x \to \infty} \frac{\ln x}{x} \{\frac{\infty}{\infty}\} = \lim_{x \to \infty} \frac{1/x}{1} = 0$, and hence $\lim_{n \to \infty} \frac{\ln n}{n} = 0$.

Thus, by (11.8), L $= 0$; C.

[24] Using (11.5)(ii), $\lim_{x \to \infty} \frac{x^2}{\ln(x+1)} \{\frac{\infty}{\infty}\} = \lim_{x \to \infty} \frac{2x}{1/(x+1)} = \lim_{x \to \infty} 2x(x+1) = \infty$.

Thus, L $= \infty$; D.

[25] Using (11.5)(ii), $\lim_{x \to \infty} \frac{4x^4 + 1}{2x^2 - 1} \{\frac{\infty}{\infty}\} = \lim_{x \to \infty} \frac{16x^3}{4x} = \lim_{x \to \infty} 4x^2 = \infty$.

Thus, L $= \infty$; D.

[26] $-\frac{1}{n} \le \frac{\cos n}{n} \le \frac{1}{n} \Rightarrow$ L $= 0$ by (11.7); C.

[27] $\lim_{x \to \infty} \frac{e^x}{x^4} = \infty$ {see Exercise 32, §10.1}. Thus, L $= \infty$; D.

[28] $\lim_{x \to \infty} e^{-x} \ln x \{0 \cdot \infty\} = \lim_{x \to \infty} \frac{\ln x}{e^x} \{\frac{\infty}{\infty}\} = \lim_{x \to \infty} \frac{1}{xe^x} = 0$. Thus, L $= 0$; C.

[29] By (7.32)(ii), L $= e$; C.

[30] $\lim_{x \to \infty} x^3 3^{-x} \{\infty \cdot 0\} = \lim_{x \to \infty} \frac{x^3}{3^x} \{\frac{\infty}{\infty}\} = \lim_{x \to \infty} \frac{3x^2}{3^x(\ln 3)} \{\frac{\infty}{\infty}\} =$

$\lim_{x \to \infty} \frac{6x}{3^x(\ln 3)^2} \{\frac{\infty}{\infty}\} = \lim_{x \to \infty} \frac{6}{3^x(\ln 3)^3} = 0$. Thus, by (11.8), L $= 0$; C.

[31] $-\frac{1}{2^n} \le \frac{\sin n}{2^n} \le \frac{1}{2^n} \Rightarrow$ L $= 0$ by (11.7); C.

[32] L $= \lim_{n \to \infty} \frac{4n^3/n^3 + 5n/n^3 + 1/n^3}{2n^3/n^3 - n^2/n^3 + 5/n^3} = 2$; C.

[33] L $= \lim_{n \to \infty} \frac{n^2(2n+1) - n^2(2n-1)}{(2n-1)(2n+1)} = \lim_{n \to \infty} \frac{2n^2}{4n^2 - 1} = \lim_{n \to \infty} \frac{2n^2/n^2}{4n^2/n^2 - 1/n^2} = \frac{1}{2}$;

C.

[34] $\lim_{x \to \infty} x \sin\frac{1}{x} \{\infty \cdot 0\} = \lim_{x \to \infty} \frac{\sin 1/x}{1/x} \{\frac{0}{0}\} = \lim_{x \to \infty} \frac{(-1/x^2)\cos(1/x)}{-1/x^2} = \cos(0) = 1$.

Thus, L $= 1$; C.

[35] Since $\{\cos \pi n\} = -1, 1, -1, 1, -1, 1, \ldots$; L DNE; D.

[36] Since $\{4 + \sin(\frac{1}{2}\pi n)\} = 5, 4, 3, 4, 5, 4, 3, 4, \ldots$; L DNE; D.

[37] L $= 1$; C {see Exercise 17, §10.2}.

[38] $\lim_{x \to \infty} \frac{x^2}{2^x} \{\frac{\infty}{\infty}\} = \lim_{x \to \infty} \frac{2x}{2^x(\ln 2)} \{\frac{\infty}{\infty}\} = \lim_{x \to \infty} \frac{2}{2^x(\ln 2)^2} = 0$. Thus, L $= 0$; C.

[39] $-\frac{1}{n^{10}} \le \frac{\cos n}{n^{10}} = \frac{n^{-10}}{\sec n} \le \frac{1}{n^{10}} \Rightarrow$ L $= 0$ by (11.7); C.

[40] Since $\lim_{n \to \infty} \frac{n^2}{1 + n^2} = 1$, even numbered terms approach 1 and

odd numbered terms approach -1. Thus, L DNE; D.

[41] $\mathrm{L} = \lim_{n\to\infty}\left(\frac{\sqrt{n+1}-\sqrt{n}}{1}\cdot\frac{\sqrt{n+1}+\sqrt{n}}{\sqrt{n+1}+\sqrt{n}}\right) = \lim_{n\to\infty}\frac{1}{\sqrt{n+1}+\sqrt{n}} = 0$; C.

[42] $\mathrm{L} = \lim_{n\to\infty}\left(\frac{\sqrt{n^2+n}-n}{1}\cdot\frac{\sqrt{n^2+n}+n}{\sqrt{n^2+n}+n}\right) = \lim_{n\to\infty}\frac{n}{\sqrt{n^2+n}+n} =$

$$\lim_{n\to\infty}\frac{n/n}{\sqrt{n^2/n^2+n/n^2}+n/n} = \frac{1}{1+1} = \tfrac{1}{2}; \text{ C.}$$

[43] (a) Next year's bird population on island A is determined by the number of birds that stay (90% of its present population) and the number of birds that migrate (5% of island C's present population). Thus, $A_{n+1} = 0.90A_n + 0.05C_n$. In a similar manner, $B_{n+1} = 0.80B_n + 0.10A_n$ and $C_{n+1} = 0.95C_n + 0.20B_n$.

(b) Let $\lim_{n\to\infty} A_n = a$, $\lim_{n\to\infty} B_n = b$, and $\lim_{n\to\infty} C_n = c$. Then, as $n\to\infty$,

$a = 0.90a + 0.05c \Rightarrow 0.1a = 0.05c \Rightarrow c = 2a$,

$b = 0.80b + 0.10a \Rightarrow 0.2b = 0.1a \Rightarrow a = 2b$,

and $c = 0.95c + 0.20b \Rightarrow 0.05c = 0.2b \Rightarrow c = 4b$.

Now, $35{,}000 = a + b + c = 2b + b + 4b = 7b \Rightarrow b = 5000$.

Thus, there will be 5000 birds on B, 10,000 birds on A, and 20,000 birds on C.

[44] (a) Since 50% of the adult population is female and each female has on the average 3 kittens, $K_{n+1} = 3(\frac{1}{2}A_{n+1}) = \frac{3}{2}A_{n+1}$. Next years adult population is the sum of surviving adults $(\frac{2}{3}A_n)$ and surviving newborns $(\frac{1}{2}K_n)$.

Thus, $A_{n+1} = \frac{2}{3}A_n + \frac{1}{2}K_n$.

(b) $A_{n+1} = \frac{2}{3}A_n + \frac{1}{2}K_n = \frac{2}{3}A_n + \frac{1}{2}(\frac{3}{2}A_n) = \frac{17}{12}A_n$.

$K_{n+1} = \frac{3}{2}A_{n+1} = \frac{3}{2}(\frac{2}{3}A_n + \frac{1}{2}K_n) = A_n + \frac{3}{4}K_n = \frac{2}{3}K_n + \frac{3}{4}K_n = \frac{17}{12}K_n$.

Using the above relations recursively,

$A_n = \frac{17}{12}A_{n-1} = \frac{17}{12}(\frac{17}{12}A_{n-2}) = (\frac{17}{12})^2A_{n-2} = (\frac{17}{12})^2\frac{17}{12}A_{n-3} = (\frac{17}{12})^3A_{n-3} = \cdots$

$= (\frac{17}{12})^{n-1}A_1$. Similarly, $K_n = (\frac{17}{12})^{n-1}K_1$.

Adult and kitten populations each increase $41\frac{2}{3}\%$ $\{\frac{17}{12} = 1.41\overline{6}\}$ each year.

[45] (a) The sequence appears to converge to 1.

(b) $a_1 = 5^{1/2}$. Assume $a_n = 5^{1/2^n}$ for some $n \geq 1$. Then, $a_{n+1} = (a_n)^{1/2} = (5^{1/2^n})^{1/2} = 5^{1/2^{n+1}}$. By mathematical induction, the statement is true.

Now, $y = 5^{1/2^x} \Rightarrow \ln y = \frac{1}{2^x}\ln 5$. Thus, $\lim_{x\to\infty}\ln y = 0$ and $\lim_{n\to\infty} a_n = e^0 = 1$.

[46] Since the sequence generated is $n, \frac{1}{n}, n, \frac{1}{n}, \ldots$, we must have $n = \frac{1}{n} \Rightarrow$

$n^2 = 1 \Rightarrow n = 1$, since $n > 0$.

[47] (a) The sequence appears to converge to approximately 0.7390851.

(b) $a_{k+1} = \cos a_k \Rightarrow \lim_{k\to\infty} a_{k+1} = \lim_{k\to\infty} \cos a_k \Rightarrow \lim_{k\to\infty} a_{k+1} = \cos\left(\lim_{k\to\infty} a_k\right) \Rightarrow$

$L = \cos L$.

[48] (a) $x_1 = 3$, $x_2 = 3.1425465$, $x_3 = x_4 = x_5 = 3.1415927$.

The sequence appears to converge to π.

(b) $x_1 = 6$, $x_2 = 6.2910062$, $x_3 = 6.2831851$, $x_4 = x_5 = 6.2831853$.

The sequence appears to converge to 2π.

(c) $x_{k+1} = x_k - \tan x_k \Rightarrow \lim_{k\to\infty} x_{k+1} = \lim_{k\to\infty} (x_k - \tan x_k) \Rightarrow L = L - \tan L \Rightarrow$

$\tan L = 0 \Rightarrow L = \pi n$.

[49] (a) $x_2 = 3.5$, $x_3 = 3.178571429$, $x_4 = 3.162319422$, $x_5 = 3.162277660$,

$x_6 = 3.162277660$

(b) $x_{k+1} = \frac{1}{2}\left(x_k + \frac{N}{x_k}\right) \Rightarrow \lim_{k\to\infty} x_{k+1} = \lim_{k\to\infty} \frac{1}{2}\left(x_k + \frac{N}{x_k}\right) \Rightarrow L = \frac{1}{2}\left(L + \frac{N}{L}\right) \Rightarrow$

$L^2 = N \Rightarrow L = \pm\sqrt{N}$. Since x_1 is assumed to be positive, all x_k are positive and hence $L = \sqrt{N}$. *Note*: The above sequence is generated by Newton's method

using the equation $f(x) = x^2 - N = 0$.

[50] (a) $a_3 = a_2 + a_1 = 1 + 1 = 2$. 1, 1, 2, 3, 5, 8, 13, 21, 34, 55

(b) $r_1 = a_2/a_1 = \frac{1}{1} = 1$, $r_2 = \frac{2}{1} = 2$, $r_3 = \frac{3}{2} = 1.5$, $r_4 = \frac{5}{3} \approx 1.6666666$,

$r_5 = \frac{8}{5} = 1.6$, $r_6 = \frac{13}{8} = 1.625$, $r_7 = \frac{21}{13} \approx 1.6153846$, $r_8 = \frac{34}{21} \approx 1.6190476$,

$r_9 = \frac{55}{34} \approx 1.6176471$, and $r_{10} = \frac{89}{55} \approx 1.6181818$.

(c) $\tau = \lim_{k\to\infty} r_k = \lim_{k\to\infty} \frac{a_{k+1}}{a_k} = \lim_{k\to\infty} \frac{a_k + a_{k-1}}{a_k} = \lim_{k\to\infty} \left(1 + \frac{a_{k-1}}{a_k}\right) = 1 + \frac{1}{\tau} \Rightarrow$

$\tau^2 = \tau + 1 \Rightarrow \tau^2 - \tau - 1 = 0 \Rightarrow \tau = \frac{1}{2}(1 + \sqrt{5}) \approx 1.618034$.

Note that since $a_k > 0$ for all k, $\frac{1}{2}(1 - \sqrt{5}) < 0$ is not a possible value for τ.

[51] (a) $f(x) = \frac{1}{4}\sin x \cos x + 1 = \frac{1}{8}\sin 2x + 1 \Rightarrow |f'(x)| = \frac{1}{4}|\cos 2x| \le \frac{1}{4} < 1$.

Let $B = \frac{1}{4}$. Thus, the sequence converges for any a_1.

(b)

$a_1 = 1$	$a_1 = -100$	
$a_2 = 1.113662$	$a_2 = 1.109162$	
$a_3 = 1.099015$	$a_3 = 1.099697$	
$a_4 = 1.101207$	$a_4 = 1.101107$	
$a_5 = 1.100884$	$a_5 = 1.100899$	$\lim_{n\to\infty} a_n \approx 1.10$

[52] (a) $f(x) = \dfrac{x^2}{x^2 + 1} + 2 \Rightarrow f'(x) = \dfrac{2x}{(x^2 + 1)^2}$. We need to maximize $|f'|$.

$f''(x) = \dfrac{2(1 - 3x^2)}{(x^2 + 1)^3}$. $f''(x) = 0 \Rightarrow x = \pm\frac{\sqrt{3}}{3} \approx \pm 0.58$. $\left|f'(\pm\frac{\sqrt{3}}{3})\right| = \frac{3\sqrt{3}}{8} \approx$

$0.65 < 1$. Let $B = \frac{3\sqrt{3}}{8}$. Thus, the sequence converges for any a_1.

(b) $a_1 = 1$ | $a_1 = -100$

$a_2 = 2.5$ | $a_2 = 2.9999$

$a_3 = 2.862069$ | $a_3 = 2.899994$

$a_4 = 2.891203$ | $a_4 = 2.893730$

$a_5 = 2.893152$ | $a_5 = 2.893318$ | $\lim_{n\to\infty} a_n \approx 2.89$

Exercises 11.2

1 (a) $S_1 = a_1 = \frac{-2}{7\cdot 5} = -\frac{2}{35};\ S_2 = S_1 + a_2 = -\frac{2}{35} + \frac{-2}{63} = -\frac{4}{45};$

$S_3 = S_2 + a_3 = -\frac{4}{45} + \frac{-2}{99} = -\frac{6}{55}.$

(b) Using partial fractions, $a_n = \dfrac{-2}{(2n+5)(2n+3)} = \dfrac{1}{2n+5} - \dfrac{1}{2n+3}.$

$$S_n = a_1 + a_2 + a_3 + \cdots + a_n$$
$$= \left(\tfrac{1}{7} - \tfrac{1}{5}\right) + \left(\tfrac{1}{9} - \tfrac{1}{7}\right) + \left(\tfrac{1}{11} - \tfrac{1}{9}\right) + \cdots + \left(\frac{1}{2n+5} - \frac{1}{2n+3}\right)$$
$$= \frac{1}{2n+5} - \frac{1}{5} = -\frac{2n}{5(2n+5)}.$$

Note: It may be easier to compute the values in part (a)

after finding the general formula in part (b).

(c) $S = \lim_{n\to\infty} S_n = \lim_{n\to\infty}\left[-\dfrac{2n}{5(2n+5)}\right] = -\dfrac{2}{5\cdot 2} = -\dfrac{1}{5}.$

2 (a) $S_1 = a_1 = \frac{5}{7\cdot 12} = \frac{5}{84};\ S_2 = S_1 + a_2 = \frac{5}{84} + \frac{5}{204} = \frac{10}{119};$

$S_3 = S_2 + a_3 = \frac{10}{119} + \frac{5}{374} = \frac{15}{154}.$

(b) $a_n = \dfrac{5}{(5n+2)(5n+7)} = \dfrac{1}{5n+2} - \dfrac{1}{5n+7}.$

$$S_n = \left(\tfrac{1}{7} - \tfrac{1}{12}\right) + \left(\tfrac{1}{12} - \tfrac{1}{17}\right) + \left(\tfrac{1}{17} - \tfrac{1}{22}\right) + \cdots + \left(\frac{1}{5n+2} - \frac{1}{5n+7}\right)$$
$$= \frac{1}{7} - \frac{1}{5n+7} = \frac{5n}{7(5n+7)}.$$

(c) $S = \lim_{n\to\infty} S_n = \lim_{n\to\infty}\left[\dfrac{5n}{7(5n+7)}\right] = \dfrac{5}{7\cdot 5} = \dfrac{1}{7}.$

3 (a) $S_1 = a_1 = \frac{1}{4-1} = \frac{1}{3};\ S_2 = S_1 + a_2 = \frac{1}{3} + \frac{1}{15} = \frac{2}{5};$

$S_3 = S_2 + a_3 = \frac{2}{5} + \frac{1}{35} = \frac{3}{7}.$

(b) $a_n = \dfrac{1}{4n^2 - 1} = \dfrac{1}{2}\left(\dfrac{1}{2n-1} - \dfrac{1}{2n+1}\right).$

$$S_n = \tfrac{1}{2}\left(1 - \tfrac{1}{3}\right) + \tfrac{1}{2}\left(\tfrac{1}{3} - \tfrac{1}{5}\right) + \tfrac{1}{2}\left(\tfrac{1}{5} - \tfrac{1}{7}\right) + \cdots + \tfrac{1}{2}\left(\frac{1}{2n-1} - \frac{1}{2n+1}\right)$$
$$= \tfrac{1}{2}\left(1 - \frac{1}{2n+1}\right) = \frac{n}{2n+1}.$$

(c) $S = \lim_{n\to\infty} S_n = \lim_{n\to\infty}\left[\dfrac{n}{2n+1}\right] = \dfrac{1}{2}.$

4 (a) $S_1 = a_1 = \frac{-1}{9+3-2} = -\frac{1}{10};\ S_2 = S_1 + a_2 = -\frac{1}{10} + \frac{-1}{40} = -\frac{1}{8};$

$S_3 = S_2 + a_3 = -\frac{1}{8} + \frac{-1}{88} = -\frac{3}{22}.$

(b) $a_n = \dfrac{-1}{9n^2 + 3n - 2} = \frac{1}{3}\left(\dfrac{1}{3n+2} - \dfrac{1}{3n-1}\right)$.

$$S_n = \tfrac{1}{3}\left(\tfrac{1}{5} - \tfrac{1}{2}\right) + \tfrac{1}{3}\left(\tfrac{1}{8} - \tfrac{1}{5}\right) + \tfrac{1}{3}\left(\tfrac{1}{11} - \tfrac{1}{8}\right) + \cdots + \tfrac{1}{3}\left(\frac{1}{3n+2} - \frac{1}{3n-1}\right)$$

$$= \tfrac{1}{3}\left(\frac{1}{3n+2} - \tfrac{1}{2}\right) = \frac{-n}{2(3n+2)}.$$

(c) $S = \lim_{n\to\infty} S_n = \lim_{n\to\infty}\left[\dfrac{-n}{2(3n+2)}\right] = \dfrac{-1}{2\cdot 3} = -\frac{1}{6}$.

5 (a) $S_1 = a_1 = \ln\frac{1}{2} = \ln 1 - \ln 2 = 0 - \ln 2 = -\ln 2$;

$S_2 = S_1 + a_2 = -\ln 2 + \ln\frac{2}{3} = -\ln 2 + (\ln 2 - \ln 3) = -\ln 3$;

$S_3 = S_2 + a_3 = -\ln 3 + \ln\frac{3}{4} = -\ln 3 + (\ln 3 - \ln 4) = -\ln 4$.

(b) $a_n = \ln\dfrac{n}{n+1} = \ln n - \ln(n+1)$.

$$S_n = (\ln 1 - \ln 2) + (\ln 2 - \ln 3) + (\ln 3 - \ln 4) + \cdots + \left[\ln n - \ln(n+1)\right]$$

$$= \ln 1 - \ln(n+1) = -\ln(n+1).$$

(c) $S = \lim_{n\to\infty} S_n = \lim_{n\to\infty}\left[-\ln(n+1)\right] = -\infty$.

Since $\{S_n\}$ diverges, the series $\sum a_n$ diverges and has no sum.

6 (a) $S_1 = a_1 = \dfrac{1}{\sqrt{2}+1}\cdot\dfrac{\sqrt{2}-1}{\sqrt{2}-1} = \dfrac{\sqrt{2}-1}{2-1} = \sqrt{2} - 1$;

$S_2 = S_1 + a_2 = (\sqrt{2} - 1) + \dfrac{1}{\sqrt{3}+\sqrt{2}}\cdot\dfrac{\sqrt{3}-\sqrt{2}}{\sqrt{3}-\sqrt{2}} = (\sqrt{2} - 1) + (\sqrt{3} - \sqrt{2}) = \sqrt{3} - 1$;

$S_3 = S_2 + a_3 = (\sqrt{3} - 1) + \dfrac{1}{2+\sqrt{3}}\cdot\dfrac{2-\sqrt{3}}{2-\sqrt{3}} = (\sqrt{3} - 1) + (2 - \sqrt{3}) = 2 - 1 = 1$.

(b) $a_n = \dfrac{1}{\sqrt{n+1}+\sqrt{n}}\cdot\dfrac{\sqrt{n+1}-\sqrt{n}}{\sqrt{n+1}-\sqrt{n}} = \sqrt{n+1} - \sqrt{n}$.

$$S_n = (\sqrt{2} - \sqrt{1}) + (\sqrt{3} - \sqrt{2}) + (\sqrt{4} - \sqrt{3}) + \cdots + (\sqrt{n+1} - \sqrt{n})$$

$$= \sqrt{n+1} - 1.$$

(c) $S = \lim_{n\to\infty} S_n = \lim_{n\to\infty}\left[\sqrt{n+1} - 1\right] = \infty$.

Since $\{S_n\}$ diverges, the series $\sum a_n$ diverges and has no sum.

Note: From (11.15), the series converges if $|r| < 1$ and diverges is $|r| \geq 1$.

7 $a = 3,\ r = \frac{1}{4} \Rightarrow S = \dfrac{3}{1 - \frac{1}{4}} = 4$; C

8 $a = 3,\ r = -\frac{1}{4} \Rightarrow S = \dfrac{3}{1 - (-\frac{1}{4})} = \frac{12}{5}$; C

9 $a = 1,\ r = -\frac{1}{\sqrt{5}} \Rightarrow S = \dfrac{1}{1 - (-\frac{1}{\sqrt{5}})} = \dfrac{\sqrt{5}}{\sqrt{5}+1}$; C

10 $a = 1,\ r = \frac{e}{3} < 1 \Rightarrow S = \dfrac{1}{1 - \frac{e}{3}} = \dfrac{3}{3-e}$; C

11 $a = 0.37,\ r = \frac{1}{100} \Rightarrow S = \dfrac{0.37}{1 - \frac{1}{100}} = \dfrac{37}{99}$; C

[12] $a = 0.628,\ r = \frac{1}{1000} \Rightarrow S = \dfrac{0.628}{1 - \frac{1}{1000}} = \dfrac{628}{999}$; C

[13] $\sum_{n=1}^{\infty} \dfrac{3^{n-1}}{2^n} = \sum_{n=1}^{\infty} \frac{1}{2}(\frac{3}{2})^{n-1};\ r = \frac{3}{2} > 1 \Rightarrow$ D

[14] $\sum_{n=1}^{\infty} \dfrac{(-5)^{n-1}}{4^n} = \sum_{n=1}^{\infty} \frac{1}{4}(-\frac{5}{4})^{n-1};\ r = -\frac{5}{4} \Rightarrow$ D

[15] $a = 1,\ r = -1 \Rightarrow$ D

[16] $a = 1,\ r = \sqrt{2} \Rightarrow$ D

[17] S converges if $|r| < 1$. In this case, $r = -x$. Thus, S converges if $|-x| < 1$, or, equivalently, $-1 < x < 1$. $a = 1,\ r = -x \Rightarrow S = \dfrac{1}{1 - (-x)} = \dfrac{1}{1 + x}$.

[18] $|x^2| < 1 \Rightarrow |x| < 1 \Rightarrow -1 < x < 1.\ a = 1,\ r = x^2 \Rightarrow S = \dfrac{1}{1 - x^2}$.

[19] $\left|\dfrac{x-3}{2}\right| < 1 \Rightarrow |x - 3| < 2 \Rightarrow -2 < x - 3 < 2 \Rightarrow 1 < x < 5$.

$$a = \frac{1}{2},\ r = \frac{x-3}{2} \Rightarrow S = \frac{1/2}{1 - \left(\frac{x-3}{2}\right)} = \frac{1}{5 - x}.$$

[20] $\left|\dfrac{x-1}{3}\right| < 1 \Rightarrow |x - 1| < 3 \Rightarrow -3 < x - 1 < 3 \Rightarrow -2 < x < 4$.

$$a = 3,\ r = \frac{x-1}{3} \Rightarrow S = \frac{3}{1 - \left(\frac{x-1}{3}\right)} = \frac{9}{4 - x}.$$

[21] $0.\overline{23} = \frac{23}{100} + \dfrac{23}{(100)^2} + \dfrac{23}{(100)^3} + \cdots = \dfrac{\frac{23}{100}}{1 - \frac{1}{100}} = \frac{23}{99}$

[22] $5.\overline{146} = 5 + \frac{146}{1000} + \dfrac{146}{(1000)^2} + \dfrac{146}{(1000)^3} + \cdots = 5 + \dfrac{\frac{146}{1000}}{1 - \frac{1}{1000}} = 5 + \frac{146}{999} = \frac{5141}{999}$

[23] $3.2\overline{394} = 3.2 + \frac{394}{10^4} + \frac{394}{10^7} + \frac{394}{10^{10}} + \cdots = 3.2 + \dfrac{\frac{394}{10,000}}{1 - \frac{1}{1000}} = 3.2 + \frac{394}{9990} = \frac{16,181}{4995}$

[24] $2.7\overline{1828} = 2.7 + \frac{1828}{10^5} + \frac{1828}{10^9} + \frac{1828}{10^{13}} + \cdots = 2.7 + \dfrac{\frac{1828}{100,000}}{1 - \frac{1}{10,000}} =$

$$2.7 + \frac{1828}{99,990} = \frac{271,801}{99,990}$$

[25] By Example 1, $\sum_{n=1}^{\infty} \dfrac{1}{n(n+1)} = 1$. By (11.19), $\sum_{n=4}^{\infty} \dfrac{1}{n(n+1)}$ also converges.

[26] By Example 1, $\sum_{n=1}^{\infty} \dfrac{1}{n(n+1)} = 1$. By (11.19), $\sum_{n=10}^{\infty} \dfrac{1}{n(n+1)}$ also converges.

[27] By (11.20)(ii) and Example 1, $\sum_{n=1}^{\infty} \dfrac{5}{n(n+1)} = 5 \sum_{n=1}^{\infty} \dfrac{1}{n(n+1)}$ converges.

[28] By (11.20)(ii) and Example 1, $\sum_{n=1}^{\infty} \dfrac{-1}{n(n+1)} = -\sum_{n=1}^{\infty} \dfrac{1}{n(n+1)}$ converges.

[29] Since $\sum_{n=1}^{\infty} \frac{1}{n}$ diverges by (11.14), so does $\sum_{n=4}^{\infty} \frac{1}{n}$ by (11.19).

[30] Since $\sum_{n=1}^{\infty} \frac{1}{n}$ diverges, so does $\sum_{n=6}^{\infty} \frac{1}{n}$ by (11.19).

[31] $3 + \frac{3}{2} + \cdots + \frac{3}{n} + \cdots = \sum_{n=1}^{\infty} \frac{3}{n} = 3 \sum_{n=1}^{\infty} \frac{1}{n}$ and is divergent.

[32] $-4 - 2 - \frac{4}{3} - \cdots - \frac{4}{n} - \cdots = \sum_{n=1}^{\infty} \frac{-4}{n} = -4 \sum_{n=1}^{\infty} \frac{1}{n}$ and is divergent.

[33] Since $\lim_{n \to \infty} \frac{3n}{5n - 1} = \frac{3}{5} \neq 0$, the series diverges.

[34] Since $\lim_{n \to \infty} \frac{1}{1 + (0.3)^n} = 1 \neq 0$, the series diverges.

[35] Since $\lim_{n \to \infty} \frac{1}{n^2 + 3} = 0$, further investigation is necessary.

[36] Since $\lim_{n \to \infty} \frac{1}{e^n + 1} = 0$, further investigation is necessary.

[37] Since $\lim_{n \to \infty} e^{-1/n} = e^0 = 1 \neq 0$, the series diverges.

[38] Since $\lim_{n \to \infty} n \sin \frac{1}{n} = 1 \neq 0$ {see 11.1.34}, the series diverges.

[39] Since $\lim_{n \to \infty} \frac{n}{\ln(n + 1)} \left\{\frac{\infty}{\infty}\right\} = \lim_{n \to \infty} \frac{1}{1/(n + 1)} = \infty \neq 0$, the series diverges.

[40] Since $\lim_{n \to \infty} \ln\left(\frac{2n}{7n - 5}\right) = \ln \frac{2}{7} \neq 0$, the series diverges.

[41] $\sum_{n=3}^{\infty} \left[(\frac{1}{4})^n + (\frac{3}{4})^n\right] = \sum_{n=3}^{\infty} (\frac{1}{4})^n \{a = (\frac{1}{4})^3, r = \frac{1}{4}\} + \sum_{n=3}^{\infty} (\frac{3}{4})^n \{a = (\frac{3}{4})^3, r = \frac{3}{4}\}$

$$= \frac{1/4^3}{1 - \frac{1}{4}} + \frac{3^3/4^3}{1 - \frac{3}{4}} = \frac{1}{48} + \frac{27}{16} = \frac{41}{24}; \text{ C}$$

[42] Since $\sum_{n=1}^{\infty} (\frac{3}{2})^n$ diverges and $\sum_{n=1}^{\infty} (\frac{2}{3})^n$ converges, the series diverges by (11.21).

[43] $\sum_{n=1}^{\infty} (2^{-n} - 2^{-3n}) = \sum_{n=1}^{\infty} (\frac{1}{2})^n - \sum_{n=1}^{\infty} (\frac{1}{2^3})^n = \frac{1/2}{1 - \frac{1}{2}} - \frac{1/8}{1 - \frac{1}{8}} = 1 - \frac{1}{7} = \frac{6}{7}$; C

[44] $\sum_{n=1}^{\infty} \left(\frac{1}{3^n} - \frac{1}{4^n}\right) = \sum_{n=1}^{\infty} (\frac{1}{3})^n - \sum_{n=1}^{\infty} (\frac{1}{4})^n = \frac{1/3}{1 - \frac{1}{3}} - \frac{1/4}{1 - \frac{1}{4}} = \frac{1}{2} - \frac{1}{3} = \frac{1}{6}$; C

[45] Since $\sum_{n=1}^{\infty} \frac{1}{8^n} = \frac{1}{7} \{a = \frac{1}{8}, r = \frac{1}{8}\}$ and $\sum_{n=1}^{\infty} \frac{1}{n(n + 1)} = 1$,

the series converges to $\frac{1}{7} + 1 = \frac{8}{7}$.

[46] Since $\sum_{n=1}^{\infty} \frac{1}{n(n + 1)}$ converges and $\sum_{n=1}^{\infty} \frac{4}{n}$ diverges, the series diverges by (11.21).

[47] $\sum_{n=1}^{\infty} \left(\frac{5}{n + 2} - \frac{5}{n + 3}\right) = 5 \sum_{n=3}^{\infty} \left(\frac{1}{n} - \frac{1}{n + 1}\right)$ {subtract the first two terms from the sum of the series in Example 1} $= 5\left[1 - (\frac{1}{2} + \frac{1}{6})\right] = 5(\frac{1}{3}) = \frac{5}{3}$; C

[48] $\sum_{n=1}^{\infty} \left(\frac{1}{n + 1} - \frac{1}{n}\right) = -\sum_{n=1}^{\infty} \left(\frac{1}{n} - \frac{1}{n + 1}\right) = -(1) = -1$; C

[49] (a) $S_1 \approx 0.21037$, $S_2 \approx 0.26720$, $S_3 \approx 0.26940$

(b) $S_4 \approx 0.26645$, $S_5 \approx 0.26551$, $S_6 \approx 0.26544$, $S_7 \approx 0.26548$; $S \approx 0.265$

[50] (a) $S_1 \approx 0.36788$, $S_2 \approx 0.39378$, $S_3 \approx 0.39400$

(b) Since $S_4 \approx 0.39400$, $S \approx 0.394$.

[51] $S_{2^k} > (k+1)\frac{1}{2} = 3 \Rightarrow k = 5$ and $m = 2^5 = 32$.

$$S_{32} = 1 + \tfrac{1}{2} + \tfrac{1}{3} + \cdots + \tfrac{1}{32} \approx 4.058495.$$

[52] $S_{2^k} > (k+1)\frac{1}{2} = 8 \Rightarrow k = 15$ and $m = 2^{15} = 32{,}768$; $S_{32{,}768} \approx 10.974439$.

[53] $\forall\, n$, let $a_n = 1$ and $b_n = -1$. Then, both $\sum\limits_{n=1}^{\infty} a_n$ and $\sum\limits_{n=1}^{\infty} b_n$ diverge,

but $\sum\limits_{n=1}^{\infty} (a_n + b_n) = 0$ converges. The statement is false.

[54] The even partial sums are $S_{2n} = 0\ \forall\, n$ and the odd partial sums are $S_{2n-1} = 1$.

In order to have a series converge it is required that $\lim\limits_{n\to\infty} S_n$ exist.

In this case the partial sums are 1, 0, 1, 0, ..., and the limit DNE.

Note: The terms of the divergent sequence cannot be regrouped as indicated.

[55] The ball initially falls 10 m, after which, in each up and down cycle it travels:

$$\tfrac{10}{1}, \tfrac{10}{2}, \tfrac{10}{4}, \tfrac{10}{8}, \ldots \Rightarrow d = 10 + \sum_{n=0}^{\infty} 10(\tfrac{1}{2})^n = 10 + \frac{10}{1 - \frac{1}{2}} = 30 \text{ m}.$$

[56] The pendulum swings $24 + 24(\frac{5}{6}) + 24(\frac{5}{6})^2 + \cdots = \sum\limits_{n=0}^{\infty} 24(\frac{5}{6})^n = \dfrac{24}{1 - \frac{5}{6}} = 144$ cm.

[57] (a) Immediately after the first dose, there are Q units of the drug in the bloodstream. $A(1) = Q$. After the second dose, there is a new Q units plus $A(1)e^{-cT} = Qe^{-cT}$ units from the first dose. Thus, $A(2) = Q + Qe^{-cT}$. Similarly,
$A(3) = Q + A(2)e^{-cT} = Q + (Q + Qe^{-cT})e^{-cT} = Q + Qe^{-cT} + Qe^{-2cT}$,
$A(4) = Q + A(3)e^{-cT} = Q + (Q + Qe^{-cT} + Qe^{-2cT})e^{-cT} =$

$$Q + Qe^{-cT} + Qe^{-2cT} + Qe^{-3cT}, \ldots \Rightarrow A(k) = \sum_{n=0}^{k-1} Qe^{-ncT}.$$

(b) Since all terms are positive, $A(k)$ is an increasing sequence with the form

$a + ar + ar^2 + \cdots + ar^{k-1}$, where $a = Q$ and $r = e^{-cT}$.

Since $c > 0$, $|r| = |e^{-cT}| < 1$. Thus, by (11.15)(ii),

$$\text{the upper bound is } \lim_{k\to\infty} A(k) = \sum_{n=0}^{\infty} Q(e^{-cT})^n = \frac{Q}{1 - e^{-cT}}.$$

(c) $\dfrac{Q}{1 - e^{-cT}} < M \Rightarrow Q - M < -Me^{-cT} \Rightarrow T > -\frac{1}{c}\ln\dfrac{M - Q}{M}$

[58] Amount $= 1{,}000{,}000\left[0.85 + (0.85)^2 + (0.85)^3 + \cdots\right] = 1{,}000{,}000\left[\dfrac{0.85}{1 - 0.85}\right] =$

$5,666,666.67.

59 Let $P(k)$ denote the current population after k days.

(a) $P(1) = N$, $P(2) = N + 0.9\,P(1) = N + 0.9N$,

$P(3) = N + 0.9\,P(2) = N + 0.9N + (0.9)^2N, \ldots \Rightarrow$

$$P(n) = N + 0.9N + (0.9)^2N + \cdots + (0.9)^{n-1}N$$

(b) Since $\lim_{n\to\infty} P(n) = \dfrac{N}{1-0.9} = 10N$, we need $10N = 20{,}000 \Rightarrow N = 2000$ flies.

60 (a) Since the drug's half-life is 2 hours, only $\frac{1}{2}(\frac{1}{2})K = \frac{1}{4}K$ remains 4 hours after a K mg dose. Let $A(n)$ denote the amount of the drug in the bloodstream immediately after n doses. Then, $A(1) = K$, $A(2) = K + \frac{1}{4}A(1) = K + \frac{1}{4}K$,

$A(3) = K + \frac{1}{4}A(2) = K + \frac{1}{4}K + (\frac{1}{4})^2K, \ldots \Rightarrow$

$$A(n) = K + \tfrac{1}{4}K + \cdots + (\tfrac{1}{4})^{n-1}K. \quad \lim_{n\to\infty} A(n) = \frac{K}{1-\frac{1}{4}} = \tfrac{4}{3}K.$$

(b) $\frac{4}{3}K < 500 \Rightarrow K < 375$ mg.

(c) Let $a(t)$ denote the amount of drug at time t (in hours) in the bloodstream after a single dose Q at time $t = 0$. Then, $a(t) = Qe^{-ct}$ and $a(4) = Qe^{-4c}$.

But $a(4) = \frac{1}{4}Q \Rightarrow Qe^{-4c} = \frac{1}{4}Q \Rightarrow e^{-4c} = \frac{1}{4} \Rightarrow e^{-c} = (\frac{1}{4})^{1/4}$.

Using part (c) of Exercise 57, we have $\dfrac{50}{1-(\frac{1}{4})^{(1/4)T}} < 500 \Rightarrow$

$1-(\frac{1}{4})^{(1/4)T} > \frac{1}{10} \Rightarrow T > 4\dfrac{\ln(9/10)}{\ln(1/4)} \approx 0.304 \text{ hr} \approx 18.24$ min,

or not more frequently than approximately every 19 min.

61 (a) From the second figure we see that $(\frac{1}{4}a_k)^2 + (\frac{3}{4}a_k)^2 = (a_{k+1})^2 \Rightarrow$

$$\tfrac{10}{16}a_k^2 = a_{k+1}^2 \Rightarrow a_{k+1} = \tfrac{1}{4}\sqrt{10}\,a_k.$$

(b) From part (a), $a_n = (\frac{1}{4}\sqrt{10})^{n-1}a_1$.

$A_{k+1} = a_{k+1}^2 = \frac{10}{16}a_k^2 = \frac{5}{8}A_k$, hence $A_n = (\frac{5}{8})^{n-1}A_1$.

$P_{k+1} = 4a_{k+1} = 4\cdot\frac{1}{4}\sqrt{10}\,a_k = \sqrt{10}\,a_k = \sqrt{10}\,(\frac{1}{4}P_k)$, hence $P_n = (\frac{1}{4}\sqrt{10})^{n-1}P_1$.

(c) $\sum_{n=1}^{\infty} P_n$ is an infinite geometric series with first term P_1 and $r = \frac{1}{4}\sqrt{10} < 1$.

$$\sum_{n=1}^{\infty} P_n = \frac{P_1}{1-\frac{1}{4}\sqrt{10}} = \frac{4P_1}{4-\sqrt{10}} = \frac{16a_1}{4-\sqrt{10}}.$$

$$\sum_{n=1}^{\infty} A_n = \sum_{n=1}^{\infty} (\tfrac{5}{8})^{n-1}A_1 = \sum_{n=1}^{\infty} (\tfrac{5}{8})^{n-1}a_1^2 = \frac{a_1^2}{1-\frac{5}{8}} = \tfrac{8}{3}a_1^2.$$

62 (a) Let s_n denote the length of a side of the nth square and $\frac{1}{2}s_n$ the length of the radius of the inscribed circle. Now $C_n = \pi(\frac{1}{2}s_n)^2 = \frac{\pi}{4}s_n^2 = \frac{\pi}{4}S_n$, i.e., $S_n = \frac{4}{\pi}C_n$. Let r_n be the radius of the nth circle. The inscribed rectangle will have a side of length $s_{n+1} = \sqrt{r_n^2 + r_n^2} = \sqrt{2}\,r_n$.

Thus, $C_n = \pi r_n^2$ and $S_{n+1} = s_{n+1}^2 = 2r_n^2 \Rightarrow C_n = \pi(\frac{1}{2}s_{n+1}^2) = \frac{\pi}{2}S_{n+1}$.

(b) The shaded region has area $(S_1 - C_1) + (S_2 - C_2) + (S_3 - C_3) + \cdots =$

$$\sum_{n=1}^{\infty} S_n - \sum_{n=1}^{\infty} C_n = \sum_{n=1}^{\infty} S_n - \sum_{n=1}^{\infty} \tfrac{\pi}{4} S_n = \frac{4-\pi}{4} \sum_{n=1}^{\infty} S_n.$$

From part (a), $S_{n+1} = \frac{2}{\pi} C_n = \frac{2}{\pi}(\frac{\pi}{4} S_n) = \frac{1}{2} S_n$.

Hence, $\displaystyle\sum_{n=1}^{\infty} S_n = S_1 + \tfrac{1}{2}S_1 + \tfrac{1}{4}S_1 + \cdots = \frac{S_1}{1-\frac{1}{2}} = 2S_1$.

Thus, the area is $\frac{4-\pi}{4}(2S_1) = \frac{4-\pi}{2} S_1$ or approximately 42.9% of S_1.

Exercises 11.3

[1] (a) $f(n) = a_n = \dfrac{1}{(3+2n)^2}$ and $f(x) = \dfrac{1}{(3+2x)^2}$.

(i) Since $(3 + 2x)^2 > 0$ if $x \ge 1$, f is a positive-valued function.

(ii) f is continuous on $\mathbb{R} - \{-\frac{3}{2}\}$ and hence is continuous on $[1, \infty)$.

(iii) By the reciprocal rule, $f'(x) = -\dfrac{2(3+2x)\cdot 2}{\left[(3+2x)^2\right]^2} = -\dfrac{4}{(2x+3)^3}$. If $x \ge 1$, then $(2x+3)^3 > 0$ and f' is negative. Thus, f is decreasing on $[1, \infty)$.

(b) $\displaystyle\int_1^{\infty} f(x)\,dx = \lim_{t\to\infty}\left[-\frac{1}{2(3+2x)}\right]_1^t = 0 - \left(-\tfrac{1}{10}\right)$; C

Note: For Exercises 2–12, each function f is positive-valued and continuous on the interval of integration. Only the function f and its derivative are listed in (a).

[2] (a) $f(x) = \dfrac{1}{(4+x)^{3/2}} \Rightarrow f'(x) = -\dfrac{3}{2(x+4)^{5/2}} < 0$ if $x \ge 1$.

(b) $\displaystyle\int_1^{\infty} f(x)\,dx = \lim_{t\to\infty}\left[-\frac{2}{(4+x)^{1/2}}\right]_1^t = 0 - \left(-\frac{2}{\sqrt{5}}\right)$; C

[3] (a) $f(x) = \dfrac{1}{4x+7} \Rightarrow f'(x) = -\dfrac{4}{(4x+7)^2} < 0$ if $x \ge 1$.

(b) $\displaystyle\int_1^{\infty} f(x)\,dx = \lim_{t\to\infty}\left[\tfrac{1}{4}\ln|4x+7|\right]_1^t = \infty$; D

[4] (a) $f(x) = \dfrac{x}{x^2+1} \Rightarrow f'(x) = \dfrac{1-x^2}{(x^2+1)^2} < 0$ if $x > 1$.

(b) $\displaystyle\int_1^{\infty} f(x)\,dx = \lim_{t\to\infty}\left[\tfrac{1}{2}\ln(x^2+1)\right]_1^t = \infty$; D

[5] (a) $f(x) = x^2 e^{-x^3} \Rightarrow f'(x) = x(2 - 3x^3)e^{-x^3} < 0$ if $x \ge 1$.

(b) $\displaystyle\int_1^{\infty} f(x)\,dx = \lim_{t\to\infty}\left[-\tfrac{1}{3}e^{-x^3}\right]_1^t = 0 - \left(-\tfrac{1}{3}e^{-1}\right) = \frac{1}{3e}$; C

[6] (a) $f(x) = \dfrac{1}{x(2x-5)} \Rightarrow f'(x) = \dfrac{5-4x}{x^2(2x-5)^2} < 0$ if $x \ge 3$.

(b) $\displaystyle\int_3^{\infty} f(x)\,dx$ {PF} $\displaystyle= \int_3^{\infty}\left(\frac{-\frac{1}{5}}{x} + \frac{\frac{2}{5}}{2x-5}\right)dx = \lim_{t\to\infty}\left[-\tfrac{1}{5}\ln|x| + \tfrac{1}{5}\ln|2x-5|\right]_3^t =$

$\displaystyle\tfrac{1}{5}\lim_{t\to\infty}\left[\ln\frac{2x-5}{x}\right]_3^t = \tfrac{1}{5}(\ln 2 - \ln\tfrac{1}{3}) = \tfrac{1}{5}\ln 6$; C

7 (a) $f(x) = \frac{\ln x}{x} \Rightarrow f'(x) = \frac{1 - \ln x}{x^2} < 0$ if $x \geq 3$.

(b) $\int_3^\infty f(x)\,dx = \lim_{t\to\infty}\left[\frac{1}{2}(\ln x)^2\right]_3^t = \infty$; D

8 (a) $f(x) = \frac{1}{x(\ln x)^2} \Rightarrow f'(x) = -\frac{2 + \ln x}{x^2(\ln x)^3} < 0$ if $x \geq 2$.

(b) $\int_2^\infty f(x)\,dx = \lim_{t\to\infty}\left[-\frac{1}{\ln x}\right]_2^t = 0 - \left(-\frac{1}{\ln 2}\right)$; C

9 (a) $f(x) = \frac{1}{x\sqrt{x^2 - 1}} \Rightarrow f'(x) = \frac{1 - 2x^2}{x^2(x^2 - 1)^{3/2}} < 0$ if $x \geq 2$.

(b) $\int_2^\infty f(x)\,dx = \lim_{t\to\infty}\left[\sec^{-1}x\right]_2^t = \frac{\pi}{2} - \frac{\pi}{3} = \frac{\pi}{6}$; C

10 (a) $f(x) = \left(\frac{1}{n - 3} - \frac{1}{n}\right) \Rightarrow f'(x) = \frac{9 - 6x}{x^2(x - 3)^2} < 0$ if $x \geq 4$.

(b) $\int_4^\infty f(x)\,dx = \lim_{t\to\infty}\left[\ln|x - 3| - \ln|x|\right]_4^t = \lim_{t\to\infty}\left[\ln\frac{x - 3}{x}\right]_4^t =$

$\ln 1 - \ln\frac{1}{4} = \ln 4$; C

11 (a) $f(x) = \frac{\arctan x}{1 + x^2} \Rightarrow f'(x) = \frac{1 - 2x\arctan x}{(1 + x^2)^2} < 0$ if $x \geq 1$.

(b) $\int_1^\infty f(x)\,dx = \lim_{t\to\infty}\left[\frac{1}{2}(\arctan x)^2\right]_1^t = \frac{1}{2}(\frac{\pi}{2})^2 - \frac{1}{2}(\frac{\pi}{4})^2 = \frac{3\pi^2}{32}$; C

12 (a) $f(x) = \frac{1}{1 + 16x^2} \Rightarrow f'(x) = -\frac{32x}{(16x^2 + 1)^2} < 0$ if $x \geq 1$.

(b) $\int_1^\infty f(x)\,dx = \lim_{t\to\infty}\left[\frac{1}{4}\tan^{-1}4x\right]_1^t = \frac{1}{4}(\frac{\pi}{2}) - \frac{1}{4}\tan^{-1}4$; C

Note: Exer. 13–20: S (the given series) converges by (11.26)(i) or diverges by (11.26)(ii).

13 S converges since $\frac{1}{n^4 + n^2 + 1} < \frac{1}{n^4}$ and $\sum_{n=1}^{\infty}\frac{1}{n^4}$ converges by (11.25).

14 S converges since $\frac{\sqrt{n}}{n^2 + 1} < \frac{\sqrt{n}}{n^2} = \frac{1}{n^{3/2}}$ and $\sum_{n=1}^{\infty}\frac{1}{n^{3/2}}$ converges by (11.25).

15 S converges since $\frac{1}{n3^n} < \frac{1}{3^n}$ and $\sum_{n=1}^{\infty}\frac{1}{3^n}$ converges by (11.15).

16 S converges since $\frac{2 + \cos n}{n^2} < \frac{3}{n^2}$ and $3\sum_{n=1}^{\infty}\frac{1}{n^2}$ converges by (11.25).

17 S diverges since $\frac{\arctan n}{n} \geq \frac{\pi/4}{n}$ and $\frac{\pi}{4}\sum_{n=1}^{\infty}\frac{1}{n}$ diverges.

18 S diverges since $\frac{\operatorname{arcsec} n}{(0.5)^n} = 2^n \operatorname{arcsec} n \geq 2^n$ {for $n \geq 2$} and $\sum_{n=1}^{\infty} 2^n$ diverges.

19 S converges since $\frac{1}{n^n} \leq \frac{1}{n^2}$ and $\sum_{n=1}^{\infty}\frac{1}{n^2}$ converges by (11.25).

20 S converges since $\frac{1}{n!} < \frac{1}{n^2}$ {for $n \geq 4$} and $\sum_{n=1}^{\infty}\frac{1}{n^2}$ converges by (11.25).

[21] Let $a_n = \dfrac{\sqrt{n}}{n + 4}$. By deleting all terms in the numerator and denominator of a_n except those that have the greatest effect on the magnitude, we obtain $b_n = \frac{\sqrt{n}}{n} = \frac{1}{\sqrt{n}}$, which is a divergent p-series with $p = \frac{1}{2}$. Using the limit comparison test (11.27), we have $\lim\limits_{n \to \infty} \frac{a_n}{b_n} = \lim\limits_{n \to \infty} \dfrac{n}{n + 4} = 1 > 0$. Thus, S diverges.

[22] Let $a_n = \dfrac{2}{3 + \sqrt{n}}$ and $b_n = \frac{1}{\sqrt{n}}$. $\lim\limits_{n \to \infty} \frac{a_n}{b_n} = \lim\limits_{n \to \infty} \dfrac{2\sqrt{n}}{3 + \sqrt{n}} = 2 > 0$.

S diverges by (11.27) since $\sum b_n$ diverges.

[23] Let $a_n = \dfrac{1}{\sqrt{4n^3 - 5n}}$ and $b_n = \frac{1}{n^{3/2}}$. $\lim\limits_{n \to \infty} \frac{a_n}{b_n} = \lim\limits_{n \to \infty} \dfrac{n^{3/2}}{\sqrt{4n^3 - 5n}} = \frac{1}{2} > 0$.

S converges by (11.27) since $\sum b_n$ converges.

[24] Let $a_n = \dfrac{1}{\sqrt{n(n + 1)(n + 2)}}$ and $b_n = \frac{1}{n^{3/2}}$. $\lim\limits_{n \to \infty} \frac{a_n}{b_n} = \lim\limits_{n \to \infty} \dfrac{\sqrt{n^3}}{\sqrt{n^3 + 3n^2 + 2n}} =$ $1 > 0$. S converges by (11.27) since $\sum b_n$ converges.

[25] Let $a_n = \dfrac{8n^2 - 7}{e^n(n + 1)^2}$ and $b_n = \frac{1}{e^n}$. $\lim\limits_{n \to \infty} \frac{a_n}{b_n} = \lim\limits_{n \to \infty} \dfrac{8n^2 - 7}{e^n(n + 1)^2} \cdot \dfrac{e^n}{1} = 8 > 0$.

S converges by (11.27) since $\sum b_n$ converges by (11.15).

[26] Let $a_n = \dfrac{3n + 5}{n 2^n}$ and $b_n = \frac{1}{2^n}$. $\lim\limits_{n \to \infty} \frac{a_n}{b_n} = \lim\limits_{n \to \infty} \dfrac{3n + 5}{n 2^n} \cdot \dfrac{2^n}{1} = 3 > 0$.

S converges by (11.27) since $\sum b_n$ converges.

[27] Let $a_n = \dfrac{1}{\sqrt{n} + 9}$ and $b_n = \frac{1}{\sqrt{n}}$. $\lim\limits_{n \to \infty} \frac{a_n}{b_n} = \lim\limits_{n \to \infty} \dfrac{\sqrt{n}}{\sqrt{n} + 9} = 1 > 0$.

S diverges by (11.27) since $\sum b_n$ diverges.

[28] Let $a_n = \dfrac{n^2}{n^3 + 1}$ and $b_n = \frac{1}{n}$. $\lim\limits_{n \to \infty} \frac{a_n}{b_n} = \lim\limits_{n \to \infty} \dfrac{n^3}{n^3 + 1} = 1 > 0$.

S diverges by (11.27) since $\sum b_n$ diverges.

[29] Let $a_n = \dfrac{2n + n^2}{n^3 + 1}$ and $b_n = \frac{1}{n}$. $\lim\limits_{n \to \infty} \frac{a_n}{b_n} = \lim\limits_{n \to \infty} \dfrac{n^3 + 2n^2}{n^3 + 1} = 1 > 0$.

S diverges by (11.27) since $\sum b_n$ diverges.

[30] Let $a_n = \dfrac{n^5 + 4n^3 + 1}{2n^8 + n^4 + 2}$ and $b_n = \frac{1}{n^3}$. $\lim\limits_{n \to \infty} \frac{a_n}{b_n} = \lim\limits_{n \to \infty} \dfrac{n^8 + 4n^6 + n^3}{2n^8 + n^4 + 2} = \frac{1}{2} > 0$.

S converges by (11.27) since $\sum b_n$ converges.

[31] Since $\dfrac{1 + 2^n}{1 + 3^n} < \dfrac{1 + 2^n}{3^n} = \frac{1}{3^n} + \left(\frac{2}{3}\right)^n$ and $\sum\limits_{n=1}^{\infty} \left[\left(\frac{1}{3}\right)^n + \left(\frac{2}{3}\right)^n\right]$ converges,

S converges by (11.26).

[32] Let $a_n = \dfrac{3n}{2n^2 - 7}$ and $b_n = \frac{1}{n}$. $\lim\limits_{n \to \infty} \frac{a_n}{b_n} = \lim\limits_{n \to \infty} \dfrac{3n^2}{2n^2 - 7} = \frac{3}{2} > 0$.

S diverges by (11.27) since $\sum b_n$ diverges.

Note: $n \geq 4$ so that $\sum a_n$ is a positive-term series, and (11.27) applies.

[33] Let $a_n = \dfrac{1}{\sqrt[3]{5n^2+1}}$ and $b_n = \dfrac{1}{n^{2/3}}$. $\displaystyle\lim_{n\to\infty}\frac{a_n}{b_n} = \lim_{n\to\infty}\frac{n^{2/3}}{\sqrt[3]{5n^2+1}} = \frac{1}{\sqrt[3]{5}} > 0$.

S diverges by (11.27) since $\sum b_n$ diverges.

[34] Since $\dfrac{\ln n}{n^4} < \dfrac{n}{n^4} = \dfrac{1}{n^3}$ and $\displaystyle\sum_{n=1}^{\infty}\frac{1}{n^3}$ converges, S converges by (11.26).

[35] Let $a_n = \dfrac{1}{\sqrt[3]{2n+1}}$ and $b_n = \dfrac{1}{n^{1/3}}$. $\displaystyle\lim_{n\to\infty}\frac{a_n}{b_n} = \lim_{n\to\infty}\frac{n^{1/3}}{\sqrt[3]{2n+1}} = \frac{1}{\sqrt[3]{2}} > 0$.

S diverges by (11.27) since $\sum b_n$ diverges.

[36] Since $\dfrac{n+\ln n}{n^3+n+1} < \dfrac{2n}{n^3+n+1} < \dfrac{2n}{n^3} = \dfrac{2}{n^2}$ and $2\displaystyle\sum_{n=1}^{\infty}\frac{1}{n^2}$ converges,

S converges by (11.26).

[37] $f(x) = xe^{-x}$. $\displaystyle\int_1^{\infty} f(x)\,dx = \lim_{t\to\infty}\Big[-xe^{-x} - e^{-x}\Big]_1^t$ {IP} $= 0 - (-2e^{-1}) = \frac{2}{e}$; C

[38] Let $a_n = \dfrac{1}{n(n+1)(n+2)}$ and $b_n = \dfrac{1}{n^3}$. $\displaystyle\lim_{n\to\infty}\frac{a_n}{b_n} = \lim_{n\to\infty}\frac{n^3}{n(n+1)(n+2)} =$

$1 > 0$. S converges by (11.27) since $\sum b_n$ converges.

[39] Let $a_n = \sin\dfrac{1}{n^2}$ and $b_n = \dfrac{1}{n^2}$. $\displaystyle\lim_{n\to\infty}\frac{a_n}{b_n} = \lim_{n\to\infty}\frac{\sin(1/n^2)}{1/n^2}$ $\{\frac{0}{0}\}$ $=$

$\displaystyle\lim_{n\to\infty}\frac{(-2/n^3)\cos(1/n^2)}{-2/n^3} = 1 > 0$. S converges by (11.27) since $\sum b_n$ converges.

[40] Let $a_n = \tan\dfrac{1}{n}$ and $b_n = \dfrac{1}{n}$. $\displaystyle\lim_{n\to\infty}\frac{a_n}{b_n} = \lim_{n\to\infty}\frac{\tan(1/n)}{1/n}$ $\{\frac{0}{0}\}$ $=$

$\displaystyle\lim_{n\to\infty}\frac{(-1/n^2)\sec^2(1/n)}{-1/n^2} = 1 > 0$. S diverges by (11.27) since $\sum b_n$ diverges.

[41] Let $a_n = \dfrac{(2n+1)^3}{(n^3+1)^2}$ and $b_n = \dfrac{1}{n^3}$. $\displaystyle\lim_{n\to\infty}\frac{a_n}{b_n} = \lim_{n\to\infty}\frac{(2n+1)^3\cdot n^3}{(n^3+1)^2} = 8 > 0$.

S converges by (11.27) since $\sum b_n$ converges.

[42] Since $\dfrac{n+\ln n}{n^2+1} \ge \dfrac{n}{n^2+1} > \dfrac{n}{n^2+n^2} = \dfrac{1}{2n}$ and $\frac{1}{2}\displaystyle\sum_{n=1}^{\infty}\frac{1}{n}$ diverges,

S diverges by (11.26).

[43] Since $\dfrac{n^2+2^n}{n+3^n} < \dfrac{n^2+2^n}{3^n} \le \dfrac{2^n+2^n}{3^n}$ {for $n \ge 4$} $= 2(\frac{2}{3})^n$ and $2\displaystyle\sum_{n=1}^{\infty}(\tfrac{2}{3})^n$

converges, S converges by (11.26).

[44] Let $a_n = \ln\left[1 + (\frac{1}{2})^n\right]$ and $b_n = (\frac{1}{2})^n$. $\displaystyle\lim_{n\to\infty}\frac{a_n}{b_n} = \lim_{n\to\infty}\frac{\ln\left[1+(\frac{1}{2})^n\right]}{(\frac{1}{2})^n}$ $\{\frac{0}{0}\}$ $=$

$\displaystyle\lim_{n\to\infty}\frac{1}{1+(\frac{1}{2})^n} = 1 > 0$. S converges by (11.27) since $\sum b_n$ converges.

[45] Since $\dfrac{\ln n}{n^3} < \dfrac{n}{n^3} = \dfrac{1}{n^2}$ and $\displaystyle\sum_{n=1}^{\infty}\frac{1}{n^2}$ converges, S converges by (11.26).

[46] Since $\dfrac{\sin n + 2^n}{n+5^n} < \dfrac{2^n+2^n}{n+5^n} < \dfrac{2\cdot 2^n}{5^n} = 2(\frac{2}{5})^n$ and $2\displaystyle\sum_{n=1}^{\infty}(\tfrac{2}{5})^n$ converges,

S converges by (11.26).

[47] (i) $k > 1$. Since $\frac{1}{n^k \ln n} < \frac{1}{n^k}$ $\{\ln n > 1$ if $n \geq 3\}$ and $\sum_{n=2}^{\infty} \frac{1}{n^k}$ converges by (11.25), S converges.

(ii) $k = 1$. Using the integral test, $\int_2^{\infty} \frac{1}{x \ln x}\, dx = \lim_{t \to \infty} \Big[\ln|\ln x|\Big]_2^t = \infty$; D.

(iii) $0 \leq k < 1$. $u = \ln x$, $x = e^u$, and $dx = e^u\, du \Rightarrow \int_2^{\infty} \frac{1}{x^k \ln x}\, dx =$

$\int_{\ln 2}^{\infty} \frac{e^u}{e^{ku} u}\, du = \int_{\ln 2}^{\infty} \frac{e^{(1-k)u}}{u}\, du$. Since $1 - k > 0$, the integrand approaches ∞ as $u \to \infty$. Thus, the integral must diverge and S also diverges.

(iv) $k < 0$. Since $\frac{1}{n^k \ln n} > \frac{1}{\ln n} > \frac{1}{n}$ $(n \geq 2)$ and $\sum_{n=2}^{\infty} \frac{1}{n}$ diverges, S diverges.

Thus, S converges iff $k > 1$.

[48] (i) $k \neq 1$. $\int_2^{\infty} \frac{1}{x(\ln x)^k}\, dx = \lim_{t \to \infty} \left[\frac{(\ln x)^{1-k}}{1-k}\right]_2^t$. If $k > 1$, $\{1 - k < 0\}$ the integral converges. If $k < 1$, $\{1 - k > 0\}$ the integral diverges.

(ii) $k = 1$. $\int_2^{\infty} \frac{1}{x \ln x}\, dx = \lim_{t \to \infty} \Big[\ln|\ln x|\Big]_2^t = \infty$; D.

Thus, S converges iff $k > 1$.

[49] (a) From the results of the proof of (11.23) with $f(x) = \frac{1}{x+1}$ and then $f(x) = \frac{1}{x}$ we have:

(i)
$$\sum_{k=2}^{n} \frac{1}{k+1} \leq \int_1^n \frac{1}{x+1}\, dx \leq \sum_{k=1}^{n-1} \frac{1}{k+1} \Rightarrow$$

$$\tfrac{1}{3} + \tfrac{1}{4} + \cdots + \frac{1}{n+1} \leq \ln(n+1) - \ln 2 \leq \tfrac{1}{2} + \tfrac{1}{3} + \cdots + \tfrac{1}{n} \Rightarrow$$

$$\{\text{since } \ln 2 < 1\}\ \ln(n+1) < 1 + \tfrac{1}{2} + \tfrac{1}{3} + \cdots + \tfrac{1}{n}.$$

(ii)
$$\sum_{k=2}^{n} \frac{1}{k} \leq \int_1^n \frac{1}{x}\, dx \leq \sum_{k=1}^{n-1} \frac{1}{k} \Rightarrow$$

$$\tfrac{1}{2} + \tfrac{1}{3} + \cdots + \tfrac{1}{n} \leq \ln n \leq 1 + \tfrac{1}{2} + \tfrac{1}{3} + \cdots + \frac{1}{n-1} \Rightarrow$$

$$1 + \tfrac{1}{2} + \tfrac{1}{3} + \cdots + \tfrac{1}{n} \leq 1 + \ln n.$$

By (i) and (ii), $\ln(n+1) < 1 + \frac{1}{2} + \frac{1}{3} + \cdots + \frac{1}{n} < 1 + \ln n$ $(n > 1)$.

(b) From part (a), $S_n > 100$ if $\ln(n+1) > 100$.

Thus, $n + 1 > e^{100} \Rightarrow n > e^{100} - 1 \approx 2.688 \times 10^{43}$.

[50] (a) $r_n = r_{n-1}\sqrt{\frac{n-1}{n}} = r_{n-2}\sqrt{\frac{n-2}{n-1}}\sqrt{\frac{n-1}{n}} = r_{n-3}\sqrt{\frac{n-3}{n-2}}\sqrt{\frac{n-2}{n-1}}\sqrt{\frac{n-1}{n}}$

$= \cdots = r_1\sqrt{\frac{n-1}{n}}\sqrt{\frac{n-2}{n-1}}\sqrt{\frac{n-3}{n-2}} \cdots \sqrt{\frac{1}{2}} = r_1\sqrt{\frac{1}{n}}$. Since $\sum_{n=1}^{\infty} r_n =$

$\sum_{n=1}^{\infty} r_1\sqrt{\frac{1}{n}} = r_1 \sum_{n=1}^{\infty} \frac{1}{\sqrt{n}} = \infty$, the height can be made arbitrarily large.

(b) The volume of the nth ball is given by

$V_n = \frac{4}{3}\pi(r_n)^3 = \frac{4}{3}\pi\left(r_1\sqrt{\frac{1}{n}}\right)^3 = \frac{4\pi r_1^3}{3n^{3/2}} = \frac{4\pi}{3n^{3/2}}$. The maximum volume is

$V_n = \sum_{n=1}^{\infty} \frac{4\pi}{3} \cdot \frac{1}{n^{3/2}} = \frac{4\pi}{3} + \frac{4\pi}{3}\sum_{n=2}^{\infty} \frac{1}{n^{3/2}} \le \frac{4\pi}{3} + \frac{4\pi}{3}\int_1^{\infty} \frac{1}{x^{3/2}}\,dx$ {using the proof of (11.23)} $= \frac{4\pi}{3} + \frac{4\pi}{3}(2) = 4\pi$. If the material weighs 1 lb/ft^3, then the total weight is less than $4\pi \cdot 1 = 4\pi$ pounds.

[51] Since $\lim_{n\to\infty} \frac{a_n}{b_n} = 0$, then by (11.3), there is an $N \ge 1$ such that if $n > N$, then $\frac{a_n}{b_n} < 1$, or $a_n < b_n$. Since $\sum b_n$ converges and $a_n < b_n$ for all but at most a finite number of terms, $\sum a_n$ must also converge by (11.26).

[52] Since $\lim_{n\to\infty} \frac{a_n}{b_n} = \infty$, there is an $M \ge 1$ such that if $k > M$, then $\frac{a_k}{b_k} > 1$, or $a_k > b_k$. Since $\sum b_n$ diverges and $b_k < a_k$ for all but at most a finite number of terms, $\sum a_n$ must also diverge.

[53] $\sum_{n=1}^{\infty} a_n = \sum_{n=1}^{N} a_n + \sum_{n=N+1}^{\infty} a_n$. Since $a_{n+1} < \int_n^{n+1} f(x)\,dx$, it follows that the error $E = \sum_{n=N+1}^{\infty} a_n < \int_N^{\infty} f(x)\,dx$. (See Figure 11.8.)

[54] Error $< \int_N^{\infty} \frac{1}{x^2}\,dx = \lim_{t\to\infty}\left[-\frac{1}{x}\right]_N^t = \frac{1}{N} \le 0.001 \Rightarrow N \ge 1000$; 1000 terms.

[55] Error $< \int_N^{\infty} \frac{1}{x^3}\,dx = \lim_{t\to\infty}\left[-\frac{1}{2x^2}\right]_N^t = \frac{1}{2N^2} \le 0.01 \Rightarrow N^2 \ge 50 \Rightarrow N \ge \sqrt{50}$; 8 terms.

[56] Error $< \int_N^{\infty} \frac{1}{x(\ln x)^2}\,dx = \lim_{t\to\infty}\left[-\frac{1}{\ln x}\right]_N^t = \frac{1}{\ln N} \le 0.05 \Rightarrow \ln N \ge 20 \Rightarrow N \ge e^{20}$; $[\![e^{20}]\!] + 1 = 485{,}165{,}196$ terms.

[57] Since $\sum a_n$ converges, $\lim_{n\to\infty} a_n = 0$ and $\lim_{n\to\infty} \frac{1}{a_n} = \infty$. By (11.17), $\sum \frac{1}{a_n}$ diverges.

[58] $(a_n - a_{n+1})^2 \ge 0 \Rightarrow a_n^2 - 2a_n a_{n+1} + a_{n+1}^2 \ge 0 \Rightarrow a_n^2 + a_{n+1}^2 \ge 2a_n a_{n+1} \Rightarrow$ $a_n^2 + 2a_n a_{n+1} + a_{n+1}^2 \ge 4a_n a_{n+1} \Rightarrow \frac{(a_n + a_{n+1})^2}{4} \ge a_n a_{n+1} \Rightarrow$ $\frac{a_n + a_{n+1}}{2} \ge \sqrt{a_n a_{n+1}}$. Since $\sum a_n$ converges, $\frac{1}{2}\sum (a_n + a_{n+1})$ converges, and $\sum \sqrt{a_n a_{n+1}}$ converges, by (11.26).

[59] $S_1 \approx 0.367879$, $S_2 \approx 0.404511$, $S_3 \approx 0.404881$; $S \approx 0.405$

$$|\text{error}| \le \int_3^{\infty} xe^{-x^2}\,dx = \lim_{t\to\infty}\left[-\tfrac{1}{2}e^{-x^2}\right]_3^t = \tfrac{1}{2}e^{-9} \approx 0.00006 < 0.5 \times 10^{-3}$$

[60] $S_1 = 1$, $S_2 = 1.0625$, $S_3 \approx 1.074846$, $S_4 \approx 1.078752$, $S_5 \approx 1.080352$, $S_6 \approx 1.081124$, $S_7 \approx 1.081540$, $S_8 \approx 1.081784$, $S_9 \approx 1.081937$; $S \approx 1.082$

$$|\text{error}| \le \int_9^{\infty} x^{-4}\,dx = \lim_{t\to\infty}\left[-\tfrac{1}{3}x^{-3}\right]_9^t = \tfrac{1}{3}(9^{-3}) \approx 0.000457 < 0.5 \times 10^{-3}$$

61 From the graphs, it is apparent that $x \geq \ln(x^k)$ for $k = 1, 2, 3$ and $x \geq 5$.

Thus, $\frac{1}{n} \leq \frac{1}{\ln(n^k)}$ for $k = 1, 2, 3$ and $n \geq 5$. By the basic comparison test,

since $\sum_{n=1}^{\infty} \frac{1}{n}$ diverges, $\sum_{n=1}^{\infty} \frac{1}{\ln(n^k)}$ also diverges for $k = 1, 2,$ and 3.

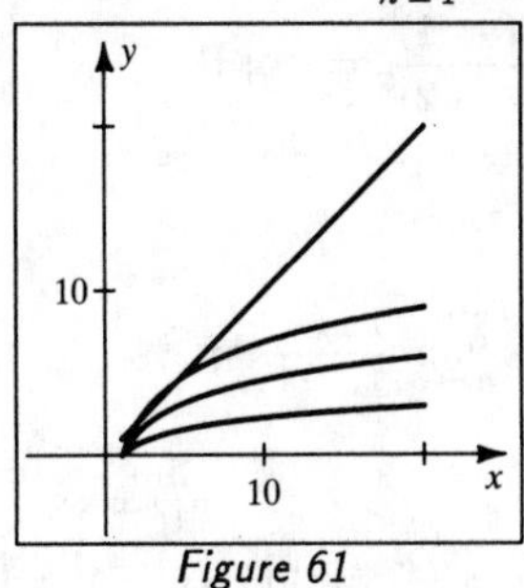

Figure 61

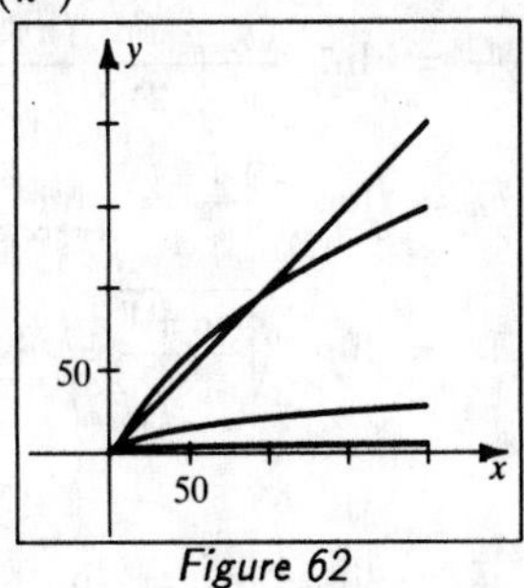

Figure 62

62 From the graphs, it is apparent that $x \geq (\ln x)^k$ for $k = 1, 2, 3$ and $x \geq 100$.

Thus, $\frac{1}{n} \leq \frac{1}{(\ln n)^k}$ for $k = 1, 2, 3$ and $n \geq 100$. By the basic comparison test,

since $\sum_{n=1}^{\infty} \frac{1}{n}$ diverges, $\sum_{n=1}^{\infty} \frac{1}{(\ln n)^k}$ also diverges for $k = 1, 2,$ and 3.

Exercises 11.4

1 $\lim_{n\to\infty} \frac{a_{n+1}}{a_n} = \lim_{n\to\infty} \left(a_{n+1} \cdot \frac{1}{a_n}\right) = \lim_{n\to\infty} \frac{3(n+1)+1}{2^{n+1}} \cdot \frac{2^n}{3n+1} =$

$\lim_{n\to\infty} \frac{3n+4}{2(3n+1)} = \frac{1}{2} < 1$; C

Note: For limits involving only powers of n, we will not divide by the highest power of n, but will merely compare leading terms to determine the limit.

2 $\lim_{n\to\infty} \frac{a_{n+1}}{a_n} = \lim_{n\to\infty} \frac{3^{n+1}}{(n+1)^2+4} \cdot \frac{n^2+4}{3^n} = \lim_{n\to\infty} \frac{3(n^2+4)}{(n+1)^2+4} = 3 > 1$; D

3 $\lim_{n\to\infty} \frac{a_{n+1}}{a_n} = \lim_{n\to\infty} \frac{5^{n+1}}{(n+1)\,3^{n+2}} \cdot \frac{n\,3^{n+1}}{5^n} = \lim_{n\to\infty} \frac{5n}{3(n+1)} = \frac{5}{3} > 1$; D

4 $\lim_{n\to\infty} \frac{a_{n+1}}{a_n} = \lim_{n\to\infty} \frac{2^n}{5^{n+1}(n+2)} \cdot \frac{5^n(n+1)}{2^{n-1}} = \lim_{n\to\infty} \frac{2(n+1)}{5(n+2)} = \frac{2}{5} < 1$; C

5 $\lim_{n\to\infty} \frac{a_{n+1}}{a_n} = \lim_{n\to\infty} \frac{100^{n+1}}{(n+1)!} \cdot \frac{n!}{100^n} = \lim_{n\to\infty} \frac{100}{n+1} = 0 < 1$; C

6 $\lim_{n\to\infty} \frac{a_{n+1}}{a_n} = \lim_{n\to\infty} \frac{(n+1)^{10}+10}{(n+1)!} \cdot \frac{n!}{n^{10}+10} = \lim_{n\to\infty} \frac{(n+1)^{10}+10}{(n+1)(n^{10}+10)} =$

$0 < 1$; C

7 $\lim_{n\to\infty} \frac{a_{n+1}}{a_n} = \lim_{n\to\infty} \frac{n+4}{(n+1)^2+2(n+1)+5} \cdot \frac{n^2+2n+5}{n+3} = 1$; the ratio test is inconclusive. Using the limit comparison test with $b_n = \frac{1}{n}$, we see that $\sum a_n$ diverges. However, the purpose here is to call attention to the condition under which the ratio test fails.

[8] $\lim_{n\to\infty} \frac{a_{n+1}}{a_n} = \lim_{n\to\infty} \frac{3(n+1)}{\sqrt{(n+1)^3+1}} \cdot \frac{\sqrt{n^3+1}}{3n} = 1$; inconclusive

[9] $\lim_{n\to\infty} \frac{a_{n+1}}{a_n} = \lim_{n\to\infty} \frac{(n+1)!}{e^{n+1}} \cdot \frac{e^n}{n!} = \lim_{n\to\infty} \frac{n+1}{e} = \infty$; D

[10] $\lim_{n\to\infty} \frac{a_{n+1}}{a_n} = \lim_{n\to\infty} \frac{(n+1)!}{(n+2)^5} \cdot \frac{(n+1)^5}{n!} = \lim_{n\to\infty} \frac{(n+1)^6}{(n+2)^5} = \infty$; D

[11] $\lim_{n\to\infty} \sqrt[n]{a_n} = \lim_{n\to\infty} \sqrt[n]{\frac{1}{n^n}} = \lim_{n\to\infty} \frac{1}{n} = 0 < 1$; C

[12] $\lim_{n\to\infty} \sqrt[n]{a_n} = \lim_{n\to\infty} \sqrt[n]{\frac{(\ln n)^n}{n^{n/2}}} = \lim_{n\to\infty} \frac{\ln n}{n^{1/2}} \left\{\frac{\infty}{\infty}\right\} = \lim_{n\to\infty} \frac{1/n}{\frac{1}{2}n^{-1/2}} =$

$\lim_{n\to\infty} \frac{2}{\sqrt{n}} = 0 < 1$; C

[13] $\lim_{n\to\infty} \sqrt[n]{a_n} = \lim_{n\to\infty} \sqrt[n]{\frac{2^n}{n^2}} = \lim_{n\to\infty} \frac{2}{n^{2/n}}$. Since $\lim_{n\to\infty} n^{2/n} = \lim_{n\to\infty} (n^{1/n})^2 =$

$(1)^2$ {see 11.1.37}, the limit is 2, which is greater than 1, and $\sum a_n$ diverges.

[14] $\lim_{n\to\infty} \sqrt[n]{a_n} = \lim_{n\to\infty} \sqrt[n]{\frac{5^{n+1}}{(\ln n)^n}} = \lim_{n\to\infty} \frac{5^{1+(1/n)}}{\ln n} = 0 < 1$; C

[15] $\lim_{n\to\infty} \sqrt[n]{a_n} = \lim_{n\to\infty} \sqrt[n]{\frac{n}{3^n}} = \lim_{n\to\infty} \frac{n^{1/n}}{3} = \frac{1}{3}$ {see 11.1.37} < 1; C

[16] $\lim_{n\to\infty} \sqrt[n]{a_n} = \lim_{n\to\infty} \sqrt[n]{\frac{n^{10}}{10^n}} = \lim_{n\to\infty} \frac{n^{10/n}}{10} = \frac{1}{10}$ {see 11.1.37} < 1; C

[17] $\lim_{n\to\infty} \sqrt[n]{a_n} = \lim_{n\to\infty} \sqrt[n]{\left(\frac{n}{2n+1}\right)^n} = \lim_{n\to\infty} \frac{n}{2n+1} = \frac{1}{2} < 1$; C

[18] $\lim_{n\to\infty} \sqrt[n]{a_n} = \lim_{n\to\infty} \sqrt[n]{\left(\frac{n}{\ln n}\right)^n} = \lim_{n\to\infty} \frac{n}{\ln n} \left\{\frac{\infty}{\infty}\right\} = \lim_{n\to\infty} \frac{1}{1/n} = \lim_{n\to\infty} n = \infty$; D

Note: The following may be solved using several methods. For a convenience, we list an abbreviation of the test used at the beginning of the problem.

NTH — nth-term test (11.17) INT — integral test (11.23)

BCT — basic comparison test (11.26) LCT — limit comparison test (11.27)

RAT — ratio test (11.28) ROT — root test (11.29)

[19] LCT Let $a_n = \frac{\sqrt{n}}{n^2+1}$ and $b_n = \frac{1}{n^{3/2}}$. $\lim_{n\to\infty} \frac{a_n}{b_n} = \lim_{n\to\infty} \frac{n^2}{n^2+1} = 1 > 0$.

S converges since $\sum b_n$ converges.

[20] LCT Let $a_n = \frac{\sqrt{n}}{3n+4}$ and $b_n = \frac{1}{n^{1/2}}$. $\lim_{n\to\infty} \frac{a_n}{b_n} = \lim_{n\to\infty} \frac{n}{3n+4} = \frac{1}{3} > 0$.

S diverges since $\sum b_n$ diverges.

[21] RAT $\lim_{n\to\infty} \frac{a_{n+1}}{a_n} = \lim_{n\to\infty} \frac{99^{n+1}\left[(n+1)^5+2\right]}{(n+1)^2\,10^{2(n+1)}} \cdot \frac{n^2\,10^{2n}}{99^n\,(n^5+2)} =$

$\lim_{n\to\infty} \frac{99n^2\left[(n+1)^5+2\right]}{10^2(n+1)^2(n^5+2)} = \frac{99}{100} < 1$; C

[22] RAT $\lim_{n\to\infty} \frac{a_{n+1}}{a_n} = \lim_{n\to\infty} \frac{(n+1)3^{2(n+1)}}{5^n} \cdot \frac{5^{n-1}}{n\,3^{2n}} = \lim_{n\to\infty} \frac{3^2(n+1)}{5n} = \frac{9}{5} > 1$; D

[23] BCT Since $\frac{2}{n^3 + e^n} < \frac{2}{n^3}$ and $2\sum_{n=1}^{\infty} \frac{1}{n^3}$ converges, S converges.

[24] LCT Let $a_n = \frac{n+1}{n^3+1}$ and $b_n = \frac{1}{n^2}$. $\lim_{n\to\infty} \frac{a_n}{b_n} = \lim_{n\to\infty} \frac{n^3+n^2}{n^3+1} = 1 > 0$.

S converges since $\sum b_n$ converges.

[25] RAT $\lim_{n\to\infty} \frac{a_{n+1}}{a_n} = \lim_{n\to\infty} \frac{2^{n+1}(n+1)!}{(n+1)^{n+1}} \cdot \frac{n^n}{2^n\,n!} = \lim_{n\to\infty} \frac{2(n+1)\,n^n}{(n+1)^{n+1}}$

$= 2\lim_{n\to\infty} \frac{n^n}{(n+1)^n} = 2\lim_{n\to\infty}\left(\frac{n}{n+1}\right)^n = 2\lim_{n\to\infty}\left(\frac{1}{\frac{n+1}{n}}\right)^n$

$= 2\lim_{n\to\infty} \frac{1^n}{\left(1+\frac{1}{n}\right)^n} = 2 \cdot \frac{1}{e} < 1$; C. *Note*: Using Example 2,

$\lim_{n\to\infty}\left(\frac{n}{n+1}\right)^n$ is the reciprocal of $\lim_{n\to\infty}\left(\frac{n+1}{n}\right)^n$, which is e.

[26] RAT From Example 2, $\lim_{n\to\infty} \frac{a_{n+1}}{a_n} = \frac{1}{e}$ {reciprocal} < 1; C

[27] ROT $\lim_{n\to\infty} \sqrt[n]{a_n} = \lim_{n\to\infty} \sqrt[n]{\frac{n^n}{10^{n+1}}} = \lim_{n\to\infty} \frac{n}{10^{1+(1/n)}} = \lim_{n\to\infty} \frac{n}{10} = \infty$; D

[28] BCT & RAT Since $\frac{10+2^n}{n!} < \frac{2^n}{n!}$ and $\lim_{n\to\infty} \frac{2^n}{(n+1)!} \cdot \frac{n!}{2^n} = \lim_{n\to\infty} \frac{2}{n+1} = 0 < 1 \Rightarrow$

$\sum_{n=1}^{\infty} \frac{2^n}{n!}$ converges, so S converges.

[29] RAT $\lim_{n\to\infty} \frac{a_{n+1}}{a_n} = \lim_{n\to\infty} \frac{[(n+1)!]^2}{(2n+2)!} \cdot \frac{(2n)!}{(n!)^2} = \lim_{n\to\infty} \frac{(n+1)^2}{(2n+2)(2n+1)} = \frac{1}{4} < 1$; C

[30] RAT $\lim_{n\to\infty} \frac{a_{n+1}}{a_n} = \lim_{n\to\infty} \frac{(2n+2)!}{2^{n+1}} \cdot \frac{2^n}{(2n)!} = \lim_{n\to\infty} \frac{(2n+2)(2n+1)}{2} = \infty$; D

[31] INT $\int_2^{\infty} \frac{1}{x\sqrt[3]{\ln x}}\,dx = \lim_{t\to\infty}\left[\frac{3}{2}(\ln x)^{2/3}\right]_2^t = \infty$; D

[32] ROT $\lim_{n\to\infty} \sqrt[n]{a_n} = \lim_{n\to\infty} \sqrt[n]{\frac{(2n)^n}{(5n+3n^{-1})^n}} = \lim_{n\to\infty} \frac{2n}{5n+\frac{3}{n}} = \frac{2}{5} < 1$; C

[33] BCT & RAT Since $\frac{\ln n}{(1.01)^n} < \frac{n}{(1.01)^n}$ and $\lim_{n\to\infty} \frac{n+1}{(1.01)^{n+1}} \cdot \frac{(1.01)^n}{n} = \frac{1}{1.01} < 1 \Rightarrow$

$\sum_{n=1}^{\infty} \frac{n}{(1.01)^n}$ converges, so S converges.

[34] NTH Since $\lim_{n\to\infty} 3^{1/n} = 3^0 = 1 \neq 0$, S diverges.

[35] NTH $\lim_{n\to\infty} n\tan\frac{1}{n}$ $\{\infty \cdot 0\}$ $= \lim_{n\to\infty} \frac{\tan(1/n)}{1/n}$ $\{\frac{0}{0}\}$ $= \lim_{n\to\infty} \frac{(-1/n^2)\sec^2(1/n)}{-1/n^2} =$

$\lim_{n\to\infty} \sec^2(1/n) = 1 \neq 0$. Thus, S diverges.

[36] BCT Since $\frac{\arctan n}{n^2} < \frac{\pi/2}{n^2}$ and $\frac{\pi}{2}\sum_{n=1}^{\infty} \frac{1}{n^2}$ converges, S converges.

[37] NTH $\lim_{n\to\infty}\left(1+\frac{1}{n}\right)^n = e$ {by (7.32)(ii)} $\neq 0$. Thus, S diverges.

[38] ROT $\lim_{n\to\infty}\sqrt[n]{a_n} = \lim_{n\to\infty}\sqrt[n]{\frac{1}{(\ln n)^n}} = \lim_{n\to\infty}\frac{1}{\ln n} = 0$; C

[39] RAT $\lim_{n\to\infty}\frac{a_{n+1}}{a_n} = \lim_{n\to\infty}\frac{1\cdot 3\cdot 5\cdot\cdots\cdot(2n-1)(2n+1)}{(n+1)!}\cdot\frac{n!}{1\cdot 3\cdot 5\cdot\cdots\cdot(2n-1)} =$

$$\lim_{n\to\infty}\frac{2n+1}{n+1} = 2 > 1;\text{ D}$$

[40] RAT $\lim_{n\to\infty}\frac{a_{n+1}}{a_n}$

$$= \lim_{n\to\infty}\frac{1\cdot 4\cdot 7\cdot\cdots\cdot(3n-2)(3n+1)}{2\cdot 4\cdot 6\cdot\cdots\cdot(2n)(2n+2)}\cdot\frac{2\cdot 4\cdot 6\cdot\cdots\cdot(2n)}{1\cdot 4\cdot 7\cdot\cdots\cdot(3n-2)}$$

$$= \lim_{n\to\infty}\frac{3n+1}{2n+2} = \frac{3}{2} > 1;\text{ D}$$

Exercises 11.5

[1] (a) $a_k = \frac{1}{k^2+7}$. $a_{k+1} = \frac{1}{(k+1)^2+7} = \frac{1}{(k^2+7)+(2k+1)}$.

Since the denominator of a_{k+1} is larger than the denominator of a_k, $a_k > a_{k+1}$.

Also, $a_{k+1} > 0$ and hence, condition (i) of (11.30) is satisfied.

Condition (ii) is satisfied since $\lim_{n\to\infty} a_n = \lim_{n\to\infty}\frac{1}{n^2+7} = 0$.

(b) The series converges by (11.30) since conditions (i) and (ii) are satisfied.

[2] (a) Condition (i) is satisfied since $\frac{a_{k+1}}{a_k} = \frac{(k+1)/5^{k+1}}{k/5^k} = \frac{k+1}{5k} = \frac{1}{5}+\frac{1}{5k} < 1$ for

every positive integer k. Condition (ii) is satisfied since

$$\lim_{n\to\infty} a_n = \lim_{n\to\infty}\frac{n}{5^n}\left\{\tfrac{\infty}{\infty}\right\} = \lim_{n\to\infty}\frac{1}{5^n(\ln 5)} = 0.$$

(b) The series converges by (11.30) since conditions (i) and (ii) are satisfied.

[3] (a) $f(x) = 1 + e^{-x} \Rightarrow f'(x) = -e^{-x} < 0$ and f is decreasing.

Hence, $a_k > a_{k+1} > 0$ and condition (i) is satisfied.

$\lim_{n\to\infty} a_n = \lim_{n\to\infty}(1+e^{-n}) = 1 \neq 0$ and condition (ii) is not satisfied.

(b) The series diverges by the nth-term test.

[4] (a) $f(x) = \frac{e^{2n}+1}{e^{2n}-1} \Rightarrow f'(x) = -\frac{4e^{2x}}{(e^{2x}-1)^2} < 0$ and f is decreasing.

Hence, $a_k > a_{k+1} > 0$ and condition (i) is satisfied. $\lim_{n\to\infty} a_n =$

$\lim_{n\to\infty}\frac{e^{2n}+1}{e^{2n}-1}\left\{\tfrac{\infty}{\infty}\right\} = \lim_{n\to\infty}\frac{2e^{2n}}{2e^{2n}} = 1 \neq 0$ and condition (ii) is not satisfied.

(b) The series diverges by the nth-term test.

Note: Let AC denote Absolutely Convergent; CC, Conditionally Convergent;

D, Divergent; AST, the Alternating Series Test; and S, the given series.

[5] $a_n = \dfrac{1}{\sqrt{2n+1}} > 0$. $\dfrac{a_{n+1}}{a_n} = \dfrac{1}{\sqrt{2(n+1)+1}} \div \dfrac{1}{\sqrt{2n+1}} = \dfrac{\sqrt{2n+1}}{\sqrt{2n+3}} < 1$ and

$\lim\limits_{n\to\infty} a_n = 0 \Rightarrow$ S converges by AST. If $b_n = \dfrac{1}{\sqrt{n}}$, $\lim\limits_{n\to\infty} \dfrac{a_n}{b_n} = \dfrac{1}{\sqrt{2}} > 0$ and

$\sum a_n$ diverges. Thus, S is CC.

[6] $a_n = \dfrac{1}{n^{2/3}} > 0$. $\dfrac{a_{n+1}}{a_n} = \dfrac{n^{2/3}}{(n+1)^{2/3}} < 1$ and $\lim\limits_{n\to\infty} a_n = 0 \Rightarrow$ S converges by AST.

However, $\sum\limits_{n=1}^{\infty} \dfrac{1}{n^{2/3}}$ diverges $\{p = \frac{2}{3} < 1\}$. Thus, S is CC.

[7] $a_n = \dfrac{1}{\ln(n+1)} > 0$. $\dfrac{a_{n+1}}{a_n} = \dfrac{\ln(n+1)}{\ln(n+2)} < 1$ and $\lim\limits_{n\to\infty} a_n = 0 \Rightarrow$ S converges by

AST. However, $\dfrac{1}{\ln(n+1)} > \dfrac{1}{n+1} > \dfrac{1}{2n}$ and $\frac{1}{2}\sum\limits_{n=1}^{\infty} \frac{1}{n}$ diverges. Thus, S is CC.

[8] $a_n = \dfrac{n}{n^2+4} > 0$. $f(x) = \dfrac{x}{x^2+4} \Rightarrow f'(x) = \dfrac{4-x^2}{(x^2+4)^2} < 0$ for $x > 2$ and

$\lim\limits_{n\to\infty} a_n = 0 \Rightarrow$ S converges by AST. If $b_n = \dfrac{1}{n}$, $\lim\limits_{n\to\infty} \dfrac{a_n}{b_n} = 1 > 0$ and

$\sum a_n$ diverges. Thus, S is CC.

[9] Since $\lim\limits_{n\to\infty} \dfrac{n}{\ln n} = \infty$, S is D.

[10] $a_n = \dfrac{\ln n}{n} > 0$. $f(x) = \dfrac{\ln x}{x} \Rightarrow f'(x) = \dfrac{1 - \ln x}{x^2} < 0$ for $x > e$ and $\lim\limits_{n\to\infty} a_n = 0 \Rightarrow$

S converges by AST. However, $\dfrac{\ln n}{n} > \dfrac{1}{n}$ for $x > e$ and $\sum\limits_{n=1}^{\infty} \dfrac{1}{n}$ diverges. Thus, S is CC.

[11] Since $\dfrac{5}{n^3+1} < \dfrac{5}{n^3}$ and $5\sum\limits_{n=1}^{\infty} \dfrac{1}{n^3}$ converges, S is AC.

[12] $\lim\limits_{n\to\infty} \left|\dfrac{a_{n+1}}{a_n}\right| = \lim\limits_{n\to\infty} \dfrac{e^{-n-1}}{e^{-n}} = e^{-1} < 1 \Rightarrow$ S is AC by (11.35)(i).

[13] $\lim\limits_{n\to\infty} \left|\dfrac{a_{n+1}}{a_n}\right| = \lim\limits_{n\to\infty} \dfrac{10^{n+1}}{(n+1)!} \cdot \dfrac{n!}{10^n} = \lim\limits_{n\to\infty} \dfrac{10}{n+1} = 0 \Rightarrow$ S is AC.

[14] $\lim\limits_{n\to\infty} \left|\dfrac{a_{n+1}}{a_n}\right| = \lim\limits_{n\to\infty} \dfrac{(n+1)!}{5^{n+1}} \cdot \dfrac{5^n}{n!} = \lim\limits_{n\to\infty} \dfrac{n+1}{5} = \infty \Rightarrow$ S is D by (11.35)(ii).

[15] $\lim\limits_{n\to\infty} |a_n| = \lim\limits_{n\to\infty} \dfrac{n^2+3}{(2n-5)^2} = \dfrac{1}{4} \neq 0 \Rightarrow$ S is D.

[16] Since $\dfrac{\sin\sqrt{n}}{\sqrt{n^3+4}} < \dfrac{1}{\sqrt{n^3+4}} < \dfrac{1}{n^{3/2}}$ and $\sum\limits_{n=1}^{\infty} \dfrac{1}{n^{3/2}}$ converges, S is AC.

[17] $a_n = \dfrac{n^{1/3}}{n+1} > 0$. $f(x) = \dfrac{x^{1/3}}{x+1} \Rightarrow f'(x) = \dfrac{1-2x}{3x^{2/3}(x+1)^2} < 0$ for $x > \dfrac{1}{2}$ and

$\lim\limits_{n\to\infty} a_n = 0 \Rightarrow$ S converges by AST. If $b_n = \dfrac{1}{n^{2/3}}$, $\lim\limits_{n\to\infty} \dfrac{a_n}{b_n} = 1 > 0$ and

$\sum a_n$ diverges. Thus, S is CC.

[18] Let $a_n = \dfrac{(n+1)^2}{n^5+1}$ and $b_n = \dfrac{1}{n^3}$. $\lim\limits_{n\to\infty} \dfrac{a_n}{b_n} = 1 > 0$ and S is AC.

[19] Since $\left|\frac{\cos\frac{\pi}{6}n}{n^2}\right| \leq \frac{1}{n^2}$ and $\sum_{n=1}^{\infty}\frac{1}{n^2}$ converges, S is AC.

[20] Since $\frac{\ln n}{(1.5)^n} < \frac{n}{(1.5)^n}$ and $\lim_{n\to\infty}\frac{n+1}{(1.5)^{n+1}}\cdot\frac{(1.5)^n}{n} = \lim_{n\to\infty}\frac{n+1}{1.5n} = \frac{1}{1.5} < 1 \Rightarrow$

$\sum_{n=1}^{\infty}\frac{n}{(1.5)^n}$ converges, S is AC.

[21] Since $\lim_{n\to\infty} n\sin\frac{1}{n}\ \{\infty\cdot 0\} = \lim_{n\to\infty}\frac{\sin(1/n)}{1/n}\ \{\frac{0}{0}\} = \lim_{n\to\infty}\frac{(-1/n^2)\cos(1/n)}{-1/n^2} =$

$\lim_{n\to\infty}\cos(1/n) = 1 \neq 0$, S is D.

[22] Since $\frac{\arctan n}{n^2} < \frac{\pi/2}{n^2}$ and $\frac{\pi}{2}\sum_{n=1}^{\infty}\frac{1}{n^2}$ converges, S is AC.

[23] $a_n = \frac{1}{n\sqrt{\ln n}} > 0$. $\frac{a_{n+1}}{a_n} = \frac{n\sqrt{\ln n}}{(n+1)\sqrt{\ln(n+1)}} < 1$ and $\lim_{n\to\infty} a_n = 0 \Rightarrow$

S converges by AST. However, S is not AC by Exercise 48, §11.3. Thus, S is CC.

[24] $\lim_{n\to\infty}\left|\frac{a_{n+1}}{a_n}\right| = \lim_{n\to\infty}\frac{2^{1/(n+1)}}{(n+1)!}\cdot\frac{n!}{2^{1/n}} = \lim_{n\to\infty}\frac{1}{(n+1)2^{1/[n(n+1)]}} = 0 < 1 \Rightarrow$ S is AC.

[25] $\lim_{n\to\infty}\sqrt[n]{|a_n|} = \lim_{n\to\infty}\sqrt[n]{\left|\frac{n^n}{(-5)^n}\right|} = \lim_{n\to\infty}\frac{n}{5} = \infty \Rightarrow$ S is D.

[26] $\lim_{n\to\infty}\sqrt[n]{|a_n|} = \lim_{n\to\infty}\sqrt[n]{\left|\frac{(n^2+1)^n}{(-n)^n}\right|} = \lim_{n\to\infty}\frac{n^2+1}{n} = \infty \Rightarrow$ S is D.

[27] Since $\lim_{x\to\infty}\frac{1+4^x}{1+3^x}\ \{\frac{\infty}{\infty}\} = \lim_{x\to\infty}\frac{4^x\ln 4}{3^x\ln 3} = \frac{\ln 4}{\ln 3}\cdot\lim_{x\to\infty}(\frac{4}{3})^x = \infty$,

$\lim_{n\to\infty} a_n \neq 0$ and S is D.

[28] $\lim_{n\to\infty}\left|\frac{a_{n+1}}{a_n}\right| = \lim_{n\to\infty}\frac{(n+1)^4}{e^{n+1}}\cdot\frac{e^n}{n^4} = \lim_{n\to\infty}\frac{(n+1)^4}{e\,n^4} = \frac{1}{e} < 1 \Rightarrow$ S is AC.

[29] $\sum_{n=1}^{\infty}(-1)^n\frac{\cos\pi n}{n} = 1 + \frac{1}{2} + \frac{1}{3} + \cdots$. This is the harmonic series. S is D.

[30] $\sum_{n=1}^{\infty}\frac{1}{n}\sin\frac{(2n-1)\pi}{2} = 1 - \frac{1}{2} + \frac{1}{3} - \cdots$. This is the alternating harmonic series.

S is CC by Example 4.

[31] Let $a_n = \frac{1}{(n-4)^2+5}$ and $b_n = \frac{1}{n^2}$. $\lim_{n\to\infty}\frac{a_n}{b_n} = 1 > 0$ and $\sum(-1)^n a_n$ is AC.

[32] $f(x) = \frac{\ln x}{x^{1/3}} \Rightarrow f'(x) = \frac{3-\ln x}{3x^{4/3}}$. $f'(x) < 0 \Rightarrow 3 - \ln x < 0 \Rightarrow x > e^3$. Hence,

condition (i) of AST is satisfied if $n > 20$. (ii) is satisfied. S converges by (11.30).

However, $\frac{\ln n}{\sqrt[3]{n}} > \frac{1}{\sqrt[3]{n}}$ for $x > e$ and $\sum_{n=1}^{\infty}\frac{1}{n^{1/3}}$ diverges. Thus, S is CC.

Note: The value of n in Exercises 33–42 was found by trial and error when the inequality could not be solved using basic algebraic operations.

Note: If the sum S_n is to be approximated to three decimal places, then $a_{n+1} < 0.5 \times 10^{-3}$. Equivalently, we use a_n and S_{n-1} to simplify the computations.

[33] $a_n = \frac{1}{n!} < 0.0005 \Rightarrow n! > 2000 \Rightarrow n \geq 7.$

$$\text{Thus, } S \approx S_6 = 1 - \tfrac{1}{1!} + \tfrac{1}{2!} - \tfrac{1}{3!} + \tfrac{1}{4!} - \tfrac{1}{5!} + \tfrac{1}{6!} \approx 0.368.$$

[34] $a_n = \frac{1}{(2n)!} < 0.0005 \Rightarrow (2n)! > 2000 \Rightarrow n \geq 4.$

$$\text{Thus, } S \approx S_3 = -1 + \tfrac{1}{2!} - \tfrac{1}{4!} + \tfrac{1}{6!} \approx -0.540.$$

[35] $a_n = \frac{1}{n^3} < 0.0005 \Rightarrow n^3 > 2000 \Rightarrow n \geq 13.$ Thus, $S \approx S_{12} =$

$$1 - \tfrac{1}{2^3} + \tfrac{1}{3^3} - \tfrac{1}{4^3} + \tfrac{1}{5^3} - \tfrac{1}{6^3} + \tfrac{1}{7^3} - \tfrac{1}{8^3} + \tfrac{1}{9^3} - \tfrac{1}{10^3} + \tfrac{1}{11^3} - \tfrac{1}{12^3} \approx 0.901.$$

[36] $a_n = \frac{1}{n^5} < 0.0005 \Rightarrow n^5 > 2000 \Rightarrow n \geq 5.$

$$\text{Thus, } S \approx S_4 = 1 - \tfrac{1}{2^5} + \tfrac{1}{3^5} - \tfrac{1}{4^5} \approx 0.972.$$

[37] $a_n = \frac{n+1}{5^n} < 0.0005 \Rightarrow 2000(n+1) < 5^n \Rightarrow n \geq 6.$

$$\text{Thus, } S \approx S_5 = \tfrac{2}{5} - \tfrac{3}{25} + \tfrac{4}{125} - \tfrac{5}{625} + \tfrac{6}{3125} \approx 0.306.$$

[38] $a_n = \frac{1}{n}(\frac{1}{2})^n < 0.0005 \Rightarrow 2000 < n2^n \Rightarrow n \geq 8.$ Thus, $S \approx S_7 =$

$$-\tfrac{1}{1}(\tfrac{1}{2}) + \tfrac{1}{2}(\tfrac{1}{2})^2 - \tfrac{1}{3}(\tfrac{1}{2})^3 + \tfrac{1}{4}(\tfrac{1}{2})^4 - \tfrac{1}{5}(\tfrac{1}{2})^5 + \tfrac{1}{6}(\tfrac{1}{2})^6 - \tfrac{1}{7}(\tfrac{1}{2})^7 \approx -0.4058.$$

Note: If we round to -0.406 (in this case), $|S - (-0.406)| > 0.0005$ because of this rounding error. However, $|S - S_n| < 0.0005$ as (11.31) predicts.

[39] $a_{n+1} = \frac{1}{(n+1)^2} < 0.00005 \Rightarrow (n+1)^2 > 20{,}000 \Rightarrow n > \sqrt{20{,}000} - 1 \Rightarrow n \geq 141.$

[40] $a_{n+1} = \frac{1}{\sqrt{n+1}} < 0.00005 \Rightarrow \sqrt{n+1} > 20{,}000 \Rightarrow n > 4 \times 10^8 - 1 \Rightarrow$

$$n \geq 4 \times 10^8.$$

[41] $a_{n+1} = \frac{1}{(n+1)^{n+1}} < 0.00005 \Rightarrow (n+1)^{n+1} > 20{,}000 \Rightarrow n+1 \geq 6 \Rightarrow n \geq 5.$

[42] $a_{n+1} = \frac{1}{(n+1)^3 + 1} < 0.00005 \Rightarrow (n+1)^3 + 1 > 20{,}000 \Rightarrow n > \sqrt[3]{19{,}999} - 1 \Rightarrow$

$$n \geq 27.$$

[43] (i) $a_n = \frac{(\ln n)^k}{n} > 0 \ (n \geq 3).$ $f(x) = \frac{(\ln x)^k}{x} \Rightarrow f'(x) = \frac{(\ln x)^{k-1}(k - \ln x)}{x^2} < 0$ for

$x > e^k \Rightarrow a_{n+1} < a_n$ for all but a finite number of terms.

(ii) $\lim_{x\to\infty} \frac{(\ln x)^k}{x} \{\frac{\infty}{\infty}\} = \lim_{x\to\infty} \frac{k(\ln x)^{k-1}}{x} \{\frac{\infty}{\infty}\} = \lim_{x\to\infty} \frac{k(k-1)(\ln x)^{k-2}}{x} \{\frac{\infty}{\infty}\}$

$= \cdots = \lim_{x\to\infty} \frac{k!}{x} = 0.$ Thus, by AST, S converges.

[44] (i) $a_n = \frac{1}{\sqrt[k]{n}} > 0.$ $\frac{a_{n+1}}{a_n} = \frac{\sqrt[k]{n}}{\sqrt[k]{n+1}} < 1 \Rightarrow a_{n+1} < a_n.$

(ii) $\lim_{n\to\infty} a_n = \lim_{n\to\infty} \frac{1}{\sqrt[k]{n}} = 0.$ Thus, by AST, S converges.

[45] No. If $a_n = b_n = (-1)^n/\sqrt{n}$, then both $\sum a_n$ and $\sum b_n$ converge by the alternating series test. However, $\sum a_n b_n = \sum 1/n$, which diverges.

[46] No. If $a_n = b_n = 1/n$, then both $\sum a_n$ and $\sum b_n$ diverge.

However, $\sum a_n b_n = \sum 1/n^2$, which converges.

Exercises 11.6

Note: Let u_n denote the nth term of the power series. Let AC denote Absolutely Convergent; C, Convergent; D, Divergent; and AST, the Alternating Series Test.

1 $u_n = \dfrac{x^n}{n+4} \Rightarrow \lim_{n\to\infty}\left|\dfrac{u_{n+1}}{u_n}\right| = \lim_{n\to\infty}\left|\dfrac{x^{n+1}}{n+5}\cdot\dfrac{n+4}{x^n}\right| = \lim_{n\to\infty}\left(\dfrac{n+4}{n+5}\right)|x| = |x|.$

$|x| < 1 \Leftrightarrow -1 < x < 1$. If $x = 1$, $\sum_{n=0}^{\infty}\dfrac{1}{n+4}(1)^n = \sum_{n=0}^{\infty}\dfrac{1}{n+4}$ is D {use LCT with $b_n = \frac{1}{n}$}. If $x = -1$, $\sum_{n=0}^{\infty}(-1)^n\dfrac{1}{n+4}$ is C by AST. ★ $[-1, 1)$

2 $u_n = \dfrac{x^n}{n^2+4} \Rightarrow \lim_{n\to\infty}\left|\dfrac{u_{n+1}}{u_n}\right| = \lim_{n\to\infty}\left|\dfrac{x^{n+1}}{(n+1)^2+4}\cdot\dfrac{n^2+4}{x^n}\right| =$

$\lim_{n\to\infty}\dfrac{n^2+4}{n^2+2n+5}|x| = |x|$. $|x| < 1 \Leftrightarrow -1 < x < 1$. If $x = 1$, $\sum_{n=0}^{\infty}\dfrac{1}{n^2+4} < \sum_{n=0}^{\infty}\dfrac{1}{n^2}$ is C. If $x = -1$, $\sum_{n=0}^{\infty}(-1)^n\dfrac{1}{n^2+4}$ is C by AST. ★ $[-1, 1]$

3 $u_n = \dfrac{n^2x^n}{2^n} \Rightarrow \lim_{n\to\infty}\left|\dfrac{u_{n+1}}{u_n}\right| = \lim_{n\to\infty}\left|\dfrac{(n+1)^2x^{n+1}}{2^{n+1}}\cdot\dfrac{2^n}{n^2x^n}\right| = \frac{1}{2}|x|.$

$\frac{1}{2}|x| < 1 \Leftrightarrow -2 < x < 2$. If $x = 2$, $\sum_{n=0}^{\infty} n^2$ is D {nth-term}.

If $x = -2$, $\sum_{n=0}^{\infty}(-1)^n n^2$ is D. ★ $(-2, 2)$

4 $u_n = \dfrac{(-3)^nx^{n+1}}{n} \Rightarrow \lim_{n\to\infty}\left|\dfrac{u_{n+1}}{u_n}\right| = \lim_{n\to\infty}\left|\dfrac{(-3)^{n+1}x^{n+2}}{n+1}\cdot\dfrac{n}{(-3)^nx^{n+1}}\right| = 3|x|.$

$3|x| < 1 \Leftrightarrow -\frac{1}{3} < x < \frac{1}{3}$. If $x = \frac{1}{3}$, $\sum_{n=1}^{\infty}\dfrac{(-3)^n}{n3^{n+1}} = \sum_{n=1}^{\infty}(-1)^n\dfrac{1}{3n}$ is C by AST.

If $x = -\frac{1}{3}$, $\sum_{n=1}^{\infty}\dfrac{(-3)^n}{n(-3)^{n+1}} = \sum_{n=1}^{\infty}-\dfrac{1}{3n}$ is D. ★ $(-\frac{1}{3}, \frac{1}{3}]$

5 $u_n = (-1)^{n-1}\dfrac{x^n}{\sqrt{n}} \Rightarrow \lim_{n\to\infty}\left|\dfrac{u_{n+1}}{u_n}\right| = \lim_{n\to\infty}\left|\dfrac{x^{n+1}}{\sqrt{n+1}}\cdot\dfrac{\sqrt{n}}{x^n}\right| = |x|.$

$|x| < 1 \Leftrightarrow -1 < x < 1$. If $x = 1$, $\sum_{n=1}^{\infty}(-1)^{n-1}\dfrac{1}{\sqrt{n}}$ is C by AST.

If $x = -1$, $\sum_{n=1}^{\infty}(-1)^{n-1}\dfrac{(-1)^n}{\sqrt{n}} = \sum_{n=1}^{\infty}\dfrac{(-1)^{2n-1}}{\sqrt{n}} =$

$\sum_{n=1}^{\infty}-\dfrac{1}{\sqrt{n}}$ {since $2n-1$ is odd} is D. ★ $(-1, 1]$

6 $u_n = \dfrac{x^n}{\ln(n+1)} \Rightarrow \lim_{n\to\infty}\left|\dfrac{u_{n+1}}{u_n}\right| = \lim_{n\to\infty}\left|\dfrac{x^{n+1}}{\ln(n+2)}\cdot\dfrac{\ln(n+1)}{x^n}\right| = |x|.$

$|x| < 1 \Leftrightarrow -1 < x < 1$. If $x = 1$, $\sum_{n=1}^{\infty}\dfrac{1}{\ln(n+1)} \geq \sum_{n=1}^{\infty}\dfrac{1}{n}$ is D.

If $x = -1$, $\sum_{n=1}^{\infty}(-1)^n\dfrac{1}{\ln(n+1)}$ is C by AST. ★ $[-1, 1)$

[7] $u_n = \dfrac{nx^n}{n^2+1} \Rightarrow \lim\limits_{n\to\infty}\left|\dfrac{u_{n+1}}{u_n}\right| = \lim\limits_{n\to\infty}\left|\dfrac{(n+1)\,x^{n+1}}{(n+1)^2+1}\cdot\dfrac{n^2+1}{n\,x^n}\right| = |x|.$

$|x| < 1 \Leftrightarrow -1 < x < 1.$ If $x = 1$, $\sum\limits_{n=2}^{\infty}\dfrac{n}{n^2+1}$ is D { LCT with $b_n = \frac{1}{n}$ }.

If $x = -1$, $\sum\limits_{n=2}^{\infty}(-1)^n\dfrac{n}{n^2+1}$ converges by AST. ★ $[-1, 1)$

[8] $u_n = \dfrac{x^n}{4^n\sqrt{n}} \Rightarrow \lim\limits_{n\to\infty}\left|\dfrac{u_{n+1}}{u_n}\right| = \lim\limits_{n\to\infty}\left|\dfrac{x^{n+1}}{4^{n+1}\sqrt{n+1}}\cdot\dfrac{4^n\sqrt{n}}{x^n}\right| = \frac{1}{4}|x|.$

$\frac{1}{4}|x| < 1 \Leftrightarrow -4 < x < 4.$ If $x = 4$, $\sum\limits_{n=1}^{\infty}\dfrac{1}{\sqrt{n}}$ is D.

If $x = -4$, $\sum\limits_{n=1}^{\infty}(-1)^n\dfrac{1}{\sqrt{n}}$ is C by AST. ★ $[-4, 4)$

[9] $u_n = \dfrac{(\ln n)\,x^n}{n^3} \Rightarrow \lim\limits_{n\to\infty}\left|\dfrac{u_{n+1}}{u_n}\right| = \lim\limits_{n\to\infty}\left|\dfrac{[\ln(n+1)]\,x^{n+1}}{(n+1)^3}\cdot\dfrac{n^3}{(\ln n)\,x^n}\right| = |x|.$

$|x| < 1 \Leftrightarrow -1 < x < 1.$ If $x = 1$, $\sum\limits_{n=2}^{\infty}\dfrac{\ln n}{n^3} \le \sum\limits_{n=2}^{\infty}\dfrac{n}{n^3} = \sum\limits_{n=2}^{\infty}\dfrac{1}{n^2}$ is C.

If $x = -1$, $\sum\limits_{n=2}^{\infty}(-1)^n\dfrac{\ln n}{n^3}$ is C by AST. ★ $[-1, 1]$

[10] $u_n = \dfrac{10^{n+1}\,x^n}{3^{2n}} = \dfrac{10^{n+1}\,x^n}{9^n} \Rightarrow \lim\limits_{n\to\infty}\left|\dfrac{u_{n+1}}{u_n}\right| = \lim\limits_{n\to\infty}\left|\dfrac{10^{n+2}\,x^{n+1}}{9^{n+1}}\cdot\dfrac{9^n}{10^{n+1}\,x^n}\right| =$

$\frac{10}{9}|x|.$ $\frac{10}{9}|x| < 1 \Leftrightarrow -\frac{9}{10} < x < \frac{9}{10}.$ If $x = \frac{9}{10}$, $\sum\limits_{n=0}^{\infty}10$ is D { nth-term }.

If $x = -\frac{9}{10}$, $\sum\limits_{n=0}^{\infty}(-1)^n\,10$ is D. ★ $(-\frac{9}{10}, \frac{9}{10})$

[11] $u_n = \dfrac{(n+1)(x-4)^n}{10^n} \Rightarrow \lim\limits_{n\to\infty}\left|\dfrac{u_{n+1}}{u_n}\right| = \lim\limits_{n\to\infty}\left|\dfrac{(n+2)(x-4)^{n+1}}{10^{n+1}}\cdot\dfrac{10^n}{(n+1)(x-4)^n}\right|$

$= \frac{1}{10}|x-4|.$ $\frac{1}{10}|x-4| < 1 \Leftrightarrow -6 < x < 14.$ If $x = 14$, $\sum\limits_{n=0}^{\infty}(n+1)$ is D { nth-

term }. If $x = -6$, $\sum\limits_{n=0}^{\infty}(-1)^n(n+1)$ is D. ★ $(-6, 14)$

[12] $u_n = \dfrac{(x-2)^n}{n(n+1)} \Rightarrow \lim\limits_{n\to\infty}\left|\dfrac{u_{n+1}}{u_n}\right| = \lim\limits_{n\to\infty}\left|\dfrac{(x-2)^{n+1}}{(n+1)(n+2)}\cdot\dfrac{n(n+1)}{(x-2)^n}\right| = |x-2|.$

$|x-2| < 1 \Leftrightarrow 1 < x < 3.$ If $x = 3$, $\sum\limits_{n=1}^{\infty}\dfrac{1}{n(n+1)} \le \sum\limits_{n=1}^{\infty}\dfrac{1}{n^2}$ is C.

If $x = 1$, $\sum\limits_{n=1}^{\infty}(-1)^n\dfrac{1}{n(n+1)}$ is C by AST. ★ $[1, 3]$

[13] $u_n = \dfrac{n!\,x^n}{100^n} \Rightarrow \lim\limits_{n\to\infty}\left|\dfrac{u_{n+1}}{u_n}\right| = \lim\limits_{n\to\infty}\left|\dfrac{(n+1)!\,x^{n+1}}{100^{n+1}}\cdot\dfrac{100^n}{n!\,x^n}\right| = \lim\limits_{n\to\infty}\dfrac{n+1}{100}|x|$

$= \infty.$ Converges only for $x = 0$.

[14] $u_n = \dfrac{(3n)!\,x^n}{(2n)!} \Rightarrow \lim\limits_{n\to\infty}\left|\dfrac{u_{n+1}}{u_n}\right| = \lim\limits_{n\to\infty}\left|\dfrac{(3n+3)!\,x^{n+1}}{(2n+2)!}\cdot\dfrac{(2n)!}{(3n)!\,x^n}\right| =$

$\lim\limits_{n\to\infty}\dfrac{(3n+3)(3n+2)(3n+1)}{(2n+2)(2n+1)}|x| = \infty$. Converges only for $x = 0$.

[15] $u_n = \dfrac{x^{2n+1}}{(-4)^n} \Rightarrow \lim\limits_{n\to\infty}\left|\dfrac{u_{n+1}}{u_n}\right| = \lim\limits_{n\to\infty}\left|\dfrac{x^{2n+3}}{(-4)^{n+1}}\cdot\dfrac{(-4)^n}{x^{2n+1}}\right| = \frac{1}{4}x^2$.

$\frac{1}{4}x^2 < 1 \Leftrightarrow -2 < x < 2$. If $x = \pm 2$, $|u_n| = 2\ \forall n$ and both series are D. ★ $(-2, 2)$

[16] $u_n = (-1)^{n-1}\dfrac{x^n}{\sqrt[3]{n}\,3^n} \Rightarrow \lim\limits_{n\to\infty}\left|\dfrac{u_{n+1}}{u_n}\right| = \lim\limits_{n\to\infty}\left|\dfrac{x^{n+1}}{\sqrt[3]{n+1}\,3^{n+1}}\cdot\dfrac{\sqrt[3]{n}\,3^n}{x^n}\right|$

$= \frac{1}{3}|x|$. $\frac{1}{3}|x| < 1 \Leftrightarrow -3 < x < 3$. If $x = 3$, $\sum\limits_{n=1}^{\infty}(-1)^{n-1}\dfrac{1}{\sqrt[3]{n}}$ is C by AST.

If $x = -3$, $\sum\limits_{n=1}^{\infty}\dfrac{-1}{\sqrt[3]{n}}$ is D. ★ $(-3, 3]$

[17] $u_n = \dfrac{2^n x^{2n}}{(2n)!} \Rightarrow \lim\limits_{n\to\infty}\left|\dfrac{u_{n+1}}{u_n}\right| = \lim\limits_{n\to\infty}\left|\dfrac{2^{n+1}x^{2n+2}}{(2n+2)!}\cdot\dfrac{(2n)!}{2^n x^{2n}}\right| =$

$\lim\limits_{n\to\infty}\dfrac{2x^2}{(2n+2)(2n+1)} = 0$. Converges $\forall x$. ★ $(-\infty, \infty)$

[18] $u_n = \dfrac{10^n x^n}{n!} \Rightarrow \lim\limits_{n\to\infty}\left|\dfrac{u_{n+1}}{u_n}\right| = \lim\limits_{n\to\infty}\left|\dfrac{10^{n+1}x^{n+1}}{(n+1)!}\cdot\dfrac{n!}{10^n x^n}\right| = \lim\limits_{n\to\infty}\dfrac{10}{n+1}|x| = 0$.

Converges $\forall x$. ★ $(-\infty, \infty)$

[19] $u_n = \dfrac{3^{2n}(x-2)^n}{n+1} = \dfrac{9^n(x-2)^n}{n+1} \Rightarrow \lim\limits_{n\to\infty}\left|\dfrac{u_{n+1}}{u_n}\right| =$

$\lim\limits_{n\to\infty}\left|\dfrac{9^{n+1}(x-2)^{n+1}}{n+2}\cdot\dfrac{n+1}{9^n(x-2)^n}\right| = 9|x-2|$. $9|x-2| < 1 \Leftrightarrow \frac{17}{9} < x < \frac{19}{9}$.

If $x = \frac{19}{9}$, $\sum\limits_{n=0}^{\infty}\dfrac{1}{n+1}$ is D. If $x = \frac{17}{9}$, $\sum\limits_{n=0}^{\infty}(-1)^n\dfrac{1}{n+1}$ is C by AST. ★ $[\frac{17}{9}, \frac{19}{9})$

[20] $u_n = \dfrac{(x-5)^n}{n5^n} \Rightarrow \lim\limits_{n\to\infty}\left|\dfrac{u_{n+1}}{u_n}\right| = \lim\limits_{n\to\infty}\left|\dfrac{(x-5)^{n+1}}{(n+1)5^{n+1}}\cdot\dfrac{n5^n}{(x-5)^n}\right| = \frac{1}{5}|x-5|$.

$\frac{1}{5}|x-5| < 1 \Leftrightarrow 0 < x < 10$. If $x = 10$, $\sum\limits_{n=1}^{\infty}\dfrac{1}{n}$ is D.

If $x = 0$, $\sum\limits_{n=1}^{\infty}(-1)^n\dfrac{1}{n}$ is C by AST. ★ $[0, 10)$

[21] $u_n = \dfrac{n^2(x+4)^n}{2^{3n}} = \dfrac{n^2(x+4)^n}{8^n} \Rightarrow \lim\limits_{n\to\infty}\left|\dfrac{u_{n+1}}{u_n}\right| =$

$\lim\limits_{n\to\infty}\left|\dfrac{(n+1)^2(x+4)^{n+1}}{8^{n+1}}\cdot\dfrac{8^n}{n^2(x+4)^n}\right| = \frac{1}{8}|x+4|$. $\frac{1}{8}|x+4| < 1 \Leftrightarrow -12 < x < 4$.

If $x = 4$, $\sum\limits_{n=0}^{\infty}n^2$ is D. If $x = -12$, $\sum\limits_{n=0}^{\infty}(-1)\,n^2$ is D. ★ $(-12, 4)$

22 $u_n = \dfrac{(x+3)^n}{2n+1} \Rightarrow \lim\limits_{n\to\infty}\left|\dfrac{u_{n+1}}{u_n}\right| = \lim\limits_{n\to\infty}\left|\dfrac{(x+3)^{n+1}}{2n+3}\cdot\dfrac{2n+1}{(x+3)^n}\right| = |x+3|.$

$|x+3| < 1 \Leftrightarrow -4 < x < -2.$ If $x = -2$, $\sum\limits_{n=0}^{\infty}\dfrac{1}{2n+1}$ is D.

If $x = -4$, $\sum\limits_{n=0}^{\infty}(-1)^n\dfrac{1}{2n+1}$ is C by AST. ★ $[-4, -2)$

23 $u_n = (-1)^n\dfrac{n^n(x-3)^n}{n+1} \Rightarrow \lim\limits_{n\to\infty}\left|\dfrac{u_{n+1}}{u_n}\right| = \lim\limits_{n\to\infty}\left|\dfrac{(n+1)^{n+1}(x-3)^{n+1}}{n+2}\cdot\dfrac{n+1}{n^n(x-3)^n}\right|$

$= \lim\limits_{n\to\infty}\left(\dfrac{n+1}{n}\right)^n(n+1)|x-3| = \lim\limits_{n\to\infty}e(n+1)|x-3| = \infty.$

Converges only for $x = 3$.

24 $u_n = (-1)^n\dfrac{n!\,(x+2)^n}{n^3} \Rightarrow \lim\limits_{n\to\infty}\left|\dfrac{u_{n+1}}{u_n}\right| = \lim\limits_{n\to\infty}\left|\dfrac{(n+1)!\,(x+2)^{n+1}}{(n+1)^3}\cdot\dfrac{n^3}{n!\,(x+2)^n}\right| =$

$\lim\limits_{n\to\infty}(n+1)|x+2| = \infty.$ Converges only for $x = -2$.

25 $u_n = \dfrac{\ln n\,(x-e)^n}{e^n} \Rightarrow \lim\limits_{n\to\infty}\left|\dfrac{u_{n+1}}{u_n}\right| = \lim\limits_{n\to\infty}\left|\dfrac{[\ln(n+1)](x-e)^{n+1}}{e^{n+1}}\cdot\dfrac{e^n}{(\ln n)(x-e)^n}\right|$

$= \frac{1}{e}|x-e|.$ $\frac{1}{e}|x-e| < 1 \Leftrightarrow 0 < x < 2e.$ If $x = 2e$, $\sum\limits_{n=1}^{\infty}\ln n$ is D {nth-term}.

If $x = 0$, $\sum\limits_{n=1}^{\infty}(-1)^n\ln n$ is D. ★ $(0, 2e)$

26 $u_n = \dfrac{n(x-1)^{2n}}{3^{2n-1}} \Rightarrow \lim\limits_{n\to\infty}\left|\dfrac{u_{n+1}}{u_n}\right| = \lim\limits_{n\to\infty}\left|\dfrac{(n+1)(x-1)^{2n+2}}{3^{2n+1}}\cdot\dfrac{3^{2n-1}}{n(x-1)^{2n}}\right| =$

$\frac{1}{9}(x-1)^2.$ $\frac{1}{9}(x-1)^2 < 1 \Leftrightarrow -2 < x < 4.$ If $x = -2$ or 4, $\sum\limits_{n=0}^{\infty}3n$ is D. ★ $(-2, 4)$

27 $u_n = (-1)^n\dfrac{(2x-1)^n}{n6^n} \Rightarrow \lim\limits_{n\to\infty}\left|\dfrac{u_{n+1}}{u_n}\right| = \lim\limits_{n\to\infty}\left|\dfrac{(2x-1)^{n+1}}{(n+1)6^{n+1}}\cdot\dfrac{n6^n}{(2x-1)^n}\right| =$

$\frac{1}{6}|2x-1|.$ $\frac{1}{6}|2x-1| < 1 \Leftrightarrow -\frac{5}{2} < x < \frac{7}{2}.$ If $x = \frac{7}{2}$, $\sum\limits_{n=1}^{\infty}(-1)^n\frac{1}{n}$ is C by AST.

If $x = -\frac{5}{2}$, $\sum\limits_{n=1}^{\infty}\frac{1}{n}$ is D. ★ $(-\frac{5}{2}, \frac{7}{2}]$

28 $u_n = \dfrac{(3x+4)^n}{\sqrt{3n+4}} \Rightarrow \lim\limits_{n\to\infty}\left|\dfrac{u_{n+1}}{u_n}\right| = \lim\limits_{n\to\infty}\left|\dfrac{(3x+4)^{n+1}}{\sqrt{3n+7}}\cdot\dfrac{\sqrt{3n+4}}{(3x+4)^n}\right| = |3x+4|.$

$|3x+4| < 1 \Leftrightarrow -\frac{5}{3} < x < -1.$ If $x = -1$, $\sum\limits_{n=0}^{\infty}\dfrac{1}{\sqrt{3n+4}}$ is D {LCT, $b_n = \frac{1}{\sqrt{n}}$}.

If $x = -\frac{5}{3}$, $\sum\limits_{n=0}^{\infty}(-1)^n\dfrac{1}{\sqrt{3n+4}}$ is C by AST. ★ $[-\frac{5}{3}, -1)$

29 $u_n = (-1)^n\dfrac{3^n(x-4)^n}{n!} \Rightarrow \lim\limits_{n\to\infty}\left|\dfrac{u_{n+1}}{u_n}\right| = \lim\limits_{n\to\infty}\left|\dfrac{3^{n+1}(x-4)^{n+1}}{(n+1)!}\cdot\dfrac{n!}{3^n(x-4)^n}\right| =$

$\lim\limits_{n\to\infty}\dfrac{3}{n+1}|x-4| = 0.$ Converges $\forall x$. ★ $(-\infty, \infty)$

[30] $u_n = (-1)^n \dfrac{e^{n+1}(x-1)^n}{n^n} \Rightarrow \lim\limits_{n\to\infty}\left|\dfrac{u_{n+1}}{u_n}\right| = \lim\limits_{n\to\infty}\left|\dfrac{e^{n+2}(x-1)^{n+1}}{(n+1)^{n+1}}\cdot\dfrac{n^n}{e^{n+1}(x-1)^n}\right|$

$= \lim\limits_{n\to\infty}\dfrac{e|x-1|}{e(n+1)} = 0$. Converges $\forall x$. ★ $(-\infty, \infty)$

[31] $\lim\limits_{n\to\infty}\left|\dfrac{u_{n+1}}{u_n}\right| =$

$\lim\limits_{n\to\infty}\left|\dfrac{1\cdot3\cdot5\cdot\cdots\cdot(2n-1)(2n+1)\,x^{n+1}}{3\cdot6\cdot9\cdot\cdots\cdot(3n)(3n+3)}\cdot\dfrac{3\cdot6\cdot9\cdot\cdots\cdot(3n)}{1\cdot3\cdot5\cdot\cdots\cdot(2n-1)\,x^n}\right| =$

$\lim\limits_{n\to\infty}\left(\dfrac{2n+1}{3n+3}\right)|x| = \frac{2}{3}|x|$. $\frac{2}{3}|x| < 1 \Leftrightarrow |x| < \frac{3}{2}$. ★ $r = \frac{3}{2}$.

Note: The convergence/divergence at the end points does not affect the radius of convergence.

[32] $\lim\limits_{n\to\infty}\left|\dfrac{u_{n+1}}{u_n}\right| =$

$\lim\limits_{n\to\infty}\left|\dfrac{2\cdot4\cdot6\cdot\cdots\cdot(2n)(2n+2)\,x^{n+1}}{4\cdot7\cdot10\cdot\cdots\cdot(3n+1)(3n+4)}\cdot\dfrac{4\cdot7\cdot10\cdot\cdots\cdot(3n+1)}{2\cdot4\cdot6\cdot\cdots\cdot(2n)\,x^n}\right| =$

$\lim\limits_{n\to\infty}\left(\dfrac{2n+2}{3n+4}\right)|x| = \frac{2}{3}|x|$. $\frac{2}{3}|x| < 1 \Leftrightarrow |x| < \frac{3}{2}$. ★ $r = \frac{3}{2}$.

[33] $\lim\limits_{n\to\infty}\left|\dfrac{u_{n+1}}{u_n}\right| = \lim\limits_{n\to\infty}\left|\dfrac{(n+1)^{n+1}x^{n+1}}{(n+1)!}\cdot\dfrac{n!}{n^n x^n}\right| =$

$\lim\limits_{n\to\infty}\left|\dfrac{(n+1)(n+1)^n x}{(n+1)n^n}\right| = \lim\limits_{n\to\infty}\left(\dfrac{n+1}{n}\right)^n|x| = \lim\limits_{n\to\infty}\left(1+\frac{1}{n}\right)^n|x| = e|x|$.

$e|x| < 1 \Leftrightarrow |x| < \frac{1}{e}$. ★ $r = \frac{1}{e}$.

[34] $\lim\limits_{n\to\infty}\left|\dfrac{u_{n+1}}{u_n}\right| = \lim\limits_{n\to\infty}\left|\dfrac{(n+2)!\,(x-5)^{n+1}}{10^{n+1}}\cdot\dfrac{10^n}{(n+1)!\,(x-5)^n}\right| =$

$\lim\limits_{n\to\infty}\dfrac{n+2}{10}|x-5| = \infty$. ★ $r = 0$.

[35] $\lim\limits_{n\to\infty}\left|\dfrac{u_{n+1}}{u_n}\right| = \lim\limits_{n\to\infty}\left|\dfrac{(n+c+1)!\,x^{n+1}}{(n+1)!\,(n+d+1)!}\cdot\dfrac{n!\,(n+d)!}{(n+c)!\,x^n}\right| =$

$\lim\limits_{n\to\infty}\dfrac{n+c+1}{(n+1)(n+d+1)}|x| = 0$. ★ $r = \infty$.

[36] $\lim\limits_{n\to\infty}\left|\dfrac{u_{n+1}}{u_n}\right| = \lim\limits_{n\to\infty}\left|\dfrac{(cn+c)!\,x^{n+1}}{[(n+1)!]^c}\cdot\dfrac{(n!)^c}{(cn)!\,x^n}\right| = \lim\limits_{n\to\infty}\dfrac{(cn+c)(cn+2)\cdots(cn+1)}{(n+1)^c}|x|$

$= \lim\limits_{n\to\infty}\dfrac{c^c n^c + \text{terms of lower degree}}{n^c + \text{terms of lower degree}}|x| = c^c|x|$. $c^c|x| < 1 \Leftrightarrow |x| < \frac{1}{c^c}$. ★ $r = \dfrac{1}{c^c}$

[37] $\lim\limits_{n\to\infty}\left|\dfrac{u_{n+1}}{u_n}\right| = \lim\limits_{n\to\infty}\left|\dfrac{(\frac{1}{2}x)^{2n+\alpha+2}}{(n+1)!\,(n+\alpha+1)!}\cdot\dfrac{n!\,(n+\alpha)!}{(\frac{1}{2}x)^{2n+\alpha}}\right| =$

$\lim\limits_{n\to\infty}\dfrac{(\frac{1}{2}x)^2}{(n+1)(n+\alpha+1)} = 0$. Thus, $r = \infty$.

[38] From (11.31), the error involved in approximating the sum using the first four terms in an alternating series is less than or equal to the fifth term. $\alpha = 0$, $0 \le x \le 1$, and $n = 4 \Rightarrow a_5 = \dfrac{(-1)^4 x^8}{4!\,4!\,2^8} \le \dfrac{1^8}{147{,}456} \approx 0.0000068 < 0.0000100$.

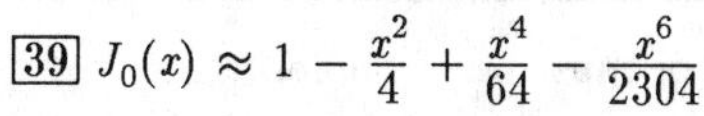

[39] $J_0(x) \approx 1 - \frac{x^2}{4} + \frac{x^4}{64} - \frac{x^6}{2304}$ [40] $J_1(x) \approx \frac{x}{2} - \frac{x^3}{16} + \frac{x^5}{384} - \frac{x^7}{18{,}432}$

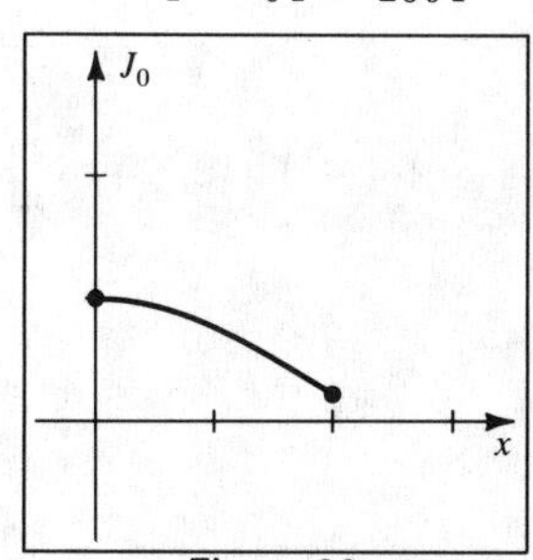

Figure 39

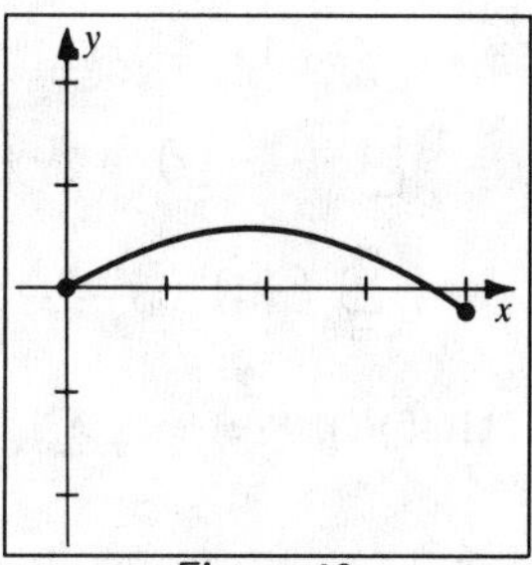

Figure 40

[41] $\lim_{n\to\infty} \left|\frac{u_{n+1}}{u_n}\right| = \lim_{n\to\infty} \left|\frac{a_{n+1}\, x^{n+1}}{a_n\, x^n}\right| = k|x|.$ $k|x| < 1 \Leftrightarrow |x| < \frac{1}{k}.$ ★ $r = \frac{1}{k}$

[42] Applying the root test, $\lim_{n\to\infty} \sqrt[n]{|a_n\, x^n|} = \lim_{n\to\infty} \left[\sqrt[n]{|a_n|}\, \sqrt[n]{|x^n|}\right] = k|x|.$

$k|x| < 1 \Leftrightarrow |x| < \frac{1}{k}.$ ★ $r = \frac{1}{k}$

[43] Since $\sum a_n\, x^n$ is convergent for $|x| < r$ and is divergent for $|x| > r$, it follows that $\sum a_n\, x^{2n} = \sum a_n\, (x^2)^n$ is convergent for $x^2 < r$ or $|x| < \sqrt{r}$ and is divergent for $x^2 > r$ or $|x| > \sqrt{r}$. Thus, the radius of convergence is $\sqrt{r}$.

[44] Since $|x| \leq 1$ and $\sum a_n$ is absolutely convergent,

$\sum |a_n x^n| \leq \sum |a_n|$ and thus $\sum a_n x^n$ is also absolutely convergent.

[45] Assume that $\sum a_n x^n$ is absolutely convergent at $x = r$. Let $x = -r$.

Then $\sum |a_n(-r)^n| = \sum |a_n r^n|$ is absolutely convergent,

which implies that $\sum a_n(-r)^n$ is convergent by (11.34). This is a contradiction.

[46] Since $\sum |a_n(-x)^n| = \sum |a_n x^n|$, the series must be absolutely convergent at $x = \pm r$.

Exercises 11.7

[1] (a) By (11.15)(i) with $a = 1$ and $r = 3x$,

$$f(x) = \frac{1}{1 - 3x} = 1 + (3x) + (3x)^2 + (3x)^3 + \cdots + (3x)^n + \cdots = \sum_{n=0}^{\infty} 3^n x^n.$$

(b) By (11.40)(i), $f'(x) = \sum_{n=1}^{\infty} n3^n x^{n-1}$. By (11.40)(ii), $\int_0^x f(t)\, dt = \sum_{n=0}^{\infty} \frac{3^n}{n+1} x^{n+1}$.

[2] (a) By (11.15)(i) with $a = 1$ and $r = -5x$, $f(x) = \frac{1}{1 + 5x} =$

$$1 + (-5x) + (-5x)^2 + (-5x)^3 + \cdots + (-1)^n(5x)^n + \cdots = \sum_{n=0}^{\infty} (-1)^n 5^n x^n.$$

(b) By (11.40)(i), $f'(x) = \sum_{n=1}^{\infty} (-1)^n n5^n x^{n-1}$.

By (11.40)(ii), $\int_0^x f(t)\, dt = \sum_{n=0}^{\infty} (-1)^n \frac{5^n}{n+1} x^{n+1}$.

3 (a) $f(x) = \dfrac{1}{2+7x} = \dfrac{1}{2}\cdot\dfrac{1}{1-(-\frac{7}{2}x)}$ {the denominator must be of the form $1-r$}.

With $a = \frac{1}{2}$ and $r = -\frac{7}{2}x$,

$$f(x) = \tfrac{1}{2}\left[1 + (-\tfrac{7}{2}x) + (-\tfrac{7}{2}x)^2 + (-\tfrac{7}{2}x)^3 + \cdots + (-1)^n(\tfrac{7}{2}x)^n + \cdots\right]$$

$$= \tfrac{1}{2}\sum_{n=0}^{\infty}(-1)^n(\tfrac{7}{2})^n x^n.$$

(b) By (11.40)(i), $f'(x) = \frac{1}{2}\sum_{n=1}^{\infty}(-1)^n\dfrac{n7^n}{2^n}x^{n-1}$.

By (11.40)(ii), $\displaystyle\int_0^x f(t)\,dt = \tfrac{1}{2}\sum_{n=0}^{\infty}(-1)^n\frac{7^n}{(n+1)2^n}x^{n+1}$.

4 (a) $f(x) = \dfrac{1}{3-2x} = \dfrac{1}{3}\cdot\dfrac{1}{1-\frac{2}{3}x} =$

$$\tfrac{1}{3}\left[1 + (\tfrac{2}{3}x) + (\tfrac{2}{3}x)^2 + (\tfrac{2}{3}x)^3 + \cdots + (\tfrac{2}{3}x)^n + \cdots\right] = \tfrac{1}{3}\sum_{n=0}^{\infty}(\tfrac{2}{3})^n x^n.$$

(b) By (11.40)(i), $f'(x) = \frac{1}{3}\sum_{n=1}^{\infty}\dfrac{n2^n}{3^n}x^{n-1}$.

By (11.40)(ii), $\displaystyle\int_0^x f(t)\,dt = \tfrac{1}{3}\sum_{n=0}^{\infty}\frac{2^n}{(n+1)3^n}x^{n+1}$.

5 $\dfrac{x^2}{1-x^2} = x^2\cdot\dfrac{1}{1-x^2} = x^2\left[1 + (x^2) + (x^2)^2 + (x^2)^3 + \cdots\right] =$

$$x^2\sum_{n=0}^{\infty}(x^2)^n = \sum_{n=0}^{\infty}x^{2n+2}.\quad |x^2| < 1 \Rightarrow |x| < 1 \Rightarrow r = 1.$$

6 $\dfrac{x}{1-x^4} = x\cdot\dfrac{1}{1-x^4} = x\left[1 + (x^4) + (x^4)^2 + (x^4)^3 + \cdots\right] =$

$$x\sum_{n=0}^{\infty}(x^4)^n = \sum_{n=0}^{\infty}x^{4n+1}.\quad |x^4| < 1 \Rightarrow |x| < 1 \Rightarrow r = 1.$$

7 $\dfrac{x}{2-3x} = x\cdot\dfrac{\frac{1}{2}}{1-\frac{3}{2}x} = \frac{1}{2}x\left[1 + (\frac{3}{2}x) + (\frac{3}{2}x)^2 + (\frac{3}{2}x)^3 + \cdots\right] =$

$$\tfrac{1}{2}x\sum_{n=0}^{\infty}(\tfrac{3}{2}x)^n = \sum_{n=0}^{\infty}\frac{3^n}{2^{n+1}}x^{n+1}.\quad \left|\tfrac{3}{2}x\right| < 1 \Rightarrow |x| < \tfrac{2}{3} \Rightarrow r = \tfrac{2}{3}.$$

8 $\dfrac{x^3}{4-x^3} = x^3\cdot\dfrac{\frac{1}{4}}{1-\frac{1}{4}x^3} = \frac{1}{4}x^3\left[1 + (\frac{1}{4}x^3) + (\frac{1}{4}x^3)^2 + (\frac{1}{4}x^3)^3 + \cdots\right] =$

$$\tfrac{1}{4}x^3\sum_{n=0}^{\infty}(\tfrac{1}{4}x^3)^n = \sum_{n=0}^{\infty}\frac{1}{4^{n+1}}x^{3n+3}.\quad \left|\tfrac{1}{4}x^3\right| < 1 \Rightarrow |x| < \sqrt[3]{4} \Rightarrow r = \sqrt[3]{4}.$$

9 $\dfrac{x^2+1}{x-1} = x + 1 + \dfrac{2}{x-1} = 1 + x - \dfrac{2}{1-x} = 1 + x - 2(1 + x + x^2 + x^3 + \cdots)$

$$= 1 + x - 2 - 2x - 2(x^2 + x^3 + x^4 + x^5 + \cdots) = -1 - x - 2\sum_{n=2}^{\infty}x^n.$$

$|x| < 1 \Rightarrow r = 1.$

[10] $\dfrac{x^2-3}{x-2} = x + 2 - \dfrac{1}{2-x} = 2 + x - \dfrac{\frac{1}{2}}{1-\frac{1}{2}x} =$

$2 + x - \frac{1}{2}\left[1 + (\frac{1}{2}x) + (\frac{1}{2}x)^2 + (\frac{1}{2}x)^3 + \cdots\right] = 2 + x - \frac{1}{2} - \frac{1}{4}x - \frac{1}{2}\sum_{n=2}^{\infty}(\frac{1}{2}x)^n =$

$\frac{3}{2} + \frac{3}{4}x - \frac{1}{2}\sum_{n=2}^{\infty}(\frac{1}{2}x)^n$. $\left|\frac{1}{2}x\right| < 1 \Rightarrow |x| < 2 \Rightarrow r = 2.$

[11] (a) $f(x) = \ln(1-x) \Rightarrow f'(x) = -\dfrac{1}{1-x} = -(1 + x + x^2 + \cdots + x^n + \cdots) =$

$-\sum_{n=0}^{\infty} x^n \Rightarrow f(x) = -\sum_{n=0}^{\infty}\dfrac{x^{n+1}}{n+1} = -\sum_{n=1}^{\infty}\dfrac{x^n}{n}.$

(b) Using (11.31), $\ln(1.2) = \ln\left[1-(-0.2)\right] \approx -(-0.2) - \dfrac{(-0.2)^2}{2} - \dfrac{(-0.2)^3}{3} \approx$

0.183, with |error| $< \dfrac{(-0.2)^4}{4} = 0.4 \times 10^{-3}$. Calculator value ≈ 0.182321557.

[12] $\ln(0.9) = \ln(1-0.1) = -(0.1) - \dfrac{(0.1)^2}{2} - \dfrac{(0.1)^3}{3} \approx -0.1053.$

Calculator value ≈ -0.105360516.

[13] $\arctan x = \sum_{n=0}^{\infty}(-1)^n\dfrac{x^{2n+1}}{2n+1} \Rightarrow \arctan\dfrac{1}{\sqrt{3}} = \sum_{n=0}^{\infty}(-1)^n\dfrac{\left(\frac{1}{\sqrt{3}}\right)^{2n+1}}{2n+1} \Rightarrow$

$\dfrac{\pi}{6} = \sum_{n=0}^{\infty}(-1)^n\dfrac{1}{(\sqrt{3})^{2n}(\sqrt{3})(2n+1)} \Rightarrow \dfrac{\pi}{6} = \dfrac{1}{\sqrt{3}}\sum_{n=0}^{\infty}(-1)^n\dfrac{1}{3^n(2n+1)}.$

[14] (a) $\frac{\pi}{4} = \arctan 1 = 1 - \frac{1}{3} + \frac{1}{5} - \frac{1}{7} + \frac{1}{9} = \frac{263}{315} \approx 0.8349.$

(b) $|\text{Error}| < a_{n+1} = a_6 = \frac{1}{11} \approx 0.0909.$

[15] Using (11.41),

$f(x) = xe^{3x} = x\left[1 + (3x) + \dfrac{(3x)^2}{2!} + \dfrac{(3x)^3}{3!} + \cdots\right] = x\sum_{n=0}^{\infty}\dfrac{(3x)^n}{n!} = \sum_{n=0}^{\infty}\dfrac{3^n}{n!}x^{n+1}.$

[16] Using (11.41), $f(x) = x^2e^{(x^2)} =$

$x^2\left[1 + (x^2) + \dfrac{(x^2)^2}{2!} + \dfrac{(x^2)^3}{3!} + \cdots\right] = x^2\sum_{n=0}^{\infty}\dfrac{(x^2)^n}{n!} = \sum_{n=0}^{\infty}\dfrac{1}{n!}x^{2n+2}.$

[17] Using (11.41), $f(x) = x^3e^{-x} =$

$x^3\left[1 + (-x) + \dfrac{(-x)^2}{2!} + \dfrac{(-x)^3}{3!} + \cdots\right] = x^3\sum_{n=0}^{\infty}\dfrac{(-x)^n}{n!} = \sum_{n=0}^{\infty}(-1)^n\dfrac{1}{n!}x^{n+3}.$

[18] Using (11.41), $f(x) = xe^{-3x} =$

$x\left[1 + (-3x) + \dfrac{(-3x)^2}{2!} + \dfrac{(-3x)^3}{3!} + \cdots\right] = x\sum_{n=0}^{\infty}\dfrac{(-3x)^n}{n!} = \sum_{n=0}^{\infty}(-1)^n\dfrac{3^n}{n!}x^{n+1}.$

[19] Using Example 3, $f(x) = x^2\ln(1+x^2) = x^2\left[(x^2) - \dfrac{(x^2)^2}{2} + \dfrac{(x^2)^3}{3} - \dfrac{(x^2)^4}{4} + \cdots\right]$

$= x^2\sum_{n=0}^{\infty}(-1)^n\dfrac{(x^2)^{n+1}}{n+1} = \sum_{n=0}^{\infty}(-1)^n\dfrac{1}{n+1}x^{2n+4}.$

[20] Using Example 3, $f(x) = x\ln(1-x) = x\left[(-x) - \frac{(-x)^2}{2} + \frac{(-x)^3}{3} - \frac{(-x)^4}{4} + \cdots\right]$

$$= x\sum_{n=0}^{\infty}(-1)^n\frac{(-x)^{n+1}}{n+1} = \sum_{n=0}^{\infty}(-1)^{2n+1}\frac{x^{n+2}}{n+1} = -\sum_{n=0}^{\infty}\frac{1}{n+1}x^{n+2}.$$

Alternatively, from Exercise 11, $x\ln(1-x) = x\left(-\sum_{n=1}^{\infty}\frac{x^n}{n}\right) = -\sum_{n=0}^{\infty}\frac{1}{n+1}x^{n+2}$.

[21] Using Example 5, $f(x) = \arctan\sqrt{x} = \left[(x^{1/2}) - \frac{(x^{1/2})^3}{3} + \frac{(x^{1/2})^5}{5} - \frac{(x^{1/2})^7}{7} + \cdots\right]$

$$= \sum_{n=0}^{\infty}(-1)^n\frac{(x^{1/2})^{2n+1}}{2n+1} = \sum_{n=0}^{\infty}(-1)^n\frac{1}{2n+1}x^{(2n+1)/2}.$$

[22] Using Example 5, $f(x) = x^4\arctan(x^4) = x^4\left[(x^4) - \frac{(x^4)^3}{3} + \frac{(x^4)^5}{5} - \frac{(x^4)^7}{7} + \cdots\right]$

$$= x^4\sum_{n=0}^{\infty}(-1)^n\frac{(x^4)^{2n+1}}{2n+1} = \sum_{n=0}^{\infty}(-1)^n\frac{1}{2n+1}x^{8n+8}.$$

[23] Using the series for $\sinh x$ (just before Example 6),

$$f(x) = \sinh(-5x) = \left[(-5x) + \frac{(-5x)^3}{3!} + \frac{(-5x)^5}{5!} + \frac{(-5x)^7}{7!} + \cdots\right] =$$

$$\sum_{n=0}^{\infty}\frac{(-5x)^{2n+1}}{(2n+1)!} = \sum_{n=0}^{\infty}\frac{-5^{2n+1}}{(2n+1)!}x^{2n+1}.$$

[24] Using the series for $\sinh x$ (just before Example 6),

$$f(x) = \sinh(x^2) = \left[(x^2) + \frac{(x^2)^3}{3!} + \frac{(x^2)^5}{5!} + \frac{(x^2)^7}{7!} + \cdots\right] =$$

$$\sum_{n=0}^{\infty}\frac{(x^2)^{2n+1}}{(2n+1)!} = \sum_{n=0}^{\infty}\frac{1}{(2n+1)!}x^{4n+2}.$$

[25] Using the series for $\cosh x$ (just before Example 6),

$$f(x) = x^2\cosh(x^3) = x^2\left[1 + \frac{(x^3)^2}{2!} + \frac{(x^3)^4}{4!} + \frac{(x^3)^6}{6!} + \cdots\right] =$$

$$x^2\sum_{n=0}^{\infty}\frac{(x^3)^{2n}}{(2n)!} = \sum_{n=0}^{\infty}\frac{1}{(2n)!}x^{6n+2}.$$

[26] Using the series for $\cosh x$ (just before Example 6), $f(x) = \cosh(-2x) =$

$$\left[1 + \frac{(-2x)^2}{2!} + \frac{(-2x)^4}{4!} + \frac{(-2x)^6}{6!} + \cdots\right] = \sum_{n=0}^{\infty}\frac{(-2x)^{2n}}{(2n)!} = \sum_{n=0}^{\infty}\frac{2^{2n}}{(2n)!}x^{2n}.$$

[27] $\displaystyle\int_0^{1/3}\frac{1}{1+x^6}\,dx = \int_0^{1/3}(1 - x^6 + x^{12} - x^{18} + \cdots)\,dx$

$$= \left[x - \tfrac{1}{7}x^7 + \tfrac{1}{13}x^{13} - \tfrac{1}{19}x^{19} + \cdots\right]_0^{1/3}.$$ Using the first two terms,

$$\mathrm{I} \approx \tfrac{1}{3} - \tfrac{1}{7}(\tfrac{1}{3})^7 \approx 0.3333, \text{ with } |\text{error}| < \tfrac{1}{13}(\tfrac{1}{3})^{13} < 0.5 \times 10^{-7}.$$

[28] $\displaystyle\int_0^{1/2}\arctan x^2\,dx = \int_0^{1/2}(x^2 - \tfrac{1}{3}x^6 + \tfrac{1}{5}x^{10} - \tfrac{1}{7}x^{14} + \cdots)\,dx =$

$$\left[\tfrac{1}{3}x^3 - \tfrac{1}{21}x^7 + \tfrac{1}{55}x^{11} - \tfrac{1}{105}x^{15} + \cdots\right]_0^{1/2}.$$ Using the first two terms, (cont.)

$\text{I} \approx \frac{1}{3}(\frac{1}{2})^3 - \frac{1}{21}(\frac{1}{2})^7 \approx 0.0413$, with $|\text{error}| < \frac{1}{55}(\frac{1}{2})^{11} < 0.9 \times 10^{-5}$.

[29] Dividing the series for $\arctan x$ by x, we have

$$\int_{0.1}^{0.2} (1 - \tfrac{1}{3}x^2 + \tfrac{1}{5}x^4 - \tfrac{1}{7}x^6 + \cdots)\,dx = \left[x - \tfrac{1}{9}x^3 + \tfrac{1}{25}x^5 - \tfrac{1}{49}x^7 + \cdots\right]_{0.1}^{0.2}.$$

Using the first two terms, $\text{I} \approx \left[x - \frac{1}{9}x^3\right]_{0.1}^{0.2} \approx 0.0992$.

Since the series satisfies (11.30), (11.31) applies.

Our estimate of $\arctan 0.2$ is $\left[0.2 - \frac{1}{9}(0.2)^3\right]$ and has an error of at most $\frac{1}{25}(0.2)^5$.

Our estimate of $\arctan 0.1$ is $\left[0.1 - \frac{1}{9}(0.1)^3\right]$ and has an error of at most $\frac{1}{25}(0.1)^5$.

If we *subtract* these two values, the error will not exceed the *sum* of the two errors.

Thus, $|\text{error}| < \frac{1}{25}(0.2)^5 + \frac{1}{25}(0.1)^5 = 0.0000132 < 0.5 \times 10^{-4}$.

[30] Since $\dfrac{x^3}{1 + x^5} = x^3(1 - x^5 + x^{10} - x^{15} + \cdots)$,

$$\text{I} = \int_0^{0.2} (x^3 - x^8 + x^{13} - x^{18} + \cdots)\,dx = \left[\tfrac{1}{4}x^4 - \tfrac{1}{9}x^9 + \tfrac{1}{14}x^{14} - \tfrac{1}{19}x^{19} + \cdots\right]_0^{0.2}.$$

Using only the first term, $\text{I} \approx \frac{1}{4}(0.2)^4 = 0.0004$ with $|\text{error}| < \frac{1}{9}(0.2)^9 < 0.6 \times 10^{-7}$.

[31] $\displaystyle\int_0^1 e^{-x^2/10}\,dx = \int_0^1 \left(1 - \frac{x^2}{10} + \frac{x^4}{2! \cdot 100} - \frac{x^6}{3! \cdot 1000} + \cdots\right) dx =$

$\left[x - \frac{1}{30}x^3 + \frac{1}{1000}x^5 - \frac{1}{42{,}000}x^7 + \cdots\right]_0^1$. Using the first three terms,

$\text{I} \approx 1 - \frac{1}{30} + \frac{1}{1000} \approx 0.9677$, with $|\text{error}| < \frac{1}{42{,}000} < 0.3 \times 10^{-4}$.

[32] $\displaystyle\int_0^{0.5} e^{-x^3}\,dx = \int_0^{0.5} \left(1 - x^3 + \frac{x^6}{2!} - \frac{x^9}{3!} + \cdots\right) dx =$

$\left[x - \frac{1}{4}x^4 + \frac{1}{14}x^7 - \frac{1}{60}x^{10} + \cdots\right]_0^{0.5}$. Using the first three terms,

$\text{I} \approx 0.5 - \frac{1}{4}(0.5)^4 + \frac{1}{14}(0.5)^7 \approx 0.4849$, with $|\text{error}| < \dfrac{(0.5)^{10}}{60} < 0.2 \times 10^{-4}$.

[33] $\dfrac{1}{1 - x^2} = \displaystyle\sum_{n=0}^{\infty} x^{2n}$. Since $\dfrac{d}{dx}\left(\dfrac{1}{1 - x^2}\right) = \dfrac{2x}{(1 - x^2)^2}$,

we have $\dfrac{2x}{(1 - x^2)^2} = \displaystyle\sum_{n=1}^{\infty} (2n)x^{2n-1}$ by (11.40)(i).

[34] $\dfrac{d}{dx}\left[\ln(3 + 2x)\right] = \dfrac{2}{3 + 2x} = \dfrac{\frac{2}{3}}{1 - (-\frac{2}{3}x)} = \frac{2}{3}\left[1 + (-\frac{2}{3}x) + (-\frac{2}{3}x)^2 + \cdots\right] =$

$\displaystyle\sum_{n=0}^{\infty} (-1)^n (\tfrac{2}{3})^{n+1} x^n$. By (5.30), $\ln\left[3 + 2(x)\right] - \ln\left[3 + 2(0)\right] = \displaystyle\int_0^x \frac{2}{3 + 2t}\,dt \Rightarrow$

$\ln(3 + 2x) = \ln 3 + \displaystyle\sum_{n=0}^{\infty} (-1)^n \frac{1}{n + 1} (\tfrac{2}{3})^{n+1} x^{n+1}$.

[35] (a) $D_x\left[J_0(x)\right] = D_x \displaystyle\sum_{n=0}^{\infty} \frac{(-1)^n}{n!\,n!}\left(\frac{x}{2}\right)^{2n} = \sum_{n=1}^{\infty} \frac{(-1)^n}{n!\,n!}(2n)\left(\frac{x}{2}\right)^{2n-1}\left(\frac{1}{2}\right) =$

$\displaystyle\sum_{n=0}^{\infty} \frac{(-1)^{n+1}}{(n + 1)!\,(n + 1)!}(n + 1)\left(\frac{x}{2}\right)^{2n+1} = \sum_{n=0}^{\infty} \frac{(-1)^n(-1)}{n!\,(n + 1)!}\left(\frac{x}{2}\right)^{2n+1} = -J_1(x)$.

(b) $\int x^3 J_2(x)\,dx = \int \left[\sum_{n=0}^{\infty} \frac{(-1)^n}{n!\,(n+2)!} \left(\frac{x}{2}\right)^{2n+2} x^3 \right] dx =$

$$\int \left[\sum_{n=0}^{\infty} \frac{(-1)^n}{n!\,(n+2)!} (8) \left(\frac{x}{2}\right)^{2n+5} \right] dx = \left[\sum_{n=0}^{\infty} \frac{(-1)^n}{n!\,(n+2)!} \cdot \frac{8 \cdot 2}{2(n+3)} \cdot \left(\frac{x}{2}\right)^{2n+6} \right] + C$$

$$= \left[\sum_{n=0}^{\infty} \frac{(-1)^n}{n!\,(n+3)!} \left(\frac{x}{2}\right)^{2n+3} x^3 \right] + C = x^3 J_3(x) + C.$$

36 (a) $\sum_{n=0}^{\infty} p_n = \sum_{n=0}^{\infty} e^{-\lambda} \frac{\lambda^n}{n!} = e^{-\lambda} \sum_{n=0}^{\infty} \frac{\lambda^n}{n!} = e^{-\lambda} e^{\lambda} = e^0 = 1$

(b) Let P denote the probability that two or more photons are absorbed.

$$P = \sum_{n=2}^{\infty} p_n = \left(\sum_{n=0}^{\infty} p_n\right) - (p_1 + p_0) = 1 - \left(\frac{e^{-\lambda}\lambda^1}{1!} + \frac{e^{-\lambda}\lambda^0}{0!}\right) =$$

$$1 - (\lambda e^{-\lambda} + e^{-\lambda}) = 1 - e^{-\lambda}(\lambda + 1).$$

37 Using Exercise 11, $\frac{\ln(1-t)}{t} = -\sum_{n=1}^{\infty} \frac{t^{n-1}}{n}.$

$$f(x) = \int_0^x \left(-\sum_{n=1}^{\infty} \frac{t^{n-1}}{n}\right) dt = \left[-\sum_{n=1}^{\infty} \frac{t^n}{n^2}\right]_0^x = -\sum_{n=1}^{\infty} \frac{1}{n^2} x^n,\ |x| < 1.$$

38 Using (11.41), $\frac{e^t - 1}{t} = \frac{\left(1 + t + \frac{t^2}{2!} + \frac{t^3}{3!} + \cdots\right) - 1}{t} = 1 + \frac{1}{2!}t + \frac{1}{3!}t^2 + \cdots$

$$= \sum_{n=1}^{\infty} \frac{t^{n-1}}{n!}.\quad f(x) = \int_0^x \left(\sum_{n=1}^{\infty} \frac{t^{n-1}}{n!}\right) dt = \left[\sum_{n=1}^{\infty} \frac{t^n}{n\,(n!)}\right]_0^x = \sum_{n=1}^{\infty} \frac{1}{n(n!)} x^n.$$

Exercises 11.8

Note: Let *DERIV* denote the beginning of the sequence:

$f(x)$, $f(c)$‡ $f'(x)$, $f'(c)$‡ $f''(x)$, $f''(c)$‡ .... For a Maclaurin series, $c = 0$.

We have used the double dagger symbol (‡) to separate the terms.

1 Let $f(x) = e^{3x}$. *DERIV*: e^{3x}, 1‡ $3e^{3x}$, 3‡ $3^2 e^{3x}$, 3^2‡ $3^3 e^{3x}$, 3^3‡ ....

$f^{(n)}(x) = 3^n e^{3x}$ and $f^{(n)}(0) = 3^n$. Thus, $a_n = \frac{f^{(n)}(0)}{n!} = \frac{3^n}{n!}$.

2 Let $f(x) = e^{-2x}$. *DERIV*: e^{-2x}, 1‡ $-2e^{-2x}$, -2‡ $(-2)^2 e^{-2x}$, $(-2)^2$‡ $(-2)^3 e^{-2x}$, $(-2)^3$‡ .... $f^{(n)}(x) = (-2)^n e^{-2x}$ and $f^{(n)}(0) = (-2)^n$.

Thus, $a_n = \frac{f^{(n)}(0)}{n!} = \frac{(-2)^n}{n!} = (-1)^n \frac{2^n}{n!}$.

3 Let $f(x) = \sin 2x$. *DERIV*: $\sin 2x$, 0‡ $2\cos 2x$, 2‡ $-2^2 \sin 2x$, 0‡ $-2^3 \cos 2x$, -2^3‡ .... If $k = 0, 1, 2, \ldots$, then $f^{(2k)}(0) = 0$ and $f^{(2k+1)}(0) = (-1)^k 2^{2k+1}$.

Thus, $a_n = 0$ if $n = 2k$, and $a_n = \frac{f^{(n)}(0)}{n!} = (-1)^k \frac{2^{2k+1}}{(2k+1)!}$ if $n = 2k + 1$.

4 Let $f(x) = \cos 3x$. *DERIV*: $\cos 3x$, 1‡ $-3\sin 3x$, 0‡ $-3^2\cos 3x$, -3^2‡ $3^3\sin 3x$, 0‡ ... .

If $k = 0, 1, 2, \ldots$, then $f^{(2k+1)}(0) = 0$ and $f^{(2k)}(0) = (-1)^k\, 3^{2k}$.

$$\text{Thus, } a_n = 0 \text{ if } n = 2k+1 \text{ and } a_n = \frac{f^{(n)}(0)}{n!} = (-1)^k \frac{3^{2k}}{(2k)!} \text{ if } n = 2k.$$

5 Let $f(x) = \dfrac{1}{1+3x}$. *DERIV*: $(1+3x)^{-1}$, 1‡ $-3(1+3x)^{-2}$, -3‡ $2\cdot 3^2(1+3x)^{-3}$, 18‡ $-2\cdot 3\cdot 3^3(1+3x)^{-4}$, -162‡ ... . $f^{(n)}(x) = (-1)^n\, n!\, 3^n (1+3x)^{-n-1}$ and

$$f^{(n)}(0) = (-1)^n\, n!\, 3^n. \text{ Thus, } a_n = \frac{f^{(n)}(0)}{n!} = \frac{(-1)^n\, n!\, 3^n}{n!} = (-1)^n\, 3^n.$$

6 Let $f(x) = \dfrac{1}{1-2x}$. *DERIV*: $(1-2x)^{-1}$, 1‡ $(-1)(-2)(1-2x)^{-2}$, 2‡ $(-1)(-2)(-2)^2(1-2x)^{-3}$, 8‡ $(-1)(-2)(-3)(-2)^3(1-2x)^{-4}$, 48‡ ... .

$f^{(n)}(x) = (-1)^n\, n!\, (-2)^n (1-2x)^{-n-1} = n!\, 2^n (1-2x)^{-n-1}$ and $f^{(n)}(0) = n!\, 2^n$.

$$\text{Thus, } a_n = \frac{f^{(n)}(0)}{n!} = \frac{n!\, 2^n}{n!} = 2^n.$$

7 Let $f(x) = \cos x$. *DERIV*: $\cos x$, 1‡ $-\sin x$, 0‡ $-\cos x$, -1‡ $\sin x$, 0‡ $\cos x$, 1‡ ... .

(a) $\left|f^{(n+1)}(z)\right| \le 1 \Rightarrow |R_n(x)| = \left|\dfrac{f^{(n+1)}(z)}{(n+1)!}\,x^{n+1}\right| \le \left|\dfrac{x^{n+1}}{(n+1)!}\right| \to 0$ as $n \to \infty$ by (11.47).

(b) $f(x) = f(0) + f'(0)x + \dfrac{f''(0)}{2!}x^2 + \cdots + \dfrac{f^{(n)}(0)}{n!}x^n + \cdots \Rightarrow$

$$\cos x = 1 - \frac{x^2}{2!} + \frac{x^4}{4!} - \frac{x^6}{6!} + \cdots = \sum_{n=0}^{\infty} (-1)^n \frac{1}{(2n)!} x^{2n}.$$

8 Let $f(x) = e^{-x}$. *DERIV*: e^{-x}, 1‡ $-e^{-x}$, -1‡ e^{-x}, 1‡ ... .

(a) $\left|f^{(n+1)}(z)\right| = |\pm e^{-z}| \le e^{|x|}$ for z between 0 and $x \Rightarrow$

$$|R_n(x)| = \left|\frac{f^{(n+1)}(z)}{(n+1)!}\,x^{n+1}\right| \le e^{|x|}\left|\frac{x^{n+1}}{(n+1)!}\right| \to 0 \text{ as } n \to \infty \text{ by (11.47).}$$

(b) $f(x) = f(0) + f'(0)x + \dfrac{f''(0)}{2!}x^2 + \cdots + \dfrac{f^{(n)}(0)}{n!}x^n + \cdots \Rightarrow$

$$e^{-x} = 1 - x + \frac{x^2}{2!} - \frac{x^3}{3!} + \cdots = \sum_{n=0}^{\infty} (-1)^n \frac{1}{n!} x^n.$$

9 Using (11.48)(a),

$$f(x) = x\sin 3x = x\sum_{n=0}^{\infty} (-1)^n \frac{(3x)^{2n+1}}{(2n+1)!} = \sum_{n=0}^{\infty} (-1)^n \frac{3^{2n+1}}{(2n+1)!} x^{2n+2}.$$

10 Using (11.48)(a),

$$f(x) = x^2\sin x = x^2\sum_{n=0}^{\infty} (-1)^n \frac{x^{2n+1}}{(2n+1)!} = \sum_{n=0}^{\infty} (-1)^n \frac{1}{(2n+1)!} x^{2n+3}.$$

11 Using (11.48)(b), $f(x) = \cos(-2x) = \displaystyle\sum_{n=0}^{\infty} (-1)^n \frac{(-2x)^{2n}}{(2n)!} = \sum_{n=0}^{\infty} (-1)^n \frac{2^{2n}}{(2n)!} x^{2n}.$

12 Using (11.48)(b), $f(x) = \cos(x^2) = \displaystyle\sum_{n=0}^{\infty} (-1)^n \frac{(x^2)^{2n}}{(2n)!} = \sum_{n=0}^{\infty} (-1)^n \frac{1}{(2n)!} x^{4n}.$

[13] Using (11.48)(b), $f(x) = \cos^2 x = \frac{1}{2} + \frac{1}{2}\cos 2x = \frac{1}{2} + \frac{1}{2}\sum_{n=0}^{\infty}(-1)^n \frac{(2x)^{2n}}{(2n)!} =$

$$\frac{1}{2} + \sum_{n=0}^{\infty}(-1)^n \frac{2^{2n-1}}{(2n)!}x^{2n} = 1 + \sum_{n=1}^{\infty}(-1)^n \frac{2^{2n-1}}{(2n)!}x^{2n}.$$

[14] $\sin^2 x = 1 - \cos^2 x = 1 - \left(1 + \sum_{n=1}^{\infty}(-1)^n \frac{2^{2n-1}}{(2n)!}x^{2n}\right)$ {from Exercise 13} $=$

$$-\sum_{n=1}^{\infty}(-1)^n \frac{2^{2n-1}}{(2n)!}x^{2n} = \sum_{n=1}^{\infty}(-1)^{n+1} \frac{2^{2n-1}}{(2n)!}x^{2n}.$$

[15] *DERIV*: 10^x, 1‡ $10^x \ln 10$, $\ln 10$‡ $10^x(\ln 10)^2$, $(\ln 10)^2$‡ $10^x(\ln 10)^3$, $(\ln 10)^3$‡ ... .

$f^{(n)}(x) = 10^x(\ln 10)^n$ and $f^{(n)}(0) = (\ln 10)^n$. Thus, $10^x =$

$$\sum_{n=0}^{\infty}\frac{f^{(n)}(0)}{n!}x^n = 1 + (\ln 10)x + \frac{(\ln 10)^2}{2!}x^2 + \frac{(\ln 10)^3}{3!}x^3 + \cdots = \sum_{n=0}^{\infty}\frac{(\ln 10)^n}{n!}x^n.$$

[16] *DERIV*: $\ln(3 + x)$, $\ln 3$‡ $(3 + x)^{-1}$, $\frac{1}{3}$‡ $-1(3 + x)^{-2}$, $-\frac{1}{3^2}$‡ $2(3 + x)^{-3}$, $\frac{2}{3^3}$‡

$-6(3 + x)^{-4}$, $-\frac{6}{3^4}$‡ ... . $f^{(n)}(x) = (-1)^{n+1}(n - 1)!\,(3 + x)^{-n}$ and

$f^{(n)}(0) = (-1)^{n+1}(n - 1)!\,3^{-n}$ for $n \geq 1$. Thus, $\ln(3 + x) = \sum_{n=0}^{\infty}\frac{f^{(n)}(0)}{n!}x^n =$

$$\ln 3 + \frac{1}{3}x - \frac{1}{3^2 \cdot 2}x^2 + \frac{1}{3^3 \cdot 3}x^3 - \frac{1}{3^4 \cdot 4}x^4 + \cdots = \ln 3 + \sum_{n=1}^{\infty}(-1)^{n+1}\frac{1}{3^n n}x^n.$$

[17] *DERIV*: $\sin x$, $\frac{1}{\sqrt{2}}$‡ $\cos x$, $\frac{1}{\sqrt{2}}$‡ $-\sin x$, $-\frac{1}{\sqrt{2}}$‡ $-\cos x$, $-\frac{1}{\sqrt{2}}$‡ ... .

$$\sin x = \frac{1}{\sqrt{2}} + \frac{1}{\sqrt{2}}(x - \tfrac{\pi}{4}) - \frac{1}{\sqrt{2}\cdot 2!}(x - \tfrac{\pi}{4})^2 - \frac{1}{\sqrt{2}\cdot 3!}(x - \tfrac{\pi}{4})^3 + \cdots$$

$$= \sum_{n=0}^{\infty}(-1)^n \frac{1}{\sqrt{2}\,(2n + 1)!}(x - \tfrac{\pi}{4})^{2n+1} + \sum_{n=0}^{\infty}(-1)^n \frac{1}{\sqrt{2}\,(2n)!}(x - \tfrac{\pi}{4})^{2n}.$$

[18] *DERIV*: $\cos x$, $\frac{1}{2}$‡ $-\sin x$, $-\frac{\sqrt{3}}{2}$‡ $-\cos x$, $-\frac{1}{2}$‡ $\sin x$, $\frac{\sqrt{3}}{2}$‡ ... .

$$\cos x = \frac{1}{2} - \frac{\sqrt{3}}{2}(x - \tfrac{\pi}{3}) - \frac{1}{2!\,2}(x - \tfrac{\pi}{3})^2 + \frac{\sqrt{3}}{3!\,2}(x - \tfrac{\pi}{3})^3 + \cdots$$

$$= \sum_{n=0}^{\infty}(-1)^n \frac{1}{2(2n)!}(x - \tfrac{\pi}{3})^{2n} + \sum_{n=0}^{\infty}(-1)^{n+1}\frac{\sqrt{3}}{2(2n + 1)!}(x - \tfrac{\pi}{3})^{2n+1}.$$

[19] *DERIV*: x^{-1}, $\frac{1}{2}$‡ $-x^{-2}$, $-\frac{1}{2^2}$‡ $2x^{-3}$, $\frac{2}{2^3}$‡ $-6x^{-4}$, $-\frac{6}{2^4}$‡ ... .

$f^{(n)}(x) = (-1)^n\, n!\, x^{-n-1}$ and $f^{(n)}(2) = (-1)^n\, n!/2^{n+1}$.

Thus, $\frac{1}{x} = \frac{1}{2} - \frac{1}{2^2}(x - 2) + \frac{2!}{2^3 \cdot 2!}(x - 2)^2 - \frac{3!}{2^4 \cdot 3!}(x - 2)^3 + \cdots =$

$$\sum_{n=0}^{\infty}(-1)^n \frac{1}{2^{n+1}}(x - 2)^n.$$

[20] *DERIV*: e^x, e^{-3}‡ e^x, e^{-3}‡ ... . $f^{(n)}(x) = e^x$ and $f^{(n)}(-3) = e^{-3} = 1/e^3$.

$$\text{Thus, } e^x = \frac{1}{e^3} + \frac{1}{e^3}(x + 3) + \frac{1}{e^3\,2!}(x + 3)^2 + \cdots = \sum_{n=0}^{\infty}\frac{1}{e^3\, n!}(x + 3)^n.$$

[21] Let $c = -1$. *DERIV*: e^{2x}, e^{-2}‡ $2e^{2x}$, $2e^{-2}$‡ 2^2e^{2x}, 2^2e^{-2}‡ ....

$f^{(n)}(x) = 2^n e^{2x}$ and $f^{(n)}(-1) = 2^n e^{-2}$. Thus, $e^{2x} =$

$$e^{-2} + 2e^{-2}(x+1) + \frac{2^2 e^{-2}}{2!}(x+1)^2 + \frac{2^3 e^{-2}}{3!}(x+1)^3 + \cdots = \sum_{n=0}^{\infty} \frac{2^n}{e^2\, n!}(x+1)^n.$$

[22] Let $c = 1$. *DERIV*: $\ln x$, 0‡ x^{-1}, 1‡ $-x^{-2}$, -1‡ $2x^{-3}$, 2‡ ....

$f^{(n)}(x) = (-1)^{n+1}(n-1)!\,x^{-n}$ and $f^{(n)}(1) = (-1)^{n+1}(n-1)!$.

$$\ln x = 0 + 1(x-1) - \tfrac{1}{2!}(x-1)^2 + \tfrac{2!}{3!}(x-1)^3 + \cdots = \sum_{n=1}^{\infty} (-1)^{n+1}\tfrac{1}{n}(x-1)^n.$$

[23] *DERIV*: $\sec x$, 2‡ $\sec x \tan x$, $2\sqrt{3}$‡ $\sec^3 x + \sec x \tan^2 x$, 14.

$$\sec x = 2 + 2\sqrt{3}(x - \tfrac{\pi}{3}) + 7(x - \tfrac{\pi}{3})^2 + \cdots.$$

[24] *DERIV*: $\tan x$, 1‡ $\sec^2 x$, 2‡ $2\sec^2 x \tan x$, 4. $\tan x = 1 + 2(x - \tfrac{\pi}{4}) + 2(x - \tfrac{\pi}{4})^2 + \cdots$.

[25] *DERIV*: $\sin^{-1} x$, $\frac{\pi}{6}$‡ $\frac{1}{(1-x^2)^{1/2}}$, $\frac{2}{\sqrt{3}}$‡ $\frac{x}{(1-x^2)^{3/2}}$, $\frac{4}{3\sqrt{3}}$.

$$\sin^{-1} x = \frac{\pi}{6} + \frac{2}{\sqrt{3}}(x - \tfrac{1}{2}) + \frac{2}{3\sqrt{3}}(x - \tfrac{1}{2})^2 + \cdots.$$

[26] *DERIV*: $\tan^{-1} x$, $\frac{\pi}{4}$‡ $\frac{1}{1+x^2}$, $\frac{1}{2}$‡ $-\frac{2x}{(1+x^2)^2}$, $-\frac{1}{2}$.

$$\tan^{-1} x = \tfrac{\pi}{4} + \tfrac{1}{2}(x-1) - \tfrac{1}{4}(x-1)^2 + \cdots.$$

[27] *DERIV*: xe^x, $-e^{-1}$‡ $(x+1)e^x$, 0‡ $(x+2)e^x$, e^{-1}‡ $(x+3)e^x$, $2e^{-1}$.

$$xe^x = -\frac{1}{e} + \frac{1}{2e}(x+1)^2 + \frac{1}{3e}(x+1)^3 + \cdots.$$

[28] *DERIV*: $\csc x$, $\frac{2}{\sqrt{3}}$‡ $-\csc x \cot x$, $\frac{2}{3}$‡ $\csc^3 x + \csc x \cot^2 x$, $\frac{10}{3\sqrt{3}}$.

$$\csc x = \tfrac{2}{\sqrt{3}} + \tfrac{2}{3}(x - \tfrac{2\pi}{3}) + \tfrac{5}{3\sqrt{3}}(x - \tfrac{2\pi}{3})^2 + \cdots.$$

Note: In Exercises 29–38, all series are alternating, so (11.31) applies. S denotes the sum of the first two nonzero terms and E the absolute value of the maximum error.

[29] $e^{-x} = \sum_{n=0}^{\infty} (-1)^n \frac{x^n}{n!}$. $\frac{1}{\sqrt{e}} = e^{-1/2} \approx 1 - (\frac{1}{2}) + \frac{(\frac{1}{2})^2}{2!}$.

$$S = 1 - \tfrac{1}{2} = 0.5.\quad E = \frac{(\frac{1}{2})^2}{2!} = \tfrac{1}{8} = 0.125.$$

[30] $e^{-x} = \sum_{n=0}^{\infty} (-1)^n \frac{x^n}{n!}$. $\frac{1}{e} = e^{-1} \approx 1 - (1) + \frac{(1)^2}{2!}$.

$$S = 1 - 1 = 0.\quad E = \frac{(1)^2}{2!} = \tfrac{1}{2} = 0.5.$$

[31] $\cos x = \sum_{n=0}^{\infty} (-1)^n \frac{x^{2n}}{(2n)!}$. $\cos 3^\circ = \cos \frac{\pi}{60} \approx 1 - \frac{(\frac{\pi}{60})^2}{2!} + \frac{(\frac{\pi}{60})^4}{4!}$.

$$S = 1 - \frac{(\frac{\pi}{60})^2}{2!} \approx 0.9986.\quad E = \frac{(\frac{\pi}{60})^4}{4!} \approx 3.13 \times 10^{-7}.$$

[32] $\sin x = \sum_{n=0}^{\infty} (-1)^n \frac{x^{2n+1}}{(2n+1)!}$. $\sin 1^\circ = \sin \frac{\pi}{180} \approx \frac{\pi}{180} - \frac{(\frac{\pi}{180})^3}{3!} + \frac{(\frac{\pi}{180})^5}{5!}$.

$$S = \frac{\pi}{180} - \frac{(\frac{\pi}{180})^3}{3!} \approx 0.0175.\quad E = \frac{(\frac{\pi}{180})^5}{5!} \approx 1.35 \times 10^{-11}.$$

33 $\tan^{-1}x = \sum_{n=0}^{\infty}(-1)^n \frac{x^{2n+1}}{2n+1}$. $\tan^{-1}0.1 \approx (0.1) - \frac{(0.1)^3}{3} + \frac{(0.1)^5}{5}$.

$$S = (0.1) - \frac{(0.1)^3}{3} \approx 0.0997.\ E = \frac{(0.1)^5}{5} = 2 \times 10^{-6}.$$

34 $\ln(1+x) = \sum_{n=0}^{\infty}(-1)^n \frac{x^{n+1}}{n+1}$. $\ln 1.5 = \ln[1 + (0.5)] \approx (0.5) - \frac{(0.5)^2}{2} + \frac{(0.5)^3}{3}$.

$$S = (0.5) - \frac{(0.5)^2}{2} = 0.375.\ E = \frac{(0.5)^3}{3} \approx 0.0417.$$

35 $e^{-x^2} = \sum_{n=0}^{\infty}\frac{(-x^2)^n}{n!} = \sum_{n=0}^{\infty}(-1)^n \frac{x^{2n}}{n!} \Rightarrow$

$$\int_0^1 e^{-x^2}\,dx = \left[\sum_{n=0}^{\infty}(-1)^n \frac{x^{2n+1}}{(2n+1)\,n!}\right]_0^1.$$

$\mathrm{I} \approx 1 - \frac{1}{3} + \frac{1}{10}$. $S = 1 - \frac{1}{3} = \frac{2}{3} \approx 0.6667$. $E = \frac{1}{10} = 0.1$.

36 $x\cos(x^3) = x\sum_{n=0}^{\infty}(-1)^n \frac{(x^3)^{2n}}{(2n)!} = \sum_{n=0}^{\infty}(-1)^n \frac{x^{6n+1}}{(2n)!}$.

$$\int_0^{1/2} x\cos(x^3)\,dx = \left[\sum_{n=0}^{\infty}(-1)^n \frac{x^{6n+2}}{(6n+2)(2n)!}\right]_0^{1/2}.\ \mathrm{I} \approx \frac{(\frac{1}{2})^2}{2(0!)} - \frac{(\frac{1}{2})^8}{8(2!)} + \frac{(\frac{1}{2})^{14}}{14(4!)}.$$

$$S = \frac{(\frac{1}{2})^2}{2(0!)} - \frac{(\frac{1}{2})^8}{8(2!)} \approx 0.1248.\ E = \frac{(\frac{1}{2})^{14}}{14(4!)} \approx 1.82 \times 10^{-7}.$$

37 $\cos(x^2) = \sum_{n=0}^{\infty}(-1)^n \frac{(x^2)^{2n}}{(2n)!} = \sum_{n=0}^{\infty}(-1)^n \frac{x^{4n}}{(2n)!}$.

$$\int_0^{0.5}\cos(x^2)\,dx = \left[\sum_{n=0}^{\infty}(-1)^n \frac{x^{4n+1}}{(4n+1)(2n)!}\right]_0^{0.5}.\ \mathrm{I} \approx \frac{(0.5)^1}{1(0!)} - \frac{(0.5)^5}{5(2!)} + \frac{(0.5)^9}{9(4!)}.$$

$$S = \frac{(0.5)^1}{1(0!)} - \frac{(0.5)^5}{5(2!)} = 0.496875.\ E = \frac{(0.5)^9}{9(4!)} \approx 9.04 \times 10^{-6}.$$

38 $\tan^{-1}(x^2) = \sum_{n=0}^{\infty}(-1)^n \frac{(x^2)^{2n+1}}{2n+1} = \sum_{n=0}^{\infty}(-1)^n \frac{x^{4n+2}}{2n+1}$.

$$\int_0^{0.1}\tan^{-1}(x^2)\,dx = \left[\sum_{n=0}^{\infty}(-1)^n \frac{x^{4n+3}}{(4n+3)(2n+1)}\right]_0^{0.1}.$$

$\mathrm{I} \approx \frac{(0.1)^3}{(3)(1)} - \frac{(0.1)^7}{(7)(3)} + \frac{(0.1)^{11}}{(11)(5)}$. $S = \frac{(0.1)^3}{(3)(1)} - \frac{(0.1)^7}{(7)(3)} \approx 0.0003$.

$$E = \frac{(0.1)^{11}}{(11)(5)} \approx 1.82 \times 10^{-13}.$$

Note: In Exercises 39–42, all series are alternating and satisfy (11.30), so (11.31) applies.

39 $\frac{1-\cos x}{x^2} = \frac{1 - \left(1 - \frac{x^2}{2!} + \frac{x^4}{4!} - \frac{x^6}{6!} + \cdots\right)}{x^2} = \frac{1}{2!} - \frac{x^2}{4!} + \frac{x^4}{6!} - \cdots =$

$\sum_{n=0}^{\infty}(-1)^n \frac{x^{2n}}{(2n+2)!}$. $\int_0^1 \frac{1-\cos x}{x^2}\,dx = \left[\sum_{n=0}^{\infty}(-1)^n \frac{x^{2n+1}}{(2n+1)(2n+2)!}\right]_0^1$.

$|\text{Error}| < \frac{1}{(2n+1)(2n+2)!} < 0.5 \times 10^{-4}$ for $n \geq 3$.

$$\text{Thus, } \mathrm{I} \approx \frac{1}{(1)(2!)} - \frac{1}{(3)(4!)} + \frac{1}{(5)(6!)} \approx 0.4864.$$

40 $\dfrac{\sin x}{x} = \dfrac{1}{x} \sum_{n=0}^{\infty} (-1)^n \dfrac{x^{2n+1}}{(2n+1)!} = \sum_{n=0}^{\infty} (-1)^n \dfrac{x^{2n}}{(2n+1)!}$.

$$\int_0^1 \frac{\sin x}{x}\,dx = \left[\sum_{n=0}^{\infty} (-1)^n \frac{x^{2n+1}}{(2n+1)(2n+1)!}\right]_0^1$$

$|\text{Error}| < \dfrac{1}{(2n+1)(2n+1)!} < 0.5 \times 10^{-4}$ for $n \geq 3$.

Thus, $\text{I} \approx \dfrac{1}{1(1!)} - \dfrac{1}{3(3!)} + \dfrac{1}{5(5!)} \approx 0.9461$.

41 $\dfrac{\ln(1+x)}{x} = \dfrac{1}{x} \sum_{n=0}^{\infty} (-1)^n \dfrac{x^{n+1}}{n+1}$. $\sum_{n=0}^{\infty} (-1)^n \dfrac{x^n}{n+1}$.

$$\int_0^{1/2} \frac{\ln(1+x)}{x}\,dx = \left[\sum_{n=0}^{\infty} (-1)^n \frac{x^{n+1}}{(n+1)^2}\right]_0^{1/2}.$$

$|\text{Error}| < \dfrac{(\frac{1}{2})^{n+1}}{(n+1)^2} < 0.5 \times 10^{-4}$ for $n \geq 8$. Thus,

$$\text{I} \approx \tfrac{1}{2} - \frac{(1/2)^2}{4} + \frac{(1/2)^3}{9} - \frac{(1/2)^4}{16} + \frac{(1/2)^5}{25} - \frac{(1/2)^6}{36} + \frac{(1/2)^7}{49} - \frac{(1/2)^8}{64} \approx 0.4484.$$

42 $\dfrac{1 - e^{-x}}{x} = \dfrac{1 - \left[1 + (-x) + \dfrac{(-x)^2}{2!} + \dfrac{(-x)^3}{3!} + \cdots\right]}{x} = 1 - \dfrac{x}{2!} + \dfrac{x^2}{3!} - \cdots =$

$\sum_{n=0}^{\infty} (-1)^n \dfrac{x^n}{(n+1)!}$. $\displaystyle\int_0^1 \frac{1 - e^{-x}}{x}\,dx = \left[\sum_{n=0}^{\infty} (-1)^n \frac{x^{n+1}}{(n+1)(n+1)!}\right]_0^1$.

$|\text{Error}| < \dfrac{1}{(n+1)(n+1)!} < 0.5 \times 10^{-4}$ for $n \geq 6$.

Thus, $\text{I} \approx 1 - \dfrac{1}{2(2!)} + \dfrac{1}{3(3!)} - \dfrac{1}{4(4!)} + \dfrac{1}{5(5!)} - \dfrac{1}{6(6!)} \approx 0.7966$.

43 (a) $\displaystyle\int_0^1 \sin(x^2)\,dx \approx \int_0^1 (x^2 - \tfrac{1}{6}x^6)\,dx = \left[\tfrac{1}{3}x^3 - \tfrac{1}{42}x^7\right]_0^1 = \tfrac{13}{42} \approx 0.309524.$

$\displaystyle\int_1^2 \sin(x^2)\,dx \approx \int_1^2 (x^2 - \tfrac{1}{6}x^6)\,dx = \left[\tfrac{1}{3}x^3 - \tfrac{1}{42}x^7\right]_1^2 = -\tfrac{29}{42} \approx -0.690476.$

(b) $g(x)$ becomes a worse approximation for $f(x)$ the further x is from zero.

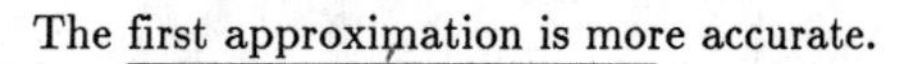
The first approximation is more accurate.

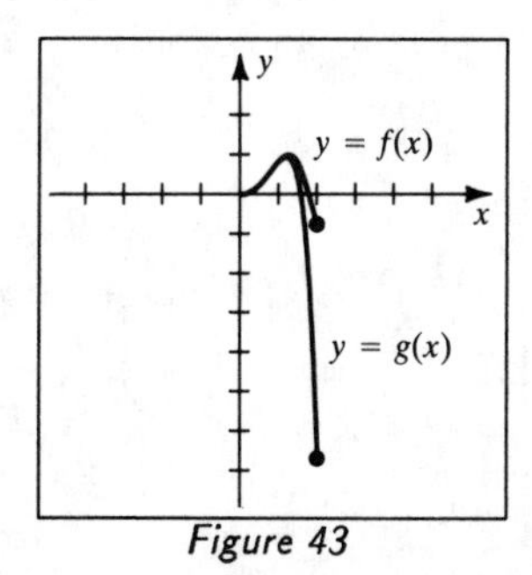

Figure 43

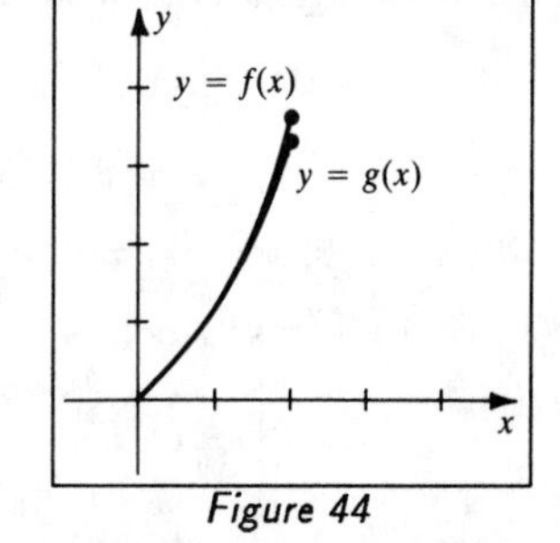

Figure 44

44 (a) $\displaystyle\int_0^1 \sinh x\,dx \approx \int_0^1 (x + \tfrac{1}{6}x^3)\,dx = \left[\tfrac{1}{2}x^2 + \tfrac{1}{24}x^4\right]_0^1 = \tfrac{13}{24} \approx 0.541667.$ (cont.)

$$\int_1^2 \sinh x\,dx \approx \int_1^2 (x + \tfrac{1}{6}x^3)\,dx = \left[\tfrac{1}{2}x^2 + \tfrac{1}{24}x^4\right]_1^2 = \tfrac{17}{8} = 2.125.$$

(b) $g(x)$ becomes a worse approximation for $f(x)$ the further x is from zero.

The first approximation is more accurate. See *Figure 44*.

[45] $\ln\left(\frac{1+x}{1-x}\right) = \ln(1+x) - \ln(1-x)$. $\ln(1-x) = \ln\left[1 + (-x)\right] =$

$$\sum_{n=0}^{\infty} (-1)^n \frac{(-x)^{n+1}}{n+1} = \sum_{n=0}^{\infty} (-1)^{2n+1} \frac{x^{n+1}}{n+1} = -\sum_{n=0}^{\infty} \frac{x^{n+1}}{n+1}.$$

Thus, $\ln(1+x) - \ln(1-x) = \sum_{n=0}^{\infty} (-1)^n \frac{x^{n+1}}{n+1} + \sum_{n=0}^{\infty} \frac{x^{n+1}}{n+1} =$

$$\left(x - \frac{x^2}{2} + \frac{x^3}{3} + \cdots\right) + \left(x + \frac{x^2}{2} + \frac{x^3}{3} + \cdots\right) = 2\sum_{n=0}^{\infty} \frac{x^{2n+1}}{2n+1}.$$

Since the first series is valid for $-1 < x \le 1$ and the second series is valid for

$-1 < -x \le 1 \Leftrightarrow -1 \le x < 1$, the final series is valid for $|x| < 1$.

[46] $\frac{1+x}{1-x} = 2 \Rightarrow x = \frac{1}{3}$. $\ln 2 = \ln\left(\frac{1+\frac{1}{3}}{1-\frac{1}{3}}\right) \approx$

$$2\left[\frac{(\frac{1}{3})}{1} + \frac{(\frac{1}{3})^3}{3} + \frac{(\frac{1}{3})^5}{5} + \frac{(\frac{1}{3})^7}{7} + \frac{(\frac{1}{3})^9}{9}\right] \approx 0.69314605. \text{ Calculator value} \approx 0.69314718.$$

[47] (a) $\pi = 4\tan^{-1}1 = 4\left[1 - \frac{1}{3} + \frac{1}{5} - \frac{1}{7} + \cdots + (-1)^n \frac{1}{2n+1} + \cdots\right]$.

(b) $\pi \approx 4 - \frac{4}{3} + \frac{4}{5} - \frac{4}{7} + \frac{4}{9} \approx 3.34$, with an error of less than $\frac{4}{11}$.

(c) For 4 decimal places we need $\frac{4}{2n+1} < 0.5 \times 10^{-4}$, or $n \ge 40{,}000$ terms.

[48] (a) $\pi = 4(\tan^{-1}\frac{1}{2} + \tan^{-1}\frac{1}{3}) = 4\left[\sum_{n=0}^{\infty} (-1)^n \frac{(\frac{1}{2})^{2n+1}}{2n+1} + \sum_{n=0}^{\infty} (-1)^n \frac{(\frac{1}{3})^{2n+1}}{2n+1}\right]$.

(b) $\pi \approx 4\Big(\left[\frac{1}{2} - \frac{1}{3}(\frac{1}{2})^3 + \frac{1}{5}(\frac{1}{2})^5 - \frac{1}{7}(\frac{1}{2})^7 + \frac{1}{9}(\frac{1}{2})^9\right]$

$+\left[\frac{1}{3} - \frac{1}{3}(\frac{1}{3})^3 + \frac{1}{5}(\frac{1}{3})^5 - \frac{1}{7}(\frac{1}{3})^7 + \frac{1}{9}(\frac{1}{3})^9\right]\Big) \approx 3.14174,$

much closer to the value of π than the result in Exercise 47.

[49] (a) The central angle θ of a circle of radius R subtended by an arc of length s is

$\theta = \frac{s}{R}$. From the figure in the text, $\cos\theta = \frac{R}{R+C} \Rightarrow \sec\theta = \frac{R+C}{R} \Rightarrow$

$R\sec\theta = R + C \Rightarrow C = R(\sec\theta - 1) = R\left[\sec(s/R) - 1\right]$.

(b) *DERIV*: $\sec x$, 1‡ $\sec x \tan x$, 0‡ $\sec^3 x + \sec x \tan^2 x$, 1‡ $5\sec^3 x \tan x + \sec x \tan^3 x$,

0‡ $18\sec^3 x \tan^2 x + 5\sec^5 x + \sec x \tan^4 x$, 5. $f(x) = \sec x \approx 1 + \frac{1}{2!}x^2 + \frac{5}{4!}x^4$.

$$C = R\left[\sec(s/R) - 1\right] \approx R\left[\left(1 + \frac{s^2}{2R^2} + \frac{5s^4}{24R^4}\right) - 1\right] = \frac{s^2}{2R} + \frac{5s^4}{24R^3}.$$

(c) $R = 3959$, $s = 5 \Rightarrow C \approx \frac{5^2}{2(3959)} + \frac{5(5)^4}{24(3959)^3} \approx 0.003157 \text{ mi} \approx 16.7 \text{ ft}.$

[50] (a) Let $f(x) = \tanh x$ and $c = 0$. *DERIV*: $\tanh x$, 0‡ $\operatorname{sech}^2 x$, 1‡ $-2\operatorname{sech}^2 x \tanh x$, 0‡

$4\operatorname{sech}^2 x \tanh^2 x - 2\operatorname{sech}^4 x$, -2. $\tanh x \approx x - \frac{2}{3!}x^3 = x - \frac{1}{3}x^3$

(b) $\tanh x \approx x \Rightarrow v^2 = \frac{gL}{2\pi}\tanh\frac{2\pi h}{L} \approx \frac{gL}{2\pi}\left(\frac{2\pi h}{L}\right) = gh.$

(c) $v^2 = \frac{gL}{2\pi}\tanh\frac{2\pi h}{L} \approx \frac{gL}{2\pi}\left[\frac{2\pi h}{L} - \frac{1}{3}\left(\frac{2\pi h}{L}\right)^3\right] = gh - \frac{1}{3}\,gL(2\pi)^2\left(\frac{h}{L}\right)^3.$

Since $L > 20h$, $h/L < \frac{1}{20}$. Error $= \left|v^2 - gh\right| < \frac{1}{3}gL(2\pi)^2(\frac{1}{20})^3 < 0.002\,gL.$

[51] $\delta = x(\sec kL - 1) \approx x\left[[1 + \frac{1}{2}(kL)^2] - 1\right]$ {by Exer. 49(b)} $= \frac{1}{2}xk^2L^2 = \frac{1}{2}PxL^2/R.$

[52] By (11.31), |error| $< \frac{x^8}{8!} \leq \frac{(\frac{\pi}{4})^8}{8!} \approx 3.6 \times 10^{-6} < 0.5 \times 10^{-5}.$

Exercises 11.9

Note: In the following exercises, $\sin x$ and $\cos x$ have been bound by 1.

In certain intervals, it may be possible to bound them by a smaller value.

[1] (a) Use (11.48)(a) to find $P_n(x)$, which consists of terms through x^n.

Thus, $P_1(x) = P_2(x) = x$, and $P_3(x) = x - \frac{1}{6}x^3$.

(c) $f(x) = \sin x$ and $f(0.05) \approx P_3(0.05) = 0.05 - \frac{1}{6}(0.05)^3 \approx 0.0500.$

$f^{(4)}(x) = \sin x$ and $\left|f^{(4)}(z)\right| \leq 1$ for all z.

$$\text{Error} \leq \left|R_3(0.05)\right| = \left|\frac{f^{(4)}(z)}{4!}(0.05)^4\right| \leq \left|\frac{1}{4!}(0.05)^4\right| \approx 2.6 \times 10^{-7}.$$

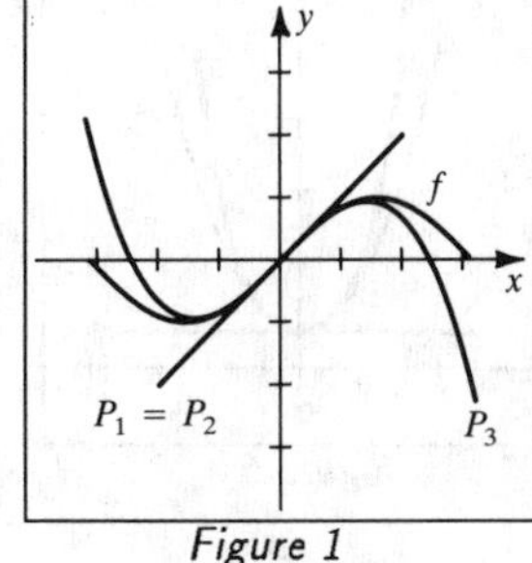

Figure 1

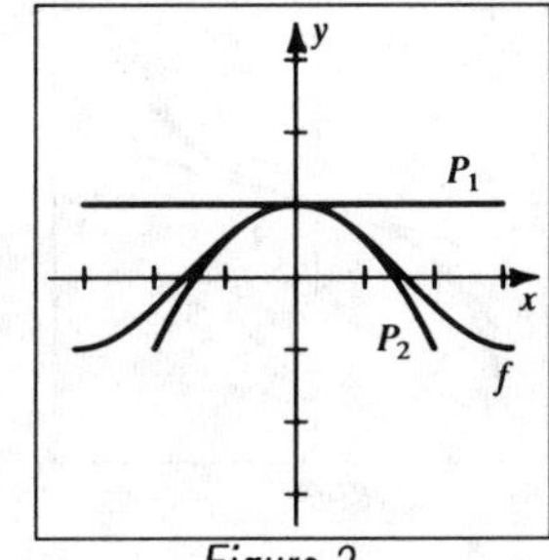

Figure 2

[2] (a) From (11.48)(b), $P_1(x) = 1$, $P_2(x) = P_3(x) = 1 - \frac{1}{2}x^2$.

(c) $f(x) = \cos x$ and $f(0.2) \approx P_3(0.2) = 1 - \frac{1}{2}(0.2)^2 \approx 0.9800.$

$f^{(4)}(x) = \cos x$ and $\left|f^{(4)}(z)\right| \leq 1$ for all z.

$$\text{Error} \leq \left|R_3(0.2)\right| = \left|\frac{f^{(4)}(z)}{4!}(0.2)^4\right| \leq \left|\frac{1}{4!}(0.2)^4\right| \approx 6.7 \times 10^{-5}.$$

[3] (a) From (11.48)(d), $P_1(x) = x$, $P_2(x) = x - \frac{1}{2}x^2$, and $P_3(x) = x - \frac{1}{2}x^2 + \frac{1}{3}x^3$.

(c) $f(x) = \ln(x + 1)$ and $f(0.9) \approx P_3(0.9) = 0.9 - \frac{1}{2}(0.9)^2 + \frac{1}{3}(0.9)^3 \approx 0.7380.$

$$\text{Error} \leq \left|R_3(0.9)\right| = \left|\frac{f^{(4)}(z)}{4!}(0.9)^4\right| = \left|\frac{-6(1 + z)^{-4}}{24}(0.9)^4\right| = \left|\frac{-(0.9)^4}{4(1 + z)^4}\right|.$$

To maximize this value, we minimize the denominator by letting $z = 0$.

Thus, error ≈ 0.164. See *Figure 3*.

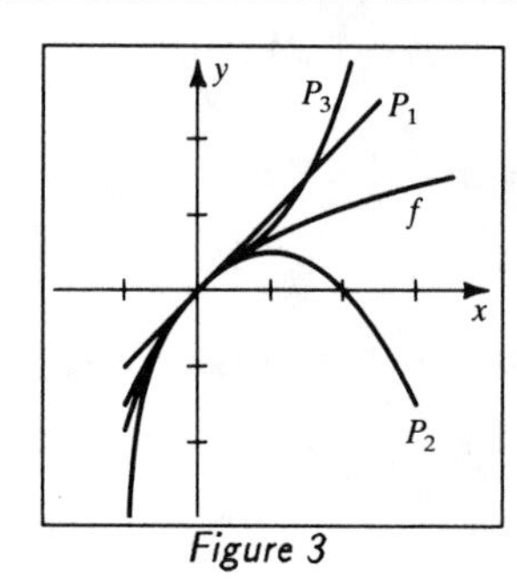

Figure 3

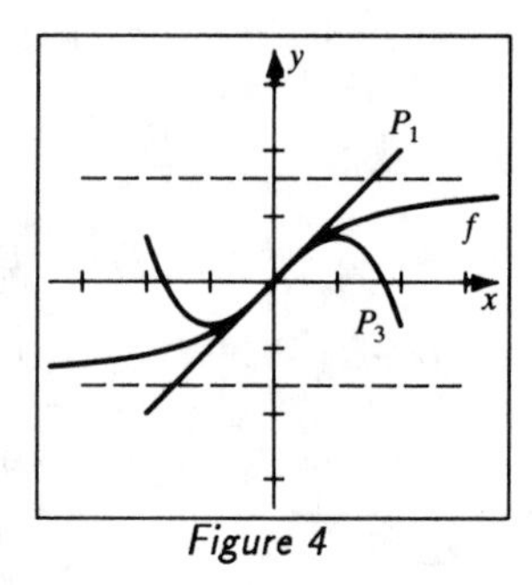

Figure 4

[4] (a) From (11.48)(e), $P_1(x) = P_2(x) = x$ and $P_3(x) = x - \frac{1}{3}x^3$.

(c) $f(x) = \tan^{-1}x$ and $f(0.1) \approx P_3(0.1) = 0.1 - \frac{1}{3}(0.1)^3 \approx 0.0997$.

$$\text{Error} \le |R_3(0.1)| = \left|\frac{f^{(4)}(z)}{4!}(0.1)^4\right| = \left|\frac{-24z(z^2-1)}{24(z^2+1)^4}(0.1)^4\right|.$$

To maximize this value, let $z = 0.1$ in the numerator, and $z^2 = 0$ in both the numerator and the denominator. Thus, error $\approx 1 \times 10^{-5}$.

[5] $f(x) = \sinh x$, $P_1(x) = x$, $P_3(x) = x + \frac{x^3}{6}$, and $P_5(x) = x + \frac{x^3}{6} + \frac{x^5}{120}$.

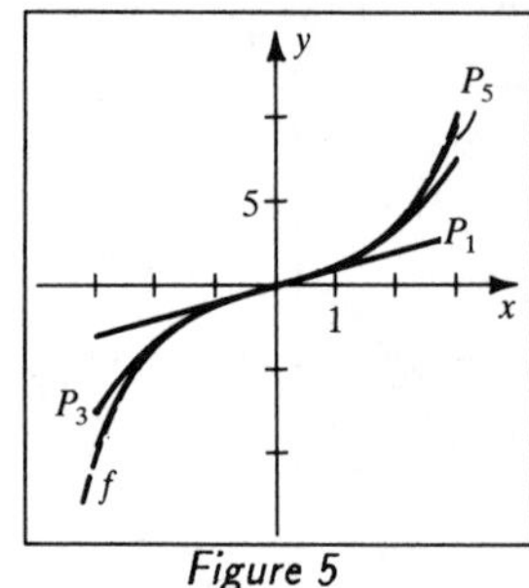

Figure 5

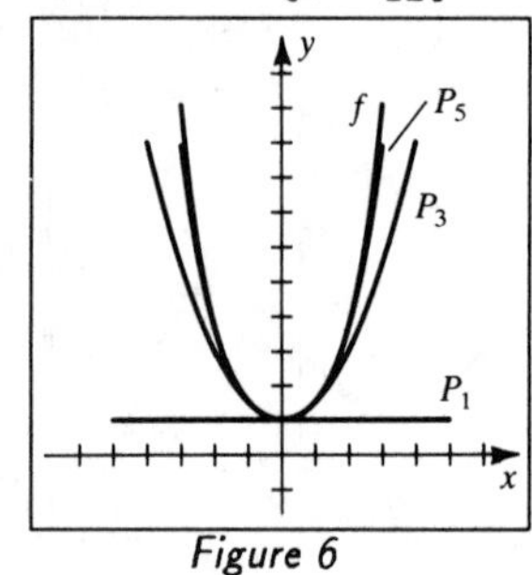

Figure 6

[6] $f(x) = \cosh x$, $P_1(x) = 1$, $P_3(x) = 1 + \frac{x^2}{2}$, and $P_5(x) = 1 + \frac{x^2}{2} + \frac{x^4}{24}$.

Note: Let *DERIV* denote the beginning of the sequence:

$$f(x),\ f(c)\ddagger\ f'(x),\ f'(c)\ddagger\ \ldots\ddagger\ f^{(n)}(x),\ f^{(n)}(c)\ddagger\ f^{(n+1)}(z).$$

[7] *DERIV*: $\sin x,\ 1\ddagger\ \cos x,\ 0\ddagger\ -\sin x,\ -1\ddagger\ -\cos x,\ 0\ddagger\ \sin z$.

$\underline{\sin x} = 1 - \frac{1}{2}(x - \frac{\pi}{2})^2 + \frac{\sin z}{24}(x - \frac{\pi}{2})^4$, z is between x and $\frac{\pi}{2}$.

[8] *DERIV*: $\cos x,\ \frac{1}{\sqrt{2}}\ddagger\ -\sin x,\ -\frac{1}{\sqrt{2}}\ddagger\ -\cos x,\ -\frac{1}{\sqrt{2}}\ddagger\ \sin x,\ \frac{1}{\sqrt{2}}\ddagger\ \cos z$.

$$\underline{\cos x} = \frac{1}{\sqrt{2}} - \frac{1}{\sqrt{2}}(x - \tfrac{\pi}{4}) - \frac{1}{2\sqrt{2}}(x - \tfrac{\pi}{4})^2 + \frac{1}{6\sqrt{2}}(x - \tfrac{\pi}{4})^3 + \frac{\cos z}{24}(x - \tfrac{\pi}{4})^4,$$

z is between x and $\frac{\pi}{4}$.

[9] *DERIV*: $x^{1/2},\ 2\ddagger\ \frac{1}{2}x^{-1/2},\ \frac{1}{4}\ddagger\ -\frac{1}{4}x^{-3/2},\ -\frac{1}{32}\ddagger\ \frac{3}{8}x^{-5/2},\ \frac{3}{256}\ddagger\ -\frac{15}{16}z^{-7/2}$.

$$\underline{\sqrt{x}} = 2 + \tfrac{1}{4}(x - 4) - \tfrac{1}{64}(x - 4)^2 + \tfrac{1}{512}(x - 4)^3 - \tfrac{5}{128}z^{-7/2}(x - 4)^4,$$

z is between x and 4.

[10] *DERIV*: e^{-x}, e^{-1}‡ $-e^{-x}$, $-e^{-1}$‡ e^{-x}, e^{-1}‡ $-e^{-x}$, $-e^{-1}$‡ e^{-z}.

$$\underline{e^{-x}} = e^{-1} - e^{-1}(x-1) + \frac{e^{-1}}{2}(x-1)^2 - \frac{e^{-1}}{6}(x-1)^3 + \frac{e^{-z}}{24}(x-1)^4,$$

z is between x and 1.

[11] $f(x) = \tan x$, $f(\frac{\pi}{4}) = 1$; $f'(x) = \sec^2 x$, $f'(\frac{\pi}{4}) = 2$;

$f''(x) = 2\sec^2 x \tan x = 2\tan^3 x + 2\tan x$, $f''(\frac{\pi}{4}) = 4$;

$f'''(x) = 6\tan^2 x \sec^2 x + 2\sec^2 x = 6\tan^4 x + 6\tan^2 x + 2\tan^2 x + 2 =$

$6\tan^4 x + 8\tan^2 x + 2.$

$$\underline{\tan x} = 1 + 2(x - \tfrac{\pi}{4}) + 2(x - \tfrac{\pi}{4})^2 + \tfrac{1}{3}(3\tan^4 z + 4\tan^2 z + 1)(x - \tfrac{\pi}{4})^3,$$

z is between x and $\frac{\pi}{4}$.

[12] *DERIV*: $(x-1)^{-2}$, 1‡ $-2(x-1)^{-3}$, -2‡ $6(x-1)^{-4}$, 6‡ $-24(x-1)^{-5}$, -24‡ $120(x-1)^{-6}$, 120‡ $-720(x-1)^{-7}$, -720‡ $5040(z-1)^{-8}$.

$$\underline{\frac{1}{(x-1)^2}} = 1 - 2(x-2) + 3(x-2)^2 - 4(x-2)^3 + 5(x-2)^4 - 6(x-2)^5$$

$$+ 7(z-1)^{-8}(x-2)^6,\ z \text{ is between } x \text{ and } 2.$$

[13] *DERIV*: x^{-1}, $-\frac{1}{2}$‡ $-x^{-2}$, $-\frac{1}{4}$‡ $2x^{-3}$, $-\frac{1}{4}$‡ $-6x^{-4}$, $-\frac{3}{8}$‡ $24x^{-5}$, $-\frac{3}{4}$‡ $-120x^{-6}$, $-\frac{15}{8}$‡ $720z^{-7}$.

$$\underline{\frac{1}{x}} = -\tfrac{1}{2} - \tfrac{1}{4}(x+2) - \tfrac{1}{8}(x+2)^2 - \tfrac{1}{16}(x+2)^3 - \tfrac{1}{32}(x+2)^4 - \tfrac{1}{64}(x+2)^5 +$$

$$z^{-7}(x+2)^6,\ z \text{ is between } x \text{ and } -2.$$

[14] *DERIV*: $x^{1/3}$, -2‡ $\frac{1}{3}x^{-2/3}$, $\frac{1}{12}$‡ $-\frac{2}{9}x^{-5/3}$, $\frac{1}{144}$‡ $\frac{10}{27}x^{-8/3}$, $\frac{5}{3456}$‡ $-\frac{80}{81}z^{-11/3}$.

$$\underline{\sqrt[3]{x}} = -2 + \tfrac{1}{12}(x+8) + \tfrac{1}{288}(x+8)^2 + \tfrac{5}{20{,}736}(x+8)^3 - \tfrac{10}{243}z^{-11/3}(x+8)^4,$$

z is between x and -8.

[15] *DERIV*: $\tan^{-1}x$, $\frac{\pi}{4}$‡ $(1+x^2)^{-1}$, $\frac{1}{2}$‡ $-2x(1+x^2)^{-2}$, $-\frac{1}{2}$‡ $(6z^2-2)(1+z^2)^{-3}$.

$$\underline{\tan^{-1}x} = \tfrac{\pi}{4} + \tfrac{1}{2}(x-1) - \tfrac{1}{4}(x-1)^2 + \frac{3z^2-1}{3(1+z^2)^3}(x-1)^3,\ z \text{ is between } x \text{ and } 1.$$

[16] *DERIV*: $\ln\sin x$, $\ln\frac{1}{2}$‡ $\cot x$, $\sqrt{3}$‡ $-\csc^2 x$, -4‡ $2\csc^2 x \cot x$, $8\sqrt{3}$‡ $-2\csc^4 z - 4\cot^2 z\csc^2 z$.

$$\underline{\ln\sin x} = -\ln 2 + \sqrt{3}(x - \tfrac{\pi}{6}) - 2(x - \tfrac{\pi}{6})^2 + \tfrac{4}{3}\sqrt{3}(x - \tfrac{\pi}{6})^3$$

$$- \frac{\csc^4 z + 2\cot^2 z\csc^2 z}{12}(x - \tfrac{\pi}{6})^4,\ z \text{ is between } x \text{ and } \tfrac{\pi}{6}.$$

[17] *DERIV*: xe^x, $-e^{-1}$‡ $xe^x + e^x$, 0‡ $xe^x + 2e^x$, e^{-1}‡ $xe^x + 3e^x$, $2e^{-1}$‡ $xe^x + 4e^x$, $3e^{-1}$‡ $ze^z + 5e^z$.

$$\underline{xe^x} = -\frac{1}{e} + \frac{1}{2e}(x+1)^2 + \frac{1}{3e}(x+1)^3 + \frac{1}{8e}(x+1)^4 + \frac{ze^z + 5e^z}{120}(x+1)^5,$$

z is between x and -1.

[18] *DERIV*: $\log x$, 1‡ $\frac{1}{x\ln 10}$, $\frac{1}{10\ln 10}$‡ $-\frac{1}{x^2\ln 10}$, $-\frac{1}{100\ln 10}$‡ $\frac{2}{z^3\ln 10}$.

$$\underline{\log x} = 1 + \frac{1}{10\ln 10}(x-10) - \frac{1}{200\ln 10}(x-10)^2 + \frac{1}{3z^3\ln 10}(x-10)^3,$$

z is between x and 10.

Note: Exer. 19-30: Since $c = 0$, z is between x and 0.

[19] *DERIV*: $\ln(x + 1)$, 0‡ $(x + 1)^{-1}$, 1‡ $-1(x + 1)^{-2}$, -1‡ $2(x + 1)^{-3}$, 2‡

$-6(x + 1)^{-4}$, -6‡ $24(z + 1)^{-5}$. $\underline{\ln(x + 1)} = x - \frac{1}{2}x^2 + \frac{1}{3}x^3 - \frac{1}{4}x^4 + \dfrac{x^5}{5(z + 1)^5}$

[20] *DERIV*: $\sin x$, 0‡ $\cos x$, 1‡ $-\sin x$, 0‡ $-\cos x$, -1‡ $\sin x$, 0‡ $\cos x$, 1‡ $-\sin x$, 0‡

$-\cos x$, -1‡ $\sin z$. $\underline{\sin x} = x - \dfrac{x^3}{3!} + \dfrac{x^5}{5!} - \dfrac{x^7}{7!} + \dfrac{\sin z}{8!}x^8$

[21] *DERIV*: $\cos x$, 1‡ $-\sin x$, 0‡ $-\cos x$, -1‡ $\sin x$, 0‡ $\cos x$, 1‡ $-\sin x$, 0‡ $-\cos x$, -1‡

$\sin x$, 0‡ $\cos x$, 1‡ $-\sin z$. $\underline{\cos x} = 1 - \dfrac{x^2}{2!} + \dfrac{x^4}{4!} - \dfrac{x^6}{6!} + \dfrac{x^8}{8!} - \dfrac{\sin z}{9!}x^9$

[22] *DERIV*: $\tan^{-1}x$, 0‡ $(1 + x^2)^{-1}$, 1‡ $-2x(1 + x^2)^{-2}$, 0‡ $(6x^2 - 2)(1 + x^2)^{-3}$, -2‡

$24(z - z^3)(1 + z^2)^{-4}$. $\underline{\tan^{-1}x} = x - \frac{1}{3}x^3 + \dfrac{z - z^3}{(1 + z^2)^4}x^4$

[23] $f^{(k)}(x) = 2^k e^{2x}$, $f^{(k)}(0) = 2^k$‡ $f^{(6)}(z) = 64e^{2z}$.

$\underline{e^{2x}} = 1 + 2x + 2x^2 + \frac{4}{3}x^3 + \frac{2}{3}x^4 + \frac{4}{15}x^5 + \frac{4}{45}e^{2z}x^6$

[24] *DERIV*: $\sec x$, 1‡ $\sec x \tan x$, 0‡ $\sec x \tan^2 x + \sec^3 x$, 1‡ $\sec x \tan^3 x + 5\sec^3 x \tan x$, 0‡

$\sec z \tan^4 z + 18\sec^3 z \tan^2 z + 5\sec^5 z$.

$\underline{\sec x} = 1 + \frac{1}{2}x^2 + \frac{1}{24}(\sec z \tan^4 z + 18\sec^3 z \tan^2 z + 5\sec^5 z)\,x^4$

[25] *DERIV*: $(x - 1)^{-2}$, 1‡ $-2(x - 1)^{-3}$, 2‡ $6(x - 1)^{-4}$, 6‡ $-24(x - 1)^{-5}$, 24‡

$120(x - 1)^{-6}$, 120‡ $-720(x - 1)^{-7}$, 720‡ $5040(z - 1)^{-8}$.

$\underline{\dfrac{1}{(x - 1)^2}} = 1 + 2x + 3x^2 + 4x^3 + 5x^4 + 6x^5 + \dfrac{7x^6}{(z - 1)^8}$

[26] *DERIV*: $(4 - x)^{1/2}$, 2‡ $-\frac{1}{2}(4 - x)^{-1/2}$, $-\frac{1}{4}$‡ $-\frac{1}{4}(4 - x)^{-3/2}$, $-\frac{1}{32}$‡

$-\frac{3}{8}(4 - x)^{-5/2}$, $-\frac{3}{256}$‡ $-\frac{15}{16}(4 - z)^{-7/2}$.

$\underline{\sqrt{4 - x}} = 2 - \frac{1}{4}x - \frac{1}{64}x^2 - \frac{1}{512}x^3 - \dfrac{5x^4}{128(4 - z)^{7/2}}$

[27] *DERIV*: $\arcsin x$, 0‡ $(1 - x^2)^{-1/2}$, 1‡ $x(1 - x^2)^{-3/2}$, 0‡ $(1 + 2z^2)(1 - z^2)^{-5/2}$.

$\underline{\arcsin x} = x + \dfrac{1 + 2z^2}{6(1 - z^2)^{5/2}}x^3$

[28] *DERIV*: e^{-x^2}, 1‡ $-2xe^{-x^2}$, 0‡ $4x^2 e^{-x^2} - 2e^{-x^2}$, -2‡ $12xe^{-x^2} - 8x^3 e^{-x^2}$, 0‡

$16z^4 e^{-z^2} - 48z^2 e^{-z^2} + 12e^{-z^2}$. $\underline{e^{-x^2}} = 1 - x^2 + \dfrac{e^{-z^2}}{6}(3 - 12z^2 + 4z^4)\,x^4$

[29] *DERIV*: $2x^4 - 5x^3$, 0‡ $8x^3 - 15x^2$, 0‡ $24x^2 - 30x$, 0‡ $48x - 30$, -30‡ 48, 48‡ 0.

For both values of n, $\underline{f(x)} = -5x^3 + 2x^4$

[30] *DERIV*: $\cosh x$, 1‡ $\sinh x$, 0‡ $\cosh x$, 1‡ $\sinh x$, 0‡ $\cosh x$, 1‡ $\sinh x$, 0‡ $\sinh z$.

If $n = 4$, $\underline{\cosh x} = 1 + \dfrac{x^2}{2} + \dfrac{x^4}{24} + \dfrac{\sinh z}{120}x^5$.

If $n = 5$, $\underline{\cosh x} = 1 + \dfrac{x^2}{2} + \dfrac{x^4}{24} + \dfrac{\cosh z}{720}x^6$.

[31] Let $x = \frac{\pi}{2} - \frac{\pi}{180}$. Then, $\sin 89^\circ \approx 1 - \frac{1}{2}(-\frac{\pi}{180})^2 \approx 0.9998$.

Since $|\sin z| \le 1$, $\left|R_3(x)\right| \le \left|\frac{1}{24}(-\frac{\pi}{180})^4\right| \approx 4 \times 10^{-9}$.

[32] Let $x = \frac{\pi}{4} + \frac{\pi}{90}$. Then, $\cos 47^\circ \approx \frac{1}{\sqrt{2}} - \frac{1}{\sqrt{2}}(\frac{\pi}{90}) - \frac{1}{2\sqrt{2}}(\frac{\pi}{90})^2 + \frac{1}{6\sqrt{2}}(\frac{\pi}{90})^3 \approx 0.6820$.

Since $|\cos z| \le 1$, $|R_3(x)| \le \left|\frac{1}{24}(\frac{\pi}{90})^4\right| < 7 \times 10^{-8}$.

[33] Let $x = 4 + 0.03$. Then, $\sqrt{4.03} \approx 2 + \frac{1}{4}(0.03) - \frac{1}{64}(0.03)^2 + \frac{1}{512}(0.03)^3 \approx 2.0075$.

Since $\left|\frac{1}{z^{7/2}}\right| \le \left|\frac{1}{4^{7/2}}\right| \approx 0.0078$ on $(4, 4.03)$,

$|R_3(x)| \le \left|\frac{5}{128}(0.0078)(0.03)^4\right| < 3 \times 10^{-10}$.

[34] Let $x = 1 + 0.02$. Then, $e^{-1.02} \approx e^{-1} - e^{-1}(0.02) + \frac{1}{2}e^{-1}(0.02)^2 - \frac{1}{6}e^{-1}(0.02)^3 \approx$ 0.3606. Since $|e^{-z}| \le |e^{-1}| \approx 0.368$ on $(1, 1.02)$,

$|R_3(x)| \le \left|\frac{1}{24}(0.368)(0.02)^4\right| < 3 \times 10^{-9}$.

[35] Let $x = -2 - 0.2$. Then,

$-\frac{1}{2.2} \approx -\frac{1}{2} - \frac{1}{4}(-0.2) - \frac{1}{8}(-0.2)^2 - \frac{1}{16}(-0.2)^3 - \frac{1}{32}(-0.2)^4 - \frac{1}{64}(-0.2)^5 \approx$

-0.454545. Thus, $|R_5(x)| \le \left|-2^{-7}(-0.2)^6\right| = 5 \times 10^{-7}$.

[36] Let $x = -8 + -0.5$.

Then, $\sqrt[3]{-8.5} \approx -2 + \frac{1}{12}(-0.5) + \frac{1}{288}(-0.5)^2 + \frac{5}{20{,}736}(-0.5)^3 \approx -2.04083$.

Thus, $|R_3(x)| \le \left|\frac{10}{243} \cdot 8^{-11/3}(-0.5)^4\right| \approx 1.3 \times 10^{-6}$.

[37] Let $x = 0.25$. Then, $\ln 1.25 \approx 0.25 - \frac{1}{2}(0.25)^2 + \frac{1}{3}(0.25)^3 - \frac{1}{4}(0.25)^4 \approx 0.22298$.

Thus, $|R_4(x)| \le \left|\frac{(0.25)^5}{5(0 + 1)^5}\right| \approx 2 \times 10^{-4}$.

[38] Let $x = 0.1$. Then, $\sin 0.1 \approx 0.1 - \frac{(0.1)^3}{6} + \frac{(0.1)^5}{120} - \frac{(0.1)^7}{5040} \approx 0.0998334$.

Thus, $|R_7(x)| \le \left|\frac{1}{8!}(0.1)^8\right| \approx 2.5 \times 10^{-13}$.

[39] Let $x = \frac{\pi}{6}$. Then, $\cos 30^\circ \approx 1 - \frac{(\frac{\pi}{6})^2}{2} + \frac{(\frac{\pi}{6})^4}{24} - \frac{(\frac{\pi}{6})^6}{720} + \frac{(\frac{\pi}{6})^8}{40{,}320} \approx 0.8660254$.

Thus, $|R_8(x)| \le \left|\frac{1}{9!}(\frac{\pi}{6})^9\right| \approx 8.2 \times 10^{-9}$.

[40] Let $x = 10 + 0.01$. Then, $\log 10.01 \approx 1 + \frac{1}{10 \ln 10}(0.01) - \frac{1}{200 \ln 10}(0.01)^2 \approx$

1.0004341. Thus, $|R_2(x)| \le \left|\frac{1}{3(10)^3 \ln 10}(0.01)^3\right| \approx 1.5 \times 10^{-10}$.

[41] Using Exercise 21 with $n = 3$, $\underline{\cos x} = 1 - \frac{x^2}{2!} + \frac{\cos z}{4!}x^4$.

$|R_3(x)| \le \left|\frac{(1)(0.1)^4}{24}\right| \approx 4.2 \times 10^{-6} < 0.5 \times 10^{-5} \Rightarrow$ <u>five decimal places.</u>

[42] *DERIV*: $(1 + x)^{1/3}$, $1‡\ \frac{1}{3}(1 + x)^{-2/3}$, $\frac{1}{3}‡\ -\frac{2}{9}(1 + z)^{-5/3}$.

$\underline{\sqrt[3]{1 + x}} = 1 + \frac{1}{3}x - \frac{x^2}{9(1 + z)^{5/3}}$. $|R_1(x)| \le \left|\frac{(-0.1)^2}{9(0.9)^{5/3}}\right| \approx 0.0013 < 0.5 \times 10^{-2} \Rightarrow$

<u>two decimal places.</u>

[43] *DERIV*: e^x, 1‡ e^x, 1‡ e^x, 1‡ e^z. $\underline{e^x} = 1 + x + \frac{x^2}{2!} + \frac{e^z x^3}{3!}$.

$$|R_2(x)| \leq \left|\frac{e^{0.1}(0.1)^3}{6}\right| \approx 0.00018 < 0.5 \times 10^{-3} \Rightarrow \underline{\text{three decimal places}}.$$

[44] Using Exercise 20 with $n = 4$, $\underline{\sin x} = x - \frac{x^3}{6} + \frac{\cos z}{120}x^5$.

$$|R_4(x)| \leq \left|\frac{(1)(0.1)^5}{120}\right| \approx 8.3 \times 10^{-8} < 0.5 \times 10^{-6} \Rightarrow \underline{\text{six decimal places}}.$$

[45] Using Exercise 19 with $n = 3$, $\underline{\ln(x + 1)} = x - \frac{1}{2}x^2 + \frac{1}{3}x^3 - \frac{x^4}{4(z + 1)^4}$.

$$|R_3(x)| \leq \left|\frac{(-0.1)^4}{4(0.9)^4}\right| \approx 0.000038 < 0.5 \times 10^{-4} \Rightarrow \underline{\text{four decimal places}}.$$

[46] Using Exercise 30 with $n = 3$, $\underline{\cosh x} = 1 + \frac{1}{2}x^2 + \frac{\cosh z}{4!}x^4$.

$$|R_3(x)| \leq \left|\frac{\cosh(0.1)(0.1)^4}{24}\right| \approx 0.0000042 < 0.5 \times 10^{-5} \Rightarrow \underline{\text{five decimal places}}.$$

[47] If f is a polynomial of degree n, then the Taylor remainder

$R_n(x) = 0$, since $f^{(n+1)}(x) = 0$. By (11.45), we have $f(x) = P_n(x)$.

Exercises 11.10

[1] (a) $(1 + x)^{1/2} = 1 + \frac{1}{2}x + \frac{(\frac{1}{2})(-\frac{1}{2})}{2!}x^2 + \frac{(\frac{1}{2})(-\frac{1}{2})(-\frac{3}{2})}{3!}x^3 + \cdots =$

$$1 + \tfrac{1}{2}x - \tfrac{1}{8}x^2 + \sum_{n=3}^{\infty}(-1)^{n-1}\frac{1 \cdot 3 \cdot 5 \cdot \cdots \cdot (2n - 3)}{2^n\, n!}x^n;\ r = 1.$$

(b) Substituting $-x^3$ for x in part (a) yields

$$(1 - x^3)^{1/2} = 1 - \tfrac{1}{2}x^3 - \tfrac{1}{8}x^6 - \sum_{n=3}^{\infty}\frac{1 \cdot 3 \cdot 5 \cdot \cdots \cdot (2n - 3)}{2^n\, n!}x^{3n};\ r = 1.$$

[2] (a) $(1 + x)^{-1/3} = 1 - \frac{1}{3}x + \frac{(-\frac{1}{3})(-\frac{4}{3})}{2!}x^2 + \frac{(-\frac{1}{3})(-\frac{4}{3})(-\frac{7}{3})}{3!}x^3 + \cdots =$

$$1 - \tfrac{1}{3}x + \tfrac{2}{9}x^2 + \sum_{n=3}^{\infty}(-1)^n\frac{1 \cdot 4 \cdot 7 \cdot \cdots \cdot (3n - 2)}{3^n\, n!}x^n;\ r = 1.$$

(b) Substituting $-x^2$ for x in part (a) yields

$$(1 - x^2)^{1/3} = 1 + \tfrac{1}{3}x^2 + \tfrac{2}{9}x^4 + \sum_{n=3}^{\infty}\frac{1 \cdot 4 \cdot 7 \cdot \cdots \cdot (3n - 2)}{3^n\, n!}x^{2n};\ r = 1.$$

[3] $(1 + x)^{-2/3} = 1 - \frac{2}{3}x + \frac{(-\frac{2}{3})(-\frac{5}{3})}{2!}x^2 + \frac{(-\frac{2}{3})(-\frac{5}{3})(-\frac{8}{3})}{3!}x^3 + \cdots =$

$$1 - \tfrac{2}{3}x + \tfrac{5}{9}x^2 + \sum_{n=3}^{\infty}\frac{(-2)(-5)(-8)\cdots(1 - 3n)}{3^n\, n!}x^n;\ r = 1.$$

[4] $(1 + x)^{1/4} = 1 + \frac{1}{4}x + \frac{(\frac{1}{4})(-\frac{3}{4})}{2!}x^2 + \frac{(\frac{1}{4})(-\frac{3}{4})(-\frac{7}{4})}{3!}x^3 + \cdots =$

$$1 + \tfrac{1}{4}x - \tfrac{3}{32}x^2 + \sum_{n=3}^{\infty}\frac{(1)(-3)(-7)\cdots(5 - 4n)}{4^n\, n!}x^n;\ r = 1.$$

5 $(1-x)^{3/5} = 1 + \frac{3}{5}(-x) + \frac{(\frac{3}{5})(-\frac{2}{5})}{2!}(-x)^2 + \frac{(\frac{3}{5})(-\frac{2}{5})(-\frac{7}{5})}{3!}(-x)^3 + \cdots =$

$$1 - \tfrac{3}{5}x - \tfrac{3}{25}x^2 + \sum_{n=3}^{\infty} \frac{(3)(-2)(-7)\cdots(8-5n)}{5^n\, n!}(-x)^n;\ r = 1.$$

6 $(1-x)^{2/3} = 1 + \frac{2}{3}(-x) + \frac{(\frac{2}{3})(-\frac{1}{3})}{2!}(-x)^2 + \frac{(\frac{2}{3})(-\frac{1}{3})(-\frac{4}{3})}{3!}(-x)^3 + \cdots =$

$$1 - \tfrac{2}{3}x - \tfrac{1}{9}x^2 + \sum_{n=3}^{\infty} \frac{(2)(-1)(-4)\cdots(5-3n)}{3^n\, n!}(-x)^n;\ r = 1.$$

7 $(1+x)^{-2} = 1 - 2x + \frac{(-2)(-3)}{2!}x^2 + \frac{(-2)(-3)(-4)}{3!}x^3 + \cdots$

$$= 1 - 2x + 3x^2 + \sum_{n=3}^{\infty} \frac{(-2)(-3)(-4)\cdots(-1-n)}{n!}x^n$$

$$= 1 - 2x + 3x^2 + \sum_{n=3}^{\infty} (-1)^n (n+1)\, x^n;\ r = 1.$$

8 $(1+x)^{-4} = 1 - 4x + \frac{(-4)(-5)}{2!}x^2 + \frac{(-4)(-5)(-6)}{3!}x^3 + \cdots$

$$= 1 - 4x + 10x^2 + \sum_{n=3}^{\infty} \frac{(-4)(-5)(-6)\cdots(-3-n)}{n!}x^n$$

$$= 1 - 4x + 10x^2 + \sum_{n=3}^{\infty} (-1)^n \tfrac{1}{6}(n+1)(n+2)(n+3)\, x^n;\ r = 1.$$

9 $(1+x)^{-3} = 1 - 3x + \frac{(-3)(-4)}{2!}x^2 + \frac{(-3)(-4)(-5)}{3!}x^3 + \cdots$

$$= 1 - 3x + 6x^2 + \sum_{n=3}^{\infty} \frac{(-3)(-4)(-5)\cdots(-2-n)}{n!}x^n$$

$$= 1 - 3x + 6x^2 + \sum_{n=3}^{\infty} (-1)^n \tfrac{1}{2}(n+1)(n+2)\, x^n;\ r = 1.$$

10 $x(1+2x)^{-2} = x\left[1 + (-2)(2x) + \frac{(-2)(-3)}{2!}(2x)^2 + \frac{(-2)(-3)(-4)}{3!}(2x)^3 + \cdots\right]$

$$= x\left[1 - 4x + 12x^2 + \sum_{n=3}^{\infty} \frac{(-2)(-3)(-4)\cdots(-1-n)}{n!}2^n x^n\right]$$

$$= x - 4x^2 + 12x^3 + \sum_{n=3}^{\infty} (-1)^n (n+1)\, 2^n x^{n+1};\ |2x| < 1 \Rightarrow r = \tfrac{1}{2}.$$

11 $(8+x)^{1/3} = 2(1 + \frac{1}{8}x)^{1/3}$

$$= 2\left[1 + \tfrac{1}{3}(\tfrac{1}{8}x) + \frac{(\frac{1}{3})(-\frac{2}{3})}{2!}(\tfrac{1}{8}x)^2 + \frac{(\frac{1}{3})(-\frac{2}{3})(-\frac{5}{3})}{3!}(\tfrac{1}{8}x)^3 + \cdots\right]$$

$$= 2\left[1 + \tfrac{1}{24}x - \tfrac{1}{576}x^2 + \sum_{n=3}^{\infty} \frac{(-2)(-5)\cdots(4-3n)}{3^n\, n!\, 8^n}x^n\right]$$

$$= 2 + \tfrac{1}{12}x - \tfrac{1}{288}x^2 + 2\sum_{n=3}^{\infty} (-1)^{n-1} \frac{2\cdot 5\cdot \cdots \cdot (3n-4)}{24^n\, n!}x^n;$$

$$\left|\tfrac{1}{8}x\right| < 1 \Rightarrow r = 8.$$

[12] $(4+x)^{3/2} = 8(1+\frac{1}{4}x)^{3/2}$

$$= 8\left[1 + \tfrac{3}{2}(\tfrac{1}{4}x) + \frac{(\frac{3}{2})(\frac{1}{2})}{2!}(\tfrac{1}{4}x)^2 + \frac{(\frac{3}{2})(\frac{1}{2})(-\frac{1}{2})}{3!}(\tfrac{1}{4}x)^3 + \cdots\right]$$

$$= 8\left[1 + \tfrac{3}{8}x + \tfrac{3}{128}x^2 + \sum_{n=3}^{\infty}(3)\frac{(-1)(-3)(-5)\cdots(5-2n)}{2^n\, n!\, 4^n}x^n\right]$$

$$= 8 + 3x + \tfrac{3}{16}x^2 + 24\sum_{n=3}^{\infty}(-1)^n\frac{1\cdot 3\cdot 5\cdot\cdots\cdot(2n-5)}{8^n\, n!}x^n;$$

$$\left|\tfrac{1}{4}x\right| < 1 \Rightarrow r = 4.$$

[13] (a) $(1-t^2)^{-1/2} = 1 + (-\frac{1}{2})(-t^2) + \frac{(-\frac{1}{2})(-\frac{3}{2})}{2!}(-t^2)^2 + \frac{(-\frac{1}{2})(-\frac{3}{2})(-\frac{5}{2})}{3!}(-t^2)^3$

$$+ \cdots = 1 + \sum_{n=1}^{\infty}\frac{1\cdot 3\cdot 5\cdot\cdots\cdot(2n-1)}{2^n\, n!}t^{2n} \Rightarrow$$

$$\int_0^x (1-t^2)^{-1/2}\,dt = x + \sum_{n=1}^{\infty}\frac{1\cdot 3\cdot 5\cdot\cdots\cdot(2n-1)}{2^n\, n!\,(2n+1)}x^{2n+1}.$$

(b) $t^2 < 1 \Rightarrow |x| < 1 \Rightarrow r = 1.$

[14] (a) $(1+t^2)^{-1/2} = 1 + (-\frac{1}{2})(t^2) + \frac{(-\frac{1}{2})(-\frac{3}{2})}{2!}(t^2)^2 + \frac{(-\frac{1}{2})(-\frac{3}{2})(-\frac{5}{2})}{3!}(t^2)^3 + \cdots$

$$= 1 + \sum_{n=1}^{\infty}(-1)^n\frac{1\cdot 3\cdot 5\cdot\cdots\cdot(2n-1)}{2^n\, n!}t^{2n} \Rightarrow$$

$$\int_0^x (1+t^2)^{-1/2}\,dt = x + \sum_{n=1}^{\infty}(-1)^n\frac{1\cdot 3\cdot 5\cdot\cdots\cdot(2n-1)}{2^n\, n!\,(2n+1)}x^{2n+1}.$$

(b) $t^2 < 1 \Rightarrow |x| < 1 \Rightarrow r = 1.$

[15] Substituting x^3 for x in Exercise 1(a), $(1+x^3)^{1/2} = 1 + \frac{1}{2}x^3 - \frac{1}{8}x^6 + \cdots \Rightarrow$

$$\int_0^{1/2}(1+x^3)^{1/2}\,dx \approx \left[x + \tfrac{1}{8}x^4\right]_0^{1/2} \approx 0.508.\quad |\text{Error}| \le \tfrac{1}{8(7)}(\tfrac{1}{2})^7 < 0.5\times 10^{-3}.$$

[16] Substituting x^2 for x in Exercise 2(a), $(1+x^2)^{-1/3} = 1 - \frac{1}{3}x^2 + \frac{2}{9}x^4 - \frac{14}{81}x^6 + \cdots$

$$\Rightarrow \int_0^{1/2}(1+x^2)^{-1/3}\,dx \approx \left[x - \tfrac{1}{9}x^3 + \tfrac{2}{45}x^5\right]_0^{1/2} \approx 0.488.$$

$$|\text{Error}| \le \tfrac{14}{81(7)}(\tfrac{1}{2})^7 < 0.5\times 10^{-3}.$$

Note: In Exercises 17–18, since the series is not alternating, we cannot estimate the error by evaluating the next nonzero term. We could try to justify the error estimate by using the method in §11.9, but this was not the original intention.

[17] Substituting x^2 for x in Exercise 5, $(1-x^2)^{3/5} = 1 - \frac{3}{5}x^2 - \frac{3}{25}x^4 - \cdots \Rightarrow$

$$\int_0^{0.2}(1-x^2)^{3/5}\,dx \approx \left[x - \tfrac{1}{5}x^3\right]_0^{0.2} \approx 0.198.$$

The actual value is approximately 0.198392.

[18] Substituting x^3 for x in Exercise 6, $(1-x^3)^{2/3} = 1 - \frac{2}{3}x^3 - \frac{1}{9}x^6 - \cdots \Rightarrow$

$$\int_0^{0.4}(1-x^3)^{2/3}\,dx \approx \left[x - \tfrac{1}{6}x^4\right]_0^{0.4} \approx 0.396.$$

The actual value is approximately 0.395706.

[19] Substituting x^3 for x in Exercise 7, $(1 + x^3)^{-2} = 1 - 2x^3 + 3x^6 - \cdots \Rightarrow$

$$\int_0^{0.3} (1 + x^3)^{-2}\,dx = \left[x - \tfrac{1}{2}x^4\right]_0^{0.3} \approx 0.296. \quad |\text{Error}| \le \tfrac{3}{7}(0.3)^7 < 0.5 \times 10^{-3}.$$

[20] Substituting $5x^2$ for x in Exercise 8, $(1 + 5x^2)^{-4} = 1 - 4(5x^2) + 10(5x^2)^2 - \cdots \Rightarrow$

$$\int_0^{0.1} (1 + 5x^2)^{-4}\,dx = \left[x - \tfrac{20}{3}x^3\right]_0^{0.1} \approx 0.093. \quad |\text{Error}| \le 50(0.1)^5 = 0.5 \times 10^{-3}.$$

[21] From the graph, $|f(x) - g(x)| \le 0.1$ for approximately $-0.7 \le x \le 1$.

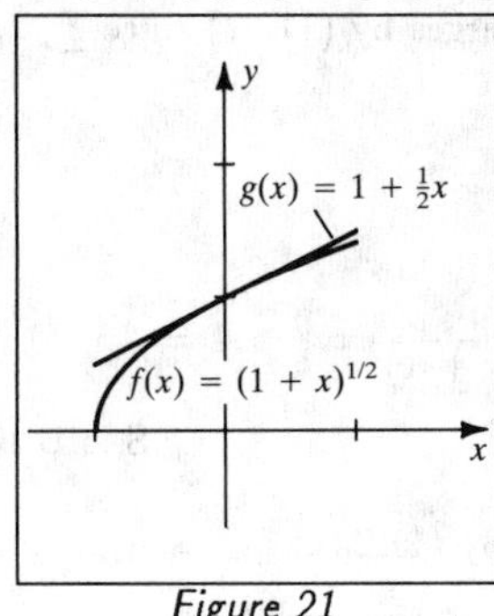

Figure 21

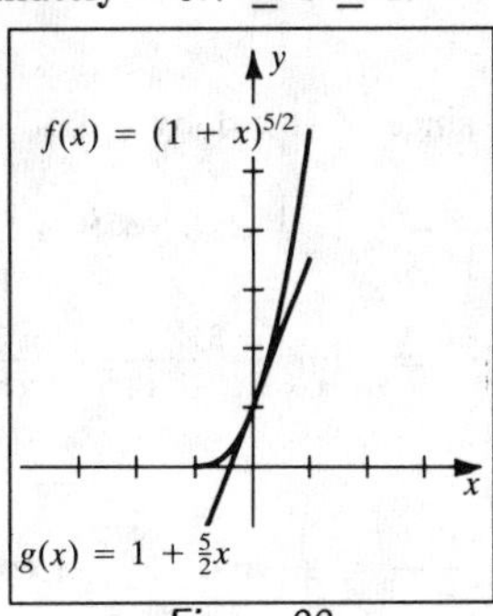

Figure 22

[22] From the graph, $|f(x) - g(x)| \le 0.1$ for approximately $-0.2 \le x \le 0.2$.

[23] (a) Since $(1 - x)^{-1/2} \approx 1 - \frac{1}{2}(-x) = 1 + \frac{1}{2}x$, let $(1 - k^2\sin^2 u)^{-1/2} \approx$

$1 + \frac{1}{2}k^2\sin^2 u$. Then, $T \approx 4\sqrt{\frac{L}{g}}\int_0^{\pi/2}(1 + \frac{1}{2}k^2\sin^2 u)\,du =$

$$4\sqrt{\frac{L}{g}}\int_0^{\pi/2}(1 + \tfrac{1}{4}k^2 - \tfrac{1}{4}k^2\cos 2u)\,du = 4\sqrt{\frac{L}{g}}\left[u + \tfrac{1}{4}k^2u - \tfrac{1}{8}k^2\sin 2u\right]_0^{\pi/2} =$$

$$4\sqrt{\frac{L}{g}}\left[\frac{\pi}{2} + \tfrac{1}{4}k^2\left(\frac{\pi}{2}\right)\right] = 2\pi\sqrt{\frac{L}{g}}\left(1 + \tfrac{1}{4}k^2\right).$$

(b) $\theta_0 = \frac{\pi}{6} \Rightarrow k = \sin(\frac{1}{2}\cdot\frac{\pi}{6}) = \sin\frac{\pi}{12}$. $T \approx 2\pi\sqrt{L/g}\,(1 + \frac{1}{4}\sin^2\frac{\pi}{12}) \approx 6.39\sqrt{L/g}$.

11.11 Review Exercises

Note: Let AC denote Absolutely Convergent; CC, Conditionally Convergent; D, Divergent; C, Convergent; and AST, the Alternating Series Test.

Note: In Exercises 1–6, let L denote the limit of the sequence if it exists.

[1] $\lim_{x\to\infty}\frac{\ln(x^2+1)}{x}\left\{\frac{\infty}{\infty}\right\} = \lim_{x\to\infty}\frac{2x}{x^2+1} = 0$. Thus, L = 0; C.

[2] $L = 100\lim_{n\to\infty}(0.99)^n = 0$; C.

[3] $\lim_{x\to\infty}\frac{10^x}{x^{10}} = \lim_{x\to\infty}\frac{10^x(\ln 10)}{10x^9} = \cdots = \lim_{x\to\infty}\frac{10^x(\ln 10)^{10}}{10!} = \infty$. Thus, L = ∞; D.

[4] L DNE, since $(-2)^n$ oscillates in sign with increasing magnitude; D.

[5] $L = \lim_{n\to\infty}\frac{5n}{(\sqrt{n}+4)(\sqrt{n}+9)} = \lim_{n\to\infty}\frac{5n}{n + 13\sqrt{n} + 36} = 5$; C.

[6] Let $u = \frac{2}{n}$. Then, $L = \lim_{u\to 0^+}\left[(1+u)^{1/u}\right]^4 = e^4$; C.

Note: In Exercises 7–32, let S denote the given series.

[7] Let $a_n = \dfrac{1}{\sqrt[3]{n(n+1)(n+2)}}$ and $b_n = \frac{1}{n}$. $\lim\limits_{n\to\infty} \frac{a_n}{b_n} =$

$$\lim_{n\to\infty} \frac{n}{\sqrt[3]{n(n+1)(n+2)}} = 1 > 0.$$ S diverges by (11.27) since $\sum b_n$ diverges.

[8] Let $a_n = \dfrac{(2n+3)^2}{(n+1)^3}$ and $b_n = \frac{1}{n}$. $\lim\limits_{n\to\infty} \frac{a_n}{b_n} = \lim\limits_{n\to\infty} \dfrac{n(2n+3)^2}{(n+1)^3} = 4 > 0.$

S diverges by (11.27) since $\sum b_n$ diverges.

[9] S is AC since $\left|-\frac{2}{3}\right| < 1$ by (11.15).

[10] $\lim\limits_{n\to\infty} \dfrac{1}{2 + (\frac{1}{2})^n} = \frac{1}{2} \neq 0 \Rightarrow$ S is D by (11.17).

[11] $\lim\limits_{n\to\infty} \frac{a_{n+1}}{a_n} = \lim\limits_{n\to\infty} \dfrac{3^{2n+3}}{(n+1)5^n} \cdot \dfrac{n5^{n-1}}{3^{2n+1}} = \lim\limits_{n\to\infty} \dfrac{9n}{5(n+1)} = \frac{9}{5} > 1 \Rightarrow$

S is D by (11.28).

[12] Since $\dfrac{1}{3^n + 2} < \dfrac{1}{3^n}$ and $\sum\limits_{n=1}^{\infty} \dfrac{1}{3^n}$ converges, S is C by (11.26).

[13] $\lim\limits_{n\to\infty} \frac{a_{n+1}}{a_n} = \lim\limits_{n\to\infty} \dfrac{(n+1)!}{\ln(n+2)} \cdot \dfrac{\ln(n+1)}{n!} = \lim\limits_{n\to\infty} \dfrac{(n+1)\ln(n+1)}{\ln(n+2)} = \infty \Rightarrow$

S is D by (11.28).

[14] $\lim\limits_{n\to\infty} \dfrac{n^2 - 1}{n^2 + 1} = 1 \neq 0 \Rightarrow$ S is D by (11.17).

[15] $\lim\limits_{n\to\infty} \left|\frac{a_{n+1}}{a_n}\right| = \lim\limits_{n\to\infty} \left|\dfrac{(n+1)^2 + 9}{(-2)^n} \cdot \dfrac{(-2)^{n-1}}{n^2 + 9}\right| = \frac{1}{2} < 1 \Rightarrow$ S is AC by (11.35).

[16] Since $\dfrac{n + \cos n}{n^3 + 1} < \dfrac{n+1}{n^3+1} < \dfrac{n+n}{n^3} = 2(\frac{1}{n^2})$ and $2\sum\limits_{n=1}^{\infty} \frac{1}{n^2}$ converges, S is C.

[17] $\lim\limits_{n\to\infty} \frac{a_{n+1}}{a_n} = \lim\limits_{n\to\infty} \dfrac{e^{n+1}}{(n+1)^e} \cdot \dfrac{n^e}{e^n} = e > 1 \Rightarrow$ S is D.

[18] $a_n = \dfrac{n}{n^2+1} > 0$. $f(x) = \dfrac{x}{x^2+1} \Rightarrow f'(x) = \dfrac{1 - x^2}{(x^2+1)^2} < 0$ for $x > 1$ and

$\lim\limits_{n\to\infty} a_n = 0 \Rightarrow$ S converges by AST. If $b_n = \frac{1}{n}$, $\lim\limits_{n\to\infty} \frac{a_n}{b_n} = 1 > 0$ and

$\sum a_n$ diverges. Thus, S is CC.

[19] $\lim\limits_{n\to\infty} \left|(-1)^n \dfrac{1}{\sqrt[n]{n}}\right| = \lim\limits_{n\to\infty} \dfrac{1}{n^{1/n}}$ {11.1.37} $= 1 \neq 0 \Rightarrow$ S is D by (11.17).

[20] $\lim\limits_{n\to\infty} \left|\frac{a_{n+1}}{a_n}\right| = \lim\limits_{n\to\infty} \left|\dfrac{(0.9)^{n+1}}{\ln(n+1)} \cdot \dfrac{\ln n}{(0.9)^n}\right| = 0.9 < 1 \Rightarrow$ S is AC.

[21] Since $\left|\dfrac{\sin(\frac{5\pi}{3}n)}{n^{5\pi/3}}\right| < \left|\dfrac{1}{n^{5\pi/3}}\right|$ and $\sum\limits_{n=1}^{\infty} \dfrac{1}{n^{5\pi/3}}$ converges $\{p = \frac{5\pi}{3} > 1\}$, S is AC.

[22] Let $a_n = \dfrac{\sqrt[3]{n-1}}{n^2 - 1}$ and $b_n = \dfrac{1}{n^{5/3}}$. $\lim\limits_{n\to\infty} \frac{a_n}{b_n} = \lim\limits_{n\to\infty} \dfrac{n^{5/3}\sqrt[3]{n-1}}{n^2 - 1} = 1 > 0.$

Thus, S is AC.

[23] $a_n = \dfrac{\sqrt{n}}{n+1} > 0.$ $f(x) = \dfrac{\sqrt{x}}{x+1} \Rightarrow f'(x) = \dfrac{1-x}{2\sqrt{x}(x+1)^2} < 0$ for $x > 1$ and

$\lim\limits_{n\to\infty} a_n = 0 \Rightarrow$ S converges by AST. If $b_n = \dfrac{1}{\sqrt{n}}$, $\lim\limits_{n\to\infty} \dfrac{a_n}{b_n} = 1 > 0$ and

$\sum a_n$ diverges. Thus, S is CC.

[24] $\lim\limits_{n\to\infty} \left|\dfrac{a_{n+1}}{a_n}\right| = \lim\limits_{n\to\infty} \left|\dfrac{2n+5}{(n+1)!} \cdot \dfrac{n!}{2n+3}\right| = \lim\limits_{n\to\infty} \left|\dfrac{2n+5}{(2n+3)(n+1)}\right| = 0 < 1 \Rightarrow$

S is AC.

[25] $\dfrac{1-\cos n}{n^2} \le \dfrac{2}{n^2}$ and $2\sum\limits_{n=1}^{\infty} \dfrac{1}{n^2}$ converges $\Rightarrow$ S is C.

[26] $\lim\limits_{n\to\infty} \left|(-1)^{n-1}\dfrac{2 \cdot 4 \cdot \cdots \cdot (2n)}{n!}\right| = \lim\limits_{n\to\infty} \dfrac{2^n\, n!}{n!} = \infty \neq 0 \Rightarrow$ S is D.

[27] $\lim\limits_{n\to\infty} \sqrt[n]{a_n} = \lim\limits_{n\to\infty} \sqrt[n]{\dfrac{(2n)^n}{n^{2n}}} = \lim\limits_{n\to\infty} \dfrac{2n}{n^2} = 0 < 1 \Rightarrow$ S is C by (11.29).

[28] $\lim\limits_{n\to\infty} \dfrac{3^{n-1}}{n^2+9} = \infty \neq 0 \Rightarrow$ S is D.

[29] $\lim\limits_{n\to\infty} \dfrac{a_{n+1}}{a_n} = \lim\limits_{n\to\infty} \dfrac{e^{2n+2}}{(2n+1)!} \cdot \dfrac{(2n-1)!}{e^{2n}} = \lim\limits_{n\to\infty} \dfrac{e^2}{(2n+1)(2n)} = 0 < 1 \Rightarrow$ S is C.

[30] Since $\sum\limits_{n=1}^{\infty} \dfrac{1}{3^n}$ converges and $-5\sum\limits_{n=1}^{\infty} \dfrac{1}{\sqrt{n}}$ diverges, S is D by (11.21).

[31] $a_n = \dfrac{\sqrt{\ln n}}{n} > 0$ for $n > 1$. $f(x) = \dfrac{\sqrt{\ln x}}{x} \Rightarrow f'(x) = \dfrac{1 - 2\ln x}{2x^2\sqrt{\ln x}} < 0$ for

$x > e^{1/2}$ $\{\approx 1.65\}$ and $\lim\limits_{n\to\infty} a_n = 0 \Rightarrow$ S converges by AST.

However, $\dfrac{\sqrt{\ln n}}{n} > \dfrac{1}{n}$ for $n > e$ and $\sum\limits_{n=2}^{\infty} \dfrac{1}{n}$ diverges. Thus, S is CC.

[32] $\dfrac{\tan^{-1} n}{\sqrt{1+n^2}} \ge \dfrac{\pi/4}{\sqrt{1+n^2}} \ge \dfrac{\pi/4}{\sqrt{n^2+n^2}} = \dfrac{\pi/4}{\sqrt{2}\,n}$ and $\dfrac{\pi}{4\sqrt{2}}\sum\limits_{n=1}^{\infty} \dfrac{1}{n}$ diverges $\Rightarrow$ S is D.

Note: In Exer. 33–38, each $f(x)$ is positive, continuous, and decreasing on the interval of integration.

[33] Let $f(x) = \dfrac{1}{(3x+2)^3}$. $\displaystyle\int_1^{\infty} f(x)\,dx = \lim_{t\to\infty}\left[-\frac{1}{6(3x+2)^2}\right]_1^t = 0 - \left(-\frac{1}{150}\right)$; C

[34] Let $f(x) = \dfrac{x}{\sqrt{x^2-1}}$. $\displaystyle\int_2^{\infty} f(x)\,dx = \lim_{t\to\infty}\left[\sqrt{x^2-1}\right]_2^t = \infty$; D

[35] Let $f(x) = \dfrac{e^{1/x}}{x^2}$. $\displaystyle\int_1^{\infty} f(x)\,dx = \lim_{t\to\infty}\left[-e^{1/x}\right]_1^t = -1 - (-e)$; C

[36] Let $f(x) = \dfrac{1}{x(\ln x)^3}$. $\displaystyle\int_2^{\infty} f(x)\,dx = \lim_{t\to\infty}\left[-\frac{1}{2(\ln x)^2}\right]_2^t = 0 - \left(-\frac{1}{2(\ln 2)^2}\right)$; C

[37] Let $f(x) = \dfrac{10}{\sqrt[3]{x+8}}$. $\displaystyle\int_1^{\infty} f(x)\,dx = \lim_{t\to\infty}\left[15(x+8)^{2/3}\right]_1^t = \infty$; D

[38] Let $f(x) = \dfrac{1}{x(x-4)} = \dfrac{\frac{1}{4}}{x-4} - \dfrac{\frac{1}{4}}{x}$ {PF}. $\displaystyle\int_5^\infty f(x)\,dx =$

$$\lim_{t\to\infty}\left[\tfrac{1}{4}\ln|x-4| - \tfrac{1}{4}\ln|x|\right]_5^t = \lim_{t\to\infty}\left[\tfrac{1}{4}\ln\left|\frac{x-4}{x}\right|\right]_5^t = \tfrac{1}{4}(\ln 1 - \ln\tfrac{1}{5});\ \text{C}$$

[39] $a_n = \dfrac{1}{(2n+1)!} < 0.0005 \Rightarrow (2n+1)! > 2000 \Rightarrow 2n+1 \geq 7 \Rightarrow n \geq 3.$

Thus, $S \approx S_2 = \dfrac{1}{3!} - \dfrac{1}{5!} \approx 0.158.$

[40] $a_n = \dfrac{1}{n^2(n^2+1)} < 0.0005 \Rightarrow n^4 + n^2 > 2000 \Rightarrow n \geq 7.$

Thus, $S = S_6 \approx \frac{1}{1(2)} - \frac{1}{4(5)} + \frac{1}{9(10)} - \frac{1}{16(17)} + \frac{1}{25(26)} - \frac{1}{36(37)} \approx 0.458.$

Note: In Exer. 41–46, let u_n denote the nth term of the power series.

[41] $u_n = \dfrac{(n+1)x^n}{(-3)^n} \Rightarrow \displaystyle\lim_{n\to\infty}\left|\frac{u_{n+1}}{u_n}\right| = \lim_{n\to\infty}\left|\frac{(n+2)\,x^{n+1}}{(-3)^{n+1}} \cdot \frac{(-3)^n}{(n+1)\,x^n}\right| = \tfrac{1}{3}|x|.$

$\tfrac{1}{3}|x| < 1 \Leftrightarrow -3 < x < 3.$ If $x = 3$, $\displaystyle\sum_{n=0}^{\infty}(-1)^n(n+1)$ is D.

If $x = -3$, $\displaystyle\sum_{n=0}^{\infty}(n+1)$ is D. ★ $(-3, 3)$

[42] $u_n = (-1)^n\dfrac{4^{2n}x^n}{\sqrt{n+1}} \Rightarrow \displaystyle\lim_{n\to\infty}\left|\frac{u_{n+1}}{u_n}\right| = \lim_{n\to\infty}\left|\frac{4^{2n+2}\,x^{n+1}}{\sqrt{n+2}} \cdot \frac{\sqrt{n+1}}{4^{2n}\,x^n}\right| = 16|x|.$

$16|x| < 1 \Leftrightarrow -\frac{1}{16} < x < \frac{1}{16}.$ If $x = \frac{1}{16}$, $\displaystyle\sum_{n=0}^{\infty}(-1)^n\frac{1}{\sqrt{n+1}}$ converges by AST.

If $x = -\frac{1}{16}$, $\displaystyle\sum_{n=0}^{\infty}\frac{1}{\sqrt{n+1}}$ diverges. ★ $(-\frac{1}{16}, \frac{1}{16}]$

[43] $u_n = \dfrac{1}{n\cdot 2^n}(x+10)^n \Rightarrow$

$$\lim_{n\to\infty}\left|\frac{u_{n+1}}{u_n}\right| = \lim_{n\to\infty}\left|\frac{(x+10)^{n+1}}{(n+1)\,2^{n+1}} \cdot \frac{n\,2^n}{(x+10)^n}\right| = \tfrac{1}{2}|x+10|.$$

$\tfrac{1}{2}|x+10| < 1 \Leftrightarrow -12 < x < -8.$ If $x = -8$, $\displaystyle\sum_{n=1}^{\infty}\frac{1}{n}$ diverges.

If $x = -12$, $\displaystyle\sum_{n=1}^{\infty}(-1)^n\frac{1}{n}$ converges. ★ $[-12, -8)$

[44] $u_n = \dfrac{1}{n(\ln n)^2}(x-1)^n \Rightarrow$

$$\lim_{n\to\infty}\left|\frac{u_{n+1}}{u_n}\right| = \lim_{n\to\infty}\left|\frac{(x-1)^{n+1}}{(n+1)\left[\ln(n+1)\right]^2} \cdot \frac{n(\ln n)^2}{(x-1)^n}\right| = |x-1|.$$

$|x-1| < 1 \Leftrightarrow 0 < x < 2.$ If $x = 2$, $\displaystyle\sum_{n=2}^{\infty}\frac{1}{n(\ln n)^2}$ converges by Exercise 8,

§11.3. If $x = 0$, $\displaystyle\sum_{n=2}^{\infty}(-1)^n\frac{1}{n(\ln n)^2}$ converges by AST. ★ $[0, 2]$

[45] $\displaystyle\lim_{n\to\infty}\left|\frac{u_{n+1}}{u_n}\right| = \lim_{n\to\infty}\left|\frac{(2n+2)!\,x^{n+1}}{(n+1)!\,(n+1)!} \cdot \frac{n!\,n!}{(2n)!\,x^n}\right| =$

$\displaystyle\lim_{n\to\infty}\frac{(2n+2)(2n+1)}{(n+1)(n+1)}|x| = 4|x|.$ $4|x| < 1 \Leftrightarrow |x| < \frac{1}{4}.$ ★ $r = \frac{1}{4}.$

[46] $\lim_{n\to\infty}\left|\frac{u_{n+1}}{u_n}\right| = \lim_{n\to\infty}\left|\frac{(x+5)^{n+1}}{(n+6)!}\cdot\frac{(n+5)!}{(x+5)^n}\right| = \lim_{n\to\infty}\frac{|x+5|}{n+6} = 0.$ ★ $r = \infty$.

[47] If $x \neq 0$, using (11.48)(b), we have $\frac{1-\cos x}{x} = \left(\frac{1}{x}\right)\left(\frac{x^2}{2!} - \frac{x^4}{4!} + \frac{x^6}{6!} - \cdots\right) =$

$\sum_{n=1}^{\infty}(-1)^{n+1}\frac{x^{2n-1}}{(2n)!}$. At $x = 0$, the series equals 0. $r = \infty$.

[48] By (11.48)(c), $xe^{-2x} = x\sum_{n=0}^{\infty}\frac{(-2x)^n}{n!} = \sum_{n=0}^{\infty}(-1)^n\frac{2^n x^{n+1}}{n!}$. $r = \infty$.

[49] By (11.48)(a), $\sin x\cos x = \frac{1}{2}\sin 2x = \frac{1}{2}\sum_{n=0}^{\infty}(-1)^n\frac{(2x)^{2n+1}}{(2n+1)!} = \sum_{n=0}^{\infty}(-1)^n\frac{2^{2n}x^{2n+1}}{(2n+1)!}$.

$r = \infty$.

[50] $\frac{d}{dx}\left[\ln(2+x)\right] = \frac{1}{2+x} = \frac{\frac{1}{2}}{1-(-\frac{1}{2}x)} = \frac{1}{2}\sum_{n=0}^{\infty}(-1)^n\left(\frac{x}{2}\right)^n$ for $\left|\frac{x}{2}\right| < 1$.

By (5.30), $\ln(2+x) - \ln(2+0) = \int_0^x \frac{1}{2+t}\,dt = \sum_{n=0}^{\infty}(-1)^n\frac{x^{n+1}}{2^{n+1}(n+1)} \Rightarrow$

$\ln(2+x) = \ln 2 + \sum_{n=1}^{\infty}(-1)^{n-1}\frac{x^n}{2^n n}$.

Since integration does not change the radius of convergence, $r = 2$.

[51] Using (11.50), $(1+x)^{2/3} = 1 + \frac{2}{3}x + \frac{(\frac{2}{3})(-\frac{1}{3})}{2!}x^2 + \frac{(\frac{2}{3})(-\frac{1}{3})(-\frac{4}{3})}{3!}x^3 + \cdots =$

$1 + \frac{2}{3}x + 2\sum_{n=2}^{\infty}(-1)^{n-1}\frac{1\cdot 4\cdot 7\cdot\cdots\cdot(3n-5)}{3^n\, n!}x^n$; $r = 1$.

[52] By Exercise 13, §11.10, $(1-x^2)^{-1/2} = 1 + \sum_{n=1}^{\infty}\frac{1\cdot 3\cdot 5\cdot\cdots\cdot(2n-1)}{2^n\, n!}x^{2n}$; $r = 1$.

[53] Let $c = -2$. *DERIV*: e^{-x}, e^2‡ $-e^{-x}$, $-e^2$‡ e^{-x}, e^2‡ ....

Thus, $e^{-x} = e^2\sum_{n=0}^{\infty}(-1)^n\frac{1}{n!}(x+2)^n$.

[54] Let $c = \frac{\pi}{2}$. *DERIV*: $\cos x$, 0‡ $-\sin x$, -1‡ $-\cos x$, 0‡ $\sin x$, 1‡ ....

Thus, $\cos x = \sum_{n=0}^{\infty}(-1)^{n+1}\frac{1}{(2n+1)!}(x-\frac{\pi}{2})^{2n+1}$. Note that $\cos x = \sin(\frac{\pi}{2} - x) =$

$-\sin(x - \frac{\pi}{2})$. The above series could be obtained using this fact and (11.48)(a).

[55] $\sqrt{x} = \sqrt{4 + (x-4)} = 2\left[1 + \frac{1}{4}(x-4)\right]^{1/2}$. Using the binomial series,

$$\sqrt{x} = 2\left[1 + \tfrac{1}{8}(x-4) + \frac{(\frac{1}{2})(-\frac{1}{2})}{2!}\left[\tfrac{1}{4}(x-4)\right]^2 + \frac{(\frac{1}{2})(-\frac{1}{2})(-\frac{3}{2})}{3!}\left[\tfrac{1}{4}(x-4)\right]^3 + \cdots\right]$$

$$= 2 + \tfrac{1}{4}(x-4) + 2\sum_{n=2}^{\infty}(-1)^{n-1}\frac{1\cdot 3\cdot 5\cdot\cdots\cdot(2n-3)}{2^n\, n!\, 4^n}(x-4)^n$$

$$= 2 + \tfrac{1}{4}(x-4) + \sum_{n=2}^{\infty}(-1)^{n-1}\frac{1\cdot 3\cdot 5\cdot\cdots\cdot(2n-3)}{2^{3n-1}\, n!}(x-4)^n.$$

[56] $e^{-1/3} \approx 1 - \frac{1}{3} + \frac{(\frac{1}{3})^2}{2!} - \frac{(\frac{1}{3})^3}{3!} + \frac{(\frac{1}{3})^4}{4!} \approx 0.717,$

with $|\text{error}| \le \frac{(\frac{1}{3})^5}{5!} \approx 3.4 \times 10^{-5} < 0.5 \times 10^{-3}$.

[57] $\displaystyle\int_0^1 x^2 e^{-x^2}\,dx = \int_0^1 x^2\left[1 - x^2 + \frac{x^4}{2!} - \frac{x^6}{3!} + \frac{x^8}{4!} - \frac{x^{10}}{5!} + \frac{x^{12}}{6!} - \cdots\right]dx$

$$= \left[\tfrac{1}{3}x^3 - \tfrac{1}{5}x^5 + \frac{x^7}{7(2!)} - \frac{x^9}{9(3!)} + \frac{x^{11}}{11(4!)} - \frac{x^{13}}{13(5!)} + \frac{x^{15}}{15(6!)} - \cdots\right]_0^1.$$

Using the first 6 terms, $I \approx \frac{1}{3} - \frac{1}{5} + \frac{1}{14} - \frac{1}{54} + \frac{1}{264} - \frac{1}{1560} \approx 0.189$,

with $|\text{error}| \le \frac{1}{10{,}800} \approx 9.26 \times 10^{-5} < 0.5 \times 10^{-3}$.

[58] $\displaystyle\int_0^1 \frac{\sin x}{\sqrt{x}}\,dx = \int_0^1\left[x^{1/2} - \frac{x^{5/2}}{3!} + \frac{x^{9/2}}{5!} - \frac{x^{13/2}}{7!} + \cdots\right]dx$

$= \left[\tfrac{2}{3}x^{3/2} - \frac{2x^{7/2}}{7(3!)} + \frac{2x^{11/2}}{11(5!)} - \frac{2x^{15/2}}{15(7!)} + \cdots\right]_0^1$. Using the first 3 terms,

$I \approx \frac{2}{3} - \frac{2}{42} + \frac{2}{1320} \approx 0.621$, with $|\text{error}| \le \frac{1}{37{,}800} \approx 2.65 \times 10^{-5} < 0.5 \times 10^{-3}$.

Note that the series expansion for f equals 0 at $x = 0$.

[59] $(1 + x)^{1/5} = 1 + \frac{1}{5}x + \frac{(\frac{1}{5})(-\frac{4}{5})}{2!}x^2 + \cdots$. Thus, $(1 + 0.01)^{1/5} \approx$

$1 + \frac{1}{5}(0.01) = 1.002$, with $|\text{error}| \le \frac{4}{50}(0.01)^2 = 8 \times 10^{-6} < 0.5 \times 10^{-3}$.

[60] $e^{-0.25} = 1 - (0.25) + \frac{(0.25)^2}{2!} - \frac{(0.25)^3}{3!} + \frac{(0.25)^4}{4!} - \cdots \approx 1 - (0.25) +$

$\frac{(0.25)^2}{2!} - \frac{(0.25)^3}{3!} \approx 0.779$, with $|\text{error}| \le \frac{(0.25)^4}{4!} \approx 1.63 \times 10^{-4} < 0.5 \times 10^{-3}$.

[61] *DERIV*: $\ln\cos x$, $\ln(\frac{\sqrt{3}}{2})$‡ $-\tan x$, $-\frac{\sqrt{3}}{3}$‡ $-\sec^2 x$, $-\frac{4}{3}$‡ $-2\sec^2 x\tan x$, $-\frac{8\sqrt{3}}{9}$‡

$-2\sec^4 z - 4\sec^2 z\tan^2 z$.

$\underline{\ln\cos x} = \ln(\frac{\sqrt{3}}{2}) - \frac{\sqrt{3}}{3}(x - \frac{\pi}{6}) - \frac{2}{3}(x - \frac{\pi}{6})^2 - \frac{4\sqrt{3}}{27}(x - \frac{\pi}{6})^3$

$- \frac{1}{12}(\sec^4 z + 2\sec^2 z\tan^2 z)(x - \frac{\pi}{6})^4$, z is between x and $\frac{\pi}{6}$.

[62] *DERIV*: $(x - 1)^{1/2}$, 1‡ $\frac{1}{2}(x - 1)^{-1/2}$, $\frac{1}{2}$‡ $-\frac{1}{4}(x - 1)^{-3/2}$, $-\frac{1}{4}$‡ $\frac{3}{8}(x - 1)^{-5/2}$, $\frac{3}{8}$‡

$-\frac{15}{16}(x - 1)^{-7/2}$, $-\frac{15}{16}$‡ $\frac{105}{32}(z - 1)^{-9/2}$.

$\underline{\sqrt{x - 1}} = 1 + \frac{1}{2}(x - 2) - \frac{1}{8}(x - 2)^2 + \frac{1}{16}(x - 2)^3 - \frac{5}{128}(x - 2)^4 +$

$\frac{7}{256}(z - 1)^{-9/2}(x - 2)^5$, z is between x and 2.

[63] *DERIV*: e^{-x^2}, 1‡ $-2xe^{-x^2}$, 0‡ $(4x^2 - 2)e^{-x^2}$, -2‡ $(-8x^3 + 12x)e^{-x^2}$, 0‡

$(16z^4 - 48z^2 + 12)e^{-z^2}$.

$\underline{e^{-x^2}} = 1 - x^2 + \frac{1}{6}(4z^4 - 12z^2 + 3)e^{-z^2}x^4$, z is between x and 0.

[64] *DERIV*: $(1-x)^{-1}$, 1‡ $1(1-x)^{-2}$, 1‡ $2(1-x)^{-3}$, 2‡ $6(1-x)^{-4}$, 6‡
$24(1-x)^{-5}$, 24‡ $120(1-x)^{-6}$, 120‡ $720(1-x)^{-7}$, 720‡ $5040(1-z)^{-8}$.

$$\frac{1}{1-x} = 1 + x + x^2 + x^3 + x^4 + x^5 + x^6 + \frac{x^7}{(1-z)^8},$$

z is between x and 0.

[65] Using Exercise 8 in §11.9 with $x = \frac{\pi}{4} - \frac{\pi}{90}$,

$\cos x = \frac{1}{\sqrt{2}} - \frac{1}{\sqrt{2}}(-\frac{\pi}{90}) - \frac{1}{2\sqrt{2}}(-\frac{\pi}{90})^2 + \frac{\sin z}{6}(-\frac{\pi}{90})^3$, z is between x and $\frac{\pi}{4}$.

Using the first three terms, $\cos 43^\circ \approx 0.7314$,

$$\text{with } |\text{error}| \le \left|R_3(x)\right| = \left|\frac{\sin z}{6}(-\frac{\pi}{90})^3\right| < \frac{\pi^3}{6 \cdot 90^3} \approx 7.09 \times 10^{-6} < 0.5 \times 10^{-4}.$$

[66] Using Exercise 20, §11.9, $\sin x = x - \frac{1}{6}x^3 + \frac{1}{120}x^5 - \frac{1}{5040}(\cos z)\,x^7$,

z is between 0 and x. $\left|R_6(x)\right| \le \left|\frac{1}{5040}(1)(\frac{\pi}{4})^7\right| \approx 3.66 \times 10^{-5} < 0.5 \times 10^{-4}$.

Chapter 12: Topics From Analytic Geometry

Exercises 12.1

Note: Let V, F, and l denote the vertex, focus, and directrix, respectively.

[1] $y = -\frac{1}{12}x^2 \Rightarrow a = -\frac{1}{12}$. $p = \frac{1}{4a} = \frac{1}{4(-\frac{1}{12})} = -3$. $V(0, 0)$; $F(0, -3)$; $y = 3$

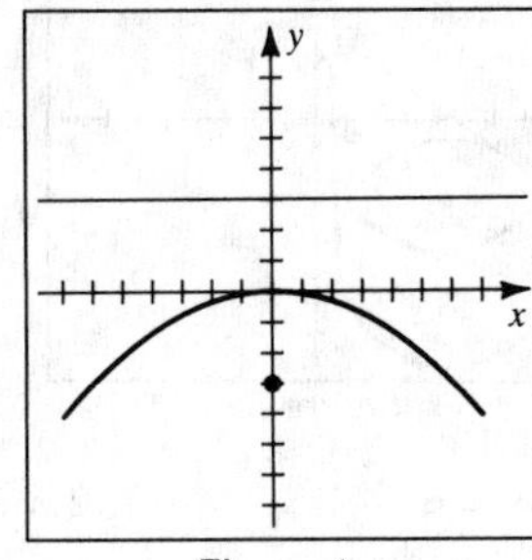

Figure 1

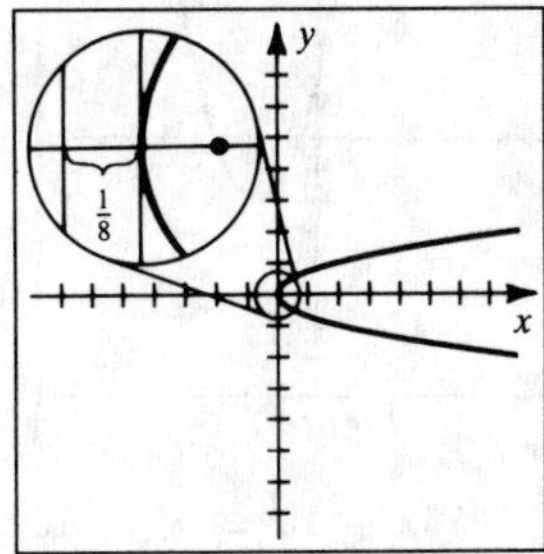

Figure 2

[2] $x = 2y^2 \Rightarrow a = 2$. $p = \frac{1}{4a} = \frac{1}{4(2)} = \frac{1}{8}$. $V(0, 0)$; $F(\frac{1}{8}, 0)$; $x = -\frac{1}{8}$

[3] $2y^2 = -3x \Rightarrow x = -\frac{2}{3}y^2 \Rightarrow a = -\frac{2}{3}$. $p = \frac{1}{4(-\frac{2}{3})} = -\frac{3}{8}$.

$V(0, 0)$; $F(-\frac{3}{8}, 0)$; $x = \frac{3}{8}$

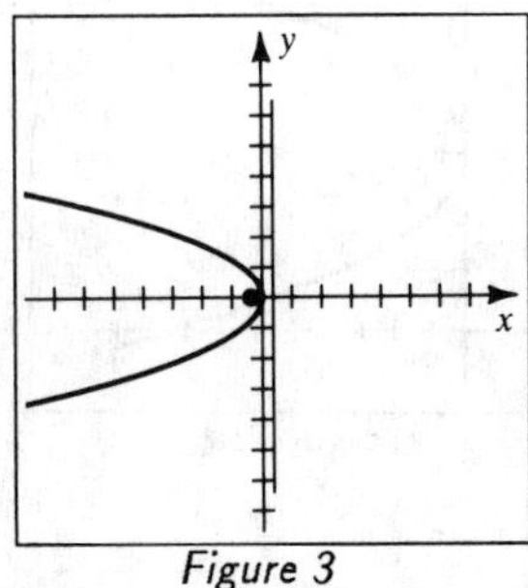

Figure 3

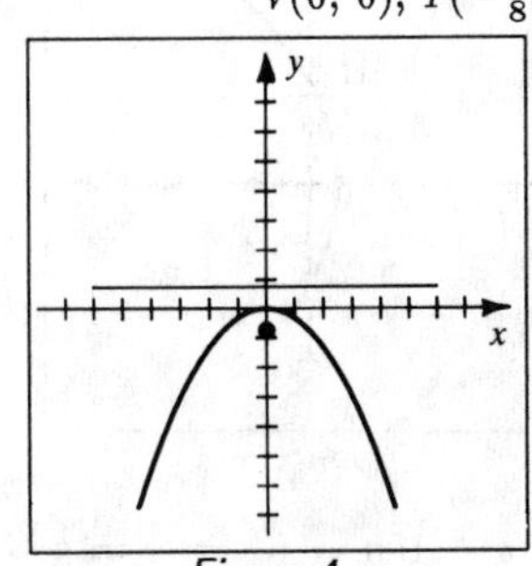

Figure 4

[4] $x^2 = -3y \Rightarrow y = -\frac{1}{3}x^2 \Rightarrow a = -\frac{1}{3}$. $p = \frac{1}{4(-\frac{1}{3})} = -\frac{3}{4}$. $V(0, 0)$; $F(0, -\frac{3}{4})$; $y = \frac{3}{4}$

[5] $y = 8x^2 \Rightarrow a = 8$. $p = \frac{1}{4(8)} = \frac{1}{32}$. $V(0, 0)$; $F(0, \frac{1}{32})$; $y = -\frac{1}{32}$

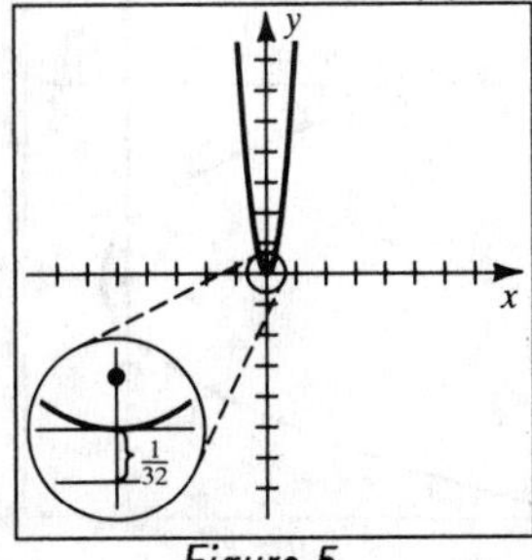

Figure 5

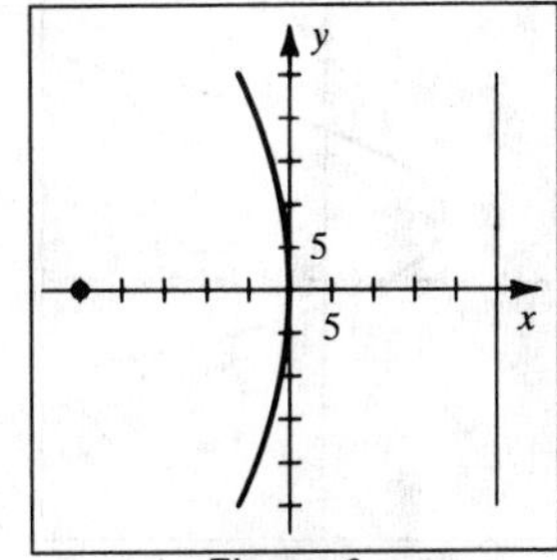

Figure 6

[6] $y^2 = -100x \Rightarrow x = -\frac{1}{100}y^2 \Rightarrow a = -\frac{1}{100}$. $p = \frac{1}{4(-\frac{1}{100})} = -25$.

$V(0, 0)$; $F(-25, 0)$; $x = 25$

[7] $y = x^2 - 4x + 2 \Rightarrow y' = 2x - 4$. $y' = 0 \Rightarrow x = 2$ and hence $y = -2$.

The vertex is $V(2, -2)$. $y = x^2 - 4x + 2 \Rightarrow a = 1$. $p = \frac{1}{4(1)} = \frac{1}{4}$.

The focus is p units from the vertex, i.e., $F(2, -2 + \frac{1}{4}) = F(2, -\frac{7}{4})$.

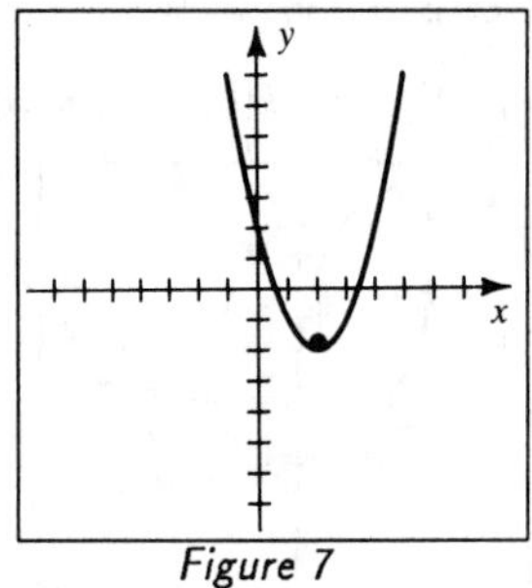

Figure 7

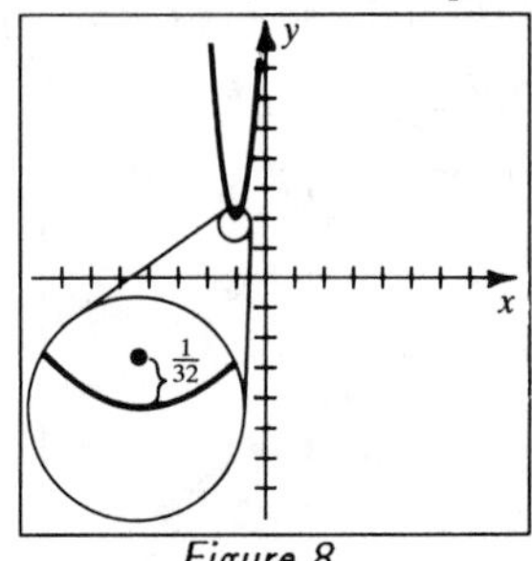

Figure 8

[8] $y = 8x^2 + 16x + 10 \Rightarrow y' = 16x + 16$. $y' = 0 \Rightarrow x = -1 \Rightarrow V(-1, 2)$.

$a = 8 \Rightarrow p = \frac{1}{32} \Rightarrow F(-1, \frac{65}{32})$.

[9] $y^2 - 12 = 12x \Rightarrow x = \frac{1}{12}y^2 - 1 \Rightarrow x' = \frac{1}{6}y$. $x' = 0 \Rightarrow y = 0 \Rightarrow V(-1, 0)$.

$a = \frac{1}{12} \Rightarrow p = 3 \Rightarrow F(2, 0)$.

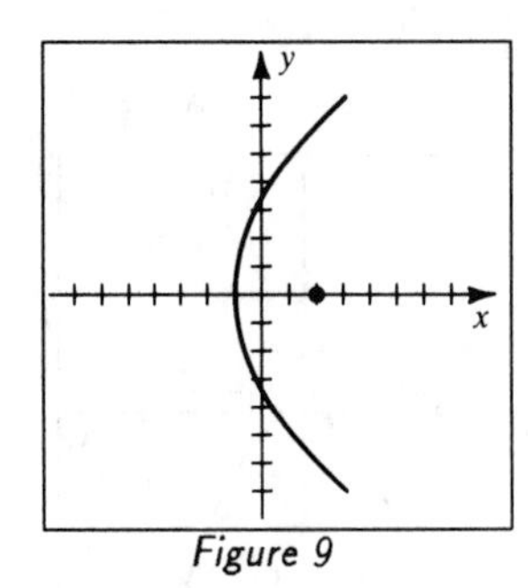

Figure 9

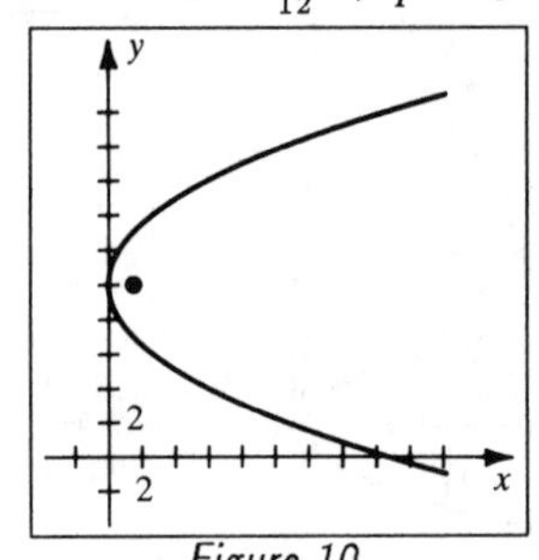

Figure 10

[10] $y^2 - 20y + 100 = 6x \Rightarrow x = \frac{1}{6}y^2 - \frac{10}{3}y + \frac{50}{3} \Rightarrow x' = \frac{1}{3}y - \frac{10}{3}$.

$x' = 0 \Rightarrow y = 10 \Rightarrow V(0, 10)$. $a = \frac{1}{6} \Rightarrow p = \frac{3}{2} \Rightarrow F(\frac{3}{2}, 10)$.

[11] $y^2 - 4y - 2x - 4 = 0 \Rightarrow x = \frac{1}{2}y^2 - 2y - 2 \Rightarrow x' = y - 2$.

$x' = 0 \Rightarrow y = 2 \Rightarrow V(-4, 2)$. $a = \frac{1}{2} \Rightarrow p = \frac{1}{2} \Rightarrow F(-\frac{7}{2}, 2)$.

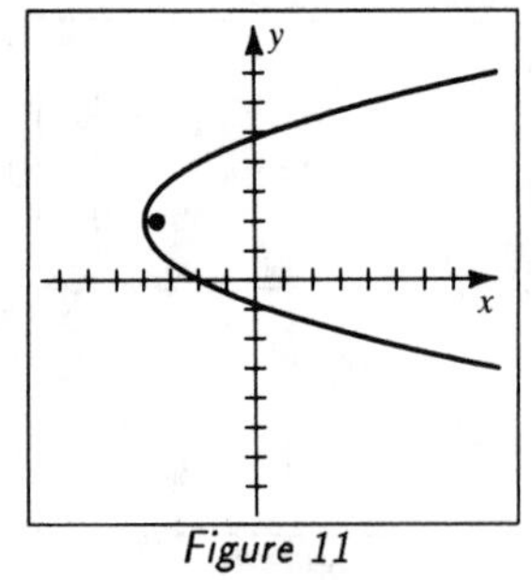

Figure 11

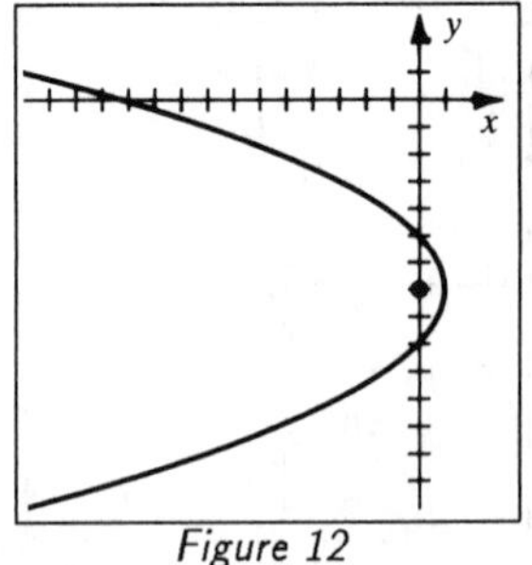

Figure 12

[12] $y^2 + 14y + 4x + 45 = 0 \Rightarrow x = -\frac{1}{4}y^2 - \frac{7}{2}y - \frac{45}{4} \Rightarrow x' = -\frac{1}{2}y - \frac{7}{2}$.

$x' = 0 \Rightarrow y = -7 \Rightarrow V(1, -7)$. $a = -\frac{1}{4} \Rightarrow p = -1 \Rightarrow F(0, -7)$.

[13] $4x^2 + 40x + y + 106 = 0 \Rightarrow y = -4x^2 - 40x - 106 \Rightarrow y' = -8x - 40.$

$y' = 0 \Rightarrow x = -5 \Rightarrow V(-5, -6).\ a = -4 \Rightarrow p = -\frac{1}{16} \Rightarrow F(-5, -\frac{97}{16}).$

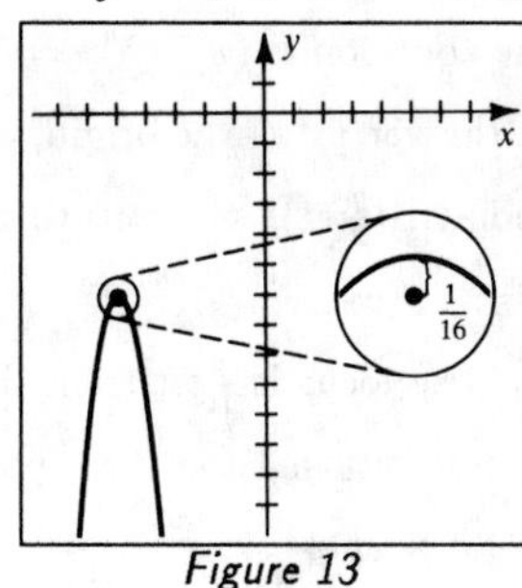

Figure 13

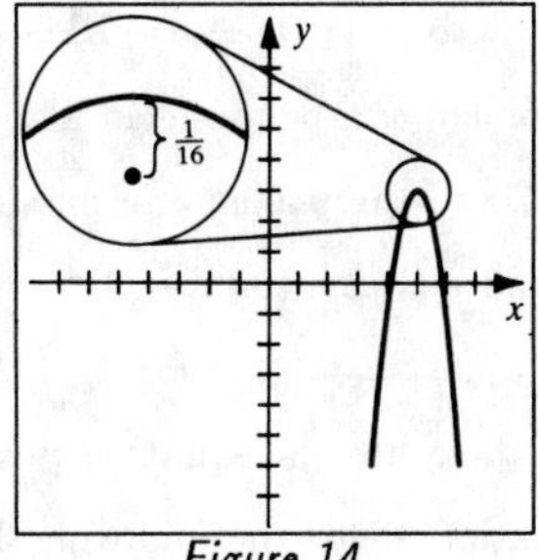

Figure 14

[14] $y = 40x - 97 - 4x^2 = -4x^2 + 40x - 97 \Rightarrow y' = -8x + 40.$

$y' = 0 \Rightarrow x = 5 \Rightarrow V(5, 3).\ a = -4 \Rightarrow p = -\frac{1}{16} \Rightarrow F(5, \frac{47}{16}).$

[15] $x^2 + 20y = 10 \Rightarrow y = -\frac{1}{20}x^2 + \frac{1}{2} \Rightarrow y' = -\frac{1}{10}x.\ y' = 0 \Rightarrow x = 0 \Rightarrow V(0, \frac{1}{2}).$

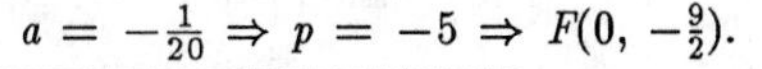

$a = -\frac{1}{20} \Rightarrow p = -5 \Rightarrow F(0, -\frac{9}{2}).$

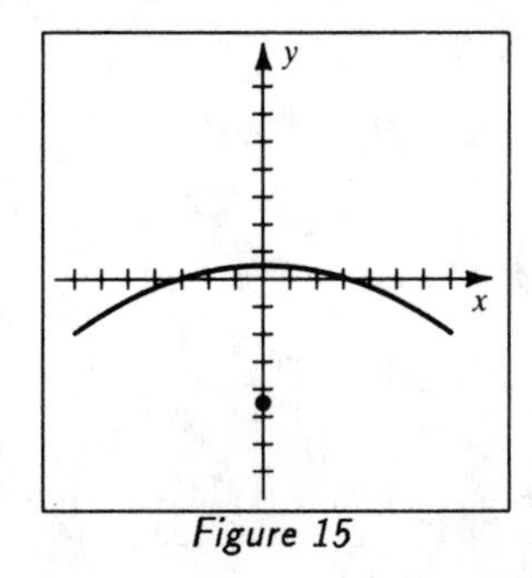

Figure 15

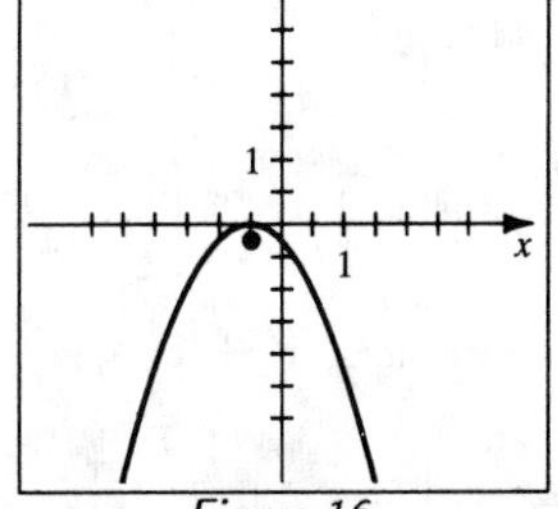

Figure 16

[16] $4x^2 + 4x + 4y + 1 = 0 \Rightarrow y = -x^2 - x - \frac{1}{4} \Rightarrow y' = -2x - 1.$

$y' = 0 \Rightarrow x = -\frac{1}{2} \Rightarrow V(-\frac{1}{2}, 0).\ a = -1 \Rightarrow p = -\frac{1}{4} \Rightarrow F(-\frac{1}{2}, -\frac{1}{4}).$

[17] $F(2, 0)$ and $l: x = -2 \Rightarrow p = 2$ and $V(0, 0)$; $(y - 0)^2 = 4p(x - 0) \Rightarrow y^2 = 8x$

[18] $F(0, -4)$ and $l: y = 4 \Rightarrow p = -4$ and $V(0, 0)$;

$(x - 0)^2 = 4p(y - 0) \Rightarrow x^2 = -16y$

[19] $V(3, -5)$ and $l: x = 2 \Rightarrow p = 1.\ (y + 5)^2 = 4p(x - 3) \Rightarrow (y + 5)^2 = 4(x - 3)$

[20] $V(-2, 3)$ and $l: y = 5 \Rightarrow p = -2.$

$(x + 2)^2 = 4p(y - 3) \Rightarrow (x + 2)^2 = -8(y - 3)$

[21] $V(-1, 0)$ and $F(-4, 0) \Rightarrow p = -3.\ (y - 0)^2 = 4p(x + 1) \Rightarrow y^2 = -12(x + 1)$

[22] $V(1, -2)$ and $F(1, 0) \Rightarrow p = 2.\ (x - 1)^2 = 4p(y + 2) \Rightarrow (x - 1)^2 = 8(y + 2)$

[23] The vertex at the origin and symmetric to the y-axis imply that the equation is of the form $y = ax^2$. Substituting $x = 2$ and $y = -3$ into this equation yields

$-3 = a \cdot 4 \Rightarrow a = -\frac{3}{4}$. Thus, the equation is $y = -\frac{3}{4}x^2$ or $3x^2 = -4y$.

[24] The vertex at $(-3, 5)$ and axis parallel to the x-axis imply that the equation is of the form $(y-5)^2 = 4p(x+3)$. Substituting $x = 5$ and $y = 9$ into this equation yields $16 = 4p \cdot 8 \Rightarrow p = \frac{1}{2}$. Thus, the equation is $(y-5)^2 = 2(x+3)$.

[25] This problem can be modeled like Exercise 23 with the vertex at the origin, symmetric to the y-axis, and passing through the point $A(\frac{3}{2}, 1)$. Substituting $x = \frac{3}{2}$ and $y = 1$ into $(x-0)^2 = 4p(y-0) \Rightarrow \frac{9}{4} = 4p \Rightarrow p = \frac{9}{16}$.

The focus is $\frac{9}{16}$ ft from the vertex.

[26] The parabola has the equation $y = ax^2 + bx + c$. Substituting the three points into this equation yields the following 3 equations: $5 = 4a + 2b + c$, $-3 = 4a - 2b + c$, and $6 = a + b + c$. Solving the third equation for c we obtain $c = 6 - a - b$. Substituting that expression in the first two equations yields $-1 = 3a + b$ and $-9 = 3a - 3b$.

The solution is $a = -1$, $b = 2$, $c = 5$ and the equation is $y = -x^2 + 2x + 5$.

[27] $x^2 = 4y \Rightarrow p = 1$ and the focus is $F(0, 1)$. Thus, the line l is $y = 1$.

(a) $A = \int_{-2}^{2} (1 - \frac{1}{4}x^2)\,dx = 2\int_{0}^{2} (1 - \frac{1}{4}x^2)\,dx = 2\Big[x - \frac{1}{12}x^3\Big]_0^2 = \frac{8}{3}$.

(b) Using shells, $V = 2\pi \int_0^2 x(1 - \frac{1}{4}x^2)\,dx = 2\pi\Big[\frac{1}{2}x^2 - \frac{1}{16}x^4\Big]_0^2 = 2\pi$.

(c) Using washers, $V = \pi\int_{-2}^{2} (1^2 - \frac{1}{16}x^4)\,dx = 2\pi\Big[x - \frac{1}{80}x^5\Big]_0^2 = \frac{16\pi}{5}$.

[28] (a) $A = \int_{-2}^{2} \Big[5 - \frac{1}{2}(y^2 + 6)\Big]\,dy = 2\int_0^2 (2 - \frac{1}{2}y^2)\,dy = 2\Big[2y - \frac{1}{6}y^3\Big]_0^2 = \frac{16}{3}$.

(b) Using washers,

$$V = \pi\int_{-2}^{2} \Big[5^2 - \tfrac{1}{4}(y^2 + 6)^2\Big]\,dy = 2\pi\Big[-\tfrac{1}{20}y^5 - y^3 + 16y\Big]_0^2 = \tfrac{224\pi}{5}.$$

(c) Using shells, $V = 2\pi\int_0^2 y\Big[5 - \frac{1}{2}(y^2 + 6)\Big]\,dy = 2\pi\Big[-\frac{1}{8}y^4 + y^2\Big]_0^2 = 4\pi$.

[29] (a) Let the parabola have the equation $x^2 = 4py$. Since the point $(r, -h)$ is on the parabola, $r^2 = 4p(-h)$, or $p = -\dfrac{r^2}{4h}$. The focal length is $|p| = \dfrac{r^2}{4h}$.

(b) Using disks with radius $= x = \sqrt{4py}$,

$$V = \pi\int_{-h}^{0} (4py)\,dy = 4p\pi\Big[\tfrac{1}{2}y^2\Big]_{-h}^{0} = 4\Big(-\frac{r^2}{4h}\Big)\pi\Big(-\frac{h^2}{2}\Big) = \tfrac{1}{2}\pi r^2 h.$$

[30] (a) $x^2 = 4py \Rightarrow y = \dfrac{1}{4p}x^2 \Rightarrow y' = \dfrac{1}{2p}x$. Using Exercise 41 of §6.5,

$$S = \int_0^a 2\pi x\sqrt{1 + \Big(\frac{1}{2p}x\Big)^2}\,dx = \frac{2\pi}{2p}\int_0^a x\sqrt{4p^2 + x^2}\,dx \ \{u = 4p^2 + x^2,\ du = 2x\,dx\}$$

$$= \frac{\pi}{2p}\int_{4p^2}^{4p^2+a^2} \sqrt{u}\,du = \frac{\pi}{2p}\Big[\frac{2}{3}u^{3/2}\Big]_{4p^2}^{4p^2+a^2}$$

(cont.)

$$= \frac{\pi}{3p}\left[(4p^2 + a^2)^{3/2} - (4p^2)^{3/2}\right]$$

$$= \frac{8p^3\pi}{3p}\left[\left(1 + \frac{a^2}{4p^2}\right)^{3/2} - 1\right] \{\text{multiply and divide by } (4p^2)^{3/2} = 8p^3\}$$

$$= \frac{8\pi p^2}{3}\left[\left(1 + \frac{a^2}{4p^2}\right)^{3/2} - 1\right].$$

(b) With $a = 125$ and $p = 50$, $S \approx 64{,}968 \text{ ft}^2$.

[31] (a) The parabola is of the form $(x - h)^2 = a(y - k)$. Since the vertex is $V(0, 10)$, we have $x^2 = a(y - 10)$. The points $(\pm 200, 90)$ on the parabola imply that $(200)^2 = a(80)$ and hence $a = 500$. Thus, $x^2 = 500(y - 10)$.

(b) Let $y = f(x) = \frac{1}{500}x^2 + 10$. Then, $f'(x) = \frac{1}{250}x$ and $L = \int_{-200}^{200} \sqrt{1 + (\frac{1}{250}x)^2}\, dx$.

(c) The spacing between each support is $\frac{400}{10} = 40$ ft. Total length =

$$f(0) + 2\sum_{n=1}^{5} f(40n) = 10 + 2(13.2 + 22.8 + 38.8 + 61.2) = 282 \text{ ft.}$$

[32] $x = ay^2 \Rightarrow y = \sqrt{\frac{x}{a}} = \frac{1}{\sqrt{a}}\sqrt{x} \Rightarrow y' = \frac{1}{2\sqrt{ax}}$. Using (6.19),

$$S = \int_0^p 2\pi\left(\frac{1}{\sqrt{a}}\sqrt{x}\right)\sqrt{1 + \left(\frac{1}{2\sqrt{ax}}\right)^2}\, dx = \frac{2\pi}{\sqrt{a}}\int_0^{1/(4a)} \sqrt{x}\sqrt{1 + \frac{1}{4ax}}\, dx$$

$$= \frac{\pi}{a}\int_0^{1/(4a)} \sqrt{4ax + 1}\, dx \; \{u = 4ax + 1,\; du = 4a\, dx\}$$

$$= \frac{\pi}{4a^2}\int_1^2 \sqrt{u}\, du = \frac{\pi}{4a^2}\left[\tfrac{2}{3}u^{3/2}\right]_1^2 = \frac{\pi}{6a^2}(2\sqrt{2} - 1).$$

[33] Let the arch have the equation $y = ax^2 + k$, where $a < 0$. The area of the rectangle with vertices $(\pm x, ax^2 + k)$, $x > 0$, is $A(x) = 2x(ax^2 + k) = 2ax^3 + 2xk$.

$A'(x) = 6ax^2 + 2k = 0 \Rightarrow x^2 = -\frac{k}{3a}$. Thus, the height of the rectangle is $y = a\left(-\frac{k}{3a}\right) + k = \frac{2}{3}k$. This is a maximum for $x \in \left[0, \sqrt{-k/a}\right]$.

[34] Without loss of generality, let the parabola have the equation $x^2 = 4py$ and hence have focus $F(0, p)$. Let $P(x, y)$ be a point on the parabola and S the square of the distance from P to F. $S = \left[d(P, F)\right]^2 = (x - 0)^2 + (y - p)^2$.

$\frac{dS}{dx} = 2x + 2(y - p)y'$ and since $y = \frac{1}{4p}x^2$ and $y' = \frac{1}{2p}x$, we have $\frac{dS}{dx} = 0 \Rightarrow$

$x + \left(\frac{1}{4p}x^2 - p\right)\frac{1}{2p}x = 0 \Rightarrow \frac{1}{8p^2}x^3 + \frac{1}{2}x = 0 \Rightarrow x(x^2 + 4p^2) = 0 \Rightarrow x = 0$, which is the x-coordinate of the vertex.

[35] (a) Let the path be described by $y(x) = ax^2 + bx + c$. $y(0) = 0 \Rightarrow c = 0$.

Also, $y'(x) = 2ax + b$ and $y'(0) = 1 \Rightarrow b = 1$. Thus, $y(x) = ax^2 + x$.

Now, let $P(x_0, y_0)$ be the point where the ball strikes the ground and $y_0 = -\frac{3}{4}x_0$.

Then $x_0^2 + y_0^2 = 50^2 \Rightarrow x_0^2 + \frac{9}{16}x_0^2 = 2500 \Rightarrow x_0 = 40$ and $y_0 = -30$.

Thus, $y(40) = a(40)^2 + 40 = -30 \Rightarrow a = -\frac{70}{1600}$ and $y(x) = -\frac{7}{160}x^2 + x$.

(b) Let h represent the height of the ball *off the ground.*

Then, $h(x) = (-\frac{7}{160}x^2 + x) - (-\frac{3}{4}x) = -\frac{7}{160}x^2 + \frac{7}{4}x.$

$h'(x) = -\frac{7}{80}x + \frac{7}{4} = 0 \Rightarrow x = 20.$ $h(20) = 17.5$ ft, which is a maximum.

[36] (a) $y' = \frac{\omega^2}{g}x \Rightarrow y = \frac{\omega^2}{2g}x^2 + C.$ $y(0) = C$ and $y(0) = f(0) \Rightarrow f(0) = C.$

$g = 32 \Rightarrow y = \frac{1}{64}\omega^2 x^2 + f(0).$

(b) $x^2 = \frac{64}{\omega^2}(y - f(0)) \Rightarrow 4p = \frac{64}{\omega^2} \Rightarrow p = \frac{16}{\omega^2} = 2 \Rightarrow \omega^2 = 8 \Rightarrow$

$\omega = 2\sqrt{2}$ rad/sec ≈ 0.45 rev/sec.

[37] (a) Note that the value of p completely determines the parabola.

If (x_1, y_1) is on the parabola, then $y_1^2 = 4p(x_1 + p) \Rightarrow$

$4p^2 + 4x_1p - y_1^2 = 0 \Rightarrow p = \dfrac{-x_1 \pm \sqrt{x_1^2 + y_1^2}}{2}.$ If $y_1 \neq 0,$

then there are exactly two values for p and hence, exactly two parabolas.

(b) $y^2 = 4p(x + p) \Rightarrow 2yy' = 4p \Rightarrow y' = \dfrac{2p}{y}.$

Calculating the value of y' and (x_1, y_1) for each value of p and multiplying these together gives $\dfrac{-x_1 + \sqrt{x_1^2 + y_1^2}}{y_1} \cdot \dfrac{-x_1 - \sqrt{x_1^2 + y_1^2}}{y_1} = \dfrac{-y_1^2}{y_1^2} = -1.$

Thus, the tangent lines are perpendicular.

[38] (a) Refer to *Figure 38.* Without loss of generality, let $x^2 = 4py.$ (Two lines remain perpendicular under translations and rotations.) Focal chord $\overline{T_1T_2}$ passes through $F(0, p)$ with slope m and y-intercept p. Thus, $mx + p = y = \dfrac{x^2}{4p} \Rightarrow$

$x^2 - 4pmx - 4p^2 = 0 \Rightarrow x_1 = 2p(m - \sqrt{m^2 + 1})$ and $x_2 = 2p(m + \sqrt{m^2 + 1}).$

Now, $y' = \dfrac{x}{2p} \Rightarrow \dfrac{x_1}{2p} \cdot \dfrac{x_2}{2p} = m^2 - (m^2 + 1) = -1 \Rightarrow$

the tangent lines at T_1 and T_2 are perpendicular.

(b) If $T_1 = (x_1, y_1)$ and $T_2 = (x_2, y_2)$, then the equations of the two tangent lines are $y = \dfrac{x_1}{2p}(x - x_1) + y_1$ and $y = \dfrac{x_2}{2p}(x - x_2) + y_2.$ Since $y_1 = \dfrac{x_1^2}{4p}$ and

$y_2 = \dfrac{x_2^2}{4p}$, we see that $\dfrac{x_1}{2p}(x - x_1) + \dfrac{x_1^2}{4p} = y = \dfrac{x_2}{2p}(x - x_2) + \dfrac{x_2^2}{4p} \Rightarrow$

$\dfrac{x_1}{2p}x - \dfrac{x_1^2}{4p} = \dfrac{x_2}{2p}x - \dfrac{x_2^2}{4p} \Rightarrow x\left(\dfrac{x_1 - x_2}{2p}\right) = \dfrac{x_1^2 - x_2^2}{4p} \Rightarrow x = \dfrac{x_1 + x_2}{2}$ is where the

tangent lines intersect. Evaluating a tangent line at $x = \dfrac{x_1 + x_2}{2}$ yields

$y = \dfrac{x_1}{2p}\left(\dfrac{x_2 - x_1}{2}\right) + \dfrac{x_1^2}{4p} = \dfrac{x_1 x_2}{4p} = -\dfrac{4p^2}{4p}$ $\{x_1, x_2$ from part (a)$\}$ $= -p,$

which is on the directrix.

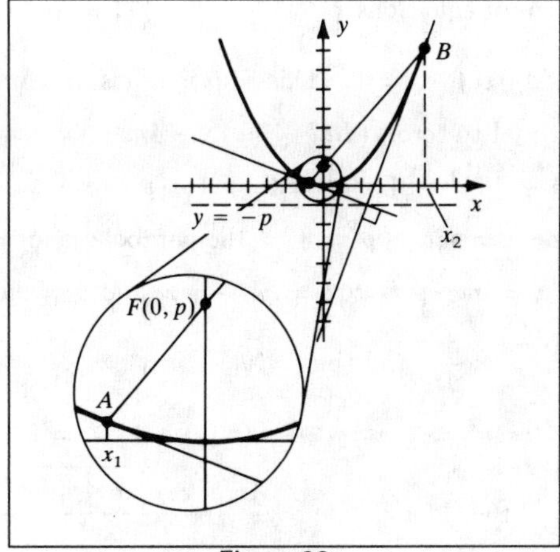

Figure 38

[39] $x^2 = 4p(y + p) \Rightarrow y = \frac{1}{4p}x^2 - p$. Plot $y = \frac{x^2}{5.2} - 1.3$ and $y = -\frac{x^2}{4} + 1$.

The points of intersection are approximately $(\pm 2.280, -0.3)$. $y' = \frac{1}{2p}x$.

$p = 1.3$ and $x = 2.280 \Rightarrow y' \approx 0.877$, $p = -1$ and $x = 2.280 \Rightarrow y' \approx -1.14$.

Now, $0.877 \times -1.14 = -0.99978 \approx -1$. Similarly, $p = 1.3$ and $x = -2.280 \Rightarrow$

$y' \approx -0.877$, $p = -1$ and $x = -2.280 \Rightarrow y' \approx 1.14$.

Now, $-0.877 \times 1.14 = -0.99978 \approx -1$.

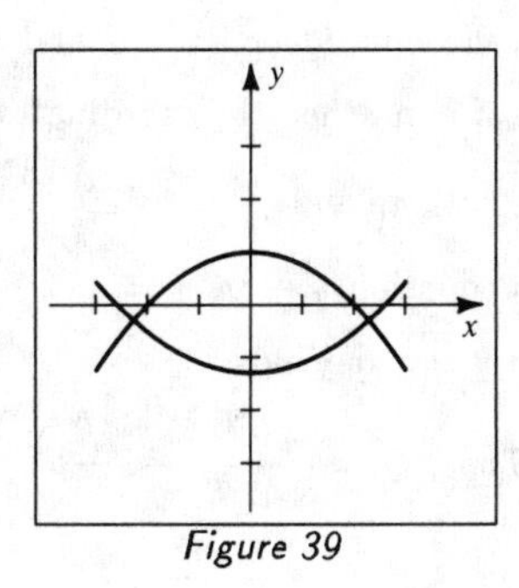

Figure 39

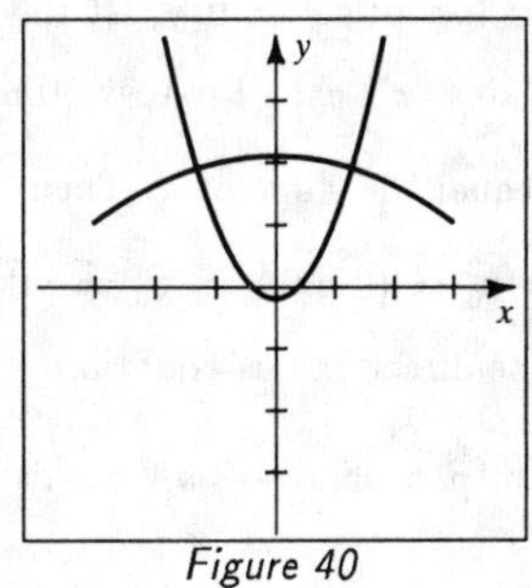

Figure 40

[40] $x^2 = 4p(y + p) \Rightarrow y = \frac{1}{4p}x^2 - p$. Plot $y = \frac{x^2}{0.8} - 0.2$ and $y = -\frac{x^2}{8.4} + 2.1$.

The points of intersection are approximately $(\pm 1.296, 1.9)$. $y' = \frac{1}{2p}x$.

$p = 0.2$ and $x = 1.296 \Rightarrow y' \approx 3.24$, $p = -2.1$ and $x = 1.296 \Rightarrow y' \approx -0.309$.

Now, $3.24 \times -0.309 \approx -1.001 \approx -1$. Similarly, $p = 0.2$ and $x = -1.296 \Rightarrow$

$y' \approx -3.24$, $p = -2.1$ and $x = -1.296 \Rightarrow y' \approx 0.309$.

Now, $-3.24 \times 0.309 \approx -1.001 \approx -1$.

[41] (a) Solving the system of equations $\begin{cases} y^2 &= 4px \\ y &= mx + b \end{cases}$, we obtain $(mx + b)^2 = 4px \Rightarrow$

$m^2x^2 + (2mb - 4p)x + b^2 = 0$. This equation has only one solution iff its discriminant is equal to zero. $(2mb - 4p)^2 - 4m^2b^2 = 0 \Rightarrow$

$4m^2b^2 - 16mbp + 16p^2 - 4m^2b^2 = 0 \Rightarrow 16p(p - mb) = 0 \Rightarrow$

$p = 0$ or $p = mb$. $\{p = 0 \Rightarrow$ the parabola is identical to the directrix.$\}$

(b) Using $P(x_1, y_1)$, $y = mx + b \Rightarrow b = y_1 - mx_1$ and $p = mb = m(y_1 - mx_1)$.

But $y^2 = 4px \Rightarrow p = \dfrac{y_1^2}{4x_1}$ and thus $m(y_1 - mx_1) = \dfrac{y_1^2}{4x_1} \Rightarrow$

$4mx_1y_1 - 4m^2x_1^2 = y_1^2 \Rightarrow (4x_1^2)m^2 - (4x_1y_1)m + y_1^2 = 0$. This is a quadratic in the slope m. $m = \dfrac{4x_1y_1 \pm \sqrt{16x_1^2y_1^2 - 16x_1^2y_1^2}}{8x_1^2} = \dfrac{4x_1y_1}{8x_1^2} = \dfrac{y_1}{2x_1}$.

[42] Note that $\beta = \angle PQF$. To show that $\alpha = \beta$, we will show that $d(Q, F) = d(F, P)$ and thus $\triangle QFP$ is isosceles. By Exercise 41, $m_{PQ} = \dfrac{y_1}{2x_1}$. If Q has coordinates $(x, 0)$, then $m_{PQ} = \dfrac{0 - y_1}{x - x_1} \Rightarrow \dfrac{y_1}{2x_1} = \dfrac{-y_1}{x - x_1} \Rightarrow x - x_1 = -2x_1 \Rightarrow x = -x_1$.

Thus $Q = (-x_1, 0)$ and $d(Q, F) = x_1 + p$. $d(F, P) = \sqrt{(x_1 - p)^2 + (y_1 - 0)^2} = \sqrt{x_1^2 - 2px_1 + p^2 + 4px_1}$ $\{$since $y_1^2 = 4px_1\}$ $= \sqrt{x_1^2 + 2px_1 + p^2} = \sqrt{(x_1 + p)^2} = x_1 + p$.

[43] Without loss of generality, let the parabola have the equation $x^2 = 4py$ and P have coordinates (x_0, y_0). First, we find the equation of l and then the coordinates of Q.

Let m equal the slope of l. Then, $y' = \dfrac{x}{2p} \Rightarrow x_0 = 2mp$ and

$4py_0 = x_0^2 = (2mp)^2 \Rightarrow y_0 = m^2p$. Thus, l has equation $y - m^2p = m(x - 2mp)$.

Since the directrix has equation $y = -p$, setting $y = -p$ yields:

$-p - m^2p = m(x - 2mp) \Rightarrow x = \dfrac{pm^2 - p}{m}$. Thus, $Q = \left(\dfrac{pm^2 - p}{m}, -p\right)$.

Since F has coordinates $(0, p)$, the slope of segment QF is $-\dfrac{2pm}{pm^2 - p}$.

Since P has coordinates $(x_0, y_0) = (2mp, pm^2)$, the slope of segment PF is $\dfrac{pm^2 - p}{2pm}$.

Their product equals -1 and hence, the two segments are perpendicular.

Exercises 12.2

Note: Let C, V, F, and M denote the center, the vertices, the foci, and the end points of the minor axis, respectively.

[1] $c^2 = 9 - 4 \Rightarrow c = \pm\sqrt{5}$; $V(\pm 3, 0)$; $F(\pm\sqrt{5}, 0)$; $M(0, \pm 2)$

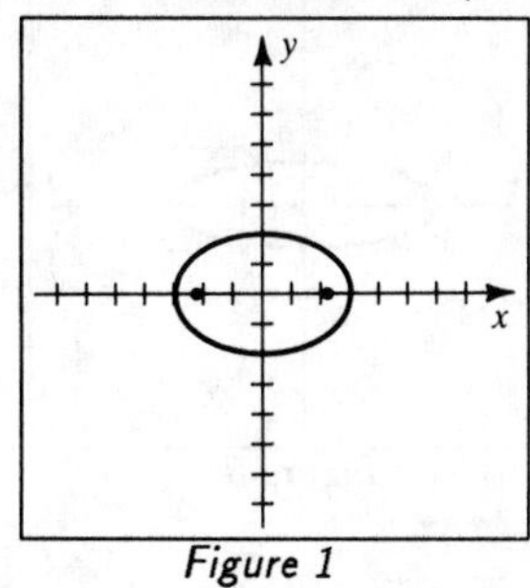

Figure 1

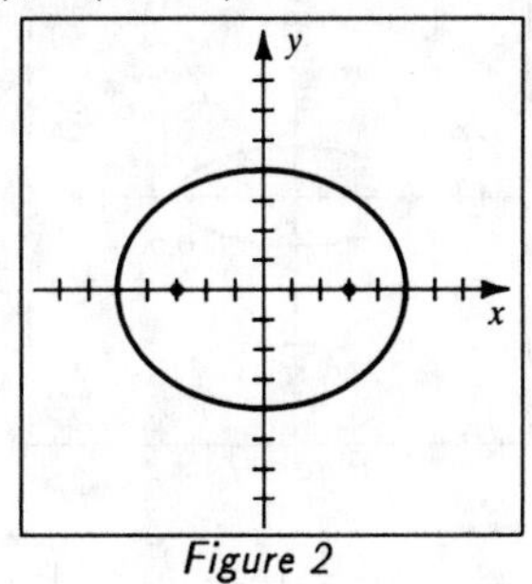

Figure 2

[2] $c^2 = 25 - 16 \Rightarrow c = \pm 3$; $V(\pm 5, 0)$; $F(\pm 3, 0)$; $M(0, \pm 4)$

[3] $4x^2 + y^2 = 16 \Rightarrow \frac{x^2}{4} + \frac{y^2}{16} = 1$; $c^2 = 16 - 4 \Rightarrow c = \pm 2\sqrt{3}$;

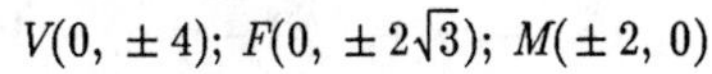
$V(0, \pm 4)$; $F(0, \pm 2\sqrt{3})$; $M(\pm 2, 0)$

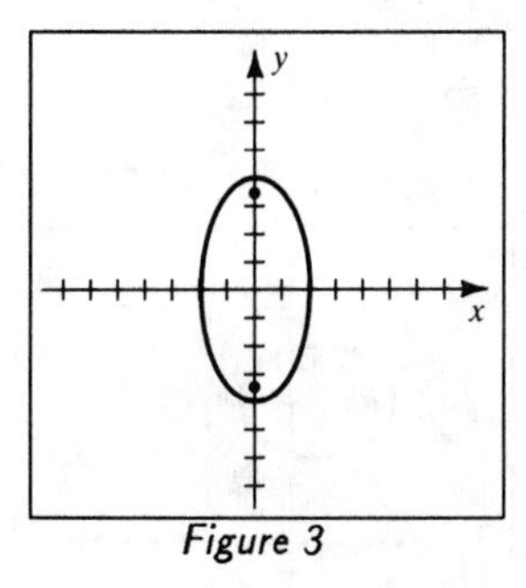

Figure 3

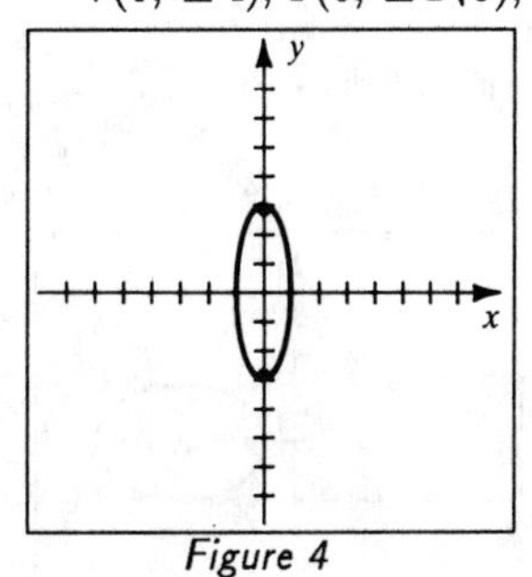

Figure 4

[4] $y^2 + 9x^2 = 9 \Rightarrow \frac{x^2}{1} + \frac{y^2}{9} = 1$; $c^2 = 9 - 1 \Rightarrow c = \pm 2\sqrt{2}$;

$V(0, \pm 3)$; $F(0, \pm 2\sqrt{2})$; $M(\pm 1, 0)$

[5] $5x^2 + 2y^2 = 10 \Rightarrow \frac{x^2}{2} + \frac{y^2}{5} = 1$; $c^2 = 5 - 2 \Rightarrow c = \pm\sqrt{3}$;

$V(0, \pm\sqrt{5})$; $F(0, \pm\sqrt{3})$; $M(\pm\sqrt{2}, 0)$

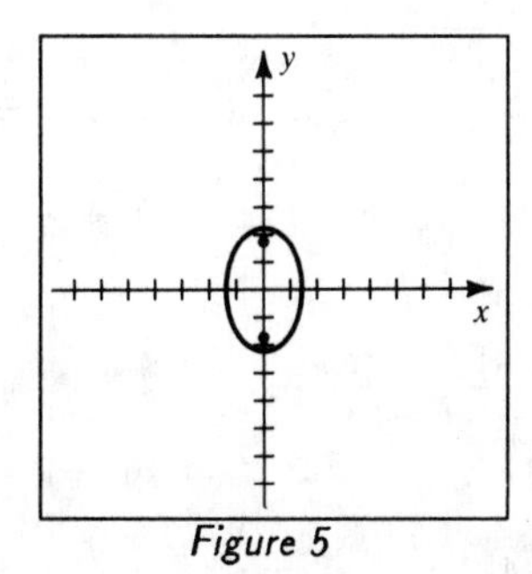

Figure 5

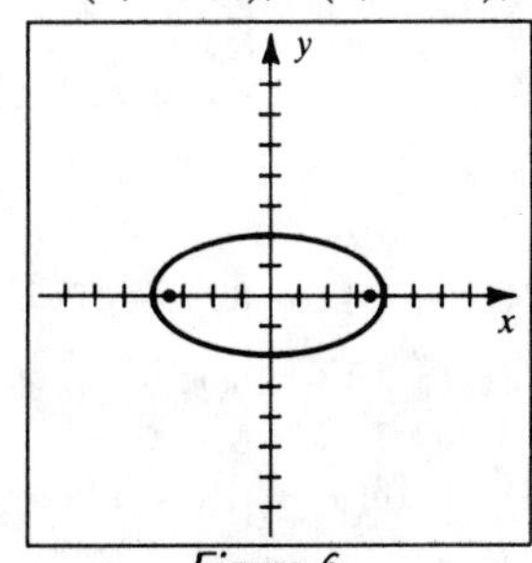

Figure 6

[6] $\frac{1}{2}x^2 + 2y^2 = 8 \Rightarrow \frac{x^2}{16} + \frac{y^2}{4} = 1$; $c^2 = 16 - 4 \Rightarrow c = \pm 2\sqrt{3}$;

$V(\pm 4, 0)$; $F(\pm 2\sqrt{3}, 0)$; $M(0, \pm 2)$

[7] $4x^2 + 25y^2 = 1 \Rightarrow \frac{x^2}{\frac{1}{4}} + \frac{y^2}{\frac{1}{25}} = 1$; $c^2 = \frac{1}{4} - \frac{1}{25} = \frac{21}{100} \Rightarrow c = \pm\frac{1}{10}\sqrt{21}$;

$V(\pm\frac{1}{2}, 0)$; $F(\pm\frac{1}{10}\sqrt{21}, 0)$; $M(0, \pm\frac{1}{5})$

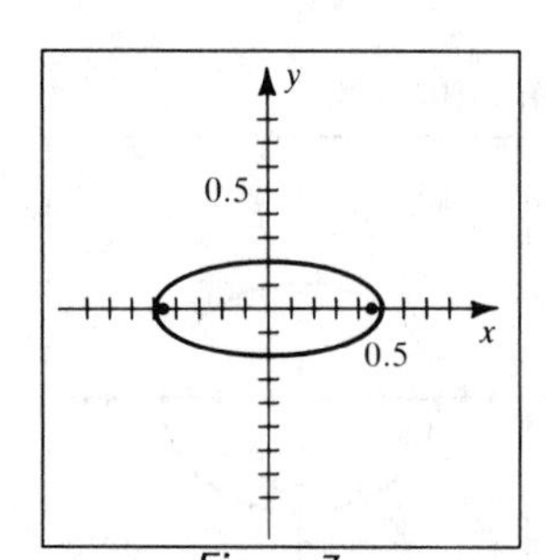

Figure 7

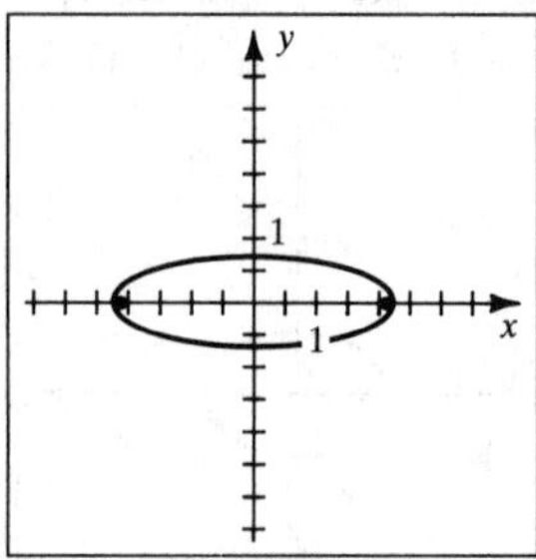

Figure 8

[8] $10y^2 + x^2 = 5 \Rightarrow \frac{x^2}{5} + \frac{y^2}{\frac{1}{2}} = 1$; $c^2 = 5 - \frac{1}{2} = \frac{9}{2} \Rightarrow c = \pm\frac{3}{2}\sqrt{2}$;

$V(\pm\sqrt{5}, 0)$; $F(\pm\frac{3}{2}\sqrt{2}, 0)$; $M(0, \pm\frac{1}{2}\sqrt{2})$

[9] $4x^2 + 9y^2 - 32x - 36y + 64 = 0 \Rightarrow$

$4(x^2 - 8x + \underline{16}) + 9(y^2 - 4y + \underline{4}) = -64 + \underline{64} + \underline{36} \Rightarrow$

$4(x - 4)^2 + 9(y - 2)^2 = 36 \Rightarrow \frac{(x - 4)^2}{9} + \frac{(y - 2)^2}{4} = 1$;

$c^2 = 9 - 4 \Rightarrow c = \pm\sqrt{5}$; $C(4, 2)$; $V(4 \pm 3, 2)$; $F(4 \pm \sqrt{5}, 2)$; $M(4, 2 \pm 2)$

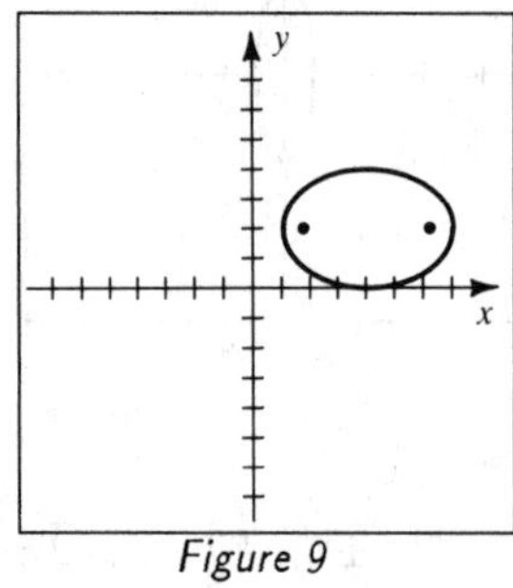

Figure 9

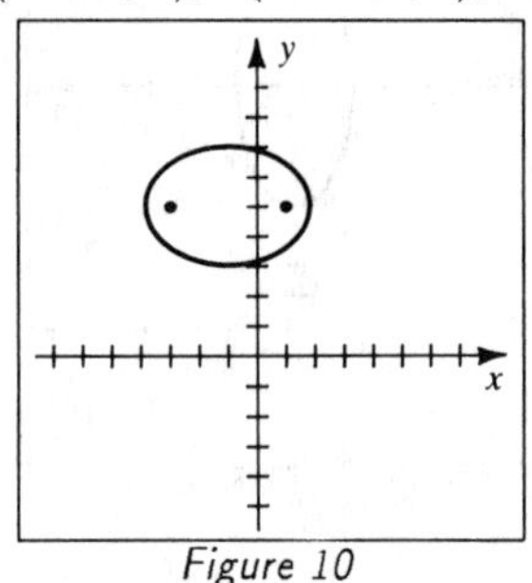

Figure 10

[10] $x^2 + 2y^2 + 2x - 20y + 43 = 0 \Rightarrow$

$(x^2 + 2x + \underline{1}) + 2(y^2 - 10y + \underline{25}) = -43 + \underline{1} + \underline{50} \Rightarrow$

$(x + 1)^2 + 2(y - 5)^2 = 8 \Rightarrow \frac{(x + 1)^2}{8} + \frac{(y - 5)^2}{4} = 1$; $c^2 = 8 - 4 \Rightarrow c = \pm 2$;

$C(-1, 5)$; $V(-1 \pm 2\sqrt{2}, 5)$; $F(-1 \pm 2, 5)$; $M(-1, 5 \pm 2)$

[11] $9x^2 + 16y^2 + 54x - 32y - 47 = 0 \Rightarrow$

$9(x^2 + 6x + \underline{9}) + 16(y^2 - 2y + \underline{1}) = 47 + \underline{81} + \underline{16} \Rightarrow$

$9(x + 3)^2 + 16(y - 1)^2 = 144 \Rightarrow \frac{(x + 3)^2}{16} + \frac{(y - 1)^2}{9} = 1$; $c^2 = 16 - 9 \Rightarrow$

$c = \pm\sqrt{7}$; $C(-3, 1)$; $V(-3 \pm 4, 1)$; $F(-3 \pm \sqrt{7}, 1)$; $M(-3, 1 \pm 3)$

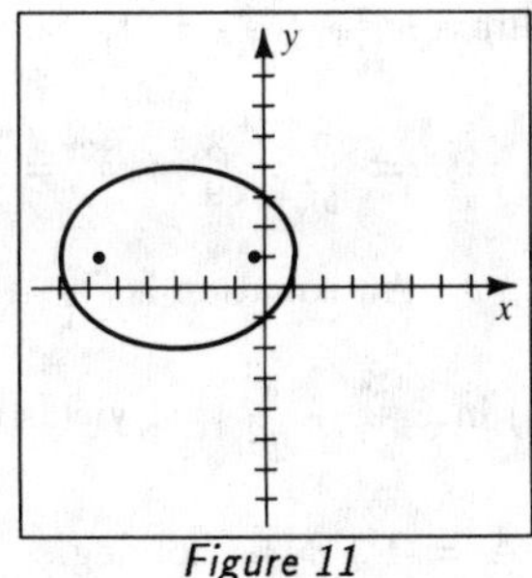

Figure 11

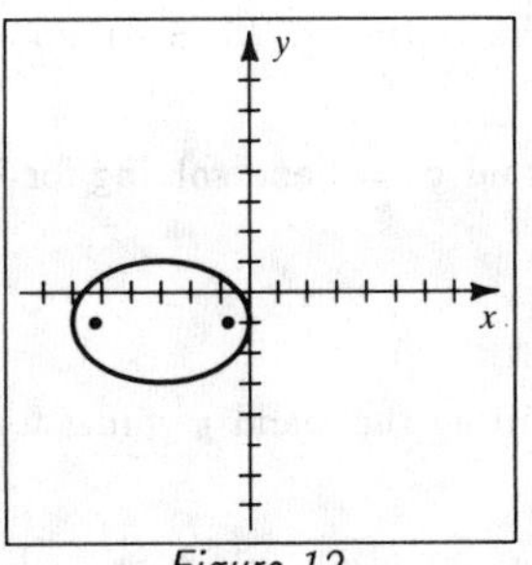

Figure 12

[12] $4x^2 + 9y^2 + 24x + 18y + 9 = 0 \Rightarrow$

$4(x^2 + 6x + \underline{9}) + 9(y^2 + 2y + \underline{1}) = -9 + \underline{36} + \underline{9} \Rightarrow$

$4(x+3)^2 + 9(y+1)^2 = 36 \Rightarrow \dfrac{(x+3)^2}{9} + \dfrac{(y+1)^2}{4} = 1;\ c^2 = 9 - 4 \Rightarrow$

$c = \pm\sqrt{5};\ C(-3, -1);\ V(-3 \pm 3, -1);\ F(-3 \pm \sqrt{5}, -1);\ M(-3, -1 \pm 2)$

[13] $25x^2 + 4y^2 - 250x - 16y + 541 = 0 \Rightarrow$

$25(x^2 - 10x + \underline{25}) + 4(y^2 - 4y + \underline{4}) = -541 + \underline{625} + \underline{16} \Rightarrow$

$25(x-5)^2 + 4(y-2)^2 = 100 \Rightarrow \dfrac{(x-5)^2}{4} + \dfrac{(y-2)^2}{25} = 1;\ c^2 = 25 - 4 \Rightarrow$

$c = \pm\sqrt{21};\ C(5, 2);\ V(5, 2 \pm 5);\ F(5, 2 \pm \sqrt{21});\ M(5 \pm 2, 2)$

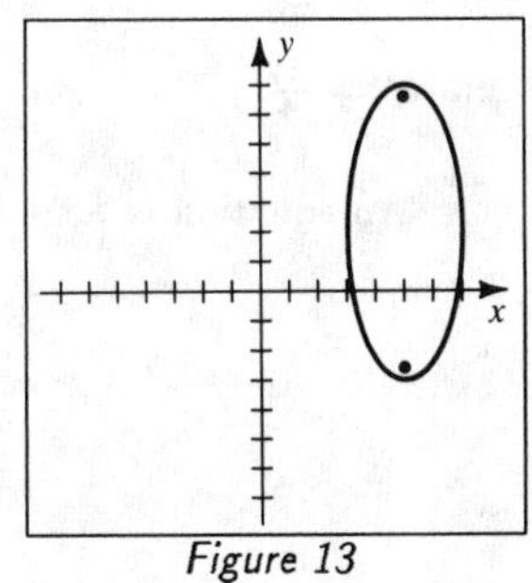

Figure 13

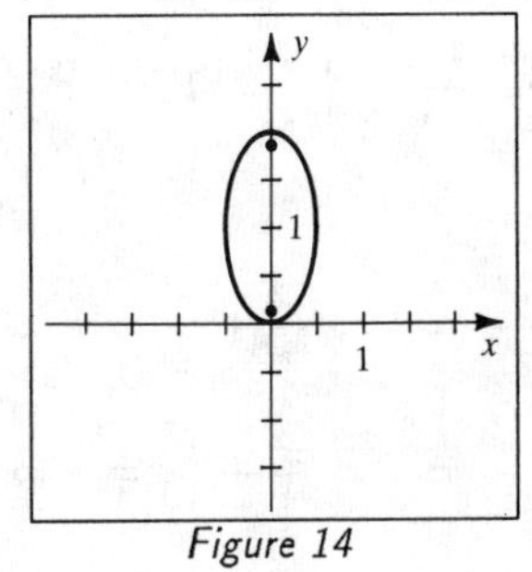

Figure 14

[14] $4x^2 + y^2 = 2y \Rightarrow 4x^2 + y^2 - 2y + \underline{1} = \underline{1} \Rightarrow$

$\dfrac{x^2}{\frac{1}{4}} + \dfrac{(y-1)^2}{1} = 1;\ c^2 = 1 - \frac{1}{4} \Rightarrow c = \pm\frac{1}{2}\sqrt{3};$

$C(0, 1);\ V(0, 1 \pm 1);\ F(0, 1 \pm \frac{1}{2}\sqrt{3});\ M(0 \pm \frac{1}{2}, 1)$

[15] $b^2 = 8^2 - 5^2 = 39$. An equation is $\dfrac{x^2}{64} + \dfrac{y^2}{39} = 1$.

[16] $b^2 = 7^2 - 2^2 = 45$. An equation is $\dfrac{x^2}{45} + \dfrac{y^2}{49} = 1$.

[17] If the length of the minor axis is 3, then $b = \frac{3}{2}$. An equation is $\dfrac{4x^2}{9} + \dfrac{y^2}{25} = 1$.

[18] If the length of the minor axis is 2, then $b = 1$. $a^2 = 3^2 + 1^2 = 10$.

An equation is $\dfrac{x^2}{10} + \dfrac{y^2}{1} = 1$.

[19] With the vertices at $(0, \pm 6)$, an equation of the ellipse is $\frac{x^2}{b^2} + \frac{y^2}{36} = 1$. Substituting $x = 3$ and $y = 2$ and solving for b^2 yields $\frac{9}{b^2} + \frac{4}{36} = 1 \Rightarrow \frac{9}{b^2} = \frac{8}{9} \Rightarrow b^2 = \frac{81}{8}$.

An equation is $\frac{8x^2}{81} + \frac{y^2}{36} = 1$.

[20] Substituting the x and y values for $(2, 3)$ and $(6, 1)$ into $\frac{x^2}{a^2} + \frac{y^2}{b^2} = 1$ yields the system of equations: $\frac{4}{a^2} + \frac{9}{b^2} = 1$ $\{E_1\}$ and $\frac{36}{a^2} + \frac{1}{b^2} = 1$ $\{E_2\}$. Solving, $E_2 - 9E_1 \Rightarrow -\frac{80}{b^2} = -8 \Rightarrow b^2 = 10$. Substituting into E_1 gives $a^2 = 40$.

An equation is $\frac{x^2}{40} + \frac{y^2}{10} = 1$.

[21] With vertices $V(0, \pm 4)$, an equation of the ellipse is $\frac{x^2}{b^2} + \frac{y^2}{16} = 1$. $e = \frac{c}{a} = \frac{3}{4}$ and $a = 4 \Rightarrow c = 3$. Thus, $b^2 = 16 - 9 = 7$. An equation is $\frac{x^2}{7} + \frac{y^2}{16} = 1$.

[22] An equation of the ellipse is $\frac{x^2}{a^2} + \frac{y^2}{b^2} = 1$. $(1, 3)$ on the ellipse $\Rightarrow \frac{1}{a^2} + \frac{9}{b^2} = 1 \Rightarrow$ $b^2 = \frac{9a^2}{a^2 - 1}$. $e = \frac{c}{a} = \frac{1}{2} \Rightarrow c = \frac{1}{2}a$. $b^2 = a^2 - c^2 = a^2 - \frac{1}{4}a^2 = \frac{3}{4}a^2$.

Thus, $\frac{9a^2}{a^2 - 1} = \frac{3}{4}a^2 \Rightarrow 12 = a^2 - 1 \Rightarrow a^2 = 13$ and $b^2 = \frac{39}{4}$.

An equation is $\frac{x^2}{13} + \frac{4y^2}{39} = 1$.

[23] $\frac{x^2}{2^2} + \frac{y^2}{(\frac{1}{3})^2} = 1 \Rightarrow \frac{x^2}{4} + \frac{y^2}{\frac{1}{9}} = 1 \Rightarrow \frac{x^2}{4} + 9y^2 = 1$

[24] $\frac{x^2}{(\frac{1}{2})^2} + \frac{y^2}{4^2} = 1 \Rightarrow \frac{x^2}{\frac{1}{4}} + \frac{y^2}{16} = 1 \Rightarrow 4x^2 + \frac{y^2}{16} = 1$

[25] Model this problem as an ellipse with $V(\pm 15, 0)$ and $M(0, \pm 10)$.

Substituting $x = 6$ into $\frac{x^2}{15^2} + \frac{y^2}{10^2} = 1$ yields $\frac{y^2}{100} = \frac{189}{225} \Rightarrow y^2 = 84$.

The desired height is $\sqrt{84} = 2\sqrt{21} \approx 9.165$ ft.

[26] $e = \frac{c}{a} = 0.017 \Rightarrow c = 0.017a = 0.017(\frac{1}{2} \cdot 186{,}000{,}000) = 1{,}581{,}000$. As in Example 7, the maximum distance is $a + c = 93{,}000{,}000 + 1{,}581{,}000 = 94{,}581{,}000$ mi.

The minimum distance is $a - c = 93{,}000{,}000 - 1{,}581{,}000 = 91{,}419{,}000$ mi.

[27] Refer to *Figure 27*. $\sin\theta = \frac{|y|}{a}$ and $\cos\theta = \frac{|x|}{b}$.

Thus, $\sin^2\theta + \cos^2\theta = 1 = \frac{y^2}{a^2} + \frac{x^2}{b^2}$, which is the equation of an ellipse. Therefore, the point $P(x, y)$ always lies on the ellipse. $\{$If $a = b$, $P(x, y)$ traces a circle.$\}$

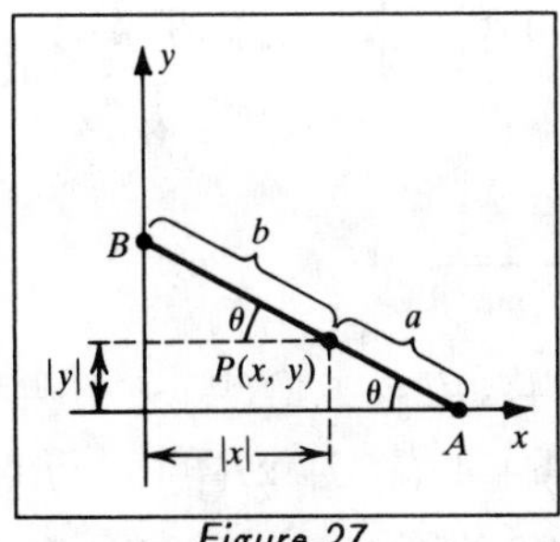

Figure 27

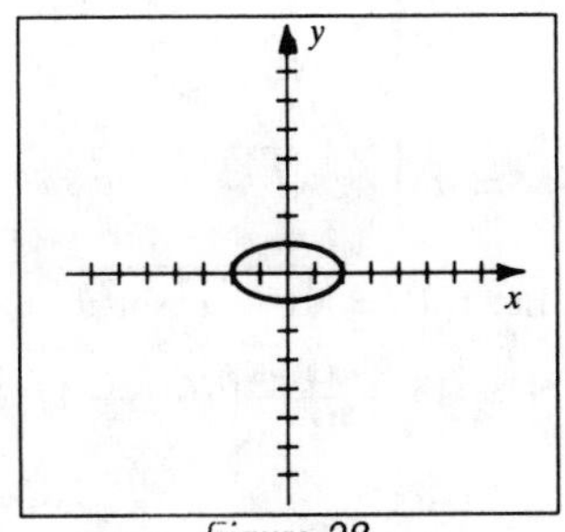

Figure 28

[28] $x^2 + y^2 = 4 \Rightarrow y = \pm\sqrt{4 - x^2}$.

The set of all midpoints have coordinates $(x, y) = (x, \pm\frac{1}{2}\sqrt{4 - x^2})$ for $-2 \le x \le 2$.

$y = \pm\frac{1}{2}\sqrt{4 - x^2} \Rightarrow 2y = \pm\sqrt{4 - x^2} \Rightarrow x^2 + 4y^2 = 4$.

The graph of $x^2 + 4y^2 = 4$ is an ellipse with $C(0, 0)$, $V(\pm 2, 0)$, and $M(0, \pm 1)$.

[29] (a) $E = \frac{1}{2}mv^2 + \frac{1}{2}kx^2 \Rightarrow \dfrac{v^2}{2E/m} + \dfrac{x^2}{2E/k} = 1$, which is the equation of an ellipse.

The lengths of the axes are $a = \sqrt{2E/m}$ and $b = \sqrt{2E/k}$.

If $m < k$, then a is the length of the major axis, and b is if $k < m$.

(b) Area $= \pi ab$ {by Example 6} $\Rightarrow A = \dfrac{2E\pi}{\sqrt{mk}}$ {from (a)} $\Rightarrow A\sqrt{mk} = 2\pi E$

[30] Since $a = p + c$, $b^2 = a^2 - c^2 = (p + c)^2 - c^2 = p^2 + 2pc = p(p + 2c)$.

Thus, the ellipse has the equation $\dfrac{[x - (p + c)]^2}{(p + c)^2} + \dfrac{y^2}{p(p + 2c)} = 1 \Rightarrow$

$$y^2 = p(p + 2c)\left[1 - \frac{(x - p - c)^2}{(p + c)^2}\right] = \frac{p(p + 2c)(2xp + 2xc - x^2)}{(p + c)^2}.$$

Consider the expression to be a rational function of c with $4px$ as the coefficient of c^2 in the numerator and 1 as the coefficient of c^2 in the denominator.

Hence, as $c \to \infty$, $y^2 \to 4px$.

[31] Let (x_1, y_1) be a point of tangency on $9x^2 + 4y^2 = 36$. By Example 5(c), the tangent line has equation $\frac{x_1 x}{4} + \frac{y_1 y}{9} = 1$. $(0, 6)$ on this line $\Rightarrow 0 + \frac{2}{3}y_1 = 1 \Rightarrow$

$y_1 = \frac{3}{2}$. $9x_1^2 + 4y_1^2 = 36 \Rightarrow x_1^2 = \frac{1}{9}(36 - 4y_1^2) = \frac{1}{9}(36 - 9) = 3$. $(\pm\sqrt{3}, \frac{3}{2})$

[32] (a) $(x^2/a^2) + (y^2/b^2) = 1 \Rightarrow y = f(x) = \frac{b}{a}(a^2 - x^2)^{1/2}$ $\{y \ge 0\}$.

Using (6.14) with $f'(x) = -\frac{bx}{a}(a^2 - x^2)^{-1/2}$ yields

$$C = 4\int_0^a \sqrt{1 + \frac{b^2x^2}{a^2(a^2 - x^2)}}\,dx \ \{x = a\sin\theta \text{ and } dx = a\cos\theta\,d\theta\}$$

$$= 4\int_0^{\pi/2} \sqrt{1 + \frac{b^2a^2\sin^2\theta}{a^2(a^2\cos^2\theta)}}\,a\cos\theta\,d\theta = 4a\int_0^{\pi/2} \sqrt{\cos^2\theta + \frac{b^2}{a^2}\sin^2\theta}\,d\theta \qquad \text{(cont.)}$$

$$= 4a\int_0^{\pi/2}\sqrt{\cos^2\theta + \frac{a^2 - c^2}{a^2}\sin^2\theta}\,d\theta = 4a\int_0^{\pi/2}\sqrt{\cos^2\theta + \sin^2\theta - \frac{c^2}{a^2}\sin^2\theta}\,d\theta$$

$$= 4a\int_0^{\pi/2}\sqrt{1 - e^2\sin^2\theta}\,d\theta$$

(b) Let $f(\theta) = \sqrt{1 - e^2\sin^2\theta} = \sqrt{1 - (0.206)^2\sin^2\theta}$.

$$S = 4a\cdot\frac{(\pi/2)-0}{3(10)}\Big(f(0) + 4f(\tfrac{\pi}{20}) + 2f(\tfrac{2\pi}{20}) + 4f(\tfrac{3\pi}{20}) + 2f(\tfrac{4\pi}{20}) + 4f(\tfrac{5\pi}{20}) + 2f(\tfrac{6\pi}{20}) + 4f(\tfrac{7\pi}{20}) + 2f(\tfrac{8\pi}{20}) + 4f(\tfrac{9\pi}{20}) + f(\tfrac{\pi}{2})\Big)$$

$$\approx a\cdot\tfrac{\pi}{15}\Big(1 + 4(0.9995) + 2(0.9980) + 4(0.9956) + 2(0.9926) + 4(0.9893) + 2(0.9860) + 4(0.9830) + 2(0.9806) + 4(0.9791) + (0.9786)\Big)$$

$$\approx (0.387)\tfrac{\pi}{15}(29.689) \approx 2.406 \text{ AU}.$$

(c) $c = ae = (0.387)(0.206) \approx 0.080$. As in Example 7,

the maximum and minimum distances are $a + c \approx 0.387 + 0.080 = 0.467$ AU

and $a - c \approx 0.387 - 0.080 = 0.307$ AU, respectively.

[33] $5x^2 + 4y^2 = 56 \Rightarrow 10x + 8yy' = 0 \Rightarrow y' = -\frac{5x}{4y}$. At $P(-2, 3)$,

$y' = \frac{5}{6}$ and an equation of the tangent line is $y - 3 = \frac{5}{6}(x + 2)$, or $5x - 6y = -28$.

The normal line is $y - 3 = -\frac{6}{5}(x + 2)$, or $6x + 5y = 3$.

[34] $9x^2 + 4y^2 = 72 \Rightarrow 18x + 8yy' = 0 \Rightarrow y' = -\frac{9x}{4y}$. At $P(2, 3)$, $y' = -\frac{3}{2}$

and an equation of the tangent line is $y - 3 = -\frac{3}{2}(x - 2)$, or $3x + 2y = 12$.

The normal line is $y - 3 = \frac{2}{3}(x - 2)$, or $2x - 3y = -5$.

[35] The upper half of the ellipse has equation $y = f(x) = \frac{b}{a}\sqrt{a^2 - x^2}$ for $-a \le x \le a$.

Using disks, $V = \pi\int_{-a}^{a}\frac{b^2}{a^2}(a^2 - x^2)\,dx = \frac{2b^2\pi}{a^2}\Big[a^2x - \frac{1}{3}x^3\Big]_0^a = \frac{4}{3}\pi ab^2$.

[36] The right half of the ellipse has equation $x = f(y) = \frac{a}{b}\sqrt{b^2 - y^2}$ for $-b \le y \le b$.

Using disks, $V = \pi\int_{-b}^{b}\frac{a^2}{b^2}(b^2 - y^2)\,dy = \frac{2a^2\pi}{b^2}\Big[b^2y - \frac{1}{3}y^3\Big]_0^b = \frac{4}{3}\pi a^2b$.

[37] Placing the major axis along the x-axis, the equation is $\frac{x^2}{8^2} + \frac{y^2}{(4.5)^2} = 1 \Rightarrow$

$y = \pm(4.5)\sqrt{1 - \frac{x^2}{64}}$. The base of the cross section has length $2 \times (4.5)\sqrt{1 - \frac{x^2}{64}}$.

$A(x) = \Big(9\sqrt{1 - \frac{x^2}{64}}\Big)^2 \Rightarrow V = \int_{-8}^{8} 81\Big(1 - \frac{x^2}{64}\Big)\,dx = 2\cdot 81\Big[x - \frac{1}{192}x^3\Big]_0^8 = 864$.

See *Figure 37.*

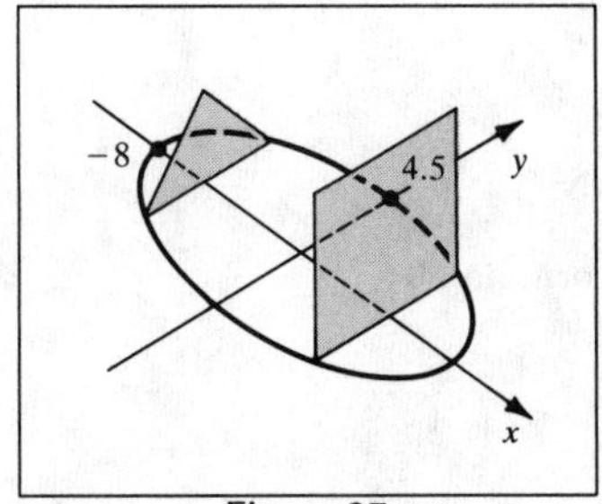

Figure 37

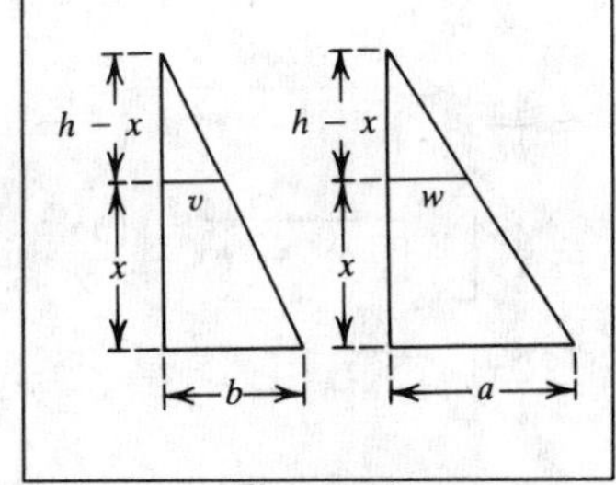

Figure 40

[38] $A(x) = \frac{\sqrt{3}}{4}\left(9 - \sqrt{1 - \frac{x^2}{64}}\right)^2 \Rightarrow V = \frac{\sqrt{3}}{4}(864)$ {from #37} $= 216\sqrt{3}$

[39] Let any cross section of the elliptical frustum be represented by $\frac{x^2}{a^2} + \frac{y^2}{b^2} = 1$, where a and b are functions of the distance h that the cross section is from the left end of the frustum in the figure. The area of this ellipse is πab by Example 6. Writing a as a linear function of h, $a = \frac{a_2 - a_1}{L}h + a_1$, where $h \in [0, L]$. Also, $\frac{a}{b} = k \Rightarrow b = \frac{a}{k}$.

Thus, $V = \int_0^L (\pi ab)\, dh = \frac{\pi}{k}\int_0^L \left(\frac{a_2 - a_1}{L}h + a_1\right)^2 dh$

$$= \frac{\pi}{k}\int_0^L \left[\left(\frac{a_2 - a_1}{L}\right)^2 h^2 + 2a_1\left(\frac{a_2 - a_1}{L}\right)h + a_1^2\right] dh$$

$$= \frac{\pi}{k}\left[\left(\frac{a_2 - a_1}{L}\right)^2 \frac{h^3}{3} + 2a_1\left(\frac{a_2 - a_1}{L}\right)\frac{h^2}{2} + a_1^2 h\right]_0^L = \frac{\pi L}{3k}(a_1^2 + a_1 a_2 + a_2^2).$$

[40] Place the x-axis with $x = 0$ at the base of the cone, $x = h$ at the top. Every cross section is an ellipse of semi-major axis length w and semi-minor axis length v.

By similar triangles in *Figure 40*, $w = \frac{a(h - x)}{h}$ and $v = \frac{b(h - x)}{h}$.

By Example 6, the area of this cross section is $\pi wv = \frac{\pi ab(h - x)^2}{h^2}$. Thus,

$$V = \int_0^h \frac{\pi ab(h - x)^2}{h^2}\, dx \; \{u = h - x,\ du = -dx\} = -\frac{\pi ab}{h^2}\int_h^0 u^2\, du = -\frac{\pi ab}{h^2}\left[\tfrac{1}{3}u^3\right]_h^0$$

$$= \tfrac{1}{3}\pi abh.$$

[41] $f(x) = \sqrt{b^2 - (b^2/a^2)x^2} \Rightarrow f'(x) = -\frac{bx}{a(a^2 - x^2)^{1/2}} \Rightarrow$

$1 + [f'(x)]^2 = 1 + \frac{b^2x^2}{a^2(a^2 - x^2)}$. Let $u = \frac{b}{a}x$ and $\frac{a}{b}\,du = dx$. Then,

$$S = \int_{-a}^a 2\pi f(x)\sqrt{1 + [f'(x)]^2}\, dx = 2\pi\int_{-b}^b \sqrt{b^2 - u^2}\sqrt{1 + \frac{u^2}{b^2 - u^2}\left(\frac{b^2}{a^2}\right)}\,\frac{a}{b}\, du$$

$$= 4\pi\int_0^b \sqrt{b^2 - u^2}\sqrt{\frac{a^2}{b^2} + \frac{u^2}{b^2 - u^2}}\, du = 4\pi\int_0^b \sqrt{\frac{a^2(b^2 - u^2)}{b^2} + u^2}\, du$$ (cont.)

$= \frac{4\pi}{b}\int_0^b \sqrt{a^2b^2 + (b^2 - a^2)u^2}\,du = \frac{4\pi}{b}\int_0^b \sqrt{a^2b^2 - c^2u^2}\,du\{\,a^2 - b^2 = c^2\,\}$

$= \frac{4\pi}{bc}\int_0^{bc} \sqrt{(ab)^2 - v^2}\,dv\{\,v = cu,\ dv = c\,du\,\}$

$= \frac{4\pi}{bc}\left[\frac{v}{2}\sqrt{(ab)^2 - v^2} + \frac{(ab)^2}{2}\sin^{-1}\frac{v}{ab}\right]_0^{bc}$ {Formula 30}

$= \frac{2\pi}{bc}\left[bc\sqrt{a^2b^2 - b^2c^2} + a^2b^2\sin^{-1}\frac{c}{a}\right]$

$= \frac{2\pi}{bc}\left[b^3c + a^2b^2\cos^{-1}\frac{b}{a}\right] = 2\pi b^2 + \frac{2\pi a^2 b}{c}\cos^{-1}\frac{b}{a}.$

If $a = 6378$ km and $b = 6356$ km, then $S \approx 509 \times 10^6$ km^2.

42 Let the ellipse have the equation $\frac{x^2}{a^2} + \frac{y^2}{b^2} = 1$. If (x, y) is the vertex shown,

$A = 4xy = \frac{4b}{a}x\sqrt{a^2 - x^2},\ -a \le x \le a.\ A' = \frac{4b}{a}\left[\frac{-x^2}{\sqrt{a^2 - x^2}} + \sqrt{a^2 - x^2}\right] = 0 \Rightarrow$

$2x^2 = a^2 \Rightarrow x = \frac{a}{\sqrt{2}}$ and $y = \frac{b}{\sqrt{2}}$. The dimensions are $\frac{2a}{\sqrt{2}}$ by $\frac{2b}{\sqrt{2}}$, or, equivalently,

$\sqrt{2}\,a$ by $\sqrt{2}\,b$. This is a maximum area since $A(\pm a) = 0$.

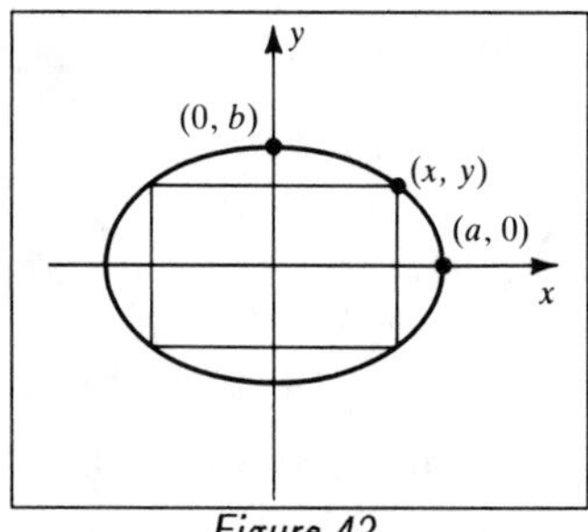

Figure 42

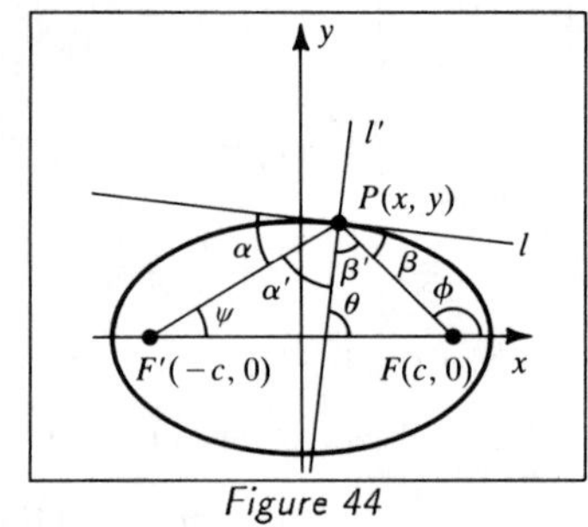

Figure 44

43 Since rotations and translations of the ellipse will not affect the result,

let its equation be $\frac{x^2}{a^2} + \frac{y^2}{b^2} = 1$. Thus, $\frac{2x}{a^2} + \frac{2y}{b^2}y' = 0 \Rightarrow y' = -\frac{b^2x}{a^2y}$.

The slope of the normal line at (x_1, y_1) is $\frac{a^2y_1}{b^2x_1}$.

Since the normal line passes through $(0, 0)$ and (x_1, y_1), its slope must be $\frac{y_1}{x_1}$.

Now, $\frac{a^2y_1}{b^2x_1} = \frac{y_1}{x_1} \Rightarrow a^2 = b^2$. Thus, the ellipse is a circle.

44 See *Figure 44*. Let $x^2/a^2 + y^2/b^2 = 1$ be the equation of the ellipse. Let ℓ' be the normal line through P with the inclination θ. By implicit differentiation,

$y' = m_\ell = -\frac{b^2x}{a^2y} \Rightarrow m_{\ell'} = \tan\theta = \frac{a^2y}{b^2x}$. Let α' and β' be as shown.

Let ψ be the inclination of $F'P$. Then, $\alpha' + \psi + (\pi - \theta) = \pi \Rightarrow \alpha' = \theta - \psi \Rightarrow$

$$\tan\alpha' = \frac{\tan\theta - \tan\psi}{1 + \tan\theta\tan\psi} = \frac{\dfrac{a^2y}{b^2x} - \dfrac{y}{x+c}}{1 + \dfrac{a^2y^2}{b^2x^2 + b^2cx}} = \frac{a^2xy + a^2cy - b^2xy}{b^2x^2 + b^2cx + a^2y^2}.$$

But, $a^2 - b^2 = c^2$ and $b^2x^2 + a^2y^2 = a^2b^2 \Rightarrow \tan\alpha' = \dfrac{c^2xy + a^2cy}{a^2b^2 + b^2cx} =$

$\dfrac{cy(cx + a^2)}{b^2(a^2 + cx)} = \dfrac{cy}{b^2}$. Now let φ be the inclination of FP so that

$\beta' + \theta + (\pi - \varphi) = \pi \Rightarrow \beta' = \varphi - \theta \Rightarrow$

$$\tan\beta' = \frac{\tan\varphi - \tan\theta}{1 + \tan\varphi\tan\theta} = \frac{\dfrac{y}{x-c} - \dfrac{a^2y}{b^2x}}{1 + \dfrac{a^2y^2}{b^2x^2 - cb^2x}} = \frac{b^2xy - a^2xy + ca^2y}{b^2x^2 - cb^2x + a^2y^2} =$$

$\dfrac{-c^2xy + ca^2y}{a^2b^2 - cb^2x} = \dfrac{cy}{b^2} = \tan\alpha' \Rightarrow \alpha' = \beta'$ since both are acute. Thus, $\alpha = \beta$.

Note: Depending on the type of software used, you may need to solve the given equation for y in order to graph the ellipses.

45 (a) $\dfrac{x^2}{2.9} + \dfrac{y^2}{2.1} = 1 \Rightarrow 2.1x^2 + 2.9y^2 = 6.09 \Rightarrow y = \pm\sqrt{\dfrac{1}{2.9}(6.09 - 2.1x^2)}$.

$\dfrac{x^2}{4.3} + \dfrac{(y - 2.1)^2}{4.9} = 1 \Rightarrow 4.9x^2 + 4.3(y - 2.1)^2 = 21.07 \Rightarrow$

$y = 2.1 \pm \sqrt{\dfrac{1}{4.3}(21.07 - 4.9x^2)}$.

From the graph, the points of intersection are approximately $(\pm 1.540, 0.618)$.

(b) Area $\approx 2\displaystyle\int_0^{1.54}\left[\sqrt{\frac{1}{2.9}(6.09 - 2.1x^2)} - \left(2.1 - \sqrt{\frac{1}{4.3}(21.07 - 4.9x^2)}\right)\right]dx$

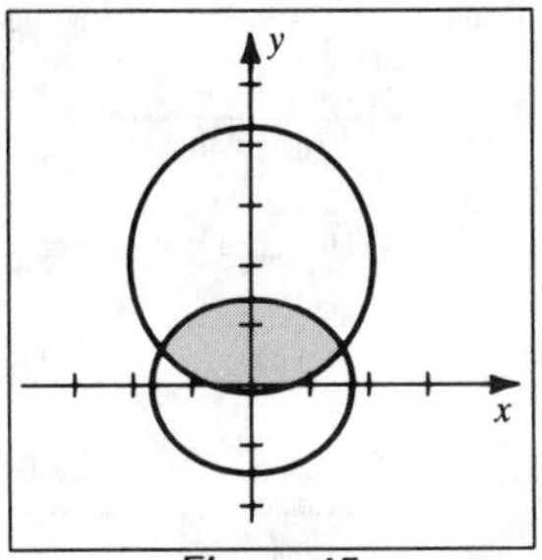

Figure 45

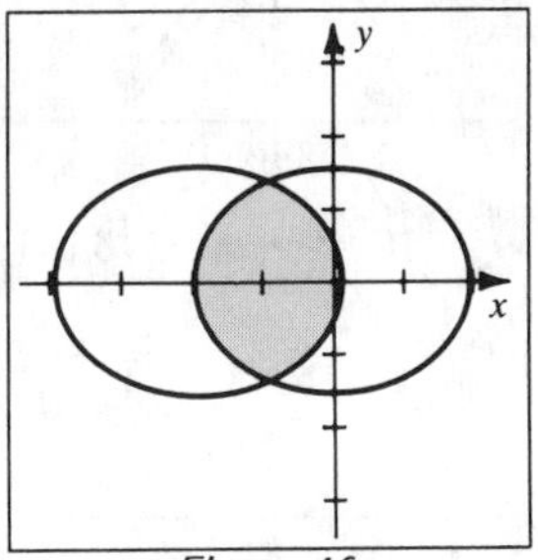

Figure 46

46 (a) $\dfrac{x^2}{3.9} + \dfrac{y^2}{2.4} = 1 \Rightarrow 2.4x^2 + 3.9y^2 = 9.36 \Rightarrow y = \pm\sqrt{\dfrac{1}{3.9}(9.36 - 2.4x^2)}$.

Since we will need to integrate with respect to y, we will also solve for x.

Thus, $x = \pm\sqrt{\dfrac{1}{2.4}(9.36 - 3.9y^2)}$. $\dfrac{(x + 1.9)^2}{4.1} + \dfrac{y^2}{2.5} = 1 \Rightarrow$

$2.5(x + 1.9)^2 + 4.1y^2 = 10.25 \Rightarrow y = \pm\sqrt{\dfrac{1}{4.1}\left[10.25 - 2.5(x + 1.9)^2\right]}$. (cont.)

Similarly, $x = -1.9 \pm \sqrt{\frac{1}{2.5}(10.25 - 4.1y^2)}$.

From the graph, the points of intersection are approximately $(-0.905, \pm 1.377)$.

(b) Area $\approx 2\int_0^{1.38}\left[-1.9 + \sqrt{\frac{1}{2.5}(10.25 - 4.1y^2)} - \left(-\sqrt{\frac{1}{2.4}(9.36 - 3.9y^2)}\right)\right]dy$

Exercises 12.3

Note: Let C, V, F, and W denote the center, the vertices, the foci, and the end points of the conjugate axis, respectively.

[1] $c^2 = 9 + 4 \Rightarrow c = \pm\sqrt{13}$; $V(\pm 3, 0)$; $F(\pm\sqrt{13}, 0)$; $W(0, \pm 2)$; $y = \pm\frac{2}{3}x$

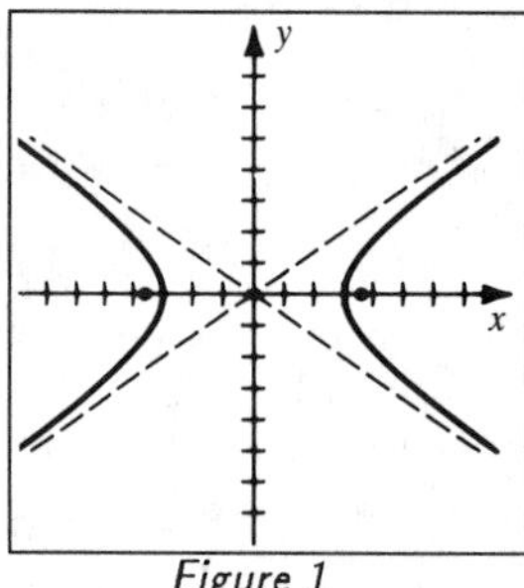

Figure 1

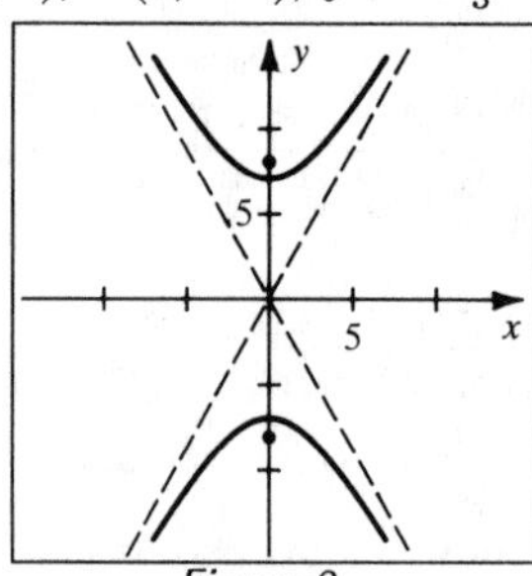

Figure 2

[2] $c^2 = 49 + 16 \Rightarrow c = \pm\sqrt{65}$; $V(0, \pm 7)$; $F(0, \pm\sqrt{65})$; $W(\pm 4, 0)$; $y = \pm\frac{7}{4}x$

[3] $c^2 = 9 + 4 \Rightarrow c = \pm\sqrt{13}$; $V(0, \pm 3)$; $F(0, \pm\sqrt{13})$; $W(\pm 2, 0)$; $y = \pm\frac{3}{2}x$

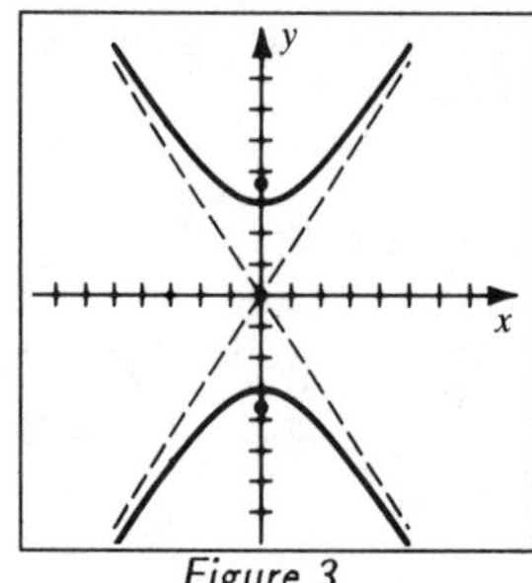

Figure 3

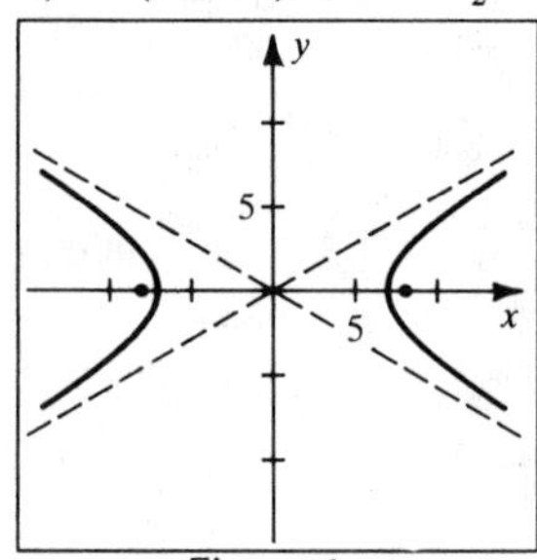

Figure 4

[4] $c^2 = 49 + 16 \Rightarrow c = \pm\sqrt{65}$; $V(\pm 7, 0)$; $F(\pm\sqrt{65}, 0)$; $W(0, \pm 4)$; $y = \pm\frac{4}{7}x$

[5] $y^2 - 4x^2 = 16 \Rightarrow \frac{y^2}{16} - \frac{x^2}{4} = 1$; $c^2 = 16 + 4 \Rightarrow c = \pm 2\sqrt{5}$;

$V(0, \pm 4)$; $F(0, \pm 2\sqrt{5})$; $W(\pm 2, 0)$; $y = \pm 2x$

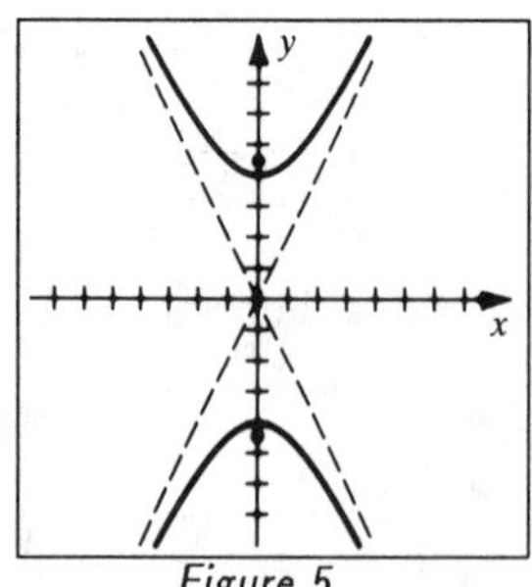

Figure 5

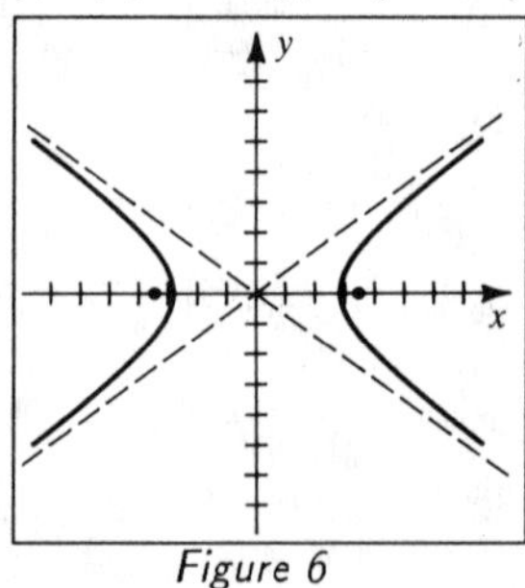

Figure 6

6 $x^2 - 2y^2 = 8 \Rightarrow \frac{x^2}{8} - \frac{y^2}{4} = 1;\ c^2 = 8 + 4 \Rightarrow c = \pm 2\sqrt{3};$

$V(\pm 2\sqrt{2}, 0);\ F(\pm 2\sqrt{3}, 0);\ W(0, \pm 2);\ y = \pm \frac{1}{2}\sqrt{2}\,x$

7 $c^2 = 1 + 1 \Rightarrow c = \pm\sqrt{2};\ V(\pm 1, 0);\ F(\pm\sqrt{2}, 0);\ W(0, \pm 1);\ y = \pm x$

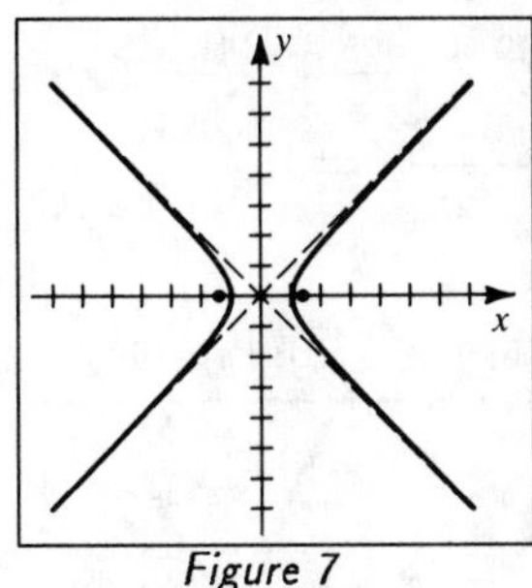

Figure 7

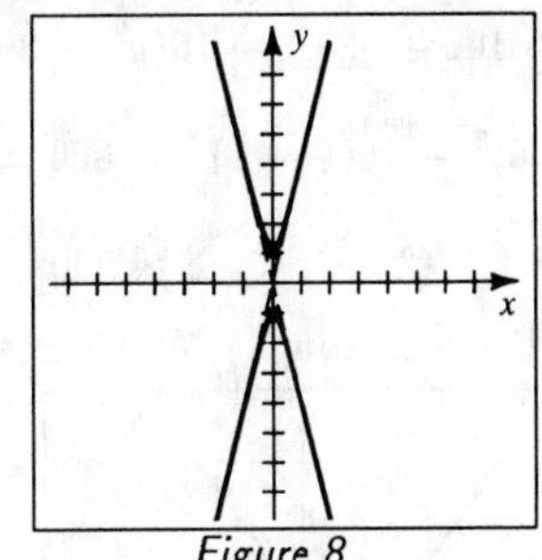

Figure 8

8 $y^2 - 16x^2 = 1 \Rightarrow \frac{y^2}{1} - \frac{x^2}{\frac{1}{16}} = 1;\ c^2 = 1 + \frac{1}{16} \Rightarrow c = \pm \frac{1}{4}\sqrt{17};$

$V(0, \pm 1);\ F(0, \pm \frac{1}{4}\sqrt{17});\ W(\pm \frac{1}{4}, 0);\ y = \pm 4x$

9 $x^2 - 5y^2 = 25 \Rightarrow \frac{x^2}{25} - \frac{y^2}{5} = 1;\ c^2 = 25 + 5 \Rightarrow c = \pm\sqrt{30};$

$V(\pm 5, 0);\ F(\pm\sqrt{30}, 0);\ W(0, \pm\sqrt{5});\ y = \pm \frac{1}{5}\sqrt{5}\,x$

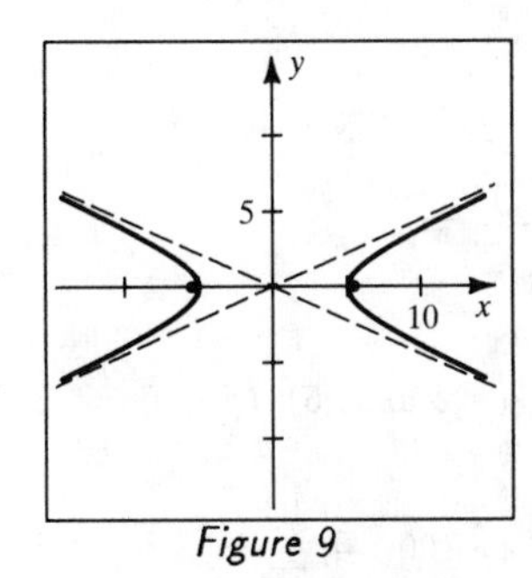

Figure 9

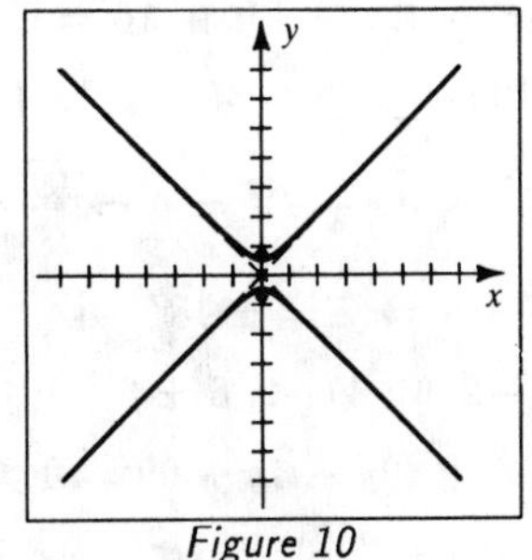

Figure 10

10 $4y^2 - 4x^2 = 1 \Rightarrow \frac{y^2}{\frac{1}{4}} - \frac{x^2}{\frac{1}{4}} = 1;\ c^2 = \frac{1}{4} + \frac{1}{4} \Rightarrow c = \pm \frac{1}{2}\sqrt{2};$

$V(0, \pm \frac{1}{2});\ F(0, \pm \frac{1}{2}\sqrt{2});\ W(\pm \frac{1}{2}, 0);\ y = \pm x$

11 $3x^2 - y^2 = -3 \Rightarrow \frac{y^2}{3} - \frac{x^2}{1} = 1;\ c^2 = 3 + 1 \Rightarrow c = \pm 2;$

$V(0, \pm\sqrt{3});\ F(0, \pm 2);\ W(\pm 1, 0);\ y = \pm\sqrt{3}\,x$

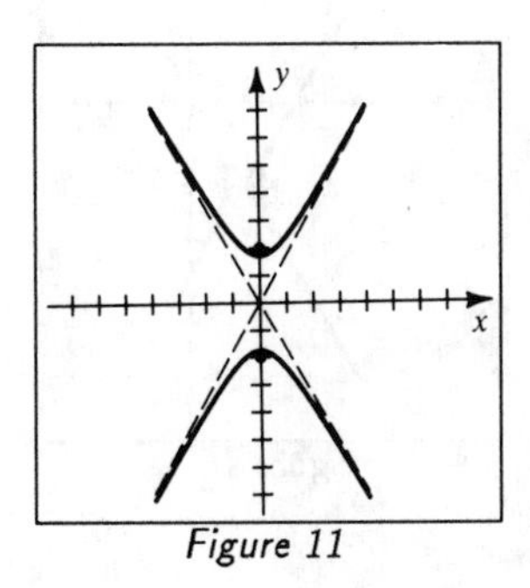

Figure 11

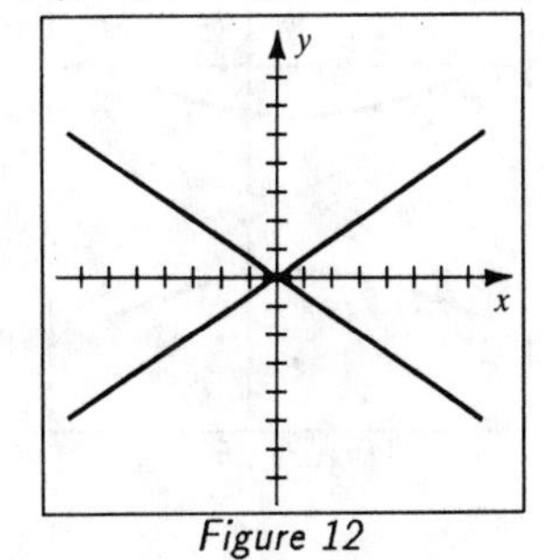

Figure 12

[12] $16x^2 - 36y^2 = 1 \Rightarrow \dfrac{x^2}{\frac{1}{16}} - \dfrac{y^2}{\frac{1}{36}} = 1$; $c^2 = \frac{1}{16} + \frac{1}{36} \Rightarrow c = \pm\frac{1}{12}\sqrt{13}$;

$V(\pm\frac{1}{4}, 0)$; $F(\pm\frac{1}{12}\sqrt{13}, 0)$; $W(0, \pm\frac{1}{6})$; $y = \pm\frac{2}{3}x$. See *Figure 12.*

[13] $25x^2 - 16y^2 + 250x + 32y + 109 = 0 \Rightarrow$

$25(x^2 + 10x + \underline{25}) - 16(y^2 - 2y + \underline{1}) = -109 + \underline{625} - \underline{16} \Rightarrow$

$25(x+5)^2 - 16(y-1)^2 = 500 \Rightarrow \dfrac{(x+5)^2}{20} - \dfrac{(y-1)^2}{\frac{125}{4}} = 1$;

$c^2 = 20 + \frac{125}{4} \Rightarrow c = \pm\frac{1}{2}\sqrt{205}$; $C(-5, 1)$;

$V(-5 \pm 2\sqrt{5}, 1)$; $F(-5 \pm \frac{1}{2}\sqrt{205}, 1)$; $W(-5, 1 \pm \frac{5}{2}\sqrt{5})$; $(y - 1) = \pm\frac{5}{4}(x + 5)$

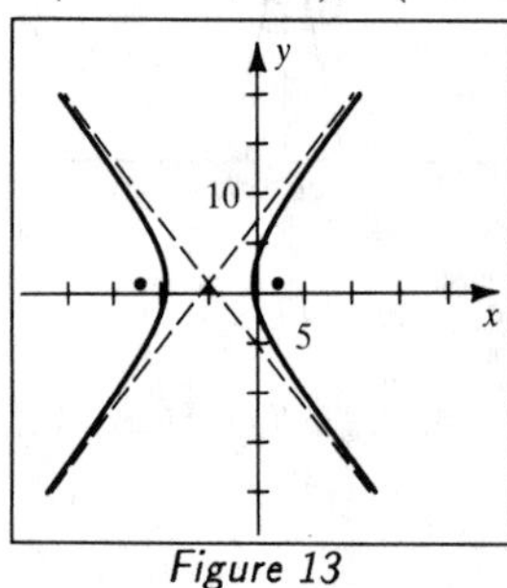

Figure 13

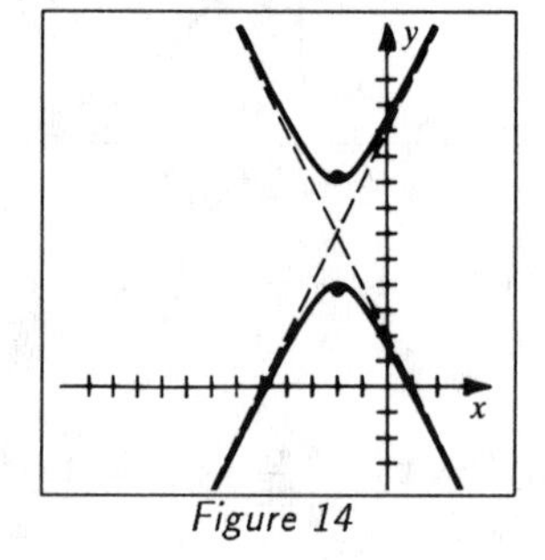

Figure 14

[14] $y^2 - 4x^2 - 12y - 16x + 16 = 0 \Rightarrow$

$(y^2 - 12y + \underline{36}) - 4(x^2 + 4x + \underline{4}) = -16 + \underline{36} - \underline{16} \Rightarrow$

$(y-6)^2 - 4(x+2)^2 = 4 \Rightarrow \dfrac{(y-6)^2}{4} - \dfrac{(x+2)^2}{1} = 1$;

$c^2 = 4 + 1 \Rightarrow c = \pm\sqrt{5}$;

$C(-2, 6)$; $V(-2, 6 \pm 2)$; $F(-2, 6 \pm \sqrt{5})$; $W(-2 \pm 1, 6)$; $(y - 6) = \pm 2(x + 2)$

[15] $4y^2 - x^2 + 40y - 4x + 60 = 0 \Rightarrow$

$4(y^2 + 10y + \underline{25}) - 1(x^2 + 4x + \underline{4}) = -60 + \underline{100} - \underline{4} \Rightarrow$

$4(y+5)^2 - (x+2)^2 = 36 \Rightarrow \dfrac{(y+5)^2}{9} - \dfrac{(x+2)^2}{36} = 1$;

$c^2 = 9 + 36 \Rightarrow c = \pm 3\sqrt{5}$; $C(-2, -5)$;

$V(-2, -5 \pm 3)$; $F(-2, -5 \pm 3\sqrt{5})$; $W(-2 \pm 6, -5)$; $(y + 5) = \pm\frac{1}{2}(x + 2)$

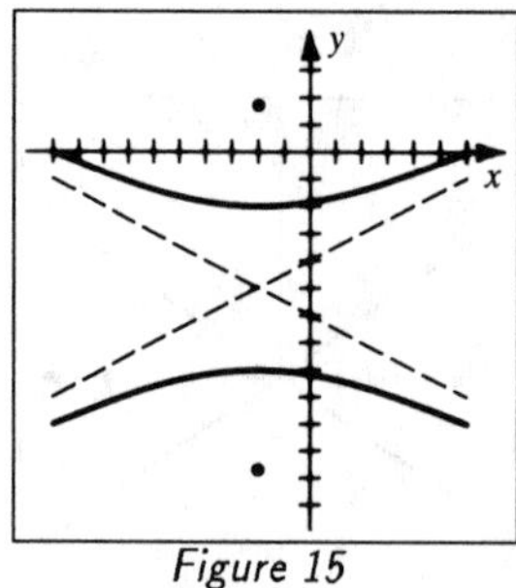

Figure 15

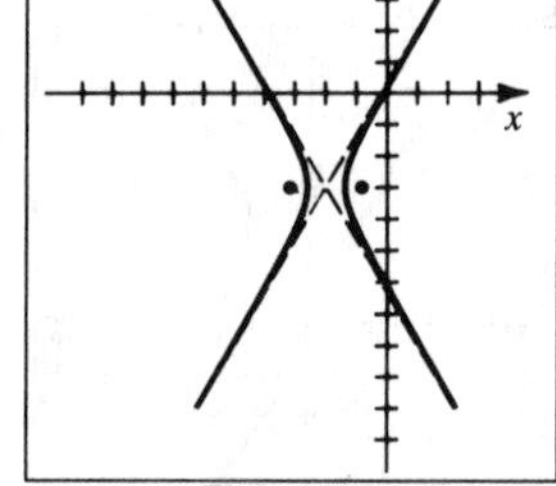

Figure 16

[16] $25x^2 - 9y^2 + 100x - 54y + 10 = 0 \Rightarrow$

$25(x^2 + 4x + \underline{4}) - 9(y^2 + 6y + \underline{9}) = -10 + \underline{100} - \underline{81} \Rightarrow$

$25(x+2)^2 - 9(y+3)^2 = 9 \Rightarrow \dfrac{(x+2)^2}{\frac{9}{25}} - \dfrac{(y+3)^2}{1} = 1;$

$c^2 = \frac{9}{25} + 1 \Rightarrow c = \pm\frac{1}{5}\sqrt{34};$ $C(-2, -3);$

$V(-2 \pm \frac{3}{5}, -3);\ F(-2 \pm \frac{1}{5}\sqrt{34}, -3);\ W(-2, -3 \pm 1);\ (y+3) = \pm\frac{5}{3}(x+2)$

[17] $9y^2 - x^2 - 36y + 12x - 36 = 0 \Rightarrow$

$9(y^2 - 4y + \underline{4}) - (x^2 - 12x + \underline{36}) = 36 + \underline{36} - \underline{36} \Rightarrow$

$9(y-2)^2 - (x-6)^2 = 36 \Rightarrow \dfrac{(y-2)^2}{4} - \dfrac{(x-6)^2}{36} = 1;$

$c^2 = 4 + 36 \Rightarrow c = \pm 2\sqrt{10};$

$C(6, 2);\ V(6, 2 \pm 2);\ F(6, 2 \pm 2\sqrt{10});\ W(6 \pm 6, 2);\ (y-2) = \pm\frac{1}{3}(x-6)$

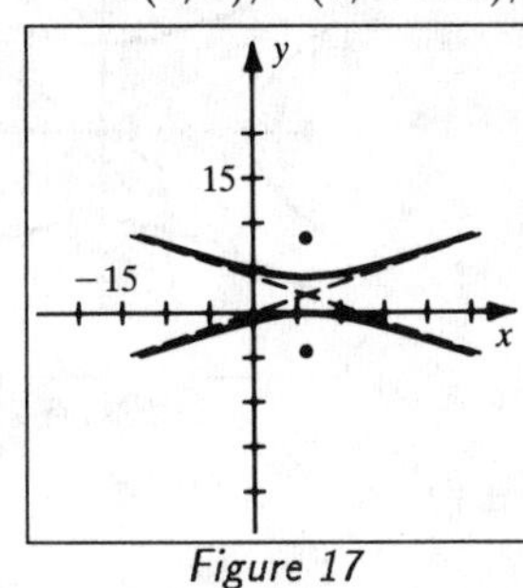

Figure 17

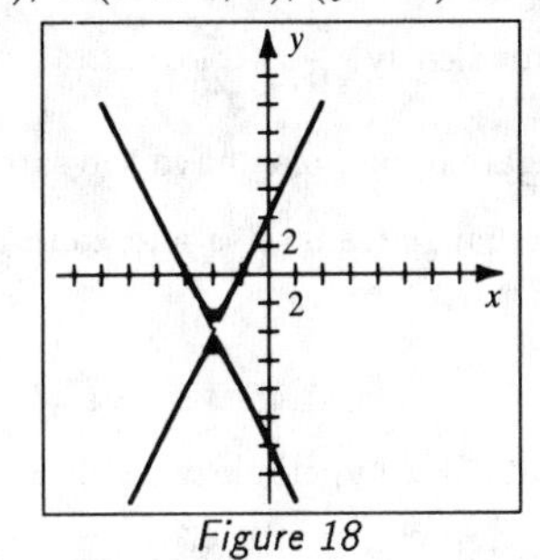

Figure 18

[18] $4x^2 - y^2 + 32x - 8y + 49 = 0 \Rightarrow$

$4(x^2 + 8x + \underline{16}) - (y^2 + 8y + \underline{16}) = -49 + \underline{64} - \underline{16} \Rightarrow$

$4(x+4)^2 - (y+4)^2 = -1 \Rightarrow \dfrac{(y+4)^2}{1} - \dfrac{(x+4)^2}{\frac{1}{4}} = 1;$

$c^2 = 1 + \frac{1}{4} \Rightarrow c = \pm\frac{1}{2}\sqrt{5};$ $C(-4, -4);$

$V(-4, -4 \pm 1);\ F(-4, -4 \pm \frac{1}{2}\sqrt{5});\ W(-4 \pm \frac{1}{2}, -4);\ (y+4) = \pm 2(x+4)$

[19] $F(0, \pm 4)$ and $V(0, \pm 1) \Rightarrow W(\pm\sqrt{15}, 0)$. An equation is $\dfrac{y^2}{1} - \dfrac{x^2}{15} = 1$.

[20] $F(\pm 8, 0)$ and $V(\pm 5, 0) \Rightarrow W(0, \pm\sqrt{39})$. An equation is $\dfrac{x^2}{25} - \dfrac{y^2}{39} = 1$.

[21] $F(\pm 5, 0)$ and $V(\pm 3, 0) \Rightarrow W(0, \pm 4)$. An equation is $\dfrac{x^2}{9} - \dfrac{y^2}{16} = 1$.

[22] $F(0, \pm 3)$ and $V(0, \pm 2) \Rightarrow W(\pm\sqrt{5}, 0)$. An equation is $\dfrac{y^2}{4} - \dfrac{x^2}{5} = 1$.

[23] Conjugate axis of length 4 and $F(0, \pm 5) \Rightarrow W(\pm 2, 0)$ and $V(0, \pm\sqrt{21})$.

An equation is $\dfrac{y^2}{21} - \dfrac{x^2}{4} = 1$.

[24] An equation of a hyperbola with vertices at $(\pm 4, 0)$ is $\dfrac{x^2}{16} - \dfrac{y^2}{b^2} = 1$. Substituting

$x = 8$ and $y = 2$ yields $4 - \dfrac{4}{b^2} = 1 \Rightarrow b^2 = \frac{4}{3}$. An equation is $\dfrac{x^2}{16} - \dfrac{3y^2}{4} = 1$.

[25] Asymptote equations of $y = \pm 2x$ and $V(\pm 3, 0) \Rightarrow W(0, \pm 6)$.

An equation is $\frac{x^2}{9} - \frac{y^2}{36} = 1$.

[26] Let the y value of V equal a, and the x value of W equal b.

Now $a = \frac{1}{3}b$ {from the asymptote equation} and $a^2 + b^2 = 10^2$ {from the foci} $\Rightarrow$ $(\frac{1}{3}b)^2 + b^2 = 10^2 \Rightarrow \frac{10}{9}b^2 = 100 \Rightarrow b^2 = 90$ and $a^2 = 10$.

An equation is $\frac{y^2}{10} - \frac{x^2}{90} = 1$.

[27] $a = 5$, $b = 2(5) = 10$. $\frac{x^2}{5^2} - \frac{y^2}{10^2} = 1 \Rightarrow \frac{x^2}{25} - \frac{y^2}{100} = 1$.

[28] $a = 2$, $2 = \frac{1}{4}(b) \Rightarrow b = 8$. $\frac{y^2}{2^2} - \frac{x^2}{8^2} = 1 \Rightarrow \frac{y^2}{4} - \frac{x^2}{64} = 1$.

[29] Their equations are $\frac{x^2}{25} - \frac{y^2}{9} = 1$ and $\frac{x^2}{25} - \frac{y^2}{9} = -1$,

or, equivalently, $\frac{y^2}{9} - \frac{x^2}{25} = 1$.

Conjugate hyperbolas have the same asymptotes and exchange transverse and conjugate axes.

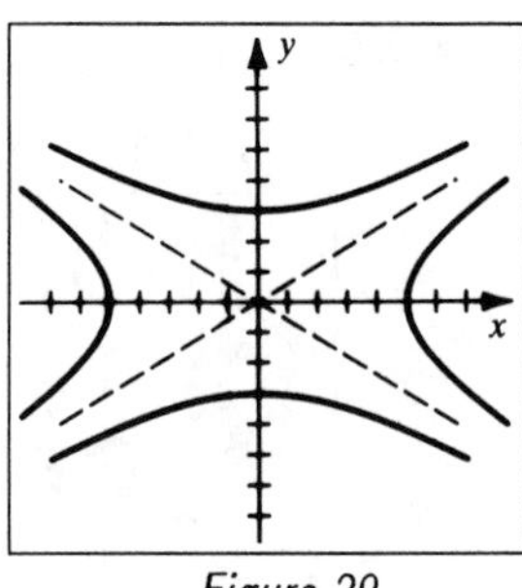

Figure 29

[30] The path is a hyperbola with $V(\pm 3, 0)$ and $W(0, \pm \frac{3}{2})$.

An equation is $\frac{x^2}{(3)^2} - \frac{y^2}{(\frac{3}{2})^2} = 1$ or equivalently, $x^2 - 4y^2 = 9$.

If only the right branch is considered, then $x = \sqrt{9 + 4y^2}$ is an equation of the path.

[31] Set up a coordinate system like Example 5. Then, $d_1 - d_2 = 2a = 160 \Rightarrow a = 80$. $b^2 = c^2 - a^2 = 100^2 - 80^2 \Rightarrow b = 60$. The equation of the hyperbola with focus A, passing through the coordinates of the ship at $P(x, y)$ is $\frac{x^2}{80^2} - \frac{y^2}{60^2} = 1$. Now,

$y = 100 \Rightarrow x = \frac{80}{3}\sqrt{34}$. The ship's coordinates are $(\frac{80}{3}\sqrt{34}, 100) \approx (155.5, 100)$.

[32] $\frac{x^2}{a^2} - \frac{y^2}{b^2} = 1 \Rightarrow \frac{2x}{a^2} - \frac{2y}{b^2}y' = 0 \Rightarrow y' = \frac{b^2x}{a^2y}$. At $P(x_1, y_1)$, the tangent line is

$$y - y_1 = \frac{b^2x_1}{a^2y_1}(x - x_1) \Rightarrow \frac{y_1y - y_1^2}{b^2} = \frac{x_1x - x_1^2}{a^2} \Rightarrow$$

$$\frac{x_1x}{a^2} - \frac{y_1y}{b^2} = \frac{x_1^2}{a^2} - \frac{y_1^2}{b^2} = 1,$$ since $P(x_1, y_1)$ is on the hyperbola.

[33] By Exercise 32, the tangent lines have the equation $\frac{x_1x}{4} - \frac{y_1y}{36} = 1$, where (x_1, y_1) is a point of tangency. $(0, 6)$ is on this line so $y_1 = -6$. Thus, $9x_1^2 - y_1^2 = 36 \Rightarrow$

$x_1^2 = \frac{1}{9}(y_1^2 + 36) = \frac{1}{9}(36 + 36) = 8$. Points of tangency are $(\pm 2\sqrt{2}, -6)$.

[34] By Exercise 32, the tangent line has the equation $\frac{x_1 x}{16} - \frac{y_1 y}{4} = 1$, where (x_1, y_1) is the point of tangency. $P(2, -1)$ is on the line, so $\frac{x_1}{8} + \frac{y_1}{4} = 1 \Rightarrow x_1 = 8 - 2y_1$. Substituting this expression for x_1 into $x^2 - 4y^2 = 16$ yields $y_1 = \frac{3}{2}$ and $x_1 = 5$.

The tangent line is $\frac{5}{16}x - \frac{3}{8}y = 1$, or $5x - 6y = 16$.

[35] $2x^2 - 5y^2 = 3 \Rightarrow 4x - 10yy' = 0 \Rightarrow y' = \frac{2x}{5y}$. At $P(-2, 1)$, $y' = -\frac{4}{5}$.

Tangent line equation: $y - 1 = -\frac{4}{5}(x + 2)$, or $4x + 5y = -3$.

Normal line equation: $y - 1 = \frac{5}{4}(x + 2)$, or $5x - 4y = -14$.

[36] $3y^2 - 2x^2 = 40 \Rightarrow 6yy' - 4x = 0 \Rightarrow y' = \frac{2x}{3y}$. At $P(2, -4)$, $y' = -\frac{1}{3}$.

Tangent line equation: $y + 4 = -\frac{1}{3}(x - 2)$, or $x + 3y = -10$.

Normal line equation: $y + 4 = 3(x - 2)$, or $3x - y = 10$.

[37] The upper and lower boundaries of the region are $y = \pm\frac{b}{a}\sqrt{x^2 - a^2}$ for $a \le x \le c$.

Thus, $A = \frac{2b}{a}\int_a^c \sqrt{x^2 - a^2}\,dx$

$= \frac{2b}{a}\left[\frac{x}{2}\sqrt{x^2 - a^2} - \frac{a^2}{2}\ln\left|x + \sqrt{x^2 - a^2}\right|\right]_a^c$ {Formula 39 or letting $x = a\sec\theta$}

$= \frac{b}{a}\left[bc - a^2\ln(b + c) + a^2\ln a\right]$ since $\sqrt{c^2 - a^2} = b$.

[38] Using shells, $V = 2\pi\int_a^c (x)\frac{2b}{a}\sqrt{x^2 - a^2}\,dx = \frac{2\pi b}{a}\left[\frac{2}{3}(x^2 - a^2)^{3/2}\right]_a^c =$

$\frac{4\pi b}{3a}(c^2 - a^2)^{3/2} = \frac{4\pi b}{3a}(b^2)^{3/2} = \frac{4b^4\pi}{3a}$.

[39] $x^2 - y^2 = 8 \Rightarrow y = \sqrt{x^2 - 8} \Rightarrow y' = \frac{x}{\sqrt{x^2 - 8}}$. $V(\sqrt{8}, 0)$ and $F(4, 0) \Rightarrow$

$S = 2\pi\int_{\sqrt{8}}^4 \sqrt{x^2 - 8}\sqrt{1 + \frac{x^2}{x^2 - 8}}\,dx = 2\pi\sqrt{2}\int_{\sqrt{8}}^4 \sqrt{x^2 - 4}\,dx$

$= 2\pi\sqrt{2}\left[\frac{x}{2}\sqrt{x^2 - 4} - 2\ln\left|x + \sqrt{x^2 - 4}\right|\right]_{\sqrt{8}}^4$ {Formula 39}

$= 2\pi\sqrt{2}\left[2\sqrt{12} - 2\ln\left|4 + \sqrt{12}\right| - 2\sqrt{2} + 2\ln\left|2 + \sqrt{8}\right|\right]$

$= 4\pi\sqrt{2}\left[2\sqrt{3} - \sqrt{2} + \ln\left(\frac{1 + \sqrt{2}}{2 + \sqrt{3}}\right)\right] \approx 28.69$.

[40] Without loss of generality, let the equation of the hyperbola be $\frac{x^2}{a^2} - \frac{y^2}{b^2} = 1$ with interior focus $F(c, 0)$. Then the square of the distance between $P(x, y)$ and $F(c, 0)$ is $f(x) = (x - c)^2 + (y - 0)^2 = (x - c)^2 + b^2\left(\frac{x^2}{a^2} - 1\right)$, $x \ge a$. Now,

$f'(x) = 2(x - c) + \frac{2xb^2}{a^2} = 2\left[\left(1 + \frac{b^2}{a^2}\right)x - c\right] = 2\left[\frac{c^2}{a^2}x - c\right]$. $f'(x) = 0 \Rightarrow$

$x = \frac{a^2}{c} = \left(\frac{a}{c}\right)\cdot a < a$. Since $f'(x) > 0$ for $x > \frac{a^2}{c}$, $f(x)$ is $\uparrow$ and its minimum must occur at the end point of its domain, i.e., at $x = a$,

which is the x-coordinate of the vertex $V(a, 0)$.

[41] $12x^2 + 24x - 4y^2 + 9 = 0 \Rightarrow 12(x+1)^2 - 4y^2 = 3 \Rightarrow \dfrac{(x+1)^2}{\frac{1}{4}} - \dfrac{y^2}{\frac{3}{4}} = 1 \Rightarrow$

$V(-1 \pm \frac{1}{2}, 0)$. By Exercise 40, the comet is closest to the sun when it is at

$V(-\frac{1}{2}, 0)$. Distance $= \frac{1}{2}$AU.

[42] Refer to *Figure 42*. Without loss of generality, let the equation of the hyperbola be $\dfrac{x^2}{a^2} - \dfrac{y^2}{b^2} = 1$. Also let $P(x, y)$ be the point of tangency in quadrant I and segment MP be parallel to the x-axis. We wish to show that $\theta_1 = \theta_2$. Now, $\theta_1 = \beta_2 - \alpha$ and $\theta_2 = \alpha - \beta_1$. Since $y' = \dfrac{b^2x}{a^2y}$, $\tan\alpha = \dfrac{b^2x}{a^2y}$. Then, $\tan\theta_1 = \tan(\beta_2 - \alpha) =$

$$\frac{\tan\beta_2 - \tan\alpha}{1 + \tan\beta_2\tan\alpha} = \frac{\dfrac{y-0}{x-c} - \dfrac{b^2x}{a^2y}}{1 + \left(\dfrac{y-0}{x-c}\right)\left(\dfrac{b^2x}{a^2y}\right)} = \frac{a^2y^2 - b^2x(x-c)}{a^2y(x-c) + b^2xy} =$$

$$\frac{b^2cx - (b^2x^2 - a^2y^2)}{y\left[(a^2+b^2)x - a^2c\right]} = \frac{b^2cx - a^2b^2}{y(c^2x - a^2c)} = \frac{b^2}{cy}.$$ In a similar manner with

$\tan\beta_1 = \dfrac{y-0}{x+c}$, $\tan\theta_2 = \tan(\alpha - \beta_1) = \dfrac{b^2}{cy}$. Since both angles are acute, $\theta_1 = \theta_2$.

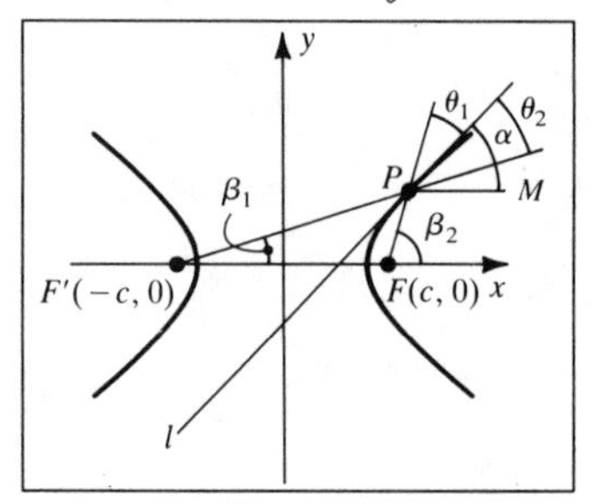

Figure 42

[43] By the reflective property of parabolic mirrors (Exercise 42 of Section 12.1) parallel rays striking the parabolic mirror will be reflected toward F_1.
By the reflective property of hyperbolic mirrors (see page 631),
these rays will be reflected toward the exterior focus of the hyperbolic mirror,
which is located below the parabolic mirror in the figure.

[44] Without loss of generality, let the hyperbola have the equation $\dfrac{x^2}{a^2} - \dfrac{y^2}{b^2} = 1$, the coordinates of P be (x_1, y_1), and the equations of the asymptotes be $y = \pm\frac{b}{a}x$.
From Exercise 32, the equation of the tangent line is $\dfrac{x_1x}{a^2} - \dfrac{y_1y}{b^2} = 1$. First, consider the point of intersection with the asymptote $y = -\frac{b}{a}x$ at $Q(x_Q, y_Q)$. Solving the equations simultaneously gives $\dfrac{x_1x}{a^2} - \left(-\frac{b}{a}x\right)\left(\dfrac{y_1}{b^2}\right) = 1 \Rightarrow x_Q = \dfrac{a^2b}{bx_1 + ay_1}$. Then, $y_Q = -\frac{b}{a}x_Q = \dfrac{-ab^2}{bx_1 + ay_1}$. Second, consider the point of intersection with the asymptote $y = \frac{b}{a}x$ at $R(x_R, y_R)$. In a similar manner, $x_R = \dfrac{a^2b}{bx_1 - ay_1}$ and

$y_R = \dfrac{ab^2}{bx_1 - ay_1}$. The first coordinate for the midpoint of QR is $\frac{1}{2}(x_Q + x_R) =$

$$\frac{1}{2}\left[\frac{a^2b^2\,2x_1}{b^2x_1^2 - a^2y_1^2}\right] = \frac{a^2b^2x_1}{a^2b^2}\ \{\text{since } (x_1,\, y_1) \text{ is on the hyperbola}\} = x_1.$$

In a similar manner, $\frac{1}{2}(y_Q + y_R) = y_1$. Hence, the midpoint is $P(x_1,\, y_1)$.

Note: Depending on the type of software used, you may need to solve the given equation for y in order to graph the hyperbolas.

[45] (a) $\dfrac{(y - 0.1)^2}{1.6} - \dfrac{(x + 0.2)^2}{0.5} = 1 \Rightarrow 0.5(y - 0.1)^2 - 1.6(x + 0.2)^2 = 0.8 \Rightarrow$

$$y = 0.1 \pm \sqrt{1.6 + 3.2(x + 0.2)^2}$$

$\dfrac{(y - 0.5)^2}{2.7} - \dfrac{(x - 0.1)^2}{5.3} = 1 \Rightarrow 5.3(y - 0.5)^2 - 2.7(x - 0.1)^2 = 14.31 \Rightarrow$

$$y = 0.5 \pm \sqrt{\tfrac{1}{5.3}\left[14.31 + 2.7(x - 0.1)^2\right]}$$

From the graph,

the point of intersection in the first quadrant is approximately (0.741, 2.206).

(b) Area $\approx$

$$\int_0^{0.74} \left\{0.5 + \sqrt{\tfrac{1}{5.3}\left[14.31 + 2.7(x - 0.1)^2\right]} - \left[0.1 + \sqrt{1.6 + 3.2(x + 0.2)^2}\right]\right\} dx$$

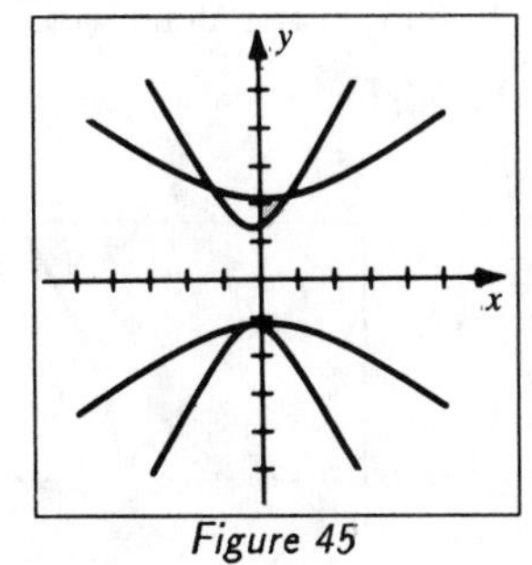

Figure 45

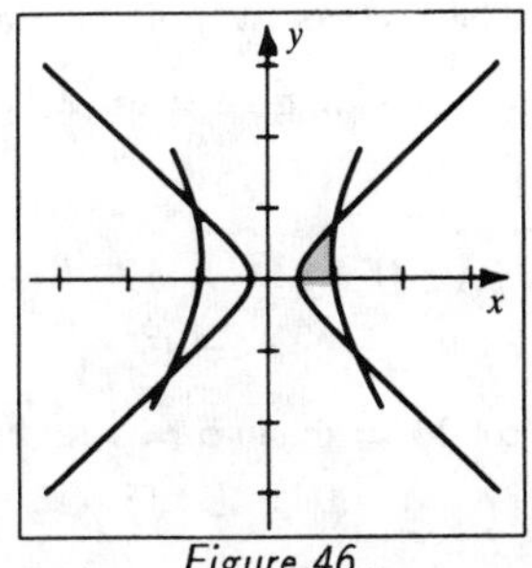

Figure 46

[46] (a) $\dfrac{(x - 0.1)^2}{0.12} - \dfrac{y^2}{0.1} = 1 \Rightarrow 0.1(x - 0.1)^2 - 0.12y^2 = 0.012 \Rightarrow$

$$y = \pm\sqrt{\tfrac{1}{0.12}\left[0.1(x - 0.1)^2 - 0.012\right]}$$

Since we will need to integrate with respect to y, we will also solve for x.

Thus, $x = 0.1 \pm \sqrt{1.2y^2 + 0.12}$.

$\dfrac{x^2}{0.9} - \dfrac{(y - 0.3)^2}{2.1} = 1 \Rightarrow 2.1x^2 - 0.9(y - 0.3)^2 = 1.89 \Rightarrow$

$$y = 0.3 \pm \sqrt{\tfrac{1}{0.9}\left[2.1x^2 - 1.89\right]}$$

Similarly, $x = \pm\sqrt{\tfrac{1}{2.1}\left[0.9(y - 0.3)^2 + 1.89\right]}$. From the graph,

the point of intersection in the first quadrant is approximately (0.994, 0.752).

(b) Area $\approx \displaystyle\int_0^{0.75} \left\{\sqrt{\tfrac{1}{2.1}\left[0.9(y - 0.3)^2 + 1.89\right]} - \left[0.1 + \sqrt{1.2y^2 + 0.12}\right]\right\} dy$

Exercises 12.4

Note: Let D denote the value of the discriminant $B^2 - 4AC$. The following is a general outline of the solutions for Exercises 1-13 in this section and 39–40 in §12.5.

(a) Discriminant value and conic type are given.

(b) The 5 steps in part (b) are :

1) $\cot 2\phi = \dfrac{A - C}{B} \Rightarrow \phi = \frac{1}{2}\cot^{-1}\left(\dfrac{A - C}{B}\right)$, where the range of $\cot^{-1}$ is $0°$ to $180°$. Note that the range of ϕ is $0° < \phi < 90°$.

2) $\cos 2\phi = \dfrac{\pm(A - C)}{\sqrt{(A - C)^2 + B^2}}$, $\sin\phi = \sqrt{\dfrac{1 - \cos 2\phi}{2}}$, and $\cos\phi = \sqrt{\dfrac{1 + \cos 2\phi}{2}}$;

Note that $\cos 2\phi$ will have the same sign as $\cot 2\phi$ since $\cot 2\phi = \dfrac{\cos 2\phi}{\sin 2\phi}$ and $\sin 2\phi$ is positive. Since ϕ is acute, $\sin\phi$ and $\cos\phi$ are positive.

3) The rotation of axes formulas are given. $\begin{cases} x = x'\cos\phi - y'\sin\phi \\ y = x'\sin\phi + y'\cos\phi \end{cases}$

4) The rotation of axes formulas are substituted into the original equation to obtain an equation in x' and y'. This equation is then simplified into a standard form.

5) The vertices (V') of the graph on the $x'y'$-plane are listed along with the corresponding vertices (V) of the graph of the original equation on the xy-plane.

[1] (a) $D = (-2)^2 - 4(1)(1) = 0$, parabola

(b) 1) $\cot 2\phi = 0$; $\phi = 45°$

2) $\cos 2\phi = 0$; $\sin\phi = \frac{1}{2}\sqrt{2}$; $\cos\phi = \frac{1}{2}\sqrt{2}$

3) $\begin{cases} x = \frac{1}{2}\sqrt{2}\,x' - \frac{1}{2}\sqrt{2}\,y' = \frac{1}{2}\sqrt{2}\,(x' - y') \\ y = \frac{1}{2}\sqrt{2}\,x' + \frac{1}{2}\sqrt{2}\,y' = \frac{1}{2}\sqrt{2}\,(x' + y') \end{cases}$

4) $2(y')^2 - 4x' = 0 \Rightarrow (y')^2 = 2(x')$

5) $V'(0, 0) \rightarrow V(0, 0)$ { x & y int. @ $2\sqrt{2}$ }

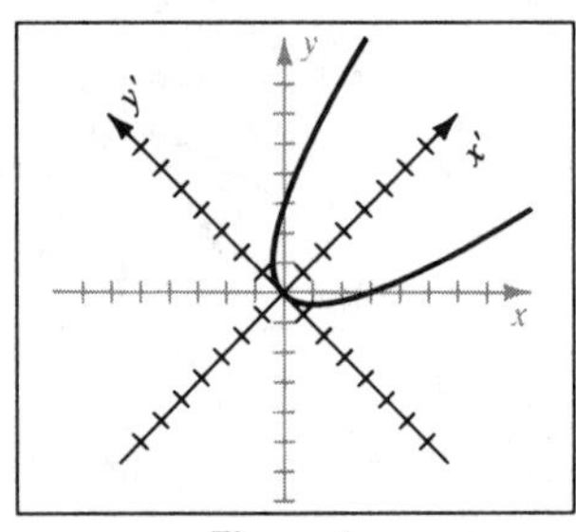

Figure 1

[2] (a) $D = (-2)^2 - 4(1)(1) = 0$, parabola

(b) 1) $\cot 2\phi = 0$; $\phi = 45°$

2) $\cos 2\phi = 0$; $\sin\phi = \frac{1}{2}\sqrt{2}$; $\cos\phi = \frac{1}{2}\sqrt{2}$

3) $\begin{cases} x = \frac{1}{2}\sqrt{2}\,x' - \frac{1}{2}\sqrt{2}\,y' = \frac{1}{2}\sqrt{2}\,(x' - y') \\ y = \frac{1}{2}\sqrt{2}\,x' + \frac{1}{2}\sqrt{2}\,y' = \frac{1}{2}\sqrt{2}\,(x' + y') \end{cases}$

4) $2(y')^2 + 4\sqrt{2}(x') = 0 \Rightarrow (y')^2 = -2\sqrt{2}(x')$

5) $V'(0, 0) \rightarrow V(0, 0)$ { x & y int. @ -4 }

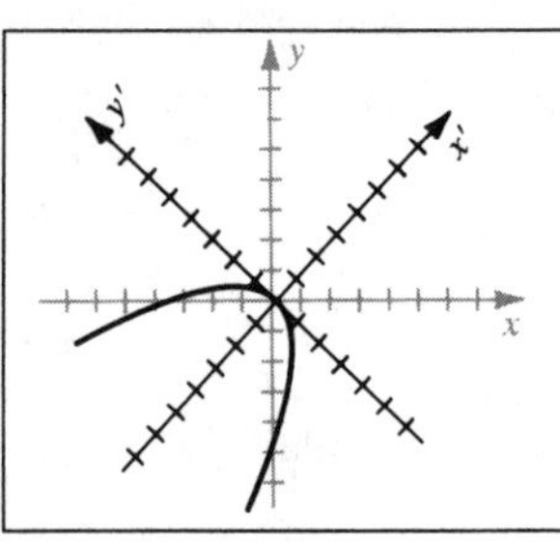

Figure 2

3 (a) $D = (-8)^2 - 4(5)(5) = -36 < 0$, ellipse

(b) 1) $\cot 2\phi = 0$; $\phi = 45^\circ$

2) $\cos 2\phi = 0$; $\sin\phi = \frac{1}{2}\sqrt{2}$; $\cos\phi = \frac{1}{2}\sqrt{2}$

3) $\begin{cases} x = \frac{1}{2}\sqrt{2}\,x' - \frac{1}{2}\sqrt{2}\,y' = \frac{1}{2}\sqrt{2}\,(x' - y') \\ y = \frac{1}{2}\sqrt{2}\,x' + \frac{1}{2}\sqrt{2}\,y' = \frac{1}{2}\sqrt{2}\,(x' + y') \end{cases}$

4) $1(x')^2 + 9(y')^2 = 9 \Rightarrow \dfrac{(x')^2}{9} + \dfrac{(y')^2}{1} = 1$

5) $V'(\pm 3, 0) \to V(\pm \frac{3}{2}\sqrt{2}, \pm \frac{3}{2}\sqrt{2})$

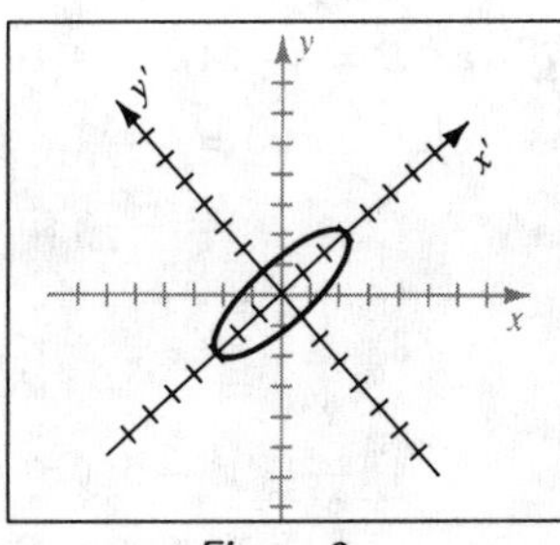

Figure 3

4 (a) $D = (-1)^2 - 4(1)(1) = -3 < 0$, ellipse

(b) 1) $\cot 2\phi = 0$; $\phi = 45^\circ$

2) $\cos 2\phi = 0$; $\sin\phi = \frac{1}{2}\sqrt{2}$; $\cos\phi = \frac{1}{2}\sqrt{2}$

3) $\begin{cases} x = \frac{1}{2}\sqrt{2}\,x' - \frac{1}{2}\sqrt{2}\,y' = \frac{1}{2}\sqrt{2}\,(x' - y') \\ y = \frac{1}{2}\sqrt{2}\,x' + \frac{1}{2}\sqrt{2}\,y' = \frac{1}{2}\sqrt{2}\,(x' + y') \end{cases}$

4) $\frac{1}{2}(x')^2 + \frac{3}{2}(y')^2 = 3 \Rightarrow \dfrac{(x')^2}{6} + \dfrac{(y')^2}{2} = 1$

5) $V'(\pm\sqrt{6}, 0) \to V(\pm\sqrt{3}, \pm\sqrt{3})$

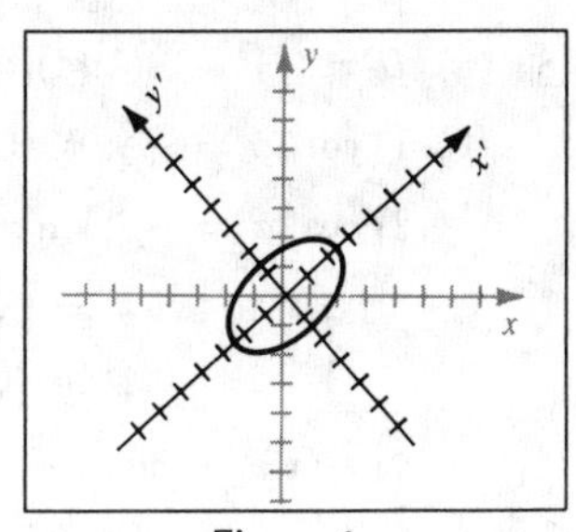

Figure 4

5 (a) $D = (10\sqrt{3})^2 - 4(11)(1) = 256 > 0$, hyperbola

(b) 1) $\cot 2\phi = \frac{1}{3}\sqrt{3}$; $\phi = 30^\circ$

2) $\cos 2\phi = \frac{1}{2}$; $\sin\phi = \frac{1}{2}$; $\cos\phi = \frac{1}{2}\sqrt{3}$

3) $\begin{cases} x = \frac{1}{2}\sqrt{3}\,x' - \frac{1}{2}y' = \frac{1}{2}(\sqrt{3}\,x' - y') \\ y = \frac{1}{2}x' + \frac{1}{2}\sqrt{3}\,y' = \frac{1}{2}(x' + \sqrt{3}\,y') \end{cases}$

4) $\frac{64}{4}(x')^2 - \frac{16}{4}(y')^2 = 4 \Rightarrow \dfrac{(x')^2}{\frac{1}{4}} - \dfrac{(y')^2}{1} = 1$

5) $V'(\pm\frac{1}{2}, 0) \to V(\pm\frac{1}{4}\sqrt{3}, \pm\frac{1}{4})$

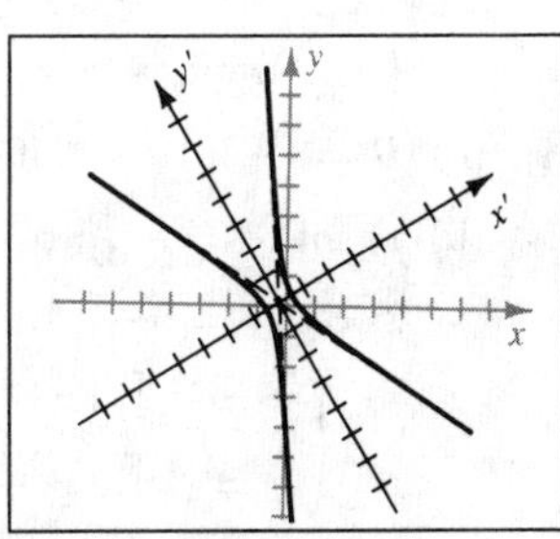

Figure 5

6 (a) $D = (-48)^2 - 4(7)(-7) = 2500 > 0$, hyperbola

(b) 1) $\cot 2\phi = -\frac{7}{24}$; $\phi \approx 53.13^\circ$

2) $\cos 2\phi = -\frac{7}{25}$; $\sin\phi = \frac{4}{5}$; $\cos\phi = \frac{3}{5}$

3) $\begin{cases} x = \frac{3}{5}x' - \frac{4}{5}y' = \frac{1}{5}(3x' - 4y') \\ y = \frac{4}{5}x' + \frac{3}{5}y' = \frac{1}{5}(4x' + 3y') \end{cases}$

4) $-\frac{625}{25}(x')^2 + \frac{625}{25}(y')^2 = 225 \Rightarrow \dfrac{(y')^2}{9} - \dfrac{(x')^2}{9} = 1$

5) $V'(0, \pm 3) \to V(\mp\frac{12}{5}, \pm\frac{9}{5})$

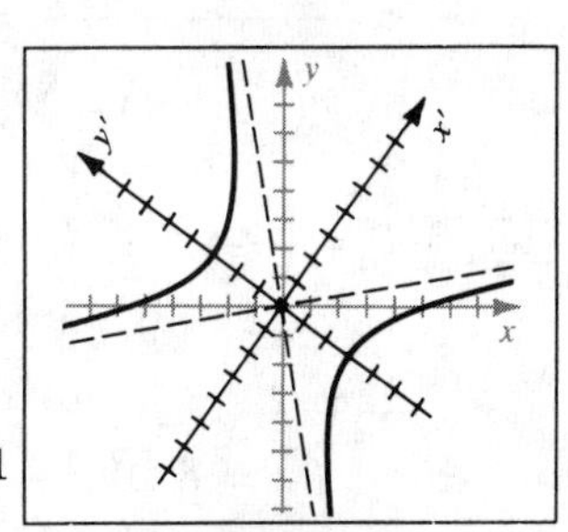

Figure 6

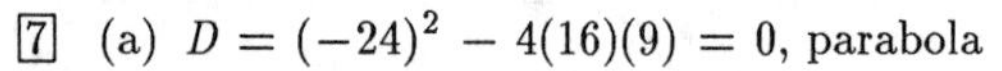

7 (a) $D = (-24)^2 - 4(16)(9) = 0$, parabola

(b) 1) $\cot 2\phi = -\frac{7}{24}$; $\phi \approx 53.13^\circ$

2) $\cos 2\phi = -\frac{7}{25}$; $\sin\phi = \frac{4}{5}$; $\cos\phi = \frac{3}{5}$

3) $\begin{cases} x = \frac{3}{5}x' - \frac{4}{5}y' = \frac{1}{5}(3x' - 4y') \\ y = \frac{4}{5}x' + \frac{3}{5}y' = \frac{1}{5}(4x' + 3y') \end{cases}$

4) $\frac{625}{25}(y')^2 - 100\,x' + 100 = 0 \Rightarrow$

$(y')^2 = 4(x' - 1)$

5) $V'(1, 0) \rightarrow V(\frac{3}{5}, \frac{4}{5})$

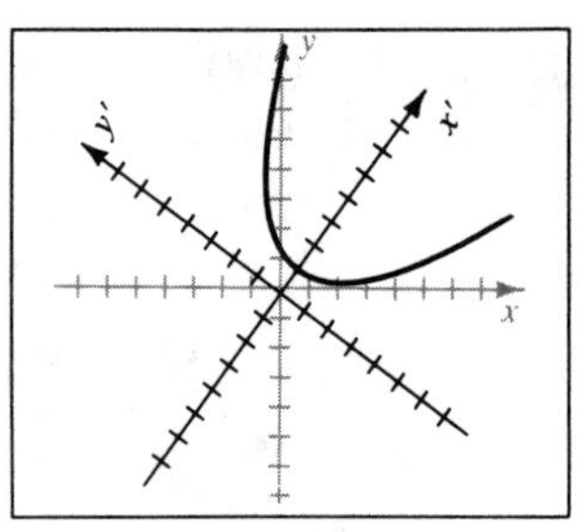

Figure 7

8 (a) $D = (4)^2 - 4(1)(4) = 0$, parabola

(b) 1) $\cot 2\phi = -\frac{3}{4}$; $\phi \approx 63.43^\circ$

2) $\cos 2\phi = -\frac{3}{5}$; $\sin\phi = \frac{2}{5}\sqrt{5}$; $\cos\phi = \frac{1}{5}\sqrt{5}$

3) $\begin{cases} x = \frac{1}{5}\sqrt{5}\,x' - \frac{2}{5}\sqrt{5}\,y' = \frac{1}{5}\sqrt{5}\,(x' - 2y') \\ y = \frac{2}{5}\sqrt{5}\,x' + \frac{1}{5}\sqrt{5}\,y' = \frac{1}{5}\sqrt{5}\,(2x' + y') \end{cases}$

4) $\frac{25}{5}(x')^2 - 30\,x' - 30\,y' + 45 = 0 \Rightarrow$

$(x' - 3)^2 = 6y'$

5) $V'(3, 0) \rightarrow V(\frac{3}{5}\sqrt{5}, \frac{6}{5}\sqrt{5})$

{The graph is tangent to the x-axis at $x = -3\sqrt{5}$.}

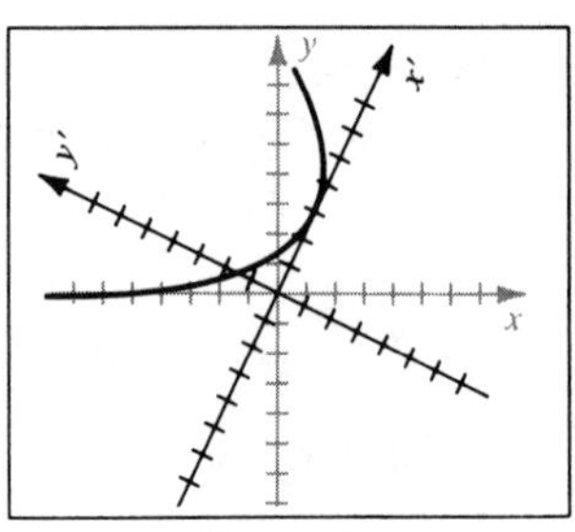

Figure 8

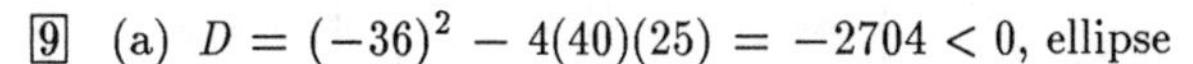

9 (a) $D = (-36)^2 - 4(40)(25) = -2704 < 0$, ellipse

(b) 1) $\cot 2\phi = -\frac{5}{12}$; $\phi \approx 56.31^\circ$

2) $\cos 2\phi = -\frac{5}{13}$; $\sin\phi = \frac{3}{13}\sqrt{13}$; $\cos\phi = \frac{2}{13}\sqrt{13}$

3) $\begin{cases} x = \frac{2}{13}\sqrt{13}\,x' - \frac{3}{13}\sqrt{13}\,y' = \frac{1}{13}\sqrt{13}\,(2x' - 3y') \\ y = \frac{3}{13}\sqrt{13}\,x' + \frac{2}{13}\sqrt{13}\,y' = \frac{1}{13}\sqrt{13}(3x' + 2y') \end{cases}$

4) $\frac{169}{13}(x')^2 + \frac{676}{13}(y')^2 - 52\,x' = 0 \Rightarrow$

$\frac{(x' - 2)^2}{4} + \frac{(y')^2}{1} = 1$

5) $V'(2 \pm 2, 0) \rightarrow V([\frac{4}{13} \pm \frac{4}{13}]\sqrt{13}, [\frac{6}{13} \pm \frac{6}{13}]\sqrt{13})$

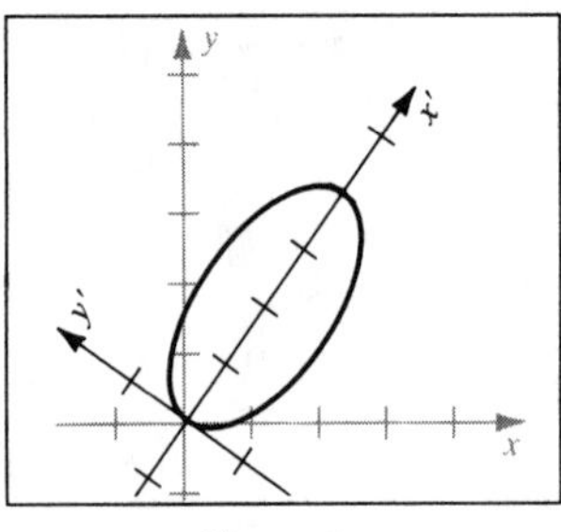

Figure 9

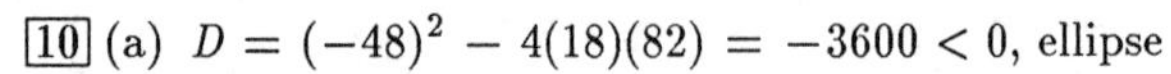

10 (a) $D = (-48)^2 - 4(18)(82) = -3600 < 0$, ellipse

(b) 1) $\cot 2\phi = \frac{4}{3}$; $\phi \approx 18.43^\circ$

2) $\cos 2\phi = \frac{4}{5}$; $\sin\phi = \frac{1}{10}\sqrt{10}$; $\cos\phi = \frac{3}{10}\sqrt{10}$

3) $\begin{cases} x = \frac{3}{10}\sqrt{10}\,x' - \frac{1}{10}\sqrt{10}\,y' = \frac{1}{10}\sqrt{10}\,(3x' - y') \\ y = \frac{1}{10}\sqrt{10}\,x' + \frac{3}{10}\sqrt{10}\,y' = \frac{1}{10}\sqrt{10}\,(x' + 3y') \end{cases}$

4) $\frac{100}{10}(x')^2 + \frac{900}{10}(y')^2 + 20x' - 80 = 0 \Rightarrow$

$\frac{(x' + 1)^2}{9} + \frac{(y')^2}{1} = 1$

5) $V'(-1 \pm 3, 0) \rightarrow V([-\frac{3}{10} \pm \frac{9}{10}]\sqrt{10}, [-\frac{1}{10} \pm \frac{3}{10}]\sqrt{10})$

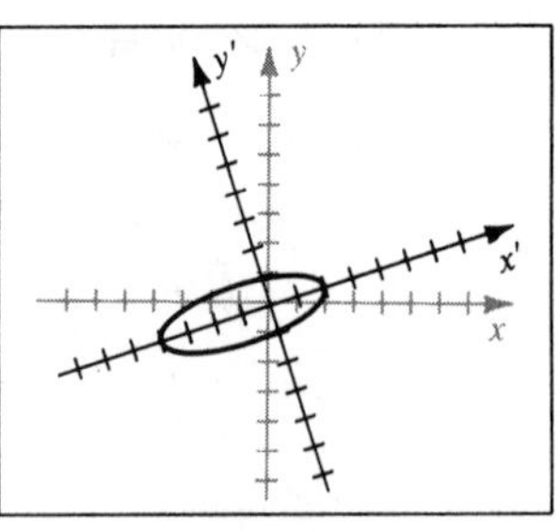

Figure 10

11 (a) $D = (6\sqrt{3})^2 - 4(5)(-1) = 128 > 0$, hyperbola

(b) 1) $\cot 2\phi = \frac{1}{3}\sqrt{3}$; $\phi = 30°$

2) $\cos 2\phi = \frac{1}{2}$; $\sin\phi = \frac{1}{2}$; $\cos\phi = \frac{1}{2}\sqrt{3}$

3) $\begin{cases} x = \frac{1}{2}\sqrt{3}\,x' - \frac{1}{2}y' = \frac{1}{2}(\sqrt{3}\,x' - y') \\ y = \frac{1}{2}x' + \frac{1}{2}\sqrt{3}\,y' = \frac{1}{2}(x' + \sqrt{3}\,y') \end{cases}$

4) $\frac{32}{4}(x')^2 - \frac{16}{4}(y')^2 - 16\,y' - 12 = 0 \Rightarrow$

$\dfrac{(y'+2)^2}{1} - \dfrac{(x')^2}{\frac{1}{2}} = 1$

5) $V'(0, -2 \pm 1) \to V(1 \mp \frac{1}{2}, -\sqrt{3} \pm \frac{1}{2}\sqrt{3})$

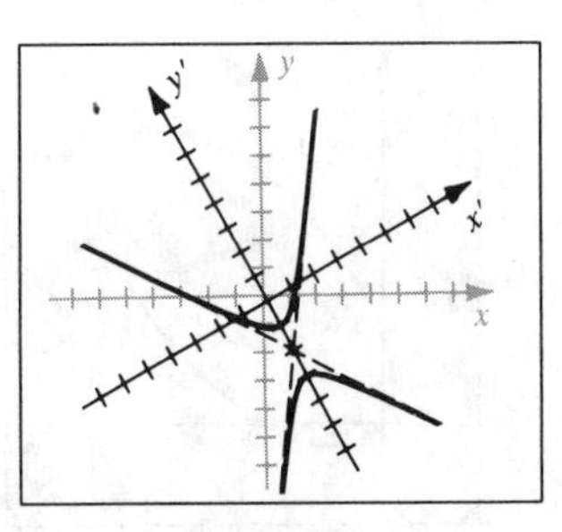

Figure 11

12 (a) $D = (20)^2 - 4(15)(0) = 400 > 0$, hyperbola

(b) 1) $\cot 2\phi = \frac{3}{4}$; $\phi \approx 26.57°$

2) $\cos 2\phi = \frac{3}{5}$; $\sin\phi = \frac{1}{5}\sqrt{5}$; $\cos\phi = \frac{2}{5}\sqrt{5}$

3) $\begin{cases} x = \frac{2}{5}\sqrt{5}\,x' - \frac{1}{5}\sqrt{5}\,y' = \frac{1}{5}\sqrt{5}\,(2x' - y') \\ y = \frac{1}{5}\sqrt{5}\,x' + \frac{2}{5}\sqrt{5}\,y' = \frac{1}{5}\sqrt{5}\,(x' + 2y') \end{cases}$

4) $\frac{100}{5}(x')^2 - \frac{25}{5}(y')^2 + 20\,y' - 100 = 0 \Rightarrow$

$\dfrac{(x')^2}{4} - \dfrac{(y'-2)^2}{16} = 1$

5) $V'(\pm 2, 2) \to V([\pm\frac{4}{5} - \frac{2}{5}]\sqrt{5}, [\pm\frac{2}{5} + \frac{4}{5}]\sqrt{5})$

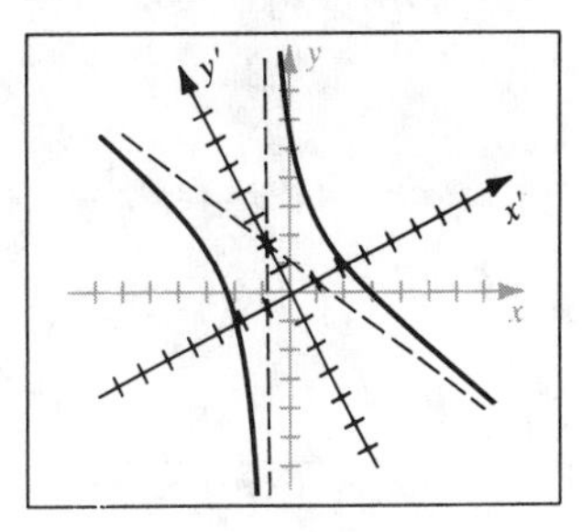

Figure 12

13 (a) $D = (-72)^2 - 4(32)(53) = -1600 < 0$, ellipse

(b) 1) $\cot 2\phi = \frac{7}{24}$; $\phi \approx 36.87°$

2) $\cos 2\phi = \frac{7}{25}$; $\sin\phi = \frac{3}{5}$; $\cos\phi = \frac{4}{5}$

3) $\begin{cases} x = \frac{4}{5}x' - \frac{3}{5}y' = \frac{1}{5}(4x' - 3y') \\ y = \frac{3}{5}x' + \frac{4}{5}y' = \frac{1}{5}(3x' + 4y') \end{cases}$

4) $\frac{125}{25}(x')^2 + \frac{2000}{25}(y')^2 = 80 \Rightarrow \dfrac{(x')^2}{16} + \dfrac{(y')^2}{1} = 1$

5) $V'(\pm 4, 0) \to V(\pm\frac{16}{5}, \pm\frac{9}{5})$

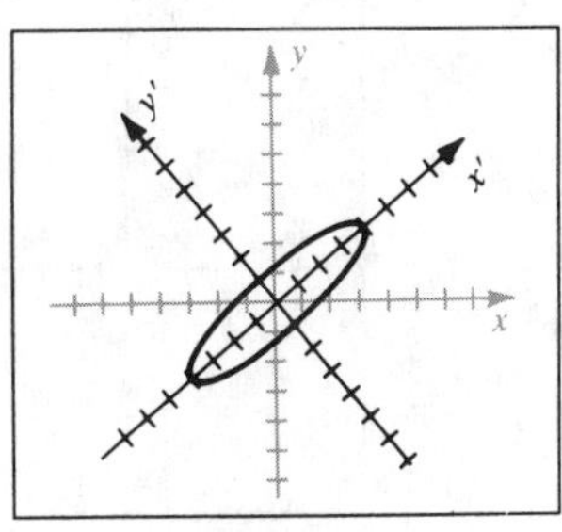

Figure 13

Note: Depending on the type of software used, you may need to solve the given equation for y.

14 $1.1x^2 - 1.3xy + y^2 - 2.9x - 1.9y = 0 \Rightarrow$

$y^2 - (1.3x + 1.9)y + (1.1x^2 - 2.9x) = 0 \Rightarrow$

$$y = \frac{(1.3x + 1.9) \pm \sqrt{(1.3x + 1.9)^2 - 4(1.1x^2 - 2.9x)}}{2}.$$ See *Figure 14*.

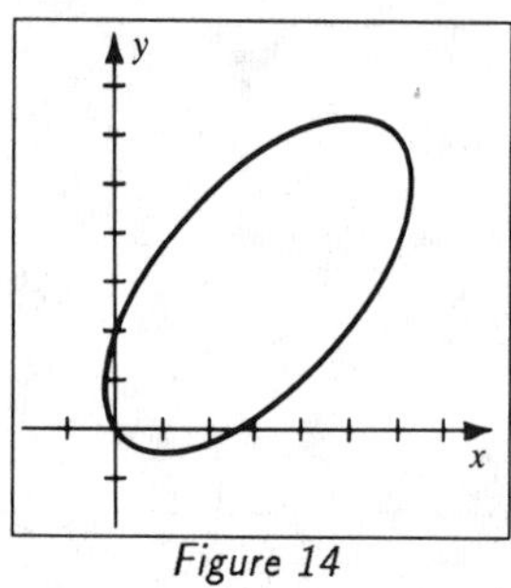

Figure 14

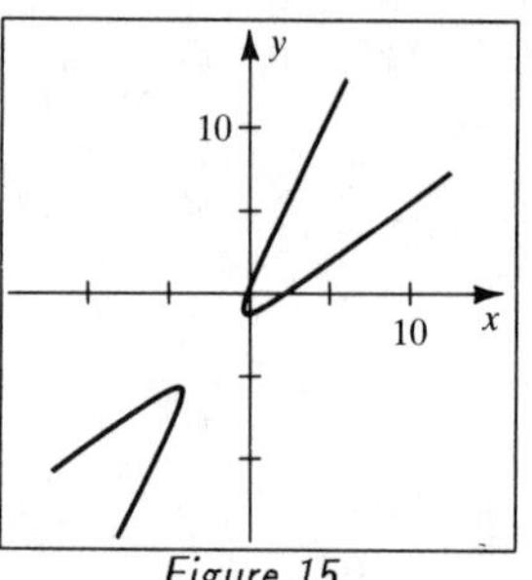

Figure 15

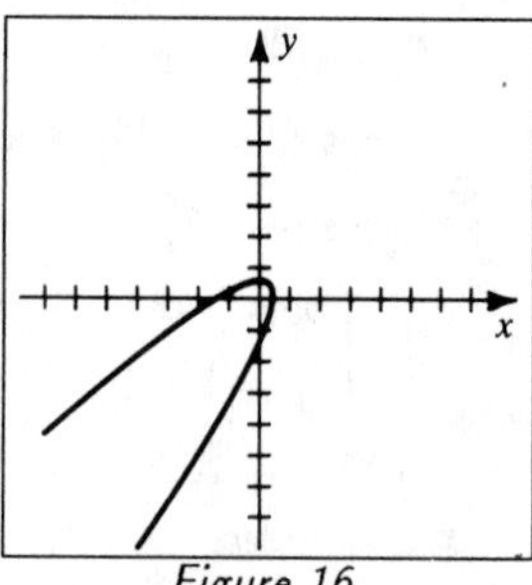

Figure 16

[15] $2.1x^2 - 4xy + 1.5y^2 - 4x + y - 1 = 0 \Rightarrow$

$1.5y^2 - (4x - 1)y + (2.1x^2 - 4x - 1) = 0 \Rightarrow$

$$y = \frac{(4x - 1) \pm \sqrt{(4x - 1)^2 - 6(2.1x^2 - 4x - 1)}}{3}$$

[16] $3.2x^2 - 4\sqrt{2}xy + 2.5y^2 + 2.1y + 3x - 2.1 = 0 \Rightarrow$

$2.5y^2 - (4\sqrt{2}x - 2.1)y + (3.2x^2 + 3x - 2.1) \Rightarrow$

$$y = \frac{(4\sqrt{2}x - 2.1) \pm \sqrt{(4\sqrt{2}x - 2.1)^2 - 10(3.2x^2 + 3x - 2.1)}}{5}$$

12.5 Review Exercises

Note: Let the notation be the same as in §12.1-12.4.

[1] $y^2 = 64x \Rightarrow x = \frac{1}{64}y^2 \Rightarrow a = \frac{1}{64}$. $p = \frac{1}{4(\frac{1}{64})} = 16$. $V(0, 0)$; $F(16, 0)$; $\ell\colon x = -16$

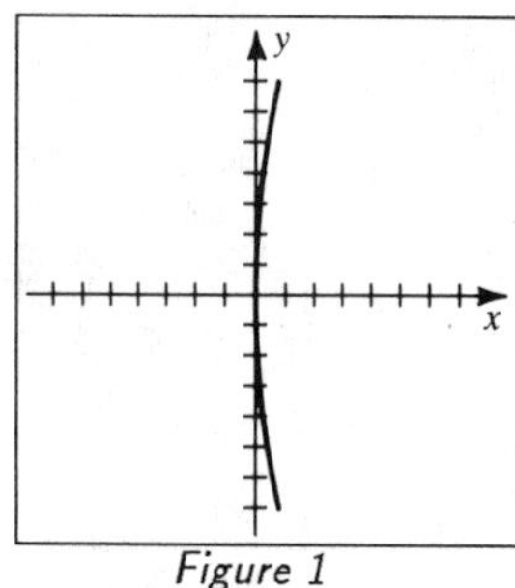

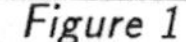
Figure 1

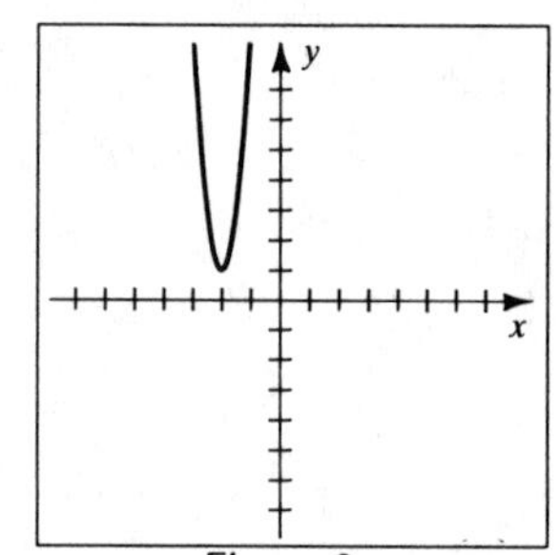

Figure 2

[2] $y = 8x^2 + 32x + 33 \Rightarrow a = 8$. $p = \frac{1}{4(8)} = \frac{1}{32}$. $V(-2, 1)$; $F(-2, \frac{33}{32})$; $\ell\colon y = \frac{31}{32}$

[3] $9y^2 = 144 - 16x^2 \Rightarrow \frac{x^2}{9} + \frac{y^2}{16} = 1$; $c^2 = 16 - 9 \Rightarrow c = \pm\sqrt{7}$;

$V(0, \pm 4)$; $F(0, \pm\sqrt{7})$; $M(\pm 3, 0)$

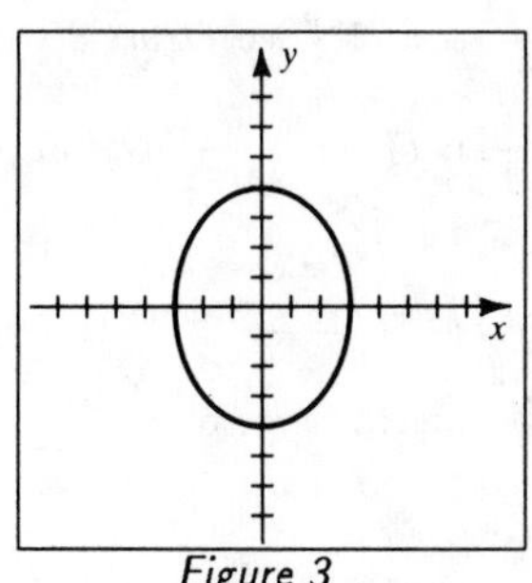

Figure 3

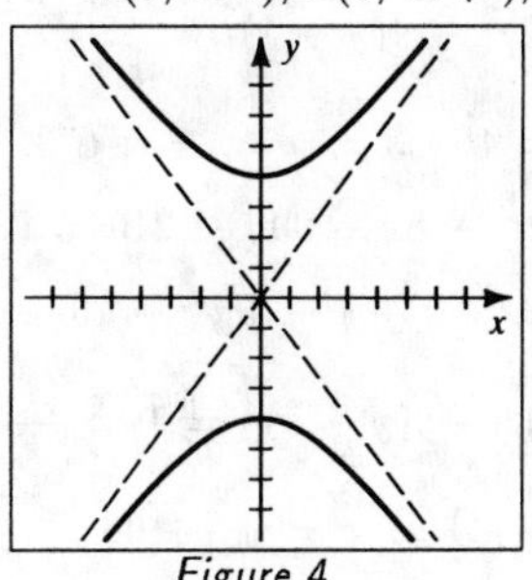

Figure 4

[4] $9y^2 = 144 + 16x^2 \Rightarrow \frac{y^2}{16} - \frac{x^2}{9} = 1$; $c^2 = 16 + 9 \Rightarrow c = \pm 5$;

$V(0, \pm 4)$; $F(0, \pm 5)$; $W(\pm 3, 0)$; $y = \pm\frac{4}{3}x$

[5] $x^2 - y^2 - 4 = 0 \Rightarrow \frac{x^2}{4} - \frac{y^2}{4} = 1$; $c^2 = 4 + 4 \Rightarrow c = \pm 2\sqrt{2}$;

$V(\pm 2, 0)$; $F(\pm 2\sqrt{2}, 0)$; $W(0, \pm 2)$; $y = \pm x$

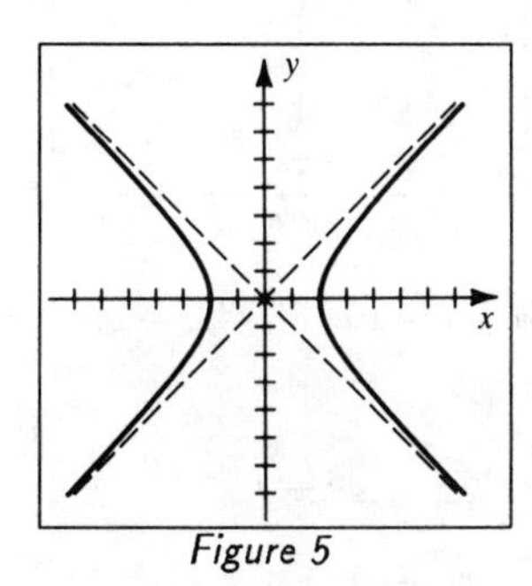

Figure 5

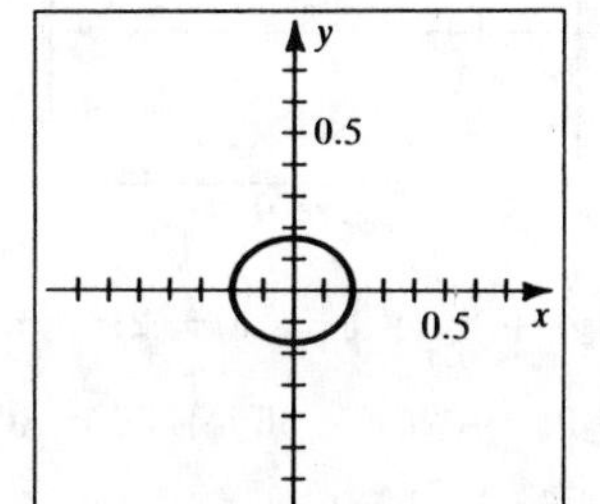

Figure 6

[6] $\frac{x^2}{\frac{1}{25}} + \frac{y^2}{\frac{1}{36}} = 1$; $c^2 = \frac{1}{25} - \frac{1}{36} \Rightarrow c = \pm\frac{1}{30}\sqrt{11}$; $V(\pm\frac{1}{5}, 0)$; $F(\pm\frac{1}{30}\sqrt{11}, 0)$; $M(0, \pm\frac{1}{6})$

[7] $25y = 100 - x^2 \Rightarrow y = 4 - \frac{1}{25}x^2 \Rightarrow a = -\frac{1}{25}$. $p = \frac{1}{4(-\frac{1}{25})} = -\frac{25}{4}$.

$V(0, 4)$; $F(0, -\frac{9}{4})$; ℓ: $y = \frac{41}{4}$

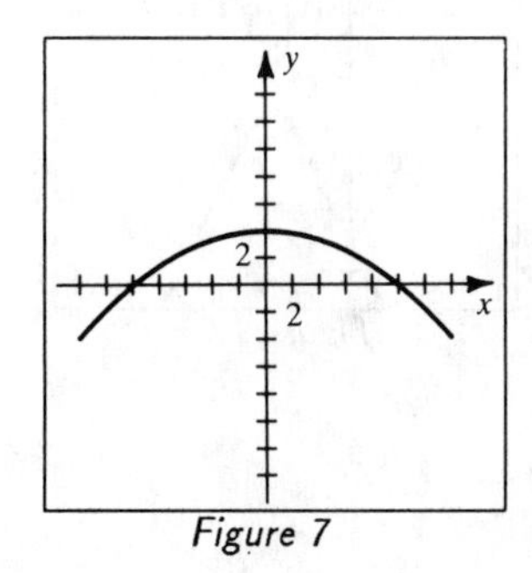

Figure 7

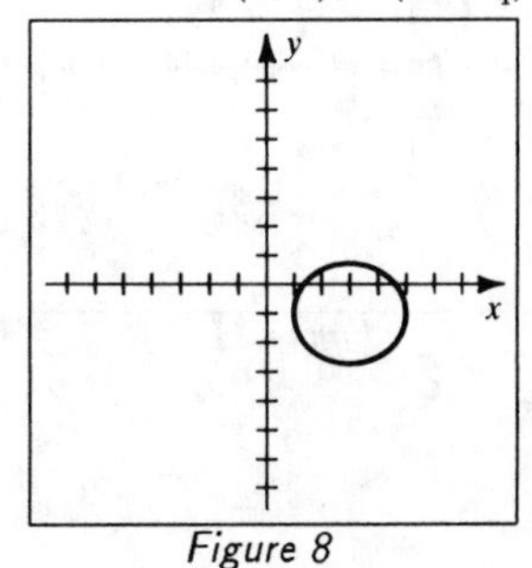

Figure 8

[8] $3x^2 + 4y^2 - 18x + 8y + 19 = 0 \Rightarrow$

$3(x^2 - 6x + \underline{\ 9\ }) + 4(y^2 + 2y + \underline{\ 1\ }) = -19 + \underline{\ 27\ } + \underline{\ 4\ } \Rightarrow$

$3(x - 3)^2 + 4(y + 1)^2 = 12 \Rightarrow \dfrac{(x - 3)^2}{4} + \dfrac{(y + 1)^2}{3} = 1.$ See *Figure 8.*

$c^2 = 4 - 3 \Rightarrow c = \pm 1$; $C(3, -1)$; $V(3 \pm 2, -1)$; $F(3 \pm 1, -1)$; $M(3, -1 \pm \sqrt{3})$

[9] $x^2 - 9y^2 + 8x + 90y - 210 = 0 \Rightarrow$

$(x^2 + 8x + \underline{\ 16\ }) - 9(y^2 - 10y + \underline{\ 25\ }) = 210 + \underline{\ 16\ } - \underline{\ 225\ } \Rightarrow$

$(x + 4)^2 - 9(y - 5)^2 = 1 \Rightarrow \dfrac{(x + 4)^2}{1} - \dfrac{(y - 5)^2}{\frac{1}{9}} = 1;$

$c^2 = 1 + \frac{1}{9} \Rightarrow c = \pm \frac{1}{3}\sqrt{10};$

$C(-4, 5)$; $V(-4 \pm 1, 5)$; $F(-4 \pm \frac{1}{3}\sqrt{10}, 5)$; $W(-4, 5 \pm \frac{1}{3})$; $(y - 5) = \pm \frac{1}{3}(x + 4)$

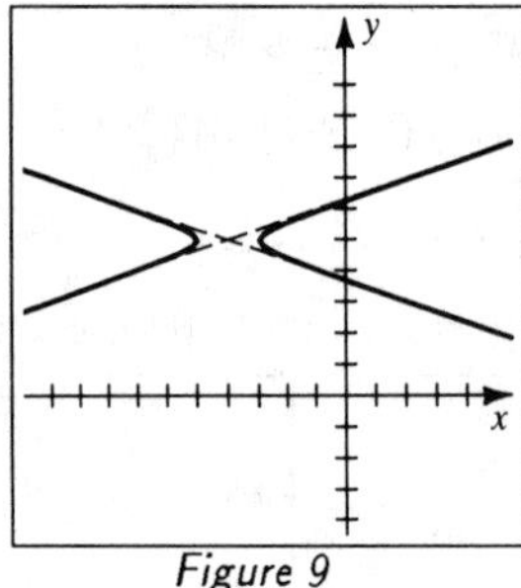

Figure 9

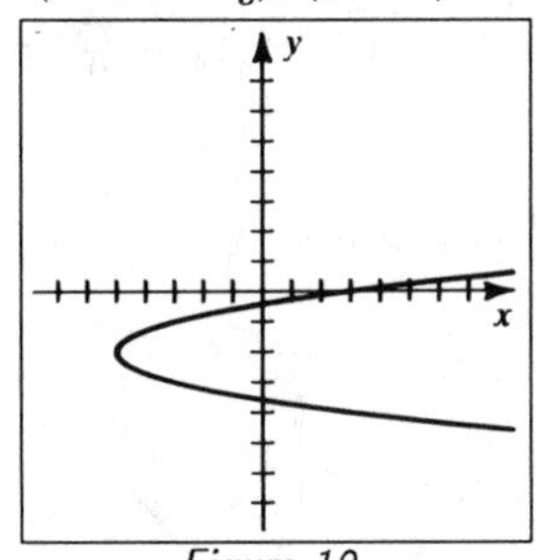

Figure 10

[10] $x = 2y^2 + 8y + 3 \Rightarrow a = 2.$ $p = \dfrac{1}{4(2)} = \frac{1}{8}.$ $V(-5, -2)$; $F(-\frac{39}{8}, -2)$; $\ell\colon x = -\frac{41}{8}$

[11] $4x^2 + 9y^2 + 24x - 36y + 36 = 0 \Rightarrow$

$4(x^2 + 6x + \underline{\ 9\ }) + 9(y^2 - 4y + \underline{\ 4\ }) = -36 + \underline{\ 36\ } + \underline{\ 36\ } \Rightarrow$

$4(x + 3)^2 + 9(y - 2)^2 = 36 \Rightarrow \dfrac{(x + 3)^2}{9} + \dfrac{(y - 2)^2}{4} = 1;$

$c^2 = 9 - 4 \Rightarrow c = \pm \sqrt{5}$; $C(-3, 2)$; $V(-3 \pm 3, 2)$; $F(-3 \pm \sqrt{5}, 2)$; $M(-3, 2 \pm 2)$

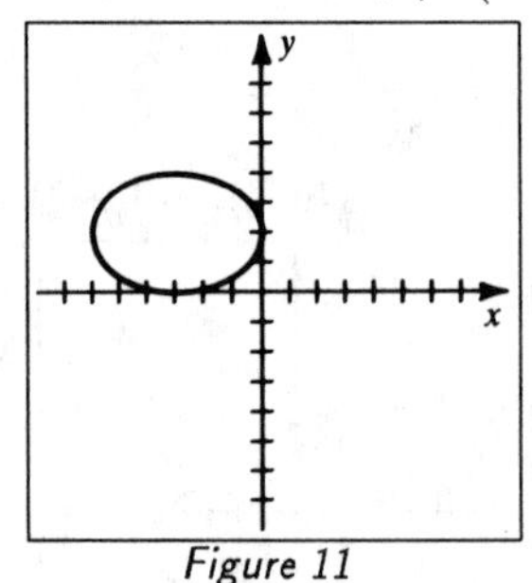

Figure 11

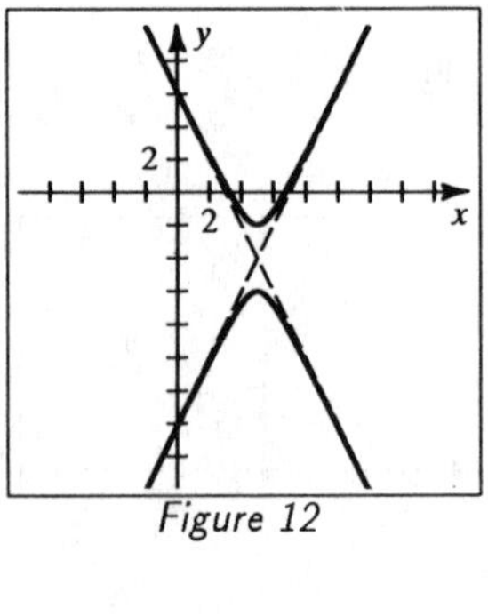

Figure 12

[12] $4x^2 - y^2 - 40x - 8y + 88 = 0 \Rightarrow$

$4(x^2 - 10x + \underline{\ 25\ }) - (y^2 + 8y + \underline{\ 16\ }) = -88 + \underline{\ 100\ } - \underline{\ 16\ } \Rightarrow$

$4(x-5)^2 - (y+4)^2 = -4 \Rightarrow \dfrac{(y+4)^2}{4} - \dfrac{(x-5)^2}{1} = 1;$

$c^2 = 4 + 1 \Rightarrow c = \pm\sqrt{5};$

$C(5, -4);\ V(5, -4 \pm 2);\ F(5, -4 \pm \sqrt{5});\ W(5 \pm 1, -4);\ (y+4) = \pm 2(x-5)$

[13] $y^2 - 8x + 8y + 32 = 0 \Rightarrow x = \frac{1}{8}y^2 + y + 4 \Rightarrow a = \frac{1}{8}.\ \ p = \dfrac{1}{4(\frac{1}{8})} = 2.$

$V(2, -4);\ F(4, -4);\ \ell\colon x = 0$

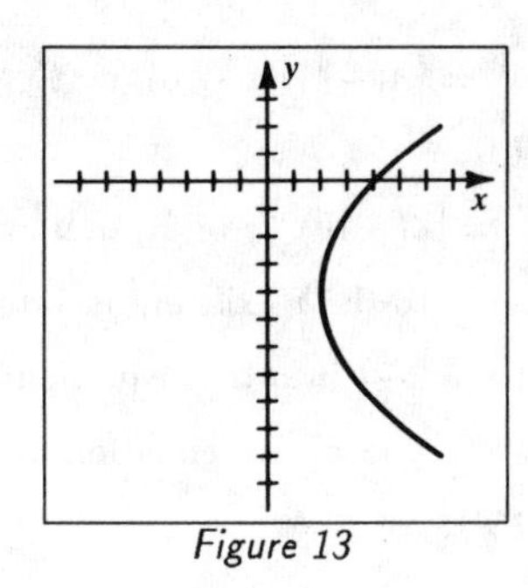

Figure 13

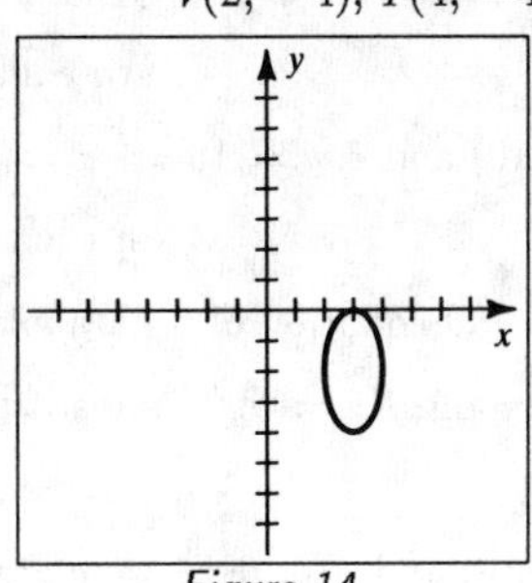

Figure 14

[14] $4x^2 + y^2 - 24x + 4y + 36 = 0 \Rightarrow$

$4(x^2 - 6x + \underline{\ 9\ }) + (y^2 + 4y + \underline{\ 4\ }) = -36 + \underline{\ 36\ } + \underline{\ 4\ } \Rightarrow$

$4(x-3)^2 + (y+2)^2 = 4 \Rightarrow \dfrac{(x-3)^2}{1} + \dfrac{(y+2)^2}{4} = 1;$

$c^2 = 4 - 1 \Rightarrow c = \pm\sqrt{3};\ C(3, -2);\ V(3, -2 \pm 2);\ F(3, -2 \pm \sqrt{3});\ M(3 \pm 1, -2)$

[15] $x^2 - 9y^2 + 8x + 7 = 0 \Rightarrow$

$(x^2 + 8x + \underline{\ 16\ }) - 9(y^2) = -7 + \underline{\ 16\ } \Rightarrow (x+4)^2 - 9(y^2) = 9 \Rightarrow$

$\dfrac{(x+4)^2}{9} - \dfrac{y^2}{1} = 1;\ c^2 = 9 + 1 \Rightarrow c = \pm\sqrt{10};$

$C(-4, 0);\ V(-4 \pm 3, 0);\ F(-4 \pm \sqrt{10}, 0);\ W(-4, 0 \pm 1);\ y = \pm\frac{1}{3}(x+4)$

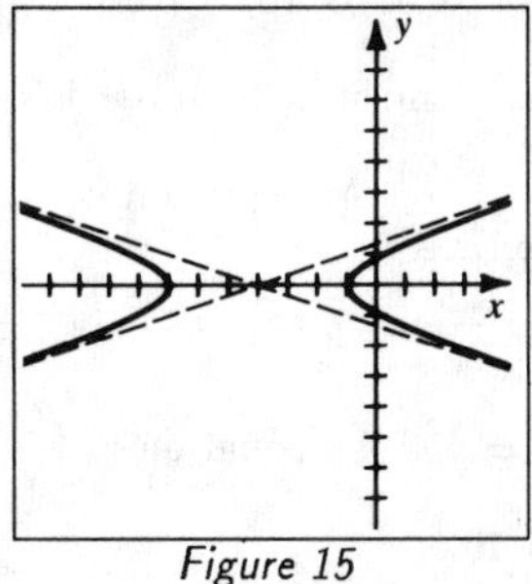

Figure 15

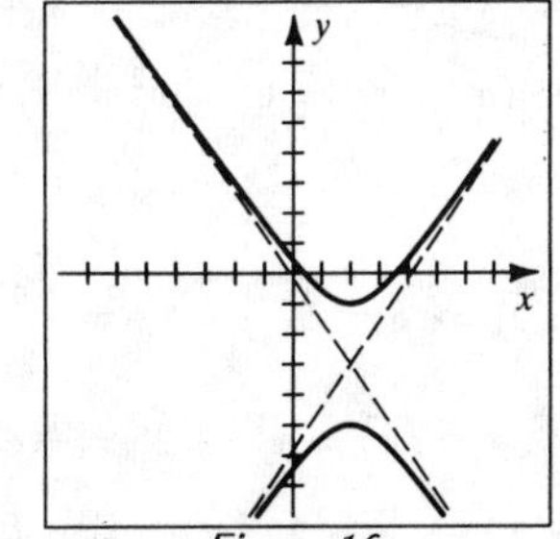

Figure 16

[16] See *Figure 16.* $y^2 - 2x^2 + 6y + 8x - 3 = 0 \Rightarrow$

$(y^2 + 6y + \underline{9}) - 2(x^2 - 4x + \underline{4}) = 3 + \underline{9} - \underline{8} \Rightarrow$

$(y+3)^2 - 2(x-2)^2 = 4 \Rightarrow \dfrac{(y+3)^2}{4} - \dfrac{(x-2)^2}{2} = 1;$

$c^2 = 4 + 2 \Rightarrow c = \pm\sqrt{6};$

$C(2, -3)$; $V(2, -3 \pm 2)$; $F(2, -3 \pm \sqrt{6})$; $W(2 \pm \sqrt{2}, -3)$; $(y + 3) = \pm\sqrt{2}\,(x - 2)$

[17] An equation is $\dfrac{y^2}{7^2} - \dfrac{x^2}{3^2} = 1$ or $\dfrac{y^2}{49} - \dfrac{x^2}{9} = 1$.

[18] $F(-4, 0)$ and $\ell\colon x = 4 \Rightarrow p = -4$ and $V(0, 0)$.

An equation is $(y - 0)^2 = [4(-4)](x - 0)$, or $y^2 = -16x$.

[19] $F(0, -10)$ and $\ell\colon y = 10 \Rightarrow p = -10$ and $V(0, 0)$.

An equation is $(x - 0)^2 = [4(-10)](y - 0)$, or $x^2 = -40y$.

[20] The general equation of a parabola that is symmetric to the x-axis and has its vertex at the origin is $x = ay^2$. Substituting $x = 5$ and $y = -1$ into that equation yields

$a = 5$. An equation is $x = 5y^2$.

[21] $V(0, \pm 10)$ and $F(0, \pm 5) \Rightarrow b^2 = 10^2 - 5^2 = 75$.

An equation is $\dfrac{x^2}{75} + \dfrac{y^2}{10^2} = 1$ or $\dfrac{x^2}{75} + \dfrac{y^2}{100} = 1$.

[22] $F(\pm 10, 0)$ and $V(\pm 5, 0) \Rightarrow b^2 = 10^2 - 5^2 = 75$.

An equation is $\dfrac{x^2}{5^2} - \dfrac{y^2}{75} = 1$ or $\dfrac{x^2}{25} - \dfrac{y^2}{75} = 1$.

[23] Asymptote equations of $y = \pm 9x$ and $V(0, \pm 6) \Rightarrow b = \frac{6}{9} = \frac{2}{3}$.

An equation is $\dfrac{y^2}{6^2} - \dfrac{x^2}{(\frac{2}{3})^2} = 1$ or $\dfrac{y^2}{36} - \dfrac{x^2}{\frac{4}{9}} = 1$.

[24] $F(\pm 2, 0) \Rightarrow c^2 = 4$. Now $\dfrac{x^2}{a^2} + \dfrac{y^2}{b^2} = 1$ can be written as $\dfrac{x^2}{a^2} + \dfrac{y^2}{a^2 - 4} = 1$ since

$b^2 = a^2 - c^2$. Substituting $x = 2$ and $y = \sqrt{2}$ into that equation yields

$\dfrac{4}{a^2} + \dfrac{2}{a^2 - 4} = 1 \Rightarrow 4a^2 - 16 + 2a^2 = a^4 - 4a^2 \Rightarrow a^4 - 10a^2 + 16 = 0 \Rightarrow$

$(a^2 - 2)(a^2 - 8) = 0 \Rightarrow a^2 = 2, 8$. Since $a > c$, a^2 must be 8 and b^2 is equal to 4.

An equation is $\dfrac{x^2}{8} + \dfrac{y^2}{4} = 1$.

[25] $M(\pm 5, 0) \Rightarrow b = 5$. $e = \dfrac{c}{a} = \dfrac{\sqrt{a^2 - b^2}}{a} = \dfrac{\sqrt{a^2 - 25}}{a} = \frac{2}{3} \Rightarrow \frac{2}{3}a = \sqrt{a^2 - 25} \Rightarrow$

$\frac{4}{9}a^2 = a^2 - 25 \Rightarrow \frac{5}{9}a^2 = 25 \Rightarrow a^2 = 45$. An equation is $\dfrac{x^2}{25} + \dfrac{y^2}{45} = 1$.

[26] $F(\pm 12, 0) \Rightarrow c = 12$. $e = \dfrac{c}{a} = \dfrac{12}{a} = \frac{3}{4} \Rightarrow a = 16$.

$b^2 = a^2 - c^2 = 16^2 - 12^2 = 112$. An equation is $\dfrac{x^2}{256} + \dfrac{y^2}{112} = 1$.

[27] (a) This problem can be modeled as an ellipse with $V(\pm 100, 0)$ and passing through the point $(25, 30)$. Substituting $x = 25$ and $y = 30$ into $\frac{x^2}{100^2} + \frac{y^2}{b^2} = 1$ yields $\frac{30^2}{b^2} = \frac{15}{16} \Rightarrow b^2 = 960$. An equation for the ellipse is $\frac{x^2}{10{,}000} + \frac{y^2}{960} = 1$.

An equation for the top half of the ellipse is $y = \sqrt{960\left(1 - \frac{x^2}{10{,}000}\right)}$.

(b) The height in the middle of the bridge is $\sqrt{960} = 8\sqrt{15} \approx 31$ ft.

[28] $P(x, y)$ is a distance of $(2 + d)$ from $(0, 0)$ and a distance of d from $(4, 0)$. The difference of these distances is $(2 + d) - d = 2$, a positive constant. By the definition of a hyperbola, $P(x, y)$ lies on the right branch of the hyperbola with foci $(0, 0)$ and $(4, 0)$. The center of the hyperbola is halfway between the foci, i.e., $(2, 0)$. The vertex is halfway from $(2, 0)$ to $(4, 0)$ since the distance from the circle to P equals the distance from P to $(4, 0)$. Thus, the vertex is $(3, 0)$ and $a = 1$.
$b^2 = c^2 - a^2 = 2^2 - 1^2 = 3$ and an equation of the right branch of the hyperbola

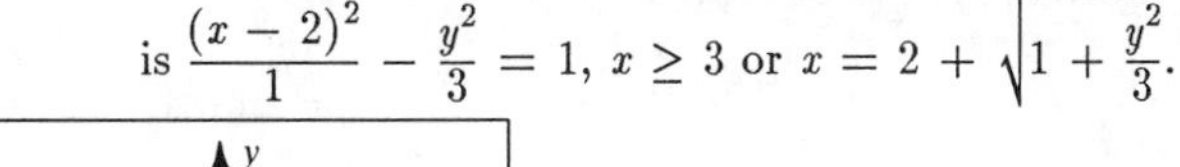

is $\frac{(x - 2)^2}{1} - \frac{y^2}{3} = 1$, $x \geq 3$ or $x = 2 + \sqrt{1 + \frac{y^2}{3}}$.

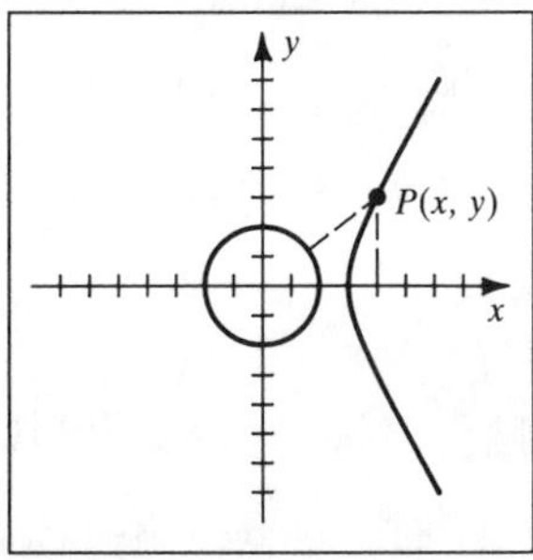

Figure 28

[29] $4x^2 - 9y^2 - 8x + 6y - 36 = 0 \Rightarrow 8x - 18yy' - 8 + 6y' = 0 \Rightarrow$
$y' = \frac{8 - 8x}{6 - 18y} = -\frac{16}{15}$ at $P(-3, 2)$.

Tangent line: $y - 2 = -\frac{16}{15}(x + 3)$; Normal line: $y - 2 = \frac{15}{16}(x + 3)$.

[30] At (x_1, y_1) on the parabola, the equation of the tangent line is
$y - y_1 = (4x_1 + 3)(x - x_1)$. $P(2, -1)$ on this line $\Rightarrow$
$-1 - y_1 = (4x_1 + 3)(2 - x_1) \Rightarrow -1 - (2x_1^2 + 3x_1 + 1) = (4x_1 + 3)(2 - x_1) \Rightarrow$

$x_1^2 - 4x_1 - 4 = 0 \Rightarrow x_1 = 2(1 \pm \sqrt{2})$.

[31] $x^2 = 4py \Rightarrow 2x = 4py' \Rightarrow y' = \frac{x}{2p}$.
Now, $y' = m \Rightarrow x = 2pm$ is the unique solution.

The equation of the tangent line is $y - \frac{(2pm)^2}{4p} = m(x - 2pm)$, or $y = mx - pm^2$.

[32] $px^2 + qy^2 = pq \Rightarrow 2px + 2qyy' = 0 \Rightarrow y' = -\frac{px}{qy}$. If (x_1, y_1) is the point of tangency, then $m = -\frac{px_1}{qy_1} \Rightarrow x_1 = -\frac{qmy_1}{p} \Rightarrow x_1^2 = \frac{q^2m^2}{p^2}(p - \frac{p}{q}x_1^2) \Rightarrow$

$x_1^2 = \dfrac{q^2m^2}{p + qm^2} \Rightarrow x_1 = \pm\dfrac{qm}{\sqrt{p + qm^2}}$. Thus, for a given m, there are two points on the ellipse with slope m. Since $y_1 = -\frac{px_1}{qm}$, $x_1 > 0$, $m > 0 \Rightarrow y_1 < 0$ and $x_1 > 0$, $m < 0 \Rightarrow y_1 > 0$. Similarly, $x_1 < 0$, $m > 0 \Rightarrow y_1 > 0$ and $x_1 < 0$, $m < 0 \Rightarrow y_1 < 0$.

Thus, $y_1 = \mp\sqrt{p - \frac{p}{q}x_1^2} = \mp\dfrac{p}{\sqrt{p + qm^2}}$ when $x_1 = \pm\dfrac{qm}{\sqrt{p + qm^2}}$.

The equations of the tangent lines are $y - y_1 = m(x - x_1) \Rightarrow$

$$y - \left(\mp\frac{p}{\sqrt{p + qm^2}}\right) = m\left[x - \left(\pm\frac{qm}{\sqrt{p + qm^2}}\right)\right] \Rightarrow y = mx - \left(\pm\frac{p + qm^2}{\sqrt{q + qm^2}}\right) \Rightarrow$$

$$y = mx \pm \sqrt{p + qm^2}.$$

[33] Without loss of generality, let $x^2 = 4py$.

Then $x = \pm\sqrt{4py}$ and $A = 2\displaystyle\int_0^p \sqrt{4py}\,dy = 4\sqrt{p}\left[\tfrac{2}{3}y^{3/2}\right]_0^p = \tfrac{8}{3}p^2$.

[34] The asymptotes of $9y^2 - 4x^2 = 36$ are $y = \pm\frac{2}{3}x$.

$y = -\frac{2}{3}x$ never intersects the parabola. $y = \frac{2}{3}x$ intersects at $x = 3, 6$

(a) $A = \displaystyle\int_3^6 \left[\tfrac{2}{3}x - (\tfrac{5}{9}x^2 - \tfrac{39}{9}x + 10)\right]dx = \tfrac{5}{2}$.

(b) Using shells, $V = 2\pi\displaystyle\int_3^6 x\left[\tfrac{2}{3}x - (\tfrac{5}{9}x^2 - \tfrac{39}{9}x + 10)\right]dx = \tfrac{45\pi}{2}$.

[35] Without loss of generality, let the ellipse have the equation $x^2/a^2 + y^2/b^2 = 1$. Using Exercise 35 of §12.2 with $a = 4$ and $b = 2$, we get $V = \frac{4}{3}\pi(4)(2)^2 = \frac{64\pi}{3}$.

[36] Using Example 6 of §12.2, the area of the elliptical cross section will be $\frac{1}{2}(\pi ab) = \frac{1}{2}\pi\cdot 5\cdot y = \frac{5\pi}{2}\sqrt{16 - x^2}$. Thus, $V = \frac{5\pi}{2}\displaystyle\int_{-4}^4 \sqrt{16 - x^2}\,dx = \tfrac{5\pi}{2}(\tfrac{1}{2}\pi\cdot 4^2) = 20\pi^2$.

Note that the integral represents the area of the upper half of a circle with radius 4.

[37] Let $\rho = 1$. $x = \pm\frac{b}{a}\sqrt{y^2 - a^2} \Rightarrow m = \displaystyle\int_a^c \tfrac{2b}{a}\sqrt{y^2 - a^2}\,dy$

$= \frac{2b}{a}\left[\frac{y}{2}\sqrt{y^2 - a^2} - \frac{a^2}{2}\ln\left|y + \sqrt{y^2 - a^2}\right|\right]_a^c$ {Formula 39}

$= \frac{b}{a}\left[cb - a^2\ln(c + b) + a^2\ln a\right]$. Also, $M_x = \displaystyle\int_a^c y\left(\tfrac{2b}{a}\sqrt{y^2 - a^2}\right)dy =$

$\frac{2b}{a}\left[\frac{1}{3}(y^2 - a^2)^{3/2}\right]_a^c = \dfrac{2b^4}{3a}$. Thus, $\bar{y} = \dfrac{M_x}{m} = \dfrac{\frac{2}{3}b^3}{bc - a^2\left[\ln(c + b) - \ln a\right]}$ and

$\bar{x} = 0$ by symmetry.

[38] (a) $D = B^2 - 4AC = (-3)^2 - 4(2)(4) = -23 < 0$, ellipse

(b) $D = B^2 - 4AC = (2)^2 - 4(3)(-1) = 16 > 0$, hyperbola

(c) $D = B^2 - 4AC = (-6)^2 - 4(1)(9) = 0$, parabola

Note: See §12.4 for a discussion of the general solution outline for Exercises 39–40.

[39] $D = (-8)^2 - 4(1)(16) = 0$, parabola

1) $\cot 2\phi = \frac{15}{8}$; $\phi \approx 14.04°$

2) $\cos 2\phi = \frac{15}{17}$; $\sin\phi = \frac{1}{17}\sqrt{17}$; $\cos\phi = \frac{4}{17}\sqrt{17}$

3) $\begin{cases} x = \frac{4}{17}\sqrt{17}\,x' - \frac{1}{17}\sqrt{17}\,y' = \frac{1}{17}\sqrt{17}\,(4x' - y') \\ y = \frac{1}{17}\sqrt{17}\,x' + \frac{4}{17}\sqrt{17}\,y' = \frac{1}{17}\sqrt{17}(x' + 4y') \end{cases}$

4) $\frac{289}{17}(y')^2 - 51(x') = 0 \Rightarrow (y')^2 = 3(x')$

5) $V'(0, 0) \rightarrow V(0, 0)$

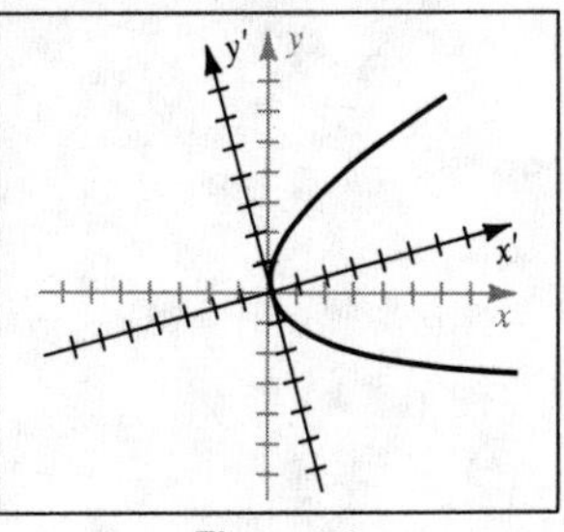

Figure 39

[40] $D = (12)^2 - 4(8)(17) = -400 < 0$, ellipse

1) $\cot 2\phi = -\frac{3}{4}$; $\phi \approx 63.43°$

2) $\cos 2\phi = -\frac{3}{5}$; $\sin\phi = \frac{2}{5}\sqrt{5}$; $\cos\phi = \frac{1}{5}\sqrt{5}$

3) $\begin{cases} x = \frac{1}{5}\sqrt{5}\,x' - \frac{2}{5}\sqrt{5}\,y' = \frac{1}{5}\sqrt{5}\,(x' - 2y') \\ y = \frac{2}{5}\sqrt{5}\,x' + \frac{1}{5}\sqrt{5}\,y' = \frac{1}{5}\sqrt{5}\,(2x' + y') \end{cases}$

4) $\frac{100}{5}(x')^2 + \frac{25}{5}(y')^2 - 40\,x' + 20\,y' = 0 \Rightarrow$
$\dfrac{(x' - 1)^2}{2} + \dfrac{(y' + 2)^2}{8} = 1$

5) $V'(1, -2 \pm 2\sqrt{2}) \rightarrow V([\frac{5}{5} \mp \frac{4}{5}\sqrt{2}]\sqrt{5}, [\pm\frac{2}{5}\sqrt{2}]\sqrt{5})$

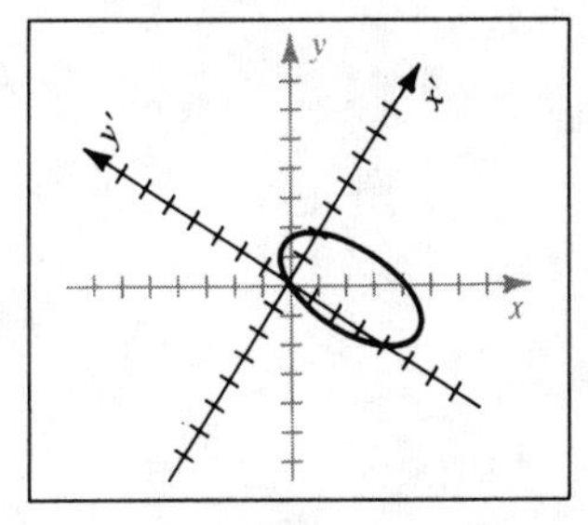

Figure 40

Chapter 13: Plane Curves and Polar Coordinates

Exercises 13.1

[1] $x = t - 2 \Rightarrow t = x + 2.\ \ y = 2t + 3 = 2(x + 2) + 3 = 2x + 7.$

As t varies from 0 to 5, (x, y) varies from $(-2, 3)$ to $(3, 13)$.

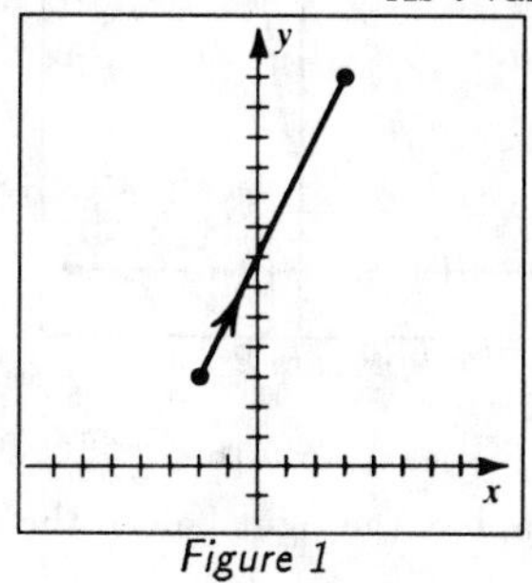

Figure 1

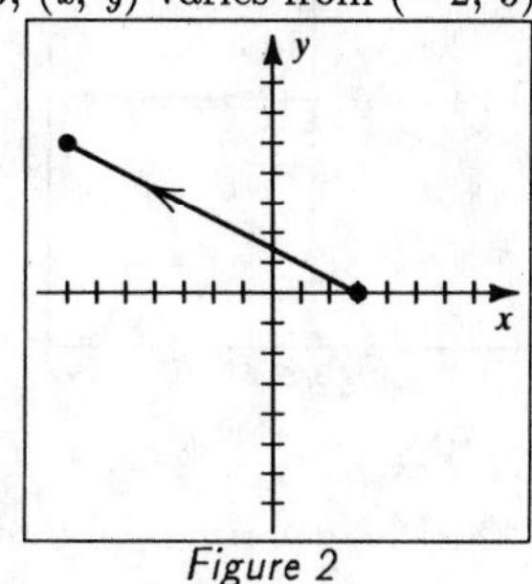

Figure 2

[2] $y = 1 + t \Rightarrow t = y - 1.\ \ x = 1 - 2t = 1 - 2(y - 1) = -2y + 3.$

As t varies from -1 to 4, (x, y) varies from $(3, 0)$ to $(-7, 5)$.

[3] $x = t^2 + 1 \Rightarrow t^2 = x - 1.\ \ y = t^2 - 1 = x - 2.$ As t varies from -2 to 2,

(x, y) varies from $(5, 3)$ to $(1, -1)$ $\{$when $t = 0\}$ and back to $(5, 3)$.

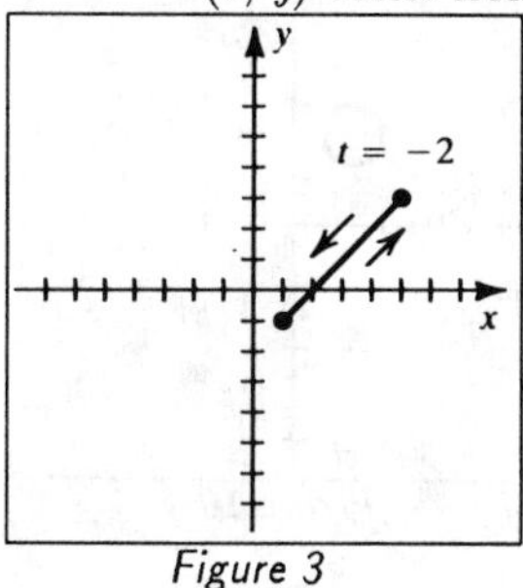

Figure 3

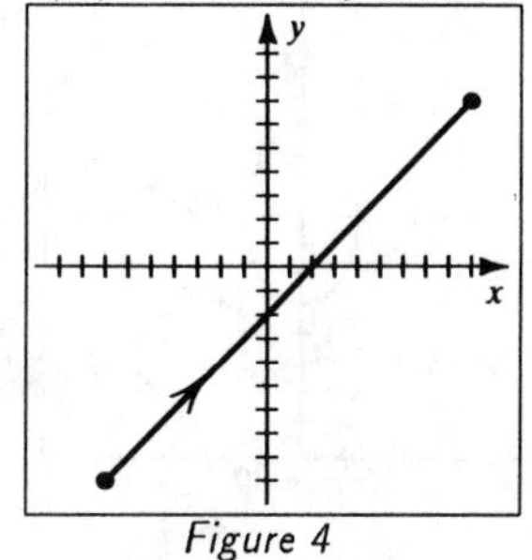

Figure 4

[4] $x = t^3 + 1 \Rightarrow t^3 = x - 1.\ \ y = t^3 - 1 = x - 2.$

As t varies from -2 to 2, (x, y) varies from $(-7, -9)$ to $(9, 7)$.

[5] $y = 2t + 3 \Rightarrow t = \frac{1}{2}(y - 3).\ \ x = 4\left[\frac{1}{2}(y - 3)\right]^2 - 5 \Rightarrow (y - 3)^2 = x + 5.$

This is a parabola with vertex at $(-5, 3)$.

Since t takes on all real values, so does y, and the curve C is the entire parabola.

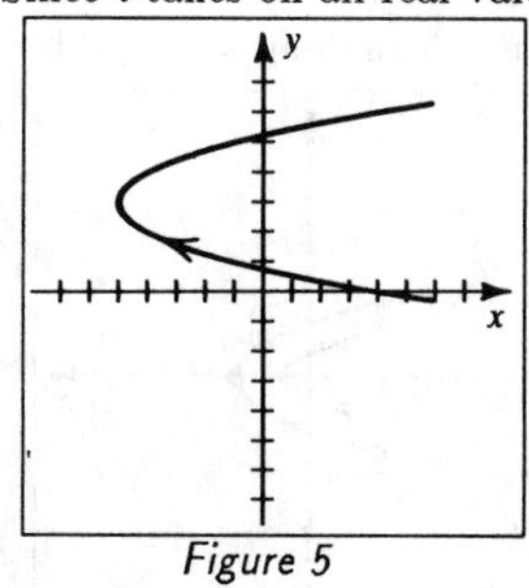

Figure 5

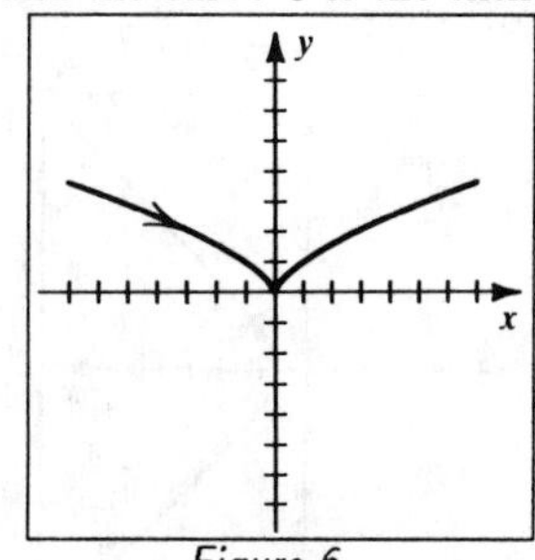

Figure 6

[6] $x = t^3 \Rightarrow t = \sqrt[3]{x}.\ \ y = t^2 = x^{2/3}.$ x takes on all real values.

[7] $y = e^{-2t} = (e^t)^{-2} = x^{-2} = \frac{1}{x^2}$.

As t varies from $-\infty$ to ∞, x varies from 0 to ∞, excluding 0.

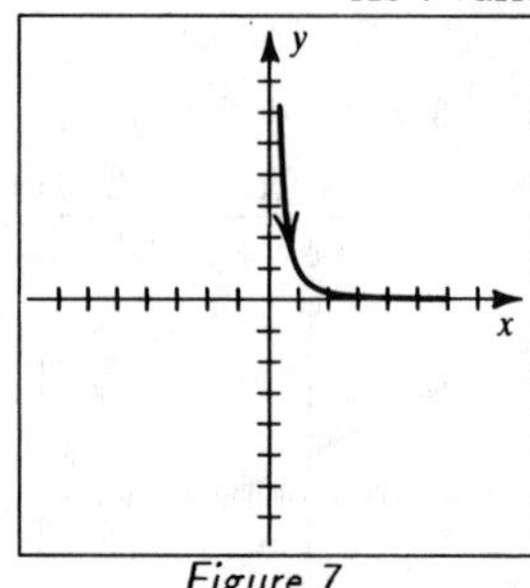

Figure 7

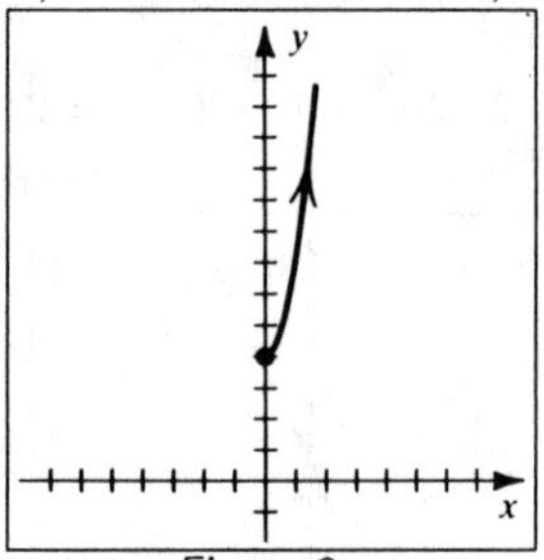

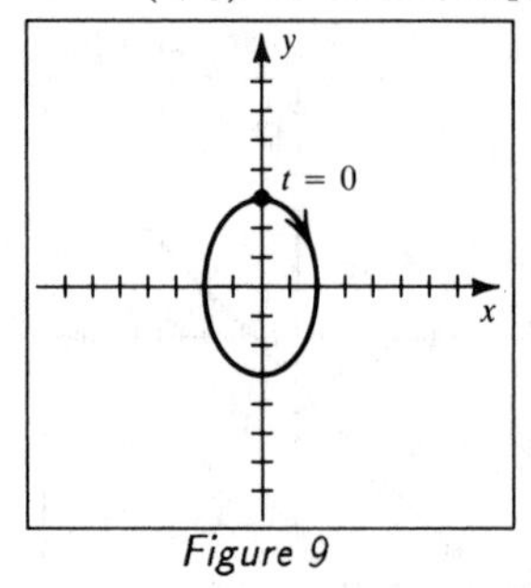

Figure 8

[8] $x = \sqrt{t} \Rightarrow t = x^2$. $y = 3t + 4 = 3x^2 + 4$. As t varies from 0 to ∞,

x varies from 0 to ∞ and the graph is the right half of the parabola.

[9] $x = 2\sin t$ and $y = 3\cos t \Rightarrow \frac{x}{2} = \sin t$ and $\frac{y}{3} = \cos t \Rightarrow$

$\frac{x^2}{4} + \frac{y^2}{9} = \sin^2 t + \cos^2 t = 1$. As t varies from 0 to 2π,

(x, y) traces the ellipse from (0, 3) in a clockwise direction back to (0, 3).

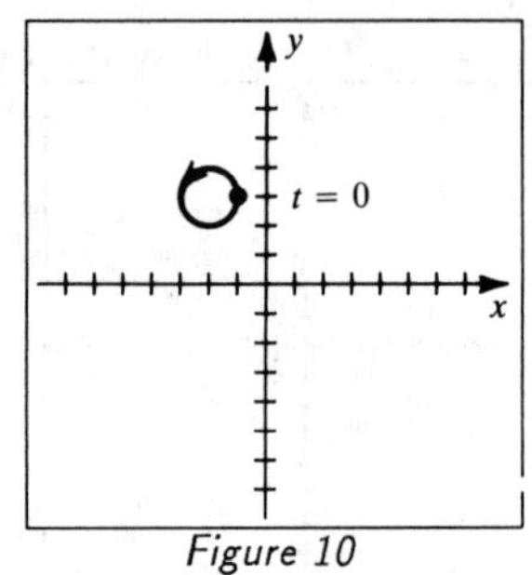

Figure 9

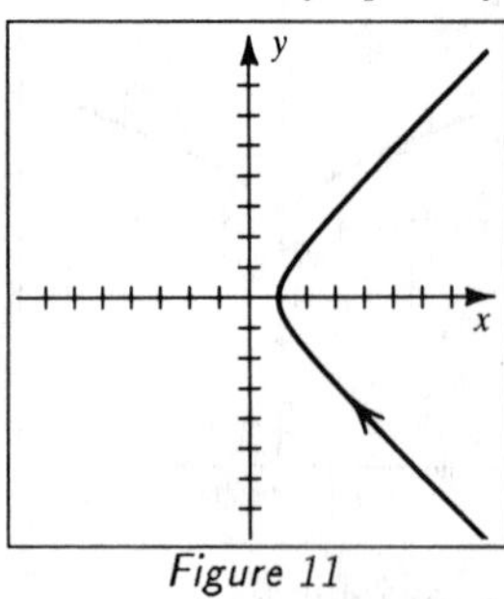

Figure 10

[10] $x = \cos t - 2$ and $y = \sin t + 3 \Rightarrow x + 2 = \cos t$ and $y - 3 = \sin t \Rightarrow$

$(x + 2)^2 + (y - 3)^2 = \cos^2 t + \sin^2 t = 1$. As t varies from 0 to 2π,

(x, y) traces the circle from $(-1, 3)$ in a counterclockwise direction back to $(-1, 3)$.

[11] $x = \sec t$ and $y = \tan t \Rightarrow x^2 - y^2 = \sec^2 t - \tan^2 t = 1$.

As t varies from $-\frac{\pi}{2}$ to $\frac{\pi}{2}$, (x, y) traces the right branch of the hyperbola along the asymptote $y = -x$ to (1, 0) and then along the asymptote $y = x$.

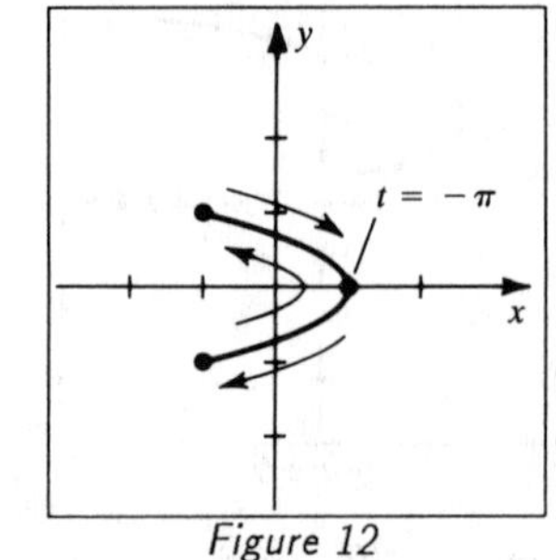

Figure 11

Figure 12

[12] $x = \cos 2t = 1 - 2\sin^2 t = 1 - 2y^2$. As t varies from $-\pi$ to π,

(x, y) varies from $(1, 0)$ {the vertex} down to $(-1, -1)$ {when $t = -\frac{\pi}{2}$}, back to the vertex when $t = 0$, up to $(-1, 1)$ {when $t = \frac{\pi}{2}$}, and finally back to the vertex.

[13] $y = 2\ln t = \ln t^2$ {since $t > 0$} $= \ln x$.

As t varies from 0 to ∞, so does x, and y varies from $-\infty$ to ∞.

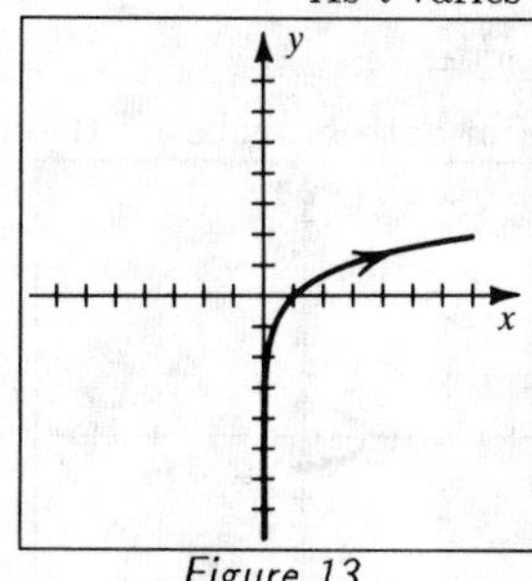

Figure 13

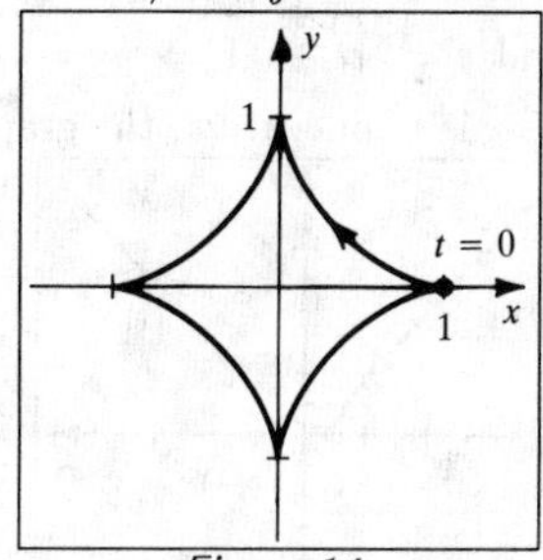

Figure 14

[14] $x = \cos^3 t$ and $y = \sin^3 t \Rightarrow x^{2/3} = \cos^2 t$ and $y^{2/3} = \sin^2 t \Rightarrow x^{2/3} + y^{2/3} = 1$ or $y = \pm(1 - x^{2/3})^{3/2}$. As t varies from 0 to 2π,

(x, y) traces the astroid from $(1, 0)$ in a counterclockwise direction back to $(1, 0)$.

[15] $y = \csc t = \dfrac{1}{\sin t} = \dfrac{1}{x}$.

As t varies from 0 to $\frac{\pi}{2}$, (x, y) varies asymptotically from the positive y-axis to $(1, 1)$.

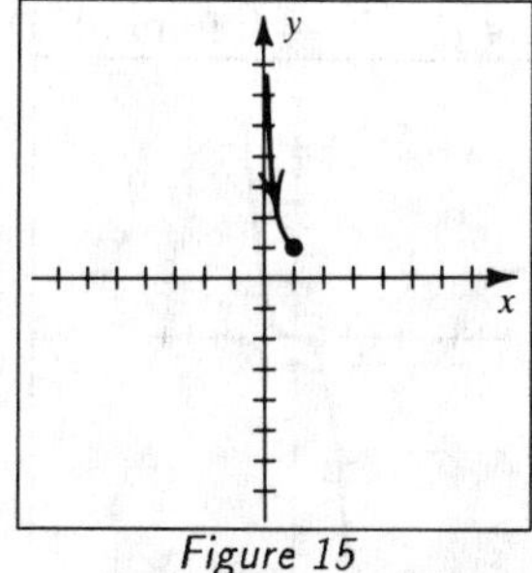

Figure 15

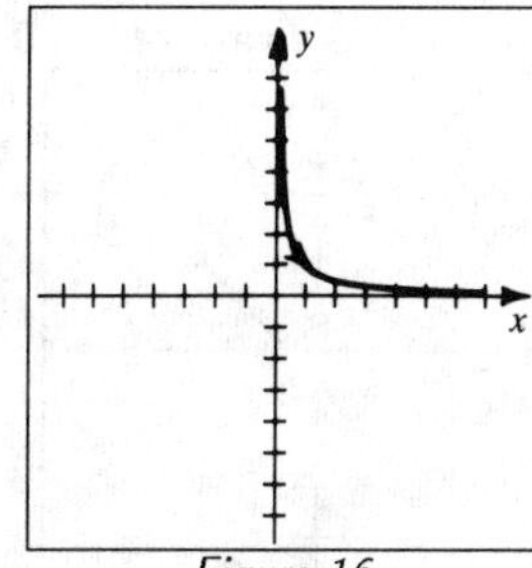

Figure 16

[16] $y = e^{-t} = (e^t)^{-1} = x^{-1} = \frac{1}{x}$. As t varies from

$-\infty$ to ∞, (x, y) varies asymptotically from the positive y-axis to the positive x-axis.

[17] $x = \cosh t$ and $y = \sinh t \Rightarrow x^2 - y^2 = \cosh^2 t - \sinh^2 t = 1$ $(x \geq 1)$.

As t varies from $-\infty$ to ∞, (x, y) traces the right branch of the hyperbola along the asymptote $y = -x$ to $(1, 0)$ and then along the asymptote $y = x$.

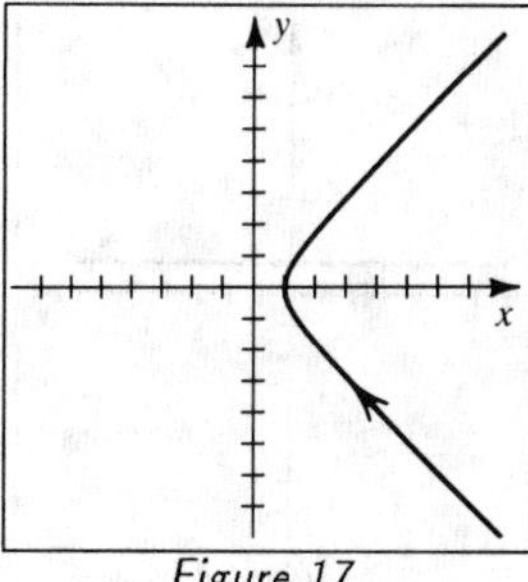

Figure 17

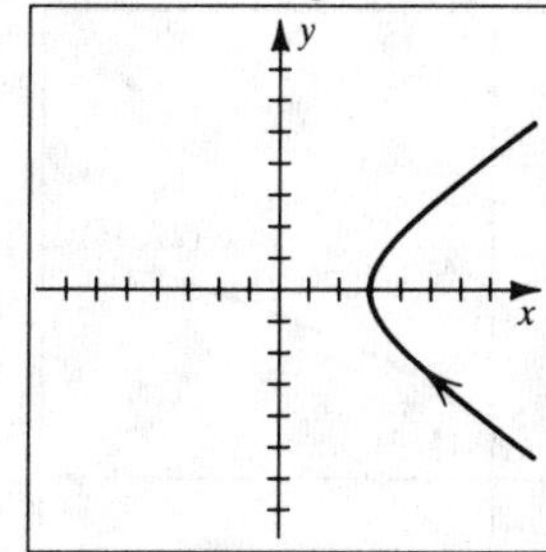

Figure 18

[18] $x = 3\cosh t$ and $y = 2\sinh t \Rightarrow \frac{x}{3} = \cosh t$ and $\frac{y}{2} = \sinh t \Rightarrow$

$\frac{x^2}{9} - \frac{y^2}{4} = \cosh^2 t - \sinh^2 t = 1$ $(x \geq 3)$.

As t varies from $-\infty$ to ∞, (x, y) traces the right branch of the hyperbola along the asymptote $y = -\frac{2}{3}x$ and then along the asymptote $y = \frac{2}{3}x$. See *Figure 18.*

[19] $x = t$ and $y = \sqrt{t^2 - 1} \Rightarrow y = \sqrt{x^2 - 1} \Rightarrow x^2 - y^2 = 1$.

Since y is nonnegative, the graph is the top half of both branches of the hyperbola.

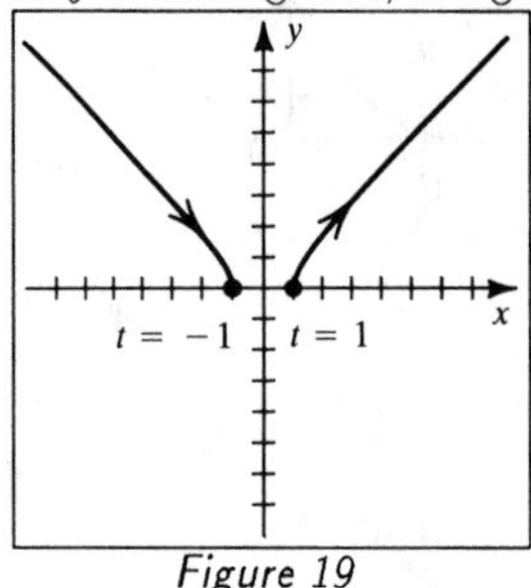

Figure 19

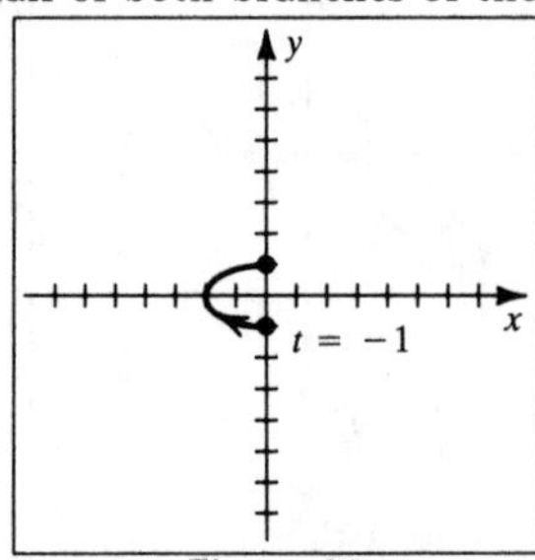

Figure 20

[20] $y = t$ and $x = -2\sqrt{1 - t^2} \Rightarrow x = -2\sqrt{1 - y^2} \Rightarrow x^2 = 4 - 4y^2 \Rightarrow x^2 + 4y^2 = 4$.

As t varies from -1 to 1, (x, y) traces the ellipse from $(0, -1)$ to $(0, 1)$.

[21] $x = t$ and $y = \sqrt{t^2 - 2t + 1} \Rightarrow y = \sqrt{x^2 - 2x + 1} = \sqrt{(x - 1)^2} = |x - 1|$.

As t varies from 0 to 4, (x, y) traces $y = |x - 1|$ from $(0, 1)$ to $(4, 3)$.

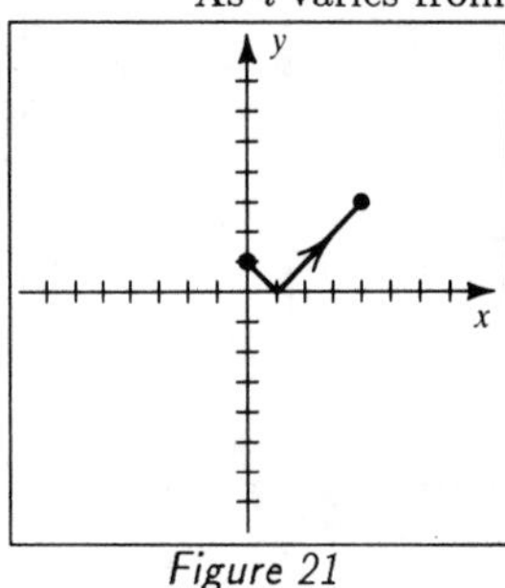

Figure 21

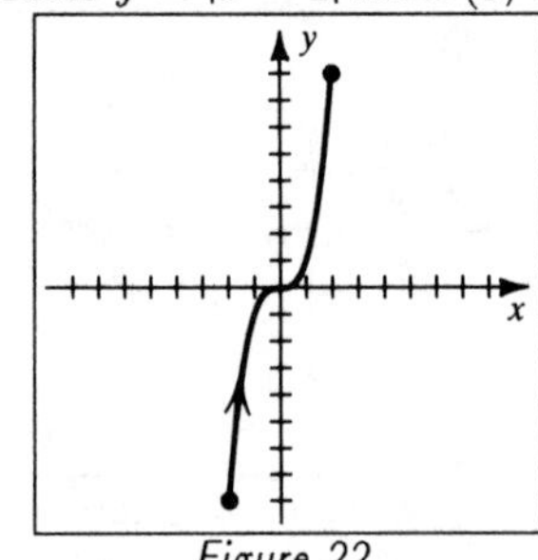

Figure 22

[22] $y = 8t^3 = (2t)^3 = x^3$.

As t varies from -1 to 1, (x, y) varies from $(-2, -8)$ to $(2, 8)$.

[23] $x = (t + 1)^3 \Rightarrow t = x^{1/3} - 1$. $y = (t + 2)^2 = (x^{1/3} + 1)^2$.

As t varies from 0 to 2, (x, y) varies from $(1, 4)$ to $(27, 16)$.

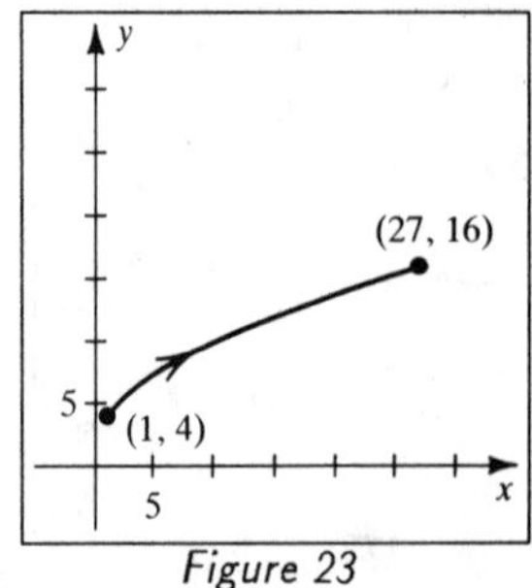

Figure 23

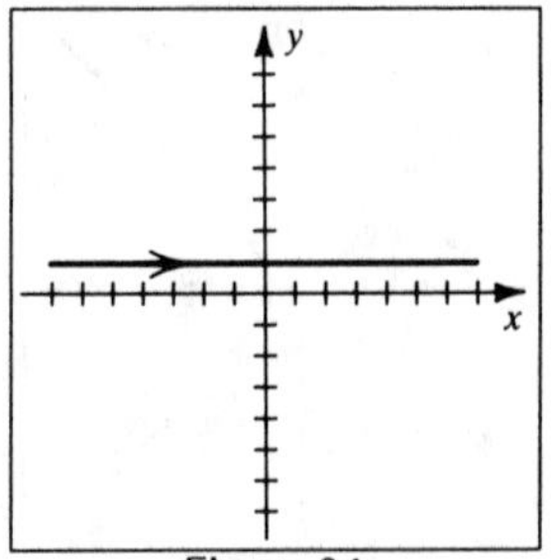

Figure 24

[24] As t varies from $-\frac{\pi}{2}$ to $\frac{\pi}{2}$, x varies from $-\infty$ to ∞.

y is always 1 so we have the graph of $y = 1$.

[25] All of the curves are a portion of the parabola $x = y^2$.

C_1: $x = t^2 = y^2$. y takes on all real values and we have the entire parabola.

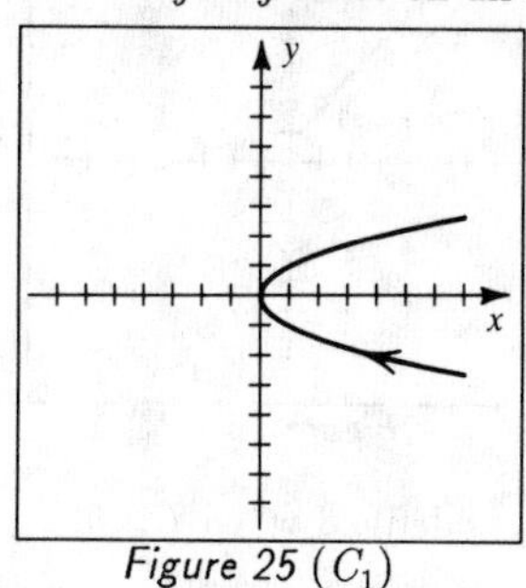

Figure 25 (C_1)

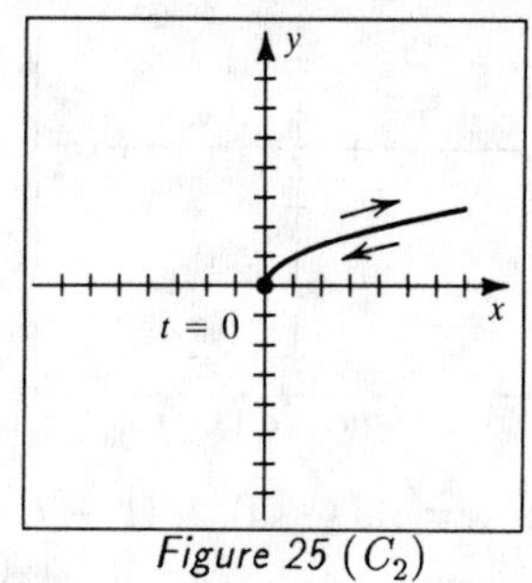

Figure 25 (C_2)

C_2: $x = t^4 = (t^2)^2 = y^2$. C_2 is only the top half since $y = t^2$ is nonnegative.

As t varies from $-\infty$ to ∞, the top portion is traced twice.

C_3: $x = \sin^2 t = (\sin t)^2 = y^2$. C_3 is the portion of the curve from $(1, -1)$ to $(1, 1)$.

The point $(1, 1)$ is reached at $t = \frac{\pi}{2} + 2\pi n$ and the point $(1, -1)$ when

$t = \frac{3\pi}{2} + 2\pi n$.

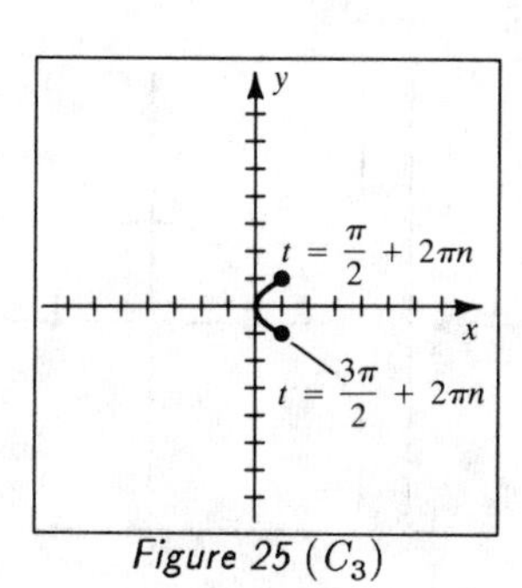

Figure 25 (C_3)

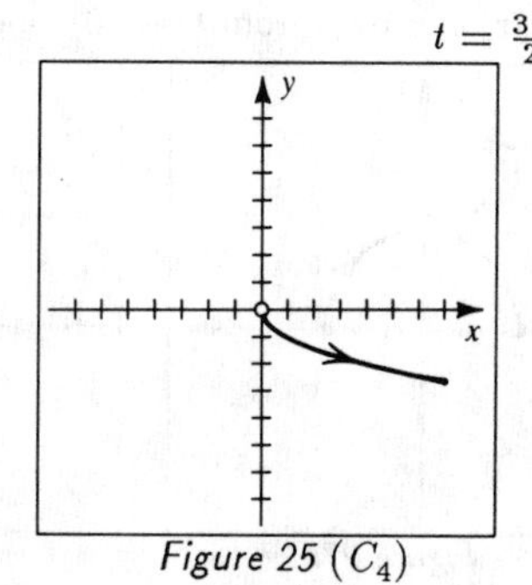

Figure 25 (C_4)

C_4: $x = e^{2t} = (e^t)^2 = (-e^t)^2 = y^2$. C_4 is the bottom half of the parabola since y

is negative. As t approaches $-\infty$, the parabola approaches the origin.

[26] All of the curves are a portion of the line $x + y = 1$.

C_1: $x + y = t + (1 - t) = 1$. C_1 is the entire line since x takes on all real values.

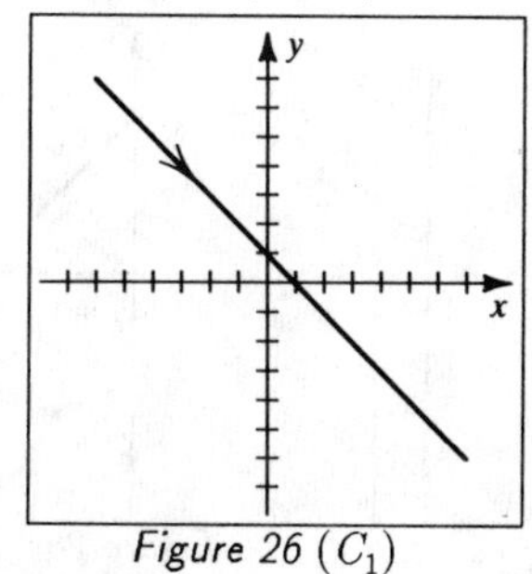

Figure 26 (C_1)

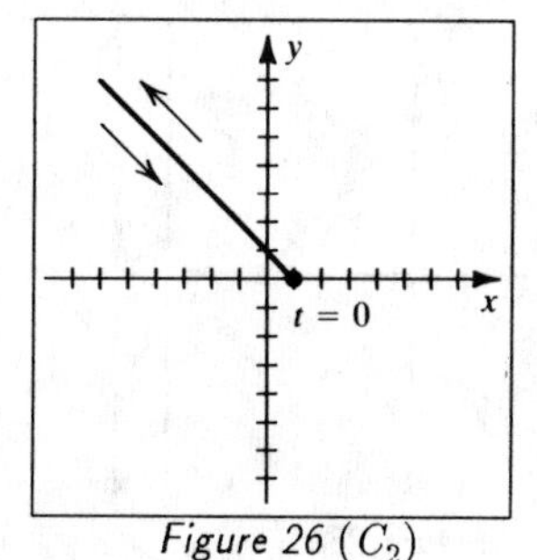

Figure 26 (C_2)

C_2: $x + y = (1 - t^2) + t^2 = 1$.

C_2 is the portion of the line where $y \geq 0$ since $y = t^2$.

C_3: $x + y = \cos^2 t + \sin^2 t = 1$. C_3 is only the portion from (0, 1) to (1, 0) since $\sin^2 t$ and $\cos^2 t$ are bounded by 0 and 1.

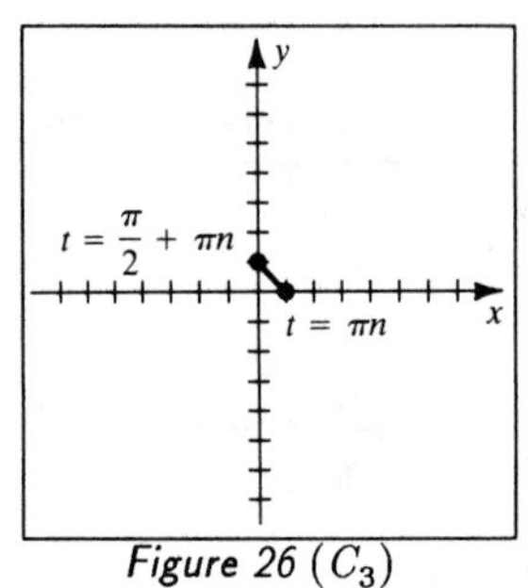

Figure 26 (C_3)

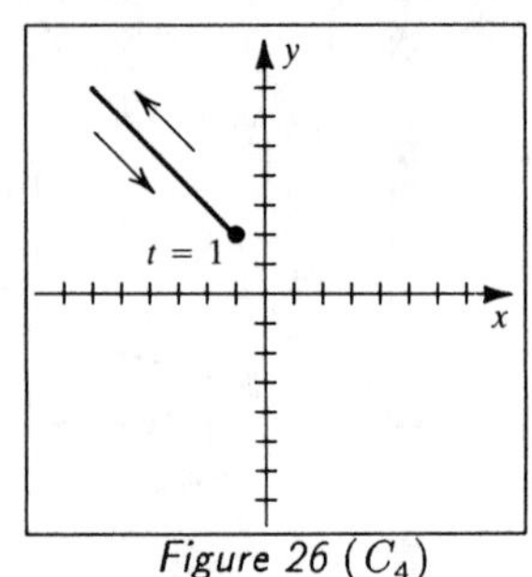

Figure 26 (C_4)

C_4: $x + y = (\ln t - t) + (1 + t - \ln t) = 1$. C_4 is defined when $t > 0$. When $t = 1$, $(x, y) = (-1, 2)$. As t approaches 0 or ∞, x approaches $-\infty$.

27 In each part, the motion is on the unit circle since $x^2 + y^2 = 1$.

(a) $P(x, y)$ moves from (1, 0) counterclockwise to (−1, 0).

(b) $P(x, y)$ moves from (0, 1) clockwise to (0, −1).

(c) $P(x, y)$ moves from (−1, 0) clockwise to (1, 0).

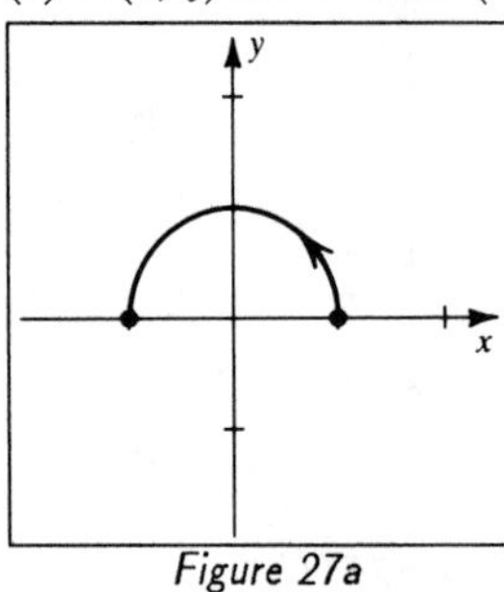

Figure 27a

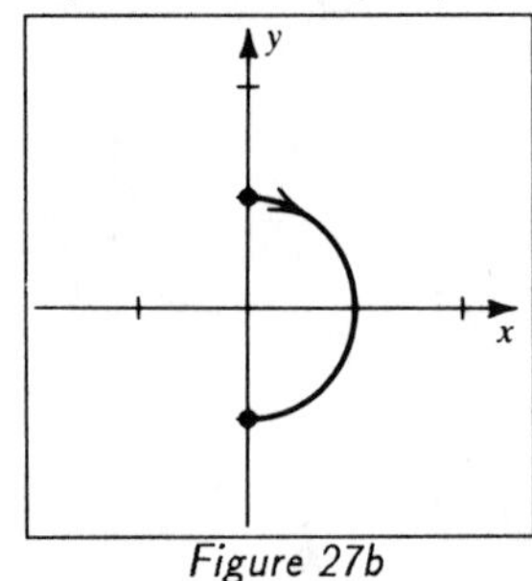

Figure 27b

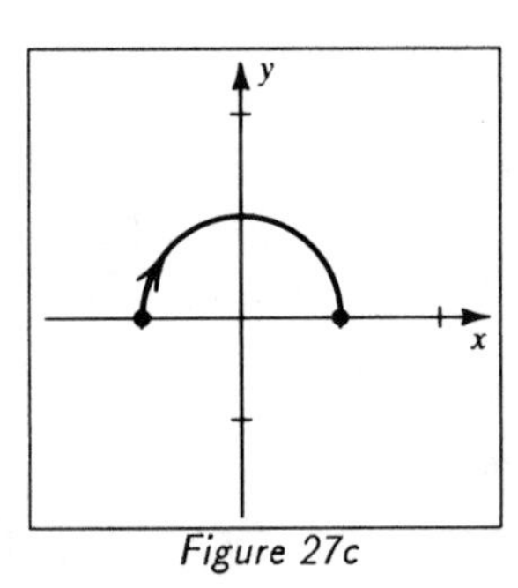

Figure 27c

28 In each part, the motion is on a line since $x + y = 1$.

(a) $P(x, y)$ moves from (0, 1) to (1, 0).

(b) $P(x, y)$ moves from (1, 0) to (0, 1).

(c) $P(x, y)$ moves from (1, 0) to (0, 1) twice in each direction.

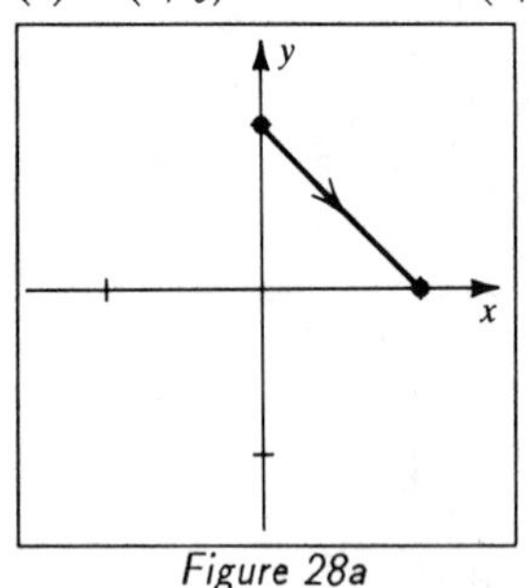

Figure 28a

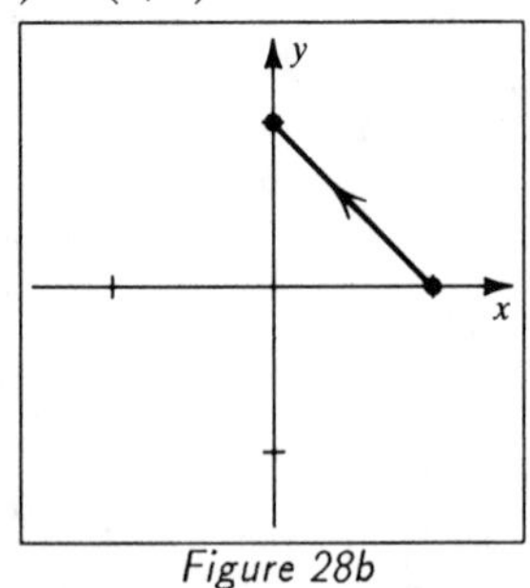

Figure 28b

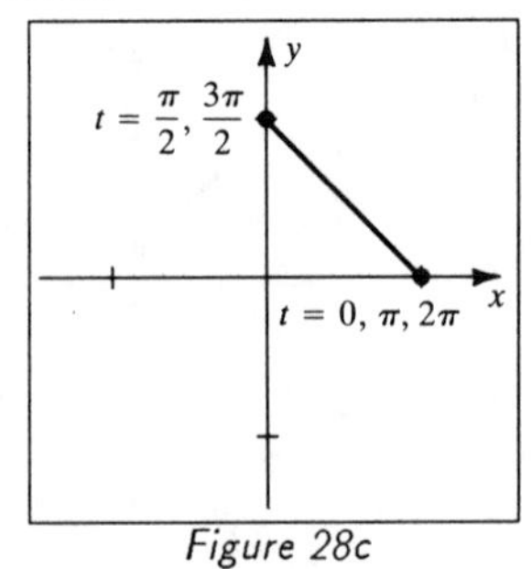

Figure 28c

[29] $x = a\cos t + h$ and $y = b\sin t + k \Rightarrow \dfrac{x-h}{a} = \cos t$ and $\dfrac{y-k}{b} = \sin t \Rightarrow$

$\dfrac{(x-h)^2}{a^2} + \dfrac{(y-k)^2}{b^2} = \cos^2 t + \sin^2 t = 1$. This is the equation of an ellipse with center (h, k) and semiaxes of lengths a and b (axes of lengths $2a$ and $2b$).

[30] $x = a\sec t + h$, $y = b\tan t + k \Rightarrow \dfrac{x-h}{a} = \sec t$, $\dfrac{y-k}{b} = \tan t \Rightarrow$

$\dfrac{(x-h)^2}{a^2} - \dfrac{(y-k)^2}{b^2} = \sec^2 t - \tan^2 t = 1$. This is an equation of a hyperbola with center (h, k), vertices $V(h \pm a, k)$ and $W(h, k \pm b)$. The transverse axis has length $2a$ and the conjugate axis has length $2b$. The right branch corresponds to $-\frac{\pi}{2} < t < \frac{\pi}{2}$ and the left branch corresponds to $\frac{\pi}{2} < t < \frac{3\pi}{2}$.

[31] Solving the first equation for t we have $t = \dfrac{x - x_1}{x_2 - x_1}$.
Substituting this expression into the second equation yields

$$y = (y_2 - y_1)\left(\frac{x - x_1}{x_2 - x_1}\right) + y_1, \quad \text{or} \quad y - y_1 = \left(\frac{y_2 - y_1}{x_2 - x_1}\right)(x - x_1),$$

which is the point-slope formula for the equation of a line through P_1 and P_2.

[32] The parametric equations give only one branch of the hyperbola since $\forall t$,
$x = a\cosh t < 0$ if $a < 0$ and $x = a\cosh t > 0$ if $a > 0$.

[33] Let $\theta = \angle FDP$ and $\alpha = \angle GDP = \angle EDP$. Then $\angle ODG = (\frac{\pi}{2} - t)$ and $\alpha = \theta - (\frac{\pi}{2} - t) = \theta + t - \frac{\pi}{2}$. Arcs AF and PF are equal in length since each is the distance rolled. Thus, $at = b\theta$ or $\theta = (\frac{a}{b})t$ and $\alpha = \dfrac{a+b}{b}t - \frac{\pi}{2}$.

Note that $\cos\alpha = \sin\left(\dfrac{a+b}{b}t\right)$ and $\sin\alpha = -\cos\left(\dfrac{a+b}{b}t\right)$.

For the location of the points as illustrated, the coordinates of P are:

$$x = d(O, G) + d(G, B) = d(O, G) + d(E, P) = (a + b)\cos t + b\sin\alpha$$
$$= (a + b)\cos t - b\cos\left(\frac{a+b}{b}t\right).$$

$$y = d(B, P) = d(G, D) - d(D, E) = (a + b)\sin t - b\cos\alpha$$
$$= (a + b)\sin t - b\sin\left(\frac{a+b}{b}t\right).$$

It can be verified that these equations are valid for all locations of the points.

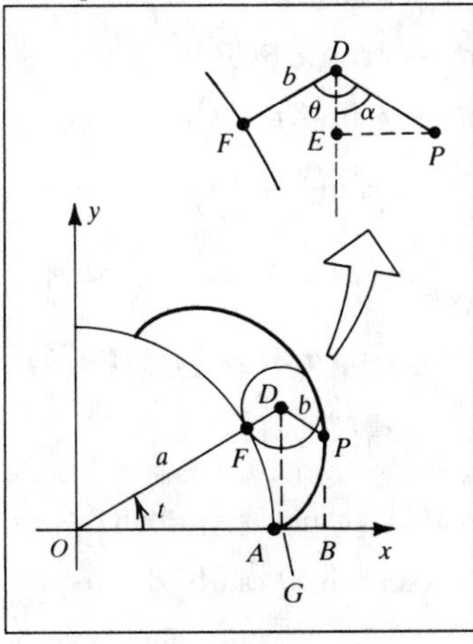

Figure 33

34 See *Figures 34a & 34b*. The line from O to C, the center of the smaller circle, must also pass through B, the common point of tangency as the smaller circle rolls inside the larger. Note that:

(i) $t = \alpha$ by the properties of parallel lines

(ii) $\overline{OB} = \overline{OA} = a$

(iii) $\overline{CB} = \overline{CP} = b$

(iv) $\overline{OC} = \overline{OB} - \overline{CB} = a - b$

(v) $\angle CPD = \beta$ by the properties of parallel lines

(vi) $\angle BCP = \alpha + \beta = t + \beta$

(vii) in ΔOCE, $\overline{OE} = \overline{OC}\cos t$ and $\overline{EC} = \overline{OC}\sin t$

(viii) in ΔDCP, $\overline{DP} = \overline{CP}\cos\angle CPD$ and $\overline{DC} = \overline{CP}\sin\angle CPD$

(ix) $\overline{DP} = \overline{EF}$

(x) in the smaller circle, $\widehat{BP} = b(\angle BCP)$

(xi) in the larger circle, $\widehat{BA} = a(\angle BOA)$

(xii) since $\widehat{BP} = \widehat{BA}$ (each is the distance rolled), $b(\angle BCP) = at \Rightarrow \angle BCP = \frac{a}{b}t$

$$
\begin{aligned}
\text{Now, } x &= \overline{OE} + \overline{EF} \\
&= \overline{OC}\cos t + \overline{CP}\cos\angle CPD \\
&= (a-b)\cos t + b\cos\beta \\
&= (a-b)\cos t + b\cos(\angle BCP - \alpha) \\
&= (a-b)\cos t + b\cos\left(\tfrac{a}{b}t - t\right) \\
&= (a-b)\cos t + b\cos\left(\frac{a-b}{b}t\right).
\end{aligned}
$$

$$
\begin{aligned}
\text{Similarly, } y &= \overline{EC} - \overline{DC} \\
&= \overline{OC}\sin t - \overline{CP}\sin\angle CPD \\
&= (a-b)\sin t - b\sin\beta \\
&= (a-b)\sin t - b\sin(\angle BCP - \alpha) \\
&= (a-b)\sin t - b\sin\left(\tfrac{a}{b}t - t\right) \\
&= (a-b)\sin t - b\sin\left(\frac{a-b}{b}t\right).
\end{aligned}
$$

If $b = \frac{1}{4}a$, then $x = (a - \frac{1}{4}a)\cos t + \frac{1}{4}a\cos\left(\dfrac{a - \frac{1}{4}a}{\frac{1}{4}a}t\right)$

$= \frac{3}{4}a\cos t + \frac{1}{4}a\cos 3t = \frac{3}{4}a\cos t + \frac{1}{4}a(4\cos^3 t - 3\cos t) = a\cos^3 t.$

Also, $y = (a - \frac{1}{4}a)\sin t - \frac{1}{4}a\sin\left(\dfrac{a - \frac{1}{4}a}{\frac{1}{4}a}t\right)$

$= \frac{3}{4}a\sin t - \frac{1}{4}a\sin 3t = \frac{3}{4}a\sin t - \frac{1}{4}a(3\sin t - 4\sin^3 t) = a\sin^3 t.$

The identities used for $\cos 3t$ and $\sin 3t$ can be derived by applying the addition, double angle, and fundamental identities.

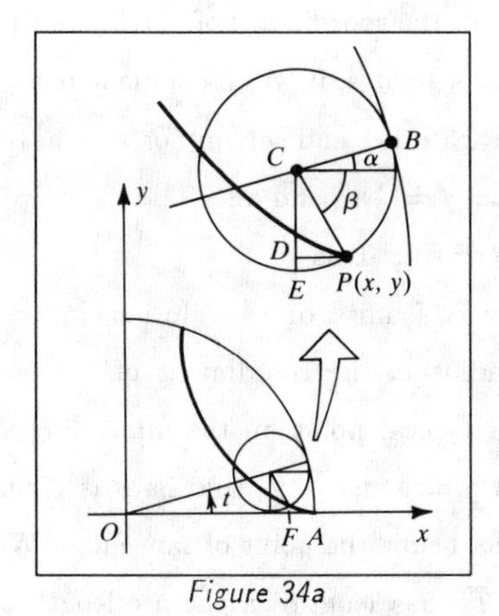

Figure 34a

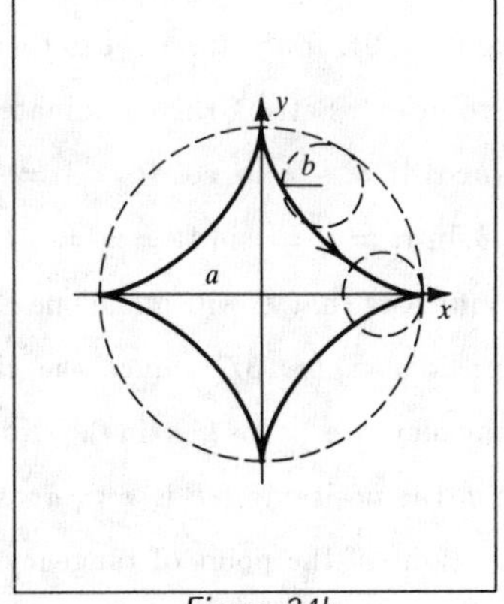

Figure 34b

[35] $b = \frac{1}{3}a \Rightarrow a = 3b$. Substituting into the equations from Exercise 33 yields:

$$x = (3b + b)\cos t - b\cos\left(\frac{3b + b}{b}t\right) = 4b\cos t - b\cos 4t$$

$$y = (3b + b)\sin t - b\sin\left(\frac{3b + b}{b}t\right) = 4b\sin t - b\sin 4t$$

As an aid in graphing, to determine where the path of the smaller circle will intersect the path of the larger circle (for the original starting point of intersection at $A(a, 0)$), we can solve $x^2 + y^2 = a^2$ for t.

$$x^2 + y^2 = 16b^2\cos^2 t - 8b^2\cos t\cos 4t + b^2\cos^2 4t + 16b^2\sin^2 t - 8b^2\sin t\sin 4t + b^2\sin^2 4t$$

$$= 17b^2 - 8b^2(\cos t\cos 4t + \sin t\sin 4t)$$

$$= 17b^2 - 8b^2[\cos(t - 4t)] = 17b^2 - 8b^2\cos 3t.$$

Thus, $x^2 + y^2 = a^2 \Rightarrow 17b^2 - 8b^2\cos 3t = a^2 = 9b^2 \Rightarrow 8b^2 = 8b^2\cos 3t \Rightarrow$ $1 = \cos 3t \Rightarrow 3t = 2\pi n \Rightarrow t = \frac{2\pi}{3}n$.

It follows that the intersection points are at $t = \frac{2\pi}{3}, \frac{4\pi}{3}$, and 2π.

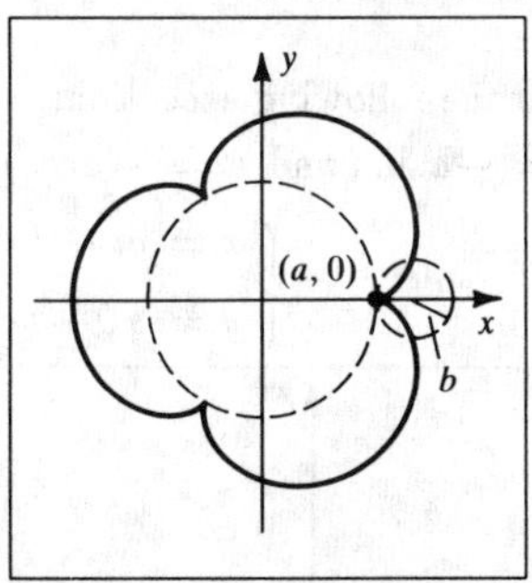

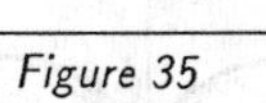
Figure 35

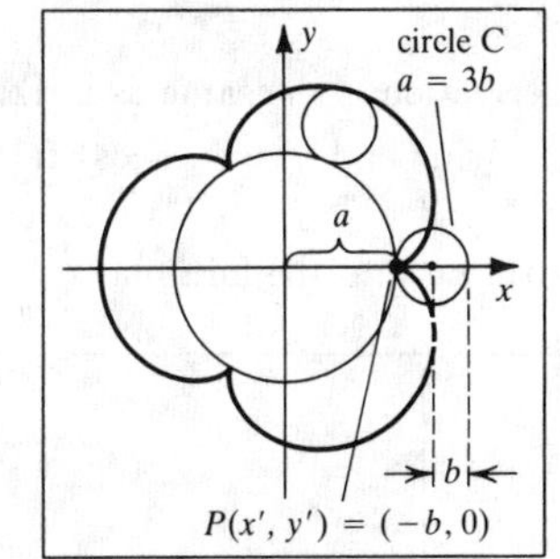

Figure 36

[36] Let $C'(h, k)$ be the center of the circle C and $a = 3b$. Its coordinates are always $x = 4b\cos t$ and $y = 4b\sin t$ since it can be thought of as being on a circle of radius $4b$. Since the equations given in Exercise 35 are for a point P relative to the origin, we can see that the coordinates of P relative to C' are $P(x', y') =$ (cont.)

$P(-b\cos 4t,\ -b\sin 4t)$. At the starting point A, the coordinates of P relative to C' are $(-b,\ 0)$. Each time P has these relative coordinates, it will have made one revolution. Setting the coordinates equal to each other and solving for t we have:

$-b\cos 4t = -b \Rightarrow \cos 4t = 1 \Rightarrow 4t = 2\pi n \Rightarrow t = \frac{\pi}{2}n$ and also

$-b\sin 4t = 0 \Rightarrow \sin 4t = 0 \Rightarrow 4t = \pi n \Rightarrow t = \frac{\pi}{4}n$. This

indicates that C will make one revolution every $\frac{\pi}{2}$ units, or 4 revolutions in 2π units.

[37] Consider *Figure 37*. Since the circle has radius a, the coordinates of the point of tangency are $(a\cos t,\ a\sin t)$. Now if P is a typical point on the unraveling string, then the position of P is $x = a\cos t + x'$ and $y = a\sin t - y'$; that is, P is x' units to the right of the point of tangency and y' units below the point of tangency. We now seek to determine x' and y' in terms of t. $\overline{TP}$ has length ta, the arc length on the circle. Since the tangent line TP is perpendicular to the radius OT, $\theta = \frac{\pi}{2} - t$. Hence, in the right triangle TAP,

$$\cos\theta = \frac{x'}{ta} \Rightarrow x' = at\cos\left(\tfrac{\pi}{2} - t\right) = at\sin t$$

and $$\sin\theta = \frac{y'}{ta} \Rightarrow y' = at\sin\left(\tfrac{\pi}{2} - t\right) = at\cos t.$$

Thus, $\quad x = a\cos t + at\sin t \quad$ and $\quad y = a\sin t - at\cos t.$

Factoring, $\quad x = a(\cos t + t\sin t) \quad$ and $\quad y = a(\sin t - t\cos t).$

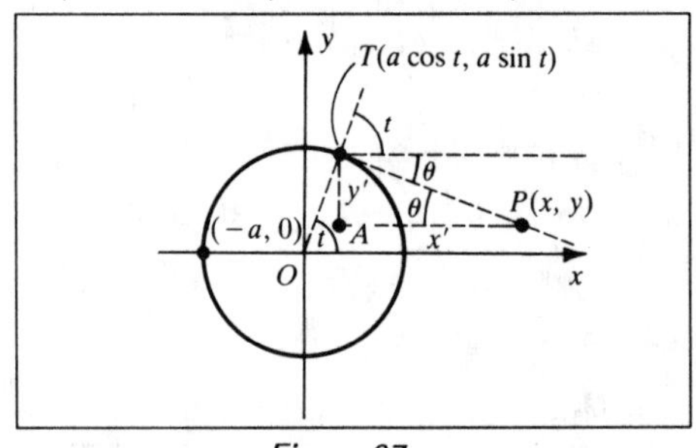

Figure 37

[38] The derivation is the same as in Example 6 with the following exception: $x' = -b\sin t$ and $y' = -b\cos t$ rather than $x' = -a\sin t$ and $y' = -a\cos t$.

Substitution into the formulas $\begin{cases} x = at + x' \\ y = a + y' \end{cases}$ yields $\begin{cases} x = at - b\sin t \\ y = a - b\cos t \end{cases}$

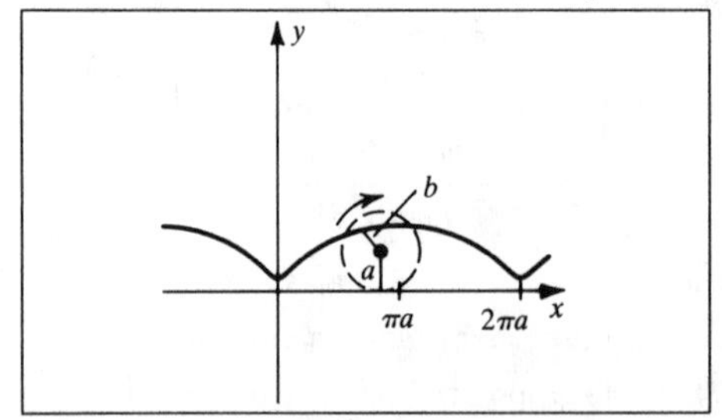

Figure 38a

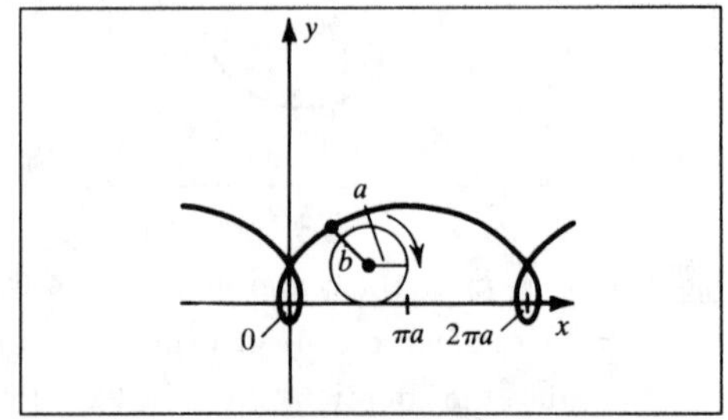

Figure 38b

[39] (a) $x = a\sin\omega t$ and $y = b\cos\omega t \Rightarrow \frac{x}{a} = \sin\omega t$ and $\frac{y}{b} = \cos\omega t \Rightarrow \frac{x^2}{a^2} + \frac{y^2}{b^2} = 1.$

The figure is an ellipse with center (0, 0) and axes of lengths $2a$ and $2b$.

(b) $f(t+p) = a\sin\left[\omega_1(t+p)\right] = a\sin\left[\omega_1 t + \omega_1 p\right] = a\sin\left[\omega_1 t + 2\pi n\right] = a\sin\omega_1 t = f(t).$

$g(t+p) = b\cos\left[\omega_2(t+p)\right] = b\cos\left[\omega_2 t + \frac{\omega_2}{\omega_1}2\pi n\right] = b\cos\left[\omega_2 t + \frac{m}{n}2\pi n\right] = b\cos\left[\omega_2 t + 2\pi m\right] = b\cos\omega_2 t = g(t).$

Since f and g are periodic with period p,

the curve retraces itself every p units of time.

[40] (a) Since $x = 2\sin 3t$ has period $\frac{2\pi}{3}$ and $y = 3\sin(1.5t)$ has period $\frac{4\pi}{3}$,

the curve will repeat itself every $\frac{4\pi}{3}$ units of time.

(b) Let $D(t)$ denote the square of the distance from the origin to a point on the graph. $D(t) = x^2 + y^2 = 4\sin^2 3t + 9\sin^2(1.5t)$.

$D'(t) = 24\sin 3t\cos 3t + 27\sin(1.5t)\cos(1.5t)$

$= 24\sin 3t\cos 3t + \frac{27}{2}\sin 3t = 3\sin 3t(8\cos 3t + \frac{9}{2}).$

$D'(t) = 0 \Rightarrow \sin 3t = 0$ or $\cos 3t = -\frac{9}{16}$. $\sin 3t = 0$ gives a minimum value.

$\cos 3t = -\frac{9}{16} \Rightarrow 3t = \cos^{-1}(-\frac{9}{16}) = k \approx 2.168$. $0 \le t \le \frac{4\pi}{3} \Rightarrow 0 \le 3t \le 4\pi$.

Thus, other solutions are $k + 2\pi \approx 8.451$, $2\pi - k \approx 4.115$, and $4\pi - k \approx 10.398$. Each value yields $D(t) \approx 9.766$,

or a maximum distance ≈ 3.125.

[41] $x = 3\sin^5 t$ and $y = 3\cos^5 t \Rightarrow (\frac{x}{3})^{2/5} + (\frac{y}{3})^{2/5} = \sin^2 t + \cos^2 t = 1.$

The graph traces an astroid.

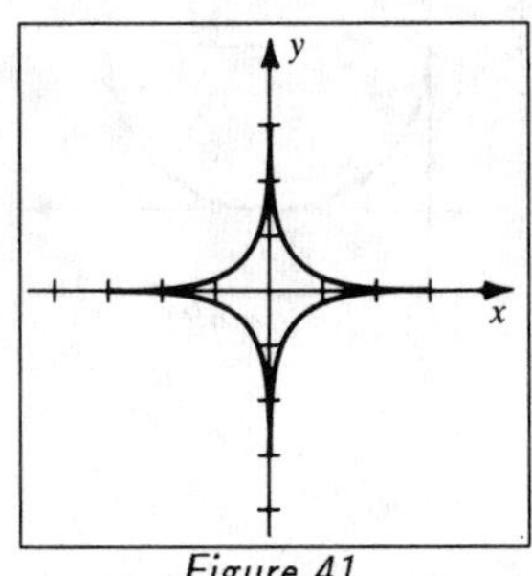

Figure 41

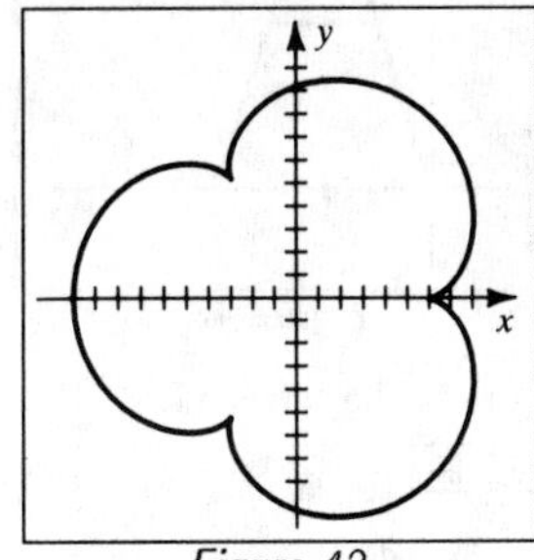

Figure 42

[42] The graph traces an epicycloid. See Exercise 35 with $b = 2$.

43 The graph traces a curtate cycloid. See Exercise 38 with $a = 3$ and $b = 2$.

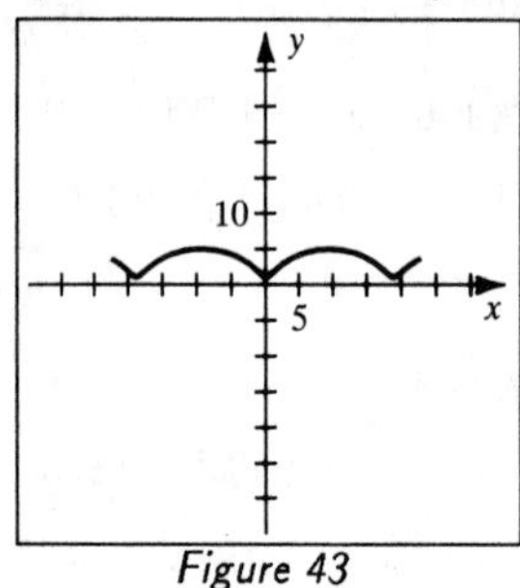

Figure 43

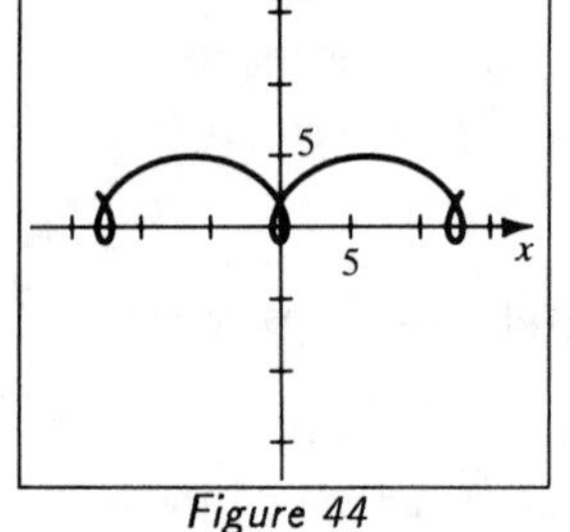

Figure 44

44 The graph traces a prolate cycloid. See Exercise 38 with $a = 2$ and $b = 3$.

45 The figure is a mask with a mouth, nose, and eyes.

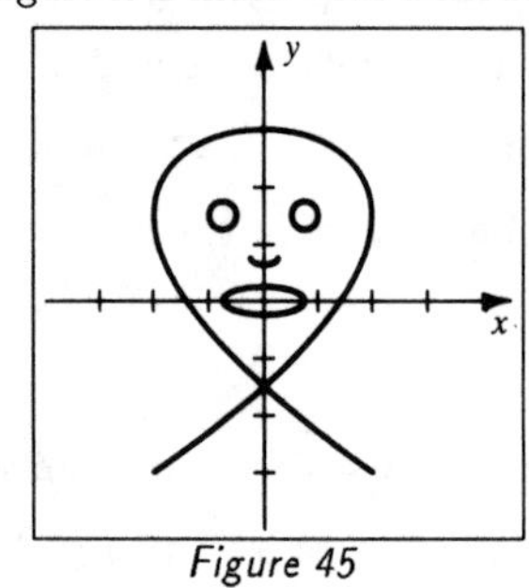

Figure 45

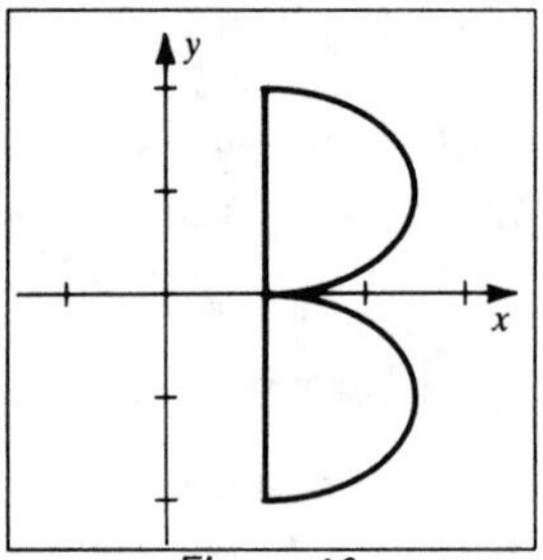

Figure 46

46 The figure is the letter B.

47 The figure is the letter A.

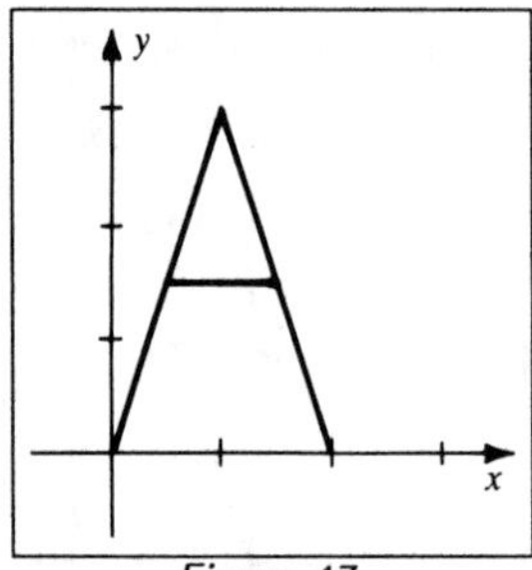

Figure 47

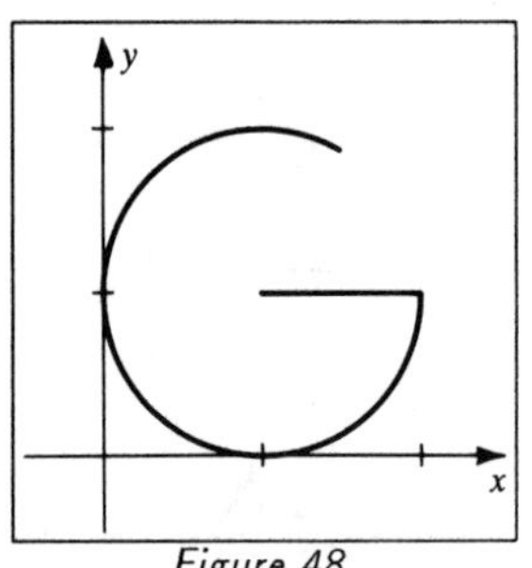

Figure 48

48 The figure is the letter G.

Exercises 13.2

1 $\frac{dx}{dt} = 2t$ and $\frac{dy}{dt} = 2t \Rightarrow \frac{dy}{dx} = \frac{dy/dt}{dx/dt} = \frac{2t}{2t} = 1$ for all $t \neq 0$. Normal: -1

2 $\frac{dx}{dt} = 3t^2$ and $\frac{dy}{dt} = 3t^2 \Rightarrow \frac{dy}{dx} = \frac{dy/dt}{dx/dt} = \frac{3t^2}{3t^2} = 1$ for all $t \neq 0$. Normal: -1

3 $\frac{dx}{dt} = 8t$ and $\frac{dy}{dt} = 2 \Rightarrow \frac{dy}{dx} = \frac{2}{8t} = \frac{1}{4}$ at $t = 1$. Normal: -4

4 $\frac{dx}{dt} = 3t^2$ and $\frac{dy}{dt} = 2t \Rightarrow \frac{dy}{dx} = \frac{2t}{3t^2} = \frac{2}{3}$ at $t = 1$. Normal: $-\frac{3}{2}$

5 $\frac{dx}{dt} = e^t$ and $\frac{dy}{dt} = -2e^{-2t} \Rightarrow \frac{dy}{dx} = \frac{-2e^{-2t}}{e^t} = -\frac{2}{e^3}$ at $t = 1$. Normal: $\frac{e^3}{2}$

6 $\frac{dx}{dt} = \frac{1}{2\sqrt{t}}$ and $\frac{dy}{dt} = 3 \Rightarrow \frac{dy}{dx} = 6\sqrt{t} = 6$ at $t = 1$. Normal: $-\frac{1}{6}$

7 $\frac{dx}{dt} = 2\cos t$ and $\frac{dy}{dt} = -3\sin t \Rightarrow \frac{dy}{dx} = \frac{-3\sin t}{2\cos t} = -\frac{3}{2}\tan 1 \approx -2.34$ at $t = 1$.

Normal: $\frac{2}{3}\cot 1 \approx 0.43$

8 $\frac{dx}{dt} = -\sin t$ and $\frac{dy}{dt} = \cos t \Rightarrow \frac{dy}{dx} = \frac{\cos t}{-\sin t} = -\cot 1 \approx -0.64$ at $t = 1$.

Normal: $\tan 1 \approx 1.56$

9 $\frac{dx}{dt} = -3t^2$ and $\frac{dy}{dt} = -12t - 18 \Rightarrow \frac{dy}{dx} = \frac{-12t - 18}{-3t^2}$. $\frac{dy}{dx} = 2 \Rightarrow$

$6t^2 - 12t - 18 = 0 \Rightarrow t = 3, -1$. $t = 3 \Rightarrow x = -27$ and $y = -108$.

$t = -1 \Rightarrow x = 1$ and $y = 12$. ★ $(-27, -108), (1, 12)$

10 $\frac{dx}{dt} = 2t + 1$ and $\frac{dy}{dt} = 10t \Rightarrow \frac{dy}{dx} = \frac{10t}{2t + 1}$. $\frac{dy}{dx} = 4 \Rightarrow 10t = 8t + 4 \Rightarrow t = 2$.

$t = 2 \Rightarrow x = 6$ and $y = 17$. ★ $(6, 17)$

Note: In Exer. 11–20, let *HT* denote Horizontal Tangent and *VT* denote Vertical Tangent.

11 (a) $\frac{dx}{dt} = 8t$ and $\frac{dy}{dt} = 3t^2 - 12 \Rightarrow \frac{dy}{dx} = \frac{3t^2 - 12}{8t} = 0$ at $t = \pm 2$.

HT at $(16, \pm 16)$; *VT* at $t = 0$ or $(0, 0)$.

(b) $\frac{d^2y}{dx^2} = \frac{dy'/dt}{dx/dt} = \frac{D_t\left[\frac{3}{8}t - \frac{3}{2t}\right]}{8t} = \frac{\frac{3}{8} + \frac{3}{2t^2}}{8t} = \frac{3t^2 + 12}{64t^3}$.

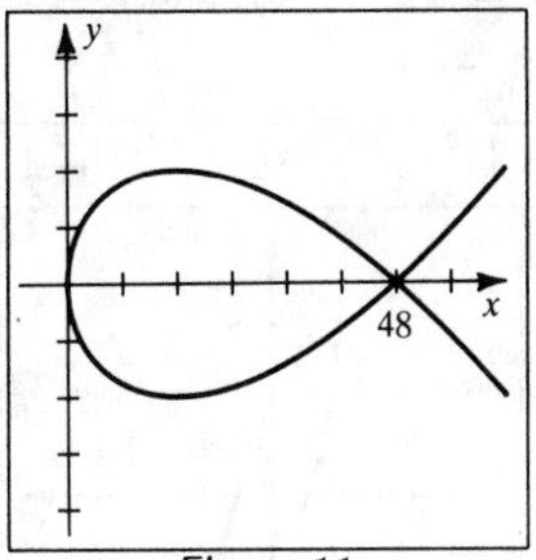

Figure 11

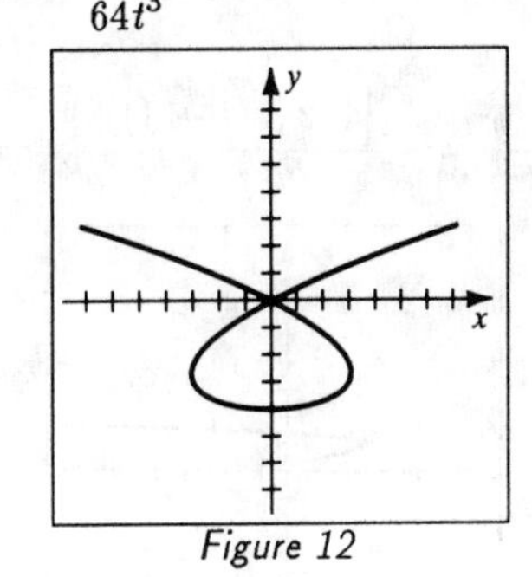

Figure 12

12 (a) $\frac{dx}{dt} = 3t^2 - 4$ and $\frac{dy}{dt} = 2t \Rightarrow \frac{dy}{dx} = \frac{2t}{3t^2 - 4} = 0$ at $t = 0$.

HT at $(0, -4)$; *VT* at $t = \pm\frac{2}{\sqrt{3}}$ or $(\mp\frac{16}{9}\sqrt{3}, -\frac{8}{3})$.

(b) $\frac{d^2y}{dx^2} = \frac{dy'/dt}{dx/dt} = \frac{D_t\left[\frac{2t}{3t^2 - 4}\right]}{3t^2 - 4} = \frac{-6t^2 - 8}{(3t^2 - 4)^3}$.

13 (a) $\frac{dx}{dt} = 3t^2$ and $\frac{dy}{dt} = 2t - 2 \Rightarrow \frac{dy}{dx} = \frac{2t-2}{3t^2} = 0$ at $t = 1$.

HT at $(2, -1)$; *VT* at $t = 0$ or $(1, 0)$.

(b) $$\frac{d^2y}{dx^2} = \frac{D_t\left[\frac{2}{3t} - \frac{2}{3t^2}\right]}{3t^2} = \frac{-\frac{2}{3t^2} + \frac{4}{3t^3}}{3t^2} = \frac{-2t+4}{9t^5}.$$

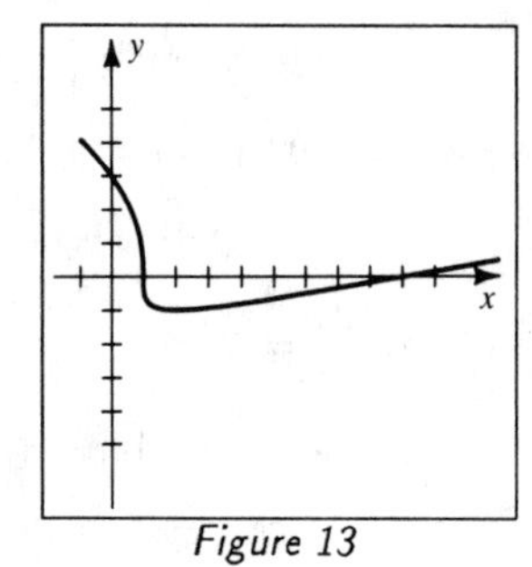

Figure 13

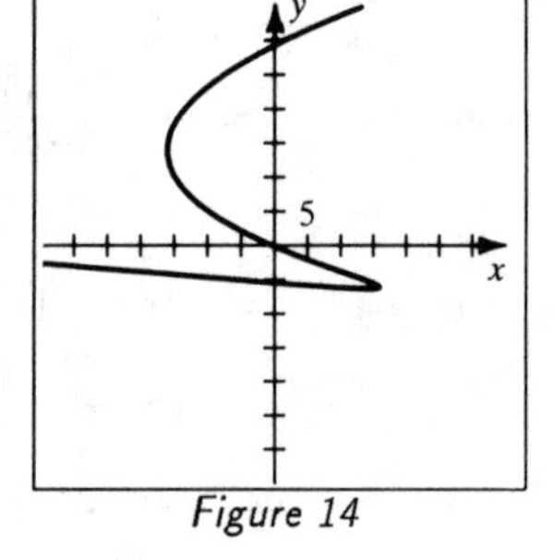

Figure 14

14 (a) $\frac{dx}{dt} = 12 - 3t^2$ and $\frac{dy}{dt} = 2t - 5 \Rightarrow \frac{dy}{dx} = \frac{2t-5}{12-3t^2} = 0$ at $t = \frac{5}{2}$.

HT at $(\frac{115}{8}, -\frac{25}{4}) = (14.375, -6.25)$;

VT at $t = -2$ or $(-16, 14)$, and $t = 2$ or $(16, -6)$.

(b) $$\frac{d^2y}{dx^2} = \frac{D_t\left[\frac{2t-5}{12-3t^2}\right]}{12-3t^2} = \frac{2t^2 - 10t + 8}{9(4-t^2)^3} = \frac{2(t-1)(t-4)}{9(4-t^2)^3}.$$

15 (a) $\frac{dx}{dt} = 6t - 6$ and $\frac{dy}{dt} = \frac{1}{2\sqrt{t}} \Rightarrow \frac{dy}{dx} = \frac{1}{12\sqrt{t}\,(t-1)} \neq 0$.

No *HT*; *VT* at $t = 0$ or $(0, 0)$, and $t = 1$ or $(-3, 1)$.

(b) $$\frac{d^2y}{dx^2} = \frac{D_t\left[(12\sqrt{t}\,(t-1))^{-1}\right]}{6t-6} = \frac{-(18t^{1/2} - 6t^{-1/2})}{864t(t-1)^3} = \frac{1-3t}{144t^{3/2}(t-1)^3}.$$

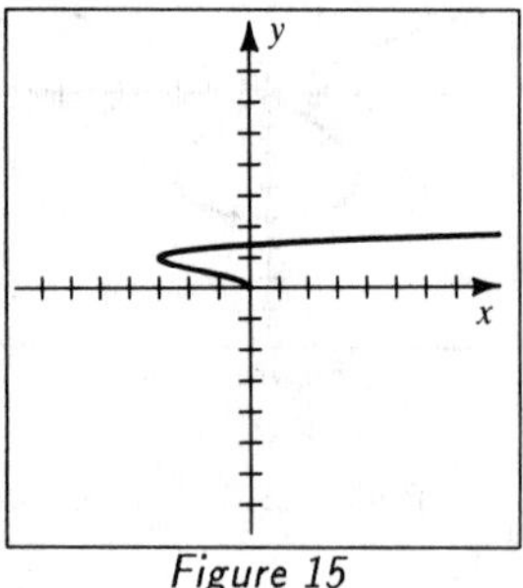

Figure 15

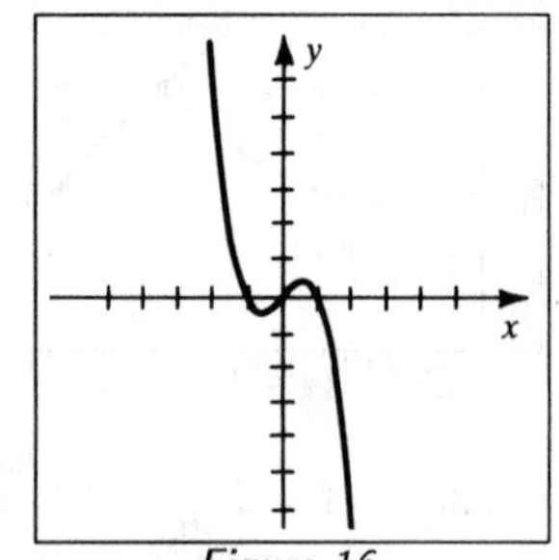

Figure 16

16 (a) $\frac{dx}{dt} = \frac{1}{3}t^{-2/3}$ and $\frac{dy}{dt} = \frac{1}{3}t^{-2/3} - 1 \Rightarrow \frac{dy}{dx} = 1 - 3t^{2/3} = 0$ at $t = \pm\frac{1}{\sqrt{27}}$.

HT at $(\frac{1}{\sqrt{3}}, \frac{2}{3\sqrt{3}})$ and $(-\frac{1}{\sqrt{3}}, -\frac{2}{3\sqrt{3}})$; no *VT*.

(b) $$\frac{d^2y}{dx^2} = \frac{D_t\left[1 - 3t^{2/3}\right]}{\frac{1}{3}t^{-2/3}} = \frac{-2t^{-1/3}}{\frac{1}{3}t^{-2/3}} = -6t^{1/3}.$$

17 (a) $\frac{dx}{dt} = -3\cos^2 t \sin t$ and $\frac{dy}{dt} = 3\sin^2 t \cos t \Rightarrow \frac{dy}{dx} = -\frac{\sin t}{\cos t} = -\tan t = 0$ at

$t = 0, \pi, 2\pi$. *HT* at $(\pm 1, 0)$; *VT* at $t = \frac{\pi}{2}, \frac{3\pi}{2}$ or $(0, \pm 1)$.

(b) $\frac{d^2y}{dx^2} = \frac{D_t[-\tan t]}{-3\cos^2 t \sin t} = \frac{-\sec^2 t}{-3\cos^2 t \sin t} = \frac{1}{3}\sec^4 t \csc t$.

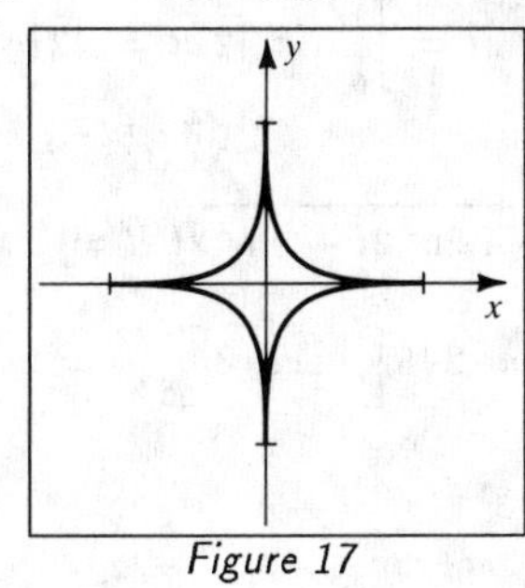

Figure 17

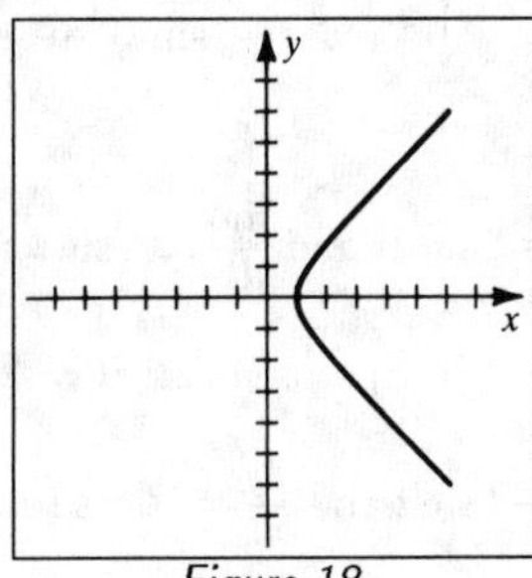

Figure 18

18 (a) $\frac{dx}{dt} = \sinh t$ and $\frac{dy}{dt} = \cosh t \Rightarrow \frac{dy}{dx} = \coth t \neq 0$. No *HT*; *VT* at $t = 0$ or $(1, 0)$.

(b) $\frac{d^2y}{dx^2} = \frac{D_t[\coth t]}{\sinh t} = \frac{-\operatorname{csch}^2 t}{\sinh t} = -\operatorname{csch}^3 t$.

19 $\frac{dx}{dt} = 8\cos 2t$ and $\frac{dy}{dt} = -6\sin 3t \Rightarrow \frac{dy}{dx} = -\frac{3\sin 3t}{4\cos 2t} = 0$ when $3t = \pi n$ or

$t = \frac{\pi}{3}n$. Since $\sin 2t$ has period π and $\cos 3t$ has period $\frac{2\pi}{3}$, the curve will retrace itself

every 2π units; that is, the period will be the $\text{lcm}(\pi, \frac{2\pi}{3}) = 2\pi$.

On $[0, 2\pi)$, $t = 0, \frac{\pi}{3}, \frac{2\pi}{3}, \pi, \frac{4\pi}{3}, \frac{5\pi}{3}$ gives *HT* at $(0, \pm 2)$, $(2\sqrt{3}, \pm 2)$, $(-2\sqrt{3}, \pm 2)$.

VT occur when $\cos 2t = 0 \Rightarrow 2t = \frac{\pi}{2} + \pi n \Rightarrow t = \frac{\pi}{4} + \frac{\pi}{2}n$.

On $[0, 2\pi)$, $t = \frac{\pi}{4}, \frac{3\pi}{4}, \frac{5\pi}{4}, \frac{7\pi}{4}$ gives *VT* at $(4, \pm\sqrt{2})$, $(-4, \pm\sqrt{2})$.

20 $\frac{dx}{dt} = \frac{5}{3}\cos\frac{1}{3}t$ and $\frac{dy}{dt} = 8\cos 2t \Rightarrow \frac{dy}{dx} = \frac{24\cos 2t}{5\cos\frac{1}{3}t} = 0$ when $2t = \frac{\pi}{2} + \pi n$ or

$t = \frac{\pi}{4} + \frac{\pi}{2}n$. Since $\sin\frac{1}{3}t$ has period 6π and $\sin 2t$ has period π, the curve retraces

itself every 6π units; that is, the period will be the $\text{lcm}(\pi, 6\pi) = 6\pi$.

On $[0, 6\pi)$, $\cos 2t = 0$ at $t = \frac{\pi}{4}, \frac{3\pi}{4}, \frac{5\pi}{4}, \ldots, \frac{21\pi}{4}, \frac{23\pi}{4}$.

HT at $(x, \pm y)$ and $(-x, \pm y)$, where (i) $x = 5\sin\frac{\pi}{12} \approx 1.29$; $y = 4$;

(ii) $x = \frac{5}{\sqrt{2}} \approx 3.54$; $y = 4$;

(iii) $x = 5\sin\frac{5\pi}{12} \approx 4.83$; $y = 4$.

VT occur when $\cos\frac{1}{3}t = 0 \Rightarrow \frac{1}{3}t = \frac{\pi}{2} + \pi n \Rightarrow t = \frac{3\pi}{2} + 3\pi n$.

On $[0, 6\pi)$, $t = \frac{3\pi}{2}, \frac{9\pi}{2}$ gives *VT* at $(\pm 5, 0)$.

21 $\frac{dx}{dt} = 10t$ and $\frac{dy}{dt} = 6t^2 \Rightarrow L = \int_0^1 \sqrt{(10t)^2 + (6t^2)^2}\,dt = \int_0^1 2t\sqrt{25 + 9t^2}\,dt =$

$\frac{1}{9}\left[\frac{2}{3}(25 + 9t^2)^{3/2}\right]_0^1 = \frac{2}{27}(34^{3/2} - 125) \approx 5.43$

[22] $\frac{dx}{dt} = 3$ and $\frac{dy}{dt} = 3t^{1/2} \Rightarrow L = \int_0^4 \sqrt{9 + 9t}\,dt = 3\Big[\frac{2}{3}(1 + t)^{3/2}\Big]_0^4 =$

$2(5^{3/2} - 1) \approx 20.36$

[23] $\frac{dx}{dt} = e^t(\cos t - \sin t)$ and $\frac{dy}{dt} = e^t(\cos t + \sin t) \Rightarrow$

$$L = \int_0^{\pi/2} \sqrt{e^{2t}(\cos t - \sin t)^2 + e^{2t}(\cos t + \sin t)^2}\,dt = \int_0^{\pi/2} e^t\sqrt{2}\,dt = \sqrt{2}\Big[e^t\Big]_0^{\pi/2} =$$

$\sqrt{2}\,(e^{\pi/2} - 1) \approx 5.39$

[24] $\frac{dx}{dt} = -2\sin 2t$ and $\frac{dy}{dt} = 2\sin t\cos t \Rightarrow L = \int_0^{\pi} \sqrt{4\sin^2 2t + \sin^2 2t}\,dt =$

$$\int_0^{\pi} \sqrt{5}\,|\sin 2t|\,dt = 2\sqrt{5}\int_0^{\pi/2} \sin 2t\,dt = 2\sqrt{5}\Big[-\tfrac{1}{2}\cos 2t\Big]_0^{\pi/2} = 2\sqrt{5} \approx 4.47$$

[25] $\frac{dx}{dt} = -t\sin t$ and $\frac{dy}{dt} = t\cos t \Rightarrow$

$$L = \int_0^{\pi/2} \sqrt{t^2\sin^2 t + t^2\cos^2 t}\,dt = \int_0^{\pi/2} t\,dt = \Big[\tfrac{1}{2}t^2\Big]_0^{\pi/2} = \tfrac{1}{8}\pi^2 \approx 1.23$$

[26] $\frac{dx}{dt} = -3\cos^2 t\sin t$ and $\frac{dy}{dt} = 3\sin^2 t\cos t \Rightarrow$

$$L = \int_0^{\pi/2} \sqrt{9\cos^4 t\sin^2 t + 9\sin^4 t\cos^2 t}\,dt = 3\int_0^{\pi/2} \sin t\cos t\sqrt{\cos^2 t + \sin^2 t}\,dt =$$

$$3\int_0^{\pi/2} \sin t\cos t\,dt = 3\Big[\tfrac{1}{2}\sin^2 t\Big]_0^{\pi/2} = \tfrac{3}{2}$$

[27] $\frac{dx}{dt} = -2\sin t$ and $\frac{dy}{dt} = 3\cos t$. $L = \int_0^{2\pi} \sqrt{4\sin^2 t + 9\cos^2 t}\,dt.$

Let $f(t) = \sqrt{4\sin^2 t + 9\cos^2 t}$. $L \approx S =$

$$\tfrac{2\pi - 0}{3(6)}\Big[f(0) + 4f(\tfrac{\pi}{3}) + 2f(\tfrac{2\pi}{3}) + 4f(\pi) + 2f(\tfrac{4\pi}{3}) + 4f(\tfrac{5\pi}{3}) + f(2\pi)\Big] \approx 15.881.$$

[28] $\frac{dx}{dt} = 12t^2 - 1$ and $\frac{dy}{dt} = 4t$.

$L = \int_0^1 \sqrt{(12t^2 - 1)^2 + (4t)^2}\,dt$. Let $f(t) = \sqrt{(12t^2 - 1)^2 + (4t)^2}$.

$$L \approx S = \tfrac{1-0}{3(6)}\Big[f(0) + 4f(\tfrac{1}{6}) + 2f(\tfrac{1}{3}) + 4f(\tfrac{1}{2}) + 2f(\tfrac{2}{3}) + 4f(\tfrac{5}{6}) + f(1)\Big] \approx 4.052.$$

[29] $\frac{dx}{dt} = 2t$ and $\frac{dy}{dt} = 2 \Rightarrow$

$$S = 2\pi\int_0^4 2t\sqrt{4t^2 + 4}\,dt = 4\pi\Big[\tfrac{2}{3}(t^2 + 1)^{3/2}\Big]_0^4 = \tfrac{8\pi}{3}(17^{3/2} - 1) \approx 578.83$$

[30] $\frac{dx}{dt} = 4$ and $\frac{dy}{dt} = 3t^2 \Rightarrow$

$$S = 2\pi\int_1^2 t^3\sqrt{16 + 9t^4}\,dt = \tfrac{\pi}{18}\Big[\tfrac{2}{3}(16 + 9t^4)^{3/2}\Big]_1^2 = \tfrac{\pi}{27}(160^{3/2} - 125) \approx 220.94$$

[31] $\frac{dx}{dt} = 2t$ and $\frac{dy}{dt} = 1 - t^2 \Rightarrow S = 2\pi\int_0^1 (t - \tfrac{1}{3}t^3)\sqrt{4t^2 + (1 - t^2)^2}\,dt =$

$$2\pi\int_0^1 (t - \tfrac{1}{3}t^3)(1 + t^2)\,dt = 2\pi\Big[\tfrac{1}{2}t^2 + \tfrac{1}{6}t^4 - \tfrac{1}{18}t^6\Big]_0^1 = \tfrac{11\pi}{9} \approx 3.84$$

[32] $\frac{dx}{dt} = 8t$ and $\frac{dy}{dt} = -2 \Rightarrow S = 2\pi\int_{-2}^{0}(3 - 2t)\sqrt{64t^2 + 4}\,dt$

$= 48\pi\int_{-2}^{0}\sqrt{t^2 + \frac{1}{16}}\,dt - 4\pi\int_{-2}^{0} t\sqrt{64t^2 + 4}\,dt$

$= 48\pi\Big[\frac{t}{2}\sqrt{t^2 + \frac{1}{16}} + \frac{1}{32}\ln\Big|t + \sqrt{t^2 + \frac{1}{16}}\Big|\Big]_{-2}^{0} - \frac{4\pi}{128}\Big[\frac{2}{3}(64t^2 + 4)^{3/2}\Big]_{-2}^{0}$

{ Formula 21 }

$= \Big[(\frac{48\pi}{32}\ln\frac{1}{4}) - 48\pi\Big(-\sqrt{4 + \frac{1}{16}} + \frac{1}{32}\ln\Big|-2 + \sqrt{4 + \frac{1}{16}}\Big|\Big)\Big] - \frac{\pi}{48}\Big[4^{3/2} - 260^{3/2}\Big]$

$= -\frac{3\pi}{2}\ln 4 + 12\pi\sqrt{65} - \frac{3\pi}{2}\ln\Big(\frac{\sqrt{65} - 8}{4}\Big) - \frac{\pi}{6} + \frac{\pi}{6}(65\sqrt{65})$

$= \frac{\pi}{6}(137\sqrt{65} - 1) - \frac{3\pi}{2}\ln(\sqrt{65} - 8) \approx 590.89$

[33] $\frac{dx}{dt} = 1 - \cos t$ and $\frac{dy}{dt} = \sin t \Rightarrow$

$S = 2\pi\int_{0}^{2\pi}(1 - \cos t)\sqrt{(1 - \cos t)^2 + \sin^2 t}\,dt$

$= 2\pi\int_{0}^{2\pi}(1 - \cos t)\sqrt{2 - 2\cos t}\,dt = 2\sqrt{2}\pi\int_{0}^{2\pi}(1 - \cos t)^{3/2}\,dt$

$= 2\sqrt{2}\pi\int_{0}^{2\pi}2^{3/2}\sin^3(\frac{1}{2}t)\,dt$ $\{\cos t = 1 - 2\sin^2(\frac{1}{2}t)\}$

$= 8\pi\int_{0}^{2\pi}\Big[1 - \cos^2(\frac{1}{2}t)\Big]\sin(\frac{1}{2}t)\,dt$ $\{u = \cos(\frac{1}{2}t),\ -2\,du = \sin(\frac{1}{2}t)\,dt\}$

$= 8\pi(-2)\int_{1}^{-1}(1 - u^2)\,du = 32\pi\Big[u - \frac{1}{3}u^3\Big]_{0}^{1} = \frac{64\pi}{3} \approx 67.02$

[34] $\frac{dx}{dt} = 1$ and $\frac{dy}{dt} = t^2 - \frac{1}{4}t^{-2} \Rightarrow$

$S = 2\pi\int_{1}^{2}(\frac{1}{3}t^3 + \frac{1}{4}t^{-1})\sqrt{1 + (t^2 - \frac{1}{4}t^{-2})^2}\,dt$

$= 2\pi\int_{1}^{2}(\frac{1}{3}t^3 + \frac{1}{4}t^{-1})\sqrt{(t^2 + \frac{1}{4}t^{-2})^2}\,dt$

$= 2\pi\int_{1}^{2}(\frac{1}{3}t^5 + \frac{1}{3}t + \frac{1}{16}t^{-3})\,dt = 2\pi\Big[\frac{1}{18}t^6 + \frac{1}{6}t^2 - \frac{1}{32}t^{-2}\Big]_{1}^{2} = \frac{515\pi}{64} \approx 25.28$

[35] $\frac{dx}{dt} = 2t^{-1/2}$ and $\frac{dy}{dt} = t - t^{-2} \Rightarrow$

$S = 2\pi\int_{1}^{4}4t^{1/2}\sqrt{4t^{-1} + (t - t^{-2})^2}\,dt = 8\pi\int_{1}^{4}t^{1/2}\sqrt{(t + t^{-2})^2}\,dt =$

$8\pi\int_{1}^{4}(t^{3/2} + t^{-3/2})\,dt = 8\pi\Big[\frac{2}{5}t^{5/2} - 2t^{-1/2}\Big]_{1}^{4} = 8\pi(\frac{67}{5}) = \frac{536\pi}{5} \approx 336.78$

[36] $\frac{dx}{dt} = 3$ and $\frac{dy}{dt} = 1 \Rightarrow S = 2\pi\int_{0}^{5}3t\sqrt{9 + 1}\,dt = 6\sqrt{10}\,\pi\Big[\frac{1}{2}t^2\Big]_{0}^{5} = 75\sqrt{10}\,\pi \approx 745.09$

[37] $\frac{dx}{dt} = e^t(\cos t + \sin t)$ and $\frac{dy}{dt} = e^t(\cos t - \sin t) \Rightarrow S = 2\pi\int_{0}^{\pi/2}e^t\sin t\sqrt{2e^{2t}}\,dt =$

$2\sqrt{2}\pi\int_{0}^{\pi/2}e^{2t}\sin t\,dt = 2\sqrt{2}\pi\Big[\frac{1}{5}e^{2t}(2\sin t - \cos t)\Big]_{0}^{\pi/2}$ { Formula 98 or IP } $=$

$\frac{2}{5}\sqrt{2}\,\pi(2e^{\pi} + 1) \approx 84.03$

[38] $\frac{dx}{dt} = 6t$ and $\frac{dy}{dt} = 6t^2 \Rightarrow S = 2\pi \int_0^1 3t^2 \sqrt{36t^2 + 36t^4}\, dt = 36\pi \int_0^1 t^3 \sqrt{1 + t^2}\, dt.$

Now, $u = 1 + t^2 \Rightarrow du = 2t\,dt$ and $t^2 = u - 1 \Rightarrow S = 18\pi \int_1^2 (u - 1)u^{1/2}\, du =$

$$18\pi\left[\tfrac{2}{5}u^{5/2} - \tfrac{2}{3}u^{3/2}\right]_1^2 = 18\pi\left[\tfrac{4}{15}\sqrt{2} - (-\tfrac{4}{15})\right] = \tfrac{24\pi}{5}(\sqrt{2} + 1) \approx 36.41$$

[39] $\frac{dx}{dt} = -2t \sin(t^2)$ and $\frac{dy}{dt} = 2\sin t \cos t = \sin 2t.$

$S = 2\pi \int_0^1 \sin^2 t \sqrt{\left[-2t\sin(t^2)\right]^2 + (\sin 2t)^2}\, dt.$ Let $f(t)$ equal the integrand.

$$S \approx \mathrm{S} = (2\pi)\tfrac{1-0}{3(4)}\left[f(0) + 4f(\tfrac{1}{4}) + 2f(\tfrac{1}{2}) + 4f(\tfrac{3}{4}) + f(1)\right] \approx 2.226.$$

[40] $\frac{dx}{dt} = 2t + 2$ and $\frac{dy}{dt} = 4t^3.$ $S = 2\pi \int_0^1 (t^2 + 2t)\sqrt{(2t + 2)^2 + (4t^3)^2}\, dt.$

Let $f(t)$ equal the integrand.

$$S \approx \mathrm{S} = (2\pi)\tfrac{1-0}{3(4)}\left[f(0) + 4f(\tfrac{1}{4}) + 2f(\tfrac{1}{2}) + 4f(\tfrac{3}{4}) + f(1)\right] \approx 32.5975.$$

Exercises 13.3

Note: For the following exercises, the substitutions $y = r\sin\theta$, $x = r\cos\theta$, $r^2 = x^2 + y^2$, and $\tan\theta = \frac{y}{x}$ are used without mention. We have found it helpful to find the "pole" values {when the graph intersects the pole} to determine which values of θ should be used in the construction of an r-θ chart. The numbers listed on each line of the r-θ chart correspond to the numbers labeled on the figures.

[1] $r = 5 \Rightarrow r^2 = 25 \Rightarrow x^2 + y^2 = 25$, a circle centered at the origin with radius 5.

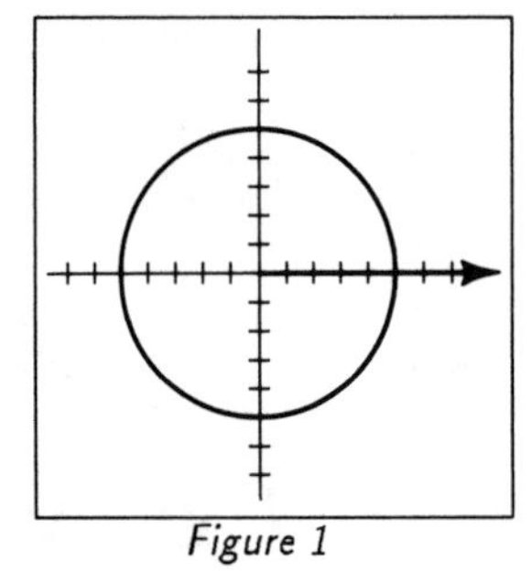
Figure 1

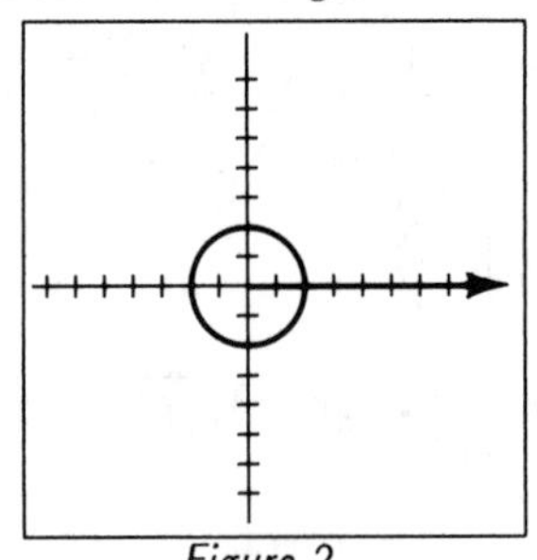
Figure 2

[2] $r = -2 \Rightarrow r^2 = 4 \Rightarrow x^2 + y^2 = 4$, a circle centered at the origin with radius 2.

[3] $\theta = -\frac{\pi}{6}$ and $r \in \mathbb{R}$. The line is $y = (\tan\theta)\,x$ or $y = -\frac{1}{3}\sqrt{3}\,x.$

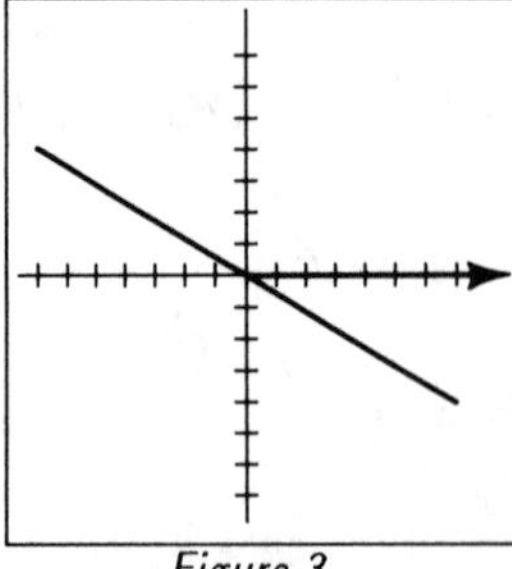
Figure 3

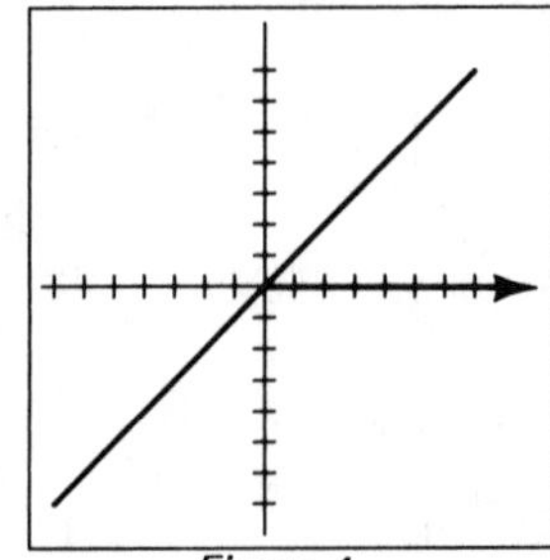
Figure 4

[4] $\theta = \frac{\pi}{4}$ and $r \in \mathbb{R}$. The line is $y = (\tan\theta)\,x$ or $y = x.$

[5] $r = 3\cos\theta \Rightarrow r^2 = 3r\cos\theta \Rightarrow x^2 + y^2 = 3x \Rightarrow$

$(x^2 - 3x + \underline{\tfrac{9}{4}}) + y^2 = \underline{\tfrac{9}{4}} \Rightarrow (x - \tfrac{3}{2})^2 + y^2 = \tfrac{9}{4}.$

	Range of θ	Range of r
1)	$0 \rightarrow \frac{\pi}{2}$	$3 \rightarrow 0$
2)	$\frac{\pi}{2} \rightarrow \pi$	$0 \rightarrow -3$

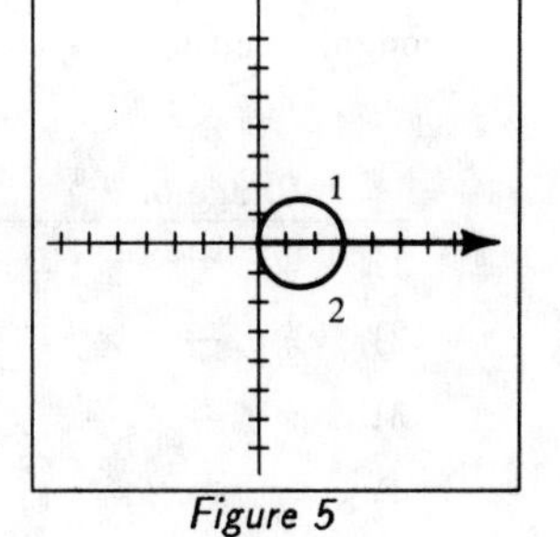

Figure 5

[6] $r = -2\sin\theta \Rightarrow r^2 = -2r\sin\theta \Rightarrow x^2 + y^2 = -2y \Rightarrow$

$x^2 + y^2 + 2y + \underline{1} = \underline{1} \Rightarrow x^2 + (y+1)^2 = 1.$

	Range of θ	Range of r
1)	$0 \rightarrow \frac{\pi}{2}$	$0 \rightarrow -2$
2)	$\frac{\pi}{2} \rightarrow \pi$	$-2 \rightarrow 0$

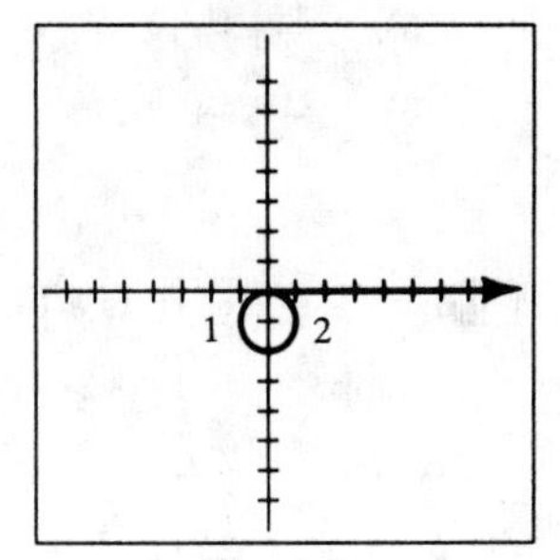

Figure 6

[7] $r = 4 - 4\sin\theta$ is a cardioid since the coefficient of $\sin\theta$ has the same magnitude as the constant term.

$0 = 4 - 4\sin\theta \Rightarrow \sin\theta = 1 \Rightarrow \theta = \frac{\pi}{2} + 2\pi n.$

	Range of θ	Range of r
1)	$0 \rightarrow \frac{\pi}{2}$	$4 \rightarrow 0$
2)	$\frac{\pi}{2} \rightarrow \pi$	$0 \rightarrow 4$
3)	$\pi \rightarrow \frac{3\pi}{2}$	$4 \rightarrow 8$
4)	$\frac{3\pi}{2} \rightarrow 2\pi$	$8 \rightarrow 4$

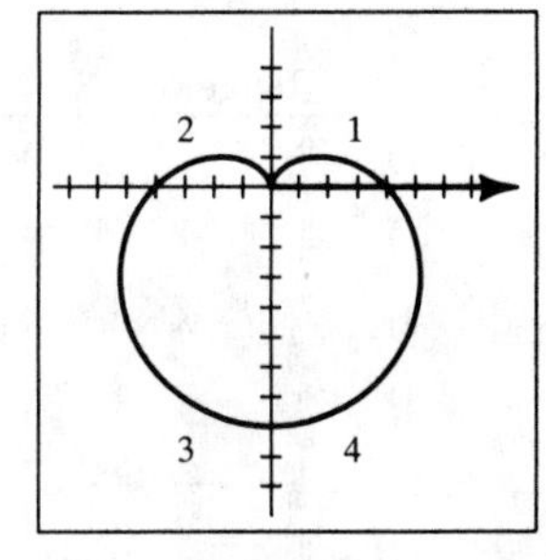

Figure 7

[8] $r = -6(1 + \cos\theta)$ is a cardioid.

$0 = -6(1 + \cos\theta) \Rightarrow \cos\theta = -1 \Rightarrow \theta = \pi + 2\pi n.$

	Range of θ	Range of r
1)	$0 \rightarrow \frac{\pi}{2}$	$-12 \rightarrow -6$
2)	$\frac{\pi}{2} \rightarrow \pi$	$-6 \rightarrow 0$
3)	$\pi \rightarrow \frac{3\pi}{2}$	$0 \rightarrow -6$
4)	$\frac{3\pi}{2} \rightarrow 2\pi$	$-6 \rightarrow -12$

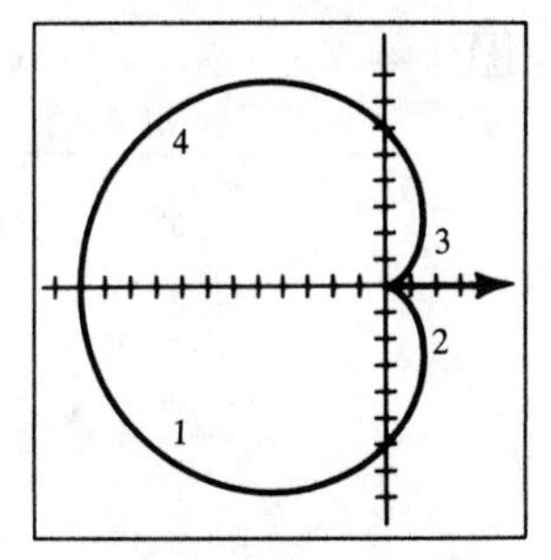

Figure 8

[9] $r = 2 + 4\sin\theta$ is a limaçon with a loop since the constant term has a smaller magnitude than the coefficient of $\sin\theta$.

$0 = 2 + 4\sin\theta \Rightarrow \sin\theta = -\frac{1}{2} \Rightarrow \theta = \frac{7\pi}{6} + 2\pi n, \frac{11\pi}{6} + 2\pi n.$

	Range of θ	Range of r
1)	$0 \rightarrow \frac{\pi}{2}$	$2 \rightarrow 6$
2)	$\frac{\pi}{2} \rightarrow \pi$	$6 \rightarrow 2$
3)	$\pi \rightarrow \frac{7\pi}{6}$	$2 \rightarrow 0$
4)	$\frac{7\pi}{6} \rightarrow \frac{3\pi}{2}$	$0 \rightarrow -2$
5)	$\frac{3\pi}{2} \rightarrow \frac{11\pi}{6}$	$-2 \rightarrow 0$
6)	$\frac{11\pi}{6} \rightarrow 2\pi$	$0 \rightarrow 2$

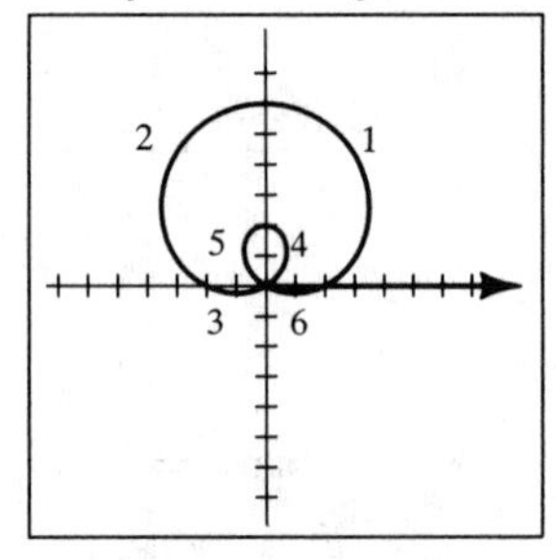

Figure 9

[10] $r = 1 + 2\cos\theta$ is a limaçon with a loop. $0 = 1 + 2\cos\theta \Rightarrow \cos\theta = -\frac{1}{2} \Rightarrow \theta = \frac{2\pi}{3} + 2\pi n, \frac{4\pi}{3} + 2\pi n.$

	Range of θ	Range of r
1)	$0 \rightarrow \frac{\pi}{2}$	$3 \rightarrow 1$
2)	$\frac{\pi}{2} \rightarrow \frac{2\pi}{3}$	$1 \rightarrow 0$
3)	$\frac{2\pi}{3} \rightarrow \pi$	$0 \rightarrow -1$
4)	$\pi \rightarrow \frac{4\pi}{3}$	$-1 \rightarrow 0$
5)	$\frac{4\pi}{3} \rightarrow \frac{3\pi}{2}$	$0 \rightarrow 1$
6)	$\frac{3\pi}{2} \rightarrow 2\pi$	$1 \rightarrow 3$

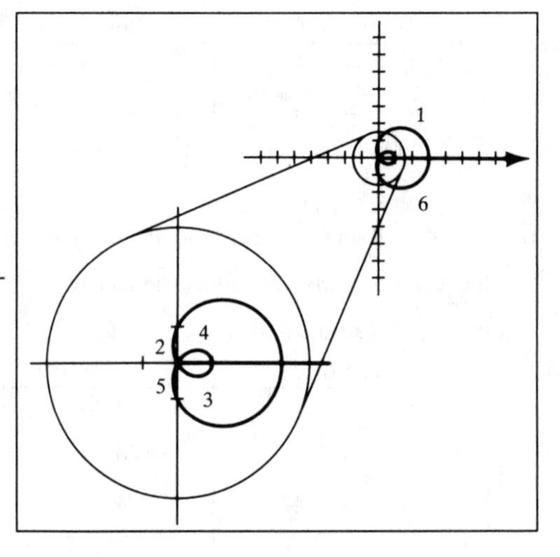

Figure 10

[11] $0 = 2 - \cos\theta \Rightarrow \cos\theta = 2 \Rightarrow$ no pole values.

	Range of θ	Range of r
1)	$0 \rightarrow \frac{\pi}{2}$	$1 \rightarrow 2$
2)	$\frac{\pi}{2} \rightarrow \pi$	$2 \rightarrow 3$
3)	$\pi \rightarrow \frac{3\pi}{2}$	$3 \rightarrow 2$
4)	$\frac{3\pi}{2} \rightarrow 2\pi$	$2 \rightarrow 1$

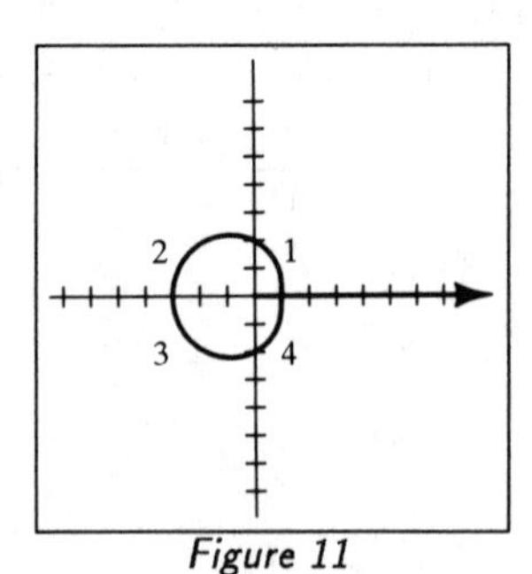

Figure 11

[12] $0 = 5 + 3\sin\theta \Rightarrow \sin\theta = -\frac{5}{3} \Rightarrow$ no pole values.

	Range of θ	Range of r
1)	$0 \rightarrow \frac{\pi}{2}$	$5 \rightarrow 8$
2)	$\frac{\pi}{2} \rightarrow \pi$	$8 \rightarrow 5$
3)	$\pi \rightarrow \frac{3\pi}{2}$	$5 \rightarrow 2$
4)	$\frac{3\pi}{2} \rightarrow 2\pi$	$2 \rightarrow 5$

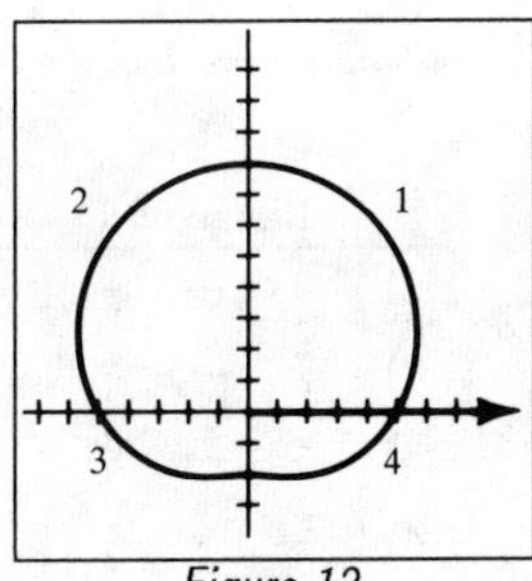

Figure 12

[13] $r = 4\csc\theta \Rightarrow r\sin\theta = 4 \Rightarrow y = 4$. r is undefined at $\theta = \pi n$.

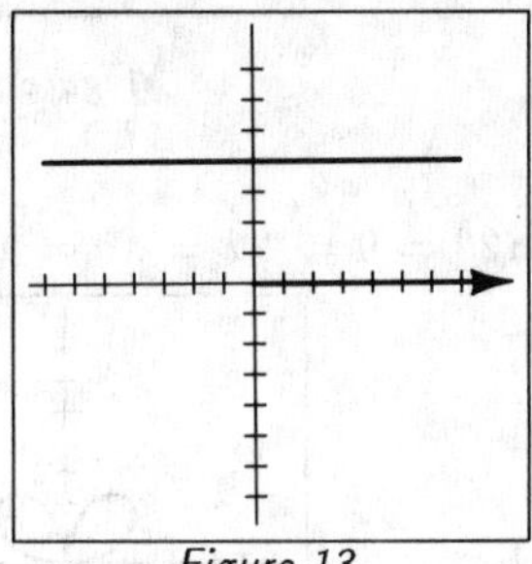
Figure 13

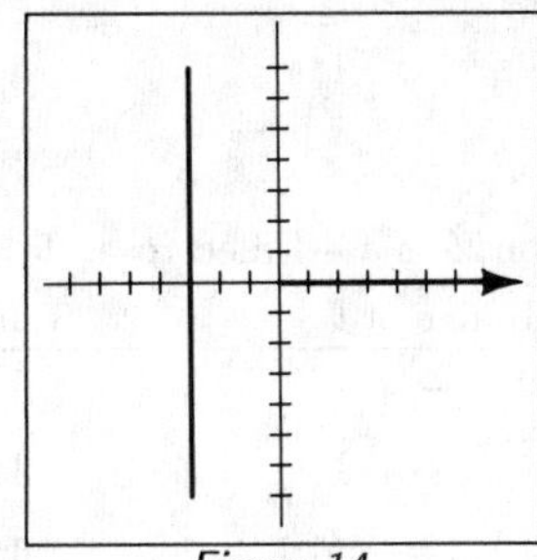
Figure 14

[14] $r = -3\sec\theta \Rightarrow r\cos\theta = -3 \Rightarrow x = -3$. r is undefined at $\theta = \frac{\pi}{2} + \pi n$.

[15] $r = 8\cos 3\theta$ is a 3-leafed rose since 3 is odd.

$0 = 8\cos 3\theta \Rightarrow \cos 3\theta = 0 \Rightarrow 3\theta = \frac{\pi}{2} + \pi n \Rightarrow \theta = \frac{\pi}{6} + \frac{\pi}{3}n.$

	Range of θ	Range of r
1)	$0 \rightarrow \frac{\pi}{6}$	$8 \rightarrow 0$
2)	$\frac{\pi}{6} \rightarrow \frac{\pi}{3}$	$0 \rightarrow -8$
3)	$\frac{\pi}{3} \rightarrow \frac{\pi}{2}$	$-8 \rightarrow 0$
4)	$\frac{\pi}{2} \rightarrow \frac{2\pi}{3}$	$0 \rightarrow 8$
5)	$\frac{2\pi}{3} \rightarrow \frac{5\pi}{6}$	$8 \rightarrow 0$
6)	$\frac{5\pi}{6} \rightarrow \pi$	$0 \rightarrow -8$

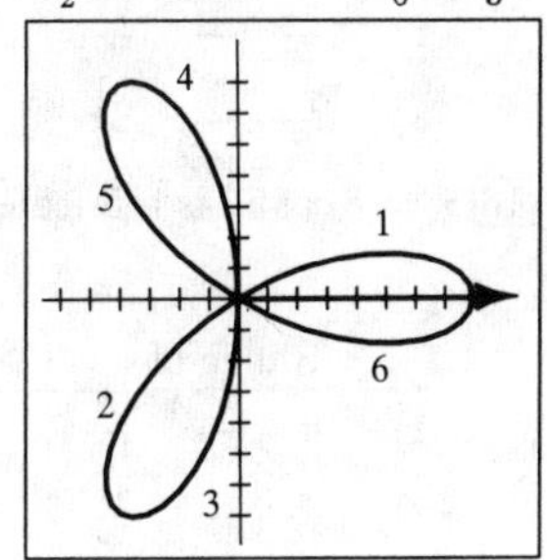

Figure 15

[16] $r = 2\sin 4\theta$ is an 8-leafed rose since 4 is even. $0 = 2\sin 4\theta \Rightarrow \sin 4\theta = 0 \Rightarrow$ $4\theta = \pi n \Rightarrow \theta = \frac{\pi}{4}n$. Steps 9 through 16 follow a similar pattern to steps 1 through 8 and are labeled in the correct order.

	Range of θ	Range of r
1)	$0 \rightarrow \frac{\pi}{8}$	$0 \rightarrow 2$
2)	$\frac{\pi}{8} \rightarrow \frac{\pi}{4}$	$2 \rightarrow 0$
3)	$\frac{\pi}{4} \rightarrow \frac{3\pi}{8}$	$0 \rightarrow -2$
4)	$\frac{3\pi}{8} \rightarrow \frac{\pi}{2}$	$-2 \rightarrow 0$
5)	$\frac{\pi}{2} \rightarrow \frac{5\pi}{8}$	$0 \rightarrow 2$
6)	$\frac{5\pi}{8} \rightarrow \frac{3\pi}{4}$	$2 \rightarrow 0$
7)	$\frac{3\pi}{4} \rightarrow \frac{7\pi}{8}$	$0 \rightarrow -2$
8)	$\frac{7\pi}{8} \rightarrow \pi$	$-2 \rightarrow 0$

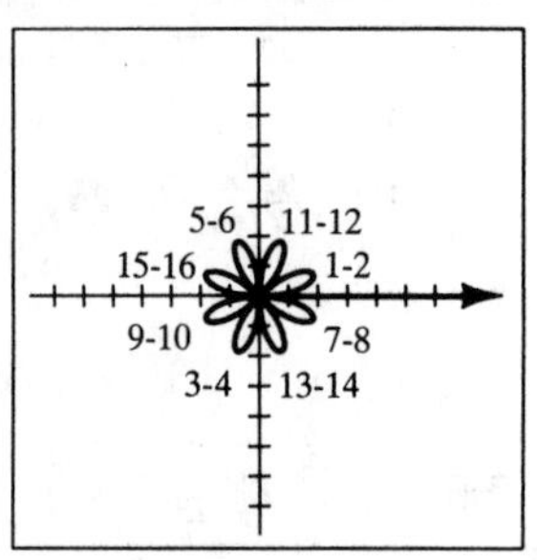

Figure 16

[17] $r = 3\sin 2\theta$ is a 4-leafed rose. $0 = 3\sin 2\theta \Rightarrow \sin 2\theta = 0 \Rightarrow 2\theta = \pi n \Rightarrow \theta = \frac{\pi}{2}n$.

	Range of θ	Range of r
1)	$0 \rightarrow \frac{\pi}{4}$	$0 \rightarrow 3$
2)	$\frac{\pi}{4} \rightarrow \frac{\pi}{2}$	$3 \rightarrow 0$
3)	$\frac{\pi}{2} \rightarrow \frac{3\pi}{4}$	$0 \rightarrow -3$
4)	$\frac{3\pi}{4} \rightarrow \pi$	$-3 \rightarrow 0$
5)	$\pi \rightarrow \frac{5\pi}{4}$	$0 \rightarrow 3$
6)	$\frac{5\pi}{4} \rightarrow \frac{3\pi}{2}$	$3 \rightarrow 0$
7)	$\frac{3\pi}{2} \rightarrow \frac{7\pi}{4}$	$0 \rightarrow -3$
8)	$\frac{7\pi}{4} \rightarrow 2\pi$	$-3 \rightarrow 0$

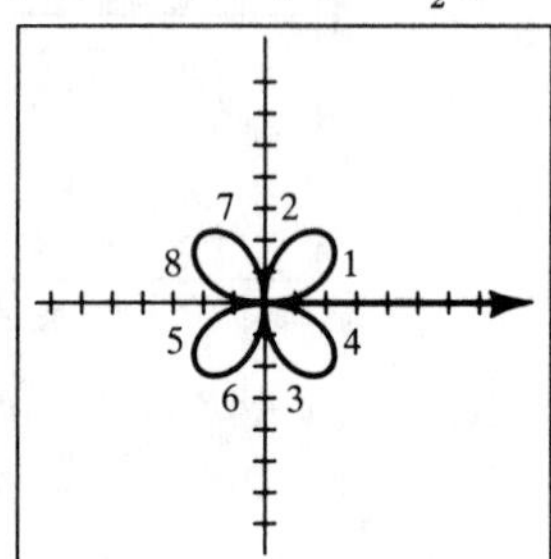

Figure 17

[18] $r = 8\cos 5\theta$ is a 5-leafed rose.

$0 = 8\cos 5\theta \Rightarrow \cos 5\theta = 0 \Rightarrow 5\theta = \frac{\pi}{2} + \pi n \Rightarrow \theta = \frac{\pi}{10} + \frac{\pi}{5}n$.

	Range of θ	Range of r
1)	$0 \rightarrow \frac{\pi}{10}$	$8 \rightarrow 0$
2)	$\frac{\pi}{10} \rightarrow \frac{2\pi}{10}$	$0 \rightarrow -8$
3)	$\frac{2\pi}{10} \rightarrow \frac{3\pi}{10}$	$-8 \rightarrow 0$
4)	$\frac{3\pi}{10} \rightarrow \frac{4\pi}{10}$	$0 \rightarrow 8$
5)	$\frac{4\pi}{10} \rightarrow \frac{5\pi}{10}$	$8 \rightarrow 0$
6)	$\frac{5\pi}{10} \rightarrow \frac{6\pi}{10}$	$0 \rightarrow -8$
7)	$\frac{6\pi}{10} \rightarrow \frac{7\pi}{10}$	$-8 \rightarrow 0$
8)	$\frac{7\pi}{10} \rightarrow \frac{8\pi}{10}$	$0 \rightarrow 8$
9)	$\frac{8\pi}{10} \rightarrow \frac{9\pi}{10}$	$8 \rightarrow 0$
10)	$\frac{9\pi}{10} \rightarrow \pi$	$0 \rightarrow -8$

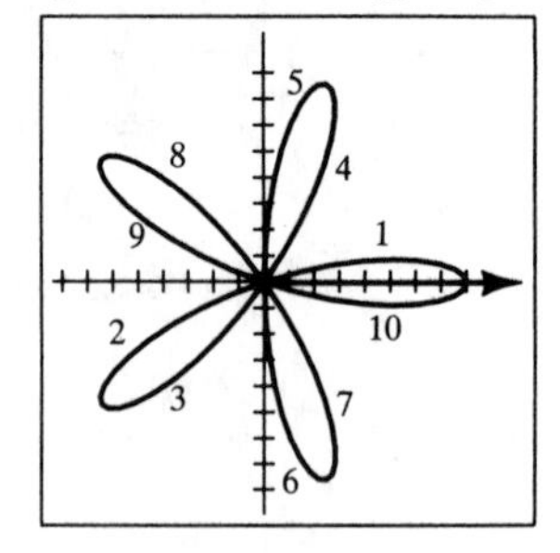

Figure 18

[19] $0 = 4\cos 2\theta \Rightarrow \cos 2\theta = 0 \Rightarrow 2\theta = \frac{\pi}{2} + \pi n \Rightarrow \theta = \frac{\pi}{4} + \frac{\pi}{2}n.$

	Range of θ	Range of r
1)	$0 \rightarrow \frac{\pi}{4}$	$\pm 2 \rightarrow 0$
2)	$\frac{\pi}{4} \rightarrow \frac{\pi}{2}$	undefined
3)	$\frac{\pi}{2} \rightarrow \frac{3\pi}{4}$	undefined
4)	$\frac{3\pi}{4} \rightarrow \pi$	$0 \rightarrow \pm 2$

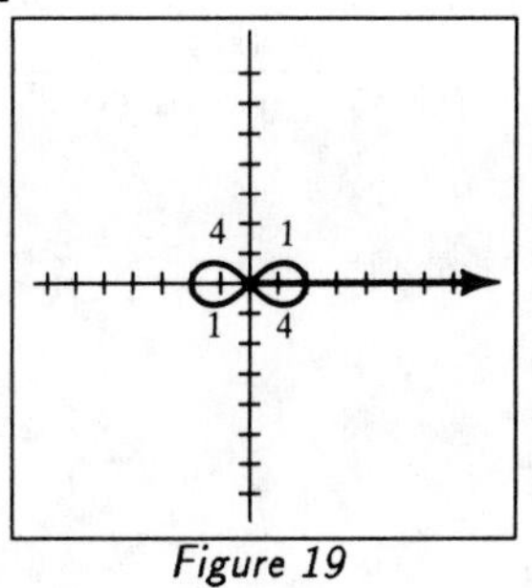

Figure 19

[20] $0 = -16\sin 2\theta \Rightarrow \sin 2\theta = 0 \Rightarrow 2\theta = \pi n \Rightarrow \theta = \frac{\pi}{2}n.$

	Range of θ	Range of r
1)	$0 \rightarrow \frac{\pi}{4}$	undefined
2)	$\frac{\pi}{4} \rightarrow \frac{\pi}{2}$	undefined
3)	$\frac{\pi}{2} \rightarrow \frac{3\pi}{4}$	$0 \rightarrow \pm 4$
4)	$\frac{3\pi}{4} \rightarrow \pi$	$\pm 4 \rightarrow 0$

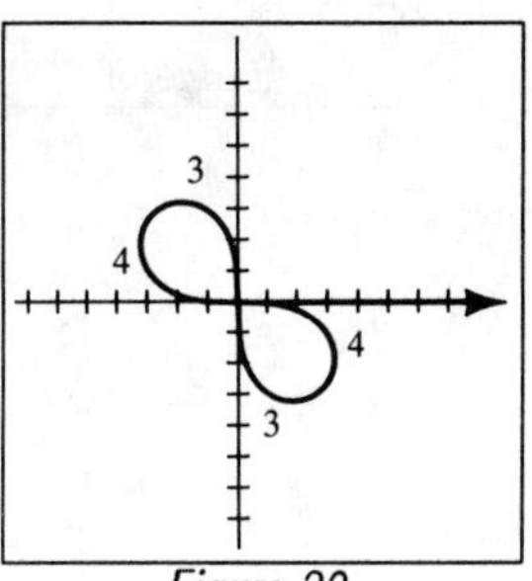

Figure 20

[21] $r = e^{\theta},\ \theta \geq 0$

	Range of θ	Range of r
1)	$0 \rightarrow \frac{\pi}{2}$	$1 \rightarrow 4.81$
2)	$\frac{\pi}{2} \rightarrow \pi$	$4.81 \rightarrow 23.14$
3)	$\pi \rightarrow \frac{3\pi}{2}$	$23.14 \rightarrow 111.3$
4)	$\frac{3\pi}{2} \rightarrow 2\pi$	$111.3 \rightarrow 535.5$

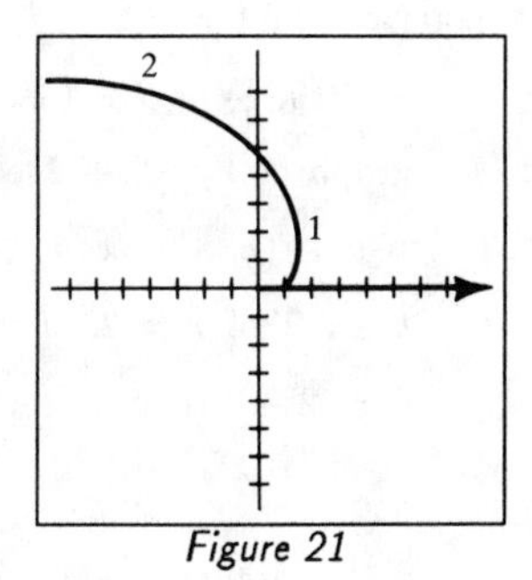

Figure 21

[22] $r = 6\sin^2(\frac{\theta}{2}) = 6\left(\frac{1 - \cos\theta}{2}\right) = 3(1 - \cos\theta)$ is a cardioid.

$0 = 3(1 - \cos\theta) \Rightarrow \cos\theta = 1 \Rightarrow \theta = 2\pi n.$

	Range of θ	Range of r
1)	$0 \rightarrow \frac{\pi}{2}$	$0 \rightarrow 3$
2)	$\frac{\pi}{2} \rightarrow \pi$	$3 \rightarrow 6$
3)	$\pi \rightarrow \frac{3\pi}{2}$	$6 \rightarrow 3$
4)	$\frac{3\pi}{2} \rightarrow 2\pi$	$3 \rightarrow 0$

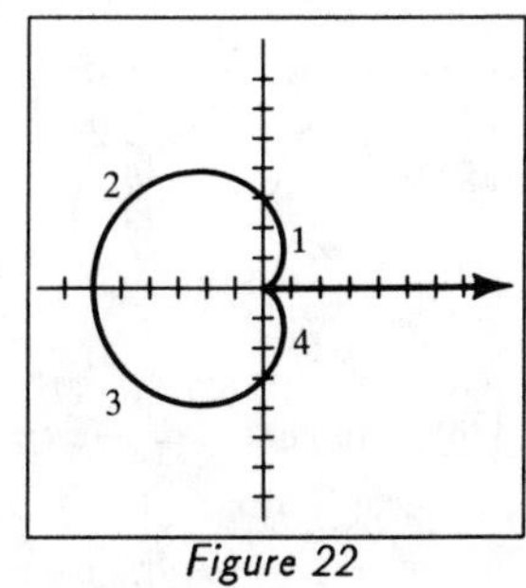

Figure 22

[23] $r = 2\theta,\ \theta \geq 0$

	Range of θ	Range of r
1)	$0 \rightarrow \frac{\pi}{2}$	$0 \rightarrow \pi$
2)	$\frac{\pi}{2} \rightarrow \pi$	$\pi \rightarrow 2\pi$
3)	$\pi \rightarrow \frac{3\pi}{2}$	$2\pi \rightarrow 3\pi$

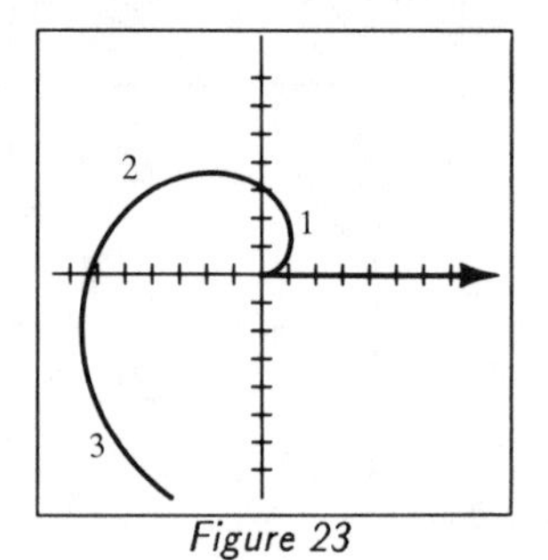

Figure 23

[24] $r\theta = 1 \Rightarrow r = \dfrac{1}{\theta}$. r is undefined at $\theta = \pi n$.

	Range of θ	Range of r
1)	$0 \rightarrow \frac{\pi}{2}$	$+\infty \rightarrow 0.63$
2)	$\frac{\pi}{2} \rightarrow \pi$	$0.63 \rightarrow 0.32$
3)	$\pi \rightarrow \frac{3\pi}{2}$	$0.32 \rightarrow 0.21$
4)	$\frac{3\pi}{2} \rightarrow 2\pi$	$0.21 \rightarrow 0.16$

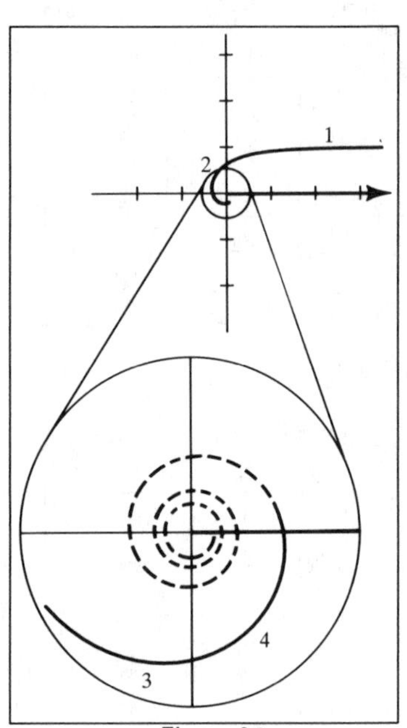

Figure 24

[25] Note that $r = 2\sec\theta$ is equivalent to $x = 2$. If $\theta \in (0, \frac{\pi}{2})$ or $\theta \in (\frac{3\pi}{2}, 2\pi)$, then $\sec\theta > 0$ and the graph of $r = 2 + 2\sec\theta$ is to the right of $x = 2$. If $\theta \in (\frac{\pi}{2}, \frac{3\pi}{2})$, $\sec\theta < 0$ and $r = 2 + 2\sec\theta$ is to the left of $x = 2$. r is undefined at $\theta = \frac{\pi}{2} + \pi n$. $0 = 2 + 2\sec\theta \Rightarrow \sec\theta = -1 \Rightarrow \theta = \pi + 2\pi n$.

	Range of θ	Range of r
1)	$0 \rightarrow \frac{\pi}{2}$	$4 \rightarrow \infty$
2)	$\frac{\pi}{2} \rightarrow \pi$	$-\infty \rightarrow 0$
3)	$\pi \rightarrow \frac{3\pi}{2}$	$0 \rightarrow -\infty$
4)	$\frac{3\pi}{2} \rightarrow 2\pi$	$\infty \rightarrow 4$

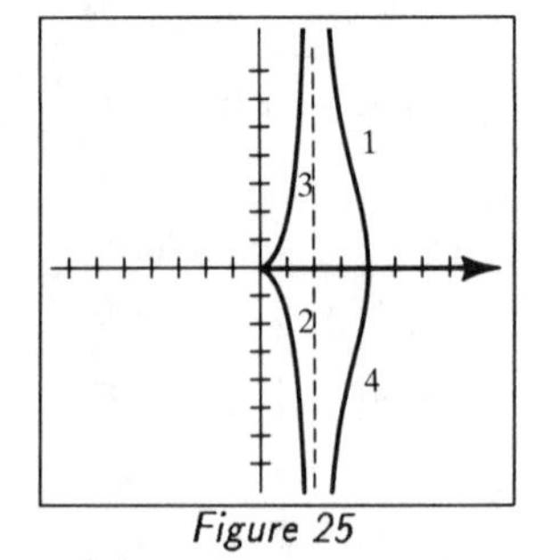

Figure 25

[26] Note that $r = -\csc\theta$ is equivalent to $y = -1$. If $\theta \in (0, \pi)$, then $\csc\theta > 0$ and the graph of $r = 1 - \csc\theta$ is above $y = -1$. If $\theta \in (\pi, 2\pi)$, $\csc\theta < 0$ and $r = 1 - \csc\theta$ is below $y = -1$. r is undefined at $\theta = \pi n$.

$0 = 1 - \csc\theta \Rightarrow \csc\theta = 1 \Rightarrow \theta = \frac{\pi}{2} + 2\pi n$. See *Figure 26*.

	Range of θ	Range of r
1)	$0 \to \frac{\pi}{2}$	$-\infty \to 0$
2)	$\frac{\pi}{2} \to \pi$	$0 \to -\infty$
3)	$\pi \to \frac{3\pi}{2}$	$\infty \to 2$
4)	$\frac{3\pi}{2} \to 2\pi$	$2 \to \infty$

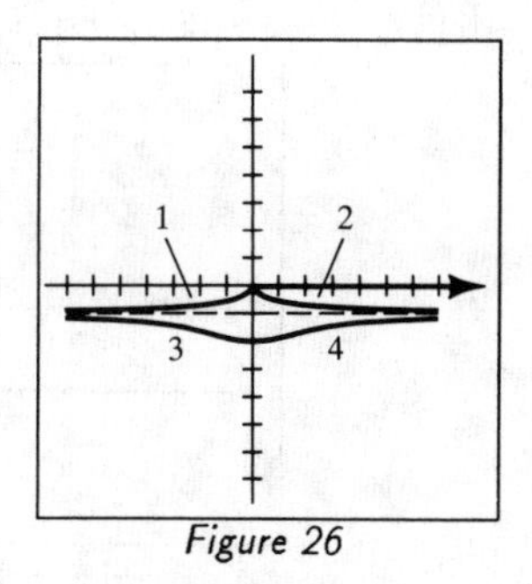

Figure 26

[27] $x = -3 \Rightarrow r\cos\theta = -3 \Rightarrow r = \dfrac{-3}{\cos\theta} \Rightarrow r = -3\sec\theta$

[28] $y = 2 \Rightarrow r\sin\theta = 2 \Rightarrow r = \dfrac{2}{\sin\theta} \Rightarrow r = 2\csc\theta$

[29] $x^2 + y^2 = 16 \Rightarrow r^2 = 16 \Rightarrow r = \pm 4$ {both are circles with radius 4}.

[30] $x^2 = 8y \Rightarrow r^2\cos^2\theta = 8r\sin\theta \Rightarrow r = \dfrac{8r\sin\theta}{r\cos^2\theta} = 8\cdot\dfrac{\sin\theta}{\cos\theta}\cdot\dfrac{1}{\cos\theta} = 8\tan\theta\sec\theta.$

[31] $2y = -x \Rightarrow \frac{y}{x} = -\frac{1}{2} \Rightarrow \tan\theta = -\frac{1}{2} \Rightarrow \theta = \tan^{-1}(-\frac{1}{2})$

[32] $y = 6x \Rightarrow \frac{y}{x} = 6 \Rightarrow \tan\theta = 6 \Rightarrow \theta = \tan^{-1}6$

[33] $y^2 - x^2 = 4 \Rightarrow r^2\sin^2\theta - r^2\cos^2\theta = 4 \Rightarrow -r^2(\cos^2\theta - \sin^2\theta) = 4 \Rightarrow$

$$-r^2\cos 2\theta = 4 \Rightarrow r^2 = \frac{-4}{\cos 2\theta} \Rightarrow r^2 = -4\sec 2\theta$$

[34] $xy = 8 \Rightarrow (r\cos\theta)(r\sin\theta) = 8 \Rightarrow r^2(\frac{1}{2})(2\sin\theta\cos\theta) = 8 \Rightarrow r^2\sin 2\theta = 16 \Rightarrow$

$$r^2 = \frac{16}{\sin 2\theta} \Rightarrow r^2 = 16\csc 2\theta$$

[35] $(x^2 + y^2)\tan^{-1}(\frac{y}{x}) = ay \Rightarrow r^2\theta = ar\sin\theta \Rightarrow r\theta = a\sin\theta$ or $r = 0 \Rightarrow r\theta = a\sin\theta$

since $\theta = 0$ gives $r = 0$ as one solution. Note that $\theta = \tan^{-1}(\frac{y}{x}) \Rightarrow \theta \in (-\frac{\pi}{2}, \frac{\pi}{2})$.

[36] $x^3 + y^3 - 3axy = 0 \Rightarrow r^3\cos^3\theta + r^3\sin^3\theta - 3ar^2\cos\theta\sin\theta = 0 \Rightarrow$

$r\cos^3\theta + r\sin^3\theta - 3a\cos\theta\sin\theta = 0$ or $r = 0 \Rightarrow$

$$r = \frac{3a\cos\theta\sin\theta}{\cos^3\theta + \sin^3\theta} \text{ since } \theta = 0 \text{ gives } r = 0.$$

[37] $r\cos\theta = 5 \Rightarrow x = 5.$

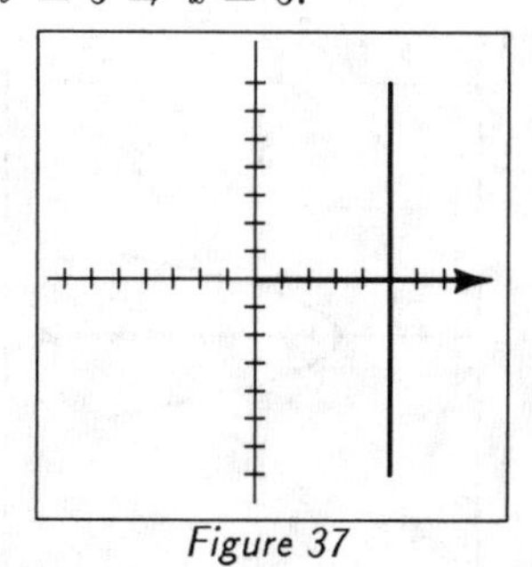
Figure 37

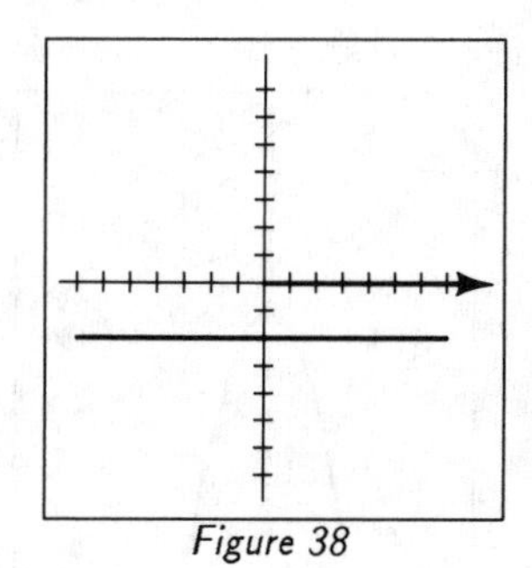
Figure 38

[38] $r\sin\theta = -2 \Rightarrow y = -2.$

[39] $r = -3\csc\theta \Rightarrow r\sin\theta = -3 \Rightarrow y = -3$. θ is undefined at πn.

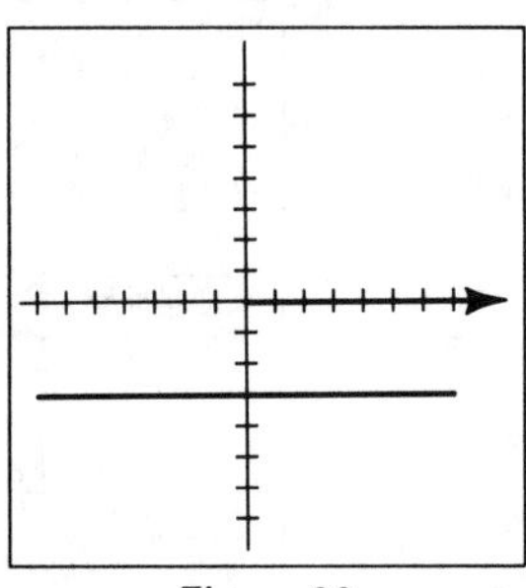

Figure 39

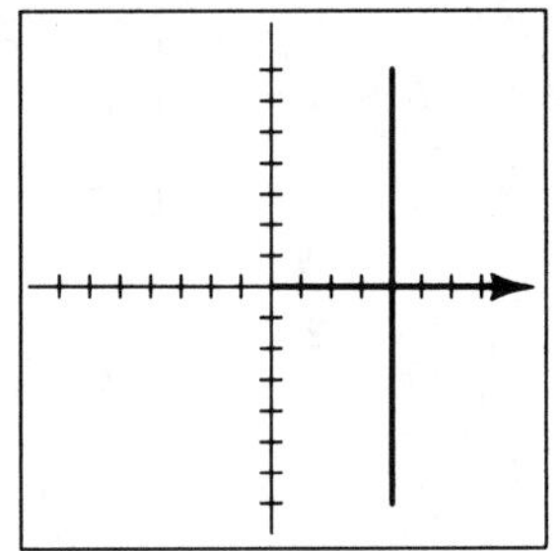

Figure 40

[40] $r = 4\sec\theta \Rightarrow r\cos\theta = 4 \Rightarrow x = 4$. θ is undefined at $\frac{\pi}{2} + \pi n$.

[41] $r^2\cos 2\theta = 1 \Rightarrow r^2(\cos^2\theta - \sin^2\theta) = 1 \Rightarrow r^2\cos^2\theta - r^2\sin^2\theta = 1 \Rightarrow x^2 - y^2 = 1$.

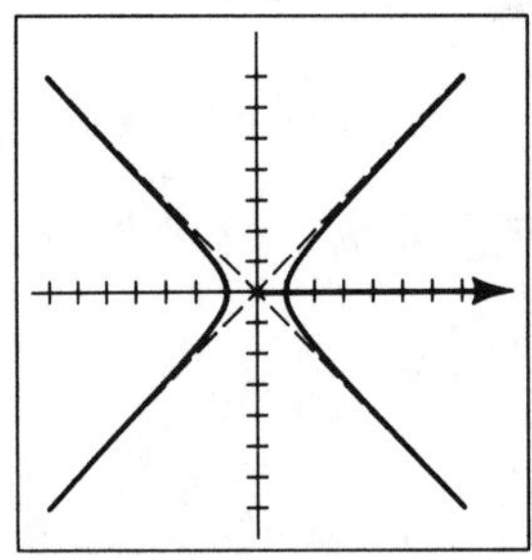

Figure 41

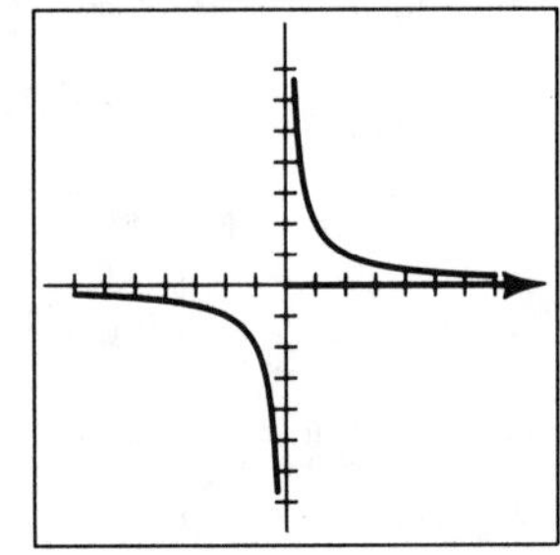

Figure 42

[42] $r^2\sin 2\theta = 4 \Rightarrow r^2(2\sin\theta\cos\theta) = 4 \Rightarrow (r\sin\theta)(r\cos\theta) = 2 \Rightarrow xy = 2$.

[43] $r(\sin\theta - 2\cos\theta) = 6 \Rightarrow r\sin\theta - 2r\cos\theta = 6 \Rightarrow y - 2x = 6$.

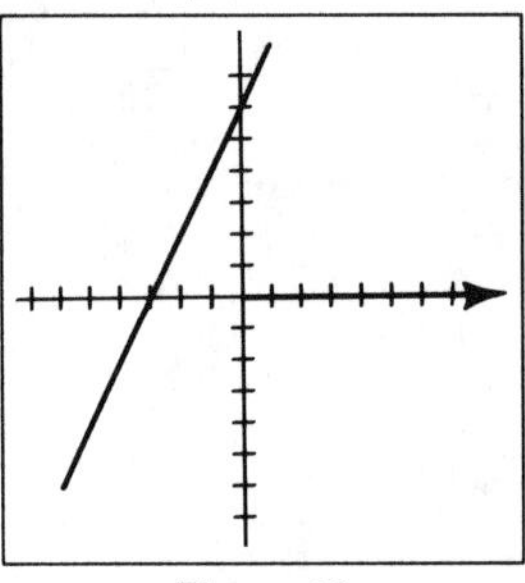

Figure 43

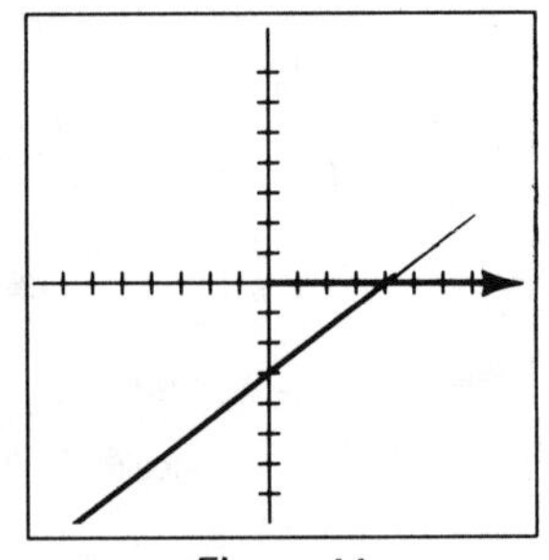

Figure 44

[44] $r(3\cos\theta - 4\sin\theta) = 12 \Rightarrow 3r\cos\theta - 4r\sin\theta = 12 \Rightarrow 3x - 4y = 12$.

[45] $r(\sin\theta + r\cos^2\theta) = 1 \Rightarrow r\sin\theta + r^2\cos^2\theta = 1 \Rightarrow y + x^2 = 1$ or $y = -x^2 + 1$.

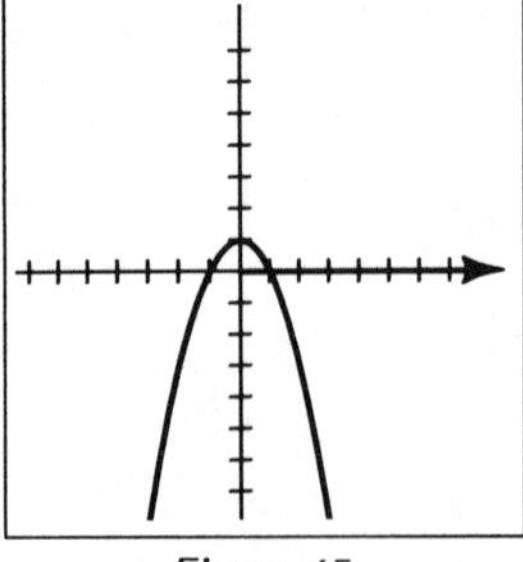

Figure 45

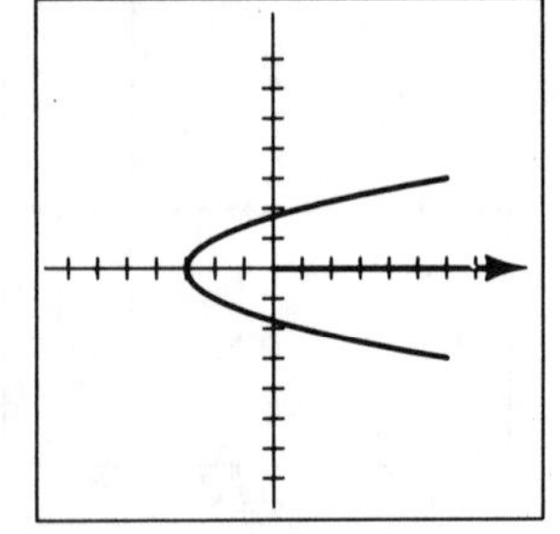

Figure 46

[46] $r(r\sin^2\theta - \cos\theta) = 3 \Rightarrow r^2\sin^2\theta - r\cos\theta = 3 \Rightarrow y^2 - x = 3 \Rightarrow x = y^2 - 3.$

[47] $r = 8\sin\theta - 2\cos\theta \Rightarrow r^2 = 8r\sin\theta - 2r\cos\theta \Rightarrow x^2 + y^2 = 8y - 2x \Rightarrow$

$$x^2 + 2x + \underline{1} + y^2 - 8y + \underline{16} = \underline{1} + \underline{16} \Rightarrow (x+1)^2 + (y-4)^2 = 17.$$

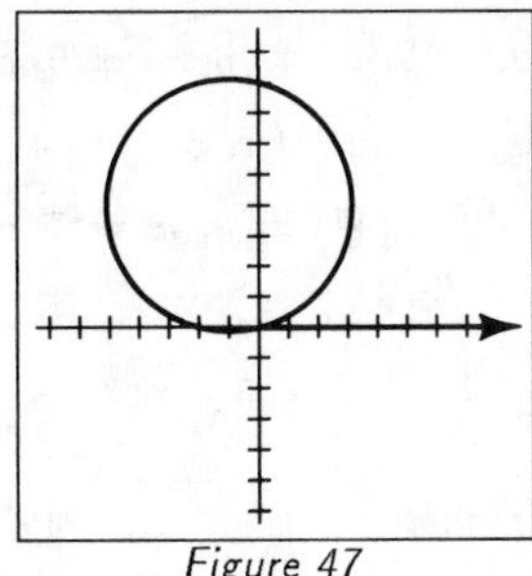
Figure 47

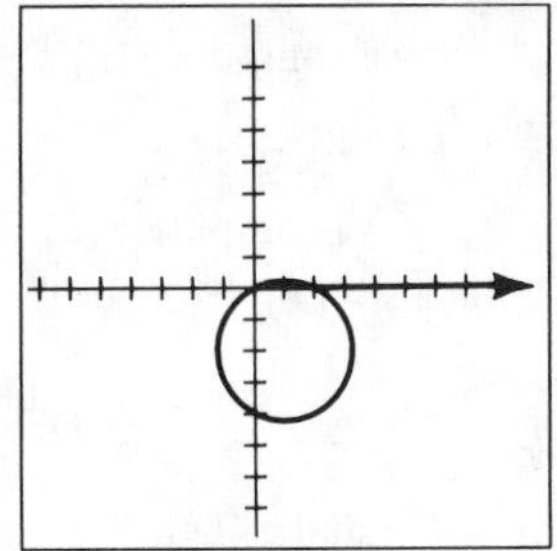
Figure 48

[48] $r = 2\cos\theta - 4\sin\theta \Rightarrow r^2 = 2r\cos\theta - 4r\sin\theta \Rightarrow x^2 + y^2 = 2x - 4y \Rightarrow$

$$x^2 - 2x + \underline{1} + y^2 + 4y + \underline{4} = \underline{1} + \underline{4} \Rightarrow (x-1)^2 + (y+2)^2 = 5.$$

[49] $r = \tan\theta \Rightarrow r^2 = \tan^2\theta \Rightarrow x^2 + y^2 = \dfrac{y^2}{x^2} \Rightarrow x^4 + x^2y^2 = y^2 \Rightarrow$

$$y^2 - x^2y^2 = x^4 \Rightarrow y^2(1 - x^2) = x^4 \Rightarrow y^2 = \frac{x^4}{1 - x^2}$$

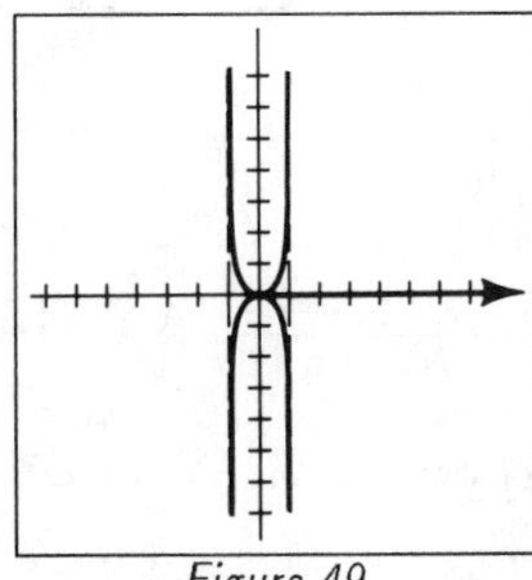
Figure 49

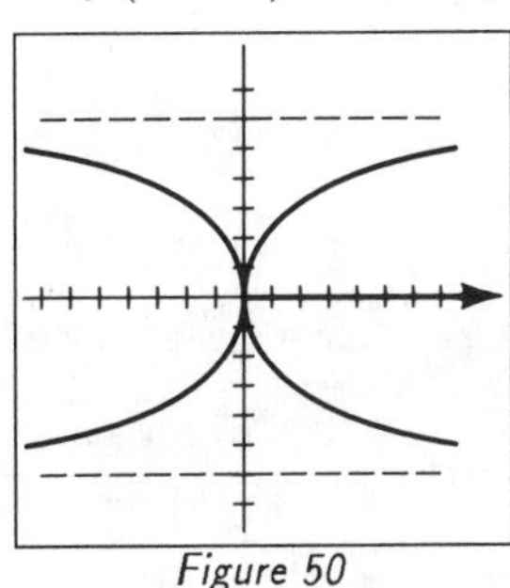
Figure 50

[50] $r = 6\cot\theta \Rightarrow r^2 = 36\cot^2\theta \Rightarrow x^2 + y^2 = 36\left(\dfrac{x^2}{y^2}\right) \Rightarrow x^2y^2 + y^4 = 36x^2 \Rightarrow$

$$36x^2 - x^2y^2 = y^4 \Rightarrow x^2(36 - y^2) = y^4 \Rightarrow x^2 = \frac{y^4}{36 - y^2}$$

Note: Exer. 51–60: With $r = f(\theta)$ and θ given, we will list r', r' evaluated at θ, and r evaluated at θ. Substitute these into the formula $m = \dfrac{r'\sin\theta + r\cos\theta}{r'\cos\theta - r\sin\theta}$ and simplify to obtain the value of m given.

[51] $r' = -2\sin\theta,\ r'(\frac{\pi}{3}) = -\sqrt{3},\ r(\frac{\pi}{3}) = 1,\ m = \dfrac{(-\sqrt{3})(\sqrt{3}/2) + (1)(1/2)}{(-\sqrt{3})(1/2) - (1)(\sqrt{3}/2)} = \sqrt{3}/3$

[52] $r' = -2\cos\theta,\ r'(\frac{\pi}{6}) = -\sqrt{3},\ r(\frac{\pi}{6}) = -1,\ m = \dfrac{(-\sqrt{3})(1/2) + (-1)(\sqrt{3}/2)}{(-\sqrt{3})(\sqrt{3}/2) - (-1)(1/2)} = \sqrt{3}$

[53] $r' = -4\cos\theta,\ r'(0) = -4,\ r(0) = 4,\ m = \dfrac{(-4)(0) + (4)(1)}{(-4)(1) - (4)(0)} = -1$

[54] $r' = -2\sin\theta,\ r'(\frac{\pi}{2}) = -2,\ r(\frac{\pi}{2}) = 1,\ m = \dfrac{(-2)(1) + (1)(0)}{(-2)(0) - (1)(1)} = 2$

55 $r' = -24\sin 3\theta$, $r'(\frac{\pi}{4}) = -12\sqrt{2}$, $r(\frac{\pi}{4}) = -4\sqrt{2}$, $m = 2$

56 $r' = 8\cos 4\theta$, $r'(\frac{\pi}{4}) = -8$, $r(\frac{\pi}{4}) = 0$, $m = 1$

57 $2rr' = -8\sin 2\theta \Rightarrow r' = \dfrac{-4\sin 2\theta}{r}$; {since r' is in terms of r, we compute $r(\theta)$ first} $r(\frac{\pi}{6}) = \pm\sqrt{2}$, $r'(\frac{\pi}{6}) = \mp\sqrt{6}$, $m = 0$ in either case

58 $2rr' = -4\cos 2\theta \Rightarrow r' = \dfrac{-2\cos 2\theta}{r}$; {since r' is in terms of r, we compute $r(\theta)$ first} $r(\frac{3\pi}{4}) = \pm\sqrt{2}$, $r'(\frac{3\pi}{4}) = 0$, $m = 1$ in either case

59 $r' = 2^{\theta}(\ln 2)$, $r'(\pi) = 2^{\pi}(\ln 2)$, $r(\pi) = 2^{\pi}$, $m = 1/\ln 2$

60 $r = \dfrac{1}{\theta} \Rightarrow r' = -\dfrac{1}{\theta^2}$, $r'(2\pi) = -\dfrac{1}{4\pi^2}$, $r(2\pi) = \dfrac{1}{2\pi}$, $m = -2\pi$

61 Let $P_1(r_1, \theta_1)$ and $P_2(r_2, \theta_2)$ be points in an $r\theta$-plane. Let $a = r_1$, $b = r_2$, $c = d(P_1, P_2)$, and $\gamma = \theta_2 - \theta_1$. Substituting into the law of cosines, $c^2 = a^2 + b^2 - 2ab\cos\gamma$, gives us the formula.

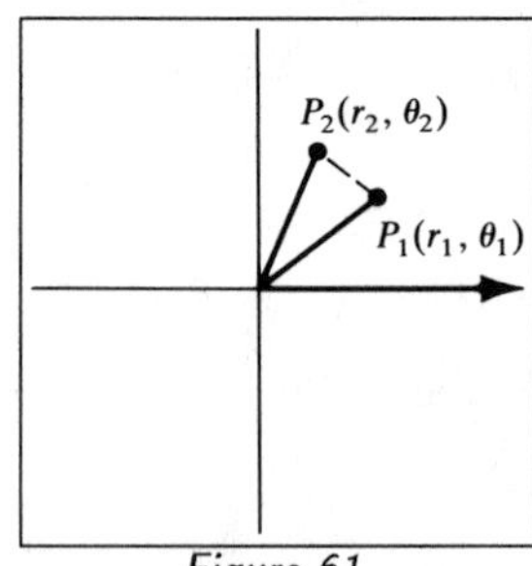

Figure 61

62 $r = a\sin\theta + b\cos\theta \Rightarrow r^2 = ar\sin\theta + br\cos\theta \Rightarrow x^2 - bx + y^2 - ay = 0 \Rightarrow$
$(x - \frac{1}{2}b)^2 + (y - \frac{1}{2}a)^2 = \frac{1}{4}b^2 + \frac{1}{4}a^2$. $C(\frac{1}{2}b, \frac{1}{2}a)$; $r = \frac{1}{2}\sqrt{b^2 + a^2}$

63 The slope of one tangent line is $m_1 = \dfrac{f'(\theta)\sin\theta + f(\theta)\cos\theta}{f'(\theta)\cos\theta - f(\theta)\sin\theta} = \dfrac{\text{A}}{\text{B}}$ and the slope of the other is $m_2 = \dfrac{g'(\theta)\sin\theta + g(\theta)\cos\theta}{g'(\theta)\cos\theta - g(\theta)\sin\theta} = \dfrac{\text{C}}{\text{D}}$. Now, $m_1 \cdot m_2 = -1 \Leftrightarrow$

$\dfrac{\text{AC}}{\text{BD}} = -1 \Leftrightarrow \dfrac{\text{AC} + \text{BD}}{\text{BD}} = 0 \Leftrightarrow \text{AC} + \text{BD} = 0 \Leftrightarrow$

$f'(\theta)\,g'(\theta)\left[\sin^2\theta + \cos^2\theta\right] + f(\theta)\,g(\theta)\left[\sin^2\theta + \cos^2\theta\right] = 0 \Leftrightarrow$

$$f'(\theta)\,g'(\theta) + f(\theta)\,g(\theta) = 0.$$

64 (a) $f(\theta) = a\sin\theta \Rightarrow f'(\theta) = a\cos\theta$. $g(\theta) = a\cos\theta \Rightarrow g'(\theta) = -a\sin\theta$.
$f'(\theta)\,g'(\theta) + f(\theta)\,g(\theta) = -a^2\sin\theta\cos\theta + a^2\sin\theta\cos\theta = 0$, for any point of intersection.

(b) $f(\theta) = a\theta \Rightarrow f'(\theta) = a$. $g(\theta) = \frac{a}{\theta} \Rightarrow g'(\theta) = -\frac{a}{\theta^2}$. $f'(\theta)\,g'(\theta) + f(\theta)\,g(\theta) = -\frac{a^2}{\theta^2} + a^2$. At a point of intersection, $a\theta = \frac{a}{\theta} \Rightarrow \theta^2 = 1$, so $-\frac{a^2}{\theta^2} + a^2 = 0$.

65 Dividing the numerator and denominator by $\cos\theta$ in (13.10) gives the desired result.

66 (a) $A(r, \theta)$ and $B(r, \theta + \frac{\pi}{2})$ have rectangular coordinates $(r\cos\theta, r\sin\theta)$ and $(r\cos[\theta + \frac{\pi}{2}], r\sin[\theta + \frac{\pi}{2}])$, respectively. Using the slope formula,

$$m = \frac{r\sin(\theta + \frac{\pi}{2}) - r\sin\theta}{r\cos(\theta + \frac{\pi}{2}) - r\cos\theta} = \frac{\cos\theta - \sin\theta}{-\sin\theta - \cos\theta} = \frac{\sin\theta - \cos\theta}{\sin\theta + \cos\theta}.$$

(b) $m = \dfrac{r'\tan\theta + r}{r' - r\tan\theta}$ {Exer. 65} and $m = \dfrac{\sin\theta - \cos\theta}{\sin\theta + \cos\theta}$ {part (a)} $= \dfrac{\tan\theta - 1}{\tan\theta + 1} \Rightarrow$

$(r'\tan\theta + r)(\tan\theta + 1) = (r' - r\tan\theta)(\tan\theta - 1) \Rightarrow$

$$r'(\tan^2\theta + 1) = -r(\tan^2\theta + 1) \Rightarrow r' = -r$$

(c) $\frac{dr}{d\theta} = -r \Rightarrow \frac{dr}{r} = -d\theta \Rightarrow \ln r = -\theta + K \Rightarrow r = e^{-\theta + K} = e^K e^{-\theta}$,

where K is a constant. Let the initial position of bug A be $(r_0, \frac{\pi}{4})$.

Then, $r_0 = e^K e^{-\pi/4} \Rightarrow e^K = r_0 e^{\pi/4}$. Thus, $r = r_0 e^{\pi/4} e^{-\theta}$,

which is a logarithmic spiral with $a = r_0 e^{\pi/4}$ and $b = -1$.

[67] The graph is symmetric with respect to the polar axis.

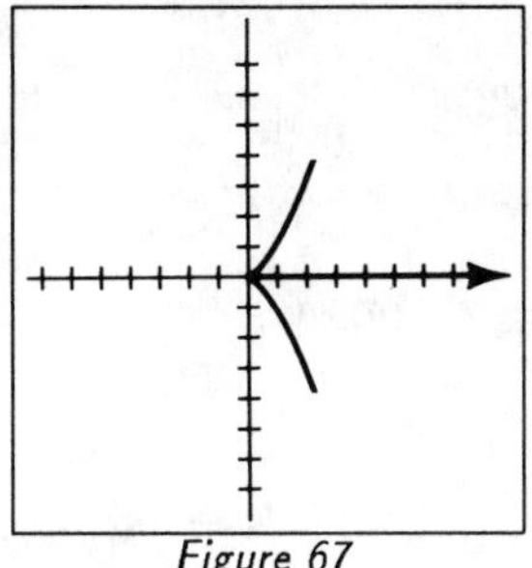
Figure 67

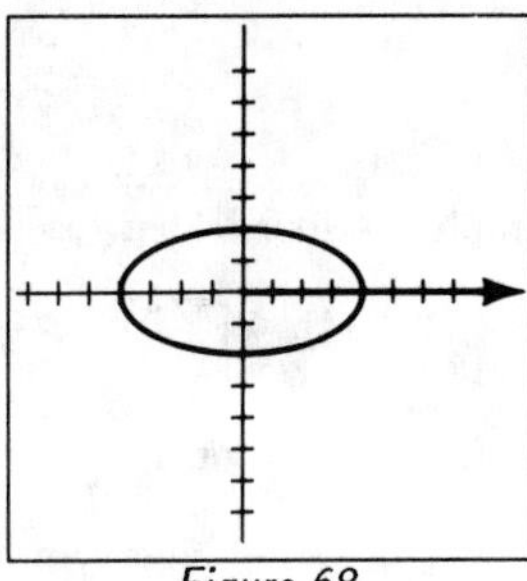
Figure 68

[68] The graph is symmetric with respect to the polar axis, the line $\theta = \frac{\pi}{2}$, and the pole.

[69] From the graph, there are six points of intersection. The approximate polar coordinates are $(1.75, \pm 0.45)$, $(4.49, \pm 1.77)$, and $(5.76, \pm 2.35)$.

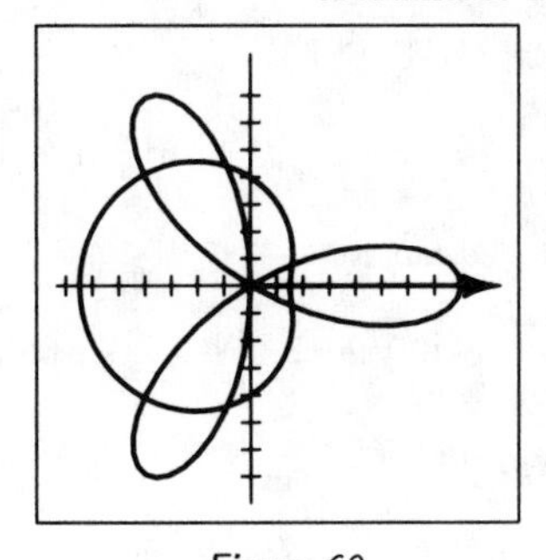
Figure 69

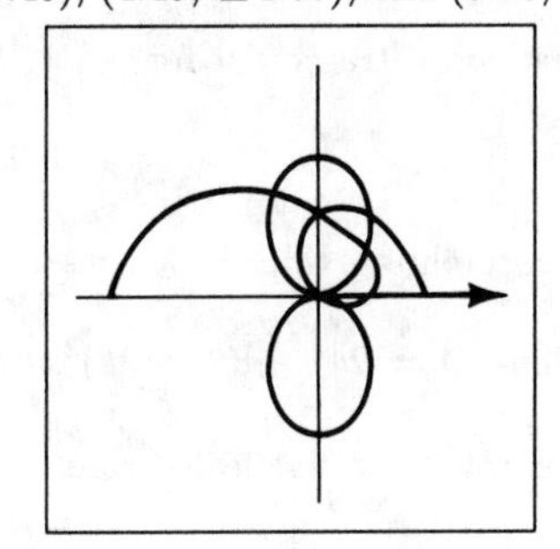
Figure 70

[70] From the graph, there are five points of intersection. The approximate polar coordinates are $(0, 0)$, $(0.32, -0.41)$, $(0.96, 0.77)$, $(1.39, 0.99)$, and $(1.64, 2.01)$.

Be sure to plot $r = \frac{3}{4}(\theta + \cos^2\theta)$ for both positive and negative values of θ.

Figure 70 was plotted using $-\pi \leq \theta \leq \pi$.

Exercises 13.4

[1] $A = 2\int_0^{\pi/2} \frac{1}{2}(2\cos\theta)^2\,d\theta = 2\int_0^{\pi/2}(1 + \cos 2\theta)\,d\theta = 2\Big[\theta + \frac{1}{2}\sin 2\theta\Big]_0^{\pi/2} = \pi$

[2] $A = 2\int_0^{\pi/2} \frac{1}{2}(5\sin\theta)^2\,d\theta = 25\int_0^{\pi/2} \frac{1}{2}(1 - \cos 2\theta)\,d\theta = \frac{25}{2}\Big[\theta - \frac{1}{2}\sin 2\theta\Big]_0^{\pi/2} = \frac{25\pi}{4}$

[3] $A = 2\int_0^{\pi} \frac{1}{2}(1 - \cos\theta)^2\,d\theta = \int_0^{\pi}(1 - 2\cos\theta + \cos^2\theta)\,d\theta =$

$$\int_0^{\pi}\left(\tfrac{3}{2} - 2\cos\theta + \tfrac{1}{2}\cos 2\theta\right)d\theta = \Big[\tfrac{3}{2}\theta - 2\sin\theta + \tfrac{1}{4}\sin 2\theta\Big]_0^{\pi} = \tfrac{3\pi}{2}$$

[4] $A = 2\int_{-\pi/2}^{\pi/2} \frac{1}{2}(6 - 6\sin\theta)^2\,d\theta = 36\int_{-\pi/2}^{\pi/2}(1 - 2\sin\theta + \sin^2\theta)\,d\theta =$

$$72\int_0^{\pi/2}\left(\tfrac{3}{2} - \tfrac{1}{2}\cos 2\theta\right)d\theta = 36\Big[3\theta - \tfrac{1}{2}\sin 2\theta\Big]_0^{\pi/2} = 54\pi$$

[5] The graph is a four-leafed rose. Using symmetry,

$$A = 4\int_0^{\pi/2} \tfrac{1}{2}\sin^2 2\theta\,d\theta = \int_0^{\pi/2}(1 - \cos 4\theta)\,d\theta = \Big[\theta - \tfrac{1}{4}\sin 4\theta\Big]_0^{\pi/2} = \tfrac{\pi}{2}.$$

[6] See Exer. 19, §13.3 for a similar graph.

$$\text{Using symmetry, } A = 4\int_0^{\pi/4} \tfrac{1}{2}(9\cos 2\theta)\,d\theta = 18\Big[\tfrac{1}{2}\sin 2\theta\Big]_0^{\pi/4} = 9.$$

[7] $A = \int_0^{\pi/2} \frac{1}{2}e^{2\theta}\,d\theta = \Big[\frac{1}{4}e^{2\theta}\Big]_0^{\pi/2} = \frac{1}{4}(e^{\pi} - 1) \approx 5.54$

[8] $A = \int_0^{\pi} \frac{1}{2}(2\theta)^2\,d\theta = \Big[\frac{2}{3}\theta^3\Big]_0^{\pi} = \frac{2}{3}\pi^3 \approx 20.67$

[9] One loop is traced out for $-\frac{\pi}{4} \le \theta \le \frac{\pi}{4}$ {see Exer. 19, §13.3}.

$$\text{Thus, } A = 2\int_0^{\pi/4} \tfrac{1}{2}(4\cos 2\theta)\,d\theta = 4\Big[\tfrac{1}{2}\sin 2\theta\Big]_0^{\pi/4} = 2.$$

[10] The graph is a three-leafed rose. One leaf is traced out for $-\frac{\pi}{6} \le \theta \le \frac{\pi}{6}$.

$$\text{Thus, } A = 2\int_0^{\pi/6} \tfrac{1}{2}(2\cos 3\theta)^2\,d\theta = 2\int_0^{\pi/6}(1 + \cos 6\theta)\,d\theta = 2\Big[\theta + \tfrac{1}{6}\sin 6\theta\Big]_0^{\pi/6} = \tfrac{\pi}{3}.$$

[11] The graph is a five-leafed rose. One leaf is traced out for $-\frac{\pi}{10} \le \theta \le \frac{\pi}{10}$. Thus,

$$A = 2\int_0^{\pi/10} \tfrac{1}{2}(3\cos 5\theta)^2\,d\theta = \tfrac{9}{2}\int_0^{\pi/10}(1 + \cos 10\theta)\,d\theta = \tfrac{9}{2}\Big[\theta + \tfrac{1}{10}\sin 10\theta\Big]_0^{\pi/10} = \tfrac{9\pi}{20}.$$

[12] The graph is a 12-leafed rose. One leaf is traced out for $0 \le \theta \le \frac{\pi}{6}$.

$$\text{Thus, } A = \int_0^{\pi/6} \tfrac{1}{2}(\sin 6\theta)^2\,d\theta = \tfrac{1}{4}\int_0^{\pi/6}(1 - \cos 12\theta)\,d\theta = \tfrac{1}{4}\Big[\theta - \tfrac{1}{12}\sin 12\theta\Big]_0^{\pi/6} = \tfrac{\pi}{24}.$$

[13] $x = 4 \Leftrightarrow r\cos\theta = 4 \Leftrightarrow r = 4\sec\theta$. $x^2 + y^2 = 25 \Leftrightarrow r = 5$.

If $0 \le \theta \le \arctan\frac{3}{4}$, then $r = 4\sec\theta$. If $\arctan\frac{3}{4} \le \theta \le \frac{\pi}{2}$, then $r = 5$.

$$A = \int_0^{\arctan(3/4)} \tfrac{1}{2}(4\sec\theta)^2\,d\theta + \int_{\arctan(3/4)}^{\pi/2} \tfrac{1}{2}(5)^2\,d\theta.$$

[14] $y = 2x - 3 \Rightarrow r\sin\theta = 2r\cos\theta - 3 \Rightarrow r(2\cos\theta - \sin\theta) = 3 \Rightarrow$

$r = \dfrac{3}{2\cos\theta - \sin\theta}$. $x^2 + y^2 = 18 \Leftrightarrow r = \sqrt{18}$.

If $-\frac{\pi}{2} \leq \theta \leq \frac{\pi}{4}$, then $r = \dfrac{3}{2\cos\theta - \sin\theta}$. If $\frac{\pi}{4} \leq \theta \leq \frac{\pi}{2}$, then $r = \sqrt{18}$.

$$A = \int_{-\pi/2}^{\pi/4} \frac{1}{2}\left(\frac{3}{2\cos\theta - \sin\theta}\right)^2 d\theta + \int_{\pi/4}^{\pi/2} \tfrac{1}{2}(\sqrt{18})^2\, d\theta.$$

[15] $x^2 + y^2 = 4 \Leftrightarrow r = 2$. $y = x \Rightarrow \frac{y}{x} = 1 \Rightarrow \tan^{-1}\frac{y}{x} = \tan^{-1}1 \Rightarrow \theta = \frac{\pi}{4}$.

$y = 4 \Leftrightarrow r\sin\theta = 4 \Leftrightarrow r = 4\csc\theta$. $y = 3x \Rightarrow \theta = \arctan 3$.

If $\frac{\pi}{4} \leq \theta \leq \arctan 3$, then the outer radius is $4\csc\theta$ and the inner radius is 2.

$$A = \int_{\pi/4}^{\arctan 3} \tfrac{1}{2}\left[(4\csc\theta)^2 - (2)^2\right] d\theta.$$

[16] $y = \frac{1}{3}x \Rightarrow \frac{y}{x} = \frac{1}{3} \Rightarrow \theta = \arctan\frac{1}{3}$. $(x - 2)^2 + y^2 = 4 \Leftrightarrow r = 4\cos\theta$.

If $0 \leq \theta \leq \arctan\frac{1}{3}$, then $r = 4\cos\theta$.

Using symmetry, $A = 2\displaystyle\int_0^{\arctan(1/3)} \tfrac{1}{2}(4\cos\theta)^2\, d\theta$.

[17] (a) Consider only the blue region in the first quadrant bounded below by the polar axis. $4\cos 2\theta = 2 \Rightarrow \cos 2\theta = \frac{1}{2} \Rightarrow 2\theta = \frac{\pi}{3} \Rightarrow \theta = \frac{\pi}{6}$.

If $0 \leq \theta \leq \frac{\pi}{6}$, then the outer radius is $4\cos 2\theta$ and the inner radius is 2.

Using symmetry, $A = 8\displaystyle\int_0^{\pi/6} \tfrac{1}{2}\left[(4\cos 2\theta)^2 - (2)^2\right] d\theta$.

(b) Consider only the green region in the first quadrant bounded below by the polar axis. If $0 \leq \theta \leq \frac{\pi}{6}$, then $r = 2$. If $\frac{\pi}{6} \leq \theta \leq \frac{\pi}{4}$, then $r = 4\cos 2\theta$.

Using symmetry, $A = 8\left[\displaystyle\int_0^{\pi/6} \tfrac{1}{2}(2)^2\, d\theta + \int_{\pi/6}^{\pi/4} \tfrac{1}{2}(4\cos 2\theta)^2\, d\theta\right]$.

[18] (a) Consider the uppermost blue region. $6\cos\theta = 2 + 2\cos\theta \Rightarrow \cos\theta = \frac{1}{2} \Rightarrow$ $\theta = \frac{\pi}{3}$. If $\frac{\pi}{3} \leq \theta \leq \frac{\pi}{2}$, then the outer radius is $2 + 2\cos\theta$ and the inner radius is $6\cos\theta$. If $\frac{\pi}{2} \leq \theta \leq \pi$, then $r = 2 + 2\cos\theta$ (the region to the left of the $\theta = \frac{\pi}{2}$ axis). Combining integrals and using symmetry,

$$A = 2\left[\int_{\pi/3}^{\pi} \tfrac{1}{2}(2 + 2\cos\theta)^2\, d\theta - \int_{\pi/3}^{\pi/2} \tfrac{1}{2}(6\cos\theta)^2\, d\theta\right].$$

(b) Consider the uppermost green region. If $0 \leq \theta \leq \frac{\pi}{3}$, then $r = 2 + 2\cos\theta$. If $\frac{\pi}{3} \leq \theta \leq \frac{\pi}{2}$, then $r = 6\cos\theta$.

Using symmetry, $A = 2\left[\displaystyle\int_0^{\pi/3} \tfrac{1}{2}(2 + 2\cos\theta)^2\, d\theta + \int_{\pi/3}^{\pi/2} \tfrac{1}{2}(6\cos\theta)^2\, d\theta\right]$.

[19] See Figure 13.35. Using symmetry, $A = 2\displaystyle\int_{\pi/3}^{\pi} \tfrac{1}{2}\left[3^2 - (2 + 2\cos\theta)^2\right] d\theta =$

$$\int_{\pi/3}^{\pi} (3 - 8\cos\theta - 2\cos 2\theta)\, d\theta = \Big[3\theta - 8\sin\theta - \sin 2\theta\Big]_{\pi/3}^{\pi} = 2\pi + \tfrac{9}{2}\sqrt{3} \approx 14.08.$$

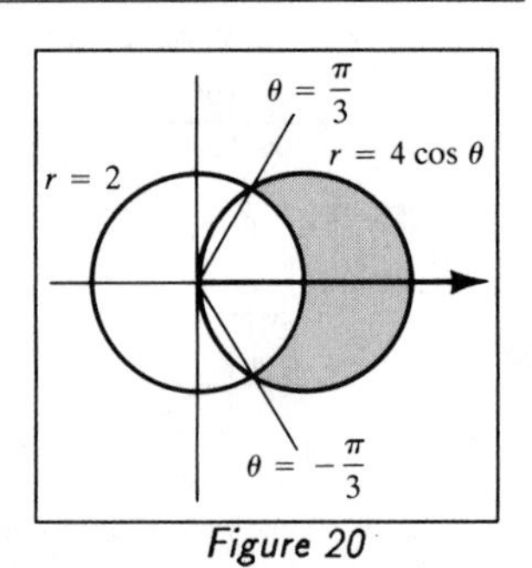

Figure 20

[20] $2 = 4\cos\theta \Rightarrow \cos\theta = \frac{1}{2}$ at $\theta = \pm\frac{\pi}{3}$.

$$A = 2\int_0^{\pi/3} \tfrac{1}{2}\Big[(4\cos\theta)^2 - 2^2\Big]\,d\theta$$
$$= 4\int_0^{\pi/3} (1 + 2\cos 2\theta)\,d\theta$$
$$= 4\Big[\theta + \sin 2\theta\Big]_0^{\pi/3} = \tfrac{4\pi}{3} + 2\sqrt{3} \approx 7.65$$

[21] $2^2 = 8\sin 2\theta \Rightarrow \sin 2\theta = \frac{1}{2}$ at $2\theta = \frac{\pi}{6}, \frac{5\pi}{6}, \frac{13\pi}{6}, \frac{17\pi}{6} \Rightarrow \theta = \frac{\pi}{12}, \frac{5\pi}{12}, \frac{13\pi}{12}, \frac{17\pi}{12}$.

Using symmetry,

$$A = 4\int_{\pi/12}^{\pi/4} \tfrac{1}{2}(8\sin 2\theta - 2^2)\,d\theta = 2\Big[-4\cos 2\theta - 4\theta\Big]_{\pi/12}^{\pi/4} = 4\sqrt{3} - \tfrac{4\pi}{3} \approx 2.74.$$

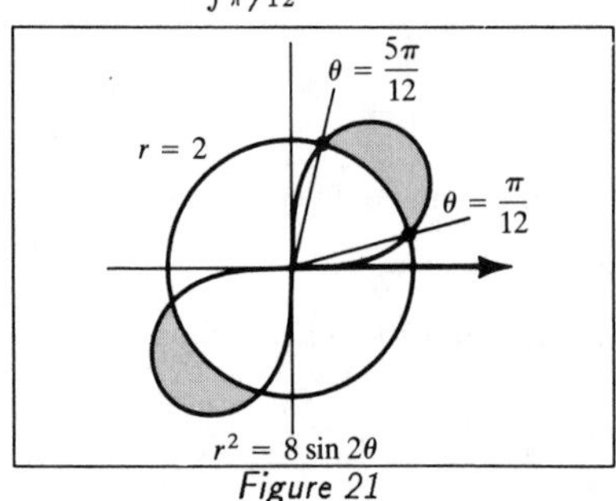

Figure 21

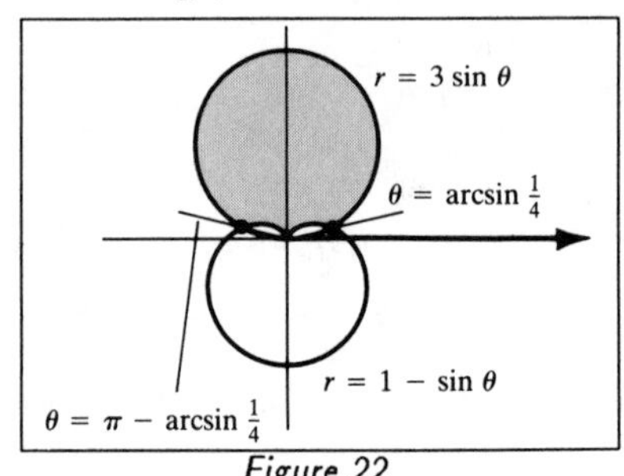

Figure 22

[22] $1 - \sin\theta = 3\sin\theta \Rightarrow \sin\theta = \frac{1}{4}$ at $\theta = \arcsin\frac{1}{4}$ {call this α} and $(\pi - \alpha)$.

Using symmetry, $A = 2\int_\alpha^{\pi/2} \frac{1}{2}\Big[(3\sin\theta)^2 - (1 - \sin\theta)^2\Big]\,d\theta =$

$$\int_\alpha^{\pi/2} (2\sin\theta - 4\cos 2\theta + 3)\,d\theta = \Big[-2\cos\theta - 2\sin 2\theta + 3\theta\Big]_\alpha^{\pi/2} =$$

$$\Big[(\tfrac{3\pi}{2}) - (-\tfrac{1}{2}\sqrt{15} - \tfrac{1}{4}\sqrt{15} + 3\alpha)\Big] = \tfrac{3\pi}{2} + \tfrac{3}{4}\sqrt{15} - 3\arcsin\tfrac{1}{4} \approx 6.86.$$

[23] $\sin\theta = \sqrt{3}\cos\theta \Rightarrow \tan\theta = \sqrt{3}$ at $\theta = \frac{\pi}{3}$. For $0 \le \theta \le \frac{\pi}{3}$, the boundary is $r = \sin\theta$ and for $\frac{\pi}{3} \le \theta \le \frac{\pi}{2}$, the boundary is $r = \sqrt{3}\cos\theta$. Thus,

$$A = \tfrac{1}{2}\int_0^{\pi/3} \sin^2\theta\,d\theta + \tfrac{1}{2}\int_{\pi/3}^{\pi/2} 3\cos^2\theta\,d\theta$$
$$= \tfrac{1}{4}\Big[\theta - \tfrac{1}{2}\sin 2\theta\Big]_0^{\pi/3} + \tfrac{3}{4}\Big[\theta + \tfrac{1}{2}\sin 2\theta\Big]_{\pi/3}^{\pi/2}$$
$$= \tfrac{1}{4}\Big[\tfrac{\pi}{3} - \tfrac{\sqrt{3}}{4}\Big] + \tfrac{3}{4}\Big[\tfrac{\pi}{2} - (\tfrac{\pi}{3} + \tfrac{\sqrt{3}}{4})\Big] = \tfrac{5\pi}{24} - \tfrac{\sqrt{3}}{4} \approx 0.22.$$

See *Figure 23*.

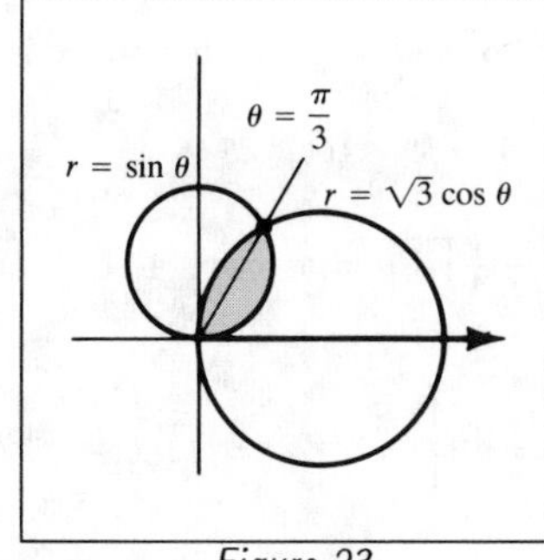

Figure 23

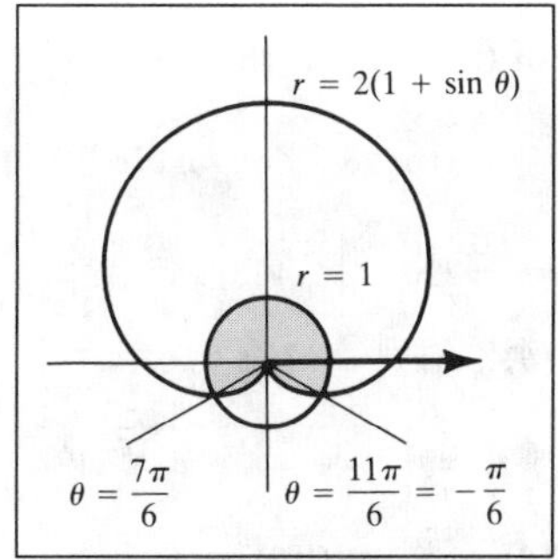

Figure 24

[24] $2(1 + \sin\theta) = 1 \Rightarrow \sin\theta = -\frac{1}{2}$ at $\theta = \frac{7\pi}{6}, \frac{11\pi}{6}$. For $-\frac{\pi}{2} \leq \theta \leq -\frac{\pi}{6}$, the boundary is $r = 2(1 + \sin\theta)$ and for $-\frac{\pi}{6} \leq \theta \leq \frac{\pi}{2}$, the boundary is $r = 1$. Using symmetry,

$$A = 2\int_{-\pi/2}^{-\pi/6} \tfrac{1}{2}(2^2)(1 + \sin\theta)^2\, d\theta + 2\int_{-\pi/6}^{\pi/2} \tfrac{1}{2}(1)^2\, d\theta$$

$$= 4\int_{-\pi/2}^{-\pi/6} (\tfrac{3}{2} + 2\sin\theta - \tfrac{1}{2}\cos 2\theta)\, d\theta + \int_{-\pi/6}^{\pi/2} 1\, d\theta$$

$$= 4\left[\tfrac{3}{2}\theta - 2\cos\theta - \tfrac{1}{4}\sin 2\theta\right]_{-\pi/2}^{-\pi/6} + \left[\,\theta\,\right]_{-\pi/6}^{\pi/2}$$

$$= (2\pi - \tfrac{7}{2}\sqrt{3}) + \tfrac{2\pi}{3} = \tfrac{8\pi}{3} - \tfrac{7}{2}\sqrt{3} \approx 2.32.$$

[25] $1 + \sin\theta = 5\sin\theta \Rightarrow \sin\theta = \frac{1}{4}$ at $\theta = \arcsin\frac{1}{4}$ {call this α} and $(\pi - \alpha)$.

For $0 \leq \theta \leq \alpha$, the boundary is $r = 5\sin\theta$ and for $\alpha \leq \theta \leq \frac{\pi}{2}$,

the boundary is $r = 1 + \sin\theta$.

$$A = 2\int_0^{\alpha} \tfrac{1}{2}(5\sin\theta)^2\, d\theta + 2\int_{\alpha}^{\pi/2} \tfrac{1}{2}(1 + \sin\theta)^2\, d\theta$$

$$= \tfrac{25}{2}\int_0^{\alpha} (1 - \cos 2\theta)\, d\theta + \int_{\alpha}^{\pi/2} (\tfrac{3}{2} + 2\sin\theta - \tfrac{1}{2}\cos 2\theta)\, d\theta$$

$$= \tfrac{25}{2}\left[\theta - \tfrac{1}{2}\sin 2\theta\right]_0^{\alpha} + \left[\tfrac{3}{2}\theta - 2\cos\theta - \tfrac{1}{4}\sin 2\theta\right]_{\alpha}^{\pi/2}$$

$$= \tfrac{25}{2}\left[(\alpha - \tfrac{1}{2}\cdot\tfrac{1}{8}\sqrt{15}) - (0)\right] + \left[(\tfrac{3\pi}{4}) - (\tfrac{3}{2}\alpha - 2\cdot\tfrac{1}{4}\sqrt{15} - \tfrac{1}{4}\cdot\tfrac{1}{8}\sqrt{15})\right] =$$

$$11\arcsin\tfrac{1}{4} - \tfrac{1}{4}\sqrt{15} + \tfrac{3\pi}{4} \approx 4.17$$

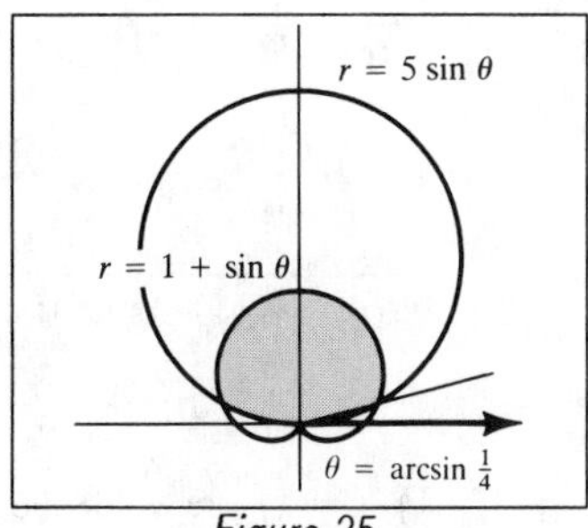

Figure 25

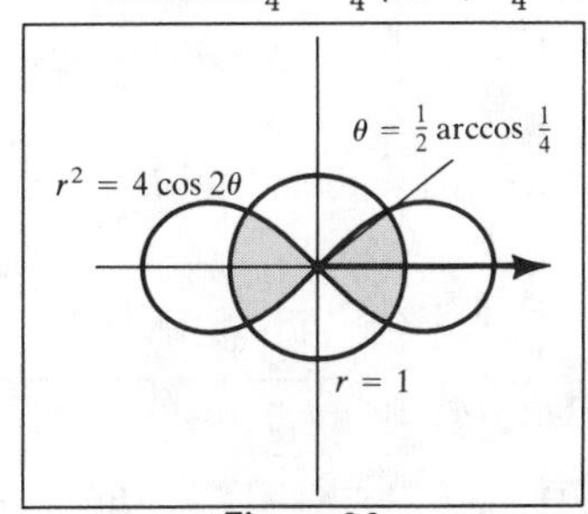

Figure 26

26 $4\cos 2\theta = 1^2 \Rightarrow \cos 2\theta = \frac{1}{4}$ at $\theta = \frac{1}{2}\arccos\frac{1}{4}$ {call this α} and $(2\pi - \alpha)$.

For $0 \le \theta \le \alpha$, the boundary is $r = 1$ and for $\alpha \le \theta \le \frac{\pi}{4}$, the boundary is $r^2 = 4\cos 2\theta$. See *Figure 26*. By symmetry, $A = 4\int_0^{\alpha} \frac{1}{2}(1)^2\, d\theta + 4\int_{\alpha}^{\pi/4} \frac{1}{2}(4\cos 2\theta)\, d\theta$

$$= 2\Big[\theta\Big]_0^{\alpha} + 8\Big[\tfrac{1}{2}\sin 2\theta\Big]_{\alpha}^{\pi/4} = 2\alpha + 4(1 - \tfrac{1}{4}\sqrt{15}) = \arccos\tfrac{1}{4} + 4 - \sqrt{15} \approx 1.45.$$

27 $f(\theta) = e^{-\theta} \Rightarrow f'(\theta) = -e^{-\theta}$.

$$L = \int_0^{2\pi} \sqrt{e^{-2\theta} + e^{-2\theta}}\, d\theta = \int_0^{2\pi} e^{-\theta}\sqrt{2}\, d\theta = \sqrt{2}\Big[-e^{-\theta}\Big]_0^{2\pi} = \sqrt{2}(1 - e^{-2\pi}) \approx 1.41.$$

28 $f(\theta) = \theta \Rightarrow f'(\theta) = 1$.

$$L = \int_0^{4\pi} \sqrt{\theta^2 + 1}\, d\theta = \Big[\tfrac{1}{2}\theta\sqrt{1 + \theta^2} + \tfrac{1}{2}\ln\Big|\theta + \sqrt{1 + \theta^2}\Big|\Big]_0^{4\pi} \{\text{Formula 21}\} =$$

$$\tfrac{1}{2}\Big[4\pi\sqrt{1 + 16\pi^2} + \ln(4\pi + \sqrt{1 + 16\pi^2})\Big] \approx 80.82.$$

29 $f(\theta) = \cos^2(\frac{1}{2}\theta) \Rightarrow f'(\theta) = -\cos(\frac{1}{2}\theta)\sin(\frac{1}{2}\theta)$.

$$L = \int_0^{\pi} \sqrt{\cos^4(\tfrac{1}{2}\theta) + \cos^2(\tfrac{1}{2}\theta)\sin^2(\tfrac{1}{2}\theta)}\, d\theta = \int_0^{\pi} \Big|\cos(\tfrac{1}{2}\theta)\Big|\sqrt{\cos^2(\tfrac{1}{2}\theta) + \sin^2(\tfrac{1}{2}\theta)}\, d\theta$$

$$= \int_0^{\pi} \Big|\cos(\tfrac{1}{2}\theta)\Big|\, d\theta = \int_0^{\pi} \cos(\tfrac{1}{2}\theta)\, d\theta = \Big[2\sin\tfrac{1}{2}\theta\Big]_0^{\pi} = 2.$$

30 $f(\theta) = 2^{\theta} \Rightarrow f'(\theta) = 2^{\theta}\ln 2$. $L = \int_0^{\pi} \sqrt{2^{2\theta} + 2^{2\theta}(\ln 2)^2}\, d\theta = \int_0^{\pi} 2^{\theta}\sqrt{1 + (\ln 2)^2}\, d\theta =$

$$\sqrt{1 + (\ln 2)^2}\Big[\frac{2^{\theta}}{\ln 2}\Big]_0^{\pi} = \frac{\sqrt{1 + (\ln 2)^2}}{\ln 2}(2^{\pi} - 1) \approx 13.74.$$

31 $r = \sin^3(\frac{1}{3}\theta)$ has a period of 6π. However, since $\sin^3\Big[\frac{1}{3}(\theta + 3\pi)\Big] = \sin^3(\frac{1}{3}\theta + \pi) = -\sin^3(\frac{1}{3}\theta) = -r$, the graph will be traced out twice on $[0, 6\pi]$. Note that (r, θ) and $(-r, \theta + 3\pi)$ denote the same point. Thus, $f(\theta) = \sin^3(\frac{1}{3}\theta) \Rightarrow f'(\theta) = \sin^2(\frac{1}{3}\theta)\cos(\frac{1}{3}\theta)$ and $L = \int_0^{3\pi} \sqrt{\sin^6(\frac{1}{3}\theta) + \sin^4(\frac{1}{3}\theta)\cos^2(\frac{1}{3}\theta)}\, d\theta = \int_0^{3\pi} \sin^2(\frac{1}{3}\theta)\, d\theta$

$$= \tfrac{1}{2}\int_0^{3\pi} \Big[1 - \cos\tfrac{2}{3}\theta\Big]\, d\theta = \tfrac{1}{2}\Big[\theta - \tfrac{3}{2}\sin\tfrac{2}{3}\theta\Big]_0^{3\pi} = \tfrac{3\pi}{2}.$$

32 $f(\theta) = 2 - 2\cos\theta \Rightarrow f'(\theta) = 2\sin\theta$.

$$L = \int_0^{2\pi} \sqrt{(2 - 2\cos\theta)^2 + (2\sin\theta)^2}\, d\theta = \sqrt{8}\int_0^{2\pi} \sqrt{1 - \cos\theta}\, d\theta$$

$$= \sqrt{8}\int_0^{2\pi} \sqrt{2\sin^2(\tfrac{1}{2}\theta)}\, d\theta = 4\int_0^{2\pi} \sin(\tfrac{1}{2}\theta)\, d\theta = 4\Big[-2\cos(\tfrac{1}{2}\theta)\Big]_0^{2\pi} = -8(-2) = 16.$$

33 $r = \theta + \cos\theta \Rightarrow \dfrac{dr}{d\theta} = 1 - \sin\theta$. $L = \int_0^{\pi/2} \sqrt{(\theta + \cos\theta)^2 + (1 - \sin\theta)^2}\, d\theta$.

Let $f(\theta) = \sqrt{(\theta + \cos\theta)^2 + (1 - \sin\theta)^2}$.

Then, $L \approx S = \frac{\pi/2 - 0}{3(4)}\Big[f(0) + 4f(\frac{\pi}{8}) + 2f(\frac{\pi}{4}) + 4f(\frac{3\pi}{8}) + f(\frac{\pi}{2})\Big] \approx 2.368495$.

[34] $r = \sin\theta + \cos^2\theta \Rightarrow \dfrac{dr}{d\theta} = \cos\theta - 2\cos\theta\,\sin\theta = \cos\theta - \sin 2\theta.$

$$L = \int_0^{\pi} \sqrt{(\sin\theta + \cos^2\theta)^2 + (\cos\theta - \sin 2\theta)^2}\,d\theta.$$

Let $f(\theta) = \sqrt{(\sin\theta + \cos^2\theta)^2 + (\cos\theta - \sin 2\theta)^2}$.

Then, $L \approx \mathrm{S} = \frac{\pi - 0}{3(4)}\left[f(0) + 4f(\frac{\pi}{4}) + 2f(\frac{\pi}{2}) + 4f(\frac{3\pi}{4}) + f(\pi)\right] \approx 3.865596.$

[35] $f(\theta) = 2 + 2\cos\theta \Rightarrow f'(\theta) = -2\sin\theta.$

$$S = \int_0^{\pi} 2\pi(2 + 2\cos\theta)\sin\theta\sqrt{(2 + 2\cos\theta)^2 + (-2\sin\theta)^2}\,d\theta$$

$$= 4\pi\int_0^{\pi}(1 + \cos\theta)\sqrt{8(1 + \cos\theta)}\sin\theta\,d\theta = 8\pi\sqrt{2}\int_0^{\pi}(1 + \cos\theta)^{3/2}\sin\theta\,d\theta$$

$$= 8\pi\sqrt{2}\left[-\tfrac{2}{5}(1 + \cos\theta)^{5/2}\right]_0^{\pi} = -\tfrac{16}{5}\pi\sqrt{2}(-2^{5/2}) = \tfrac{128\pi}{5} \approx 80.42.$$

[36] $r^2 = 4\cos 2\theta \Rightarrow 2rr' = -8\sin 2\theta \Rightarrow r' = \dfrac{-2\sin 2\theta}{\sqrt{\cos 2\theta}}$ on $[0, \frac{\pi}{4}]$. Using symmetry,

$$S = 2\int_0^{\pi/4}(2\pi)\,2\sqrt{\cos 2\theta}\sin\theta\sqrt{4\cos 2\theta + \left(\frac{-2\sin 2\theta}{\sqrt{\cos 2\theta}}\right)^2}\,d\theta$$

$$= 8\pi\int_0^{\pi/4}\sqrt{\cos 2\theta}\sqrt{4\cos 2\theta + \left(\frac{4\sin^2 2\theta}{\cos 2\theta}\right)}\sin\theta\,d\theta$$

$$= 8\pi\int_0^{\pi/4}\sqrt{4\cos^2 2\theta + 4\sin^2 2\theta}\sin\theta\,d\theta$$

$$= 16\pi\int_0^{\pi/4}\sin\theta\,d\theta = 16\pi\Big[-\cos\theta\Big]_0^{\pi/4} = -16\pi(\tfrac{\sqrt{2}}{2} - 1) = 8\pi(2 - \sqrt{2}) \approx 14.72.$$

[37] $f(\theta) = 2a\sin\theta \Rightarrow f'(\theta) = 2a\cos\theta$. The entire circle is traced for $0 \le \theta \le \pi$. Thus,

$$S = \int_0^{\pi} 2\pi(2a\sin\theta)\sin\theta\sqrt{4a^2\sin^2\theta + 4a^2\cos^2\theta}\,d\theta = 8\pi a^2\int_0^{\pi}\sin^2\theta\,d\theta$$

$$= 4\pi a^2\int_0^{\pi}(1 - \cos 2\theta)\,d\theta = 4\pi a^2\left[\theta - \tfrac{1}{2}\sin 2\theta\right]_0^{\pi} = 4\pi^2 a^2.$$

[38] $f(\theta) = 2a\cos\theta \Rightarrow f'(\theta) = -2a\sin\theta.$

The upper half of the circle (above the polar axis) is traced for $0 \le \theta \le \frac{\pi}{2}$. Thus,

$$S = \int_0^{\pi/2} 2\pi(2a\cos\theta)\sin\theta\sqrt{4a^2\cos^2\theta + 4a^2\sin^2\theta}\,d\theta$$

$$= 8\pi a^2\int_0^{\pi/2}\sin\theta\,\cos\theta\,d\theta = 8\pi a^2\left[\tfrac{1}{2}\sin^2\theta\right]_0^{\pi/2} = 4\pi a^2.$$

[39] The right half of the curve is traced out for $-\frac{\pi}{2} \le \theta \le \frac{\pi}{2}$.

Using symmetry and (13.14), $S = 2\displaystyle\int_0^{\pi/2} 2\pi\sin^2\theta\,\cos\theta\sqrt{\sin^4\theta + \sin^2 2\theta}\,d\theta.$

Let $f(\theta) = \sin^2\theta\,\cos\theta\sqrt{\sin^4\theta + \sin^2 2\theta}$.

Then, $\mathrm{T} \approx (4\pi)\frac{\pi/2 - 0}{2(4)}\left[f(0) + 2f(\frac{\pi}{8}) + 2f(\frac{\pi}{4}) + 2f(\frac{3\pi}{8}) + f(\frac{\pi}{2})\right] \approx 4.219428.$

[40] The right half of the curve is traced out for $-\frac{\pi}{2} \le \theta \le \frac{\pi}{2}$.

Using symmetry and (13.14), $S = 2\int_0^{\pi/2} 2\pi \cos^2\theta \cos\theta \sqrt{\cos^4\theta + \sin^2 2\theta}\, d\theta$.

Let $f(\theta) = \cos^3\theta \sqrt{\cos^4\theta + \sin^2 2\theta}$.

Then, $T \approx (4\pi)\frac{\pi/2 - 0}{2(4)}\left[f(0) + 2f(\frac{\pi}{8}) + 2f(\frac{\pi}{4}) + 2f(\frac{3\pi}{8}) + f(\frac{\pi}{2})\right] \approx 8.931091$.

[41] $f(\theta) = a \Rightarrow f'(\theta) = 0$. Since $x = a\cos\theta$, the average radius of the frustum generated by revolving a small section of the circle about the line $x = b$ is $b - a\cos\theta$. Thus,

$$S = \int_0^{2\pi} (2\pi)(b - a\cos\theta)\sqrt{a^2 + 0^2}\, d\theta = 2\pi a\left[b\theta - a\sin\theta\right]_0^{2\pi} = 4\pi^2 ab.$$

[42] $A = \int_{2\pi}^{4\pi} \frac{1}{2}r^2\, d\theta - \int_0^{2\pi} \frac{1}{2}r^2\, d\theta = \frac{1}{2}\int_{2\pi}^{4\pi} a^2\theta^2\, d\theta - \frac{1}{2}\int_0^{2\pi} a^2\theta^2\, d\theta =$

$$\frac{1}{2}a^2\left[\frac{1}{3}\theta^3\right]_{2\pi}^{4\pi} - \frac{1}{2}a^2\left[\frac{1}{3}\theta^3\right]_0^{2\pi} = \frac{1}{6}a^2\left[(4\pi)^3 - (2\pi)^3 - (2\pi)^3\right] = 8a^2\pi^3.$$

[43] $f(\theta) = e^{-\theta} \Rightarrow f'(\theta) = -e^{-\theta}$.

$$S = \int_0^{\pi/2} (2\pi)\, e^{-\theta}\cos\theta \sqrt{(e^{-\theta})^2 + (-e^{-\theta})^2}\, d\theta = 2\pi\sqrt{2}\int_0^{\pi/2} e^{-2\theta}\cos\theta\, d\theta$$

$$= 2\pi\sqrt{2}\left[\frac{e^{-2\theta}}{5}(\sin\theta - 2\cos\theta)\right]_0^{\pi/2} \{\text{Formula 99}\} = \frac{2}{5}\pi\sqrt{2}(e^{-\pi} + 2) \approx 3.63.$$

Exercises 13.5

Note: For the ellipse, the major axis is vertical if the denominator contains $\sin\theta$, horizontal if the denominator contains $\cos\theta$. For the hyperbola, the transverse axis is vertical if the denominator contains $\sin\theta$, horizontal if the denominator contains $\cos\theta$. The focus at the pole is called F and V is the vertex associated with (or closest to) F. $d(V, F)$ denotes the distance from the vertex to the focus. The foci are not asked for in the directions, but are listed. For the parabola, the directrix is on the right, on the left, above, or below the focus depending on the term "+cos", "−cos", "+sin", or "−sin", respectively, appearing in the denominator.

[1] Divide the numerator and denominator by the constant term in the denominator, i.e., 6. $r = \dfrac{12}{6 + 2\sin\theta} = \dfrac{2}{1 + \frac{1}{3}\sin\theta} \Rightarrow e = \frac{1}{3} < 1$, ellipse. From the last note, we see that the denominator has $\sin\theta$ and we have vertices at $\theta = \frac{\pi}{2}$ and $\frac{3\pi}{2}$.

$V(\frac{3}{2}, \frac{\pi}{2})$ and $V'(3, \frac{3\pi}{2})$. $d(V, F) = \frac{3}{2} \Rightarrow F' = (\frac{3}{2}, \frac{3\pi}{2})$.

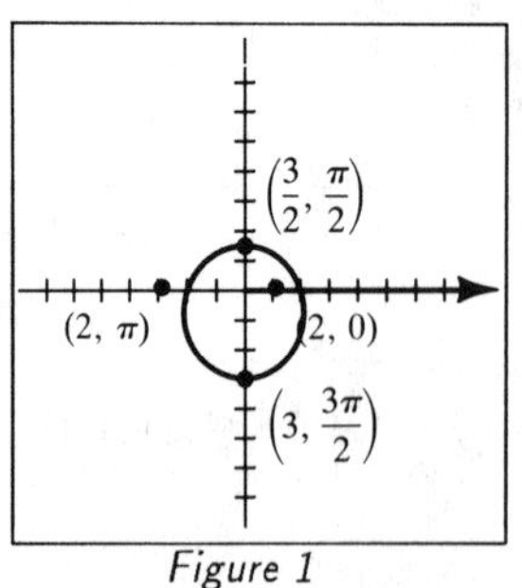

Figure 1

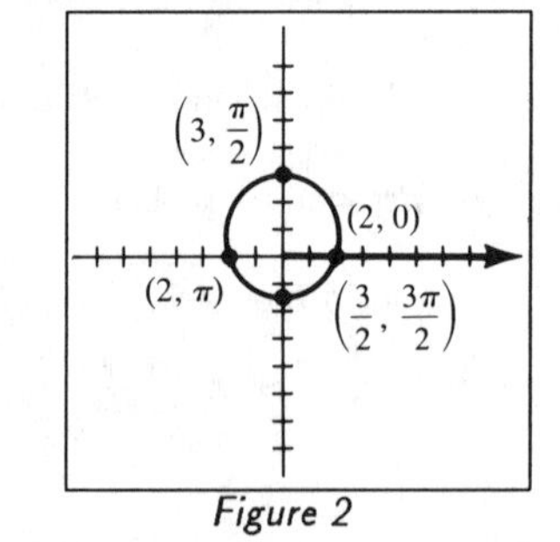

Figure 2

[2] $r = \dfrac{12}{6 - 2\sin\theta} = \dfrac{2}{1 - \frac{1}{3}\sin\theta} \Rightarrow e = \frac{1}{3} < 1$, ellipse.

$V(\frac{3}{2}, \frac{3\pi}{2})$ and $V'(3, \frac{\pi}{2})$. $d(V, F) = \frac{3}{2} \Rightarrow F' = (\frac{3}{2}, \frac{\pi}{2})$.

[3] $r = \dfrac{12}{2 - 6\cos\theta} = \dfrac{6}{1 - 3\cos\theta} \Rightarrow e = 3 > 1$, hyperbola.

$V(\frac{3}{2}, \pi)$ and $V'(-3, 0)$. $d(V, F) = \frac{3}{2} \Rightarrow F' = (-\frac{9}{2}, 0)$.

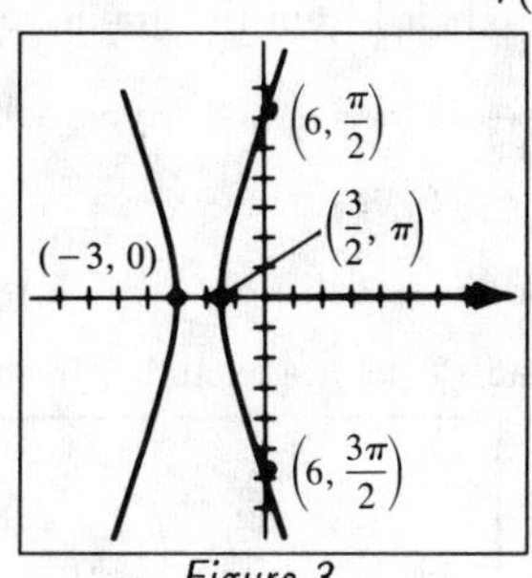

Figure 3

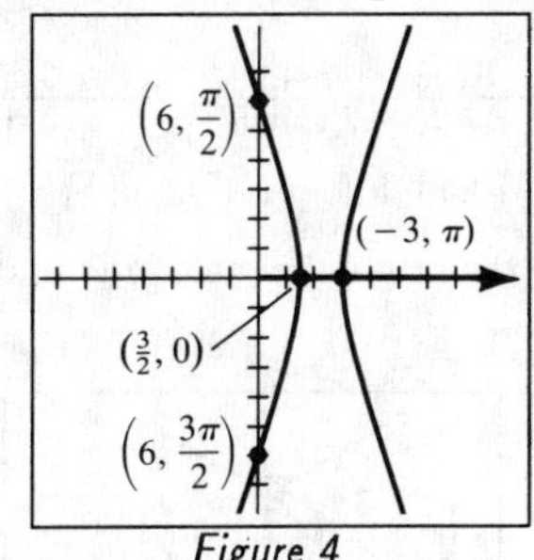

Figure 4

[4] $r = \dfrac{12}{2 + 6\cos\theta} = \dfrac{6}{1 + 3\cos\theta} \Rightarrow e = 3 > 1$, hyperbola.

$V(\frac{3}{2}, 0)$ and $V'(-3, \pi)$. $d(V, F) = \frac{3}{2} \Rightarrow F' = (-\frac{9}{2}, \pi)$.

[5] $r = \dfrac{3}{2 + 2\cos\theta} = \dfrac{\frac{3}{2}}{1 + 1\cos\theta} \Rightarrow e = 1$, parabola. Note that the expression is undefined in the $\theta = \pi$ direction. The vertex is in the $\theta = 0$ direction, $V(\frac{3}{4}, 0)$.

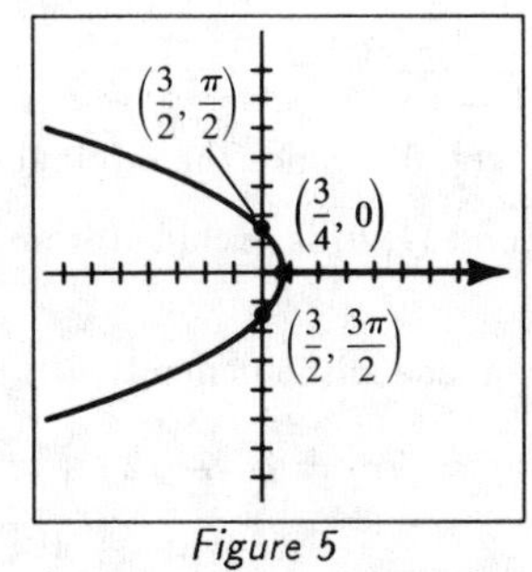

Figure 5

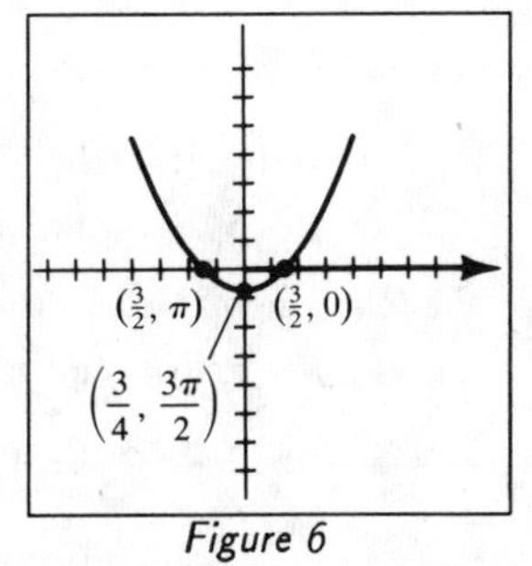

Figure 6

[6] $r = \dfrac{3}{2 - 2\sin\theta} = \dfrac{\frac{3}{2}}{1 - 1\sin\theta} \Rightarrow e = 1$, parabola.

The vertex is in the $\theta = \frac{3\pi}{2}$ direction, $V(\frac{3}{4}, \frac{3\pi}{2})$.

[7] $r = \dfrac{4}{\cos\theta - 2} = \dfrac{-2}{1 - \frac{1}{2}\cos\theta} \Rightarrow e = \frac{1}{2} < 1$, ellipse.

$V(-\frac{4}{3}, \pi)$ and $V'(-4, 0)$. $d(V, F) = \frac{4}{3} \Rightarrow F' = (-\frac{8}{3}, 0)$

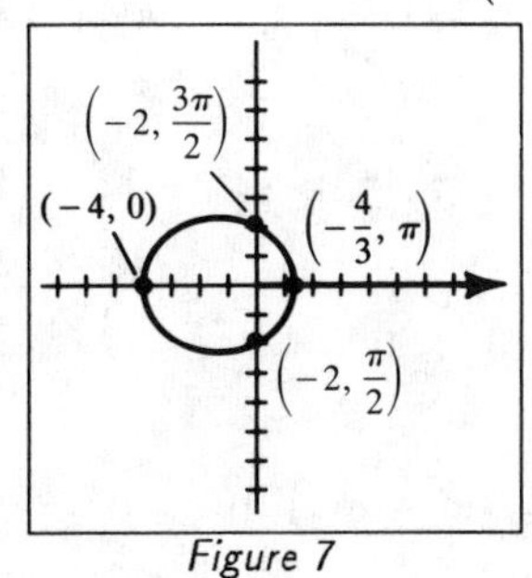

Figure 7

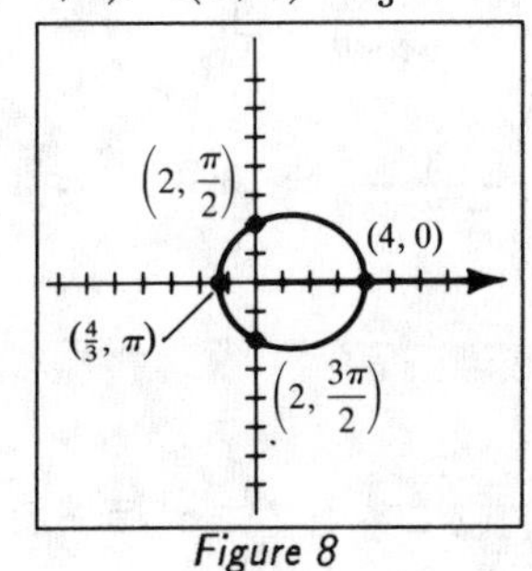

Figure 8

[8] $r = \dfrac{4\sec\theta}{2\sec\theta - 1}\cdot\dfrac{\cos\theta}{\cos\theta} = \dfrac{4}{2 - 1\cos\theta} = \dfrac{2}{1 - \frac{1}{2}\cos\theta} \Rightarrow e = \frac{1}{2} < 1$, ellipse.

$V(\frac{4}{3}, \pi)$ and $V'(4, 0)$. $d(V, F) = \frac{4}{3} \Rightarrow F' = (\frac{8}{3}, 0)$.

Since the original equation is undefined when $\sec\theta$ is undefined,

the points $(2, \frac{\pi}{2})$ and $(2, \frac{3\pi}{2})$ are excluded from the graph. See *Figure 8.*

[9] $r = \dfrac{6\csc\theta}{2\csc\theta + 3}\cdot\dfrac{\sin\theta}{\sin\theta} = \dfrac{6}{2 + 3\sin\theta} = \dfrac{3}{1 + \frac{3}{2}\sin\theta} \Rightarrow e = \frac{3}{2} > 1$, hyperbola.

$V(\frac{6}{5}, \frac{\pi}{2})$ and $V'(-6, \frac{3\pi}{2})$. $d(V, F) = \frac{6}{5} \Rightarrow F' = (-\frac{36}{5}, \frac{3\pi}{2})$.

Since the original equation is undefined when $\csc\theta$ is undefined,

the points $(3, 0)$ and $(3, \pi)$ are excluded from the graph.

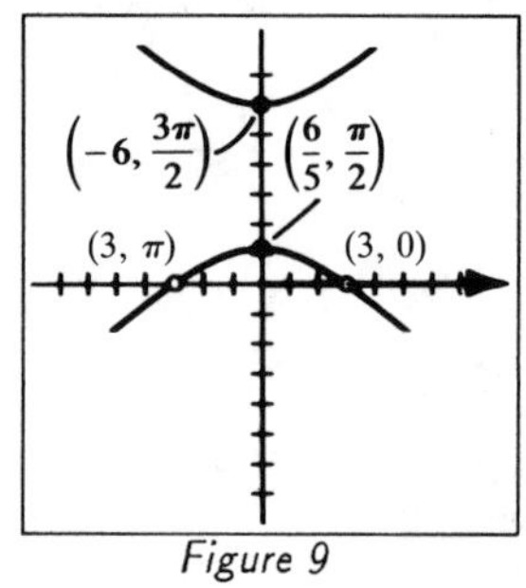

Figure 9

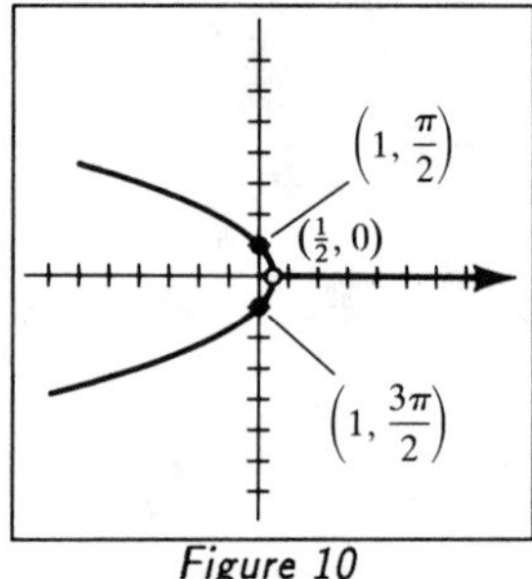

Figure 10

[10] $r = \csc\theta(\csc\theta - \cot\theta) = \dfrac{1}{\sin\theta}\left(\dfrac{1 - \cos\theta}{\sin\theta}\right) = \dfrac{1 - \cos\theta}{1 - \cos^2\theta} = \dfrac{1}{1 + 1\cos\theta} \Rightarrow e = 1$,

parabola. The vertex is in the $\theta = 0$ direction, $V(\frac{1}{2}, 0)$. Since the original equation

is undefined when $\csc\theta$ is undefined, the point $(\frac{1}{2}, 0)$ is excluded from the graph.

Note: For the following exercises, the substitutions

$x = r\cos\theta$, $y = r\sin\theta$, and $r^2 = x^2 + y^2$ are made without mention.

[11] $r = \dfrac{12}{6 + 2\sin\theta} \Rightarrow 6r + 2y = 12 \Rightarrow 3r = 6 - y \Rightarrow 9r^2 = 36 - 12y + y^2 \Rightarrow$

$9x^2 + 8y^2 + 12y - 36 = 0$

[12] $r = \dfrac{12}{6 - 2\sin\theta} \Rightarrow 6r - 2y = 12 \Rightarrow 3r = y + 6 \Rightarrow 9r^2 = y^2 + 12y + 36 \Rightarrow$

$9x^2 + 8y^2 - 12y - 36 = 0$

[13] $r = \dfrac{12}{2 - 6\cos\theta} \Rightarrow 2r - 6x = 12 \Rightarrow r = 3x + 6 \Rightarrow r^2 = 9x^2 + 36x + 36 \Rightarrow$

$8x^2 - y^2 + 36x + 36 = 0$

[14] $r = \dfrac{12}{2 + 6\cos\theta} \Rightarrow 2r + 6x = 12 \Rightarrow r = 6 - 3x \Rightarrow r^2 = 36 - 36x + 9x^2 \Rightarrow$

$8x^2 - y^2 - 36x + 36 = 0$

[15] $r = \dfrac{3}{2 + 2\cos\theta} \Rightarrow 2r + 2x = 3 \Rightarrow 2r = 3 - 2x \Rightarrow 4r^2 = 4x^2 - 12x + 9 \Rightarrow$

$4y^2 + 12x - 9 = 0$

[16] $r = \dfrac{3}{2 - 2\sin\theta} \Rightarrow 2r - 2y = 3 \Rightarrow 2r = 2y + 3 \Rightarrow 4r^2 = 4y^2 + 12y + 9 \Rightarrow$

$4x^2 - 12y - 9 = 0$

[17] $r = \dfrac{4}{\cos\theta - 2} \Rightarrow x - 2r = 4 \Rightarrow x - 4 = 2r \Rightarrow x^2 - 8x + 16 = 4r^2 \Rightarrow$

$3x^2 + 4y^2 + 8x - 16 = 0$

[18] $r = \dfrac{4\sec\theta}{2\sec\theta - 1}\cdot\dfrac{\cos\theta}{\cos\theta} = \dfrac{4}{2 - 1\cos\theta} \Rightarrow 2r - x = 4 \Rightarrow 2r = x + 4 \Rightarrow$

$4r^2 = x^2 + 8x + 16 \Rightarrow 3x^2 + 4y^2 - 8x - 16 = 0.$

r is undefined when $\theta = \frac{\pi}{2}$ or $\frac{3\pi}{2}$. For the rectangular equation, these points correspond to $x = 0$ (or $r\cos\theta = 0$). Substituting $x = 0$ into the above rectangular equation yields $4y^2 = 16$ or $y = \pm 2$. $\therefore$ exclude $(0, \pm 2)$

[19] $r = \dfrac{6\csc\theta}{2\csc\theta + 3}\cdot\dfrac{\sin\theta}{\sin\theta} = \dfrac{6}{2 + 3\sin\theta} \Rightarrow 2r + 3y = 6 \Rightarrow 2r = 6 - 3y \Rightarrow$

$4r^2 = 36 - 36y + 9y^2 \Rightarrow 4x^2 - 5y^2 + 36y - 36 = 0.$

r is undefined when $\theta = 0$ or π. For the rectangular equation, these points correspond to $y = 0$ (or $r\sin\theta = 0$). Substituting $y = 0$ into the above rectangular equation yields $4x^2 = 36$ or $x = \pm 3$. $\therefore$ exclude $(\pm 3, 0)$

[20] $r = \csc\theta(\csc\theta - \cot\theta) = \dfrac{1}{\sin\theta}\left(\dfrac{1 - \cos\theta}{\sin\theta}\right) = \dfrac{1 - \cos\theta}{1 - \cos^2\theta} = \dfrac{1}{1 + \cos\theta} \Rightarrow$

$r + x = 1 \Rightarrow r = 1 - x \Rightarrow r^2 = 1 - 2x + x^2 \Rightarrow y^2 + 2x - 1 = 0.$

r is undefined when $\theta = 0$ or π. For the rectangular equation, this point corresponds to $y = 0$ (or $r\sin\theta = 0$). Substituting $y = 0$ into the above rectangular equation yields $2x - 1 = 0$ or $x = \frac{1}{2}$. $\therefore$ exclude $(\frac{1}{2}, 0)$

[21] $r = 2\sec\theta \Rightarrow r\cos\theta = 2 \Rightarrow x = 2$. Thus, $d = 2$ and since the directrix is on the right of the focus at the pole, we use "$+\cos\theta$". $r = \dfrac{2(\frac{1}{3})}{1 + \frac{1}{3}\cos\theta}\cdot\dfrac{3}{3} = \dfrac{2}{3 + \cos\theta}$.

[22] $r = 4\csc\theta \Rightarrow r\sin\theta = 4 \Rightarrow y = 4$. Thus, $d = 4$ and since the directrix is above the focus at the pole, we use "$+\sin\theta$". $r = \dfrac{4(\frac{2}{5})}{1 + \frac{2}{5}\sin\theta}\cdot\dfrac{5}{5} = \dfrac{8}{5 + 2\sin\theta}$.

[23] $r = -3\csc\theta \Rightarrow r\sin\theta = -3 \Rightarrow y = -3 \Rightarrow d = 3$ and use "$-\sin\theta$".

$$r = \frac{3(4)}{1 - 4\sin\theta} = \frac{12}{1 - 4\sin\theta}.$$

[24] $r = -4\sec\theta \Rightarrow r\cos\theta = -4 \Rightarrow x = -4 \Rightarrow d = 4$ and use "$-\cos\theta$".

$$r = \frac{4(3)}{1 - 3\cos\theta} = \frac{12}{1 - 3\cos\theta}.$$

[25] $r\cos\theta = 5 \Rightarrow x = 5 \Rightarrow d = 5$ and use "$+\cos\theta$". $r = \dfrac{5(1)}{1 + 1\cos\theta} = \dfrac{5}{1 + \cos\theta}$.

[26] $r\sin\theta = -2 \Rightarrow y = -2$. Thus, $d = 2$ and since the directrix is under the focus at the pole, we use "$-\sin\theta$". $r = \dfrac{2(1)}{1 - 1\sin\theta} = \dfrac{2}{1 - \sin\theta}$.

[27] For a parabola, $e = 1$. The vertex is 4 units on top of the focus at the pole so $d = 2(4)$ and we should use "$+\sin\theta$" in the denominator. $r = \dfrac{8}{1 + \sin\theta}$

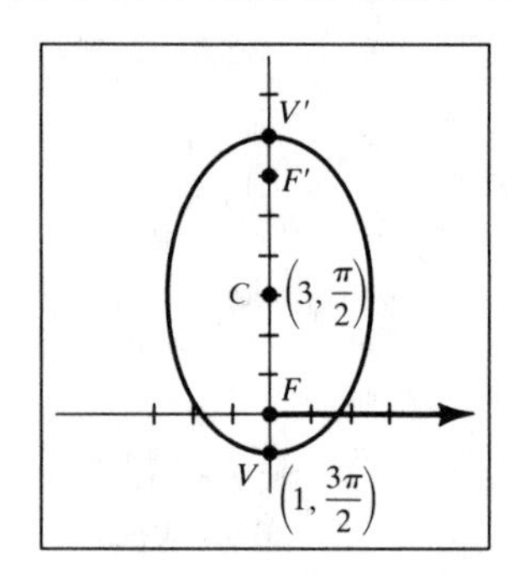

Figure 28

[28] (a) See *Figure 28.* $e = \frac{c}{a} = \dfrac{d(C, F)}{d(C, V)} = \frac{3}{4}$.

(b) Since the vertex is under the focus at the pole, use "$-\sin\theta$". $r = \dfrac{d(\frac{3}{4})}{1 - \frac{3}{4}\sin\theta}$ and $r = 1$ when $\theta = \frac{3\pi}{2} \Rightarrow 1 = \dfrac{d(\frac{3}{4})}{1 - \frac{3}{4}(-1)} \Rightarrow 1 = \dfrac{\frac{3}{4}d}{\frac{7}{4}} \Rightarrow$ $d = \frac{7}{3}$. Thus, $r = \dfrac{(\frac{7}{3})(\frac{3}{4})}{1 - \frac{3}{4}\sin\theta} \cdot \dfrac{4}{4} = \dfrac{7}{4 - 3\sin\theta}$.

$\left\{\dfrac{x^2}{7} + \dfrac{(y-3)^2}{16} = 1\right\}$

Note: Exer. 29–32: With $r = f(\theta)$ and θ given, we will list r', r' evaluated at θ, and r evaluated at θ. Substitute these into the formula $m = \dfrac{r'\sin\theta + r\cos\theta}{r'\cos\theta - r\sin\theta}$ and simplify to obtain the value of m given.

[29] $r' = \dfrac{-24\cos\theta}{(6 + 2\sin\theta)^2}$, $r'(\frac{\pi}{6}) = -\frac{12}{49}\sqrt{3}$, $r(\frac{\pi}{6}) = \frac{12}{7}$, $m = -\frac{3}{5}\sqrt{3}$

[30] $r' = \dfrac{24\cos\theta}{(6 - 2\sin\theta)^2}$, $r'(0) = \frac{2}{3}$, $r(0) = 2$, $m = 3$

[31] $r' = \dfrac{-72\sin\theta}{(2 - 6\cos\theta)^2}$, $r'(\frac{\pi}{2}) = -18$, $r(\frac{\pi}{2}) = 6$, $m = 3$

[32] $r' = \dfrac{72\sin\theta}{(2 + 6\cos\theta)^2}$, $r'(\frac{\pi}{3}) = \frac{36}{25}\sqrt{3}$, $r(\frac{\pi}{3}) = \frac{12}{5}$, $m = -\frac{7}{3}\sqrt{3}$

[33] $r = 2\sec\theta$ is the line $x = 2$. $A = \displaystyle\int_{\pi/6}^{\pi/3} \tfrac{1}{2}(2\sec\theta)^2\, d\theta$.

[34] Since $r = \csc\theta\cot\theta \Leftrightarrow r = \dfrac{1}{\sin\theta}\cdot\dfrac{\cos\theta}{\sin\theta} \Leftrightarrow r\sin^2\theta = \cos\theta \Leftrightarrow r^2\sin^2\theta = r\cos\theta$, $r = \csc\theta\cot\theta$ is the parabola $y^2 = x$. $A = \displaystyle\int_{\pi/6}^{\pi/4} \tfrac{1}{2}(\csc\theta\cot\theta)^2\, d\theta$.

[35] $r = \dfrac{4}{1 - \cos\theta}$ is a parabola similar to the one shown in Figure 13.45. By letting $\theta = \frac{\pi}{2}$, π, and $\frac{3\pi}{2}$, we see that it includes the points $(4, \frac{\pi}{2})$, $(2, \pi)$, and $(4, \frac{3\pi}{2})$.

$$A = \int_{\pi/4}^{5\pi/4} \frac{1}{2}\left(\frac{4}{1 - \cos\theta}\right)^2 d\theta.$$

[36] $r = \dfrac{2}{1 + \sin\theta}$ is a parabola that has the shape of the curve in Figure 13.42(i). By letting $\theta = 0$, $\frac{\pi}{2}$, and π, we see that it includes the points $(2, 0)$, $(1, \frac{\pi}{2})$, and $(2, \pi)$.

$$A = \int_{\pi/3}^{4\pi/3} \frac{1}{2}\left(\frac{2}{1 + \sin\theta}\right)^2 d\theta.$$

[37] (a) Let V and C denote the vertex closest to the sun and the center of the ellipse, respectively. Let s denote the distance from V to the directrix to the left of V. $d(O, V) = d(C, V) - d(C, O) = a - c = a - ea = a(1 - e)$.

Also, by (13.15), $\dfrac{d(O, V)}{s} = e \Rightarrow s = \dfrac{d(O, V)}{e} = \dfrac{a(1 - e)}{e}$. Now,

$d = s + d(O, V) = \dfrac{a(1 - e)}{e} + a(1 - e) = \dfrac{a(1 - e^2)}{e}$ and $de = a(1 - e^2)$.

Thus, the equation of the orbit is $r = \dfrac{(1 - e^2)a}{1 - e\cos\theta}$.

(b) The minimum distance occurs when $\theta = \pi$. $r_{\text{per}} = \dfrac{(1 - e^2)a}{1 - e(-1)} = a(1 - e)$.

The maximum distance occurs when $\theta = 0$. $r_{\text{aph}} = \dfrac{(1 - e^2)a}{1 - e(1)} = a(1 + e)$.

[38] $r = \dfrac{(1 - e^2)a}{1 - e\cos\theta} = \dfrac{(1 + e)[(1 - e)a]}{1 - e\cos\theta} = \dfrac{(1 + 0.249)(29.62)}{1 - 0.249\cos\theta}$ $\{\text{since } r_{\text{per}} = a(1 - e)\}$

$\approx \dfrac{37.00}{1 - 0.249\cos\theta}$ is an equation of Pluto's orbit.

$r_{\text{aph}} = a(1 + e) = \left(\dfrac{r_{\text{per}}}{1 - e}\right)(1 + e) = \left(\dfrac{29.62}{1 - 0.249}\right)(1 + 0.249) \approx 49.26$ AU.

[39] (a) Using (9.6), $\displaystyle\int \frac{1}{1 - e\cos\theta}\,d\theta = \int \frac{1}{1 - e\left(\dfrac{1 - u^2}{1 + u^2}\right)}\left(\frac{2}{1 + u^2}\right)du$

$$= \int \frac{2}{(1 - e) + (1 + e)u^2}\,du = \frac{2}{1 + e}\int \frac{1}{\left(\dfrac{1 - e}{1 + e}\right) + u^2}\,du$$

$$= \left(\frac{2}{1 + e}\right)\sqrt{\frac{1 + e}{1 - e}}\tan^{-1}\sqrt{\frac{1 + e}{1 - e}}\,u + C.$$ As $\theta \to \pi$,

$u = \tan\left(\frac{1}{2}\theta\right) \to \infty$ and as $\theta \to -\pi$, $u = \tan\left(\frac{1}{2}\theta\right) \to -\infty$. Thus, $\displaystyle\int_{-\pi}^{\pi} \frac{1}{1 - e\cos\theta}\,d\theta$

$$= \lim_{t\to-\infty}\left[\left(\frac{2}{\sqrt{1 - e^2}}\right)\tan^{-1}\sqrt{\frac{1 + e}{1 - e}}\,u\right]_t^0 + \lim_{s\to\infty}\left[\left(\frac{2}{\sqrt{1 - e^2}}\right)\tan^{-1}\sqrt{\frac{1 + e}{1 - e}}\,u\right]_0^s$$

$$= -\frac{2}{\sqrt{1 - e^2}}\left(-\frac{\pi}{2}\right) + \frac{2}{\sqrt{1 - e^2}}\left(\frac{\pi}{2}\right) = \frac{2\pi}{\sqrt{1 - e^2}}.$$

(b) Using the expression for r from Exercise 37(a), the average radius (distance) will

be given by $\displaystyle\frac{1}{\pi - (-\pi)}\int_{-\pi}^{\pi} \frac{(1 - e^2)a}{1 - e\cos\theta}\,d\theta = \frac{(1 - e^2)a}{2\pi}\cdot\frac{2\pi}{\sqrt{1 - e^2}} = a\sqrt{1 - e^2}$.

[40] $1.46 = a\sqrt{1 - e^2}$ and $e = 0.223 \Rightarrow a \approx 1.50$ AU.

$r = \dfrac{(1 - e^2)a}{1 - e\cos\theta} \approx \dfrac{1.42}{1 - 0.223\cos\theta}$ is the equation of the orbit.

The minimum value of r occurs when $r = a(1 - e) \approx 1.16$ AU.

13.6 Review Exercises

[1] $x = \frac{1}{t} + 1 \Rightarrow t = \dfrac{1}{x - 1}$ and $y = 2(x - 1) - \left(\dfrac{1}{x - 1}\right) = \dfrac{2(x^2 - 2x + 1) - 1}{x - 1} =$

$\dfrac{2x^2 - 4x + 1}{x - 1}$. This is a rational function with a vertical asymptote at $x = 1$ and an oblique asymptote of $y = 2x - 2$. The graph has a minimum point at $\left(\frac{5}{4}, -\frac{7}{2}\right)$ when $t = 4$ and then approaches the oblique asymptote as t approaches 0. See *Figure 1*.

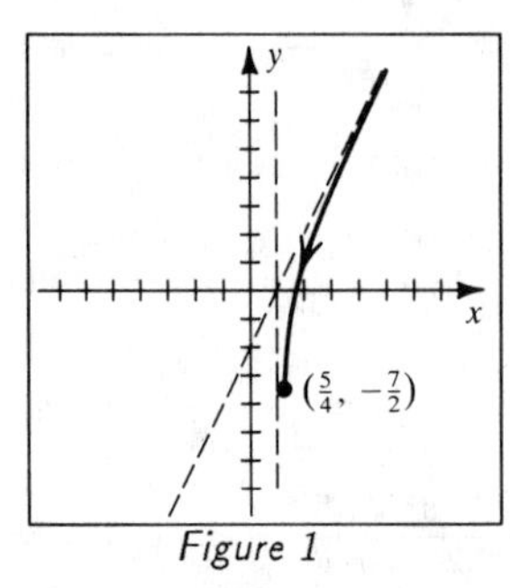

Figure 1

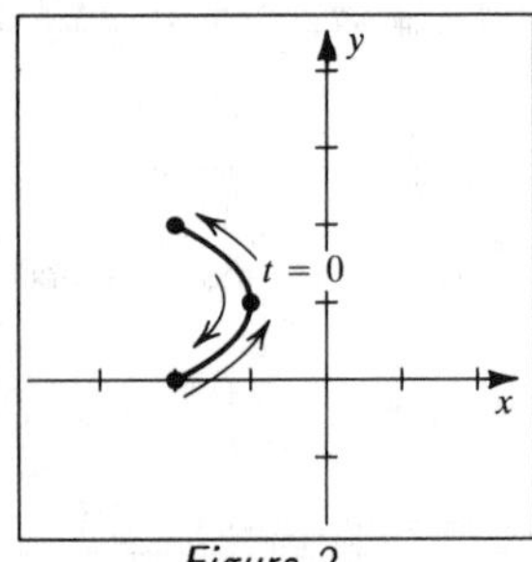

Figure 2

[2] $x = \cos^2 t - 2 \Rightarrow x + 2 = \cos^2 t.$ $y = \sin t + 1 \Rightarrow (y - 1)^2 = \sin^2 t.$
$\sin^2 t + \cos^2 t = 1 = x + 2 + (y - 1)^2 \Rightarrow (y - 1)^2 = -(x + 1).$
This is a parabola with vertex at $(-1, 1)$ and opening to the left. $t = 0$ corresponds to the vertex and as t varies from 0 to 2π, the point (x, y) moves to $(-2, 2)$ at $t = \frac{\pi}{2}$, back to the vertex at $t = \pi$, down to $(-2, 0)$ at $t = \frac{3\pi}{2}$,
and finishes at the vertex at $t = 2\pi$.

[3] $x = \sqrt{t} \Rightarrow x^2 = t$ and $y = 2^{-x^2}$. The graph is a bell-shaped curve with a maximum point at $t = 0$ or $(0, 1)$. As t increases, x increases, and y gets close to 0.

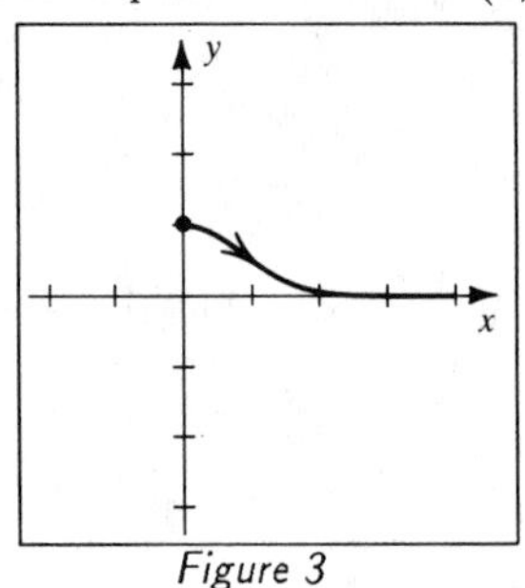

Figure 3

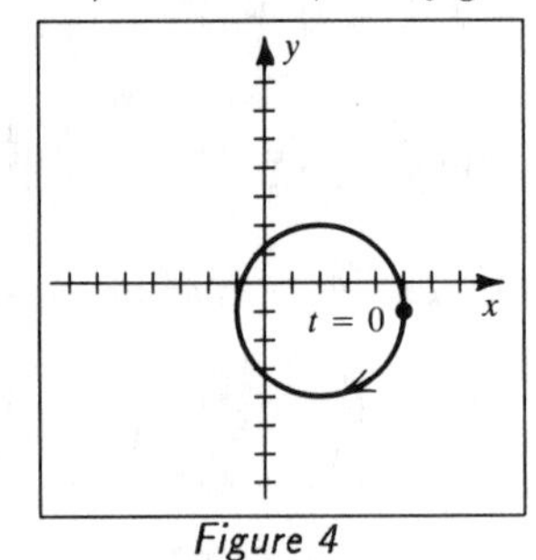

Figure 4

[4] $x = 3\cos t + 2$ and $y = -3\sin t - 1 \Rightarrow x - 2 = 3\cos t$ and $y + 1 = -3\sin t \Rightarrow$
$(x - 2)^2 + (y + 1)^2 = 9\cos^2 t + 9\sin^2 t = 9.$ As t varies from 0 to 2π, (x, y) traces
the circle from $(5, -1)$ in a clockwise direction back to $(5, -1)$.

[5] All of the curves are a portion of the circle $x^2 + y^2 = 16$.
C_1: $y = \sqrt{16 - t^2} = \sqrt{16 - x^2}$. Since $y = \sqrt{16 - t^2}$,
y must be nonnegative and we have the top half of the circle.

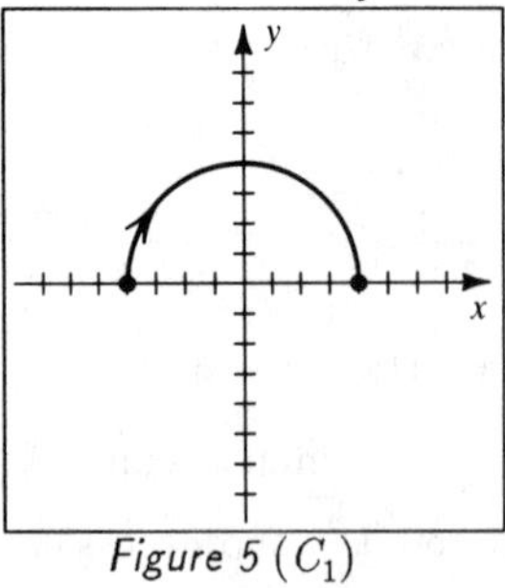

Figure 5 (C_1)

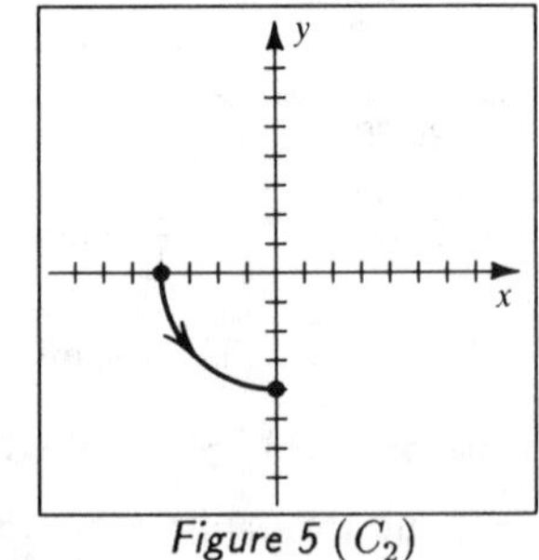

Figure 5 (C_2)

C_2: $x = -\sqrt{16 - t} = -\sqrt{16 - (-\sqrt{t})^2} = -\sqrt{16 - y^2}$.

This is the left half of the circle. Since $y = -\sqrt{t}$, y can only be nonpositive.

Hence we have only the third quadrant portion of the circle.

C_3: $x = 4\cos t,\ y = 4\sin t \Rightarrow \frac{x}{4} = \cos t,\ \frac{y}{4} = \sin t \Rightarrow \frac{x^2}{16} = \cos^2 t,\ \frac{y^2}{16} = \sin^2 t \Rightarrow$

$\frac{x^2}{16} + \frac{y^2}{16} = \cos^2 t + \sin^2 t = 1 \Rightarrow x^2 + y^2 = 16$. This is the entire circle.

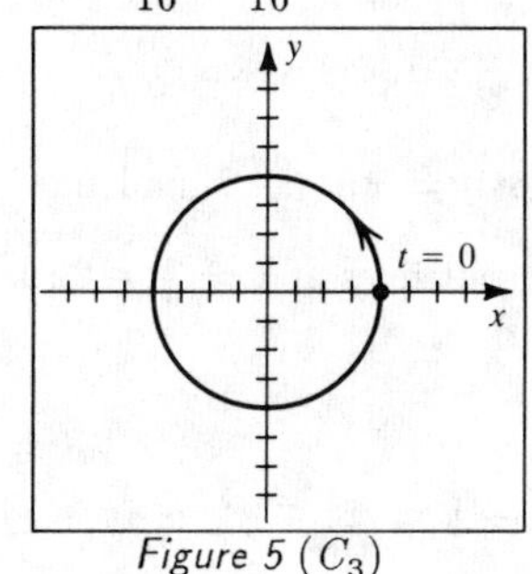

Figure 5 (C_3)

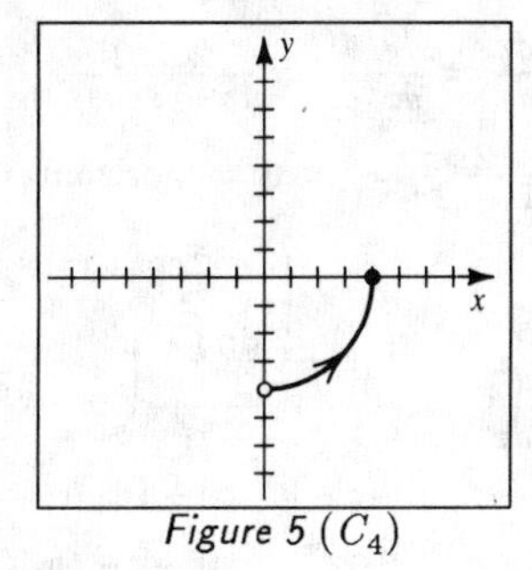

Figure 5 (C_4)

C_4: $y = -\sqrt{16 - e^{2t}} = -\sqrt{16 - (e^t)^2} = -\sqrt{16 - x^2}$.

This is the bottom half of the circle. Since e^t is positive, x takes on all positive real values. Note that $(0, -4)$ is *not* included on the graph since $x \neq 0$.

6 C_1: $x = t^2,\ y = t^3 \Rightarrow y = (\pm x^{1/2})^3 = \pm x^{3/2};\ x \geq 0$.

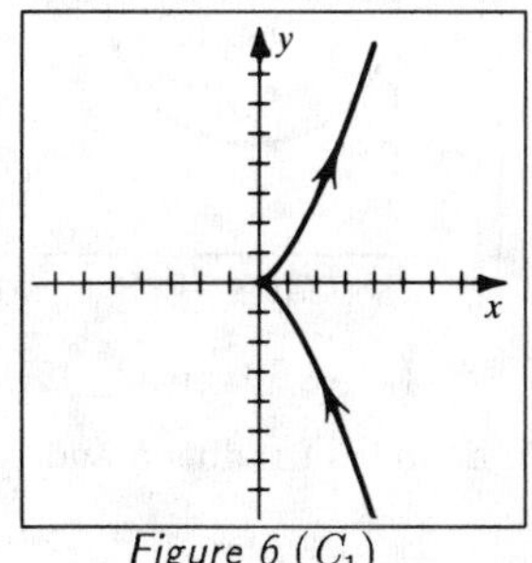

Figure 6 (C_1)

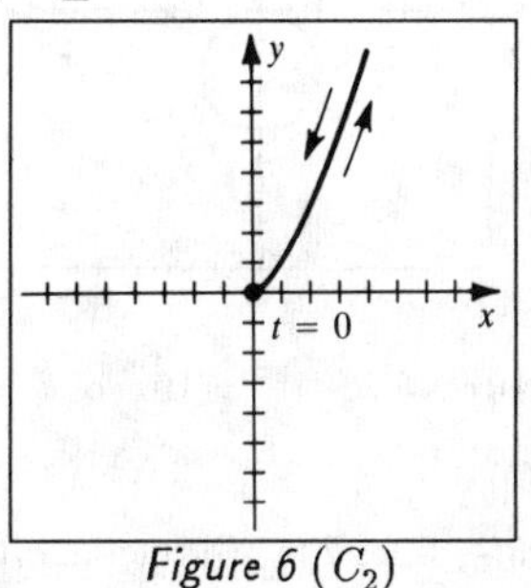

Figure 6 (C_2)

C_2: $x = t^4,\ y = t^6 \Rightarrow y = (\pm x^{1/4})^6 = x^{3/2};\ x, y \geq 0$.

C_3: $x = e^{2t},\ y = e^{3t} \Rightarrow y = (x^{1/2})^3 = x^{3/2};\ x, y > 0$.

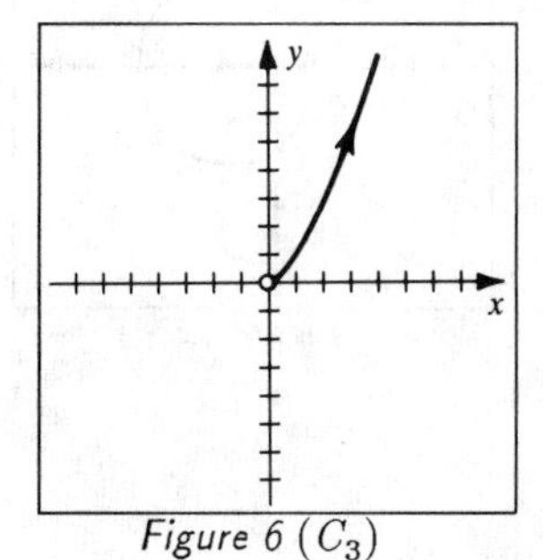

Figure 6 (C_3)

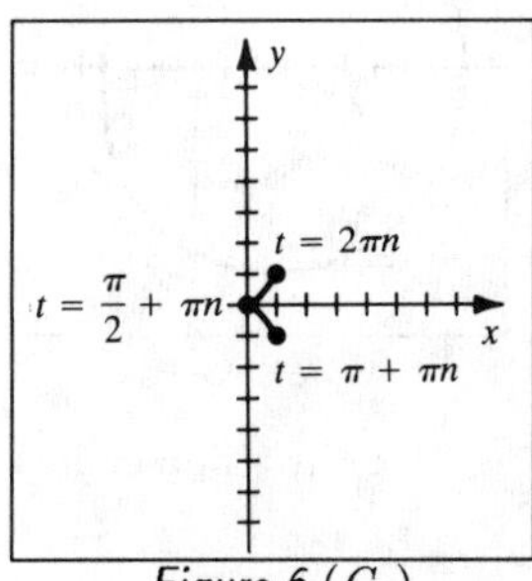

Figure 6 (C_4)

C_4: $x = 1 - \sin^2 t = \cos^2 t,\ y = \cos^3 t \Rightarrow y = (\pm x^{1/2})^3 = \pm x^{3/2}$;

$0 \leq x \leq 1$ and $-1 \leq y \leq 1$.

7 (a) $\frac{dy}{dx} = \frac{dy/dt}{dx/dt} = \frac{6t^2 + 4}{2t} = \frac{3t^2 + 2}{t}$.

(b) Since $\frac{3t^2 + 2}{t} \neq 0$, there are no horizontal tangents.

There is a vertical tangent at $t = 0$.

(c) $\frac{d^2y}{dx^2} = \frac{D_t\left[3t + \frac{2}{t}\right]}{2t} = \frac{3 - \frac{2}{t^2}}{2t} = \frac{3t^2 - 2}{2t^3}$.

8 (a) $\frac{dy}{dx} = \frac{dy/dt}{dx/dt} = \frac{2\sin t}{1 - 2\cos t}$.

(b) $\frac{2\sin t}{1 - 2\cos t} = 0 \Rightarrow$ horizontal tangent lines at $t = \pi n$. $\frac{dy}{dx}$ is undefined when

$1 - 2\cos t = 0 \Rightarrow$ vertical tangent lines at $t = \frac{\pi}{3} + 2\pi n, \frac{5\pi}{3} + 2\pi n$.

(c) $\frac{d^2y}{dx^2} = \frac{D_t\left[\frac{2\sin t}{1 - 2\cos t}\right]}{1 - 2\cos t} = \frac{2\cos t - 4}{(1 - 2\cos t)^3}$.

9 $r = -4\sin\theta \Rightarrow r^2 = -4r\sin\theta \Rightarrow x^2 + y^2 = -4y \Rightarrow x^2 + y^2 + 4y + \underline{4} = \underline{4}$

$\Rightarrow x^2 + (y + 2)^2 = 4$, a circle of radius 2 and center $(0, -2)$.

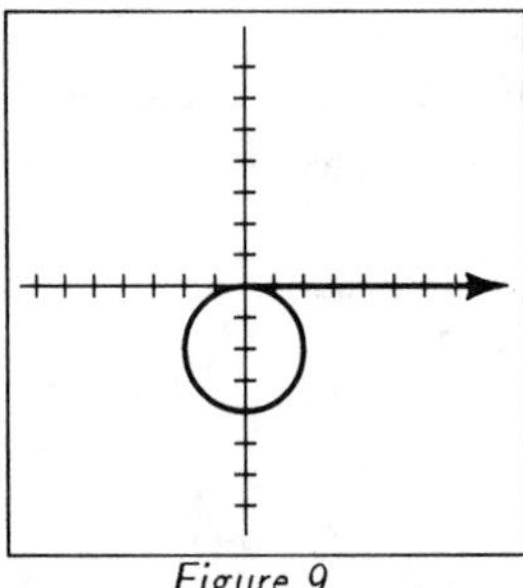

Figure 9

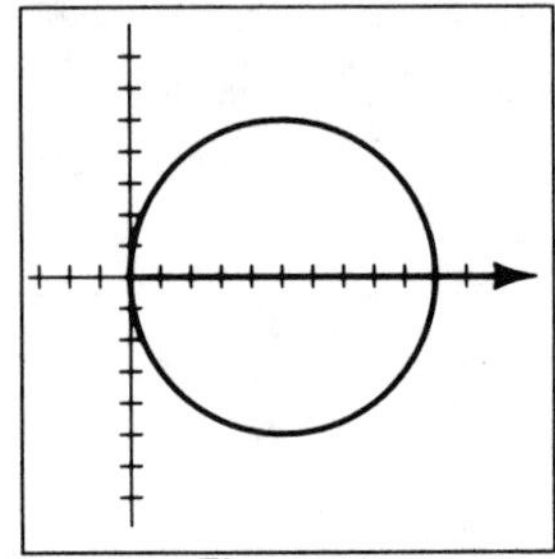

Figure 10

10 $r = 10\cos\theta \Rightarrow r^2 = 10r\cos\theta \Rightarrow x^2 + y^2 = 10x \Rightarrow (x^2 - 10x + \underline{25}) + y^2 = \underline{25}$

$\Rightarrow (x - 5)^2 + y^2 = 25$, a circle of radius 5 and center $(5, 0)$.

11 $r = 6 - 3\cos\theta$ is a limaçon without a loop.

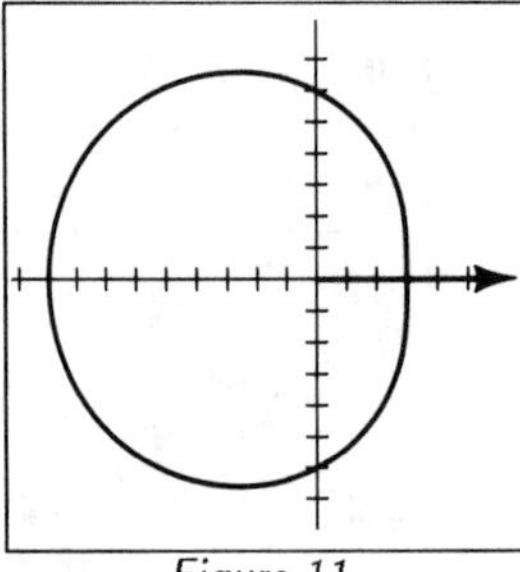

Figure 11

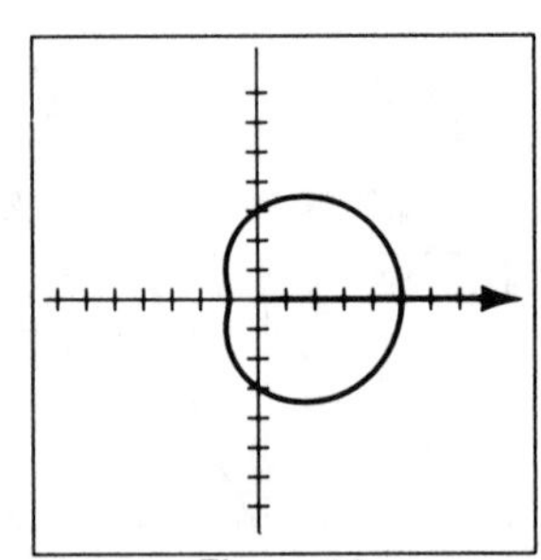

Figure 12

12 $r = 3 + 2\cos\theta$ is a limaçon without a loop.

[13] $r^2 = 9\sin 2\theta$ is a lemniscate with loops in QI and QIII.

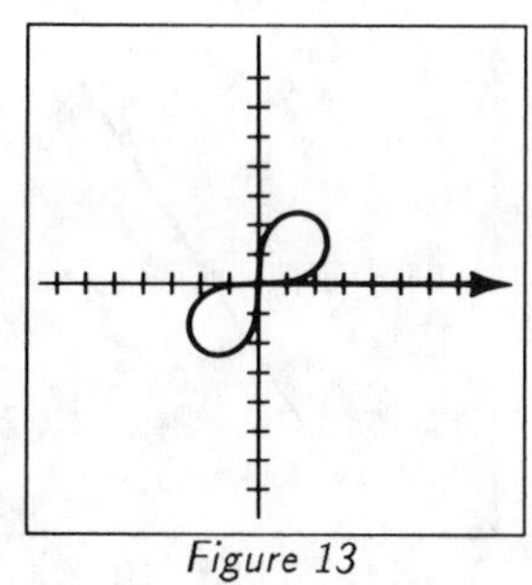
Figure 13

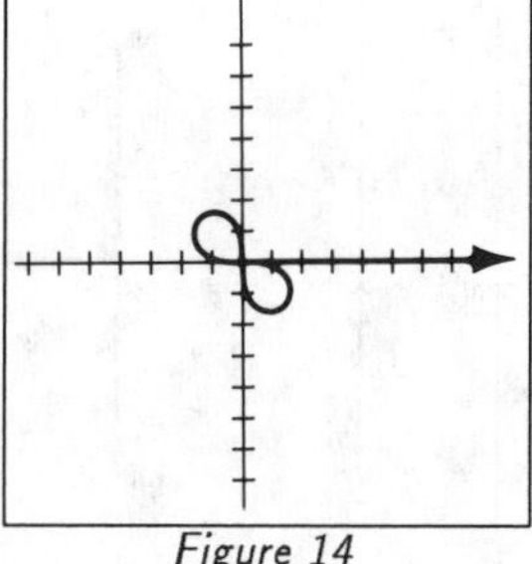
Figure 14

[14] $r^2 = -4\sin 2\theta$ is a lemniscate with loops in QII and QIV.

[15] $r = 3\sin 5\theta$ is a 5-leafed rose.

One leaf is centered on the line $\theta = \frac{\pi}{2}$ and the others are equally spaced 72° apart.

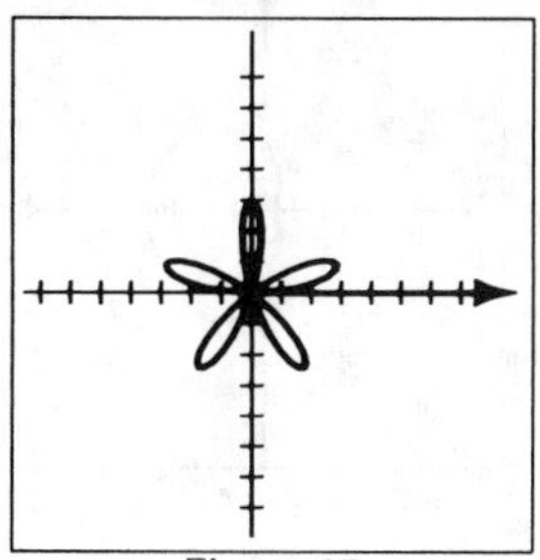
Figure 15

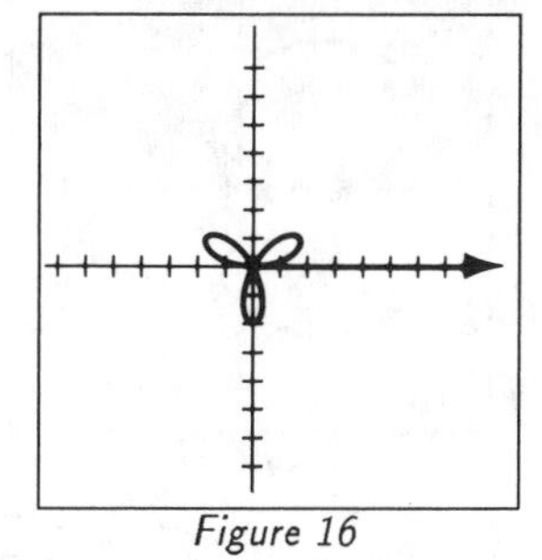
Figure 16

[16] $r = 2\sin 3\theta$ is a three-leafed rose.

One leaf is centered on the line $\theta = \frac{3\pi}{2}$ and the others are equally spaced 120° apart.

[17] $2r = \theta \Rightarrow r = \frac{1}{2}\theta$. Positive values of θ yield the "counterclockwise spiral" while the "clockwise spiral" is obtained from the negative values of θ.

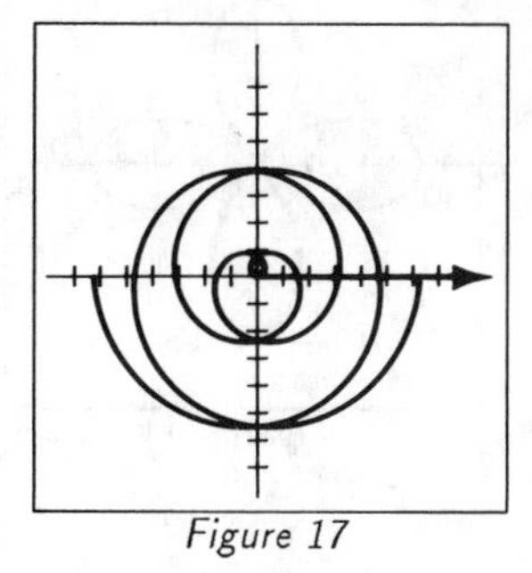
Figure 17

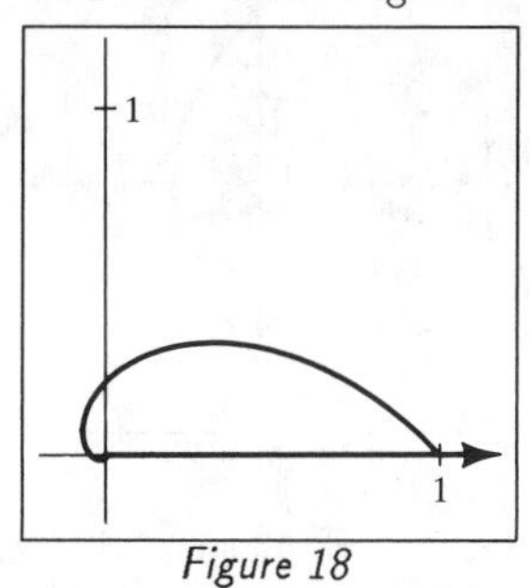

Figure 18

[18] $r = e^{-\theta}$, $\theta \geq 0$, is a spiral starting at (1, 0) that approaches the pole as $\theta \to \infty$.

[19] $r = 8\sec\theta \Rightarrow r\cos\theta = 8 \Rightarrow x = 8$, a vertical line.

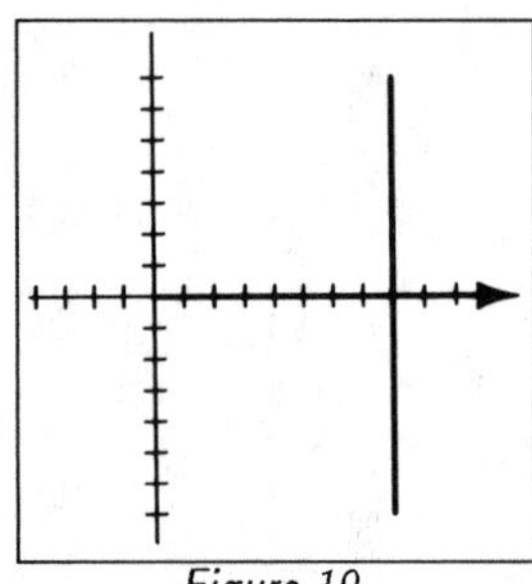

Figure 19

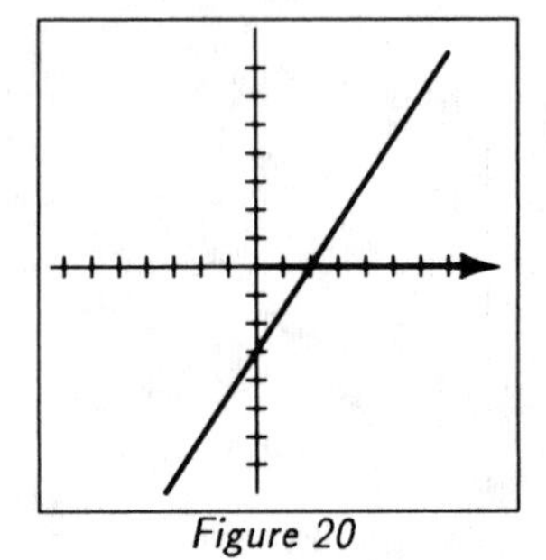

Figure 20

[20] $r(3\cos\theta - 2\sin\theta) = 6 \Rightarrow 3x - 2y = 6$, a line.

[21] $r = 4 - 4\cos\theta$ is a cardioid.

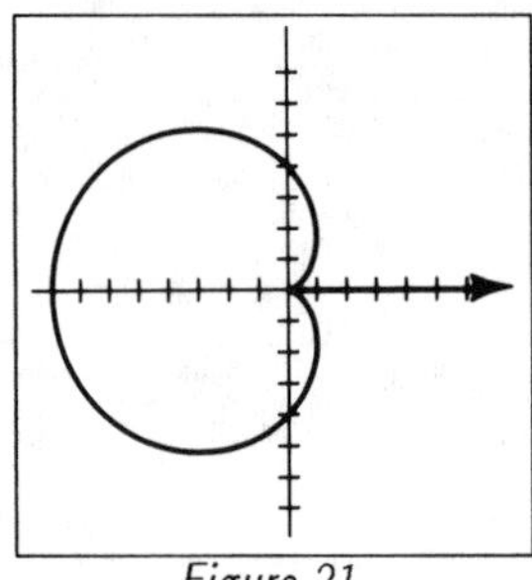

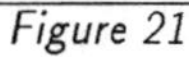

Figure 21

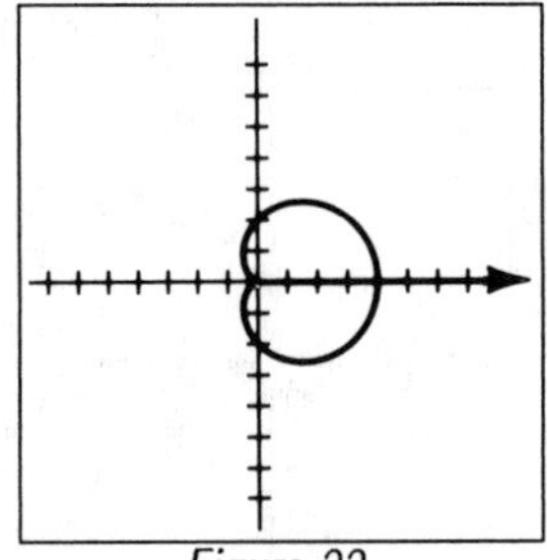

Figure 22

[22] $r = 4\cos^2(\frac{1}{2}\theta) = 4\left(\frac{1 + \cos\theta}{2}\right) = 2(1 + \cos\theta)$, a cardioid.

[23] $r = 6 - r\cos\theta \Rightarrow r + r\cos\theta = 6 \Rightarrow r(1 + \cos\theta) = 6 \Rightarrow r = \frac{6}{1 + 1\cos\theta} \Rightarrow$

$e = 1$, a parabola. The vertex is in the $\theta = 0$ direction, $V(3, 0)$.

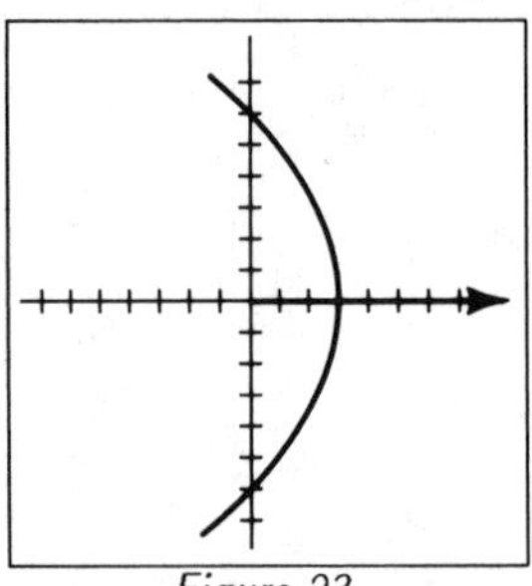

Figure 23

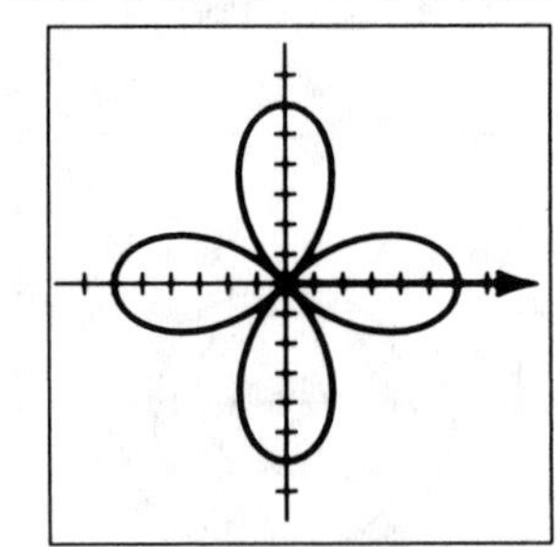

Figure 24

[24] $r = 6\cos 2\theta$ is a four-leafed rose.

One leaf is centered on the polar axis and the others are equally spaced 90° apart.

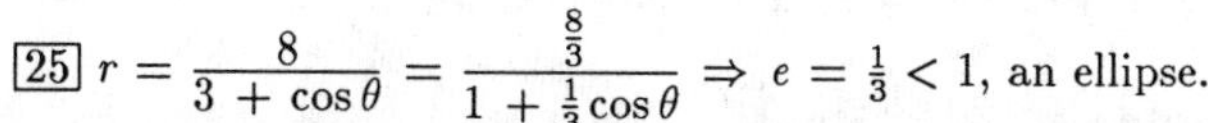

[25] $r = \dfrac{8}{3 + \cos\theta} = \dfrac{\frac{8}{3}}{1 + \frac{1}{3}\cos\theta} \Rightarrow e = \frac{1}{3} < 1$, an ellipse.

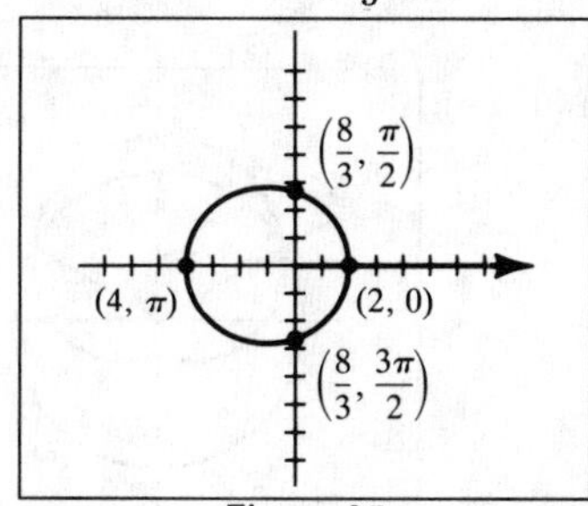

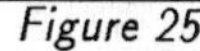
Figure 25

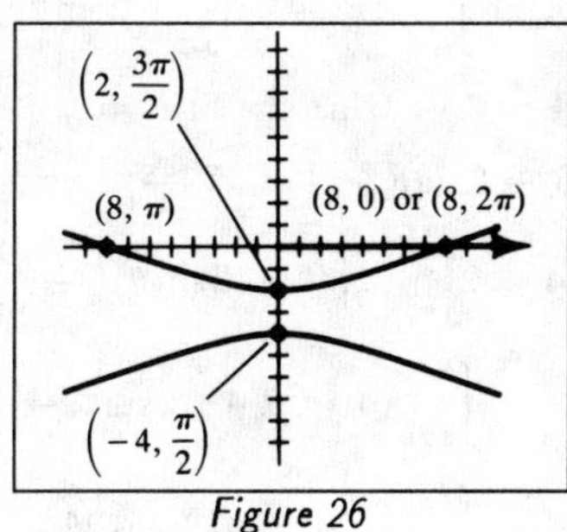

Figure 26

[26] $r = \dfrac{8}{1 - 3\sin\theta} \Rightarrow e = 3 > 1$, a hyperbola.

[27] $y^2 = 4x \Rightarrow r^2\sin^2\theta = 4r\cos\theta \Rightarrow r = \dfrac{4r\cos\theta}{r\sin^2\theta} = 4 \cdot \dfrac{\cos\theta}{\sin\theta} \cdot \dfrac{1}{\sin\theta} \Rightarrow r = 4\cot\theta\csc\theta$.

[28] $x^2 + y^2 - 3x + 4y = 0 \Rightarrow r^2 - 3r\cos\theta + 4r\sin\theta = 0 \Rightarrow$

$r - 3\cos\theta + 4\sin\theta = 0 \Rightarrow r = 3\cos\theta - 4\sin\theta$.

[29] $2x - 3y = 8 \Rightarrow 2r\cos\theta - 3r\sin\theta = 8 \Rightarrow r(2\cos\theta - 3\sin\theta) = 8$.

[30] $x^2 + y^2 = 2xy \Rightarrow r^2 = 2r^2\cos\theta\sin\theta \Rightarrow 1 = 2\sin\theta\cos\theta \Rightarrow \sin 2\theta = 1 \Rightarrow$

$2\theta = \frac{\pi}{2} + 2\pi n \Rightarrow \theta = \frac{\pi}{4}, \frac{5\pi}{4}$ on $[0, 2\pi)$, which are the same lines.

In rectangular coordinates: $x^2 + y^2 = 2xy \Rightarrow x^2 - 2xy + y^2 = 0 \Rightarrow$

$(x - y)^2 = 0 \Rightarrow x - y = 0$ or $y = x$.

[31] $y^2 = x^2 - 2x \Rightarrow r^2\sin^2\theta = r^2\cos^2\theta - 2r\cos\theta \Rightarrow$

$r(\cos^2\theta - \sin^2\theta) = 2\cos\theta \Rightarrow r\cos 2\theta = 2\cos\theta \Rightarrow r = 2\cos\theta\sec 2\theta$.

[32] $x^2 = y^2 + 3y \Rightarrow r^2\cos^2\theta = r^2\sin^2\theta + 3r\sin\theta \Rightarrow$

$r(\cos^2\theta - \sin^2\theta) = 3\sin\theta \Rightarrow r\cos 2\theta = 3\sin\theta \Rightarrow r = 3\sin\theta\sec 2\theta$.

[33] $r^2 = \tan\theta \Rightarrow x^2 + y^2 = \frac{y}{x} \Rightarrow x^3 + xy^2 = y$.

[34] $r = 2\cos\theta + 3\sin\theta \Rightarrow r^2 = 2r\cos\theta + 3r\sin\theta \Rightarrow x^2 + y^2 = 2x + 3y$.

[35] $r^2 = 4\sin 2\theta \Rightarrow r^2 = 4(2\sin\theta\cos\theta) \Rightarrow r^2 = 8\sin\theta\cos\theta \Rightarrow$

$r^2 \cdot r^2 = 8(r\sin\theta)(r\cos\theta) \Rightarrow (x^2 + y^2)^2 = 8xy$.

[36] $r^2 = \sec 2\theta \Rightarrow r^2\cos 2\theta = 1 \Rightarrow r^2(\cos^2\theta - \sin^2\theta) = 1 \Rightarrow$

$r^2\cos^2\theta - r^2\sin^2\theta = 1 \Rightarrow x^2 - y^2 = 1$.

[37] $\theta = \sqrt{3} \Rightarrow \tan^{-1}(\frac{y}{x}) = \sqrt{3} \Rightarrow \frac{y}{x} = \tan\sqrt{3} \Rightarrow y = (\tan\sqrt{3})\,x$.

Note that $\tan\sqrt{3} \approx -6.15$. This is a line through the origin making an angle of

approximately 99.24° with the positive x-axis. The line is *not* $y = \frac{\pi}{3}x$.

[38] $r = -6 \Rightarrow r^2 = 36 \Rightarrow x^2 + y^2 = 36$.

[39] $r' = \dfrac{6\sin\theta}{(2 + 2\cos\theta)^2}$, $r'(\frac{\pi}{2}) = \frac{3}{2}$, $r(\frac{\pi}{2}) = \frac{3}{2}$, $m = -1$ using (13.10).

[40] $r' = 3e^{3\theta}$, $r'(\frac{\pi}{4}) = 3e^{3\pi/4}$, $r(\frac{\pi}{4}) = e^{3\pi/4}$, $m = 2$ using (13.10).

[41] $A = \int_0^{\pi/2} \frac{1}{2}r^2\,d\theta = \int_0^{\pi/2} 2\sin 2\theta\,d\theta = \Big[-\cos 2\theta\Big]_0^{\pi/2} = 2.$

[42] $4 = 3 + 2\sin\theta \Rightarrow \sin\theta = \frac{1}{2} \Rightarrow \theta = \frac{\pi}{6}, \frac{5\pi}{6}$ on $[0, 2\pi)$.

$3 + 2\sin\theta \geq 4$ on $[\frac{\pi}{6}, \frac{5\pi}{6}]$. Using symmetry,

$$A = 2\int_{\pi/6}^{\pi/2} \tfrac{1}{2}\Big[(3 + 2\sin\theta)^2 - 4^2\Big]\,d\theta$$

$$= \int_{\pi/6}^{\pi/2} (4\sin^2\theta + 12\sin\theta - 7)\,d\theta$$

$$= \int_{\pi/6}^{\pi/2} (12\sin\theta - 2\cos 2\theta - 5)\,d\theta$$

$$= \Big[-12\cos\theta - \sin 2\theta - 5\theta\Big]_{\pi/6}^{\pi/2} = \tfrac{13}{2}\sqrt{3} - \tfrac{5\pi}{3} \approx 6.02.$$

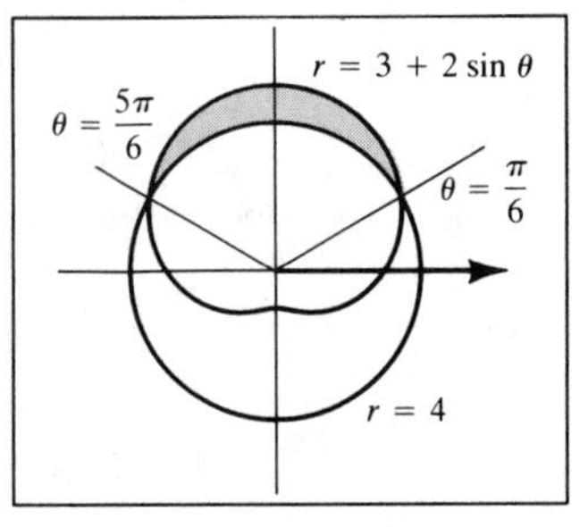

Figure 42

[43] $\frac{dx}{dt} = 2\cos t$ and $\frac{dy}{dt} = 2\sin t\cos t \Rightarrow L = \int_0^{\pi/2} \sqrt{4\cos^2 t + 4\sin^2 t\cos^2 t}\,dt$

$$= \int_0^{\pi/2} \sqrt{1 + \sin^2 t}\,(2\cos t)\,dt = 2\int_0^1 \sqrt{1 + u^2}\,du \;\{u = \sin t\}$$

$$= 2\Big[\tfrac{u}{2}\sqrt{1 + u^2} + \tfrac{1}{2}\ln\big|u + \sqrt{1 + u^2}\big|\Big]_0^1 \;\{\text{Formula 21}\} = \sqrt{2} + \ln(1 + \sqrt{2}) \approx 2.30.$$

[44] $f(\theta) = \frac{1}{\theta} \Rightarrow f'(\theta) = -\frac{1}{\theta^2}$. $L = \int_1^2 \sqrt{(1/\theta^2) + (1/\theta^4)}\,d\theta = \int_1^2 \frac{1}{\theta^2}\sqrt{\theta^2 + 1}\,d\theta$

$$= \Big[-\frac{\sqrt{\theta^2 + 1}}{\theta} + \ln\big|\theta + \sqrt{\theta^2 + 1}\big|\Big]_1^2 \;\{\text{Formula 24}\}$$

$$= -\tfrac{1}{2}\sqrt{5} + \ln(2 + \sqrt{5}) + \sqrt{2} - \ln(1 + \sqrt{2}) \approx 0.86.$$

[45] $\frac{dx}{dt} = 4t$ and $\frac{dy}{dt} = 4 \Rightarrow$

$$S = 2\pi\int_0^1 (2t^2 + 1)\sqrt{16t^2 + 16}\,dt = 16\pi\int_0^1 t^2\sqrt{t^2 + 1}\,dt + 8\pi\int_0^1 \sqrt{t^2 + 1}\,dt$$

$$= 16\pi\Big[\tfrac{t}{8}(1 + 2t^2)\sqrt{1 + t^2} - \tfrac{1}{8}\ln\big|t + \sqrt{1 + t^2}\big|\Big]_0^1 \;\{\text{Formula 22}\} +$$

$$8\pi\Big[\tfrac{t}{2}\sqrt{1 + t^2} + \tfrac{1}{2}\ln\big|t + \sqrt{1 + t^2}\big|\Big]_0^1 \;\{\text{Formula 21}\}$$

$$= 16\pi\Big[\tfrac{3}{8}\sqrt{2} - \tfrac{1}{8}\ln(1 + \sqrt{2})\Big] + 8\pi\Big[\tfrac{1}{2}\sqrt{2} + \tfrac{1}{2}\ln(1 + \sqrt{2})\Big]$$

$$= 2\pi\Big[5\sqrt{2} + \ln(1 + \sqrt{2})\Big] \approx 49.97.$$

[46] $f(\theta) = e^\theta \Rightarrow f'(\theta) = e^\theta$. Using (13.14), $S = \int_0^1 2\pi\,e^\theta\cos\theta\sqrt{e^{2\theta} + e^{2\theta}}\,d\theta =$

$$2\pi\sqrt{2}\int_0^1 e^{2\theta}\cos\theta\,d\theta = 2\pi\sqrt{2}\Big[\frac{e^{2\theta}}{5}(2\cos\theta + \sin\theta)\Big]_0^1 \;\{\text{Formula 99}\} =$$

$$\tfrac{2}{5}\pi\sqrt{2}\Big[e^2(2\cos 1 + \sin 1) - 2\Big] \approx 21.69.$$

[47] See Exercise 19, §13.3 for a graph of this form with $a = 2$. $r^2 = a^2 \cos 2\theta \Rightarrow$

$2rr' = -2a^2 \sin 2\theta \Rightarrow r' = -\dfrac{a \sin 2\theta}{\sqrt{\cos 2\theta}}$ on $[0, \frac{\pi}{4}]$. Using symmetry,

$$S = 2\int_0^{\pi/4} 2\pi\, a\sqrt{\cos 2\theta}\sin\theta \sqrt{a^2\cos 2\theta + \left(-\frac{a\sin 2\theta}{\sqrt{\cos 2\theta}}\right)^2}\, d\theta$$

$$= 4\pi a\int_0^{\pi/4} \sqrt{\cos 2\theta}\sqrt{a^2\cos 2\theta + \frac{a^2\sin^2 2\theta}{\cos 2\theta}}\sin\theta\, d\theta$$

$$= 4\pi a\int_0^{\pi/4} \sqrt{a^2\cos^2 2\theta + a^2\sin^2 2\theta}\sin\theta\, d\theta = 4\pi a^2\int_0^{\pi/4} \sin\theta\, d\theta$$

$$= 4\pi a^2\Big[-\cos\theta\Big]_0^{\pi/4} = -4\pi a^2(\tfrac{\sqrt{2}}{2} - 1) = 2\pi a^2(2 - \sqrt{2}) \approx 3.68a^2.$$

[48] Since t is always an interior angle of the right triangle AOB, it must be acute.

From the figure, $\cos t = \frac{|x|}{a}$ and $\sin t = \frac{|y|}{b} \Rightarrow |x| = a\cos t$ and $|y| = b\sin t \Rightarrow$

(i) $x = a\cos t,\ y = b\sin t$ in Quadrant I

(ii) $x = -a\cos t,\ y = b\sin t$ in Quadrant II

(iii) $x = -a\cos t,\ y = -b\sin t$ in Quadrant III

(iv) $x = a\cos t,\ y = -b\sin t$ in Quadrant IV

Since $1 = \sin^2 t + \cos^2 t = \frac{x^2}{a^2} + \frac{y^2}{b^2}$, the curve is an ellipse. {see Exercise 27, §12.3}

Chapter 14: Vectors and Surfaces

Exercises 14.1

[1] $\mathrm{a} = <2, 5> \Rightarrow \|\mathrm{a}\| = \sqrt{2^2 + 5^2} = \sqrt{29}$

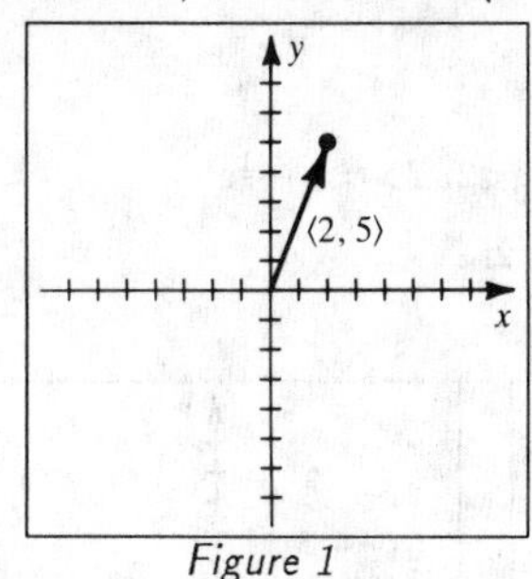

Figure 1

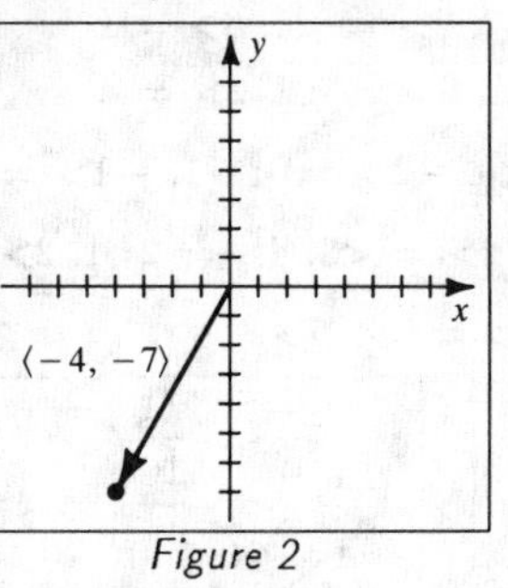

Figure 2

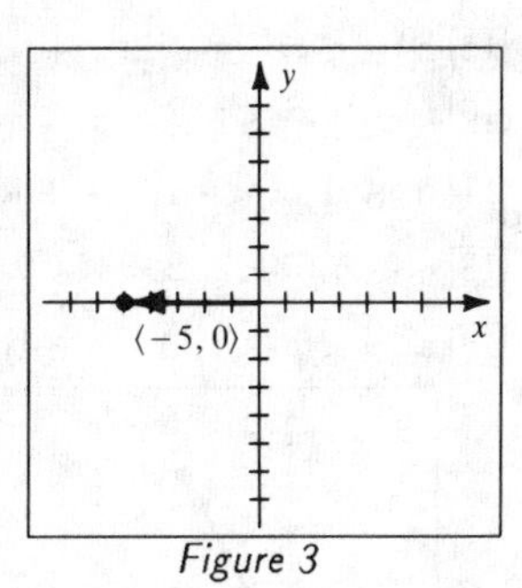

Figure 3

[2] $\mathrm{a} = <-4, -7> \Rightarrow \|\mathrm{a}\| = \sqrt{(-4)^2 + (-7)^2} = \sqrt{65}$

[3] $\mathrm{a} = <-5, 0> \Rightarrow \|\mathrm{a}\| = \sqrt{(-5)^2 + 0^2} = \sqrt{(-5)^2} = |-5| = 5$

[4] $\mathrm{a} = -18\mathrm{j} \Rightarrow \|\mathrm{a}\| = \sqrt{0^2 + (-18)^2} = \sqrt{(-18)^2} = |-18| = 18$

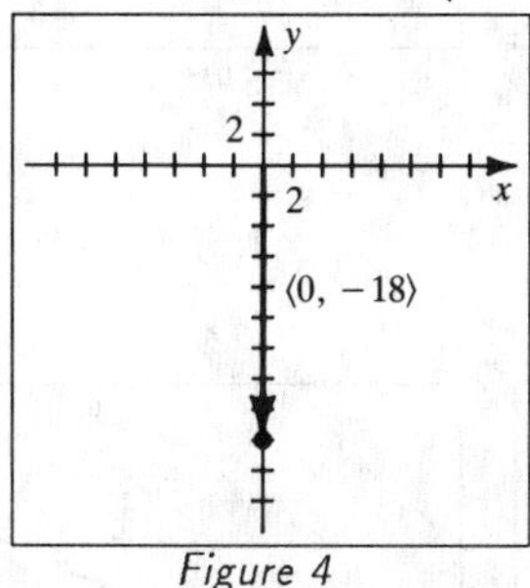

Figure 4

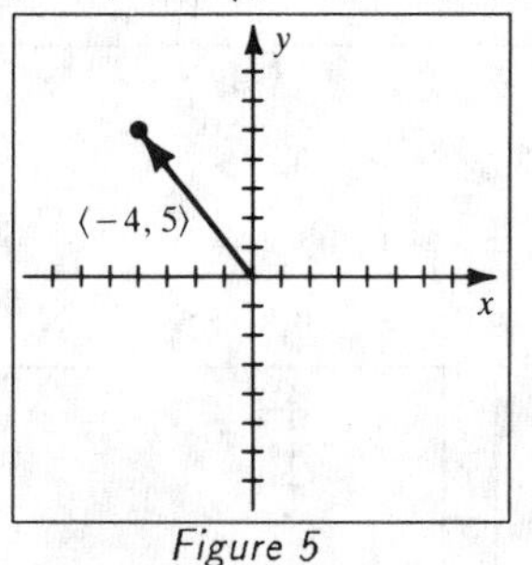

Figure 5

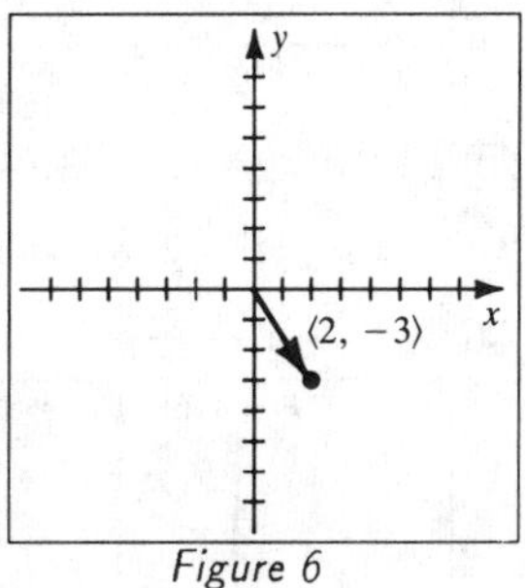

Figure 6

[5] $\mathrm{a} = -4\mathrm{i} + 5\mathrm{j} \Rightarrow \|\mathrm{a}\| = \sqrt{(-4)^2 + 5^2} = \sqrt{41}$

[6] $\mathrm{a} = 2\mathrm{i} - 3\mathrm{j} \Rightarrow \|\mathrm{a}\| = \sqrt{2^2 + (-3)^2} = \sqrt{13}$

[7] $\mathrm{a} + \mathrm{b} = <2, -3> + <1, 4> = <2 + 1, -3 + 4> = <3, 1>$.

$\mathrm{a} - \mathrm{b} = <2, -3> - <1, 4> = <2 - 1, -3 - 4> = <1, -7>$.

$2\mathrm{a} = 2<2, -3> = <2 \cdot 2, 2 \cdot (-3)> = <4, -6>$.

$-3\mathrm{b} = -3<1, 4> = <-3 \cdot 1, -3 \cdot 4> = <-3, -12>$.

$4\mathrm{a} - 5\mathrm{b} = 4<2, -3> - 5<1, 4> = <8, -12> - <5, 20> =$

$<8 - 5, -12 - 20> = <3, -32>$.

Note: In Exercises 8–12, the answers are given in the following order:

	a + b,	a − b,	2a,	−3b,	4a − 5b
[8]	$<-5, -1>$,	$<1, -9>$,	$<-4, -10>$,	$<9, -12>$,	$<7, -40>$
[9]	$<-15, 6>$,	$<1, -2>$,	$<-14, 4>$,	$<24, -12>$,	$<12, -12>$
[10]	$<5, 22>$,	$<-1, -2>$,	$<4, 20>$,	$<-9, -36>$,	$<-7, -20>$
[11]	$2\mathrm{i} + 7\mathrm{j}$,	$4\mathrm{i} - 3\mathrm{j}$,	$6\mathrm{i} + 4\mathrm{j}$,	$3\mathrm{i} - 15\mathrm{j}$,	$17\mathrm{i} - 17\mathrm{j}$
[12]	$-4\mathrm{i} - \mathrm{j}$,	$-6\mathrm{i} + 5\mathrm{j}$,	$-10\mathrm{i} + 4\mathrm{j}$,	$-3\mathrm{i} + 9\mathrm{j}$,	$-25\mathrm{i} + 23\mathrm{j}$

[13] $\mathbf{a} + \mathbf{b} = \langle 2, 0\rangle + \langle -1, 0\rangle = \langle 1, 0\rangle = -\langle -1, 0\rangle = -\mathbf{b}$

{ Alternate answer: $\frac{1}{2}\mathbf{a}$ }

[14] $\mathbf{c} - \mathbf{d} = \langle 0, 2\rangle - \langle 0, -1\rangle = \langle 0, 3\rangle = -3\langle 0, -1\rangle = -3\mathbf{d}$

{ Alternate answer: $\frac{3}{2}\mathbf{c}$ }

[15] $\mathbf{b} + \mathbf{e} = \langle -1, 0\rangle + \langle 2, 2\rangle = \langle 1, 2\rangle = \mathbf{f}$

[16] $\mathbf{f} - \mathbf{b} = \langle 1, 2\rangle - \langle -1, 0\rangle = \langle 2, 2\rangle = \mathbf{e}$

[17] $\mathbf{b} + \mathbf{d} = \langle -1, 0\rangle + \langle 0, -1\rangle = \langle -1, -1\rangle = -\frac{1}{2}\langle 2, 2\rangle = -\frac{1}{2}\mathbf{e}$

[18] $\mathbf{e} + \mathbf{c} = \langle 2, 2\rangle + \langle 0, 2\rangle = \langle 2, 4\rangle = 2\langle 1, 2\rangle = 2\mathbf{f}$

[19] $\mathbf{a} = \langle 5 - 1, 3 - (-4)\rangle = \langle 4, 7\rangle$

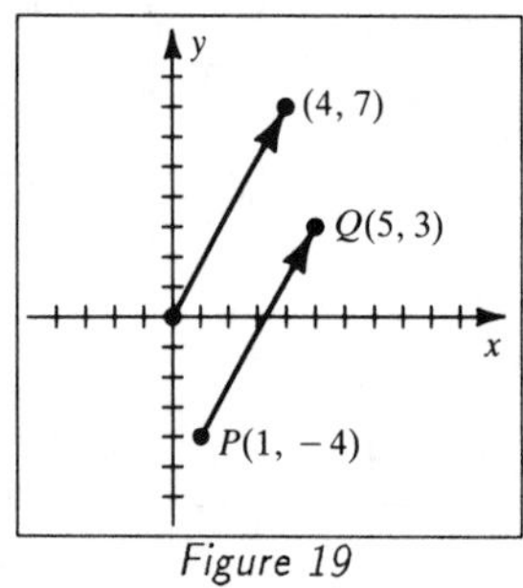

Figure 19

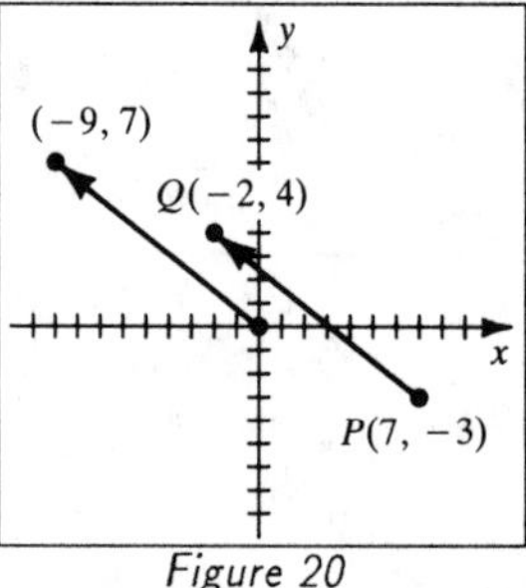

Figure 20

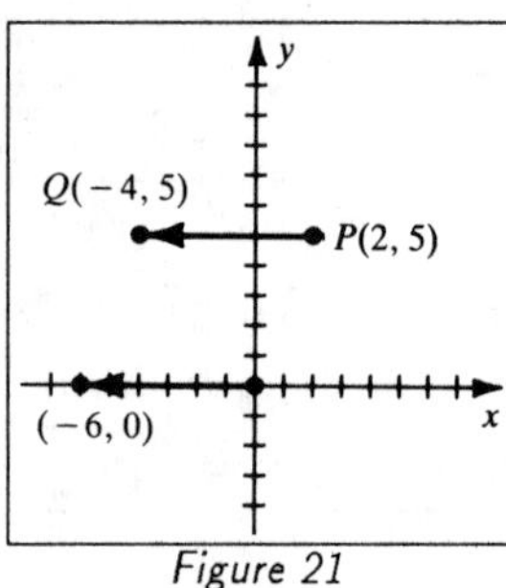

Figure 21

[20] $\mathbf{a} = \langle -2 - 7, 4 - (-3)\rangle = \langle -9, 7\rangle$

[21] $\mathbf{a} = \langle -4 - 2, 5 - 5\rangle = \langle -6, 0\rangle$

[22] $\mathbf{a} = \langle -4 - (-4), -2 - 6\rangle = \langle 0, -8\rangle$

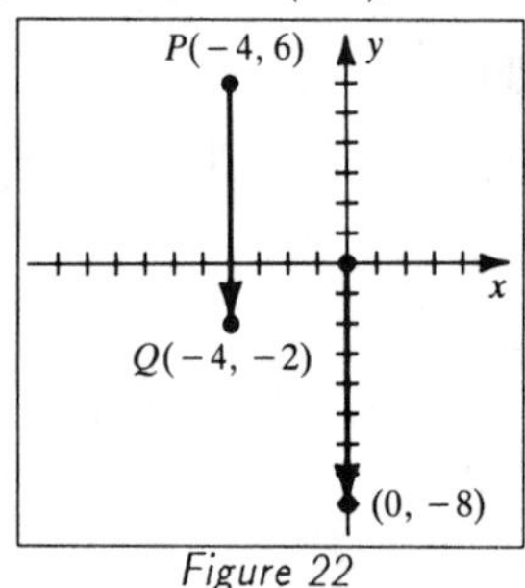

Figure 22

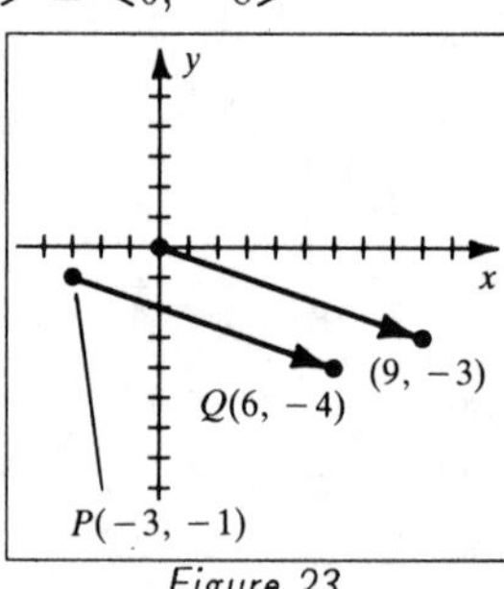

Figure 23

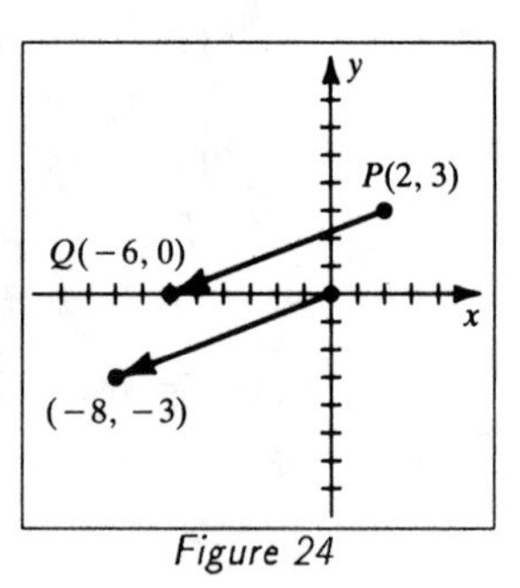

Figure 24

[23] $\mathbf{a} = \langle 6 - (-3), -4 - (-1)\rangle = \langle 9, -3\rangle$

[24] $\mathbf{a} = \langle -6 - 2, 0 - 3\rangle = \langle -8, -3\rangle$

Note: In Exercises 25–28, let $\mathbf{u}$ denote the unit vector in the direction of $\mathbf{a}$.

[25] (a) $\|\mathbf{a}\| = \sqrt{(-8)^2 + 15^2} = \sqrt{289} = 17$; $\quad \mathbf{u} = \frac{\mathbf{a}}{\|\mathbf{a}\|} = -\frac{8}{17}\mathbf{i} + \frac{15}{17}\mathbf{j}$

(b) $-\mathbf{u} = -(-\frac{8}{17}\mathbf{i} + \frac{15}{17}\mathbf{j}) = \frac{8}{17}\mathbf{i} - \frac{15}{17}\mathbf{j}$

[26] (a) $\|\mathbf{a}\| = \sqrt{5^2 + (-3)^2} = \sqrt{34}$; $\quad \mathbf{u} = \frac{\mathbf{a}}{\|\mathbf{a}\|} = \frac{5}{\sqrt{34}}\mathbf{i} - \frac{3}{\sqrt{34}}\mathbf{j}$

(b) $-\mathbf{u} = -(\frac{5}{\sqrt{34}}\mathbf{i} - \frac{3}{\sqrt{34}}\mathbf{j}) = -\frac{5}{\sqrt{34}}\mathbf{i} + \frac{3}{\sqrt{34}}\mathbf{j}$

[27] (a) $\|\mathbf{a}\| = \sqrt{2^2 + (-5)^2} = \sqrt{29}$; $\quad \mathbf{u} = \frac{\mathbf{a}}{\|\mathbf{a}\|} = \left\langle \frac{2}{\sqrt{29}}, -\frac{5}{\sqrt{29}} \right\rangle$

(b) $-\mathbf{u} = -(\left\langle \frac{2}{\sqrt{29}}, -\frac{5}{\sqrt{29}} \right\rangle) = \left\langle -\frac{2}{\sqrt{29}}, \frac{5}{\sqrt{29}} \right\rangle$

[28] (a) $\|\mathbf{a}\| = \sqrt{0^2 + 6^2} = 6$; $\mathbf{u} = \frac{\mathbf{a}}{\|\mathbf{a}\|} = <\frac{0}{6}, \frac{6}{6}> = <0, 1>$

(b) $-\mathbf{u} = -(<0, 1>) = <0, -1>$

[29] (a) $2<-6, 3> = <-12, 6>$ (b) $\frac{1}{2}<-6, 3> = <-3, \frac{3}{2}>$

[30] (a) $-3(8\mathbf{i} - 5\mathbf{j}) = -24\mathbf{i} + 15\mathbf{j}$ (b) $-\frac{1}{3}(8\mathbf{i} - 5\mathbf{j}) = -\frac{8}{3}\mathbf{i} + \frac{5}{3}\mathbf{j}$

Note: In Exercises 31–32, let $\mathbf{v}$ denote the desired vector.

[31] $\mathbf{a} = 4\mathbf{i} - 7\mathbf{j} \Rightarrow \|\mathbf{a}\| = \sqrt{65}$. $\mathbf{v} = 6\left(\frac{\mathbf{a}}{\|\mathbf{a}\|}\right) = \frac{24}{\sqrt{65}}\mathbf{i} - \frac{42}{\sqrt{65}}\mathbf{j}$.

[32] $\mathbf{a} = <2, -5> \Rightarrow \|\mathbf{a}\| = \sqrt{29}$. $\mathbf{v} = -4\left(\frac{\mathbf{a}}{\|\mathbf{a}\|}\right) = \left<-\frac{8}{\sqrt{29}}, \frac{20}{\sqrt{29}}\right>$.

[33] (a) $\mathbf{a} = 3\mathbf{i} - 4\mathbf{j} \Rightarrow \|\mathbf{a}\| = 5$. $\|c\mathbf{a}\| = |c|\,\|\mathbf{a}\|\,\{(14.9)\} = |c|(5)$. $5|c| = 3 \Rightarrow c = \pm\frac{3}{5}$.

(b) None, since $\|c\mathbf{a}\| \geq 0$ for all $c\mathbf{a}$.

(c) $\|c\mathbf{a}\| = 0 \Rightarrow c\mathbf{a} = \mathbf{0} \Rightarrow c = 0$, since $\mathbf{a} \neq \mathbf{0}$.

[34] (a), (b), & (c) are the same as in Exercise 33 except $\|\mathbf{a}\| = 13$.

[35] Horizontal $= 50\cos 35^\circ \approx 40.96$. Vertical $= 50\sin 35^\circ \approx 28.68$.

[36] Horizontal $= 20\cos 40^\circ \approx 15.32$. Vertical $= 20\sin 40^\circ \approx 12.86$.

[37] Horizontal $= 20\cos 108^\circ \approx -6.18$. Vertical $= 20\sin 108^\circ \approx 19.02$.

[38] Horizontal $= 160\cos 7.5^\circ \approx 158.63$. Vertical $= 160\sin 7.5^\circ \approx 20.88$.

Note: In Exercises 39–40, let $\mathbf{v}$ denote the plane's air velocity,

$\mathbf{r}$ the plane's ground velocity, and $\mathbf{w}$ the wind velocity.

[39] $\mathbf{v} = <300\cos(-60^\circ), 300\sin(-60^\circ)> = <150, -150\sqrt{3}>$.

$\mathbf{w} = <30\cos 30^\circ, 30\sin 30^\circ> = <15\sqrt{3}, 15>$.

Thus, $\mathbf{r} = \mathbf{v} + \mathbf{w} = <150 + 15\sqrt{3}, 15 - 150\sqrt{3}>$ and $\|\mathbf{r}\| = \sqrt{90{,}900} \approx 301.5$ mi/hr.

Since $\tan^{-1}\left(\frac{15 - 150\sqrt{3}}{150 + 15\sqrt{3}}\right) \approx -54.3^\circ$, the true course is $90^\circ + 54.3^\circ = 144.3^\circ$.

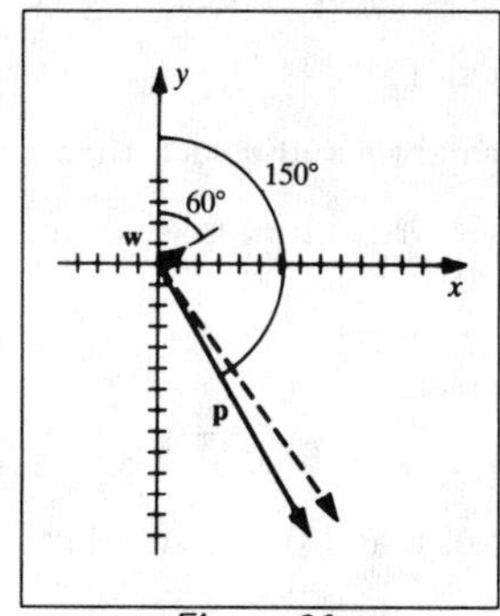

Figure 39

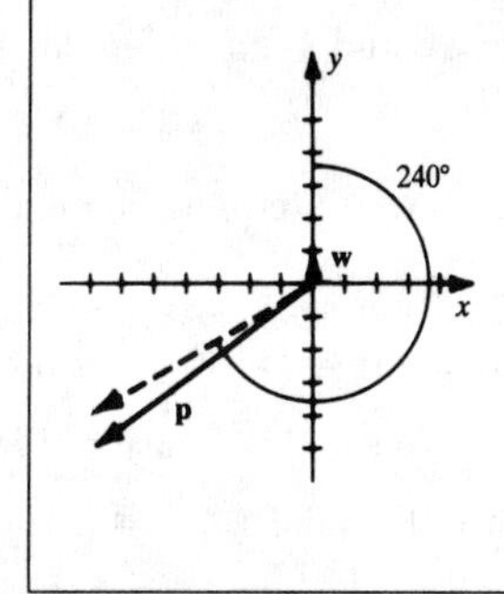

Figure 40

[40] $\mathbf{r} = <400\cos(-150^\circ), 400\sin(-150^\circ)> = <-200\sqrt{3}, -200>$. $\mathbf{w} = <0, 50>$.

Thus, $\mathbf{v} + \mathbf{w} = \mathbf{r} \Rightarrow \mathbf{v} = \mathbf{r} - \mathbf{w} \Rightarrow \mathbf{v} = <-200\sqrt{3}, -250> \Rightarrow$

$\|\mathbf{v}\| = \sqrt{182{,}500} \approx 427.2$ mi/hr. $\text{Tan}^{-1}\left(\frac{-250}{-200\sqrt{3}}\right) \approx 35.8^\circ \Rightarrow$ the true compass

heading is 35.8° south of due east, or in a direction of $270^\circ - 35.8^\circ = 234.2^\circ$.

[41] The vertical components of the forces must add up to zero for the large ship to move along the line segment AB. The vertical component of the smaller tug is $3200\sin(-30^\circ) = -1600$. The vertical component of the larger tug is $4000\sin\theta$.

$$4000\sin\theta = 1600 \Rightarrow \theta = \sin^{-1}(0.4) \approx 23.6^\circ.$$

[42] (a) Consider the force of 160 pounds to be the resultant vector of two vectors whose initial point is at the astronaut's feet, one along the positive x-axis and the other along the negative y-axis. The angle formed by the resultant vector and the positive x-axis is the complement of θ, $90^\circ - \theta$.

Now $\quad \cos(90^\circ - \theta) = \dfrac{x\text{-component}}{160} \Rightarrow x\text{-component} = 160\sin\theta$

and $\quad \sin(90^\circ - \theta) = \dfrac{y\text{-component}}{160} \Rightarrow y\text{-component} = 160\cos\theta.$

(b) $27 = 160\cos\theta \Rightarrow \theta = \cos^{-1}(\frac{27}{160}) \approx 80.28^\circ$ on the moon.

$60 = 160\cos\theta \Rightarrow \theta = \cos^{-1}(\frac{3}{8}) \approx 67.98^\circ$ on Mars.

[43] $p\langle 3, -1\rangle + q\langle 4, 3\rangle = \langle -6, -11\rangle \Rightarrow \begin{cases} 3p + 4q = -6 & (E_1) \\ -p + 3q = -11 & (E_2) \end{cases}$

$$3E_2 + E_1 \Rightarrow p = 2 \text{ and } q = -3.$$

[44] Let $\mathbf{c} = \langle c_1, c_2\rangle$. Then $\mathbf{c} = p\mathbf{a} + q\mathbf{b} \Leftrightarrow c_1 = pa_1 + qb_1$ and $c_2 = pa_2 + qb_2$.

There are three cases to consider:

(a) $b_1 \neq 0$ and $b_2 \neq 0 \Rightarrow \frac{a_1}{b_1} \neq \frac{a_2}{b_2}$ {since $\mathbf{a}$ and $\mathbf{b}$ are nonparallel} $\Rightarrow$

$a_1b_2 \neq a_2b_1 \Rightarrow a_1b_2 - a_2b_1 \neq 0 \Rightarrow$

the system of equations has a unique solution p and q by Cramer's rule.

(b) $b_1 = 0$ and $b_2 \neq 0$: $c_1 = pa_1$ and $c_2 = pa_2 + qb_2$.

Since $\mathbf{a} \neq \mathbf{0}$, $a_1 \neq 0$ and hence $p = c_1/a_1$. Thus, $q = (c_2 - pa_2)/b_2$.

(c) $b_1 \neq 0$ and $b_2 = 0$: Similar to the argument in (b),

we can find scalars p and q such that $\mathbf{c} = p\mathbf{a} + q\mathbf{b}$.

The geometric interpretation is that any $\mathbf{c}$ in V_2 is the sum of two vectors parallel to $\mathbf{a}$ and $\mathbf{b}$.

[45] $\|\mathbf{a} + \mathbf{b}\| = \|\mathbf{a}\| + \|\mathbf{b}\|$ if $\mathbf{a}$ and $\mathbf{b}$ have the same directions.

To see this, consider Figure 14.4 with $\mathbf{a} = \overrightarrow{AB}$, $\mathbf{b} = \overrightarrow{BC}$, and $\mathbf{a} + \mathbf{b} = \overrightarrow{AC}$.

[46] (a) Since the point (a_1, a_2) is the terminal point of $\mathbf{a}$ and $\|\mathbf{a}\|$ is the length of $\mathbf{a}$, we know from trigonometry that $\cos\theta = \dfrac{a_1}{\|\mathbf{a}\|}$ and $\sin\theta = \dfrac{a_2}{\|\mathbf{a}\|}$.

Thus, $\langle a_1, a_2\rangle = \langle \|\mathbf{a}\|\cos\theta, \|\mathbf{a}\|\sin\theta\rangle = \|\mathbf{a}\|(\cos\theta\,\mathbf{i} + \sin\theta\,\mathbf{j})$.

(b) Let $\mathbf{u}$ be any unit vector.

Then by part (a), $\mathbf{u} = \|\mathbf{u}\|(\cos\theta\,\mathbf{i} + \sin\theta\,\mathbf{j}) = \cos\theta\,\mathbf{i} + \sin\theta\,\mathbf{j}$, since $\|\mathbf{u}\| = 1$.

47 $\|\mathbf{r} - \mathbf{r}_0\| = c \Leftrightarrow \|\mathbf{r} - \mathbf{r}_0\|^2 = c^2 \Leftrightarrow (x - x_0)^2 + (y - y_0)^2 = c^2$.

This is a circle with center (x_0, y_0) and radius c.

48 $\mathbf{r} - \mathbf{r}_0 = c\mathbf{a} \Leftrightarrow \langle x - x_0, y - y_0 \rangle = c\langle a_1, a_2 \rangle \Leftrightarrow$

$(x - x_0) = ca_1$ and $y - y_0 = ca_2 \Leftrightarrow x = a_1 c + x_0$ and $y = a_2 c + y_0$,

which is a parametric equation of a line through (x_0, y_0).

49 $\mathbf{a} + (\mathbf{b} + \mathbf{c}) = \langle a_1, a_2 \rangle + (\langle b_1, b_2 \rangle + \langle c_1, c_2 \rangle)$

$= \langle a_1, a_2 \rangle + \langle b_1 + c_1, b_2 + c_2 \rangle$

$= \langle a_1 + b_1 + c_1, a_2 + b_2 + c_2 \rangle$

$= \langle a_1 + b_1, a_2 + b_2 \rangle + \langle c_1, c_2 \rangle$

$= (\langle a_1, a_2 \rangle + \langle b_1, b_2 \rangle) + \langle c_1, c_2 \rangle = (\mathbf{a} + \mathbf{b}) + \mathbf{c}$

50 $\mathbf{a} + \mathbf{0} = \langle a_1, a_2 \rangle + \langle 0, 0 \rangle = \langle a_1 + 0, a_2 + 0 \rangle = \langle a_1, a_2 \rangle = \mathbf{a}$

51 $0\mathbf{a} = 0\langle a_1, a_2 \rangle = \langle 0a_1, 0a_2 \rangle = \langle 0, 0 \rangle = \mathbf{0}$

Also, $p\mathbf{0} = p\langle 0, 0 \rangle = \langle p0, p0 \rangle = \langle 0, 0 \rangle = \mathbf{0}$

52 $1\mathbf{a} = 1\langle a_1, a_2 \rangle = \langle 1a_1, 1a_2 \rangle = \langle a_1, a_2 \rangle = \mathbf{a}$

53 $(p + q)\mathbf{a} = (p + q)\langle a_1, a_2 \rangle$

$= \langle (p + q)a_1, (p + q)a_2 \rangle$

$= \langle pa_1 + qa_1, pa_2 + qa_2 \rangle$

$= \langle pa_1, pa_2 \rangle + \langle qa_1, qa_2 \rangle$

$= p\langle a_1, a_2 \rangle + q\langle a_1, a_2 \rangle = p\mathbf{a} + q\mathbf{a}$

54 $p(\mathbf{a} - \mathbf{b}) = p\langle a_1 - b_1, a_2 - b_2 \rangle$

$= \langle pa_1 - pb_1, pa_2 - pb_2 \rangle$

$= \langle pa_1, pa_2 \rangle - \langle pb_1, pb_2 \rangle = p\mathbf{a} - p\mathbf{b}$

55 If $p\mathbf{a} = \mathbf{0}$, then $\langle pa_1, pa_2 \rangle = \langle 0, 0 \rangle$ and hence $pa_1 = 0$ and $pa_2 = 0$.

Since $p \neq 0$, $a_1 = a_2 = 0$, or, equivalently, $\mathbf{a} = \mathbf{0}$.

56 If $p\mathbf{a} = \mathbf{0}$, then $\langle pa_1, pa_2 \rangle = \langle 0, 0 \rangle$ and hence $pa_1 = 0$ and $pa_2 = 0$.

Since $\mathbf{a} \neq \mathbf{0}$, either $a_1 \neq 0$ or $a_2 \neq 0$, and hence $p = 0$.

Exercises 14.2

Note: In Exercises 1–6, use (14.13), (14.14), and the 3-D generalization of (14.7).

1 (a) $d(A, B) = \sqrt{(4 - 2)^2 + (-2 - 4)^2 + (3 + 5)^2} = \sqrt{4 + 36 + 64} = \sqrt{104}$

(b) $M(A, B) = \left(\frac{2 + 4}{2}, \frac{4 + (-2)}{2}, \frac{-5 + 3}{2}\right) = (3, 1, -1)$

(c) The vector in V_3 that corresponds to $\overrightarrow{AB}$ is

$\langle 4 - 2, -2 - 4, 3 - (-5) \rangle = \langle 2, -6, 8 \rangle$.

2 (a) $\sqrt{101}$ (b) $(\frac{3}{2}, 1, 3)$ (c) $\langle 1, 6, -8 \rangle$

3 (a) $\sqrt{53}$ (b) $(-\frac{1}{2}, -1, 1)$ (c) $\langle 7, -2, 0 \rangle$

4 (a) $\sqrt{33}$ (b) $(\frac{1}{2}, 3, -2)$ (c) $\langle 1, -4, 4 \rangle$

[5] (a) $\sqrt{3}$ (b) $(\frac{1}{2}, \frac{1}{2}, \frac{1}{2})$ (c) $<-1, 1, 1>$

[6] (a) 9 (b) $(-4, -\frac{1}{2}, 2)$ (c) $<-8, -1, 4>$

[7] (a) $<1, 3, 0>$ (b) $<-5, 9, 2>$ (c) $<-22, 42, 9>$ (d) $\sqrt{41}$ (e) $3\sqrt{41}$

[8] (a) $<-3, 2, -2>$ (b) $<5, 2, -4>$ (c) $<21, 10, -19>$ (d) $\sqrt{14}$ (e) $3\sqrt{14}$

[9] (a) $4i - 2j - 3k$ (b) $2i - 6j + 7k$ (c) $11i - 28j + 30k$ (d) $\sqrt{29}$ (e) $3\sqrt{29}$

[10] (a) $3i - j + 3k$ (b) $i - j + 5k$ (c) $6i - 5j + 24k$ (d) $\sqrt{21}$ (e) $3\sqrt{21}$

[11] (a) $i + k$ (b) $i + 2j - k$ (c) $5i + 9j - 4k$ (d) $\sqrt{2}$ (e) $3\sqrt{2}$

[12] (a) $2i + 3k$ (b) $2i - 3k$ (c) $10i - 12k$ (d) 2 (e) 6

[13] $2a = <4, 6, 8>$; $-3b = <-3, 6, -6>$; $a + b = <3, 1, 6>$; $a - b = <1, 5, 2>$

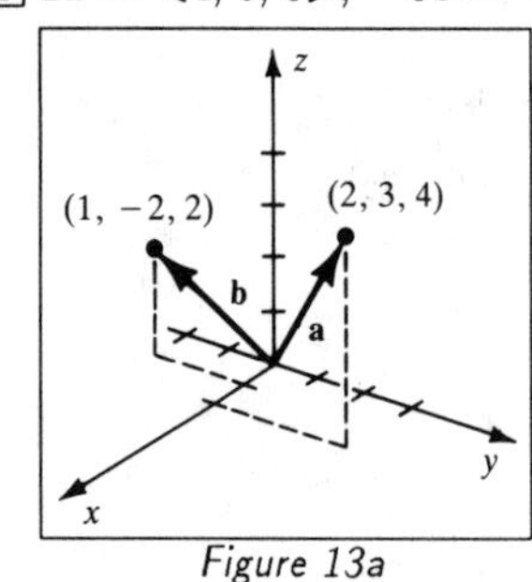

Figure 13a

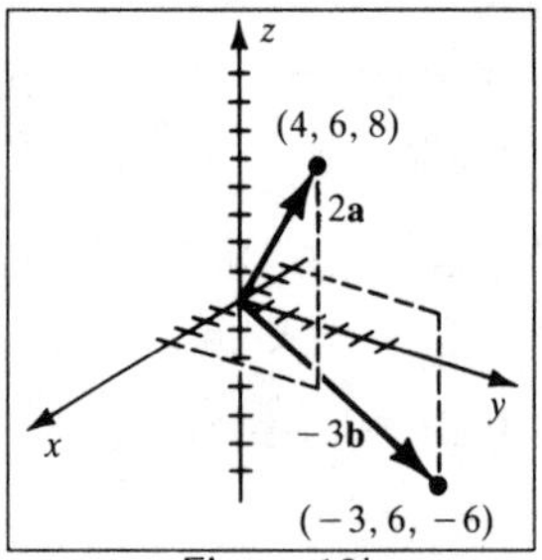

Figure 13b

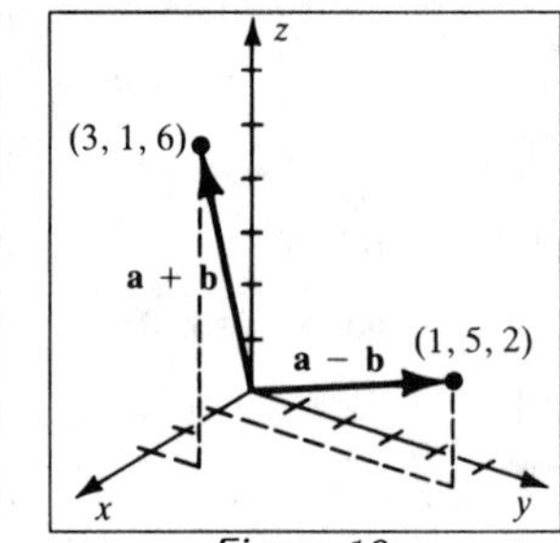

Figure 13c

[14] $2a = -2i + 4j + 6k$; $-3b = 6j - 3k$; $a + b = -i + 4k$; $a - b = -i + 4j + 2k$

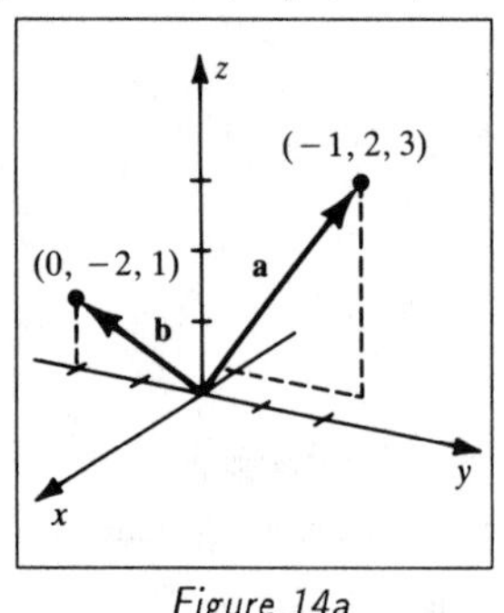

Figure 14a

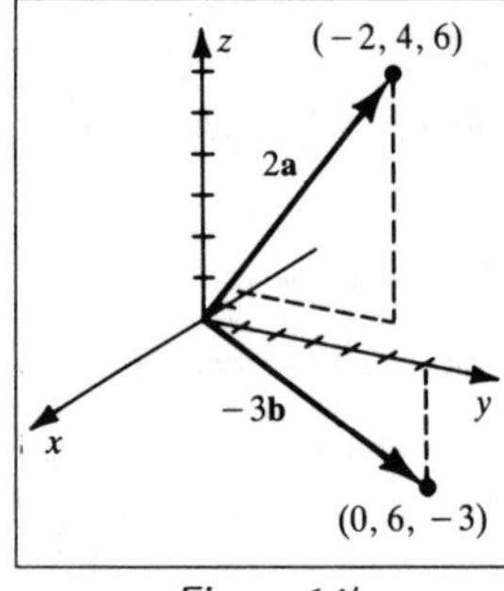

Figure 14b

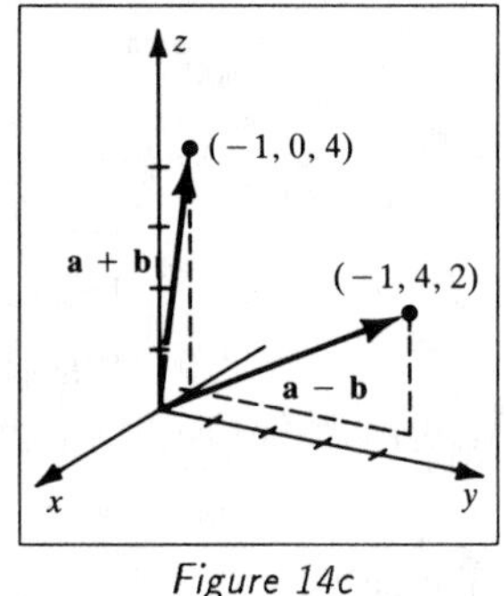

Figure 14c

[15] $a = <-4, 10, -2> \Rightarrow \|a\| = \sqrt{120} = 2\sqrt{30}$. $u = \frac{a}{\|a\|} = \frac{1}{\sqrt{30}}<-2, 5, -1>$.

[16] $a = 3i - 7j + 2k \Rightarrow \|a\| = \sqrt{62}$. $u = \frac{a}{\|a\|} = \frac{1}{\sqrt{62}}(3i - 7j + 2k)$.

[17] (a) $2a = 28i - 30j + 12k$ (b) $-\frac{1}{3}a = -\frac{14}{3}i + 5j - 2k$

(c) $\frac{2a}{\|a\|} = \frac{2}{\sqrt{457}}(14i - 15j + 6k)$

[18] (a) $2a = <-12, -6, 12>$ (b) $-\frac{1}{3}a = <2, 1, -2>$

(c) $\frac{2a}{\|a\|} = \frac{2}{9}<-6, -3, 6> = \frac{2}{3}<-2, -1, 2>$

[19] $C(3, -1, 2)$ and $r = 3 \Rightarrow (x - 3)^2 + (y + 1)^2 + (z - 2)^2 = 3^2 = 9$

[20] $C(4, -5, 1)$ and $r = 5 \Rightarrow (x - 4)^2 + (y + 5)^2 + (z - 1)^2 = 5^2 = 25$

[21] $C(-5, 0, 1)$ and $r = \frac{1}{2} \Rightarrow (x + 5)^2 + y^2 + (z - 1)^2 = (\frac{1}{2})^2 = \frac{1}{4}$

[22] $C(0, -3, -6)$ and $r = \sqrt{3} \Rightarrow x^2 + (y + 3)^2 + (z + 6)^2 = (\sqrt{3})^2 = 3$

[23] (a) Tangent to xy-plane $\Rightarrow r = 6$; $(x + 2)^2 + (y - 4)^2 + (z + 6)^2 = 6^2 = 36$

(b) Tangent to xz-plane $\Rightarrow r = 4$; $(x + 2)^2 + (y - 4)^2 + (z + 6)^2 = 4^2 = 16$

(c) Tangent to yz-plane $\Rightarrow r = 2$; $(x + 2)^2 + (y - 4)^2 + (z + 6)^2 = 2^2 = 4$

[24] (a) Tangent to xy-plane $\Rightarrow r = 2$; $(x - 3)^2 + (y + 1)^2 + (z - 2)^2 = 2^2 = 4$

(b) Tangent to xz-plane $\Rightarrow r = 1$; $(x - 3)^2 + (y + 1)^2 + (z - 2)^2 = 1^2 = 1$

(c) Tangent to yz-plane $\Rightarrow r = 3$; $(x - 3)^2 + (y + 1)^2 + (z - 2)^2 = 3^2 = 9$

[25] The center of the sphere is the midpoint of AB, namely, $(-3, \frac{5}{2}, 0)$.

The radius is $\frac{1}{2} \cdot d(A, B) = \frac{1}{2}\sqrt{89}$. $(x + 3)^2 + (y - \frac{5}{2})^2 + z^2 = \frac{89}{4}$

[26] The center of the sphere is the midpoint of AB, namely, $(2, 0, -3)$.

The radius is $\frac{1}{2} \cdot d(A, B) = \frac{1}{2}\sqrt{248}$. $(x - 2)^2 + y^2 + (z + 3)^2 = 62$

[27] The center of the sphere must be $(1, 1, 1)$ and

an equation is $(x - 1)^2 + (y - 1)^2 + (z - 1)^2 = 1$.

[28] $C(2, 3, -1) \Rightarrow (x - 2)^2 + (y - 3)^2 + (z + 1)^2 = r^2$ is an equation of the sphere.

Substituting $x = 1$, $y = 7$, and $z = -9$ into that equation yields $1 + 16 + 64 = r^2$,

or $r^2 = 81$.

[29] $(x^2 + 4x + 4) + (y^2 - 2y + 1) + (z^2 + 2z + 1) = -2 + 4 + 1 + 1 \Rightarrow$

$(x + 2)^2 + (y - 1)^2 + (z + 1)^2 = 4$; $C(-2, 1, -1)$, $r = 2$

[30] $(x^2 - 6x + 9) + (y^2 - 10y + 25) + (z^2 + 6z + 9) = -34 + 9 + 25 + 9 \Rightarrow$

$(x - 3)^2 + (y - 5)^2 + (z + 3)^2 = 9$; $C(3, 5, -3)$, $r = 3$

[31] $(x^2 - 8x + 16) + y^2 + (z^2 + 8z + 16) = -16 + 16 + 16 \Rightarrow$

$(x - 4)^2 + y^2 + (z + 4)^2 = 16$; $C(4, 0, -4)$, $r = 4$

[32] $4(x^2 - x + \frac{1}{4}) + 4(y^2 + 2y + 1) + 4z^2 = 3 + 1 + 4 \Rightarrow$

$(x - \frac{1}{2})^2 + (y + 1)^2 + z^2 = 2$; $C(\frac{1}{2}, -1, 0)$, $r = \sqrt{2}$

[33] $x^2 + (y^2 + 4y + 4) + z^2 = 4 \Rightarrow x^2 + (y + 2)^2 + z^2 = 4$; $C(0, -2, 0)$, $r = 2$

[34] $x^2 + y^2 + (z^2 - z + \frac{1}{4}) = \frac{1}{4} \Rightarrow x^2 + y^2 + (z - \frac{1}{2})^2 = \frac{1}{4}$; $C(0, 0, \frac{1}{2})$, $r = \frac{1}{2}$

[35] All points inside or on the sphere of radius 1 with center at the origin

[36] All points outside the sphere of radius 1 with center at the origin

[37] All points inside or on a rectangular box with center at the origin and having edges

of lengths 2, 4, and 6 in the x, y, and z directions, respectively

[38] All points outside or on a rectangular box with center at the origin and having edges

of lengths 2, 4, and 6 in the x, y, and z directions, respectively

[39] All points inside or on a cylindrical region of radius 5 and altitude 6 with center at

the origin and axis along the z-axis

[40] All points whose projection onto the xy-plane lies outside or

on the ellipse $(x^2/4) + (y^2/9) \geq 1$

[41] All points not on a coordinate plane

[42] All points between the spheres of radii 2 and 3 with centers at the origin

[43] Following the hint, let $A = (x_1, y_1, 0)$, $B = (x_2, y_2, 0)$, $C = (x_3, y_3, 0)$, and $D = (0, 0, z_4)$. Then,

$$P_1 = \left(\frac{x_1 + x_2}{2}, \frac{y_1 + y_2}{2}, 0\right),\ P_2 = \left(\frac{x_2}{2}, \frac{y_2}{2}, \frac{z_4}{2}\right),\ P_3 = \left(\frac{x_2 + x_3}{2}, \frac{y_2 + y_3}{2}, 0\right),$$

$$P_1' = \left(\frac{x_3}{2}, \frac{y_3}{2}, \frac{z_4}{2}\right),\ P_2' = \left(\frac{x_1 + x_3}{2}, \frac{y_1 + y_3}{2}, 0\right), \text{ and } P_3' = \left(\frac{x_1}{2}, \frac{y_1}{2}, \frac{z_4}{2}\right).$$

P_1P_1', P_2P_2', and P_3P_3' all have the common midpoint,

$$P = \left(\frac{x_1 + x_2 + x_3}{4}, \frac{y_1 + y_2 + y_3}{4}, \frac{z_4}{4}\right).$$

[44] (a) In the xy-plane: $V_1 = (k/2, k/2, 0)$; xz-plane: $V_2 = (k/2, 0, k/2)$;

yz-plane: $V_3(0, k/2, k/2)$; $z = k$ plane: $V_4(k/2, k/2, k)$;

$y = k$ plane: $V_5(k/2, k, k/2)$; $x = k$ plane: $V_6(k, k/2, k/2)$

(b) Using symmetry, the length of each edge will be equal. $d(V_5, V_6) = k/\sqrt{2}$.

[45] The magnitude of the force is $|\mathrm{F}| = \dfrac{|k(+q)(-1)|}{\|\overrightarrow{BA}\|^2}$ for some positive constant k.

The direction of the force is in the same direction as

the unit vector $\dfrac{\overrightarrow{BA}}{\|\overrightarrow{BA}\|}$. Thus, $\mathrm{F} = \dfrac{kq}{\|\overrightarrow{BA}\|^3}\overrightarrow{BA}$.

[46] (a) The force caused by the charge at (1, 0, 0) is $\mathrm{F}_1 = \dfrac{kq}{\|\mathrm{v} - \mathrm{i}\|^3}(\mathrm{v} - \mathrm{i})$,

at (0, 1, 0) is $\mathrm{F}_2 = \dfrac{kq}{\|\mathrm{v} - \mathrm{j}\|^3}(\mathrm{v} - \mathrm{j})$, and at (0, 0, 1) is $\mathrm{F}_3 = \dfrac{kq}{\|\mathrm{v} - \mathrm{k}\|^3}(\mathrm{v} - \mathrm{k})$.

$$\text{Now, } \mathrm{F} = \mathrm{F}_1 + \mathrm{F}_2 + \mathrm{F}_3 = kq\left(\frac{\mathrm{v} - \mathrm{i}}{\|\mathrm{v} - \mathrm{i}\|^3} + \frac{\mathrm{v} - \mathrm{j}}{\|\mathrm{v} - \mathrm{j}\|^3} + \frac{\mathrm{v} - \mathrm{k}}{\|\mathrm{v} - \mathrm{k}\|^3}\right).$$

(b) Since the particle is to be placed equidistant from the three points,

$\|\mathrm{v} - \mathrm{i}\| = \|\mathrm{v} - \mathrm{j}\| = \|\mathrm{v} - \mathrm{k}\|$. Thus, $\mathrm{F} = kq\left[\dfrac{3\mathrm{v} - (\mathrm{i} + \mathrm{j} + \mathrm{k})}{\|\mathrm{v} - \mathrm{i}\|^3}\right] = 0 \Rightarrow$

$\mathrm{v} = \frac{1}{3}(\mathrm{i} + \mathrm{j} + \mathrm{k})$ or $P(x, y, z) = (\frac{1}{3}, \frac{1}{3}, \frac{1}{3})$.

Exercises 14.3

[1] $\mathrm{a} \cdot \mathrm{b} = (-2)(7) + (3)(4) + (1)(5) = -14 + 12 + 5 = 3$

[2] $\mathrm{b} \cdot \mathrm{c} = (7)(1) + (4)(-5) + (5)(2) = 7 - 20 + 10 = -3$

[3] (a) $\mathrm{a} \cdot (\mathrm{b} + \mathrm{c}) = \langle -2, 3, 1 \rangle \cdot \langle 8, -1, 7 \rangle = -12$

(b) $\mathrm{a} \cdot \mathrm{b} + \mathrm{a} \cdot \mathrm{c} = 3 + (-15) = -12$

[4] (a) $(\mathrm{a} - \mathrm{c}) \cdot \mathrm{b} = \langle -3, 8, -1 \rangle \cdot \langle 7, 4, 5 \rangle = 6$

(b) $\mathrm{a} \cdot \mathrm{b} - \mathrm{c} \cdot \mathrm{b} = 3 - (-3) = 6$

[5] $(2\mathrm{a} + \mathrm{b}) \cdot 3\mathrm{c} = (\langle -4, 6, 2 \rangle + \langle 7, 4, 5 \rangle) \cdot \langle 3, -15, 6 \rangle$

$= \langle 3, 10, 7 \rangle \cdot \langle 3, -15, 6 \rangle = -99$

[6] $(\mathrm{a} - \mathrm{b}) \cdot (\mathrm{b} + \mathrm{c}) = \langle -9, -1, -4 \rangle \cdot \langle 8, -1, 7 \rangle = -99$

[7] $\mathrm{comp}_{\mathrm{c}}\,\mathrm{b} = \mathrm{b} \cdot \frac{1}{\|\mathrm{c}\|}\mathrm{c} = \langle 7, 4, 5 \rangle \cdot \frac{1}{\sqrt{30}}\langle 1, -5, 2 \rangle = -\frac{3}{\sqrt{30}}$

[8] $\text{comp}_{\mathbf{b}}\,\mathbf{c} = \mathbf{c}\cdot\frac{1}{\|\mathbf{b}\|}\mathbf{b} = \frac{1}{\|\mathbf{b}\|}(\mathbf{c}\cdot\mathbf{b}) = \frac{1}{\sqrt{90}}(-3) = -\frac{1}{\sqrt{10}}$

[9] $\text{comp}_{\mathbf{b}}\,(\mathbf{a}+\mathbf{c}) = (\mathbf{a}+\mathbf{c})\cdot\frac{1}{\|\mathbf{b}\|}\mathbf{b} = \langle -1, -2, 3\rangle\cdot\frac{1}{\sqrt{90}}\langle 7, 4, 5\rangle = 0$

[10] $\text{comp}_{\mathbf{c}}\,\mathbf{c} = \mathbf{c}\cdot\frac{1}{\|\mathbf{c}\|}\mathbf{c} = \frac{\mathbf{c}\cdot\mathbf{c}}{\|\mathbf{c}\|} = \frac{\|\mathbf{c}\|^2}{\|\mathbf{c}\|} = \|\mathbf{c}\| = \sqrt{30}$

[11] Using (14.19), $\cos\theta = \frac{-3}{\sqrt{89}\sqrt{6}} = \frac{-3}{\sqrt{534}} \Rightarrow \theta = \arccos\frac{-3}{\sqrt{534}} \approx 97.5^\circ$.

[12] $\cos\theta = \frac{1}{\sqrt{66}\sqrt{26}} = \frac{1}{\sqrt{1716}} \Rightarrow \theta = \arccos\frac{1}{\sqrt{1716}} \approx 88.6^\circ$

[13] $\cos\theta = \frac{12}{\sqrt{13}\sqrt{52}} = \frac{6}{13} \Rightarrow \theta = \arccos\frac{6}{13} \approx 62.5^\circ$

[14] $\cos\theta = \frac{4}{\sqrt{35}\sqrt{14}} = \frac{4}{7\sqrt{10}} \Rightarrow \theta = \arccos\frac{4}{7\sqrt{10}} \approx 79.6^\circ$

[15] Since $\mathbf{a}\cdot\mathbf{b} = 0$, **a** and **b** are orthogonal by Theorem (14.21).

[16] Since $\mathbf{a}\cdot\mathbf{b} = 0$, **a** and **b** are orthogonal by Theorem (14.21).

[17] $\mathbf{a}\cdot\mathbf{b} = c^2 - 2c - 15$. $\mathbf{a}\cdot\mathbf{b} = 0 \Rightarrow (c+3)(c-5) = 0 \Rightarrow c = -3, 5$.

[18] $\mathbf{a}\cdot\mathbf{b} = 4 + 44 - 3c^2$. $\mathbf{a}\cdot\mathbf{b} = 0 \Rightarrow 48 = 3c^2 \Rightarrow c = \pm 4$.

[19] $\overrightarrow{PQ} = \langle -2, 7, 5\rangle$ and $\overrightarrow{RS} = \langle -6, 1, 11\rangle \Rightarrow \overrightarrow{PQ}\cdot\overrightarrow{RS} = 74$

[20] $\overrightarrow{QS} = \langle -5, -4, 1\rangle$ and $\overrightarrow{RP} = \langle 1, -2, 5\rangle \Rightarrow \overrightarrow{QS}\cdot\overrightarrow{RP} = 8$

[21] $\cos\theta = \frac{74}{\sqrt{78}\sqrt{158}} \Rightarrow \theta = \cos^{-1}\frac{37}{\sqrt{3081}} \approx 48.2^\circ$

[22] $\cos\theta = \frac{8}{\sqrt{42}\sqrt{30}} \Rightarrow \theta = \cos^{-1}\frac{4}{\sqrt{315}} \approx 77.0^\circ$

[23] $\text{comp}_{\overrightarrow{QR}}\,\overrightarrow{PS} = \overrightarrow{PS}\cdot\frac{1}{\|\overrightarrow{QR}\|}\overrightarrow{QR} = \langle -7, 3, 6\rangle\cdot\frac{1}{\sqrt{126}}\langle 1, -5, -10\rangle = \frac{-82}{\sqrt{126}}$

[24] $\text{comp}_{\overrightarrow{PS}}\,\overrightarrow{QR} = \overrightarrow{QR}\cdot\frac{1}{\|\overrightarrow{PS}\|}\overrightarrow{PS} = \langle 1, -5, -10\rangle\cdot\frac{1}{\sqrt{94}}\langle -7, 3, 6\rangle = \frac{-82}{\sqrt{94}}$

Note: In Exercises 25–30, W denotes work.

[25] $W = \mathbf{a}\cdot\overrightarrow{PQ} = \langle -1, 5, -3\rangle\cdot\langle -2, 4, 7\rangle = 1$

[26] $W = \mathbf{a}\cdot\overrightarrow{PQ} = \langle 8, 0, -4\rangle\cdot\langle 5, -1, -5\rangle = 60$

[27] Since the force vector is $\frac{4\mathbf{a}}{\|\mathbf{a}\|} = \frac{4}{\sqrt{3}}(\mathbf{i}+\mathbf{j}+\mathbf{k})$, $W = \frac{4}{\sqrt{3}}(\mathbf{i}+\mathbf{j}+\mathbf{k})\cdot(-3\mathbf{j}) = -4\sqrt{3}$ ft-lb.

[28] Since the force vector is $5\mathbf{k}$, $W = 5\mathbf{k}\cdot(\mathbf{i}+2\mathbf{j}+3\mathbf{k}) = 15$ joules.

[29] Since the force vector is $20(\cos\theta\,\mathbf{i} + \sin\theta\,\mathbf{j})$ {Exer. 46, §14.1} $= 20(\frac{1}{2}\sqrt{3}\,\mathbf{i} + \frac{1}{2}\,\mathbf{j})$,
$W = (10\sqrt{3}\,\mathbf{i} + 10\,\mathbf{j})\cdot(100\mathbf{i}) = 1000\sqrt{3}$ ft-lb.

[30] The force vector is $20(\cos 60^\circ\,\mathbf{i} + \sin 60^\circ\,\mathbf{j}) = 20(\frac{1}{2}\,\mathbf{i} + \frac{1}{2}\sqrt{3}\,\mathbf{j})$.

The direction vector is $100(\cos 30^\circ\,\mathbf{i} + \sin 30^\circ\,\mathbf{j}) = 100(\frac{1}{2}\sqrt{3}\,\mathbf{i} + \frac{1}{2}\mathbf{j})$. Thus,

$W = 20(\frac{1}{2}\,\mathbf{i} + \frac{1}{2}\sqrt{3}\,\mathbf{j})\cdot 100(\frac{1}{2}\sqrt{3}\,\mathbf{i} + \frac{1}{2}\,\mathbf{j}) = 1000\sqrt{3}$ ft-lb. Note that the force in relation to the direction of movement is exactly the same as in Exercise 29.

[31] Following the hint, $\overrightarrow{PA} = \mathbf{v}_1 - \mathbf{v}_2$ and $\overrightarrow{PB} = -\mathbf{v}_1 - \mathbf{v}_2$. Now,

$$\begin{aligned}\overrightarrow{PA}\cdot\overrightarrow{PB} &= (\mathbf{v}_1 - \mathbf{v}_2)\cdot(-\mathbf{v}_1 - \mathbf{v}_2) = -\mathbf{v}_1\cdot\mathbf{v}_1 - \mathbf{v}_1\cdot\mathbf{v}_2 + \mathbf{v}_2\cdot\mathbf{v}_1 + \mathbf{v}_2\cdot\mathbf{v}_2\\ &= -\|\mathbf{v}_1\|^2 + \|\mathbf{v}_2\|^2 = -r^2 + r^2 = 0.\end{aligned}$$

Thus, $\overrightarrow{PA}$ and $\overrightarrow{PB}$ are orthogonal and APB is a right triangle.

[32] $A(a,\,0,\,0)$, $B(0,\,b,\,0)$, and $P(\frac{1}{2}a,\,\frac{1}{2}b,\,\frac{1}{2}c) \Rightarrow$

$\overrightarrow{PA} = (\frac{1}{2}a,\,-\frac{1}{2}b,\,-\frac{1}{2}c)$ and $\overrightarrow{PB} = (-\frac{1}{2}a,\,\frac{1}{2}b,\,-\frac{1}{2}c)$. Now, $\cos\theta = \dfrac{\overrightarrow{PA}\cdot\overrightarrow{PB}}{\|\overrightarrow{PA}\|\|\overrightarrow{PB}\|} =$

$$\frac{\frac{1}{4}c^2 - \frac{1}{4}a^2 - \frac{1}{4}b^2}{\frac{1}{4}a^2 + \frac{1}{4}b^2 + \frac{1}{4}c^2} \Rightarrow \theta = \cos^{-1}\frac{c^2 - a^2 - b^2}{a^2 + b^2 + c^2}.$$

[33] From the figure, $a = b = c = 1 \Rightarrow \theta = \cos^{-1}(\frac{-1}{3}) \approx 109.5^\circ$.

[34] If $\overrightarrow{BA} = \langle x_1,\,y_1,\,z_1\rangle$ and $\overrightarrow{CD} = \langle x_2,\,y_2,\,z_2\rangle$, then their projections onto the xy-plane are $\langle x_1,\,y_1,\,0\rangle$ and $\langle x_2,\,y_2,\,0\rangle$, respectively.

θ is the angle between these projections and $\cos\theta = \dfrac{x_1x_2 + y_1y_2}{\sqrt{x_1^2 + y_1^2}\,\sqrt{x_2^2 + y_2^2}}$.

[35] (a) $a_1 = \mathbf{a}\cdot\mathbf{i} = \|\mathbf{a}\|\|\mathbf{i}\|\cos\alpha = \|\mathbf{a}\|(1)\cos\alpha \Rightarrow \cos\alpha = \dfrac{a_1}{\|\mathbf{a}\|}$

$a_2 = \mathbf{a}\cdot\mathbf{j} = \|\mathbf{a}\|\|\mathbf{j}\|\cos\beta = \|\mathbf{a}\|(1)\cos\beta \Rightarrow \cos\beta = \dfrac{a_2}{\|\mathbf{a}\|}$

$a_3 = \mathbf{a}\cdot\mathbf{k} = \|\mathbf{a}\|\|\mathbf{k}\|\cos\gamma = \|\mathbf{a}\|(1)\cos\gamma \Rightarrow \cos\gamma = \dfrac{a_3}{\|\mathbf{a}\|}$

(b) $\cos^2\alpha + \cos^2\beta + \cos^2\gamma = \dfrac{a_1^2 + a_2^2 + a_3^2}{\|\mathbf{a}\|^2} = 1$

[36] (a) Using Exercise 35(a), $\cos\alpha = \frac{-2}{\sqrt{30}}$, $\cos\beta = \frac{1}{\sqrt{30}}$, and $\cos\gamma = \frac{5}{\sqrt{30}}$.

(b) The direction angles of $\mathbf{i}$ are 0°, 90°, 90°; $\mathbf{j}$ are 90°, 0°, 90°; $\mathbf{k}$ are 90°, 90°, 0°.

The direction cosines of $\mathbf{i}$ are 1, 0, 0; $\mathbf{j}$ are 0, 1, 0; $\mathbf{k}$ are 0, 0, 1.

(c) Using Exercise 35(a), let $a_1 = a_2 = a_3 = 1$, and hence $\mathbf{a} = \langle 1,\,1,\,1\rangle$.

Two unit vectors are $\pm\frac{1}{\sqrt{3}}(\mathbf{i} + \mathbf{j} + \mathbf{k})$ since $\|\mathbf{a}\| = \sqrt{3}$.

[37] $d = (l^2 + m^2 + n^2)^{1/2} = (k^2\cos^2\alpha + k^2\cos^2\beta + k^2\cos^2\gamma)^{1/2}$

$= \left[k^2(\cos^2\alpha + \cos^2\beta + \cos^2\gamma)\right]^{1/2} = k$ by Exercise 35.

Since $l = k\cos\alpha$, $\cos\alpha = l/k = l/d$. Similarly, $\cos\beta = m/d$, and $\cos\gamma = n/d$.

[38] (a) The unit vector in the direction of $\mathbf{a} = \langle a_1,\,a_2,\,a_3\rangle$ is $\frac{1}{\|\mathbf{a}\|}\langle a_1,\,a_2,\,a_3\rangle =$ $\langle\cos\alpha_1,\,\cos\beta_1,\,\cos\gamma_1\rangle = \langle l_1/d_1,\,m_1/d_1,\,n_1/d_1\rangle$, where $d_1 = (l_1^2 + m_1^2 + n_1^2)^{1/2}$. Similarly, the unit vector in the direction of $\mathbf{b}$ is $\langle l_2/d_2,\,m_2/d_2,\,n_2/d_2\rangle$. Thus, $\mathbf{a}$ and $\mathbf{b}$ are orthogonal $\Leftrightarrow \frac{\mathbf{a}}{\|\mathbf{a}\|}\cdot\frac{\mathbf{b}}{\|\mathbf{b}\|} = 0 \Leftrightarrow$

$$\frac{l_1l_2 + m_1m_2 + n_1n_2}{d_1d_2} = 0 \Leftrightarrow l_1l_2 + m_1m_2 + n_1n_2 = 0.$$

(b) $\mathbf{a}$ and $\mathbf{b}$ are parallel iff $\frac{\mathbf{a}}{\|\mathbf{a}\|} = c\,\frac{\mathbf{b}}{\|\mathbf{b}\|}$ for some nonzero constant c.

Thus, $\dfrac{l_1}{d_1} = c\,\dfrac{l_2}{d_2} \Rightarrow l_1 = \dfrac{c\,d_1}{d_2}\,l_2 \Rightarrow l_1 = k\,l_2$, where $k = \dfrac{c\,d_1}{d_2} > 0$.

Similarly, $m_1 = k\,m_2$ and $n_1 = k\,n_2$.

[39] $\mathbf{a}\cdot\mathbf{a} = \langle a_1,\,a_2,\,a_3\rangle\cdot\langle a_1,\,a_2,\,a_3\rangle$

$= a_1^2 + a_2^2 + a_3^2 = \left(\sqrt{a_1^2 + a_2^2 + a_3^2}\right)^2 = \|\mathbf{a}\|^2$

[40] $\mathbf{a}\cdot\mathbf{b} = \langle a_1, a_2, a_3\rangle \cdot \langle b_1, b_2, b_3\rangle$

$= a_1b_1 + a_2b_2 + a_3b_3$

$= b_1a_1 + b_2a_2 + b_3a_3$

$= \langle b_1, b_2, b_3\rangle \cdot \langle a_1, a_2, a_3\rangle = \mathbf{b}\cdot\mathbf{a}$

[41] $(c\mathbf{a})\cdot\mathbf{b} = (c\langle a_1, a_2, a_3\rangle)\cdot\langle b_1, b_2, b_3\rangle$

$= \langle ca_1, ca_2, ca_3\rangle \cdot \langle b_1, b_2, b_3\rangle$

$= ca_1b_1 + ca_2b_2 + ca_3b_3$

$= c(a_1b_1 + a_2b_2 + a_3b_3) = c(\mathbf{a}\cdot\mathbf{b})$

$c(\mathbf{a}\cdot\mathbf{b}) = c(\langle a_1, a_2, a_3\rangle \cdot \langle b_1, b_2, b_3\rangle)$

$= c(a_1b_1 + a_2b_2 + a_3b_3)$

$= ca_1b_1 + ca_2b_2 + ca_3b_3$

$= a_1(cb_1) + a_2(cb_2) + a_3(cb_3)$

$= \langle a_1, a_2, a_3\rangle \cdot \langle cb_1, cb_2, cb_3\rangle$

$= \mathbf{a}\cdot(c\langle b_1, b_2, b_3\rangle) = \mathbf{a}\cdot(c\mathbf{b})$

[42] $\mathbf{0}\cdot\mathbf{a} = \langle 0, 0, 0\rangle \cdot \langle a_1, a_2, a_3\rangle = 0(a_1) + 0(a_2) + 0(a_3) = 0 + 0 + 0 = 0$

[43] If $c\mathbf{a} + d\mathbf{b} = p\mathbf{a} + q\mathbf{b}$, then $(c - p)\mathbf{a} = (q - d)\mathbf{b}$. Assume $c \neq p$.

Then $\mathbf{a} = \dfrac{q-d}{c-p}\mathbf{b}$, that is, $\mathbf{a}$ and $\mathbf{b}$ are parallel, a contradiction. Hence $c = p$.

Similarly, $q = d$.

[44] If $\mathbf{c}$ is orthogonal to $\mathbf{a}$ and $\mathbf{b}$, then $\mathbf{c}\cdot\mathbf{a} = 0$ and $\mathbf{c}\cdot\mathbf{b} = 0$ and

hence $\mathbf{c}\cdot(p\mathbf{a} + q\mathbf{b}) = \mathbf{c}\cdot(p\mathbf{a}) + \mathbf{c}\cdot(q\mathbf{b}) = p(\mathbf{c}\cdot\mathbf{a}) + q(\mathbf{c}\cdot\mathbf{b}) = p(0) + q(0) = 0$.

Thus, $\mathbf{c}$ is orthogonal to $p\mathbf{a} + q\mathbf{b}$ for all scalars p and q.

[45] $\|\mathbf{a}+\mathbf{b}\|^2 = (\mathbf{a}+\mathbf{b})\cdot(\mathbf{a}+\mathbf{b})$

$= \mathbf{a}\cdot\mathbf{a} + \mathbf{a}\cdot\mathbf{b} + \mathbf{b}\cdot\mathbf{a} + \mathbf{b}\cdot\mathbf{b} = \|\mathbf{a}\|^2 + 2(\mathbf{a}\cdot\mathbf{b}) + \|\mathbf{b}\|^2$

[46] Using Exercise 45, $\|\mathbf{a}+\mathbf{b}\|^2 + \|\mathbf{a}-\mathbf{b}\|^2 =$

$\left[\|\mathbf{a}\|^2 + 2(\mathbf{a}\cdot\mathbf{b}) + \|\mathbf{b}\|^2\right] + \left[\|\mathbf{a}\|^2 - 2(\mathbf{a}\cdot\mathbf{b}) + \|\mathbf{b}\|^2\right] = 2(\|\mathbf{a}\|^2 + \|\mathbf{b}\|^2)$

[47] $|\mathbf{a}\cdot\mathbf{b}| = \|\mathbf{a}\|\|\mathbf{b}\||\cos\theta| = \|\mathbf{a}\|\|\mathbf{b}\| \Leftrightarrow \cos\theta = \pm 1$.

Thus, $\mathbf{a}$ and $\mathbf{b}$ have the same or opposite direction.

[48] Using Exercise 45, $\|\mathbf{a}+\mathbf{b}\|^2 = \|\mathbf{a}\|^2 + 2(\mathbf{a}\cdot\mathbf{b}) + \|\mathbf{b}\|^2$.

Also, $(\|\mathbf{a}\| + \|\mathbf{b}\|)^2 = \|\mathbf{a}\|^2 + 2\|\mathbf{a}\|\|\mathbf{b}\| + \|\mathbf{b}\|^2$. Thus, the equality holds

iff $\mathbf{a}\cdot\mathbf{b} = \|\mathbf{a}\|\|\mathbf{b}\| \Leftrightarrow \cos\theta = 1 \Leftrightarrow$ $\mathbf{a}$ and $\mathbf{b}$ have the same direction.

[49] $\mathbf{a} = \mathbf{b} + (\mathbf{a}-\mathbf{b}) \Rightarrow \|\mathbf{a}\| \leq \|\mathbf{b}\| + \|\mathbf{a}-\mathbf{b}\| \Rightarrow \|\mathbf{a}\| - \|\mathbf{b}\| \leq \|\mathbf{a}-\mathbf{b}\|$

[50] $(\mathbf{a}+\mathbf{b})\cdot(\mathbf{a}-\mathbf{b}) = \mathbf{a}\cdot(\mathbf{a}-\mathbf{b}) + \mathbf{b}\cdot(\mathbf{a}-\mathbf{b})$

$= \mathbf{a}\cdot\mathbf{a} - \mathbf{a}\cdot\mathbf{b} + \mathbf{b}\cdot\mathbf{a} - \mathbf{b}\cdot\mathbf{b} = \mathbf{a}\cdot\mathbf{a} - \mathbf{b}\cdot\mathbf{b}$

[51] Using Exercise 45, $\frac{1}{4}(\|\mathbf{a}+\mathbf{b}\|^2 - \|\mathbf{a}-\mathbf{b}\|^2) =$

$\frac{1}{4}\left[\|\mathbf{a}\|^2 + 2(\mathbf{a}\cdot\mathbf{b}) + \|\mathbf{b}\|^2 - \|\mathbf{a}\|^2 + 2(\mathbf{a}\cdot\mathbf{b}) - \|\mathbf{b}\|^2\right] = \frac{1}{4}\left[4(\mathbf{a}\cdot\mathbf{b})\right] = \mathbf{a}\cdot\mathbf{b}$.

[52] $\text{comp}_{\mathbf{c}}(\mathbf{a} + \mathbf{b}) = (\mathbf{a} + \mathbf{b}) \cdot \frac{1}{\|\mathbf{c}\|}\mathbf{c}$

$$= \mathbf{a} \cdot \left(\frac{1}{\|\mathbf{c}\|}\mathbf{c}\right) + \mathbf{b} \cdot \left(\frac{1}{\|\mathbf{c}\|}\mathbf{c}\right)$$

$$= \text{comp}_{\mathbf{c}}\,\mathbf{a} + \text{comp}_{\mathbf{c}}\,\mathbf{b}$$

Exercises 14.4

[1] $\mathbf{a} \times \mathbf{b} = \begin{vmatrix} \mathbf{i} & \mathbf{j} & \mathbf{k} \\ 1 & -2 & 3 \\ 2 & 1 & -4 \end{vmatrix} = \begin{vmatrix} -2 & 3 \\ 1 & -4 \end{vmatrix}\mathbf{i} - \begin{vmatrix} 1 & 3 \\ 2 & -4 \end{vmatrix}\mathbf{j} + \begin{vmatrix} 1 & -2 \\ 2 & 1 \end{vmatrix}\mathbf{k} =$

$5\mathbf{i} + 10\mathbf{j} + 5\mathbf{k} = <5, 10, 5>$

[2] $\mathbf{a} \times \mathbf{b} = \begin{vmatrix} \mathbf{i} & \mathbf{j} & \mathbf{k} \\ -5 & 1 & -1 \\ 3 & 6 & -2 \end{vmatrix} = \begin{vmatrix} 1 & -1 \\ 6 & -2 \end{vmatrix}\mathbf{i} - \begin{vmatrix} -5 & -1 \\ 3 & -2 \end{vmatrix}\mathbf{j} + \begin{vmatrix} -5 & 1 \\ 3 & 6 \end{vmatrix}\mathbf{k} =$

$4\mathbf{i} - 13\mathbf{j} - 33\mathbf{k} = <4, -13, -33>$

[3] $\mathbf{a} \times \mathbf{b} = \begin{vmatrix} \mathbf{i} & \mathbf{j} & \mathbf{k} \\ 0 & 1 & 2 \\ 1 & 2 & 0 \end{vmatrix} = \begin{vmatrix} 1 & 2 \\ 2 & 0 \end{vmatrix}\mathbf{i} - \begin{vmatrix} 0 & 2 \\ 1 & 0 \end{vmatrix}\mathbf{j} + \begin{vmatrix} 0 & 1 \\ 1 & 2 \end{vmatrix}\mathbf{k} =$

$-4\mathbf{i} + 2\mathbf{j} - \mathbf{k} = <-4, 2, -1>$

[4] $\mathbf{a} \times \mathbf{b} = \begin{vmatrix} \mathbf{i} & \mathbf{j} & \mathbf{k} \\ 0 & 0 & 4 \\ -7 & 1 & 0 \end{vmatrix} = \begin{vmatrix} 0 & 4 \\ 1 & 0 \end{vmatrix}\mathbf{i} - \begin{vmatrix} 0 & 4 \\ -7 & 0 \end{vmatrix}\mathbf{j} + \begin{vmatrix} 0 & 0 \\ -7 & 1 \end{vmatrix}\mathbf{k} =$

$-4\mathbf{i} - 28\mathbf{j} = <-4, -28, 0>$

[5] $\mathbf{a} \times \mathbf{b} = \begin{vmatrix} \mathbf{i} & \mathbf{j} & \mathbf{k} \\ 5 & -6 & -1 \\ 3 & 0 & 1 \end{vmatrix} = \begin{vmatrix} -6 & -1 \\ 0 & 1 \end{vmatrix}\mathbf{i} - \begin{vmatrix} 5 & -1 \\ 3 & 1 \end{vmatrix}\mathbf{j} + \begin{vmatrix} 5 & -6 \\ 3 & 0 \end{vmatrix}\mathbf{k} =$

$-6\mathbf{i} - 8\mathbf{j} + 18\mathbf{k}$

[6] $\mathbf{a} \times \mathbf{b} = \begin{vmatrix} \mathbf{i} & \mathbf{j} & \mathbf{k} \\ 2 & 1 & 0 \\ 0 & -5 & 2 \end{vmatrix} = \begin{vmatrix} 1 & 0 \\ -5 & 2 \end{vmatrix}\mathbf{i} - \begin{vmatrix} 2 & 0 \\ 0 & 2 \end{vmatrix}\mathbf{j} + \begin{vmatrix} 2 & 1 \\ 0 & -5 \end{vmatrix}\mathbf{k} = 2\mathbf{i} - 4\mathbf{j} - 10\mathbf{k}$

[7] $\mathbf{a} \times \mathbf{b} = \begin{vmatrix} \mathbf{i} & \mathbf{j} & \mathbf{k} \\ -3 & 1 & 2 \\ 9 & -3 & -6 \end{vmatrix} = \begin{vmatrix} 1 & 2 \\ -3 & -6 \end{vmatrix}\mathbf{i} - \begin{vmatrix} -3 & 2 \\ 9 & -6 \end{vmatrix}\mathbf{j} + \begin{vmatrix} -3 & 1 \\ 9 & -3 \end{vmatrix}\mathbf{k} =$

$0\mathbf{i} + 0\mathbf{j} + 0\mathbf{k} = \mathbf{0}$

[8] $\mathbf{a} \times \mathbf{b} = \begin{vmatrix} \mathbf{i} & \mathbf{j} & \mathbf{k} \\ 3 & -1 & 8 \\ 0 & 5 & 0 \end{vmatrix} = \begin{vmatrix} -1 & 8 \\ 5 & 0 \end{vmatrix}\mathbf{i} - \begin{vmatrix} 3 & 8 \\ 0 & 0 \end{vmatrix}\mathbf{j} + \begin{vmatrix} 3 & -1 \\ 0 & 5 \end{vmatrix}\mathbf{k} = -40\mathbf{i} + 15\mathbf{k}$

[9] $\mathbf{a} \times \mathbf{b} = \begin{vmatrix} \mathbf{i} & \mathbf{j} & \mathbf{k} \\ 4 & -6 & 2 \\ -2 & 3 & -1 \end{vmatrix} = \begin{vmatrix} -6 & 2 \\ 3 & -1 \end{vmatrix}\mathbf{i} - \begin{vmatrix} 4 & 2 \\ -2 & -1 \end{vmatrix}\mathbf{j} + \begin{vmatrix} 4 & -6 \\ -2 & 3 \end{vmatrix}\mathbf{k} =$

$0\mathbf{i} + 0\mathbf{j} + 0\mathbf{k} = \mathbf{0}$

[10] $\mathbf{a} \times \mathbf{b} = \begin{vmatrix} \mathbf{i} & \mathbf{j} & \mathbf{k} \\ 3 & 0 & 0 \\ 0 & 0 & 4 \end{vmatrix} = \begin{vmatrix} 0 & 0 \\ 0 & 4 \end{vmatrix}\mathbf{i} - \begin{vmatrix} 3 & 0 \\ 0 & 4 \end{vmatrix}\mathbf{j} + \begin{vmatrix} 3 & 0 \\ 0 & 0 \end{vmatrix}\mathbf{k} = -12\mathbf{j}$

[11] Since $\mathbf{a} \times \mathbf{b} = \mathbf{0}$, $\mathbf{a}$ and $\mathbf{b}$ are parallel by Corollary (14.31).

[12] Since $\mathbf{a} \times \mathbf{b} = \mathbf{0}$, $\mathbf{a}$ and $\mathbf{b}$ are parallel by Corollary (14.31).

[13] $\mathbf{b} \times \mathbf{c} = <4, 12, 5>$ and $\mathbf{a} \times (\mathbf{b} \times \mathbf{c}) = <12, -14, 24>$;

$\mathbf{a} \times \mathbf{b} = <1, 3, 2>$ and $(\mathbf{a} \times \mathbf{b}) \times \mathbf{c} = <16, -2, -5>$

[14] $\mathbf{b} - \mathbf{c} = <-4, 3, -4>$ and $\mathbf{a} \times (\mathbf{b} - \mathbf{c}) = <3, 12, 6>$;

$\mathbf{a} \times \mathbf{b} = <1, 3, 2>$, $\mathbf{a} \times \mathbf{c} = <-2, -9, -4>$, and $(\mathbf{a} \times \mathbf{b}) - (\mathbf{a} \times \mathbf{c}) = <3, 12, 6>$

Note: In Exercises 15–18, let c be a nonzero scalar.

15 (a) $\overrightarrow{PQ} = \langle -1, 4, -3 \rangle$ and $\overrightarrow{PR} = \langle 2, -3, -1 \rangle$.

A vector perpendicular to the plane is $c(\overrightarrow{PR} \times \overrightarrow{PQ}) = c\langle 13, 7, 5 \rangle$.

(b) $A = \frac{1}{2}\left\|\overrightarrow{PR} \times \overrightarrow{PQ}\right\| = \frac{1}{2}\sqrt{243} = \frac{9}{2}\sqrt{3}$

16 (a) $\overrightarrow{PQ} = \langle 5, -1, -8 \rangle$ and $\overrightarrow{PR} = \langle 7, 1, -6 \rangle$.

A vector perpendicular to the plane is $c(\overrightarrow{PR} \times \overrightarrow{PQ}) = c\langle -14, 26, -12 \rangle$.

(b) $A = \frac{1}{2}\left\|\overrightarrow{PR} \times \overrightarrow{PQ}\right\| = \frac{1}{2}\sqrt{1016} = \sqrt{254}$

17 (a) $\overrightarrow{PQ} = \langle -4, 5, 0 \rangle$ and $\overrightarrow{PR} = \langle -4, 0, 2 \rangle$.

A vector perpendicular to the plane is $c(\overrightarrow{PR} \times \overrightarrow{PQ}) = c\langle -10, -8, -20 \rangle$.

(b) $A = \frac{1}{2}\left\|\overrightarrow{PR} \times \overrightarrow{PQ}\right\| = \frac{1}{2}\sqrt{564} = \sqrt{141}$

18 (a) $\overrightarrow{PQ} = \langle 1, 0, -3 \rangle$ and $\overrightarrow{PR} = \langle 6, -2, 1 \rangle$.

A vector perpendicular to the plane is $c(\overrightarrow{PR} \times \overrightarrow{PQ}) = c\langle 6, 19, 2 \rangle$.

(b) $A = \frac{1}{2}\left\|\overrightarrow{PR} \times \overrightarrow{PQ}\right\| = \frac{1}{2}\sqrt{401}$

19 $$d = \frac{\left\|\overrightarrow{QR} \times \overrightarrow{QP}\right\|}{\left\|\overrightarrow{QR}\right\|} = \frac{\|\langle -3, -1, 1 \rangle \times \langle 1, -4, -3 \rangle\|}{\|\langle -3, -1, 1 \rangle\|} = \frac{\|\langle 7, -8, 13 \rangle\|}{\|\langle -3, -1, 1 \rangle\|} = \frac{\sqrt{282}}{\sqrt{11}} \approx 5.06$$

20 $$d = \frac{\left\|\overrightarrow{QR} \times \overrightarrow{QP}\right\|}{\left\|\overrightarrow{QR}\right\|} = \frac{\|\langle -2, 7, -7 \rangle \times \langle -5, 6, -3 \rangle\|}{\|\langle -2, 7, -7 \rangle\|} = \frac{\|\langle 21, 29, 23 \rangle\|}{\|\langle -2, 7, -7 \rangle\|} = \frac{\sqrt{1811}}{\sqrt{102}} \approx 4.21$$

21 $\mathbf{a} \times \mathbf{b} = \langle a_2b_3 - a_3b_2, b_1a_3 - a_1b_3, a_1b_2 - a_2b_1 \rangle$ and

$\mathbf{b} \times \mathbf{c} = \langle b_2c_3 - b_3c_2, c_1b_3 - b_1c_3, b_1c_2 - b_2c_1 \rangle$.

$\mathbf{a} \cdot (\mathbf{b} \times \mathbf{c}) = a_1b_2c_3 - a_1b_3c_2 + a_2c_1b_3 - a_2b_1c_3 + a_3b_1c_2 - a_3b_2c_1$ and

$(\mathbf{a} \times \mathbf{b}) \cdot \mathbf{c} = a_2b_3c_1 - a_3b_2c_1 + b_1a_3c_2 - a_1b_3c_2 + a_1b_2c_3 - a_2b_1c_3$.

Comparing terms, we see that $\mathbf{a} \cdot (\mathbf{b} \times \mathbf{c}) = (\mathbf{a} \times \mathbf{b}) \cdot \mathbf{c}$. Also, $\begin{vmatrix} a_1 & a_2 & a_3 \\ b_1 & b_2 & b_3 \\ c_1 & c_2 & c_3 \end{vmatrix} =$

$$\begin{vmatrix} b_2 & b_3 \\ c_2 & c_3 \end{vmatrix} a_1 - \begin{vmatrix} b_1 & b_3 \\ c_1 & c_3 \end{vmatrix} a_2 + \begin{vmatrix} b_1 & b_2 \\ c_1 & c_2 \end{vmatrix} a_3 =$$

$$(b_2c_3 - c_2b_3)a_1 - (b_1c_3 - c_1b_3)a_2 + (b_1c_2 - c_1b_2)a_3 = \mathbf{a} \cdot (\mathbf{b} \times \mathbf{c}) = (\mathbf{a} \times \mathbf{b}) \cdot \mathbf{c}.$$

22 $\overrightarrow{AB} = \langle 1, -1, 2 \rangle$, $\overrightarrow{AC} = \langle 0, 3, -1 \rangle$, and $\overrightarrow{AD} = \langle 3, -4, 1 \rangle$.

Thus, $V = \left|(\overrightarrow{AB} \times \overrightarrow{AC}) \cdot \overrightarrow{AD}\right| = \begin{vmatrix} 1 & -1 & 2 \\ 0 & 3 & -1 \\ 3 & -4 & 1 \end{vmatrix} = \left|1(-1) + 1(3) + 2(-9)\right| = 16.$

23 $\overrightarrow{AB} = \langle 1, -1, 3 \rangle$, $\overrightarrow{AC} = \langle 2, -3, 2 \rangle$, and $\overrightarrow{AD} = \langle 3, -4, 1 \rangle$.

Thus, $V = \left|(\overrightarrow{AB} \times \overrightarrow{AC}) \cdot \overrightarrow{AD}\right| = \begin{vmatrix} 1 & -1 & 3 \\ 2 & -3 & 2 \\ 3 & -4 & 1 \end{vmatrix} = \left|1(5) + 1(-4) + 3(1)\right| = 4.$

[24] $\Rightarrow$: b × c is orthogonal to both b and c. $\mathbf{a}\cdot(\mathbf{b}\times\mathbf{c}) = 0 \Rightarrow$

a is also orthogonal to b × c.

Since a, b, and c all have a common initial point, they lie in the same plane.

$\Leftarrow$: If a, b, and c are coplanar, then b × c is orthogonal to a, b, and c.

Hence, $\mathbf{a}\cdot(\mathbf{b}\times\mathbf{c}) = 0$.

[25] By (14.29), a × b is orthogonal to b. By (14.21), $(\mathbf{a}\times\mathbf{b})\cdot\mathbf{b} = 0$.

[26] No. $\mathbf{a}\times\mathbf{b} = \mathbf{a}\times\mathbf{c} \Rightarrow (\mathbf{a}\times\mathbf{b}) - (\mathbf{a}\times\mathbf{c}) = \mathbf{0} \Rightarrow \mathbf{a}\times(\mathbf{b}-\mathbf{c}) = \mathbf{0} \Rightarrow$ a and $(\mathbf{b}-\mathbf{c})$ are parallel. Thus, it is not necessary that $\mathbf{b}-\mathbf{c} = \mathbf{0}$, but only that $\mathbf{b}-\mathbf{c} = k\mathbf{a}$ for some scalar k. For example, let $\mathbf{a} = \mathbf{i}$, $\mathbf{b} = 2\mathbf{i}+\mathbf{j}$, and $\mathbf{c} = \mathbf{i}+\mathbf{j}$.

[27] $\mathbf{a}\times\mathbf{b} = \mathbf{a}\times\mathbf{c} \Rightarrow \mathbf{a}\times(\mathbf{b}-\mathbf{c}) = \mathbf{0}$ and $\mathbf{a}\cdot\mathbf{b} = \mathbf{a}\cdot\mathbf{c} \Rightarrow \mathbf{a}\cdot(\mathbf{b}-\mathbf{c}) = 0$. It follows that $\|\mathbf{a}\|\|\mathbf{b}-\mathbf{c}\|\sin\theta = 0$ and $\|\mathbf{a}\|\|\mathbf{b}-\mathbf{c}\|\cos\theta = 0$. Since $\|\mathbf{a}\| \neq 0$ and $\sin\theta$ and $\cos\theta$ cannot be zero at the same time, we must have $\|\mathbf{b}-\mathbf{c}\| = 0$, or $\mathbf{b} = \mathbf{c}$.

[28] $$(m\mathbf{a})\times\mathbf{b} = \begin{vmatrix} \mathbf{i} & \mathbf{j} & \mathbf{k} \\ ma_1 & ma_2 & ma_3 \\ b_1 & b_2 & b_3 \end{vmatrix}$$

$$= (ma_2b_3 - ma_3b_2)\mathbf{i} - (ma_1b_3 - ma_3b_1)\mathbf{j} + (ma_1b_2 - ma_2b_1)\mathbf{k}$$

$$= m\left[(a_2b_3 - a_3b_2)\mathbf{i} - (a_1b_3 - a_3b_1)\mathbf{j} + (a_1b_2 - a_2b_1)\mathbf{k}\right] = m(\mathbf{a}\times\mathbf{b})$$

$$= (a_2\,mb_3 - a_3\,mb_2)\mathbf{i} - (a_1\,mb_3 - a_3\,mb_1)\mathbf{j} + (a_1\,mb_2 - a_2\,mb_1)\mathbf{k}$$

$$= \mathbf{a}\times(m\mathbf{b})$$

[29] $$(\mathbf{a}+\mathbf{b})\times\mathbf{c} = \begin{vmatrix} \mathbf{i} & \mathbf{j} & \mathbf{k} \\ a_1+b_1 & a_2+b_2 & a_3+b_3 \\ c_1 & c_2 & c_3 \end{vmatrix} =$$

$$\left[(a_2+b_2)c_3 - (a_3+b_3)c_2\right]\mathbf{i} - \left[(a_1+b_1)c_3 - (a_3+b_3)c_1\right]\mathbf{j} + \left[(a_1+b_1)c_2 - (a_2+b_2)c_1\right]\mathbf{k}.$$

$$\mathbf{a}\times\mathbf{c} = \begin{vmatrix} \mathbf{i} & \mathbf{j} & \mathbf{k} \\ a_1 & a_2 & a_3 \\ c_1 & c_2 & c_3 \end{vmatrix} = (a_2c_3 - c_2a_3)\mathbf{i} - (a_1c_3 - c_1a_3)\mathbf{j} + (a_1c_2 - c_1a_2)\mathbf{k}.$$

$$\mathbf{b}\times\mathbf{c} = \begin{vmatrix} \mathbf{i} & \mathbf{j} & \mathbf{k} \\ b_1 & b_2 & b_3 \\ c_1 & c_2 & c_3 \end{vmatrix} = (b_2c_3 - c_2b_3)\mathbf{i} - (b_1c_3 - c_1b_3)\mathbf{j} + (b_1c_2 - c_1b_2)\mathbf{k}.$$

Hence, $(\mathbf{a}\times\mathbf{c}) + (\mathbf{b}\times\mathbf{c}) = (\mathbf{a}+\mathbf{b})\times\mathbf{c}$.

[30] $\mathbf{a}\times\mathbf{b} = (a_2b_3 - a_3b_2)\mathbf{i} - (a_1b_3 - a_3b_1)\mathbf{j} + (a_1b_2 - a_2b_1)\mathbf{k}$ and

$$(\mathbf{a}\times\mathbf{b})\cdot\mathbf{c} = (a_2b_3c_1 - a_3b_2c_1) - (a_1b_3c_2 - a_3b_1c_2) + (a_1b_2c_3 - a_2b_1c_3).$$

$\mathbf{b}\times\mathbf{c} = (b_2c_3 - c_2b_3)\mathbf{i} - (b_1c_3 - c_1b_3)\mathbf{j} + (b_1c_2 - c_1b_2)\mathbf{k}$ and

$$\mathbf{a}\cdot(\mathbf{b}\times\mathbf{c}) = (a_1b_2c_3 - a_1c_2b_3) - (a_2b_1c_3 - a_2c_1b_3) + (a_3b_1c_2 - a_3c_1b_2).$$

Comparing terms, we see that the result follows.

31 $\mathbf{a} \times (\mathbf{b} \times \mathbf{c}) = \begin{vmatrix} \mathbf{i} & \mathbf{j} & \mathbf{k} \\ a_1 & a_2 & a_3 \\ b_2c_3 - c_2b_3 & b_3c_1 - b_1c_3 & b_1c_2 - c_1b_2 \end{vmatrix} =$

$$\left[a_2(b_1c_2 - c_1b_2) - a_3(b_3c_1 - b_1c_3)\right]\mathbf{i} - \left[a_1(b_1c_2 - c_1b_2) - a_3(b_2c_3 - c_2b_3)\right]\mathbf{j} + \left[a_1(b_3c_1 - b_1c_3) - a_2(b_2c_3 - c_2b_3)\right]\mathbf{k}.$$

$$\begin{aligned}(\mathbf{a}\cdot\mathbf{c})\mathbf{b} &= (a_1c_1 + a_2c_2 + a_3c_3)(b_1\mathbf{i} + b_2\mathbf{j} + b_3\mathbf{k}) \\ &= (a_1c_1b_1 + a_2c_2b_1 + a_3c_3b_1)\mathbf{i} + (a_1c_1b_2 + a_2c_2b_2 + a_3c_3b_2)\mathbf{j} + (a_1c_1b_3 + a_2c_2b_3 + a_3c_3b_3)\mathbf{k}.\end{aligned}$$

$$\begin{aligned}(\mathbf{a}\cdot\mathbf{b})\mathbf{c} &= (a_1b_1 + a_2b_2 + a_3b_3)(c_1\mathbf{i} + c_2\mathbf{j} + c_3\mathbf{k}) \\ &= (a_1b_1c_1 + a_2b_2c_1 + a_3b_3c_1)\mathbf{i} + (a_1b_1c_2 + a_2b_2c_2 + a_3b_3c_2)\mathbf{j} + (a_1b_1c_3 + a_2b_2c_3 + a_3b_3c_3)\mathbf{k}.\end{aligned}$$

Taking $(\mathbf{a}\cdot\mathbf{c})\mathbf{b} - (\mathbf{a}\cdot\mathbf{b})\mathbf{c}$ and comparing terms, the result follows.

32
$$\begin{aligned}(\mathbf{a}+\mathbf{b}) \times (\mathbf{a}-\mathbf{b}) &= (\mathbf{a}+\mathbf{b}) \times \mathbf{a} - (\mathbf{a}+\mathbf{b}) \times \mathbf{b} && \{14.33(\text{iii})\} \\ &= \mathbf{a}\times\mathbf{a} + \mathbf{b}\times\mathbf{a} - \mathbf{a}\times\mathbf{b} - \mathbf{b}\times\mathbf{b} && \{14.33(\text{iv})\} \\ &= \mathbf{0} + \mathbf{b}\times\mathbf{a} + \mathbf{b}\times\mathbf{a} + \mathbf{0} = 2(\mathbf{b}\times\mathbf{a}) && \{14.33(\text{i})\}\end{aligned}$$

33 $\mathbf{a} \times (\mathbf{b} \times \mathbf{c}) + \mathbf{b} \times (\mathbf{c} \times \mathbf{a}) + \mathbf{c} \times (\mathbf{a} \times \mathbf{b}) =$

$(\mathbf{a}\cdot\mathbf{c})\mathbf{b} - (\mathbf{a}\cdot\mathbf{b})\mathbf{c} + (\mathbf{b}\cdot\mathbf{a})\mathbf{c} - (\mathbf{b}\cdot\mathbf{c})\mathbf{a} + (\mathbf{c}\cdot\mathbf{b})\mathbf{a} - (\mathbf{c}\cdot\mathbf{a})\mathbf{b} =$

$(\mathbf{a}\cdot\mathbf{c} - \mathbf{c}\cdot\mathbf{a})\mathbf{b} + (\mathbf{b}\cdot\mathbf{a} - \mathbf{a}\cdot\mathbf{b})\mathbf{c} + (\mathbf{c}\cdot\mathbf{b} - \mathbf{b}\cdot\mathbf{c})\mathbf{a} = 0\mathbf{b} + 0\mathbf{c} + 0\mathbf{a} = \mathbf{0}$

34
$$\begin{aligned}(\mathbf{a}\times\mathbf{b})\times\mathbf{c} = -\mathbf{c}\times(\mathbf{a}\times\mathbf{b}) &= (-\mathbf{c}\cdot\mathbf{b})\mathbf{a} - (-\mathbf{c}\cdot\mathbf{a})\mathbf{b} \\ &= (\mathbf{c}\cdot\mathbf{a})\mathbf{b} - (\mathbf{c}\cdot\mathbf{b})\mathbf{a} = (\mathbf{a}\cdot\mathbf{c})\mathbf{b} - (\mathbf{b}\cdot\mathbf{c})\mathbf{a}\end{aligned}$$

35 Using (14.33)(v), $(\mathbf{a}\times\mathbf{b})\cdot(\mathbf{c}\times\mathbf{d}) = \mathbf{a}\cdot\left[\mathbf{b}\times(\mathbf{c}\times\mathbf{d})\right] = \mathbf{a}\cdot\left[(\mathbf{b}\cdot\mathbf{d})\mathbf{c} - (\mathbf{b}\cdot\mathbf{c})\mathbf{d}\right] =$

$\mathbf{a}\cdot(\mathbf{b}\cdot\mathbf{d})\mathbf{c} - \mathbf{a}\cdot(\mathbf{b}\cdot\mathbf{c})\mathbf{d} = (\mathbf{b}\cdot\mathbf{d})(\mathbf{a}\cdot\mathbf{c}) - (\mathbf{b}\cdot\mathbf{c})(\mathbf{a}\cdot\mathbf{d})$ by (14.17)(iv) $= \begin{vmatrix} \mathbf{a}\cdot\mathbf{c} & \mathbf{b}\cdot\mathbf{c} \\ \mathbf{a}\cdot\mathbf{d} & \mathbf{b}\cdot\mathbf{d} \end{vmatrix}$.

36 Using (14.33)(vi), $(\mathbf{a}\times\mathbf{b})\times(\mathbf{c}\times\mathbf{d}) = \left[(\mathbf{a}\times\mathbf{b})\cdot\mathbf{d}\right]\mathbf{c} - \left[(\mathbf{a}\times\mathbf{b})\cdot\mathbf{c}\right]\mathbf{d}$.

37
$$\begin{aligned}(\mathbf{a}\times\mathbf{b})\cdot(\mathbf{b}\times\mathbf{c})\times(\mathbf{c}\times\mathbf{a}) &= (\mathbf{a}\times\mathbf{b})\cdot\left[(\mathbf{b}\times\mathbf{c}\cdot\mathbf{a})\mathbf{c} - (\mathbf{b}\times\mathbf{c}\cdot\mathbf{c})\mathbf{a}\right] \ \{\text{Exer. 36}\} \\ &= (\mathbf{b}\times\mathbf{c}\cdot\mathbf{a})(\mathbf{a}\times\mathbf{b}\cdot\mathbf{c}) - 0 \ \{\text{since } \mathbf{b}\times\mathbf{c}\cdot\mathbf{c} = 0\} \\ &= (\mathbf{b}\times\mathbf{c}\cdot\mathbf{a})(\mathbf{a}\cdot\mathbf{b}\times\mathbf{c}) \ \{\text{by (14.33)(v)}\} \\ &= (\mathbf{a}\cdot\mathbf{b}\times\mathbf{c})(\mathbf{a}\cdot\mathbf{b}\times\mathbf{c}) = (\mathbf{a}\cdot\mathbf{b}\times\mathbf{c})^2\end{aligned}$$

Exercises 14.5

1 By (14.34), a parametric equation for the line through $P(4, 2, -3)$ and parallel to $\mathbf{a} = \langle \frac{1}{3}, 2, \frac{1}{2} \rangle$ is $x = 4 + \frac{1}{3}t$, $y = 2 + 2t$, and $z = -3 + \frac{1}{2}t$.

2 $P(5, 0, -2)$; $\mathbf{a} = \langle -1, -4, 1 \rangle \Rightarrow x = 5 - t,\ y = -4t,\ z = -2 + t$

3 $P(0, 0, 0)$; $\mathbf{a} = \mathbf{j} \Rightarrow x = 0,\ y = t,\ z = 0$

4 $P(1, 2, 3)$; $\mathbf{a} = \mathbf{i} + 2\mathbf{j} + 3\mathbf{k} \Rightarrow x = 1 + t,\ y = 2 + 2t,\ z = 3 + 3t$

5 $\mathbf{a} = \overrightarrow{P_1P_2} = <-3, 8, -3>$. Thus, $x = 5 - 3t$, $y = -2 + 8t$, $z = 4 - 3t$.

Now, $z = 0 \Rightarrow t = \frac{4}{3} \Rightarrow x = 1$ and $y = \frac{26}{3}$. $(1, \frac{26}{3}, 0)$ is the point of intersection with the xy-plane. Similarly, $y = 0 \Rightarrow t = \frac{1}{4} \Rightarrow xz$-plane intersection point is $(\frac{17}{4}, 0, \frac{13}{4})$, and $x = 0 \Rightarrow t = \frac{5}{3} \Rightarrow yz$-plane intersection point is $(0, \frac{34}{3}, -1)$.

6 $\mathbf{a} = \overrightarrow{P_1P_2} = <10, 10, -7>$. Thus, $x = -3 + 10t$, $y = 1 + 10t$, $z = -1 - 7t$.

Now, $z = 0 \Rightarrow t = -\frac{1}{7} \Rightarrow x = -\frac{31}{7}$ and $y = -\frac{3}{7}$. $(-\frac{31}{7}, -\frac{3}{7}, 0)$ is the point of intersection with the xy-plane. Similarly, $y = 0 \Rightarrow t = -\frac{1}{10} \Rightarrow xz$-plane intersection point is $(-4, 0, -\frac{3}{10})$,

and $x = 0 \Rightarrow t = \frac{3}{10} \Rightarrow yz$-plane intersection point is $(0, 4, -\frac{31}{10})$.

7 $\mathbf{a} = \overrightarrow{P_1P_2} = <-8, 0, -2>$. Thus, $x = 2 - 8t$, $y = 0 + 0t$, $z = 5 - 2t$.

Now, $z = 0 \Rightarrow t = \frac{5}{2} \Rightarrow x = -18$ and $y = 0$. $(-18, 0, 0)$ is the point of intersection with the xy-plane. y is always equal to 0, so the line lies in xz-plane.

Also, $x = 0 \Rightarrow t = \frac{1}{4} \Rightarrow yz$-plane intersection point is $(0, 0, \frac{9}{2})$.

8 $\mathbf{a} = \overrightarrow{P_1P_2} = <0, 0, -7>$. Thus, $x = 2$, $y = -2$, $z = 4 - 7t$.

Now, $z = 0 \Rightarrow t = \frac{4}{7} \Rightarrow x = 2$ and $y = -2$. $(2, -2, 0)$ is the point of intersection with the xy-plane. $y = 0$ and $x = 0$ have no solution,

so the line never intersects the xz-plane or the yz-plane.

9 Using the point $P(-6, 4, -3)$, the coefficients of t $\{-3, 1, \text{and } 9\}$ for l,

and the parameter s, we have $x = -6 - 3s$, $y = 4 + s$, and $z = -3 + 9s$.

10 $\mathbf{a} = \overrightarrow{P_1P_2} = <8, -2, -1>$ and $P(4, -1, 0) \Rightarrow x = 4 + 8t$, $y = -1 - 2t$, $z = -t$

Note: In Exercises 11–14, if the two lines intersect, let t_0 and v_0 denote the values of t and v for the intersection point.

11

$$\begin{aligned} 1 + 2t_0 &= 4 - v_0 & 2t_0 + v_0 &= 3 \\ 1 - 4t_0 &= -1 + 6v_0 \Rightarrow & 4t_0 + 6v_0 &= 2 \Rightarrow t_0 = 2,\ v_0 = -1 \\ 5 - t_0 &= 4 + v_0 & t_0 + v_0 &= 1 \end{aligned}$$

Lines intersect at $(5, -7, 3)$.

12

$$\begin{aligned} 1 - 6t_0 &= 2 + 2v_0 & 6t_0 + 2v_0 &= -1 \\ 3 + 2t_0 &= 6 + v_0 \Rightarrow & 2t_0 - v_0 &= 3 \Rightarrow t_0 = \tfrac{1}{2},\ v_0 = -2 \\ 1 - 2t_0 &= 2 + v_0 & 2t_0 + v_0 &= -1 \end{aligned}$$

Lines intersect at $(-2, 4, 0)$.

13

$$\begin{aligned} 3 + t_0 &= 4 - v_0 & t_0 + v_0 &= 1 \\ 2 - 4t_0 &= 3 + v_0 \Rightarrow & 4t_0 + v_0 &= -1 \Rightarrow \text{no solution} \\ t_0 &= -2 + 3v_0 & t_0 - 3v_0 &= -2 \end{aligned}$$

Lines do not intersect.

[14] $\begin{aligned} 2 - 5t_0 &= 4 - 3v_0 \\ 6 + 2t_0 &= 7 + 5v_0 \\ -3 - 2t_0 &= 1 + 4v_0 \end{aligned} \Rightarrow \begin{aligned} 5t_0 - 3v_0 &= -2 \\ 2t_0 - 5v_0 &= 1 \\ 2t_0 + 4v_0 &= -4 \end{aligned} \Rightarrow$ no solution

Lines do not intersect.

[15] a = <−2, 3, 5>, b = <4, 4, 1> ⇒

$$\cos\theta = \frac{\mathbf{a}\cdot\mathbf{b}}{\|\mathbf{a}\|\|\mathbf{b}\|} = \frac{-8 + 12 + 5}{\sqrt{38}\sqrt{33}} \Rightarrow \theta = \cos^{-1}\left(\frac{9}{\sqrt{38}\sqrt{33}}\right) \approx 75^\circ \text{ and } 180^\circ - \theta.$$

Note: The answer in the first printing was for $x = -1 + t$ instead of $x = -1 + 4t$.

[16] a = <3, −1, 2>, b = <−1, −2, 1> ⇒

$$\cos\theta = \frac{\mathbf{a}\cdot\mathbf{b}}{\|\mathbf{a}\|\|\mathbf{b}\|} = \frac{-3 + 2 + 2}{\sqrt{14}\sqrt{6}} \Rightarrow \theta = \cos^{-1}\left(\frac{1}{\sqrt{14}\sqrt{6}}\right) \approx 84^\circ \text{ and } 180^\circ - \theta.$$

[17] a = <−3, 8, −3>, b = <10, 10, −7> ⇒

$$\cos\theta = \frac{\mathbf{a}\cdot\mathbf{b}}{\|\mathbf{a}\|\|\mathbf{b}\|} = \frac{-30 + 80 + 21}{\sqrt{82}\sqrt{249}} \Rightarrow \theta = \cos^{-1}\left(\frac{71}{\sqrt{82}\sqrt{249}}\right) \approx 60^\circ \text{ and } 180^\circ - \theta.$$

[18] a = <3, 3, 1>, b = <4, −3, −9> ⇒

$$\cos\theta = \frac{\mathbf{a}\cdot\mathbf{b}}{\|\mathbf{a}\|\|\mathbf{b}\|} = \frac{12 - 9 - 9}{\sqrt{19}\sqrt{106}} \Rightarrow \theta = \cos^{-1}\left(\frac{-6}{\sqrt{19}\sqrt{106}}\right) \approx 98^\circ \text{ and } 180^\circ - \theta.$$

[19] (a) $z = 4$ (b) $x = 6$ (c) $y = -7$

[20] (a) $x = -2$ (b) $y = 5$ (c) $z = -8$

[21] By (14.36), an equation of the plane through $P(-11, 4, -2)$ with normal vector $\mathbf{a} = 6\mathbf{i} - 5\mathbf{j} - \mathbf{k}$ is $6(x + 11) - 5(y - 4) - 1(z + 2) = 0$, or $6x - 5y - z = -84$.

[22] $4(x - 4) + 2(y - 2) - 9(z + 9) = 0$, or $4x + 2y - 9z = 101$.

[23] By (14.37), the plane $3x - y + 2z = 10$ has normal vector <3, −1, 2>. An equation of the plane is $3(x - 2) - 1(y - 5) + 2(z + 6) = 0$, or $3x - y + 2z = -11$.

[24] $1(x - 0) - 6(y - 0) + 4(z - 0) = 0$, or $x - 6y + 4z = 0$.

[25] Letting $y = 0$ to find the trace of $x + 4y - 5z = 8$ in the xz-plane gives us $x - 5z = 8$. The equation of the plane is $x + ky - 5z = 8$. Substituting $x = -4$, $y = 1$, and $z = 6$ yields $k = 42$. ★ $x + 42y - 5z = 8$

[26] $\overrightarrow{OP}$ = <0, 2, 5>, $\overrightarrow{OQ}$ = <1, 4, 0>, and a = $\overrightarrow{OP} \times \overrightarrow{OQ}$ = <−20, 5, −2>. a is normal to the plane, so an equation is $-20(x - 0) + 5(y - 0) - 2(z - 0) = 0$, or $20x - 5y + 2z = 0$.

[27] $\overrightarrow{PQ}$ = <−2, 2, −1>, $\overrightarrow{PR}$ = <0, −2, −1>, and a = $\overrightarrow{PQ} \times \overrightarrow{PR}$ = <−4, −2, 4>. a is normal to the plane, so an equation is $-4(x - 1) - 2(y - 1) + 4(z - 3) = 0$, or $2x + y - 2z = -3$.

[28] $\overrightarrow{PQ}$ = <−4, −1, −3>, $\overrightarrow{PR}$ = <0, −6, 0>, and a = $\overrightarrow{PQ} \times \overrightarrow{PR}$ = <−18, 0, 24>. a is normal to the plane, so an equation is

$-18(x - 3) + 0(y - 2) + 24(z - 1) = 0$, or $3x - 4z = 5$.

[29] (a) $x = 3$ (b) $y = -2$ (c) $z = 5$ •

[30] (a) $x = -4$ (b) $y = 0$ (c) $z = -\frac{2}{3}$ •

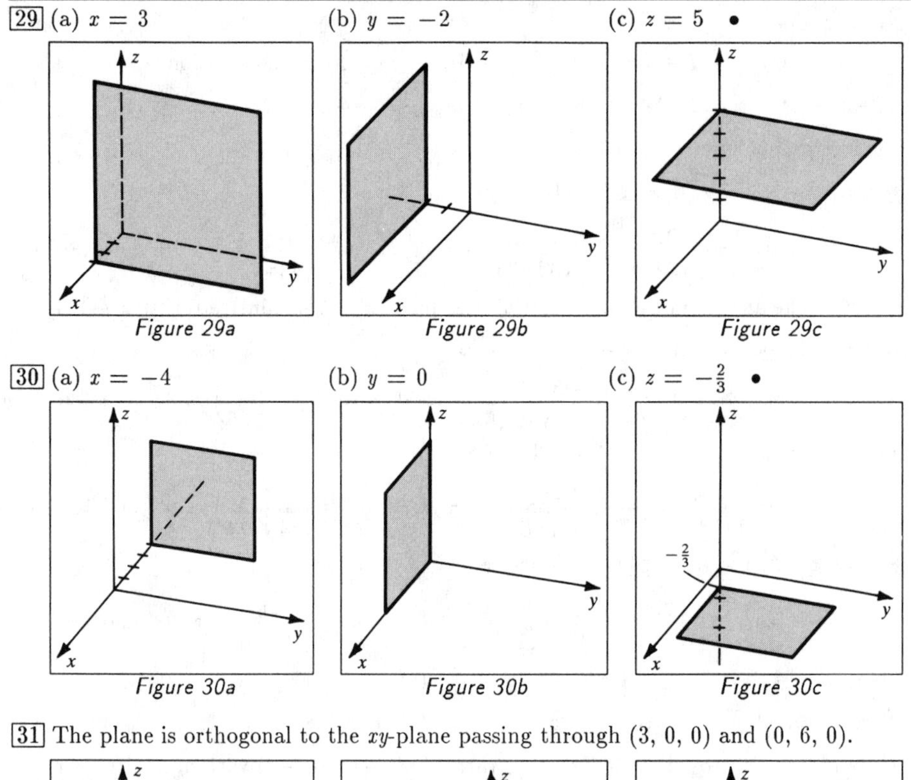

Figure 29a Figure 29b Figure 29c

Figure 30a Figure 30b Figure 30c

[31] The plane is orthogonal to the xy-plane passing through (3, 0, 0) and (0, 6, 0).

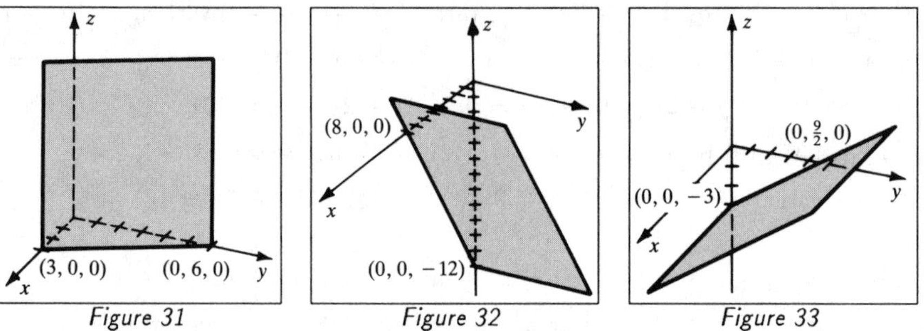

Figure 31 Figure 32 Figure 33

[32] The plane is orthogonal to the xz-plane passing through (8, 0, 0) and (0, 0, −12).

[33] The plane is orthogonal to the yz-plane passing through (0, $\frac{9}{2}$, 0) and (0, 0, −3).

[34] The plane passes through the points $(-4, 0, 0)$, $(0, -20, 0)$, and $(0, 0, 5)$.

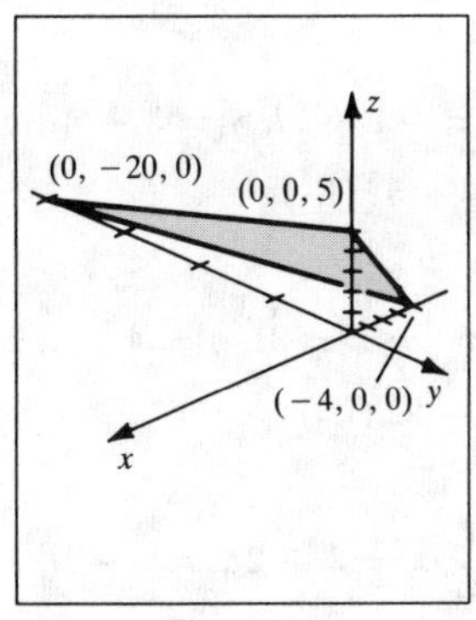

Figure 34

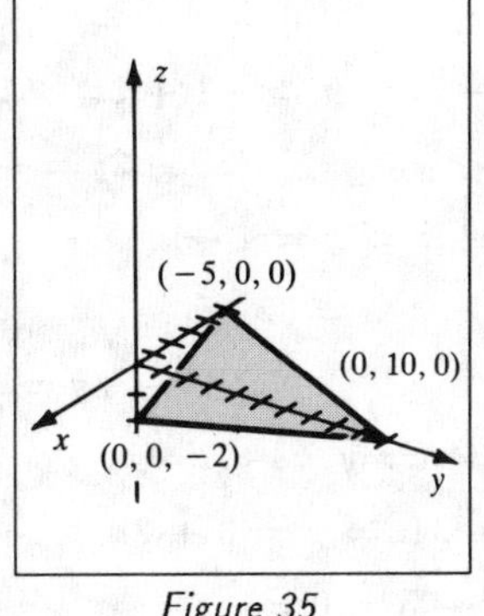

Figure 35

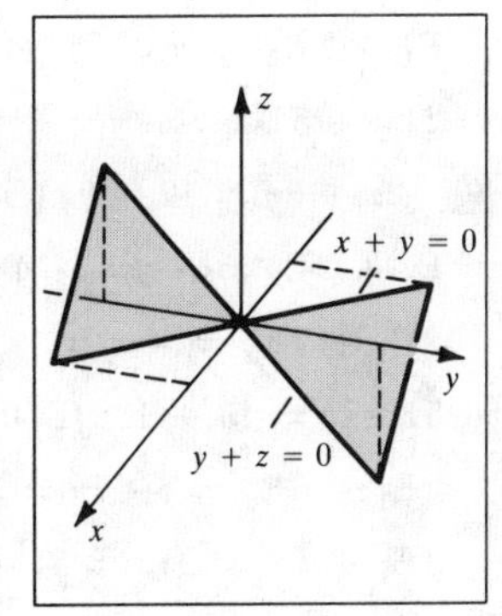

Figure 36

[35] The plane passes through the points $(-5, 0, 0)$, $(0, 10, 0)$, and $(0, 0, -2)$.

[36] The point of intersection with the three coordinate axes is $(0, 0, 0)$.

The trace in the xy-plane is $x + y = 0$, in the xz-plane is $x + z = 0$,

and in the yz-plane is $y + z = 0$.

[37] The trace of the plane in the xz-plane is $x + z = 5$.

Since the plane is orthogonal to the xz-plane, the y-coordinate coefficient is zero.

[38] The trace of the plane in the yz-plane is $y + z = 4$.

Since the plane is orthogonal to the yz-plane, the x-coordinate coefficient is zero.

[39] The trace of the plane in the xy-plane is $3x + 2y = 6$.

Since the plane is orthogonal to the xy-plane, the z-coordinate coefficient is zero.

[40] The trace of the plane in the xz-plane is $x + 2z = 4$.

Since the plane is orthogonal to the xz-plane, the y-coordinate coefficient is zero.

[41] A normal vector for the plane is $<4, -1, 3>$ and an equation for the plane is

$$4(x - 1) - 1(y - 2) + 3(z + 3) = 0, \text{ or } 4x - y + 3z + 7 = 0.$$

[42] A normal vector for the plane is $<-2, 3, -1>$ and an equation for the plane is

$$-2(x - 3) + 3(y + 2) - 1(z - 4) = 0, \text{ or } -2x + 3y - z + 16 = 0.$$

[43] $P_1(5, -2, 4)$, $P_2(2, 6, 1) \Rightarrow \overrightarrow{P_1P_2} = <-3, 8, -3>$ and

$\dfrac{x - 5}{-3} = \dfrac{y + 2}{8} = \dfrac{z - 4}{-3}$ is a symmetric form for the line through P_1 and P_2.

[44] $P_1(-3, 1, -1)$, $P_2(7, 11, -8) \Rightarrow \overrightarrow{P_1P_2} = <10, 10, -7>$; $\dfrac{x + 3}{10} = \dfrac{y - 1}{10} = \dfrac{z + 1}{-7}$

[45] $P_1(4, 2, -3)$, $P_2(-3, 2, 5) \Rightarrow \overrightarrow{P_1P_2} = <-7, 0, 8>$; $\dfrac{x - 4}{-7} = \dfrac{z + 3}{8}$; $y = 2$

[46] $P_1(5, -7, 4)$, $P_2(-2, -1, 4) \Rightarrow \overrightarrow{P_1P_2} = <-7, 6, 0>$; $\dfrac{x - 5}{-7} = \dfrac{y + 7}{6}$; $z = 4$

[47] $x + 2y = 7 + 9z\ (\mathrm{E}_1)$; $\qquad 2x - 3y = -17z\ (\mathrm{E}_2)$

$2\,\mathrm{E}_1 - \mathrm{E}_2 \Rightarrow 7y = 14 + 35z \Rightarrow y = 2 + 5z$

$3\,\mathrm{E}_1 + 2\,\mathrm{E}_2 \Rightarrow 7x = 21 - 7z \Rightarrow x = 3 - z$ $\qquad$ ★ $x = 3 - t$, $y = 2 + 5t$, $z = t$

48 $2x + 5y = 13 - 16z\ (E_1)$; $-x - 2y = -5 + 6z\ (E_2)$

$E_1 + 2E_2 \Rightarrow y = 3 - 4z$

$2E_1 + 5E_2 \Rightarrow -x = 1 - 2z \Rightarrow x = -1 + 2z$ ★ $x = -1 + 2t,\ y = 3 - 4t,\ z = t$

49 $-2x + 3y = 12 - 9z\ (E_1)$; $x - 2y = -8 + 5z\ (E_2)$

$E_1 + 2E_2 \Rightarrow -y = -4 + z \Rightarrow y = 4 - z$

$2E_1 + 3E_2 \Rightarrow -x = -3z \Rightarrow x = 3z$ ★ $x = 3t,\ y = 4 - t,\ z = t$

50 $5x - y = 15 + 12z\ (E_1)$; $2x + 3y = 6 - 2z\ (E_2)$

$2E_1 - 5E_2 \Rightarrow -17y = 34z \Rightarrow y = -2z$

$3E_1 + E_2 \Rightarrow 17x = 51 + 34z \Rightarrow x = 3 + 2z$ ★ $x = 3 + 2t,\ y = -2t,\ z = t$

51 $P(1, -1, 2)$; $3x - 7y + z - 5 = 0 \Rightarrow$

$$h = \frac{|3(1) - 7(-1) + 1(2) - 5|}{\sqrt{3^2 + (-7)^2 + 1^2}} = \frac{7}{\sqrt{59}} \approx 0.91.$$

52 $P(3, 1, -2)$; $2x + 4y - 5z + 1 = 0 \Rightarrow$

$$h = \frac{|2(3) + 4(1) - 5(-2) + 1|}{\sqrt{2^2 + 4^2 + (-5)^2}} = \frac{21}{\sqrt{45}} = \frac{7}{\sqrt{5}} \approx 3.13.$$

53 The normal vectors are $\mathbf{a} = \langle 4, -2, 6\rangle$ and $\mathbf{b} = \langle -6, 3, -9\rangle$. They are parallel since $\mathbf{b} = -\frac{3}{2}\mathbf{a}$ and hence, the planes are parallel. Since $(0, 0, \frac{1}{2})$ is on the first plane, its distance from the second plane is $h = \dfrac{|-6(0) + 3(0) - 9(\frac{1}{2}) - 4|}{\sqrt{(-6)^2 + 3^2 + (-9)^2}} = \dfrac{17}{6\sqrt{14}} \approx 0.76.$

54 The normal vectors are $\mathbf{a} = \langle 3, 12, -6\rangle$ and $\mathbf{b} = \langle 5, 20, -10\rangle$. They are parallel since $\mathbf{b} = \frac{5}{3}\mathbf{a}$ and hence, the planes are parallel. Since $(0, 0, \frac{1}{3})$ is on the first plane, its distance from the second plane is $h = \dfrac{|5(0) + 20(0) - 10(\frac{1}{3}) - 7|}{\sqrt{5^2 + 20^2 + (-10)^2}} = \dfrac{31}{15\sqrt{21}} \approx 0.45.$

55 $\mathbf{a} = \overrightarrow{AB} = \langle 1, 2, 2\rangle$ and $\mathbf{b} = \overrightarrow{CD} = \langle -6, 2, 5\rangle$.

$\mathbf{a} \times \mathbf{b} = \langle 6, -17, 14\rangle$ and $\|\mathbf{a} \times \mathbf{b}\| = \sqrt{521}$. Also, $\mathbf{c} = \overrightarrow{AC} = \langle 3, 3, -4\rangle$.

$$\text{Thus, } d = \frac{|(\mathbf{a} \times \mathbf{b}) \cdot \mathbf{c}|}{\|\mathbf{a} \times \mathbf{b}\|} = \frac{89}{\sqrt{521}} \approx 3.90.$$

56 $\mathbf{a} = \overrightarrow{AB} = \langle -1, 1, 5\rangle$ and $\mathbf{b} = \overrightarrow{CD} = \langle 7, 2, -2\rangle$.

$\mathbf{a} \times \mathbf{b} = \langle -12, 33, -9\rangle$ and $\|\mathbf{a} \times \mathbf{b}\| = \sqrt{1314}$. Also, $\mathbf{c} = \overrightarrow{AC} = \langle -3, -4, 2\rangle$.

$$\text{Thus, } d = \frac{|(\mathbf{a} \times \mathbf{b}) \cdot \mathbf{c}|}{\|\mathbf{a} \times \mathbf{b}\|} = \frac{114}{\sqrt{1314}} = \frac{38}{\sqrt{146}} \approx 3.14.$$

57 The points $Q(1, 4, -3)$ $\{t = 0\}$ and $R(4, 2, -2)$ $\{t = 1\}$ lie on ℓ and in the plane. $\overrightarrow{QP} = \langle 4, -4, 5\rangle$ and $\overrightarrow{QR} = \langle 3, -2, 1\rangle$. $\overrightarrow{QP} \times \overrightarrow{QR} = \langle 6, 11, 4\rangle$.

The plane is $6(x - 5) + 11(y - 0) + 4(z - 2) = 0$, or $6x + 11y + 4z = 38$.

58 The points $Q(5, -1, 7)$ $\{t = 0\}$ and $R(6, 1, 6)$ $\{t = 1\}$ lie on ℓ and in the plane. $\overrightarrow{QP} = \langle -1, -2, -7\rangle$ and $\overrightarrow{QR} = \langle 1, 2, -1\rangle$. $\overrightarrow{QP} \times \overrightarrow{QR} = \langle 16, -8, 0\rangle$.

The plane is $16(x - 4) - 8(y + 3) + 0(z - 0) = 0$, or $2x - y = 11$.

[59] Let D denote the projection of A onto the line through B and C.

Let $\mathbf{a} = \overrightarrow{BA} = \langle -1, -10, 3\rangle$ and $\mathbf{c} = \overrightarrow{BC} = \langle 4, -5, 7\rangle$. $\|\mathbf{a}\|$, $\|\overrightarrow{BD}\|$,

and $\overline{AD}$ form a right triangle. $\|\mathbf{a}\| = \sqrt{110}$ and $\|\overrightarrow{BD}\| = \operatorname{comp}_{\mathbf{c}}\mathbf{a} = \frac{\mathbf{a}\cdot\mathbf{c}}{\|\mathbf{c}\|} = \frac{67}{\sqrt{90}}$.

Thus, $\overline{AD} = \sqrt{(\sqrt{110})^2 - (67/\sqrt{90})^2} = \sqrt{5411/90} \approx 7.75$.

[60] Using the same notation as in Exercise 59, $\mathbf{a} = \langle 3, 4, 4\rangle$ and $\mathbf{c} = \langle 2, -4, 6\rangle$.

$\|\mathbf{a}\| = \sqrt{41}$ and $\operatorname{comp}_{\mathbf{c}}\mathbf{a} = \frac{14}{\sqrt{56}} = \frac{\sqrt{14}}{2}$.

Thus, $\overline{AD} = \sqrt{(\sqrt{41})^2 - (\sqrt{14}/2)^2} = \frac{5}{2}\sqrt{6} \approx 6.12$.

[61] Refer to Example 3 of Section 14.4. To obtain two points on the line, let $t = 0$ and $t = 1$ to get $A(3, -4, 1)$ and $B(1, -1, 3)$. $\mathbf{a} = \overrightarrow{AB} = \langle -2, 3, 2\rangle$ and

$$\mathbf{b} = \overrightarrow{AP} = \langle -1, 5, -3\rangle.\quad d = \frac{\|\mathbf{a}\times\mathbf{b}\|}{\|\mathbf{a}\|} = \frac{\|\langle -19, -8, -7\rangle\|}{\sqrt{17}} = \sqrt{\frac{474}{17}} \approx 5.28.$$

[62] As in Exercise 61, we have $A(1, 3, 0)$ and $B(5, 2, 3)$. $\mathbf{a} = \overrightarrow{AB} = \langle 4, -1, 3\rangle$ and

$$\mathbf{b} = \overrightarrow{AP} = \langle 2, -2, -1\rangle.\quad d = \frac{\|\mathbf{a}\times\mathbf{b}\|}{\|\mathbf{a}\|} = \frac{\|\langle 7, 10, -6\rangle\|}{\sqrt{26}} = \sqrt{\frac{185}{26}} \approx 2.67.$$

[63] $(10x - 15y + 6z = 30)\cdot\frac{1}{30} \Leftrightarrow \frac{x}{3} + \frac{y}{-2} + \frac{z}{5} = 1$

[64] $(12x + 15y - 20z = 60)\cdot\frac{1}{60} \Leftrightarrow \frac{x}{5} + \frac{y}{4} + \frac{z}{-3} = 1$

[65] Using the intercept form, an equation is $\frac{x}{2} + \frac{y}{3} + \frac{z}{4} = 1$, or $6x + 4y + 3z = 12$.

[66] Using the intercept form, an equation is $\frac{x}{4} + \frac{y}{5} + \frac{z}{2} = 1$, or $5x + 4y + 10z = 20$.

[67] (a) From the graph, $P \approx (0.55, 0.30)$.

(b) $f'(x) = 2x\cos x^2 \Rightarrow f'(0.55) \approx 1.05$. $g'(x) = -\sin x - 1 \Rightarrow g'(0.55) \approx -1.52$.

The vectors $\mathbf{v} = \mathbf{i} + 1.05\mathbf{j}$ and $\mathbf{w} = \mathbf{i} - 1.52\mathbf{j}$ are approximately parallel to the tangent lines. $\cos\theta = \frac{\mathbf{v}\cdot\mathbf{w}}{\|\mathbf{v}\|\|\mathbf{w}\|} \approx \frac{-0.60}{(1.45)(1.82)} \approx -0.23 \Rightarrow \theta \approx 103^\circ$.

The angles are 103° and 77°.

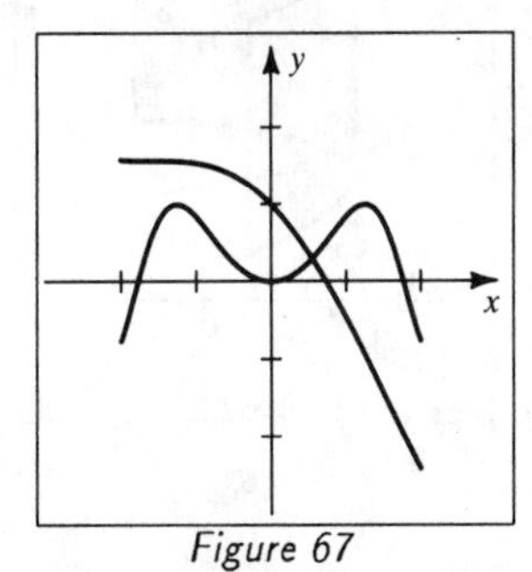

Figure 67

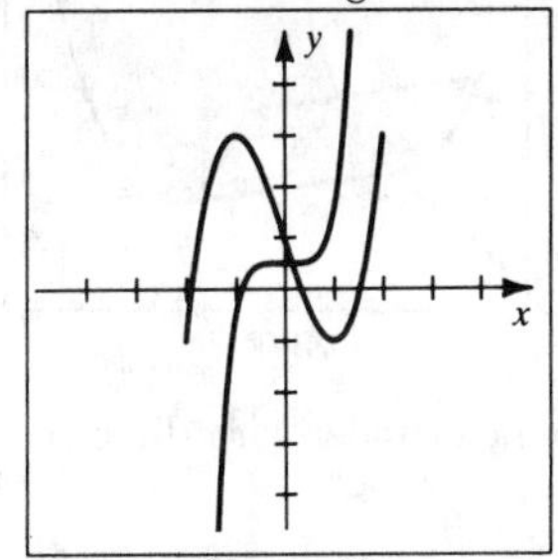

Figure 68

[68] (a) From the graph, $P \approx (0.17, 0.50)$.

(b) $f'(x) = -3 + 3x^2 \Rightarrow f'(0.17) \approx -2.91$. $g'(x) = 5x^4 \Rightarrow g'(0.17) \approx 0.004$.

The vectors $\mathbf{v} = \mathbf{i} - 2.91\mathbf{j}$ and $\mathbf{w} = \mathbf{i} + 0.004\mathbf{j}$ are approximately parallel to the tangent lines. $\cos\theta = \frac{\mathbf{v}\cdot\mathbf{w}}{\|\mathbf{v}\|\|\mathbf{w}\|} \approx \frac{0.99}{(3.08)(1.00)} \approx 0.32 \Rightarrow \theta \approx 71^\circ$.

The angles are 71° and 109°.

Exercises 14.6

[1] Circular cylinder with directrix $x^2 + y^2 = 9$ and rulings parallel to the z-axis.

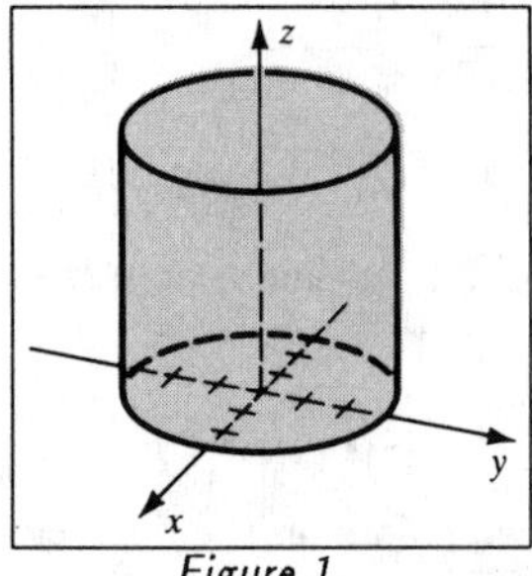

Figure 1

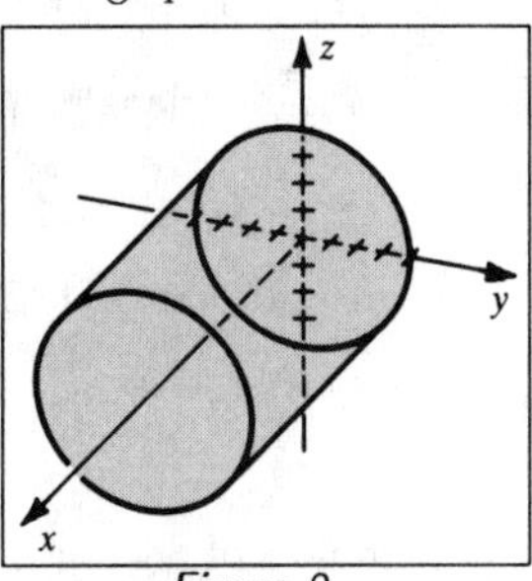

Figure 2

[2] Circular cylinder with directrix $y^2 + z^2 = 16$ and rulings parallel to the x-axis.

[3] Elliptic cylinder with directrix $4y^2 + 9z^2 = 36$ and rulings parallel to the x-axis.

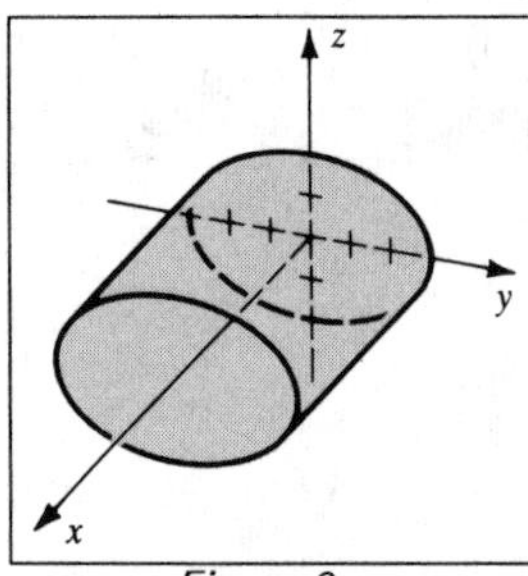

Figure 3

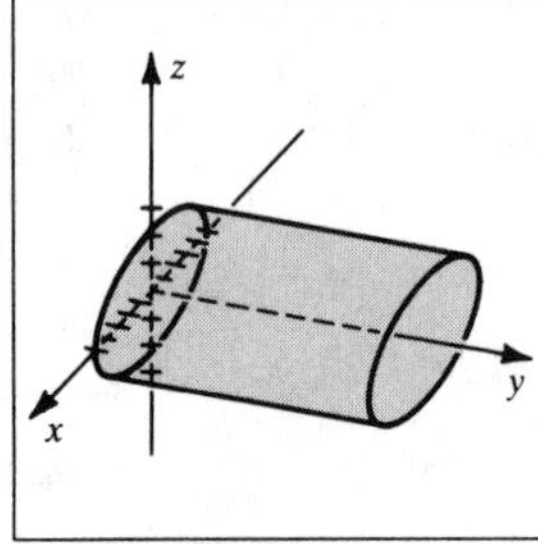

Figure 4

[4] Elliptic cylinder with directrix $x^2 + 5z^2 = 25$ and rulings parallel to the y-axis.

[5] Parabolic cylinder with directrix $x^2 = 9z$ and rulings parallel to the y-axis.

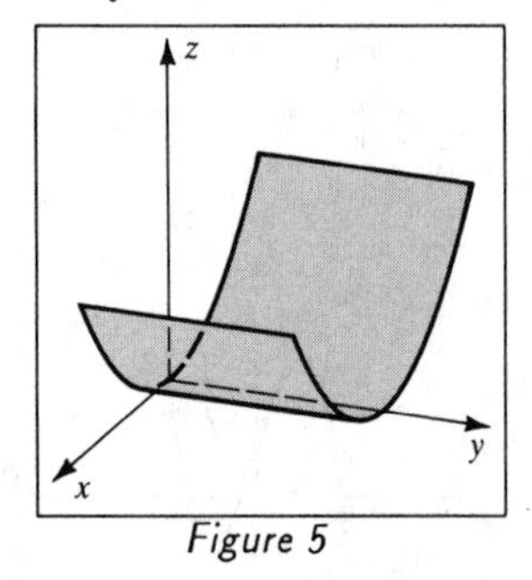

Figure 5

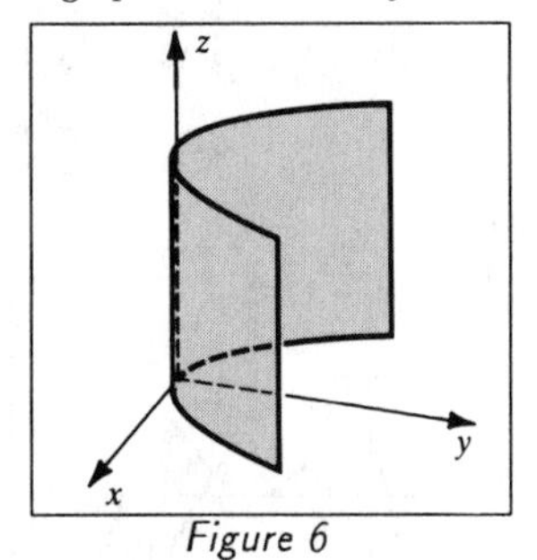

Figure 6

[6] Parabolic cylinder with directrix $x^2 = 4y$ and rulings parallel to the z-axis.

[7] Hyperbolic cylinder with directrix $y^2 - x^2 = 16$ and rulings parallel to the z-axis.

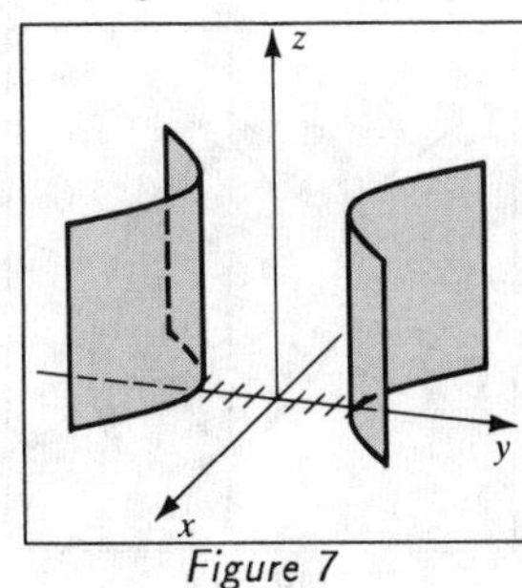

Figure 7

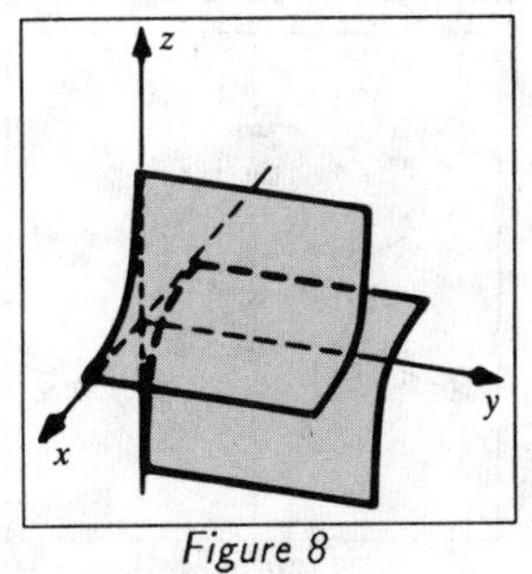

Figure 8

[8] Hyperbolic cylinder with directrix $xz = 1$ and rulings parallel to the y-axis.

[9] The surface is an elliptic paraboloid with z positive—choice K.

[10] The surface is a circular paraboloid with y positive—choice N.

[11] The surface is a hyperboloid of one sheet with the x^2 term being negative—choice C.

[12] The surface is a hyperboloid of one sheet with the y^2 term being negative—choice O.

[13] The surface is a hyperboloid of two sheets with the y^2 term being positive—choice Q.

[14] The surface is a hyperboloid of two sheets with the x^2 term being positive—choice H.

[15] The surface is a cone opening in the positive and negative y direction and having the form $y^2 = Ax^2 + Bz^2$—choice P.

[16] The surface is a cone opening in the positive and negative x direction and having the form $x^2 = Ay^2 + Bz^2$—choice G.

[17] The surface is an ellipsoid with x- and y-intercepts larger than the z-intercept—choice A.

[18] The surface is an ellipsoid with z- and y-intercepts larger than the x-intercept—choice J.

[19] The surface is a hyperbolic paraboloid with z positive in the xz-plane, z negative in the yz-plane, and "facing along the x-axis"—choice E.

[20] The surface is a hyperbolic paraboloid with y positive in the xy-plane, y negative in the yz-plane, and "facing along the x-axis"—choice I.

Note: It may be an interesting review exercise to orally quiz students by having them identify all of the choices A–R. The other choices are:

B. Elliptic paraboloid opening in the positive x direction

D. Cone opening in the positive and negative z direction

F. Hyperboloid of two sheets with the z^2 term positive (opening in the positive and negative z direction)

L. Ellipsoid with y-intercepts being the largest and x-intercepts the smallest

M. Hyperbolic paraboloid with y positive in the yz-plane, y negative in the xy-plane, and "facing along the z-axis"

R. Hyperboloid of one sheet with the z^2 term negative (opening in the positive and negative z direction)

[21] Ellipsoid with x-, y-, and z-intercepts at ± 2, ± 3, and ± 4.

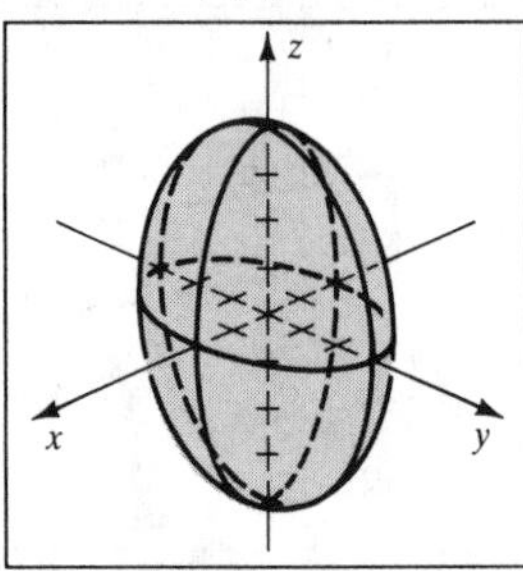

Figure 21

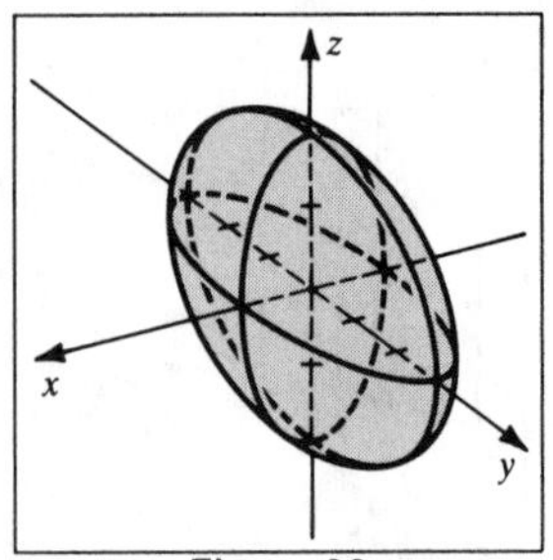

Figure 22

[22] Ellipsoid with x-, y-, and z-intercepts at ± 1, ± 3, and ± 2.

[23] (a) $\frac{x^2}{4} + y^2 - z^2 = 1$ (b) $x^2 + \frac{z^2}{4} - y^2 = 1$ •

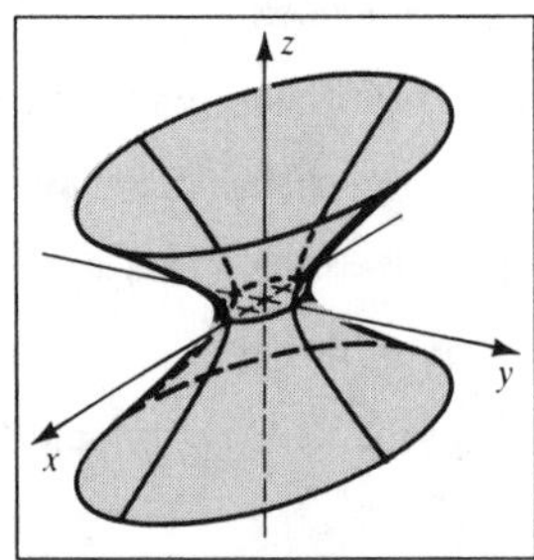

Figure 23a

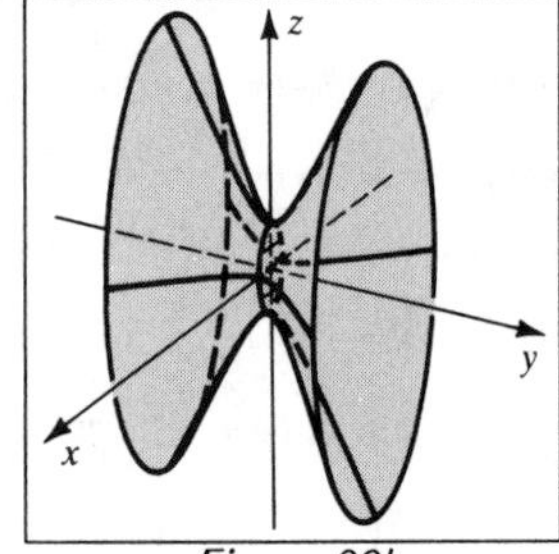

Figure 23b

[24] (a) $z^2 + x^2 - y^2 = 1$ (b) $y^2 + \frac{z^2}{4} - x^2 = 1$ •

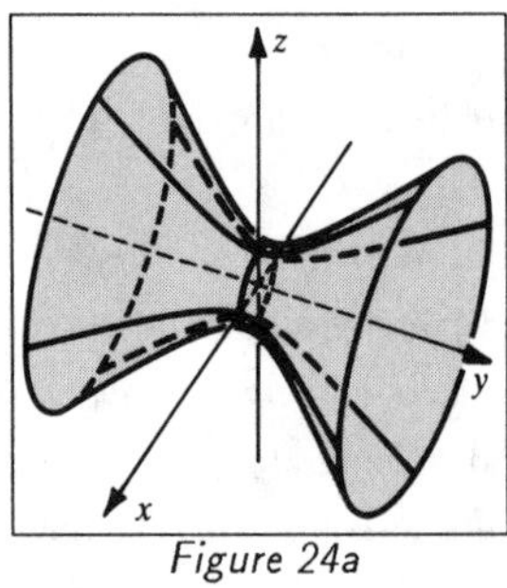

Figure 24a

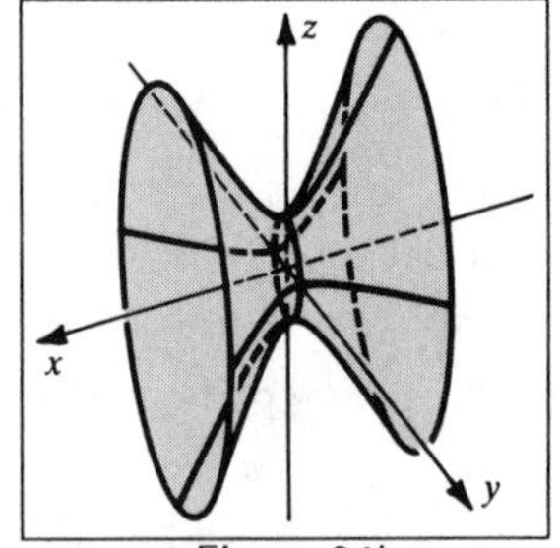

Figure 24b

[25] (a) $x^2 - \frac{y^2}{4} - z^2 = 1$ (b) $\frac{z^2}{4} - y^2 - x^2 = 1$ •

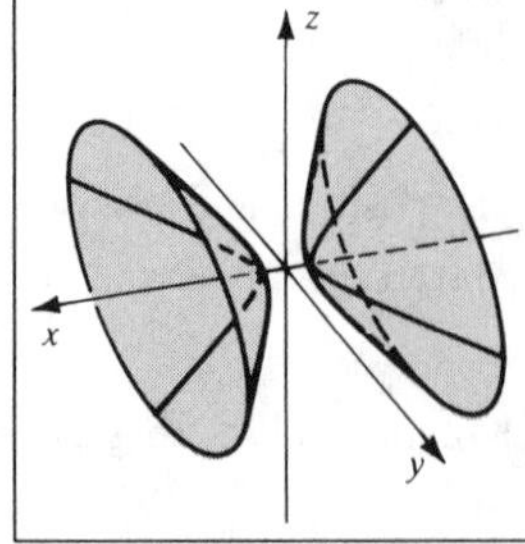

Figure 25a

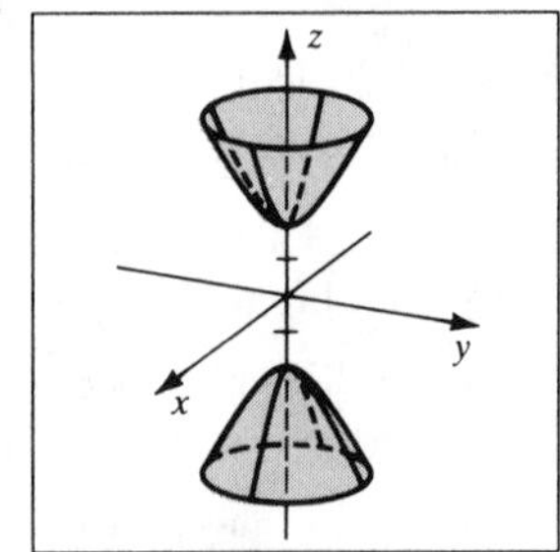

Figure 25b

[26] (a) $z^2 - \frac{x^2}{4} - \frac{y^2}{4} = 1$

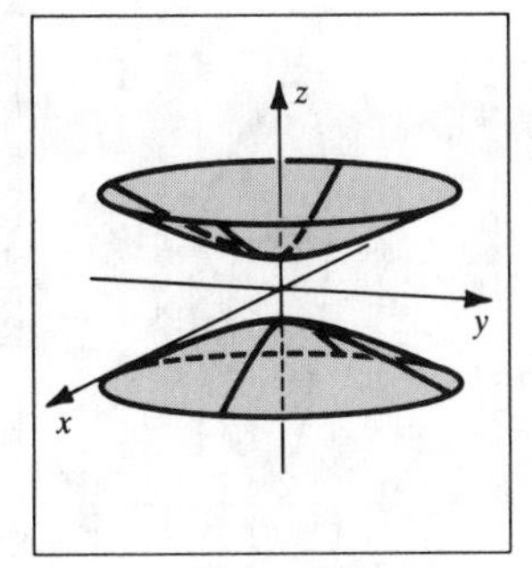

Figure 26a

(b) $\frac{y^2}{4} - x^2 - \frac{z^2}{9} = 1$ •

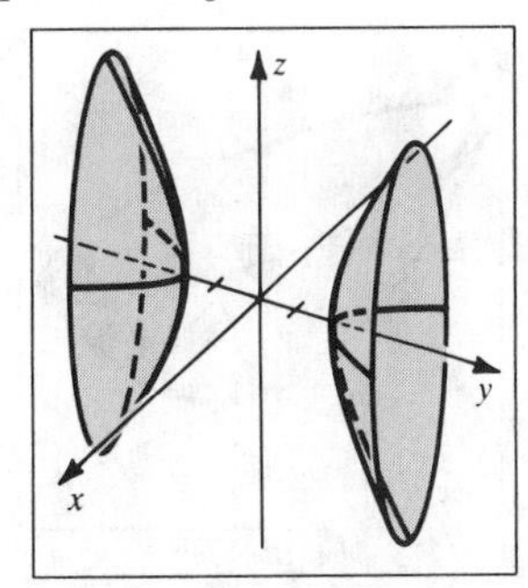

Figure 26b

[27] (a) $\frac{x^2}{9} + \frac{y^2}{4} = \frac{z^2}{4}$

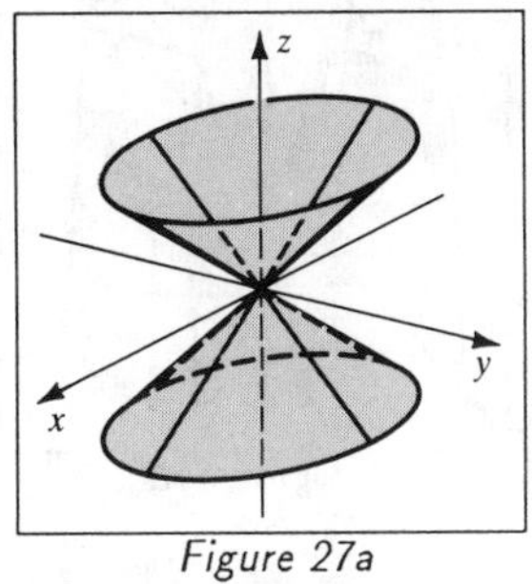

Figure 27a

(b) $\frac{x^2}{4} - y^2 + \frac{z^2}{9} = 0$ •

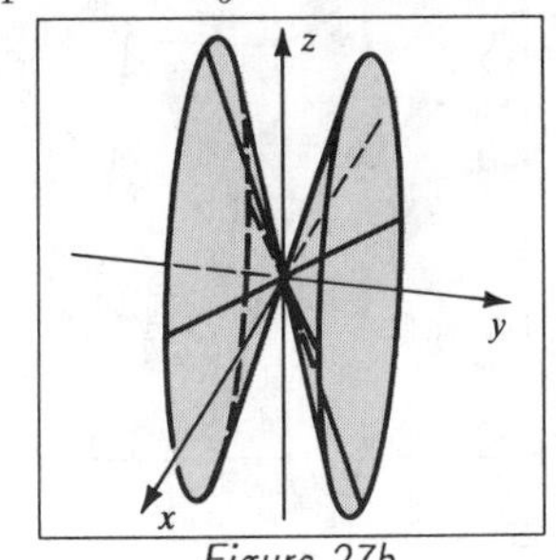

Figure 27b

[28] (a) $\frac{x^2}{25} + \frac{y^2}{9} - z^2 = 0$

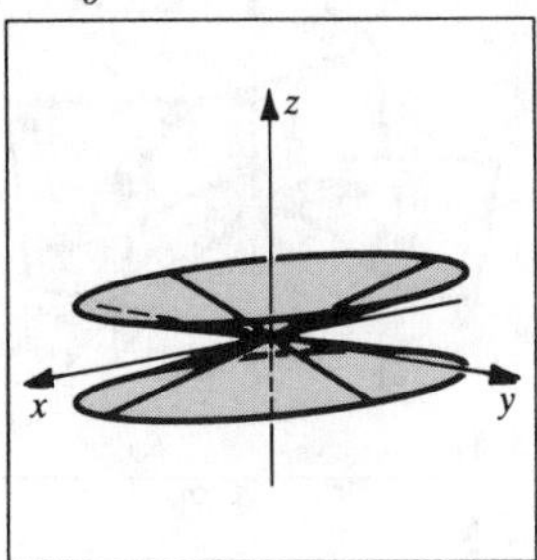

Figure 28a

(b) $x^2 = 4y^2 + z^2$ •

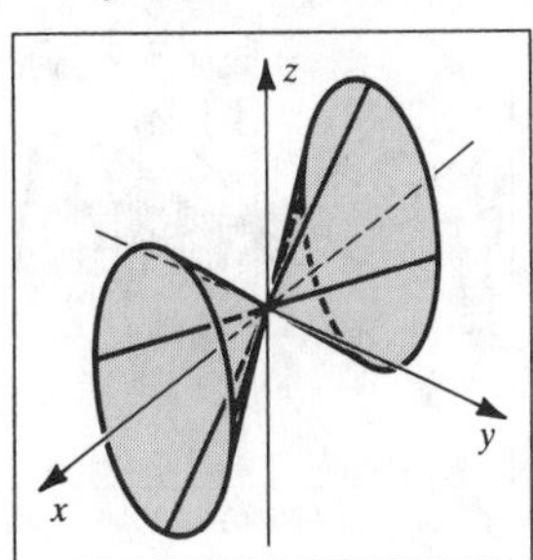

Figure 28b

[29] (a) $y = \frac{x^2}{4} + \frac{z^2}{9}$

Figure

(b) $x^2 + \frac{y^2}{4} - z = 0$ •

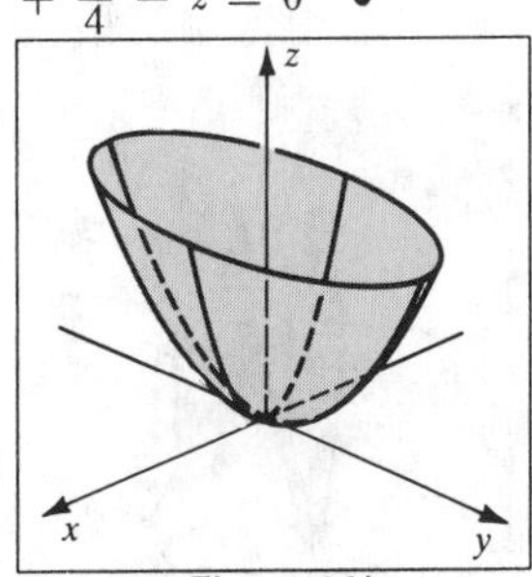

Figure 29b

[30] (a) $z = x^2 + \frac{y^2}{9}$ (b) $\frac{z^2}{25} + \frac{y^2}{9} - x = 0$ •

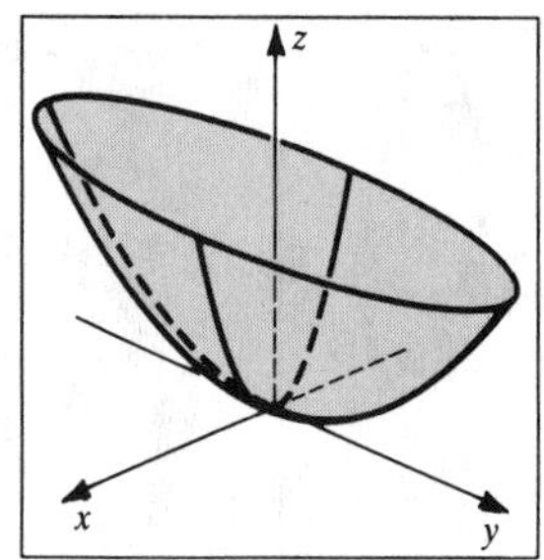

Figure 30a

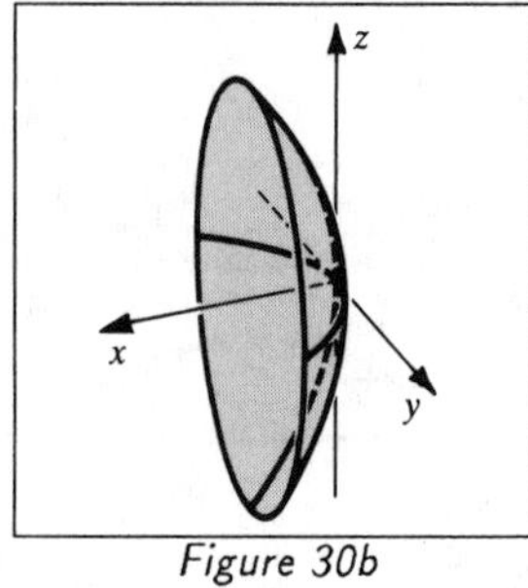

Figure 30b

[31] (a) $z = x^2 - y^2$ (b) $z = y^2 - x^2$ •

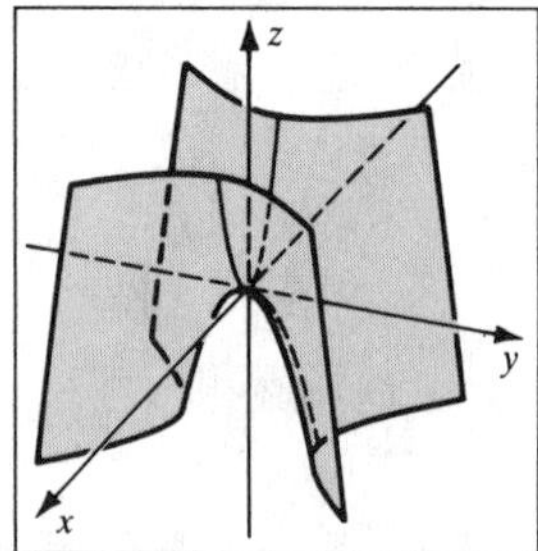

Figure 31a

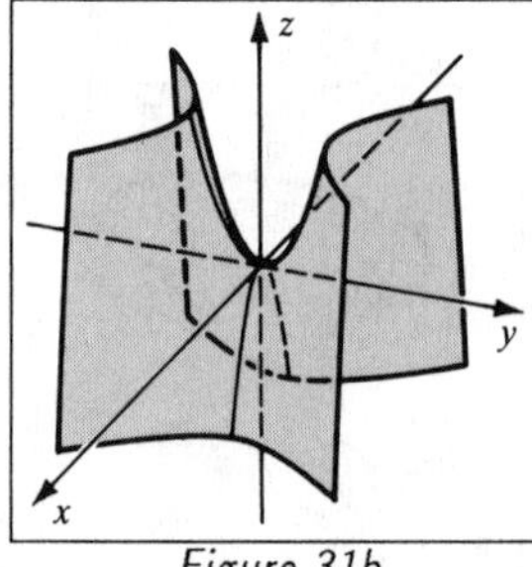

Figure 31b

[32] (a) $z = \frac{y^2}{9} - \frac{x^2}{4}$ (b) $z = \frac{x^2}{4} - \frac{y^2}{9}$ •

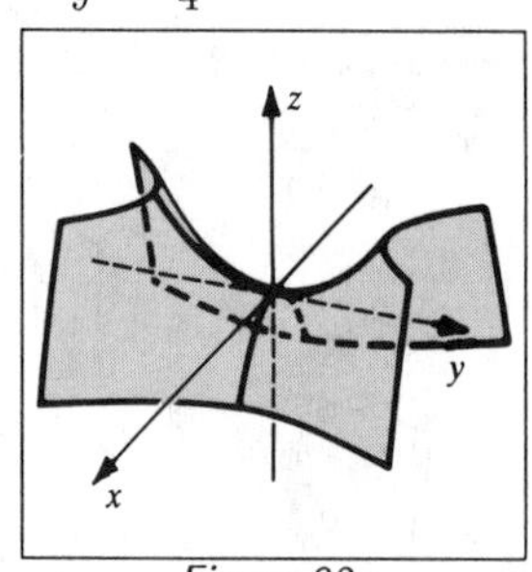

Figure 32a

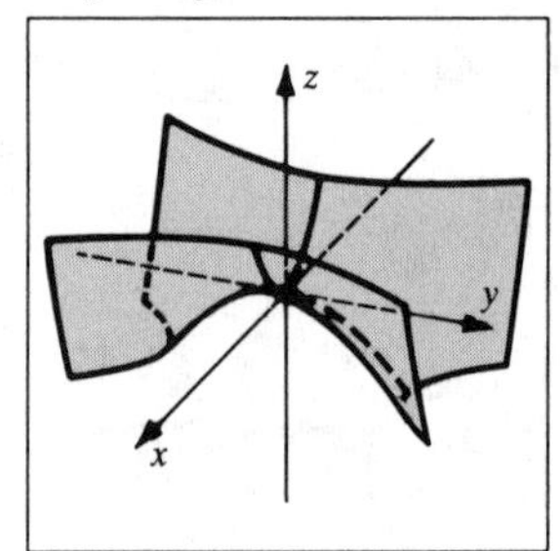

Figure 32b

[33] $16x^2 - 4y^2 - z^2 + 1 = 0 \Leftrightarrow z^2 + \frac{y^2}{1/4} - \frac{x^2}{1/16} = 1$;

a hyperboloid of one sheet with axis along the x-axis.

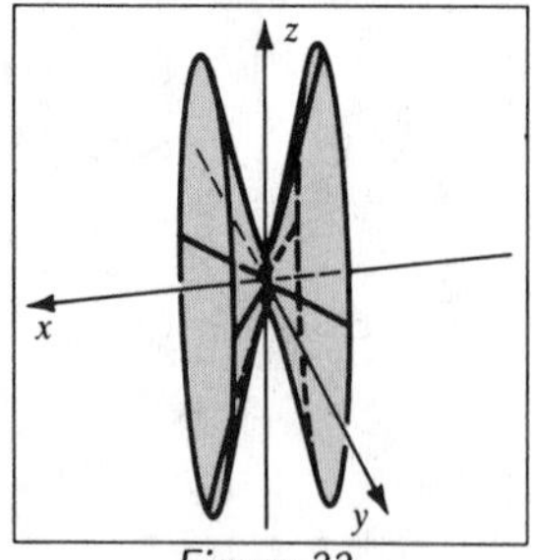

Figure 33

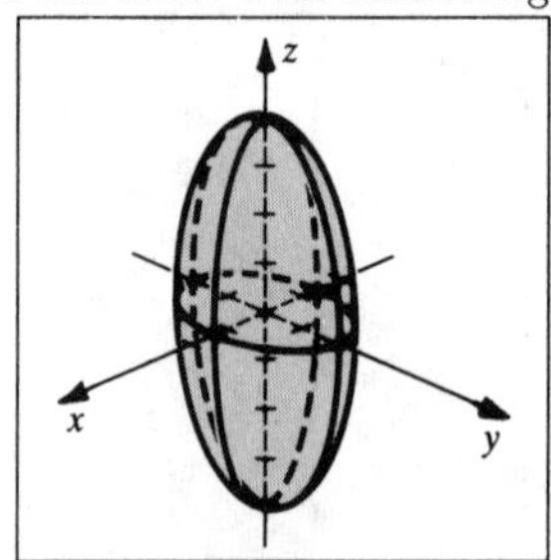

Figure 34

[34] $8x^2 + 4y^2 + z^2 = 16 \Leftrightarrow \frac{x^2}{2} + \frac{y^2}{4} + \frac{z^2}{16} = 1$;

an ellipsoid with x-, y-, and z-intercepts at $\pm\sqrt{2}$, ± 2, and ± 4.

[35] $36x = 9y^2 + z^2 \Leftrightarrow x = \frac{y^2}{4} + \frac{z^2}{36}$; a paraboloid with axis along the x-axis.

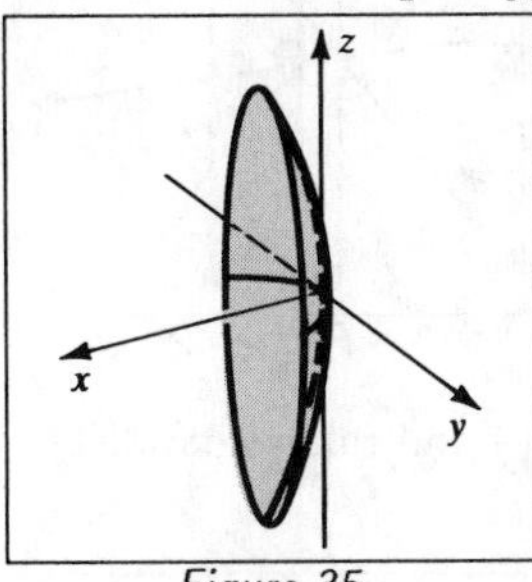

Figure 35

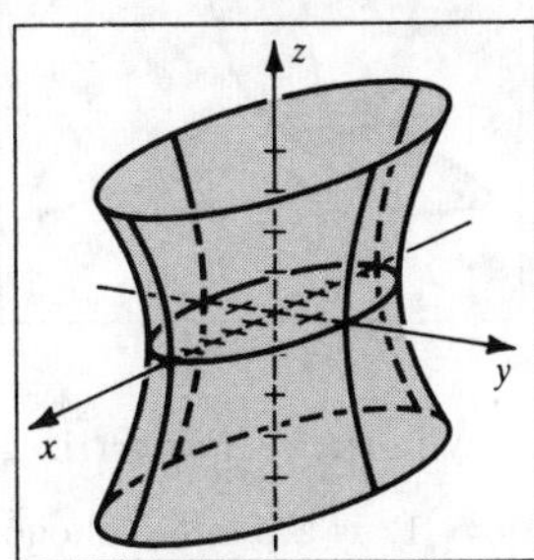

Figure 36

[36] $16x^2 + 100y^2 - 25z^2 = 400 \Leftrightarrow \frac{x^2}{25} + \frac{y^2}{4} - \frac{z^2}{16} = 1$;

a hyperboloid of one sheet with axis along the z-axis.

[37] $x^2 - 16y^2 = 4z^2 \Leftrightarrow x^2 = \frac{y^2}{1/16} + \frac{z^2}{1/4}$; a cone with axis along the x-axis.

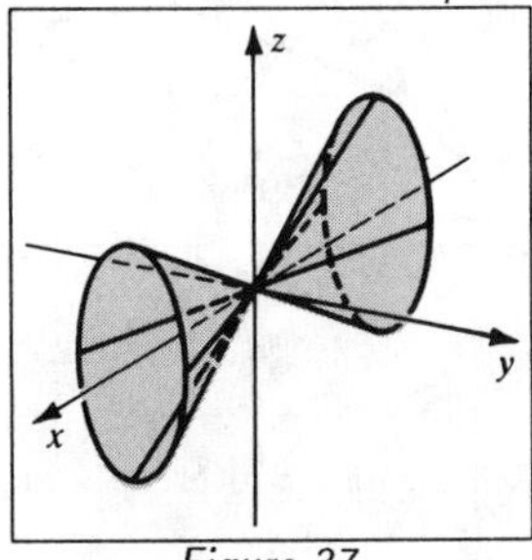

Figure 37

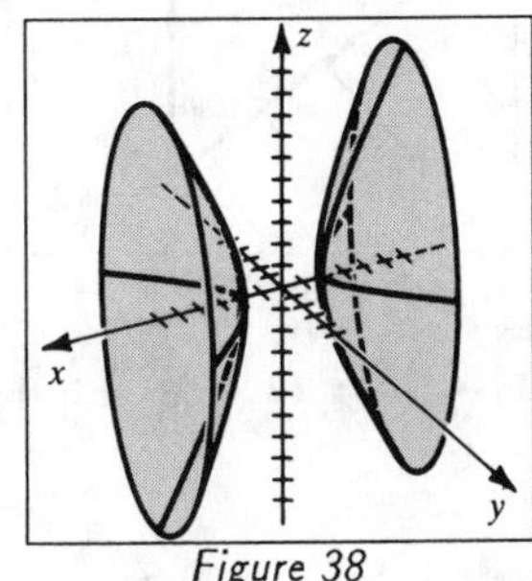

Figure 38

[38] $3x^2 - 4y^2 - z^2 = 12 \Leftrightarrow \frac{x^2}{4} - \frac{y^2}{3} - \frac{z^2}{12} = 1$;

a hyperboloid of two sheets with axis along the x-axis.

[39] $9x^2 + 4y^2 + z^2 = 36 \Leftrightarrow \frac{x^2}{4} + \frac{y^2}{9} + \frac{z^2}{36} = 1$;

an ellipsoid with x-, y-, and z-intercepts at ± 2, ± 3, and ± 6.

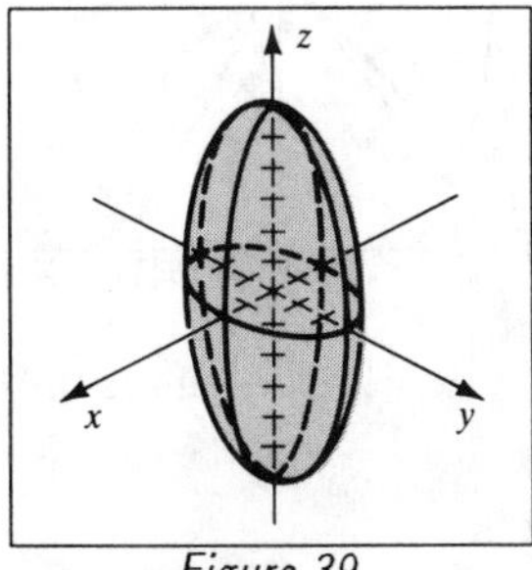

Figure 39

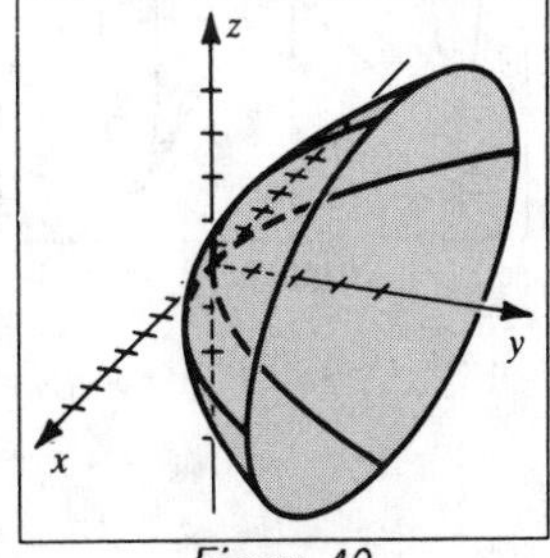

Figure 40

[40] $16y = x^2 + 4z^2 \Leftrightarrow y = \frac{x^2}{16} + \frac{z^2}{4}$; a paraboloid with axis along the y-axis.

[41] Cylinder with directrix $z = e^y$ and rulings parallel to the x-axis.

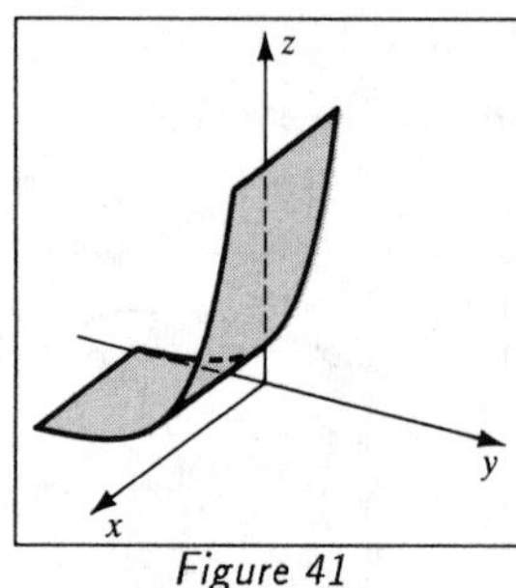

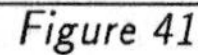
Figure 41

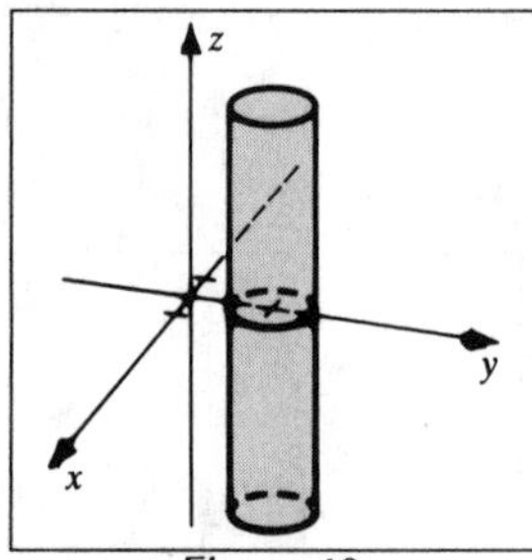

Figure 42

[42] Circular cylinder with directrix $x^2 + (y - 2)^2 = 1$ and rulings parallel to the z-axis.

[43] $4x - 3y = 12$ is a plane orthogonal to the xy-plane.

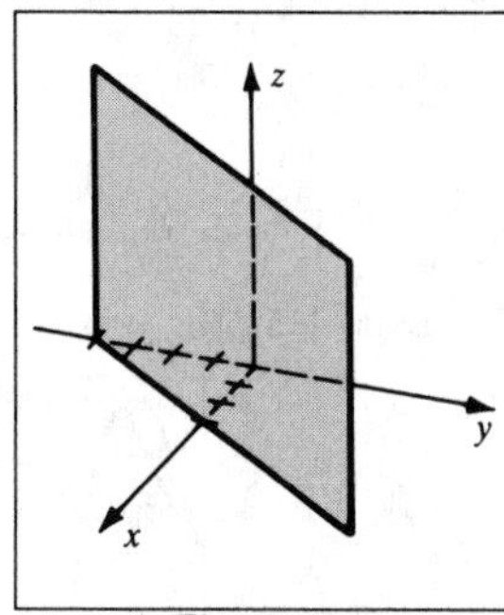

Figure 43

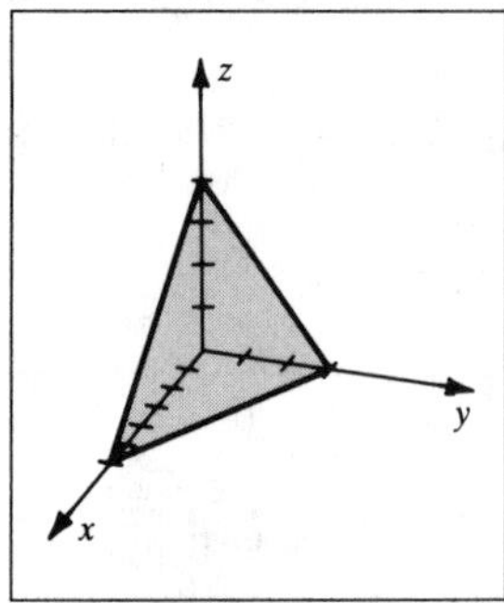

Figure 44

[44] $2x + 4y + 3z = 12 \Leftrightarrow \frac{x}{6} + \frac{y}{3} + \frac{z}{4} = 1$;

a plane with x-, y-, and z-intercepts at 6, 3, and 4.

[45] $y^2 - 9x^2 - z^2 - 9 = 0 \Leftrightarrow \frac{y^2}{9} - x^2 - \frac{z^2}{9} = 1$;

a hyperboloid of two sheets with axis on the y-axis.

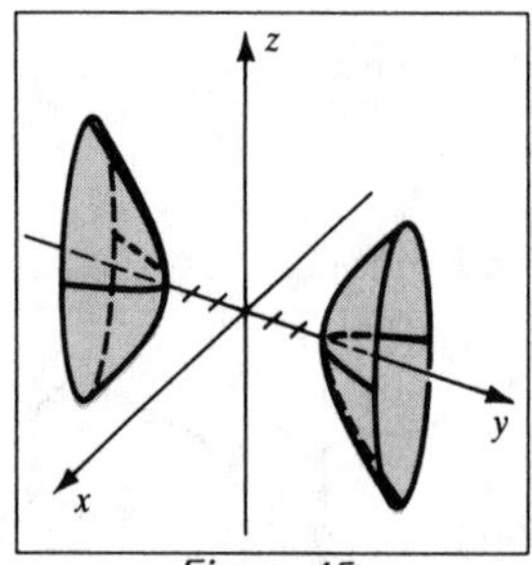

Figure 45

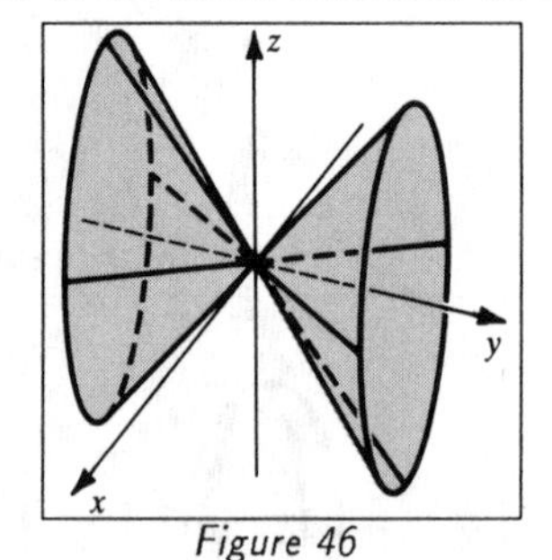

Figure 46

[46] $36x^2 - 16y^2 + 9z^2 = 0 \Leftrightarrow y^2 = \frac{x^2}{4/9} + \frac{z^2}{16/9}$; a cone with axis along the y-axis.

[47] $z = \frac{1}{5}y^2 - 3|x|$ •

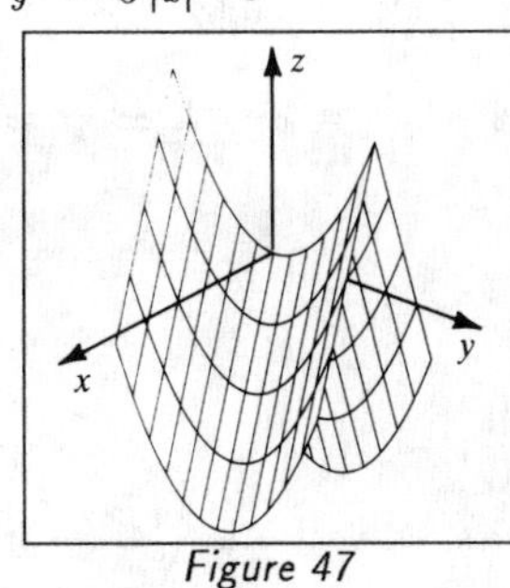

Figure 47

[48] $z = x^2 + 3xy + 4y^2$ •

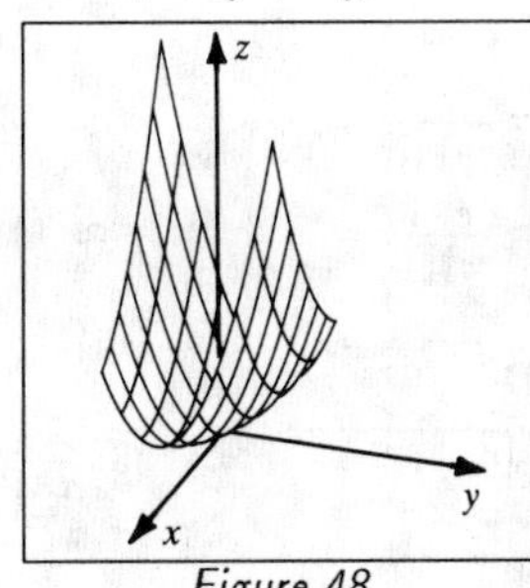

Figure 48

[49] $z = xy + x^2$ •

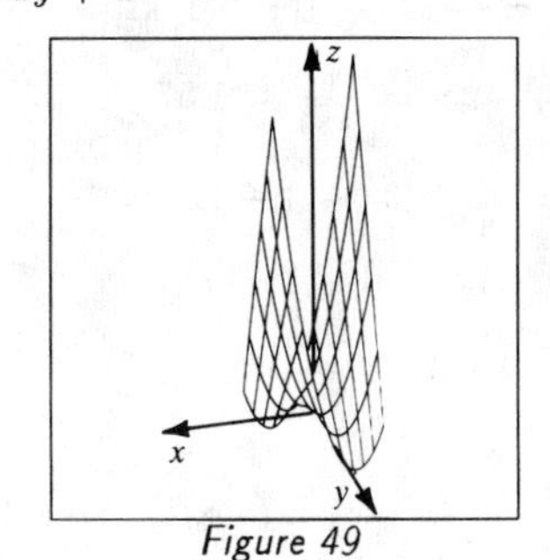

Figure 49

[50] $z = \dfrac{y^2}{16} - \dfrac{xy}{15} - \dfrac{x^2}{9}$ •

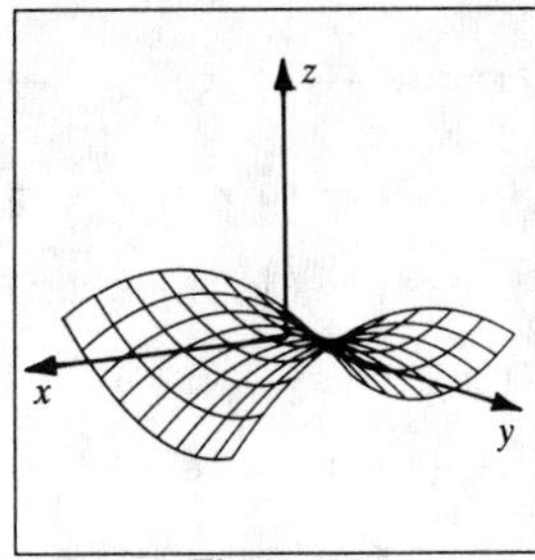

Figure 50

[51] Substituting $x^2 + z^2$ for x^2 gives $x^2 + z^2 + 4y^2 = 16$, which is an ellipsoid.

[52] Substituting $y^2 + z^2$ for y^2 gives $y^2 + z^2 = 4x$, which is a paraboloid.

[53] Substituting $x^2 + y^2$ for y^2 gives $z = 4 - x^2 - y^2$, which is a paraboloid.

[54] Substituting $\sqrt{x^2 + z^2}$ for z $\{z > 0\}$ gives $\sqrt{x^2 + z^2} = e^{-y^2} \Rightarrow x^2 + z^2 = e^{-2y^2}$.

[55] Substituting $y^2 + z^2$ for z^2 gives $y^2 + z^2 - x^2 = 1$,

which is a hyperboloid of one sheet.

[56] Substituting $\pm\sqrt{x^2 + z^2}$ for x gives $(\pm\sqrt{x^2 + y^2})\,z = 1 \Rightarrow (x^2 + y^2)\,z^2 = 1$.

[57] (a) The Clarke ellipsoid is flatter at the north and south poles.

(b) Since $a = b$, these curves will be circles, or more generally, ellipses.

(c) $y = mx \Rightarrow \dfrac{x^2}{a^2} + \dfrac{m^2x^2}{b^2} + \dfrac{z^2}{c^2} = 1 \Rightarrow \left(\dfrac{1 + m^2}{a^2}\right)x^2 + \dfrac{z^2}{c^2} = 1$ $\{$since $a = b\}$,

which are ellipses.

14.7 Review Exercises

[1] $3\mathbf{a} - 2\mathbf{b} = (9\mathbf{i} - 3\mathbf{j} - 12\mathbf{k}) - (4\mathbf{i} + 10\mathbf{j} - 4\mathbf{k}) = 5\mathbf{i} - 13\mathbf{j} - 8\mathbf{k}$

[2] $\mathbf{a}\cdot(\mathbf{b} - \mathbf{c}) = (3\mathbf{i} - \mathbf{j} - 4\mathbf{k})\cdot(3\mathbf{i} + 5\mathbf{j} - 8\mathbf{k}) = 3(3) - (5) - 4(-8) = 36$

[3] $\|-3\mathbf{b}\| = \|-6\mathbf{i} - 15\mathbf{j} + 6\mathbf{k}\| = \sqrt{(-6)^2 + (-15)^2 + 6^2} = \sqrt{297} = 3\sqrt{33}$

[4] $\|\mathbf{b} + \mathbf{c}\| = \|(\mathbf{i} + 5\mathbf{j} + 4\mathbf{k})\| = \sqrt{1^2 + 5^2 + 4^2} = \sqrt{42}$

[5] $\|\mathbf{a}\|^2 = \mathbf{a}\cdot\mathbf{a} = 3^2 + (-1)^2 + (-4)^2 = 9 + 1 + 16 = 26$

[6] $\|\mathbf{a} \times \mathbf{a}\| = \|\mathbf{0}\| = 0$

7 $\cos\theta = \dfrac{\mathbf{a}\cdot\mathbf{c}}{\|\mathbf{a}\|\|\mathbf{c}\|} = \dfrac{-27}{\sqrt{26}\sqrt{37}} = \dfrac{-27}{\sqrt{962}} \Rightarrow \theta = \arccos\dfrac{-27}{\sqrt{962}} \approx 150.52°$

8 $\cos\alpha = \dfrac{\mathbf{a}\cdot\mathbf{i}}{\|\mathbf{a}\|\|\mathbf{i}\|} = \frac{3}{\sqrt{26}} \Rightarrow \alpha \approx 53.96°$; $\cos\beta = \dfrac{\mathbf{b}\cdot\mathbf{i}}{\|\mathbf{b}\|\|\mathbf{i}\|} = \frac{-1}{\sqrt{26}} \Rightarrow \beta \approx 101.31°$;

$\cos\gamma = \dfrac{\mathbf{c}\cdot\mathbf{i}}{\|\mathbf{c}\|\|\mathbf{i}\|} = \frac{-4}{\sqrt{26}} \Rightarrow \gamma \approx 141.67°$

9 $\mathbf{u} = \frac{\mathbf{a}}{\|\mathbf{a}\|} = \frac{1}{\sqrt{26}}(3\mathbf{i} - \mathbf{j} - 4\mathbf{k})$

10 $\|\mathbf{b}\| = \sqrt{4 + 25 + 4} = \sqrt{33}$; $\frac{-\mathbf{a}}{\|\mathbf{a}\|} = \dfrac{-3\mathbf{i} + \mathbf{j} + 4\mathbf{k}}{\sqrt{26}}$;

the desired vector is $\frac{2\sqrt{33}}{\sqrt{26}}(-3\mathbf{i} + \mathbf{j} + 4\mathbf{k})$

11 $\mathbf{a} \times \mathbf{b} = 22\mathbf{i} - 2\mathbf{j} + 17\mathbf{k}$

12 $(\mathbf{b} + \mathbf{c}) \times \mathbf{a} = (\mathbf{i} + 5\mathbf{j} + 4\mathbf{k}) \times (3\mathbf{i} - \mathbf{j} - 4\mathbf{k}) = -16\mathbf{i} + 16\mathbf{j} - 16\mathbf{k}$

13 $\text{comp}_{\mathbf{b}}\,\mathbf{a} = \mathbf{a}\cdot\dfrac{\mathbf{b}}{\|\mathbf{b}\|} = \langle 3, -1, -4\rangle \cdot \frac{1}{\sqrt{33}}\langle 2, 5, -2\rangle = \frac{9}{\sqrt{33}} \approx 1.57$

14 $\text{comp}_{\mathbf{a}}(\mathbf{b} \times \mathbf{c}) = (\mathbf{b} \times \mathbf{c})\cdot\frac{\mathbf{a}}{\|\mathbf{a}\|} = \langle 30, -10, 5\rangle \cdot \frac{1}{\sqrt{26}}\langle 3, -1, -4\rangle = \frac{80}{\sqrt{26}} \approx 15.69$

15 $(2\mathbf{a})\cdot(3\mathbf{a}) = (2\cdot 3)(\mathbf{a}\cdot\mathbf{a}) = 6(26) = 156$

16 $(2\mathbf{a}) \times (3\mathbf{a}) = (2\cdot 3)(\mathbf{a} \times \mathbf{a}) = 6(\mathbf{0}) = \mathbf{0}$

17 $(\mathbf{a} \times \mathbf{c}) + (\mathbf{c} \times \mathbf{a}) = (\mathbf{a} \times \mathbf{c}) - (\mathbf{a} \times \mathbf{c}) = \mathbf{0}$

18 $(\mathbf{a} \times \mathbf{c}) \times (\mathbf{c} \times \mathbf{a}) = (\mathbf{a} \times \mathbf{c}) \times (-1)(\mathbf{a} \times \mathbf{c}) = -1\big[(\mathbf{a} \times \mathbf{c}) \times (\mathbf{a} \times \mathbf{c})\big] = -1(\mathbf{0}) = \mathbf{0}$

19 $V = |(\mathbf{a} \times \mathbf{b})\cdot\mathbf{c}| = |\langle 22, -2, 17\rangle \cdot \langle -1, 0, 6\rangle| = |80| = 80$

20 $A = \frac{1}{2}\|\mathbf{a} \times \mathbf{b}\| = \frac{1}{2}\|\langle 22, -2, 17\rangle\| = \frac{1}{2}\sqrt{777} \approx 13.94$

21 Since $\mathbf{a}\cdot\mathbf{b} = 0$, $\mathbf{a}$ and $\mathbf{b}$ are orthogonal by (14.21).

22 The force vector is $\frac{18\mathbf{a}}{\|\mathbf{a}\|}$.

The work done is $\frac{18\mathbf{a}}{\|\mathbf{a}\|}\cdot\overrightarrow{PQ} = \frac{18}{9}(\langle 4, 7, 4\rangle \cdot \langle 2, 4, 3\rangle) = 2(48) = 96$ ft-lb.

23 (a) $\sqrt{(-6)^2 + (-1)^2 + 1^2} = \sqrt{38}$

(b) $M(A, B) = \left(\dfrac{5 - 1}{2}, \dfrac{-3 - 4}{2}, \dfrac{2 + 3}{2}\right) = (2, -\frac{7}{2}, \frac{5}{2})$

(c) $(x + 1)^2 + (y + 4)^2 + (z - 3)^2 = 4^2 = 16$

(d) $y = -4$ (e) $\overrightarrow{BA} = \langle 6, 1, -1\rangle$; $x = 5 + 6t$, $y = -3 + t$, $z + 2 - t$

(f) $\overrightarrow{AB} = \langle -6, -1, 1\rangle$;

$-6(x - 5) - 1(y + 3) + 1(z - 2) = 0$, or $6x + y - z = 25$

24 Since the normal vector $\mathbf{n}$ for the plane must be orthogonal to both $\overrightarrow{BA} = \langle 0, 7, 2\rangle$ and $\mathbf{i}$, $\mathbf{n} = \overrightarrow{BA} \times \mathbf{i} = \langle 0, 2, -7\rangle$. Using $A(0, 4, 9)$ and $\mathbf{n}$,

an equation of the plane is $2(y - 4) - 7(z - 9) = 0$, or $2y - 7z = -55$.

25 $\left(\frac{x}{5} + \frac{y}{-2} + \frac{z}{6} = 1\right)\cdot 30 \Leftrightarrow 6x - 15y + 5z = 30$

26 Solving each equation for t yields $\dfrac{x - 3}{4} = 2 - y = \frac{z}{2}$.

[27] $2x + y = 8 - 4z$ (E_1); $x + 3y = -1 + z$ (E_2)

$E_1 - 2E_2 \Rightarrow -5y = 10 - 6z \Rightarrow y = -2 + \frac{6}{5}z$

$3E_1 - E_2 \Rightarrow 5x = 25 - 13z \Rightarrow x = 5 - \frac{13}{5}z$

If we let $z = 5t$, then $x = -13t + 5$ and $y = 6t - 2$.

[28] The plane's equation has the form $5x - 4y + 3z = k$.

Substituting $x = 1$, $y = 3$, and $z = -2$ yields $k = -13$.

[29] The plane's equation has the form $ax + 3y - 4z = 11$.

Substituting $x = 4$, $y = 1$, and $z = 2$ yields $4a = 16$, or $a = 4$.

[30] $(x - 4)^2 + (y + 3)^2 = 25$

[31] $\frac{x^2}{8^2} + \frac{y^2}{3^2} + \frac{z^2}{1^2} = 1$, or $\frac{x^2}{64} + \frac{y^2}{9} + z^2 = 1$

[32] Substituting $\pm\sqrt{x^2 + y^2}$ for x gives $z = \pm\sqrt{x^2 + y^2}$, or $z^2 = x^2 + y^2$.

[33] (a) $\overrightarrow{PQ} = \langle -5, 3, -1 \rangle$ and $\overrightarrow{PR} = \langle 2, -4, 2 \rangle$;

$$\overrightarrow{PQ} \times \overrightarrow{PR} = \langle 2, 8, 14 \rangle \Rightarrow \mathbf{u} = \frac{1}{2\sqrt{66}}\langle 2, 8, 14 \rangle = \frac{1}{\sqrt{66}}\langle 1, 4, 7 \rangle$$

(b) Using $P(2, -1, 1)$ and the vector $\langle 1, 4, 7 \rangle$ from part (a) gives us

$$1(x - 2) + 4(y + 1) + 7(z - 1) = 0, \text{ or } x + 4y + 7z = 5.$$

(c) $\overrightarrow{QR} = \langle 7, -7, 3 \rangle$; $x = 2 + 7t$, $y = -1 - 7t$, $z = 1 + 3t$

(d) $\overrightarrow{QP} \cdot \overrightarrow{QR} = \langle 5, -3, 1 \rangle \cdot \langle 7, -7, 3 \rangle = 59$

(e) $\cos\theta = \dfrac{\overrightarrow{QP} \cdot \overrightarrow{QR}}{\|\overrightarrow{QP}\|\|\overrightarrow{QR}\|} = \dfrac{59}{\sqrt{35}\sqrt{107}} \Rightarrow \theta = \arccos\dfrac{59}{\sqrt{3745}} \approx 15.40^\circ$

(f) area $= \frac{1}{2}\|\overrightarrow{PQ} \times \overrightarrow{PR}\| = \frac{1}{2}(2\sqrt{66}) = \sqrt{66} \approx 8.12$

(g) $d = \dfrac{\|\overrightarrow{PQ} \times \overrightarrow{PR}\|}{\|\overrightarrow{PQ}\|} = \dfrac{2\sqrt{66}}{\sqrt{35}} \approx 2.75$

[34] Direction vectors for the given lines are $\mathbf{a}_1 = \langle 2, -4, 8 \rangle$ and $\mathbf{a}_2 = \langle 7, -2, -2 \rangle$.

$$\cos\theta = \frac{\mathbf{a}_1 \cdot \mathbf{a}_2}{\|\mathbf{a}_1\|\|\mathbf{a}_2\|} = \frac{6}{\sqrt{84}\sqrt{57}} \Rightarrow \theta = \arccos\frac{1}{\sqrt{133}} \approx 85.03^\circ \text{ and } 180^\circ - \theta$$

[35] Setting each factor equal to t and solving for x, y, and z yields $x = 3 + 2t$,

$$y = -1 - 4t,\ z = 5 + 8t \quad \text{and} \quad x = -1 + 7t,\ y = 6 - 2t,\ z = -\tfrac{7}{2} - 2t.$$

[36]
$$\begin{aligned} 2 + t_0 &= -4 + 5v_0 & \quad t_0 - 5v_0 &= -6 \\ 1 + t_0 &= 2 - 2v_0 \Rightarrow & \quad t_0 + 2v_0 &= 1 \Rightarrow t_0 = -1,\ v_0 = 1 \\ 4 + 7t_0 &= 1 - 4v_0 & \quad 7t_0 + 4v_0 &= -3 \end{aligned}$$

Lines intersect at $(1, 0, -3)$.

[37] Direction vectors for the given lines are $\mathbf{a}_1 = \langle 1, 1, 7 \rangle$ and $\mathbf{a}_2 = \langle 5, -2, -4 \rangle$.

$$\cos\theta = \frac{\mathbf{a}_1 \cdot \mathbf{a}_2}{\|\mathbf{a}_1\|\|\mathbf{a}_2\|} = \frac{-25}{\sqrt{51}\sqrt{45}} \Rightarrow \theta = \arccos\frac{-25}{\sqrt{2295}} \approx 121.46^\circ \text{ and } 180^\circ - \theta$$

[38] Let $P = (1, 3, 0)$ and $Q = (3, 2, 5)$ be on the line.

Then, $\overrightarrow{PA} = \langle 2, -2, -1 \rangle$ and $\overrightarrow{PQ} = \langle 2, -1, 5 \rangle$.

$$\text{Using Example 3, §14.4, } d = \frac{\left\| \overrightarrow{PQ} \times \overrightarrow{PA} \right\|}{\left\| \overrightarrow{PQ} \right\|} = \frac{\| \langle 11, 12, -2 \rangle \|}{\| \langle 2, -1, 5 \rangle \|} = \sqrt{\frac{269}{30}} \approx 2.99$$

[39] $x^2 + y^2 + z^2 - 14x + 6y - 8z + 10 = 0 \Leftrightarrow$

$(x - 7)^2 + (y + 3)^2 + (z - 4)^2 = 8^2$; a sphere with center $(7, -3, 4)$ and radius 8.

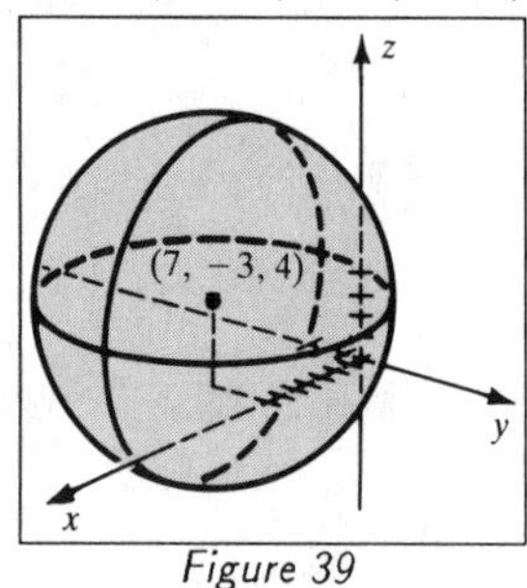

Figure 39

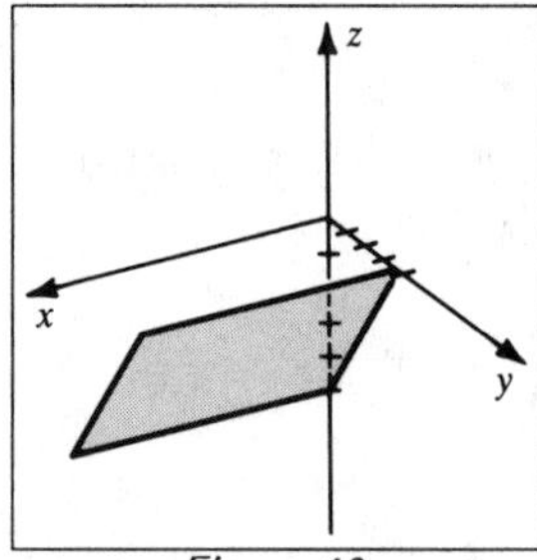

Figure 40

[40] $4y - 3z - 15 = 0$ is a plane orthogonal to the yz-plane.

The trace in the yz-plane is $y = \frac{15}{4}$.

[41] $3x - 5y + 2z = 10$ is a plane passing through the points $(\frac{10}{3}, 0, 0)$, $(0, -2, 0)$, and $(0, 0, 5)$.

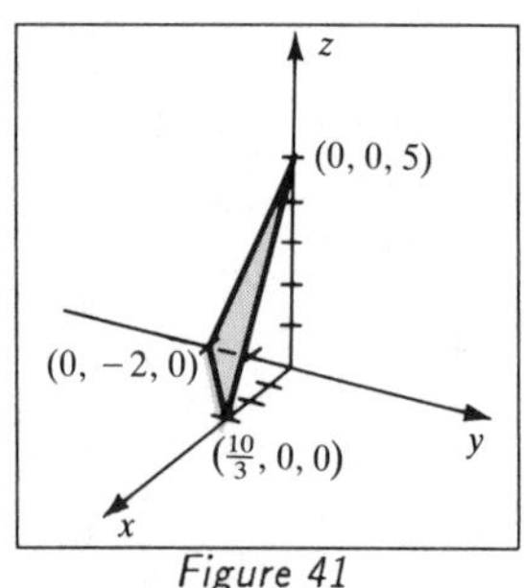

Figure 41

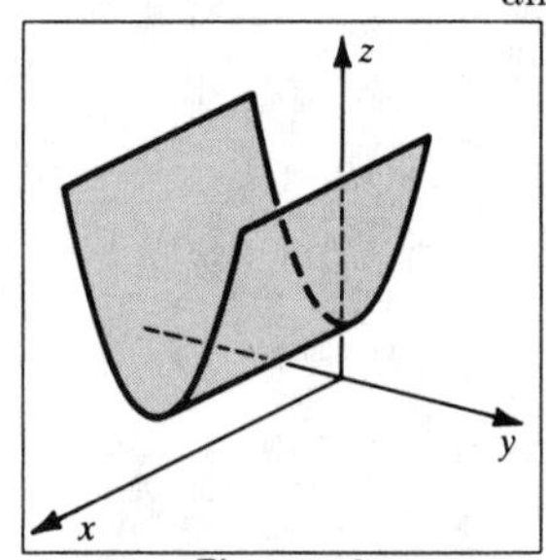

Figure 42

[42] Parabolic cylinder with directrix $y = z^2 + 1$ and rulings parallel to the x-axis.

[43] Elliptic cylinder with directrix $9x^2 + 4z^2 = 36$ and rulings parallel to the y-axis.

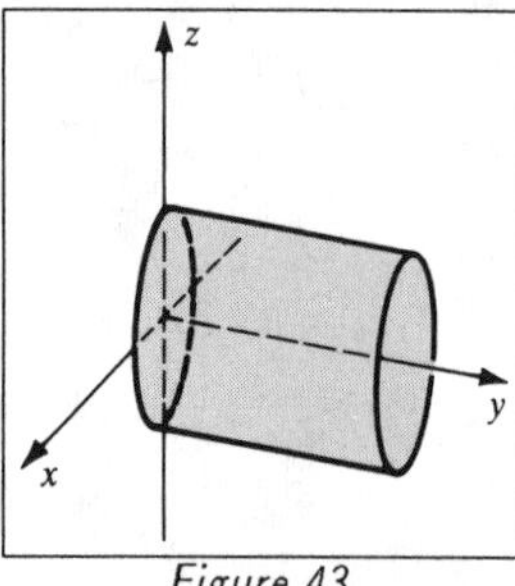

Figure 43

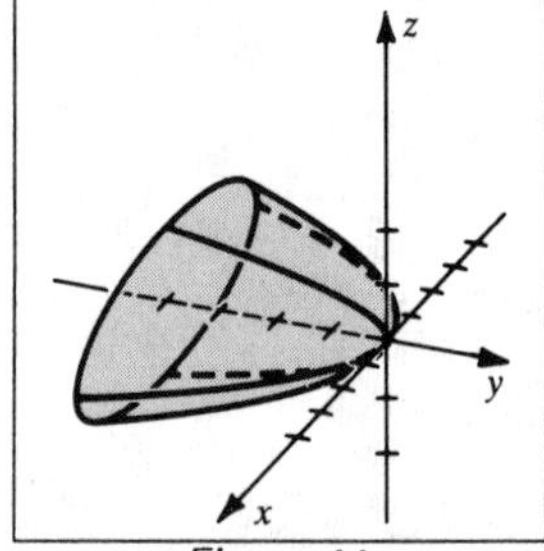

Figure 44

[44] $x^2 + 4y + 9z^2 = 0 \Leftrightarrow y = -\frac{x^2}{4} - \frac{z^2}{4/9}$;

a paraboloid with axis along the negative y-axis.

[45] $z^2 - 4x^2 = 9 - 4y^2 \Leftrightarrow \frac{z^2}{9} + \frac{y^2}{9/4} - \frac{x^2}{9/4} = 1$;

a hyperboloid of one sheet with axis along the x-axis.

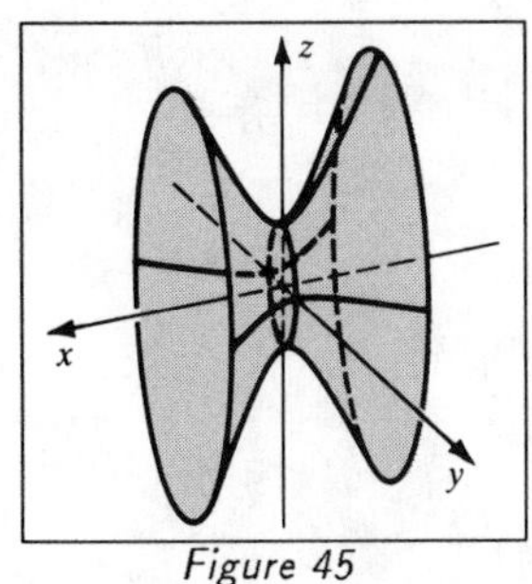

Figure 45

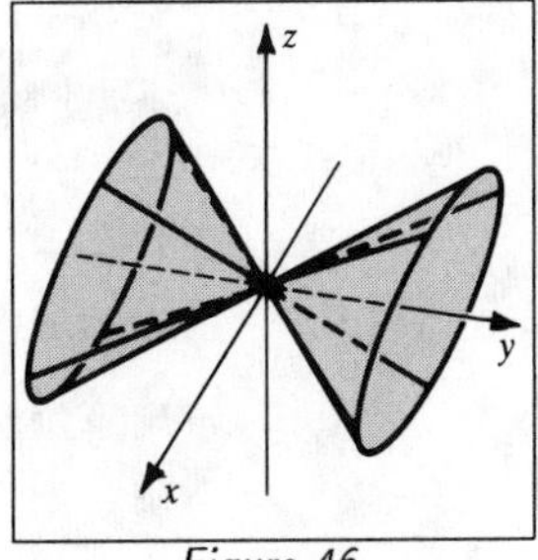

Figure 46

[46] $2x^2 + 4z^2 - y^2 = 0 \Leftrightarrow y^2 = \frac{x^2}{1/2} + \frac{z^2}{1/4}$; a cone with axis along the y-axis.

[47] $z^2 - 4x^2 - y^2 = 4 \Leftrightarrow \frac{z^2}{4} - x^2 - \frac{y^2}{4} = 1$;

a hyperboloid of two sheets with axis along the z-axis.

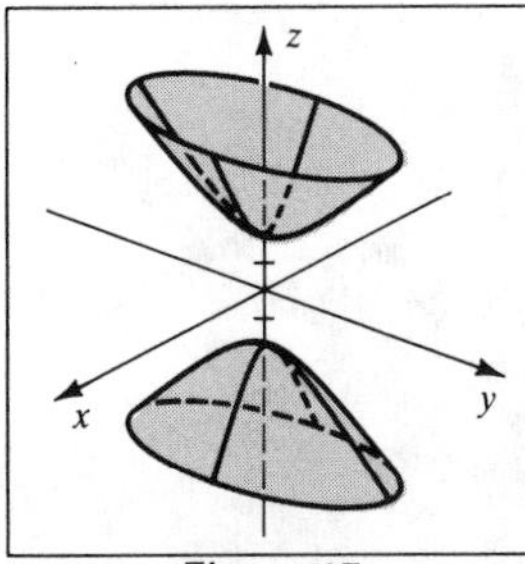

Figure 47

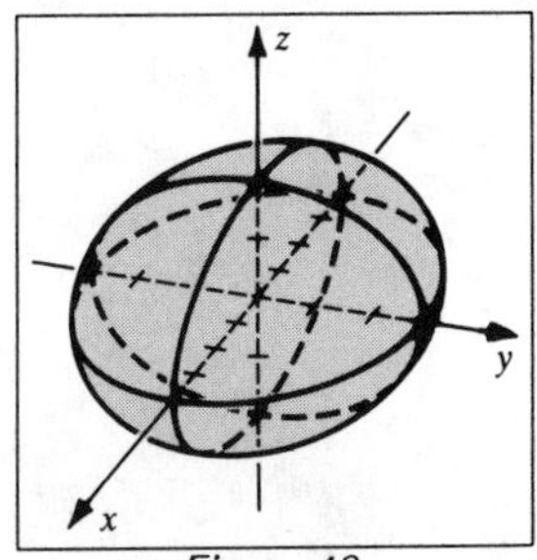

Figure 48

[48] $x^2 + 2y^2 + 4z^2 = 16 \Leftrightarrow \frac{x^2}{16} + \frac{y^2}{8} + \frac{z^2}{4} = 1$;

an ellipsoid with x-, y-, and z-intercepts at ± 4, $\pm 2\sqrt{2}$, and ± 2.

[49] $x^2 - 4y^2 = 4z \Leftrightarrow z = \frac{x^2}{4} - y^2$; a hyperbolic paraboloid with z positive in the xz-plane, z negative in the yz-plane, and "facing along the x-axis".

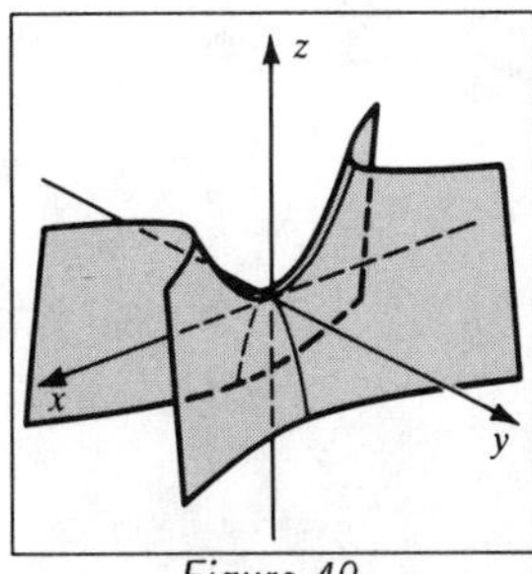

Figure 49

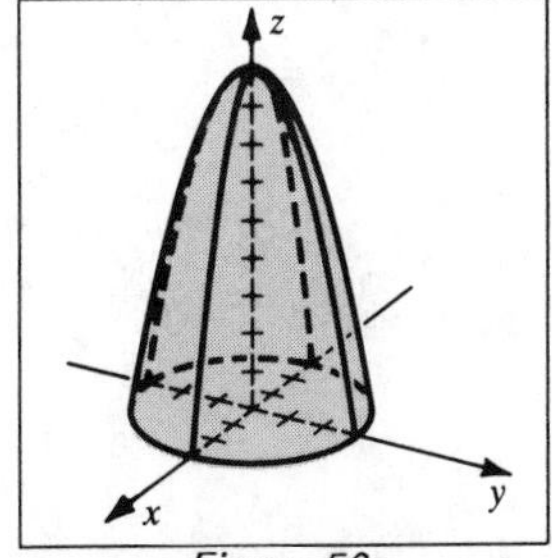

Figure 50

[50] $z = 9 - x^2 - y^2 \Leftrightarrow z = 9 - (x^2 + y^2)$;

a circular paraboloid with vertex (0, 0, 9) and opening downward.

Chapter 15: Vector-Valued Functions

Exercises 15.1

1 $\mathbf{r}(0) = \langle 0, 1 \rangle$. $\mathbf{r}(1) = \langle 3, -8 \rangle$.

$x = 3t$ and $y = 1 - 9t^2 \Rightarrow y = 1 - (3t)^2 = 1 - x^2$.

To determine the orientation, examine $\mathbf{r}(t)$ for increasing values of t.

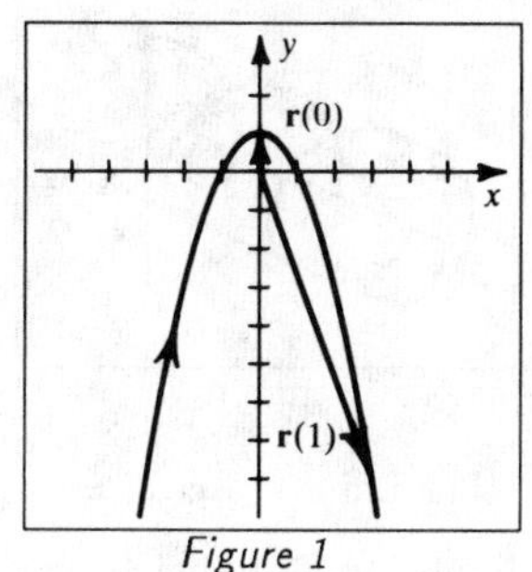

Figure 1

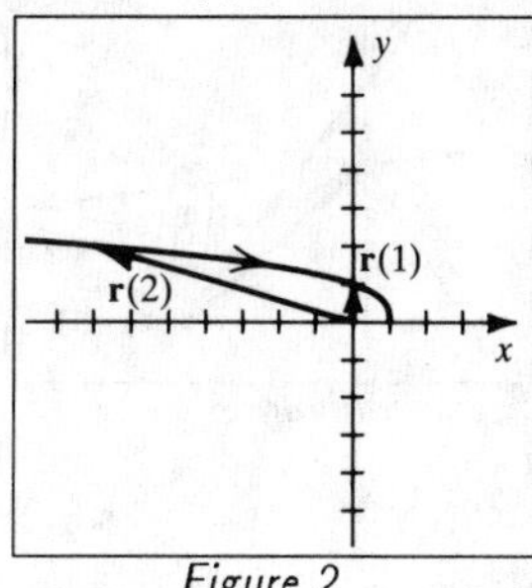

Figure 2

2 $\mathbf{r}(1) = \langle 0, 1 \rangle$. $\mathbf{r}(2) = \langle -7, 2 \rangle$. $x = 1 - t^3$ and $y = t \Rightarrow x = 1 - y^3$.

Since $y = t \geq 0$, $x \leq 1$.

3 $\mathbf{r}(1) = \langle 0, 3 \rangle$. $\mathbf{r}(2) = \langle 7, 6 \rangle$. $x = t^3 - 1$ and $y = t^2 + 2 \Rightarrow$

$t = (x + 1)^{1/3}$ and $y = (x + 1)^{2/3} + 2$ for $-9 \leq x \leq 7$.

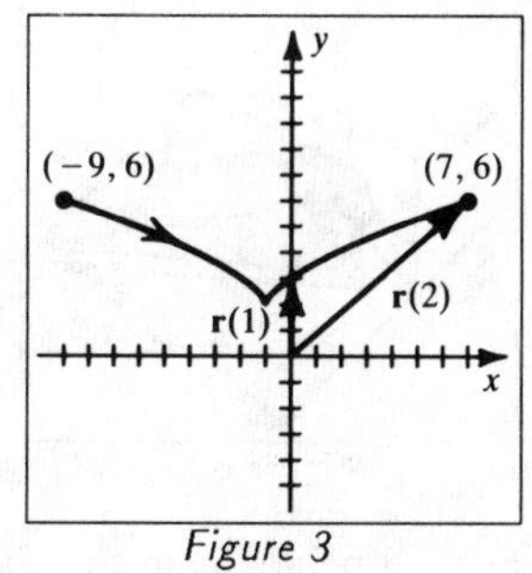

Figure 3

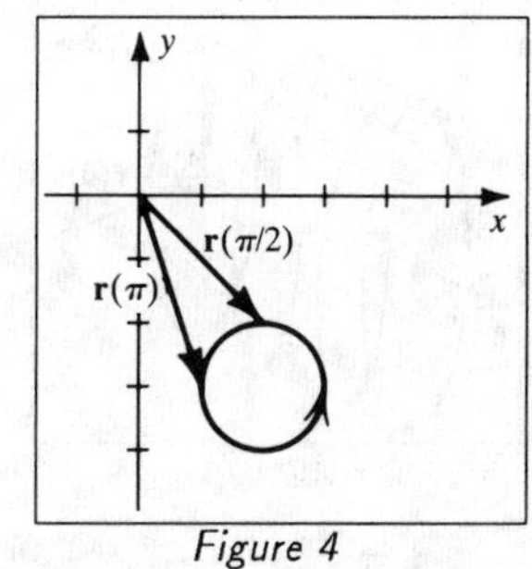

Figure 4

4 $\mathbf{r}(\frac{\pi}{2}) = \langle 2, -2 \rangle$. $\mathbf{r}(\pi) = \langle 1, -3 \rangle$. $x = 2 + \cos t$ and $y = \sin t - 3 \Rightarrow$

$\cos t = x - 2$ and $\sin t = y + 3 \Rightarrow (x - 2)^2 + (y + 3)^2 = \cos^2 t + \sin^2 t = 1$.

Starting at $(3, -3)$ $\{t = 0\}$, this circle is traversed in a counterclockwise direction.

[5] $\mathbf{r}(-1) = \langle 2, 3, -1 \rangle$. $\mathbf{r}(0) = \langle 3, 2, 1 \rangle$.

$x = 3 + t$, $y = 2 - t$, and $z = 1 + 2t \Rightarrow \frac{x-3}{1} = \frac{2-y}{1} = \frac{z-1}{2}$,

the symmetric form of a half-line with end point $(2, 3, -1)$ $\{t = -1\}$.

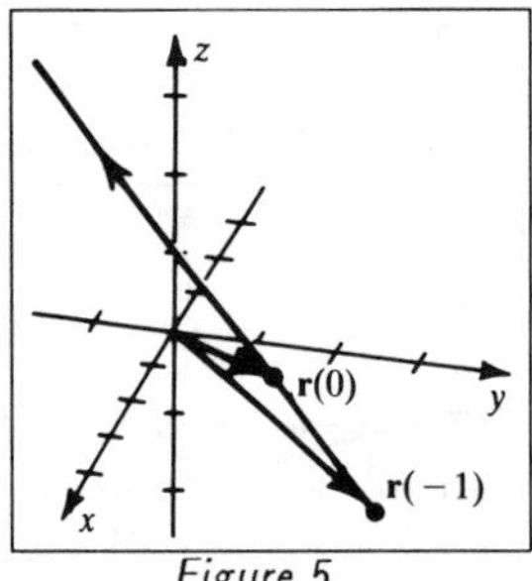

Figure 5

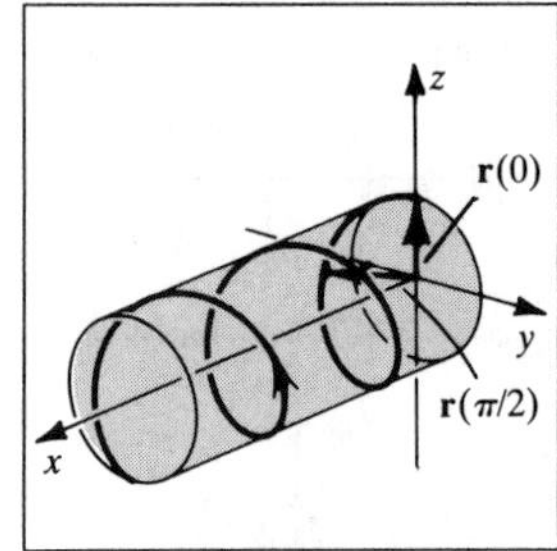

Figure 6

[6] $\mathbf{r}(0) = \langle 0, 0, 3 \rangle$. $\mathbf{r}(\frac{\pi}{2}) = \langle \frac{\pi}{2}, -3, 0 \rangle$. $y = -3\sin t$ and $z = 3\cos t \Rightarrow$

$y^2 = 9\sin^2 t$ and $z^2 = 9\cos^2 t \Rightarrow y^2 + z^2 = 9$. Similar to Example 4,

this is a circular helix along the positive x-axis with end point $(0, 0, 3)$.

[7] $\mathbf{r}(0) = \langle 0, 4, 0 \rangle$. $\mathbf{r}(\frac{\pi}{2}) = \langle \frac{\pi}{2}, 0, 9 \rangle$. $y = 4\cos t$ and $z = 9\sin t \Rightarrow \frac{y^2}{16} + \frac{z^2}{81} = 1$.

C is an elliptic helix along the positive x-axis with end point $(0, 4, 0)$.

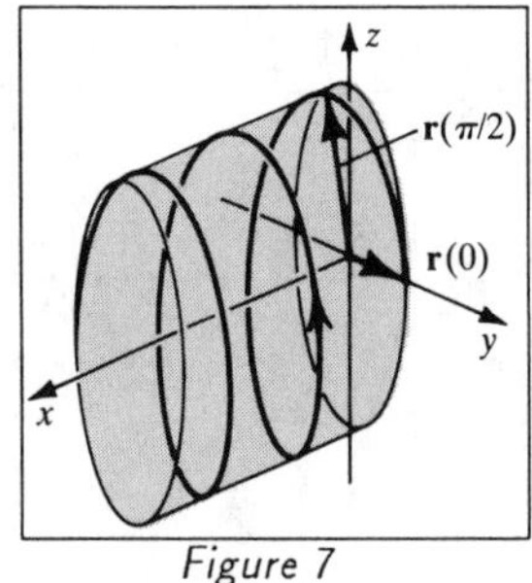

Figure 7

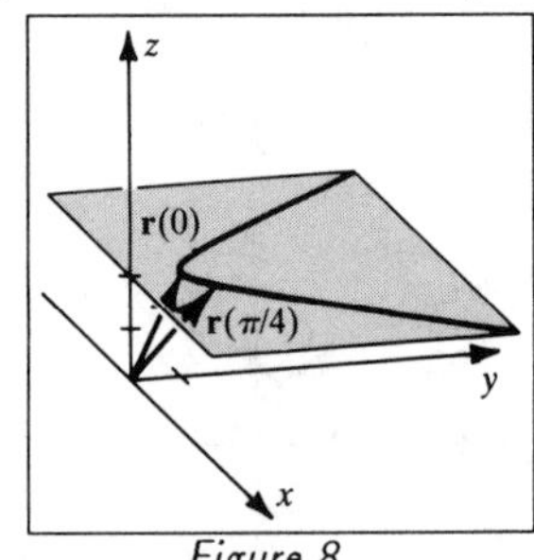

Figure 8

[8] $\mathbf{r}(0) = \langle 0, 1, 2 \rangle$. $\mathbf{r}(\frac{\pi}{4}) = \langle 1, \sqrt{2}, 2 \rangle$. Since $z = 2$, all points lie in the plane

$z = 2$. $x = \tan t$ and $y = \sec t \Rightarrow y^2 - x^2 = 1$. Since $-\frac{\pi}{2} < t < \frac{\pi}{2}$, $y > 0$.

[9] $x = e^t \cos t$ and $y = e^t \sin t \Rightarrow x^2 + y^2 = e^{2t} \Rightarrow r^2 = e^{2t} \Rightarrow r = e^t$.

Since $0 \le t \le \pi$, $y = e^t \sin t \ge 0$ and C is in quadrants I and II.

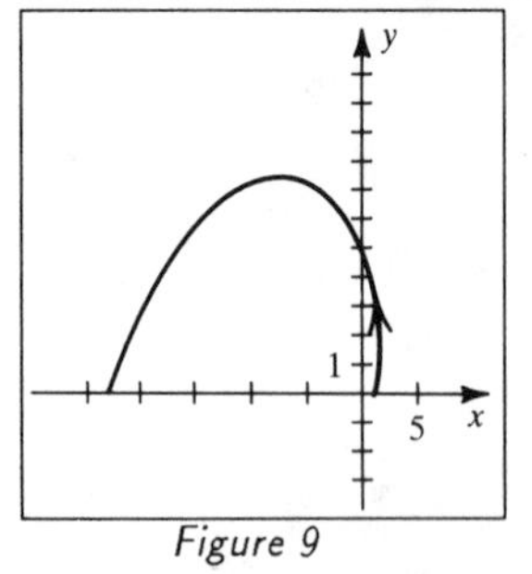

Figure 9

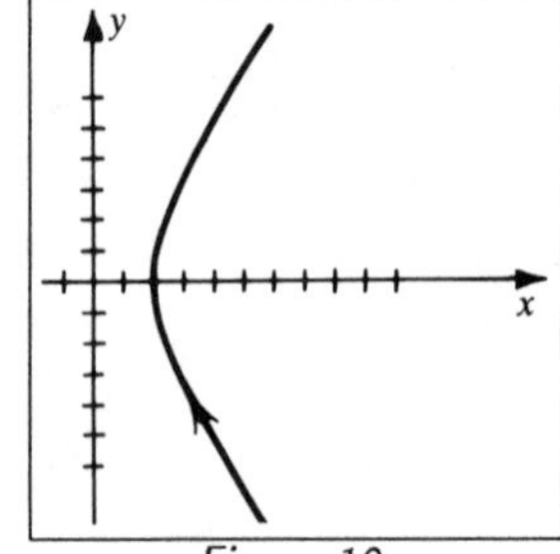

Figure 10

[10] $x = 2\cosh t$ and $y = 3\sinh t \Rightarrow \frac{x^2}{4} - \frac{y^2}{9} = 1$. $x = 2\cosh t \ge 2$ since $\cosh t \ge 1$.

[11] This is a twisted cubic. See Example 3. $x = t$, $y = 2t^2$, and $z = 3t^3 \Rightarrow$

$y = 2x^2$ and $z = 3x^3$. $t \in \mathbb{R} \Rightarrow x \in \mathbb{R}$, $z \in \mathbb{R}$, and $y \geq 0$.

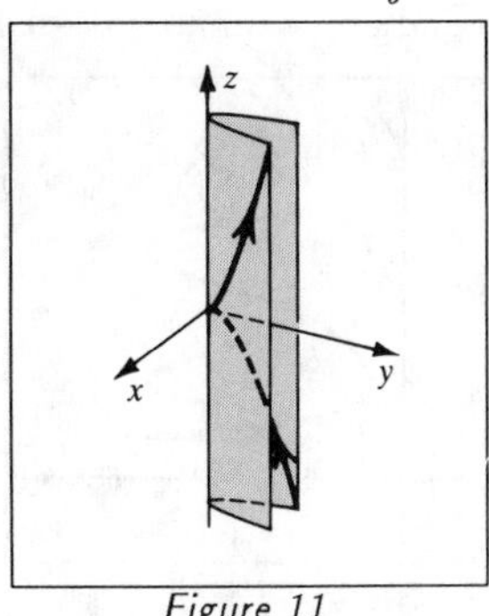

Figure 11

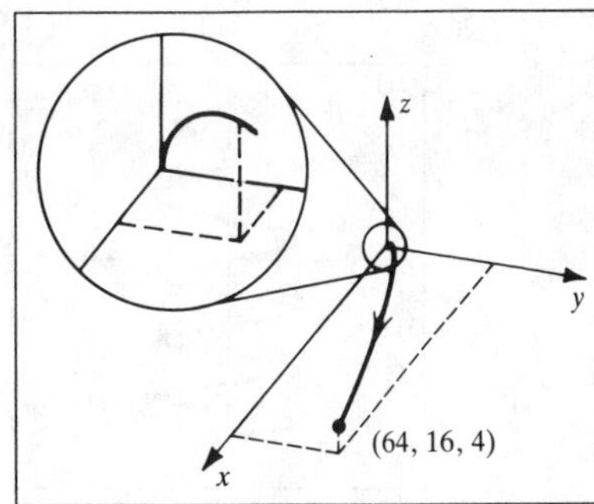

Figure 12

[12] This is a twisted cubic like the one in Example 3 except the roles of x and z have been interchanged and $0 \leq t \leq 4$. The end points are (0, 0, 0) and (64, 16, 4).

[13] Since $z = 3$, all points lie in the plane $z = 3$.

$x = t^2 + 1$ and $y = t \Rightarrow x = y^2 + 1$.

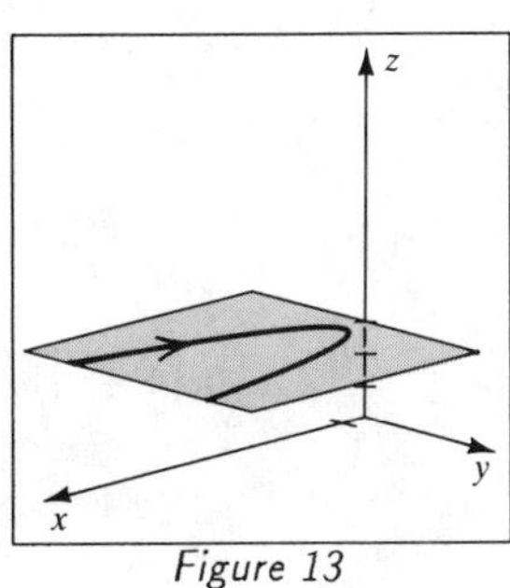

Figure 13

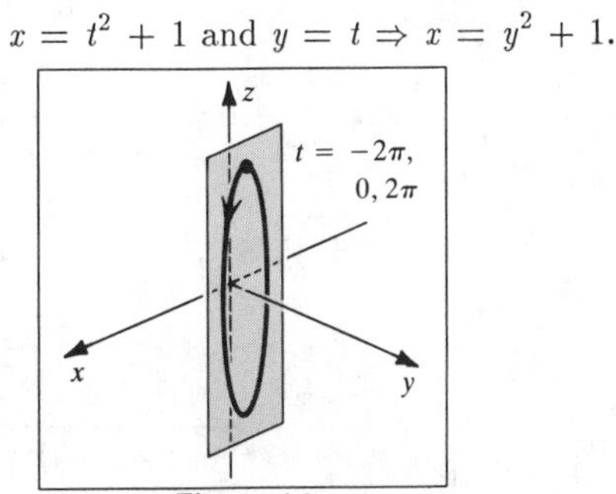

Figure 14

[14] Since $y = 4$, all points lie in the plane $y = 4$.

$x = 6\sin t$ and $z = 25\cos t \Rightarrow \frac{x^2}{36} + \frac{z^2}{625} = 1$.

[15] $x = t$, $y = t$, and $z = \sin t \Rightarrow$

$y = x$ and C is a sine curve along the line $y = x$, orthogonal to the xy-plane.

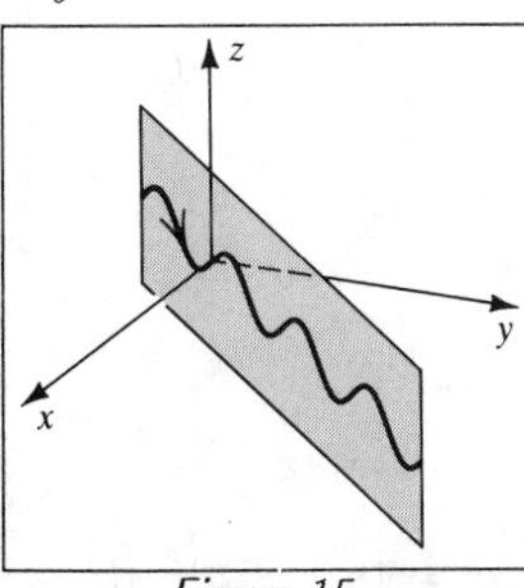

Figure 15

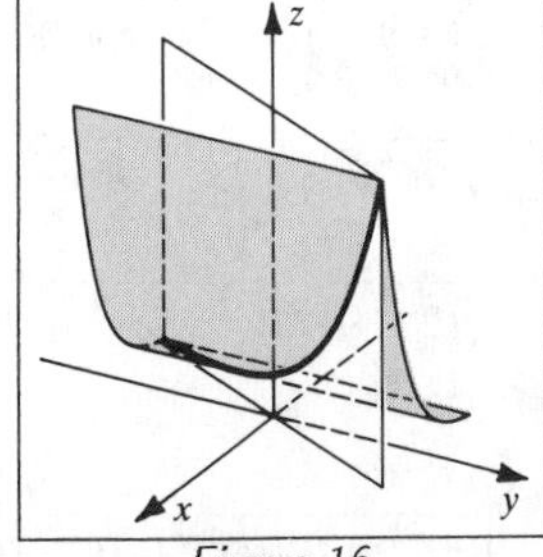

Figure 16

[16] $x = t$, $y = 2t$, and $z = e^t \Rightarrow$

all points lie in the plane $y = 2x$ and on the cylinder $z = e^x$.

[17] $\mathbf{r}(t) = 3\sin(t^2)\mathbf{i} + (4 - t^{3/2})\mathbf{j}$, with $0 \le t \le 5$,

has end points $(0, 4)$ and approximately $(-0.40, -7.18)$.

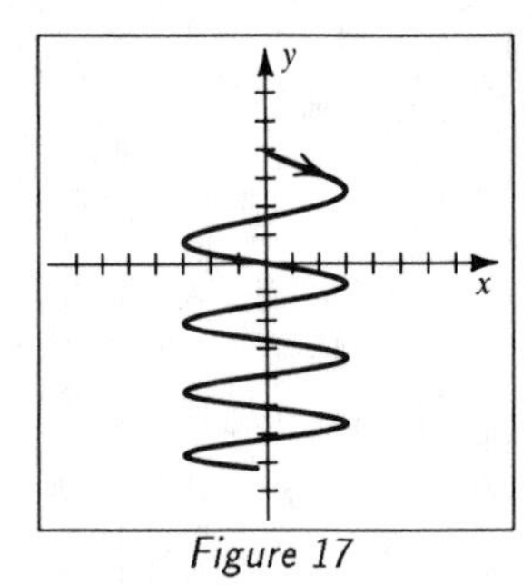

Figure 17

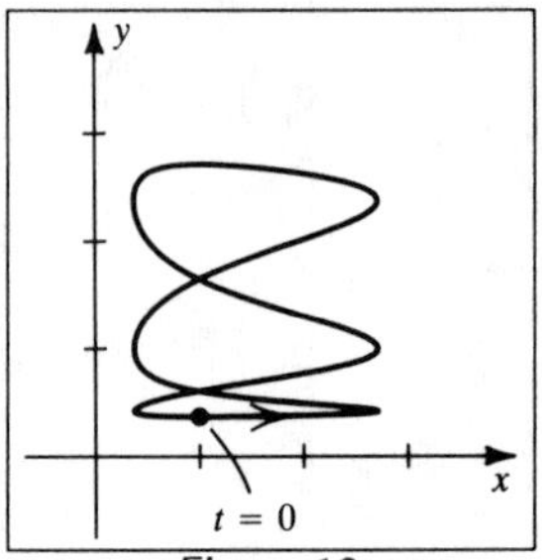

Figure 18

[18] $\mathbf{r}(t) = e^{\sin 3t}\mathbf{i} + e^{-\cos t}\mathbf{j}$, with $0 \le t \le 2\pi$,

goes through the points $(1, e^{-1}) \approx (1, 0.37)$ when $t = 0, 2\pi$.

[19] $$L = \int_0^2 \sqrt{\left(\frac{dx}{dt}\right)^2 + \left(\frac{dy}{dt}\right)^2 + \left(\frac{dz}{dt}\right)^2}\,dt = \int_0^2 \sqrt{(5)^2 + (8t)^2 + (6t)^2}\,dt$$

$$= 5\int_0^2 \sqrt{4t^2 + 1}\,dt = \tfrac{5}{2}\int_0^4 \sqrt{1 + u^2}\,du \; \{u = 2t\}$$

$$= \tfrac{5}{2}\left[\tfrac{u}{2}\sqrt{1 + u^2} + \tfrac{1}{2}\ln\left|u + \sqrt{1 + u^2}\right|\right]_0^4 \; \{\text{Formula 21}\} =$$

$$\tfrac{5}{4}\left[4\sqrt{17} + \ln(4 + \sqrt{17})\right] \approx 23.23$$

[20] $$L = \int_0^1 \sqrt{(2t)^2 + (t\cos t + \sin t)^2 + (-t\sin t + \cos t)^2}\,dt$$

$$= \int_0^1 \sqrt{1 + 5t^2}\,dt = \frac{1}{\sqrt{5}}\int_0^{\sqrt{5}} \sqrt{1 + u^2}\,du \; \{u = \sqrt{5}\,t\}$$

$$= \frac{1}{\sqrt{5}}\left[\tfrac{u}{2}\sqrt{1 + u^2} + \tfrac{1}{2}\ln\left|u + \sqrt{1 + u^2}\right|\right]_0^{\sqrt{5}} \; \{\text{Formula 21}\}$$

$$= \tfrac{1}{2}\sqrt{6} + \tfrac{1}{2\sqrt{5}}\ln(\sqrt{5} + \sqrt{6}) \approx 1.57$$

[21] $$L = \int_0^{2\pi} \sqrt{(-e^t\sin t + e^t\cos t)^2 + (e^t)^2 + (e^t\cos t + e^t\sin t)^2}\,dt$$

$$= \int_0^{2\pi} \sqrt{3e^{2t}}\,dt = \sqrt{3}\int_0^{2\pi} e^t\,dt = \sqrt{3}(e^{2\pi} - 1) \approx 925.77$$

[22] $$L = \int_0^{2\pi} \sqrt{(2)^2 + (12\cos 3t)^2 + (-12\sin 3t)^2}\,dt = \int_0^{2\pi} \sqrt{148}\,dt =$$

$$\sqrt{148}(2\pi - 0) = 4\pi\sqrt{37} \approx 76.44$$

[23] $$L = \int_0^1 \sqrt{(6t)^2 + (3t^2)^2 + (6)^2}\,dt = \int_0^1 \sqrt{9(t^4 + 4t^2 + 4)}\,dt = 3\int_0^1 \sqrt{(t^2 + 2)^2}\,dt =$$

$$3\int_0^1 (t^2 + 2)\,dt = 7$$

[24] $L = \int_0^2 \sqrt{(-4t)^2 + (4)^2 + (4t)^2}\,dt = 4\int_0^2 \sqrt{2t^2 + 1}\,dt$

$= \frac{4}{\sqrt{2}}\int_0^{2\sqrt{2}} \sqrt{1 + u^2}\,du \;\{u = \sqrt{2}\,t\} = \frac{4}{\sqrt{2}}\left[\frac{u}{2}\sqrt{1 + u^2} + \frac{1}{2}\ln\left|u + \sqrt{1 + u^2}\right|\right]_0^{2\sqrt{2}}$

$\{\text{Formula 21}\} = 12 + \sqrt{2}\ln(2\sqrt{2} + 3) \approx 14.49$

[25] (a) $b^2(x^2 + y^2) = b^2(a^2 e^{2\mu t}\cos^2 t + a^2 e^{2\mu t}\sin^2 t)$

$= b^2 a^2 e^{2\mu t}(\cos^2 t + \sin^2 t)$

$= a^2(b^2 e^{2\mu t}) = a^2 z^2$

(b) The curve C starts at the point $(4, 0, 4)$ $\{t = 0\}$ and "spirals" on the cone $z^2 = x^2 + y^2$ toward the vertex of the cone.

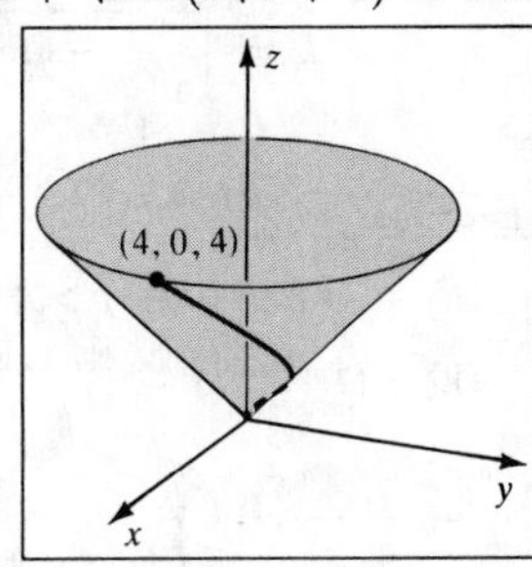

Figure 25

(c) $x = 4e^{-t}\cos t$, $y = 4e^{-t}\sin t$, and $z = 4e^{-t} \Rightarrow \sqrt{\left(\frac{dx}{dt}\right)^2 + \left(\frac{dy}{dt}\right)^2 + \left(\frac{dz}{dt}\right)^2} =$

$\sqrt{\left[-4e^{-t}(\sin t + \cos t)\right]^2 + \left[4e^{-t}(\cos t - \sin t)\right]^2 + \left[-4e^{-t}\right]^2} = \sqrt{48e^{-2t}} =$

$4\sqrt{3}\,e^{-t}$. Thus, $L = \int_0^\infty 4\sqrt{3}\,e^{-t}\,dt = 4\sqrt{3}\lim_{t\to\infty}\left[-e^{-t}\right]_0^t = 4\sqrt{3}$.

[26] (a) $\frac{x^2}{a^2} + \frac{y^2}{b^2} + \frac{z^2}{c^2} = \sin^2 t\sin^2\alpha + \sin^2 t\cos^2\alpha + \cos^2 t$

$= \sin^2 t(\sin^2\alpha + \cos^2\alpha) + \cos^2 t = \sin^2 t + \cos^2 t = 1$

(b) Planes that contain the z-axis are of the form $y = kx$. Thus, we may show that y/x is a constant. $\frac{y}{x} = \frac{b\sin t\cos\alpha}{a\sin t\sin\alpha} = \frac{b}{a}\tan\alpha \Rightarrow y = \left(\frac{b}{a}\tan\alpha\right)x$, where $\frac{b}{a}\tan\alpha$ is a *constant*. Thus, C lies in a plane which contains the z-axis.

(c) C is the intersection of the ellipsoid $\frac{x^2}{a^2} + \frac{y^2}{b^2} + \frac{z^2}{c^2} = 1$ with the plane $y = \left(\frac{b}{a}\tan\alpha\right)x$, an ellipse.

[27] (a) Let $Ax + By + Cz = D$ be the equation of an arbitrary plane.
Then $x = at$, $y = bt^2$, and $z = ct^3 \Rightarrow A(at) + B(bt^2) + C(ct^3) = D$, which is a cubic equation in t. This equation has at most 3 roots.

(b) $L = \int_0^1 \sqrt{(6)^2 + (6t)^2 + (3t^2)^2}\,dt = 7$ {see Exercise 23 solution}

[28] (a) In the first figure, let the horizontal axis be associated with the variable t. Since the slope of the line segment is m, $z = mt$. Now, when the rectangle is transformed into a cylinder, the line segment $0 \le t \le 2\pi$ $\{z = 0\}$ is transformed into the unit circle $x^2 + y^2 = 1$, where $t = 0, 2\pi$ correspond to the point $(1, 0)$. Moreover, $x = \cos t$ and $y = \sin t$ is a parametrization of this circle.

(b) Let $t = t_0$ correspond to the point P. In rectangle $ABCD$, P has coordinates (t_0, mt_0) and its distance from A is $d(A, P) = \sqrt{t_0^2 + m^2 t_0^2} = \sqrt{1 + m^2}\, t_0$. {The shortest distance in a plane between A and P is a straight line distance.} On the cylinder, the length of the helix from A to P is

$$L = \int_0^{t_0} \sqrt{(-\sin t)^2 + (\cos t)^2 + (m)^2}\, dt = \int_0^{t_0} \sqrt{1 + m^2}\, dt = \sqrt{1 + m^2}\, t_0.$$

The distance L is equal to the minimum distance $d(A, P)$.

Exercises 15.2

[1] (a) $t - 1 \geq 0 \Rightarrow t \geq 1$ and $2 - t \geq 0 \Rightarrow t \leq 2$. Domain $D = [1, 2]$.

(b) $\mathbf{r}'(t) = \frac{1}{2}(t - 1)^{-1/2}\mathbf{i} - \frac{1}{2}(2 - t)^{-1/2}\mathbf{j}$;

$$\mathbf{r}''(t) = -\tfrac{1}{4}(t - 1)^{-3/2}\mathbf{i} - \tfrac{1}{4}(2 - t)^{-3/2}\mathbf{j}$$

[2] (a) $D = \{t : t \neq 0\}$.

(b) $\mathbf{r}'(t) = -\dfrac{1}{t^2}\mathbf{i} + 3\cos 3t\mathbf{j}$; $\mathbf{r}''(t) = \dfrac{2}{t^3}\mathbf{i} - 9\sin 3t\mathbf{j}$

[3] (a) $D = \{t : t \neq \frac{\pi}{2} + \pi n\}$.

(b) $\mathbf{r}'(t) = \sec^2 t\mathbf{i} + (2t + 8)\mathbf{j}$; $\mathbf{r}''(t) = 2\sec^2 t \tan t\mathbf{i} + 2\mathbf{j}$

[4] (a) Since the domain of $\sin^{-1}$ is $[-1, 1]$, $D = [-1, 1]$.

(b) $\mathbf{r}'(t) = 2te^{(t^2)}\mathbf{i} + (1 - t^2)^{-1/2}\mathbf{j}$;

$$\mathbf{r}''(t) = (4t^2 + 2)\, e^{(t^2)}\mathbf{i} + t(1 - t^2)^{-3/2}\mathbf{j}$$

[5] (a) $D = \{t : t \neq \frac{\pi}{2} + \pi n\}$.

(b) $\mathbf{r}'(t) = 2t\mathbf{i} + \sec^2 t\mathbf{j}$; $\mathbf{r}''(t) = 2\mathbf{i} + 2\sec^2 t \tan t\mathbf{j}$

[6] (a) $D = \{t : t \neq 0\}$.

(b) $\mathbf{r}'(t) = \frac{1}{3}t^{-2/3}\mathbf{i} - t^{-2}\mathbf{j} - e^{-t}\mathbf{k}$; $\mathbf{r}''(t) = -\frac{2}{9}t^{-5/3}\mathbf{i} + 2t^{-3}\mathbf{j} + e^{-t}\mathbf{k}$

[7] (a) $D = \{t : t \geq 0\}$.

(b) $\mathbf{r}'(t) = \dfrac{1}{2\sqrt{t}}\mathbf{i} + 2e^{2t}\mathbf{j} + \mathbf{k}$; $\mathbf{r}''(t) = -\dfrac{1}{4t\sqrt{t}}\mathbf{i} + 4e^{2t}\mathbf{j}$

[8] (a) Since the domain of $\ln(1 - t)$ is $t < 1$, $D = \{t : t < 1\}$.

(b) $\mathbf{r}'(t) = -\dfrac{1}{1 - t}\mathbf{i} + \cos t\mathbf{j} + 2t\mathbf{k}$; $\mathbf{r}''(t) = -\dfrac{1}{(1 - t)^2}\mathbf{i} - \sin t\mathbf{j} + 2\mathbf{k}$

[9] (a) $x = -\frac{1}{4}t^4$ and $y = t^2 \Rightarrow x = -\frac{1}{4}y^2$; $y \geq 0$.

(b) $\mathbf{r}(2) = -4\mathbf{i} + 4\mathbf{j}$; $\mathbf{r}'(t) = -t^3\mathbf{i} + 2t\mathbf{j}$; $\mathbf{r}'(2) = -8\mathbf{i} + 4\mathbf{j}$

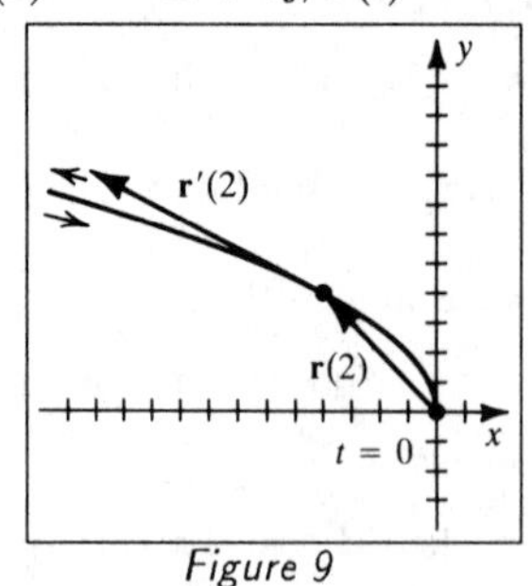

Figure 9

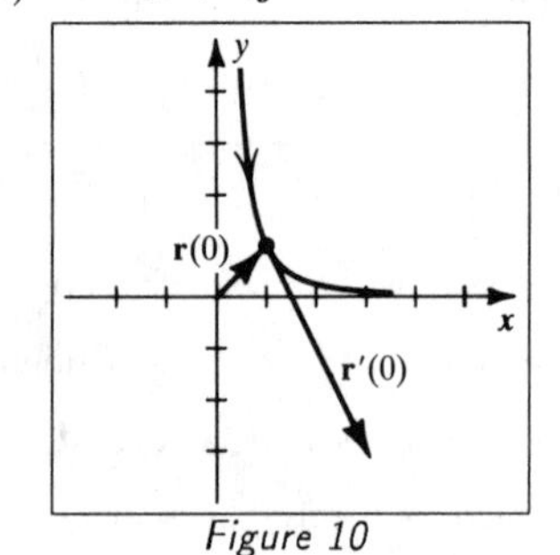

Figure 10

[10] (a) $x = e^{2t}$ and $y = e^{-4t} \Rightarrow y = \dfrac{1}{(e^{2t})^2} = \dfrac{1}{x^2}$; $x \geq 0$

(b) $\mathbf{r}(0) = \mathbf{i} + \mathbf{j}$; $\mathbf{r}'(t) = 2e^{2t}\,\mathbf{i} - 4e^{-4t}\,\mathbf{j}$; $\mathbf{r}'(0) = 2\mathbf{i} - 4\mathbf{j}$

[11] (a) $x = 4\cos t$ and $y = 2\sin t \Rightarrow \dfrac{x^2}{16} + \dfrac{y^2}{4} = 1$

(b) $\mathbf{r}(\frac{3\pi}{4}) = -2\sqrt{2}\,\mathbf{i} + \sqrt{2}\,\mathbf{j}$; $\mathbf{r}'(t) = -4\sin t\,\mathbf{i} + 2\cos t\,\mathbf{j}$; $\mathbf{r}'(\frac{3\pi}{4}) = -2\sqrt{2}\,\mathbf{i} - \sqrt{2}\,\mathbf{j}$

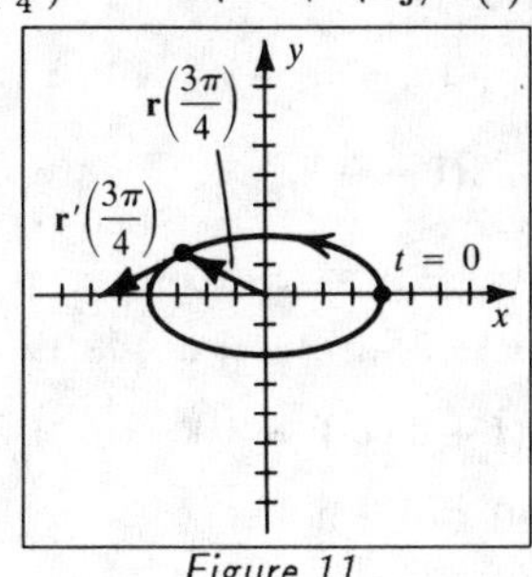

Figure 11

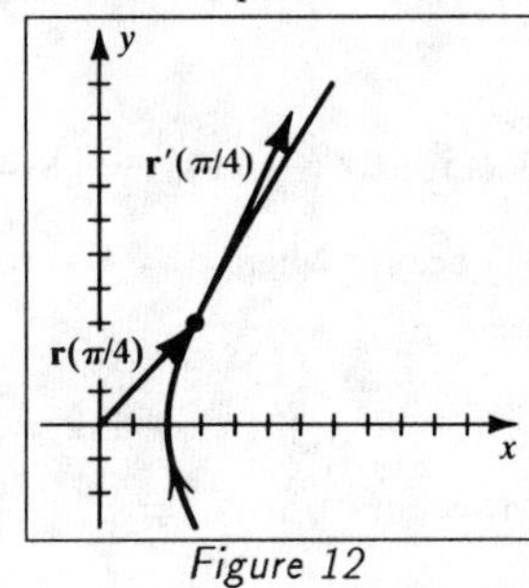

Figure 12

[12] (a) $x = 2\sec t$ and $y = 3\tan t \Rightarrow \dfrac{x^2}{4} - \dfrac{y^2}{9} = 1$, $x \geq 2$ since $|t| < \frac{\pi}{2}$

(b) $\mathbf{r}(\frac{\pi}{4}) = 2\sqrt{2}\,\mathbf{i} + 3\mathbf{j}$; $\mathbf{r}'(t) = 2\sec t\tan t\,\mathbf{i} + 3\sec^2 t\,\mathbf{j}$; $\mathbf{r}'(\frac{\pi}{4}) = 2\sqrt{2}\,\mathbf{i} + 6\mathbf{j}$

[13] (a) $x = t^3$ and $y = t^{-3} \Rightarrow y = \frac{1}{x}$, $x > 0$ since $t > 0$

(b) $\mathbf{r}(1) = \mathbf{i} + \mathbf{j}$; $\mathbf{r}'(t) = 3t^2\,\mathbf{i} - 3t^{-4}\,\mathbf{j}$; $\mathbf{r}'(1) = 3\mathbf{i} - 3\mathbf{j}$

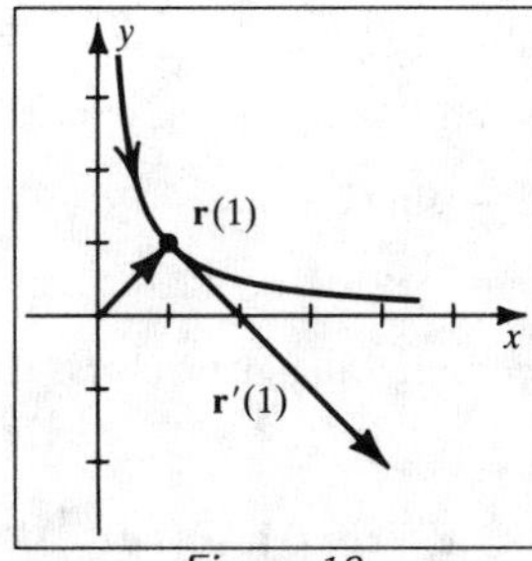

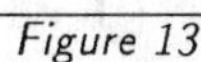
Figure 13

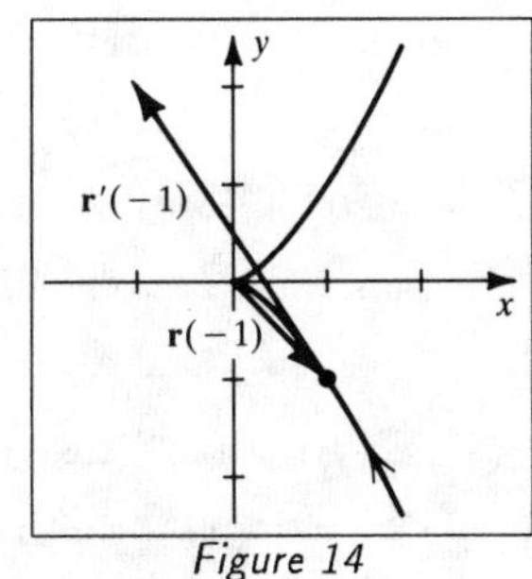

Figure 14

[14] (a) $x = t^2$ and $y = t^3 \Rightarrow x = y^{2/3}$

(b) $\mathbf{r}(-1) = \mathbf{i} - \mathbf{j}$; $\mathbf{r}'(t) = 2t\,\mathbf{i} + 3t^2\,\mathbf{j}$; $\mathbf{r}'(-1) = -2\mathbf{i} + 3\mathbf{j}$

[15] (a) $x = 2t - 1$ and $y = 4 - t \Rightarrow y = -\frac{1}{2}x + \frac{7}{2}$

(b) $\mathbf{r}(3) = 5\mathbf{i} + \mathbf{j}$; $\mathbf{r}'(t) = 2\mathbf{i} - \mathbf{j}$; $\mathbf{r}'(3) = 2\mathbf{i} - \mathbf{j}$

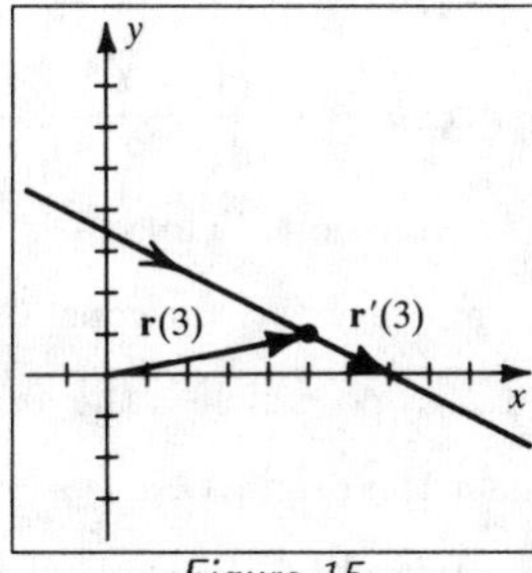

Figure 15

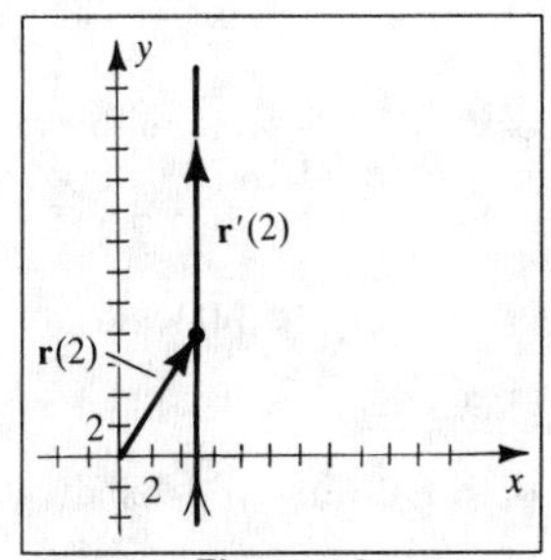

Figure 16

16 (a) $x = 5$ and $y = t^3 \Rightarrow x = 5$ since $y \in \mathbb{R}$.

(b) $\mathbf{r}(2) = 5\mathbf{i} + 8\mathbf{j}$; $\mathbf{r}'(t) = 3t^2\mathbf{j}$; $\mathbf{r}'(2) = 12\mathbf{j}$

17 $\mathbf{r}(t) = (2t^3 - 1)\mathbf{i} + (-5t^2 + 3)\mathbf{j} + (8t + 2)\mathbf{k} \Rightarrow \mathbf{r}'(t) = 6t^2\mathbf{i} - 10t\mathbf{j} + 8\mathbf{k}$.

$P(1, -2, 10)$ occurs when $z = 8t + 2 = 10$, or $t = 1$. $\mathbf{r}'(1) = 6\mathbf{i} - 10\mathbf{j} + 8\mathbf{k}$.

A parametric equation for the tangent line to C at P is

$$x = 1 + 6t,\ y = -2 - 10t,\ z = 10 + 8t.$$

18 $\mathbf{r}(t) = 4\sqrt{t}\,\mathbf{i} + (t^2 - 10)\mathbf{j} + \frac{4}{t}\mathbf{k} \Rightarrow \mathbf{r}'(t) = \dfrac{2}{\sqrt{t}}\mathbf{i} + 2t\mathbf{j} - \dfrac{4}{t^2}\mathbf{k}$.

$P(8, 6, 1)$ occurs when $z = 4/t = 1$, or $t = 4$. $\mathbf{r}'(4) = \mathbf{i} + 8\mathbf{j} - \frac{1}{4}\mathbf{k}$.

An equation is $x = 8 + t,\ y = 6 + 8t,\ z = 1 - \frac{1}{4}t$.

19 $\mathbf{r}(t) = e^t\mathbf{i} + te^t\mathbf{j} + (t^2 + 4)\mathbf{k} \Rightarrow \mathbf{r}'(t) = e^t\mathbf{i} + (t + 1)e^t\mathbf{j} + 2t\mathbf{k}$.

$P(1, 0, 4)$ occurs when $x = e^t = 1$, or $t = 0$. $\mathbf{r}'(0) = \mathbf{i} + \mathbf{j}$.

An equation is $x = 1 + t,\ y = t,\ z = 4$.

20 $\mathbf{r}(t) = t\sin t\,\mathbf{i} + t\cos t\,\mathbf{j} + t\mathbf{k} \Rightarrow \mathbf{r}'(t) = (t\cos t + \sin t)\mathbf{i} + (-t\sin t + \cos t)\mathbf{j} + \mathbf{k}$.

$P(\frac{\pi}{2}, 0, \frac{\pi}{2})$ occurs when $z = t = \frac{\pi}{2}$. $\mathbf{r}'(\frac{\pi}{2}) = \mathbf{i} - \frac{\pi}{2}\mathbf{j} + \mathbf{k}$.

An equation is $x = \frac{\pi}{2} + t,\ y = -\frac{\pi}{2}t,\ z = \frac{\pi}{2} + t$.

21 $\mathbf{r}(t) = e^{2t}\mathbf{i} + e^{-t}\mathbf{j} + (t^2 + 4)\mathbf{k} \Rightarrow \mathbf{r}'(t) = 2e^{2t}\mathbf{i} - e^{-t}\mathbf{j} + 2t\mathbf{k}$.

$P(1, 1, 4)$ occurs when $x = e^{2t} = 1$, or $t = 0$. $\mathbf{r}'(0) = 2\mathbf{i} - \mathbf{j}$ and $\|\mathbf{r}'(0)\| = \sqrt{5}$.

Two unit tangent vectors: $\pm\frac{1}{\sqrt{5}}(2\mathbf{i} - \mathbf{j})$

22 $\mathbf{r}(t) = (2 + \sin t)\mathbf{i} + \cos t\,\mathbf{j} + t\mathbf{k} \Rightarrow \mathbf{r}'(t) = \cos t\,\mathbf{i} - \sin t\,\mathbf{j} + \mathbf{k}$.

$P(2, 1, 0)$ occurs when $z = t = 0$. $\mathbf{r}'(0) = \mathbf{i} + \mathbf{k}$ and $\|\mathbf{r}'(0)\| = \sqrt{2}$.

Two unit tangent vectors: $\pm\frac{1}{\sqrt{2}}(\mathbf{i} + \mathbf{k})$

23 $\mathbf{r}(t) = (ae^{\mu t}\cos t)\mathbf{i} + (ae^{\mu t}\sin t)\mathbf{j} + (be^{\mu t})\mathbf{k} \Rightarrow$

$\mathbf{r}'(t) = (a\mu e^{\mu t}\cos t - ae^{\mu t}\sin t)\mathbf{i} + (a\mu e^{\mu t}\sin t + ae^{\mu t}\cos t)\mathbf{j} + (b\mu e^{\mu t})\mathbf{k} \Rightarrow$

$\|\mathbf{r}'(t)\| = e^{\mu t}\sqrt{a^2\mu^2 + a^2 + b^2\mu^2}$. Now, $\mathbf{r}'(t)\cdot\mathbf{k} = b\mu e^{\mu t} = \|\mathbf{r}'(t)\|\|\mathbf{k}\|\cos\theta \Rightarrow$

$$\cos\theta = \frac{\mathbf{r}'(t)\cdot\mathbf{k}}{\|\mathbf{r}'(t)\|} = \frac{b\mu}{\sqrt{a^2\mu^2 + a^2 + b^2\mu^2}},\ \text{which is a constant.}$$

24 $\mathbf{r}(t) = (3t - t^3)\mathbf{i} + (3t^2)\mathbf{j} + (3t + t^3)\mathbf{k} \Rightarrow$

$\mathbf{r}'(t) = (3 - 3t^2)\mathbf{i} + 6t\mathbf{j} + 3(1 + t^2)\mathbf{k} \Rightarrow \|\mathbf{r}'(t)\| = \sqrt{18}\,(t^2 + 1)$.

Since $\cos\theta = \dfrac{\mathbf{r}'(t)\cdot\mathbf{u}}{\|\mathbf{r}'(t)\|}$, it follows that if $\mathbf{u} = \mathbf{k}$, then $\cos\theta = \dfrac{3(1 + t^2)}{\sqrt{18}\,(t^2 + 1)} = \dfrac{3}{\sqrt{18}}$,

a constant. Thus, $\mathbf{u} = \langle 0, 0, 1\rangle$.

25 Since $\mathbf{r}(t) = \mathbf{r}'(t)$, it follows that $x = ae^t$, $y = be^t$, and $z = ce^t$ for constants a, b, and c. Let $s = e^t > 0$. Then $x = as$, $y = bs$, and $z = cs$, which is the graph of all points on a half-line in a fixed octant that approaches the origin O.

The octant is determined by the signs of a, b, and c.

[26] $D_t\|\mathbf{r}(t)\|^2 = D_t\left[\mathbf{r}(t)\cdot\mathbf{r}(t)\right] = \mathbf{r}(t)\cdot\mathbf{r}'(t) + \mathbf{r}'(t)\cdot\mathbf{r}(t) = 2\left[\mathbf{r}(t)\cdot\mathbf{r}'(t)\right] = 2(0) = 0.$

Thus, $\|\mathbf{r}(t)\|^2$ is a constant and C lies on

a sphere with radius $\|\mathbf{r}(t)\|$ and center at the origin.

[27] $\int_0^2 (6t^2\mathbf{i} - 4t\mathbf{j} + 3\mathbf{k})\,dt = \left[2t^3\mathbf{i} - 2t^2\mathbf{j} + 3t\mathbf{k}\right]_0^2 = 16\mathbf{i} - 8\mathbf{j} + 6\mathbf{k}$

[28] $\int_{-1}^1 (-5t\mathbf{i} + 8t^3\mathbf{j} - 3t^2\mathbf{k})\,dt = 2\int_0^1 (-3t^2\mathbf{k})\,dt = -2\left[t^3\,\mathbf{k}\right]_0^1 = -2\mathbf{k}$

[29] $\int_0^{\pi/4} (\sin t\mathbf{i} - \cos t\mathbf{j} + \tan t\mathbf{k})\,dt = \left[-\cos t\mathbf{i} - \sin t\mathbf{j} + \ln|\sec t|\mathbf{k}\right]_0^{\pi/4} =$

$\left(1 - \frac{1}{\sqrt{2}}\right)\mathbf{i} - \frac{1}{\sqrt{2}}\mathbf{j} + (\ln\sqrt{2})\mathbf{k}$

[30] $\int_0^1 [te^{(t^2)}\mathbf{i} + \sqrt{t}\mathbf{j} + (t^2 + 1)^{-1}\mathbf{k}]\,dt = \left[\frac{1}{2}e^{(t^2)}\mathbf{i} + \frac{2}{3}t^{3/2}\mathbf{j} + \tan^{-1}t\mathbf{k}\right]_0^1 =$

$(\frac{1}{2}e - \frac{1}{2})\mathbf{i} + \frac{2}{3}\mathbf{j} + \frac{\pi}{4}\mathbf{k}$

[31] $\mathbf{r}'(t) = t^2\mathbf{i} + (6t + 1)\mathbf{j} + 8t^3\mathbf{k} \Rightarrow \mathbf{r}(t) = \frac{1}{3}t^3\mathbf{i} + (3t^2 + t)\mathbf{j} + 2t^4\mathbf{k} + \mathbf{c}.$

$\mathbf{r}(0) = 2\mathbf{i} - 3\mathbf{j} + \mathbf{k}$ and $\mathbf{r}(0) = 0\mathbf{i} + 0\mathbf{j} + 0\mathbf{k} + \mathbf{c} \Rightarrow \mathbf{c} = \mathbf{r}(0).$

Thus, $\mathbf{r}(t) = (\frac{1}{3}t^3 + 2)\mathbf{i} + (3t^2 + t - 3)\mathbf{j} + (2t^4 + 1)\mathbf{k}.$

[32] $\mathbf{r}'(t) = 2\mathbf{i} - 4t^3\mathbf{j} + 6\sqrt{t}\mathbf{k} \Rightarrow \mathbf{r}(t) = 2t\mathbf{i} - t^4\mathbf{j} + 4t^{3/2}\mathbf{k} + \mathbf{c}.$

$\mathbf{r}(0) = \mathbf{i} + 5\mathbf{j} + 3\mathbf{k}$ and $\mathbf{r}(0) = 0\mathbf{i} + 0\mathbf{j} + 0\mathbf{k} + \mathbf{c} \Rightarrow \mathbf{c} = \mathbf{r}(0).$

Thus, $\mathbf{r}(t) = (2t + 1)\mathbf{i} + (5 - t^4)\mathbf{j} + (4t^{3/2} + 3)\mathbf{k}.$

[33] $\mathbf{r}''(t) = 6t\mathbf{i} - 12t^2\mathbf{j} + \mathbf{k} \Rightarrow \mathbf{r}'(t) = 3t^2\mathbf{i} - 4t^3\mathbf{j} + t\mathbf{k} + \mathbf{c}_1.$

$\mathbf{r}'(0) = \mathbf{i} + 2\mathbf{j} - 3\mathbf{k}$ and $\mathbf{r}'(0) = 0\mathbf{i} + 0\mathbf{j} + 0\mathbf{k} + \mathbf{c}_1 \Rightarrow \mathbf{c}_1 = \mathbf{r}'(0).$

Hence, $\mathbf{r}'(t) = (3t^2 + 1)\mathbf{i} + (2 - 4t^3)\mathbf{j} + (t - 3)\mathbf{k}$ and

$\mathbf{r}(t) = (t^3 + t)\mathbf{i} + (2t - t^4)\mathbf{j} + (\frac{1}{2}t^2 - 3t)\mathbf{k} + \mathbf{c}_2.$

$\mathbf{r}(0) = 7\mathbf{i} + \mathbf{k}$ and $\mathbf{r}(0) = 0\mathbf{i} + 0\mathbf{j} + 0\mathbf{k} + \mathbf{c}_2 \Rightarrow \mathbf{c}_2 = \mathbf{r}(0).$

Thus, $\mathbf{r}(t) = (t^3 + t + 7)\mathbf{i} + (2t - t^4)\mathbf{j} + (\frac{1}{2}t^2 - 3t + 1)\mathbf{k}.$

[34] $\mathbf{r}''(t) = 6t\mathbf{i} + 3\mathbf{j} \Rightarrow \mathbf{r}'(t) = 3t^2\mathbf{i} + 3t\mathbf{j} + \mathbf{c}_1.$

$\mathbf{r}'(0) = 4\mathbf{i} - \mathbf{j} + \mathbf{k}$ and $\mathbf{r}'(0) = 0\mathbf{i} + 0\mathbf{j} + 0\mathbf{k} + \mathbf{c}_1 \Rightarrow \mathbf{c}_1 = \mathbf{r}'(0).$

Hence, $\mathbf{r}'(t) = (3t^2 + 4)\mathbf{i} + (3t - 1)\mathbf{j} + \mathbf{k}$ and

$\mathbf{r}(t) = (t^3 + 4t)\mathbf{i} + (\frac{3}{2}t^2 - t)\mathbf{j} + t\mathbf{k} + \mathbf{c}_2.$

$\mathbf{r}(0) = 5\mathbf{j}$ and $\mathbf{r}(0) = 0\mathbf{i} + 0\mathbf{j} + 0\mathbf{k} + \mathbf{c}_2 \Rightarrow \mathbf{c}_2 = \mathbf{r}(0).$

Thus, $\mathbf{r}(t) = (t^3 + 4t)\mathbf{i} + (\frac{3}{2}t^2 - t + 5)\mathbf{j} + t\mathbf{k}.$

[35] From Exercise 19, $\mathbf{r}'(0) = \mathbf{i} + \mathbf{j}$. An equation of the plane is

$1(x - 1) + 1(y - 0) + 0(z - 4) = 0$, or $x + y = 1$.

[36] From Exercise 20, $\mathbf{r}'(\frac{\pi}{2}) = \mathbf{i} - \frac{\pi}{2}\mathbf{j} + \mathbf{k}$. An equation of the plane is

$1(x - \frac{\pi}{2}) - \frac{\pi}{2}(y - 0) + 1(z - \frac{\pi}{2}) = 0$, or $x - \frac{\pi}{2}y + z = \pi$.

[37] (1) $\mathbf{u}(t) \cdot \mathbf{v}(t) = t \sin t + t^2 \cos t + 2t^3 \sin t \Rightarrow$

$$D_t\big[\mathbf{u}(t) \cdot \mathbf{v}(t)\big] = t \cos t + \sin t - t^2 \sin t + 2t \cos t + 2t^3 \cos t + 6t^2 \sin t$$
$$= (1 + 5t^2)\sin t + (2t^3 + 3t)\cos t.$$

(2) $\mathbf{u}(t) \times \mathbf{v}(t) = (2t^2 \sin t - t^3 \cos t)\,\mathbf{i} - (2t \sin t - t^3 \sin t)\,\mathbf{j} + (t \cos t - t^2 \sin t)\,\mathbf{k}$

$$\Rightarrow D_t\big[\mathbf{u}(t) \times \mathbf{v}(t)\big] = (2t^2 \cos t + 4t \sin t + t^3 \sin t - 3t^2 \cos t)\,\mathbf{i}$$
$$- (2t \cos t + 2 \sin t - t^3 \cos t - 3t^2 \sin t)\,\mathbf{j}$$
$$+ (-t \sin t + \cos t - t^2 \cos t - 2t \sin t)\,\mathbf{k} =$$
$$\big[(t^3 + 4t)\sin t - t^2 \cos t\big]\mathbf{i} + \big[(3t^2 - 2)\sin t + (t^3 - 2t)\cos t\big]\mathbf{j} +$$
$$\big[-3t \sin t + (1 - t^2)\cos t\big]\mathbf{k}.$$

[38] (1) $\mathbf{u}(t) \cdot \mathbf{v}(t) = 2te^{-t} - 6te^{-t} + t^2 \Rightarrow$

$$D_t\big[\mathbf{u}(t) \cdot \mathbf{v}(t)\big] = -2te^{-t} + 2e^{-t} + 6te^{-t} - 6e^{-t} + 2t = 2t + 4te^{-t} - 4e^{-t}.$$

(2) $\mathbf{u}(t) \times \mathbf{v}(t) = (6t + t^2 e^{-t})\,\mathbf{i} + (t^2 e^{-t} - 2t)\,\mathbf{j} - (8te^{-t})\,\mathbf{k} \Rightarrow D_t\big[\mathbf{u}(t) \times \mathbf{v}(t)\big] =$

$$\big[6 + (2t - t^2)\,e^{-t}\big]\mathbf{i} + \big[(2t - t^2)\,e^{-t} - 2\big]\mathbf{j} + \big[(8t - 8)\,e^{-t}\big]\mathbf{k}.$$

Note: For many remaining exercises, we will use the notation

$\mathbf{u}(t) = \langle u_1, u_2, u_3 \rangle$, where $u_n = f_n(t)$ for $n = 1, 2, 3$.

Similarly, $\mathbf{v}(t) = \langle v_1, v_2, v_3 \rangle$, where $v_n = g_n(t)$ for $n = 1, 2, 3$.

[39] (a) $\lim_{t \to a}\big[\mathbf{u}(t) + \mathbf{v}(t)\big] = \lim_{t \to a} \langle u_1 + v_1, u_2 + v_2, u_3 + v_3 \rangle$

$$= \langle \lim_{t \to a} u_1 + \lim_{t \to a} v_1, \lim_{t \to a} u_2 + \lim_{t \to a} v_2, \lim_{t \to a} u_3 + \lim_{t \to a} v_3 \rangle$$

$$= \langle \lim_{t \to a} u_1, \lim_{t \to a} u_2, \lim_{t \to a} u_3 \rangle + \langle \lim_{t \to a} v_1, \lim_{t \to a} v_2, \lim_{t \to a} v_3 \rangle$$

$$= \lim_{t \to a} \langle u_1, u_2, u_3 \rangle + \lim_{t \to a} \langle v_1, v_2, v_3 \rangle = \lim_{t \to a} \mathbf{u}(t) + \lim_{t \to a} \mathbf{v}(t).$$

(b) $\lim_{t \to a}\big[\mathbf{u}(t) \cdot \mathbf{v}(t)\big] = \lim_{t \to a}\big[u_1 v_1 + u_2 v_2 + u_3 v_3\big]$

$$= \lim_{t \to a} u_1 \lim_{t \to a} v_1 + \lim_{t \to a} u_2 \lim_{t \to a} v_2 + \lim_{t \to a} u_3 \lim_{t \to a} v_3$$

$$= \langle \lim_{t \to a} u_1, \lim_{t \to a} u_2, \lim_{t \to a} u_3 \rangle \cdot \langle \lim_{t \to a} v_1, \lim_{t \to a} v_2, \lim_{t \to a} v_3 \rangle$$

$$= \lim_{t \to a} \mathbf{u}(t) \cdot \lim_{t \to a} \mathbf{v}(t)$$

(c) $\lim_{t \to a} c\mathbf{u}(t) = \lim_{t \to a} \langle cu_1, cu_2, cu_3 \rangle = \langle c \lim_{t \to a} u_1, c \lim_{t \to a} u_2, c \lim_{t \to a} u_3 \rangle =$

$$c \lim_{t \to a} \langle u_1, u_2, u_3 \rangle = c \lim_{t \to a} \mathbf{u}(t)$$

[40] $\lim_{t \to a} f(t)\,\mathbf{u}(t) = \lim_{t \to a} \langle f(t)\,u_1, f(t)\,u_2, f(t)\,u_3 \rangle$

$$= \langle \lim_{t \to a} f(t) \lim_{t \to a} u_1, \lim_{t \to a} f(t) \lim_{t \to a} u_2, \lim_{t \to a} f(t) \lim_{t \to a} u_3 \rangle$$

$$= \lim_{t \to a} f(t) \langle \lim_{t \to a} u_1, \lim_{t \to a} u_2, \lim_{t \to a} u_3 \rangle$$

$$= \lim_{t \to a} f(t) \lim_{t \to a} \langle u_1, u_2, u_3 \rangle = \lim_{t \to a} f(t) \lim_{t \to a} \mathbf{u}(t).$$

[41] Let $\mathbf{u}(t) = \langle u_1, u_2, u_3\rangle$. Now, $\lim_{t\to a} \mathbf{u}(t) = \mathbf{b} = \langle b_1, b_2, b_3\rangle \Rightarrow$

$\lim_{t\to a} u_1 = b_1$, $\lim_{t\to a} u_2 = b_2$, and $\lim_{t\to a} u_3 = b_3$. Thus, $\forall \epsilon > 0$, $\exists \delta_1$, δ_2, and δ_3,

respectively, such that $|t - a| < \delta_1 \Rightarrow |u_1 - b_1| < \frac{\epsilon}{\sqrt{3}}$, $|t - a| < \delta_2 \Rightarrow$

$|u_2 - b_2| < \frac{\epsilon}{\sqrt{3}}$, and $|t - a| < \delta_3 \Rightarrow |u_3 - b_3| < \frac{\epsilon}{\sqrt{3}}$.

Let $\delta = \min(\delta_1, \delta_2, \delta_3)$. Then if $|t - a| < \delta$, $\|\mathbf{u}(t) - \mathbf{b}\| =$

$\sqrt{(u_1 - b_1)^2 + (u_2 - b_2)^2 + (u_3 - b_3)^2} \le \sqrt{\left(\frac{\epsilon}{\sqrt{3}}\right)^2 + \left(\frac{\epsilon}{\sqrt{3}}\right)^2 + \left(\frac{\epsilon}{\sqrt{3}}\right)^2} = \epsilon.$

Conversely, if $\forall \epsilon > 0$, $\exists \delta > 0$ such that $\|\mathbf{u}(t) - \mathbf{b}\| < \epsilon$ whenever $0 < |t - a| < \delta$,

then, for the (ϵ, δ) pair, $\sqrt{(u_1 - b_1)^2 + (u_2 - b_2)^2 + (u_3 - b_3)^2} < \epsilon \Rightarrow$

$|u_1 - b_1| < \epsilon$, $|u_2 - b_2| < \epsilon$, and $|u_3 - b_3| < \epsilon \Rightarrow$

$\lim_{t\to a} u_1 = b_1$, $\lim_{t\to a} u_2 = b_2$, and $\lim_{t\to a} u_3 = b_3 \Rightarrow \lim_{t\to a} \mathbf{u}(t) = \mathbf{b}$. Graphically, this

means that as $t \to a$, the terminal point of $\mathbf{u}(t)$ approaches the terminal point of $\mathbf{b}$.

[42] $\lim_{t\to a}\left[\mathbf{u}(t) \times \mathbf{v}(t)\right] = \lim_{t\to a}\left[(u_2v_3 - v_2u_3)\mathbf{i} - (u_1v_3 - v_1u_3)\mathbf{j} + (u_1v_2 - v_1u_2)\mathbf{k}\right]$

$= \left[\lim_{t\to a} u_2 \lim_{t\to a} v_3 - \lim_{t\to a} v_2 \lim_{t\to a} u_3\right]\mathbf{i}$

$- \left[\lim_{t\to a} u_1 \lim_{t\to a} v_3 - \lim_{t\to a} v_1 \lim_{t\to a} u_3\right]\mathbf{j}$

$+ \left[\lim_{t\to a} u_1 \lim_{t\to a} v_2 - \lim_{t\to a} v_1 \lim_{t\to a} u_2\right]\mathbf{k} = \lim_{t\to a} \mathbf{u}(t) \times \lim_{t\to a} \mathbf{v}(t)$

[43] $D_t\left[\mathbf{u}(t) + \mathbf{v}(t)\right] = D_t\langle u_1 + v_1, u_2 + v_2, u_3 + v_3\rangle$

$= \langle u_1' + v_1', u_2' + v_2', u_3' + v_3'\rangle$

$= \langle u_1', u_2', u_3'\rangle + \langle v_1', v_2', v_3'\rangle = \mathbf{u}'(t) + \mathbf{v}'(t).$

[44] $\mathbf{u}(t) \times \mathbf{v}(t) = \langle u_2 v_3 - v_2 u_3, v_1 u_3 - u_1 v_3, u_1 v_2 - v_1 u_2\rangle$.

$D_t\left[\mathbf{u}(t) \times \mathbf{v}(t)\right] = \langle u_2 v_3' + u_2' v_3 - v_2 u_3' - v_2' u_3,$

$v_1 u_3' + v_1' u_3 - u_1 v_3' - u_1' v_3,$

$u_1 v_2' + u_1' v_2 - v_1 u_2' - v_1' u_2\rangle$

$= \langle u_2 v_3' - v_2' u_3, v_1' u_3 - u_1 v_3', u_1 v_2' - v_1' u_2\rangle +$

$\langle u_2' v_3 - v_2 u_3', v_1 u_3' - u_1' v_3, u_1' v_2 - v_1 u_2'\rangle$

$= \langle u_1, u_2, u_3\rangle \times \langle v_1', v_2', v_3'\rangle +$

$\langle u_1', u_2', u_3'\rangle \times \langle v_1, v_2, v_3\rangle = \mathbf{u}(t) \times \mathbf{v}'(t) + \mathbf{u}'(t) \times \mathbf{v}(t).$

[45] $D_t\left[f(t)\,\mathbf{u}(t)\right] = D_t\langle f(t)\,u_1, f(t)\,u_2, f(t)\,u_3\rangle$

$= \langle f(t)\,u_1' + f'(t)\,u_1, f(t)\,u_2' + f'(t)\,u_2, f(t)\,u_3' + f'(t)\,u_3\rangle$

$= \langle f(t)\,u_1', f(t)\,u_2', f(t)\,u_3'\rangle + \langle f'(t)\,u_1, f'(t)\,u_2, f'(t)\,u_3\rangle$

$= f(t)\,\mathbf{u}'(t) + f'(t)\,\mathbf{u}(t)$

[46] Let $\mathbf{u}(t) = \langle m(t),\ n(t),\ p(t)\rangle$. $D_t\,\mathbf{u}(f(t)) = D_t \langle m(f(t)),\ n(f(t)),\ p(f(t))\rangle$

$$= \langle m'(f(t))f'(t),\ n'(f(t))f'(t),\ p'(f(t))f'(t)\rangle$$
$$= f'(t)\langle m'(f(t)),\ n'(f(t)),\ p'(f(t))\rangle = f'(t)\ \mathbf{u}'(f(t))$$

[47] Using (15.8), $D_t\big[\mathbf{u}(t)\cdot\mathbf{v}(t)\times\mathbf{w}(t)\big] = \mathbf{u}(t)\cdot D_t\big[\mathbf{v}(t)\times\mathbf{w}(t)\big] + \mathbf{u}'(t)\cdot\big[\mathbf{v}(t)\times\mathbf{w}(t)\big]$

$$= \mathbf{u}(t)\cdot\big[\mathbf{v}(t)\times\mathbf{w}'(t) + \mathbf{v}'(t)\times\mathbf{w}(t)\big] + \mathbf{u}'(t)\cdot\big[\mathbf{v}(t)\times\mathbf{w}(t)\big]$$
$$= \big[\mathbf{u}(t)\cdot\mathbf{v}(t)\times\mathbf{w}'(t)\big] + \big[\mathbf{u}(t)\cdot\mathbf{v}'(t)\times\mathbf{w}(t)\big] + \big[\mathbf{u}'(t)\cdot\mathbf{v}(t)\times\mathbf{w}(t)\big]$$

[48] $D_t\big[\mathbf{u}(t)\times\mathbf{u}'(t)\big] = \mathbf{u}(t)\times\mathbf{u}''(t) + \mathbf{u}'(t)\times\mathbf{u}'(t)$

$$= \mathbf{u}(t)\times\mathbf{u}''(t) + \mathbf{0} = \mathbf{u}(t)\times\mathbf{u}''(t).$$

[49] (a) $\displaystyle\int_a^b \big[\mathbf{u}(t) + \mathbf{v}(t)\big]\,dt$

$$= \left(\int_a^b (u_1 + v_1)\,dt\right)\mathbf{i} + \left(\int_a^b (u_2 + v_2)\,dt\right)\mathbf{j} + \left(\int_a^b (u_3 + v_3)\,dt\right)\mathbf{k}$$
$$= \int_a^b (u_1\mathbf{i} + u_2\mathbf{j} + u_3\mathbf{k})\,dt + \int_a^b (v_1\mathbf{i} + v_2\mathbf{j} + v_3\mathbf{k})\,dt = \int_a^b \mathbf{u}(t)\,dt + \int_a^b \mathbf{v}(t)\,dt$$

(b) $\displaystyle\int_a^b c\,\mathbf{u}(t)\,dt = \left(c\int_a^b u_1\,dt\right)\mathbf{i} + \left(c\int_a^b u_2\,dt\right)\mathbf{j} + \left(c\int_a^b u_3\,dt\right)\mathbf{k}$

$$= c\left[\left(\int_a^b u_1\,dt\right)\mathbf{i} + \left(\int_a^b u_2\,dt\right)\mathbf{j} + \left(\int_a^b u_3\,dt\right)\mathbf{k}\right] = c\int_a^b \mathbf{u}(t)\,dt$$

[50] Let $\mathbf{R}'(t) = \mathbf{u}(t)$. RHS $= \mathbf{c}\cdot\displaystyle\int_a^b \mathbf{u}(t)\,dt = \mathbf{c}\cdot\big[\mathbf{R}(b) - \mathbf{R}(a)\big]$ (†).

$D_t\big[\mathbf{c}\cdot\mathbf{R}(t)\big] = D_t\,\mathbf{c}\cdot\mathbf{R}(t) + \mathbf{c}\cdot\mathbf{R}'(t) = \mathbf{c}\cdot\mathbf{R}'(t)$ {c is constant} $= \mathbf{c}\cdot\mathbf{u}(t)$. Thus,

LHS $= \displaystyle\int_a^b \mathbf{c}\cdot\mathbf{u}(t)\,dt = \big[\mathbf{c}\cdot\mathbf{R}(b) - \mathbf{c}\cdot\mathbf{R}(a)\big] = \mathbf{c}\cdot\big[\mathbf{R}(b) - \mathbf{R}(a)\big] = \mathbf{c}\cdot\int_a^b \mathbf{u}(t)\,dt$ from (†).

Exercises 15.3

[1] $y = 4t^2 + 1 = (2t)^2 + 1 = x^2 + 1$. $\mathbf{v}(t) = \mathbf{r}'(t) = 2\mathbf{i} + 8t\mathbf{j}$; $\mathbf{v}(1) = 2\mathbf{i} + 8\mathbf{j}$.

$\mathbf{a}(t) = \mathbf{r}''(t) = 8\mathbf{j}$; $\mathbf{a}(1) = 8\mathbf{j}$.

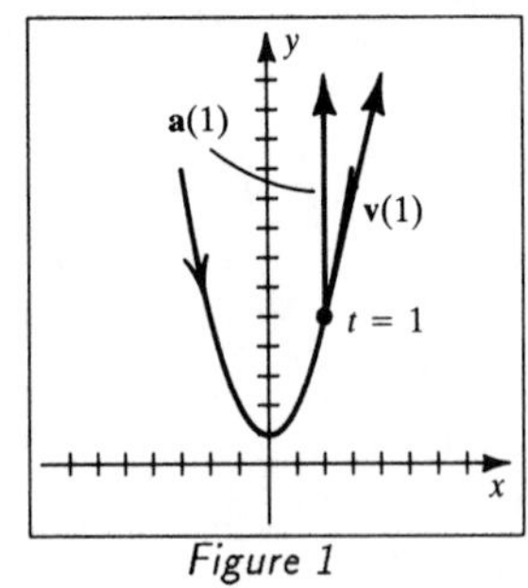

Figure 1

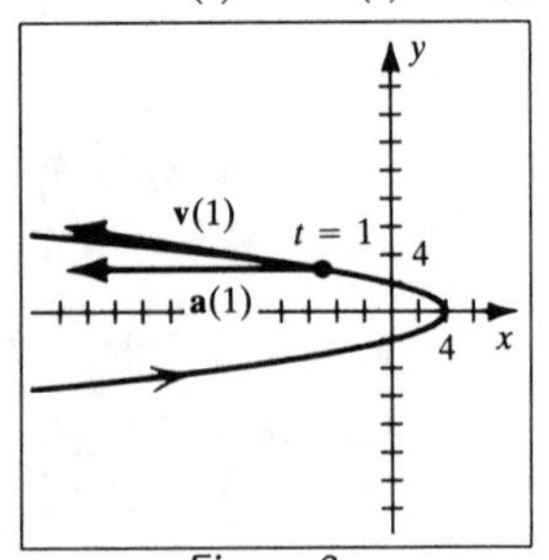

Figure 2

[2] $x = 4 - 9t^2 = 4 - (3t)^2 = 4 - y^2$.

$\mathbf{v}(t) = \mathbf{r}'(t) = -18t\mathbf{i} + 3\mathbf{j}$; $\mathbf{v}(1) = -18\mathbf{i} + 3\mathbf{j}$. $\mathbf{a}(t) = \mathbf{r}''(t) = -18\mathbf{i}$; $\mathbf{a}(1) = -18\mathbf{i}$.

[3] $y = 4\cos 2t = 4(1 - 2\sin^2 t) = 4 - 8x^2$. $\mathbf{v}(t) = \cos t\mathbf{i} - 8\sin 2t\mathbf{j}$;

$\mathbf{v}(\frac{\pi}{6}) = \frac{1}{2}\sqrt{3}\,\mathbf{i} - 4\sqrt{3}\,\mathbf{j}$. $\mathbf{a}(t) = -\sin t\mathbf{i} - 16\cos 2t\mathbf{j}$; $\mathbf{a}(\frac{\pi}{6}) = -\frac{1}{2}\mathbf{i} - 8\mathbf{j}$.

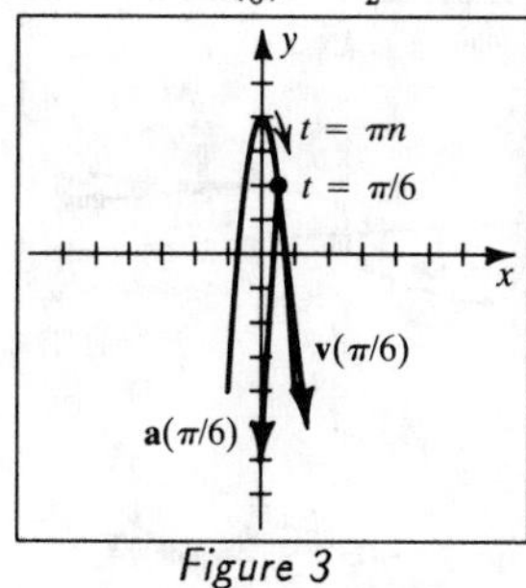

Figure 3

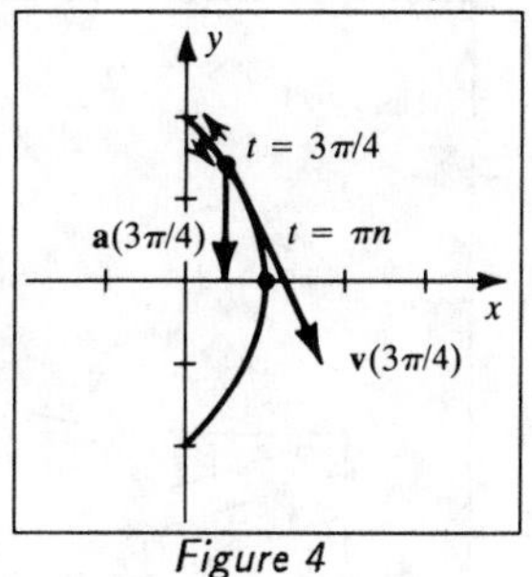

Figure 4

[4] $x = \cos^2 t = 1 - \sin^2 t = 1 - \frac{1}{4}y^2$, $|y| \le 2$.

$\mathbf{v}(t) = -2\cos t\sin t\mathbf{i} + 2\cos t\mathbf{j} = -\sin 2t\mathbf{i} + 2\cos t\mathbf{j}$; $\mathbf{v}(\frac{3\pi}{4}) = \mathbf{i} - \sqrt{2}\,\mathbf{j}$.

$\mathbf{a}(t) = -2\cos 2t\mathbf{i} - 2\sin t\mathbf{j}$; $\mathbf{a}(\frac{3\pi}{4}) = -\sqrt{2}\,\mathbf{j}$.

[5] This is a circular helix along the z-axis. $t = 0$ corresponds to $(1, 0, 0)$.

$\mathbf{v}(t) = -\sin t\mathbf{i} + \cos t\mathbf{j} + \mathbf{k}$; $\mathbf{v}(\frac{\pi}{2}) = -\mathbf{i} + \mathbf{k}$. $\mathbf{a}(t) = -\cos t\mathbf{i} - \sin t\mathbf{j}$; $\mathbf{a}(\frac{\pi}{2}) = -\mathbf{j}$.

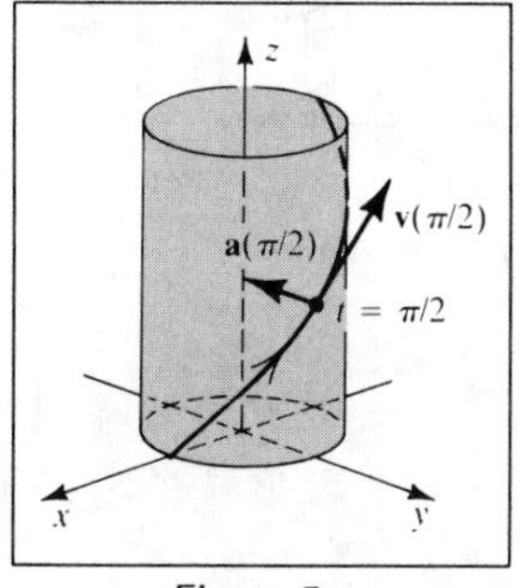

Figure 5

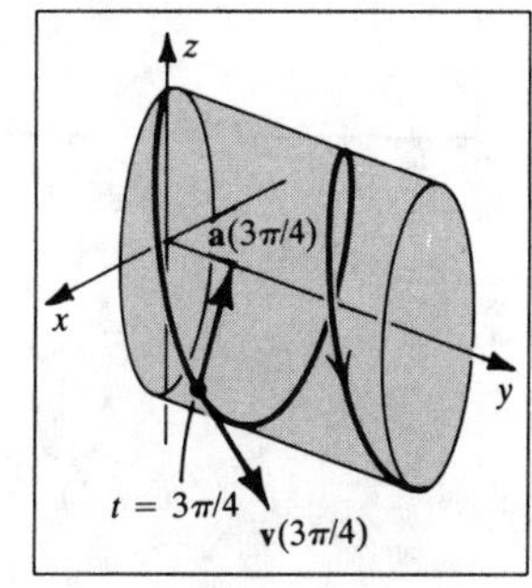

Figure 6

[6] This is an elliptic helix along the y-axis. $\mathbf{v}(t) = 4\cos t\mathbf{i} + 2\mathbf{j} - 9\sin t\mathbf{k}$;

$\mathbf{v}(\frac{3\pi}{4}) = -2\sqrt{2}\,\mathbf{i} + 2\mathbf{j} - \frac{9}{2}\sqrt{2}\,\mathbf{k}$. $\mathbf{a}(t) = -4\sin t\mathbf{i} - 9\cos t\mathbf{k}$; $\mathbf{a}(\frac{3\pi}{4}) = -2\sqrt{2}\,\mathbf{i} + \frac{9}{2}\sqrt{2}\,\mathbf{k}$.

[7] $y^2 = t^2$, $z^2 = 4t^2 \Rightarrow y^2 + z^2 = 5t^2 = 5x$.

The curve lies on the circular paraboloid $5x = y^2 + z^2$.

$\mathbf{v}(t) = 2t\mathbf{i} + \mathbf{j} + 2\mathbf{k}$; $\mathbf{v}(1) = 2\mathbf{i} + \mathbf{j} + 2\mathbf{k}$.

$\mathbf{a}(t) = 2\mathbf{i}$; $\mathbf{a}(1) = 2\mathbf{i}$.

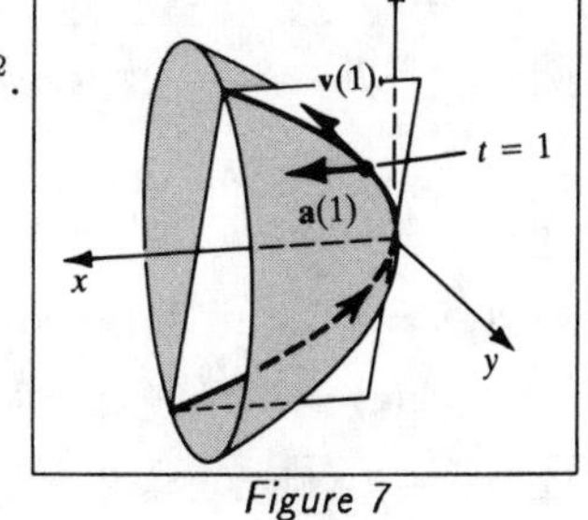

Figure 7

[8] This is a twisted cubic. $\mathbf{v}(t) = 2t\mathbf{i} + 3t^2\mathbf{j} + \mathbf{k}$; $\mathbf{v}(2) = 4\mathbf{i} + 12\mathbf{j} + \mathbf{k}$.

$\mathbf{a}(t) = 2\mathbf{i} + 6t\mathbf{j}$; $\mathbf{a}(2) = 2\mathbf{i} + 12\mathbf{j}$.

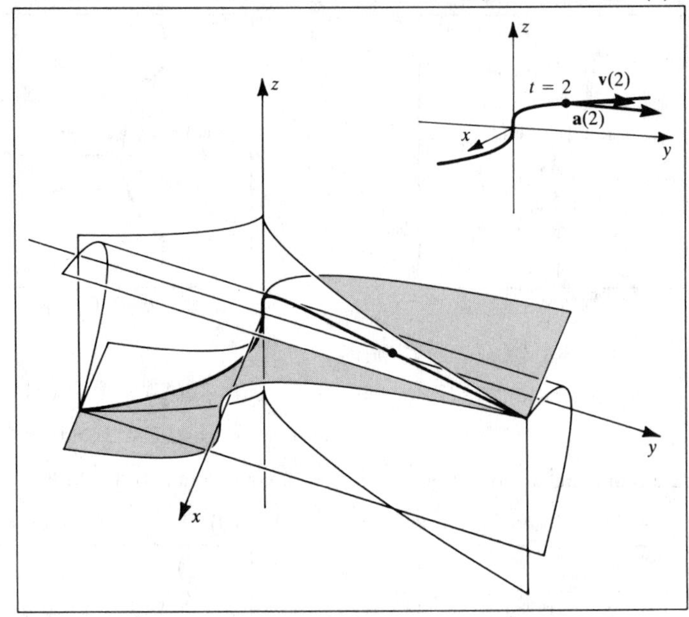

Figure 8

[9] $\mathbf{v}(t) = -2t^{-2}\mathbf{i} - 3(t+1)^{-2}\mathbf{j}$; $\mathbf{v}(2) = -\frac{1}{2}\mathbf{i} - \frac{1}{3}\mathbf{j}$.

$\mathbf{a}(t) = 4t^{-3}\mathbf{i} + 6(t+1)^{-3}\mathbf{j}$; $\mathbf{a}(2) = \frac{1}{2}\mathbf{i} + \frac{2}{9}\mathbf{j}$.

$\|\mathbf{v}(2)\| = \sqrt{(-\frac{1}{2})^2 + (-\frac{1}{3})^2} = \frac{1}{6}\sqrt{13} \approx 0.60$.

[10] $\mathbf{v}(t) = \frac{1}{2}t^{-1/2}\mathbf{i} + \frac{1}{2}t^{-1/2}\mathbf{j}$; $\mathbf{v}(4) = \frac{1}{4}\mathbf{i} + \frac{1}{4}\mathbf{j}$.

$\mathbf{a}(t) = -\frac{1}{4}t^{-3/2}\mathbf{i} - \frac{1}{4}t^{-3/2}\mathbf{j}$; $\mathbf{a}(4) = -\frac{1}{32}\mathbf{i} - \frac{1}{32}\mathbf{j}$.

$\|\mathbf{v}(4)\| = \sqrt{(\frac{1}{4})^2 + (\frac{1}{4})^2} = \frac{1}{2\sqrt{2}} \approx 0.35$.

[11] $\mathbf{v}(t) = 2e^{2t}\mathbf{i} - e^{-t}\mathbf{j}$; $\mathbf{v}(0) = 2\mathbf{i} - \mathbf{j}$. $\mathbf{a}(t) = 4e^{2t}\mathbf{i} + e^{-t}\mathbf{j}$; $\mathbf{a}(0) = 4\mathbf{i} + \mathbf{j}$.

$\|\mathbf{v}(0)\| = \sqrt{(2)^2 + (-1)^2} = \sqrt{5} \approx 2.24$.

[12] $\mathbf{v}(t) = 2\mathbf{i} - 2te^{-t^2}\mathbf{j}$; $\mathbf{v}(1) = 2\mathbf{i} - \frac{2}{e}\mathbf{j}$. $\mathbf{a}(t) = (4t^2 - 2)\,e^{-t^2}\mathbf{j}$; $\mathbf{a}(1) = \frac{2}{e}\mathbf{j}$.

$\|\mathbf{v}(1)\| = \sqrt{(2)^2 + (-2/e)^2} = 2\sqrt{1 + (1/e^2)} \approx 2.13$.

[13] $\mathbf{v}(t) = (e^t\cos t - e^t\sin t)\mathbf{i} + (e^t\sin t + e^t\cos t)\mathbf{j} + e^t\mathbf{k}$;

$\mathbf{v}(\frac{\pi}{2}) = e^{\pi/2}(-\mathbf{i} + \mathbf{j} + \mathbf{k})$. $\mathbf{a}(t) = -2e^t\sin t\mathbf{i} + 2e^t\cos t\mathbf{j} + e^t\mathbf{k}$;

$\mathbf{a}(\frac{\pi}{2}) = e^{\pi/2}(-2\mathbf{i} + \mathbf{k})$. $\|\mathbf{v}(\frac{\pi}{2})\| = e^{\pi/2}\sqrt{(-1)^2 + (1)^2 + (1)^2} = \sqrt{3}\,e^{\pi/2} \approx 8.33$.

[14] $\mathbf{v}(t) = (\cos t - t\sin t)\mathbf{i} + (\sin t + t\cos t)\mathbf{j} + 2t\mathbf{k}$; $\mathbf{v}(\frac{\pi}{2}) = -\frac{\pi}{2}\mathbf{i} + \mathbf{j} + \pi\mathbf{k}$.

$\mathbf{a}(t) = (-2\sin t - t\cos t)\mathbf{i} + (2\cos t - t\sin t)\mathbf{j} + 2\mathbf{k}$; $\mathbf{a}(\frac{\pi}{2}) = -2\mathbf{i} - \frac{\pi}{2}\mathbf{j} + 2\mathbf{k}$.

$\|\mathbf{v}(\frac{\pi}{2})\| = \sqrt{(-\frac{\pi}{2})^2 + (1)^2 + (\pi)^2} = \sqrt{1 + \frac{5}{4}\pi^2} \approx 3.65$.

[15] For all t, $\mathbf{v}(t) = \mathbf{i} + 2\mathbf{j} + 3\mathbf{k}$, $\mathbf{a}(t) = 0$, and $\|\mathbf{v}(t)\| = \sqrt{14} \approx 3.74$.

16 $\mathbf{v}(t) = 2\mathbf{i} + 18t\mathbf{k}$; $\mathbf{v}(2) = 2\mathbf{i} + 36\mathbf{k}$. $\mathbf{a}(t) = 18\mathbf{k}$ for all t.

$$\|\mathbf{v}(2)\| = \sqrt{(2)^2 + (0)^2 + (36)^2} = 10\sqrt{13} \approx 36.06.$$

17 Let $\mathbf{r}(t)$ be the position vector of the particle.

By Theorem (15.9), §15.2, $\|\mathbf{r}'(t)\|$ constant $\Rightarrow$ $\mathbf{r}''(t)$ is orthogonal to $\mathbf{r}'(t)$ for every t; that is, the velocity and acceleration are orthogonal.

18 $\mathbf{r}''(t) = \mathbf{0} \Rightarrow \mathbf{r}'(t) = a\mathbf{i} + b\mathbf{j} + c\mathbf{k} \Rightarrow \mathbf{r}(t) = (at + d)\mathbf{i} + (bt + e)\mathbf{j} + (ct + f)\mathbf{k}$,

where a, b, c, d, e, and f are all constants. Thus, $x = at + d$, $y = bt + e$, and

$z = ct + f$. These are parametric equations of a line.

Note: If the acceleration is zero, then the velocity and hence, the direction of the particle is constant. Thus, it must move in a straight line.

19 (a) $\|\mathbf{a}(t)\| = 32 \text{ ft/sec}^2 = \frac{32}{5280} \text{ mi/sec}^2$.

$$\|\mathbf{a}(t)\| = \frac{v^2}{k} \Rightarrow \frac{32}{5280} = \frac{v^2}{4000 + 150} \Rightarrow v \approx 5.015 \text{ mi/sec} \approx 18{,}054 \text{ mi/hr}.$$

(b) $\omega = \frac{v}{k} \approx \dfrac{18{,}054}{4150} \approx 4.35 \text{ rad/hr} \Rightarrow T = \frac{2\pi}{\omega} \approx 1.4443 \text{ hr} \approx 86.7 \text{ min}.$

20 $T = \frac{2\pi}{\omega} = \frac{2\pi k}{v}\ \{\omega = \frac{v}{k}\} = \dfrac{2\pi k}{\sqrt{gk}}\ \{g = \frac{v^2}{k} \text{ or } v = \sqrt{gk}\,\} = 2\pi\sqrt{k/g} \Rightarrow$

$$k = \frac{T^2 g}{4\pi^2} \approx \frac{(88 \times 60)^2(32)}{4\pi^2} \approx 22{,}597{,}380 \text{ ft} \approx 4280 \text{ mi. Altitude} \approx 280 \text{ mi.}$$

Note: Since the time units for g are in seconds, T must also be in seconds.

21 $\mathbf{r}(t) = t(v_0 \cos\alpha)\mathbf{i} + (-\frac{1}{2}gt^2 + tv_0 \sin\alpha + h_0)\mathbf{j} =$

$$(1500\cos 30^\circ)\,t\mathbf{i} + (-\tfrac{1}{2}gt^2 + 1500\sin 30^\circ\, t)\mathbf{j} = 750\sqrt{3}\,t\mathbf{i} + (-\tfrac{1}{2}gt^2 + 750t)\mathbf{j}$$

(a) $\mathbf{v}(t) = \mathbf{r}'(t) = 750\sqrt{3}\,\mathbf{i} + (-gt + 750)\mathbf{j}$

(b) $h = \dfrac{v_0^2 \sin^2\alpha}{2g} = \dfrac{(1500)^2 \sin^2 30^\circ}{2g} = \dfrac{(1500)^2}{8g} \approx 8789 \text{ ft. } \{g \approx 32\}$

(c) $d = \dfrac{v_0^2 \sin 2\alpha}{g} = \dfrac{(1500)^2 \sin 60^\circ}{g} = \dfrac{(1500)^2\sqrt{3}}{2g} \approx 60{,}892 \text{ ft.}$

(d) $t = \dfrac{2v_0 \sin\alpha}{g} = \dfrac{2(1500)\sin 30^\circ}{g} = \dfrac{1500}{g}$.

$$\left\|\mathbf{v}(\tfrac{1500}{g})\right\| \approx \sqrt{(750\sqrt{3})^2 + (-750)^2} = 1500 \text{ ft/sec, which is the initial velocity.}$$

22 $\mathbf{r}(t) = (1500\cos 60^\circ)\,t\mathbf{i} + (-\frac{1}{2}gt^2 + 1500\sin 60^\circ\, t)\mathbf{j} = 750\,t\mathbf{i} + (-\frac{1}{2}gt^2 + 750\sqrt{3}\,t)\mathbf{j}$

(a) $\mathbf{v}(t) = \mathbf{r}'(t) = 750\,\mathbf{i} + (-gt + 750\sqrt{3})\mathbf{j}$

(b) $h = \dfrac{v_0^2 \sin^2\alpha}{2g} = \dfrac{(1500)^2 \sin^2 60^\circ}{2g} = \dfrac{(1500)^2(3)}{8g} \approx 26{,}367 \text{ ft.}$

(c) $d = \dfrac{v_0^2 \sin 2\alpha}{g} = \dfrac{(1500)^2 \sin 120^\circ}{g} = \dfrac{(1500)^2\sqrt{3}}{2g} \approx 60{,}892 \text{ ft.}$

(d) $t = \dfrac{2v_0 \sin\alpha}{g} = \dfrac{2(1500)\sin 60^\circ}{g} = \dfrac{1500\sqrt{3}}{g}$.

$$\left\|\mathbf{v}(\tfrac{1500\sqrt{3}}{g})\right\| \approx \sqrt{(750)^2 + (-750\sqrt{3})^2} = 1500 \text{ ft/sec.}$$

[23] $d = \dfrac{v_0^2 \sin 2\alpha}{g} \Rightarrow v_0 = \sqrt{(dg)/\sin 2\alpha} = \sqrt{250g} \approx 40\sqrt{5} \approx 89.4$ ft/sec.

[24] The initial height of 1000 ft must be included in the $\mathbf{j}$-component of $\mathbf{r}(t)$.

$\mathbf{r}(t) = t(v_0 \cos\alpha)\mathbf{i} + (-\frac{1}{2}gt^2 + tv_0 \sin\alpha + h_0)\mathbf{j} = 1800t\mathbf{i} + (-\frac{1}{2}gt^2 + 1000)\mathbf{j}$.

$y = -\frac{1}{2}gt^2 + 1000 = 0 \Rightarrow t = \sqrt{2000/g} \approx 7.9$ seconds after firing. $\mathbf{r}(\sqrt{2000/g}) =$

$1800\sqrt{2000/g}\,\mathbf{i} \approx 14{,}230$ ft from a point directly under the firing position.

[25] From Example 2,

$$\|\mathbf{a}(t)\| = \omega^2 k \Rightarrow \omega^2 = \frac{8g}{k} \Rightarrow \omega \approx \sqrt{\tfrac{256}{30}} \approx 2.92 \text{ rad/sec} \approx 0.46 \text{ rev/sec}.$$

[26] $T = \frac{2\pi}{\omega} \Rightarrow \omega = \frac{2\pi}{T}$. $v = \omega k = \frac{2\pi k}{T}$ in km/day. Divide by 24×60^2 for km/sec.

(1) Earth: $v = \dfrac{2\pi(149.6 \times 10^6)}{365.3} \approx 2.5731 \times 10^6$ km/day ≈ 29.78 km/sec.

(2) Venus: $v = \dfrac{2\pi(108.2 \times 10^6)}{224.7} \approx 3.0255 \times 10^6$ km/day ≈ 35.02 km/sec.

(3) Neptune: $v = \dfrac{2\pi(4498 \times 10^6)}{60{,}188} \approx 0.4696 \times 10^6$ km/day ≈ 5.43 km/sec.

[27] $mg = \|\mathbf{F}\| = G\dfrac{mM}{(R+d)^2} \Rightarrow g = \dfrac{GM}{(R+d)^2}$. From Exercise 20,

$$k = \frac{T^2 g}{4\pi^2} \Rightarrow T^2 = \frac{k4\pi^2}{g}\ \{k = R + d\} \Rightarrow T^2 = \frac{(R+d)4\pi^2}{g} = \frac{4\pi^2}{GM}(R+d)^3.$$

[28] (a) $T^2 = 0.00346\left[1 + \frac{1000}{3959}\right]^3 \approx 0.00680 \Rightarrow T \approx 0.08246$ days ≈ 1.98 hours.

(b) $1^2 = 0.00346\left[1 + \dfrac{d}{3959}\right]^3 \Rightarrow d = 3959\left(\sqrt[3]{\frac{1}{0.00346}} - 1\right) \approx 22{,}216$ mi.

[29] $\mathbf{r}(t) = t(v_0 \cos\alpha)\mathbf{i} + (-\frac{1}{2}gt^2 + tv_0 \sin\alpha + h_0)\mathbf{j}$,

where $v_0 = 100$ mi/hr $= 146\frac{2}{3}$ ft/sec and $\tan\alpha = -\frac{2}{58} \Rightarrow \alpha \approx -1.975°$.

Now, $tv_0 \cos\alpha = 58 \Rightarrow t = \dfrac{58}{v_0 \cos\alpha} \approx 0.3957$ sec.

The drop caused by gravity is given by $d = -\frac{1}{2}gt^2 \approx (-16)(0.3957)^2 \approx -2.51$ ft.

[30] From Example 6(b), $d = \dfrac{v_0^2 \sin 2\alpha}{g} \Rightarrow v_0^2 = \dfrac{dg}{\sin 2\alpha} \approx \dfrac{(150)(32)}{\sin 60°} \approx 5542.6 \Rightarrow$

$v_0 \approx 74.45$ ft/sec ≈ 50.76 mi/hr.

[31] $D_h\,\mathbf{v} = \frac{d}{dh}\left[(12 + 0.006h^{3/2})\mathbf{i} + (10 + 0.005h^{3/2})\mathbf{j}\right] = 0.009\sqrt{h}\,\mathbf{i} + 0.0075\sqrt{h}\,\mathbf{j}$.

When $h = 150$, $\|D_h\,\mathbf{v}\| = \sqrt{(0.009)^2(150) + (0.0075)^2(150)} \approx 0.14$ (mi/hr)/ft.

Exercises 15.4

[1] (a) $\mathbf{r}(t) = t\mathbf{i} - \frac{1}{2}t^2\mathbf{j} \Rightarrow \mathbf{r}'(t) = \mathbf{i} - t\mathbf{j} \Rightarrow \|\mathbf{r}'(t)\| = (1 + t^2)^{1/2}$.

$$\mathbf{T}(t) = \frac{1}{\|\mathbf{r}'(t)\|}\mathbf{r}'(t) = \frac{1}{(1+t^2)^{1/2}}\mathbf{i} - \frac{t}{(1+t^2)^{1/2}}\mathbf{j} \Rightarrow$$

$$\mathbf{T}'(t) = -\frac{t}{(1+t^2)^{3/2}}\mathbf{i} - \frac{1}{(1+t^2)^{3/2}}\mathbf{j} \Rightarrow \|\mathbf{T}'(t)\| = \frac{1}{1+t^2}.$$

$$\mathbf{N}(t) = \frac{1}{\|\mathbf{T}'(t)\|}\mathbf{T}'(t) = -\frac{t}{(1+t^2)^{1/2}}\mathbf{i} - \frac{1}{(1+t^2)^{1/2}}\mathbf{j}.$$

(b) $x = t$ and $y = -\frac{1}{2}t^2 \Rightarrow y = -\frac{1}{2}x^2$.

$$T(1) = \tfrac{1}{\sqrt{2}}\mathbf{i} - \tfrac{1}{\sqrt{2}}\mathbf{j} \text{ and } N(1) = -\tfrac{1}{\sqrt{2}}\mathbf{i} - \tfrac{1}{\sqrt{2}}\mathbf{j}.$$

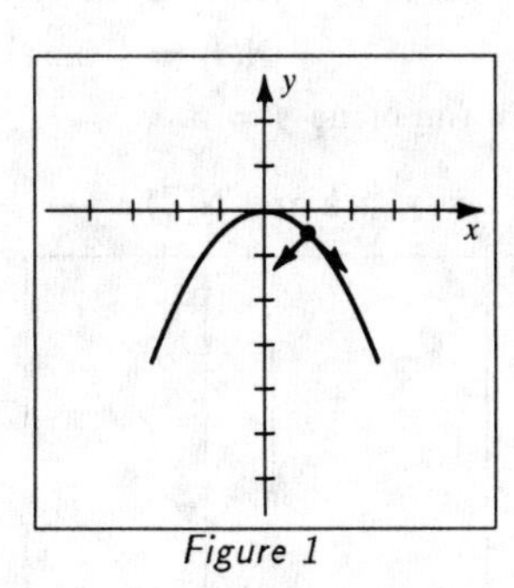

Figure 1

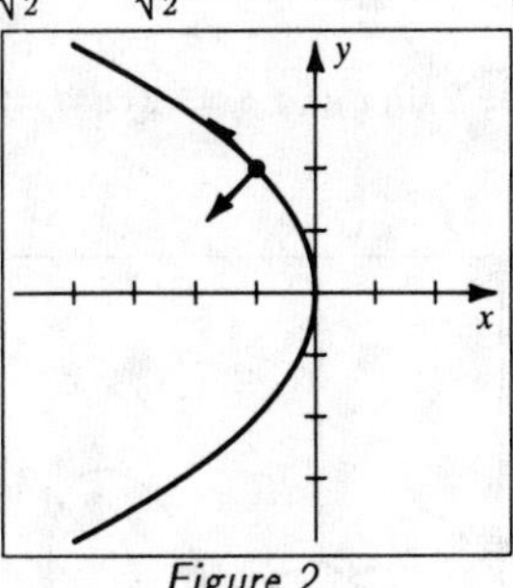

Figure 2

2 (a) $\mathbf{r}(t) = -t^2\mathbf{i} + 2t\mathbf{j} \Rightarrow \mathbf{r}'(t) = -2t\mathbf{i} + 2\mathbf{j} \Rightarrow \|\mathbf{r}'(t)\| = 2(t^2+1)^{1/2}$.

$$T(t) = \frac{1}{\|\mathbf{r}'(t)\|}\mathbf{r}'(t) = -\frac{t}{(t^2+1)^{1/2}}\mathbf{i} + \frac{1}{(t^2+1)^{1/2}}\mathbf{j} \Rightarrow$$

$$T'(t) = -\frac{1}{(1+t^2)^{3/2}}\mathbf{i} - \frac{t}{(1+t^2)^{3/2}}\mathbf{j} \Rightarrow \|T'(t)\| = \frac{1}{1+t^2}.$$

$$N(t) = \frac{1}{\|T'(t)\|}T'(t) = -\frac{1}{(1+t^2)^{1/2}}\mathbf{i} - \frac{t}{(1+t^2)^{1/2}}\mathbf{j}.$$

(b) $x = -t^2$ and $y = 2t \Rightarrow x = -\frac{1}{4}y^2$.

$$T(1) = -\tfrac{1}{\sqrt{2}}\mathbf{i} + \tfrac{1}{\sqrt{2}}\mathbf{j} \text{ and } N(1) = -\tfrac{1}{\sqrt{2}}\mathbf{i} - \tfrac{1}{\sqrt{2}}\mathbf{j}.$$

3 (a) $\mathbf{r}(t) = t^3\mathbf{i} + 3t\mathbf{j} \Rightarrow \mathbf{r}'(t) = 3t^2\mathbf{i} + 3\mathbf{j} \Rightarrow \|\mathbf{r}'(t)\| = 3(t^4+1)^{1/2}$.

$$T(t) = \frac{t^2}{(t^4+1)^{1/2}}\mathbf{i} + \frac{1}{(t^4+1)^{1/2}}\mathbf{j} \Rightarrow T'(t) = \frac{2t}{(t^4+1)^{3/2}}\mathbf{i} - \frac{2t^3}{(t^4+1)^{3/2}}\mathbf{j}$$

$$\Rightarrow \|T'(t)\| = \left|\frac{2t}{t^4+1}\right|. \quad N(t) = \frac{1}{(t^4+1)^{1/2}}\mathbf{i} - \frac{t^2}{(t^4+1)^{1/2}}\mathbf{j}.$$

(b) $x = t^3$ and $y = 3t \Rightarrow x = \frac{1}{27}y^3$. $T(1) = \frac{1}{\sqrt{2}}\mathbf{i} + \frac{1}{\sqrt{2}}\mathbf{j}$ and $N(1) = \frac{1}{\sqrt{2}}\mathbf{i} - \frac{1}{\sqrt{2}}\mathbf{j}$.

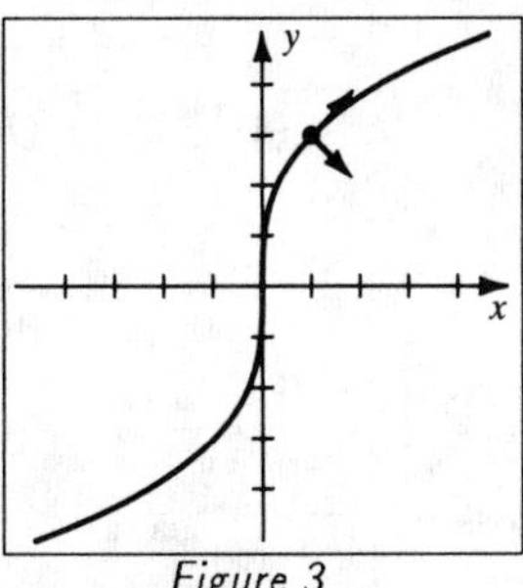

Figure 3

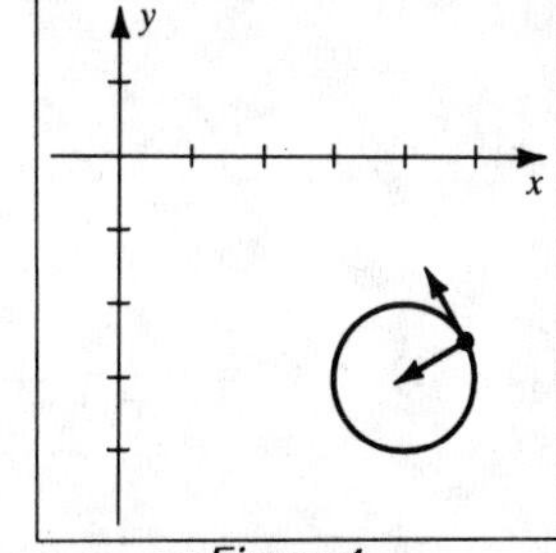

Figure 4

4 (a) $\mathbf{r}(t) = (4+\cos t)\mathbf{i} - (3-\sin t)\mathbf{j} \Rightarrow \mathbf{r}'(t) = -\sin t\,\mathbf{i} + \cos t\,\mathbf{j} \Rightarrow \|\mathbf{r}'(t)\| = 1$.

$T(t) = -\sin t\,\mathbf{i} + \cos t\,\mathbf{j} \Rightarrow T'(t) = -\cos t\,\mathbf{i} - \sin t\,\mathbf{j} \Rightarrow \|T'(t)\| = 1$.

$$N(t) = -\cos t\,\mathbf{i} - \sin t\,\mathbf{j}.$$

(b) $x - 4 = \cos t$ and $y + 3 = \sin t \Rightarrow (x-4)^2 + (y+3)^2 = 1$.

$$T(\tfrac{\pi}{6}) = -\tfrac{1}{2}\mathbf{i} + \tfrac{1}{2}\sqrt{3}\,\mathbf{j} \text{ and } N(\tfrac{\pi}{6}) = -\tfrac{1}{2}\sqrt{3}\,\mathbf{i} - \tfrac{1}{2}\mathbf{j}.$$

5 (a) $\mathbf{r}(t) = 2\sin t\mathbf{i} + 3\mathbf{j} + 2\cos t\mathbf{k} \Rightarrow \mathbf{r}'(t) = 2\cos t\mathbf{i} - 2\sin t\mathbf{k} \Rightarrow \|\mathbf{r}'(t)\| = 2.$

$\mathbf{T}(t) = \cos t\mathbf{i} - \sin t\mathbf{k} \Rightarrow \mathbf{T}'(t) = -\sin t\mathbf{i} - \cos t\mathbf{k} \Rightarrow \|\mathbf{T}'(t)\| = 1.$

$\mathbf{N}(t) = -\sin t\mathbf{i} - \cos t\mathbf{k}.$

(b) $x = 2\sin t$ and $z = 2\cos t \Rightarrow x^2 + z^2 = 4$ in the plane $y = 3$.

$\mathbf{T}(\frac{\pi}{4}) = \frac{1}{\sqrt{2}}\mathbf{i} - \frac{1}{\sqrt{2}}\mathbf{k}$ and $\mathbf{N}(\frac{\pi}{4}) = -\frac{1}{\sqrt{2}}\mathbf{i} - \frac{1}{\sqrt{2}}\mathbf{k}.$

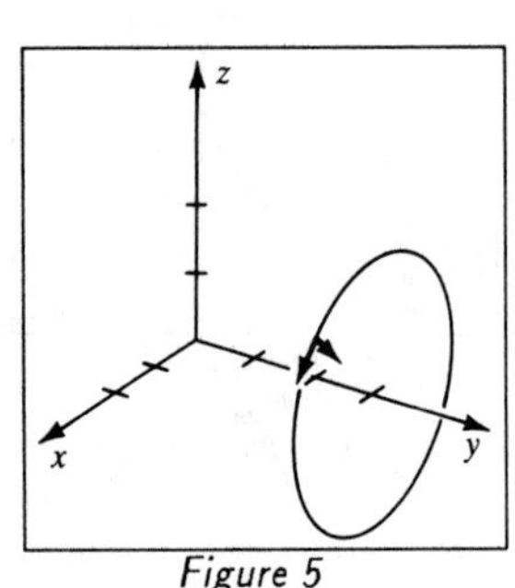

Figure 5

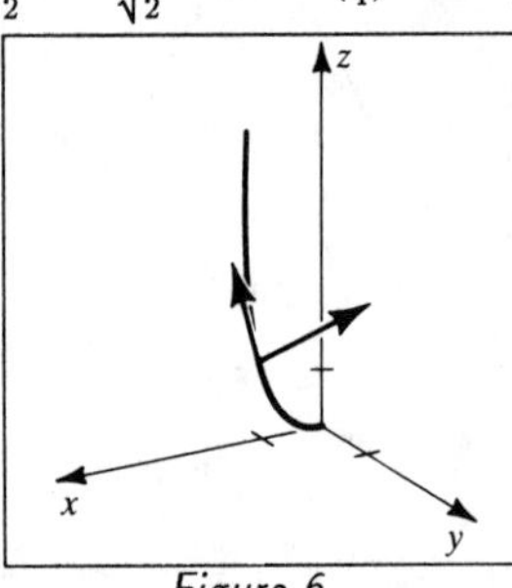

Figure 6

6 (a) $\mathbf{r}(t) = t\mathbf{i} + \frac{1}{2}t^2\mathbf{j} + t^2\mathbf{k} \Rightarrow \mathbf{r}'(t) = \mathbf{i} + t\mathbf{j} + 2t\mathbf{k} \Rightarrow \|\mathbf{r}'(t)\| = (1 + 5t^2)^{1/2}.$

$$\mathbf{T}(t) = \frac{1}{(1 + 5t^2)^{1/2}}(\mathbf{i} + t\mathbf{j} + 2t\mathbf{k}) \Rightarrow \mathbf{T}'(t) = \frac{1}{(1 + 5t^2)^{3/2}}(-5t\mathbf{i} + \mathbf{j} + 2\mathbf{k})$$

$$\Rightarrow \|\mathbf{T}'(t)\| = \frac{\sqrt{5}}{1 + 5t^2}. \quad \mathbf{N}(t) = \frac{1}{(5 + 25t^2)^{1/2}}(-5t\mathbf{i} + \mathbf{j} + 2\mathbf{k}).$$

(b) $x = \mathrm{t}$, $y = \frac{1}{2}t^2$, and $z = t^2 \Rightarrow y = \frac{1}{2}x^2$ and $z = x^2$.

$\mathbf{T}(1) = \frac{1}{\sqrt{6}}(\mathbf{i} + \mathbf{j} + 2\mathbf{k})$ and $\mathbf{N}(1) = \frac{1}{\sqrt{30}}(-5\mathbf{i} + \mathbf{j} + 2\mathbf{k}).$

Note: Exercises 7-12 use (15.18) and Exercises 13-18 use (15.19).

7 $y = 2 - x^3 \Rightarrow y' = -3x^2 \Rightarrow y'' = -6x$. At $P(1, 1)$, $y' = -3$, $y'' = -6$,

$$\text{and } K = \frac{|y''|}{\left[1 + (y')^2\right]^{3/2}} = \frac{6}{\left[1 + (-3)^2\right]^{3/2}} = \frac{6}{10^{3/2}} \approx 0.19.$$

8 $y = x^4 \Rightarrow y' = 4x^3 \Rightarrow y'' = 12x^2$.

$$\text{At } P(1, 1),\ K = \frac{12}{\left[1 + (4)^2\right]^{3/2}} = \frac{12}{17^{3/2}} \approx 0.17.$$

9 $y = e^{(x^2)} \Rightarrow y' = 2xe^{(x^2)} \Rightarrow y'' = (4x^2 + 2)\,e^{(x^2)}$.

$$\text{At } P(0, 1),\ K = \frac{2}{\left[1 + (0)^2\right]^{3/2}} = 2.$$

10 $y = \ln(x - 1) \Rightarrow y' = \frac{1}{x - 1} \Rightarrow y'' = -\frac{1}{(x - 1)^2}$.

$$\text{At } P(2, 0),\ K = \frac{1}{\left[1 + (1)^2\right]^{3/2}} = \frac{1}{2^{3/2}} \approx 0.35.$$

11 $y = \cos 2x \Rightarrow y' = -2\sin 2x \Rightarrow y'' = -4\cos 2x$.

$$\text{At } P(0, 1),\ K = \frac{4}{\left[1 + (0)^2\right]^{3/2}} = 4.$$

12 $y = \sec x \Rightarrow y' = \sec x\tan x \Rightarrow y'' = \sec^3 x + \sec x\tan^2 x$.

$$\text{At } P(\tfrac{\pi}{3}, 2),\ K = \frac{14}{\left[1 + (2\sqrt{3})^2\right]^{3/2}} = \frac{14}{13^{3/2}} \approx 0.30.$$

[13] $x = f(t) = t - 1 \Rightarrow f'(t) = 1$ and $f''(t) = 0$. $y = g(t) = \sqrt{t} \Rightarrow$

$g'(t) = \frac{1}{2}t^{-1/2}$ and $g''(t) = -\frac{1}{4}t^{-3/2}$. At $P(3, 2)$, $t = x + 1 = 4$ and

$$K = \frac{|f'(t)g''(t) - g'(t)f''(t)|}{\left[(f'(t))^2 + (g'(t))^2\right]^{3/2}} = \frac{\left|(1)(-\frac{1}{32}) - (\frac{1}{4})(0)\right|}{\left[(1)^2 + (\frac{1}{4})^2\right]^{3/2}} = \frac{2}{17^{3/2}} \approx 0.03.$$

[14] $f'(t) = 1$, $f''(t) = 0$, $g'(t) = 2t + 4$, and $g''(t) = 2$.

$$\text{At } P(1, 3),\ t = x - 1 = 0 \text{ and } K = \frac{|(1)(2) - (4)(0)|}{\left[(1)^2 + (4)^2\right]^{3/2}} = \frac{2}{17^{3/2}} \approx 0.03.$$

[15] $f'(t) = 1 - 2t$, $f''(t) = -2$, $g'(t) = -3t^2$, and $g''(t) = -6t$.

$$\text{At } P(0, 1),\ t = \sqrt[3]{1 - y} = 0 \text{ and } K = \frac{|(1)(0) - (0)(-2)|}{\left[(1)^2 + (0)^2\right]^{3/2}} = \frac{0}{1} = 0.$$

[16] $f'(t) = 1 - \cos t$, $f''(t) = \sin t$, $g'(t) = \sin t$, and $g''(t) = \cos t$.

$$\text{At } P(\tfrac{\pi}{2} - 1, 1),\ t = \cos^{-1}(1 - y) = \tfrac{\pi}{2} \text{ and } K = \frac{|(1)(0) - (1)(1)|}{\left[(1)^2 + (1)^2\right]^{3/2}} = \frac{1}{2^{3/2}} \approx 0.35.$$

[17] $f'(t) = 2\cos t$, $f''(t) = -2\sin t$, $g'(t) = -3\sin t$, and $g''(t) = -3\cos t$. At

$$P(1, \tfrac{3}{2}\sqrt{3}),\ t = \sin^{-1}(\tfrac{1}{2}x) = \tfrac{\pi}{6} \text{ and } K = \frac{\left|(\sqrt{3})(-\frac{3}{2}\sqrt{3}) - (-\frac{3}{2})(-1)\right|}{\left[(\sqrt{3})^2 + (-\frac{3}{2})^2\right]^{3/2}} = \frac{48}{21^{3/2}} \approx 0.50.$$

[18] $f'(t) = -3\cos^2 t \sin t$, $f''(t) = 3(2\cos t \sin^2 t - \cos^3 t)$,

$g'(t) = 3\sin^2 t \cos t$, and $g''(t) = 3(2\sin t \cos^2 t - \sin^3 t)$. At $P(\frac{1}{4}\sqrt{2}, \frac{1}{4}\sqrt{2})$,

$t = \cos^{-1}(\sqrt[3]{x}) = \cos^{-1}\left(\sqrt[3]{\tfrac{1}{4}\sqrt{2}}\right) = \cos^{-1}\left(\sqrt[3]{\tfrac{1}{8}(\sqrt{2})^3}\right) = \cos^{-1}(\tfrac{1}{2}\sqrt{2}) = \tfrac{\pi}{4}$ and

$$K = \{\text{with } a = \tfrac{3}{4}\sqrt{2}\}\ \frac{|(-a)(a) - (a)(a)|}{\left[(-a)^2 + (a)^2\right]^{3/2}} = \frac{\left|-\frac{9}{4}\right|}{(\frac{9}{4})^{3/2}} = \frac{2}{3}.$$

[19] (a) $y' = \cos x$ and $y'' = -\sin x$. At $P(\frac{\pi}{2}, 1)$, $K = \dfrac{1}{\left[1 + 0^2\right]^{3/2}} = 1 \Rightarrow \rho = \dfrac{1}{K} = 1$.

(b) Since $y' = 0$ and the graph is concave down at $x = \frac{\pi}{2}$, it follows that the center of curvature is on a vertical line 1 unit below P, that is, $(\frac{\pi}{2}, 0)$.

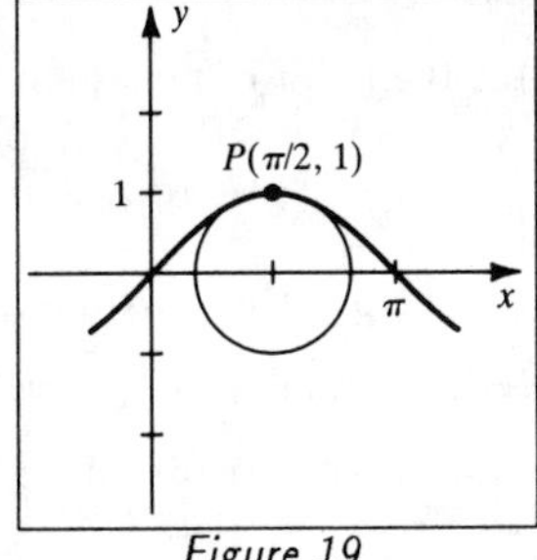

Figure 19

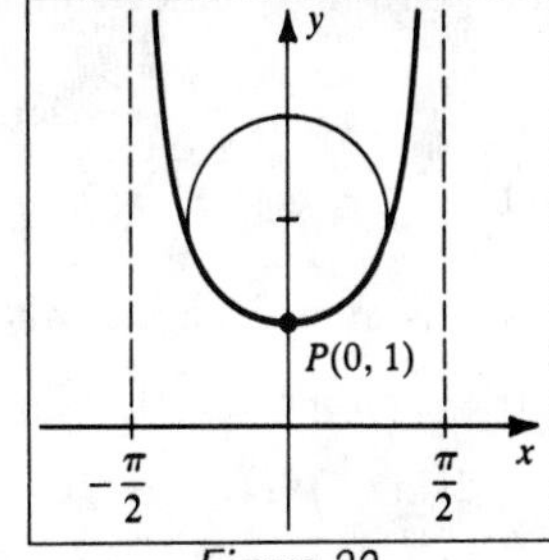

Figure 20

[20] (a) $y' = \sec x \tan x$ and $y'' = \sec^3 x - \sec x \tan^2 x$.

$$\text{At } P(0, 1),\ K = \frac{1}{\left[1 + 0^2\right]^{3/2}} = 1 \Rightarrow \rho = \frac{1}{K} = 1.$$

(b) Since $y' = 0$ and the graph is concave up at $x = 0$, it follows that the center of curvature is on a vertical line 1 unit above P, that is, $(0, 2)$. *Note*: The graph of the secant is actually *above* the graph of the circle of curvature on $0 < |x| < 1$!

[21] (a) $y' = e^x$ and $y'' = e^x$. At $P(0, 1)$, $K = \dfrac{1}{\left[1 + 1^2\right]^{3/2}} = \dfrac{1}{2^{3/2}} \Rightarrow \rho = \dfrac{1}{K} = 2\sqrt{2}$.

(b) Since $y' = 1$ at $x = 0$, it follows that the center of curvature is on the line with slope -1 passing through $(0, 1)$, or, $y = -x + 1$. The center C is a distance of $2\sqrt{2}$ from $(0, 1)$ on the concave side. If a is the side of the isosceles right triangle with hypotenuse $CP = 2\sqrt{2}$, then $a^2 + a^2 = (2\sqrt{2})^2 \Rightarrow a = 2$ and hence, the center of curvature is $(-2, 3)$.

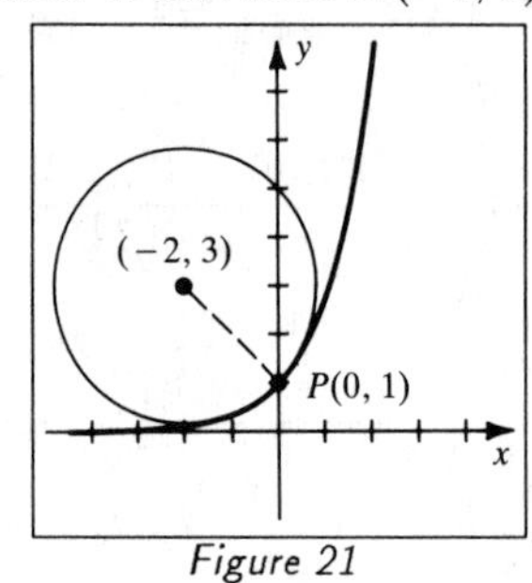

Figure 21

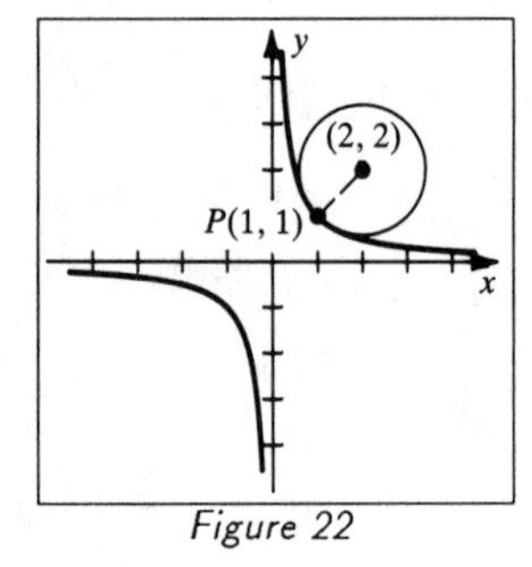

Figure 22

[22] (a) $y' = -\dfrac{1}{x^2}$ and $y'' = \dfrac{2}{x^3}$. At $P(1, 1)$, $K = \dfrac{2}{\left[1 + (-1)^2\right]^{3/2}} = \dfrac{1}{\sqrt{2}} \Rightarrow$
$\rho = 1/K = \sqrt{2}$.

(b) Since $y' = -1$ at $x = 1$, it follows that the center of curvature is on the line with slope 1 passing through $(1, 1)$, or, $y = x$. The center is a distance of $\sqrt{2}$ from $(1, 1)$ on the concave side. Thus, the center of curvature is $(2, 2)$.

[23] $y' = -e^{-x}$, $y'' = e^{-x} \Rightarrow K(x) = \dfrac{e^{-x}}{\left[1 + e^{-2x}\right]^{3/2}}$.

$$K'(x) = \frac{(1 + e^{-2x})^{3/2}(-e^{-x}) - (e^{-x})(\frac{3}{2})(1 + e^{-2x})^{1/2}(-2e^{-2x})}{(1 + e^{-2x})^3} = 0 \Rightarrow$$

$(-e^{-x})(1 + e^{-2x})^{1/2}\left[(1 + e^{-2x}) - 3e^{-2x}\right] = 0 \Rightarrow 1 - 2e^{-2x} = 0 \Rightarrow x = \ln\sqrt{2}$.

Since $K'(x) > 0$ if $x < \ln\sqrt{2}$ and $K'(x) < 0$ if $x > \ln\sqrt{2}$,

$(\ln\sqrt{2}, 1/\sqrt{2})$ is the point of maximum curvature.

[24] $y' = \sinh x$, $y'' = \cosh x \Rightarrow K(x) = \dfrac{\cosh x}{\left[1 + \sinh^2 x\right]^{3/2}} = \dfrac{1}{\cosh^2 x}$. $K(x)$ is maximum when $\cosh x$ is minimum, i.e., when $x = 0$. $(0, 1)$ is the point of maximum curvature.

[25] Parametric equations for the ellipse are $x = f(t) = 2\cos t$ and $y = g(t) = 3\sin t$.

$f'(t) = -2\sin t$, $f''(t) = -2\cos t$, $g'(t) = 3\cos t$, and $g''(t) = -3\sin t$.

$K(t) = \dfrac{\left|6\sin^2 t + 6\cos^2 t\right|}{\left[4\sin^2 t + 9\cos^2 t\right]^{3/2}} = \dfrac{6}{\left[4 + 5\cos^2 t\right]^{3/2}}$. $K(t)$ is a maximum when

$\cos^2 t = 0$, i.e., when $t = \frac{\pi}{2}, \frac{3\pi}{2}$. $(0, \pm 3)$ are the points of maximum curvature.

[26] Parametric equations for the right branch of the hyperbola are

$x = f(t) = 2\cosh t$ and $y = g(t) = 3\sinh t$.

$f'(t) = 2\sinh t$, $f''(t) = 2\cosh t$, $g'(t) = 3\cosh t$, and $g''(t) = 3\sinh t$. (cont.)

$$K(t) = \frac{|6\sinh^2 t - 6\cosh^2 t|}{\left[4\sinh^2 t + 9\cosh^2 t\right]^{3/2}} = \frac{6}{\left[4\sinh^2 t + 9\cosh^2 t\right]^{3/2}}.$$

$K(t)$ is maximum when $\sinh^2 t$ and $\cosh^2 t$ are minimum, i.e., when $t = 0$. Thus, K is maximum at $(2, 0)$. By symmetry, $(-2, 0)$ is also a point of maximum curvature.

27 $y' = \frac{1}{x}$, $y'' = -\frac{1}{x^2} \Rightarrow K(x) = \dfrac{1}{x^2\left[1 + \left(\frac{1}{x}\right)^2\right]^{3/2}} = \dfrac{x}{\left[1 + x^2\right]^{3/2}}.$

$$K'(x) = \frac{(1 + x^2)^{3/2}(1) - (x)(\frac{3}{2})(1 + x^2)^{1/2}(2x)}{(1 + x^2)^3} = 0 \Rightarrow$$

$(1 + x^2)^{1/2}\left[(1 + x^2) - 3x^2\right] = 0 \Rightarrow 1 - 2x^2 = 0 \Rightarrow x = \frac{1}{\sqrt{2}}$, $(x > 0)$.

Since $K'(x) > 0$ when $x < \frac{1}{\sqrt{2}}$ and $K'(x) < 0$ when $x > \frac{1}{\sqrt{2}}$, $\left(\frac{1}{\sqrt{2}}, -\frac{1}{2}\ln 2\right)$ is the point of maximum curvature.

28 $y' = \cos x$, $y'' = -\sin x \Rightarrow K(x) = \dfrac{|\sin x|}{\left[1 + \cos^2 x\right]^{3/2}}.$

$K(x)$ is maximum when $\sin x$ is maximum and $\cos x$ is minimum, i.e., when $x = \frac{\pi}{2} + \pi n$. $(\frac{\pi}{2} + \pi n, (-1)^n)$ are the points of maximum curvature.

Note: In Exercises 29-32, the curvature is 0 only when $y'' = 0$ from (15.18).

29 $y = x^4 - 12x^2 \Rightarrow y' = 4x^3 - 24x \Rightarrow y'' = 12x^2 - 24 = 0 \Rightarrow x = \pm\sqrt{2}$. $(\pm\sqrt{2}, -20)$

30 $y = \tan x \Rightarrow y' = \sec^2 x \Rightarrow y'' = 2\sec^2 x \tan x = 0 \Rightarrow x = \pi n$. $(\pi n, 0)$

31 $y = \sinh x \Rightarrow y' = \cosh x \Rightarrow y'' = \sinh x = 0 \Rightarrow x = 0$. $(0, 0)$

32 $y = e^{-x^2} \Rightarrow y' = -2xe^{-x^2} \Rightarrow y'' = (4x^2 - 2)\,e^{-x^2} = 0 \Rightarrow x = \pm\frac{1}{\sqrt{2}}$. $\left(\pm\frac{1}{\sqrt{2}}, e^{-1/2}\right)$

33 Since $r = f(\theta)$, if we let $x = r\cos\theta$ and $y = r\sin\theta$, then x and y are also functions of θ. $x' = -r\sin\theta + r'\cos\theta$. $(x')^2 = r^2\sin^2\theta - 2rr'\sin\theta\cos\theta + (r')^2\cos^2\theta$.

$x'' = -r\cos\theta - r'\sin\theta - r'\sin\theta + r''\cos\theta$.

$y' = r\cos\theta + r'\sin\theta$. $(y')^2 = r^2\cos^2\theta + 2rr'\sin\theta\cos\theta + (r')^2\sin^2\theta$.

$y'' = -r\sin\theta + r'\cos\theta + r'\cos\theta + r''\sin\theta$.

Simplifying $K = \dfrac{|x'y'' - y'x''|}{\left[(x')^2 + (y')^2\right]^{3/2}}$ gives us $\dfrac{\left|2(r')^2 - rr'' + r^2\right|}{\left[(r')^2 + r^2\right]^{3/2}}.$

34 $r = a(1 - \cos\theta) \Rightarrow r' = a\sin\theta \Rightarrow r'' = a\cos\theta$.

$$K = \frac{\left|2(a^2\sin^2\theta) - a^2(\cos\theta - \cos^2\theta) + a^2(1 - \cos\theta)^2\right|}{\left[a^2\sin^2\theta + a^2(1 - \cos\theta)^2\right]^{3/2}} =$$

$$\frac{|3a^2 - 3a^2\cos\theta|}{(2a^2 - 2a^2\cos\theta)^{3/2}} = \frac{3a^2|1 - \cos\theta|}{(2a^2)^{3/2}(1 - \cos\theta)^{3/2}} = \frac{3}{2\sqrt{2}\,a(1 - \cos\theta)^{1/2}}.$$

[35] $r = \sin 2\theta \Rightarrow r' = 2\cos 2\theta \Rightarrow r'' = -4\sin 2\theta$.

$$K = \frac{\left|8\cos^2 2\theta + 4\sin^2 2\theta + \sin^2 2\theta\right|}{\left[4\cos^2 2\theta + \sin^2 2\theta\right]^{3/2}} = \frac{8(\cos^2 2\theta + \sin^2 2\theta) - 3\sin^2\theta}{\left[(\cos^2 2\theta + \sin^2 2\theta) + 3\cos^2 2\theta\right]^{3/2}}$$

$$= \frac{8 - 3\sin^2 2\theta}{(1 + 3\cos^2 2\theta)^{3/2}}.$$

[36] $r = e^{a\theta} \Rightarrow r' = ae^{a\theta} \Rightarrow r'' = a^2e^{a\theta}$.

$$K = \frac{\left|2a^2e^{2a\theta} - a^2e^{2a\theta} + e^{2a\theta}\right|}{\left[a^2e^{2a\theta} + e^{2a\theta}\right]^{3/2}} = \frac{e^{2a\theta}(1 + a^2)}{e^{3a\theta}(1 + a^2)^{3/2}} = \frac{1}{e^{a\theta}(1 + a^2)^{1/2}}.$$

[37] The equation of the circle of curvature is

[1] $(x - h)^2 + (y - k)^2 = r^2 = \left(\frac{1}{K}\right)^2 = \frac{\left[1 + (y')^2\right]^3}{(y'')^2}$ since $K = \frac{|y''|}{\left[1 + (y')^2\right]^{3/2}}$.

The center of curvature must be on the normal line at the point $P(x, y)$, i.e.,

[2] $y - k = -\frac{1}{y'}(x - h)$. We must solve these two equations simultaneously for h and k. Squaring [2] results in [3] $(x - h)^2 = (y')^2(y - k)^2$. Substituting [3] into [1] yields $\left[1 + (y')^2\right](y - k)^2 = \frac{\left[1 + (y')^2\right]^3}{(y'')^2} \Rightarrow k = y \pm \frac{\left[1 + (y')^2\right]}{y''}$.

The positive solution is the correct one since if the graph of f is concave up $(y'' > 0)$, $k > y$ and if the graph of f is concave down $(y'' < 0)$, $k < y$. Substituting $(y - k)^2 = \frac{(x - h)^2}{(y')^2}$ into [1] yields $\left[1 + (y')^{-2}\right](x - h)^2 = \frac{\left[1 + (y')^2\right]^3}{(y'')^2} \Rightarrow$

$(x - h)^2 = \frac{\left[1 + (y')^2\right]^3}{(y'')^2} \cdot \frac{(y')^2}{\left[1 + (y')^2\right]} \Rightarrow h = x \pm \frac{y'\left[1 + (y')^2\right]}{y''}$. The negative solution is correct in this case. To verify this, there are four cases to consider, depending on the signs of y' and y''. For example, if $y' < 0$ and $y'' > 0$, then f is decreasing and its graph is concave up—so $h > x$ is required and the negative solution is necessary. The other three cases are done in a similar manner.

[38] At $P(1, 1)$, $y' = -3$ and $y'' = -6$. $h = x - \frac{y'[1 + (y')^2]}{y''} =$

$$1 - \frac{-3[1 + 9]}{-6} = -4 \text{ and } k = y + \frac{[1 + (y')^2]}{y''} = 1 + \frac{[1 + 9]}{-6} = -\tfrac{2}{3}.$$

[39] At $P(1, 1)$, $y' = 4$ and $y'' = 12$.

$$h = 1 - \frac{4[1 + 16]}{12} = -\tfrac{14}{3} \text{ and } k = 1 + \frac{[1 + 16]}{12} = \tfrac{29}{12}.$$

[40] At $P(0, 1)$, $y' = 0$ and $y'' = 2$. $h = 0$ and $k = 1 + \frac{[1 + 0]}{2} = \frac{3}{2}$.

[41] At $P(2, 0)$, $y' = 1$ and $y'' = -1$.

$$h = 2 - \frac{1[1 + 1]}{-1} = 4 \text{ and } k = 0 + \frac{[1 + 1]}{-1} = -2.$$

[42] At $P(0, 1)$, $y' = 0$ and $y'' = -4$. $h = 0$ and $k = 1 + \frac{[1 + 0]}{-4} = \frac{3}{4}$.

[43] We need to find the center of curvature for $y = -\frac{1}{27}x^3$ at $P(3, -1)$.
If the circular arc has this as its center, the curvature will be continuous at $x = 3$.
$y' = -\frac{1}{9}x^2$ and $y'' = -\frac{2}{9}x$ when $x = 3$, $y' = -1$ and $y'' = -\frac{2}{3}$. From Exercise 37,

$$h = 3 - \frac{(-1)\left[1 + (-1)^2\right]}{-\frac{2}{3}} = 0 \text{ and } k = -1 + \frac{\left[1 + (-1)^2\right]}{-\frac{2}{3}} = -4.$$

[44] $y' = m \Rightarrow y'' = 0 \Rightarrow K(x) = \dfrac{0}{\left[1 + m^2\right]^{3/2}} = 0.$

[45] Without loss of generality, let $y = ax^2 + bx + c$. Thus, $y' = 2ax + b$ and $y'' = 2a$.

$K(x) = \dfrac{|2a|}{\left[1 + (2ax + b)^2\right]^{3/2}}$. $K(x)$ is maximum when $2ax + b = 0$, or $x = -\dfrac{b}{2a}$,

i.e., when x is the x-coordinate of the vertex of the parabola.

[46] Without loss of generality, let $\dfrac{x^2}{a^2} + \dfrac{y^2}{b^2} = 1$ with $a > b > 0$.

Parametric equations for the ellipse are $x = f(t) = a \cos t$ and $y = g(t) = b \sin t$.

Hence $f'(t) = -a \sin t$, $f''(t) = -a \cos t$, $g'(t) = b \cos t$, $g''(t) = -b \sin t$, and

$$K(t) = \frac{\left|ab \sin^2 t + ab \cos^2 t\right|}{\left[a^2 \sin^2 t + b^2 \cos^2 t\right]^{3/2}} = \frac{|ab|}{\left[(a^2 - b^2)\sin^2 t + b^2\right]^{3/2}}.$$

$K(t)$ is maximum when $\sin^2 t = 0$ or when $t = 0, \pi$, i.e., $(\pm a, 0)$.

$K(t)$ is minimum when $\sin^2 t = 1$, or when $t = \frac{\pi}{2}, \frac{3\pi}{2}$, i.e., $(0, \pm b)$.

Thus, since $a > b$, maximum curvature occurs at the ends of the major axis and

minimum curvature occurs at the ends of the minor axis.

[47] Without loss of generality, let $\dfrac{x^2}{a^2} - \dfrac{y^2}{b^2} = 1$ with $a, b > 0$. Parametric equations for

the hyperbola are $x = f(t) = a \cosh t$ and $y = g(t) = b \sinh t$.

Hence $f'(t) = a \sinh t$, $f''(t) = a \cosh t$, $g'(t) = b \cosh t$, $g''(t) = b \sinh t$, and

$$K(t) = \frac{\left|ab \sinh^2 t - ab \cosh^2 t\right|}{\left[a^2 \sinh^2 t + b^2 \cosh^2 t\right]^{3/2}} = \frac{ab}{\left[(a^2 + b^2)\sinh^2 t + b^2\right]^{3/2}}.$$

$K(t)$ is maximum when $\sinh^2 t = 0$, or $t = 0$. Thus, maximum curvature occurs at

$(a, 0)$ and by symmetry at $(-a, 0)$, that is, at the ends of the transverse axis.

[48] Let the curve have an arc length parametrization and position vector $\mathbf{r}(s)$.

From the text, $\mathbf{r}'(s) = \mathbf{T}(s) = \cos\theta\, \mathbf{i} + \sin\theta\, \mathbf{j}$.

Case 1: If $\dfrac{d\theta}{ds} = a$, where a is a nonzero constant, it follows that $\theta = as + b$, where b

is a constant. Thus, $\mathbf{r}'(s) = \cos(as + b)\mathbf{i} + \sin(as + b)\mathbf{j}$ and hence

$\mathbf{r}(s) = \frac{1}{a}\sin(as + b)\mathbf{i} - \frac{1}{a}\cos(as + b)\mathbf{j} + \mathbf{c}$, where $\mathbf{c}$ is a constant vector.

$\left\|\mathbf{r}(s) - \mathbf{c}\right\| = \frac{1}{|a|}$, $a \neq 0$. This is an equation of a circular arc with center at the

terminal point of $\mathbf{c}$ and radius $\frac{1}{|a|}$.

Case 2: If $a = 0$, then θ is constant and so $\mathbf{r}'(s)$ is a constant vector.

Hence, $\mathbf{r}(s)$ does not change direction and $\mathbf{r}(s)$ must trace out a straight line.

[49] $f'(t) = 4$ and $g'(t) = 3$. $s = \displaystyle\int_0^t \sqrt{(4)^2 + (3)^2}\, du = 5t \Rightarrow t = \frac{1}{5}s$.

Thus, $x = \frac{4}{5}s - 3$ and $y = \frac{3}{5}s + 5$; $s \geq 0$.

[50] $f'(t) = 6t$ and $g'(t) = 6t^2$. $s = \int_0^t \sqrt{36u^2 + 36u^4}\, du = \int_0^t 6u\sqrt{1 + u^2}\, du =$

$\left[2(1 + u^2)^{3/2}\right]_0^t = 2(1 + t^2)^{3/2} - 2 \Rightarrow t = \left[\left(\frac{s+2}{2}\right)^{2/3} - 1\right]^{1/2}$.

Thus, $x = 3\left[\left(\frac{s+2}{2}\right)^{2/3} - 1\right]$ and $y = 2\left[\left(\frac{s+2}{2}\right)^{2/3} - 1\right]^{3/2}$; $s \geq 0$.

[51] $f'(t) = -4\sin t$ and $g'(t) = 4\cos t$. $s = \int_0^t \sqrt{16\sin^2 u + 16\cos^2 u}\, du = \int_0^t 4\, du = 4t$

$\Rightarrow t = \frac{1}{4}s$. Thus, $x = 4\cos\frac{1}{4}s$ and $y = 4\sin\frac{1}{4}s$; $0 \leq s \leq 8\pi$.

[52] $f'(t) = e^t(\cos t - \sin t)$ and $g'(t) = e^t(\sin t + \cos t)$.

$$s = \int_0^t \sqrt{e^{2u}(\cos^2 u - 2\cos u \sin u + \sin^2 u) + e^{2u}(\sin^2 u + 2\sin u \cos u + \cos^2 u)}\, du$$

$= \int_0^t \sqrt{2}\, e^u\, du = \sqrt{2}(e^t - 1) \Rightarrow t = \ln\left(1 + \frac{s}{\sqrt{2}}\right)$. Thus,

$x = \left(1 + \frac{s}{\sqrt{2}}\right)\cos\left[\ln\left(1 + \frac{s}{\sqrt{2}}\right)\right]$ and $y = \left(1 + \frac{s}{\sqrt{2}}\right)\sin\left[\ln\left(1 + \frac{s}{\sqrt{2}}\right)\right]$; $s \geq 0$.

[53] Let $h(s) = \int_0^s k(t)\, dt$, $x = f(s) = \int_0^s \cos h(t)\, dt$, and $y = g(s) = \int_0^s \sin h(t)\, dt$.

Let C be the curve defined parametrically by $x = f(s)$ and $y = g(s)$. We will show that the curvature of C is given by $k(s)$. $x' = \cos[h(s)]$, $x'' = -\sin[h(s)]\, h'(s)$, $y' = \sin[h(s)]$, and $y'' = \cos[h(s)]\, h'(s)$. Using the fact that $h'(s) = k(s)$, we have

$$K = \frac{|x'y'' - x''y'|}{\left[(x')^2 + (y')^2\right]^{3/2}} = \frac{\left|k(s)\cos^2[h(s)] + k(s)\sin^2[h(s)]\right|}{\left[\cos^2[h(s)] + \sin^2[h(s)]\right]^{3/2}} =$$

$|k(s)| = k(s)$ since k is nonnegative.

[54] (1) Let $k = K$, a nonzero constant. Using Exercise 53, we let $h(s) = \int_0^s K\, dt = Ks$.

Parametric equations for C are $x = \int_0^s \cos(Kt)\, dt = \left[\frac{1}{K}\sin(Kt)\right]_0^s = \frac{1}{K}\sin Ks$

and $y = \int_0^s \sin(Kt)\, dt = \left[-\frac{1}{K}\cos(Kt)\right]_0^s = -\frac{1}{K}\cos(Ks) + \frac{1}{K}$.

Thus, $Kx = \sin(Ks)$ and $Ky - 1 = -\cos(Ks)$ give us $(Kx)^2 + (Ky - 1)^2 = 1$, which is a circle.

(2) If $k = K = 0$, then $h(s) = \int_0^s 0\, dt = 0$,

$x = \int_0^s \cos 0\, dt = s$, and $y = \int_0^s \sin 0\, dt = 0$, which is a line.

Exercises 15.5

Note: In Exercises 1-8, we use (15.22) for $a_{\mathbf{T}}$, (15.23) for $a_{\mathbf{N}}$, and (15.25) for K.

You could also use (15.24) for finding $a_{\mathbf{N}}$.

[1] $\mathbf{r}(t) = t^2\mathbf{i} + (3t + 2)\mathbf{j}$, $\mathbf{r}'(t) = 2t\mathbf{i} + 3\mathbf{j}$, $\mathbf{r}''(t) = 2\mathbf{i}$, and $\|\mathbf{r}'(t)\| = \sqrt{4t^2 + 9}$.

Using (15.22), $a_{\mathrm{T}} = \dfrac{\mathbf{r}'(t)\cdot\mathbf{r}''(t)}{\|\mathbf{r}'(t)\|} = \dfrac{4t}{(4t^2+9)^{1/2}}$.

Using (15.23), $a_{\mathrm{N}} = \dfrac{\|\mathbf{r}'(t)\times\mathbf{r}''(t)\|}{\|\mathbf{r}'(t)\|} = \dfrac{\|-6\mathbf{k}\|}{(4t^2+9)^{1/2}} = \dfrac{6}{(4t^2+9)^{1/2}}$.

Using (15.25), $K = a_{\mathrm{N}}\dfrac{1}{\|\mathbf{r}'(t)\|^2} = \dfrac{6}{(4t^2+9)^{1/2}}\cdot\dfrac{1}{4t^2+9} = \dfrac{6}{(4t^2+9)^{3/2}}$.

[2] $\mathbf{r}(t) = (2t^2 - 1)\mathbf{i} + 5t\mathbf{j}$, $\mathbf{r}'(t) = 4t\mathbf{i} + 5\mathbf{j}$, $\mathbf{r}''(t) = 4\mathbf{i}$, and $\|\mathbf{r}'(t)\| = \sqrt{16t^2 + 25}$.

Using (15.22), $a_{\mathrm{T}} = \dfrac{\mathbf{r}'(t)\cdot\mathbf{r}''(t)}{\|\mathbf{r}'(t)\|} = \dfrac{16t}{(16t^2+25)^{1/2}}$.

Using (15.23), $a_{\mathrm{N}} = \dfrac{\|\mathbf{r}'(t)\times\mathbf{r}''(t)\|}{\|\mathbf{r}'(t)\|} = \dfrac{\|-20\mathbf{k}\|}{(16t^2+25)^{1/2}} = \dfrac{20}{(16t^2+25)^{1/2}}$.

Using (15.25), $K = a_{\mathrm{N}}\dfrac{1}{\|\mathbf{r}'(t)\|^2} = \dfrac{20}{(16t^2+25)^{1/2}}\cdot\dfrac{1}{16t^2+25} = \dfrac{20}{(16t^2+25)^{3/2}}$.

[3] $\mathbf{r}(t) = 3t\mathbf{i} + t^3\mathbf{j} + 3t^2\mathbf{k}$, $\mathbf{r}'(t) = 3\mathbf{i} + 3t^2\mathbf{j} + 6t\mathbf{k}$, $\mathbf{r}''(t) = 6t\mathbf{j} + 6\mathbf{k}$, and

$\|\mathbf{r}'(t)\| = 3\sqrt{t^4 + 4t^2 + 1}$. $a_{\mathrm{T}} = \dfrac{18t^3 + 36t}{3(t^4+4t^2+1)^{1/2}} = \dfrac{6t^3+12t}{(t^4+4t^2+1)^{1/2}}$.

$a_{\mathrm{N}} = \dfrac{\|-18t^2\mathbf{i} - 18\mathbf{j} + 18t\mathbf{k}\|}{3(t^4+4t^2+1)^{1/2}} = \dfrac{6(t^4+t^2+1)^{1/2}}{(t^4+4t^2+1)^{1/2}}$.

$K = a_{\mathrm{N}}\cdot\dfrac{1}{9(t^4+4t^2+1)} = \dfrac{2(t^4+t^2+1)^{1/2}}{3(t^4+4t^2+1)^{3/2}}$.

[4] $\mathbf{r}(t) = 4t\mathbf{i} + t^2\mathbf{j} + 2t^2\mathbf{k}$, $\mathbf{r}'(t) = 4\mathbf{i} + 2t\mathbf{j} + 4t\mathbf{k}$, $\mathbf{r}''(t) = 2\mathbf{j} + 4\mathbf{k}$, and

$\|\mathbf{r}'(t)\| = 2\sqrt{4 + 5t^2}$. $a_{\mathrm{T}} = \dfrac{20t}{2(4+5t^2)^{1/2}} = \dfrac{10t}{(4+5t^2)^{1/2}}$.

$a_{\mathrm{N}} = \dfrac{\|-16\mathbf{j} + 8\mathbf{k}\|}{2(4+5t^2)^{1/2}} = \dfrac{4\sqrt{5}}{(4+5t^2)^{1/2}}$. $K = a_{\mathrm{N}}\cdot\dfrac{1}{4(4+5t^2)} = \dfrac{\sqrt{5}}{(4+5t^2)^{3/2}}$.

[5] $\mathbf{r}(t) = t(\cos t\,\mathbf{i} + \sin t\,\mathbf{j})$, $\mathbf{r}'(t) = (\cos t - t\sin t)\mathbf{i} + (\sin t + t\cos t)\mathbf{j}$,

$\mathbf{r}''(t) = (-2\sin t - t\cos t)\mathbf{i} + (2\cos t - t\sin t)\mathbf{j}$, and $\|\mathbf{r}'(t)\| = \sqrt{1 + t^2}$.

$a_{\mathrm{T}} = \dfrac{t}{(1+t^2)^{1/2}}$. $a_{\mathrm{N}} = \dfrac{\|(2+t^2)\mathbf{k}\|}{(1+t^2)^{1/2}} = \dfrac{2+t^2}{(1+t^2)^{1/2}}$.

$K = a_{\mathrm{N}}\cdot\dfrac{1}{1+t^2} = \dfrac{2+t^2}{(1+t^2)^{3/2}}$.

[6] $\mathbf{r}(t) = \cosh t\,\mathbf{i} + \sinh t\,\mathbf{j}$, $\mathbf{r}'(t) = \sinh t\,\mathbf{i} + \cosh t\,\mathbf{j}$, $\mathbf{r}''(t) = \cosh t\,\mathbf{i} + \sinh t\,\mathbf{j}$, and

$\|\mathbf{r}'(t)\| = \sqrt{\sinh^2 t + \cosh^2 t}$. $a_{\mathrm{T}} = \dfrac{2\cosh t\sinh t}{(\sinh^2 t + \cosh^2 t)^{1/2}}$. $a_{\mathrm{N}} = \dfrac{\|-\mathbf{k}\|}{(\sinh^2 t + \cosh^2 t)^{1/2}}$

$= \dfrac{1}{(\sinh^2 t + \cosh^2 t)^{1/2}}$. $K = a_{\mathrm{N}}\cdot\dfrac{1}{\sinh^2 t + \cosh^2 t} = \dfrac{1}{(\sinh^2 t + \cosh^2 t)^{3/2}}$.

[7] $\mathbf{r}(t) = 4\cos t\,\mathbf{i} + 9\sin t\,\mathbf{j} + t\,\mathbf{k}$, $\mathbf{r}'(t) = -4\sin t\,\mathbf{i} + 9\cos t\,\mathbf{j} + \mathbf{k}$,

$\mathbf{r}''(t) = -4\cos t\,\mathbf{i} - 9\sin t\,\mathbf{j}$, and $\|\mathbf{r}'(t)\| = \sqrt{16\sin^2 t + 81\cos^2 t + 1}$.

$$a_{\mathrm{T}} = \frac{-65\sin t\cos t}{(16\sin^2 t + 81\cos^2 t + 1)^{1/2}}.$$

$$a_{\mathrm{N}} = \frac{\|9\sin t\,\mathbf{i} - 4\cos t\,\mathbf{j} + 36\mathbf{k}\|}{(16\sin^2 t + 81\cos^2 t + 1)^{1/2}} = \frac{(81\sin^2 t + 16\cos^2 t + 1296)^{1/2}}{(16\sin^2 t + 81\cos^2 t + 1)^{1/2}}.$$

$$K = a_{\mathrm{N}} \cdot \frac{1}{16\sin^2 t + 81\cos^2 t + 1} = \frac{(81\sin^2 t + 16\cos^2 t + 1296)^{1/2}}{(16\sin^2 t + 81\cos^2 t + 1)^{3/2}}.$$

[8] $\mathbf{r}(t) = e^t(\sin t\,\mathbf{i} + \cos t\,\mathbf{j} + \mathbf{k})$, $\mathbf{r}'(t) = e^t(\sin t + \cos t)\mathbf{i} + e^t(\cos t - \sin t)\mathbf{j} + e^t\mathbf{k}$,

$\mathbf{r}''(t) = 2e^t\cos t\,\mathbf{i} - 2e^t\sin t\,\mathbf{j} + e^t\mathbf{k}$, and $\|\mathbf{r}'(t)\| = \sqrt{3}\,e^t$. $a_{\mathrm{T}} = \dfrac{3e^{2t}}{\sqrt{3}\,e^t} = \sqrt{3}\,e^t$.

$\mathbf{r}'(t) \times \mathbf{r}''(t) = e^{2t}\big[(\sin t + \cos t)\mathbf{i} + (\cos t - \sin t)\mathbf{j} - 2\mathbf{k}\big]$ and

$$a_{\mathrm{N}} = \frac{\sqrt{6}\,e^{2t}}{\sqrt{3}\,e^t} = \sqrt{2}\,e^t. \quad K = a_{\mathrm{N}} \cdot \frac{1}{3e^{2t}} = \frac{\sqrt{2}}{3}e^{-t}.$$

[9] Let $\mathbf{r}(t) = f(t)\mathbf{i} + g(t)\mathbf{j}$, where $g(t) = [f(t)]^2$ $\{y = x^2\}$ and $f'(t) = 3$.

Thus, $f(t) = 3t + C$. Since $f(1) = 1$, $f(t) = 3t - 2$ and $g(t) = (3t - 2)^2$.

Hence $\mathbf{r}'(t) = 3\mathbf{i} + 6(3t - 2)\mathbf{j}$, $\mathbf{r}''(t) = 18\mathbf{j}$, and $\|\mathbf{r}'(t)\| = \sqrt{9 + 36(3t - 2)^2}$.

Let $t = 1$. $\mathbf{r}'(1) = 3\mathbf{i} + 6\mathbf{j}$, $\mathbf{r}''(1) = 18\mathbf{j}$, and $\|\mathbf{r}'(1)\| = \sqrt{45}$.

$$a_{\mathrm{T}} = \frac{108}{\sqrt{45}} = \frac{36}{\sqrt{5}} \approx 16.10. \quad a_{\mathrm{N}} = \frac{\|54\,\mathbf{k}\|}{\sqrt{45}} = \frac{54}{\sqrt{45}} = \frac{18}{\sqrt{5}} \approx 8.05.$$

[10] Let $\mathbf{r}(t) = f(t)\mathbf{i} + g(t)\mathbf{j}$, where $g(t) = 2[f(t)]^3 - f(t)$ $\{y = 2x^3 - x\}$ and $f'(t) = 3$.

Thus, $f(t) = 3t + C$. Since $f(1) = 1$, $f(t) = 3t - 2$ and

$g(t) = 2(3t - 2)^3 - (3t - 2)$. Hence, $\mathbf{r}'(t) = 3\mathbf{i} + \big[18(3t - 2)^2 - 3\big]\mathbf{j}$,

$\mathbf{r}''(t) = 108(3t - 2)\mathbf{j}$, and $\|\mathbf{r}'(t)\| = \sqrt{9 + \big[18(3t - 2)^2 - 3\big]^2}$.

Let $t = 1$. $\mathbf{r}'(1) = 3\mathbf{i} + 15\mathbf{j}$, $\mathbf{r}''(1) = 108\mathbf{j}$, and $\|\mathbf{r}'(1)\| = \sqrt{234}$.

$$a_{\mathrm{T}} = \frac{1620}{\sqrt{234}} \approx 105.90. \quad a_{\mathrm{N}} = \frac{\|324\,\mathbf{k}\|}{\sqrt{234}} = \frac{324}{\sqrt{234}} \approx 21.18.$$

[11] Since $v = \dfrac{ds}{dt}$ is a constant, $a_{\mathrm{T}} = \dfrac{d^2s}{dt^2} = 0$. But a_{T} is also equal to $\dfrac{\mathbf{r}'(t)}{\|\mathbf{r}'(t)\|} \cdot \mathbf{r}''(t)$,

which is $\mathbf{T}(t) \cdot \mathbf{r}''(t)$. Thus, the acceleration $\{\mathbf{r}''(t)\}$ is normal to the unit tangent vector $\{\mathbf{T}(t)\}$, and therefore normal to C.

[12] If $\mathbf{r}''(t) = 0$, then $\|\mathbf{r}'(t) \times \mathbf{r}''(t)\| = 0$, $K = \left|\dfrac{d\theta}{ds}\right| = 0$, and $\theta = c$, a constant.

Thus, the motion is on a line.

[13] Since the speed is constant, $\dfrac{dv}{dt} = 0$. Thus, $\mathbf{a}(t) = \dfrac{dv}{dt}\mathbf{T}(s) + Kv^2\,\mathbf{N}(s)$ reduces to

$Kv^2\,\mathbf{N}(s)$. Hence, $\|\mathbf{a}(t)\| = Kv^2 = cK$, where $c = v^2$ is a constant and $\|\mathbf{N}(s)\| = 1$.

[14] From Exercise 13, $c = v^2$. Thus, if the speed of Q is twice that of P, then c is four times greater for Q than it was for P, and $\|\mathbf{a}(t)\| = cK$ is also four times greater.

[15] Assuming that the acceleration of the particle is continuous, then $f''(x) = 0$ at a point of inflection. By (15.18), $K = 0$ and thus, $a_{\mathrm{N}} = Kv^2 = 0$.

16 $\mathbf{r}(t) = f(t)\mathbf{i} + g(t)\mathbf{j} + 0\mathbf{k} \Rightarrow \mathbf{r}'(t) = f'(t)\mathbf{i} + g'(t)\mathbf{j} \Rightarrow \mathbf{r}''(t) = f''(t)\mathbf{i} + g''(t)\mathbf{j}$.

$\|\mathbf{r}'(t)\| = \left[(f'(t))^2 + (g'(t))^2\right]^{1/2}$ and $\mathbf{r}'(t) \times \mathbf{r}''(t) = \left[f'(t)\,g''(t) - g'(t)f''(t)\right]\mathbf{k}$.

Using (15.25), $K = \dfrac{\|\mathbf{r}'(t) \times \mathbf{r}''(t)\|}{\|\mathbf{r}'(t)\|^3} = \dfrac{|f'(t)\,g''(t) - g'(t)f''(t)|}{\left[(f'(t))^2 + (g'(t))^2\right]^{3/2}}$, i.e., Theorem (15.19).

17 $\mathbf{r}(t) = a\cos t\,\mathbf{i} + a\sin t\,\mathbf{j} + bt\,\mathbf{k} \Rightarrow \mathbf{r}'(t) = -a\sin t\,\mathbf{i} + a\cos t\,\mathbf{j} + b\,\mathbf{k} \Rightarrow$

$\mathbf{r}''(t) = -a\cos t\,\mathbf{i} - a\sin t\,\mathbf{j}$. $\|\mathbf{r}'(t)\| = \sqrt{a^2 + b^2}$ and

$\mathbf{r}'(t) \times \mathbf{r}''(t) = ab\sin t\,\mathbf{i} - ab\cos t\,\mathbf{j} + a^2\mathbf{k}$. Thus, $K = \dfrac{a(b^2 + a^2)^{1/2}}{(a^2 + b^2)^{3/2}} = \dfrac{a}{a^2 + b^2}$.

18 $\mathbf{r}(t) = a\cos t\,\mathbf{i} + b\sin t\,\mathbf{j} + ct\,\mathbf{k} \Rightarrow \mathbf{r}'(t) = -a\sin t\,\mathbf{i} + b\cos t\,\mathbf{j} + c\,\mathbf{k} \Rightarrow$

$\mathbf{r}''(t) = -a\cos t\,\mathbf{i} - b\sin t\,\mathbf{j}$. $\|\mathbf{r}'(t)\| = \sqrt{a^2\sin^2 t + b^2\cos^2 t + c^2}$ and

$\mathbf{r}'(t) \times \mathbf{r}''(t) = bc\sin t\,\mathbf{i} - ac\cos t\,\mathbf{j} + ab\,\mathbf{k}$.

Thus, $K = \dfrac{(b^2c^2\sin^2 t + a^2c^2\cos^2 t + a^2b^2)^{1/2}}{(a^2\sin^2 t + b^2\cos^2 t + c^2)^{3/2}}$.

15.7 Review Exercises

1 $y^2 - x^2 = \sec^2 t - \tan^2 t = 1$. $\mathbf{v}(t) = \mathbf{r}'(t) = \sec^2 t\,\mathbf{i} - \sec t\tan t\,\mathbf{j}$; $\mathbf{v}(\frac{\pi}{4}) = 2\mathbf{i} - \sqrt{2}\mathbf{j}$.

$\mathbf{a}(t) = \mathbf{r}''(t) = 2\sec^2 t\tan t\,\mathbf{i} - (\sec^3 t + \tan^2 t\sec t)\mathbf{j}$; $\mathbf{a}(\frac{\pi}{4}) = 4\mathbf{i} - 3\sqrt{2}\mathbf{j}$.

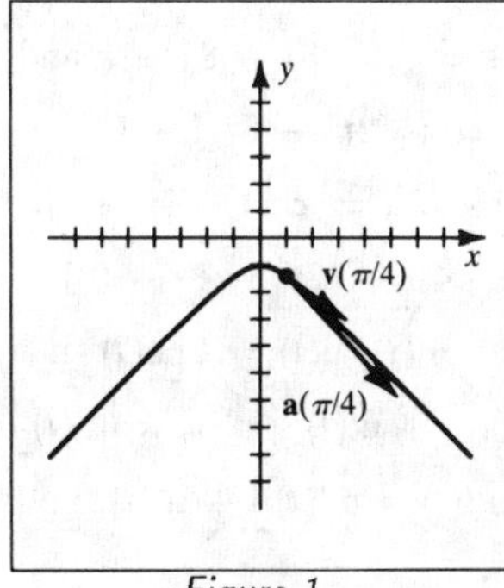

Figure 1

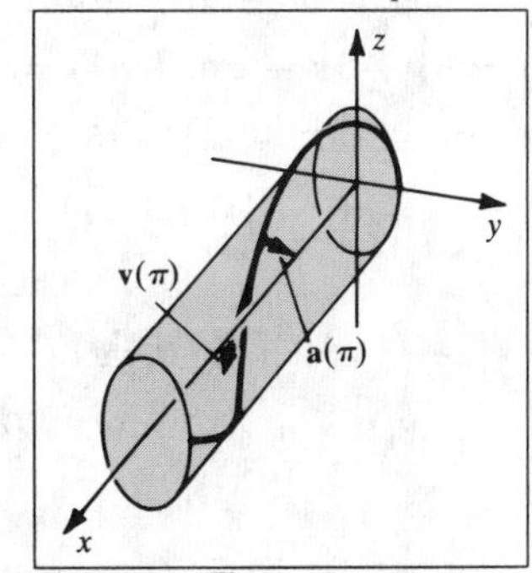

Figure 2

2 This is an elliptic helix along the x-axis. $\mathbf{v}(t) = 2t\mathbf{i} - 3\sin t\,\mathbf{j} + 5\cos t\,\mathbf{k}$;

$\mathbf{v}(\pi) = 2\pi\mathbf{i} - 5\mathbf{k}$. $\mathbf{a}(t) = 2\mathbf{i} - 3\cos t\,\mathbf{j} - 5\sin t\,\mathbf{k}$; $\mathbf{a}(\pi) = 2\mathbf{i} + 3\mathbf{j}$.

3 $\mathbf{r}(t) = t^2\mathbf{i} + (4t^2 - t^4)\mathbf{j} \Rightarrow \mathbf{r}'(t) = 2t\mathbf{i} + (8t - 4t^3)\mathbf{j}$, $\mathbf{r}''(t) = 2\mathbf{i} + (8 - 12t^2)\mathbf{j}$,

and $\|\mathbf{r}'(t)\| = 2|t|\sqrt{17 - 16t^2 + 4t^4}$.

4 $\mathbf{r}(t) = (2 - \sin t)\mathbf{i} + (1 - \cos t)\mathbf{j} \Rightarrow \mathbf{r}'(t) = -\cos t\,\mathbf{i} + \sin t\,\mathbf{j}$, $\mathbf{r}''(t) = \sin t\,\mathbf{i} + \cos t\,\mathbf{j}$,

and $\|\mathbf{r}'(t)\| = 1$.

5 (a) $\mathbf{r}'(t) = e^t(\sin t + \cos t)\mathbf{i} + e^t(\cos t - \sin t)\mathbf{j} + e^t\mathbf{k}$.

$\mathbf{r}'(0) = \mathbf{i} + \mathbf{j} + \mathbf{k}$ and $\|\mathbf{r}'(0)\| = \sqrt{3} \Rightarrow \mathbf{T}(t) = \frac{1}{\sqrt{3}}(\mathbf{i} + \mathbf{j} + \mathbf{k})$.

(b) $\mathbf{r}(0) = 0\mathbf{i} + 1\mathbf{j} + 1\mathbf{k} \Rightarrow$ the point corresponding to $t = 0$ is $P(0, 1, 1)$.

Since $\mathbf{r}'(0) = 1\mathbf{i} + 1\mathbf{j} + 1\mathbf{k}$, an equation of the tangent line to C at P is

$x = t,\ y = 1 + t,\ z = 1 + t$.

[6] $L = \int_0^1 \sqrt{e^{2t}(\sin t + \cos t)^2 + e^{2t}(\cos t - \sin t)^2 + e^{2t}}\,dt = \int_0^1 \sqrt{3}\,e^t\,dt =$

$\sqrt{3}\,(e - 1) \approx 2.98$

[7] $\mathrm{u}(t) \cdot \mathrm{v}'(t) = (t^2\mathrm{i} + 6t\mathrm{j} + t\mathrm{k}) \cdot (\mathrm{i} - 5\mathrm{j} + 8t\mathrm{k}) = t^2 - 30t + 8t^2 = 9t^2 - 30t =$

$3t(3t - 10)$. $\mathrm{u}(t) \cdot \mathrm{v}'(t) = 0 \Leftrightarrow t = 0, \frac{10}{3}$.

[8] $\mathrm{u}(t) \times \mathrm{v}(t) = (24t^3 + 5t^2)\mathrm{i} + (t^2 - 4t^4)\mathrm{j} - (5t^3 + 6t^2)\mathrm{k} \Rightarrow$

$D_t\big[\mathrm{u}(t) \times \mathrm{v}(t)\big] = (72t^2 + 10t)\mathrm{i} + (2t - 16t^3)\mathrm{j} - (15t^2 + 12t)\mathrm{k}$.

$\mathrm{u}(t) \cdot \mathrm{v}(t) = t^3 - 30t^2 + 4t^3 = 5t^3 - 30t^2 \Rightarrow D_t\big[\mathrm{u}(t) \cdot \mathrm{v}(t)\big] = 15t^2 - 60t$.

[9] $\mathrm{r}'(t) = 3\mathrm{i} + 3t^2\mathrm{j} + 4t^3\mathrm{k}$, $\mathrm{r}''(t) = 6t\mathrm{j} + 12t^2\mathrm{k}$, and $\|\mathrm{r}'(t)\| = \sqrt{9 + 9t^4 + 16t^6}$.

$\mathrm{r}'(1) = 3\mathrm{i} + 3\mathrm{j} + 4\mathrm{k}$, $\mathrm{r}''(1) = 6\mathrm{j} + 12\mathrm{k}$, $\|\mathrm{r}'(1)\| = \sqrt{9 + 9 + 16} = \sqrt{34}$.

[10] $\int (\sin 3t\mathrm{i} + e^{-2t}\mathrm{j} + \cos t\mathrm{k})\,dt = -\frac{1}{3}\cos 3t\mathrm{i} - \frac{1}{2}e^{-2t}\mathrm{j} + \sin t\mathrm{k} + \mathrm{c}$

[11] $\int_0^1 (4t\mathrm{i} + t^3\mathrm{j} - \mathrm{k})\,dt = \Big[2t^2\mathrm{i} + \frac{1}{4}t^4\mathrm{j} - t\mathrm{k}\Big]_0^1 = 2\mathrm{i} + \frac{1}{4}\mathrm{j} - \mathrm{k}$

[12] $\mathrm{u}'(t) = e^{-t}\mathrm{i} - 4\sin 2t\mathrm{j} + 3\sqrt{t}\,\mathrm{k} \Rightarrow \mathrm{u}(t) = -e^{-t}\mathrm{i} + 2\cos 2t\mathrm{j} + 2t^{3/2}\mathrm{k} + \mathrm{c}$.

$\mathrm{u}(0) = -\mathrm{i} + 2\mathrm{j} \Rightarrow \mathrm{c} = 0$, so $\mathrm{u}(t) = -e^{-t}\mathrm{i} + 2\cos 2t\mathrm{j} + 2t^{3/2}\mathrm{k}$.

[13] $\mathrm{a}(t) = 12t\mathrm{j} + 5\mathrm{k} \Rightarrow \mathrm{v}(t) = 6t^2\mathrm{j} + 5t\mathrm{k} + \mathrm{c}$.

$\mathrm{v}(1) = 6\mathrm{j} + 5\mathrm{k} + \mathrm{c}$ and $\mathrm{v}(1) = 3\mathrm{i} - 2\mathrm{j} + 4\mathrm{k} \Rightarrow \mathrm{c} = 3\mathrm{i} - 8\mathrm{j} - \mathrm{k}$ and hence

$\mathrm{v}(t) = 3\mathrm{i} + (6t^2 - 8)\mathrm{j} + (5t - 1)\mathrm{k}$. $\mathrm{r}(t) = 3t\mathrm{i} + (2t^3 - 8t)\mathrm{j} + (\frac{5}{2}t^2 - t)\mathrm{k} + \mathrm{c}$.

$\mathrm{r}(1) = 3\mathrm{i} - 6\mathrm{j} + \frac{3}{2}\mathrm{k} + \mathrm{c}$ and $\mathrm{r}(1) = -\mathrm{i} + 3\mathrm{j} + \frac{3}{2}\mathrm{k} \Rightarrow \mathrm{c} = -4\mathrm{i} + 9\mathrm{j}$.

Thus, $\mathrm{r}(t) = (3t - 4)\mathrm{i} + (2t^3 - 8t + 9)\mathrm{j} + (\frac{5}{2}t^2 - t)\mathrm{k}$.

[14] $\|\mathrm{u}(t)\|^2 = \mathrm{u}(t) \cdot \mathrm{u}(t) \Rightarrow D_t\|\mathrm{u}(t)\|^2 = \mathrm{u}(t) \cdot \mathrm{u}'(t) + \mathrm{u}'(t) \cdot \mathrm{u}(t) = 2\big[\mathrm{u}(t) \cdot \mathrm{u}'(t)\big]$.

[15] $D_t\big[\mathrm{u}(t) \cdot \mathrm{u}'(t) \times \mathrm{u}''(t)\big] = \mathrm{u}(t) \cdot D_t\big[\mathrm{u}'(t) \times \mathrm{u}''(t)\big] + \mathrm{u}'(t) \cdot \big[\mathrm{u}'(t) \times \mathrm{u}''(t)\big]$

$= \mathrm{u}(t) \cdot \big[\mathrm{u}'(t) \times \mathrm{u}'''(t) + \mathrm{u}''(t) \times \mathrm{u}''(t)\big] + \mathrm{u}'(t) \cdot \big[\mathrm{u}'(t) \times \mathrm{u}''(t)\big]$

$= \mathrm{u}(t) \cdot \mathrm{u}'(t) \times \mathrm{u}'''(t)$

since $\mathrm{u}''(t) \times \mathrm{u}''(t) = \mathbf{0}$ and $\mathrm{u}'(t) \times \mathrm{u}''(t)$ is orthogonal to $\mathrm{u}'(t)$.

[16] $y' = (1 + x)\,e^x$, $y'' = (2 + x)\,e^x$. At $P(0, 0)$, $K = \dfrac{2}{\big[1 + (1)^2\big]^{3/2}} = \dfrac{1}{\sqrt{2}}$.

[17] $f'(t) = -(1 + t)^{-2}$, $f''(t) = 2(1 + t)^{-3}$, $g'(t) = (1 - t)^{-2}$, $g''(t) = 2(1 - t)^{-3}$.

At $P(\frac{2}{3}, 2)$, $t = \frac{1}{2}$ and $K = \dfrac{\left|(-\frac{4}{9})(16) - (4)(\frac{16}{27})\right|}{\big[(-\frac{4}{9})^2 + (4)^2\big]^{3/2}} = \dfrac{108}{82^{3/2}} \approx 0.15$

[18] $\mathrm{r}(t) = 2t^2\mathrm{i} + t^4\mathrm{j} + 4t\mathrm{k} \Rightarrow \mathrm{r}'(t) = 4t\mathrm{i} + 4t^3\mathrm{j} + 4\mathrm{k} \Rightarrow \mathrm{r}''(t) = 4\mathrm{i} + 12t^2\mathrm{j}$.

$\|\mathrm{r}'(t)\| = 4\sqrt{t^6 + t^2 + 1}$ and $\mathrm{r}'(t) \times \mathrm{r}''(t) = -48t^2\mathrm{i} + 16\mathrm{j} + 32t^3\mathrm{k}$.

Thus, $K = \dfrac{\|-48t^2\mathrm{i} + 16\mathrm{j} + 32t^3\mathrm{k}\|}{(4\sqrt{t^6 + t^2 + 1})^3} = \dfrac{(4t^6 + 9t^4 + 1)^{1/2}}{4(t^6 + t^2 + 1)^{3/2}}$.

[19] $y' = 3x^2 - 3$, $y'' = 6x \Rightarrow K(x) = \dfrac{|6x|}{\left[1 + (3x^2 - 3)^2\right]^{3/2}}$. If $x > 0$, then

$$K'(x) = \frac{(9x^4 - 18x^2 + 10)^{3/2}(6) - (6x)(\tfrac{3}{2})(9x^4 - 18x^2 + 10)^{1/2}(36x^3 - 36x)}{(9x^4 - 18x^2 + 10)^3} \text{ and}$$

$K'(x) = 0 \Rightarrow (9x^4 - 18x^2 + 10)(6) - (9x)(36x^3 - 36x) = 0 \Rightarrow$

$45x^4 - 36x^2 - 10 = 0 \Rightarrow x^2 = \dfrac{36 \pm \sqrt{3096}}{90} \Rightarrow x \approx 1.009$ since $x > 0$.

This gives a maximum since $K'(x) > 0$ on $[0, 1.009)$ and $K'(x) < 0$ on $(1.009, \infty)$.

By symmetry, another maximum exists at $x \approx -1.009$.

[20] $y' = \sinh x$, $y'' = \cosh x$. At $P(0, 1)$, $K = \dfrac{1}{\left[1 + 0^2\right]^{3/2}} = 1 \Rightarrow \rho = \dfrac{1}{K} = 1$.

Since $y' = 0$ and the graph is concave up at $x = 0$, it follows that the center of curvature is on a vertical line 1 unit above P, i.e., $(0, 2)$.

The circle of curvature is $x^2 + (y - 2)^2 = 1$.

[21] $r = 2 + \sin\theta \Rightarrow r' = \cos\theta \Rightarrow r'' = -\sin\theta$. Use Exercise 33 in §15.4. At $P(2, \pi)$,

$r' = -1$, $r'' = 0$, and $K = \dfrac{\left|2(-1)^2 - (2)(0) + (2)^2\right|}{\left[(-1)^2 + (2)^2\right]^{3/2}} = \dfrac{6}{5\sqrt{5}} \Rightarrow \rho = \tfrac{5}{6}\sqrt{5} \approx 1.86$.

[22] (a) $\mathbf{r}(t) = (t^2 + 1)\mathbf{i} + 4t\mathbf{j} \Rightarrow \mathbf{r}'(t) = 2t\mathbf{i} + 4\mathbf{j} \Rightarrow \|\mathbf{r}'(t)\| = 2\sqrt{t^2 + 4}$.

$$\mathbf{T}(t) = \frac{1}{\|\mathbf{r}'(t)\|}\mathbf{r}'(t) = \frac{t}{(t^2 + 4)^{1/2}}\mathbf{i} + \frac{2}{(t^2 + 4)^{1/2}}\mathbf{j} \Rightarrow$$

$$\mathbf{T}'(t) = \frac{4}{(t^2 + 4)^{3/2}}\mathbf{i} - \frac{2t}{(t^2 + 4)^{3/2}}\mathbf{j} \text{ and}$$

$\|\mathbf{T}'(t)\| = \dfrac{2}{t^2 + 4}$. Thus,

$$\mathbf{N}(t) = \frac{1}{\|\mathbf{T}'(t)\|}\mathbf{T}'(t) = \frac{2}{(t^2 + 4)^{1/2}}\mathbf{i} - \frac{t}{(t^2 + 4)^{1/2}}\mathbf{j}.$$

(b) $x = t^2 + 1$ and $y = 4t \Rightarrow x = \frac{1}{16}y^2 + 1$.

$\mathbf{T}(1) = \frac{1}{\sqrt{5}}\mathbf{i} + \frac{2}{\sqrt{5}}\mathbf{j}$ and $\mathbf{N}(1) = \frac{2}{\sqrt{5}}\mathbf{i} - \frac{1}{\sqrt{5}}\mathbf{j}$.

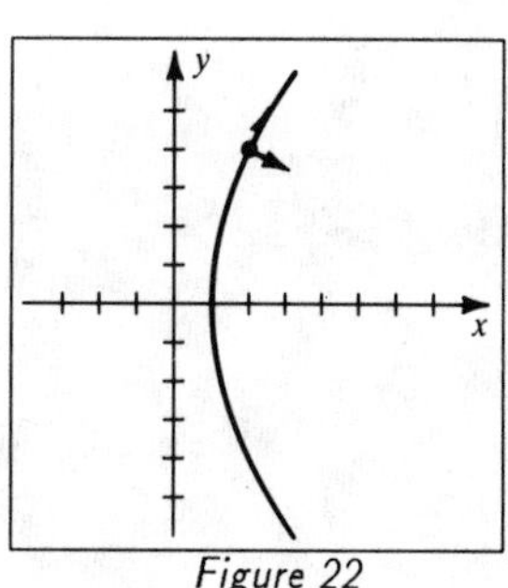

Figure 22

[23] $\mathbf{r}(t) = \sin 2t\mathbf{i} + \cos t\mathbf{j}$, $\mathbf{r}'(t) = 2\cos 2t\mathbf{i} - \sin t\mathbf{j}$, $\mathbf{r}''(t) = -4\sin 2t\mathbf{i} - \cos t\mathbf{j}$,

and $\|\mathbf{r}'(t)\| = \sqrt{4\cos^2 2t + \sin^2 t}$. $a_{\mathrm{T}} = \dfrac{-8\cos 2t\sin 2t + \sin t\cos t}{(4\cos^2 2t + \sin^2 t)^{1/2}}$.

$$a_{\mathrm{N}} = \frac{\|-(2\cos 2t\cos t + 4\sin 2t\sin t)\,\mathbf{k}\|}{(4\cos^2 2t + \sin^2 t)^{1/2}} = \frac{2|\cos 2t\cos t + 2\sin 2t\sin t|}{(4\cos^2 2t + \sin^2 t)^{1/2}}.$$

[24] $\mathbf{r}(t) = 3t\mathbf{i} + t^3\mathbf{j} + t\mathbf{k}$, $\mathbf{r}'(t) = 3\mathbf{i} + 3t^2\mathbf{j} + \mathbf{k}$, $\mathbf{r}''(t) = 6t\mathbf{j}$, and $\|\mathbf{r}'(t)\| = \sqrt{9t^4 + 10}$.

$$a_{\mathrm{T}} = \frac{18t^3}{(9t^4 + 10)^{1/2}}.\quad a_{\mathrm{N}} = \frac{\|-6t\mathbf{i} + 18t\mathbf{k}\|}{(9t^4 + 10)^{1/2}} = \frac{6\sqrt{10}\,|t|}{(9t^4 + 10)^{1/2}}.$$

Chapter 16: Partial Differentiation

Exercises 16.1

Note: In Exercises 1–6, let D denote the domain of f.

[1] f is defined for all ordered pairs $(x, y) \Rightarrow D = \mathbb{R}^2$. $f(x, y) = 2x - y^2 \Rightarrow$

$$f(-2, 5) = 2(-2) - 5^2 = -29,\ f(5, -2) = 6, \text{ and } f(0, -2) = -4.$$

[2] Since we can't divide by 0, $D = \{(x, y): x \neq 0\}$.

$$f(x, y) = \frac{y + 2}{x} \Rightarrow f(3, 1) = \tfrac{3}{3} = 1,\ f(1, 3) = \tfrac{5}{1} = 5, \text{ and } f(2, 0) = \tfrac{2}{2} = 1.$$

[3] Since we can't divide by 0, $u - 2v \neq 0$, and $D = \{(u, v): u \neq 2v\}$. $f(u, v) =$

$$\frac{uv}{u - 2v} \Rightarrow f(2, 3) = \tfrac{6}{-4} = -\tfrac{3}{2},\ f(-1, 4) = \tfrac{-4}{-9} = \tfrac{4}{9}, \text{ and } f(0, 1) = \tfrac{0}{-2} = 0.$$

[4] $1 - r \geq 0 \Rightarrow r \leq 1$ and $D = \{(r, s): r \leq 1,\ s \neq 0\}$.

$$f(r, s) = \sqrt{1 - r} - e^{r/s} \Rightarrow f(1, 1) = -e,\ f(0, 4) = 0, \text{ and } f(-3, 3) = 2 - e^{-1}.$$

[5] $25 - x^2 - y^2 - z^2 \geq 0 \Rightarrow x^2 + y^2 + z^2 \leq 25$.

$D = \{(x, y, z): x^2 + y^2 + z^2 \leq 25\}$.

$$f(x, y, z) = \sqrt{25 - x^2 - y^2 - z^2} \Rightarrow f(1, -2, 2) = 4 \text{ and } f(-3, 0, 2) = 2\sqrt{3}.$$

[6] Since $\tan x$ is undefined at $x = \frac{\pi}{2} + \pi n$, $D = \{(x, y, z): x \neq \frac{\pi}{2} + \pi n\}$.

$$f(x, y, z) = 2 + \tan x + y \sin z \Rightarrow f(\tfrac{\pi}{4}, 4, \tfrac{\pi}{6}) = 5 \text{ and } f(0, 0, 0) = 2.$$

Note: In Exercises 7–14, let $z = f(x, y)$.

[7] $z = \sqrt{1 - x^2 - y^2} \Rightarrow x^2 + y^2 + z^2 = 1,\ z \geq 0$.

The top half of the sphere with radius 1 and center $(0, 0, 0)$

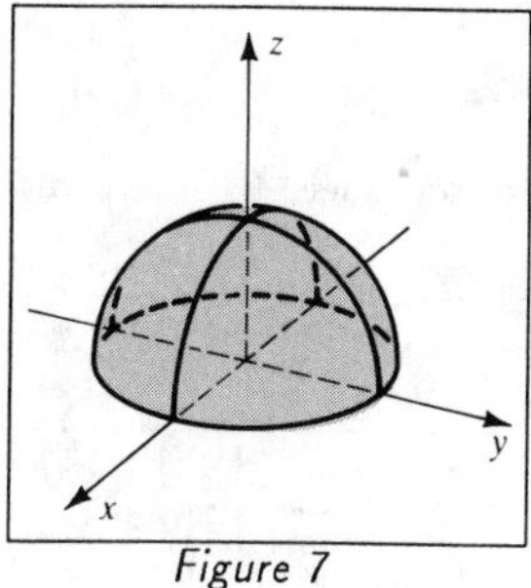

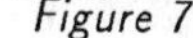
Figure 7

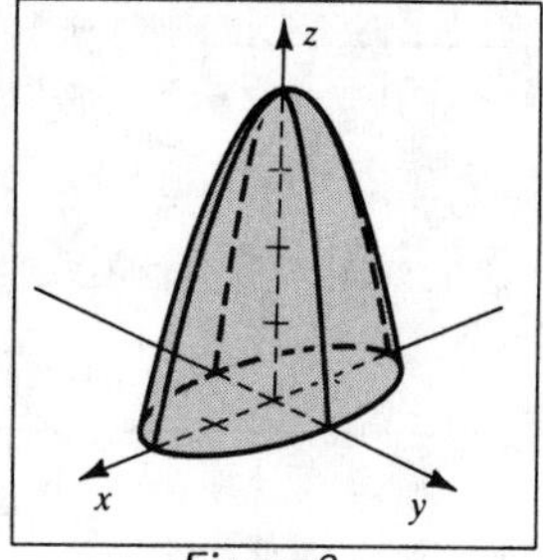

Figure 8

[8] $z = 4 - x^2 - 4y^2 = 4 - \left(x^2 + \dfrac{y^2}{1/4}\right)$. An elliptic paraboloid opening down with

z-intercept 4, x-intercepts ± 2, and y-intercepts ± 1

[9] $z = x^2 + y^2 - 1 \Rightarrow x^2 + y^2 = z + 1.$

A circular paraboloid opening upwards with vertex $(0, 0, -1)$

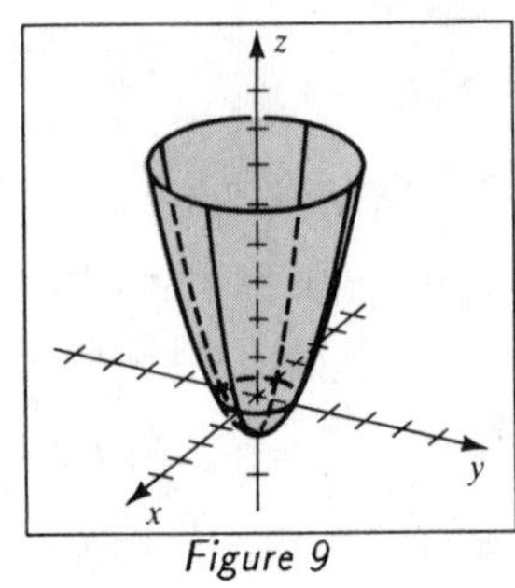

Figure 9

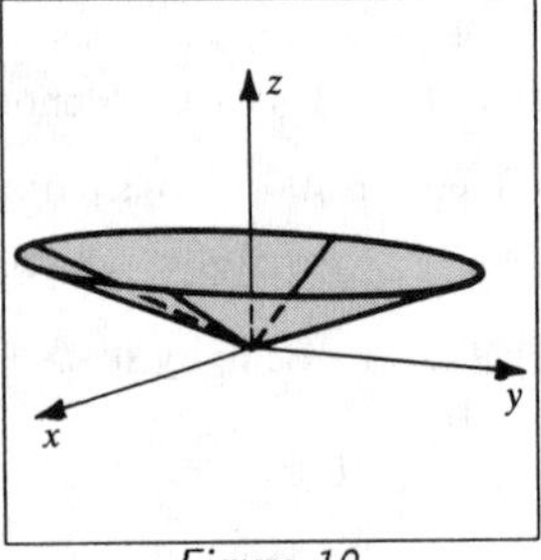

Figure 10

[10] $z = \frac{1}{6}\sqrt{9x^2 + 4y^2} \Rightarrow 36z^2 = 9x^2 + 4y^2.$

The upper portion of a cone with vertex at $(0, 0, 0)$

[11] $z = 6 - 2x - 3y \Rightarrow 2x + 3y + z = 6.$

A plane with intercepts $x = 3$, $y = 2$, and $z = 6$

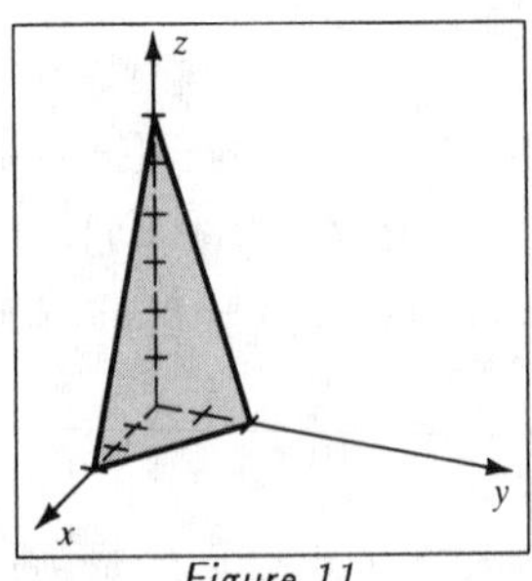

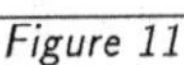
Figure 11

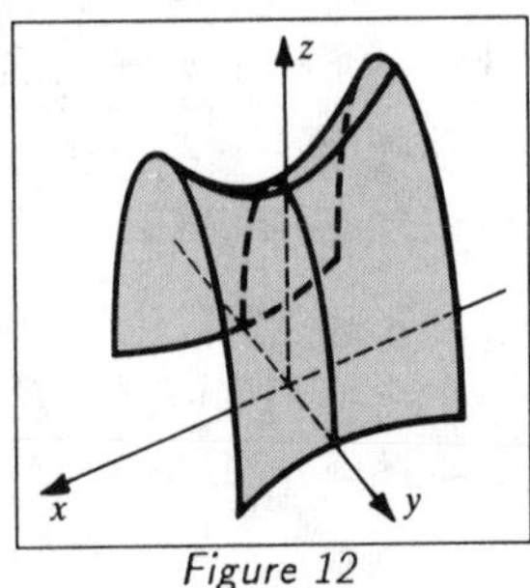

Figure 12

[12] $z = \sqrt{72 + 4x^2 - 9y^2} \Rightarrow \frac{y^2}{8} + \frac{z^2}{72} - \frac{x^2}{18} = 1,\ z \geq 0.$

The upper half of a hyperboloid of one sheet with axis along the x-axis

[13] $z = \sqrt{y^2 - 4x^2 - 16} \Rightarrow \frac{y^2}{16} - \frac{x^2}{4} - \frac{z^2}{16} = 1,\ z \geq 0.$

The upper half of a hyperboloid of two sheets with axis along the y-axis

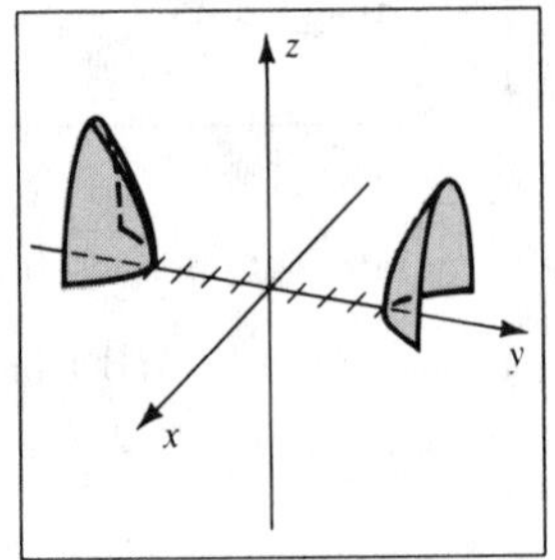

Figure 13

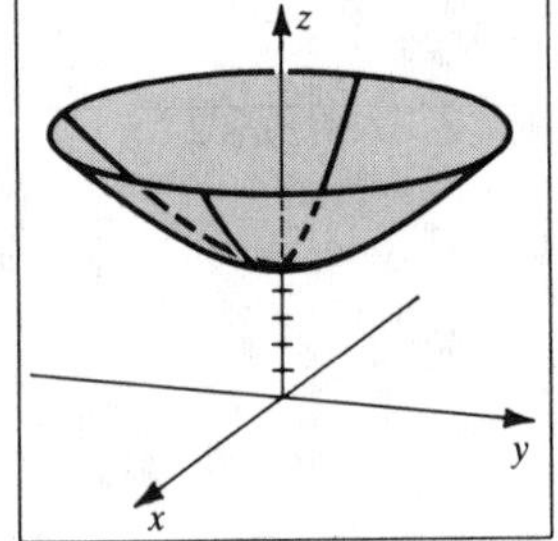

Figure 14

[14] $z = \sqrt{x^2 + 4y^2 + 25} \Rightarrow \frac{z^2}{25} - \frac{x^2}{25} - \frac{4y^2}{25} = 1,\ z \geq 0.$

The upper half of a hyperboloid of two sheets with axis along the z-axis

[15] $y^2 - x^2 = k$ are hyperbolas with asymptotes $y = \pm x$ and a vertical (horizontal) transverse axis if $k > 0$ $(k < 0)$. If $k = 0$, we have $y = \pm x$.

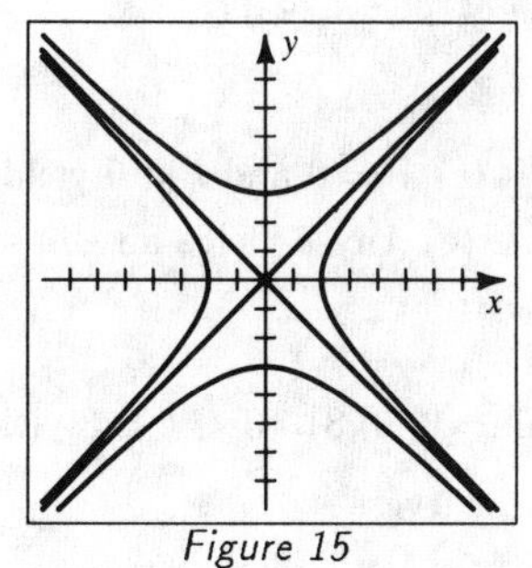

Figure 15

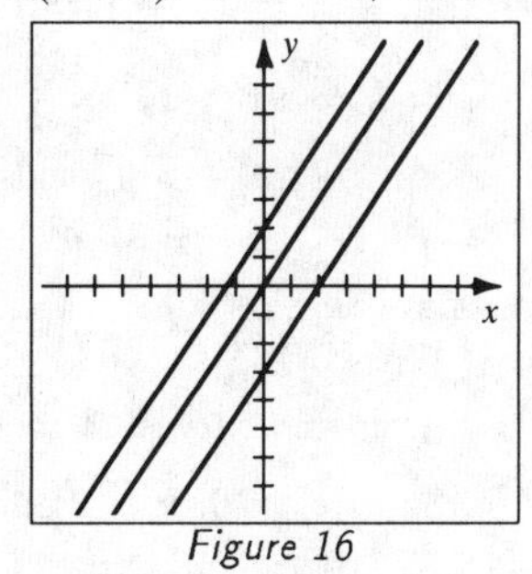

Figure 16

[16] $3x - 2y = k \Rightarrow y = \frac{3}{2}x - \frac{1}{2}k$. These are lines with slope $\frac{3}{2}$ and y-intercept $-\frac{1}{2}k$.

[17] $x^2 - y = k \Rightarrow y = x^2 - k$. These are parabolas with vertices $(0, -k)$.

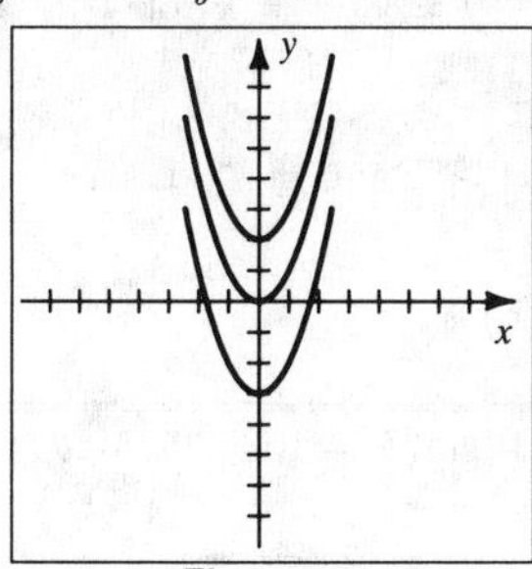

Figure 17

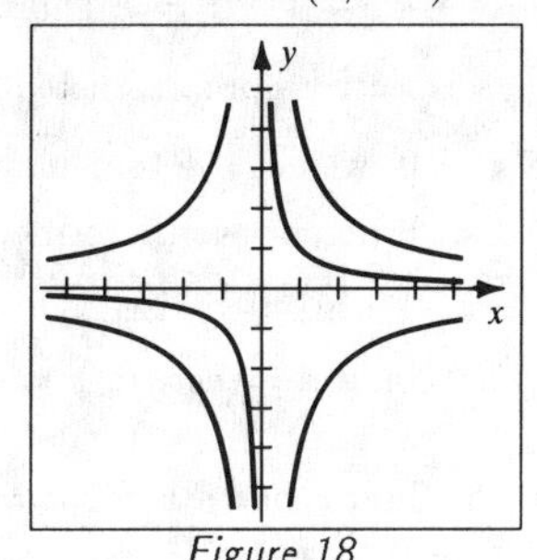

Figure 18

[18] $xy = k \Rightarrow y = \frac{k}{x}$. These are hyperbolas with $x = 0$ and $y = 0$ as asymptotes, and $y = x$ $(y = -x)$ as the transverse axis if $k > 0$ $(k < 0)$.

If $k = 0$, we have the x-axis and the y-axis.

[19] $(x - 2)^2 + (y + 3)^2 = k$.

These are circles with center $(2, -3)$ and radius $\sqrt{k}$, where $k > 0$.

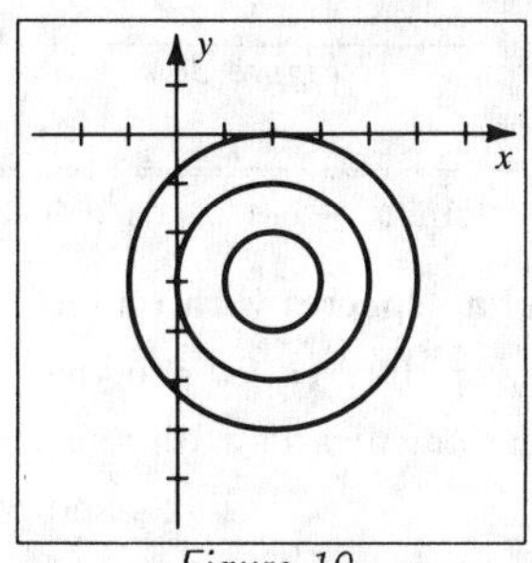

Figure 19

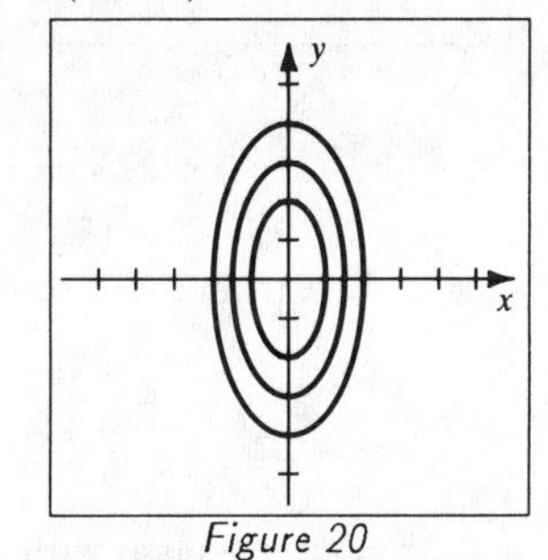

Figure 20

[20] $4x^2 + y^2 = k \Rightarrow \dfrac{x^2}{k/4} + \dfrac{y^2}{k} = 1$. These are ellipses centered at the origin with intercepts $x = \pm\sqrt{k}/2$ and $y = \pm\sqrt{k}$, where $k > 0$.

[21] Since $f(1, 4) = 4\arctan 1 = 4(\frac{\pi}{4}) = \pi$, an equation is $y \arctan x = \pi$.

[22] Since $f(0, 2) = 4e^0 = 4$, an equation is $(2x + y^2)e^{xy} = 4$.

23 Since $f(2, -1, 3) = 4 + 4 - 9 = -1$, an equation is $x^2 + 4y^2 - z^2 = -1$.

24 Since $f(1, 4, -2) = 16 + 1 = 17$, an equation is $z^2y + x = 17$.

25 $\dfrac{1}{9x^2 + y^2} = k \Rightarrow 9x^2 + y^2 = \dfrac{1}{k}$.

(a) $k > 0$ yields ellipses. (b) & (c) $k = 0$ and $k < 0$ yield no curves.

26 $\ln(2x^2 + y^2) = k \Rightarrow 2x^2 + y^2 = e^k$. (a), (b), & (c) Any k yields an ellipse.

27 $8e^{-(x^2+y^2)/4} = k \Rightarrow x^2 + y^2 = -4\ln\frac{k}{8}$.

(a) $k > 8$ yields no curves. $k = 8$ yields the point $(0, 0, 8)$. $0 < k < 8$ yields circles.

(b) & (c) Since $e^x > 0$, $k = 0$ and $k < 0$ yield no curves.

28 (a) If $k > 1$, there is no curve. If $k = 1$, we get the lines $x = t$, $y = 0$, $z = 1$ and $x = 0$, $y = t$, $z = 1$ for t in $\mathbb{R}$. If $0 < k < 1$, we get the lines $x = t$, $y = \pm\sqrt{1 - k^2}$, $z = k$ and $x = \pm\sqrt{1 - k^2}$, $y = t$, $z = k$ for t in $\mathbb{R}$ and using the respective domains listed in the text.

(b) If $k = 0$, we get the four planes formed by $|x| \geq 1$ and $|y| \geq 1$.

(c) If $k < 0$, no curves are formed.

29 (d) 30 (e) 31 (a) 32 (c) 33 (f) 34 (b)

35 $f(x, y) = k$ and $x = r\cos\theta$, $y = r\sin\theta \Rightarrow r = \dfrac{k^{1/4}}{\sqrt[4]{\sin^4\theta + 8\sin^2\theta\cos^2\theta + \cos^4\theta}}$

Note: The figure shown is for $k = 1$, 16, and 81.

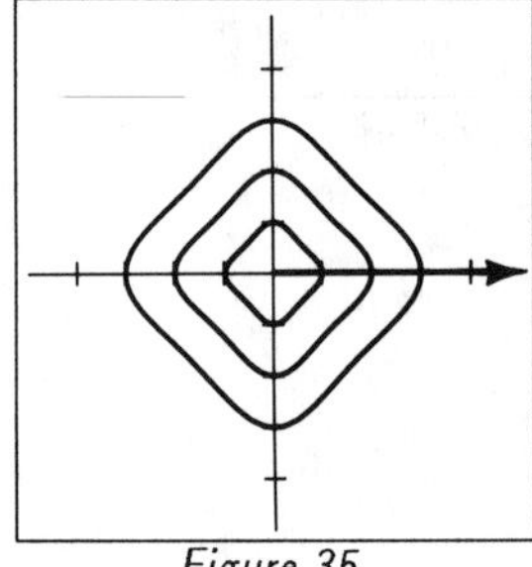

Figure 35

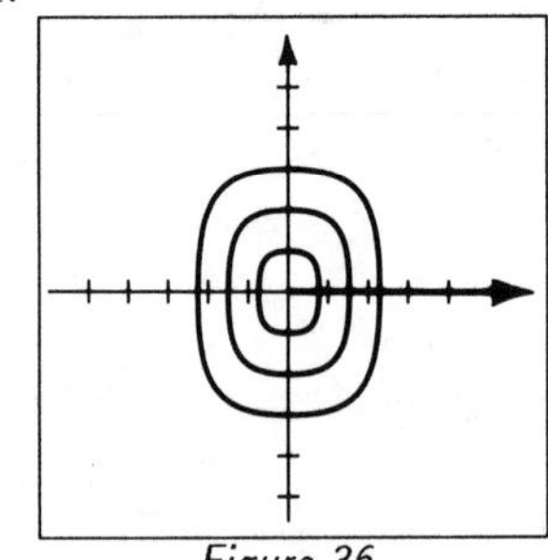

Figure 36

36 $f(x, y) = k$ and $x = r\cos\theta$, $y = r\sin\theta \Rightarrow r = \dfrac{k^{1/4}}{\sqrt[4]{\sin^4\theta + \sin^2\theta\cos^2\theta + 3\cos^4\theta}}$

37 $x^2 + y^2 + z^2 = k$ include the origin ($k = 0$) and all spheres with center $(0, 0, 0)$ and radius $\sqrt{k}$. $k < 0$ yields no surface.

38 $z + x^2 + 4y^2 = k$ are elliptic paraboloids opening downwards with axis along the z-axis and z-intercept k.

39 $x + 2y + 3z = k$ are planes with x-intercept k, y-intercept $\frac{k}{2}$, and z-intercept $\frac{k}{3}$.

40 $x^2 + y^2 - z^2 = k$ are hyperboloids of one sheet if $k > 0$, hyperboloids of two sheets if $k < 0$, and a cone if $k = 0$. Their axes are along the z-axis.

41 $x^2 + y^2 = k$ are right circular cylinders with axis along the z-axis and radius $\sqrt{k}$. $k = 0$ gives us the z-axis and $k < 0$ yields no surface.

42 $z = k$ are planes parallel to the xy-plane with z-intercept k.

[43] The temperature is given by $T = \dfrac{c}{\sqrt{x^2 + y^2}}$, where c is a constant.

(a) $T = k \Rightarrow x^2 + y^2 = (\frac{c}{k})^2$. These are circles with center $(0, 0)$ and radius $\frac{c}{k}$.

(b) $x = 4,\ y = 3,\ T = 40 \Rightarrow 40 = \dfrac{c}{\sqrt{4^2 + 3^2}} \Rightarrow c = 200.$

At $T = 20$, the equation in part (a) becomes $x^2 + y^2 = (\frac{200}{20})^2 = 100.$

[44] (a) $k = \dfrac{6}{(x^2 + 4y^2 + 9z^2)^{1/2}} \Rightarrow x^2 + 4y^2 + 9z^2 = (\frac{6}{k})^2.$

These are ellipsoids centered at the origin.

(b) $V = 120 \Rightarrow x^2 + 4y^2 + 9z^2 = (\frac{6}{120})^2 \Rightarrow 400x^2 + 1600y^2 + 3600z^2 = 1$

[45] There are five: m_0, m, x, y, and z. m_0 and m constant $\Rightarrow$

$x^2 + y^2 + z^2 = \dfrac{Gm_0m}{k}$, which are spheres centered at the origin.

The force F is constant if (x, y, z) moves along a level surface.

[46] $PV = kT \Rightarrow P = \frac{kT}{V}$. If $\frac{kT}{V} = c$, where c is a constant, then $V = \frac{k}{c}\,T$ are lines in the VT-plane passing through the origin. The pressure is constant as a point moves on a level curve.

[47] (a) $P = kAv^3$, where $k > 0$ is a constant.

(b) $c = kAv^3 \Rightarrow A = \frac{c}{kv^3}$, where $c \geq 0$ is a constant.

A typical level curve (see *Figure 47*) shows the combinations of areas and wind velocities that result in a fixed power $P = c$.

(c) $d = 10 \Rightarrow r = 5$. $3000 = k(\pi\, 5^2)(20^3) \Rightarrow$

$k = \frac{3}{200\pi}$. Thus, $4000 = \frac{3}{200\pi}Av^3 \Rightarrow Av^3 = \frac{8}{3}\pi \times 10^5.$

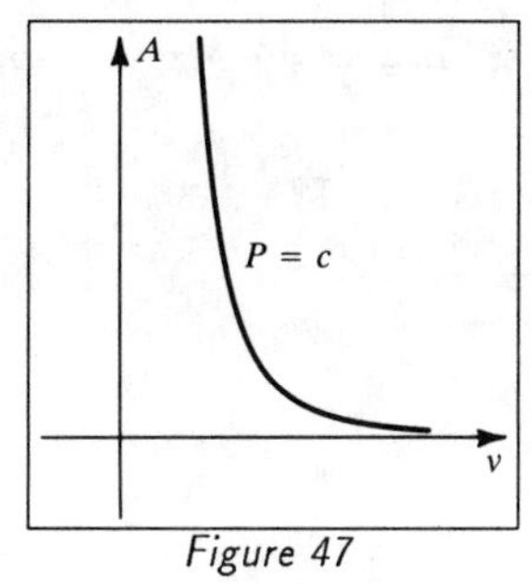

Figure 47

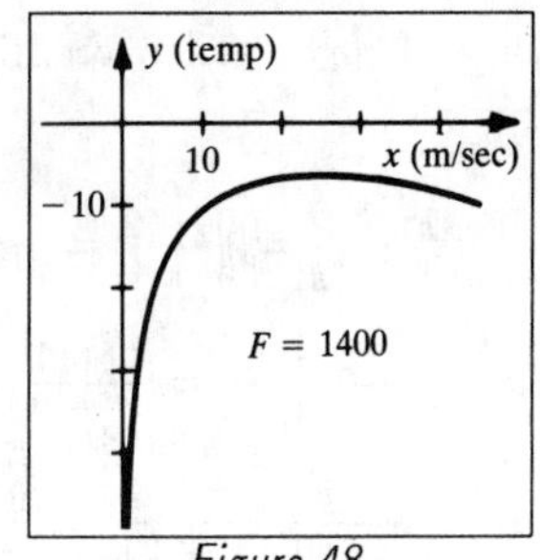

Figure 48

[48] (a) $F = 0 \Rightarrow y = 33$ or $10\sqrt{x} - x + 10.5 = 0$ (†). If it's 33 °C, human skin is neither warmed nor chilled by the wind. Skin temperature is approximately 33 °C. Solving (†), we get $x = (5 + \frac{1}{2}\sqrt{142})^2 \approx 120.1 \notin [0, 50]$.

Hence, $F = 0$ only when $y = 33$.

(b) $1400 = (33 - y)(10\sqrt{x} - x + 10.5) \Rightarrow$

$$y = 33 - \frac{1400}{10\sqrt{x} - x + 10.5} = \frac{33x - 330\sqrt{x} + 1053.5}{\left[\sqrt{x} - (5 + \frac{1}{2}\sqrt{142})\right]\left[\sqrt{x} - (5 - \frac{1}{2}\sqrt{142})\right]}.$$

By using the techniques in Chapter 4, we obtain *Figure 48*.

[49] Example: 5′11″ and 175 lb are approximately 180 cm and 80 kg.

From the graph, we have a surface area of approximately 2.0 m^2.

Using the formula, we obtain $S = 0.007184(80)^{0.425}(180)^{0.725} \approx 1.996\ \mathrm{m}^2$.

[50] $c = \frac{20}{\pi}\tan^{-1}\left(\frac{2y}{1 - x^2 - y^2}\right) \Rightarrow$

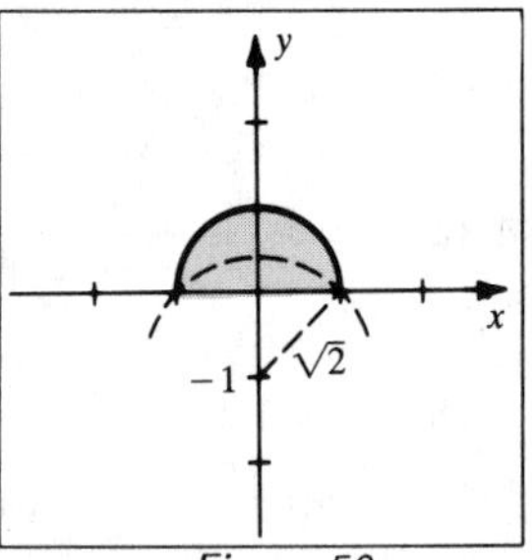

Figure 50

$\tan\left(\frac{c\pi}{20}\right) = \frac{2y}{1 - x^2 - y^2}$. If $d = \tan\left(\frac{c\pi}{20}\right)$, then

$d - dx^2 - dy^2 = 2y \Rightarrow 1 = x^2 + y^2 + \frac{2}{d}y \Rightarrow$

$1 + \frac{1}{d^2} = x^2 + \left(y + \frac{1}{d}\right)^2$.

The center of the circle is $\left(0, -\frac{1}{d}\right)$, which lies on the negative y-axis. The points $(\pm 1, 0)$ satisfy the last equation.

When $c = 5$, $d = 1$, and the equation of the circle becomes $x^2 + (y + 1)^2 = 2$.

[51] Without loss of generality, let the two cities be located at $F'(-c, 0)$ and $F(c, 0)$. If the plant is located at $P(x, y)$, then the sum of the distances from P to each city is equal to the constant M. By (12.5), this is the definition of an ellipse. See Figure 12.14.

[52] (a) $ax^2 + by^2 + c = k \Rightarrow \frac{x^2}{(k - c)/a} + \frac{y^2}{(k - c)/b} = 1\ (k > c)$. They are ellipses.

(b) A region of low pressure will occur when its atmospheric pressure is less than that of the area surrounding the region. As we move away from the origin, the atmospheric pressure always increases. p is minimum at $(0, 0)$ and increases with each level curve. Therefore, there is an area of low pressure near the origin.

Exercises 16.2

[1] $\lim_{(x, y)\to(0, 0)} \frac{x^2 - 2}{3 + xy} = \frac{0 - 2}{3 + 0} = -\frac{2}{3}$

[2] $\lim_{(x, y)\to(2, 1)} \frac{4 + x}{2 - y} = \frac{4 + 2}{2 - 1} = 6$

[3] $\lim_{(x, y)\to(\pi/2, 1)} \frac{y + 1}{2 - \cos x} = \frac{1 + 1}{2 - 0} = 1$

[4] $\lim_{(x, y)\to(-1, 3)} \frac{y^2 + x}{(x - 1)(y + 2)} = \frac{9 - 1}{(-2)(5)} = -\frac{4}{5}$

[5] $\lim_{(x, y)\to(0, 0)} \frac{x^4 - y^4}{x^2 + y^2} = \lim_{(x, y)\to(0, 0)} \frac{(x^2 + y^2)(x^2 - y^2)}{x^2 + y^2} =$

$\lim_{(x, y)\to(0, 0)} (x^2 - y^2) = 0$.

[6] $\lim_{(x, y)\to(1, 2)} \frac{xy - y}{x^2 - x + 2xy - 2y} = \lim_{(x, y)\to(1, 2)} \frac{y(x - 1)}{x(x - 1) + 2y(x - 1)} =$

$\lim_{(x, y)\to(1, 2)} \frac{y(x - 1)}{(x + 2y)(x - 1)} = \lim_{(x, y)\to(1, 2)} \frac{y}{x + 2y} = \frac{2}{1 + 4} = \frac{2}{5}$

[7] $\displaystyle\lim_{(x,y)\to(0,0)} \frac{3x^3 - 2x^2y + 3y^2x - 2y^3}{x^2 + y^2} = \lim_{(x,y)\to(0,0)} \frac{x^2(3x - 2y) + y^2(3x - 2y)}{x^2 + y^2} =$

$$\lim_{(x,y)\to(0,0)} \frac{(x^2 + y^2)(3x - 2y)}{x^2 + y^2} = \lim_{(x,y)\to(0,0)} (3x - 2y) = 0$$

[8] $\displaystyle\lim_{(x,y)\to(0,0)} \frac{x^3 - x^2y + xy^2 - y^3}{x^2 + y^2} = \lim_{(x,y)\to(0,0)} \frac{x^2(x - y) + y^2(x - y)}{x^2 + y^2} =$

$$\lim_{(x,y)\to(0,0)} \frac{(x^2 + y^2)(x - y)}{x^2 + y^2} = \lim_{(x,y)\to(0,0)} (x - y) = 0$$

[9] $\displaystyle\lim_{(x,y,z)\to(2,3,1)} \frac{y^2 - 4y + 3}{x^2z(y - 3)} = \lim_{(x,y,z)\to(2,3,1)} \frac{(y - 3)(y - 1)}{x^2z(y - 3)} =$

$$\lim_{(x,y,z)\to(2,3,1)} \frac{y - 1}{x^2z} = \frac{3 - 1}{4(1)} = \frac{1}{2}$$

[10] $\displaystyle\lim_{(x,y,z)\to(2,1,2)} \frac{x^2 - z^2}{x^3 - z^3} = \lim_{(x,y,z)\to(2,1,2)} \frac{(x + z)(x - z)}{(x - z)(x^2 + xz + z^2)} =$

$$\lim_{(x,y,z)\to(2,1,2)} \frac{x + z}{x^2 + xz + z^2} = \frac{2 + 2}{4 + 4 + 4} = \frac{1}{3}$$

Note: In future exercises, L denotes the indicated limit.

[11] Along the path $x = 0$, $L = \dfrac{-y^2}{2y^2} = -\dfrac{1}{2}$, and along the path $y = 0$, $L = \dfrac{2x^2}{x^2} = 2$.

Thus, L DNE.

[12] Along the path $x = 0$, $L = \dfrac{5y^2}{4y^2} = \dfrac{5}{4}$, and along the path $y = 0$, $L = \dfrac{x^2}{3x^2} = \dfrac{1}{3}$.

Thus, L DNE.

[13] $L = \displaystyle\lim_{(x,y)\to(1,2)} \frac{(x - 1)(y - 2)}{(x - 1)^2 + (y - 2)^2}$. Along the path $(y - 2) = m(x - 1)$,

$$\frac{(x - 1)(y - 2)}{(x - 1)^2 + (y - 2)^2} = \frac{m(x - 1)^2}{(x - 1)^2 + m^2(x - 1)^2} = \frac{m}{1 + m^2}.$$

Since different values for L can be obtained using different values of m, L DNE.

[14] $L = \displaystyle\lim_{(x,y)\to(2,1)} \frac{(x - 2)^2}{(y - 1)(x - 2)}$.

Along the path $(y - 1) = m(x - 2)$, $\dfrac{(x - 2)^2}{(y - 1)(x - 2)} = \dfrac{(x - 2)^2}{m(x - 2)^2} = \dfrac{1}{m}$.

Since different values for L can be obtained using different values of m, L DNE.

[15] Along the path $y = 2x$,

$L = \dfrac{24x^5}{62x^5} = \dfrac{12}{31}$. Along the path $y = 0$, $L = \dfrac{0}{-2x^5} = 0$. Thus, L DNE.

[16] Along the path $y = x$, as $x \to 0$,

$L = \dfrac{3x^2}{7x^4} \to \infty$. Along the path $y = -x$, as $x \to 0$, $L = -\dfrac{3x^2}{7x^4} \to -\infty$. Thus, L DNE.

[17] Along the z-axis ($x = y = 0$),

$L = \dfrac{0}{z^2} = 0$, and along the path $x = y = z$, $L = \dfrac{3z^2}{3z^2} = 1$. Thus, L DNE.

[18] Along the y-axis $(x = z = 0)$,

$$L = \frac{3y^2}{y^2} = 3, \text{ and along the path } x = y = z,\ L = \frac{6z^2}{3z^2} = 2. \text{ Thus, L DNE.}$$

[19] Along the path $x = at$, $y = bt$, and $z = ct$,

$$L = \frac{a^3t^3 + b^3t^3 + c^3t^3}{abct^3} = \frac{(a^3 + b^3 + c^3)t^3}{abct^3} = \frac{a^3 + b^3 + c^3}{abc}.$$ Since different values for L can be obtained using different values of a, b, and c, L DNE.

[20] Along the path $x = 2 + at$, $y = 1 + bt$, and $z = ct$,

$$L = \frac{(at + bt + ct)^5}{c^3t^3(at)(bt)} = \frac{(a + b + c)^5t^5}{abc^3t^5} = \frac{(a + b + c)^3}{abc^3}.$$ Since different values for L can be obtained using different values of a, b, and c, L DNE.

Note: In Exercises 21–24 and 33–34, we use the fact that as $(x, y) \to (0, 0)$, $r \to 0$.

[21] $\dfrac{xy^2}{x^2 + y^2} = \dfrac{r^3 \cos\theta \sin^2\theta}{r^2} = r\cos\theta \sin^2\theta \to 0$ for all θ. Thus, L = 0.

[22] $\dfrac{x^3 - y^3}{x^2 + y^2} = \dfrac{r^3 \cos^3\theta - r^3 \sin^3\theta}{r^2} = r(\cos^3\theta - \sin^3\theta) \to 0$ for all θ. Thus, L = 0.

[23] $\dfrac{x^2 + y^2}{\sin(x^2 + y^2)} = \dfrac{r^2}{\sin r^2} \to 1$ for all θ. Thus, L = 1.

[24] $\dfrac{\sinh(x^2 + y^2)}{x^2 + y^2} = \dfrac{\sinh r^2}{r^2} \to 1 \left\{\text{since } \lim\limits_{w \to 0} \dfrac{\sinh w}{w} \left\{\tfrac{0}{0}\right\} = \lim\limits_{w \to 0} \dfrac{\cosh w}{1} = 1\right\}.$

Thus, L = 1.

[25] $x + y - 1 > 0 \Rightarrow x + y > 1$. f is continuous on $\{(x, y): x + y > 1\}$.

[26] $x^2 - y^2 \neq 0 \Rightarrow x^2 \neq y^2$. f is continuous on $\{(x, y): x^2 \neq y^2\}$.

[27] $1 - y^2 \geq 0 \Rightarrow |y| \leq 1$. f is continuous on $\{(x, y): x \geq 0 \text{ and } |y| \leq 1\}$.

[28] $25 - x^2 - y^2 \geq 0 \Rightarrow x^2 + y^2 \leq 25$. f is continuous on $\{(x, y): x^2 + y^2 \leq 25\}$.

[29] $x^2 + y^2 - z^2 \neq 0 \Rightarrow z^2 \neq x^2 + y^2$. f is continuous on $\{(x, y, z): z^2 \neq x^2 + y^2\}$.

[30] $\tan z$ is defined if $\tan z \neq \frac{\pi}{2} + \pi n$.

f is continuous on $\{(x, y, z): xy \geq 0 \text{ and } z \neq \frac{\pi}{2} + \pi n\}$.

[31] $x - 2 \geq 0 \Rightarrow x \geq 2$. f is continuous on $\{(x, y, z): x \geq 2,\ yz > 0\}$.

[32] $4 - x^2 - y^2 - z^2 \geq 0 \Rightarrow x^2 + y^2 + z^2 \leq 4$.

f is continuous on $\{(x, y, z): x^2 + y^2 + z^2 \leq 4\}$.

[33] $L = \lim\limits_{r \to 0} \dfrac{r^2}{\ln r^2} = 0$.

[34] $L = \lim\limits_{r \to 0} \dfrac{\sin(r^2 + r^2\cos^2\theta)}{r^2} \left\{\tfrac{0}{0}\right\} = \lim\limits_{r \to 0} \dfrac{(2r + 2r\cos^2\theta)\cos(r^2 + r^2\cos^2\theta)}{2r} =$

$\lim\limits_{r \to 0} (1 + \cos^2\theta)\cos(r^2 + r^2\cos^2\theta) = 1 + \cos^2\theta$. Since L depends on θ, L DNE.

[35] $h(x, y) = \dfrac{(x^2 - y^2)^2 - 4}{x^2 - y^2} = \dfrac{x^4 - 2x^2y^2 + y^4 - 4}{x^2 - y^2}$; continuous on $\{(x, y): x^2 \neq y^2\}$.

[36] $h(x, y) = \ln(3x + 2y - 4 + 5) = \ln(3x + 2y + 1)$;

continuous on $\{(x, y): 3x + 2y > -1\}$.

[37] $h(x, y) = (x + \tan y)^2 + 1 = x^2 + 2x\tan y + \tan^2 y + 1$;

continuous on $\{(x, y): y \neq \frac{\pi}{2} + \pi n\}$.

[38] $h(x, y) = e^{y\ln x} = (e^{\ln x})^y = x^y$; continuous on $\{(x, y): x > 0\}$.

[39] $g(f(x, y)) = e^{f(x, y)} = e^{x^2 + 2y}$; $h(f(x, y)) = (x^2 + 2y)^2 - 3(x^2 + 2y) =$

$(x^2 + 2y)(x^2 + 2y - 3)$; $f(g(t), h(t)) = (e^t)^2 + 2(t^2 - 3t) = e^{2t} + 2(t^2 - 3t)$

[40] $g(f(x, y, z)) = [f(x, y, z)]^2 = (2x + ye^z)^2$

[41] $f(g(x, y), k(x, y)) = (x - 2y)(2x + y) - 3(x - 2y) + (2x + y) =$

$2x^2 - 3xy - 2y^2 - x + 7y$

[42] $f(f(x, y), f(x, y)) = 2(2x + y) + 2x + y = 6x + 3y$

[43] The statement $\lim_{(x,y,z,w)\to(a,b,c,d)} f(x, y, z, w) = L$ means that

for every $\epsilon > 0$ there is a $\delta > 0$ such that if

$0 < \sqrt{(x - a)^2 + (y - b)^2 + (z - c)^2 + (w - d)^2} < \delta$, *then* $|f(x, y, z, w) - L| < \epsilon$.

[44] Let $\epsilon = f(a, b) > 0$. Since f is continuous, $\exists\delta > 0$ such that if

$0 < \sqrt{(x - a)^2 + (y - b)^2} < \delta$, then $|f(x, y) - f(a, b)| < \epsilon = f(a, b) \Rightarrow$

$-f(a, b) < f(x, y) - f(a, b) < f(a, b) \Rightarrow 0 < f(x, y) < 2f(a, b)$.

Thus, f is positive inside the circle of radius δ, centered at (a, b).

[45] (a) Let $f(x, y) = x$ and $\epsilon > 0$ be given. Choose $\delta = \epsilon$.

If $0 < \sqrt{(x - a)^2 + (y - b)^2} < \delta$,

then $\sqrt{(x - a)^2} = |x - a| = |f(x, y) - a| < \delta = \epsilon$. Thus, $\lim_{(x, y)\to(a, b)} x = a$.

(b) Let $f(x, y) = y$ and $\epsilon > 0$ be given. Choose $\delta = \epsilon$.

If $0 < \sqrt{(x - a)^2 + (y - b)^2} < \delta$,

then $\sqrt{(y - b)^2} = |y - b| = |f(x, y) - b| < \delta = \epsilon$. Thus, $\lim_{(x, y)\to(a, b)} y = b$.

[46] If $c = 0$, then $\lim_{(x, y)\to(a, b)} cf(x, y) = \lim_{(x, y)\to(a, b)} 0 = 0 = 0 \cdot L$.

If $c \neq 0$, then $\lim_{(x, y)\to(a, b)} f(x, y) = L$ means for any positive number $\frac{\epsilon}{|c|}$,

$\exists\delta > 0$ such that if $0 < \sqrt{(x - a)^2 + (y - b)^2} < \delta$, then $|f(x, y) - L| < \frac{\epsilon}{|c|} \Rightarrow$

$|cf(x, y) - cL| < \epsilon$. That is, given $\epsilon > 0$, $\exists\delta > 0$ such that if

$0 < \sqrt{(x - a)^2 + (y - b)^2} < \delta$, then $|cf(x, y) - cL| < \epsilon$ or

$\lim_{(x, y)\to(a, b)} cf(x, y) = cL$.

Exercises 16.3

Note: We sometimes denote $f_x(x, y)$ with f_x, $g_r(r, s)$ with g_r, etc.

[1] $f(x, y) = 2x^4y^3 - xy^2 + 3y + 1 \Rightarrow f_x = 8x^3y^3 - y^2$; $f_y = 6x^4y^2 - 2xy + 3$

[2] $f(x,\,y) = (x^3 - y^2)^5 \Rightarrow f_x = 5(x^3 - y^2)^4(3x^2) = 15x^2(x^3 - y^2)^4;$

$$f_y = 5(x^3 - y^2)^4(-2y) = -10y(x^3 - y^2)^4$$

[3] $f(r,\,s) = \sqrt{r^2 + s^2} \Rightarrow f_r = \frac{1}{2}(r^2 + s^2)^{-1/2}(2r) = \dfrac{r}{(r^2 + s^2)^{1/2}};\; f_s = \dfrac{s}{(r^2 + s^2)^{1/2}}$

[4] $f(s,\,t) = \frac{t}{s} - \frac{s}{t} \Rightarrow f_s = -\frac{t}{s^2} - \frac{1}{t};\; f_t = \frac{1}{s} + \frac{s}{t^2}$

[5] $f(x,\,y) = xe^y + y \sin x \Rightarrow f_x = e^y + y \cos x;\; f_y = xe^y + \sin x$

[6] $f(x,\,y) = e^x \ln xy \Rightarrow$

$$f_x = e^x \ln xy + e^x \cdot \frac{1}{xy} \cdot y = e^x \ln xy + \frac{e^x}{x};\; f_y = e^x \cdot \frac{1}{xy} \cdot x = \frac{e^x}{y}$$

[7] $f(t,\,v) = \ln \sqrt{\dfrac{t + v}{t - v}} = \frac{1}{2}\ln(t + v) - \frac{1}{2}\ln(t - v) \Rightarrow$

$$f_t(t,\,v) = \frac{1}{2(t + v)} - \frac{1}{2(t - v)} = \frac{-2v}{2(t^2 - v^2)} = -\frac{v}{t^2 - v^2};$$

$$f_v(t,\,v) = \frac{1}{2(t + v)} + \frac{1}{2(t - v)} = \frac{2t}{2(t^2 - v^2)} = \frac{t}{t^2 - v^2}$$

[8] $f(u,\,w) = \arctan \frac{u}{w} \Rightarrow$

$$f_u(u,\,w) = \frac{1}{1 + (u/w)^2} \cdot \frac{1}{w} = \frac{w}{w^2 + u^2};\; f_w(u,\,w) = \frac{1}{1 + (u/w)^2} \cdot -\frac{u}{w^2} = -\frac{u}{w^2 + u^2}$$

[9] $f(x,\,y) = x \cos \frac{x}{y} \Rightarrow f_x = \cos \frac{x}{y} - \frac{x}{y} \sin \frac{x}{y};\; f_y = x(-\sin \frac{x}{y})(-\frac{x}{y^2}) = (\frac{x}{y})^2 \sin \frac{x}{y}$

[10] $f(x,\,y) = \sqrt{4x^2 - y^2}\, \sec x \Rightarrow$

$$f_x(x,\,y) = (4x^2 - y^2)^{1/2} \sec x \tan x + \frac{4x \sec x}{(4x^2 - y^2)^{1/2}};\; f_y(x,\,y) = \frac{-y \sec x}{(4x^2 - y^2)^{1/2}}$$

[11] $f(x,\,y,\,z) = 3x^2 z + xy^2 \Rightarrow f_x(x,\,y,\,z) = 6xz + y^2;\; f_y(x,\,y,\,z) = 2xy;\; f_z(x,\,y,\,z) = 3x^2$

[12] $f(x,\,y,\,z) = x^2 y^3 z^4 + 2x - 5yz \Rightarrow f_x(x,\,y,\,z) = 2xy^3 z^4 + 2;$

$$f_y(x,\,y,\,z) = 3x^2 y^2 z^4 - 5z;\; f_z(x,\,y,\,z) = 4x^2 y^3 z^3 - 5y$$

[13] $f(r,\,s,\,t) = r^2 e^{2s} \cos t \Rightarrow$

$$f_r(r,\,s,\,t) = 2re^{2s} \cos t;\; f_s(r,\,s,\,t) = 2r^2 e^{2s} \cos t;\; f_t(r,\,s,\,t) = -r^2 e^{2s} \sin t$$

[14] $f(x,\,y,\,t) = \dfrac{x^2 - t^2}{1 + \sin 3y} \Rightarrow f_x(x,\,y,\,t) = \dfrac{1}{1 + \sin 3y}(2x) = \dfrac{2x}{1 + \sin 3y};$

$$f_y(x,\,y,\,t) = -\frac{3 \cos 3y\,(x^2 - t^2)}{(1 + \sin 3y)^2};\; f_t(x,\,y,\,t) = \frac{-2t}{1 + \sin 3y}$$

[15] $f(x,\,y,\,z) = xe^z - ye^x + ze^{-y} \Rightarrow$

$$f_x(x,\,y,\,z) = e^z - ye^x;\; f_y(x,\,y,\,z) = -e^x - ze^{-y};\; f_z(x,\,y,\,z) = xe^z + e^{-y}$$

[16] $f(r,\,s,\,v,\,p) = r^3 \tan s + \sqrt{s}\, e^{(v2)} - v \cos 2p \Rightarrow f_r(r,\,s,\,v,\,p) = 3r^2 \tan s;$

$$f_s(r,\,s,\,v,\,p) = r^3 \sec^2 s + \frac{e^{(v2)}}{2\sqrt{s}};\; f_v(r,\,s,\,v,\,p) = 2v\sqrt{s}\, e^{(v2)} - \cos 2p;$$

$$f_p(r,\,s,\,v,\,p) = 2v \sin 2p$$

[17] $f(q,\,v,\,w) = \sin^{-1}\sqrt{qv} + \sin vw \Rightarrow f_q(q,\,v,\,w) = \dfrac{1}{\sqrt{1 - (\sqrt{qv})^2}} \cdot \frac{1}{2}(qv)^{-1/2} \cdot v =$

$$\frac{v}{2\sqrt{qv}\,\sqrt{1 - qv}};\; f_v(q,\,v,\,w) = \frac{q}{2\sqrt{qv}\,\sqrt{1 - qv}} + w \cos vw;\; f_w(q,\,v,\,w) = v \cos vw$$

[18] $f(x, y, z) = xyze^{xyz} \Rightarrow f_x(x, y, z) = xyz \cdot e^{xyz} \cdot yz + yze^{xyz} = yz(xyz + 1)e^{xyz};$

$f_y(x, y, z) = xz(xyz + 1)e^{xyz}; \ f_z(x, y, z) = xy(xyz + 1)e^{xyz}$

[19] $w = xy^4 - 2x^2y^3 + 4x^2 - 3y \Rightarrow w_x = y^4 - 4xy^3 + 8x \Rightarrow w_{xy} = 4y^3 - 12xy^2;$

$w_y = 4xy^3 - 6x^2y^2 - 3 \Rightarrow w_{yx} = 4y^3 - 12xy^2$

[20] $w = \dfrac{x^2}{x + y} \Rightarrow w_x = \dfrac{(x + y)(2x) - (x^2)(1)}{(x + y)^2} = \dfrac{x^2 + 2xy}{(x + y)^2} \Rightarrow$

$w_{xy} = \dfrac{(x + y)^2(2x) - (x^2 + 2xy)(2)(x + y)}{(x + y)^4} = -\dfrac{2xy}{(x + y)^3};$

$w_y = -\dfrac{x^2}{(x + y)^2} \Rightarrow w_{yx} = -\dfrac{(x + y)^2(2x) - (x^2)(2)(x + y)}{(x + y)^4} = -\dfrac{2xy}{(x + y)^3}$

[21] $w = x^3e^{-2y} + y^{-2}\cos x \Rightarrow$

$w_x = 3x^2e^{-2y} - y^{-2}\sin x \Rightarrow w_{xy} = -6x^2e^{-2y} + 2y^{-3}\sin x;$

$w_y = -2x^3e^{-2y} - 2y^{-3}\cos x \Rightarrow w_{yx} = -6x^2e^{-2y} + 2y^{-3}\sin x$

[22] $w = y^2e^{(x^2)} + \dfrac{1}{x^2y^3} \Rightarrow$

$w_x = 2xy^2e^{(x^2)} - 2x^{-3}y^{-3} \Rightarrow w_{xy} = 4xye^{(x^2)} + 6x^{-3}y^{-4};$

$w_y = 2ye^{(x^2)} - 3x^{-2}y^{-4} \Rightarrow w_{yx} = 4xye^{(x^2)} + 6x^{-3}y^{-4}$

[23] $w = x^2\cosh\frac{z}{y} \Rightarrow w_x = 2x\cosh\frac{z}{y} \Rightarrow w_{xy} = -\frac{2xz}{y^2}\sinh\frac{z}{y};$

$w_y = x^2\sinh\frac{z}{y}\cdot\left(-\frac{z}{y^2}\right) = -\frac{x^2z}{y^2}\sinh\frac{z}{y} \Rightarrow w_{yx} = -\frac{2xz}{y^2}\sinh\frac{z}{y}$

[24] $w = \sqrt{x^2 + y^2 + z^2} = (x^2 + y^2 + z^2)^{1/2} \Rightarrow$

$w_x = \frac{1}{2}(x^2 + y^2 + z^2)^{-1/2}(2x) \Rightarrow w_{xy} = -xy(x^2 + y^2 + z^2)^{-3/2};$

$w_y = \frac{1}{2}(x^2 + y^2 + z^2)^{-1/2}(2y) \Rightarrow w_{yx} = -xy(x^2 + y^2 + z^2)^{-3/2}$

[25] $w = 3x^2y^3z + 2xy^4z^2 - yz \Rightarrow$

$w_x = 6xy^3z + 2y^4z^2 \Rightarrow w_{xy} = 18xy^2z + 8y^3z^2 \Rightarrow w_{xyz} = 18xy^2 + 16y^3z$

[26] $w = u^4vt^2 - 3uv^2t^3 \Rightarrow$

$w_t = 2u^4vt - 9uv^2t^2 \Rightarrow w_{tu} = 8u^3vt - 9v^2t^2 \Rightarrow w_{tut} = 8u^3v - 18v^2t$

[27] $u = v\sec(rt) \Rightarrow u_r = tv\sec(rt)\tan(rt) \Rightarrow u_{rv} = t\sec(rt)\tan(rt) \Rightarrow$

$u_{rvr} = t\left[t\sec(rt)\sec^2(rt) + t\sec(rt)\tan^2(rt)\right] = t^2(\sec rt)(\sec^2 rt + \tan^2 rt)$

[28] $v = y\ln(x^2 + z^4) \Rightarrow v_z = \dfrac{4yz^3}{x^2 + z^4} \Rightarrow v_{zz} = \dfrac{(x^2 + z^4)(12yz^2) - (4yz^3)(4z^3)}{(x^2 + z^4)^2} =$

$\dfrac{12yz^2x^2 - 4yz^6}{(x^2 + z^4)^2} = \dfrac{4z^2(3yx^2 - z^4)}{(x^2 + z^4)^2} \Rightarrow v_{zzy} = \dfrac{4z^2(3x^2 - z^4)}{(x^2 + z^4)^2}$

[29] $w = \sin xyz \Rightarrow w_x = yz\cos xyz \Rightarrow w_{xy} = z\cos xyz - xyz^2\sin xyz \Rightarrow$

$w_{xyz} = \cos xyz - xyz\sin xyz - 2xyz\sin xyz - x^2y^2z^2\cos xyz =$

$(1 - x^2y^2z^2)\cos xyz - 3xyz\sin xyz$

[30] $w = \dfrac{x^2}{y^2 + z^2} = x^2(y^2 + z^2)^{-1} \Rightarrow w_y = -2x^2y(y^2 + z^2)^{-2} \Rightarrow$

$w_{yy} = -2x^2(y^2 + z^2)^{-2} + 8x^2y^2(y^2 + z^2)^{-3} \Rightarrow$

$w_{yyz} = 8x^2z(y^2 + z^2)^{-3} - 48x^2y^2z(y^2 + z^2)^{-4} =$

$$8x^2z(y^2 + z^2)^{-4}\left[(y^2 + z^2) - 6y^2\right] = \frac{8x^2z(z^2 - 5y^2)}{(y^2 + z^2)^4}$$

[31] $w = r^4s^3t - 3s^2e^{rt} \Rightarrow w_r = 4r^3s^3t - 3s^2te^{rt} \Rightarrow w_{rr} = 12r^2s^3t - 3s^2t^2e^{rt} \Rightarrow$

$w_{rrs} = 36r^2s^2t - 6st^2e^{rt}$; $w_{rs} = 12r^3s^2t - 6ste^{rt} \Rightarrow w_{rsr} = 36r^2s^2t - 6st^2e^{rt}$;

$w_s = 3r^4s^2t - 6se^{rt} \Rightarrow w_{sr} = 12r^3s^2t - 6ste^{rt} \Rightarrow w_{srr} = 36r^2s^2t - 6st^2e^{rt}$

[32] $w = \tan uv + 2\ln(u + v) \Rightarrow w_u = v\sec^2(uv) + 2(u + v)^{-1} \Rightarrow$

$w_{uv} = 2uv\sec^2(uv)\tan(uv) + \sec^2(uv) - 2(u + v)^{-2} \Rightarrow$

$w_{uvv} = 4u\sec^2(uv)\tan(uv) + 4u^2v\sec^2(uv)\tan^2(uv) + 2u^2v\sec^4(uv) + 4(u + v)^{-3}$;

$w_v = u\sec^2(uv) + 2(u + v)^{-1} \Rightarrow$

$w_{vu} = \sec^2(uv) + 2uv\sec^2(uv)\tan(uv) - 2(u + v)^{-2} \Rightarrow$

$w_{vuv} = w_{uvv}$ since $w_{vu} = w_{uv}$

$w_{vv} = 2u^2\sec^2(uv)\tan(uv) - 2(u + v)^{-2} \Rightarrow w_{vvu} =$

$4u\sec^2(uv)\tan(uv) + 4u^2v\sec^2(uv)\tan^2(uv) + 2u^2v\sec^4(uv) + 4(u + v)^{-3} = w_{uvv}$

Note: In Exercises 33–36, let $w = f(x, y)$. In each exercise, $w_{xx} + w_{yy} = 0$.

[33] $w = \ln\sqrt{x^2 + y^2} = \frac{1}{2}\ln(x^2 + y^2) \Rightarrow$

$$w_x = \frac{x}{x^2 + y^2} \Rightarrow w_{xx} = \frac{y^2 - x^2}{(x^2 + y^2)^2};\ w_y = \frac{y}{x^2 + y^2} \Rightarrow w_{yy} = \frac{x^2 - y^2}{(x^2 + y^2)^2}.$$

[34] $w = \arctan\frac{y}{x} \Rightarrow w_x = \dfrac{-y/x^2}{\left[1 + (y/x)^2\right]} = \dfrac{-y}{x^2 + y^2} \Rightarrow w_{xx} = \dfrac{2xy}{(x^2 + y^2)^2}$;

$$w_y = \frac{1/x}{\left[1 + (y/x)^2\right]} = \frac{x}{x^2 + y^2} \Rightarrow w_{yy} = \frac{-2xy}{(x^2 + y^2)^2}.$$

[35] $w = \cos x\sinh y + \sin x\cosh y \Rightarrow$

$w_x = -\sin x\sinh y + \cos x\cosh y \Rightarrow w_{xx} = -\cos x\sinh y - \sin x\cosh y$;

$w_y = \cos x\cosh y + \sin x\sinh y \Rightarrow w_{yy} = \cos x\sinh y + \sin x\cosh y.$

[36] $w = e^{-x}\cos y + e^{-y}\cos x \Rightarrow$

$w_x = -e^{-x}\cos y - e^{-y}\sin x \Rightarrow w_{xx} = e^{-x}\cos y - e^{-y}\cos x$;

$w_y = -e^{-x}\sin y - e^{-y}\cos x \Rightarrow w_{yy} = -e^{-x}\cos y + e^{-y}\cos x.$

[37] $w = \cos(x - y) + \ln(x + y) \Rightarrow w_x = -\sin(x - y) + \dfrac{1}{x + y} \Rightarrow$

$w_{xx} = -\cos(x - y) - \dfrac{1}{(x + y)^2}$; $w_y = \sin(x - y) + \dfrac{1}{x + y} \Rightarrow$

$w_{yy} = -\cos(x - y) - \dfrac{1}{(x + y)^2}$. Thus, $w_{xx} - w_{yy} = 0$.

[38] $w = (y - 2x)^3 - \sqrt{y - 2x} \Rightarrow w_x = -6(y - 2x)^2 + (y - 2x)^{-1/2} \Rightarrow$

$w_{xx} = 24(y - 2x) + (y - 2x)^{-3/2}$; $w_y = 3(y - 2x)^2 - \frac{1}{2}(y - 2x)^{-1/2} \Rightarrow$

$w_{yy} = 6(y - 2x) + \frac{1}{4}(y - 2x)^{-3/2}$. Thus, $w_{xx} - 4\,w_{yy} = 0$.

[39] $w = e^{-c^2 t} \sin cx \Rightarrow w_x = ce^{-c^2 t} \cos cx \Rightarrow$

$$w_{xx} = -c^2 e^{-c^2 t} \sin cx;\ w_t = -c^2 e^{-c^2 t} \sin cx. \text{ Thus, } w_{xx} = w_t.$$

[40] $PV = knT \Rightarrow \dfrac{\partial V}{\partial T} = \dfrac{kn}{P}, \dfrac{\partial T}{\partial P} = \dfrac{V}{kn}$, and $\dfrac{\partial P}{\partial V} = -\dfrac{knT}{V^2}$.

$$\text{Thus, } \frac{\partial V}{\partial T}\frac{\partial T}{\partial P}\frac{\partial P}{\partial V} = \frac{kn}{P} \cdot \frac{V}{kn} \cdot \frac{-knT}{V^2} = \frac{-knT}{PV} = -1 \text{ since } PV = knT.$$

[41] $v = (\sin akt)(\sin kx) \Rightarrow v_x = k(\sin akt)(\cos kx) \Rightarrow v_{xx} = -k^2(\sin akt)(\sin kx)$;

$$v_t = ak(\cos akt)(\sin kx) \Rightarrow v_{tt} = -a^2k^2(\sin akt)(\sin kx). \text{ Thus, } v_{tt} = a^2 v_{xx}.$$

[42] $v = (x - at)^4 + \cos(x + at) \Rightarrow v_x = 4(x - at)^3 - \sin(x + at) \Rightarrow$

$v_{xx} = 12(x - at)^2 - \cos(x + at);\ v_t = -4a(x - at)^3 - a \sin(x + at) \Rightarrow$

$$v_{tt} = 12a^2(x - at)^2 - a^2 \cos(x + at). \text{ Thus, } v_{tt} = a^2 v_{xx}.$$

[43] $u(x, y) = x^2 - y^2;\ v(x, y) = 2xy \Rightarrow$

$u_x = 2x$ and $u_y = -2y;\ v_x = 2y = -u_y$ and $v_y = 2x = u_x$.

[44] $u(x, y) = \dfrac{y}{x^2 + y^2};\ v(x, y) = \dfrac{x}{x^2 + y^2} \Rightarrow u_x = \dfrac{-2xy}{(x^2 + y^2)^2}$ and

$$u_y = \frac{x^2 - y^2}{(x^2 + y^2)^2};\ v_x = \frac{y^2 - x^2}{(x^2 + y^2)^2} = -u_y \text{ and } v_y = \frac{-2xy}{(x^2 + y^2)^2}.$$

[45] $u(x, y) = e^x \cos y;\ v(x, y) = e^x \sin y \Rightarrow u_x = e^x \cos y$ and $u_y = -e^x \sin y$;

$$v_x = e^x \sin y = -u_y \text{ and } v_y = e^x \cos y = u_x.$$

[46] $u(x, y) = \cos x \cosh y + \sin x \sinh y;\ v(x, y) = \cos x \cosh y - \sin x \sinh y \Rightarrow$

$u_x = -\sin x \cosh y + \cos x \sinh y$ and $u_y = \cos x \sinh y + \sin x \cosh y$;

$v_x = -\sin x \cosh y - \cos x \sinh y = -u_y$ and $v_y = \cos x \sinh y - \sin x \cosh y = u_x$.

[47] There are 9 second partial derivatives: $w_{xx}, w_{xy}, w_{xz}, w_{yx}, w_{yy}, w_{yz}, w_{zx}, w_{zy}, w_{zz}$

[48] $w_t = \displaystyle\lim_{h \to 0} \frac{f(x, y, z, t + h, v) - f(x, y, z, t, v)}{h}$

[49] (a) $T = 10(x^2 + y^2)^2 \Rightarrow T_x(x, y) = 40x(x^2 + y^2) \Rightarrow T_x(1, 2) = 200$ deg/cm

(b) $T_y(x, y) = 40y(x^2 + y^2) \Rightarrow T_y(1, 2) = 400$ deg/cm

[50] (a) $f(x, y) = 300 - 2x^2 - 3y^2 \Rightarrow$

$$f_x(x, y) = -4x \Rightarrow f_x(4, 9) = -16 \text{ (depth in ft)/(distance in ft)}$$

(b) $f_y(x, y) = -6y \Rightarrow f_y(4, 9) = -54$ (depth in ft)/(distance in ft)

[51] (a) $V = 100/(x^2 + y^2 + z^2) \Rightarrow$

$$V_x(x, y, z) = -200x(x^2 + y^2 + z^2)^{-2} \Rightarrow V_x(2, -1, 1) = -\tfrac{100}{9} \text{ volts/in}$$

(b) $V_y(x, y, z) = -200y(x^2 + y^2 + z^2)^{-2} \Rightarrow V_y(2, -1, 1) = \frac{50}{9}$ volts/in

(c) $V_z(x, y, z) = -200z(x^2 + y^2 + z^2)^{-2} \Rightarrow V_z(2, -1, 1) = -\frac{50}{9}$ volts/in

[52] (a) $T = 4x^2 - y^2 + 16z^2 \Rightarrow T_x(x, y, z) = 8x \Rightarrow T_x(4, -2, 1) = 32$ deg/cm

(b) $T_y(x, y, z) = -2y \Rightarrow T_y(4, -2, 1) = 4$ deg/cm

(c) $T_z(x, y, z) = 32z \Rightarrow T_z(4, -2, 1) = 32$ deg/cm

[53] $C(x,\, y) = \dfrac{200}{x^2}\Big[e^{-0.02(y-10)^2/x^2} + e^{-0.02(y+10)^2/x^2}\Big] \Rightarrow$

$$\frac{\partial C}{\partial x} = -\frac{400}{x^3}\Big[e^{-0.02(y-10)^2/x^2} + e^{-0.02(y+10)^2/x^2}\Big] +$$

$$\frac{200}{x^2}\Bigg[e^{-0.02(y-10)^2/x^2}\Big(\frac{0.04(y-10)^2}{x^3}\Big) + e^{-0.02(y+10)^2/x^2}\Big(\frac{0.04(y+10)^2}{x^3}\Big)\Bigg]$$

$\partial C/\partial x \approx -36.58\ (\mu\text{g/m}^3)/\text{m}$

is the rate at which the concentration changes in the horizontal direction at (2, 5).

$$\frac{\partial C}{\partial y} = \frac{200}{x^2}\Bigg[e^{-0.02(y-10)^2/x^2}\Big(\frac{-0.04(y-10)}{x^2}\Big) + e^{-0.02(y+10)^2/x^2}\Big(\frac{-0.04(y+10)}{x^2}\Big)\Bigg]$$

$\partial C/\partial y \approx -0.229\ (\mu\text{g/m}^3)/\text{m}$

is the rate at which the concentration changes in the vertical direction at (2, 5).

[54] $I = V/\sqrt{R^2 + L^2\omega^2} \Rightarrow \dfrac{\partial I}{\partial R} = -\dfrac{VR}{(R^2 + L^2\omega^2)^{3/2}}.$

$\partial I/\partial R$ is the rate of change in the current w.r.t. resistance.

$\dfrac{\partial I}{\partial L} = -\dfrac{\omega^2 VL}{(R^2 + L^2\omega^2)^{3/2}}$ is the rate of change in the current w.r.t. inductance.

[55] (a) $S(p,\, q) = \dfrac{p}{q + p(1-q)} \Rightarrow S(10,\, 0.8) = \frac{25}{7} \approx 3.57,$

$S(100,\, 0.8) = \frac{125}{26} \approx 4.81$, and $S(1000,\, 0.8) = \frac{1250}{251} \approx 4.98.$

$$\lim_{p\to\infty} S(p,\, 0.8) = \lim_{p\to\infty}\frac{p}{0.8 + 0.2p} = \lim_{p\to\infty}\frac{5p}{p+4} = 5$$

(b) $\dfrac{\partial S}{\partial q} = -p\cdot\dfrac{1-p}{[q + p(1-q)]^2} = \dfrac{p(p-1)}{[q + p(1-q)]^2}$

(c) $q = 1 \Rightarrow \dfrac{\partial S}{\partial q} = p(p-1) = p^2 - p.$ Since $\dfrac{\partial S}{\partial q} \geq 0$ for $p \geq 1$ when $q = 1$,

as the number of processors increases, the rate of change of the speedup increases.

[56] $E = \dfrac{S(p,\, q)}{p} = \dfrac{1}{q + p(1-q)} \Rightarrow \dfrac{\partial E}{\partial p} = \dfrac{q-1}{[q + p(1-q)]^2} < 0$ since $0 \leq q < 1.$

[57] (a) $T = T_0 e^{-\lambda x}\sin(\omega t - \lambda x) \Rightarrow \dfrac{\partial T}{\partial t} = T_0\omega e^{-\lambda x}\cos(\omega t - \lambda x)$

is the rate of change of temperature w.r.t. time at a fixed depth x and

$$\frac{\partial T}{\partial x} = -T_0\lambda e^{-\lambda x}\big[\cos(\omega t - \lambda x) + \sin(\omega t - \lambda x)\big]$$

is the rate of change of temperature w.r.t. the depth at a fixed time t.

(b) $\dfrac{\partial^2 T}{\partial x^2} = T_0\lambda^2 e^{-\lambda x}\big[\cos(\omega t - \lambda x) + \sin(\omega t - \lambda x)\big] +$

$$T_0\lambda^2 e^{-\lambda x}\big[-\sin(\omega t - \lambda x) + \cos(\omega t - \lambda x)\big] = 2\lambda^2 T_0 e^{-\lambda x}\cos(\omega t - \lambda x).$$

If $k = \dfrac{2\lambda^2}{\omega}$, then $\dfrac{\partial T}{\partial t} = k\dfrac{\partial^2 T}{\partial x^2}.$

[58] $w = (\sin ax)(\cos by)e^{-\sqrt{a^2+b^2}\,z} \Rightarrow w_x = a(\cos ax)(\cos by)\,e^{-\sqrt{a^2+b^2}\,z} \Rightarrow$

$w_{xx} = -a^2(\sin ax)(\cos by)\,e^{-\sqrt{a^2+b^2}\,z}.\quad w_y = -b(\sin ax)(\sin by)\,e^{-\sqrt{a^2+b^2}\,z} \Rightarrow$

$w_{yy} = -b^2(\sin ax)(\cos by)\, e^{-\sqrt{a^2+b^2}\, z}$. $w_z = -\sqrt{a^2 + b^2}\,(\sin ax)(\cos by)\, e^{-\sqrt{a^2+b^2}\, z} \Rightarrow$

$w_{zz} = (a^2 + b^2)(\sin ax)(\cos by)\, e^{-\sqrt{a^2+b^2}\, z}$. Thus, $w_{xx} + w_{yy} + w_{zz} = 0$.

[59] (a) $V = 27.63y - 0.112xy \Rightarrow \partial V/\partial x = -0.112y$ ml/yr

is the rate at which lung capacity decreases with age for an adult male.

(b) $\partial V/\partial y = 27.63 - 0.112x$ ml/cm is difficult to interpret because we usually think of adult height y as fixed instead of a function of age x.

[60] $I(x, t) = 1000e^{-0.10x}\sin^3(\frac{\pi}{12}t) \Rightarrow \dfrac{\partial I}{\partial t} = 1000e^{-0.10x}(3)\sin^2(\frac{\pi}{12}t)\cos(\frac{\pi}{12}t)(\frac{\pi}{12})$.

$\partial I/\partial t = 0$ ft-candles/hr is the rate at which light intensity changes with respect to time at a fixed depth of 5 meters when $t = 6$. $\dfrac{\partial I}{\partial x} = -100e^{-0.10x}\sin^3(\frac{\pi}{12}t)$.

$\partial I/\partial x = -100e^{-1/2} \approx -60.65$ ft-candles/m is the rate at which light intensity changes with respect to depth at a fixed depth of 5 meters when $t = 6$.

[61] $e_k = \dfrac{p_k}{q_k}\dfrac{\partial q_k}{\partial p_k} = \dfrac{p_k}{q_k}\left(b_k p_1^{-a_{k1}} p_2^{-a_{k2}} \cdots p_{k-1}^{-a_{k,k-1}} p_{k+1}^{-a_{k,k+1}} \cdots p_n^{-a_{kn}}\right)\left(-a_{kk}p_k^{-a_{kk}-1}\right)$

$= -a_{kk}\dfrac{q_k}{q_k} = -a_{kk}$ for every k

[62] $\dfrac{\partial q_j}{\partial p_k} = \left(b_j p_1^{-a_{j1}} p_2^{-a_{j2}} \cdots p_{k-1}^{-a_{j,k-1}} p_{k+1}^{-a_{j,k+1}} \cdots p_n^{-a_{jn}}\right)\left(-a_{jk}p_k^{-a_{jk}-1}\right) = 0 \Leftrightarrow$

$a_{jk} = 0$ since $b_j > 0$.

[63] $z = 9 - x^2 - y^2 \Rightarrow \dfrac{\partial z}{\partial y} = -2y \Rightarrow$ slope of l at $P(1, 2, 4)$ is -4,

and an equation of l is $(z - 4) = -4(y - 2)$, or, $z = -4y + 12$.

Thus, parametric equations are $x = 1$, $y = t$, $z = -4t + 12$.

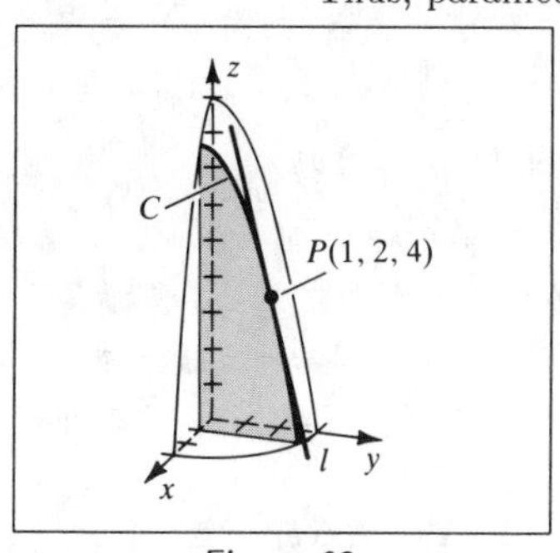

Figure 63

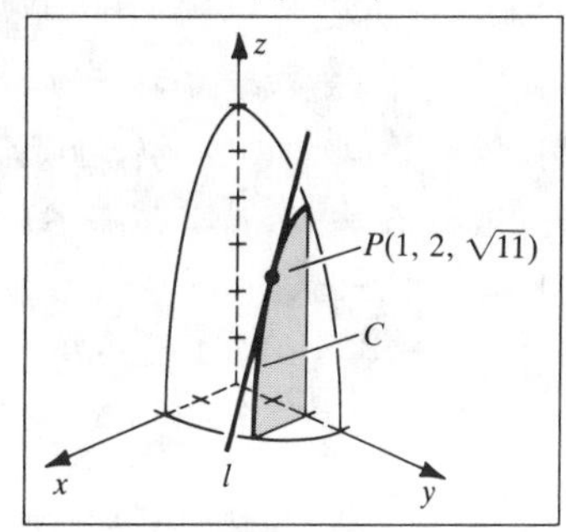

Figure 64

[64] C is the intersection of the plane $y = 2$ with the upper half of the ellipsoid

$\dfrac{x^2}{4} + \dfrac{y^2}{9} + \dfrac{z^2}{36} = 1$ and l is the tangent line to C at the point $P(1, 2, \sqrt{11})$.

$z = \sqrt{36 - 9x^2 - 4y^2} \Rightarrow \dfrac{\partial z}{\partial x} = \dfrac{-9x}{(36 - 9x^2 - 4y^2)^{1/2}} \Rightarrow$ slope of l at $P(1, 2, \sqrt{11})$ is

$-\frac{9}{\sqrt{11}}$, and an equation of l is $z - \sqrt{11} = -\frac{9}{\sqrt{11}}(x - 1)$, or, $z = -\frac{9}{\sqrt{11}}x + \frac{20}{\sqrt{11}}$.

Thus, parametric equations are $x = t$, $y = 2$, $z = \frac{1}{\sqrt{11}}(-9t + 20)$.

[65] $f_x(0.5,\ 0.2) \approx \dfrac{f(0.48,\ 0.2) - 4f(0.49,\ 0.2) + 3f(0.5,\ 0.2)}{2(0.01)} \approx 0.0079600438$

$f_y(0.5,\ 0.2) \approx \dfrac{f(0.5,\ 0.18) - 4f(0.5,\ 0.19) + 3f(0.5,\ 0.2)}{2(0.01)} \approx 0.0597349919$

$f(x,\ y) = y^2 \sin(xy) \Rightarrow f_x(x,\ y) = y^3 \cos(xy) \Rightarrow f_x(0.5,\ 0.2) \approx 0.0079600333.$

$f_y(x,\ y) = 2y\sin(xy) + xy^2\cos(xy) \Rightarrow f_y(0.5,\ 0.2) \approx 0.0598334499.$

[66] $f_x(0.5,\ 0.2) \approx \dfrac{f(0.48,\ 0.2) - 4f(0.49,\ 0.2) + 3f(0.5,\ 0.2)}{2(0.01)} \approx 0.1279679978$

$f_y(0.5,\ 0.2) \approx \dfrac{f(0.5,\ 0.18) - 4f(0.5,\ 0.19) + 3f(0.5,\ 0.2)}{2(0.01)} = 0.2599$

$f(x,\ y) = xy^3 + 4x^3y^2 \Rightarrow f_x(x,\ y) = y^3 + 12x^2y^2 \Rightarrow f_x(0.5,\ 0.2) = 0.128.$

$f_y(x,\ y) = 3xy^2 + 8x^3y \Rightarrow f_y(0.5,\ 0.2) = 0.26.$

[67] $f_{xx}(0.6,\ 0.8) \approx \dfrac{f(0.61,\ 0.8) - 2f(0.6,\ 0.8) + f(0.59,\ 0.8)}{(0.01)^2} \approx 1.8369$

$f_{yy}(0.6,\ 0.8) \approx \dfrac{f(0.6,\ 0.81) - 2f(0.6,\ 0.8) + f(0.6,\ 0.79)}{(0.01)^2} \approx 4.1743$

[68] $f_{xx}(0.6,\ 0.8) \approx \dfrac{f(0.61,\ 0.8) - 2f(0.6,\ 0.8) + f(0.59,\ 0.8)}{(0.01)^2} \approx 0.7121$

$f_{yy}(0.6,\ 0.8) \approx \dfrac{f(0.6,\ 0.81) - 2f(0.6,\ 0.8) + f(0.6,\ 0.79)}{(0.01)^2} \approx 1.3741$

Exercises 16.4

[1] (a) $w = f(x,\ y) = 5y^2 - xy \Rightarrow \Delta w = f(x + \Delta x,\ y + \Delta y) - f(x,\ y)$

$= 5(y + \Delta y)^2 - (x + \Delta x)(y + \Delta y) - (5y^2 - xy)$

$= 5y^2 + 10y\,\Delta y + 5(\Delta y)^2 - xy - x\,\Delta y - y\,\Delta x - \Delta x\,\Delta y - 5y^2 + xy$

$= 10y\,\Delta y - x\,\Delta y - y\,\Delta x + 5(\Delta y)^2 - \Delta x\,\Delta y$

(b) $dw = f_x(x,\ y)\,dx + f_y(x,\ y)\,dy = -y\,dx + (10y - x)\,dy$

(c) Since $dx = \Delta x$ and $dy = \Delta y$, $dw - \Delta w = \Delta x\Delta y - 5(\Delta y)^2$.

[2] (a) $w = xy - y^2 + 3x \Rightarrow$

$\Delta w = (x + \Delta x)(y + \Delta y) - (y + \Delta y)^2 + 3(x + \Delta x) - (xy - y^2 + 3x)$

$= xy + x\,\Delta y + y\,\Delta x + (\Delta x)(\Delta y) - y^2 - 2y\,\Delta y - (\Delta y)^2 + 3x + 3\Delta x$
$- xy + y^2 - 3x$

$= x\,\Delta y - 2y\,\Delta y + y\,\Delta x + 3\,\Delta x + \Delta x\,\Delta y - (\Delta y)^2$

(b) $dw = f_x(x,\ y)\,dx + f_y(x,\ y)\,dy = (y + 3)dx + (x - 2y)dy$

(c) Since $dx = \Delta x$ and $dy = \Delta y$, $dw - \Delta w = \Delta x\Delta y - (\Delta y)^2$.

Note: In Exercises 3–6, the expressions for ϵ_1 and ϵ_2 are not unique. See Example 2.

[3] $\Delta w = \left[4(y + \Delta y)^2 - 3(x + \Delta x)(y + \Delta y) + 2(x + \Delta x)\right] - \left[4y^2 - 3xy + 2x\right]$

$= 8y\Delta y + 4(\Delta y)^2 - 3x\Delta y - 3y\Delta x - 3\Delta x\Delta y + 2\Delta x$

$= (-3y + 2)\Delta x + (8y - 3x)\Delta y - (3\Delta y)\Delta x + (4\Delta y)\Delta y.$

Since $f_x = -3y + 2$ and $f_y = 8y - 3x$, $\epsilon_1 = -3\Delta y$ and $\epsilon_2 = 4\Delta y$.

[4] $\Delta w = \left[2(x + \Delta x) - (y + \Delta y)\right]^2 - \left[2x - y\right]^2$

$= 4(x + \Delta x)^2 - 4(x + \Delta x)(y + \Delta y) + (y + \Delta y)^2 - (4x^2 - 4xy + y^2)$

$= 8x\Delta x + 4(\Delta x)^2 - 4x\Delta y - 4y\Delta x - 4(\Delta x\Delta y) + 2y\Delta y + (\Delta y)^2$

$= (8x - 4y)\Delta x + (-4x + 2y)\Delta y + (4\Delta x - 4\Delta y)\Delta x + (\Delta y)\Delta y.$

Since $f_x = 8x - 4y$ and $f_y = -4x + 2y$, $\epsilon_1 = 4\Delta x - 4\Delta y$ and $\epsilon_2 = \Delta y$.

[5] $\Delta w = \left[(x + \Delta x)^3 + (y + \Delta y)^3\right] - \left[x^3 + y^3\right]$

$= 3x^2\Delta x + 3x(\Delta x)^2 + (\Delta x)^3 + 3y^2\Delta y + 3y(\Delta y)^2 + (\Delta y)^3$

$= (3x^2)\Delta x + (3y^2)\Delta y + \left[3x\Delta x + (\Delta x)^2\right]\Delta x + \left[3y\Delta y + (\Delta y)^2\right]\Delta y.$

Since $f_x = 3x^2$ and $f_y = 3y^2$, $\epsilon_1 = 3x\Delta x + (\Delta x)^2$ and $\epsilon_2 = 3y\Delta y + (\Delta y)^2$.

[6] $\Delta w = \left[2(x + \Delta x)^2 - (x + \Delta x)(y + \Delta y)^2 + 3(y + \Delta y)\right] - \left[2x^2 - xy^2 + 3y\right]$

$= 4x\Delta x + 2(\Delta x)^2 - 2xy\Delta y - x(\Delta y)^2 - y^2\Delta x - 2y\Delta x\Delta y - \Delta x(\Delta y)^2 + 3\Delta y$

$= (4x - y^2)\Delta x + (-2xy + 3)\Delta y + (2\Delta x - 2y\Delta y)\Delta x + (-x\Delta y - \Delta x\Delta y)\Delta y.$

Since $f_x = 4x - y^2$ and $f_y = -2xy + 3$,

$\epsilon_1 = 2\Delta x - 2y\Delta y$ and $\epsilon_2 = -x\Delta y - \Delta x\Delta y$.

[7] $w = x^3 - x^2y + 3y^2 \Rightarrow dw = w_x\,dx + w_y\,dy = (3x^2 - 2xy)\,dx + (-x^2 + 6y)\,dy$

[8] $w = 5x^2 + 4y - 3xy^3 \Rightarrow dw = (10x - 3y^3)\,dx + (4 - 9xy^2)\,dy$

[9] $w = x^2\sin y + 2y^{3/2} \Rightarrow dw = (2x\sin y)\,dx + (x^2\cos y + 3y^{1/2})\,dy$

[10] $w = ye^{-2x} - 3x^4 \Rightarrow dw = (-2ye^{-2x} - 12x^3)\,dx + (e^{-2x})\,dy$

[11] $w = x^2e^{xy} + (1/y^2) \Rightarrow dw = xe^{xy}(xy + 2)\,dx + (x^3e^{xy} - 2y^{-3})\,dy$

[12] $w = \ln(x^2 + y^2) + x\tan^{-1}y \Rightarrow$

$$dw = \left(\frac{2x}{x^2 + y^2} + \tan^{-1}y\right)dx + \left(\frac{2y}{x^2 + y^2} + \frac{x}{1 + y^2}\right)dy$$

[13] $w = x^2\ln(y^2 + z^2) \Rightarrow dw = w_x\,dx + w_y\,dy + w_z\,dz =$

$$\left[2x\ln(y^2 + z^2)\right]dx + \left(\frac{2x^2y}{y^2 + z^2}\right)dy + \left(\frac{2x^2z}{y^2 + z^2}\right)dz$$

[14] $w = x^2y^3z + e^{-2z} \Rightarrow dw = (2xy^3z)\,dx + (3x^2y^2z)\,dy + (x^2y^3 - 2e^{-2z})\,dz$

[15] $w = \dfrac{xyz}{x + y + z} \Rightarrow$

$$dw = \left[\frac{yz(y + z)}{(x + y + z)^2}\right]dx + \left[\frac{xz(x + z)}{(x + y + z)^2}\right]dy + \left[\frac{xy(x + y)}{(x + y + z)^2}\right]dz$$

[16] $w = x^2e^{yz} + y\ln z \Rightarrow dw = (2xe^{yz})\,dx + (x^2ze^{yz} + \ln z)\,dy + (x^2ye^{yz} + \frac{y}{z})\,dz$

[17] $w = x^2z + 4yt^3 - xz^2t \Rightarrow$

$$dw = (2xz - z^2t)\,dx + (4t^3)\,dy + (x^2 - 2xzt)\,dz + (12yt^2 - xz^2)\,dt$$

[18] $w = x^2y^3zt^{-1}v^4 \Rightarrow dw = (2xy^3zt^{-1}v^4)\,dx + (3x^2y^2zt^{-1}v^4)\,dy +$

$$(x^2y^3t^{-1}v^4)\,dz + (-x^2y^3zt^{-2}v^4)\,dt + (4x^2y^3zt^{-1}v^3)\,dv$$

[19] $f(x, y) = x^2 - 3x^3y^2 + 4x - 2y^3 + 6 \Rightarrow$

$$df = (2x - 9x^2y^2 + 4)\,dx + (-6x^3y - 6y^2)\,dy.$$

$x = -2,\ y = 3,\ dx = -2.02 - (-2) = -0.02$, and $dy = 3.01 - 3 = 0.01 \Rightarrow$

$$df = (-324)(-0.02) + (90)(0.01) = 7.38.$$

[20] $f(x, y) = x^2 - 2xy + 3y \Rightarrow df = (2x - 2y)\,dx + (-2x + 3)\,dy.$ $x = 1,\ y = 2,$

$dx = 0.03$, and $dy = -0.01 \Rightarrow df = (-2)(0.03) + (1)(-0.01) = -0.07.$

[21] $f(x, y, z) = x^2z^3 - 3yz^2 + x^{-3} + 2y^{1/2}z \Rightarrow$

$$df = (2xz^3 - 3x^{-4})\,dx + (-3z^2 + y^{-1/2}z)\,dy + (3x^2z^2 - 6yz + 2y^{1/2})\,dz.$$

$x = 1,\ y = 4,\ z = 2,\ dx = 0.02,\ dy = -0.03$, and $dz = -0.04 \Rightarrow$

$$df = (13)(0.02) + (-11)(-0.03) + (-32)(-0.04) = 1.87.$$

[22] $f(x, y, z) = xy + xz + yz \Rightarrow df = (y + z)\,dx + (x + z)\,dy + (x + y)\,dz.$

$x = -1,\ y = 2,\ z = 3,\ dx = 0.02,\ dy = -0.01$, and $dz = 0.03 \Rightarrow$

$$df = (5)(0.02) + (2)(-0.01) + (1)(0.03) = 0.11.$$

[23] (a) $S(x, y, z) = 2xy + 2xz + 2yz \Rightarrow$

$dS = (2y + 2z)\,dx + (2x + 2z)\,dy + (2x + 2y)\,dz.$

$x = 3,\ y = 4,\ z = 5$, and $dx = dy = dz = \pm\frac{1}{16}$ in. $= \pm\frac{1}{192}$ ft $\Rightarrow$

$$dS = (18 + 16 + 14)(\pm\tfrac{1}{192}) = \pm\tfrac{48}{192} = \pm\tfrac{1}{4}\text{ ft}^2.$$

(b) $V(x, y, z) = xyz \Rightarrow dV = (yz)\,dx + (xz)\,dy + (xy)\,dz \Rightarrow$

$$dV = (20 + 15 + 12)(\pm\tfrac{1}{192}) = \pm\tfrac{47}{192}\text{ ft}^3.$$

[24] (a) $h(x, y) = (x^2 + y^2)^{1/2} \Rightarrow dh = \left[\dfrac{x}{(x^2 + y^2)^{1/2}}\right]dx + \left[\dfrac{y}{(x^2 + y^2)^{1/2}}\right]dy.$

$x = 3,\ y = 4$, and $dx = dy = \pm 0.02 \Rightarrow dh = (\frac{3}{5} + \frac{4}{5})(\pm 0.02) = \pm 0.028$ cm.

(b) $A(x, y) = \frac{1}{2}xy \Rightarrow dA = (\frac{1}{2}y)\,dx + (\frac{1}{2}x)\,dy = \frac{1}{2}(4 + 3)(\pm 0.02) = \pm 0.07\text{ cm}^2.$

[25] (a) $P = 15{,}700S^{5/2}RD = 15{,}700(0.54)^{5/2}(0.113/2)(2) \approx 380$ lb

(b) $dP = \dfrac{\partial P}{\partial S}\,dS + \dfrac{\partial P}{\partial R}\,dR + \dfrac{\partial P}{\partial D}\,dD$

$= 15{,}700\left[(\frac{5}{2})S^{3/2}RD\,dS + S^{5/2}D\,dR + S^{5/2}R\,dD\right] \Rightarrow$

$$\frac{dP}{P} = \frac{5}{2}\frac{dS}{S} + \frac{dR}{R} + \frac{dD}{D} = \tfrac{5}{2}(\pm 3\%) + (\pm 2\%) + (\pm 2\%) = \pm 11.5\%$$

[26] (†) $\dfrac{1}{R} = \dfrac{1}{R_1} + \dfrac{1}{R_2} + \dfrac{1}{R_3} \Rightarrow -\dfrac{1}{R^2}\,dR = -\dfrac{1}{R_1^2}\,dR_1 - \dfrac{1}{R_2^2}\,dR_2 - \dfrac{1}{R_3^2}\,dR_3 \Rightarrow$

$$dR = R^2\left(\frac{1}{R_1^2}\,dR_1 + \frac{1}{R_2^2}\,dR_2 + \frac{1}{R_3^2}\,dR_3\right) = R^2\left(\frac{1}{R_1}\frac{dR_1}{R_1} + \frac{1}{R_2}\frac{dR_2}{R_2} + \frac{1}{R_3}\frac{dR_3}{R_3}\right).$$

$R_1 = 100,\ R_2 = 200,\ R_3 = 400$, and $\dfrac{dR_1}{R_1} = \dfrac{dR_2}{R_2} = \dfrac{dR_3}{R_3} = \pm 0.01 \Rightarrow$

$R = \frac{400}{7}$ {from (†)} and $dR = (\frac{400}{7})^2(\frac{1}{100} + \frac{1}{200} + \frac{1}{400})(\pm 0.01) = \pm\frac{4}{7}$ ohm.

[27] $s = \dfrac{A}{A - W} \Rightarrow ds = s_A\, dA + s_W\, dW = \left[\dfrac{-W}{(A - W)^2}\right] dA + \left[\dfrac{A}{(A - W)^2}\right] dW.$

When making scientific measurements, we do not usually know whether a measurement is too high or too low. This is the reason we have included both the "+" and "−" signs. It is possible that one measurement can be too high and the other measurement can be too low. To obtain the maximum error in the calculated value of s, we choose the signs of dA and dW to match the signs of s_A and s_W, respectively. Hence, $A = 12$ lb, $W = 5$ lb, $dA = \pm\frac{1}{2}$ ounce $= \pm\frac{1}{32}$ lb,

$$\text{and } dW = \pm\tfrac{1}{16} \Rightarrow ds = (-\tfrac{5}{49})(-\tfrac{1}{32}) + (\tfrac{12}{49})(\tfrac{1}{16}) = \tfrac{29}{1568} = \pm\, 0.0185.$$

[28] $PV = kT \Rightarrow P = \dfrac{kT}{V} \Rightarrow dP = -\dfrac{kT}{V^2}\, dV + \dfrac{k}{V}\, dT.$ $P = 0.5$, $V = 64$, $T = 350$, $dV = 70 - 64 = 6$, and $dT = 345 - 350 = -5 \Rightarrow k = \dfrac{PV}{T} = \dfrac{16}{175}$ and

$$dP = -\frac{(16/175)(350)}{64^2}(6) + \frac{16/175}{64}(-5) = -\frac{121}{2240} \approx -0.05 \text{ lb/in.}^2.$$

[29] Using Exercise 27,

$$\frac{ds}{s} = -\frac{W}{A(A - W)}\, dA + \frac{1}{A - W}\, dW = -\frac{W}{A - W}\frac{dA}{A} + \frac{W}{A - W}\frac{dW}{W} =$$

$$\frac{W}{A - W}\left(\frac{dW}{W} - \frac{dA}{A}\right) = \frac{W}{A - W}\left[4\% - (-2\%)\right] = \pm\frac{6W}{A - W}\%.$$

[30] $PV = kT \Rightarrow V = \dfrac{kT}{P} \Rightarrow \dfrac{dV}{V} = \dfrac{(k/P)\, dT - (kT/P^2)\, dP}{kT/P} \cdot \dfrac{P^2}{P^2} = \dfrac{kP\, dT - kT\, dP}{kTP} =$

$$\frac{dT}{T} - \frac{dP}{P} = 0.8\% - (-0.5\%) = \pm\, 1.3\%$$

[31] $R = \dfrac{kL}{D^2} \Rightarrow dR = \dfrac{k}{D^2}\, dL - \dfrac{2kL}{D^3}\, dD \Rightarrow \dfrac{dR}{R} = \dfrac{k}{D^2 R}\, dL - \dfrac{2kL}{D^3 R}\, dD = \dfrac{dL}{L} - 2\dfrac{dD}{D}.$

$$\frac{dL}{L} = \pm\, 1\% \text{ and } \frac{dD}{D} = \pm\, 3\% \Rightarrow \frac{dR}{R} = 1\% - 2(-3\%) = \pm\, 7\%.$$

[32] $F = \pi P R^4/(8vl) \Rightarrow$

$$dF = \frac{\pi R^4}{8vl}\, dP + \frac{4\pi P R^3}{8vl}\, dR \Rightarrow \frac{dF}{F} = \frac{dP}{P} + 4\frac{dR}{R} = 3\% + 4(-2\%) = -5\%.$$

[33] $T = 8(2x^2 + 4y^2 + 9z^2)^{1/2} \Rightarrow dT = \dfrac{4}{(2x^2 + 4y^2 + 9z^2)^{1/2}}(4x\, dx + 8y\, dy + 18z\, dz).$

$x = 6$, $y = 3$, $z = 2$, $dx = 0.1$, $dy = 0.3$, $dz = -0.02 \Rightarrow$

$$dT = \tfrac{4}{12}\left[24(0.1) + 24(0.3) + 36(-0.02)\right] = 2.96.$$

[34] $A = \frac{1}{2}a^2 \sin\theta$, where a is the length of the equal sides and θ is the angle between them. Thus, $dA = (a \sin\theta)\, da + (\frac{1}{2}a^2 \cos\theta)\, d\theta$. $a = 100$, $\theta = 120^\circ$, $da = 1$, and

$$d\theta = -1^\circ = -\tfrac{\pi}{180} \text{ rad} \Rightarrow dA = (50\sqrt{3})(1) + (-2500)(-\tfrac{\pi}{180}) \approx 130 \text{ in.}^2.$$

[35] $h = \dfrac{x}{\cot\alpha - \cot\beta} \Rightarrow$

$$dh = -\frac{x\csc^2\alpha}{(\cot\alpha - \cot\beta)^2}\,d\alpha + \frac{x\csc^2\beta}{(\cot\alpha - \cot\beta)^2}\,d\beta + \frac{1}{(\cot\alpha - \cot\beta)^2}\,dx.$$

$d\alpha = d\beta = \pm\,30'' = \pm\frac{30\pi}{60\cdot 60\cdot 180} = \pm\frac{\pi}{21{,}600} \approx \pm\,0.000145.$

$$\alpha = 15^\circ,\ \beta = 20^\circ, \text{ and } x = 2000 \Rightarrow dh = -\frac{2000\csc^2 15^\circ}{(\cot 15^\circ - \cot 20^\circ)^2}\cdot\left(-\frac{\pi}{21{,}600}\right)$$

$$+\frac{2000\csc^2 20^\circ}{(\cot 15^\circ - \cot 20^\circ)^2}\cdot\frac{\pi}{21{,}600} + \frac{1}{(\cot 15^\circ - \cot 20^\circ)^2}\,dx.$$

Approximating, $dh \approx 7.0448 + 1.0316\,dx$. $|dh| \le 10 \Rightarrow |7.0448 + 1.0316\,dx| \le 10$

$\Rightarrow -16.527 \le dx \le 2.865$. Thus, we choose $|dx| \le 2.865$ to assure us that

$|dh| \le 10$, that is, the maximum error in x must not exceed $\pm\,2.9$ ft.

[36] $$T = \frac{xy(\ln x - \ln y)}{(x-y)\ln 2} \Rightarrow dT = \left[\frac{xy}{x-y}\cdot\frac{1}{x\ln 2} + \frac{\ln x - \ln y}{\ln 2}\cdot\frac{-y^2}{(x-y)^2}\right]dx +$$

$$\left[\frac{xy}{x-y}\cdot\frac{-1}{y\ln 2} + \frac{\ln x - \ln y}{\ln 2}\cdot\frac{x^2}{(x-y)^2}\right]dy.$$

$x = 30,\ y = 60,\ dx = (0.10)(30) = 3$, and $dy = (0.10)(60) = 6 \Rightarrow$

$$dT = \left(-\frac{2}{\ln 2} + 4\right)(3) + \left(\frac{1}{\ln 2} - 1\right)(6) = 12 - 6 = 6 \text{ min.}$$

[37] $S = 2\pi rh + 2\pi r^2 \Rightarrow dS = (2\pi h + 4\pi r)\,dr + (2\pi r)\,dh = 2\pi\left[(h + 2r)\,dr + r\,dh\right]$.

$r = 3,\ h = 8$, and $dr = dh = \pm\,0.05 \Rightarrow dS = 2\pi(14 + 3)(\pm\,0.05) = \pm\,1.7\pi$ in.^{2}.

[38] $S = 2xy + 2xz + 2yz \Rightarrow dS = (2y + 2z)\,dx + (2x + 2z)\,dy + (2x + 2y)\,dz$.

$x = 9,\ y = 6,\ z = 4,\ dx = 0.02,\ dy = -0.03$, and $dz = 0.01 \Rightarrow$

$$dS = 2(6 + 4)(0.02) + 2(9 + 4)(-0.03) + 2(9 + 6)(0.01) = -0.08 \text{ in.}^2.$$

The exact change in surface area is

$$\Delta S = 2\left[(9.02)(5.97) + (9.02)(4.01) + (5.97)(4.01)\right] - 2\left[(9)(6) + (9)(4) + (6)(4)\right] = -0.0814 \text{ in.}^2.$$

[39] $f_x(x, y) = \dfrac{4xy^2}{(x^2 + y^2)^2}$ and $f_y(x, y) = \dfrac{-4yx^2}{(x^2 + y^2)^2}$ are continuous except at $(0, 0)$ which is outside the domain of f. Thus, f is continuous on any rectangle R not containing the origin and by (16.17), f is differentiable at every point in its domain.

[40] $$f_x(x, y, z) = \frac{y^2 + z^2 - x^2 - 2xy - 2xz}{(x^2 + y^2 + z^2)^2},$$

$$f_y(x, y, z) = \frac{x^2 + z^2 - y^2 - 2yx - 2yz}{(x^2 + y^2 + z^2)^2}, \text{ and}$$

$$f_z(x, y, z) = \frac{x^2 + y^2 - z^2 - 2zx - 2zy}{(x^2 + y^2 + z^2)^2}.$$ Since f_x, f_y, and f_z are continuous except at $(0, 0, 0)$, which is outside the domain of f, by (16.17) f is differentiable on D.

Note that in this case a rectangle is actually a "box shape" in three dimensions.

41 (a) $f_x(0, 0) = \lim_{h \to 0} \frac{f(0 + h, 0) - f(0, 0)}{h} = \lim_{h \to 0} \frac{0 - 0}{h} = 0;$

$f_y(0, 0) = \lim_{h \to 0} \frac{f(0, 0 + h) - f(0, 0)}{h} = \lim_{h \to 0} \frac{0 - 0}{h} = 0.$

Thus, $f_x(0, 0)$ and $f_y(0, 0)$ exist and equal 0.

(b) Consider L $= \lim_{(x, y) \to (0, 0)} \frac{xy}{x^2 + y^2}$. Along the path $y = 0$, L $= 0$, whereas along the path $y = x$, L $= \frac{1}{2}$. Thus, L DNE, and f cannot be continuous at (0, 0).

(c) Since f is not continuous at (0, 0), it cannot be differentiable at (0, 0) by the contrapositive of (16.18).

42 (a) $f_x(0, 0, 0) = \lim_{h \to 0} \frac{f(0 + h, 0, 0) - f(0, 0, 0)}{h} = \lim_{h \to 0} \frac{0 - 0}{h} = 0.$

Similarly, $f_y(0, 0, 0) = f_z(0, 0, 0) = 0$.

Thus, $f_x(0, 0, 0)$, $f_y(0, 0, 0)$, and $f_z(0, 0, 0)$ exist and equal 0.

(b) Consider L $= \lim_{(x, y, z) \to (0, 0, 0)} \frac{xyz}{x^3 + y^3 + z^3}$. Along the path $y = 0$, L $= 0$, whereas along the path $x = y = z$, L $= \frac{1}{3}$. Thus, L DNE, and f is not continuous at (0, 0, 0). Hence, f cannot be differentiable at (0, 0, 0).

43 Let $f(x, y) = \frac{x^2}{4} + \frac{y^2}{9} - 1 = 0$ and $g(x, y) = \frac{(x - 1)^2}{10} + \frac{(y + 1)^2}{5} - 1 = 0.$

Then, $f_x(x, y) = \frac{1}{2}x$, $f_y(x, y) = \frac{2}{9}y$, $g_x(x, y) = \frac{1}{5}(x - 1)$, and $g_y(x, y) = \frac{2}{5}(y + 1)$.
For the first iteration with $(x_1, y_1) = (2, 1)$ we have:

$$\begin{bmatrix} f_x(2, 1) & f_y(2, 1) \\ g_x(2, 1) & g_y(2, 1) \end{bmatrix} \begin{bmatrix} \Delta x_1 \\ \Delta y_1 \end{bmatrix} \approx \begin{bmatrix} -f(2, 1) \\ -g(2, 1) \end{bmatrix} \Rightarrow$$

$$\begin{bmatrix} 1 & \frac{2}{9} \\ \frac{1}{5} & \frac{4}{5} \end{bmatrix} \begin{bmatrix} \Delta x_1 \\ \Delta y_1 \end{bmatrix} \approx \begin{bmatrix} -\frac{1}{9} \\ \frac{1}{10} \end{bmatrix} \Rightarrow \begin{bmatrix} \Delta x_1 \\ \Delta y_1 \end{bmatrix} \approx \begin{bmatrix} -0.1471 \\ 0.1618 \end{bmatrix}$$

Thus, $x_2 = x_1 + \Delta x_1 \approx 1.8529$ and $y_2 = y_1 + \Delta y_1 \approx 1.1618$.

In a similar manner, $(x_3, y_3) \approx (1.8460, 1.1546)$, and $(x_4, y_4) \approx (1.8460, 1.1546)$.

44 Let $f(x, y) = \frac{x^2}{9} + \frac{y^2}{4} - 1 = 0$ and $g(x, y) = \frac{(x - 1)^2}{2} - \frac{(y - 1)^2}{3} - 1 = 0.$

Then, $f_x(x, y) = \frac{2}{9}x$, $f_y(x, y) = \frac{1}{2}y$, $g_x(x, y) = x - 1$, and $g_y(x, y) = -\frac{2}{3}(y - 1)$.
For the first iteration with $(x_1, y_1) = (-1.5, -1.5)$ we have:

$$\begin{bmatrix} f_x(-1.5, -1.5) & f_y(-1.5, -1.5) \\ g_x(-1.5, -1.5) & g_y(-1.5, -1.5) \end{bmatrix} \begin{bmatrix} \Delta x_1 \\ \Delta y_1 \end{bmatrix} \approx \begin{bmatrix} -f(-1.5, -1.5) \\ -g(-1.5, -1.5) \end{bmatrix} \Rightarrow$$

$$\begin{bmatrix} -\frac{1}{3} & -\frac{3}{4} \\ -\frac{5}{2} & \frac{5}{3} \end{bmatrix} \begin{bmatrix} \Delta x_1 \\ \Delta y_1 \end{bmatrix} \approx \begin{bmatrix} \frac{3}{16} \\ -\frac{1}{24} \end{bmatrix} \Rightarrow \begin{bmatrix} \Delta x_1 \\ \Delta y_1 \end{bmatrix} \approx \begin{bmatrix} -0.1157 \\ -0.1986 \end{bmatrix}$$

Thus, $x_2 = x_1 + \Delta x_1 \approx -1.6157$ and $y_2 = y_1 + \Delta y_1 \approx -1.6986$. In a similar manner, $(x_3, y_3) \approx (-1.6105, -1.6874)$, and $(x_4, y_4) \approx (-1.6105, -1.6874)$.

[45] Using Definition (16.20)(ii) and following the technique used for two equations with first approximation (x_1, y_1, z_1) we have:

$f_x(x_1, y_1, z_1)\Delta x + f_y(x_1, y_1, z_1)\Delta y + f_z(x_1, y_1, z_1)\Delta z \approx -f(x_1, y_1, z_1)$

$g_x(x_1, y_1, z_1)\Delta x + g_y(x_1, y_1, z_1)\Delta y + g_z(x_1, y_1, z_1)\Delta z \approx -g(x_1, y_1, z_1)$

$h_x(x_1, y_1, z_1)\Delta x + h_y(x_1, y_1, z_1)\Delta y + h_z(x_1, y_1, z_1)\Delta z \approx -h(x_1, y_1, z_1)$

Writing these in matrix form and suppressing (x_1, y_1, z_1) for each function gives us:

$$\begin{bmatrix} f_x & f_y & f_z \\ g_x & g_y & g_z \\ h_x & h_y & h_z \end{bmatrix}\begin{bmatrix} \Delta x \\ \Delta y \\ \Delta z \end{bmatrix} = \begin{bmatrix} -f \\ -g \\ -h \end{bmatrix}$$

Exercises 16.5

[1] $\dfrac{\partial w}{\partial x} = \dfrac{\partial w}{\partial u}\dfrac{\partial u}{\partial x} + \dfrac{\partial w}{\partial v}\dfrac{\partial v}{\partial x} = (\sin v)(2x) + (u \cos v)(y) =$
$2x \sin(xy) + y(x^2 + y^2)\cos(xy)$

$\dfrac{\partial w}{\partial y} = \dfrac{\partial w}{\partial u}\dfrac{\partial u}{\partial y} + \dfrac{\partial w}{\partial v}\dfrac{\partial v}{\partial y} = (\sin v)(2y) + (u \cos v)(x) =$
$2y \sin(xy) + x(x^2 + y^2)\cos(xy)$

[2] $\dfrac{\partial w}{\partial x} = \dfrac{\partial w}{\partial u}\dfrac{\partial u}{\partial x} + \dfrac{\partial w}{\partial v}\dfrac{\partial v}{\partial x} = (v)(\sin y) + (u + 2v)(y \cos x) =$
$y \sin x \sin y + y(x \sin y + 2y \sin x)\cos x$

$\dfrac{\partial w}{\partial y} = \dfrac{\partial w}{\partial u}\dfrac{\partial u}{\partial y} + \dfrac{\partial w}{\partial v}\dfrac{\partial v}{\partial y} = (v)(x \cos y) + (u + 2v)(\sin x) =$
$xy \sin x \cos y + (x \sin y + 2y \sin x)\sin x$

[3] $\dfrac{\partial w}{\partial r} = \dfrac{\partial w}{\partial u}\dfrac{\partial u}{\partial r} + \dfrac{\partial w}{\partial v}\dfrac{\partial v}{\partial r} = (2u + 2v)(\ln s) + (2u)(2) =$
$2(r \ln s + 2r + s)(\ln s) + 4r \ln s = 2r(\ln s)^2 + 8r \ln s + 2s \ln s$

$\dfrac{\partial w}{\partial s} = \dfrac{\partial w}{\partial u}\dfrac{\partial u}{\partial s} + \dfrac{\partial w}{\partial v}\dfrac{\partial v}{\partial s} = (2u + 2v)(\frac{r}{s}) + (2u)(1) =$
$2(r \ln s + 2r + s)(\frac{r}{s}) + 2r \ln s = \dfrac{2r^2 \ln s}{s} + \dfrac{4r^2}{s} + 2r + 2r \ln s$

[4] $\dfrac{\partial w}{\partial r} = \dfrac{\partial w}{\partial t}\dfrac{\partial t}{\partial r} + \dfrac{\partial w}{\partial v}\dfrac{\partial v}{\partial r} = (ve^{tv})(1) + (te^{tv})(s) = e^{tv}(v + ts) = s(2r + s)e^{rs(r+s)}$

$\dfrac{\partial w}{\partial s} = \dfrac{\partial w}{\partial t}\dfrac{\partial t}{\partial s} + \dfrac{\partial w}{\partial v}\dfrac{\partial v}{\partial s} = (ve^{tv})(1) + (te^{tv})(r) = e^{tv}(v + tr) = r(2s + r)e^{rs(r+s)}$

[5] $\dfrac{\partial z}{\partial x} = \dfrac{\partial z}{\partial r}\dfrac{\partial r}{\partial x} + \dfrac{\partial z}{\partial s}\dfrac{\partial s}{\partial x} + \dfrac{\partial z}{\partial v}\dfrac{\partial v}{\partial x} = (3r^2)(e^y) + (1)(ye^x) + (2v)(2xy) =$
$3x^2e^{3y} + ye^x + 4x^3y^2$

$\dfrac{\partial z}{\partial y} = \dfrac{\partial z}{\partial r}\dfrac{\partial r}{\partial y} + \dfrac{\partial z}{\partial s}\dfrac{\partial s}{\partial y} + \dfrac{\partial z}{\partial v}\dfrac{\partial v}{\partial y} = (3r^2)(xe^y) + (1)(e^x) + (2v)(x^2) =$
$3x^3e^{3y} + e^x + 2x^4y$

[6] $\dfrac{\partial z}{\partial x} = \dfrac{\partial z}{\partial p}\dfrac{\partial p}{\partial x} + \dfrac{\partial z}{\partial q}\dfrac{\partial q}{\partial x} + \dfrac{\partial z}{\partial w}\dfrac{\partial w}{\partial x} = (q)(2) + (p + w)(1) + (q)(-2) = p + w = y$

$\dfrac{\partial z}{\partial y} = \dfrac{\partial z}{\partial p}\dfrac{\partial p}{\partial y} + \dfrac{\partial z}{\partial q}\dfrac{\partial q}{\partial y} + \dfrac{\partial z}{\partial w}\dfrac{\partial w}{\partial y} = (q)(-1) + (p + w)(-2) + (q)(2) =$
$q - 2(p + w) = x - 4y$

[7] $\dfrac{\partial r}{\partial u} = \dfrac{\partial r}{\partial x}\dfrac{\partial x}{\partial u} + \dfrac{\partial r}{\partial y}\dfrac{\partial y}{\partial u} = (\ln y)(3) + (\frac{x}{y})(vt) = 3\ln(uvt) + \dfrac{3u + vt}{u} =$
$3\ln(uvt) + 3 + \frac{vt}{u}$

$$\frac{\partial r}{\partial v} = \frac{\partial r}{\partial x}\frac{\partial x}{\partial v} + \frac{\partial r}{\partial y}\frac{\partial y}{\partial v} = (\ln y)(t) + (\tfrac{x}{y})(ut) = t\ln(uvt) + \frac{3u + vt}{v} = t\ln(uvt) + \tfrac{3u}{v} + t$$

$$\frac{\partial r}{\partial t} = \frac{\partial r}{\partial x}\frac{\partial x}{\partial t} + \frac{\partial r}{\partial y}\frac{\partial y}{\partial t} = (\ln y)(v) + (\tfrac{x}{y})(uv) = v\ln(uvt) + \frac{3u + vt}{t} = v\ln(uvt) + \tfrac{3u}{t} + v$$

[8] $$\frac{\partial r}{\partial u} = \frac{\partial r}{\partial w}\frac{\partial w}{\partial u} + \frac{\partial r}{\partial z}\frac{\partial z}{\partial u} = (2w\cos z)(2uvt) + (-w^2\sin z)(t^2) = u^3v^2t^2\left[4\cos(ut^2) - ut^2\sin(ut^2)\right]$$

$$\frac{\partial r}{\partial v} = \frac{\partial r}{\partial w}\frac{\partial w}{\partial v} + \frac{\partial r}{\partial z}\frac{\partial z}{\partial v} = (2w\cos z)(u^2t) + (-w^2\sin z)(0) = 2u^4vt^2\cos(ut^2)$$

$$\frac{\partial r}{\partial t} = \frac{\partial r}{\partial w}\frac{\partial w}{\partial t} + \frac{\partial r}{\partial z}\frac{\partial z}{\partial t} = (2w\cos z)(u^2v) + (-w^2\sin z)(2ut) = 2u^4v^2t\left[\cos(ut^2) - ut^2\sin(ut^2)\right]$$

[9] $$\frac{\partial p}{\partial r} = \frac{\partial p}{\partial u}\frac{\partial u}{\partial r} + \frac{\partial p}{\partial v}\frac{\partial v}{\partial r} + \frac{\partial p}{\partial w}\frac{\partial w}{\partial r} = (2u)(2) + (6v)(-1) + (-8w)(1) =$$
$$4(x - 3y + 2r - s) - 6(2x + y - r + 2s) - 8(-x + 2y + r + s) = -34y + 6r - 24s$$

[10] $$\frac{\partial s}{\partial y} = \frac{\partial s}{\partial t}\frac{\partial t}{\partial y} + \frac{\partial s}{\partial r}\frac{\partial r}{\partial y} + \frac{\partial s}{\partial u}\frac{\partial u}{\partial y} + \frac{\partial s}{\partial v}\frac{\partial v}{\partial y}$$
$$= (r)(2xyz) + (t)(x^2z) + (e^v)(xz^2) + (ue^v)(xz)$$
$$= 2x^3y^2z^2 + x^3y^2z^2 + xz^2e^{xyz} + x^2yz^3e^{xyz} = 3x^3y^2z^2 + xz^2(1 + xyz)e^{xyz}$$

[11] $$\frac{dw}{dt} = \frac{\partial w}{\partial x}\frac{dx}{dt} + \frac{\partial w}{\partial y}\frac{dy}{dt} = (3x^2)\left[-\frac{1}{(t+1)^2}\right] + (-3y^2)\left[\frac{1}{(t+1)^2}\right] = -\frac{3(1 + t^2)}{(t+1)^4}$$

[12] $$\frac{dw}{dt} = \frac{\partial w}{\partial u}\frac{du}{dt} + \frac{\partial w}{\partial v}\frac{dv}{dt} = \left(\frac{1}{u + v}\right)(-2e^{-2t}) + \left(\frac{1}{u + v}\right)(3t^2 - 2t) = \frac{3t^2 - 2t - 2e^{-2t}}{e^{-2t} + t^3 - t^2}$$

[13] $$\frac{dw}{dt} = \frac{\partial w}{\partial r}\frac{dr}{dt} + \frac{\partial w}{\partial s}\frac{ds}{dt} + \frac{\partial w}{\partial v}\frac{dv}{dt}$$
$$= (2r)(2\sin t\cos t) + (-\tan v)(-\sin t) + (-s\sec^2 v)(4)$$
$$= 4\sin^3 t\cos t + \tan(4t)\sin t - 4\cos t\sec^2(4t)$$

[14] $$\frac{dw}{dt} = \frac{\partial w}{\partial x}\frac{dx}{dt} + \frac{\partial w}{\partial y}\frac{dy}{dt} + \frac{\partial w}{\partial z}\frac{dz}{dt} = (2xy^3z^4)(2) + (3x^2y^2z^4)(3) + (4x^2y^3z^3)(5)$$
$$= xy^2z^3(4yz + 9xz + 20xy) = 3(2t + 1)(3t - 2)^2(5t + 4)^3(90t^2 + 35t - 12)$$

Note: Exercises 15–18 use (16.22).

[15] $$F(x, y) = 2x^3 + x^2y + y^3 - 1 = 0 \Rightarrow \frac{dy}{dx} = -\frac{F_x(x, y)}{F_y(x, y)} = -\frac{6x^2 + 2xy}{x^2 + 3y^2}$$

[16] $$F(x, y) = x^4 + 2x^2y^2 - 3xy^3 + 2x = 0 \Rightarrow$$
$$\frac{dy}{dx} = -\frac{F_x(x, y)}{F_y(x, y)} = -\frac{4x^3 + 4xy^2 - 3y^3 + 2}{4x^2y - 9xy^2}$$

[17] $$F(x, y) = 6x + \sqrt{xy} - 3y + 4 = 0 \Rightarrow$$
$$\frac{dy}{dx} = -\frac{F_x(x, y)}{F_y(x, y)} = -\frac{6 + \dfrac{y}{2\sqrt{xy}}}{\dfrac{x}{2\sqrt{xy}} - 3}\cdot\frac{2\sqrt{xy}}{2\sqrt{xy}} = \frac{12\sqrt{xy} + y}{6\sqrt{xy} - x}$$

[18] $$F(x, y) = x^{2/3} + y^{2/3} - 4 = 0 \Rightarrow \frac{dy}{dx} = -\frac{F_x(x, y)}{F_y(x, y)} = -\frac{\frac{2}{3}x^{-1/3}}{\frac{2}{3}y^{-1/3}} = -(\tfrac{y}{x})^{1/3}$$

Note: Exercises 19–22 use (16.23).

[19] $F(x,\ y,\ z) = 2xz^3 - 3yz^2 + x^2y^2 + 4z = 0 \Rightarrow \dfrac{\partial z}{\partial x} = -\dfrac{F_x(x,\ y,\ z)}{F_z(x,\ y,\ z)} =$

$$-\frac{2z^3 + 2xy^2}{6xz^2 - 6yz + 4} \text{ and } \frac{\partial z}{\partial y} = -\frac{F_y(x,\ y,\ z)}{F_z(x,\ y,\ z)} = -\frac{2x^2y - 3z^2}{6xz^2 - 6yz + 4}$$

[20] $F(x,\ y,\ z) = xz^2 + 2x^2y - 4y^2z + 3y - 2 = 0 \Rightarrow \dfrac{\partial z}{\partial x} = -\dfrac{F_x(x,\ y,\ z)}{F_z(x,\ y,\ z)} =$

$$-\frac{z^2 + 4xy}{2xz - 4y^2} \text{ and } \frac{\partial z}{\partial y} = -\frac{F_y(x,\ y,\ z)}{F_z(x,\ y,\ z)} = -\frac{2x^2 - 8yz + 3}{2xz - 4yz}$$

[21] $F(x,\ y,\ z) = xe^{yz} - 2ye^{xz} + 3ze^{xy} - 1 = 0 \Rightarrow$

$\dfrac{\partial z}{\partial x} = -\dfrac{F_x(x,\ y,\ z)}{F_z(x,\ y,\ z)} = -\dfrac{e^{yz} - 2yze^{xz} + 3yze^{xy}}{xye^{yz} - 2xye^{xz} + 3e^{xy}}$ and

$$\frac{\partial z}{\partial y} = -\frac{F_y(x,\ y,\ z)}{F_z(x,\ y,\ z)} = -\frac{xze^{yz} - 2e^{xz} + 3xze^{xy}}{xye^{yz} - 2xye^{xz} + 3e^{xy}}$$

[22] $F(x,\ y,\ z) = yx^2 + z^2 + \cos(xyz) - 4 = 0 \Rightarrow \dfrac{\partial z}{\partial x} = -\dfrac{F_x(x,\ y,\ z)}{F_z(x,\ y,\ z)} =$

$$-\frac{2xy - yz\sin(xyz)}{2z - xy\sin(xyz)} \text{ and } \frac{\partial z}{\partial y} = -\frac{F_y(x,\ y,\ z)}{F_z(x,\ y,\ z)} = -\frac{x^2 - xz\sin(xyz)}{2z - xy\sin(xyz)}$$

[23] (a) $V = \pi r^2 h \Rightarrow \dfrac{dV}{dt} = \dfrac{\partial V}{\partial r}\dfrac{dr}{dt} + \dfrac{\partial V}{\partial h}\dfrac{dh}{dt} = (2\pi rh)\dfrac{dr}{dt} + (\pi r^2)\dfrac{dh}{dt}.$

$r = 4,\ h = 7,\ \dfrac{dr}{dt} = 0.01$, and $\dfrac{dh}{dt} = 0.02 \Rightarrow$

$$\frac{dV}{dt} = (56\pi)(0.01) + (16\pi)(0.02) = 0.88\pi \approx 2.76 \text{ in.}^3/\text{min.}$$

(b) $S = 2\pi rh \Rightarrow \dfrac{dS}{dt} = \dfrac{\partial S}{\partial r}\dfrac{dr}{dt} + \dfrac{\partial S}{\partial h}\dfrac{dh}{dt} = (2\pi h)\dfrac{dr}{dt} + (2\pi r)\dfrac{dh}{dt} =$

$$(14\pi)(0.01) + (8\pi)(0.02) = 0.3\pi \approx 0.94 \text{ in.}^2/\text{min.}$$

[24] Let s denote the length of an equal side, and θ the included angle. $A = \frac{1}{2}s^2\sin\theta \Rightarrow$

$\dfrac{dA}{dt} = \dfrac{\partial A}{\partial s}\dfrac{ds}{dt} + \dfrac{\partial A}{\partial \theta}\dfrac{d\theta}{dt} = (s\sin\theta)\dfrac{ds}{dt} + (\frac{1}{2}s^2\cos\theta)\dfrac{d\theta}{dt}.$ $s = 20,\ \theta = 60°,\ \dfrac{ds}{dt} = 0.1,$

and $\dfrac{d\theta}{dt} = 2° = \frac{\pi}{90} \Rightarrow \dfrac{dA}{dt} = (10\sqrt{3})(0.1) + (100)(\frac{\pi}{90}) = \sqrt{3} + \frac{10\pi}{9} \approx 5.22 \text{ ft}^2/\text{hr.}$

[25] $T = \dfrac{PV}{k} \Rightarrow \dfrac{dT}{dt} = \dfrac{\partial T}{\partial P}\dfrac{dP}{dt} + \dfrac{\partial T}{\partial V}\dfrac{dV}{dt} = \dfrac{V}{k}\dfrac{dP}{dt} + \dfrac{P}{k}\dfrac{dV}{dt}$

[26] $V = \pi r^2 h \Rightarrow \dfrac{dV}{dt} = \dfrac{\partial V}{\partial r}\dfrac{dr}{dt} + \dfrac{\partial V}{\partial h}\dfrac{dh}{dt} = (2\pi rh)\dfrac{dr}{dt} + (\pi r^2)\dfrac{dh}{dt}$

[27] $V = \dfrac{8T}{P} \Rightarrow \dfrac{dV}{dt} = \dfrac{\partial V}{\partial T}\dfrac{dT}{dt} + \dfrac{\partial V}{\partial P}\dfrac{dP}{dt} = \left(\dfrac{8}{P}\right)\dfrac{dT}{dt} - \left(\dfrac{8T}{P^2}\right)\dfrac{dP}{dt}.$ $T = 200,\ P = 10,$

$$\frac{dT}{dt} = 2, \text{ and } \frac{dP}{dt} = \tfrac{1}{2} \Rightarrow \frac{dV}{dt} = (\tfrac{8}{10})(2) - (\tfrac{1600}{100})(\tfrac{1}{2}) = -6.4 \text{ in.}^3/\text{min.}$$

[28] $V = \frac{1}{3}\pi r^2 h \Rightarrow h = \dfrac{3V}{\pi r^2} \Rightarrow \dfrac{dh}{dt} = \dfrac{\partial h}{\partial V}\dfrac{dV}{dt} + \dfrac{\partial h}{\partial r}\dfrac{dr}{dt} = \left(\dfrac{3}{\pi r^2}\right)\dfrac{dV}{dt} + \left(-\dfrac{6V}{\pi r^3}\right)\dfrac{dr}{dt}.$

$V = 40,\ r = 5,\ \dfrac{dV}{dt} = 6$, and $\dfrac{dr}{dt} = \frac{1}{4} \Rightarrow$

$$\frac{dh}{dt} = (\tfrac{3}{25\pi})(6) + (-\tfrac{240}{125\pi})(\tfrac{1}{4}) = \tfrac{6}{25\pi} \approx 0.076 \text{ in./min.}$$

[29] $\frac{dS}{dt} = \frac{\partial S}{\partial x}\frac{dx}{dt} + \frac{\partial S}{\partial y}\frac{dy}{dt} = 0.007184\left[0.425x^{-0.575}y^{0.725}\frac{dx}{dt} + 0.725x^{0.425}y^{-0.275}\frac{dy}{dt}\right].$

$x = 13,\ y = 86,\ \frac{dx}{dt} = 2$, and $\frac{dy}{dt} = 9 \Rightarrow$

$$\frac{dS}{dt} \approx 0.007184\left[2.45687(2) + 0.63354(9)\right] \approx 0.07626 \text{ m}^2/\text{yr} = 762.6 \text{ cm}^2/\text{yr}.$$

[30] $T = \frac{1}{k}\left(P + \frac{a}{V^2}\right)(V - b) \Rightarrow$

$$\frac{dT}{dt} = \frac{\partial T}{\partial P}\frac{dP}{dt} + \frac{\partial T}{\partial V}\frac{dV}{dt} = \frac{1}{k}(V - b)\frac{dP}{dt} + \frac{1}{k}\left[\left(P + \frac{a}{V^2}\right)(1) + (V - b)\left(-\frac{2a}{V^3}\right)\right]$$

$$= \frac{1}{k}\left[(V - b)\frac{dP}{dt} + \left(P - \frac{a}{V^2} + \frac{2ab}{V^3}\right)\frac{dV}{dt}\right]$$

[31] $\frac{\partial}{\partial R_k}\left(\frac{1}{R}\right) = \frac{\partial}{\partial R_k}\left(\sum_{i=1}^{n}\frac{1}{R_i}\right) \Rightarrow -\frac{1}{R^2}\frac{\partial R}{\partial R_k} = -\frac{1}{R_k^2} \Rightarrow \frac{\partial R}{\partial R_k} = \left(\frac{R}{R_k}\right)^2$

[32] Let $u = tx$, $v = ty$, and $w = f(u, v)$. $\frac{\partial w}{\partial t} = \frac{\partial w}{\partial u}\frac{\partial u}{\partial t} + \frac{\partial w}{\partial v}\frac{\partial v}{\partial t} =$

$f_u(u, v)\cdot x + f_v(u, v)\cdot y$. Also, $w = f(u, v) = t^n f(x, y) \Rightarrow \frac{\partial w}{\partial t} = nt^{n-1}f(x, y)$.

Since t can be any real number, we let $t = 1$ and it follows that

$$xf_x(x, y) + yf_y(x, y) = nf(x, y).$$

[33] $f(tx, ty) = 2(tx)^3 + 3(tx)^2(ty) + (ty)^3 = t^3(2x^3 + 3x^2y + y^3) = t^3 f(x, y) \Rightarrow$

$n = 3$. $xf_x(x, y) + yf_y(x, y) = x(6x^2 + 6xy) + y(3x^2 + 3y^2) =$

$$3(2x^3 + 3x^2y + y^3) = 3f(x, y).$$

[34] $f(tx, ty) = \frac{(tx)^3(ty)}{(tx)^2 + (ty)^2} = \frac{t^4(x^3y)}{t^2(x^2 + y^2)} = t^2 f(x, y) \Rightarrow n = 2.$

$xf_x(x, y) + yf_y(x, y) = x\left[\frac{x^4y + 3x^2y^3}{(x^2 + y^2)^2}\right] + y\left[\frac{x^5 - x^3y^2}{(x^2 + y^2)^2}\right] =$

$$\frac{2x^5y + 2x^3y^3}{(x^2 + y^2)^2} = \frac{2x^3y(x^2 + y^2)}{(x^2 + y^2)^2} = 2\cdot\frac{x^3y}{x^2 + y^2} = 2f(x, y).$$

[35] $f(tx, ty) = \arctan\frac{ty}{tx} = \arctan\frac{y}{x} = f(x, y) = t^n f(x, y) \Rightarrow n = 0.$

$$xf_x(x, y) + yf_y(x, y) = x\left[\frac{-y}{x^2 + y^2}\right] + y\left[\frac{x}{x^2 + y^2}\right] = 0.$$

[36] $f(tx, ty) = (tx)(ty)e^{(ty)/(tx)} = t^2xye^{y/x} = t^2 f(x, y) \Rightarrow n = 2.$

$xf_x(x, y) + yf_y(x, y) = x\left[ye^{y/x} - \frac{y^2}{x}e^{y/x}\right] + y\left[xe^{y/x} + ye^{y/x}\right] =$

$$2xye^{y/x} = 2f(x, y).$$

[37] $w_r = w_x x_r + w_y y_r = w_x(\cos\theta) + w_y(\sin\theta)$ and

$w_\theta = w_x x_\theta + w_y y_\theta = w_x(-r\sin\theta) + w_y(r\cos\theta) \Rightarrow$

$w_r^2 + \frac{1}{r^2}w_\theta^2 = w_x^2\cos^2\theta + 2w_xw_y\cos\theta\sin\theta + w_y^2\sin^2\theta +$

$$\frac{1}{r^2}\left[w_x^2r^2\sin^2\theta - 2w_xw_yr^2\sin\theta\cos\theta + w_y^2r^2\cos^2\theta\right] =$$

$$w_x^2(\sin^2\theta + \cos^2\theta) + w_y^2(\sin^2\theta + \cos^2\theta) = w_x^2 + w_y^2.$$

[38] $w_r = w_x x_r + w_y y_r \Rightarrow$

$$w_{rr} = (w_x x_r)_r + (w_y y_r)_r$$
$$= (w_{xx} x_r + w_{xy} y_r)\, x_r + w_x x_{rr} + (w_{yx} x_r + w_{yy} y_r)\, y_r + w_y y_{rr}$$
$$= w_{xx} x_r^2 + 2w_{xy} x_r y_r + w_{yy} y_r^2 + w_x x_{rr} + w_y y_{rr}.$$

Similarly, $w_{\theta\theta} = w_{xx} x_\theta^2 + 2w_{xy} x_\theta y_\theta + w_{yy} y_\theta^2 + w_x x_{\theta\theta} + w_y y_{\theta\theta}$.

Hence, $w_{rr} + w_{\theta\theta} = w_{xx}(x_r^2 + x_\theta^2) + 2w_{xy}(x_r y_r + x_\theta y_\theta) + w_{yy}(y_r^2 + y_\theta^2) +$
$w_x(x_{rr} + x_{\theta\theta}) + w_y(y_{rr} + y_{\theta\theta})$.

$x_r = e^r \cos\theta$, $x_\theta = -e^r \sin\theta$, $y_r = e^r \sin\theta$, and $y_\theta = e^r \cos\theta \Rightarrow$

$$w_{rr} + w_{\theta\theta} = w_{xx} e^{2r}(\cos^2\theta + \sin^2\theta) + 2w_{xy} e^{2r}(\cos\theta \sin\theta - \sin\theta \cos\theta) +$$
$$w_{yy} e^{2r}(\sin^2\theta + \cos^2\theta) + w_x(e^r \cos\theta - e^r \cos\theta) + w_y(e^r \sin\theta - e^r \sin\theta)$$
$$= e^{2r}(w_{xx} + w_{yy}).$$ Thus, $w_{xx} + w_{yy} = e^{-2r}(w_{rr} + w_{\theta\theta})$.

[39] From Exercise 38, $w_r = w_x x_r + w_y y_r$,

$w_{rr} = w_{xx} x_r^2 + 2w_{xy} x_r y_r + w_{yy} y_r^2 + w_x x_{rr} + w_y y_{rr}$, and

$w_{\theta\theta} = w_{xx} x_\theta^2 + 2w_{xy} x_\theta y_\theta + w_{yy} y_\theta^2 + w_x x_{\theta\theta} + w_y y_{\theta\theta}$.

Hence, $w_{rr} + \frac{1}{r^2} w_{\theta\theta} + \frac{1}{r} w_r = w_{xx}\left(x_r^2 + \frac{1}{r^2} x_\theta^2\right) + 2w_{xy}\left(x_r y_r + \frac{1}{r^2} x_\theta y_\theta\right) +$

$$w_{yy}\left(y_r^2 + \frac{1}{r^2} y_\theta^2\right) + w_x\left(x_{rr} + \frac{1}{r^2} x_{\theta\theta}\right) + w_y\left(y_{rr} + \frac{1}{r^2} y_{\theta\theta}\right) + \frac{1}{r}(w_x x_r + w_y y_r) =$$

$$w_{xx}(\cos^2\theta + \sin^2\theta) + 2w_{xy}(\cos\theta \sin\theta - \sin\theta \cos\theta) + w_{yy}(\sin^2\theta + \cos^2\theta) +$$

$$w_x\left(0 - \tfrac{1}{r}\cos\theta\right) + w_y\left(0 - \tfrac{1}{r}\sin\theta\right) + \tfrac{1}{r}(w_x \cos\theta + w_y \sin\theta) = w_{xx} + w_{yy}.$$

[40] Let $v = f(u) + g(w)$, where $u = x - at$ and $w = x + at$.

$v_t = f'(u)\, u_t + g'(w)\, w_t = f'(u)(-a) + g'(w)(a)$ and

$$v_{tt} = -a f''(u)\, u_t + a g''(w)\, w_t = a^2 f''(u) + a^2 g''(w) = a^2\big[f''(u) + g''(w)\big].$$

$v_x = f'(u)\, u_x + g'(w)\, w_x = f'(u) + g'(w)$ and

$$v_{xx} = f''(u)\, u_x + g''(w)\, w_x = f''(u) + g''(w).$$ Thus, $v_{tt} = a^2 v_{xx}$.

Note: f and g are both functions of a single variable—

so a derivative symbol is used rather than a partial derivative symbol.

[41] $w_x = -\sin(x + y) - \sin(x - y) \Rightarrow w_{xx} = -\cos(x + y) - \cos(x - y)$.

$w_y = -\sin(x + y) + \sin(x - y) \Rightarrow w_{yy} = -\cos(x + y) - \cos(x - y)$.

Thus, $w_{xx} - w_{yy} = 0$. Note that this is a special case of Exercise 40 with

$f(x) = \cos x$, $t = y$, and $a = 1$.

[42] Let $u = x^2 + y^2$, and hence, $w = f(u)$.

$w_x = f'(u)\, u_x = f'(u)(2x)$ and $w_y = f'(u)\, u_y = f'(u)(2y)$.

Thus, $y\, w_x - x\, w_y = 2xyf'(u) - 2xyf'(u) = 0$. See note for Exercise 40.

[43] $w_x = w_u u_x + w_v v_x \Rightarrow w_{xx} = (w_u u_x)_x + (w_v v_x)_x$

$$= \big[(w_{uu} u_x + w_{uv} v_x)\, u_x + w_u u_{xx}\big] + \big[(w_{vu} u_x + w_{vv} v_x)\, v_x + w_v v_{xx}\big]$$
$$= w_{uu}(u_x)^2 + (w_{uv} + w_{vu})\, u_x v_x + w_{vv}(v_x)^2 + w_u u_{xx} + w_v v_{xx}.$$

[44] $w_x = w_u u_x + w_v v_x \Rightarrow w_{xy} = (w_u u_x)_y + (w_v v_x)_y$

$= \left[(w_{uu} u_y + w_{uv} v_y) u_x + w_u u_{xy}\right] + \left[(w_{vu} u_y + w_{vv} v_y) v_x + w_v v_{xy}\right]$

$= w_{uu} u_x u_y + w_{uv} u_x v_y + w_{vu} u_y v_x + w_{vv} v_x v_y + w_u u_{xy} + w_v v_{xy}$.

[45] Parametric equations for the line passing through $A(x_1, y_1)$ and $B(x_2, y_2)$ are $x(t) = x_1 + (x_2 - x_1)t$ and $y(t) = y_1 + (y_2 - y_1)t$ for $0 \le t \le 1$. On this line, f can be written as a function of the single variable t. Hence, the one-dimensional mean value theorem applies and there exists $t^* \in (0, 1)$ such that if $x^* = x(t^*)$ and $y^* = y(t^*)$, then we have $f(x(1), y(1)) - f(x(0), y(0)) = \left[\frac{df}{dt}(x^*, y^*)\right](1 - 0) \Rightarrow$

$$f(x_2, y_2) - f(x_1, y_1) = \frac{\partial f}{\partial x}\frac{dx}{dt} + \frac{\partial f}{\partial y}\frac{dy}{dt} = f_x(x^*, y^*)(x_2 - x_1) + f_y(x^*, y^*)(y_2 - y_1).$$

[46] By Exercise 45, $f(x_2, y_2) - f(x_1, y_1) = 0 \Rightarrow f(x_2, y_2) = f(x_1, y_1)$ for all (x_2, y_2) and (x_1, y_1) in R. Since these points are arbitrary, f must be constant on R.

[47] Following Exercise 45, parametric equations for the line passing through $A(x_1, y_1, z_1)$ and $B(x_2, y_2, z_2)$ are $x(t) = x_1 + (x_2 - x_1)t$, $y(t) = y_1 + (y_2 - y_1)t$, and $z(t) = z_1 + (z_2 - z_1)t$ for $0 \le t \le 1$. On this line, f can be written as a function of the single variable t. Hence, the one-dimensional mean value theorem applies and there exists $t^* \in (0, 1)$ such that if $x^* = x(t^*)$, $y^* = y(t^*)$, and $z^* = z(t^*)$, then we have $f(x(1), y(1), z(1)) - f(x(0), y(0), z(0)) = \left[\frac{df}{dt}(x^*, y^*, z^*)\right](1 - 0) \Rightarrow$

$$f(x_2, y_2, z_2) - f(x_1, y_1, z_1) = \frac{\partial f}{\partial x}\frac{dx}{dt} + \frac{\partial f}{\partial y}\frac{dy}{dt} + \frac{\partial f}{\partial z}\frac{dz}{dt} =$$

$$f_x(x^*, y^*, z^*)(x_2 - x_1) + f_y(x^*, y^*, z^*)(y_2 - y_1) + f_z(x^*, y^*, z^*)(z_2 - z_1).$$

[48] By Exercise 47, $f(x_2, y_2, z_2) - f(x_1, y_1, z_1) = 0 \Rightarrow f(x_2, y_2, z_2) = f(x_1, y_1, z_1)$ for all (x_2, y_2, z_2) and (x_1, y_1, z_1) in R (which is a "box shape" in three dimensions).

Since these points are arbitrary, f must be constant on R.

[49] (1) $u_r = u_x x_r + u_y y_r = u_x \cos\theta + u_y \sin\theta$ and

$v_\theta = v_x x_\theta + v_y y_\theta = v_x(-r\sin\theta) + v_y(r\cos\theta) = r(-v_x\sin\theta + v_y\cos\theta)$.

Thus, $u_r = u_x\cos\theta + u_y\sin\theta = v_y\cos\theta - v_x\sin\theta$ (Cauchy-Riemann) $= \frac{1}{r}v_\theta$.

(2) $v_r = v_x x_r + v_y y_r = v_x\cos\theta + v_y\sin\theta$ and

$u_\theta = u_x x_\theta + u_y y_\theta = u_x(-r\sin\theta) + u_y(r\cos\theta) = r(-u_x\sin\theta + u_y\cos\theta)$.

Thus, $v_r = v_x\cos\theta + v_y\sin\theta = -u_y\cos\theta + u_x\sin\theta = -\frac{1}{r}u_\theta$.

[50] $u = \frac{r\sin\theta}{r^2} = \frac{\sin\theta}{r}$ and $v = \frac{r\cos\theta}{r^2} = \frac{\cos\theta}{r} \Rightarrow u_r = -\frac{\sin\theta}{r^2}$, $u_\theta = \frac{\cos\theta}{r}$,

$v_r = -\frac{\cos\theta}{r^2}$, and $v_\theta = -\frac{\sin\theta}{r}$. Thus, $u_r = \frac{1}{r}v_\theta$ and $v_r = -\frac{1}{r}u_\theta$.

Exercises 16.6

Note: Exercises 1–6 use (16.26) and (16.31).

[1] $\nabla f(x, y) = f_x(x, y)\mathbf{i} + f_y(x, y)\mathbf{j} = \dfrac{x}{(x^2 + y^2)^{1/2}}\mathbf{i} + \dfrac{y}{(x^2 + y^2)^{1/2}}\mathbf{j} \Rightarrow$
$\nabla f(-4, 3) = -\frac{4}{5}\mathbf{i} + \frac{3}{5}\mathbf{j}$

[2] $\nabla f(x, y) = -5\mathbf{i} + 7\mathbf{j} \Rightarrow \nabla f(2, 6) = -5\mathbf{i} + 7\mathbf{j}$

[3] $\nabla f(x, y) = 3e^{3x}\tan y\,\mathbf{i} + e^{3x}\sec^2 y\,\mathbf{j} \Rightarrow \nabla f(0, \frac{\pi}{4}) = 3\mathbf{i} + 2\mathbf{j}$

[4] $\nabla f(x, y) = \left[\ln(x - y) + \frac{x}{x - y}\right]\mathbf{i} - \frac{x}{x - y}\mathbf{j} \Rightarrow \nabla f(5, 4) = 5\mathbf{i} - 5\mathbf{j}$

[5] $\nabla f(x, y, z) = f_x(x, y, z)\mathbf{i} + f_y(x, y, z)\mathbf{j} + f_z(x, y, z)\mathbf{k} = -4x\mathbf{i} + z^3\mathbf{j} + 3yz^2\mathbf{k} \Rightarrow$
$\nabla f(2, -3, 1) = -8\mathbf{i} + \mathbf{j} - 9\mathbf{k}$

[6] $\nabla f(x, y, z) = y^2e^z\mathbf{i} + 2xye^z\mathbf{j} + xy^2e^z\mathbf{k} \Rightarrow \nabla f(2, -1, 0) = \mathbf{i} - 4\mathbf{j} + 2\mathbf{k}$

Note: Exercises 7–20 use (16.25) and (16.32).

[7] $D_{\mathbf{u}}f(x, y) = \nabla f(x, y)\cdot\mathbf{u} = f_x(x, y)\,u_1 + f_y(x, y)\,u_2$
$= (2x - 5y)(\frac{\sqrt{2}}{2}) + (-5x + 6y)(\frac{\sqrt{2}}{2}) \Rightarrow$
$D_{\mathbf{u}}f(3, -1) = (11)(\frac{\sqrt{2}}{2}) + (-21)(\frac{\sqrt{2}}{2}) = -5\sqrt{2} \approx -7.07.$

[8] $D_{\mathbf{u}}f(x, y) = (3x^2 - 6xy)(-\frac{1}{2}) + (-3x^2 - 3y^2)(\frac{1}{2}\sqrt{3}) \Rightarrow$
$D_{\mathbf{u}}f(1, -2) = -\frac{15}{2} - \frac{15}{2}\sqrt{3} = -\frac{15}{2}(1 + \sqrt{3}) \approx -20.49.$

[9] $\mathbf{a} = 2\mathbf{i} - 3\mathbf{j}$ and $\|\mathbf{a}\| = \sqrt{2^2 + (-3)^2} = \sqrt{13} \Rightarrow \mathbf{u} = \frac{2}{\sqrt{13}}\mathbf{i} - \frac{3}{\sqrt{13}}\mathbf{j}.$
$D_{\mathbf{u}}f(x, y) = \left(-\frac{y}{x^2 + y^2}\right)(\frac{2}{\sqrt{13}}) + \left(\frac{x}{x^2 + y^2}\right)(-\frac{3}{\sqrt{13}}) \Rightarrow$
$D_{\mathbf{u}}f(4, -4) = \frac{8}{32\sqrt{13}} - \frac{12}{32\sqrt{13}} = -\frac{1}{8\sqrt{13}} \approx -0.03.$

[10] $\mathbf{a} = -\mathbf{i} + 4\mathbf{j}$ and $\|\mathbf{a}\| = \sqrt{17} \Rightarrow \mathbf{u} = -\frac{1}{\sqrt{17}}\mathbf{i} + \frac{4}{\sqrt{17}}\mathbf{j}.$
$D_{\mathbf{u}}f(x, y) = (2x\ln y)(-\frac{1}{\sqrt{17}}) + \left(\frac{x^2}{y}\right)(\frac{4}{\sqrt{17}}) \Rightarrow D_{\mathbf{u}}f(5, 1) = 0 + \frac{100}{\sqrt{17}} = \frac{100}{\sqrt{17}} \approx 24.25.$

[11] $\mathbf{a} = \mathbf{i} + 5\mathbf{j}$ and $\|\mathbf{a}\| = \sqrt{26} \Rightarrow \mathbf{u} = \frac{1}{\sqrt{26}}\mathbf{i} + \frac{5}{\sqrt{26}}\mathbf{j}.$
$D_{\mathbf{u}}f(x, y) = \dfrac{9x}{\sqrt{9x^2 - 4y^2 - 1}}(\frac{1}{\sqrt{26}}) + \dfrac{-4y}{\sqrt{9x^2 - 4y^2 - 1}}(\frac{5}{\sqrt{26}}) \Rightarrow$
$D_{\mathbf{u}}f(3, -2) = \frac{27}{8}(\frac{1}{\sqrt{26}}) + \frac{8}{8}(\frac{5}{\sqrt{26}}) = \frac{67}{8\sqrt{26}} \approx 1.64.$

[12] $\mathbf{a} = 3\mathbf{i} + 4\mathbf{j} \Rightarrow \mathbf{u} = \frac{3}{5}\mathbf{i} + \frac{4}{5}\mathbf{j}.$ $D_{\mathbf{u}}f(x, y) = \dfrac{2y}{(x + y)^2}(\frac{3}{5}) + \dfrac{-2x}{(x + y)^2}(\frac{4}{5}) \Rightarrow$
$D_{\mathbf{u}}f(2, -1) = \frac{-2}{1}(\frac{3}{5}) + \frac{-4}{1}(\frac{4}{5}) = -\frac{22}{5}.$

[13] $\mathbf{a} = 5\mathbf{i} + \mathbf{j} \Rightarrow \mathbf{u} = \frac{5}{\sqrt{26}}\mathbf{i} + \frac{1}{\sqrt{26}}\mathbf{j}.$
$D_{\mathbf{u}}f(x, y) = (\cos^2 y)(\frac{5}{\sqrt{26}}) + (-2x\cos y\sin y)(\frac{1}{\sqrt{26}}) \Rightarrow$
$D_{\mathbf{u}}f(2, \frac{\pi}{4}) = (\frac{1}{2})(\frac{5}{\sqrt{26}}) + (-2)(\frac{1}{\sqrt{26}}) = \frac{1}{2\sqrt{26}} \approx 0.098.$

[14] $\mathbf{a} = -\mathbf{i} + 3\mathbf{j} \Rightarrow \mathbf{u} = -\frac{1}{\sqrt{10}}\mathbf{i} + \frac{3}{\sqrt{10}}\mathbf{j}.$ $D_{\mathbf{u}}f(x, y) = (e^{3y})(-\frac{1}{\sqrt{10}}) + (3xe^{3y})(\frac{3}{\sqrt{10}}) \Rightarrow$
$D_{\mathbf{u}}f(4, 0) = (1)(-\frac{1}{\sqrt{10}}) + (12)(\frac{3}{\sqrt{10}}) = \frac{35}{\sqrt{10}} \approx 11.07.$

15 $\mathbf{a} = \mathbf{i} + 2\mathbf{j} - 3\mathbf{k}$ and $\|\mathbf{a}\| = \sqrt{14} \Rightarrow \mathbf{u} = \frac{1}{\sqrt{14}}\mathbf{i} + \frac{2}{\sqrt{14}}\mathbf{j} - \frac{3}{\sqrt{14}}\mathbf{k}$.

$$D_{\mathbf{u}}f(x, y, z) = \nabla f(x, y, z) \cdot \mathbf{u} = f_x(x, y, z)\,u_1 + f_y(x, y, z)\,u_2 + f_z(x, y, z)\,u_3$$
$$= (y^3z^2)(\tfrac{1}{\sqrt{14}}) + (3xy^2z^2)(\tfrac{2}{\sqrt{14}}) + (2xy^3z)(-\tfrac{3}{\sqrt{14}}) \Rightarrow$$

$$D_{\mathbf{u}}f(2, -1, 4) = (-16)(\tfrac{1}{\sqrt{14}}) + (96)(\tfrac{2}{\sqrt{14}}) + (-16)(-\tfrac{3}{\sqrt{14}}) = \tfrac{224}{\sqrt{14}} = 16\sqrt{14} \approx 59.87.$$

16 $\mathbf{a} = 2\mathbf{i} - 3\mathbf{j} + \mathbf{k} \Rightarrow \mathbf{u} = \frac{2}{\sqrt{14}}\mathbf{i} - \frac{3}{\sqrt{14}}\mathbf{j} + \frac{1}{\sqrt{14}}\mathbf{k}$.

$$D_{\mathbf{u}}f(x, y, z) = (2x + 4y)(\tfrac{2}{\sqrt{14}}) + (3z + 4x)(-\tfrac{3}{\sqrt{14}}) + (3y)(\tfrac{1}{\sqrt{14}}) \Rightarrow$$

$$D_{\mathbf{u}}f(1, 0, -5) = (2)(\tfrac{2}{\sqrt{14}}) + (-11)(-\tfrac{3}{\sqrt{14}}) + (0)(\tfrac{1}{\sqrt{14}}) = \tfrac{37}{\sqrt{14}} \approx 9.89.$$

17 $\mathbf{a} = 3\mathbf{i} + \mathbf{j} - 5\mathbf{k} \Rightarrow \mathbf{u} = \frac{3}{\sqrt{35}}\mathbf{i} + \frac{1}{\sqrt{35}}\mathbf{j} - \frac{5}{\sqrt{35}}\mathbf{k}$.

$$D_{\mathbf{u}}f(x, y, z) = (yz^2e^{xy})(\tfrac{3}{\sqrt{35}}) + (xz^2e^{xy})(\tfrac{1}{\sqrt{35}}) + (2ze^{xy})(-\tfrac{5}{\sqrt{35}}) \Rightarrow$$

$$D_{\mathbf{u}}f(-1, 2, 3) = (18e^{-2})(\tfrac{3}{\sqrt{35}}) + (-9e^{-2})(\tfrac{1}{\sqrt{35}}) + (6e^{-2})(-\tfrac{5}{\sqrt{35}}) = \tfrac{15e^{-2}}{\sqrt{35}} \approx 0.34.$$

18 $\mathbf{a} = 2\mathbf{i} + 3\mathbf{j} - 2\mathbf{k} \Rightarrow \mathbf{u} = \frac{2}{\sqrt{17}}\mathbf{i} + \frac{3}{\sqrt{17}}\mathbf{j} - \frac{2}{\sqrt{17}}\mathbf{k}$.

$$D_{\mathbf{u}}f(x, y, z) = \frac{\sqrt{y}\sin z}{2\sqrt{x}}(\tfrac{2}{\sqrt{17}}) + \frac{\sqrt{x}\sin z}{2\sqrt{y}}(\tfrac{3}{\sqrt{17}}) + (\sqrt{xy}\cos z)(-\tfrac{2}{\sqrt{17}}) \Rightarrow$$

$$D_{\mathbf{u}}f(4, 9, \tfrac{\pi}{4}) = (\tfrac{3}{4\sqrt{2}})(\tfrac{2}{\sqrt{17}}) + (\tfrac{1}{3\sqrt{2}})(\tfrac{3}{\sqrt{17}}) + (\tfrac{6}{\sqrt{2}})(-\tfrac{2}{\sqrt{17}}) = -\tfrac{19}{2\sqrt{34}} \approx -1.63.$$

19 $\mathbf{a} = -3\mathbf{i} + \mathbf{k} \Rightarrow \mathbf{u} = -\frac{3}{\sqrt{10}}\mathbf{i} + \frac{1}{\sqrt{10}}\mathbf{k}$.

$$D_{\mathbf{u}}f(x, y, z) = (y + z)(-\tfrac{3}{\sqrt{10}}) + (x + y)(\tfrac{1}{\sqrt{10}}) \Rightarrow$$

$$D_{\mathbf{u}}f(5, 7, 1) = (8)(-\tfrac{3}{\sqrt{10}}) + (12)(\tfrac{1}{\sqrt{10}}) = -\tfrac{12}{\sqrt{10}} \approx -3.79.$$

20 $\mathbf{a} = 6\mathbf{i} + \mathbf{k} \Rightarrow \mathbf{u} = \frac{6}{\sqrt{37}}\mathbf{i} + \frac{1}{\sqrt{37}}\mathbf{k}$.

$$D_{\mathbf{u}}f(x, y, z) = \left[\frac{z^2}{1 + (x + y)^2}\right](\tfrac{6}{\sqrt{37}}) + \left[2z\tan^{-1}(x + y)\right](\tfrac{1}{\sqrt{37}}) \Rightarrow$$

$$D_{\mathbf{u}}f(0, 0, 4) = (16)(\tfrac{6}{\sqrt{37}}) + (0)(\tfrac{1}{\sqrt{37}}) = \tfrac{96}{\sqrt{37}} \approx 15.78.$$

Note: In Exercises 21–24, let $\mathbf{v}$ denote the unit vector in the direction of $\nabla f(x, y)$.
These exercises use (16.28) and (16.29) in parts (b) and (c), respectively.

21 (a) $\overrightarrow{PQ} = (-3 - 2)\mathbf{i} + (1 - 0)\mathbf{j} = -5\mathbf{i} + \mathbf{j} \Rightarrow \mathbf{u} = -\frac{5}{\sqrt{26}}\mathbf{i} + \frac{1}{\sqrt{26}}\mathbf{j}$.

$$D_{\mathbf{u}}f(x, y) = (2xe^{-2y})(-\tfrac{5}{\sqrt{26}}) + (-2x^2e^{-2y})(\tfrac{1}{\sqrt{26}}) \Rightarrow$$

$$D_{\mathbf{u}}f(2, 0) = (4)(-\tfrac{5}{\sqrt{26}}) + (-8)(\tfrac{1}{\sqrt{26}}) = -\tfrac{28}{\sqrt{26}},$$

which is the directional derivative of f at P in the direction from P to Q.

(b) $\nabla f(2, 0) = 4\mathbf{i} - 8\mathbf{j} \Rightarrow \mathbf{v} = \frac{4}{\sqrt{80}}\mathbf{i} - \frac{8}{\sqrt{80}}\mathbf{j} = \frac{1}{\sqrt{5}}\mathbf{i} - \frac{2}{\sqrt{5}}\mathbf{j}$,

which is a unit vector in the direction in which f increases most rapidly at P.
The rate of change of f in that direction is $\|\nabla f(2, 0)\| = \sqrt{80}$.

(c) $-\mathbf{v} = -\frac{1}{\sqrt{5}}\mathbf{i} + \frac{2}{\sqrt{5}}\mathbf{j}$ is a unit vector in the direction in which f decreases most rapidly at P. The rate of change of f in that direction is $-\|\nabla f(2, 0)\| = -\sqrt{80}$.

22 (a) $\overrightarrow{PQ} = \frac{\pi}{3}\mathbf{i} - \frac{\pi}{6}\mathbf{j} = \frac{\pi}{6}(2\mathbf{i} - \mathbf{j}) \Rightarrow \mathbf{u} = \frac{2}{\sqrt{5}}\mathbf{i} - \frac{1}{\sqrt{5}}\mathbf{j}.$

$D_{\mathbf{u}}f(x, y) = \left[2\cos(2x - y)\right](\frac{2}{\sqrt{5}}) + \left[-\cos(2x - y)\right](-\frac{1}{\sqrt{5}}) \Rightarrow$

$D_{\mathbf{u}}f(-\frac{\pi}{3}, \frac{\pi}{6}) = (-\sqrt{3})(\frac{2}{\sqrt{5}}) + (\frac{1}{2}\sqrt{3})(-\frac{1}{\sqrt{5}}) = -\frac{1}{2}\sqrt{15}.$

(b) $\nabla f(-\frac{\pi}{3}, \frac{\pi}{6}) = -\sqrt{3}\,\mathbf{i} + \frac{1}{2}\sqrt{3}\,\mathbf{j} \Rightarrow \mathbf{v} = -\frac{2\sqrt{3}}{\sqrt{15}}\mathbf{i} + \frac{\sqrt{3}}{\sqrt{15}}\mathbf{j} = -\frac{2}{\sqrt{5}}\mathbf{i} + \frac{1}{\sqrt{5}}\mathbf{j}.$

$\left\|\nabla f(-\frac{\pi}{3}, \frac{\pi}{6})\right\| = \frac{1}{2}\sqrt{15}.$

(c) $-\mathbf{v} = \frac{2}{\sqrt{5}}\mathbf{i} - \frac{1}{\sqrt{5}}\mathbf{j}.$ $-\left\|\nabla f(-\frac{\pi}{3}, \frac{\pi}{6})\right\| = -\frac{1}{2}\sqrt{15}.$

23 (a) $\overrightarrow{PQ} = 2\mathbf{i} - 8\mathbf{j} + 3\mathbf{k} \Rightarrow \mathbf{u} = \frac{2}{\sqrt{77}}\mathbf{i} - \frac{8}{\sqrt{77}}\mathbf{j} + \frac{3}{\sqrt{77}}\mathbf{k}.$

$D_{\mathbf{u}}f(x, y, z) = \left[\frac{x}{(x^2 + y^2 + z^2)^{1/2}}\right](\frac{2}{\sqrt{77}}) + \left[\frac{y}{(x^2 + y^2 + z^2)^{1/2}}\right](-\frac{8}{\sqrt{77}}) +$

$\left[\frac{z}{(x^2 + y^2 + z^2)^{1/2}}\right](\frac{3}{\sqrt{77}}) \Rightarrow D_{\mathbf{u}}f(-2, 3, 1) =$

$(-\frac{2}{\sqrt{14}})(\frac{2}{\sqrt{77}}) + (\frac{3}{\sqrt{14}})(-\frac{8}{\sqrt{77}}) + (\frac{1}{\sqrt{14}})(\frac{3}{\sqrt{77}}) = \frac{-25}{\sqrt{14}\sqrt{77}} = -\frac{25}{7\sqrt{22}}.$

(b) $\nabla f(-2, 3, 1) = -\frac{2}{\sqrt{14}}\mathbf{i} + \frac{3}{\sqrt{14}}\mathbf{j} + \frac{1}{\sqrt{14}}\mathbf{k} = \mathbf{v}.$ $\left\|\nabla f(-2, 3, 1)\right\| = 1.$

(c) $-\mathbf{v} = \frac{2}{\sqrt{14}}\mathbf{i} - \frac{3}{\sqrt{14}}\mathbf{j} - \frac{1}{\sqrt{14}}\mathbf{k}.$ $-\left\|\nabla f(-2, 3, 1)\right\| = -1.$

24 (a) $\overrightarrow{PQ} = 3\mathbf{i} + 2\mathbf{j} - 6\mathbf{k} \Rightarrow \mathbf{u} = \frac{3}{7}\mathbf{i} + \frac{2}{7}\mathbf{j} - \frac{6}{7}\mathbf{k}.$

$D_{\mathbf{u}}f(x, y, z) = \left(\frac{1}{y}\right)(\frac{3}{7}) + \left(-\frac{x}{y^2} - \frac{1}{z}\right)(\frac{2}{7}) + \left(\frac{y}{z^2}\right)(-\frac{6}{7}) \Rightarrow$

$D_{\mathbf{u}}f(0, -1, 2) = (-1)(\frac{3}{7}) + (-\frac{1}{2})(\frac{2}{7}) + (-\frac{1}{4})(-\frac{6}{7}) = -\frac{5}{14}.$

(b) $\nabla f(0, -1, 2) = -\mathbf{i} - \frac{1}{2}\mathbf{j} - \frac{1}{4}\mathbf{k} \Rightarrow \mathbf{v} = -\frac{4}{\sqrt{21}}\mathbf{i} - \frac{2}{\sqrt{21}}\mathbf{j} - \frac{1}{\sqrt{21}}\mathbf{k}.$

$\left\|\nabla f(0, -1, 2)\right\| = \frac{1}{4}\sqrt{21}.$

(c) $-\mathbf{v} = \frac{4}{\sqrt{21}}\mathbf{i} + \frac{2}{\sqrt{21}}\mathbf{j} + \frac{1}{\sqrt{21}}\mathbf{k}.$ $-\left\|\nabla f(0, -1, 2)\right\| = -\frac{1}{4}\sqrt{21}.$

25 (a) $T(x, y) = \frac{k}{(x^2 + y^2)^{1/2}}$ and $T(3, 4) = 100 \Rightarrow k = 500.$

$\nabla T(x, y) = \frac{-500x}{(x^2 + y^2)^{3/2}}\mathbf{i} + \frac{-500y}{(x^2 + y^2)^{3/2}}\mathbf{j}$ and $\mathbf{a} = \mathbf{i} + \mathbf{j} \Rightarrow$

$D_{\mathbf{u}}\,T(3, 4) = (-12\mathbf{i} - 16\mathbf{j}) \cdot (\frac{1}{\sqrt{2}}\mathbf{i} + \frac{1}{\sqrt{2}}\mathbf{j}) = \frac{-28}{\sqrt{2}},$

which is the rate of change of T at P in the direction of $\mathbf{i} + \mathbf{j}$.

(b) T increases most rapidly in the direction of $\nabla T(3, 4) = -12\mathbf{i} - 16\mathbf{j}$.

(c) T decreases most rapidly in the direction of $-\nabla T(3, 4) = 12\mathbf{i} + 16\mathbf{j}$.

(d) The rate of change will be 0 in the direction that is orthogonal to ∇T at (3, 4).

$\nabla T(3, 4) \cdot \mathbf{w} = 0 \Rightarrow (-12\mathbf{i} - 16\mathbf{j}) \cdot (w_1\mathbf{i} + w_2\mathbf{j}) = 0 \Rightarrow$

$-12w_1 - 16w_2 = 0 \Rightarrow w_1 = -\frac{4}{3}w_2.$ If $w_2 = -3$, then $w_1 = 4$. Thus,

the rate of change is 0 in the direction of $\mathbf{w} = 4\mathbf{i} - 3\mathbf{j}$, or any multiple of $\mathbf{w}$.

26 (a) $\nabla f(x, y) = -4x\mathbf{i} - 6y\mathbf{j}$. The depth of the water decreases most rapidly in the direction of $-\nabla f(4, 9) = 16\mathbf{i} + 54\mathbf{j}$.

(b) The depth remains the same if its rate of change is zero.

$\nabla f(4, 9) \cdot \mathbf{w} = 0 \Rightarrow (-16\mathbf{i} - 54\mathbf{j}) \cdot (w_1\mathbf{i} + w_2\mathbf{j}) = 0 \Rightarrow -16w_1 - 54w_2 = 0 \Rightarrow$ $w_1 = -\frac{27}{8}w_2$. If $w_2 = -8$, then $w_1 = 27$. Thus, the depth remains the same in the direction of $\mathbf{w} = 27\mathbf{i} - 8\mathbf{j}$, or any multiple of $\mathbf{w}$.

27 (a) $\nabla V(x, y, z) = 2x\mathbf{i} + 8y\mathbf{j} + 18z\mathbf{k}$.

$\mathbf{a} = \overrightarrow{PO} = -2\mathbf{i} + \mathbf{j} - 3\mathbf{k} \Rightarrow \mathbf{u} = -\frac{2}{\sqrt{14}}\mathbf{i} + \frac{1}{\sqrt{14}}\mathbf{j} - \frac{3}{\sqrt{14}}\mathbf{k}$.

$$\nabla V(2, -1, 3) \cdot \mathbf{u} = (4)(-\tfrac{2}{\sqrt{14}}) + (-8)(\tfrac{1}{\sqrt{14}}) + (54)(-\tfrac{3}{\sqrt{14}}) = -\tfrac{178}{\sqrt{14}}.$$

(b) The maximum rate of change is in the direction of

$$\nabla V(2, -1, 3) = 4\mathbf{i} - 8\mathbf{j} + 54\mathbf{k}.$$

(c) The maximum rate of change at P is

$$\|\nabla V(2, -1, 3)\| = \sqrt{4^2 + (-8)^2 + 54^2} = \sqrt{2996} \approx 54.7.$$

28 (a) $\nabla T(x, y, z) = 8x\mathbf{i} - 2y\mathbf{j} + 32z\mathbf{k}$. $\mathbf{a} = 2\mathbf{i} + 6\mathbf{j} - 3\mathbf{k} \Rightarrow \mathbf{u} = \frac{2}{7}\mathbf{i} + \frac{6}{7}\mathbf{j} - \frac{3}{7}\mathbf{k}$.

$$\nabla T(4, -2, 1) \cdot \mathbf{u} = (32)(\tfrac{2}{7}) + (4)(\tfrac{6}{7}) + (32)(-\tfrac{3}{7}) = -\tfrac{8}{7}.$$

(b) T increases most rapidly in the direction of $\nabla T(4, -2, 1) = 32\mathbf{i} + 4\mathbf{j} + 32\mathbf{k}$.

(c) The maximum rate is

$$\|\nabla T(4, -2, 1)\| = \sqrt{32^2 + 4^2 + 32^2} = \sqrt{2064} = 4\sqrt{129} \approx 45.4.$$

(d) T decreases most rapidly in the direction of

$$-\nabla T(4, -2, 1) = -32\mathbf{i} - 4\mathbf{j} - 32\mathbf{k}.$$

(e) The minimum rate is $-\|\nabla T(4, -2, 1)\| = -4\sqrt{129}$.

29 (a) Let the circle have radius r. Since a radius is always perpendicular to the tangent line, it follows that $\mathbf{a} = x\mathbf{i} + y\mathbf{j} = r\cos\theta\,\mathbf{i} + r\sin\theta\,\mathbf{j}$ is normal to the circle at the point (r, θ) and the unit normal vector is $\mathbf{n} = \cos\theta\,\mathbf{i} + \sin\theta\,\mathbf{j}$.

The boundary is insulated iff $\nabla T(x, y) \cdot \mathbf{n} = 0$, or, $\frac{\partial T}{\partial x}\cos\theta + \frac{\partial T}{\partial y}\sin\theta = 0$.

Since $\frac{\partial T}{\partial r} = \frac{\partial T}{\partial x}\frac{\partial x}{\partial r} + \frac{\partial T}{\partial y}\frac{\partial y}{\partial r} = \frac{\partial T}{\partial x}\cos\theta + \frac{\partial T}{\partial y}\sin\theta$, the result follows.

(b) Since $\frac{\partial T}{\partial r} = \nabla T(x, y) \cdot \mathbf{n}$, $\frac{\partial T}{\partial r}$ is the rate of change of temperature in the direction normal to the circular boundary.

30 (a) Let B have coordinates $(x, y) = (r\cos\theta, r\sin\theta)$.

The unit vector in the direction of $\overrightarrow{AB}$ is $\cos\theta\,\mathbf{i} + \sin\theta\,\mathbf{j}$ and the unit normal vector to $\overrightarrow{AB}$ shown in the figure is $\mathbf{n} = -\sin\theta\,\mathbf{i} + \cos\theta\,\mathbf{j}$.

The boundary is insulated iff $\nabla T(x, y) \cdot \mathbf{n} = 0$, or, $\frac{\partial T}{\partial x}(-\sin\theta) + \frac{\partial T}{\partial y}(\cos\theta) = 0$.

Since $\frac{\partial T}{\partial \theta} = \frac{\partial T}{\partial x}\frac{\partial x}{\partial \theta} + \frac{\partial T}{\partial y}\frac{\partial y}{\partial \theta} = \frac{\partial T}{\partial x}(-r\sin\theta) + \frac{\partial T}{\partial y}(r\cos\theta)$,

it follows that the boundary is insulated iff $\frac{1}{r}\frac{\partial T}{\partial \theta} = 0$, or, $\frac{\partial T}{\partial \theta} = 0$.

(b) Since $\frac{1}{r}\frac{\partial T}{\partial \theta} = \nabla T(x, y) \cdot \mathbf{n}$, $\frac{\partial T}{\partial \theta}$ is r times

the rate of change of temperature in the direction normal to the boundary AB.

31 (a) $\lim_{h \to 0} \frac{f(x + h, y) - f(x - h, y)}{2h}$

$$= \lim_{h \to 0} \frac{f(x + h, y) - f(x, y) - f(x - h, y) + f(x, y)}{2h}$$

$$= \frac{1}{2} \lim_{h \to 0} \frac{f(x + h, y) - f(x, y)}{h} - \frac{1}{2} \lim_{h \to 0} \frac{f(x - h, y) - f(x, y)}{h}$$

$$= \tfrac{1}{2} f_x(x, y) + \tfrac{1}{2} \lim_{k \to 0} \frac{f(x + k, y) - f(x, y)}{k} \;\{\text{where } k = -h\}$$

$= \frac{1}{2} f_x(x, y) + \frac{1}{2} f_x(x, y) = f_x(x, y)$. A similar solution holds for f_y.

(b) $f_x(1, 2) \approx \frac{f(1 + 0.01, 2) - f(1 - 0.01, 2)}{2(0.01)} \approx 1.00003333$ and

$f_y(1, 2) \approx \frac{f(1, 2 + 0.01) - f(1, 2 - 0.01)}{2(0.01)} \approx -0.11111235 \Rightarrow$

$\nabla f(1, 2) \approx 1.00003333\mathbf{i} - 0.11111235\mathbf{j}$.

$f_x(x, y) = \frac{3x^2}{1 + y} \Rightarrow f_x(1, 2) = 1$. $f_y(x, y) = -\frac{x^3}{(1 + y)^2} \Rightarrow f_y(1, 2) = -\frac{1}{9}$.

Thus, $\nabla f(1, 2) = \mathbf{i} - \frac{1}{9}\mathbf{j}$.

32 (a) $T_x(3, 3) \approx \frac{T(4, 3) - T(2, 3)}{2} = \frac{76 - 72}{2} = 2$ and

$T_y(3, 3) \approx \frac{T(3, 4) - T(3, 2)}{2} = \frac{69 - 80}{2} = -5.5 \Rightarrow \nabla T(3, 3) = 2\mathbf{i} - 5.5\mathbf{j}$.

(b) Maximum heat transfer from hot to cold is in the direction of

$-\nabla T(3, 3) = -2\mathbf{i} + 5.5\mathbf{j}$.

(c) $\nabla T(3, 3) \cdot \frac{\mathbf{a}}{\|\mathbf{a}\|} = (2)(-\frac{1}{\sqrt{5}}) + (-5.5)(\frac{2}{\sqrt{5}}) = -\frac{13}{\sqrt{5}} \approx -5.81$ °F/mm

33 $f_x(x, y, z) \approx \frac{f(x + h, y, z) - f(x - h, y, z)}{2h} \Rightarrow$

$f_x(1, 1, 1) \approx \frac{f(1.01, 1, 1) - f(0.99, 1, 1)}{2(0.01)} \approx 0.327669$

$f_y(x, y, z) \approx \frac{f(x, y + h, z) - f(x, y - h, z)}{2h} \Rightarrow$

$f_y(1, 1, 1) \approx \frac{f(1, 1.01, 1) - f(1, 0.99, 1)}{2(0.01)} \approx -0.234474$

$f_z(x, y, z) \approx \frac{f(x, y, z + h) - f(x, y, z - h)}{2h} \Rightarrow$

$f_z(1, 1, 1) \approx \frac{f(1, 1, 1.01) - f(1, 1, 0.99)}{2(0.01)} \approx 0.130948$

$D_{\mathbf{u}} f(x, y, z) \approx \nabla f(x, y, z) \cdot \mathbf{u} \approx (0.327669 - 0.234474 + 0.130948)/\sqrt{3} \approx 0.1294$

34 $f_x(x, y, z) \approx \dfrac{f(x+h, y, z) - f(x-h, y, z)}{2h} \Rightarrow$

$$f_x(1, 1, 1) \approx \frac{f(1.01, 1, 1) - f(0.99, 1, 1)}{2(0.01)} \approx 0.762596$$

$f_y(x, y, z) \approx \dfrac{f(x, y+h, z) - f(x, y-h, z)}{2h} \Rightarrow$

$$f_y(1, 1, 1) \approx \frac{f(1, 1.01, 1) - f(1, 0.99, 1)}{2(0.01)} \approx -1.194589$$

$f_z(x, y, z) \approx \dfrac{f(x, y, z+h) - f(x, y, z-h)}{2h} \Rightarrow$

$$f_z(1, 1, 1) \approx \frac{f(1, 1, 1.01) - f(1, 1, 0.99)}{2(0.01)} \approx 0.929418$$

$$D_{\mathbf{u}} f(x, y, z) \approx \nabla f(x, y, z) \cdot \mathbf{u} \approx \left[2(0.762596) + 1.194589 + 0.929418\right]/\sqrt{6} \approx 1.4898$$

35 $\nabla(cu) = (cu)_x \mathbf{i} + (cu)_y \mathbf{j} = cu_x \mathbf{i} + cu_y \mathbf{j} = c(u_x \mathbf{i} + u_y \mathbf{j}) = c\nabla u$

36 $\nabla(u+v) = (u+v)_x \mathbf{i} + (u+v)_y \mathbf{j} = (u_x + v_x)\mathbf{i} + (u_y + v_y)\mathbf{j} =$

$$(u_x \mathbf{i} + u_y \mathbf{j}) + (v_x \mathbf{i} + v_y \mathbf{j}) = \nabla u + \nabla v$$

37 $\nabla(uv) = (uv)_x \mathbf{i} + (uv)_y \mathbf{j} = (u_x v + u v_x)\mathbf{i} + (u_y v + u v_y)\mathbf{j} =$

$$(uv_x \mathbf{i} + uv_y \mathbf{j}) + (u_x v \mathbf{i} + u_y v \mathbf{j}) = u(v_x \mathbf{i} + v_y \mathbf{j}) + v(u_x \mathbf{i} + u_y \mathbf{j}) = u\nabla v + v\nabla u$$

38 $\nabla(\frac{u}{v}) = (\frac{u}{v})_x \mathbf{i} + (\frac{u}{v})_y \mathbf{j} = \left(\dfrac{vu_x - uv_x}{v^2}\right)\mathbf{i} + \left(\dfrac{vu_y - uv_y}{v^2}\right)\mathbf{j} =$

$$\frac{1}{v^2}\left[(vu_x \mathbf{i} + vu_y \mathbf{j}) - (uv_x \mathbf{i} + uv_y \mathbf{j})\right] = \frac{v\nabla u - u\nabla v}{v^2},\ v \neq 0$$

39 $\nabla u^n = (u^n)_x \mathbf{i} + (u^n)_y \mathbf{j} = nu^{n-1} u_x \mathbf{i} + nu^{n-1} u_y \mathbf{j} = nu^{n-1}(u_x \mathbf{i} + u_y \mathbf{j}) =$

$$nu^{n-1}\nabla u$$

40 $\nabla w = \nabla(h(u)) = (h(u))_x \mathbf{i} + (h(u))_y \mathbf{j} = h'(u)\, u_x \mathbf{i} + h'(u)\, u_y \mathbf{j} =$

$$h'(u)(u_x \mathbf{i} + u_y \mathbf{j}) = \frac{dw}{du}\nabla u$$

41 (a) $\mathbf{u} = \cos\theta\, \mathbf{i} + \sin\theta\, \mathbf{j} \Rightarrow D_{\mathbf{u}} f(x, y) = f_x(x, y)\cos\theta + f_y(x, y)\sin\theta$.

(b) $\theta = \frac{5\pi}{6} \Rightarrow \mathbf{u} = -\frac{1}{2}\sqrt{3}\,\mathbf{i} + \frac{1}{2}\mathbf{j}$. $D_{\mathbf{u}} f(x, y) = (2x + 2y)(-\frac{1}{2}\sqrt{3}) + (2x - 2y)(\frac{1}{2}) \Rightarrow$

$$D_{\mathbf{u}} f(2, -3) = (-2)(-\tfrac{1}{2}\sqrt{3}) + (10)(\tfrac{1}{2}) = 5 + \sqrt{3}.$$

42 $\theta = \frac{\pi}{3} \Rightarrow \mathbf{u} = \frac{1}{2}\mathbf{i} + \frac{1}{2}\sqrt{3}\,\mathbf{j}$.

$D_{\mathbf{u}} f(x, y) = \left[4y(xy + y^2)^3\right](\frac{1}{2}) + \left[4(x + 2y)(xy + y^2)^3\right](\frac{1}{2}\sqrt{3}) \Rightarrow$

$$D_{\mathbf{u}} f(2, -1) = (4)(\tfrac{1}{2}) + (0)(\tfrac{1}{2}\sqrt{3}) = 2.$$

43 $\nabla f(x, y) = f_x(x, y)\mathbf{i} + f_y(x, y)\mathbf{j} = 0 \Rightarrow f_x(x, y) = f_y(x, y) = 0$.

Thus, by Exercise 46 of §16.5, $f(x, y)$ is constant on R.

44 $\dfrac{dw}{dt} = \dfrac{\partial w}{\partial x}\dfrac{dx}{dt} + \dfrac{\partial w}{\partial y}\dfrac{dy}{dt} = \left(\dfrac{\partial w}{\partial x}\mathbf{i} + \dfrac{\partial w}{\partial y}\mathbf{j}\right)\cdot\left(\dfrac{dx}{dt}\mathbf{i} + \dfrac{dy}{dt}\mathbf{j}\right) = \nabla w \cdot \mathbf{r}'(t)$

Exercises 16.7

1 $F(x, y, z) = 4x^2 - y^2 + 3z^2 - 10 \Rightarrow$

$$\nabla F(x, y, z) = 8x\mathbf{i} - 2y\mathbf{j} + 6z\mathbf{k} \Rightarrow \nabla F(2, -3, 1) = 16\mathbf{i} + 6\mathbf{j} + 6\mathbf{k}.$$

Tangent plane: $16(x - 2) + 6(y + 3) + 6(z - 1) = 0$.

Normal line: $x = 2 + 16t,\ y = -3 + 6t,\ z = 1 + 6t$.

[2] $F(x, y, z) = 9x^2 - 4y^2 - 25z^2 - 40 \Rightarrow$

$$\nabla F(x, y, z) = 18x\mathbf{i} - 8y\mathbf{j} - 50z\mathbf{k} \Rightarrow \nabla F(4, 1, -2) = 72\mathbf{i} - 8\mathbf{j} + 100\mathbf{k}.$$

Tangent plane: $72(x - 4) - 8(y - 1) + 100(z + 2) = 0.$

Normal line: $x = 4 + 72t,\ y = 1 - 8t,\ z = -2 + 100t.$

[3] $F(x, y, z) = 4x^2 + 9y^2 - z \Rightarrow$

$$\nabla F(x, y, z) = 8x\mathbf{i} + 18y\mathbf{j} - \mathbf{k} \Rightarrow \nabla F(-2, -1, 25) = -16\mathbf{i} - 18\mathbf{j} - \mathbf{k}.$$

Tangent plane: $16(x + 2) + 18(y + 1) + (z - 25) = 0.$

Normal line: $x = -2 + 16t,\ y = -1 + 18t,\ z = 25 + t.$

[4] $F(x, y, z) = 4x^2 - y^2 - z \Rightarrow$

$$\nabla F(x, y, z) = 8x\mathbf{i} - 2y\mathbf{j} - \mathbf{k} \Rightarrow \nabla F(5, -8, 36) = 40\mathbf{i} + 16\mathbf{j} - \mathbf{k}.$$

Tangent plane: $40(x - 5) + 16(y + 8) - (z - 36) = 0.$

Normal line: $x = 5 + 40t,\ y = -8 + 16t,\ z = 36 - t.$

[5] $F(x, y, z) = xy + 2yz - xz^2 + 10 \Rightarrow \nabla F(x, y, z) =$

$$(y - z^2)\mathbf{i} + (x + 2z)\mathbf{j} + (2y - 2xz)\mathbf{k} \Rightarrow \nabla F(-5, 5, 1) = 4\mathbf{i} - 3\mathbf{j} + 20\mathbf{k}.$$

Tangent plane: $4(x + 5) - 3(y - 5) + 20(z - 1) = 0.$

Normal line: $x = -5 + 4t,\ y = 5 - 3t,\ z = 1 + 20t.$

[6] $F(x, y, z) = x^3 - 2xy + z^3 + 7y + 6 \Rightarrow \nabla F(x, y, z) =$

$$(3x^2 - 2y)\mathbf{i} + (-2x + 7)\mathbf{j} + 3z^2\mathbf{k} \Rightarrow \nabla F(1, 4, -3) = -5\mathbf{i} + 5\mathbf{j} + 27\mathbf{k}.$$

Tangent plane: $-5(x - 1) + 5(y - 4) + 27(z + 3) = 0.$

Normal line: $x = 1 - 5t,\ y = 4 + 5t,\ z = -3 + 27t.$

[7] $F(x, y, z) = 2e^{-x}\cos y - z \Rightarrow \nabla F(x, y, z) =$

$$(-2e^{-x}\cos y)\mathbf{i} + (-2e^{-x}\sin y)\mathbf{j} - \mathbf{k} \Rightarrow \nabla F(0, \tfrac{\pi}{3}, 1) = -\mathbf{i} - \sqrt{3}\mathbf{j} - \mathbf{k}.$$

Tangent plane: $-1(x - 0) - \sqrt{3}(y - \frac{\pi}{3}) - 1(z - 1) = 0$, or, equivalently,

$x + \sqrt{3}(y - \frac{\pi}{3}) + (z - 1) = 0.$ Normal line: $x = t,\ y = \frac{\pi}{3} + \sqrt{3}t,\ z = 1 + t.$

[8] $F(x, y, z) = \ln xy - z \Rightarrow$

$$\nabla F(x, y, z) = \frac{1}{x}\mathbf{i} + \frac{1}{y}\mathbf{j} - \mathbf{k} \Rightarrow \nabla F(\tfrac{1}{2}, 2, 0) = 2\mathbf{i} + \tfrac{1}{2}\mathbf{j} - \mathbf{k}.$$

Tangent plane: $2(x - \frac{1}{2}) + \frac{1}{2}(y - 2) - z = 0.$

Normal line: $x = \frac{1}{2} + 2t,\ y = 2 + \frac{1}{2}t,\ z = -t.$

[9] $F(x, y, z) = \ln\left(\frac{y}{2z}\right) - x \Rightarrow$

$$\nabla F(x, y, z) = -\mathbf{i} + \frac{1}{y}\mathbf{j} - \frac{1}{z}\mathbf{k} \Rightarrow \nabla F(0, 2, 1) = -\mathbf{i} + \tfrac{1}{2}\mathbf{j} - \mathbf{k}.$$

Tangent plane: $-x + \frac{1}{2}(y - 2) - (z - 1) = 0.$

Normal line: $x = -t,\ y = 2 + \frac{1}{2}t,\ z = 1 - t.$

[10] $F(x, y, z) = xyz - 4xz^3 + y^3 - 10 \Rightarrow \nabla F(x, y, z) =$

$$(yz - 4z^3)\mathbf{i} + (xz + 3y^2)\mathbf{j} + (xy - 12xz^2)\mathbf{k} \Rightarrow \nabla F(-1, 2, 1) = -2\mathbf{i} + 11\mathbf{j} + 10\mathbf{k}.$$

Tangent plane: $-2(x + 1) + 11(y - 2) + 10(z - 1) = 0.$

Normal line: $x = -1 - 2t,\ y = 2 + 11t,\ z = 1 + 10t.$

[11] $f(2, 1) = -3 = y^2 - x^2 \Rightarrow x^2 - y^2 = 3.$

$\nabla f(x, y) = -2x\mathbf{i} + 2y\mathbf{j} \Rightarrow \nabla f(2, 1) = -4\mathbf{i} + 2\mathbf{j}.$

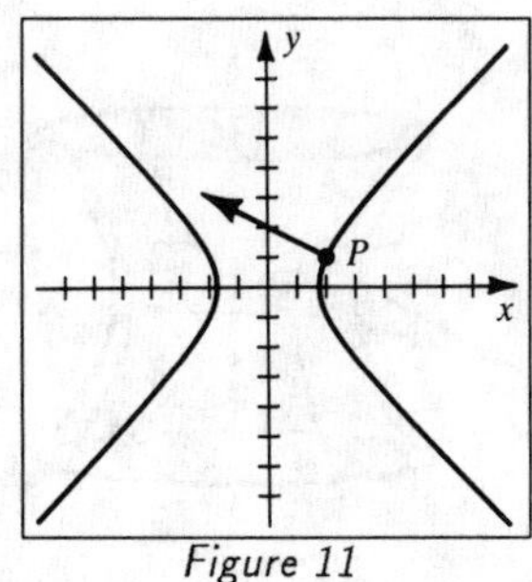

Figure 11

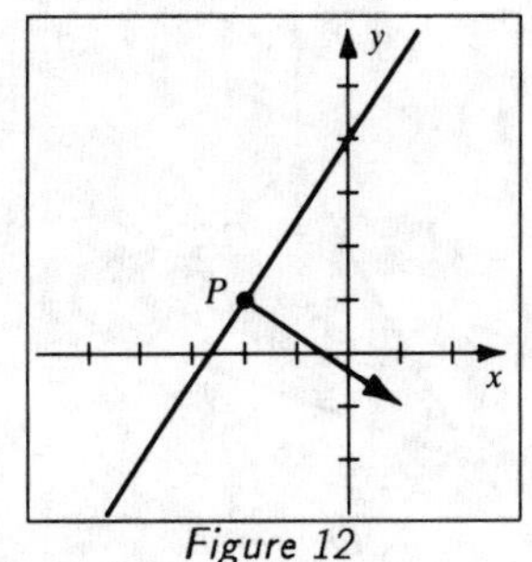

Figure 12

[12] $f(-2, 1) = -8 = 3x - 2y \Rightarrow y = \frac{3}{2}x + 4.$ $\nabla f(-2, 1) = 3\mathbf{i} - 2\mathbf{j}.$

[13] $f(-3, 5) = 4 = x^2 - y \Rightarrow y = x^2 - 4.$

$\nabla f(x, y) = 2x\mathbf{i} - \mathbf{j} \Rightarrow \nabla f(-3, 5) = -6\mathbf{i} - \mathbf{j}.$

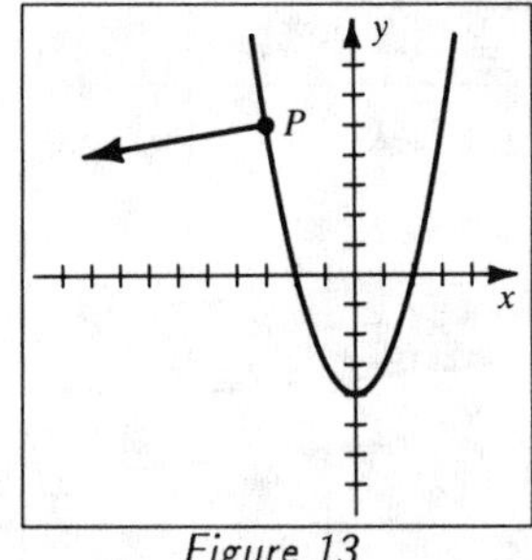

Figure 13

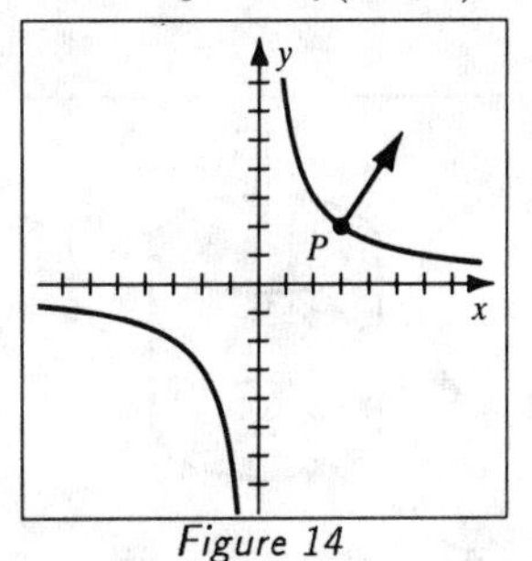

Figure 14

[14] $f(3, 2) = 6 = xy \Rightarrow y = \frac{6}{x}.$ $\nabla f(x, y) = y\mathbf{i} + x\mathbf{j} \Rightarrow \nabla f(3, 2) = 2\mathbf{i} + 3\mathbf{j}.$

[15] $F(1, 5, 2) = 30 = x^2 + y^2 + z^2$, a sphere with radius $\sqrt{30}$ centered at O.

$\nabla F(x, y, z) = 2x\mathbf{i} + 2y\mathbf{j} + 2z\mathbf{k} \Rightarrow \nabla F(1, 5, 2) = 2\mathbf{i} + 10\mathbf{j} + 4\mathbf{k}.$

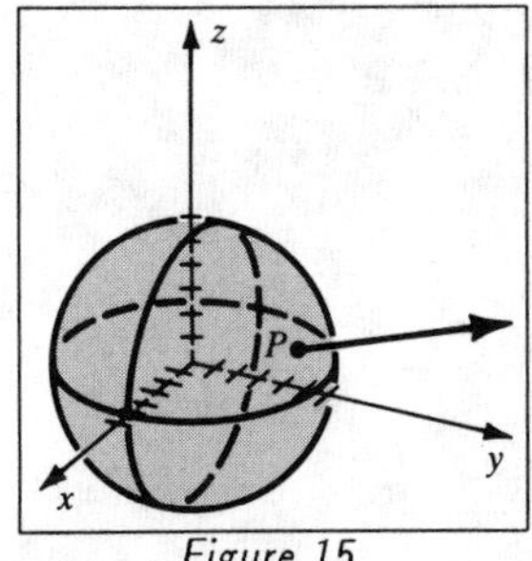

Figure 15

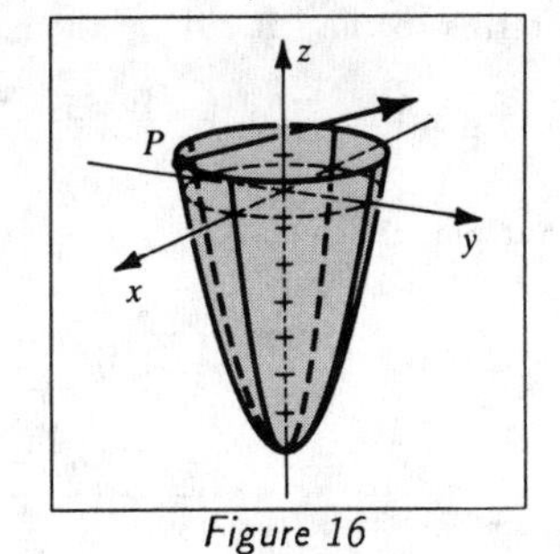

Figure 16

[16] $F(2, -2, 1) = -7 = z - x^2 - y^2 \Rightarrow$

$z + 7 = x^2 + y^2$, a circular paraboloid with vertex $(0, 0, -7)$.

$\nabla F(x, y, z) = -2x\mathbf{i} - 2y\mathbf{j} + \mathbf{k} \Rightarrow \nabla F(2, -2, 1) = -4\mathbf{i} + 4\mathbf{j} + \mathbf{k}.$

[17] $F(3, 4, 1) = 14 = x + 2y + 3z$, a plane with x-, y-, and z-intercepts of 14, 7, and $\frac{14}{3}$, respectively. $\nabla F(3, 4, 1) = \mathbf{i} + 2\mathbf{j} + 3\mathbf{k}$.

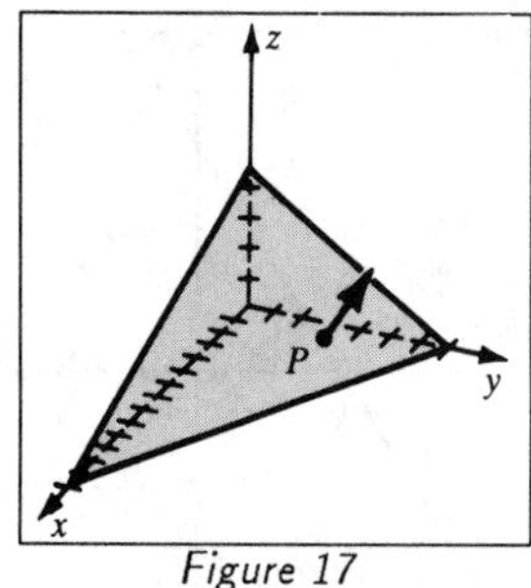

Figure 17

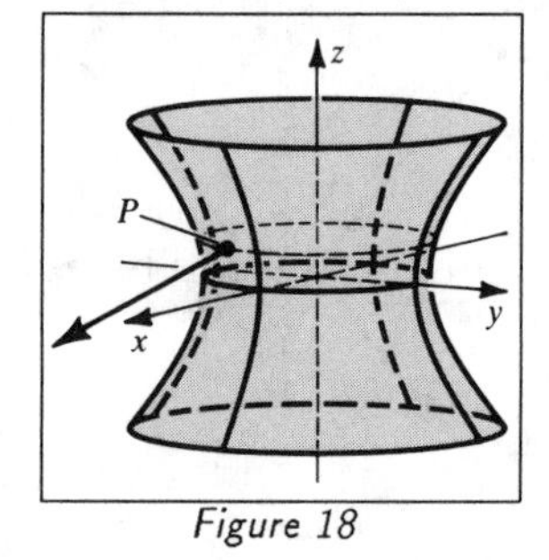

Figure 18

[18] $F(3, -1, 1) = 9 = x^2 + y^2 - z^2$, a hyperboloid of one sheet with axis along the z-axis. $\nabla F(x, y, z) = 2x\mathbf{i} + 2y\mathbf{j} - 2z\mathbf{k} \Rightarrow \nabla F(3, -1, 1) = 6\mathbf{i} - 2\mathbf{j} - 2\mathbf{k}$.

[19] $F(2, 0, 3) = 4 = x^2 + y^2$, a circular cylinder along the z-axis. $\nabla F(x, y, z) = 2x\mathbf{i} + 2y\mathbf{j} \Rightarrow \nabla F(2, 0, 3) = 4\mathbf{i}$.

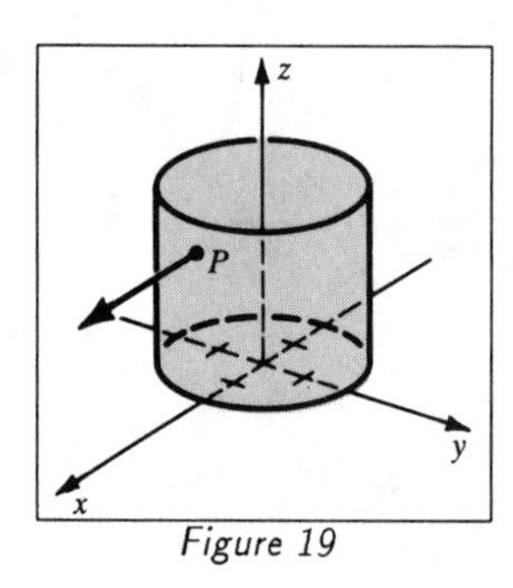

Figure 19

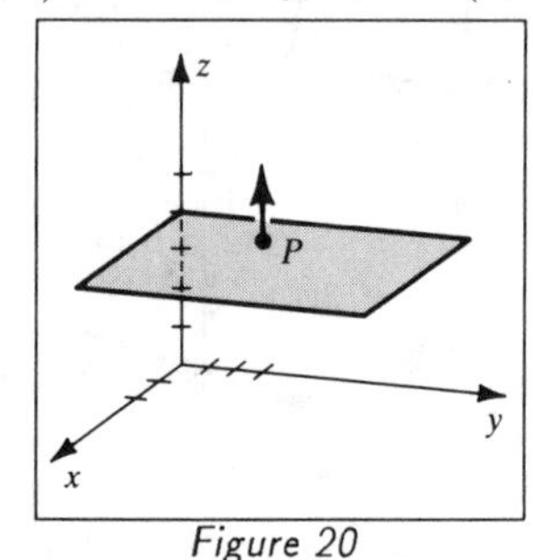

Figure 20

[20] $F(2, 3, 4) = 4 = z$, a plane parallel to the xy-plane. $\nabla F(2, 3, 4) = \mathbf{k}$.

Note: In Exercises 21–24, let $F(x, y, z)$ denote the function resulting from transposing all terms to the left side of the equation.

[21] $\nabla F(x_0, y_0, z_0) = \dfrac{2x_0}{a^2}\mathbf{i} + \dfrac{2y_0}{b^2}\mathbf{j} + \dfrac{2z_0}{c^2}\mathbf{k}$.

Tangent plane: $\dfrac{2x_0}{a^2}(x - x_0) + \dfrac{2y_0}{b^2}(y - y_0) + \dfrac{2z_0}{c^2}(z - z_0) = 0 \Rightarrow$

$$\frac{xx_0}{a^2} - \frac{x_0^2}{a^2} + \frac{yy_0}{b^2} - \frac{y_0^2}{b^2} + \frac{zz_0}{c^2} - \frac{z_0^2}{c^2} = 0 \Rightarrow$$

$$\frac{xx_0}{a^2} + \frac{yy_0}{b^2} + \frac{zz_0}{c^2} = \frac{x_0^2}{a^2} + \frac{y_0^2}{b^2} + \frac{z_0^2}{c^2},$$

which equals 1 because $P_0(x_0, y_0, z_0)$ satisfies the equation of the quadric surface.

[22] $\nabla F(x_0, y_0, z_0) = \dfrac{2x_0}{a^2}\mathbf{i} - \dfrac{2y_0}{b^2}\mathbf{j} + \dfrac{2z_0}{c^2}\mathbf{k}$.

Tangent plane: $\dfrac{2x_0}{a^2}(x - x_0) - \dfrac{2y_0}{b^2}(y - y_0) + \dfrac{2z_0}{c^2}(z - z_0) = 0 \Rightarrow$

$$\frac{xx_0}{a^2} - \frac{yy_0}{b^2} + \frac{zz_0}{c^2} = \frac{x_0^2}{a^2} - \frac{y_0^2}{b^2} + \frac{z_0^2}{c^2} = 1.$$

[23] $\nabla F(x_0, y_0, z_0) = \frac{2x_0}{a^2}\mathbf{i} - \frac{2y_0}{b^2}\mathbf{j} - \frac{2z_0}{c^2}\mathbf{k}.$

Tangent plane: $\frac{2x_0}{a^2}(x - x_0) - \frac{2y_0}{b^2}(y - y_0) - \frac{2z_0}{c^2}(z - z_0) = 0 \Rightarrow$

$$\frac{xx_0}{a^2} - \frac{yy_0}{b^2} - \frac{zz_0}{c^2} = \frac{x_0^2}{a^2} - \frac{y_0^2}{b^2} - \frac{z_0^2}{c^2} = 1.$$

[24] $\nabla F(x_0, y_0, z_0) = \frac{2x_0}{a^2}\mathbf{i} + \frac{2y_0}{b^2}\mathbf{j} - c\mathbf{k}.$

Tangent plane: $\frac{2x_0}{a^2}(x - x_0) + \frac{2y_0}{b^2}(y - y_0) - c(z - z_0) = 0 \Rightarrow$

$\frac{2xx_0}{a^2} + \frac{2yy_0}{b^2} = \frac{2x_0^2}{a^2} + \frac{2y_0^2}{b^2} + cz - cz_0 =$

$$2\left(\frac{x_0^2}{a^2} + \frac{y_0^2}{b^2} - cz_0\right) + cz_0 + cz = 2(0) + c(z_0 + z) = c(z + z_0).$$

[25] The normal to the hyperboloid is $2x\mathbf{i} - 4y\mathbf{j} - 8z\mathbf{k}$ and to the plane is $4\mathbf{i} - 2\mathbf{j} + 4\mathbf{k}$. The tangent plane of the hyperboloid is parallel to the given plane when these normal vectors are scalar multiples of each other, i.e., when $2x = 4c$, $-4y = -2c$, and $-8z = 4c$. Substituting for x, y, and z into the equation of the hyperboloid yields $(2c)^2 - 2(\frac{1}{2}c)^2 - 4(-\frac{1}{2}c)^2 = 16 \Rightarrow \frac{5}{2}c^2 = 16 \Rightarrow c = \pm\frac{4\sqrt{2}}{\sqrt{5}}$.

There are two points: $\left(\frac{8\sqrt{2}}{\sqrt{5}}, \frac{2\sqrt{2}}{\sqrt{5}}, -\frac{2\sqrt{2}}{\sqrt{5}}\right)$ and $\left(-\frac{8\sqrt{2}}{\sqrt{5}}, -\frac{2\sqrt{2}}{\sqrt{5}}, \frac{2\sqrt{2}}{\sqrt{5}}\right)$.

[26] The normal to the given surface at (x_0, y_0, z_0) is $\nabla F(x_0, y_0, z_0) = \frac{2}{3}x_0^{-1/3}\mathbf{i} + \frac{2}{3}y_0^{-1/3}\mathbf{j} + \frac{2}{3}z_0^{-1/3}\mathbf{k}$, and the tangent plane is $x_0^{-1/3}(x - x_0) + y_0^{-1/3}(y - y_0) + z_0^{-1/3}(z - z_0) = 0$. Let A, B, and C be the x-, y-, and z-intercepts, respectively. To find the x-intercept, let $y = z = 0$, and then $(A - x_0) = x_0^{1/3}(y_0^{2/3} + z_0^{2/3}) = x_0^{1/3}(a^{2/3} - x_0^{2/3}) \Rightarrow A = x_0^{1/3}a^{2/3}$. Similarly, $B = y_0^{1/3}a^{2/3}$ and $C = z_0^{1/3}a^{2/3}$.

Thus, $A^2 + B^2 + C^2 = (x_0^{2/3} + y_0^{2/3} + z_0^{2/3})a^{4/3} = (a^{2/3})a^{4/3} = a^2$.

[27] Without loss of generality, let $F(x, y, z) = x^2 + y^2 + z^2 - a^2 = 0$. $\nabla F(x, y, z) = 2x\mathbf{i} + 2y\mathbf{j} + 2z\mathbf{k}$. At (x_0, y_0, z_0), the normal line is given by $x = x_0 + 2x_0t$, $y = y_0 + 2y_0t$, $z = z_0 + 2z_0t$. $x = 0 \Rightarrow 0 = x_0 + 2x_0t \Rightarrow t = -\frac{1}{2}$.

Letting $t = -\frac{1}{2}$ gives $y = z = 0$.

[28] $F(x, y, z) = 4x^2 + 9y^2 - z = 0 \Rightarrow \nabla F(x, y, z) = 8x\mathbf{i} + 18y\mathbf{j} - \mathbf{k}$.

At (x_0, y_0, z_0), the normal line is parallel to $\mathbf{a} = 8x_0\mathbf{i} + 18y_0\mathbf{j} - \mathbf{k}$.

$\overrightarrow{PQ} = 7\mathbf{i} - 5\mathbf{j} - \mathbf{k}$ is parallel to $\mathbf{a}$ if $8x_0 = 7c$, $18y_0 = -5c$, and $-1 = -c$.

If $-1 = -c$, then $c = 1$, $x_0 = \frac{7}{8}$, $y_0 = -\frac{5}{18}$, and $z_0 = 4(\frac{7}{8})^2 + 9(-\frac{5}{18})^2 = \frac{541}{144}$.

Thus, the point $(x_0, y_0, z_0) = (\frac{7}{8}, -\frac{5}{18}, \frac{541}{144})$ satisfies the given conditions.

[29] Direction vectors for the normal lines are given by ∇F and ∇G. The vectors are orthogonal $\Leftrightarrow \nabla F(x, y, z) \cdot \nabla G(x, y, z) = 0 \Leftrightarrow F_xG_x + F_yG_y + F_zG_z = 0$.

[30] $F(x, y, z) = x^2 + y^2 + z^2 - a^2 = 0$ and $G(x, y, z) = x^2 + y^2 - z^2 = 0 \Rightarrow$

$F_xG_x + F_yG_y + F_zG_z = 4x^2 + 4y^2 - 4z^2 = 4(x^2 + y^2 - z^2) = 0,$

since $x^2 + y^2 - z^2 = 0$ for every point on the cone.

[31] Step 1: $f_x(x, y) = 2x \Rightarrow f_x(1.3, 1.1) = 2.6.$

$f_y(x, y) = -3y^2 \Rightarrow f_y(1.3, 1.1) = -3.63.$

$g_x(x, y) = 2x \Rightarrow g_x(1.3, 1.1) = 2.6.$ $g_y(x, y) = 2y \Rightarrow g_y(1.3, 1.1) = 2.2.$

$f(1.3, 1.1) = 0.359$ and $g(1.3, 1.1) = -0.1.$ Using (16.35) gives us

$z - 0.359 = 2.6(x - 1.3) - 3.63(y - 1.1)$ and

$z + 0.1 = 2.6(x - 1.3) + 2.2(y - 1.1).$

Step 2: Letting $z = 0$ yields $2.6x - 3.63y = -0.972$ and

$2.6x + 2.2y = 5.9$ as the traces in the xy-plane.

Step 3: Solving yields $y \approx 1.1787$ and $x \approx 1.2718 \Rightarrow (x_2, y_2) = (1.2718, 1.1787).$

[32] Step 1: $f_x(x, y) = \cos x \Rightarrow f_x(1.2, 0.3) \approx 0.362358.$

$f_y(x, y) = \sin y \Rightarrow f_y(1.2, 0.3) \approx 0.295520.$

$g_x(x, y) = 2x \Rightarrow g_x(1.2, 0.3) = 2.4.$

$g_y(x, y) = -2y \Rightarrow g_y(1.2, 0.3) = -0.6.$

$f(1.2, 0.3) \approx -0.0232974$ and $g(1.2, 0.3) = 0.05.$ Using (16.35) gives us

$z + 0.0232974 = 0.362358(x - 1.2) + 0.295520(y - 0.3)$ and

$z - 0.05 = 2.4(x - 1.2) - 0.6(y - 0.3).$

Step 2: Letting $z = 0$ yields $0.362358x + 0.295520y = 0.546783$ and

$2.4x - 0.6y = 2.65$ as the traces in the xy-plane.

Step 3: Solving yields $x \approx 1.1991$ and $y \approx 0.3799 \Rightarrow (x_2, y_2) = (1.1991, 0.3799).$

Exercises 16.8

Note: As in (16.39), let $D(x, y) = f_{xx}(x, y)f_{yy}(x, y) - [f_{xy}(x, y)]^2.$

Also, let f_x, f_{xx}, f_y, and f_{yy} all be evaluated at (x, y).

[1] $f_x = -2x - 4 = 0$ and $f_y = -2y + 2 = 0 \Rightarrow x = -2$ and $y = 1.$

$f_{xx} = -2$, $f_{xy} = 0$, and $f_{yy} = -2 \Rightarrow D(-2, 1) = (-2)(-2) - 0^2 = 4 > 0.$

Since $f_{xx} < 0$, $f(-2, 1) = 4$ is a *LMAX* by 16.40(i).

[2] $f_x = 2x - 2 = 0$ and $f_y = 2y - 6 = 0 \Rightarrow x = 1$ and $y = 3.$

$f_{xx} = 2$, $f_{xy} = 0$, and $f_{yy} = 2 \Rightarrow D(1, 3) = 2(2) - 0^2 = 4 > 0.$

Since $f_{xx} > 0$, $f(1, 3) = 2$ is a *LMIN* by 16.40(ii).

[3] $f_x = 2x - 1 = 0$ and $f_y = 8y + 2 = 0 \Rightarrow x = \frac{1}{2}$ and $y = -\frac{1}{4}.$

$f_{xx} = 2$, $f_{xy} = 0$, and $f_{yy} = 8 \Rightarrow D(\frac{1}{2}, -\frac{1}{4}) = 2(8) - 0^2 = 16 > 0.$

Since $f_{xx} > 0$, $f(\frac{1}{2}, -\frac{1}{4}) = -\frac{1}{2}$ is a *LMIN*.

[4] $f_x = 4 - 4x = 0$ and $f_y = 3 - 2y = 0 \Rightarrow x = 1$ and $y = \frac{3}{2}.$

$f_{xx} = -4$, $f_{xy} = 0$, and $f_{yy} = -2 \Rightarrow D(1, \frac{3}{2}) = -4(-2) - 0^2 = 8 > 0.$

Since $f_{xx} < 0$, $f(1, \frac{3}{2}) = \frac{37}{4}$ is a *LMAX*.

[5] $f_x = 2x + 2y = 0$ and $f_y = 2x + 6y = 0 \Rightarrow x = y = 0$. $f_{xx} = 2$, $f_{xy} = 2$, and $f_{yy} = 6 \Rightarrow D(0, 0) = 2(6) - 2^2 = 8 > 0$. Since $f_{xx} > 0$, $f(0, 0) = 0$ is a *LMIN*.

Note: We will use the notation SP for saddle point.

[6] $f_x = 2x - 3y - 6 = 0$ and $f_y = -3x - 2y + 2 = 0 \Rightarrow x = \frac{18}{13}$ and $y = -\frac{14}{13}$.
$f_{xx} = 2$, $f_{xy} = -3$, and $f_{yy} = -2 \Rightarrow D(\frac{18}{13}, -\frac{14}{13}) = 2(-2) - (-3)^2 = -13 < 0 \Rightarrow (\frac{18}{13}, -\frac{14}{13}, f(\frac{18}{13}, -\frac{14}{13}))$ is a SP by 16.41. There are no local extrema.

[7] $f_x = 3x^2 + 3y = 0$ and $f_y = 3x - 3y^2 = 0$. From the first equation, $y = -x^2$.
Substituting this into the second equation yields $x = x^4 \Rightarrow x(x^3 - 1) = 0 \Rightarrow$
$x = 0, 1$ and hence $y = 0, -1$. $f_{xx} = 6x$, $f_{xy} = 3$, and $f_{yy} = -6y \Rightarrow$
$D(0, 0) = 0(0) - 3^2 = -9 < 0$ and $D(1, -1) = 6(6) - 3^2 = 27 > 0$.
Thus, $(0, 0, f(0, 0))$ is a SP and since $f_{xx}(1, -1) > 0$, $f(1, -1) = -1$ is a *LMIN*.

[8] $f_x = 2x + y = 0$ and $f_y = x = 0 \Rightarrow x = 0$ and $y = -2x = 0$. $f_{xx} = 2$, $f_{xy} = 1$, and $f_{yy} = 0 \Rightarrow D(0, 0) = 2(0) - 1^2 = -1 < 0 \Rightarrow (0, 0, f(0, 0))$ is a SP.

[9] $f_x = x + 2y + 1 = 0$ and $f_y = 2x - y - 8 = 0 \Rightarrow x = 3$ and $y = -2$.
$f_{xx} = 1$, $f_{xy} = 2$, and $f_{yy} = -1 \Rightarrow D(3, -2) = 1(-1) - 2^2 = -5 < 0 \Rightarrow$
$(3, -2, f(3, -2))$ is a SP.

[10] $f_x = -4x - 2y - 14 = 0$ and $f_y = -2x - 3y - 5 = 0 \Rightarrow x = -4$ and $y = 1$.
$f_{xx} = -4$, $f_{xy} = -2$, and $f_{yy} = -3 \Rightarrow D(-4, 1) = -4(-3) - (-2)^2 = 8 > 0$.
Since $f_{xx} < 0$, $f(-4, 1) = \frac{51}{2}$ is a *LMAX*.

[11] $f_x = x^2 + x - 6 = 0$ and $f_y = -2y^2 + 32 = 0 \Rightarrow x = -3, 2$ and $y = \pm 4$.
$f_{xx} = 2x + 1$, $f_{xy} = 0$, and $f_{yy} = -4y \Rightarrow D(x, y) = -4y(2x + 1)$.
$D(2, 4) = -80 < 0 \Rightarrow (2, 4, f(2, 4))$ is a SP.
$D(-3, -4) = -80 < 0 \Rightarrow (-3, -4, f(-3, -4))$ is a SP.
$D(2, -4) = 80 > 0$ and $f_{xx}(2, -4) > 0 \Rightarrow f(2, -4) = -\frac{266}{3}$ is a *LMIN*.
$D(-3, 4) = 80 > 0$ and $f_{xx}(-3, 4) < 0 \Rightarrow f(-3, 4) = \frac{617}{6}$ is a *LMAX*.

[12] $f_x = x^2 - 3x = 0$ and $f_y = y^2 - 4 = 0 \Rightarrow x = 0, 3$ and $y = \pm 2$.
$f_{xx} = 2x - 3$, $f_{xy} = 0$, and $f_{yy} = 2y \Rightarrow D(x, y) = (2x - 3)(2y)$.
$D(0, 2) = -12 < 0 \Rightarrow (0, 2, f(0, 2))$ is a SP.
$D(3, -2) = -12 < 0 \Rightarrow (3, -2, f(3, -2))$ is a SP.
$D(3, 2) = 12 > 0$ and $f_{xx}(3, 2) > 0 \Rightarrow f(3, 2) = -\frac{59}{6}$ is a *LMIN*.
$D(0, -2) = 12 > 0$ and $f_{xx}(0, -2) < 0 \Rightarrow f(0, -2) = \frac{16}{3}$ is a *LMAX*.

[13] $f_x = 2x^3 - 6x^2 + 4y = 0$ and $f_y = 4x + 2y = 0$. $f_y = 0 \Rightarrow y = -2x$.

$f_x = 0 \Rightarrow 2x^3 - 6x^2 = -4y \Rightarrow 6x^2 - 2x^3 + 8x = 0 \Rightarrow 2x(4 - x)(1 + x) = 0 \Rightarrow$

$x = 0, 4, -1$ and hence $y = 0, -8, 2$. $f_{xx} = 6x^2 - 12x$, $f_{xy} = 4$, and $f_{yy} = 2 \Rightarrow$

$D(x, y) = (6x^2 - 12x)(2) - 4^2 = 12x(x - 2) - 16$.

$D(0, 0) = -16 < 0 \Rightarrow (0, 0, f(0, 0))$ is a SP.

Note: f_{yy} may be used in place of f_{xx} in (16.40).

$D(4, -8) = 80 > 0$ and $f_{yy} > 0 \Rightarrow f(4, -8) = -64$ is a *LMIN*.

$D(-1, 2) = 20 > 0$ and $f_{yy} > 0 \Rightarrow f(-1, 2) = -\frac{3}{2}$ is a *LMIN*.

[14] $f_x = x^2 + 4y - 9$ and $f_y = 4x - 2y$. $f_y = 0 \Rightarrow y = 2x$.

$f_x = 0 \Rightarrow x^2 + 8x - 9 = 0 \Rightarrow x = -9, 1$ and hence $y = -18, 2$.

$f_{xx} = 2x$, $f_{xy} = 4$, and $f_{yy} = -2 \Rightarrow D(x, y) = (2x)(-2) - 4^2 = -4x - 16$.

$D(1, 2) = -20 < 0 \Rightarrow (1, 2, f(1, 2))$ is a SP.

$D(-9, -18) = 20 > 0$ and $f_{yy} < 0 \Rightarrow f(-9, -18) = 162$ is a *LMAX*.

[15] $f_x = 4x^3 + 32 = 0$ and $f_y = 3y^2 - 9 = 0 \Rightarrow x = -2$ and $y = \pm\sqrt{3}$.

$f_{xx} = 12x^2$, $f_{xy} = 0$, and $f_{yy} = 6y \Rightarrow D(-2, \sqrt{3}) > 0$ and $D(-2, -\sqrt{3}) < 0$.

Thus, $(-2, -\sqrt{3}, f(-2, -\sqrt{3}))$ is a SP, and since $f_{xx}(-2, \sqrt{3}) > 0$,

$f(-2, \sqrt{3}) = -48 - 6\sqrt{3}$ is a *LMIN*.

[16] $f_x = -x^2 + y = 0$ and $f_y = x + y - 12 = 0 \Rightarrow x = -4, 3$ and hence $y = 16, 9$.

$f_{xx} = -2x$, $f_{xy} = 1$, and $f_{yy} = 1 \Rightarrow D(x, y) = (-2x)(1) - 1^2 = -2x - 1$.

$D(3, 9) = -7 < 0 \Rightarrow (3, 9, f(3, 9))$ is a SP.

$D(-4, 16) = 7 > 0$ and $f_{yy} > 0 \Rightarrow f(-4, 16) = -\frac{320}{3}$ is a *LMIN*.

[17] $f_x = e^x \sin y = 0$ and $f_y = e^x \cos y = 0 \Rightarrow \sin y = \cos y = 0$ $\{ e^x \neq 0 \}$, which has no solution since $\sin^2 y + \cos^2 y = 1$. Thus, f has no extrema or saddle points.

[18] $f_x = \sin y = 0$ and $f_y = x \cos y = 0 \Rightarrow x = 0$ and $y = \pi n$.

$\{ \cos y = \sin y = 0$ has no solution.$\}$ $f_{xx} = 0$, $f_{xy} = \cos y$, and $f_{yy} = -x \sin y \Rightarrow$

$D(x, y) = -\cos^2 y \Rightarrow D(0, \pi n) = -1 < 0$. Thus, f has SP at $(0, \pi n, f(0, \pi n))$.

[19] $f(x, y) = 4x^{-1} + xy + 8y^{-1}$. $f_x = -4x^{-2} + y = 0$ and $f_y = x - 8y^{-2} = 0 \Rightarrow$

$x^2 y = 4$ and $xy^2 = 8 \Rightarrow x = 2^{1/3}$ and $y = 2^{4/3}$.

$f_{xx} = 8x^{-3}$, $f_{xy} = 1$, and $f_{yy} = 16y^{-3} \Rightarrow D(2^{1/3}, 2^{4/3}) = 3 > 0$.

$f_{xx}(2^{1/3}, 2^{4/3}) > 0 \Rightarrow f(2^{1/3}, 2^{4/3}) = 6\sqrt[3]{4}$ is a *LMIN*.

[20] $f_x = \dfrac{y}{(x + y)^2} = 0$ and $f_y = \dfrac{-x}{(x + y)^2} = 0 \Rightarrow$ no solution since $(x, y) = (0, 0)$ is not in the domain of f. Thus, f has no extrema or saddle points.

[21] $f_x = (x^2 + 3y^2)e^{-(x^2+y^2)}(-2x) + 2xe^{-(x^2+y^2)} = 2xe^{-(x^2+y^2)}(1 - x^2 - 3y^2)$.

$f_y = (x^2 + 3y^2)e^{-(x^2+y^2)}(-2y) + 6ye^{-(x^2+y^2)} = 2ye^{-(x^2+y^2)}(3 - x^2 - 3y^2)$.

$f_x = 0$ and $f_y = 0 \Rightarrow x(1 - x^2 - 3y^2) = 0$ and $y(3 - x^2 - 3y^2) = 0$. If $x = 0$, then $y = 0, \pm 1$. If $1 - x^2 - 3y^2 = 0$, then $x^2 = 1 - 3y^2$, $y = 0$, and hence $x = \pm 1$. The five critical points are $(0, 0)$, $(0, 1)$, $(0, -1)$, $(1, 0)$, and $(-1, 0)$. Since $f(x, y) \geq 0$ for all (x, y), $f(0, 0) = 0$ is a *LMIN*. Since $f(x, y) \to 0$ as $x^2 + y^2 \to \infty$, a maximum value will occur at one or more of the critical points. $f(\pm 1, 0) = e^{-1}$ and $f(0, \pm 1) = 3e^{-1}$. Thus, $f(0, \pm 1) = 3e^{-1}$ are maximums. $(\pm 1, 0, f(\pm 1, 0))$ are SP.

22 (a) $f_x = e^{-(x^2+y^2)/4}(y^2 - \frac{1}{2}x^2y^2) = 0$ and $f_y = e^{-(x^2+y^2)/4}(2xy - \frac{1}{2}xy^3) = 0 \Rightarrow$ $y^2(2 - x^2) = 0$ and $xy(4 - y^2) = 0$. If $y = 0$, then x can be any value, and hence, there are an infinite number of critical points.

(b) If $x = \pm\sqrt{2}$, then $y = 0, \pm 2$. Thus, four critical points are $(\pm\sqrt{2}, 2)$ and $(\pm\sqrt{2}, -2)$. Any point on the line $y = 0$ { x-axis } is also a critical point.

23 $x^2 + 4y^2 = 1 \Rightarrow -1 \leq x \leq 1$ and $-\frac{1}{2}\sqrt{1 - x^2} \leq y \leq \frac{1}{2}\sqrt{1 - x^2}$.

(1) On the upper boundary $y = \frac{1}{2}\sqrt{1 - x^2}$,

$f(x, \frac{1}{2}\sqrt{1 - x^2}) = 1 - x + \sqrt{1 - x^2} = h(x)$. $h'(x) = -1 - x(1 - x^2)^{-1/2}$.

$h'(x) = 0 \Rightarrow x = -\sqrt{1 - x^2} \Rightarrow x = -\frac{1}{\sqrt{2}}$. $f(-\frac{1}{\sqrt{2}}, \frac{1}{2\sqrt{2}}) = 1 + \sqrt{2}$.

(2) On the lower boundary $y = -\frac{1}{2}\sqrt{1 - x^2}$,

$f(x, -\frac{1}{2}\sqrt{1 - x^2}) = 1 - x - \sqrt{1 - x^2} = h(x)$. $h'(x) = -1 + x(1 - x^2)^{-1/2}$.

$h'(x) = 0 \Rightarrow x = \sqrt{1 - x^2} \Rightarrow x = \frac{1}{\sqrt{2}}$. $f(\frac{1}{\sqrt{2}}, -\frac{1}{2\sqrt{2}}) = 1 - \sqrt{2}$.

At the intersection of the upper and lower boundaries, $f(-1, 0) = 2$ and $f(1, 0) = 0$. Using Exercise 3, the *MIN* is $f(\frac{1}{2}, -\frac{1}{4}) = -\frac{1}{2}$ and the *MAX* is $f(-\frac{1}{\sqrt{2}}, \frac{1}{2\sqrt{2}}) = 1 + \sqrt{2}$.

24 In R, $-2 \leq x \leq 2$ and $|x| \leq y \leq 2$.

(1) On the line $y = 2$, $f(x, 2) = -2x^2 + 4x + 7 = h(x)$. $h'(x) = 0 \Rightarrow x = 1$. $f(1, 2) = 9$.

(2) On the line $y = x$, $f(x, x) = -3x^2 + 7x + 5 = h(x)$. $h'(x) = 0 \Rightarrow x = \frac{7}{6}$. $f(\frac{7}{6}, \frac{7}{6}) = \frac{109}{12}$.

(3) On the line $y = -x$, $f(x, -x) = -3x^2 + x + 5 = h(x)$. $h'(x) = 0 \Rightarrow x = \frac{1}{6}$. $f(\frac{1}{6}, -\frac{1}{6}) = \frac{61}{12}$.

At the corners of R, $f(0, 0) = 5$, $f(2, 2) = 7$, and $f(-2, 2) = -9$.

Using Exercise 4, the *MIN* is $f(-2, 2) = -9$ and the *MAX* is $f(1, \frac{3}{2}) = \frac{37}{4}$.

25 (1) On the line $y = -1$, $f(x, -1) = x^2 - 2x + 3 = h(x)$. $h'(x) = 0 \Rightarrow x = 1$. $f(1, -1) = 2$.

(2) On the line $y = 3$, $f(x, 3) = x^2 + 6x + 27 = h(x)$. $h'(x) = 0 \Rightarrow x = -3$. This value is outside R.

(3) On the line $x = -2$, $f(-2, y) = 3y^2 - 4y + 4 = h(y)$. $h'(y) = 0 \Rightarrow y = \frac{2}{3}$. $f(-2, \frac{2}{3}) = \frac{8}{3}$. (cont.)

(4) On the line $x = 4$, $f(4, y) = 3y^2 + 8y + 16 = h(y)$. $h'(y) = 0 \Rightarrow y = -\frac{4}{3}$.

This value is outside R.

At the corners of R, $f(-2, -1) = 11$, $f(4, -1) = 11$, $f(-2, 3) = 19$, and $f(4, 3) = 67$. Using Exercise 5, the *MIN* is $f(0, 0) = 0$ and the *MAX* is $f(4, 3) = 67$.

26 (1) On the line $y = -2$, $f(x, -2) = x^2 - 8 = h(x)$. $h'(x) = 0 \Rightarrow x = 0$.

$f(0, -2) = -8$.

(2) On the line $y = 2$, $f(x, 2) = x^2 - 12x = h(x)$. $h'(x) = 0 \Rightarrow x = 6$.

This value is outside R.

(3) On the line $x = -3$, $f(-3, y) = -y^2 + 11y + 27 = h(y)$.

$h'(y) = 0 \Rightarrow y = \frac{11}{2}$. This value is outside R.

(4) On the line $x = 3$, $f(3, y) = -y^2 - 7y - 9 = h(y)$. $h'(y) = 0 \Rightarrow y = -\frac{7}{2}$.

This value is outside R.

At the corners of R, $f(-3, -2) = 1$, $f(3, -2) = 1$, $f(-3, 2) = 45$, and $f(3, 2) = -27$. Using Exercise 6, the *MIN* is $f(3, 2) = -27$ and the *MAX* is $f(-3, 2) = 45$.

27 The boundaries of the triangle are $x = 1$, $y = -2$, and $y = 2x$.

Thus, $-1 \le x \le 1$ and $-2 \le y \le 2x$.

(1) On the line $x = 1$, $f(1, y) = 1 + 3y - y^3 = h(y)$. $h'(y) = 0 \Rightarrow y = \pm 1$.

$f(1, 1) = 3$ and $f(1, -1) = -1$.

(2) On the line $y = -2$, $f(x, -2) = x^3 - 6x + 8 = h(x)$. $h'(x) = 0 \Rightarrow x = \pm\sqrt{2}$.

These values are outside R.

(3) On the line $y = 2x$, $f(x, 2x) = -7x^3 + 6x^2 = h(x)$. $h'(x) = 0 \Rightarrow x = 0, \frac{4}{7}$.

$f(0, 0) = 0$ and $f(\frac{4}{7}, \frac{8}{7}) = \frac{32}{49}$.

At the corners of R, $f(1, 2) = -1$, $f(1, -2) = 3$, and $f(-1, -2) = 13$. Using Exercise 7, the *MIN* are $f(1, 2) = f(1, -1) = -1$ and the *MAX* is $f(-1, -2) = 13$.

28 (1) On the parabola $y = x^2$, $f(x, x^2) = x^2 + x^3 = h(x)$. $h'(x) = 0 \Rightarrow x = 0, -\frac{2}{3}$.

$f(0, 0) = 0$ and $f(-\frac{2}{3}, \frac{4}{9}) = \frac{4}{27}$.

(2) On the line $y = 9$, $f(x, 9) = x^2 + 9x = h(x)$. $h'(x) = 0 \Rightarrow x = -\frac{9}{2}$.

This value is outside R.

At the corners of R, $f(-3, 9) = -18$ and $f(3, 9) = 36$.

Using Exercise 8, the *MIN* is $f(-3, 9) = -18$ and the *MAX* is $f(3, 9) = 36$.

29 Let f be the square of the distance from $P(2, 1, -1)$ to the plane $4x - 3y + z = 5$.

$f(x, y) = (x - 2)^2 + (y - 1)^2 + \left[(5 - 4x + 3y) + 1\right]^2 =$

$(x - 2)^2 + (y - 1)^2 + (6 - 4x + 3y)^2$. $f_x = 34x - 24y - 52 = 0$ and

$f_y = -24x + 20y + 34 = 0 \Rightarrow x = \frac{28}{13}$ and $y = \frac{23}{26}$. $f(\frac{28}{13}, \frac{23}{26}) = \frac{1}{26} \Rightarrow d = 1/\sqrt{26}$.

30 Let $P(0, 0, -2)$ be a point on the first plane.

As in Exercise 29, $f(x, y) = (x - 0)^2 + (y - 0)^2 + \left[(2x + 3y - 4) + 2\right]^2 =$

$x^2 + y^2 + (2x + 3y - 2)^2$. $f_x = 10x + 12y - 8 = 0$ and

$f_y = 12x + 20y - 12 = 0 \Rightarrow x = \frac{2}{7}$ and $y = \frac{3}{7}$. $f(\frac{2}{7}, \frac{3}{7}) = \frac{2}{7} \Rightarrow d = \sqrt{14}/7$.

[31] $xy^3z^2 = 16 \Rightarrow z^2 = \frac{16}{xy^3}$. Then, $f(x, y) = x^2 + y^2 + \frac{16}{xy^3}$ is the square of the distance from a point on the graph to the origin. $f_x = 2x - \frac{16}{x^2y^3} = 0$ and $f_y = 2y - \frac{48}{xy^4} = 0 \Rightarrow xy = 2$ and $xy^5 = 24 \Rightarrow y = \pm 12^{1/4}$ and $x = \pm 2/12^{1/4}$.

These solutions give 4 points but only $(2/12^{1/4}, 12^{1/4})$ and $(-2/12^{1/4}, -12^{1/4})$ are valid since $xy = 2 > 0$. Both of these points produce the same distance.

$z^2 = \frac{16}{xy^3} \Rightarrow z = \pm\frac{2\sqrt{2}}{12^{1/4}}$. Thus, the following 4 points on the graph give minimum distance. $\left(\frac{2}{\sqrt[4]{12}}, \sqrt[4]{12}, \pm\frac{2\sqrt{2}}{\sqrt[4]{12}}\right), \left(-\frac{2}{\sqrt[4]{12}}, -\sqrt[4]{12}, \pm\frac{2\sqrt{2}}{\sqrt[4]{12}}\right)$

[32] $x + y + z = 1000$ and $P = xyz \Rightarrow P = xy(1000 - x - y) = 1000xy - x^2y - xy^2$.

$P_x = 1000y - 2xy - y^2 = 0$ and $P_y = 1000x - x^2 - 2xy = 0 \Rightarrow$

$y(1000 - 2x - y) = 0$ and $x(1000 - 2y - x) = 0 \Rightarrow$

$1000 - 2x - y = 0 = 1000 - 2y - x \Rightarrow x = y = \frac{1000}{3}$. Also, $z = \frac{1000}{3}$.

[33] Since we are asked to find the relative dimensions, without loss of generality, let $V = 1$ unit and let z be the height of the box. Then, $xyz = 1$, or $z = \frac{1}{xy}$, and

$A = xy + 2xz + 2yz = xy + \frac{2}{y} + \frac{2}{x}$. $A_x = y - \frac{2}{x^2} = 0$ and $A_y = x - \frac{2}{y^2} = 0 \Rightarrow$

$x^2y = 2 = xy^2 \Rightarrow xy(x - y) = 0 \Rightarrow x = 0,\ y = 0$, or $x = y$.

Since $x, y > 0$, we must have $x = y$. Thus, $z = \frac{1}{xy} = \frac{y}{2}$ $\{x = \frac{2}{y^2}\}$.

The box should have a square base with height $\frac{1}{2}$ the length of the side of the base.

[34] As in Exercise 33, let $A = 1$ unit. Thus, $1 = xy + 2xz + 2yz \Rightarrow z = \frac{1 - xy}{2x + 2y}$ and

$V = xyz = \frac{xy(1 - xy)}{2(x + y)}$. $V_x = \frac{2(x + y)(y - 2xy^2) - xy(1 - xy)(2)}{4(x + y)^2} = 0$ and

$V_y = \frac{2(x + y)(y - 2x^2y) - xy(1 - xy)(2)}{4(x + y)^2} = 0 \Rightarrow \frac{y^2(1 - x^2 - 2xy)}{2(x + y)^2} = 0$ and

$\frac{x^2(1 - y^2 - 2xy)}{2(x + y)^2} = 0 \Rightarrow 1 - x^2 - 2xy = 0 = 1 - y^2 - 2xy$ since $x, y > 0$.

Thus, $x = y = \sqrt{1/3}$ and $z = \frac{1 - xy}{2(x + y)} = \frac{1 - \frac{1}{3}}{4(\sqrt{1/3})} = \frac{1}{2}\sqrt{1/3}$.

The box should have a square base with height $\frac{1}{2}$ the length of the side of the base—the same shape as the box in Exercise 33.

[35] Let $F(x, y, z) = 16x^2 + 4y^2 + 9z^2 - 144 = 0$.

Then, $z_x = -\frac{F_x}{F_z} = -\frac{32x}{18z}$, $z_y = -\frac{F_y}{F_z} = -\frac{8y}{18z}$, and $V = (2x)(2y)(2z) = 8xyz$.

$V_x = 8yz + 8xy(z_x) = 0$ and $V_y = 8xz + 8xy(z_y) = 0 \Rightarrow 18z^2 - 32x^2 = 0$ and $18z^2 - 8y^2 = 0 \Rightarrow x^2 = \frac{9}{16}z^2$ and $y^2 = \frac{9}{4}z^2$. $16(\frac{9}{16}z^2) + 4(\frac{9}{4}z^2) + 9z^2 - 144 = 0 \Rightarrow$ $z^2 = \frac{16}{3} \Rightarrow z = 4/\sqrt{3}$, $x = 3/\sqrt{3}$, and $y = 6/\sqrt{3}$.

Thus, the dimensions are $8/\sqrt{3}$, $6/\sqrt{3}$, and $12/\sqrt{3}$.

[36] Let $F(x, y, z) = \left(\frac{x}{a}\right)^2 + \left(\frac{y}{b}\right)^2 + \left(\frac{z}{c}\right)^2 - 1 = 0$. Then, $z_x = -\frac{F_x}{F_z} = -\frac{c^2x}{a^2z}$ and

$z_y = -\frac{F_y}{F_z} = -\frac{c^2y}{b^2z}$, and $V = 8xyz$. $V_x = 8yz + 8xy(z_x) = 0$ and

$V_y = 8xz + 8xy(z_y) \Rightarrow a^2z^2 - c^2x^2 = 0$ and $b^2z^2 - c^2y^2 = 0 \Rightarrow$

$x^2 = \frac{a^2z^2}{c^2}$ and $y^2 = \frac{b^2z^2}{c^2}$. Substituting into $\frac{x^2}{a^2} + \frac{y^2}{b^2} + \frac{z^2}{c^2} = 1$ yields

$z^2 = \frac{c^2}{3} \Rightarrow x = \frac{a}{\sqrt{3}}$, $y = \frac{b}{\sqrt{3}}$, and $z = \frac{c}{\sqrt{3}}$. The dimensions are $\frac{2a}{\sqrt{3}}, \frac{2b}{\sqrt{3}}$, and $\frac{2c}{\sqrt{3}}$.

[37] $V = xyz = xy(12 - 4x - 3y) = 12xy - 4x^2y - 3xy^2$.

$V_x = y(12 - 8x - 3y) = 0$ and $V_y = 2x(6 - 2x - 3y) = 0 \Rightarrow$

$12 - 8x - 3y = 0 = 6 - 2x - 3y$ {since $x, y > 0$} $\Rightarrow x = 1$, $y = \frac{4}{3}$, and $z = 4$.

[38] $V = xyz$ {with $\frac{x}{a} + \frac{y}{b} + \frac{z}{c} = 1$} $\Rightarrow V = cxy\left(1 - \frac{x}{a} - \frac{y}{b}\right)$.

$V_x = cy\left(1 - \frac{2x}{a} - \frac{y}{b}\right) = 0$ and $V_y = cx\left(1 - \frac{x}{a} - \frac{2y}{b}\right) = 0 \Rightarrow$

$1 - \frac{2x}{a} - \frac{y}{b} = 0 = 1 - \frac{x}{a} - \frac{2y}{b} \Rightarrow \frac{y}{b} = \frac{x}{a}$. Substituting into V_x yields

$1 - \frac{2x}{a} - \frac{x}{a} = 0 \Rightarrow x = \frac{a}{3}$. Similarly, $y = \frac{b}{3}$, and hence $z = \frac{c}{3}$.

[39] Let x = base length, y = base width, and z = height.

$C = 2(2xy) + 2xz + 2yz$ and $8 = xyz \Rightarrow C = 4xy + \frac{16}{y} + \frac{16}{x}$.

$C_x = 4y - \frac{16}{x^2} = 0$ and $C_y = 4x - \frac{16}{y^2} = 0 \Rightarrow$

$yx^2 = 4 = xy^2 \Rightarrow x = y = \sqrt[3]{4}$ ft, since $x, y > 0$. Also, $z = \frac{8}{xy} = 2\sqrt[3]{4}$ ft.

[40] The perimeter of the window is $P = x + 2y + x\sec\theta = 12$. So,

$y = \frac{1}{2}(12 - x - x\sec\theta)$ and the area is $A = xy + \frac{1}{2}x(\frac{1}{2}x\tan\theta) = xy + \frac{1}{4}x^2\tan\theta =$

$6x - \frac{1}{4}x^2(2 + 2\sec\theta - \tan\theta)$. $A_\theta = -\frac{1}{4}x^2(2\sec\theta\tan\theta - \sec^2\theta) = 0 \Rightarrow$

$2\tan\theta = \sec\theta$ {since $x^2 \neq 0$ and $\sec\theta \neq 0$} $\Rightarrow 2\sin\theta = 1 \Rightarrow \theta = \frac{\pi}{6}$.

$A_x = 6 - \frac{1}{2}x(2 + 2\sec\theta - \tan\theta) = 6 - \frac{1}{2}x(2 + \frac{4}{\sqrt{3}} - \frac{1}{\sqrt{3}}) = 6 - \frac{1}{2}x(2 + \sqrt{3}) = 0$

$\Rightarrow x = \frac{12}{2 + \sqrt{3}} = 24 - 12\sqrt{3}$ and $y = 6 - 2\sqrt{3}$.

[41] Let ℓ, w, and h denote the length, width, and height of the box. Then,

$\ell + 2w + 2h = 108 \Rightarrow \ell = 108 - 2w - 2h$. So, $V = \ell wh = wh(108 - 2w - 2h)$.

$V_w = 108h - 4hw - 2h^2 = 0$ and $V_h = 108w - 2w^2 - 4wh = 0 \Rightarrow$

$h(54 - 2w - h) = 0$ and $w(54 - 2h - w) = 0 \Rightarrow 54 - 2w - h = 0$ and

$54 - 2h - w = 0$ {$h, w \neq 0$} $\Rightarrow h = w$ and hence $3h = 54$, or $h = 18$.

Substituting gives $w = 18$ and $\ell = 36$.

[42] Let $\mathbf{a} = x\mathbf{i} + y\mathbf{j} + z\mathbf{k}$. $\|\mathbf{a}\| = \sqrt{x^2 + y^2 + z^2} = 8 \Rightarrow z^2 = 64 - x^2 - y^2$.

Since we want to maximize the sum f of the components, we assume $x, y, z \geq 0$.

$f(x, y) = x + y + \sqrt{64 - x^2 - y^2}$.

$f_x = 1 - \dfrac{x}{(64 - x^2 - y^2)^{1/2}} = 0$ and $f_y = 1 - \dfrac{y}{(64 - x^2 - y^2)^{1/2}} = 0 \Rightarrow$

$x = y = \frac{8}{\sqrt{3}}$ and $z = \frac{8}{\sqrt{3}}$. The vector is $\frac{8}{\sqrt{3}}(\mathbf{i} + \mathbf{j} + \mathbf{k})$.

[43] Consider a figure similar to Figure 16.76. The fourth device must lie on the lines $y = x$ and $y = -\tan 15^\circ(x - 4)$. Hence, $x = -\tan 15^\circ(x - 4)$ or

$$x = \frac{4 \tan 15^\circ}{1 + \tan 15^\circ} = \frac{4(2 - \sqrt{3})}{1 + (2 - \sqrt{3})} = 2 - \tfrac{2}{3}\sqrt{3}.$$

The fourth device should be placed at $(x_0, y_0) = (2 - \frac{2}{3}\sqrt{3}, 2 - \frac{2}{3}\sqrt{3})$.

[44] Without loss of generality, suppose angle $P_1P_2P_3$ is 120° or greater.

Since P_4 is located inside the triangle $P_1P_2P_3$, angle $P_1P_4P_3$ must be greater than 120°. Thus, P_4 cannot be located so that all angles are equal to 120°.

[45] $f_m(m, b) = \sum\limits_{k=1}^{n} 2(y_k - mx_k - b)(-x_k) = 0$ and

$f_b(m, b) = \sum\limits_{k=1}^{n} 2(y_k - mx_k - b)(-1) = 0 \Rightarrow$

$\sum\limits_{k=1}^{n} (mx_k^2 + bx_k) = \sum\limits_{k=1}^{n} x_k y_k$ and $\sum\limits_{k=1}^{n} (mx_k + b) = \sum\limits_{k=1}^{n} y_k \Rightarrow$

$$\left(\sum_{k=1}^{n} x_k^2\right)m + \left(\sum_{k=1}^{n} x_k\right)b = \sum_{k=1}^{n} x_k y_k \text{ and } \left(\sum_{k=1}^{n} x_k\right)m + nb = \sum_{k=1}^{n} y_k$$

[46] $\sum\limits_{k=1}^{n} d_k = \sum\limits_{k=1}^{n} \left[y_k - (mx_k + b)\right] = \sum\limits_{k=1}^{n} y_k - \left(m\sum\limits_{k=1}^{n} x_k + nb\right) = 0$ by Exercise 45.

[47] $\sum\limits_{k=1}^{3} x_k^2 = 1^2 + 4^2 + 7^2 = 66$, $\sum\limits_{k=1}^{3} x_k = 1 + 4 + 7 = 12$,

$\sum\limits_{k=1}^{3} x_k y_k = 1(3) + 4(5) + 7(6) = 65$, and $\sum\limits_{k=1}^{3} y_k = 3 + 5 + 6 = 14 \Rightarrow$

$66m + 12b = 65$ and $12m + 3b = 14 \Rightarrow m = \frac{1}{2}$ and $b = \frac{8}{3}$. Thus, $y = \frac{1}{2}x + \frac{8}{3}$.

[48] $\sum\limits_{k=1}^{4} x_k^2 = 117$, $\sum\limits_{k=1}^{4} x_k = 19$, $\sum\limits_{k=1}^{4} x_k y_k = 57$, and $\sum\limits_{k=1}^{4} y_k = 10 \Rightarrow 117m + 19b = 57$

and $19m + 4b = 10 \Rightarrow m = \frac{38}{107}$ and $b = \frac{87}{107}$. Thus, $y = \frac{38}{107}x + \frac{87}{107}$.

[49] We will fit the data to a line using the method of least squares.

$\sum\limits_{k=1}^{10} x_k^2 = 54{,}785$, $\sum\limits_{k=1}^{10} x_k = 723$, $\sum\limits_{k=1}^{10} x_k y_k = 54{,}277$, and $\sum\limits_{k=1}^{10} y_k = 708 \Rightarrow$

$54{,}785m + 723b = 54{,}277$ and $723m + 10b = 708 \Rightarrow m = \frac{30{,}886}{25{,}121} \approx 1.23$ and

$b = -\frac{454{,}491}{25{,}121} \approx -18.09$. Thus, $y \approx 1.23x - 18.09$ and $x = 70 \Rightarrow y \approx 68$.

[50] We will fit the data to a line using the method of least squares.

$\sum\limits_{k=1}^{6} x_k^2 = 38.2$, $\sum\limits_{k=1}^{6} x_k = 15$, $\sum\limits_{k=1}^{6} x_k y_k = 8.04$, and $\sum\limits_{k=1}^{6} y_k = 3 \Rightarrow$

$38.2m + 15b = 8.04$ and $15m + 6b = 3 \Rightarrow m = \frac{324}{420} \approx 0.77$ and

$b = -\frac{600}{420} \approx -1.43$. Thus, $y \approx 0.77x - 1.43$ and $x = 2.5 \Rightarrow y \approx 0.50$ in.

[51] The least squares criterion is satisfied when the sum of the squares of the distances is minimal. The sum of the squares of the distances f from P to each city is $f(x, y) = \left[(x-2)^2 + (y-3)^2\right] + \left[(x-7)^2 + (y-2)^2\right] + \left[(x-5)^2 + (y-6)^2\right]$.

Then, $f_x(x, y) = 2(x-2) + 2(x-7) + 2(x-5) = 0$ and

$f_y(x, y) = 2(y-3) + 2(y-2) + 2(y-6) = 0 \Rightarrow 6x = 28$ and $6y = 22 \Rightarrow$

$x = \frac{14}{3}$ and $y = \frac{11}{3}$.

[52] Let $f(x, y) = \sum_{k=1}^{n} \left[(x - x_k)^2 + (y - y_k)^2\right]$.

Then, $f_x(x, y) = \sum_{k=1}^{n} 2(x - x_k) = 0$ and $f_y(x, y) = \sum_{k=1}^{n} 2(y - y_k) = 0 \Rightarrow$

$nx = \sum_{k=1}^{n} x_k$ and $ny = \sum_{k=1}^{n} y_k \Rightarrow x = \frac{1}{n}\sum_{k=1}^{n} x_k$ and $y = \frac{1}{n}\sum_{k=1}^{n} y_k$.

[53] (a) $f(a, b, c) = \sum_{k=1}^{n} (z_k - ax_k - by_k - c)^2$

$$\Rightarrow\ f_a(a, b, c) = \sum_{k=1}^{n} 2(z_k - ax_k - by_k - c)(-x_k) = 0$$

$$f_b(a, b, c) = \sum_{k=1}^{n} 2(z_k - ax_k - by_k - c)(-y_k) = 0$$

$$f_c(a, b, c) = \sum_{k=1}^{n} 2(z_k - ax_k - by_k - c)(-1) = 0$$

$$\Rightarrow\ \left(\sum_{k=1}^{n} x_k^2\right)a + \left(\sum_{k=1}^{n} x_k y_k\right)b + \left(\sum_{k=1}^{n} x_k\right)c = \sum_{k=1}^{n} x_k z_k$$

$$\left(\sum_{k=1}^{n} x_k y_k\right)a + \left(\sum_{k=1}^{n} y_k^2\right)b + \left(\sum_{k=1}^{n} y_k\right)c = \sum_{k=1}^{n} y_k z_k$$

$$\left(\sum_{k=1}^{n} x_k\right)a + \left(\sum_{k=1}^{n} y_k\right)b + nc = \sum_{k=1}^{n} z_k$$

(b) The system is $a + b + c = 2$, $a + 2b + 2c = 2$, and $a + 2b + 4c = 3$.

Solving yields $a = 2$, $b = -\frac{1}{2}$, and $c = \frac{1}{2}$.

Thus, $z = 2x - \frac{1}{2}y + \frac{1}{2}$, or, $4x - y - 2z + 1 = 0$.

[54] We want to maximize P. $p + q + r = 1 \Rightarrow r = 1 - p - q$.

So, $P = 2pq + 2p(1 - p - q) + 2q(1 - p - q) = 2p - 2p^2 + 2q - 2pq - 2q^2$.

$P_p = 2 - 4p - 2q = 0$ and $P_q = 2 - 2p - 4q = 0 \Rightarrow p = q = r = \frac{1}{3}$.

$P = \frac{2}{3}$ at these values, which is a maximum. Thus, $P \leq \frac{2}{3}$ for all p, q, and r.

[55] $f_x(x, y) = 3x^2 \sin x + x^3 \cos x - y = 0 \Rightarrow y = 3x^2 \sin x + x^3 \cos x$.

$f_y(x, y) = -x + 8y + 1 = 0 \Rightarrow y = \frac{1}{8}(x - 1)$.

Their point of intersection in R is approximately $(-0.35, -0.17)$,

which is a critical point of f. See *Figure 55.*

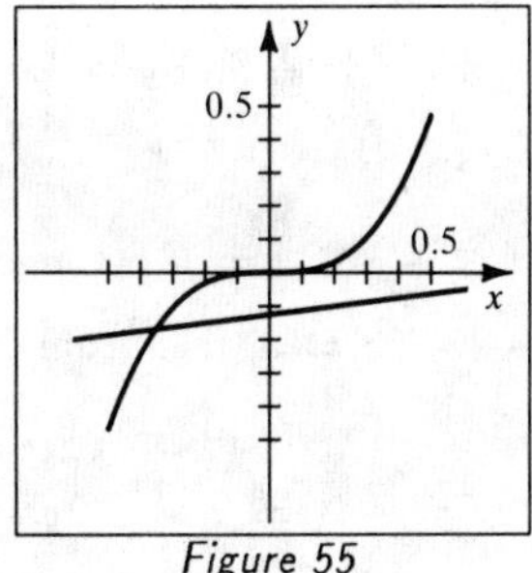

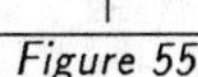
Figure 55

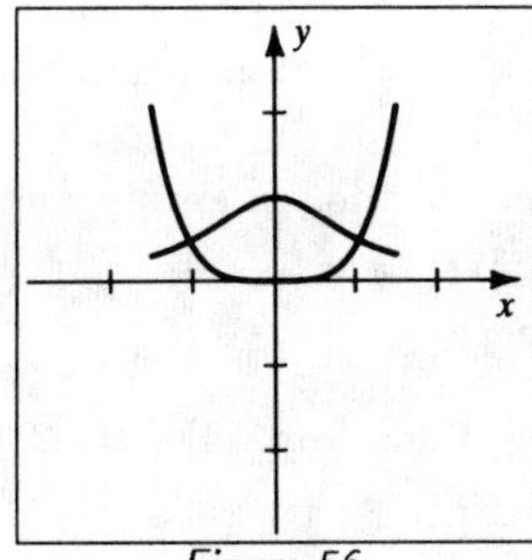

Figure 56

[56] $f_x(x,\,y) = y - \dfrac{1}{1 + x^2} = 0 \Rightarrow y = \dfrac{1}{1 + x^2}$.

$f_y(x,\,y) = x - \frac{5}{4}y^{1/4} = 0 \Rightarrow y = \frac{256}{625}x^4$.

Their points of intersection in R are approximately $(\pm\,1.04,\,0.48)$, which are critical points of f.

Exercises 16.9

Note: In Exercises 1–10, the equations listed first result from equating gradients and letting conditions equal 0. Let [[1]], [[2]], [[3]], etc., denote equation numbers.

[1] (16.43) $\Rightarrow -4y + 8x = 2x\lambda$ [[1]], $2y - 4x = 2y\lambda$ [[2]], and $x^2 + y^2 - 1 = 0$ [[3]].
Adding [[1]] plus twice [[2]] $\Rightarrow 0 = 2x\lambda + 4y\lambda = 2\lambda(x + 2y) \Rightarrow \lambda = 0$ or $x = -2y$.
If $\lambda = 0$, then from [[1]], $y = 2x$ and using [[3]] we find that
$(x,\,y) = (\frac{1}{\sqrt{5}},\,\frac{2}{\sqrt{5}})$ or $(-\frac{1}{\sqrt{5}},\,-\frac{2}{\sqrt{5}})$. If $x = -2y$, substituting into [[3]] yields
$(x,\,y) = (\frac{2}{\sqrt{5}},\,-\frac{1}{\sqrt{5}})$ or $(-\frac{2}{\sqrt{5}},\,\frac{1}{\sqrt{5}})$. $f(x,\,y) = 0$ at the first two points, which are *LMIN*, and $f(x,\,y) = 5$ at the second two points, which are *LMAX*.

[2] $4x + y = 2\lambda$ [[1]], $x - 2y + 1 = 3\lambda$ [[2]], and $2x + 3y - 1 = 0$ [[3]].
Substituting $y = 2\lambda - 4x$ {from [[1]]} into [[2]] yields $x = \frac{7}{9}\lambda - \frac{1}{9}$,
so $y = -\frac{10}{9}\lambda + \frac{4}{9}$ {from [[1]]}, and $2x + 3y - 1 = -\frac{16}{9}\lambda + \frac{1}{9}$ {from [[3]]} $= 0$.
Hence, $\lambda = \frac{1}{16}$, $x = -\frac{1}{16}$, and $y = \frac{3}{8}$. $f(-\frac{1}{16},\,\frac{3}{8}) = \frac{56}{256}$ is a *LMIN*.
Note: To convince yourself that this is a *LMIN*, pick another $(x,\,y)$ satisfying the constraint—say $(\frac{1}{2},\,0)$. Substituting into f yields $\frac{1}{2}$, a value *larger* than $\frac{56}{256}$.

[3] (16.45) $\Rightarrow 1 = 2x\lambda$ [[1]], $1 = 2y\lambda$ [[2]], $1 = 2z\lambda$ [[3]], and $x^2 + y^2 + z^2 - 25 = 0$ [[4]].
From [[1]], [[2]], and [[3]], $x = y = z$, and using [[4]], $x = y = z = \pm\frac{5}{\sqrt{3}}$.
$f(\frac{5}{\sqrt{3}},\,\frac{5}{\sqrt{3}},\,\frac{5}{\sqrt{3}}) = 5\sqrt{3}$ is a *LMAX* and $f(-\frac{5}{\sqrt{3}},\,-\frac{5}{\sqrt{3}},\,-\frac{5}{\sqrt{3}}) = -5\sqrt{3}$ is a *LMIN*.

[4] $2x = \lambda$ [[1]], $2y = \lambda$ [[2]], $2z = \lambda$ [[3]], and $x + y + z - 25 = 0$ [[4]]. From [[1]], [[2]], and [[3]], $x = y = z$, and using [[4]], $x = y = z = \frac{25}{3}$. $f(\frac{25}{3},\,\frac{25}{3},\,\frac{25}{3}) = \frac{625}{3}$ is a *LMIN*.

[5] $2x = \lambda$ [[1]], $2y = -\lambda$ [[2]], $2z = \lambda$ [[3]], and $x - y + z - 1 = 0$ [[4]]. From [[1]], [[2]], and [[3]], $x = -y = z$, and using [[4]], $x = -y = z = \frac{1}{3}$. $f(\frac{1}{3},\,-\frac{1}{3},\,\frac{1}{3}) = \frac{1}{3}$ is a *LMIN*.

6 $1 = 8x\lambda$ [[1]], $2 = 2y\lambda$ [[2]], $-3 = -\lambda$ [[3]], and $4x^2 + y^2 - z = 0$ [[4]].

From [[3]], $\lambda = 3$. Substituting into [[1]] and [[2]] yields $x = \frac{1}{24}$ and $y = \frac{1}{3}$.

Using [[4]], $z = \frac{17}{144}$. $f(\frac{1}{24}, \frac{1}{3}, \frac{17}{144}) = \frac{17}{48}$ is a *LMAX*.

7 See Example 4 for an example using 2 constraints.

$2x = \lambda$ [[1]], $2y = -\lambda + 2y\mu$ [[2]], $2z = -2z\mu$ [[3]], $x - y - 1 = 0$ [[4]], and $y^2 - z^2 - 1 = 0$ [[5]]. From [[3]], $z = 0$ or $\mu = -1$. If $z = 0$, then from [[5]], $y = 1, -1$ and from [[4]], $x = 2, 0$. This gives us $(2, 1, 0)$ and $(0, -1, 0)$.

If $\mu = -1$, then [[2]] gives $\lambda = -4y$. Using [[1]], $2x = -4y \Rightarrow x = -2y$.

Substituting into [[4]] yields $y = -\frac{1}{3}$. But substituting $y = -\frac{1}{3}$ into [[5]] gives $z^2 < 0$, which has no real solution, and hence $\mu \neq -1$. $f(0, -1, 0) = 1$ and $f(2, 1, 0) = 5$ are both *LMIN*. (Note that the intersection of $x - y = 1$ and $y^2 - z^2 = 1$ defines 2 separate curves when $y \geq 1$ or $y \leq -1$. The two points found give the square of the minimum distance between these curves and the origin. No maximum exists.)

8 $-2x = \lambda + 2x\mu$ [[1]], $-2y = \lambda + 2y\mu$ [[2]], $1 = \lambda$ [[3]], $x + y + z - 1 = 0$ [[4]], and $x^2 + y^2 - 4 = 0$ [[5]]. From [[3]], $\lambda = 1$. Substituting into [[1]] and [[2]] yields $2x(\mu + 1) = -1 = 2y(\mu + 1) \Rightarrow x = y$. From [[5]], $x = y = \pm\sqrt{2}$ and using [[4]], $z = 1 \mp 2\sqrt{2}$. $f(\sqrt{2}, \sqrt{2}, 1 - 2\sqrt{2}) = -3 - 2\sqrt{2}$ is a *LMIN* and

$f(-\sqrt{2}, -\sqrt{2}, 1 + 2\sqrt{2}) = -3 + 2\sqrt{2}$ is a *LMAX*.

9 $yzt = \lambda$ [[1]], $xzt = 2y\mu$ [[2]], $xyt = -\lambda$ [[3]], $xyz = \mu$ [[4]], $x - z - 2 = 0$ [[5]], and $y^2 + t - 4 = 0$ [[6]]. From [[1]] and [[3]], $yzt = -xyt \Rightarrow yt(x + z) = 0$.

If $yt = 0$, then $f(x, y, z, t) = 0$. If $x = -z$, then from [[5]], $z = -1$ and $x = 1$.

Multiplying [[2]] by y and [[4]] by t yields $2y^2\mu = xyzt = \mu t \Rightarrow \mu(2y^2 - t) = 0 \Rightarrow$ $\mu = 0$ or $t = 2y^2$. If $\mu = 0$, then $xyz = 0$ {from [[4]]} and $f(x, y, z, t) = 0$.

If $t = 2y^2$, then from [[6]], $y = \pm\frac{2}{\sqrt{3}}$ and $t = \frac{8}{3}$.

$f(1, \frac{2}{\sqrt{3}}, -1, \frac{8}{3}) = -\frac{16}{3\sqrt{3}}$ is a *LMIN* and $f(1, -\frac{2}{\sqrt{3}}, -1, \frac{8}{3}) = \frac{16}{3\sqrt{3}}$ is a *LMAX*.

10 $2x = 3\lambda$ [[1]], $2y = 4\lambda$ [[2]], $2z = \mu$ [[3]], $2t = \mu$ [[4]], $3x + 4y - 5 = 0$ [[5]], and $z + t - 2 = 0$ [[6]]. From [[1]] and [[2]], $x = \frac{3}{2}\lambda$ and $y = 2\lambda$. Substituting into [[5]] yields $3(\frac{3}{2}\lambda) + 4(2\lambda) - 5 = 0 \Rightarrow \lambda = \frac{2}{5}$, and hence $x = \frac{3}{5}$ and $y = \frac{4}{5}$. From [[3]] and [[4]], $z = t = \frac{1}{2}\mu$. Substituting into [[6]] gives us $\mu = 2$, and hence $z = t = 1$.

$f(\frac{3}{5}, \frac{4}{5}, 1, 1) = 3$ is a *LMIN*.

11 If f is the square of the distance from $P(x, y, z)$ to $(2, 3, 4)$, then we have $f(x, y, z) = (x - 2)^2 + (y - 3)^2 + (z - 4)^2$ and $g(x, y, z) = x^2 + y^2 + z^2 - 9 = 0$ [[1]].

$\nabla f = \lambda \nabla g \Rightarrow 2(x - 2) = 2x\lambda$ [[2]], $2(y - 3) = 2y\lambda$ [[3]], and $2(z - 4) = 2z\lambda$ [[4]].

From [[2]], [[3]], and [[4]], $\lambda = \frac{x - 2}{x} = \frac{y - 3}{y} = \frac{z - 4}{z} \Rightarrow y = \frac{3}{2}x$ and $z = 2x$.

Substituting into [[1]] yields $x = \pm\frac{6}{\sqrt{29}}$, and hence $y = \pm\frac{9}{\sqrt{29}}$, and $z = \pm\frac{12}{\sqrt{29}}$.

The positive values lead to a *minimum* distance,

whereas the negative values give a *maximum* distance.

[12] Let f be the square of the distance from the point (x, y, z) to the origin.

Thus, $f(x, y, z) = x^2 + y^2 + z^2$, $g(x, y, z) = x + 3y - 2z - 11 = 0$ ⟦1⟧,

and $h(x, y, z) = 2x - y + z - 3 = 0$ ⟦2⟧. $\nabla f = \lambda \nabla g + \mu \nabla h \Rightarrow$

$2x = \lambda + 2\mu$ ⟦3⟧, $2y = 3\lambda - \mu$ ⟦4⟧, and $2z = -2\lambda + \mu$ ⟦5⟧.

Solving ⟦3⟧, ⟦4⟧, and ⟦5⟧ for x, y, and z, respectively, and then substituting these expressions into ⟦1⟧ and ⟦2⟧ gives us the system $7\lambda - \frac{3}{2}\mu = 11$ and $-\frac{3}{2}\lambda + 3\mu = 3$.

Solving yields $\lambda = \mu = 2$, and hence $x = 3$, $y = 2$, and $z = -1$.

[13] Let x and y be the base dimensions and z the height. $g(x, y, z) = xyz - 2 = 0$ ⟦1⟧

and $C(x, y, z) = (2yz + 2xz) + 2xy + \frac{3}{2}xy = 2yz + 2xz + \frac{7}{2}xy$.

$\nabla C = \lambda \nabla g \Rightarrow 2z + \frac{7}{2}y = yz\lambda$ ⟦2⟧, $2z + \frac{7}{2}x = xz\lambda$ ⟦3⟧, and $2y + 2x = xy\lambda$ ⟦4⟧.

Multiplying ⟦2⟧ and ⟦3⟧ by x and y, respectively, and then subtracting yields

$2z(x - y) = 0$, or $x = y$. Multiplying ⟦2⟧ and ⟦4⟧ by x and z, respectively,

and then subtracting yields $\frac{7}{2}xy - 2yz = 0 \Rightarrow y(7x - 4z) = 0$, or $z = \frac{7}{4}x$.

Substituting into ⟦1⟧ gives $x = \frac{2}{\sqrt[3]{7}}$, $y = \frac{2}{\sqrt[3]{7}}$, and $z = \frac{7}{2\sqrt[3]{7}}$.

[14] Let ℓ and w be the base dimensions and h the height. The box has

fixed volume V and $V(\ell, w, h) = \ell wh$. We want to minimize the surface area

$S(\ell, w, h) = 2\ell w + 2\ell h + 2wh$. $\nabla S = \lambda \nabla V \Rightarrow 2w + 2h = wh\lambda$ ⟦1⟧,

$2\ell + 2h = \ell h\lambda$ ⟦2⟧, and $2\ell + 2w = \ell w\lambda$ ⟦3⟧. Multiplying ⟦1⟧ and ⟦2⟧ by ℓ and w, respectively, and then subtracting yields $2h(\ell - w) = 0$, or $\ell = w$. Multiplying ⟦1⟧ and ⟦3⟧ by ℓ and h, respectively, and then subtracting yields $2w(\ell - h) = 0$,

or $\ell = h$. Thus, $\ell = w = h$ and the rectangular box is a cube.

[15] $V(x, y, z) = xyz$ and $g(x, y, z) = 2x + 3y + 4z - 12 = 0$ ⟦1⟧.

$\nabla V = \lambda \nabla g \Rightarrow yz = 2\lambda$ ⟦2⟧, $xz = 3\lambda$ ⟦3⟧, and $xy = 4\lambda$ ⟦4⟧.

From ⟦2⟧, ⟦3⟧, and ⟦4⟧, $xyz = 2x\lambda = 3y\lambda = 4z\lambda \Rightarrow y = \frac{2}{3}x$ and $z = \frac{1}{2}x$.

Substituting into ⟦1⟧ yields $x = 2$, $y = \frac{4}{3}$, and $z = 1$. $V(2, \frac{4}{3}, 1) = \frac{8}{3}$.

[16] $V(x, y, z) = xyz$ and $g(x, y, z) = 2x + 3y + 5z - 90 = 0$ ⟦1⟧.

$\nabla V = \lambda \nabla g \Rightarrow yz = 2\lambda$ ⟦2⟧, $xz = 3\lambda$ ⟦3⟧, and $xy = 5\lambda$ ⟦4⟧.

From ⟦2⟧, ⟦3⟧, and ⟦4⟧, $xyz = 2x\lambda = 3y\lambda = 5z\lambda \Rightarrow y = \frac{2}{3}x$ and $z = \frac{2}{5}x$.

Substituting into ⟦1⟧ yields $x = 15$, $y = 10$, and $z = 6$.

[17] $V(r, h) = \pi r^2 h$ and $g(r, h) = 2\pi r^2 + 2\pi rh - S = 0$ ⟦1⟧.

$\nabla V = \lambda \nabla g \Rightarrow 2\pi rh = (4\pi r + 2\pi h)\lambda$ ⟦2⟧ and $\pi r^2 = 2\pi r\lambda$ ⟦3⟧. From ⟦3⟧, $\lambda = \frac{r}{2}$.

Substituting into ⟦2⟧ yields $h = 2r$, and the height is twice the radius.

18 $V(x, y, z) = (2x)(2y)(2z) = 8xyz$ and $g(x, y, z) = 4x^2 + 4y^2 + z^2 - 36 = 0$ $\llbracket 1\rrbracket$.

$\nabla V = \lambda \nabla g \Rightarrow 8yz = 8x\lambda$ $\llbracket 2\rrbracket$, $8xz = 8y\lambda$ $\llbracket 3\rrbracket$, and $8xy = 2z\lambda$ $\llbracket 4\rrbracket$.

From $\llbracket 2\rrbracket$, $\llbracket 3\rrbracket$, and $\llbracket 4\rrbracket$, $8xyz = 8x^2\lambda = 8y^2\lambda = 2z^2\lambda \Rightarrow y = x$ and $z = 2x$.

Substituting into $\llbracket 1\rrbracket$ yields $12x^2 = 36 \Rightarrow x = \sqrt{3}$, $y = \sqrt{3}$, and $z = 2\sqrt{3}$.

Thus, the dimensions are $2\sqrt{3} \times 2\sqrt{3} \times 4\sqrt{3}$.

19 $A^2 = f(x, y, z) = s(s - x)(s - y)(s - z)$ and $g(x, y, z) = x + y + z - p = 0$ $\llbracket 1\rrbracket$.

$\nabla f = \lambda \nabla g \Rightarrow -s(s - y)(s - z) = \lambda$ $\llbracket 2\rrbracket$, $-s(s - x)(s - z) = \lambda$ $\llbracket 3\rrbracket$, and

$-s(s - x)(s - y) = \lambda$ $\llbracket 4\rrbracket$. From $\llbracket 2\rrbracket$, $\llbracket 3\rrbracket$, and $\llbracket 4\rrbracket$,

$s - x = s - y = s - z \Rightarrow x = y = z$. From $\llbracket 1\rrbracket$, $x = y = z = \frac{1}{3}p$.

20 Let α, β, and γ denote the three angles of the triangle.

$f(\alpha, \beta, \gamma) = \sin\alpha \sin\beta \sin\gamma$ and $g(\alpha, \beta, \gamma) = \alpha + \beta + \gamma - \pi = 0$ $\llbracket 1\rrbracket$.

$\nabla f = \lambda \nabla g \Rightarrow \cos\alpha \sin\beta \sin\gamma = \lambda$ $\llbracket 2\rrbracket$, $\sin\alpha \cos\beta \sin\gamma = \lambda$ $\llbracket 3\rrbracket$, and

$\sin\alpha \sin\beta \cos\gamma = \lambda$ $\llbracket 4\rrbracket$. From $\llbracket 2\rrbracket$ and $\llbracket 3\rrbracket$, $\cos\alpha \sin\beta \sin\gamma = \sin\alpha \cos\beta \sin\gamma \Rightarrow$

$\cos\alpha \sin\beta - \sin\alpha \cos\beta = 0$ $\{\sin\gamma \neq 0\} \Rightarrow \sin(\beta - \alpha) = 0 \Rightarrow \beta - \alpha = \pi n$.

Since $0 < \alpha, \beta < \pi$, we have $n = 0$, so $\beta = \alpha$. Similarly, using $\llbracket 3\rrbracket$ and $\llbracket 4\rrbracket$,

$\beta = \gamma$. Since all angles are equal, the triangle is equilateral.

21 Refer to *Figure 21*. The strength of the beam is given by $f(x, y) = k(2x)(2y)^2 = 8kxy^2$, where k is a constant.

$g(x, y) = x^2/12^2 + y^2/8^2 - 1 = 0$ $\llbracket 1\rrbracket$.

$\nabla f = \lambda \nabla g \Rightarrow 8ky^2 = \frac{1}{72}x\lambda$ $\llbracket 2\rrbracket$ and

$16kxy = \frac{1}{32}y\lambda$ $\llbracket 3\rrbracket$. Multiplying $\llbracket 2\rrbracket$ by $2x$ and

$\llbracket 3\rrbracket$ by $y \Rightarrow \frac{1}{36}x^2\lambda = \frac{1}{32}y^2\lambda \Rightarrow y^2 = \frac{8}{9}x^2$.

Figure 21

Substituting into $\llbracket 1\rrbracket$ yields $3x^2 = 144 \Rightarrow x = 4\sqrt{3}$ and $y = \frac{8}{3}\sqrt{6}$. Thus, the width is $2x = 8\sqrt{3}$ in. and the depth is $2y = \frac{16}{3}\sqrt{6}$ in.

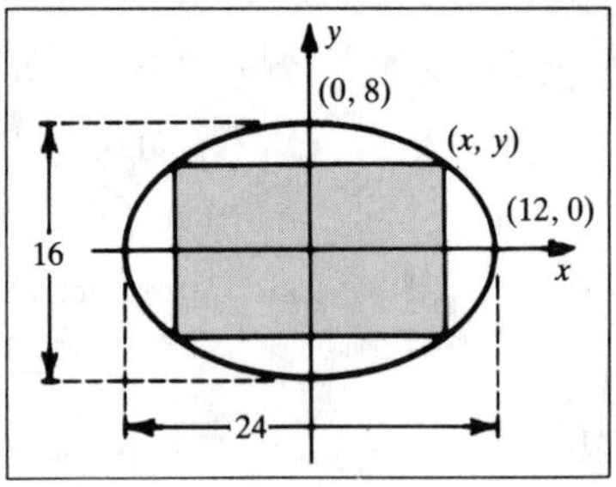

22 $f(x, y) = x^{1/5}y^{4/5}$ and $g(x, y) = xC + yL - M = 0$ $\llbracket 1\rrbracket$. $\nabla f = \lambda \nabla g \Rightarrow$

$\frac{1}{5}x^{-4/5}y^{4/5} = C\lambda$ $\llbracket 2\rrbracket$ and $\frac{4}{5}x^{1/5}y^{-1/5} = L\lambda$ $\llbracket 3\rrbracket$. Multiplying $\llbracket 2\rrbracket$ by $5L$ and $\llbracket 3\rrbracket$ by

$5C$ yields $Lx^{-4/5}y^{4/5} = 4Cx^{1/5}y^{-1/5}$. Multiplying by $x^{4/5}y^{1/5}$ gives us $Ly = 4Cx$,

or $y = \dfrac{4Cx}{L}$ $\llbracket 4\rrbracket$. Substituting into $\llbracket 1\rrbracket$ yields $x = \dfrac{M}{5C}$ and by $\llbracket 4\rrbracket$, $y = \dfrac{4M}{5L}$.

23 $f(x, y) = y - \cos x + 2x$ and $g(x, y) = x^2 + 2y^2 - 1 = 0$ $\llbracket 1\rrbracket$. $\nabla f = \lambda \nabla g \Rightarrow$

$2 + \sin x = 2x\lambda$ $\llbracket 2\rrbracket$ and $1 = 4y\lambda$ $\llbracket 3\rrbracket$. Now, $\llbracket 2\rrbracket \Rightarrow x \neq 0$ and $\llbracket 3\rrbracket \Rightarrow y \neq 0$.

Thus, all extrema must satisfy $\dfrac{1}{4y} = \lambda = \dfrac{2 + \sin x}{2x} \Rightarrow y = \dfrac{x}{4 + 2\sin x}$.

Graphing $y = \dfrac{x}{4 + 2\sin x}$ and $x^2 + 2y^2 - 1 = 0$, on the same coordinate axes,

gives the points of intersection as approximately $(0.97, 0.17)$ and $(-0.87, -0.35)$.

$f(0.97,\ 0.17) \approx 1.55$ is a maximum value and

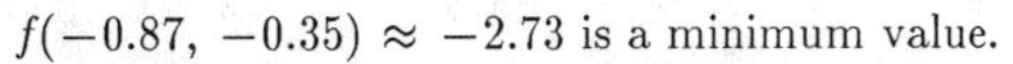

$f(-0.87,\ -0.35) \approx -2.73$ is a minimum value.

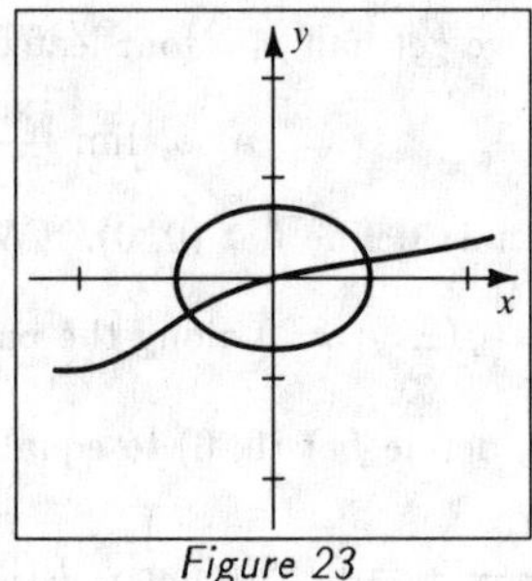

Figure 23

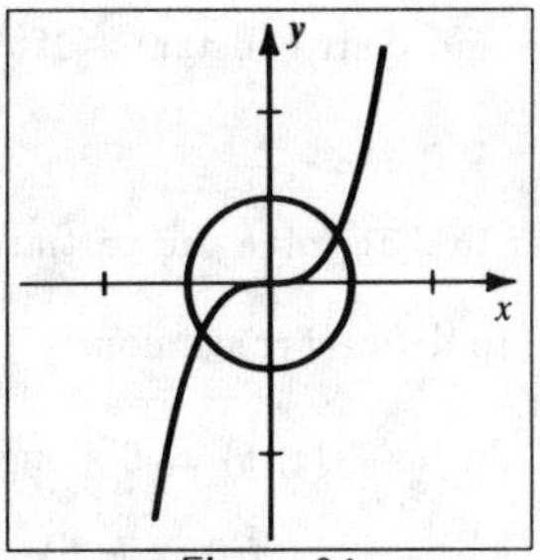

Figure 24

[24] $f(x,\ y) = \frac{1}{5}x^5 + \frac{1}{3}y^3$ and $g(x,\ y) = x^2 + y^2 - 1 = 0$ [[1]]. $\nabla f = \lambda \nabla g \Rightarrow$ $x^4 = 2x\lambda$ [[2]] and $y^2 = 2y\lambda$ [[3]]. From [[1]], $x = 0 \Rightarrow y = \pm 1$ and $y = 0 \Rightarrow$ $x = \pm 1$. If $x \neq 0$ and $y \neq 0$, then $\frac{1}{2}y = \lambda = \frac{1}{2}x^3 \Rightarrow y = x^3$. Graphing $y = x^3$ and $x^2 + y^2 - 1 = 0$, on the same coordinate axes, gives the points of intersection as approximately $(0.83,\ 0.56)$ and $(-0.83,\ -0.56)$. Evaluating these points in f gives $f(0.83,\ 0.56) \approx 0.14$, $f(-0.83,\ -0.56) \approx -0.14$, $f(0,\ 1) = \frac{1}{3}$, $f(0,\ -1) = -\frac{1}{3}$, $f(1,\ 0) = \frac{1}{5}$, $f(-1,\ 0) = -\frac{1}{5}$. Max $= \frac{1}{3}$ at $(0,\ 1)$; min $= -\frac{1}{3}$ at $(0,\ -1)$.

16.10 Review Exercises

[1] $36 - 4x^2 + 9y^2 \geq 0 \Rightarrow D = \{(x,\ y)\colon 4x^2 - 9y^2 \leq 36\}$. $f(3,\ 4) = 12 \Rightarrow$ $36 - 4x^2 + 9y^2 = 12^2 = 144$. The level curve is the hyperbola $9y^2 - 4x^2 = 108$.

[2] $D = \{(x,\ y)\colon xy > 0\}$. $f(2,\ 3) = \ln 6 \Rightarrow \ln xy = \ln 6$.

The level curve is the hyperbola $xy = 6$.

[3] $D = \{(x,\ y,\ z)\colon z^2 > x^2 + y^2\}$. $f(0,\ 0,\ 1) = 1$,

and the level surface is the hyperboloid of two sheets $z^2 - x^2 - y^2 = 1$.

[4] $D = \{(x,\ y,\ z)\colon y \neq x,\ z \neq \frac{\pi}{2} + \pi n\}$.

$f(5,\ 3,\ 0) = \frac{1}{2}$, and the level surface is the graph of $\sec z = \frac{1}{2}(x - y)$.

[5] $$\lim_{(x,\,y)\to(0,\,0)} \frac{3xy + 5}{y^2 + 4} = \frac{3(0)(0) + 5}{0^2 + 4} = \frac{5}{4}$$

[6] $$\lim_{(x,\,y,\,z)\to(1,\,3,\,-2)} \frac{z^2 + z - 2}{xyz + 2xy} = \lim_{(x,\,y,\,z)\to(1,\,3,\,-2)} \frac{(z + 2)(z - 1)}{xy(z + 2)} =$$

$$\lim_{(x,\,y,\,z)\to(1,\,3,\,-2)} \frac{z - 1}{xy} = \frac{-2 - 1}{1(3)} = \frac{-3}{3} = -1$$

[7] Along the path $y = x$, the limiting value is 0.

Along the path $y = 0$, the limiting value is 1. Thus, the limit DNE.

[8] Along the path $y = x$, the limiting value is $\dfrac{x^4}{x^4 + 2x^4} = \dfrac{1}{3}$.

Along the path $y = 0$, the limiting value is 0. Thus, the limit DNE.

[9] $f(x, y) = \dfrac{xy}{(x^2 + y^2)^{3/2}} = \dfrac{r^2 \sin\theta \cos\theta}{r^3} = \dfrac{\frac{1}{2}\sin 2\theta}{r} = k \Rightarrow r = c\sin 2\theta,$

where k and c are constants. If $c > 0$ (or $c < 0$), we get half of a four-leafed rose.

Also, $\lim\limits_{r\to 0} \dfrac{\frac{1}{2}\sin 2\theta}{r}$ DNE.

[10] In order to determine the partial derivatives, we must define f at $(0, 0)$. (We will not be able to define f continuously.) Since $\lim\limits_{(x,y)\to(0,0)} f(x, y) = 0$ along the path $y = 0$ and $\lim\limits_{(x,y)\to(0,0)} f(x, y) = 0$ along the path $x = 0$, define f at $(0, 0)$ to equal 0.

(1) $f_x(0, 0) = \lim\limits_{h\to 0} \dfrac{f(0 + h, 0) - f(0, 0)}{h} = \dfrac{0/h^4 - 0}{h} = 0.$ Similarly, $f_y(0, 0) = 0.$

(2) Along the path $y = x$, $\mathrm{L} = \lim\limits_{(x,y)\to(0,0)} \dfrac{x^2 y^2}{(x^2 + y^2)^2} = \dfrac{1}{4}$ and along the path $y = 0$, $\mathrm{L} = 0$. Thus, L DNE and f cannot be continuous at $(0, 0)$.

[11] $f(x, y) = x^3 \cos y - y^2 + 4x \Rightarrow f_x(x, y) = 3x^2 \cos y + 4;\ f_y(x, y) = -x^3 \sin y - 2y$

[12] $f(r, s) = r^2 e^{rs} \Rightarrow f_r(r, s) = 2re^{rs} + r^2 s e^{rs} = re^{rs}(2 + rs);$

$f_s(r, s) = r^2 e^{rs} \cdot r = r^3 e^{rs}$

[13] $f(x, y, z) = \dfrac{x^2 + y^2}{y^2 + z^2} \Rightarrow f_x(x, y, z) = \dfrac{(y^2 + z^2)(2x) - (x^2 + y^2)(0)}{(y^2 + z^2)^2} = \dfrac{2x}{y^2 + z^2};$

$f_y(x, y, z) = \dfrac{2y(z^2 - x^2)}{(y^2 + z^2)^2};\ f_z(x, y, z) = -\dfrac{2z(x^2 + y^2)}{(y^2 + z^2)^2}$

[14] $f(u, v, t) = u \ln \frac{v}{t} \Rightarrow f_u(u, v, t) = \ln \frac{v}{t};\ f_v(u, v, t) = u \cdot \dfrac{1}{v/t} \cdot \dfrac{1}{t} = \frac{u}{v};$

$f_t(u, v, t) = u \cdot \dfrac{1}{v/t} \cdot \left(-\frac{v}{t^2}\right) = -\frac{u}{t}$

[15] $f(x, y, z, t) = x^2 z\sqrt{2y + t} \Rightarrow f_x(x, y, z, t) = 2xz\sqrt{2y + t};\ f_y(x, y, z, t) = \dfrac{x^2 z}{\sqrt{2y + t}};$

$f_z(x, y, z, t) = x^2\sqrt{2y + t};\ f_t(x, y, z, t) = \dfrac{x^2 z}{2\sqrt{2y + t}}$

[16] $f(v, w) = v^2 \cos w + w^2 \cos v \Rightarrow$

$f_v(v, w) = 2v \cos w - w^2 \sin v;\ f_w(v, w) = -v^2 \sin w + 2w \cos v$

[17] $f(x, y) = x^3 y^2 - 3xy^3 + x^4 - 3y + 2 \Rightarrow$

$f_x = 3x^2 y^2 - 3y^3 + 4x^3$ and $f_y = 2x^3 y - 9xy^2 - 3.$ $f_{xx}(x, y) = 6xy^2 + 12x^2;$

$f_{xy}(x, y) = f_{yx}(x, y) = 6x^2 y - 9y^2;\ f_{yy}(x, y) = 2x^3 - 18xy$

[18] $f(x, y, z) = x^2 e^{y^2 - z^2} \Rightarrow f_x(x, y, z) = 2xe^{y^2 - z^2},\ f_y(x, y, z) = 2x^2 y e^{y^2 - z^2}$, and

$f_z(x, y, z) = -2x^2 z e^{y^2 - z^2}.$ $f_{xx} = 2e^{y^2 - z^2};\ f_{yy} = 2x^2(2y^2 + 1)e^{y^2 - z^2};$

$f_{zz} = 2x^2(2z^2 - 1)e^{y^2 - z^2};\ f_{xy} = f_{yx} = 4xye^{y^2 - z^2};\ f_{xz} = f_{zx} = -4xze^{y^2 - z^2};$

$f_{yz} = f_{zy} = -4x^2 yze^{y^2 - z^2}$

[19] $u = \dfrac{1}{(x^2 + y^2 + z^2)^{1/2}} \Rightarrow u_x = \dfrac{-x}{(x^2 + y^2 + z^2)^{3/2}} \Rightarrow u_{xx} = \dfrac{2x^2 - y^2 - z^2}{(x^2 + y^2 + z^2)^{5/2}}$.

Similarly, $u_{yy} = \dfrac{2y^2 - x^2 - z^2}{(x^2 + y^2 + z^2)^{5/2}}$ and $u_{zz} = \dfrac{2z^2 - x^2 - y^2}{(x^2 + y^2 + z^2)^{5/2}}$.

Thus, $u_{xx} + u_{yy} + u_{zz} = 0$.

[20] (a) $dw = \left(\dfrac{2xy^3}{1 + x^4} + 2\right) dx + (3y^2 \tan^{-1} x^2 - 1)\, dy$

(b) $dw = (2x \sin yz)\, dx + (x^2 z \cos yz)\, dy + (x^2 y \cos yz)\, dz$

[21] (a)
$$\begin{aligned} \Delta w &= w(x + \Delta x,\, y + \Delta y) - w(x,\, y) \\ &= (x + \Delta x)^2 + 3(x + \Delta x)(y + \Delta y) - (y + \Delta y)^2 - (x^2 + 3xy - y^2) \\ &= 2x\Delta x + (\Delta x)^2 + 3x\Delta y + 3y\Delta x + 3\,\Delta x\Delta y - 2y\Delta y - (\Delta y)^2 \\ &= (2x + 3y)\,\Delta x + (3x - 2y)\,\Delta y + (\Delta x)^2 + 3\,\Delta x\Delta y - (\Delta y)^2. \end{aligned}$$

$dw = w_x\, dx + w_y\, dy = (2x + 3y)\, dx + (3x - 2y)\, dy.$

(b) $x = -1$, $y = 2$, $\Delta x = dx = -1.1 - (-1) = -0.1$, and $\Delta y = dy = 2.1 - 2 = 0.1 \Rightarrow \Delta w = 4(-0.1) - 7(0.1) + (-0.1)^2 + 3(-0.1)(0.1) - (0.1)^2 = -1.13$ and $dw = 4(-0.1) - 7(0.1) = -1.1$.

[22] $dR = \dfrac{\partial R}{\partial V} dV + \dfrac{\partial R}{\partial I}\, dI = \dfrac{1}{I}\, dV - \dfrac{V}{I^2}\, dI \Rightarrow \dfrac{dR}{R} = \dfrac{dV}{V} - \dfrac{dI}{I} = 3\% - (-2\%) = 5\%.$

[23] $f(x,\, y) = \dfrac{xy}{y^2 - x^2} \Rightarrow f_x = \dfrac{y(x^2 + y^2)}{(y^2 - x^2)^2}$ and $f_y = \dfrac{-x(x^2 + y^2)}{(y^2 - x^2)^2}$. Since f_x and f_y are defined on the domain of f, f is differentiable throughout its domain.

[24] $f(x,\, y,\, z) = \dfrac{xyz}{x - y} \Rightarrow f_x = -\dfrac{y^2 z}{x - y}$, $f_y = \dfrac{x^2 z}{x - y}$, and $f_z = \dfrac{xy}{x - y}$. Since f_x, f_y, and f_z are defined on the domain of f, f is differentiable throughout its domain.

[25] $s_x = s_u u_x + s_v v_x + s_w w_x = (v - w)(2) + (u + w)(4) + (v - u)(-1) = v + 2w + 5u = 12x + 18y.$

$s_y = s_u u_y + s_v v_y + s_w w_y = (v - w)(3) + (u + w)(-1) + (v - u)(2) = 5v - 4w - 3u = 18x - 22y.$

[26] $z_r = z_x x_r + z_y y_r = (ye^x)(1) + (e^x)(2) = (y + 2)e^x = (2r + 3s - t + 2)e^{r+st}.$

$z_s = z_x x_s + z_y y_s = (ye^x)(t) + (e^x)(3) = (yt + 3)e^x = (2rt + 3st - t^2 + 3)e^{r+st}.$

$z_t = z_x x_t + z_y y_t = (ye^x)(s) + (e^x)(-1) = (ys - 1)e^x = (2rs + 3s^2 - st - 1)e^{r+st}.$

[27] $\dfrac{dw}{dt} = \dfrac{\partial w}{\partial x}\dfrac{dx}{dt} + \dfrac{\partial w}{\partial y}\dfrac{dy}{dt} + \dfrac{\partial w}{\partial z}\dfrac{dz}{dt} = (\sin yz)(-3e^{-t}) + (xz \cos yz)(2t) + (xy \cos yz)(3) = 3e^{-t}(9t^2 \cos 3t^3 - \sin 3t^3)$

[28] By (16.22), $F(x,\, y) = x^3 - 4xy^3 - 3y + x - 2 = 0$ and $y = f(x) \Rightarrow$

$$\frac{dy}{dx} = -\frac{F_x}{F_y} = -\frac{3x^2 - 4y^3 + 1}{-12xy^2 - 3} = \frac{3x^2 - 4y^3 + 1}{12xy^2 + 3}.$$

[29] By (16.23), $F(x, y, z) = x^2y + z\cos y - xz^3 = 0 \Rightarrow$

$$\frac{\partial z}{\partial x} = -\frac{F_x}{F_z} = \frac{z^3 - 2xy}{\cos y - 3xz^2} \text{ and } \frac{\partial z}{\partial y} = -\frac{F_y}{F_z} = \frac{z\sin y - x^2}{\cos y - 3xz^2}.$$

[30] (a) $\mathbf{a} = -3\mathbf{i} - 4\mathbf{j} \Rightarrow \mathbf{u} = -\frac{3}{5}\mathbf{i} - \frac{4}{5}\mathbf{j}$.

$D_{\mathbf{u}}f(x, y) = (6x + 5y)(-\frac{3}{5}) + (-2y + 5x)(-\frac{4}{5}) \Rightarrow$

$$D_{\mathbf{u}}f(2, -1) = (7)(-\tfrac{3}{5}) + (12)(-\tfrac{4}{5}) = -\tfrac{69}{5}.$$

(b) The maximum rate of increase is $\|\nabla f(2, -1)\| = \|7\mathbf{i} + 12\mathbf{j}\| = \sqrt{193}$.

[31] (a) $\overrightarrow{PQ} = -3\mathbf{i} + 4\mathbf{j} - 4\mathbf{k} \Rightarrow \mathbf{u} = -\frac{3}{\sqrt{41}}\mathbf{i} + \frac{4}{\sqrt{41}}\mathbf{j} - \frac{4}{\sqrt{41}}\mathbf{k}$.

$D_{\mathbf{u}}T(x, y, z) = (6x)(-\frac{3}{\sqrt{41}}) + (4y)(\frac{4}{\sqrt{41}}) + (-4)(-\frac{4}{\sqrt{41}}) \Rightarrow$

$$D_{\mathbf{u}}T(-1, -3, 2) = -\tfrac{14}{\sqrt{41}}.$$

(b) The maximum rate of change of T at P is

$$\|\nabla T(-1, -3, 2)\| = \|-6\mathbf{i} - 12\mathbf{j} - 4\mathbf{k}\| = \sqrt{196} = 14.$$

[32] The population density increases most rapidly in the direction of ∇Q.

(a) $$\nabla Q(x, y) = \left[ae^{-b\sqrt{x^2+y^2}}\left(\frac{-bx}{(x^2 + y^2)^{1/2}}\right)\right]\mathbf{i} + \left[ae^{-b\sqrt{x^2+y^2}}\left(\frac{-by}{(x^2 + y^2)^{1/2}}\right)\right]\mathbf{j}$$

$$= \frac{abe^{-b\sqrt{x^2+y^2}}}{(x^2 + y^2)^{1/2}}(-x\mathbf{i} - y\mathbf{j}).$$ It increases most rapidly in the direction of $-x\mathbf{i} - y\mathbf{j}$, that is, toward the center of the city.

(b) $\nabla Q(x, y) =$

$$ae^{-b\sqrt{x^2+y^2} + c(x^2+y^2)}\left[\left(\frac{-bx}{(x^2 + y^2)^{1/2}} + 2cx\right)\mathbf{i} + \left(\frac{-by}{(x^2 + y^2)^{1/2}} + 2cy\right)\mathbf{j}\right].$$

If $2c - \dfrac{b}{(x^2 + y^2)^{1/2}} > 0$, or, equivalently, $(x^2 + y^2)^{1/2} > \dfrac{b}{2c}$,

the direction is toward $x\mathbf{i} + y\mathbf{j}$, that is, away from the center of the city.

If $2c - \dfrac{b}{(x^2 + y^2)^{1/2}} < 0$, or, equivalently, $(x^2 + y^2)^{1/2} < \dfrac{b}{2c}$,

the direction is toward $-x\mathbf{i} - y\mathbf{j}$, that is, toward the center of the city.

[33] $F(x, y, z) = 4x^2 - 2y^2 - 7z = 0 \Rightarrow$

$$\nabla F(x, y, z) = 8x\mathbf{i} - 4y\mathbf{j} - 7\mathbf{k} \Rightarrow \nabla F(-2, -1, 2) = -16\mathbf{i} + 4\mathbf{j} - 7\mathbf{k}.$$

Tangent plane: $-16(x + 2) + 4(y + 1) - 7(z - 2) = 0$

Normal line: $x = -2 - 16t$, $y = -1 + 4t$, $z = 2 - 7t$

[34] Let $\mathbf{r}(t) = t\mathbf{i} + t^2\mathbf{j} + t^3\mathbf{k}$, and hence $\mathbf{r}'(t) = \mathbf{i} + 2t\mathbf{j} + 3t^2\mathbf{k}$.

$x = 2 \Rightarrow t = 2$ and $\mathbf{r}'(2) = \mathbf{i} + 4\mathbf{j} + 12\mathbf{k} \Rightarrow \mathbf{u} = \frac{1}{\sqrt{161}}(\mathbf{i} + 4\mathbf{j} + 12\mathbf{k})$.

$$D_{\mathbf{u}}f(x, y, z) = (z)(\tfrac{1}{\sqrt{161}}) + (2y)(\tfrac{4}{\sqrt{161}}) + (x)(\tfrac{12}{\sqrt{161}}) \Rightarrow D_{\mathbf{u}}f(2, 4, 8) = \tfrac{64}{\sqrt{161}}.$$

[35] $f(x,\ y,\ z) = \dfrac{x^2}{a^2} - \dfrac{y^2}{b^2} + \dfrac{z^2}{c^2} = 0 \Rightarrow \nabla f(x,\ y,\ z) = \dfrac{2x}{a^2}\mathbf{i} - \dfrac{2y}{b^2}\mathbf{j} + \dfrac{2z}{c^2}\mathbf{k}$. At $(x_0,\ y_0,\ z_0)$,

an equation of the tangent plane is $\dfrac{2x_0}{a^2}(x - x_0) - \dfrac{2y_0}{b^2}(y - y_0) + \dfrac{2z_0}{c^2}(z - z_0) = 0$.

If $x = y = z = 0$ (the origin), then the equation becomes

$$-2\left(\frac{x_0^2}{a^2} - \frac{y_0^2}{b^2} + \frac{z_0^2}{c^2}\right) = -2(0) = 0.$$

[36] $f(0,\ 5) = 1 \Rightarrow$ the level curve is $\frac{1}{4}x^2 + \frac{1}{25}y^2 = 1$, an ellipse.

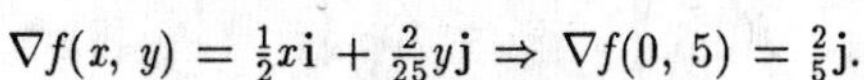

$$\nabla f(x,\ y) = \tfrac{1}{2}x\mathbf{i} + \tfrac{2}{25}y\mathbf{j} \Rightarrow \nabla f(0,\ 5) = \tfrac{2}{5}\mathbf{j}.$$

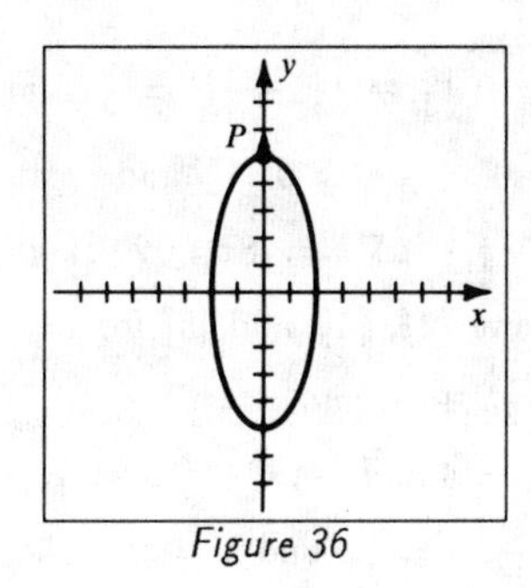

Figure 36

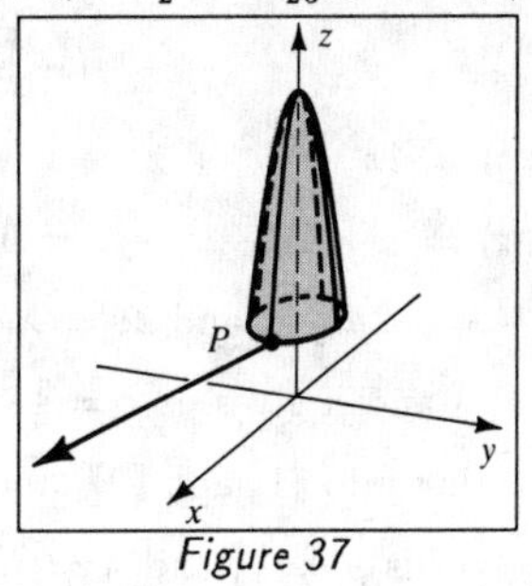

Figure 37

[37] $F(1,\ 0,\ 0) = 4 \Rightarrow$ the level surface is $z = -4x^2 - 9y^2 + 4$, a paraboloid.

$$\nabla F(x,\ y,\ z) = 8x\mathbf{i} + 18y\mathbf{j} + \mathbf{k} \Rightarrow \nabla F(1,\ 0,\ 0) = 8\mathbf{i} + \mathbf{k}.$$

[38] $f_x = -4/x^2 - y = 0$ and $f_y = -2/y^2 - x = 0 \Rightarrow y^4 + y = 0\ \{x^2 = 4/y^4\} \Rightarrow$ $y = -1\ \{y \neq 0\}$, and hence $x = -2$. $f_{xx} = 8/x^3$, $f_{xy} = -1$, and $f_{yy} = 4/y^3 \Rightarrow$ $D(x,\ y) = 32/(x^3y^3) - 1$. $D(-2,\ -1) = 3 > 0$ and $f_{xx}(-2,\ -1) = -1 < 0 \Rightarrow$

$f(-2,\ -1) = -6$ is a *LMAX*.

[39] $f_x = 2x = 0$ and $f_y = 3 - 3y^2 = 0 \Rightarrow x = 0$ and $y = \pm 1$. $f_{xx} = 2$, $f_{xy} = 0$, and $f_{yy} = -6y \Rightarrow D(x,\ y) = -12y$. $D(0,\ 1) = -12 < 0 \Rightarrow (0,\ 1,\ f(0,\ 1))$ is a SP.

$D(0,\ -1) = 12 > 0$ and $f_{xx} > 0 \Rightarrow f(0,\ -1) = -2$ is a *LMIN*.

[40] Let x and y be the base dimensions and z the height.

$C(x,\ y) = 2xy + xy + 2xz + 2yz = 3xy + 2xz + 2yz$. $V = xyz \Rightarrow z = \dfrac{V}{xy}$ and so

$C(x,\ y) = 3xy + \dfrac{2V}{y} + \dfrac{2V}{x}$. $C_x = 3y - \dfrac{2V}{x^2} = 0$ and $C_y = 3x - \dfrac{2V}{y^2} = 0 \Rightarrow$

$3xy^2 = 2V = 3x^2y \Rightarrow x = y$. From $C_x = 0$, we see that $3x^3 = 2V$, and hence

$x = y = \sqrt[3]{\frac{2}{3}V}$. $z = \dfrac{V}{xy} = (\frac{3}{2})^{2/3}\,V^{1/3}$. To relate this value of z to x,

note that $V^{1/3} = x(\frac{2}{3})^{-1/3}$. Hence, $z = (\frac{3}{2})^{2/3}(\frac{2}{3})^{-1/3}x = (\frac{3}{2})^{2/3}(\frac{3}{2})^{1/3}x = \frac{3}{2}x$.

The box should have a square base with height $\frac{3}{2}$ the length of a side of the base.

41 $f(x, y, z) = xyz$ and $g(x, y, z) = x^2 + 4y^2 + 2z^2 - 8 = 0$ [[1]]. $\nabla f = \lambda \nabla g \Rightarrow$ $yz = 2x\lambda$ [[2]], $xz = 8y\lambda$ [[3]], and $xy = 4z\lambda$ [[4]]. From [[2]], [[3]], and [[4]], $xyz = 2x^2\lambda = 8y^2\lambda = 4z^2\lambda \Rightarrow y^2 = \frac{1}{4}x^2$ and $z^2 = \frac{1}{2}x^2$. Substituting into [[1]] yields $x = \pm\sqrt{\frac{8}{3}}$, $y = \pm\sqrt{\frac{2}{3}}$, and $z = \pm\sqrt{\frac{4}{3}}$. These values represent eight points.

$f(-\sqrt{\frac{8}{3}}, -\sqrt{\frac{2}{3}}, -\sqrt{\frac{4}{3}}) = f(-\sqrt{\frac{8}{3}}, \sqrt{\frac{2}{3}}, \sqrt{\frac{4}{3}}) = f(\sqrt{\frac{8}{3}}, -\sqrt{\frac{2}{3}}, \sqrt{\frac{4}{3}}) =$

$f(\sqrt{\frac{8}{3}}, \sqrt{\frac{2}{3}}, -\sqrt{\frac{4}{3}}) = -\frac{8}{9}\sqrt{3}$ are *LMIN*.

$f(\sqrt{\frac{8}{3}}, \sqrt{\frac{2}{3}}, \sqrt{\frac{4}{3}}) = f(-\sqrt{\frac{8}{3}}, -\sqrt{\frac{2}{3}}, \sqrt{\frac{4}{3}}) =$

$f(-\sqrt{\frac{8}{3}}, \sqrt{\frac{2}{3}}, -\sqrt{\frac{4}{3}}) = f(\sqrt{\frac{8}{3}}, -\sqrt{\frac{2}{3}}, -\sqrt{\frac{4}{3}}) = \frac{8}{9}\sqrt{3}$ are *LMAX*.

42 $f(x, y, z) = 4x^2 + y^2 + z^2$, $g(x, y, z) = 2x - y + z - 4 = 0$ [[1]], and $h(x, y, z) = x + 2y - z - 1 = 0$ [[2]]. $\nabla f = \lambda \nabla g + \mu \nabla h \Rightarrow 8x = 2\lambda + \mu$ [[3]], $2y = -\lambda + 2\mu$ [[4]], and $2z = \lambda - \mu$ [[5]]. Solving [[3]], [[4]], and [[5]] for x, y, and z in terms of λ and μ gives us $x = \frac{1}{4}\lambda + \frac{1}{8}\mu$, $y = -\frac{1}{2}\lambda + \mu$, and $z = \frac{1}{2}\lambda - \frac{1}{2}\mu$. Substituting into [[1]] and [[2]] produces $\frac{3}{2}\lambda - \frac{5}{4}\mu = 4$ and $-\frac{5}{4}\lambda + \frac{21}{8}\mu = 1$. Solving yields $\mu = \frac{52}{19}$ and $\lambda = \frac{94}{19}$. Thus, $x = \frac{30}{19}$, $y = \frac{5}{19}$, and $z = \frac{21}{19}$. The two constraints define a line in three dimensions that has arbitrarily large coordinates for x, y, and z. Thus, f will have no maximum value. $f(\frac{30}{19}, \frac{5}{19}, \frac{21}{19}) = \frac{214}{19}$ is a *LMIN*.

43 From the geometry of the problem, we see that $x = 0$, and the problem can be reduced to a two-dimensional problem in the yz-plane with the parabola $z = \frac{1}{4}y^2$ and the point $(y, z) = (5, 0)$. If D is the square of the distance from (y, z) to $(5, 0)$, then $D(y) = (y - 5)^2 + (z - 0)^2 = (y - 5)^2 + \frac{1}{16}y^4$. $D'(y) = 2(y - 5) + \frac{1}{4}y^3$. $D' = 0 \Rightarrow y^3 + 8y - 40 = 0 \Rightarrow y \approx 2.66$ and $z \approx 1.76$.

Thus, the point is approximately (0, 2.66, 1.76).

44 The volume of the cylinder is $\pi r^2 h = 4\pi h$ and of the cone is $\frac{1}{3}\pi r^2 h = \frac{4}{3}\pi k$. Thus, $V = 4\pi h + \frac{4}{3}\pi k = 100$. The curved surface area of the cylinder is $2\pi rh = 4\pi h$ and of the cone is $\pi r\sqrt{r^2 + h^2} = 2\pi\sqrt{k^2 + 4}$. Using Lagrange multipliers,

$f(k, h) = 4\pi h + 2\pi\sqrt{k^2 + 4}$ and $g(k, h) = 4\pi h + \frac{4}{3}\pi k - 100 = 0$ [[1]].

$\nabla f = \lambda \nabla g \Rightarrow 4\pi = 4\pi\lambda$ [[2]] and $\dfrac{2\pi k}{\sqrt{k^2 + 4}} = \dfrac{4\pi}{3}\lambda$ [[3]]. From [[2]], $\lambda = 1$.

Substituting into [[3]] yields $k/\sqrt{k^2 + 4} = \frac{2}{3} \Rightarrow$

$k = \frac{4}{\sqrt{5}} \approx 1.79$ and using [[1]] gives $h = \frac{25}{\pi} - \frac{1}{3}k = \frac{25}{\pi} - \frac{4}{3\sqrt{5}} \approx 7.36$.

45 $P = \dfrac{s}{(T + sk_0)(C + sk_1)} \Rightarrow$

$$\frac{\partial P}{\partial s} = \frac{(T + sk_0)(C + sk_1)(1) - (s)[(T + sk_0)k_1 + (C + sk_1)k_0]}{(T + sk_0)^2(C + sk_1)^2}.$$

$\partial P/\partial s = 0 \Rightarrow TC + Tsk_1 + Csk_0 + s^2k_0k_1 = s(Tk_1 + sk_0k_1 + Ck_0 + sk_0k_1) \Rightarrow$

$$TC = s^2k_0k_1 \Rightarrow s = \sqrt{\frac{TC}{k_0k_1}}.$$

46 (a) A maximum will occur when $\sin \pi x = \sin \pi y = 1$.

Since $0 \leq x,\ y \leq 1$, this occurs when $x = y = \frac{1}{2}$.

(b) $U(x,\ y,\ t) = 20e^{-2k\pi^2t}\sin \pi x \sin \pi y \Rightarrow U_t = -40k\pi^2 e^{-2k\pi^2t}\sin \pi x \sin \pi y$,

$U_x = 20\pi e^{-2k\pi^2t}\cos \pi x \sin \pi y$, $U_{xx} = -20\pi^2 e^{-2k\pi^2t}\sin \pi x \sin \pi y$,

$U_y = 20\pi e^{-2k\pi^2t}\sin \pi x \cos \pi y$, and $U_{yy} = -20\pi^2 e^{-2k\pi^2t}\sin \pi x \sin \pi y$.

$$U_{xx} + U_{yy} = -40\pi^2 e^{-2k\pi^2t}\sin \pi x \sin \pi y \Rightarrow U_t = k(U_{xx} + U_{yy}).$$

(c) $U(x,\ y,\ t) = (20\sin \pi x \sin \pi y)\,e^{-2k\pi^2t} = Te^{-ct}$, where $c = 2k\pi^2$.

Since $|\sin \pi x| \leq 1$ and $|\sin \pi y| \leq 1$, $|T| \leq 20$.

As $t \to \infty$, $e^{-ct} \to 0$, and the temperature U approaches $0\,^\circ$C.

Chapter 17: Multiple Integrals

Exercises 17.1

[1] This is an R_x region. Recall that an R_x region lies between the graphs of two equations $y = f(x)$ and $y = g(x)$, with f and g continuous, and $f(x) \geq g(x)$ for every x in $[a, b]$, where a and b are the smallest and largest x-coordinates of the points (x, y) in the region.

[2] R_x

[3] This is an R_y region. Recall that an R_y region is a region that lies between the graphs of two equations of the form $x = f(y)$ and $x = g(y)$, with f and g continuous and $f(y) \geq g(y)$ for all y in $[c, d]$, where c and d are the smallest and largest y-coordinates of points in the region.

[4] This region could be considered as one R_x region or two R_y regions.

[5] Neither

[6] R_y

[7] This region could be considered as two R_x regions or two R_y regions.

[8] R_x and R_y

[9] Neither

[10] R_y

[11] Note that $\Delta A_k = 1$ for all k.

(a) $\sum_{k=1}^{7} f(u_k, v_k)\Delta A_k = f(0, 2) + f(1, 2) + f(0, 1) + f(1, 1) + f(0, 0) + f(1, 0) + f(2, 0) = 5 + 9 + 3 + 7 + 1 + 5 + 9 = 39$

(b) $\sum_{k=1}^{7} f(u_k, v_k)\Delta A_k = f(1, 3) + f(2, 3) + f(1, 2) + f(2, 2) + f(1, 1) + f(2, 1) + f(3, 1) = 11 + 15 + 9 + 13 + 7 + 11 + 15 = 81$

(c) $\sum_{k=1}^{7} f(u_k, v_k)\Delta A_k = f(\frac{1}{2}, \frac{5}{2}) + f(\frac{3}{2}, \frac{5}{2}) + f(\frac{1}{2}, \frac{3}{2}) + f(\frac{3}{2}, \frac{3}{2}) + f(\frac{1}{2}, \frac{1}{2}) + f(\frac{3}{2}, \frac{1}{2}) + f(\frac{5}{2}, \frac{1}{2}) = 8 + 12 + 6 + 10 + 4 + 8 + 12 = 60$

[12] (a) $\sum_{k=1}^{6} f(u_k, v_k)\Delta A_k$

$= (4)\sum_{k=1}^{6} f(u_k, v_k)$ {since each $\Delta A_k = 4$}

$= 4[f(4, 2) + f(6, 2) + f(2, 0) + f(4, 0) + f(6, 0) + f(8, 0)]$

$= 4(8 + 12) = 80$

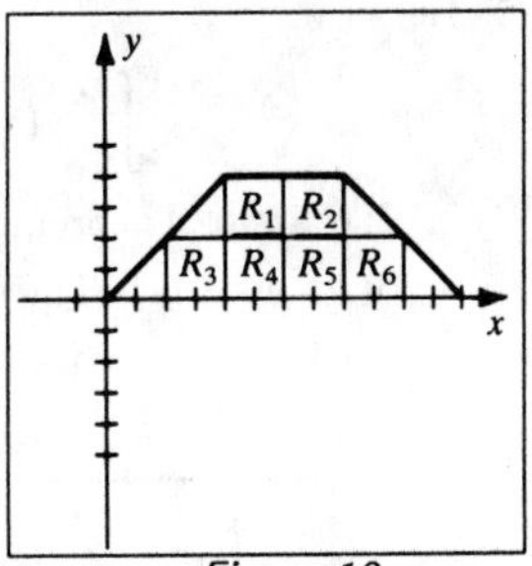

Figure 12

(b) $\sum_{k=1}^{6} f(u_k, v_k)\Delta A_k$

$= 4[f(6, 4) + f(8, 4) + f(4, 2) + f(6, 2) + f(8, 2) + f(10, 2)]$

$= 4(24 + 32 + 8 + 12 + 16 + 20) = 448$

(c) $\sum_{k=1}^{6} f(u_k, v_k)\Delta A_k$

$= 4[f(5, 3) + f(7, 3) + f(3, 1) + f(5, 1) + f(7, 1) + f(9, 1)]$

$= 4(15 + 21 + 3 + 5 + 7 + 9) = 240$

[13] $\int_1^2\int_{-1}^2 (12xy^2 - 8x^3)\,dy\,dx = \int_1^2\Big[4xy^3 - 8x^3y\Big]_{-1}^2\,dx = \int_1^2 (36x - 24x^3)\,dx =$

$\Big[18x^2 - 6x^4\Big]_1^2 = -36$

[14] $\int_0^3\int_{-2}^{-1} (4xy^3 + y)\,dx\,dy = \int_0^3\Big[2x^2y^3 + xy\Big]_{-2}^{-1}\,dy = \int_0^3 (-6y^3 + y)\,dy =$

$\Big[-\frac{3}{2}y^4 + \frac{1}{2}y^2\Big]_0^3 = -117$

[15] $\int_1^2\int_{1-x}^{\sqrt{x}} x^2y\,dy\,dx = \int_1^2\Big[\frac{1}{2}x^2y^2\Big]_{1-x}^{\sqrt{x}}\,dx = \int_1^2 \frac{1}{2}(-x^4 + 3x^3 - x^2)\,dx =$

$\frac{1}{2}\Big[-\frac{1}{5}x^5 + \frac{3}{4}x^4 - \frac{1}{3}x^3\Big]_1^2 = \frac{163}{120}$

[16] $\int_{-1}^1\int_{x^3}^{x+1} (3x + 2y)\,dy\,dx = \int_{-1}^1\Big[3xy + y^2\Big]_{x^3}^{x+1}\,dx =$

$\int_{-1}^1 (4x^2 + 5x + 1 - 3x^4 - x^6)\,dx = 2\int_0^1 (-x^6 - 3x^4 + 4x^2 + 1)\,dx$ { $5x$ is odd } $=$

$2\Big[-\frac{1}{7}x^7 - \frac{3}{5}x^5 + \frac{4}{3}x^3 + x\Big]_0^1 = \frac{334}{105}$

[17] $\int_0^2\int_{y^2}^{2y} (4x - y)\,dx\,dy = \int_0^2\Big[2x^2 - xy\Big]_{y^2}^{2y}\,dy = \int_0^2 (6y^2 + y^3 - 2y^4)\,dy =$

$\Big[2y^3 + \frac{1}{4}y^4 - \frac{2}{5}y^5\Big]_0^2 = \frac{36}{5}$

[18] $\int_0^1\int_{-y-1}^{y-1} (x^2 + y^2)\,dx\,dy = \int_0^1\Big[\frac{1}{3}x^3 + xy^2\Big]_{-y-1}^{y-1}\,dy =$

$\int_0^1\Big[\frac{1}{3}(y-1)^3 + \frac{1}{3}(y+1)^3 + 2y^3\Big]dy = \Big[\frac{1}{12}(y-1)^4 + \frac{1}{12}(y+1)^4 + \frac{1}{2}y^4\Big]_0^1 = \frac{5}{3}$

[19] $\int_1^2\int_{x^3}^{x} e^{y/x}\,dy\,dx = \int_1^2\Big[xe^{y/x}\Big]_{x^3}^{x}\,dx = \int_1^2 x(e - e^{(x^2)})\,dx =$

$\Big[\frac{1}{2}ex^2 - \frac{1}{2}e^{(x^2)}\Big]_1^2 = \frac{1}{2}(4e - e^4) \approx -21.86$

[20] $\int_0^{\pi/6}\int_0^{\pi/2} (x\cos y - y\cos x)\,dy\,dx = \int_0^{\pi/6}\Big[x\sin y - \frac{1}{2}y^2\cos x\Big]_0^{\pi/2}\,dx =$

$\int_0^{\pi/6} (x - \frac{1}{8}\pi^2\cos x)\,dx = \Big[\frac{1}{2}x^2 - \frac{1}{8}\pi^2\sin x\Big]_0^{\pi/6} = -\frac{7}{144}\pi^2 \approx -0.48$

[21] (a) To apply Theorem (17.8)(i), we first examine the y-values. In this case, $0 \le y \le \sqrt{x}$. The smallest x-value is 0 and the largest x-value is 4.

$$\int_0^4\int_0^{\sqrt{x}} f(x, y)\,dy\,dx$$

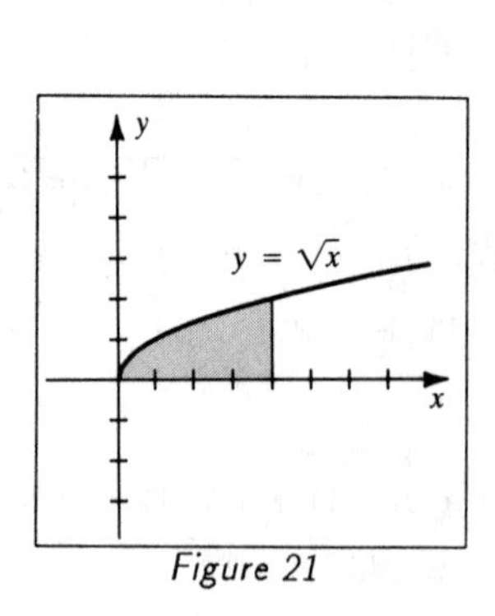

Figure 21

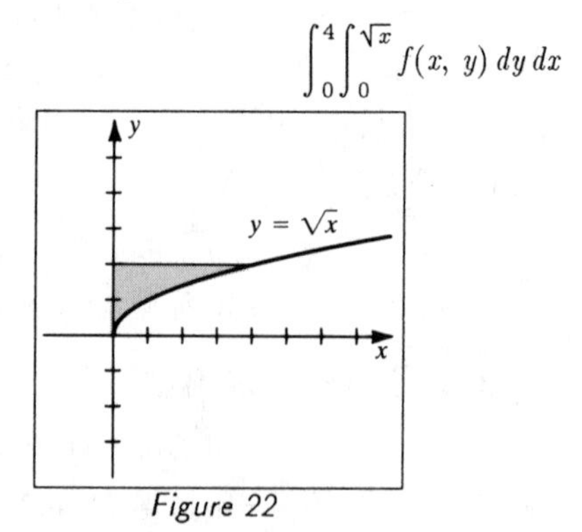

Figure 22

(b) For Theorem (17.8)(ii), we note that $y^2 \le x \le 4$.

The smallest y-value is 0 and the largest y-value is 2. $\int_0^2 \int_{y^2}^4 f(x, y)\, dx\, dy$

[22] (a) $\sqrt{x} \le y \le 2$ and $0 \le x \le 4 \Rightarrow \iint_R f(x, y)\, dA = \int_0^4 \int_{\sqrt{x}}^2 f(x, y)\, dy\, dx$

(b) $0 \le x \le y^2$ and $0 \le y \le 2 \Rightarrow \iint_R f(x, y)\, dA = \int_0^2 \int_0^{y^2} f(x, y)\, dx\, dy$

[23] (a) $x^3 \le y \le 8$ and $0 \le x \le 2 \Rightarrow \iint_R f(x, y)\, dA = \int_0^2 \int_{x^3}^8 f(x, y)\, dy\, dx$

(b) $0 \le x \le y^{1/3}$ and $0 \le y \le 8 \Rightarrow \iint_R f(x, y)\, dA = \int_0^8 \int_0^{y^{1/3}} f(x, y)\, dx\, dy$

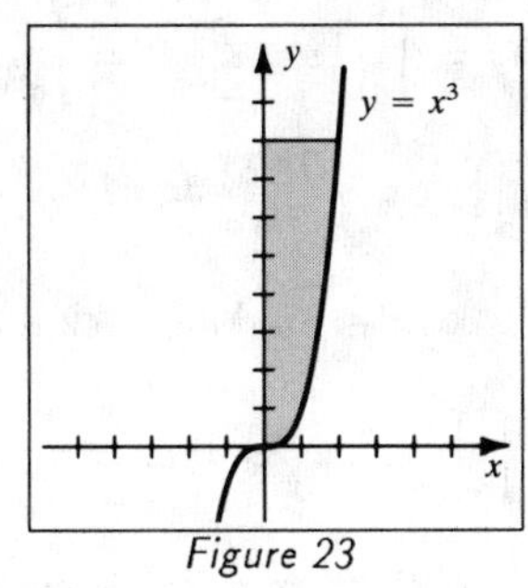

Figure 23

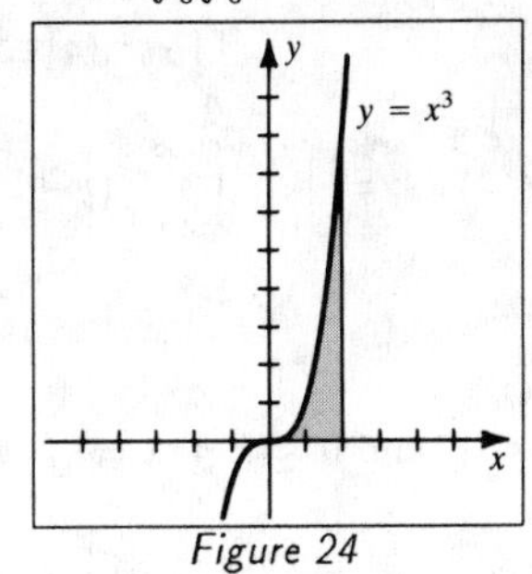

Figure 24

[24] (a) $0 \le y \le x^3$ and $0 \le x \le 2 \Rightarrow \iint_R f(x, y)\, dA = \int_0^2 \int_0^{x^3} f(x, y)\, dy\, dx$

(b) $y^{1/3} \le x \le 2$ and $0 \le x \le 8 \Rightarrow \iint_R f(x, y)\, dA = \int_0^8 \int_{y^{1/3}}^2 f(x, y)\, dx\, dy$

[25] (a) $x^3 \le y \le \sqrt{x}$ and $0 \le x \le 1 \Rightarrow \iint_R f(x, y)\, dA = \int_0^1 \int_{x^3}^{\sqrt{x}} f(x, y)\, dy\, dx$

(b) $y^2 \le x \le y^{1/3}$ and $0 \le y \le 1 \Rightarrow \iint_R f(x, y)\, dA = \int_0^1 \int_{y^2}^{y^{1/3}} f(x, y)\, dx\, dy$

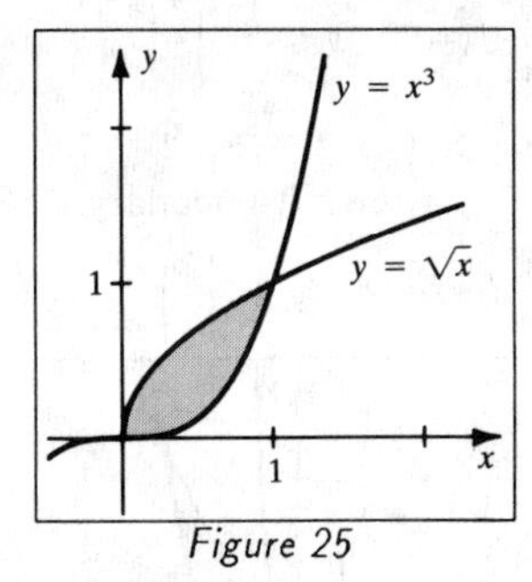

Figure 25

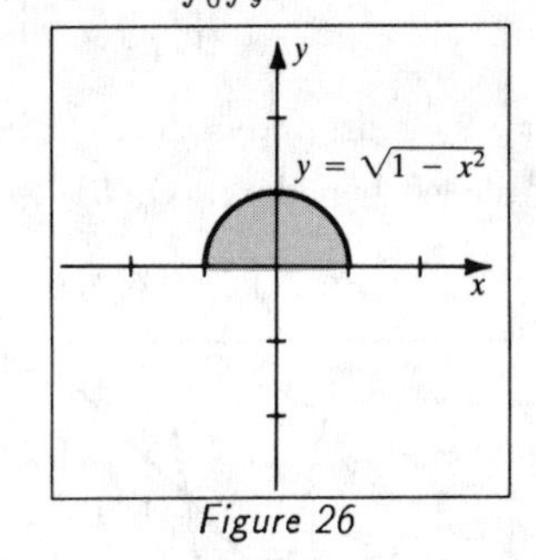

Figure 26

[26] (a) $0 \le y \le \sqrt{1 - x^2}$ and $-1 \le x \le 1 \Rightarrow \iint_R f(x, y)\, dA = \int_{-1}^1 \int_0^{\sqrt{1-x^2}} f(x, y)\, dy\, dx$

(b) $-\sqrt{1 - y^2} \le x \le \sqrt{1 - y^2}$ and $0 \le y \le 1 \Rightarrow$

$$\iint_R f(x, y)\, dA = \int_0^1 \int_{-\sqrt{1-y^2}}^{\sqrt{1-y^2}} f(x, y)\, dx\, dy$$

[27] $-1 \le x \le 2$ and $-1 \le y \le 4 \Rightarrow \iint_R (y + 2x)\,dA =$

$$\int_{-1}^{4}\int_{-1}^{2} (y + 2x)\,dx\,dy = \int_{-1}^{4}\Big[yx + x^2\Big]_{-1}^{2} dy = \int_{-1}^{4}(3y + 3)\,dy = \tfrac{75}{2}.$$

We could also use $\int_{-1}^{2}\int_{-1}^{4}(y + 2x)\,dy\,dx$.

[28] $1 \le y \le 2x + 5$ and $-2 \le x \le 2 \Rightarrow \iint_R (x - y)\,dA = \int_{-2}^{2}\int_{1}^{2x+5}(x - y)\,dy\,dx =$

$$\int_{-2}^{2}\Big[xy - \tfrac{1}{2}y^2\Big]_{1}^{2x+5} dx = \int_{-2}^{2}(-6x - 12)\,dx = 2\int_{0}^{2}(-12)\,dx = -48$$

[29] The region R is bounded by $y = 1$, $x = -2y$, and $x = 3y$.

$$\iint_R xy^2\,dA = \int_{0}^{1}\int_{-2y}^{3y} xy^2\,dx\,dy = \int_{0}^{1}\Big[\tfrac{1}{2}x^2y^2\Big]_{-2y}^{3y} = \int_{0}^{1}\tfrac{5}{2}y^4\,dy = \tfrac{1}{2}$$

[30] $\iint_R (y + 1)\,dA = \int_{0}^{\pi/4}\int_{\sin x}^{\cos x}(y + 1)\,dy\,dx = \int_{0}^{\pi/4}\Big[\tfrac{1}{2}y^2 + y\Big]_{\sin x}^{\cos x} dx =$

$$\int_{0}^{\pi/4}\left(\tfrac{1}{2}\cos 2x + \cos x - \sin x\right)dx = \sqrt{2} - \tfrac{3}{4} \approx 0.66$$

[31] $0 \le y \le x^2$ and $0 \le x \le 2 \Rightarrow \iint_R x^3 \cos xy\,dA = \int_{0}^{2}\int_{0}^{x^2} x^3 \cos xy\,dy\,dx =$

$$\int_{0}^{2}\Big[x^2 \sin xy\Big]_{0}^{x^2} dx = \int_{0}^{2} x^2 \sin x^3\,dx = \tfrac{1}{3}(1 - \cos 8) \approx 0.38$$

[32] The region R is bounded by $y = 4$, $x = \frac{1}{2}y$, and $x = -y$.

$$\iint_R e^{x/y}\,dA = \int_{0}^{4}\int_{-y}^{y/2} e^{x/y}\,dx\,dy = \int_{0}^{4}\Big[ye^{x/y}\Big]_{-y}^{y/2} dy = \int_{0}^{4} y(e^{1/2} - e^{-1})\,dy =$$

$$8(e^{1/2} - e^{-1}) \approx 10.25$$

[33] (a) $y = x + 4$ is always the upper boundary. On $1 \le x \le 2$, $y = 9 - 4x$ is the lower boundary. On $2 \le x \le 4$, $y = \frac{1}{8}x^3$ is the lower boundary.

$$\iint_R f(x, y)\,dA = \int_{1}^{2}\int_{9-4x}^{x+4} f(x, y)\,dy\,dx + \int_{2}^{4}\int_{x^3/8}^{x+4} f(x, y)\,dy\,dx$$

(b) $x = 2y^{1/3}$ is always the right boundary. On $1 \le y \le 5$, $x = (9 - y)/4$ is the left boundary. On $5 \le y \le 8$, $x = y - 4$ is the left boundary.

$$\iint_R f(x, y)\,dA = \int_{1}^{5}\int_{(9-y)/4}^{2y^{1/3}} f(x, y)\,dx\,dy + \int_{5}^{8}\int_{y-4}^{2y^{1/3}} f(x, y)\,dx\,dy$$

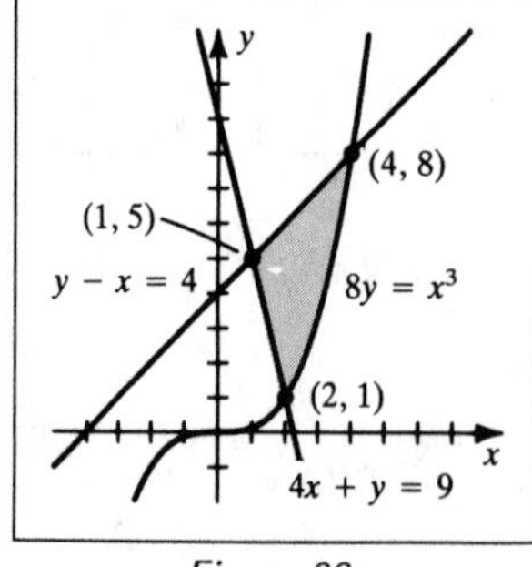

Figure 33

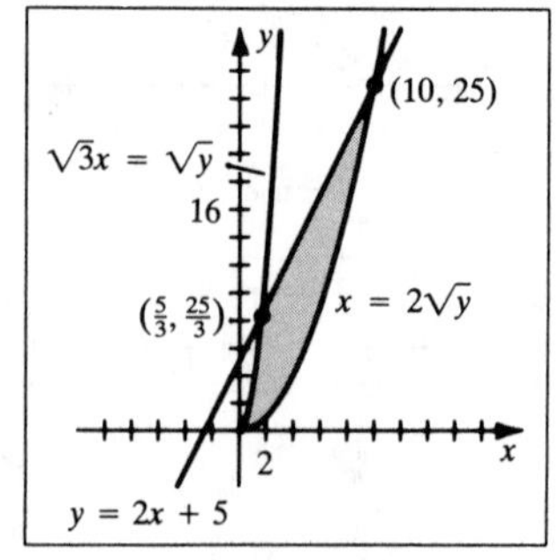

Figure 34

[34] (a) $\iint_R f(x, y)\, dA = \int_0^{5/3} \int_{x^2/4}^{3x^2} f(x, y)\, dy\, dx + \int_{5/3}^{10} \int_{x^2/4}^{2x+5} f(x, y)\, dy\, dx$

(b) $\iint_R f(x, y)\, dA = \int_0^{25/3} \int_{\sqrt{y/3}}^{2\sqrt{y}} f(x, y)\, dx\, dy + \int_{25/3}^{25} \int_{(y-5)/2}^{2\sqrt{y}} f(x, y)\, dx\, dy$

[35] (a) $\iint_R f(x, y)\, dA = \int_{-1}^{1} \int_{-x-3}^{2x} f(x, y)\, dy\, dx + \int_{1}^{3} \int_{-x-3}^{3-x^2} f(x, y)\, dy\, dx$

(b) $\iint_R f(x, y)\, dA = \int_{-6}^{-2} \int_{-y-3}^{\sqrt{3-y}} f(x, y)\, dx\, dy + \int_{-2}^{2} \int_{y/2}^{\sqrt{3-y}} f(x, y)\, dx\, dy$

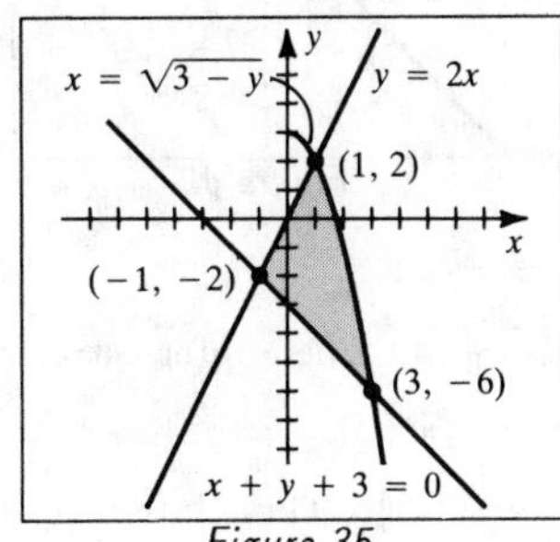

Figure 35

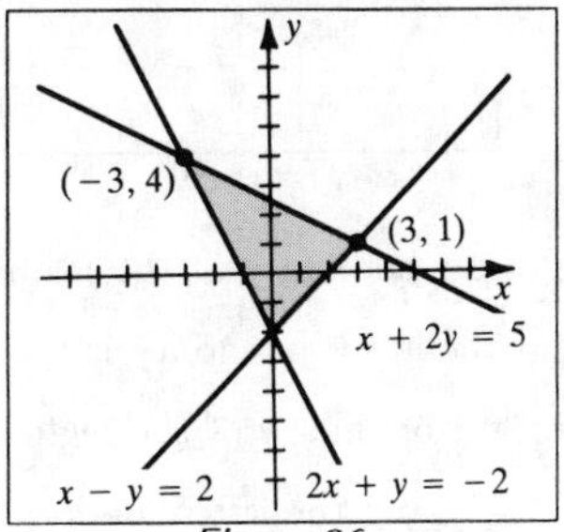

Figure 36

[36] (a) $\iint_R f(x, y)\, dA = \int_{-3}^{0} \int_{-2x-2}^{(5-x)/2} f(x, y)\, dy\, dx + \int_{0}^{3} \int_{x-2}^{(5-x)/2} f(x, y)\, dy\, dx$

(b) $\iint_R f(x, y)\, dA = \int_{-2}^{1} \int_{-(y+2)/2}^{y+2} f(x, y)\, dx\, dy + \int_{1}^{4} \int_{-(y+2)/2}^{5-2y} f(x, y)\, dx\, dy$

[37] (a) $\iint_R f(x, y)\, dA = \int_0^1 \int_{1-x}^{e^x} f(x, y)\, dy\, dx + \int_1^e \int_{\ln x}^{1+e-x} f(x, y)\, dy\, dx$

(b) $\iint_R f(x, y)\, dA = \int_0^1 \int_{1-y}^{e^y} f(x, y)\, dx\, dy + \int_1^e \int_{\ln y}^{1+e-y} f(x, y)\, dx\, dy$

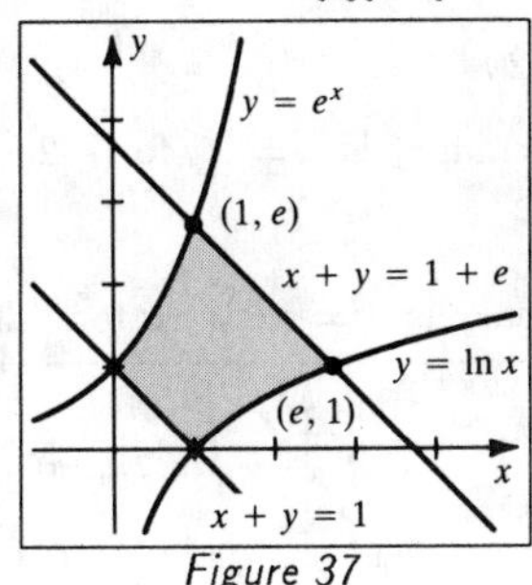

Figure 37

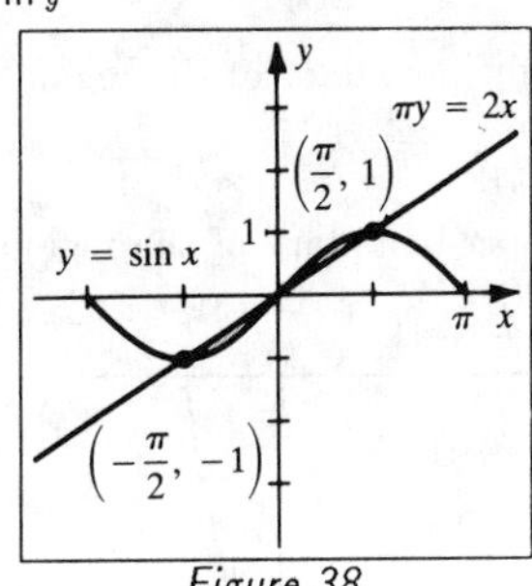

Figure 38

[38] (a) $\iint_R f(x, y)\, dA = \int_{-\pi/2}^{0} \int_{\sin x}^{2x/\pi} f(x, y)\, dy\, dx + \int_0^{\pi/2} \int_{2x/\pi}^{\sin x} f(x, y)\, dy\, dx$

(b) $\iint_R f(x, y)\, dA = \int_{-1}^{0} \int_{\pi y/2}^{\arcsin y} f(x, y)\, dx\, dy + \int_0^1 \int_{\arcsin y}^{\pi y/2} f(x, y)\, dx\, dy$

39 The upper boundary of the region is the parabola $y = 4 - x^2$

and the lower boundary is the lower half of the circle $x^2 + y^2 = 4$ for $-1 \le x \le 2$.

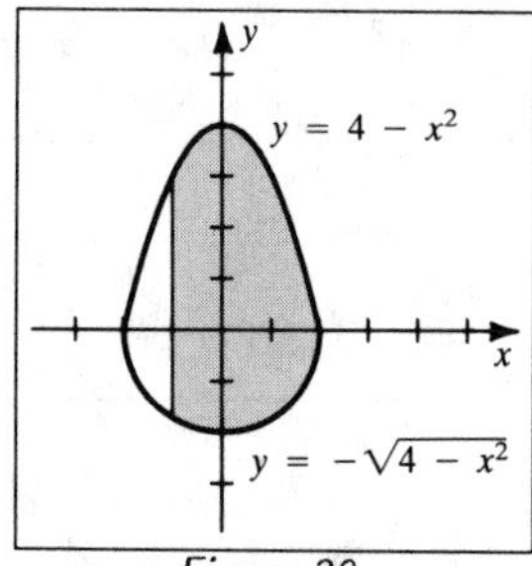

Figure 39

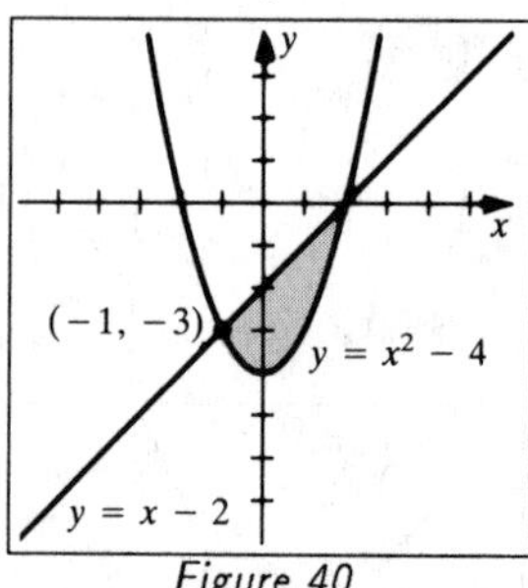

Figure 40

40 The upper boundary of the region is the line $y = x - 2$

and the lower boundary is the parabola $y = x^2 - 4$. These intersect at $x = -1, 2$.

41 The right boundary of the region is the curve $x = \sqrt[3]{y}$ $\{$or $y = x^3\}$

and the left boundary is $x = \sqrt{y}$ $\{$or $y = x^2\}$. These intersect at $y = 0, 1$.

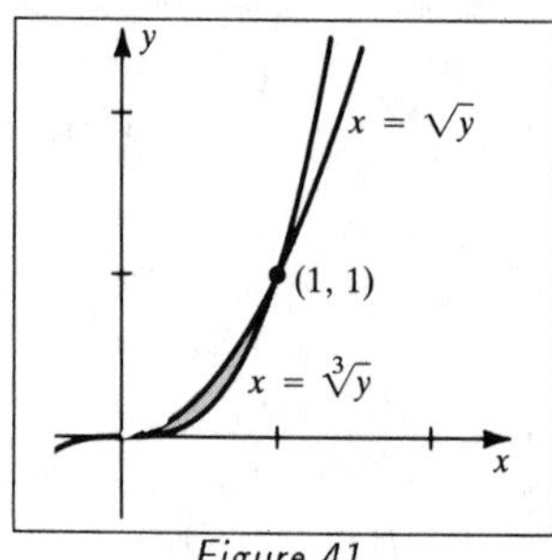

Figure 41

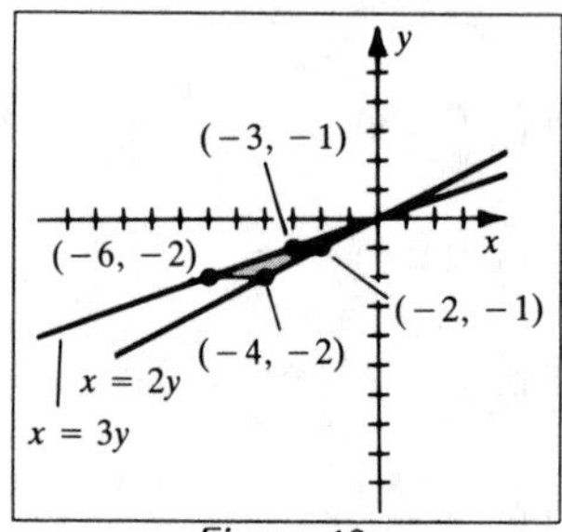

Figure 42

42 The right boundary of the region is the line $x = 2y$

and the left boundary is $x = 3y$ for $-2 \le y \le -1$.

43 The upper boundary of the region is $y = e^x$

and the lower boundary is $y = \arctan x$ for $-3 \le x \le 1$.

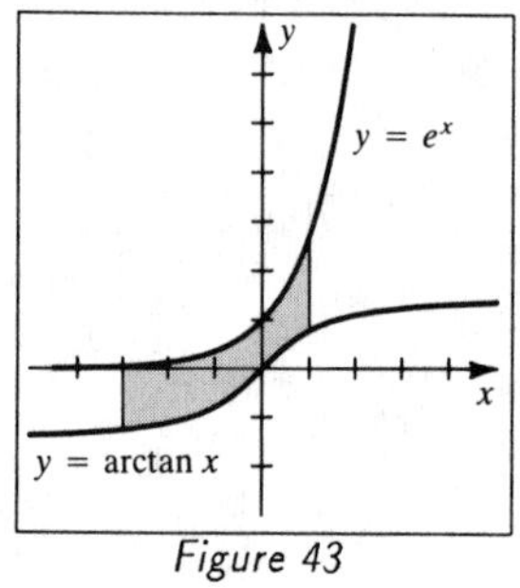

Figure 43

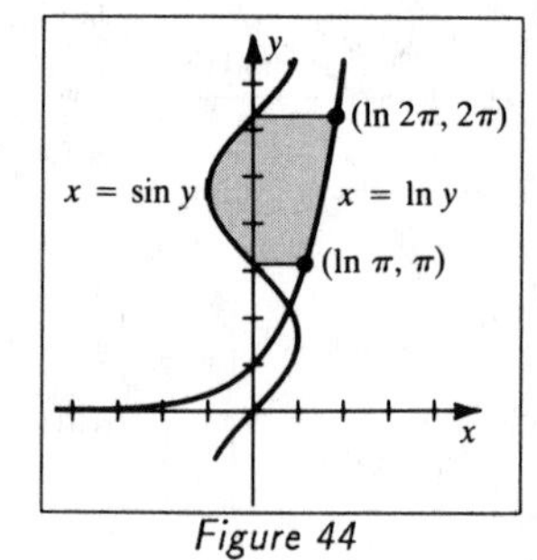

Figure 44

44 The right boundary of the region is $x = \ln y$ $\{$or $y = e^x\}$

and the left boundary is $x = \sin y$ for $\pi \le y \le 2\pi$.

45 R is the region bounded by $y = 2$, $x = 0$, and $y = 2x$.

$$\int_0^1 \int_{2x}^2 e^{(y^2)}\,dy\,dx = \int_0^2 \int_0^{y/2} e^{(y^2)}\,dx\,dy = \int_0^2 \Big[x e^{(y^2)} \Big]_0^{y/2} dy =$$

$$\tfrac{1}{2}\int_0^2 y e^{(y^2)}\,dy = \tfrac{1}{2}\Big[\tfrac{1}{2}e^{(y^2)}\Big]_0^2 = \tfrac{1}{4}(e^4 - 1) \approx 13.40.$$

46 R is the region bounded by $x = 3$, $y = 9$, and $x = \sqrt{y}$ {or $y = x^2$}.

$$\int_0^9 \int_{\sqrt{y}}^3 \sin x^3\,dx\,dy = \int_0^3 \int_0^{x^2} \sin x^3\,dy\,dx = \int_0^3 \Big[y \sin x^3 \Big]_0^{x^2} dx =$$

$$\int_0^3 x^2 \sin x^3\,dx = -\tfrac{1}{3}\Big[\cos x^3\Big]_0^3 = -\tfrac{1}{3}(\cos 27 - 1) \approx 0.43.$$

47 R is the region bounded by $y = 0$, $x = 4$, and $x = y^2$ {or $y = \sqrt{x}$}.

$$\int_0^2 \int_{y^2}^4 y \cos x^2\,dx\,dy = \int_0^4 \int_0^{\sqrt{x}} y \cos x^2\,dy\,dx = \int_0^4 \Big[\tfrac{1}{2}y^2 \cos x^2\Big]_0^{\sqrt{x}} dx =$$

$$\int_0^4 \tfrac{1}{2}x \cos x^2\,dx = \tfrac{1}{4}\Big[\sin x^2\Big]_0^4 = \tfrac{1}{4}\sin 16 \approx -0.07.$$

48 R is the region bounded by $y = 0$, $x = e$, and $y = \ln x$ {or $x = e^y$}.

$$\int_1^e \int_0^{\ln x} y\,dy\,dx = \int_0^1 \int_{e^y}^e y\,dx\,dy = \int_0^1 \Big[yx\Big]_{e^y}^e dy = \int_0^1 y(e - e^y)\,dy =$$

$$\Big[(\tfrac{1}{2}ey^2) - (ye^y - e^y)\Big]_0^1 = \tfrac{1}{2}e - 1 \approx 0.36.$$

49 R is the region bounded by $y = 0$, $x = 2$, and $x = \sqrt[3]{y}$ {or $y = x^3$}.

$$\int_0^8 \int_{\sqrt[3]{y}}^2 \frac{y}{\sqrt{16 + x^7}}\,dx\,dy = \int_0^2 \int_0^{x^3} \frac{y}{(16 + x^7)^{1/2}}\,dy\,dx = \int_0^2 \left[\frac{y^2}{2(16 + x^7)^{1/2}}\right]_0^{x^3} dx =$$

$$\int_0^2 \frac{x^6}{2(16 + x^7)^{1/2}}\,dx = \Big[\tfrac{1}{7}(16 + x^7)^{1/2}\Big]_0^2 = \tfrac{1}{7}(12 - 4) = \tfrac{8}{7}.$$

50 R is the region bounded by $x = 0$, $y = 2$, and $y = x$.

$$\int_0^2 \int_x^2 y^4 \cos(xy^2)\,dy\,dx = \int_0^2 \int_0^y y^4 \cos(xy^2)\,dx\,dy = \int_0^2 y^2\Big[\sin(xy^2)\Big]_0^y dy =$$

$$\int_0^2 y^2 \sin y^3\,dy = \Big[-\tfrac{1}{3}\cos y^3\Big]_0^2 = -\tfrac{1}{3}(\cos 8 - 1) \approx 0.38.$$

51 Let $f(x, y) = e^{x^2 y^2}$, $G(y) = \int_0^1 e^{x^2 y^2}\,dx$, and $I = \int_0^1 G(y)\,dy$.

$I \approx \frac{1-0}{2(3)}\Big[G(0) + 2G(\tfrac{1}{3}) + 2G(\tfrac{2}{3}) + G(1)\Big]$. $G(0) = \int_0^1 f(x, 0)\,dx = \int_0^1 e^0\,dx = 1.$

$$G(\tfrac{1}{3}) = \int_0^1 f(x, \tfrac{1}{3})\,dx = \int_0^1 e^{x^2/9}\,dx \approx$$

$$\tfrac{1-0}{2(3)}\Big[f(0, \tfrac{1}{3}) + 2f(\tfrac{1}{3}, \tfrac{1}{3}) + 2f(\tfrac{2}{3}, \tfrac{1}{3}) + f(1, \tfrac{1}{3})\Big] \approx 1.0406.$$

$$G(\tfrac{2}{3}) = \int_0^1 f(x, \tfrac{2}{3})\,dx = \int_0^1 e^{4x^2/9}\,dx \approx$$

$$\tfrac{1-0}{2(3)}\Big[f(0, \tfrac{2}{3}) + 2f(\tfrac{1}{3}, \tfrac{2}{3}) + 2f(\tfrac{2}{3}, \tfrac{2}{3}) + f(1, \tfrac{2}{3})\Big] \approx 1.1829. \text{ (cont.)}$$

$G(1) = \int_0^1 f(x,\,1)\,dx = \int_0^1 e^{(x^2)}\,dx \approx$

$\frac{1-0}{2(3)}\Big[f(0,\,1) + 2f(\tfrac{1}{3},\,1) + 2f(\tfrac{2}{3},\,1) + f(1,\,1)\Big] \approx 1.5121.$

Thus, $I \approx \frac{1-0}{2(3)}\Big[1 + 2(1.040601) + 2(1.182942) + 1.512094\Big] \approx 1.1599.$

[52] Let $f(x,\,y) = \sin(x^2 + y^3)$, $G(y) = \int_0^1 \sin(x^2 + y^3)\,dx$, and $I = \int_0^1 G(y)\,dy$.

$I \approx \frac{1-0}{2(3)}\Big[G(0) + 2G(\tfrac{1}{3}) + 2G(\tfrac{2}{3}) + G(1)\Big].$

$G(0) = \int_0^1 f(x,\,0)\,dx = \int_0^1 \sin(x^2)\,dx \approx$

$\frac{1-0}{2(3)}\Big[f(0,\,0) + 2f(\tfrac{1}{3},\,0) + 2f(\tfrac{2}{3},\,0) + f(1,\,0)\Big] \approx 0.3205.$

$G(\tfrac{1}{3}) = \int_0^1 f(x,\,\tfrac{1}{3})\,dx = \int_0^1 \sin\Big[x^2 + (\tfrac{1}{3})^3\Big]\,dx \approx$

$\frac{1-0}{2(3)}\Big[f(0,\,\tfrac{1}{3}) + 2f(\tfrac{1}{3},\,\tfrac{1}{3}) + 2f(\tfrac{2}{3},\,\tfrac{1}{3}) + f(1,\,\tfrac{1}{3})\Big] \approx 0.3532.$

$G(\tfrac{2}{3}) = \int_0^1 f(x,\,\tfrac{2}{3})\,dx = \int_0^1 \sin\Big[x^2 + (\tfrac{2}{3})^3\Big]\,dx \approx$

$\frac{1-0}{2(3)}\Big[f(0,\,\tfrac{2}{3}) + 2f(\tfrac{1}{3},\,\tfrac{2}{3}) + 2f(\tfrac{2}{3},\,\tfrac{2}{3}) + f(1,\,\tfrac{2}{3})\Big] \approx 0.5661.$

$G(1) = \int_0^1 f(x,\,1)\,dx = \int_0^1 \sin(x^2 + 1)\,dx \approx$

$\frac{1-0}{2(3)}\Big[f(0,\,1) + 2f(\tfrac{1}{3},\,1) + 2f(\tfrac{2}{3},\,1) + f(1,\,1)\Big] \approx 0.9212.$

Thus, $I \approx \frac{1-0}{2(3)}\Big[0.3205 + 2(0.3532) + 2(0.5661) + 0.9212\Big] \approx 0.5134.$

[53] Let $f(x,\,y) = \cos(x^2 e^y)$, $G(y) = \int_0^1 \cos(x^2 e^y)\,dx$, and $I = \int_0^1 G(y)\,dy$.

$I \approx \frac{1-0}{3(2)}\Big[G(0) + 4G(\tfrac{1}{2}) + G(1)\Big].$

$G(0) = \int_0^1 f(x,\,0)\,dx = \int_0^1 \cos(x^2)\,dx \approx \frac{1-0}{3(2)}\Big[f(0,\,0) + 4f(\tfrac{1}{2},\,0) + f(1,\,0)\Big] \approx 0.9027.$

$G(\tfrac{1}{2}) = \int_0^1 f(x,\,\tfrac{1}{2})\,dx = \int_0^1 \cos(x^2 e^{1/2})\,dx \approx$

$\frac{1-0}{3(2)}\Big[f(0,\,\tfrac{1}{2}) + 4f(\tfrac{1}{2},\,\tfrac{1}{2}) + f(1,\,\tfrac{1}{2})\Big] \approx 0.7645.$

$G(1) = \int_0^1 f(x,\,1)\,dx = \int_0^1 \cos(x^2 e)\,dx \approx$

$\frac{1-0}{3(2)}\Big[f(0,\,1) + 4f(\tfrac{1}{2},\,1) + f(1,\,1)\Big] \approx 0.5333.$

Thus, $I \approx \frac{1-0}{3(2)}\Big[0.9027 + 4(0.7645) + 0.5333\Big] \approx 0.7490.$

[54] Let $f(x,\,y) = \dfrac{1}{1 + x^4 + y^4}$, $G(y) = \int_{-1/3}^{1/3} \left[\dfrac{1}{1 + x^4 + y^4}\right] dx$, and $I = \int_0^1 G(y)\,dy$.

$I \approx \frac{1-0}{3(2)}\Big[G(0) + 4G(\tfrac{1}{2}) + G(1)\Big].$

$G(0) = \int_{-1/3}^{1/3} f(x,\,0)\,dx = \int_{-1/3}^{1/3} \left[\dfrac{1}{1 + x^4}\right] dx \approx$

$\frac{1/3-(-1/3)}{3(2)}\Big[f(-\tfrac{1}{3},\,0) + 4f(0,\,0) + f(\tfrac{1}{3},\,0)\Big] \approx 0.6640.$

$$G(\tfrac{1}{2}) = \int_{-1/3}^{1/3} f(x, \tfrac{1}{2})\,dx = \int_{-1/3}^{1/3}\left[\frac{1}{1 + x^4 + (\frac{1}{2})^4}\right]dx \approx$$

$$\tfrac{1/3-(-1/3)}{3(2)}\left[f(-\tfrac{1}{3}, \tfrac{1}{2}) + 4f(0, \tfrac{1}{2}) + f(\tfrac{1}{3}, \tfrac{1}{2})\right] \approx 0.6250.$$

$$G(1) = \int_{-1/3}^{1/3} f(x, 1)\,dx = \int_{-1/3}^{1/3}\left[\frac{1}{2 + x^4}\right]dx \approx$$

$$\tfrac{1/3-(-1/3)}{3(2)}\left[f(-\tfrac{1}{3}, 1) + 4f(0, 1) + f(\tfrac{1}{3}, 1)\right] \approx 0.3327.$$

Thus, $I \approx \frac{1-0}{3(2)}\left[0.6640 + 4(0.6250) + 0.3327\right] \approx 0.5828.$

Exercises 17.2

[1] This is an R_x region with upper boundary $y = 4x - x^2$ and lower boundary $y = -x$ on $0 \le x \le 5$. $\displaystyle\int_0^5 \int_{-x}^{4x-x^2} dy\,dx$

[2] This is an R_x region with upper boundary $y = 1/(x^2 + 1)$ and lower boundary $y = x - 3$ on $-1 \le x \le 2$. $\displaystyle\int_{-1}^2 \int_{x-3}^{1/(x^2+1)} dy\,dx$

[3] This is an R_y region with right boundary $x = (9 - y)/2$ and left boundary $x = -\sqrt{9 - y^2}$ on $-3 \le y \le 3$. $\displaystyle\int_{-3}^3 \int_{-\sqrt{9-y^2}}^{(9-y)/2} dx\,dy$

[4] This is an R_y region with right boundary $x = 4 - y^2$ and left boundary $x = -y - 2$ on $-2 \le y \le 2$. $\displaystyle\int_{-2}^2 \int_{-y-2}^{4-y^2} dx\,dy$

[5] $A = \displaystyle\int_1^2 \int_{-x^2}^{1/x^2} dy\,dx = \int_1^2 \left[\frac{1}{x^2} - (-x^2)\right]dx = \left[-\frac{1}{x} + \frac{1}{3}x^3\right]_1^2 = \frac{17}{6}$

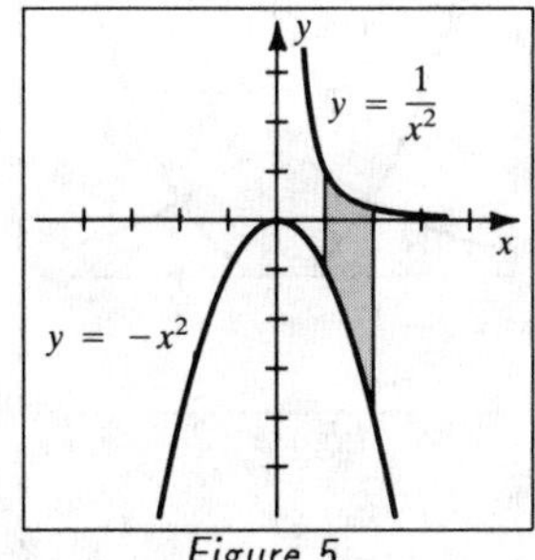

Figure 5

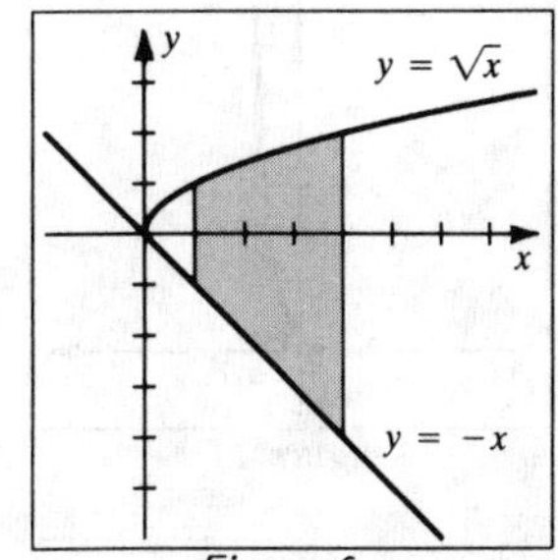

Figure 6

[6] $A = \displaystyle\int_1^4 \int_{-x}^{\sqrt{x}} dy\,dx = \int_1^4 \left[\sqrt{x} - (-x)\right]dx = \left[\frac{2}{3}x^{3/2} + \frac{1}{2}x^2\right]_1^4 = \frac{73}{6}$

[7] $A = \displaystyle\int_{-1}^2 \int_{-y^2}^{y+4} dx\,dy = \int_{-1}^2 (y + 4 + y^2)\,dy = \left[\frac{1}{3}y^3 + \frac{1}{2}y^2 + 4y\right]_{-1}^2 = \frac{33}{2}$

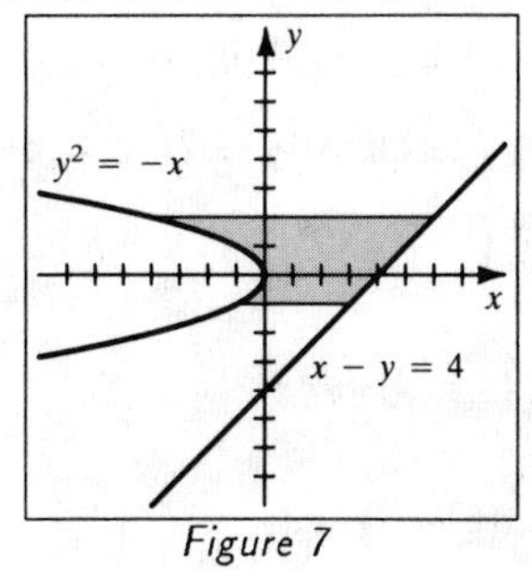

Figure 7

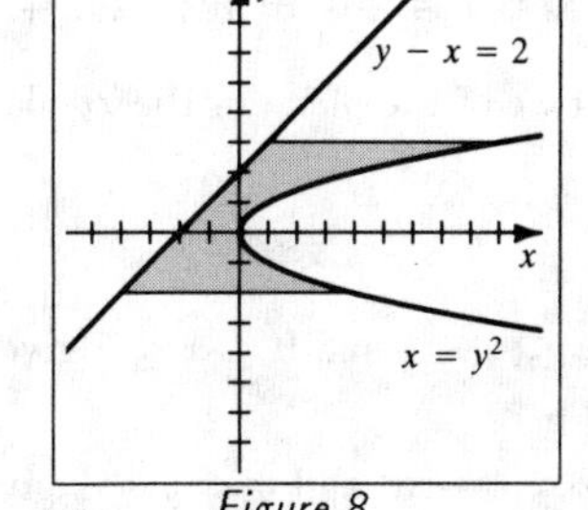

Figure 8

[8] $A = \int_{-2}^{3}\int_{y-2}^{y^2} dx\,dy = \int_{-2}^{3} (y^2 - y + 2)\,dy = \left[\frac{1}{3}y^3 - \frac{1}{2}y^2 + 2y\right]_{-2}^{3} = \frac{115}{6}$

[9] $A = \int_{0}^{1}\int_{x}^{3x} dy\,dx + \int_{1}^{2}\int_{x}^{4-x} dy\,dx = \int_{0}^{1} 2x\,dx + \int_{1}^{2} (4 - 2x)\,dx = 1 + 1 = 2$

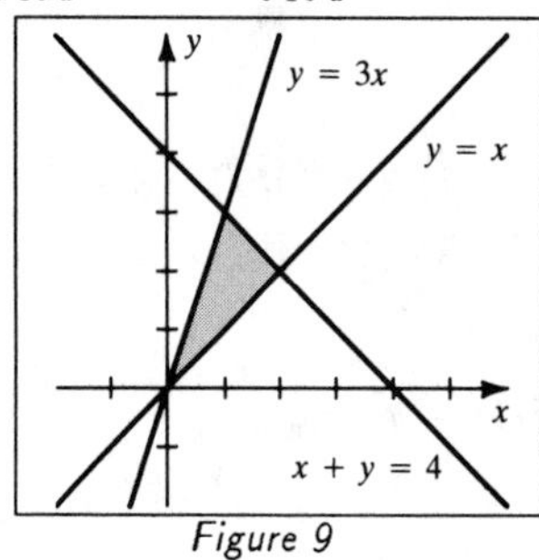

Figure 9

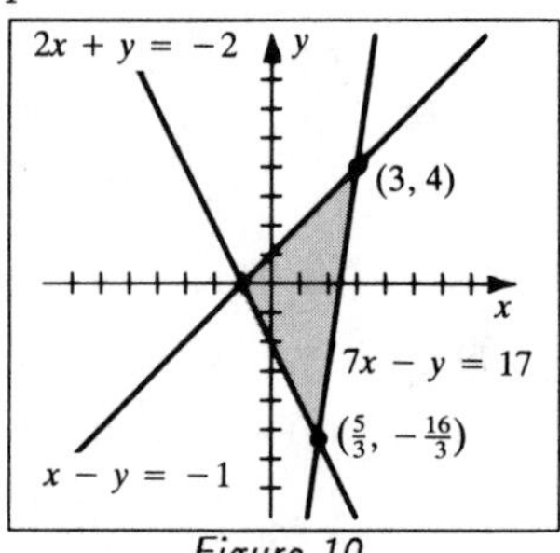

Figure 10

[10] $A = \int_{-1}^{5/3}\int_{-2x-2}^{x+1} dy\,dx + \int_{5/3}^{3}\int_{7x-17}^{x+1} dy\,dx$

$= \int_{-1}^{5/3} (3x + 3)\,dx + \int_{5/3}^{3} (18 - 6x)\,dx = \frac{32}{3} + \frac{16}{3} = 16$

[11] $A = \int_{-\pi}^{\pi}\int_{\sin x}^{e^x} dy\,dx = \int_{-\pi}^{\pi} (e^x - \sin x)\,dx = \left[e^x + \cos x\right]_{-\pi}^{\pi} = e^{\pi} - e^{-\pi} \approx 23.10$

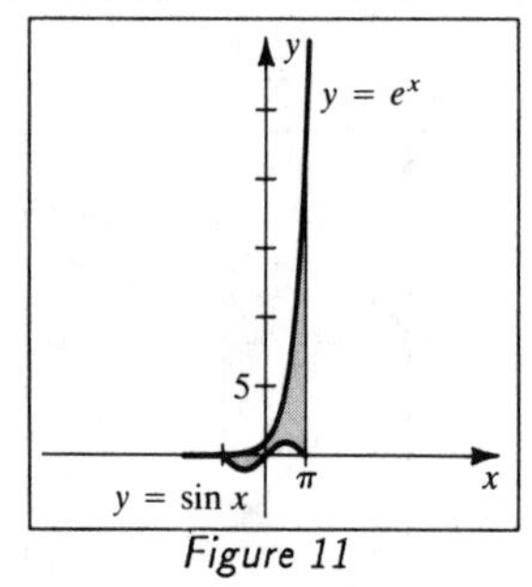

Figure 11

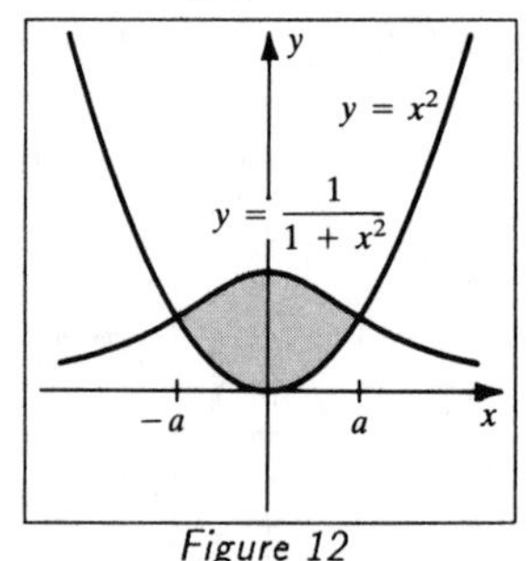

Figure 12

[12] $x^2 = \frac{1}{1 + x^2}$ at $a = \sqrt{\frac{1}{2}(\sqrt{5} - 1)} \approx 0.786$ and $-a$. $A = \int_{-a}^{a}\int_{x^2}^{1/(x^2+1)} dy\,dx =$

$2\int_{0}^{a}\int_{x^2}^{1/(x^2+1)} dy\,dx = 2\int_{0}^{a}\left[\frac{1}{x^2 + 1} - x^2\right] dx = 2\tan^{-1} a - \frac{2}{3}a^3 \approx 1.01.$

[13] The plane has x-, y-, and z-intercepts at 3, 4, and 5, respectively. An equation of the plane is $\frac{x}{3} + \frac{y}{4} + \frac{z}{5} = 1 \Leftrightarrow 20x + 15y + 12z = 60 \Leftrightarrow z = \frac{1}{12}(60 - 20x - 15y)$.

The trace of the plane in the xy-plane is $20x + 15y = 60$, or $y = \frac{1}{3}(12 - 4x)$.

$$\int_{0}^{3}\int_{0}^{(12-4x)/3} \tfrac{1}{12}(60 - 20x - 15y)\,dy\,dx$$

[14] $0 \le y \le 4 - x$ and $0 \le x \le 2$ gives us $\int_{0}^{2}\int_{0}^{4-x} (4 - x^2)\,dy\,dx.$

[15] $0 \le x \le 4 - y^2$ and $0 \le y \le 2$ gives us $\int_{0}^{2}\int_{0}^{4-y^2} (6 - x)\,dx\,dy.$

[16] Solving the equation of the sphere for z yields $z = \sqrt{25 - x^2 - y^2}$.

$0 \le x \le 4 - y$ and $0 \le y \le 4$ gives us $\int_0^4 \int_0^{4-y} \sqrt{25 - x^2 - y^2}\, dx\, dy$.

[17] S is the plane $z = 3$ over the rectangular region R in the xy-plane that has upper boundary $y = 2$ and lower boundary $y = -1$ from $x = 0$ to $x = 4$.

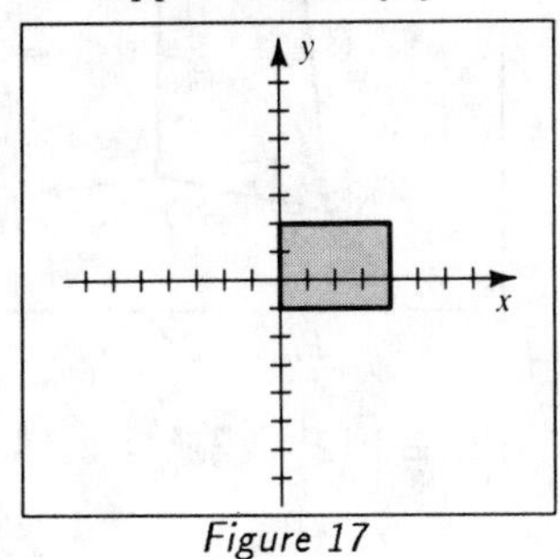

Figure 17

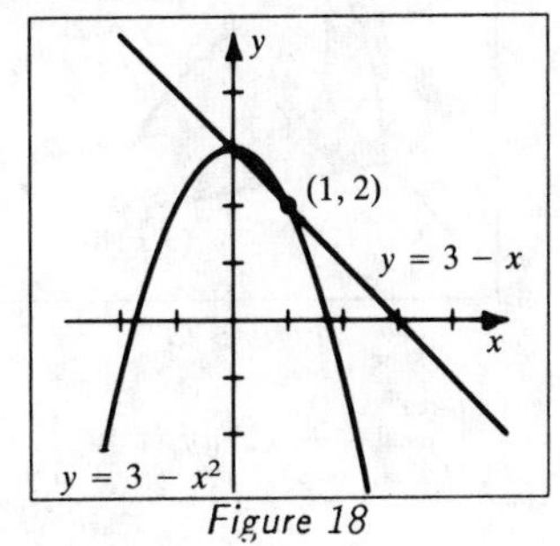

Figure 18

[18] S is the hemisphere $z = \sqrt{25 - x^2 - y^2}$ over the region R in the xy-plane that has upper boundary $y = 3 - x^2$ and lower boundary $y = 3 - x$ from $x = 0$ to $x = 1$.

[19] S is the paraboloid $z = x^2 + y^2$ over the region R in the xy-plane that has upper boundary $y = 1 - x^2$ and lower boundary $y = x - 1$ from $x = -2$ to $x = 1$.

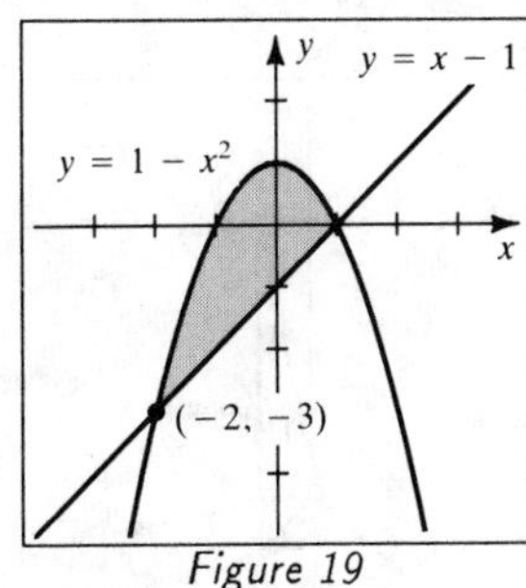

Figure 19

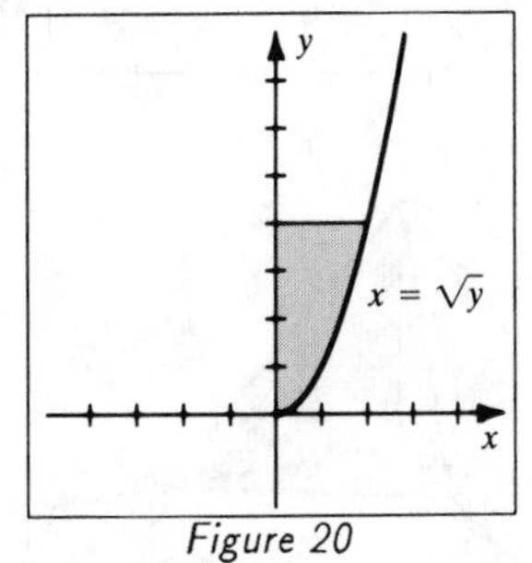

Figure 20

[20] S is the cone $z = \sqrt{x^2 + y^2}$ over the region R in the xy-plane that has right boundary $x = \sqrt{y}$ and left boundary $x = 0$ from $y = 0$ to $y = 4$.

[21] $V = \int_0^2 \int_0^1 (4x^2 + y^2)\, dy\, dx = \int_0^2 \Big[4x^2 y + \tfrac{1}{3}y^3 \Big]_0^1 dx = \int_0^2 (4x^2 + \tfrac{1}{3})\, dx =$

$\Big[\tfrac{4}{3}x^3 + \tfrac{1}{3}x \Big]_0^2 = \tfrac{34}{3}$

[22] $V = \int_0^1 \int_0^{2x} (x^2 + 4y^2)\, dy\, dx = \int_0^1 \Big[x^2 y + \tfrac{4}{3}y^3 \Big]_0^{2x} dx = \int_0^1 \tfrac{38}{3}x^3\, dx = \Big[\tfrac{19}{6}x^4 \Big]_0^1 = \tfrac{19}{6}$

[23] $V = \int_0^3\int_0^{2x} (9 - x^2)^{1/2}\, dy\, dx = \int_0^3 2x(9 - x^2)^{1/2}\, dx = \left[-\frac{2}{3}(9 - x^2)^{3/2}\right]_0^3 = 18$

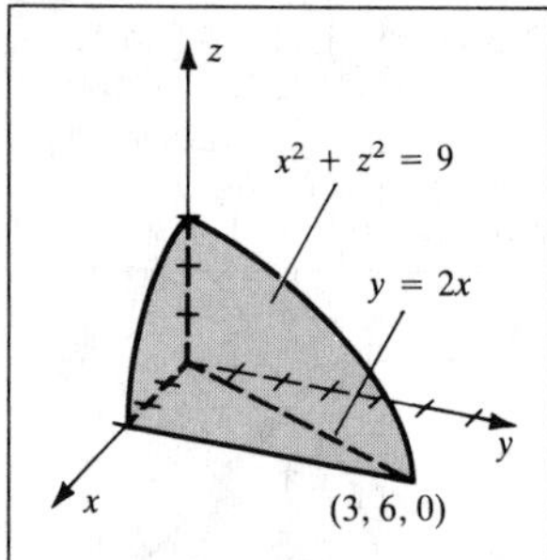

Figure 23

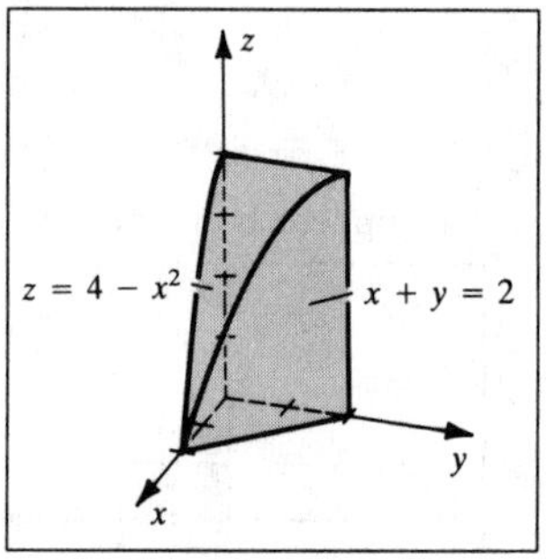

Figure 24

[24] $V = \int_0^2\int_0^{2-x} (4 - x^2)\, dy\, dx = \int_0^2 (4 - x^2)(2 - x)\, dx = \left[\frac{1}{4}x^4 - \frac{2}{3}x^3 - 2x^2 + 8x\right]_0^2$
$= \frac{20}{3}$

[25] $V = \int_0^2\int_0^{4-2x} (4 - 2x - y)\, dy\, dx = \int_0^2 \left[(4 - 2x)y - \frac{1}{2}y^2\right]_0^{4-2x} dx$

$= \int_0^2 \left[(4 - 2x)^2 - \frac{1}{2}(4 - 2x)^2\right] dx = \frac{1}{2}\int_0^2 (4 - 2x)^2\, dx = 2\int_0^2 (2 - x)^2\, dx$

$= 2\int_0^2 (x^2 - 4x + 4)\, dx = 2\left[\frac{1}{3}x^3 - 2x^2 + 4x\right]_0^2 = 2\left(\frac{8}{3}\right) = \frac{16}{3}$

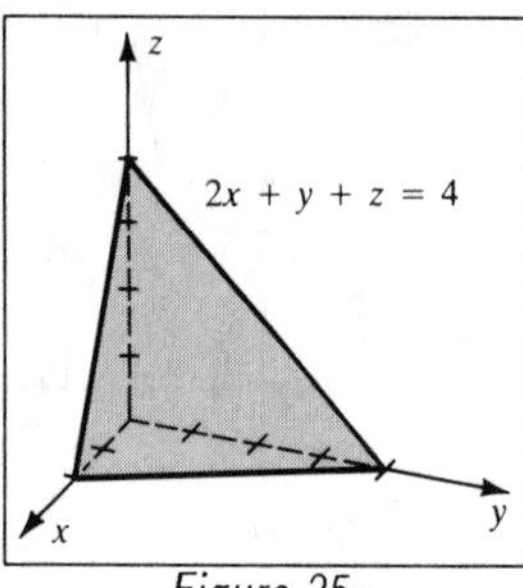

Figure 25

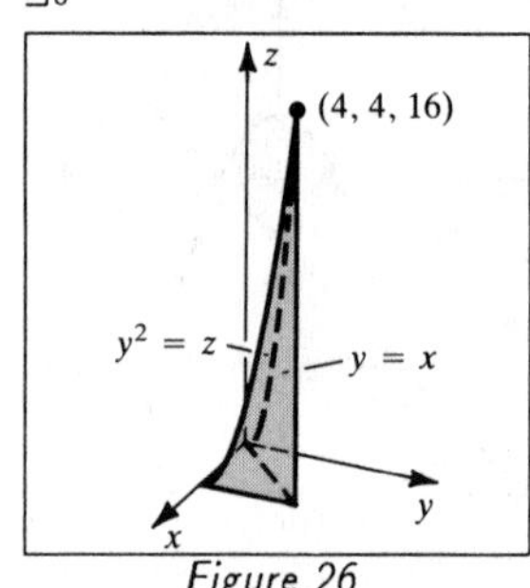

Figure 26

[26] $V = \int_0^4\int_0^x y^2\, dy\, dx = \int_0^4 \frac{1}{3}x^3\, dx = \frac{1}{3}\left[\frac{1}{4}x^4\right]_0^4 = \frac{64}{3}$

[27] $V = \int_0^2\int_0^{4-x^2} (x^2 + y^2)\, dy\, dx = \int_0^2 \left(\frac{64}{3} - 12x^2 + 3x^4 - \frac{1}{3}x^6\right) dx = \frac{832}{35}$

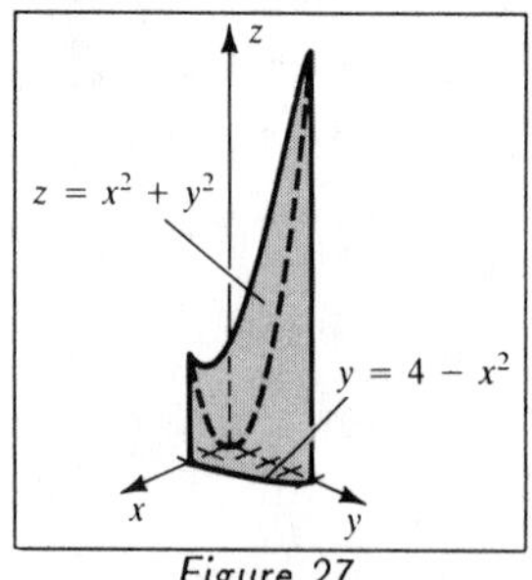

Figure 27

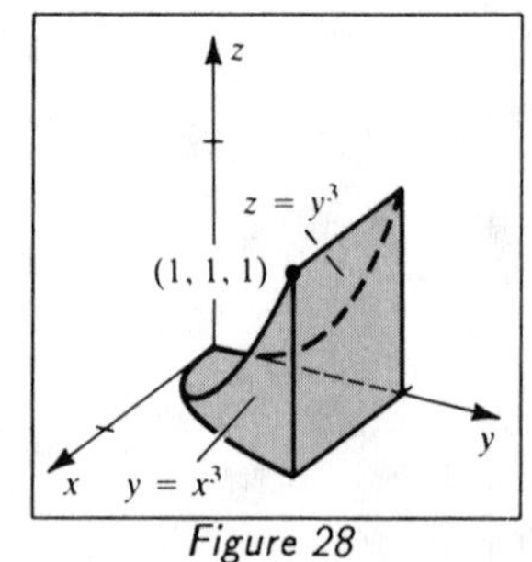

Figure 28

28 $V = \int_0^1 \int_{x^3}^1 y^3\, dy\, dx = \frac{1}{4}\int_0^1 (1 - x^{12})\, dx = \frac{1}{4}\Big[x - \frac{1}{13}x^{13}\Big]_0^1 = \frac{3}{13}$

29 $V = \int_0^4 \int_{x^2/16}^{\sqrt{x}/2} x^3\, dy\, dx = \int_0^4 (\frac{1}{2}x^{7/2} - \frac{1}{16}x^5)\, dx = \Big[\frac{1}{9}x^{9/2} - \frac{1}{96}x^6\Big]_0^4 = \frac{128}{9}$

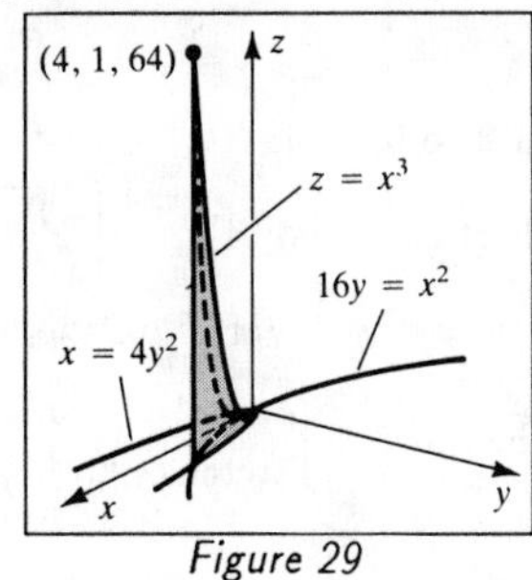

Figure 29

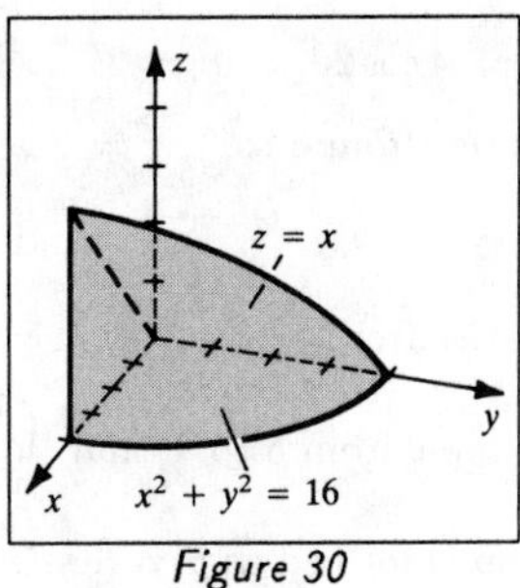

Figure 30

30 $V = \int_0^4 \int_0^{\sqrt{16-x^2}} x\, dy\, dx = \int_0^4 x(16 - x^2)^{1/2}\, dx = \Big[-\frac{1}{3}(16 - x^2)^{3/2}\Big]_0^4 = \frac{64}{3}$

31 In the xy-plane, $y = 4 - x^2$ intersects $x + y = 2$ at $(-1, 3)$ and $(2, 0)$.

$$V = \int_{-1}^2 \int_{2-x}^{4-x^2} (x^2 + 4)\, dy\, dx = \int_{-1}^2 (8 + 4x - 2x^2 + x^3 - x^4)\, dx = \frac{423}{20}$$

32 In the xy-plane, $y = x^3$ intersects $y = x^4$ at $(0, 0)$ and $(1, 1)$.

$$V = \int_0^1 \int_{x^4}^{x^3} (x + y + 4)\, dy\, dx = \int_0^1 (4x^3 - 3x^4 - x^5 + \frac{1}{2}x^6 - \frac{1}{2}x^8)\, dx = \frac{157}{630}$$

33 Note that $\Delta y_j\, \Delta x_k = \frac{1}{4}(\frac{1}{4}) = \frac{1}{16}$ for all j and k and that the (u_k, v_j) are midpoints of the rectangles in the grid. $(u_1, v_1) = (\frac{1}{4}(1) - \frac{1}{8}, \frac{1}{4}(1) - \frac{1}{8}) = (\frac{1}{8}, \frac{1}{8})$.
$(u_1, v_2) = (\frac{1}{4}(1) - \frac{1}{8}, \frac{1}{4}(2) - \frac{1}{8}) = (\frac{1}{8}, \frac{3}{8})$. $(u_2, v_1) = (\frac{1}{4}(2) - \frac{1}{8}, \frac{1}{4}(1) - \frac{1}{8}) = (\frac{3}{8}, \frac{1}{8})$.
The coordinates for the other five points are found in a similar manner.

$$V = \int_0^1 \int_0^{1/2} f(x, y)\, dy\, dx = \int_0^1 \int_0^{1/2} \sin\big[\cos(xy)\big]\, dy\, dx \approx \sum_{k=1}^{4} \sum_{j=1}^{2} f(u_k, v_j)\, \Delta y_j\, \Delta x_k =$$

$$(\tfrac{1}{16}) \sum_{k=1}^{4} \sum_{j=1}^{2} f(u_k, v_j) = \tfrac{1}{16}[f(\tfrac{1}{8}, \tfrac{1}{8}) + f(\tfrac{1}{8}, \tfrac{3}{8}) + f(\tfrac{3}{8}, \tfrac{1}{8}) + f(\tfrac{3}{8}, \tfrac{3}{8}) +$$

$$f(\tfrac{5}{8}, \tfrac{1}{8}) + f(\tfrac{5}{8}, \tfrac{3}{8}) + f(\tfrac{7}{8}, \tfrac{1}{8}) + f(\tfrac{7}{8}, \tfrac{3}{8})] \approx (\tfrac{1}{16})6.6751 \approx 0.4172.$$

34 Similar to Exercise 33 with $f(x, y) = \sqrt{x^4 + y^4}$, $V \approx (\frac{1}{16})2.8543 \approx 0.1784$.

Exercises 17.3

1 As θ varies from 0 to $\frac{\pi}{2}$, $r = 4\sin\theta$ varies from 0 to 4,
and this is one-half of the area. $2\int_0^{\pi/2} \int_0^{4\sin\theta} r\, dr\, d\theta$

2 As θ varies from 0 to $\frac{\pi}{2}$, $r = -4\cos\theta$ varies from 0 to -4,
and this is one-half of the area. $2\int_0^{\pi/2} \int_0^{-4\cos\theta} r\, dr\, d\theta$

3 $r = 0 \Rightarrow 1 + 2\cos\theta = 0 \Rightarrow \cos\theta = -\frac{1}{2} \Rightarrow \theta = \frac{2\pi}{3}, \frac{4\pi}{3}$ for $0 \le \theta < 2\pi$.
As θ varies from 0 to $\frac{2\pi}{3}$, $r = 1 + 2\cos\theta$ varies from 3 to 0,
and this is one-half of the area. $2\int_0^{2\pi/3} \int_0^{1+2\cos\theta} r\, dr\, d\theta$

4 As θ varies from $-\frac{\pi}{2}$ to $\frac{\pi}{2}$ {or $\frac{\pi}{2}$ to $\frac{3\pi}{2}$}, $r = 3 - 3\sin\theta$ varies from 6 to 0,

and this is one-half of the area. $2\int_{-\pi/2}^{\pi/2}\int_0^{3-3\sin\theta} r\,dr\,d\theta$

5 The lemniscate should be vertically centered on the polar axis.

$r = 0 \Rightarrow 4\cos 2\theta = 0 \Rightarrow 2\theta = \frac{\pi}{2} + \pi n \Rightarrow \theta = \frac{\pi}{4} + \frac{\pi}{2}n.$

As θ varies from 0 to $\frac{\pi}{4}$, $r = 2\sqrt{\cos 2\theta}$ varies from 2 to 0,

and this is one-fourth of the area. $4\int_0^{\pi/4}\int_0^{2\sqrt{\cos 2\theta}} r\,dr\,d\theta$

6 As θ varies from $\arctan\frac{4}{3}$ to $\frac{\pi}{2}$, $r = 5$ is constant, $r = 4\csc\theta$ varies on the horizontal line from 5 to 4, and this is one-half of the area. $2\int_{\arctan(4/3)}^{\pi/2}\int_{4\csc\theta}^{5} r\,dr\,d\theta$

7 One loop is formed as θ varies from 0 to $\frac{\pi}{3}$.

$$A = \int_0^{\pi/3}\int_0^{4\sin 3\theta} r\,dr\,d\theta = \int_0^{\pi/3}\left[\tfrac{1}{2}r^2\right]_0^{4\sin 3\theta} d\theta = \int_0^{\pi/3} 8\sin^2 3\theta\,d\theta =$$

$$8\int_0^{\pi/3}\tfrac{1}{2}(1 - \cos 6\theta)\,d\theta = 4\left[\theta - \tfrac{1}{6}\sin 6\theta\right]_0^{\pi/3} = \tfrac{4\pi}{3}.$$

8 One loop is formed as θ varies from $-\frac{\pi}{8}$ to $\frac{\pi}{8}$.

$$A = 2\int_0^{\pi/8}\int_0^{2\cos 4\theta} r\,dr\,d\theta = 2\int_0^{\pi/8}\left[\tfrac{1}{2}r^2\right]_0^{2\cos 4\theta} d\theta = 4\int_0^{\pi/8}\cos^2 4\theta\,d\theta =$$

$$4\int_0^{\pi/8}\tfrac{1}{2}(1 + \cos 8\theta)\,d\theta = 2\left[\theta + \tfrac{1}{8}\sin 8\theta\right]_0^{\pi/8} = \tfrac{\pi}{4}.$$

9 From *Figure 9*, we see that $2 - 2\cos\theta = 3$ when $\theta = \frac{2\pi}{3}, \frac{4\pi}{3}$.

$$A = 2\int_{2\pi/3}^{\pi}\int_3^{2-2\cos\theta} r\,dr\,d\theta = \int_{2\pi/3}^{\pi}\left[(2 - 2\cos\theta)^2 - 3^2\right]d\theta$$

$$= \int_{2\pi/3}^{\pi}\left[2\cos 2\theta - 8\cos\theta - 3\right]d\theta = \left[\sin 2\theta - 8\sin\theta - 3\theta\right]_{2\pi/3}^{\pi} =$$

$$\tfrac{9}{2}\sqrt{3} - \pi \approx 4.65.$$

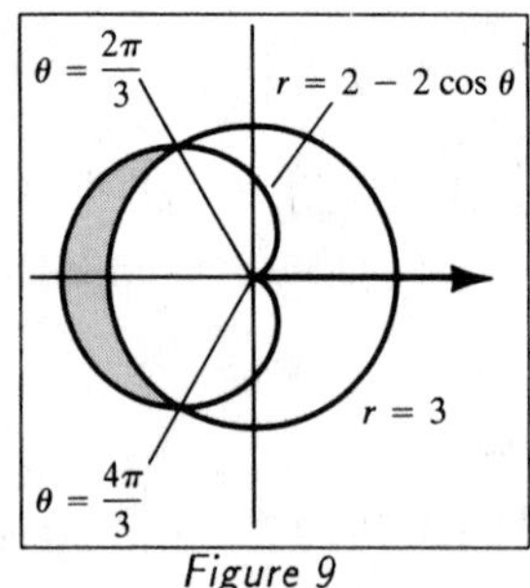

Figure 9

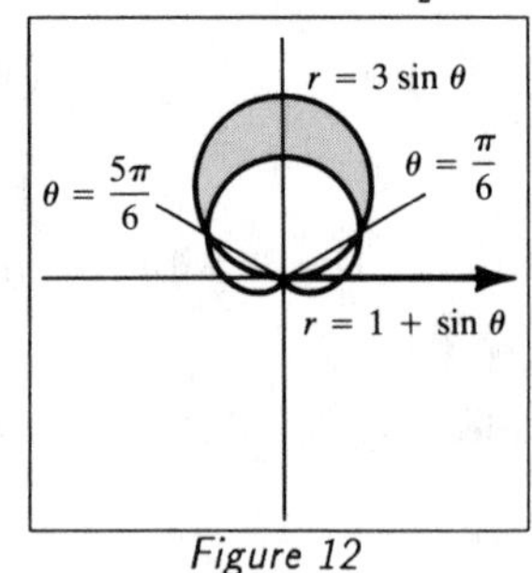

Figure 12

10 $A = \int_0^{2\pi}\int_0^{3+2\sin\theta} r\,dr\,d\theta = \int_0^{2\pi}\tfrac{1}{2}(3 + 2\sin\theta)^2\,d\theta =$

$$\tfrac{1}{2}\int_0^{2\pi}(11 + 12\sin\theta - 2\cos 2\theta)\,d\theta = \tfrac{1}{2}\left[11\theta - 12\cos\theta - \sin 2\theta\right]_0^{2\pi} = 11\pi.$$

11 $A = 2\int_0^{\pi/4}\int_0^{3\sqrt{\cos 2\theta}} r\,dr\,d\theta = \int_0^{\pi/4} 9\cos 2\theta\,d\theta = 9\Big[\tfrac{1}{2}\sin 2\theta\Big]_0^{\pi/4} = \tfrac{9}{2}.$

12 From *Figure 12*, we see that $3\sin\theta = 1 + \sin\theta$ at $\theta = \frac{\pi}{6}, \frac{5\pi}{6}$.

$$A = 2\int_{\pi/6}^{\pi/2}\int_{1+\sin\theta}^{3\sin\theta} r\,dr\,d\theta = \int_{\pi/6}^{\pi/2}\Big[9\sin^2\theta - (1+\sin\theta)^2\Big]\,d\theta$$
$$= \int_{\pi/6}^{\pi/2}\Big[3 - 4\cos 2\theta - 2\sin\theta\Big]\,d\theta = \Big[3\theta - 2\sin 2\theta + 2\cos\theta\Big]_{\pi/6}^{\pi/2} = \pi.$$

13 $\iint_R (x^2+y^2)^{3/2}\,dA = \int_0^{2\pi}\int_0^2 (r^3)\,r\,dr\,d\theta = \int_0^{2\pi}\tfrac{32}{5}\,d\theta = \tfrac{64\pi}{5}.$

14 $\iint_R x^2(x^2+y^2)^3\,dA = \int_0^{\pi}\int_0^1 (r\cos\theta)^2(r^2)^3\,r\,dr\,d\theta =$

$$\int_0^{\pi}\int_0^1 r^9\cos^2\theta\,dr\,d\theta = \tfrac{1}{10}\int_0^{\pi}\cos^2\theta\,d\theta = \tfrac{1}{20}\Big[\theta + \tfrac{1}{2}\sin 2\theta\Big]_0^{\pi} = \tfrac{\pi}{20}.$$

15 $\iint_R \frac{x^2}{x^2+y^2}\,dA = \int_0^{2\pi}\int_a^b (\cos^2\theta)\,r\,dr\,d\theta = \tfrac{1}{2}(b^2-a^2)\int_0^{2\pi}\cos^2\theta\,d\theta = \tfrac{\pi}{2}(b^2-a^2).$

16 In polar coordinates, $x^2+y^2 = 2y \Rightarrow r^2 = 2r\sin\theta \Rightarrow r = 2\sin\theta$.

This circle is traced out once as θ varies from 0 to π.

$$\iint_R (x+y)\,dA = \int_0^{\pi}\int_0^{2\sin\theta} r(\cos\theta+\sin\theta)\,r\,dr\,d\theta = \tfrac{1}{3}\int_0^{\pi}\Big[r^3(\cos\theta+\sin\theta)\Big]_0^{2\sin\theta}\,d\theta$$
$$= \tfrac{8}{3}\int_0^{\pi}\sin^3\theta\cos\theta\,d\theta + \tfrac{8}{3}\int_0^{\pi}\sin^4\theta\,d\theta$$
$$= \tfrac{8}{3}\Big[\tfrac{1}{4}\sin^4\theta\Big]_0^{\pi} + \tfrac{8}{3}\int_0^{\pi}\tfrac{1}{4}\big(\tfrac{3}{2} - 2\cos 2\theta + \tfrac{1}{2}\cos 4\theta\big)\,d\theta$$
$$= 0 + \tfrac{2}{3}\Big[\tfrac{3}{2}\theta - \sin 2\theta + \tfrac{1}{8}\sin 4\theta\Big]_0^{\pi} = \pi.$$

17 The triangle is bounded by the lines $y = 0$, $y = x$, and $x = 3$.

In polar coordinates, these are $\theta = 0$, $\theta = \frac{\pi}{4}$, and $r = 3\sec\theta$.

$$\iint_R \sqrt{x^2+y^2}\,dA = \int_0^{\pi/4}\int_0^{3\sec\theta}(r)\,r\,dr\,d\theta = \tfrac{1}{3}\int_0^{\pi/4}27\sec^3\theta\,d\theta =$$
$$\tfrac{9}{2}\Big[\sec\theta\tan\theta + \ln|\sec\theta+\tan\theta|\Big]_0^{\pi/4}\ \{\text{Formula } 71\} = \tfrac{9}{2}\Big[\sqrt{2} + \ln(\sqrt{2}+1)\Big] \approx 10.33$$

18 $y = \sqrt{2x - x^2} \Rightarrow x^2 + y^2 = 2x \Rightarrow r^2 = 2r\cos\theta \Rightarrow r = 2\cos\theta$.

$$\iint_R \sqrt{x^2+y^2}\,dA = \int_{\pi/4}^{\pi/2}\int_0^{2\cos\theta}(r)\,r\,dr\,d\theta = \tfrac{8}{3}\int_{\pi/4}^{\pi/2}\cos^3\theta\,d\theta =$$
$$\tfrac{8}{9}\Big[(2+\cos^2\theta)\sin\theta\Big]_{\pi/4}^{\pi/2}\{\text{Formula } 68\} = \tfrac{8}{9}\big(2 - \tfrac{5}{4}\sqrt{2}\big) \approx 0.21.$$

Note: In Exercises 19–24, let R denote the region in the xy-plane.

19 R is the upper half of the circle $x^2 + y^2 = a^2$.

$$\int_{-a}^{a}\int_0^{\sqrt{a^2-x^2}} e^{-(x^2+y^2)}\,dy\,dx = \int_0^{\pi}\int_0^a e^{-r^2}\,r\,dr\,d\theta = \int_0^{\pi} -\tfrac{1}{2}e^{-a^2}\,d\theta = \tfrac{\pi}{2}(1 - e^{-a^2}).$$

[20] R is the part of the circle $x^2 + y^2 = a^2$ in the first quadrant.

$$\int_0^a \int_0^{\sqrt{a^2-x^2}} (x^2 + y^2)^{3/2}\, dy\, dx = \int_0^{\pi/2}\int_0^a (r^3)\, r\, dr\, d\theta = \tfrac{1}{5}a^5 \int_0^{\pi/2} d\theta = \tfrac{\pi}{10}a^5.$$

[21] R is the region under the line $y = x$ from $x = 1$ to 2.

$$\int_1^2 \int_0^x \frac{1}{\sqrt{x^2 + y^2}}\, dy\, dx = \int_0^{\pi/4}\int_{\sec\theta}^{2\sec\theta} (\tfrac{1}{r})\, r\, dr\, d\theta = \int_0^{\pi/4} \sec\theta\, d\theta =$$

$$\Big[\ln|\sec\theta + \tan\theta|\Big]_0^{\pi/4} = \ln(\sqrt{2} + 1) - \ln 1 \approx 0.88.$$

[22] R is the part of the circle $x^2 + y^2 = 1$ in the first quadrant. $\int_0^1 \int_0^{\sqrt{1-x^2}} e^{\sqrt{x^2+y^2}}\, dy\, dx$

$$= \int_0^{\pi/2}\int_0^1 e^r\, r\, dr\, d\theta = \int_0^{\pi/2}\Big[re^r - e^r\Big]_0^1 d\theta = \int_0^{\pi/2} 1\, d\theta = \tfrac{\pi}{2}.$$

[23] R is the part of the circle $x^2 + y^2 = 4$ in the first quadrant.

$$\int_0^2 \int_0^{\sqrt{4-y^2}} \cos(x^2 + y^2)\, dx\, dy = \int_0^{\pi/2}\int_0^2 \cos(r^2)\, r\, dr\, d\theta = \int_0^{\pi/2}\Big[\tfrac{1}{2}\sin(r^2)\Big]_0^2 d\theta =$$

$$\int_0^{\pi/2} \tfrac{1}{2}\sin 4\, d\theta = \tfrac{\pi}{4}\sin 4 \approx -0.59.$$

[24] R is the region inside the circle $x^2 + (y - 1)^2 = 1$.

In polar coordinates, this region is bounded by $r = 2\sin\theta$ with $\theta = 0$ to π.

$\int_0^2 \int_{-\sqrt{2y-y^2}}^{\sqrt{2y-y^2}} x\, dx\, dy = \int_0^\pi \int_0^{2\sin\theta} (r\cos\theta)\, r\, dr\, d\theta = \frac{8}{3}\int_0^\pi \sin^3\theta \cos\theta\, d\theta = 0.$ Intuitively, we see this is true since the plane $z = x$ intersects the cylinder $x^2 + (y - 1)^2 = 1$ in a manner that would give us equal amounts of "positive" and "negative" volume.

[25] The height of the solid in the *first* octant is $z = \sqrt{25 - x^2 - y^2} = \sqrt{25 - r^2}$ and the projection of the solid onto the xy-plane is the region between the circles of radii 3 and 5 centered at the origin. Thus, by symmetry in each octant, $V =$

$$8\int_0^{\pi/2}\int_3^5 (25 - r^2)^{1/2}\, r\, dr\, d\theta = 8\int_0^{\pi/2}\Big[-\tfrac{1}{3}(25 - r^2)^{3/2}\Big]_3^5 d\theta = 8\int_0^{\pi/2} \tfrac{64}{3}\, d\theta = \tfrac{256\pi}{3}.$$

[26] The height of the solid above the xy-plane is $z = \sqrt{16 - 4x^2 - 4y^2} = 2\sqrt{4 - r^2}$. The trace of the solid in the xy-plane is the region inside the circle $x^2 + y^2 = 1$. Thus, by symmetry about the xy-plane, $V =$

$$2\int_0^{2\pi}\int_0^1 2(4 - r^2)^{1/2}\, r\, dr\, d\theta = 2\int_0^{2\pi}\Big[-\tfrac{2}{3}(3^{3/2} - 8)\Big] d\theta = \tfrac{8\pi}{3}(8 - 3\sqrt{3}) \approx 23.49.$$

[27] The upper half of the solid is the cone $z = r$ and the trace of the solid in the $r\theta$-plane is $r = 2\cos\theta$ $\{x^2 + y^2 = 2x\}$. Thus, by symmetry in 4 octants,

$$V = 4\int_0^{\pi/2}\int_0^{2\cos\theta} (r)\, r\, dr\, d\theta = \tfrac{32}{3}\int_0^{\pi/2} \cos^3\theta\, d\theta = \tfrac{32}{3}\int_0^{\pi/2} (1 - \sin^2\theta)\cos\theta\, d\theta =$$

$$\tfrac{32}{3}\int_0^1 (1 - u^2)\, du\ \{u = \sin\theta\} = \tfrac{32}{3}(\tfrac{2}{3}) = \tfrac{64}{9}.$$

[28] The height of the solid above the $r\theta$-plane is $z = 4r^2$. The trace of the solid in the $r\theta$-plane is $r = 3\sin\theta$. $V = \int_0^{\pi}\int_0^{3\sin\theta} (4r^2)\, r\, dr\, d\theta = \int_0^{\pi} 81\sin^4\theta\, d\theta =$

$$\tfrac{81}{4}\Big[\tfrac{3}{2}\theta - \sin 2\theta + \tfrac{1}{8}\sin 4\theta\Big]_0^{\pi} = \tfrac{243\pi}{8} \approx 95.43.$$

[29] The height of the solid above the $r\theta$-plane is $z = \sqrt{16 - r^2}$. The trace of the solid in the $r\theta$-plane is $r = 4\sin\theta$. Thus, by symmetry in 4 octants, $V =$

$$4\int_0^{\pi/2}\int_0^{4\sin\theta} (16 - r^2)^{1/2}\, r\, dr\, d\theta = 4\int_0^{\pi/2}\Big[-\tfrac{1}{3}(16 - r^2)^{3/2}\Big]_0^{4\sin\theta} d\theta =$$

$$-\tfrac{4}{3}\int_0^{\pi/2} 64(\cos^3\theta - 1)\, d\theta = -\tfrac{256}{3}\Big[\sin\theta - \tfrac{1}{3}\sin^3\theta - \theta\Big]_0^{\pi/2} = \tfrac{128}{9}(3\pi - 4) \approx 77.15.$$

[30] The height of the solid above the xy-plane is $z = 9 - r^2$ and the trace of the solid in the xy-plane occurs when $5 = 9 - r^2$, or $r = 2$.

$$V = \int_0^{2\pi}\int_0^{2}\Big[(9 - r^2) - 5\Big] r\, dr\, d\theta = \int_0^{2\pi}\Big[2r^2 - \tfrac{1}{4}r^4\Big]_0^2 d\theta = \int_0^{2\pi} 4\, d\theta = 8\pi.$$

[31] $\lim_{a\to\infty} \iint_{R_a} e^{-(x^2+y^2)}\, dA = \lim_{a\to\infty}\int_0^{2\pi}\int_0^{a} e^{-r^2}\, r\, dr\, d\theta =$

$$\lim_{a\to\infty}\int_0^{2\pi}\Big[-\tfrac{1}{2}e^{-r^2}\Big]_0^a d\theta = -\tfrac{1}{2}\lim_{a\to\infty}\left(e^{-a^2} - 1\right)\int_0^{2\pi} d\theta = -\tfrac{1}{2}(0 - 1)(2\pi) = \pi.$$

[32] (a) Since $\int_{-\infty}^{\infty} e^{-x^2}\, dx = \int_{-\infty}^{\infty} e^{-y^2}\, dy = \mathrm{I}$, we know from Exercise 31 that $\pi = \mathrm{I}\cdot\mathrm{I} = \mathrm{I}^2 \Rightarrow \mathrm{I} = \sqrt{\pi}$. This is the area of the unbounded region that lies under the graph of $y = e^{-x^2}$ and over the x-axis.

(b) $u = \frac{x}{\sqrt{2}}$ and $\sqrt{2}\, du = dx \Rightarrow$

$$\frac{1}{\sqrt{2\pi}}\int_{-\infty}^{\infty} e^{-x^2/2}\, dx = \frac{1}{\sqrt{\pi}}\int_{-\infty}^{\infty} e^{-u^2}\, du = \frac{1}{\sqrt{\pi}}\cdot\sqrt{\pi} = 1.$$

[33] $\mathrm{I} = \iint_R \sqrt{1 + (x^2 + y^2)^2}\, dA = \int_0^2\int_0^{\pi/2}\sqrt{1 + r^4}\, r\, d\theta\, dr = \frac{\pi}{2}\int_0^2\sqrt{1 + r^4}\, r\, dr$. Let $f(r) = r\sqrt{1 + r^4}$. $\mathrm{I} \approx (\frac{\pi}{2})\frac{2-0}{3(4)}\Big[f(0) + 4f(\frac{1}{2}) + 2f(1) + 4f(\frac{3}{2}) + f(2)\Big] \approx 7.3067.$

[34] $\mathrm{I} = \iint_R \sin\sqrt[3]{x^2 + y^2}\, dA = \int_0^1\int_{-\pi/2}^{\pi/2}\sin(r^{2/3})\, r\, d\theta\, dr = \pi\int_0^1 \sin(r^{2/3})\, r\, dr$. Let $f(r) = r\sin(r^{2/3})$. $\mathrm{I} \approx (\pi)\frac{1-0}{3(4)}\Big[f(0) + 4f(\frac{1}{4}) + 2f(\frac{1}{2}) + 4f(\frac{3}{4}) + f(1)\Big] \approx 1.0529.$

Exercises 17.4

Note: Let $\mathbb{S}$ denote surface area. Also, let $z = f(x, y)$.

1. $z = \sqrt{4 - x^2 - y^2} \Rightarrow f_x = \dfrac{-x}{\sqrt{4 - x^2 - y^2}}$ and $f_y = \dfrac{-y}{\sqrt{4 - x^2 - y^2}}$.

   Using the first octant portion and symmetry about the x- and y-axes,

   $\mathbb{S} = \iint_R \sqrt{[f_x(x, y)]^2 + [f_y(x, y)]^2 + 1}\, dA =$

   $$4\int_0^1\int_0^1 \sqrt{\left(\frac{-x}{\sqrt{4 - x^2 - y^2}}\right)^2 + \left(\frac{-y}{\sqrt{4 - x^2 - y^2}}\right)^2 + 1}\, dy\, dx.$$

2. $z = \sqrt{1 - x^2 + y^2} \Rightarrow f_x = \dfrac{-x}{\sqrt{1 - x^2 + y^2}}$ and $f_y = \dfrac{y}{\sqrt{1 - x^2 + y^2}}$.

   Using the first octant portion and symmetry about the x- and y-axes,

   $$\mathbb{S} = 4\int_0^1\int_0^{1-x} \sqrt{\left(\frac{-x}{\sqrt{1 - x^2 + y^2}}\right)^2 + \left(\frac{y}{\sqrt{1 - x^2 + y^2}}\right)^2 + 1}\, dy\, dx.$$

3. $z = \frac{1}{6}\sqrt{16x^2 + 9y^2 + 144} \Rightarrow$

   $$f_x = \frac{8x}{3\sqrt{16x^2 + 9y^2 + 144}} \text{ and } f_y = \frac{3y}{2\sqrt{16x^2 + 9y^2 + 144}}.$$

   Using the first octant portion and symmetry about the x- and y-axes,

   $$\mathbb{S} = 4\int_0^3\int_0^{\sqrt{9-x^2}} \sqrt{\left(\frac{8x}{3\sqrt{16x^2 + 9y^2 + 144}}\right)^2 + \left(\frac{3y}{2\sqrt{16x^2 + 9y^2 + 144}}\right)^2 + 1}\, dy\, dx.$$

4. $z = \sqrt{\dfrac{y^2}{25} - \dfrac{x^2}{9}} \Rightarrow f_x = \dfrac{-x}{9\sqrt{(y^2/25) - (x^2/9)}}$ and $f_y = \dfrac{y}{25\sqrt{(y^2/25) - (x^2/9)}}$.

   Using the first octant portion and symmetry about the y-axis,

   $$\mathbb{S} = 2\int_0^5\int_0^{(3/5)y} \sqrt{\left(\frac{-x}{9\sqrt{(y^2/25) - (x^2/9)}}\right)^2 + \left(\frac{y}{25\sqrt{(y^2/25) - (x^2/9)}}\right)^2 + 1}\, dx\, dy.$$

5. $z = y + \frac{1}{2}x^2 \Rightarrow f_x = x$ and $f_y = 1$. $\mathbb{S} = \int_0^1\int_0^1 \sqrt{x^2 + 1^2 + 1}\, dy\, dx =$

   $\int_0^1 \sqrt{x^2 + 2}\, dx = \frac{1}{2}\left[x\sqrt{x^2 + 2} + 2\ln\left(x + \sqrt{x^2 + 2}\right)\right]_0^1$ {Formula 21} $=$

   $$\tfrac{1}{2}\left[\sqrt{3} + 2\ln(1 + \sqrt{3}) - \ln 2\right] \approx 1.52.$$

6. R is bounded by the lines $x = 0$, $y = 2$, and $y = x$. $z = y^2 \Rightarrow f_x = 0$ and $f_y = 2y$.

   $$\mathbb{S} = \int_0^2\int_0^y \sqrt{(2y)^2 + 1}\, dx\, dy = \int_0^2 \sqrt{4y^2 + 1}\, y\, dy = \tfrac{1}{12}(17^{3/2} - 1) \approx 5.76.$$

7. $z = 1 - \frac{c}{a}x - \frac{c}{b}y \Rightarrow \mathbb{S} = \iint_R \sqrt{(-\frac{c}{a})^2 + (-\frac{c}{b})^2 + 1}\, dA =$

   $$c\sqrt{(\tfrac{1}{a})^2 + (\tfrac{1}{b})^2 + (\tfrac{1}{c})^2}\int_0^{2\pi}\int_0^k r\, dr\, d\theta = \pi ck^2\sqrt{(\tfrac{1}{a})^2 + (\tfrac{1}{b})^2 + (\tfrac{1}{c})^2}.$$

[8] $z = \sqrt{9 - y^2} \Rightarrow f_x = 0$ and $f_y = \dfrac{-y}{(9 - y^2)^{1/2}}$.

$$S = \int_0^3 \int_0^{\sqrt{9-y^2}} \sqrt{\frac{y^2}{9 - y^2} + 1}\, dx\, dy = \int_0^3 \int_0^{\sqrt{9-y^2}} \frac{3}{\sqrt{9 - y^2}}\, dx\, dy = 3\int_0^3 dy = 9.$$

[9] The plane $z = 1$ intersects $z = x^2 + y^2$ when $x^2 + y^2 = 1$.

Inside this circle is the region R. $S = \iint_R \sqrt{(2x)^2 + (2y)^2 + 1}\, dA =$

$$\int_0^{2\pi} \int_0^1 (4r^2 + 1)^{1/2}\, r\, dr\, d\theta = \tfrac{1}{8}\int_0^{2\pi} \tfrac{2}{3}(5^{3/2} - 1)\, d\theta = \tfrac{\pi}{6}(5^{3/2} - 1) \approx 5.33.$$

[10] $S = \iint_R \sqrt{0^2 + 1^2 + 1}\, dA = \sqrt{2}\int_0^{2\pi}\int_0^1 r\, dr\, d\theta = \sqrt{2}\int_0^{2\pi} \tfrac{1}{2}\, d\theta = \sqrt{2}\,\pi \approx 4.44.$

[11] The upper portion of the sphere has equation $z = \sqrt{a^2 - x^2 - y^2} \Rightarrow$

$f_x = \dfrac{-x}{(a^2 - x^2 - y^2)^{1/2}}$ and $f_y = \dfrac{-y}{(a^2 - x^2 - y^2)^{1/2}}$. Thus, by symmetry about the

xy-plane, $S = 2\iint_R \dfrac{a}{(a^2 - x^2 - y^2)^{1/2}}\, dA$. Changing to polar coordinates,

$x^2 + y^2 = ay \Leftrightarrow r = a\sin\theta$, and $S = 2a\int_0^{\pi}\int_0^{a\sin\theta} (a^2 - r^2)^{-1/2}\, r\, dr\, d\theta =$

$$2a\int_0^{\pi} \Big[-(a^2 - r^2)^{1/2}\Big]_0^{a\sin\theta} d\theta = -2a^2\int_0^{\pi} (|\cos\theta| - 1)\, d\theta =$$

$$2(-2a^2)\int_0^{\pi/2} (\cos\theta - 1)\, d\theta \;\{\text{by symmetry}\} = -4a^2\Big[\sin\theta - \theta\Big]_0^{\pi/2} =$$

$$-4a^2(1 - \tfrac{\pi}{2}) = 2a^2(\pi - 2).$$

[12] In the first octant, $z \geq 0$ so $x \geq y \geq 0$, or $0 \leq \theta \leq \frac{\pi}{4}$.

$$S = \iint_R \sqrt{(2x)^2 + (-2y)^2 + 1}\, dA = \int_0^{\pi/4}\int_0^1 \sqrt{4r^2 + 1}\, r\, dr\, d\theta =$$

$$\tfrac{1}{8}\int_0^{\pi/4} \tfrac{2}{3}(5^{3/2} - 1)\, d\theta = \tfrac{\pi}{48}(5^{3/2} - 1) \approx 0.67.$$

[13] $S_{\text{roof}} = \iint_R \sqrt{(-\frac{14}{25}x)^2 + (-\frac{14}{25}y)^2 + 1}\, dA = \iint_R \sqrt{1 + \frac{196}{625}(x^2 + y^2)}\, dA$

$$= \int_0^{2\pi}\int_0^5 \sqrt{1 + \tfrac{196}{625}r^2}\, r\, dr\, d\theta = 2\pi\Big[\tfrac{2}{3}(\tfrac{625}{2\cdot 196})(1 + \tfrac{196}{625}r^2)^{3/2}\Big]_0^5$$

$$= \tfrac{625\pi}{294}\Big[(8.84)^{3/2} - 1\Big] \approx 168.9 \text{ ft}^2.$$

The area of floor is $\pi r^2 = \pi(5)^2 = 25\pi \approx 78.5 \text{ ft}^2$.

The total amount of material is approximately $168.9 + 78.5 = 247.4 \text{ ft}^2$.

[14] $A = \text{Area}_{\text{sphere}} - \text{Area}_{\text{cutout}} = 4\pi(1)^2 - \iint_R (1 - x^2 - y^2)^{-1/2}\, dA =$

$$4\pi - \int_0^{2\pi}\int_0^{1/4} (1 - r^2)^{-1/2}\, r\, dr\, d\theta = 4\pi - 2\pi(1 - \tfrac{1}{4}\sqrt{15}) \approx 12.37 \text{ m}^2.$$

$$C = \delta A \approx (1.2 \times 10^{-5})(12.37) \approx 1.48 \times 10^{-4} \text{ coulomb}$$

[15] $z = \frac{1}{3}x^3 + \cos y \Rightarrow f_x = x^2$ and $f_y = -\sin y \Rightarrow S = \int_0^1 \int_0^1 \sqrt{x^4 + \sin^2 y + 1}\, dx\, dy.$

Let $F(x, y) = \sqrt{x^4 + \sin^2 y + 1}$. $G(y) = \int_0^1 \sqrt{x^4 + \sin^2 y + 1}\, dx$, and $S = \int_0^1 G(y)\, dy.$

$S \approx \frac{1-0}{2(2)}\left[G(0) + 2G(\frac{1}{2}) + G(1)\right].$

$G(0) = \int_0^1 F(x, 0)\, dx = \int_0^1 \sqrt{x^4 + 1}\, dx \approx$

$\frac{1-0}{2(2)}\left[F(0, 0) + 2F(\frac{1}{2}, 0) + F(1, 0)\right] \approx 1.1189.$

$G(\frac{1}{2}) = \int_0^1 F(x, \frac{1}{2})\, dx = \int_0^1 \sqrt{x^4 + \sin^2(\frac{1}{2}) + 1}\, dx \approx$

$\frac{1-0}{2(2)}\left[F(0, \frac{1}{2}) + 2F(\frac{1}{2}, \frac{1}{2}) + F(1, \frac{1}{2})\right] \approx 1.2190.$

$G(1) = \int_0^1 F(x, 1)\, dx = \int_0^1 \sqrt{x^4 + \sin^2 1 + 1}\, dx \approx$

$\frac{1-0}{2(2)}\left[F(0, 1) + 2F(\frac{1}{2}, 1) + F(1, 1)\right] \approx 1.4035.$

Thus, $S \approx \frac{1-0}{2(2)}\left[1.1189 + 2(1.2190) + 1.4035\right] \approx 1.2401.$

[16] $z = x^3 y \Rightarrow f_x = 3x^2 y$ and $f_y = x^3 \Rightarrow S = \int_{1/4}^{3/4} \int_{1/4}^{3/4} \sqrt{9x^4 y^2 + x^6 + 1}\, dx\, dy.$

Let $F(x, y) = \sqrt{9x^4 y^2 + x^6 + 1}$, $G(y) = \int_{1/4}^{3/4} \sqrt{9x^4 y^2 + x^6 + 1}\, dx$, and

$S = \int_{1/4}^{3/4} G(y)\, dy$. $S \approx \frac{3/4 - 1/4}{2(2)}\left[G(\frac{1}{4}) + 2G(\frac{1}{2}) + G(\frac{3}{4})\right].$

$G(\frac{1}{4}) = \int_{1/4}^{3/4} F(x, \frac{1}{4})\, dx = \int_{1/4}^{3/4} \sqrt{\frac{9}{16}x^4 + x^6 + 1}\, dx \approx$

$\frac{3/4 - 1/4}{2(2)}\left[F(\frac{1}{4}, \frac{1}{4}) + 2F(\frac{1}{2}, \frac{1}{4}) + F(\frac{3}{4}, \frac{1}{4})\right] \approx 0.5270.$

$G(\frac{1}{2}) = \int_{1/4}^{3/4} F(x, \frac{1}{2})\, dx = \int_{1/4}^{3/4} \sqrt{\frac{9}{4}x^4 + x^6 + 1}\, dx \approx$

$\frac{3/4 - 1/4}{2(2)}\left[F(\frac{1}{4}, \frac{1}{2}) + 2F(\frac{1}{2}, \frac{1}{2}) + F(\frac{3}{4}, \frac{1}{2})\right] \approx 0.5662.$

$G(\frac{3}{4}) = \int_{1/4}^{3/4} F(x, \frac{3}{4})\, dx = \int_{1/4}^{3/4} \sqrt{\frac{81}{16}x^4 + x^6 + 1}\, dx \approx$

$\frac{3/4 - 1/4}{2(2)}\left[F(\frac{1}{4}, \frac{3}{4}) + 2F(\frac{1}{2}, \frac{3}{4}) + F(\frac{3}{4}, \frac{3}{4})\right] \approx 0.6232.$

Thus, $S \approx \frac{3/4 - 1/4}{2(2)}\left[0.5269786 + 2(0.5662277) + 0.6231878\right] \approx 0.2853.$

Exercises 17.5

[1] $\int_0^3 \int_{-1}^0 \int_1^2 (x + 2y + 4z)\, dx\, dy\, dz = \int_0^3 \int_{-1}^0 (\frac{3}{2} + 2y + 4z)\, dy\, dz = \int_0^3 (\frac{1}{2} + 4z)\, dz = \frac{39}{2}$

[2] $\int_0^1 \int_{-1}^2 \int_1^3 (6x^2 z + 5xy^2)\, dz\, dx\, dy = \int_0^1 \int_{-1}^2 (24x^2 + 10xy^2)\, dx\, dy = \int_0^1 (72 + 15y^2)\, dy =$

77

[3] $\int_0^1 \int_{x+1}^{2x} \int_z^{x+z} x\, dy\, dz\, dx = \int_0^1 \int_{x+1}^{2x} x^2\, dz\, dx = \int_0^1 (x^3 - x^2)\, dx = -\frac{1}{12}$

[4] $\int_1^2 \int_0^{z^2} \int_{x+z}^{x-z} z\, dy\, dx\, dz = \int_1^2 \int_0^{z^2} (-2z^2)\, dx\, dz = \int_1^2 (-2z^4)\, dz = -\frac{62}{5}$

[5] $\int_{-1}^{2}\int_{1}^{x^2}\int_{0}^{x+y} 2x^2y\,dz\,dy\,dx = \int_{-1}^{2}\int_{1}^{x^2}(2x^3y + 2x^2y^2)\,dy\,dx =$

$$\int_{-1}^{2}(x^7 + \tfrac{2}{3}x^8 - x^3 - \tfrac{2}{3}x^2)\,dx = \tfrac{513}{8} = 64.125$$

[6] $\int_{2}^{3}\int_{0}^{3y}\int_{1}^{yz}(2x + y + z)\,dx\,dz\,dy = \int_{2}^{3}\int_{0}^{3y}(y^2z^2 + y^2z + yz^2 - 1 - y - z)\,dz\,dy =$

$$\int_{2}^{3}(9y^5 + \tfrac{27}{2}y^4 - \tfrac{15}{2}y^2 - 3y)\,dy = \tfrac{7561}{5} = 1512.2$$

[7] (1) <u>z integrated first</u> The trace in the xy-plane is $x + 2y = 6$.

$$\int_{0}^{6}\int_{0}^{(6-x)/2}\int_{0}^{(6-x-2y)/3} f(x,\,y,\,z)\,dz\,dy\,dx;\quad \int_{0}^{3}\int_{0}^{6-2y}\int_{0}^{(6-x-2y)/3} f(x,\,y,\,z)\,dz\,dx\,dy$$

(2) <u>x integrated first</u> The trace in the yz-plane is $2y + 3z = 6$.

$$\int_{0}^{3}\int_{0}^{(6-2y)/3}\int_{0}^{6-2y-3z} f(x,\,y,\,z)\,dx\,dz\,dy;\quad \int_{0}^{2}\int_{0}^{(6-3z)/2}\int_{0}^{6-2y-3z} f(x,\,y,\,z)\,dx\,dy\,dz$$

(3) <u>y integrated first</u> The trace in the xz-plane is $x + 3z = 6$.

$$\int_{0}^{6}\int_{0}^{(6-x)/3}\int_{0}^{(6-x-3z)/2} f(x,\,y,\,z)\,dy\,dz\,dx;\quad \int_{0}^{2}\int_{0}^{6-3z}\int_{0}^{(6-x-3z)/2} f(x,\,y,\,z)\,dy\,dx\,dz$$

[8] (1) <u>z integrated first</u> The trace in the xy-plane is $x^2 + y^2 = 9$.

$$\int_{-3}^{3}\int_{-\sqrt{9-x^2}}^{\sqrt{9-x^2}}\int_{0}^{2} f(x,\,y,\,z)\,dz\,dy\,dx;\quad \int_{-3}^{3}\int_{-\sqrt{9-y^2}}^{\sqrt{9-y^2}}\int_{0}^{2} f(x,\,y,\,z)\,dz\,dx\,dy$$

(2) <u>x integrated first</u>

The trace in the yz-plane is the rectangle $-3 \le y \le 3$ and $0 \le z \le 2$.

$$\int_{-3}^{3}\int_{0}^{2}\int_{-\sqrt{9-y^2}}^{\sqrt{9-y^2}} f(x,\,y,\,z)\,dx\,dz\,dy;\quad \int_{0}^{2}\int_{-3}^{3}\int_{-\sqrt{9-y^2}}^{\sqrt{9-y^2}} f(x,\,y,\,z)\,dx\,dy\,dz$$

(3) <u>y integrated first</u>

The trace in the xz-plane is the rectangle $-3 \le x \le 3$ and $0 \le z \le 2$.

$$\int_{-3}^{3}\int_{0}^{2}\int_{-\sqrt{9-x^2}}^{\sqrt{9-x^2}} f(x,\,y,\,z)\,dy\,dz\,dx;\quad \int_{0}^{2}\int_{-3}^{3}\int_{-\sqrt{9-x^2}}^{\sqrt{9-x^2}} f(x,\,y,\,z)\,dy\,dx\,dz$$

[9] (1) <u>z integrated first</u> The trace in the xy-plane is $4x^2 + y^2 = 9$.

$$\int_{-3/2}^{3/2}\int_{-\sqrt{9-4x^2}}^{\sqrt{9-4x^2}}\int_{0}^{9-4x^2-y^2} f(x,\,y,\,z)\,dz\,dy\,dx;$$

$$\int_{-3}^{3}\int_{-\sqrt{9-y^2}/2}^{\sqrt{9-y^2}/2}\int_{0}^{9-4x^2-y^2} f(x,\,y,\,z)\,dz\,dx\,dy$$

(2) <u>x integrated first</u> The trace in the yz-plane is $z + y^2 = 9$.

$$\int_{-3}^{3}\int_{0}^{9-y^2}\int_{-\sqrt{9-z-y^2}/2}^{\sqrt{9-z-y^2}/2} f(x,\,y,\,z)\,dx\,dz\,dy;\quad \int_{0}^{9}\int_{-\sqrt{9-z}}^{\sqrt{9-z}}\int_{-\sqrt{9-z-y^2}/2}^{\sqrt{9-z-y^2}/2} f(x,\,y,\,z)\,dx\,dy\,dz$$

(3) <u>y integrated first</u> The trace in the xz-plane is $z + 4x^2 = 9$.

$$\int_{-3/2}^{3/2}\int_{0}^{9-4x^2}\int_{-\sqrt{9-4x^2-z}}^{\sqrt{9-4x^2-z}} f(x,\,y,\,z)\,dy\,dz\,dx;$$

$$\int_{0}^{9}\int_{-\sqrt{9-z}/2}^{\sqrt{9-z}/2}\int_{-\sqrt{9-4x^2-z}}^{\sqrt{9-4x^2-z}} f(x,\,y,\,z)\,dy\,dx\,dz$$

10 (1) <u>z integrated first</u> The trace in the xy-plane is $36x^2 + 9y^2 = 36$.

$$\int_{-1}^{1}\int_{-2\sqrt{1-x^2}}^{2\sqrt{1-x^2}}\int_{-\sqrt{36-36x^2-9y^2}/2}^{\sqrt{36-36x^2-9y^2}/2} f(x,\ y,\ z)\,dz\,dy\,dx;$$

$$\int_{-2}^{2}\int_{-\sqrt{4-y^2}/2}^{\sqrt{4-y^2}/2}\int_{-\sqrt{36-36x^2-9y^2}/2}^{\sqrt{36-36x^2-9y^2}/2} f(x,\ y,\ z)\,dz\,dx\,dy$$

(2) <u>x integrated first</u> The trace in the yz-plane is $9y^2 + 4z^2 = 36$.

$$\int_{-2}^{2}\int_{-\sqrt{36-9y^2}/2}^{\sqrt{36-9y^2}/2}\int_{-\sqrt{36-9y^2-4z^2}/6}^{\sqrt{36-9y^2-4z^2}/6} f(x,\ y,\ z)\,dx\,dz\,dy;$$

$$\int_{-3}^{3}\int_{-\sqrt{36-4z^2}/3}^{\sqrt{36-4z^2}/3}\int_{-\sqrt{36-9y^2-4z^2}/6}^{\sqrt{36-9y^2-4z^2}/6} f(x,\ y,\ z)\,dx\,dy\,dz$$

(3) <u>y integrated first</u> The trace in the xz-plane is $36x^2 + 4z^2 = 36$.

$$\int_{-1}^{1}\int_{-3\sqrt{1-x^2}}^{3\sqrt{1-x^2}}\int_{-\sqrt{36-36x^2-4z^2}/3}^{\sqrt{36-36x^2-4z^2}/3} f(x,\ y,\ z)\,dy\,dz\,dx;$$

$$\int_{-3}^{3}\int_{-\sqrt{9-z^2}/3}^{\sqrt{9-z^2}/3}\int_{-\sqrt{36-36x^2-4z^2}/3}^{\sqrt{36-36x^2-4z^2}/3} f(x,\ y,\ z)\,dy\,dx\,dz$$

11 We will double the volume in the first octant. $V = 2\int_{0}^{4}\int_{0}^{4-y}\int_{0}^{\sqrt{4-z}} dx\,dz\,dy =$

$$2\int_{0}^{4}\int_{0}^{4-y} (4 - z)^{1/2}\,dz\,dy = -\tfrac{4}{3}\int_{0}^{4} (y^{3/2} - 8)\,dy = -\tfrac{4}{3}(\tfrac{64}{5} - 32) = \tfrac{128}{5}.$$

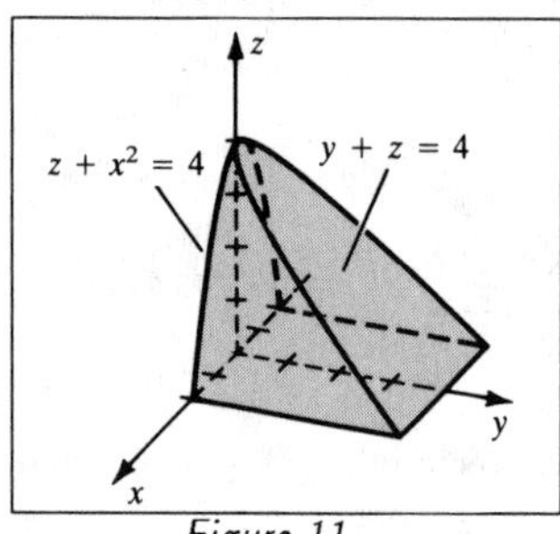

Figure 11

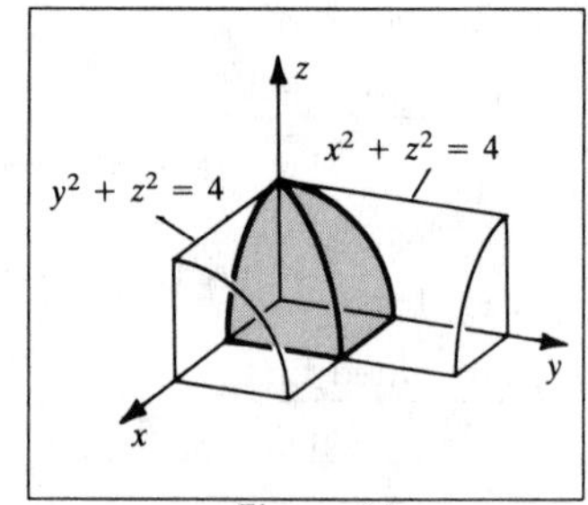

Figure 12

12 We will take 8 times the volume in the first octant.

$$V = 8\int_{0}^{2}\int_{0}^{\sqrt{4-z^2}}\int_{0}^{\sqrt{4-z^2}} dy\,dx\,dz = 8\int_{0}^{2} (4 - z^2)\,dz = \tfrac{128}{3}.$$

13 $2 - z^2 = z^2$ at $z = \pm 1$. We will double the volume over $0 \le y \le 1$.

$$V = 2\int_{-1}^{1}\int_{z^2}^{1}\int_{0}^{4-z} dx\,dy\,dz = 2\int_{-1}^{1}\int_{z^2}^{1} (4 - z)\,dy\,dz = 2\int_{-1}^{1} (4 - z - 4z^2 + z^3)\,dz =$$

$$2 \cdot 2\int_{0}^{1} (4 - 4z^2)\,dz = 4 \cdot 4\int_{0}^{1} (1 - z^2)\,dz = 16(\tfrac{2}{3}) = \tfrac{32}{3}.$$

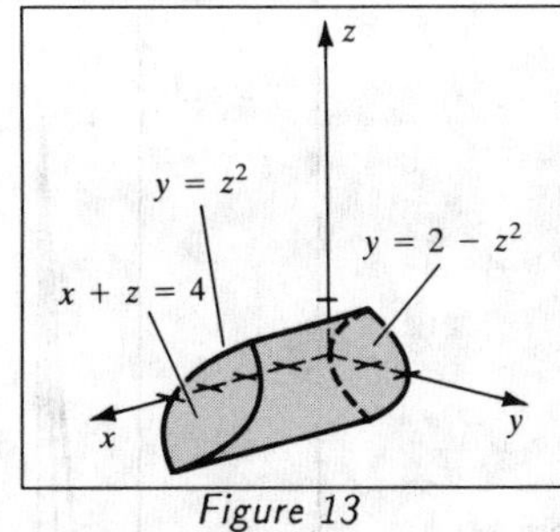

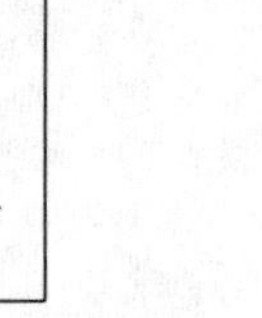

Figure 13

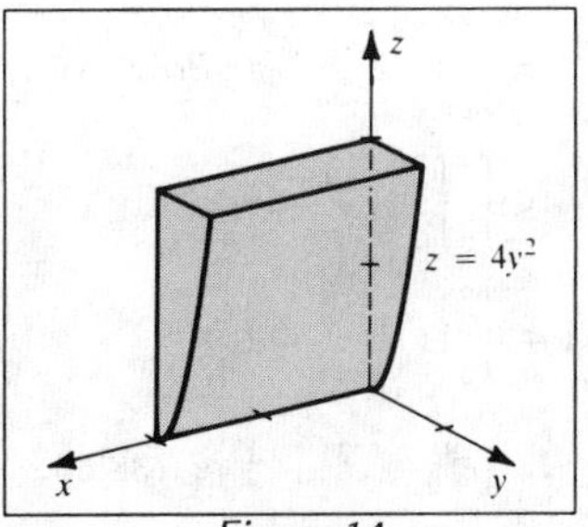

Figure 14

[14] We will double the volume in the first octant.

$$V = 2\int_0^2\int_0^{1/\sqrt{2}}\int_{4y^2}^{2} dz\,dy\,dx = 2\int_0^2\int_0^{1/\sqrt{2}}(2 - 4y^2)\,dy\,dx = 4\int_0^2 \frac{\sqrt{2}}{3}\,dx = \tfrac{8}{3}\sqrt{2} \approx 3.77.$$

[15] $$V = \int_{-1}^{1}\int_{-\sqrt{1-y^2}}^{\sqrt{1-y^2}}\int_0^{2-y-z} dx\,dz\,dy = \int_{-1}^{1}\int_{-\sqrt{1-y^2}}^{\sqrt{1-y^2}}(2 - y - z)\,dz\,dy =$$

$$4\int_{-1}^{1}\sqrt{1 - y^2}\,dy - 2\int_{-1}^{1} y\sqrt{1 - y^2}\,dy.$$

The integrand of the second integral is odd, so the integral is equal to zero. The first integral is 4 times the area of a semicircle with radius 1. Thus, $V = 4(\frac{1}{2}\pi \cdot 1^2) = 2\pi$.

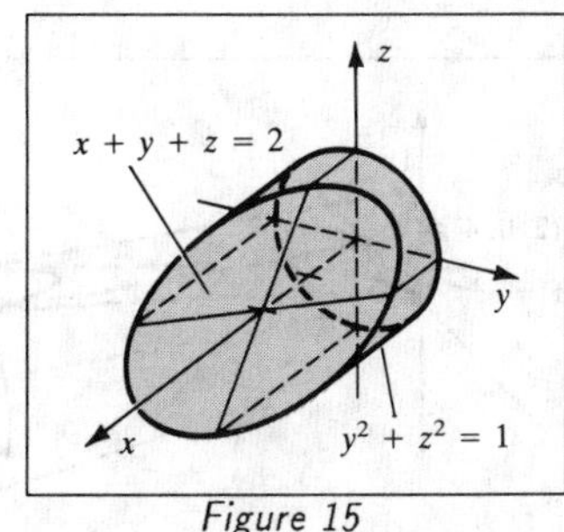

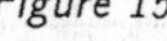

Figure 15

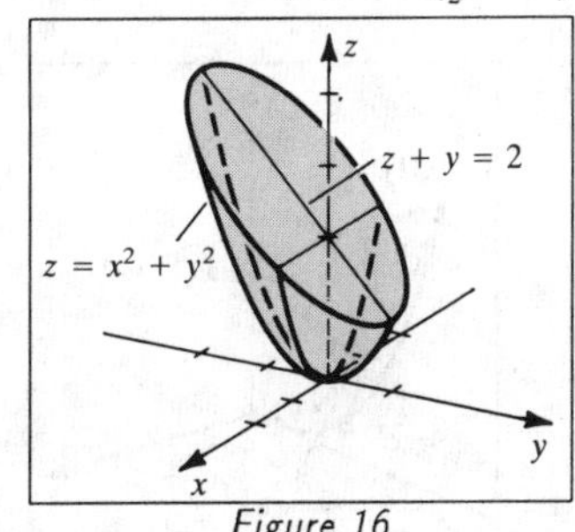

Figure 16

[16] We will double the volume of the region from the front to the yz-plane.

$$V = 2\int_{-2}^{1}\int_{y^2}^{2-y}\int_0^{\sqrt{z-y^2}} dx\,dz\,dy = 2\int_{-2}^{1}\int_{y^2}^{2-y}\sqrt{z - y^2}\,dz\,dy$$

$$= \tfrac{4}{3}\int_{-2}^{1}(2 - y - y^2)^{3/2}\,dy = \tfrac{4}{3}\int_{-2}^{1}\left[\tfrac{9}{4} - (y + \tfrac{1}{2})^2\right]^{3/2} dy.$$ Using Formula 37 with

$u = y + \frac{1}{2}$, $du = dy$, and $a = \frac{3}{2}$, $V = \frac{4}{3}\int_{-3/2}^{3/2}\left[(\frac{3}{2})^2 - u^2\right]^{3/2} du = \frac{81\pi}{32} \approx 7.95.$

[17] $$V = 2\int_0^3\int_{-1}^{2}\int_0^{9-x^2} dz\,dy\,dx$$

$$= 2\int_0^3\int_{-1}^{2}(9 - x^2)\,dy\,dx$$

$$= 6\int_0^3 (9 - x^2)\,dx = 108.$$

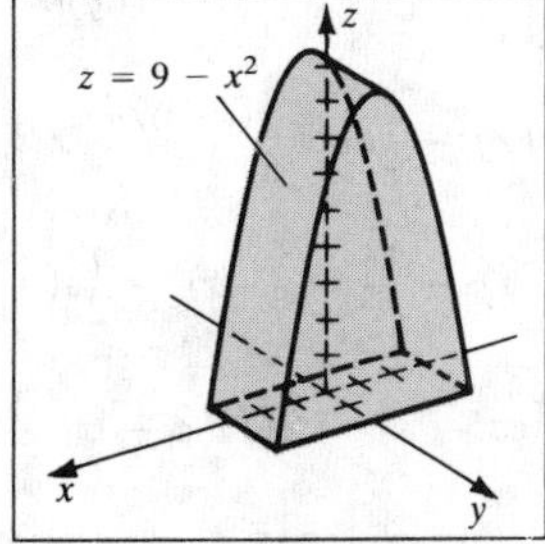

Figure 17

[18] $V = \int_0^2\int_0^{3x}\int_0^{e^{x+y}} dz\,dy\,dx$

$= \int_0^2\int_0^{3x} e^{x+y}\,dy\,dx$

$= \int_0^2 (e^{4x} - e^x)\,dx$

$= \frac{1}{4}e^8 - e^2 + \frac{3}{4} \approx 738.60.$

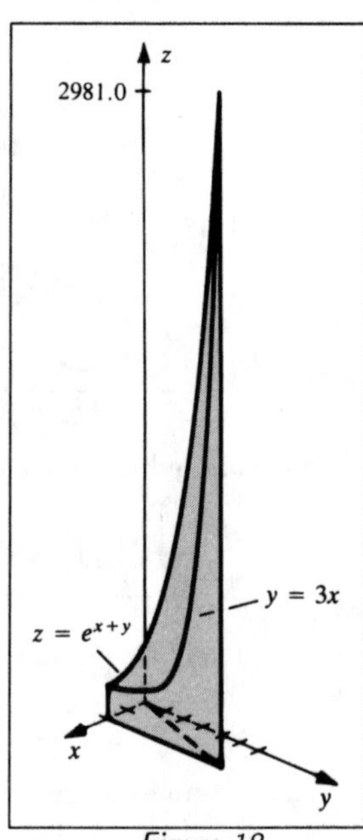

Figure 18

[19] The solid lies between the two parallel cylinders $z = x^2$ and $z = x^3$ from $y = 0$ to $y = z^2$. Thus,

$V = \int_0^1\int_{x^3}^{x^2}\int_0^{z^2} dy\,dz\,dx$

$= \int_0^1\int_{x^3}^{x^2} z^2\,dz\,dx$

$= \frac{1}{3}\int_0^1 (x^6 - x^9)\,dx = \frac{1}{70}.$

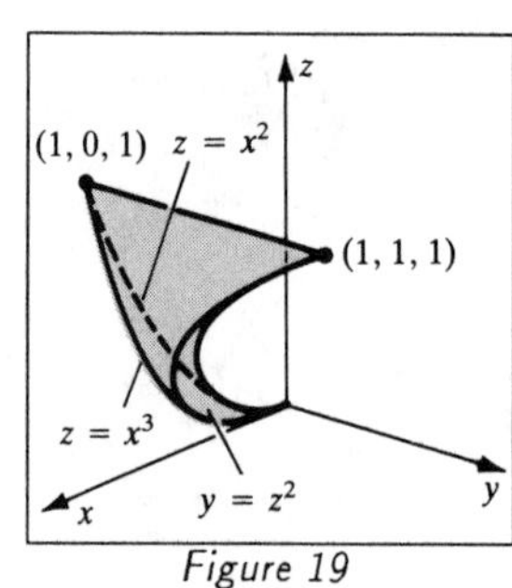

Figure 19

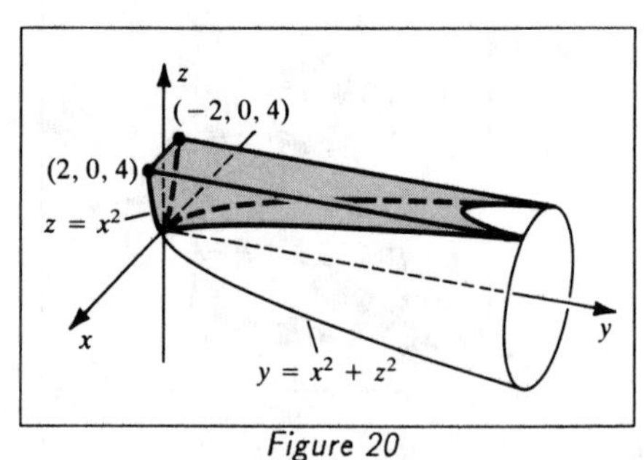

Figure 20

[20] $V = 2\int_0^2\int_{x^2}^4\int_0^{x^2+z^2} dy\,dz\,dx = 2\int_0^2\int_{x^2}^4 (x^2 + z^2)\,dz\,dx =$

$2\int_0^2 (4x^2 + \frac{64}{3} - x^4 - \frac{1}{3}x^6)\,dx = \frac{8576}{105} \approx 81.68.$

[21] $\frac{x}{a} + \frac{y}{b} + \frac{z}{c} = 1 \Rightarrow bcx + acy + abz = abc \Rightarrow z = \dfrac{c(ab - bx - ay)}{ab}.$

$V = \int_0^a\int_0^{b(a-x)/a}\int_0^{c(ab-bx-ay)/(ab)} dz\,dy\,dx = \frac{c}{ab}\int_0^a\int_0^{b(a-x)/a} (ab - bx - ay)\,dy\,dx$

$= \frac{c}{ab}\int_0^a \Big[aby - bxy - \frac{1}{2}ay^2\Big]_0^{b(a-x)/a} dx$

$= \frac{c}{ab}\int_0^a \Big[b^2(a - x) - \frac{b^2}{a}x(a - x) - \frac{b^2}{2a}(a^2 - 2ax + x^2)\Big]\,dx$

$= \frac{c}{2a^2b}\int_0^a \Big[2ab^2(a - x) - 2b^2x(a - x) - b^2(a^2 - 2ax + x^2)\Big]\,dx$

$= \frac{c}{2a^2b}\Big[a^2b^2x - ab^2x^2 + \frac{1}{3}b^2x^3\Big]_0^a = \frac{c}{2a^2b}\cdot\frac{a^3b^2}{3} = \frac{abc}{6}$

[22] $\frac{x^2}{a^2} + \frac{y^2}{b^2} + \frac{z^2}{c^2} = 1 \Rightarrow b^2c^2x^2 + a^2c^2y^2 + a^2b^2z^2 = a^2b^2c^2 \Rightarrow$

$$z^2 = \frac{1}{a^2b^2} \cdot c^2(a^2b^2 - b^2x^2 - a^2y^2) \Rightarrow z = \pm \frac{c}{ab}\sqrt{a^2b^2 - b^2x^2 - a^2y^2}.$$

$$V = 8\int_0^a \int_0^{(b/a)\sqrt{a^2-x^2}} \int_0^{[c/(ab)]\sqrt{a^2b^2-b^2x^2-a^2y^2}} dz\,dy\,dx$$

$$= \frac{8c}{ab}\int_0^a \int_0^{(b/a)\sqrt{a^2-x^2}} \sqrt{a^2b^2 - b^2x^2 - a^2y^2}\,dy\,dx$$

$$= \frac{8c}{b}\int_0^a \int_0^{(b/a)\sqrt{a^2-x^2}} \sqrt{\left(b^2 - \frac{b^2}{a^2}x^2\right) - y^2}\,dy\,dx \left\{\text{let A} = b^2 - \frac{b^2}{a^2}x^2\right\}$$

$$= \frac{8c}{b}\int_0^a \left[\frac{y}{2}\sqrt{\text{A} - y^2} + \frac{\text{A}}{2}\sin^{-1}\frac{y}{\sqrt{\text{A}}}\right]_0^{(b/a)\sqrt{a^2-x^2}} dx \text{ \{Formula 30\}}$$

$$= \frac{8c}{b}\int_0^a \left[\frac{1}{2}\left(b^2 - \frac{b^2}{a^2}x^2\right)\sin^{-1}1\right] dx$$

$$= \frac{8c}{b}\int_0^a \frac{b^2}{2a^2} \cdot \frac{\pi}{2}(a^2 - x^2)\,dx = \frac{2\pi bc}{a^2} \cdot \frac{2a^3}{3} = \frac{4}{3}\pi abc$$

[23] Q is the region bounded by the planes $z = 0$, $z = 1$, $x = 2$, $x = 3$ and the cylinders $y = \sqrt{1 - z}$ and $y = \sqrt{4 - z}$.

[24] Q is the region bounded by the planes $y = 0$ and $y = 4 - x$ and the cylinders $x = z^3$ and $x = \sqrt{z}$.

[25] Q is the region under the plane $z = x + y$ and over the region in the xy-plane bounded by the parabola $y = x^2$ and the line $y = 2x$.

[26] Q is the region under the surface $z = xy$ and over the triangular region in the xy-plane bounded by the lines $y = x$, $y = 3x$, and $x = 1$.

[27] Q is the region bounded by the paraboloid $z = x^2 + y^2$ and the planes $z = 1$ and $z = 2$.

[28] Q is the region bounded by the cone $z^2 = x^2 + y^2$ and the planes $z = 1$ and $z = 4$.

[29] $m = \iint_R \delta(x, y)\,dA = \int_0^1 \int_0^{e^{-x}} y^2\,dy\,dx.$

[30] $m = \iint_R \delta(x, y)\,dA = \int_1^2 \int_0^{1/y^2} (x^2 + y^2)\,dx\,dy$

[31] $m = \iiint_Q \delta(x, y, z)\,dV = \int_0^2 \int_0^{4-2y} \int_0^{4-x-2y} (x^2 + y^2)\,dz\,dx\,dy$

[32] *Note*: Since the density cannot be negative, we will change ky to $z + 1$.

$$m = \iiint_Q \delta(x, y, z)\,dV = \int_{-2}^2 \int_{-\sqrt{4-x^2}}^{\sqrt{4-x^2}} \int_0^{4-x^2-y^2} (z + 1)\,dz\,dy\,dx$$

[33] $m = \int_0^{1000} \int_0^{1000} \int_0^{1000} (1.225 - 0.000113z)\,dz\,dy\,dx = 1.1685 \times 10^9$ kg.

[34] $P = \frac{1}{A}\iiint_Q \delta g\,dV = \frac{1}{1}\int_0^1 \int_0^1 \int_0^{30,000} \delta g\,dz\,dy\,dx = 10{,}170\,g \approx 99{,}666$ N.

[35] $\Delta z_k = \Delta y_j = \Delta x_i = \frac{1}{2} \Rightarrow \Delta z_k\,\Delta y_j\,\Delta x_i = \frac{1}{8}$.

Now, $u_1 = v_1 = w_1 = \frac{1}{4}$ and $u_2 = v_2 = w_2 = \frac{3}{4}$.

$$m \approx \sum_{i=1}^{2}\sum_{j=1}^{2}\sum_{k=1}^{2} \delta(u_i, v_j, w_k)\,\Delta z_k\,\Delta y_j\,\Delta x_i$$
$$= \Big(\delta(\tfrac{1}{4}, \tfrac{1}{4}, \tfrac{1}{4}) + \delta(\tfrac{1}{4}, \tfrac{1}{4}, \tfrac{3}{4}) + \delta(\tfrac{1}{4}, \tfrac{3}{4}, \tfrac{1}{4}) + \delta(\tfrac{1}{4}, \tfrac{3}{4}, \tfrac{3}{4}) + \delta(\tfrac{3}{4}, \tfrac{1}{4}, \tfrac{1}{4})$$
$$+ \delta(\tfrac{3}{4}, \tfrac{1}{4}, \tfrac{3}{4}) + \delta(\tfrac{3}{4}, \tfrac{3}{4}, \tfrac{1}{4}) + \delta(\tfrac{3}{4}, \tfrac{3}{4}, \tfrac{3}{4})\Big)(\tfrac{1}{8}) \approx 6.1420(\tfrac{1}{8}) \approx 0.7678$$

[36] $\Delta z_k = \Delta y_j = \Delta x_i = \frac{1}{4} \Rightarrow \Delta z_k\,\Delta y_j\,\Delta x_i = \frac{1}{64}$. Now, $u_1 = v_1 = w_1 = \frac{1}{8}$,

$u_2 = v_2 = w_2 = \frac{3}{8}$, $u_3 = v_3 = w_3 = \frac{5}{8}$, and $u_4 = v_4 = w_4 = \frac{7}{8}$.

$$m \approx \sum_{i=1}^{4}\sum_{j=1}^{4}\sum_{k=1}^{4} \delta(u_i, v_j, w_k)\,\Delta z_k\,\Delta y_j\,\Delta x_i$$
$$= \Big(\delta(\tfrac{1}{8}, \tfrac{1}{8}, \tfrac{1}{8}) + \delta(\tfrac{1}{8}, \tfrac{1}{8}, \tfrac{3}{8}) + \delta(\tfrac{1}{8}, \tfrac{1}{8}, \tfrac{5}{8}) + \delta(\tfrac{1}{8}, \tfrac{1}{8}, \tfrac{7}{8}) + \delta(\tfrac{1}{8}, \tfrac{3}{8}, \tfrac{1}{8})$$
$$+ \cdots + \delta(\tfrac{7}{8}, \tfrac{7}{8}, \tfrac{5}{8}) + \delta(\tfrac{7}{8}, \tfrac{7}{8}, \tfrac{7}{8})\Big)(\tfrac{1}{64}) \approx 14.4353(\tfrac{1}{64}) \approx 0.2256$$

Exercises 17.6

Note: Let k be a constant of proportionality.

[1] $m = \displaystyle\int_0^9\int_0^{\sqrt{x}} (x + y)\,dy\,dx = \int_0^9 (x^{3/2} + \tfrac{1}{2}x)\,dx = \tfrac{2349}{20}$.

$M_y = \displaystyle\int_0^9\int_0^{\sqrt{x}} x(x + y)\,dy\,dx = \int_0^9 (x^{5/2} + \tfrac{1}{2}x^2)\,dx = \tfrac{31,347}{42}$.

$M_x = \displaystyle\int_0^9\int_0^{\sqrt{x}} y(x + y)\,dy\,dx = \int_0^9 (\tfrac{1}{2}x^2 + \tfrac{1}{3}x^{3/2})\,dx = \tfrac{1539}{10}$.

Thus, $\bar{x} = \dfrac{M_y}{m} = \tfrac{1290}{203}$ and $\bar{y} = \dfrac{M_x}{m} = \tfrac{38}{29}$.

[2] $m = \displaystyle\int_0^8\int_0^{x^{1/3}} (y^2)\,dy\,dx = \int_0^8 (\tfrac{1}{3}x)\,dx = \tfrac{32}{3}$.

$M_y = \displaystyle\int_0^8\int_0^{x^{1/3}} x(y^2)\,dy\,dx = \int_0^8 (\tfrac{1}{3}x^2)\,dx = \tfrac{512}{9}$.

$M_x = \displaystyle\int_0^8\int_0^{x^{1/3}} y(y^2)\,dy\,dx = \int_0^8 (\tfrac{1}{4}x^{4/3})\,dx = \tfrac{96}{7}$.

Thus, $\bar{x} = \dfrac{M_y}{m} = \tfrac{16}{3}$ and $\bar{y} = \dfrac{M_x}{m} = \tfrac{9}{7}$.

[3] By symmetry, we can double the mass in the first quadrant to find m.

$m = 2\displaystyle\int_0^2\int_{x^2}^4 kx\,dy\,dx = 2k\int_0^2 (4x - x^3)\,dx = 8k$.

$M_y = \displaystyle\int_{-2}^2\int_{x^2}^4 x\,k|x|\,dy\,dx = \int_{-2}^2 kx|x|(4 - x^2)\,dx = 0$

{the integrand is odd since $|x|$ is even, and odd · even · even is odd}.

$M_x = \displaystyle\int_{-2}^2\int_{x^2}^4 y\,k|x|\,dy\,dx = \int_{-2}^2 \tfrac{1}{2}(16 - x^4)\,k|x|\,dx = k\int_0^2 (16 - x^4)\,x\,dx = \tfrac{64}{3}k$.

Thus, $\bar{x} = \dfrac{M_y}{m} = 0$ and $\bar{y} = \dfrac{M_x}{m} = \tfrac{8}{3}$.

[4] $x^3 = 2x$ at $x = \pm\sqrt{2}$ and $\delta(x, y) = k|y|$. We can double the mass in the first quadrant to find m. $m = 2\int_0^{\sqrt{2}}\int_{x^3}^{2x}(ky)\,dy\,dx = k\int_0^{\sqrt{2}}(4x^2 - x^6)\,dx = \frac{32}{21}\sqrt{2}\,k.$

$M_y = \int_{-\sqrt{2}}^{0}\int_{2x}^{x^3} x(-ky)\,dy\,dx + \int_0^{\sqrt{2}}\int_{x^3}^{2x} x(ky)\,dy\,dx = \int_{-\sqrt{2}}^{\sqrt{2}} \frac{1}{2}k(4x^3 - x^7)\,dx = 0$

since the integrand is odd. $M_x = \int_{-\sqrt{2}}^{0}\int_{2x}^{x^3}(-ky^2)\,dy\,dx + \int_0^{\sqrt{2}}\int_{x^3}^{2x} ky^2\,dy\,dx =$

$\frac{1}{3}k\int_{-\sqrt{2}}^{\sqrt{2}}(8x^3 - x^9)\,dx = 0$ since the integrand is odd.

Thus, $\bar{x} = \frac{M_y}{m} = 0$ and $\bar{y} = \frac{M_x}{m} = 0.$

[5] *Note:* $y \geq 0 \Rightarrow \delta(x, y) = |xy| = |x|\,y.$

$m = \int_{-1}^{1}\int_0^{e^{-x^2}} |x|\,y\,dy\,dx = \int_{-1}^{1}\frac{1}{2}|x|\,e^{-2x^2}\,dx = \int_0^1 xe^{-2x^2}\,dx = \frac{1}{4}(1 - e^{-2}).$

$M_y = \int_{-1}^{1}\int_0^{e^{-x^2}} x|x|\,y\,dy = \int_{-1}^{1}\frac{1}{2}x|x|\,e^{-2x^2}\,dx = 0$ since the integrand is odd.

$M_x = \int_{-1}^{1}\int_0^{e^{-x^2}} y|x|\,y\,dy\,dx = \frac{1}{3}\int_{-1}^{1}|x|\,e^{-3x^2}\,dx = \frac{2}{3}\int_0^1 xe^{-3x^2}\,dx = \frac{1}{9}(1 - e^{-3}).$

Thus, $\bar{x} = \frac{M_y}{m} = 0$ and $\bar{y} = \frac{M_x}{m} = \frac{4(1 - e^{-3})}{9(1 - e^{-2})} \approx 0.49.$

[6] $m = \int_0^{\pi}\int_0^{\sin x} y\,dy\,dx = \frac{1}{2}\int_0^{\pi}\sin^2 x\,dx = \frac{1}{4}\int_0^{\pi}(1 - \cos 2x)\,dx = \frac{\pi}{4}.$

$M_y = \int_0^{\pi}\int_0^{\sin x} xy\,dy\,dx = \frac{1}{2}\int_0^{\pi} x\sin^2 x\,dx = \frac{1}{4}\int_0^{\pi} x\,dx - \frac{1}{4}\int_0^{\pi} x\cos 2x\,dx = \frac{\pi^2}{8}.$

$M_x = \int_0^{\pi}\int_0^{\sin x} y\cdot y\,dy\,dx = \frac{1}{3}\int_0^{\pi}\sin^3 x\,dx = \frac{1}{3}\int_0^{\pi}(1 - \cos^2 x)\sin x\,dx = \frac{4}{9}.$

Thus, $\bar{x} = \frac{M_y}{m} = \frac{\pi}{2}$ and $\bar{y} = \frac{M_x}{m} = \frac{16}{9\pi}.$

[7] $m = \int_{-\pi/4}^{\pi/4}\int_{1/2}^{\sec x} 4\,dy\,dx = \int_{-\pi/4}^{\pi/4}(4\sec x - 2)\,dx = 4\int_0^{\pi/4}(2\sec x - 1)\,dx =$

$8\ln(\sqrt{2} + 1) - \pi \approx 3.91.$

$M_x = \int_{-\pi/4}^{\pi/4}\int_{1/2}^{\sec x} 4y\,dy\,dx = \int_{-\pi/4}^{\pi/4}(2\sec^2 x - \frac{1}{2})\,dx = \int_0^{\pi/4}(4\sec^2 x - 1)\,dx =$

$4 - \frac{\pi}{4}.$

By symmetry, $\bar{x} = 0$ and $\bar{y} = \frac{M_x}{m} = \frac{4 - (\pi/4)}{8\ln(\sqrt{2} + 1) - \pi} \approx 0.82.$

[8] $m = \int_1^2\int_0^{\ln x}\frac{1}{x}\,dy\,dx = \int_1^2\frac{\ln x}{x}\,dx = \left[\frac{1}{2}(\ln x)^2\right]_1^2 = \frac{1}{2}(\ln 2)^2 \approx 0.24.$

$M_y = \int_1^2\int_0^{\ln x}\frac{1}{x}\cdot x\,dy\,dx = \int_1^2 \ln x\,dx = \left[x\ln x - x\right]_1^2 = 2\ln 2 - 1.$

$M_x = \int_1^2\int_0^{\ln x}\frac{1}{x}\cdot y\,dy\,dx = \int_1^2\frac{1}{2x}(\ln x)^2\,dx = \left[\frac{1}{6}(\ln x)^3\right]_1^2 = \frac{1}{6}(\ln 2)^3.$

Thus, $\bar{x} = \frac{M_y}{m} = \frac{2(2\ln 2 - 1)}{(\ln 2)^2} \approx 1.61$ and $\bar{y} = \frac{M_x}{m} = \frac{1}{3}\ln 2 \approx 0.23.$

9 $I_x = \int_0^9\int_0^{\sqrt{x}} y^2(x+y)\,dy\,dx = \int_0^9 (\frac{1}{3}x^{5/2} + \frac{1}{4}x^2)\,dx = 3^5(\frac{31}{28}) = \frac{7533}{28}.$

$I_y = \int_0^9\int_0^{\sqrt{x}} x^2(x+y)\,dy\,dx = \int_0^9 (x^{7/2} + \frac{1}{2}x^3)\,dx = 3^7(\frac{19}{8}) = \frac{41{,}553}{8}.$

$I_O = I_x + I_y = 3^5(\frac{1259}{56}) = \frac{305{,}937}{56}.$

10 $I_x = \int_0^8\int_0^{x^{1/3}} y^2(y^2)\,dy\,dx = \frac{1}{5}\int_0^8 x^{5/3}\,dx = \frac{96}{5}.$

$I_y = \int_0^8\int_0^{x^{1/3}} x^2(y^2)\,dy\,dx = \frac{1}{3}\int_0^8 x^3\,dx = \frac{1024}{3}.$ $I_O = I_x + I_y = \frac{5408}{15}.$

11 $I_x = \int_{-2}^2\int_{x^2}^4 y^2\,k|x|\,dy\,dx = \frac{k}{3}\int_{-2}^2 |x|(64 - x^6)\,dx = \frac{2k}{3}\int_0^2 (64x - x^7)\,dx = 64k.$

$I_y = \int_{-2}^2\int_{x^2}^4 x^2\,k|x|\,dy\,dx = \int_{-2}^2 kx^2|x|(4 - x^2)\,dx = 2k\int_0^2 (4x^3 - x^5)\,dx = \frac{32}{3}k.$

{Note that in both cases the integrand on $[-2, 2]$ is even.} $I_O = I_x + I_y = \frac{224}{3}k.$

12 $I_x = \int_{-\sqrt{2}}^0\int_{2x}^{x^3} y^2(-ky)\,dy\,dx + \int_0^{\sqrt{2}}\int_{x^3}^{2x} y^2(ky)\,dy\,dx = \frac{1}{4}k\int_{-\sqrt{2}}^{\sqrt{2}} (16x^4 - x^{12})\,dx =$

$\frac{1}{2}k\int_0^{\sqrt{2}} (16x^4 - x^{12})\,dx = \frac{256}{65}\sqrt{2}\,k.$

$I_y = \int_{-\sqrt{2}}^0\int_{2x}^{x^3} x^2(-ky)\,dy\,dx + \int_0^{\sqrt{2}}\int_{x^3}^{2x} x^2ky\,dy\,dx = \frac{1}{2}k\int_{-\sqrt{2}}^{\sqrt{2}} (4x^4 - x^8)\,dx$

$= k\int_0^{\sqrt{2}} (4x^4 - x^8)\,dx = \frac{64}{45}\sqrt{2}\,k.$ $I_O = I_x + I_y = \frac{3136}{585}\sqrt{2}\,k.$

13 (a) Position the square so that its vertices are $(0, 0)$, $(0, a)$, $(a, 0)$, and (a, a). $I_x = I_y$ are the moments of inertia with respect to a side and

$I_x = \delta\int_0^a\int_0^a y^2\,dy\,dx = \frac{1}{3}a^3\delta\int_0^a dx = \frac{1}{3}a^4\delta.$

(b) Position the square as shown in *Figure 13*.

The diagonals correspond to the x- and y-axes.

We will calculate I_y from $x = 0$ to $(a/\sqrt{2})$ and $y = 0$ to $(a/\sqrt{2}) - x$, and then quadruple it.

$I_y = 4\delta\int_0^{(a/\sqrt{2})}\int_0^{(a/\sqrt{2})-x} x^2\,dy\,dx$

$= 4\delta\int_0^{(a/\sqrt{2})}\left(\frac{a}{\sqrt{2}}x^2 - x^3\right)dx = \frac{1}{12}a^4\delta.$

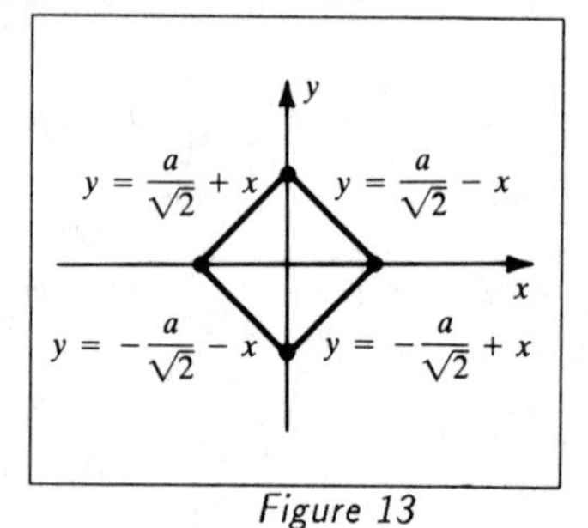

Figure 13

(c) If we position the square as in *Figure 13*, then the center of mass is at the origin and the moment of inertia with respect to the center of mass corresponds to I_O.

Since $I_x = I_y$, $I_O = 2I_y = \frac{1}{6}a^4\delta.$

14 (a) Position the triangle as in *Figure 14ab*. The altitude corresponds to the y-axis from $y = 0$ to $\frac{1}{2}\sqrt{3}\,a$. Since the figure is symmetric with respect to the y-axis and the integrand is even, we will calculate I_y for $x = 0$ to $\frac{1}{2}a$ and double it.

$$I_y = 2\delta\int_0^{\sqrt{3}\,a/2}\int_0^{-y/\sqrt{3}+a/2} x^2\,dx\,dy = \tfrac{2}{3}\delta\int_0^{\sqrt{3}\,a/2}\left(-\frac{y}{\sqrt{3}}+\tfrac{1}{2}a\right)^3 dy$$

$$= \tfrac{2}{3}\delta\int_0^{\sqrt{3}\,a/2}\left(\frac{-y^3}{3\sqrt{3}}+\frac{ay^2}{2}-\frac{3a^2y}{4\sqrt{3}}+\frac{a^3}{8}\right)dy$$

$$= \tfrac{2}{3}\delta\left[\frac{-9a^4}{(16)(4)(3\sqrt{3})}+\frac{3\sqrt{3}\,a^4}{(8)(3)(2)}-\frac{(3)(3)a^4}{(4)(2)(4\sqrt{3})}+\frac{\sqrt{3}\,a^4}{(2)(8)}\right] = \frac{2\delta}{\sqrt{3}}\left(\frac{3a^4}{192}\right) = \frac{\sqrt{3}\,a^4\delta}{96}.$$

(b) Position the triangle as in part (a). Then a side corresponds to the x-axis for $x = -\frac{1}{2}a$ to $\frac{1}{2}a$. We will calculate I_x for $x = 0$ to $\frac{1}{2}a$ and double it.

$$I_x = 2\delta\int_0^{\sqrt{3}\,a/2}\int_0^{-y/\sqrt{3}+a/2} y^2\,dx\,dy$$

$$= 2\delta\int_0^{\sqrt{3}\,a/2}\left(\frac{-y^3}{\sqrt{3}}+\frac{ay^2}{2}\right)dy = \delta\left[\frac{3\sqrt{3}\,a^4}{24}-\frac{9a^4}{32\sqrt{3}}\right] = \frac{\sqrt{3}\,a^4\delta}{32}.$$

(c) Position the triangle so that a vertex is at the origin as shown in *Figure 14c*. Then the moment of inertia with respect to a vertex corresponds to I_O.

$$I_O = \delta\int_0^{\sqrt{3}\,a/2}\int_{y/\sqrt{3}}^{(\sqrt{3}\,a-y)/\sqrt{3}} (x^2+y^2)\,dx\,dy$$

$$= \delta\int_0^{\sqrt{3}\,a/2}\left[\frac{\left[a-(y/\sqrt{3})\right]^3}{3}-\frac{y^3}{9\sqrt{3}}+y^2\left(a-\frac{y}{\sqrt{3}}-\frac{y}{\sqrt{3}}\right)\right]dy$$

$$= \delta\left[\frac{-\sqrt{3}\left[a-(y/\sqrt{3})\right]^4}{12}-\frac{y^4}{36\sqrt{3}}+\frac{ay^3}{3}-\frac{y^4}{2\sqrt{3}}\right]_0^{\sqrt{3}\,a/2}$$

$$= \delta\left[\frac{-\sqrt{3}}{12}\left(\frac{a}{2}\right)^4+\frac{\sqrt{3}\,a^4}{12}-\frac{9a^4}{(16)(36)\sqrt{3}}+\frac{3\sqrt{3}\,a^4}{(8)(3)}-\frac{9a^4}{(32)\sqrt{3}}\right] = \frac{5\sqrt{3}\,a^4\delta}{48}$$

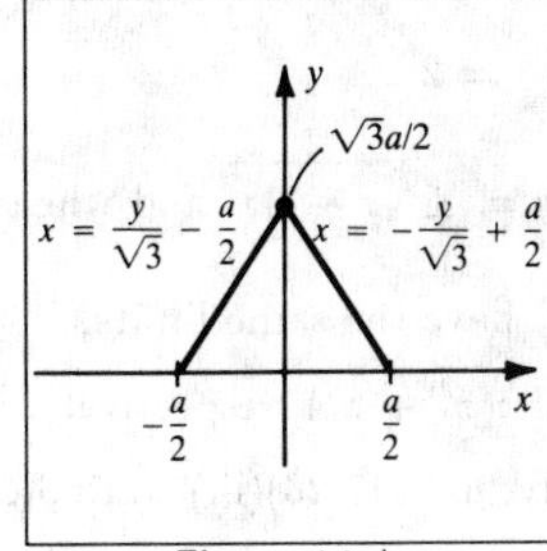

Figure 14ab

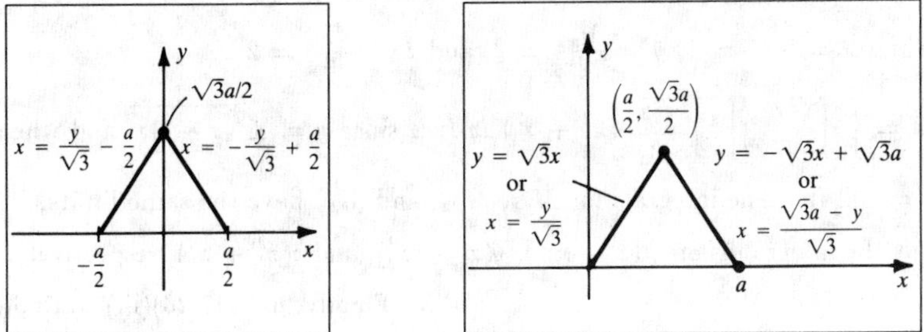

Figure 14c

15 The area of the square is a^2, and so the mass is $m = a^2\delta$.

Thus, $I = md^2 \Rightarrow d = \left(\frac{I_x}{m}\right)^{1/2} = \left(\frac{\frac{1}{3}a^4\delta}{a^2\delta}\right)^{1/2} = \frac{a}{\sqrt{3}}$.

16 The area of the triangle is $\frac{1}{4}\sqrt{3}\,a^2$, and so the mass is $m = \frac{1}{4}\sqrt{3}\,a^2\delta$.

Thus, $I = md^2 \Rightarrow d = \left(\frac{I_y}{m}\right)^{1/2} = \left(\frac{\frac{1}{96}\sqrt{3}\,a^4\delta}{\frac{1}{4}\sqrt{3}\,a^2\delta}\right)^{1/2} = \frac{a}{\sqrt{24}}$.

[17] Position the fixed corner at the origin. Then $\delta(x, y, z) = k(x^2 + y^2 + z^2)$.

By symmetry, $\bar{x} = \bar{y} = \bar{z}$. $m = \int_0^a\int_0^a\int_0^a k(x^2 + y^2 + z^2)\,dx\,dy\,dz =$

$$\int_0^a\int_0^a k(\tfrac{1}{3}a^3 + ay^2 + az^2)\,dy\,dz = \int_0^a k(\tfrac{1}{3}a^4 + \tfrac{1}{3}a^4 + a^2z^2)\,dz = ka^5.$$

$$M_{yz} = \int_0^a\int_0^a\int_0^a k(x^3 + xy^2 + xz^2)\,dx\,dy\,dz = \int_0^a\int_0^a k(\tfrac{1}{4}a^4 + \tfrac{1}{2}a^2y^2 + \tfrac{1}{2}a^2z^2)\,dy\,dz$$

$$= \int_0^a k(\tfrac{1}{4}a^5 + \tfrac{1}{6}a^5 + \tfrac{1}{2}a^3z^2)\,dz = \tfrac{7}{12}ka^6. \qquad \text{Thus, } \bar{x} = \bar{y} = \bar{z} = \frac{M_{yz}}{m} = \tfrac{7}{12}a.$$

[18] $\delta(x, y, z) = k|y| = ky$ since $y \geq 0$. $m = \int_0^5\int_0^{(10-2x)/5}\int_0^{10-2x-5y} ky\,dz\,dy\,dx =$

$$k\int_0^5\int_0^{(10-2x)/5} (10y - 2xy - 5y^2)\,dy\,dx = k\int_0^5 \tfrac{4}{75}(5 - x)^3\,dx = \tfrac{25}{3}k.$$

$$M_{yz} = \int_0^5\int_0^{(10-2x)/5}\int_0^{10-2x-5y} x\,ky\,dz\,dy\,dx$$

$$= k\int_0^5\int_0^{(10-2x)/5} x(10y - 2xy - 5y^2)\,dy\,dx = k\int_0^5 \tfrac{4}{75}x(5 - x)^3\,dx = \tfrac{25}{3}k.$$

$$M_{xz} = \int_0^5\int_0^{(10-2x)/5}\int_0^{10-2x-5y} y\,ky\,dz\,dy\,dx$$

$$= k\int_0^5\int_0^{(10-2x)/5} (10y^2 - 2xy^2 - 5y^3)\,dy\,dx = k\int_0^5 \tfrac{4}{375}(5 - x)^4\,dx = \tfrac{20}{3}k.$$

$$M_{xy} = \int_0^{10}\int_0^{(10-z)/5}\int_0^{(10-5y-z)/2} z\,ky\,dx\,dy\,dz$$

$$= k\int_0^{10}\int_0^{(10-z)/5} \tfrac{1}{2}z\left[y(10 - z) - 5y^2\right]dy\,dz = k\int_0^{10} \tfrac{1}{300}z(10 - z)^3\,dz = \tfrac{50}{3}k.$$

Thus, $\bar{x} = \frac{M_{yz}}{m} = 1$, $\bar{y} = \frac{M_{xz}}{m} = \frac{4}{5}$, and $\bar{z} = \frac{M_{xy}}{m} = 2$.

[19] $m = \int_0^4\int_{-\sqrt{x}/2}^{\sqrt{x}/2}\int_{-\sqrt{x-4z^2}}^{\sqrt{x-4z^2}} (x^2 + z^2)\,dy\,dz\,dx$ since $y = \pm\sqrt{x - 4z^2}$ and when $y = 0$,

$z = \pm\frac{1}{2}\sqrt{x}$. The integrals for M_{yz}, M_{xz}, and M_{xy} have the same limits,

but the integrands are $x(x^2 + z^2)$, $y(x^2 + z^2)$, and $z(x^2 + z^2)$, respectively.

Finally, use (17.26)(iii) to find $\bar{x}$, $\bar{y}$, and $\bar{z}$.

[20] This is a hyperboloid of two sheets with axis along the y-axis. We need the part

where $1 \leq y \leq 2$. $m = \int_{-\sqrt{3}}^{\sqrt{3}}\int_{-\sqrt{3-x^2}}^{\sqrt{3-x^2}}\int_{\sqrt{1+x^2+z^2}}^{2} x^2y^2z^2\,dy\,dz\,dx$ since for $y \geq 1$,

$y = \sqrt{1 + x^2 + z^2}$ and the trace of the hyperboloid in the plane $y = 2$ is the circle

$x^2 + z^2 = 3 \Rightarrow z = \pm\sqrt{3 - x^2}$. The integrals for M_{yz}, M_{xz}, and M_{xy} have the

same limits, but the integrands are $x^3y^2z^2$, $x^2y^3z^2$, and $x^2y^2z^3$, respectively.

Finally, use (17.26)(iii) to find $\bar{x}$, $\bar{y}$, and $\bar{z}$.

Note: In Exer. 21-24, without loss of generality let $\delta(x, y, z) = 1$.

[21] $m = \int_{-a}^{a}\int_{-\sqrt{a^2-x^2}}^{\sqrt{a^2-x^2}}\int_{0}^{\sqrt{a^2-x^2-y^2}} dz\,dy\,dx$. The integral for M_{xy} has the same limits, but the integrand is z. $\bar{z} = M_{xy}/m$ and by symmetry, $\bar{x} = \bar{y} = 0$.

[22] Let the equation of the cone be $\frac{x^2}{a^2} + \frac{y^2}{a^2} = \frac{z^2}{h^2}$ for $z \geq 0$.

Thus, $z = \frac{h}{a}\sqrt{x^2 + y^2}$ and $m = \int_{-a}^{a}\int_{-\sqrt{a^2-x^2}}^{\sqrt{a^2-x^2}}\int_{(h/a)\sqrt{x^2+y^2}}^{h} dz\,dy\,dx$.

The integral for M_{xy} has the same limits, but the integrand is z.

$\bar{z} = M_{xy}/m$ and by symmetry, $\bar{x} = \bar{y} = 0$.

[23] (a) See *Figure 23*. $m = \int_{0}^{3}\int_{0}^{9-x^2}\int_{0}^{6-2x} dy\,dz\,dx$;

the integrals for M_{yz}, M_{xz}, and M_{xy} have the same limits, but the integrands are x, y, and z, respectively.

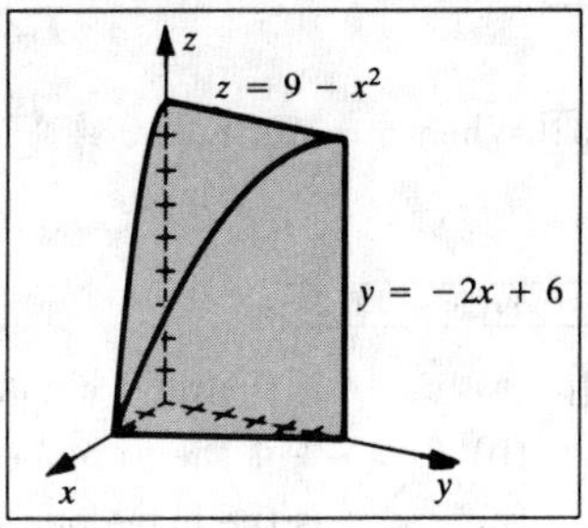

Figure 23

(b)
$$m = \int_{0}^{3}\int_{0}^{9-x^2}\int_{0}^{6-2x} dy\,dz\,dx = \int_{0}^{3}\int_{0}^{9-x^2}(6 - 2x)\,dz\,dx = \int_{0}^{3}(2x^3 - 6x^2 - 18x + 54)\,dx = \tfrac{135}{2}.$$

$$M_{xy} = \int_{0}^{3}\int_{0}^{9-x^2}\int_{0}^{6-2x} z\,dy\,dz\,dx = \int_{0}^{3}\int_{0}^{9-x^2} z(6 - 2x)\,dz\,dx = \int_{0}^{3}(-x^5 + 3x^4 + 18x^3 - 54x^2 - 81x + 243)\,dx = \tfrac{2673}{10}.$$

$$M_{yz} = \int_{0}^{3}\int_{0}^{9-x^2}\int_{0}^{6-2x} x\,dy\,dz\,dx = \int_{0}^{3}\int_{0}^{9-x^2} x(6 - 2x)\,dz\,dx = \int_{0}^{3}(2x^4 - 6x^3 - 18x^2 + 54x)\,dx = \tfrac{567}{10}.$$

$$M_{xz} = \int_{0}^{3}\int_{0}^{9-x^2}\int_{0}^{6-2x} y\,dy\,dz\,dx = \int_{0}^{3}\int_{0}^{9-x^2}(18 - 12x + 2x^2)\,dz\,dx = 2\int_{0}^{3}(-x^4 + 6x^3 - 54x + 81)\,dx = \tfrac{729}{5}.$$

Thus, $\bar{x} = \frac{M_{yz}}{m} = \frac{21}{25}$, $\bar{y} = \frac{M_{xz}}{m} = \frac{54}{25}$, and $\bar{z} = \frac{M_{xy}}{m} = \frac{99}{25}$.

[24] $m = \int_{0}^{1}\int_{x^3}^{x^2}\int_{0}^{x^2} dz\,dy\,dx$. The integrals for M_{yz}, M_{xz}, and M_{xy} have the same limits, but the integrands are x, y, and z, respectively. Then use (17.26)(iii).

[25] $I_z = \int_{-a}^{a}\int_{-\sqrt{a^2-x^2}}^{\sqrt{a^2-x^2}}\int_{-\sqrt{a^2-x^2-y^2}}^{\sqrt{a^2-x^2-y^2}}(x^2 + y^2)(x^2 + y^2 + z^2)\,dz\,dy\,dx$.

[26] $I_z = \int_{0}^{36}\int_{-z/3}^{z/3}\int_{-\sqrt{z^2-9y^2}}^{\sqrt{z^2-9y^2}}(x^2 + y^2)(x^2 + y^2)\,dx\,dy\,dz$.

27 $I_z = \int_0^a \int_0^{b(1-x/a)} \int_0^{c(1-x/a-y/b)} (x^2 + y^2)\,\delta\, dz\, dy\, dx.$

28 $I_z = \int_{-a}^{a} \int_{-b\sqrt{1-x^2/a^2}}^{b\sqrt{1-x^2/a^2}} \int_{-c\sqrt{1-x^2/a^2-y^2/b^2}}^{c\sqrt{1-x^2/a^2-y^2/b^2}} (x^2 + y^2)\,\delta\, dz\, dy\, dx.$

29 When $n = 10$, $\delta_{\text{av}} \approx \frac{1}{10}\sum_{k=1}^{10} \delta(x_k, y_k) \approx 1.2$.

Then, $m = (\text{Area of } R) \cdot \delta_{\text{av}} \approx 1^2(1.2) = 1.2$.

30 When $n = 10$, $\delta_{\text{av}} \approx \frac{1}{10}\sum_{k=1}^{10} \delta(x_k, y_k) \approx 2.7$.

Then, $m = (\text{Area of } R) \cdot \delta_{\text{av}} \approx \left[\frac{1}{4}\pi(1)^2\right](2.7) \approx 2.1$.

31 When $n = 10$, $\delta_{\text{av}} \approx \frac{1}{10}\sum_{k=1}^{10} \delta(x_k, y_k, z_k) \approx 1.3$.

Then, $m = (\text{Volume of } Q) \cdot \delta_{\text{av}} \approx 1^3(1.3) = 1.3$.

32 When $n = 10$, $\delta_{\text{av}} \approx \frac{1}{10}\sum_{k=1}^{10} \delta(x_k, y_k, z_k) \approx 1.0$.

Then, $m = (\text{Volume of } Q) \cdot \delta_{\text{av}} \approx \left[\frac{1}{8} \cdot \frac{4}{3}\pi(1)^3\right](1.0) \approx 0.52$.

Exercises 17.7

1 (a) $r = 4$ is the right circular cylinder of radius 4 with axis along the z-axis.

(b) $\theta = -\frac{\pi}{2}$ is the yz-plane. In general, this plane is $\theta = \frac{\pi}{2} + \pi n$.

(c) $z = 1$ is the plane parallel to the xy-plane with z-intercept 1.

2 (a) $r = -3$ is the right circular cylinder of radius 3 with axis along the z-axis.

(b) $\theta = \frac{\pi}{4}$ is the plane containing the z-axis and bisecting the first octant.

(c) $z = -2$ is the plane parallel to the xy-plane with z-intercept -2.

3 $r = -3\sec\theta \Rightarrow r\cos\theta = -3 \Rightarrow x = -3$,

the plane parallel to the yz-plane with x-intercept -3.

4 $r = -\csc\theta \Rightarrow r\sin\theta = -1 \Rightarrow y = -1$,

the plane parallel to the xz-plane with y-intercept -1.

5 $z = 4r^2 \Rightarrow z = 4x^2 + 4y^2$, a paraboloid with vertex $(0, 0, 0)$ and opening upward.

6 $z = 4 - r^2 \Rightarrow z = 4 - (x^2 + y^2)$,

a paraboloid with vertex $(0, 0, 4)$ and opening downward.

7 $r = 6\sin\theta \Rightarrow r^2 = 6r\sin\theta \Rightarrow x^2 + y^2 = 6y$,

the right circular cylinder with trace $x^2 + (y - 3)^2 = 9$ in the xy-plane.

8 $r\sec\theta = 4 \Rightarrow r^2 = 4r\cos\theta \Rightarrow x^2 + y^2 = 4x$,

the right circular cylinder with trace $(x - 2)^2 + y^2 = 4$ in the xy-plane.

9 $z = 2r \Rightarrow z^2 = 4r^2$, the cone $z^2 = 4x^2 + 4y^2$.

10 $3z = r \Rightarrow 9z^2 = r^2$, the cone $9z^2 = x^2 + y^2$.

11 $r^2 = 9 - z^2 \Rightarrow x^2 + y^2 + z^2 = 9$, the sphere with center at the origin and radius 3.

12 $r^2 + z^2 = 16 \Rightarrow x^2 + y^2 + z^2 = 16$,

the sphere with center at the origin and radius 4.

13 $r = 2\csc\theta\cot\theta \Rightarrow r\sin^2\theta = 2\cos\theta \Rightarrow r^2\sin^2\theta = 2r\cos\theta$,

the cylinder with trace $y^2 = 2x$ in the xy-plane and rulings parallel to the z-axis.

14 $r = \tan\theta\sec\theta \Rightarrow r\cos^2\theta = \sin\theta \Rightarrow r^2\cos^2\theta = r\sin\theta$,

the cylinder with trace $x^2 = y$ in the xy-plane and rulings parallel to the z-axis.

[15] $x^2 + y^2 + z^2 = 4 \Rightarrow (x^2 + y^2) + z^2 = 4 \Rightarrow r^2 + z^2 = 4$

[16] $x^2 + y^2 = 4z \Rightarrow r^2 = 4z$

[17] $3x + y - 4z = 12 \Rightarrow 3r\cos\theta + r\sin\theta - 4z = 12$

[18] $y = x \Rightarrow r\sin\theta = r\cos\theta \Rightarrow \tan\theta = 1$, or $\theta = \frac{\pi}{4}$ {or even $\theta = \frac{\pi}{4} + \pi n$}

[19] $x^2 = 4 - y^2 \Rightarrow x^2 + y^2 = 4 \Rightarrow r^2 = 4 \Rightarrow r = 2$ {or $r = -2$}

[20] $x^2 + (y-2)^2 = 4 \Rightarrow x^2 + y^2 - 4y = 0 \Rightarrow r^2 = 4r\sin\theta \Rightarrow r = 4\sin\theta$

[21] $x^2 - 4z^2 + y^2 = 0 \Rightarrow r^2 = 4z^2 \Rightarrow r = 2z$ {or $r = -2z$}

[22] $x^2 - y^2 - z^2 = 1 \Rightarrow r^2\cos^2\theta - r^2\sin^2\theta - z^2 = 1 \Rightarrow$

$$r^2(\cos^2\theta - \sin^2\theta) - z^2 = 1 \Rightarrow r^2\cos 2\theta - z^2 = 1$$

[23] $y^2 + z^2 = 9 \Rightarrow r^2\sin^2\theta + z^2 = 9$

[24] $x^2 + z^2 = 9 \Rightarrow r^2\cos^2\theta + z^2 = 9$

[25] The trace of $z = 16 - r^2$ in the xy-plane $\{z = 0\}$ is $r = 4$.

$$\iiint_Q f(r, \theta, z)\,dV = \int_0^{2\pi}\int_0^4\int_0^{16-r^2} f(r, \theta, z)\,r\,dz\,dr\,d\theta.$$

[26] $\iiint_Q f(r, \theta, z)\,dV = \int_0^{2\pi}\int_0^2\int_0^{4-r} f(r, \theta, z)\,r\,dz\,dr\,d\theta$, with $r \geq 0$.

[27] $z = 4$ and $z = \sqrt{25 - r^2} \Rightarrow r = 3$. $\iiint_Q f(r, \theta, z)\,dV =$

$$\int_0^{2\pi}\int_0^3\int_0^4 f(r, \theta, z)\,r\,dz\,dr\,d\theta + \int_0^{2\pi}\int_3^4\int_0^{\sqrt{25-r^2}} f(r, \theta, z)\,r\,dz\,dr\,d\theta.$$

[28] $z = r$ and $z = \sqrt{32 - r^2} \Rightarrow r = 4$.

$$\iiint_Q f(r, \theta, z)\,dV = \int_0^{2\pi}\int_0^4\int_r^{\sqrt{32-r^2}} f(r, \theta, z)\,r\,dz\,dr\,d\theta, \text{ with } r \geq 0.$$

Note: In Exercises 29–30, without loss of generality, let $\delta(x, y, z) = 1$.

[29] (a) In cylindrical coordinates: $z = r^2$, $r^2 = 4$, and $z = 0$.

$$V = \int_0^{2\pi}\int_0^2\int_0^{r^2} r\,dz\,dr\,d\theta = \int_0^{2\pi}\int_0^2 r^3\,dr\,d\theta = 4\int_0^{2\pi} d\theta = 8\pi.$$

(b) $M_{xy} = \int_0^{2\pi}\int_0^2\int_0^{r^2} zr\,dz\,dr\,d\theta = \int_0^{2\pi}\int_0^2 \frac{1}{2}r^5\,dr\,d\theta = \frac{16}{3}\int_0^{2\pi} d\theta = \frac{32\pi}{3}$.

By symmetry, $\bar{x} = \bar{y} = 0$. $\bar{z} = \frac{M_{xy}}{V} = \frac{4}{3}$, and the centroid is $(0, 0, \frac{4}{3})$.

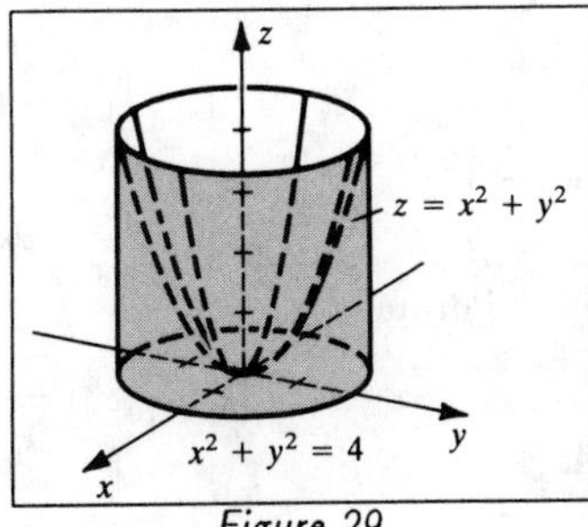

Figure 29

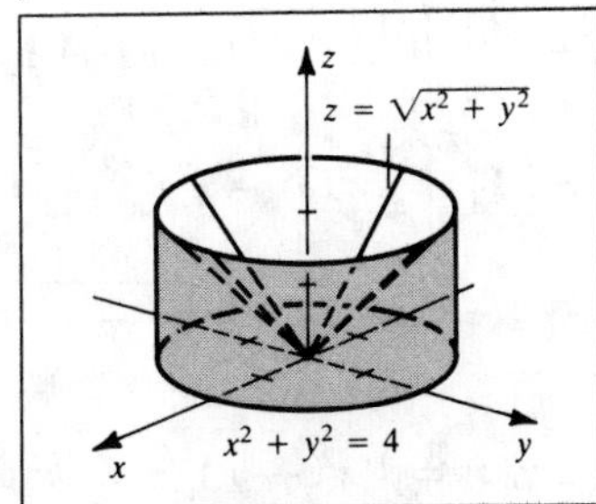

Figure 30

[30] (a) See *Figure 30.* In cylindrical coordinates: $z = r$, $r^2 = 4$, and $z = 0$.

$$V = \int_0^{2\pi}\int_0^2\int_0^r r\,dz\,dr\,d\theta = \int_0^{2\pi}\int_0^2 r^2\,dr\,d\theta = \tfrac{8}{3}\int_0^{2\pi} d\theta = \tfrac{16\pi}{3}.$$

(b) $$M_{xy} = \int_0^{2\pi}\int_0^2\int_0^r zr\,dz\,dr\,d\theta = \int_0^{2\pi}\int_0^2 \tfrac{1}{2}r^3\,dr\,d\theta = 2\int_0^{2\pi} d\theta = 4\pi.$$

By symmetry, $\bar{x} = \bar{y} = 0$. $\bar{z} = \dfrac{M_{xy}}{V} = \tfrac{3}{4}$, and the centroid is $(0, 0, \tfrac{3}{4})$.

[31] Let the cylinder have equation $r = a$ for $0 \le z \le h$.

(a) $$I_z = \delta\int_0^{2\pi}\int_0^a\int_0^h (r^2)\,r\,dz\,dr\,d\theta = h\delta\int_0^{2\pi}\int_0^a r^3\,dr\,d\theta = \tfrac{1}{4}ha^4\delta\int_0^{2\pi} d\theta = \tfrac{1}{2}\pi ha^4\delta.$$

(b) $$I_x = \delta\int_0^{2\pi}\int_0^a\int_0^h (r^2\sin^2\theta + z^2)\,r\,dz\,dr\,d\theta = h\delta\int_0^{2\pi}\int_0^a (r^3\sin^2\theta + \tfrac{1}{3}h^2 r)\,dr\,d\theta$$

$$= ha^2\delta\int_0^{2\pi} (\tfrac{1}{4}a^2\sin^2\theta + \tfrac{1}{6}h^2)\,d\theta = \pi ha^2(\tfrac{1}{4}a^2 + \tfrac{1}{3}h^2)\delta.$$

[32] In cylindrical coordinates: $z = r$ and $z = r^2$.

$r = r^2 \Rightarrow r = 0, 1$.

(a) $$m = \delta\int_0^{2\pi}\int_0^1\int_{r^2}^r r\,dz\,dr\,d\theta$$

$$= \delta\int_0^{2\pi}\int_0^1 (r^2 - r^3)\,dr\,d\theta = \tfrac{1}{12}\delta\int_0^{2\pi} d\theta$$

$$= \tfrac{1}{6}\pi\delta.$$

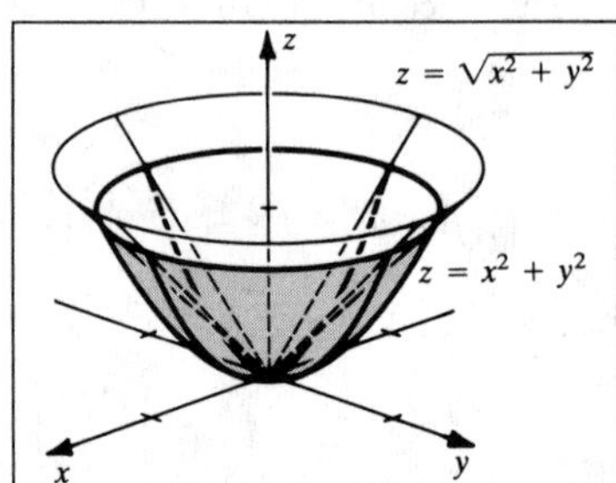

Figure 32

$$M_{xy} = \delta\int_0^{2\pi}\int_0^1\int_{r^2}^r zr\,dz\,dr\,d\theta$$

$$= \delta\int_0^{2\pi}\int_0^1 \tfrac{1}{2}(r^3 - r^5)\,dr\,d\theta = \tfrac{1}{24}\delta\int_0^{2\pi} d\theta = \tfrac{1}{12}\pi\delta.$$

By symmetry, $\bar{x} = \bar{y} = 0$. $\bar{z} = \dfrac{M_{xy}}{m} = \tfrac{1}{2}$.

(b) $$I_z = \delta\int_0^{2\pi}\int_0^1\int_{r^2}^r (r^2)\,r\,dz\,dr\,d\theta = \delta\int_0^{2\pi}\int_0^1 (r^4 - r^5)\,dr\,d\theta = \tfrac{1}{30}\delta\int_0^{2\pi} d\theta = \tfrac{1}{15}\pi\delta.$$

Note: Let k be a constant of proportionality.

[33] Let the sphere have equation $x^2 + y^2 + z^2 = a^2$ and the line l correspond to the z-axis. Then, $\delta(x, y, z) = k(x^2 + y^2)^{1/2} = kr$ and by symmetry,

$$m = 2\int_0^{2\pi}\int_0^a\int_0^{\sqrt{a^2-r^2}} (kr)\,r\,dz\,dr\,d\theta$$

$$= 2k\int_0^a\int_0^{2\pi} r^2(a^2 - r^2)^{1/2}\,d\theta\,dr = 4k\pi\int_0^a r^2(a^2 - r^2)^{1/2}\,dr$$

$$= 4k\pi\left[\tfrac{r}{8}(2r^2 - a^2)\sqrt{a^2 - r^2} + \tfrac{1}{8}a^4\arcsin\tfrac{r}{a}\right]_0^a \{\text{Formula 31}\} =$$

$$(4k\pi)(\tfrac{\pi}{16}a^4) = \tfrac{1}{4}k\pi^2a^4.$$

[34] $\delta(x, y, z) = k(x^2 + y^2)^{1/2} = kr$ and $r \le z \le 4$.

$$m = \int_0^{2\pi}\int_0^4\int_r^4 (kr)\,r\,dz\,dr\,d\theta = k\int_0^{2\pi}\int_0^4 (4r^2 - r^3)\,dr\,d\theta = \tfrac{64}{3}k\int_0^{2\pi} d\theta = \tfrac{128}{3}k\pi.$$

[35] Since l corresponds to the z-axis, we will find I_z.

$$I_z = \int_0^a \int_0^{2\pi} \int_{-\sqrt{a^2-r^2}}^{\sqrt{a^2-r^2}} (r^2)(kr)\, r\, dz\, d\theta\, dr = 2k \int_0^a \int_0^{2\pi} r^4 (a^2 - r^2)^{1/2}\, d\theta\, dr =$$

$4k\pi \int_0^a r^4(a^2 - r^2)^{1/2}\, dr$. To evaluate, let $r = a \sin\theta$ and $dr = a\cos\theta\, d\theta$.

$r = 0, a \Rightarrow \theta = 0, \frac{\pi}{2}$. Thus, $I_z = 4k\pi a^6 \int_0^{\pi/2} \sin^4\theta \cos^2\theta\, d\theta =$

$$4k\pi a^6 \Big[\tfrac{1}{16}\theta - \tfrac{1}{64}\sin 4\theta - \tfrac{1}{48}\sin^3 2\theta\Big]_0^{\pi/2} \{\text{Exercise 8, §9.2}\} = 4k\pi a^6(\tfrac{\pi}{32}) = \tfrac{1}{8}k\pi^2 a^6.$$

[36] $I_z = \int_0^{2\pi} \int_0^4 \int_r^4 (r^2)(kr)\, r\, dz\, dr\, d\theta = k \int_0^{2\pi} \int_0^4 (4r^4 - r^5)\, dr\, d\theta = \frac{4^6}{30} k \int_0^{2\pi} d\theta = \frac{4096}{15} k\pi.$

[37] $m = \int_0^{2\pi} \int_0^3 \int_0^{10{,}000} \delta\, r\, dz\, dr\, d\theta$

$$= \int_0^{2\pi} \int_0^3 \Big[1.2z - \tfrac{1}{2}(1.05 \times 10^{-4})z^2 + \tfrac{1}{3}(2.6 \times 10^{-9})z^3\Big]_0^{10{,}000} r\, dr\, d\theta$$

$$= \frac{22{,}850}{3} \int_0^{2\pi} \int_0^3 r\, dr\, d\theta = \frac{22{,}850}{3} \cdot \frac{9}{2} \int_0^{2\pi} d\theta = 68{,}550\pi \approx 215{,}356.2, \text{ or } 215{,}360 \text{ kg}$$

[38] (a) In cylindrical coordinates: $r = 2\sin\theta$ and $r^2 + z^2 = 4$. Also,

$\delta(x, y, z) = k|z| = kz$ for $z \geq 0$. By symmetry,

$$m = 2\int_0^{\pi} \int_0^{2\sin\theta} \int_0^{\sqrt{4-r^2}} (kz)\, r\, dz\, dr\, d\theta$$

$$= k \int_0^{\pi} \int_0^{2\sin\theta} (4r - r^3)\, dr\, d\theta$$

$$= k \int_0^{\pi} (8\sin^2\theta - 4\sin^4\theta)\, d\theta$$

$$= k \int_0^{\pi} \Big[\tfrac{5}{2} - 2\cos 2\theta - \tfrac{1}{2}\cos 4\theta\Big]\, d\theta = \tfrac{5}{2}k\pi.$$

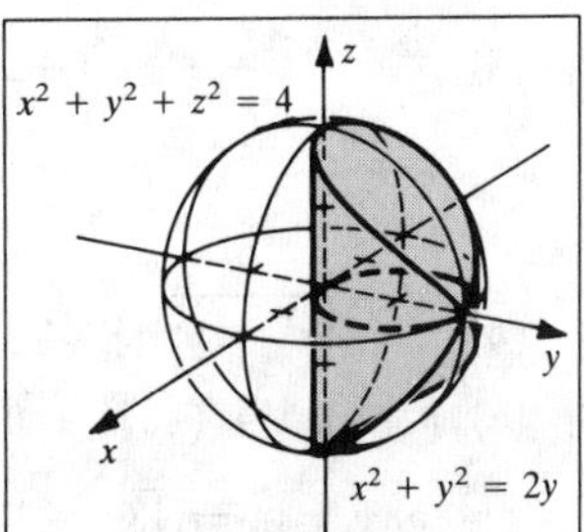

Figure 38

(b) $M_{xz} = \int_0^{\pi} \int_0^{2\sin\theta} \int_{-\sqrt{4-r^2}}^{\sqrt{4-r^2}} (r\sin\theta)(k|z|)\, r\, dz\, dr\, d\theta$

$$= 2k \int_0^{\pi} \int_0^{2\sin\theta} \int_0^{\sqrt{4-r^2}} (r\sin\theta)\, z\, r\, dz\, dr\, d\theta \ \{\text{integrand is even w.r.t. } z\}$$

$$= 2k \int_0^{\pi} \int_0^{2\sin\theta} \sin\theta \cdot \tfrac{1}{2}(4r^2 - r^4)\, dr\, d\theta$$

$$= 32k \int_0^{\pi} \Big[\tfrac{1}{3}\sin^4\theta - \tfrac{1}{5}\sin^6\theta\Big]\, d\theta = 32k(\tfrac{\pi}{16}) = 2\pi k.$$

By symmetry, $\bar{x} = \bar{z} = 0$. $\bar{y} = \frac{M_{xz}}{m} = \frac{4}{5}$, and its center of mass is $(0, \frac{4}{5}, 0)$.

[39] The region in the xy-plane is $x = \sqrt{1 - y^2}$ from $y = 0$ to 1, which is the quarter

circle $r = 1$ for $\theta = 0$ to $\frac{\pi}{2}$. $\int_0^1 \int_0^{\sqrt{1-y^2}} \int_0^{\sqrt{4-x^2-y^2}} z\, dz\, dx\, dy =$

$$\int_0^{\pi/2} \int_0^1 \int_0^{\sqrt{4-r^2}} z\, r\, dz\, dr\, d\theta = \tfrac{1}{2}\int_0^{\pi/2} \int_0^1 (4r - r^3)\, dr\, d\theta = \tfrac{1}{2} \cdot \tfrac{7}{4} \int_0^{\pi/2} d\theta = \tfrac{7\pi}{16}.$$

[40] The region in the xy-plane is $y = \pm\sqrt{2x - x^2}$ from $x = 0$ to 2, which is the circle $r = 2\cos\theta$ for $\theta = -\frac{\pi}{2}$ to $\frac{\pi}{2}$. $\int_0^2\int_{-\sqrt{2x-x^2}}^{\sqrt{2x-x^2}}\int_0^{x^2+y^2}\sqrt{x^2 + y^2}\,dz\,dy\,dx =$

$\int_{-\pi/2}^{\pi/2}\int_0^{2\cos\theta}\int_0^{r^2}(r)\,r\,dz\,dr\,d\theta = \int_{-\pi/2}^{\pi/2}\int_0^{2\cos\theta} r^4\,dr\,d\theta = \frac{1}{5}\int_{-\pi/2}^{\pi/2} 32\cos^5\theta\,d\theta =$

$\frac{64}{5}\int_0^{\pi/2}(1 - 2\sin^2\theta + \sin^4\theta)\cos\theta\,d\theta = \frac{64}{5}\int_0^1(1 - 2u^2 + u^4)\,du\ \{u = \sin\theta\} =$

$\frac{64}{5}\cdot\frac{8}{15} = \frac{512}{75}$.

Exercises 17.8

[1] (a) $x = \rho\sin\phi\cos\theta = 4\sin\frac{\pi}{6}\cos\frac{\pi}{2} = 4(\frac{1}{2})(0) = 0$,

$y = \rho\sin\phi\sin\theta = 4\sin\frac{\pi}{6}\sin\frac{\pi}{2} = 4(\frac{1}{2})(1) = 2$,

$z = \rho\cos\phi = 4\cos\frac{\pi}{6} = 4(\frac{1}{2}\sqrt{3}) = 2\sqrt{3}$; $(0, 2, 2\sqrt{3})$

(b) $r = \sqrt{x^2 + y^2} = \sqrt{0^2 + 2^2} = 2$, $\theta = \frac{\pi}{2}$, $z = 2\sqrt{3}$; $(2, \frac{\pi}{2}, 2\sqrt{3})$

[2] (a) $x = 1\sin\frac{3\pi}{4}\cos\frac{2\pi}{3} = (\frac{1}{2}\sqrt{2})(-\frac{1}{2}) = -\frac{1}{4}\sqrt{2}$, $y = 1\sin\frac{3\pi}{4}\sin\frac{2\pi}{3} =$

$(\frac{1}{2}\sqrt{2})(\frac{1}{2}\sqrt{3}) = \frac{1}{4}\sqrt{6}$, $z = 1\cos\frac{3\pi}{4} = -\frac{1}{2}\sqrt{2}$; $(-\frac{1}{4}\sqrt{2}, \frac{1}{4}\sqrt{6}, -\frac{1}{2}\sqrt{2})$

(b) $r = \sqrt{(-\frac{1}{4}\sqrt{2})^2 + (\frac{1}{4}\sqrt{6})^2} = \frac{1}{4}\sqrt{8} = \frac{1}{2}\sqrt{2}$, $\theta = \frac{2\pi}{3}$, $z = -\frac{1}{2}\sqrt{2}$; $(\frac{1}{2}\sqrt{2}, \frac{2\pi}{3}, -\frac{1}{2}\sqrt{2})$

[3] (a) $\rho^2 = 1^2 + 1^2 + (-2\sqrt{2})^2 = 10 \Rightarrow \rho = \sqrt{10}$, $\cos\phi = \frac{z}{\rho} = \frac{-2\sqrt{2}}{\sqrt{10}} = \frac{-2}{\sqrt{5}} \Rightarrow$

$\phi = \cos^{-1}(-\frac{2}{\sqrt{5}})$, $\tan\theta = \frac{y}{x} = 1 \Rightarrow \theta = \frac{\pi}{4}$ since x and y are both positive;

$(\sqrt{10}, \cos^{-1}(-\frac{2}{\sqrt{5}}), \frac{\pi}{4})$

(b) $r = \sqrt{x^2 + y^2} = \sqrt{1^2 + 1^2} = \sqrt{2}$, $\theta = \frac{\pi}{4}$, $z = -2\sqrt{2}$; $(\sqrt{2}, \frac{\pi}{4}, -2\sqrt{2})$

[4] (a) $\rho^2 = 1^2 + (\sqrt{3})^2 + 0^2 = 4 \Rightarrow \rho = 2$, $\cos\phi = \frac{z}{\rho} = 0 \Rightarrow \phi = \frac{\pi}{2}$,

$\tan\theta = \frac{y}{x} = \sqrt{3} \Rightarrow \theta = \frac{\pi}{3}$ since x and y are both positive; $(2, \frac{\pi}{2}, \frac{\pi}{3})$

(b) $r = \sqrt{1^2 + (\sqrt{3})^2} = \sqrt{4} = 2$, $\theta = \frac{\pi}{3}$, $z = 0$; $(2, \frac{\pi}{3}, 0)$

[5] (a) $\rho = 3$ is the sphere of radius 3 and center O.

(b) $\phi = \frac{\pi}{6}$ is a half-cone with vertex O and vertex angle $2\cdot\frac{\pi}{6} = \frac{\pi}{3}$.

(c) $\theta = \frac{\pi}{3}$ is a half-plane with edge on the z-axis and making an angle of $\frac{\pi}{3}$ with the xz-plane.

[6] (a) $\rho = 5$ is the sphere of radius 5 and center O.

(b) $\phi = \frac{2\pi}{3}$ is a half-cone with vertex O and vertex angle $2\cdot\frac{2\pi}{3} = \frac{4\pi}{3}$.

(c) $\theta = \frac{\pi}{4}$ is a half-plane with edge on the z-axis and making an angle of $\frac{\pi}{4}$ with the xz-plane.

[7] $\rho = 4\cos\phi \Rightarrow \rho^2 = 4\rho\cos\phi \Rightarrow x^2 + y^2 + z^2 = 4z \Rightarrow x^2 + y^2 + (z - 2)^2 = 4$, the sphere of radius 2 and center $(0, 0, 2)$.

[8] $\rho\sec\phi = 6 \Rightarrow \rho = 6\cos\phi \Rightarrow \rho^2 = 6\rho\cos\phi \Rightarrow x^2 + y^2 + z^2 = 6z \Rightarrow$

$x^2 + y^2 + (z - 3)^2 = 9$, the sphere of radius 3 and center $(0, 0, 3)$.

[9] $\rho\cos\phi = 3$ is the plane $z = 3$.

[10] $\rho = 4\sec\phi \Rightarrow \rho\cos\phi = 4$. This is the plane $z = 4$.

[11] $\rho = 6\sin\phi\cos\theta \Rightarrow \rho^2 = 6\rho\sin\phi\cos\theta \Rightarrow x^2 + y^2 + z^2 = 6x \Rightarrow$

$(x-3)^2 + y^2 + z^2 = 9$, the sphere of radius 3 and center (3, 0, 0).

[12] $\rho = 8\sin\phi\sin\theta \Rightarrow \rho^2 = 8\rho\sin\phi\sin\theta \Rightarrow x^2 + y^2 + z^2 = 8y \Rightarrow$

$x^2 + (y-4)^2 + z^2 = 16$, the sphere of radius 4 and center (0, 4, 0).

[13] $\rho = 5\csc\phi\csc\theta \Rightarrow \rho\sin\phi\sin\theta = 5$, the plane $y = 5$.

[14] $\rho = 2\csc\phi\sec\theta \Rightarrow \rho\sin\phi\cos\theta = 2$, the plane $x = 2$.

[15] $\rho = 5\csc\phi \Rightarrow \rho\sin\phi = 5 \Rightarrow \rho^2\sin^2\phi = 25 \Rightarrow$

$\rho^2\sin^2\phi(\cos^2\theta + \sin^2\theta) = 25$ {associate $\rho\sin\phi$ with $\cos\theta$ for x} $\Rightarrow$

$\rho^2\sin^2\phi\cos^2\theta + \rho^2\sin^2\phi\sin^2\theta = 25 \Rightarrow x^2 + y^2 = 25$,

the right circular cylinder of radius 5 with axis along the z-axis.

[16] $\rho\sin\phi = 3 \Rightarrow \rho^2\sin^2\phi = 9 \Rightarrow \rho^2\sin^2\phi(\cos^2\theta + \sin^2\theta) = 9 \Rightarrow$

$\rho^2\sin^2\phi\cos^2\theta + \rho^2\sin^2\phi\sin^2\theta = 9 \Rightarrow x^2 + y^2 = 9$,

the right circular cylinder of radius 3 with axis along the z-axis.

[17] $\tan\phi = 2 \Rightarrow \sin\phi = 2\cos\phi \Rightarrow \rho\sin\phi = 2\rho\cos\phi$ {associate $\cos\phi$ with ρ for z} $\Rightarrow$

$\rho^2\sin^2\phi = 4\rho^2\cos^2\phi$ {need ρ^2 for (17.31)(ii)} $\Rightarrow \rho^2\sin^2\phi(\cos^2\theta + \sin^2\theta) = 4z^2$

$\Rightarrow \rho^2\sin^2\phi\cos^2\theta + \rho^2\sin^2\phi\sin^2\theta = 4z^2$, the cone $x^2 + y^2 = 4z^2$.

[18] $\tan\theta = 4 \Rightarrow \sin\theta = 4\cos\theta \Rightarrow \rho\sin\phi\sin\theta = 4\rho\sin\phi\cos\theta$, the plane $y = 4x$.

[19] *Note:* Use $\csc\phi$, not $\csc\theta$. $\rho = 6\cot\phi\csc\phi \Rightarrow \rho\sin^2\phi = 6\cos\phi \Rightarrow$

$\rho^2\sin^2\phi(\cos^2\theta + \sin^2\theta) = 6\rho\cos\phi \Rightarrow \rho^2\sin^2\phi\cos^2\theta + \rho^2\sin^2\phi\sin^2\theta = 6z$,

the paraboloid $6z = x^2 + y^2$.

[20] $\rho^2 - 3\rho + 2 = 0 \Rightarrow (\rho - 2)(\rho - 1) = 0 \Rightarrow \rho = 1$ and $\rho = 2$,

spheres of radii 1 and 2 with centers at the origin.

[21] $x^2 + y^2 + z^2 = 4 \Rightarrow \rho^2 = 4 \Rightarrow \rho = 2$ {$\rho \geq 0$, so $\rho \neq -2$}.

[22] $x^2 + y^2 = 4z \Rightarrow \rho^2\sin^2\phi\cos^2\theta + \rho^2\sin^2\phi\sin^2\theta = 4\rho\cos\phi \Rightarrow$

$\rho^2\sin^2\phi(\cos^2\theta + \sin^2\theta) = 4\rho\cos\phi \Rightarrow \rho^2\sin^2\phi = 4\rho\cos\phi \Rightarrow \rho\sin^2\phi = 4\cos\phi$.

[23] $3x + y - 4z = 12 \Rightarrow 3\rho\sin\phi\cos\theta + \rho\sin\phi\sin\theta - 4\rho\cos\phi = 12 \Rightarrow$

$\rho(3\sin\phi\cos\theta + \sin\phi\sin\theta - 4\cos\phi) = 12$.

[24] $y = x \Rightarrow \rho\sin\phi\sin\theta = \rho\sin\phi\cos\theta \Rightarrow \sin\theta = \cos\theta \Rightarrow \tan\theta = 1$, or $\theta = \frac{\pi}{4}$.

Note: $\rho = 0$ is a solution. $\theta = \frac{\pi}{4}$ includes this solution by passing through the origin. $\sin\phi = 0$ is also a solution. $\sin\phi = 0 \Rightarrow \phi = 0, \pi$ which is the z-axis.

$\theta = \frac{\pi}{4}$ includes the z-axis.

[25] $x^2 = 4 - y^2 \Rightarrow x^2 + y^2 = 4 \Rightarrow \rho^2\sin^2\phi\cos^2\theta + \rho^2\sin^2\phi\sin^2\theta = 4 \Rightarrow$

$\rho^2\sin^2\phi(\cos^2\theta + \sin^2\theta) = 4 \Rightarrow \rho^2\sin^2\phi = 4 \Rightarrow \rho^2 = 4\csc^2\phi \Rightarrow \rho = 2\csc\phi$.

[26] $x^2 + (y-2)^2 = 4 \Rightarrow x^2 + y^2 - 4y = 0 \Rightarrow$

$\rho^2\sin^2\phi\cos^2\theta + \rho^2\sin^2\phi\sin^2\theta = 4\rho\sin\phi\sin\theta \Rightarrow \rho^2\sin^2\phi = 4\rho\sin\phi\sin\theta \Rightarrow$

$\rho\sin\phi = 4\sin\theta \Rightarrow \rho = 4\sin\theta\csc\phi$.

[27] $x^2 - 4z^2 + y^2 = 0 \Rightarrow \rho^2 \sin^2\phi = 4\rho^2 \cos^2\phi \Rightarrow \tan^2\phi = 4.$

[28] $x^2 - y^2 - z^2 = 1 \Rightarrow \rho^2 \sin^2\phi(\cos^2\theta - \sin^2\theta) - \rho^2 \cos^2\phi = 1 \Rightarrow$

$$\rho^2(\sin^2\phi \cos 2\theta - \cos^2\phi) = 1.$$

[29] $y^2 + z^2 = 9 \Rightarrow \rho^2 \sin^2\phi \sin^2\theta + \rho^2 \cos^2\phi = 9 \Rightarrow \rho^2(\sin^2\phi \sin^2\theta + \cos^2\phi) = 9.$

[30] $x^2 + z^2 = 9 \Rightarrow \rho^2 \sin^2\phi \cos^2\theta + \rho^2 \cos^2\phi = 9 \Rightarrow \rho^2(\sin^2\phi \cos^2\theta + \cos^2\phi) = 9.$

[31] Let the hemisphere have the equation $x^2 + y^2 + z^2 = a^2$ for $z \geq 0$.

Then, $\delta(x, y, z) = k(x^2 + y^2 + z^2)^{1/2} = k\rho$.

$$m = \int_0^{\pi/2}\int_0^a\int_0^{2\pi} (k\rho)(\rho^2 \sin\phi)\, d\theta\, d\rho\, d\phi = 2k\pi\int_0^{\pi/2}\int_0^a \rho^3 \sin\phi\, d\rho\, d\phi =$$

$$\tfrac{1}{2}k\pi a^4\int_0^{\pi/2} \sin\phi\, d\phi = \tfrac{1}{2}k\pi a^4.$$

Since $z = \rho\cos\phi$, $M_{xy} = \int_0^{\pi/2}\int_0^a\int_0^{2\pi} (\rho\cos\phi)(k\rho)(\rho^2 \sin\phi)\, d\theta\, d\rho\, d\phi =$

$$2k\pi\int_0^{\pi/2}\int_0^a \rho^4 \sin\phi\cos\phi\, d\rho\, d\phi = \tfrac{2}{5}k\pi a^5\int_0^{\pi/2} \sin\phi\cos\phi\, d\phi = \tfrac{1}{5}k\pi a^5\int_0^{\pi/2} \sin 2\phi\, d\phi =$$

$\tfrac{1}{5}k\pi a^5$. By symmetry, $\bar{x} = \bar{y} = 0$. $\bar{z} = \dfrac{M_{xy}}{m} = \tfrac{2}{5}a$.

[32] See *Figure 32.* θ varies from 0 to 2π and ϕ from $\frac{\pi}{4}$ to $\frac{\pi}{2}$.

From the diagram, $\rho\sin\phi = 2 \Rightarrow \rho = 2\csc\phi$. Thus,

$$V = \int_0^{2\pi}\int_{\pi/4}^{\pi/2}\int_0^{2\csc\phi} (\rho^2 \sin\phi)\, d\rho\, d\phi\, d\theta = \int_0^{2\pi}\int_{\pi/4}^{\pi/2} \tfrac{8}{3}\csc^2\phi\, d\phi\, d\theta = \tfrac{8}{3}\int_0^{2\pi} d\theta = \tfrac{16\pi}{3}.$$

$$M_{xy} = \int_0^{2\pi}\int_{\pi/4}^{\pi/2}\int_0^{2\csc\phi} (\rho\cos\phi)(\rho^2 \sin\phi)\, d\rho\, d\phi\, d\theta = \int_0^{2\pi}\int_{\pi/4}^{\pi/2} 4\cot\phi\csc^2\phi\, d\phi\, d\theta =$$

$4\int_0^{2\pi}\left[-\tfrac{1}{2}\cot^2\phi\right]_{\pi/4}^{\pi/2} d\theta = 2\int_0^{2\pi} d\theta = 4\pi$. By symmetry, $\bar{x} = \bar{y} = 0$. $\bar{z} = \dfrac{M_{xy}}{m} = \tfrac{3}{4}$.

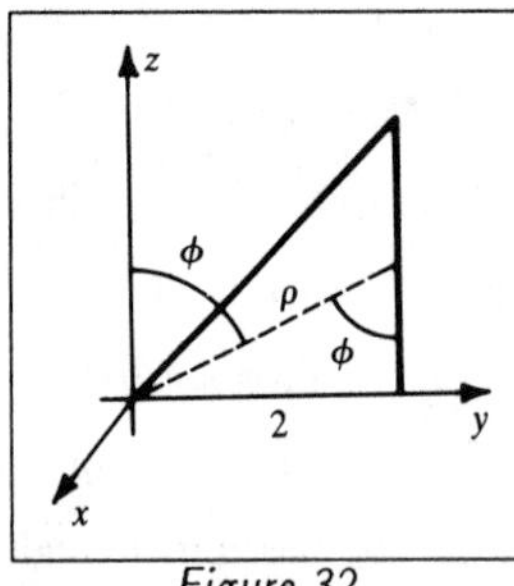

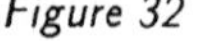
Figure 32

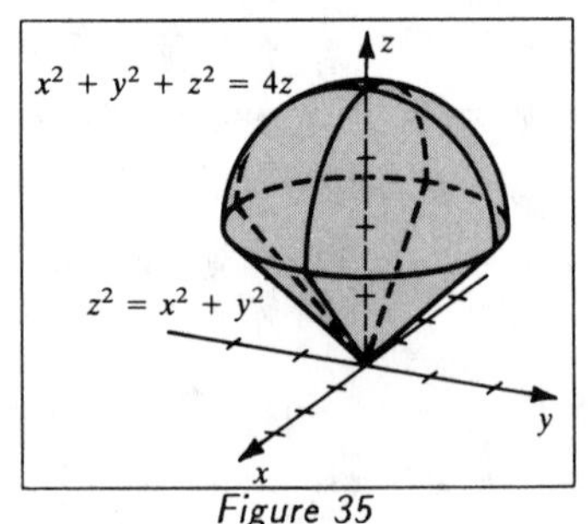

Figure 35

[33] Since $x^2 + y^2 = \rho^2\sin^2\phi$, $I_z = \int_0^{\pi/2}\int_0^a\int_0^{2\pi} (\rho^2\sin^2\phi)(k\rho)(\rho^2\sin\phi)\, d\theta\, d\rho\, d\phi =$

$$2k\pi\int_0^{\pi/2}\int_0^a \rho^5\sin^3\phi\, d\rho\, d\phi = \tfrac{1}{3}k\pi a^6\int_0^{\pi/2} (1 - \cos^2\phi)\sin\phi\, d\phi =$$

$$\tfrac{1}{3}k\pi a^6\int_{-1}^{0} (1 - u^2)\, du \;\{u = -\cos\phi\} = \tfrac{2}{9}k\pi a^6.$$

[34] Let the hemisphere have the equation $x^2 + y^2 + z^2 = a^2$ for $z \geq 0$. We choose the diameter that lies on the x-axis. Then, $y^2 + z^2 = \rho^2 \sin^2\phi \sin^2\theta + \rho^2 \cos^2\phi$.

$$I_x = \int_0^{\pi/2}\int_0^{2\pi}\int_0^a (\rho^2 \sin^2\phi \sin^2\theta + \rho^2 \cos^2\phi)(\delta)(\rho^2 \sin\phi)\, d\rho\, d\theta\, d\phi$$

$$= \tfrac{1}{5}a^5\delta \int_0^{\pi/2}\int_0^{2\pi} (\sin^3\phi \sin^2\theta + \cos^2\phi \sin\phi)\, d\theta\, d\phi$$

$$= \tfrac{1}{5}a^5\delta \int_0^{\pi/2}\Big[\tfrac{1}{2}\sin^3\phi(\theta - \tfrac{1}{2}\sin 2\theta) + (\cos^2\phi \sin\phi)\theta\Big]_0^{2\pi} d\phi$$

$$= \tfrac{1}{5}a^5\delta \int_0^{\pi/2} (\pi \sin^3\phi + 2\pi \cos^2\phi \sin\phi)\, d\phi$$

$$= \tfrac{1}{5}\pi a^5\delta \Big[-\tfrac{1}{3}(2 + \sin^2\phi)\cos\phi - \tfrac{2}{3}\cos^3\phi\Big]_0^{\pi/2} \{\text{Formula 67}\} = \tfrac{4}{15}\pi a^5\delta.$$

[35] $4z - z^2 = x^2 + y^2 = z^2 \Rightarrow z = 0, 2$. "Above the cone" means above the plane $z = 2$. See *Figure 35*. As ϕ varies between 0 and $\frac{\pi}{4}$, ρ varies between the plane $z = 2$ ($\rho\cos\phi = 2$ or $\rho = 2\sec\phi$) and the surface of the sphere ($\rho^2 = 4\rho\cos\phi$ or $\rho = 4\cos\phi$). Thus, $V = \int_0^{\pi/4}\int_0^{2\pi}\int_{2\sec\phi}^{4\cos\phi} (\rho^2 \sin\phi)\, d\rho\, d\theta\, d\phi =$

$$\tfrac{1}{3}\int_0^{\pi/4}\int_0^{2\pi} (64\cos^3\phi - 8\sec^3\phi)\sin\phi\, d\theta\, d\phi =$$

$$\tfrac{16\pi}{3}\int_0^{\pi/4}(8\cos^3\phi \sin\phi - \sec^2\phi \tan\phi)\, d\phi = \tfrac{16\pi}{3}\Big[-2\cos^4\phi - \tfrac{1}{2}\tan^2\phi\Big]_0^{\pi/4} = \tfrac{16\pi}{3}.$$

Note: The solid is a hemisphere of radius 2, which has a volume of $\frac{2}{3}\pi r^3 = \frac{16\pi}{3}$.

[36] $V = \int_0^{2\pi}\int_{\pi/4}^{3\pi/4}\int_0^1 (\rho^2 \sin\phi)\, d\rho\, d\phi\, d\theta = \frac{1}{3}\int_0^{2\pi}\int_{\pi/4}^{3\pi/4} \sin\phi\, d\phi\, d\theta = \frac{1}{3}\sqrt{2}\int_0^{2\pi} d\theta = \frac{2\pi}{3}\sqrt{2}$.

[37] $m = \int_0^{2\pi}\int_0^{\pi}\int_1^2 (k\rho^2)(\rho^2 \sin\phi)\, d\rho\, d\phi\, d\theta = \frac{31}{5}k\int_0^{2\pi}\int_0^{\pi} \sin\phi\, d\phi\, d\theta = \frac{62}{5}k\int_0^{2\pi} d\theta = \frac{124}{5}k\pi$.

[38] (a) $m = \int_0^{2\pi}\int_0^{\pi}\int_{6{,}370{,}000}^{6{,}373{,}000} \delta(\rho^2 \sin\phi)\, d\rho\, d\phi\, d\theta$

$$= \int_0^{2\pi}\int_0^{\pi} \sin\phi \Big[\frac{619.09}{3}\rho^3 - \frac{9.7 \times 10^{-5}}{4}\rho^4\Big]_{6{,}370{,}000}^{6{,}373{,}000} d\phi\, d\theta \approx$$

$$(1.28 \times 10^{17})\int_0^{2\pi}\Big[-\cos\phi\Big]_0^{\pi} d\phi\, d\theta = (1.28 \times 10^{17})(2)(2\pi) \approx 1.6 \times 10^{18}\text{ kg}.$$

(b) Using the result from part (a), $\dfrac{1.6 \times 10^{18}}{5.1 \times 10^{18}} \approx 0.31 = 31\%$.

[39] This is similar to Example 6 with $\rho = \sqrt{8}$ and $c = \frac{\pi}{4}$.

$$\int_{-2}^{2}\int_{-\sqrt{4-x^2}}^{\sqrt{4-x^2}}\int_{\sqrt{x^2+y^2}}^{\sqrt{8-x^2-y^2}} (x^2 + y^2 + z^2)\, dz\, dy\, dx = \int_0^{2\pi}\int_0^{\pi/4}\int_0^{\sqrt{8}} (\rho^2)\rho^2 \sin\phi\, d\rho\, d\phi\, d\theta$$

$$= \tfrac{128}{5}\sqrt{2}\int_0^{2\pi}\int_0^{\pi/4} \sin\phi\, d\phi\, d\theta = \tfrac{128}{5}(\sqrt{2} - 1)\int_0^{2\pi} d\theta = \tfrac{256\pi}{5}(\sqrt{2} - 1) \approx 66.63.$$

40 The region is the wedge cut from the hemisphere as indicated in *Figure 40*.

$\int_0^{\sqrt{2}} \int_y^{\sqrt{4-y^2}} \int_0^{\sqrt{4-x^2-y^2}} f(x, y, z)\, dV$

{where $f(x, y, z) = \sqrt{x^2 + y^2 + z^2}$ and $dV = dz\, dx\, dy$}

$= \int_0^{\pi/4} \int_0^{\pi/2} \int_0^2 (\rho)(\rho^2 \sin\phi)\, d\rho\, d\phi\, d\theta$

$= 4\int_0^{\pi/4} \int_0^{\pi/2} \sin\phi\, d\phi\, d\theta \;=\; 4\int_0^{\pi/4} d\theta$

$= \pi.$

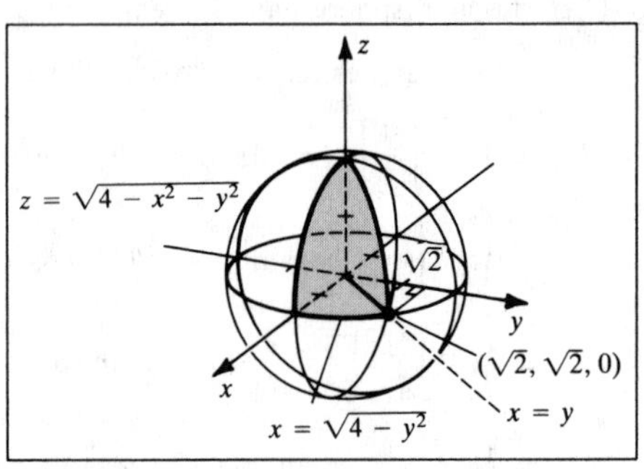

Figure 40

41 (a) $x = L\sin\phi\,\cos\theta = 12(\frac{1}{\sqrt{2}})(-\frac{1}{2}) = -\frac{6}{\sqrt{2}} = -3\sqrt{2}$ in.

$y = L\sin\phi\,\sin\theta = 12(\frac{1}{\sqrt{2}})(\frac{\sqrt{3}}{2}) = \frac{6\sqrt{3}}{\sqrt{2}} = 3\sqrt{6}$ in.

$z = L\cos\phi = 12(-\frac{1}{\sqrt{2}}) = -6\sqrt{2}$ in.

The hand is located at $(-3\sqrt{2}, 3\sqrt{6}, -6\sqrt{2})$.

(b) $(x, y, z) = (-8, -8, 8) \Rightarrow \theta = 225^\circ$, $\phi = \arctan\frac{\sqrt{(-8)^2 + (-8)^2}}{8} =$ $\arctan\frac{\sqrt{128}}{8}$ (hence the angle between the arm and the xy-plane is $\arctan\frac{8}{\sqrt{128}}$), and $L = \sqrt{(-8)^2 + (-8)^2 + 8^2} = \sqrt{192}$ in. Thus, θ should be increased by $225^\circ - 120^\circ = 105^\circ$, ϕ should be decreased by $(45^\circ + \arctan\frac{8}{\sqrt{128}}) \approx 80.26^\circ$, and L should be increased by $\sqrt{192} - 12 \approx 1.86$ in.

42 (a) $\Delta x \approx dx = \frac{\partial x}{\partial\theta}\,d\theta + \frac{\partial x}{\partial\phi}\,d\phi + \frac{\partial x}{\partial L}\,dL$

$= (-L\sin\phi\,\sin\theta)\,\Delta\theta + (L\cos\phi\,\cos\theta)\,\Delta\phi + (\sin\phi\,\cos\theta)\,\Delta L$

$\Delta y \approx dy = \frac{\partial y}{\partial\theta}\,d\theta + \frac{\partial y}{\partial\phi}\,d\phi + \frac{\partial y}{\partial L}\,dL$

$= (L\sin\phi\,\cos\theta)\,\Delta\theta + (L\cos\phi\,\sin\theta)\,\Delta\phi + (\sin\phi\,\sin\theta)\,\Delta L$

$\Delta z \approx dz = \frac{\partial z}{\partial\theta}\,d\theta + \frac{\partial z}{\partial\phi}\,d\phi + \frac{\partial z}{\partial L}\,dL = (0)\,\Delta\theta + (-L\sin\phi)\,\Delta\phi + (\cos\phi)\,\Delta L$

(b) $\Delta x \approx dx = (-12\sin\frac{\pi}{4}\,\sin\frac{3\pi}{4})(-\frac{\pi}{180}) + (12\cos\frac{\pi}{4}\,\cos\frac{3\pi}{4})(\frac{\pi}{360}) +$

$(\sin\frac{\pi}{4}\,\cos\frac{3\pi}{4})(0.1) = \frac{\pi}{60} - \frac{1}{20} \approx 0.0024$ in.

$\Delta y \approx dy = (12\sin\frac{\pi}{4}\,\cos\frac{3\pi}{4})(-\frac{\pi}{180}) + (12\cos\frac{\pi}{4}\,\sin\frac{3\pi}{4})(\frac{\pi}{360}) + (\sin\frac{\pi}{4}\,\sin\frac{3\pi}{4})(0.1)$

$= \frac{\pi + 1}{20} \approx 0.21$ in.

$\Delta z \approx dz = (-12\sin\frac{\pi}{4})(\frac{\pi}{360}) + (\cos\frac{\pi}{4})(0.1) = \frac{(3-\pi)\sqrt{2}}{60} \approx -0.0033$ in.

Exercises 17.9

1 (a) $u = c \Rightarrow 3x = c \Rightarrow x = \frac{1}{3}c$; vertical lines.

$v = d \Rightarrow 5y = d \Rightarrow y = \frac{1}{5}d$; horizontal lines.

(b) We must solve the given equations for x and y.

$u = 3x \Rightarrow x = \frac{1}{3}u$. $v = 5y \Rightarrow y = \frac{1}{5}v$.

[2] (a) $u = c \Rightarrow \frac{1}{2}y = c \Rightarrow y = 2c$; horizontal lines.

$v = d \Rightarrow \frac{1}{3}x = d \Rightarrow x = 3d$; vertical lines.

(b) $u = \frac{1}{2}y \Rightarrow y = 2u$. $v = \frac{1}{3}x \Rightarrow x = 3v$.

[3] (a) $u = c \Rightarrow x - y = c \Rightarrow y = x - c$; $v = d \Rightarrow 2x + 3y = d \Rightarrow y = -\frac{2}{3}x + \frac{1}{3}d$.

These are lines with slopes 1 and $-\frac{2}{3}$, respectively.

(b) $-2(x - y = u)$ and $2x + 3y = v \Rightarrow y = -\frac{2}{5}u + \frac{1}{5}v$.

$3(x - y = u)$ and $2x + 3y = v \Rightarrow x = \frac{3}{5}u + \frac{1}{5}v$.

[4] (a) $u = c \Rightarrow -5x + 4y = c \Rightarrow y = \frac{5}{4}x + \frac{1}{4}c$. $v = d \Rightarrow 2x - 3y = d \Rightarrow$

$y = \frac{2}{3}x - \frac{1}{3}d$. These are lines with slopes $\frac{5}{4}$ and $\frac{2}{3}$, respectively.

(b) $2(u = -5x + 4y)$ and $5(v = 2x - 3y) \Rightarrow y = -\frac{2}{7}u - \frac{5}{7}v$

$3(u = -5x + 4y)$ and $4(v = 2x - 3y) \Rightarrow x = -\frac{3}{7}u - \frac{4}{7}v$

[5] (a) $u = c \Rightarrow x^3 = c \Rightarrow x = c^{1/3}$; vertical lines.

$v = d \Rightarrow x + y = d \Rightarrow y = -x + d$; lines with slope -1.

(b) $u = x^3 \Rightarrow x = u^{1/3}$. $v = x + y \Rightarrow y = v - x = v - u^{1/3}$.

[6] (a) $u = c \Rightarrow x + 1 = c \Rightarrow x = c - 1$; vertical lines.

$v = d \Rightarrow 2 - y^3 = d \Rightarrow y = \sqrt[3]{2 - d}$; horizontal lines.

(b) $u = x + 1 \Rightarrow x = u - 1$. $v = 2 - y^3 \Rightarrow y^3 = 2 - v \Rightarrow y = \sqrt[3]{2 - v}$.

[7] (a) $u = c \Rightarrow e^x = c \Rightarrow x = \ln c$; vertical lines.

$v = d \Rightarrow e^y = d \Rightarrow y = \ln d$; horizontal lines.

(b) $u = e^x \Rightarrow x = \ln u$. $v = e^y \Rightarrow y = \ln v$.

[8] (a) $u = c \Rightarrow e^{2y} = c \Rightarrow 2y = \ln c \Rightarrow y = \frac{1}{2}\ln c$; horizontal lines.

$v = d \Rightarrow e^{-3x} = d \Rightarrow -3x = \ln d \Rightarrow x = -\frac{1}{3}\ln d$; vertical lines.

(b) $u = e^{2y} \Rightarrow y = \frac{1}{2}\ln u$. $v = e^{-3x} \Rightarrow x = -\frac{1}{3}\ln v$.

[9] (a) The vertices (0, 0), (0, 1), (2, 1), and (2, 0) are transformed into the vertices (0, 0), (0, 5), (6, 5), and (6, 0), respectively. The sides of the rectangle are segments of $x = 0$, $y = 1$, $x = 2$, and $y = 0$. In the uv-plane, these are $u = 0$, $v = 5$, $u = 6$, and $v = 0$, respectively. This is a rectangle.

(b) $x^2 + y^2 = 1 \Rightarrow \left(\frac{u}{3}\right)^2 + \left(\frac{v}{5}\right)^2 = 1 \Rightarrow \frac{u^2}{9} + \frac{v^2}{25} = 1$, an ellipse.

[10] (a) The vertices (0, 0), (3, 6), and (9, 4) are transformed into the vertices (0, 0), (3, 1), and (2, 3), respectively. The sides of the triangle are segments of $y = 2x$, $y = -\frac{1}{3}x + 7$, and $y = \frac{4}{9}x$. In the uv-plane, these are $v = \frac{1}{3}u$, $v = -2u + 7$, and $v = \frac{3}{2}u$, respectively. This is a triangle.

(b) $3x - 2y = 4 \Rightarrow 3(3v) - 2(2u) = 4 \Rightarrow 9v - 4u = 4$, a line.

11 (a) The vertices $(0, 0)$, $(0, 1)$, and $(2, 0)$ are transformed into the vertices $(0, 0)$, $(-1, 3)$, and $(2, 4)$, respectively. The sides of the triangle are segments of $x = 0$, $y = 0$, and $y = -\frac{1}{2}x + 1$. From Exercise 3, we have $x = \frac{3}{5}u + \frac{1}{5}v$ and $y = -\frac{2}{5}u + \frac{1}{5}v$. In the uv-plane, these are $v = -3u$, $v = 2u$, and $v = \frac{1}{3}u + \frac{10}{3}$, respectively. This is a triangle.

(b) $x + 2y = 1 \Rightarrow (\frac{3}{5}u + \frac{1}{5}v) + 2(-\frac{2}{5}u + \frac{1}{5}v) = 1 \Rightarrow -u + 3v = 5$, a line.

12 (a) The vertices $(0, 0)$, $(1, -1)$, $(2, 0)$, and $(1, 1)$ are transformed into the vertices $(0, 0)$, $(-9, 5)$, $(-10, 4)$, and $(-1, -1)$, respectively. The sides of the rectangle are segments of $y = -x$, $y = x - 2$, $y = -x + 2$, and $y = x$. From Exercise 4, we have $x = -\frac{3}{7}u - \frac{4}{7}v$ and $y = -\frac{2}{7}u - \frac{5}{7}v$. In the uv-plane, these are $v = -\frac{5}{9}u$, $v = u + 14$, $v = -\frac{5}{9}u - \frac{14}{9}$, and $v = u$, respectively. This is a rectangle.

(b) $x^2 + y^2 = 1 \Rightarrow \left[-\frac{1}{7}(3u + 4v)\right]^2 + \left[-\frac{1}{7}(2u + 5v)\right]^2 = 1 \Rightarrow$
$\frac{1}{49}(9u^2 + 24uv + 16v^2) + \frac{1}{49}(4u^2 + 20uv + 25v^2) = 1 \Rightarrow$
$13u^2 + 44uv + 41v^2 = 49$, an ellipse.

Note: Let J denote the indicated Jacobian.

13 $x = u^2 - v^2$ and $y = 2uv \Rightarrow J = \dfrac{\partial(x, y)}{\partial(u, v)} = \begin{vmatrix} \dfrac{\partial x}{\partial u} & \dfrac{\partial x}{\partial v} \\ \dfrac{\partial y}{\partial u} & \dfrac{\partial y}{\partial v} \end{vmatrix} = \begin{vmatrix} 2u & -2v \\ 2v & 2u \end{vmatrix} =$
$(2u)(2u) - (2v)(-2v) = 4u^2 + 4v^2$.

14 $x = e^u \sin v$ and $y = e^u \cos v \Rightarrow J = \begin{vmatrix} e^u \sin v & e^u \cos v \\ e^u \cos v & -e^u \sin v \end{vmatrix} =$
$-e^{2u} \sin^2 v - e^{2u} \cos^2 v = -e^{2u}(\sin^2 v + \cos^2 v) = -e^{2u}$.

15 $x = ve^{-2u}$ and $y = u^2 e^{-v} \Rightarrow J = \begin{vmatrix} -2ve^{-2u} & e^{-2u} \\ 2ue^{-v} & -u^2 e^{-v} \end{vmatrix} =$
$2u^2 ve^{-2u} e^{-v} - 2ue^{-2u} e^{-v} = 2u(uv - 1)e^{-(2u+v)}$.

16 $x = \dfrac{u}{u^2 + v^2}$, $y = \dfrac{v}{u^2 + v^2} \Rightarrow J = \begin{vmatrix} \dfrac{v^2 - u^2}{(u^2 + v^2)^2} & \dfrac{-2uv}{(u^2 + v^2)^2} \\ \dfrac{-2uv}{(u^2 + v^2)^2} & \dfrac{u^2 - v^2}{(u^2 + v^2)^2} \end{vmatrix}$
$= \dfrac{(v^2 - u^2)(u^2 - v^2) - (-2uv)(-2uv)}{(u^2 + v^2)^4} = \dfrac{-v^4 - 2u^2v^2 - u^4}{(u^2 + v^2)^4} = -\dfrac{(v^2 + u^2)^2}{(u^2 + v^2)^4} =$
$-1/(u^2 + v^2)^2$.

17 $x = 2u + 3v - w$, $y = v - 5w$, and $z = u + 4w \Rightarrow J = \dfrac{\partial(x, y, z)}{\partial(u, v, w)} =$

$$\begin{vmatrix} \dfrac{\partial x}{\partial u} & \dfrac{\partial x}{\partial v} & \dfrac{\partial x}{\partial w} \\ \dfrac{\partial y}{\partial u} & \dfrac{\partial y}{\partial v} & \dfrac{\partial y}{\partial w} \\ \dfrac{\partial z}{\partial u} & \dfrac{\partial z}{\partial v} & \dfrac{\partial z}{\partial w} \end{vmatrix} = \begin{vmatrix} 2 & 3 & -1 \\ 0 & 1 & -5 \\ 1 & 0 & 4 \end{vmatrix} = (2)(4) - (3)(5) + (-1)(-1) = -6.$$

[18] $x = u^2 + vw$, $y = 2v + u^2w$, and $z = uvw \Rightarrow$

$$J = \begin{vmatrix} 2u & w & v \\ 2uw & 2 & u^2 \\ vw & uw & uv \end{vmatrix} = (2u)(2uv - u^3w) - w(2u^2vw - u^2vw) + v(2u^2w^2 - 2vw)$$

$$= 4u^2v - 2u^4w + u^2vw^2 - 2v^2w.$$

[19] $y = 0 \Rightarrow v = 0$; $x = 2 \Rightarrow u + v = 2$; $y = 2x \Rightarrow x = v \Rightarrow u = 0$. $J = 2$.

As (u, v) traverses the vertices $(0, 0)$, $(2, 0)$, and $(0, 2)$ in the positive direction, (x, y) traverses the vertices $(0, 0)$, $(2, 0)$, and $(2, 4)$ in a positive direction.

Thus, using (17.35), $\iint_R (y - x)\,dx\,dy = \int_0^2 \int_0^{2-u} (v - u)(2)\,dv\,du.$

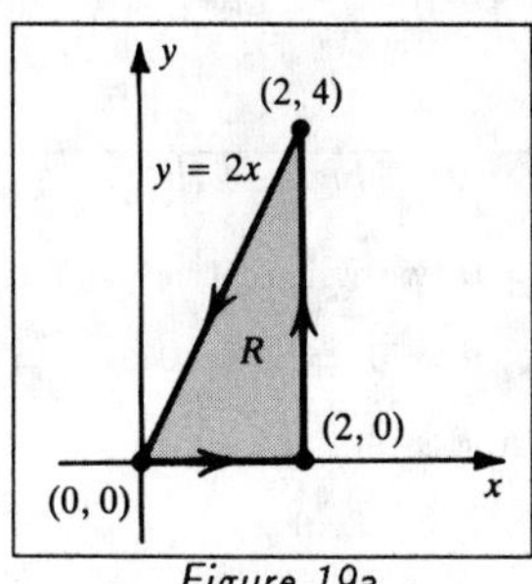

Figure 19a

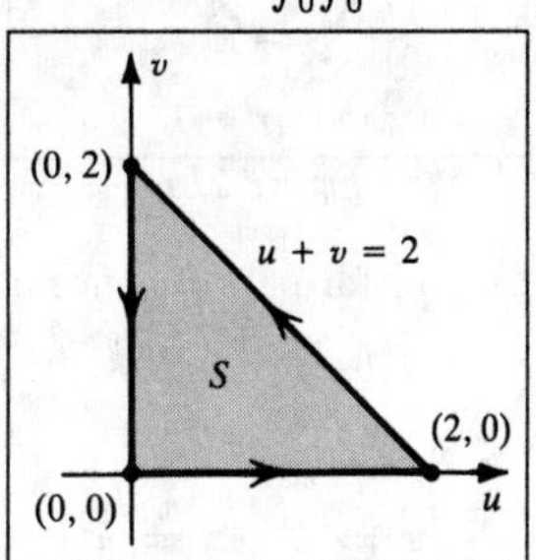

Figure 19b

[20] $y = 3x \Rightarrow u - \frac{1}{3}v = u - 2v \Rightarrow v = 0$; $y = \frac{1}{2}x \Rightarrow 6u - 2v = u - 2v \Rightarrow u = 0$; $x = 4 \Rightarrow u - 2v = 4$. $J = 5$. As (u, v) traverses the vertices $(0, 0)$, $(0, -2)$, and $(4, 0)$ in a positive direction, (x, y) traverses the vertices $(0, 0)$, $(4, 2)$, and $(4, 12)$ in a positive direction. Thus, $\iint_R (3x - 4y)\,dx\,dy = \int_{-2}^0 \int_0^{2v+4} (-9u - 2v)(5)\,du\,dv.$

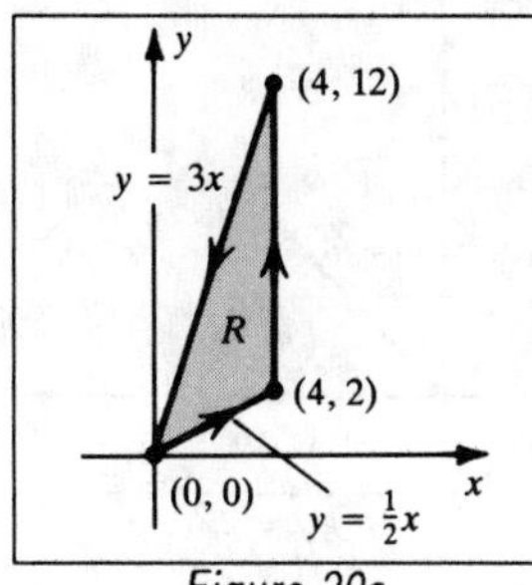

Figure 20a

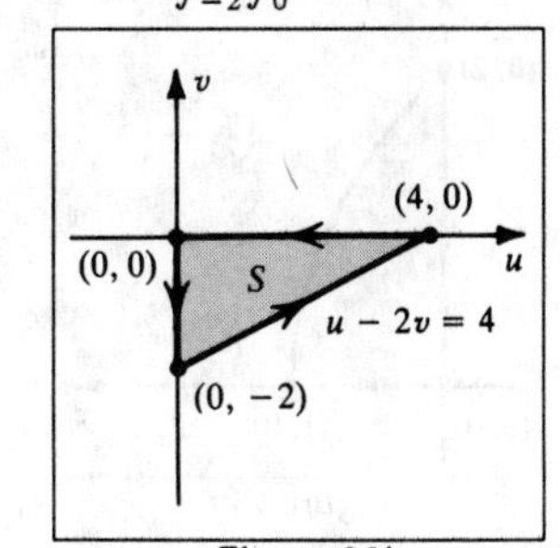

Figure 20b

21 $\frac{1}{4}x^2 + \frac{1}{9}y^2 = 1 \Rightarrow \frac{1}{4}(4u^2) + \frac{1}{9}(9v^2) = 1 \Rightarrow u^2 + v^2 = 1$; $J = 6$.

As (u, v) traverses the points $(1, 0)$, $(0, 1)$, and $(-1, 0)$ in the positive direction, (x, y) traverses the points $(2, 0)$, $(0, 3)$, and $(-2, 0)$ in the positive direction.

Thus, $\iint_R (\frac{1}{4}x^2 + \frac{1}{9}y^2)\,dx\,dy = \int_{-1}^{1}\int_{-\sqrt{1-v^2}}^{\sqrt{1-v^2}} (u^2 + v^2)(6)\,du\,dv.$

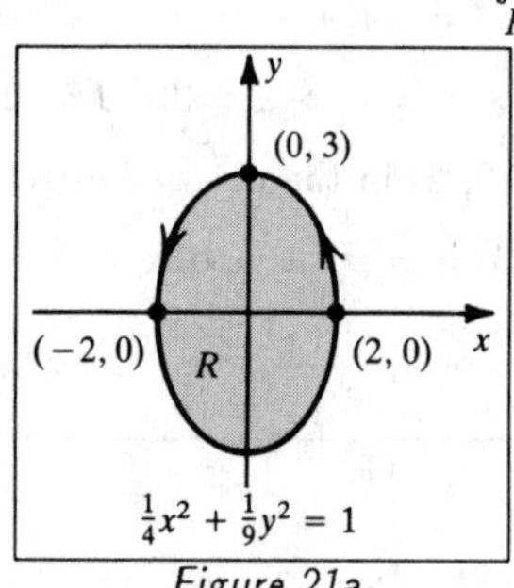

Figure 21a

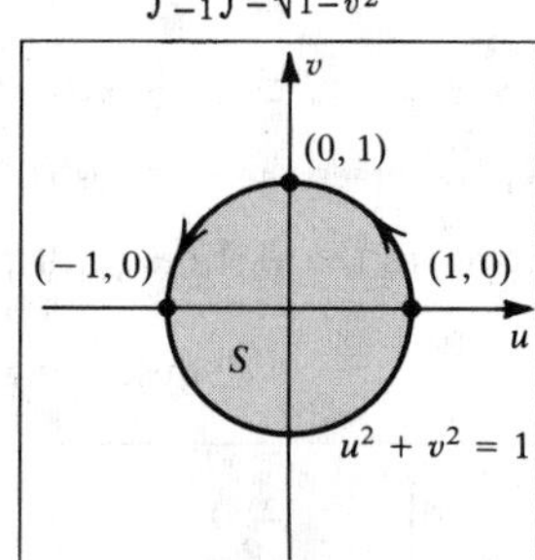

Figure 21b

22 Since the transformation is not one-to-one and $y \geq 0$, we will restrict u, v to be positive. $x = 0 \Rightarrow v^2 = u^2 \Rightarrow v = u$ $(u, v \geq 0)$. $y = 0 \Rightarrow u = 0$ or $v = 0$. Since $0 \leq x \leq 1 \Rightarrow 0 \leq u^2 - v^2 \leq 1$, $u = 0$ is impossible.

$y = 2\sqrt{1 - x} \Rightarrow 1 - x = u^2v^2 \Rightarrow u^2 + u^2v^2 - v^2 - 1 = 0 \Rightarrow$

$(u^2 - 1)(v^2 + 1) = 0 \Rightarrow u = \pm 1 \Rightarrow u = 1$ $(u \geq 0)$. $J = 4u^2 + 4v^2$.

As (u, v) traverses the vertices $(0, 0)$, $(1, 0)$, and $(1, 1)$ in a positive direction, (x, y) traverses the vertices $(0, 0)$, $(1, 0)$, and $(0, 2)$ in a positive direction.

Thus, $\iint_R xy\,dx\,dy = \int_0^1\int_0^u (u^2 - v^2)(2uv)(4u^2 + 4v^2)\,dv\,du.$

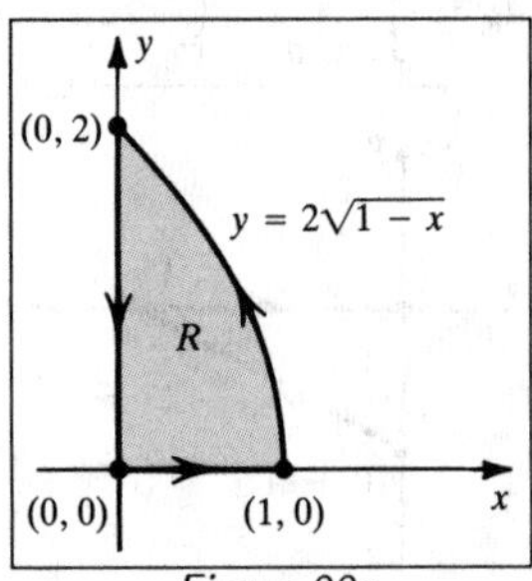

Figure 22a

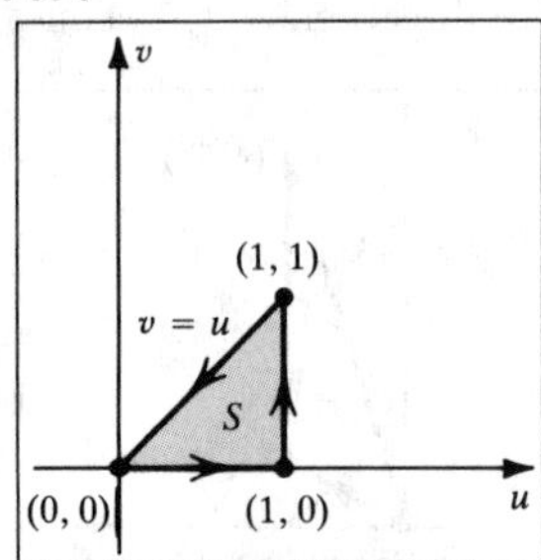

Figure 22b

23 The vertices (0, 1), (1, 0), (2, 1), and (1, 2) are transformed into the vertices (−1, 1), (1, 1), (1, 3), and (−1, 3) in a positive direction. Solving for x and y gives us $x = \frac{1}{2}(u + v)$ and $y = \frac{1}{2}(v - u)$. The line segments in the xy-plane are transformed into the following boundaries in the uv-plane: $y = -x + 1$, $0 \leq x \leq 1 \Rightarrow$ $v = 1$, $-1 \leq u \leq 1$. $y = x - 1$, $1 \leq x \leq 2 \Rightarrow u = 1$, $1 \leq v \leq 3$. $y = -x + 3$, $1 \leq x \leq 2 \Rightarrow v = 3$, $-1 \leq u \leq 1$. $y = x + 1$, $0 \leq x \leq 1 \Rightarrow u = -1$, $1 \leq v \leq 3$. $J = \frac{1}{2}$. Thus, $\iint_R (x - y)^2 \cos^2(x + y)\,dx\,dy = \int_1^3\int_{-1}^1 (u^2 \cos^2 v)(\frac{1}{2})\,du\,dv =$

$$\frac{1}{3}\int_1^3 \cos^2 v\,dv = \frac{1}{3} + \frac{1}{12}\sin 6 - \frac{1}{12}\sin 2 \approx 0.23.$$

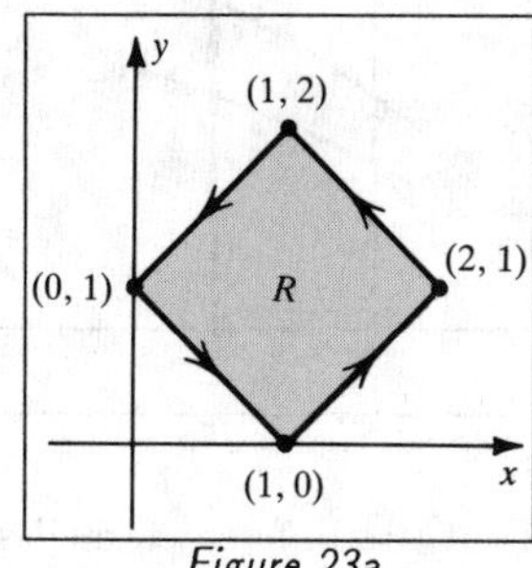

Figure 23a

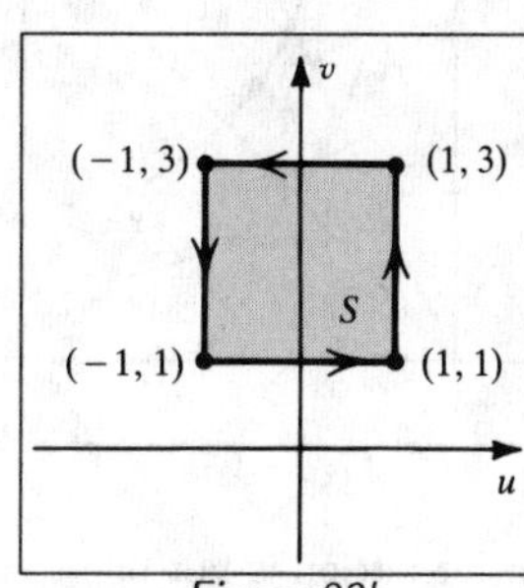

Figure 23b

24 The vertices (1, 1), (2, 0), (4, 0), and (2, 2) are transformed into the vertices (0, 2), (−2, 2), (−4, 4), and (0, 4) in a negative direction. Solving for x and y gives us $x = \frac{1}{2}(v - u)$ and $y = \frac{1}{2}(u + v)$. $y = 0$, $2 \leq x \leq 4 \Rightarrow v = -u$, $-4 \leq u \leq -2$. $y = -x + 4$, $2 \leq x \leq 4 \Rightarrow v = 4$, $-4 \leq u \leq 0$. $y = x$, $1 \leq x \leq 2 \Rightarrow$ $u = 0$, $2 \leq v \leq 4$. $y = -x + 2$, $1 \leq x \leq 2 \Rightarrow v = 2$, $-2 \leq u \leq 0$. $J = -\frac{1}{2}$. Thus, $\iint_R \sin\left(\frac{y - x}{y + x}\right) dx\,dy = -\int_2^4\int_{-v}^0 \sin(\frac{u}{v})(-\frac{1}{2})\,du\,dv = \frac{1}{2}\int_2^4 v\left[\cos(-1) - 1\right] dv =$

$$\tfrac{1}{2}(\cos 1 - 1)(6) = 3(\cos 1 - 1) \approx -1.38.$$

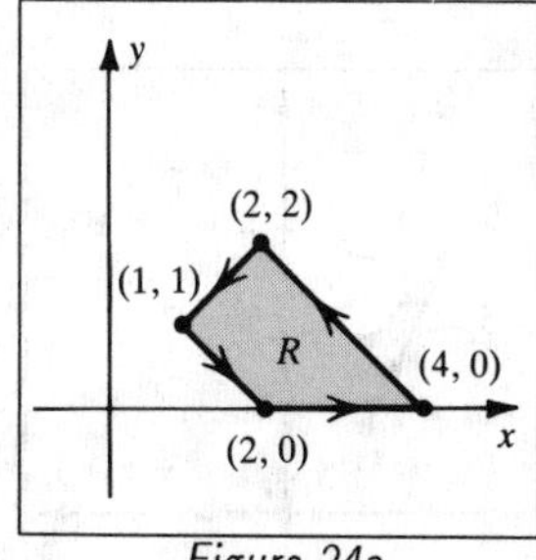

Figure 24a

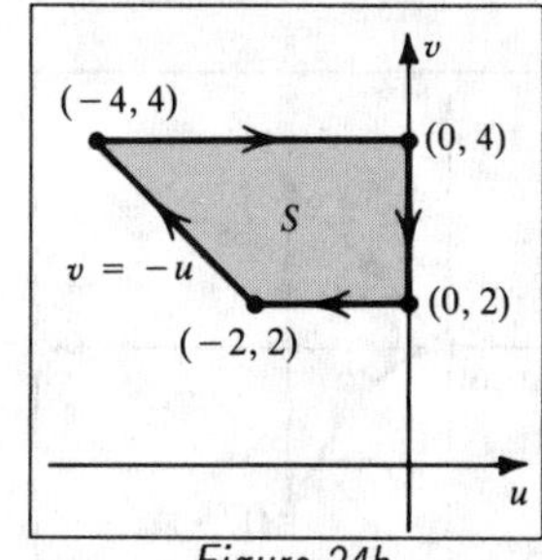

Figure 24b

25 The vertices $(\frac{1}{2}\sqrt{2}, \sqrt{2})$, $(1, 1)$, $(\sqrt{2}, \sqrt{2})$, and $(1, 2)$ are transformed into the vertices $(1, \sqrt{2})$, $(1, 1)$, $(2, \sqrt{2})$, and $(2, 2)$ in a positive direction. $x = u/v$, $y = v \Rightarrow u = xy$, $v = y$, and u and v are positive since x and y are. $xy = 1$ and $xy = 2 \Rightarrow$ $u = 1$ and $u = 2$. $y = x \Rightarrow v = u/v \Rightarrow u = v^2$, or $v = \sqrt{u}$. $y = 2x \Rightarrow$ $v = 2(u/v) \Rightarrow v = \sqrt{2u}$. Both curves are traced out in a positive direction. $J = \frac{1}{v}$. Thus, $\iint_R (x^2 + 2y^2)\, dx\, dy =$

$$\int_1^2 \int_{\sqrt{u}}^{\sqrt{2u}} \left(\frac{u^2}{v^2} + 2v^2\right)\left(\frac{1}{v}\right) dv\, du = \int_1^2 \left[-\frac{u^2}{2v^2} + v^2\right]_{\sqrt{u}}^{\sqrt{2u}} du = \frac{5}{4}\int_1^2 u\, du = \frac{15}{8}.$$

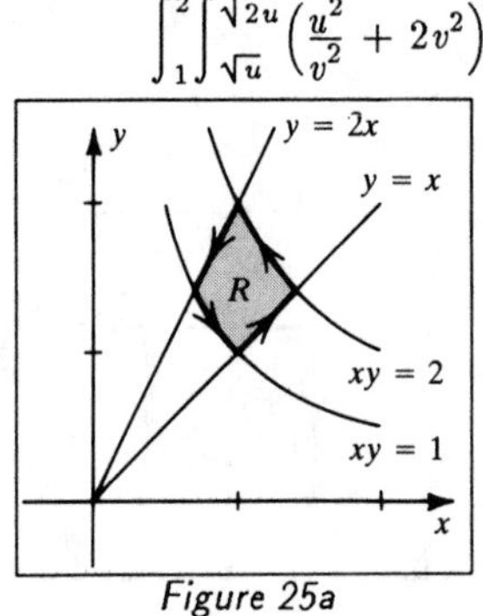

Figure 25a

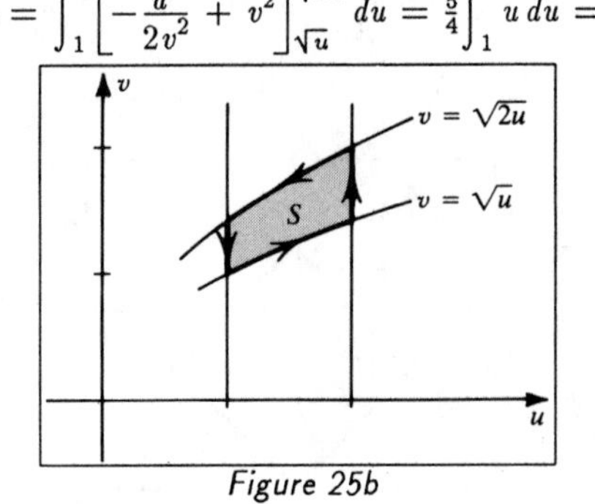

Figure 25b

26 The transformation is not one-to-one and it traces out two different triangles. $x = \sqrt{-y} \Rightarrow u + v = \sqrt{u^2 - v} \Rightarrow v^2 + 2uv + v = 0 \Rightarrow v = -2u - 1$ or $v = 0$. $x = y \Rightarrow u + v = v - u^2 \Rightarrow u(u + 1) = 0 \Rightarrow u = 0$ or $u = -1$. $x = 1 \Rightarrow$ $u + v = 1$. See the figures for the vertices and orientations of the curves. Notice that S_1 is oriented in a positive direction and S_2 is oriented in a negative direction. We choose to integrate over S_1. $J = 1 + 2u$. Thus, $\iint_R (4x - 4y + 1)^{-2}\, dx\, dy =$

$$\int_0^1 \int_0^{1-v} \frac{1}{(4u^2 + 4u + 1)^2}(1 + 2u)\, du\, dv = \frac{1}{4}\int_0^1 \left[-\frac{1}{(4u^2 + 4u + 1)}\right]_0^{1-v} dv =$$

$$\frac{1}{4}\int_0^1 \left[1 - \frac{1}{4v^2 - 12v + 9}\right] dv = \frac{1}{4}\int_0^1 dv - \frac{1}{4}\int_0^1 \frac{1}{(2v - 3)^2}\, dv = \frac{1}{4} - \frac{1}{12} = \frac{1}{6}.$$

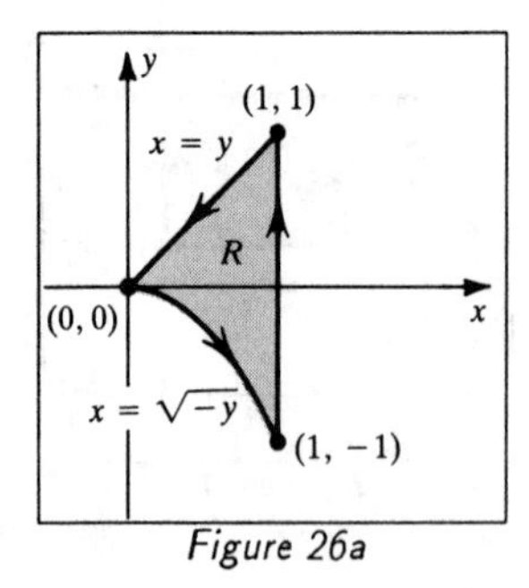

Figure 26a

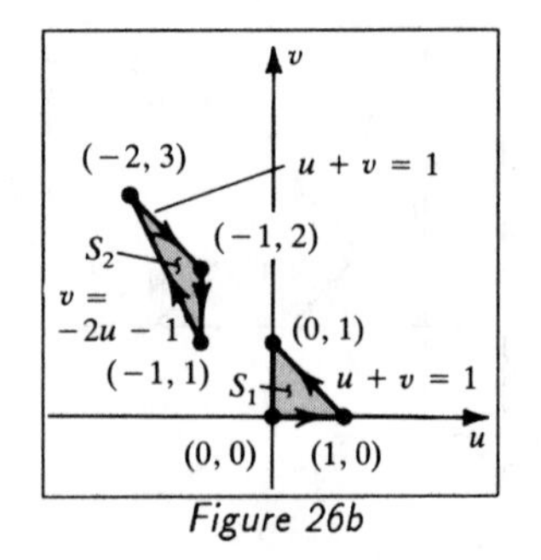

Figure 26b

[27] The vertices $(-2, 0)$, $(-1, 0)$, $(0, 2)$, and $(0, 4)$ are transformed into the vertices $(4, -2)$, $(2, -1)$, $(2, 4)$, and $(4, 8)$ in a negative direction. The four line segments in the xy-plane are transformed into four line segments in the uv-plane connecting the vertices as shown in the figures. Since $x = \frac{1}{5}v - \frac{2}{5}u$ and $y = \frac{1}{5}u + \frac{2}{5}v$, $J = -\frac{1}{5}$.

$$\text{Thus, } \iint_R \frac{2y + x}{y - 2x}\, dx\, dy = -\int_2^4 \int_{-u/2}^{2u} \left(\tfrac{v}{u}\right)\left(-\tfrac{1}{5}\right) dv\, du = \tfrac{1}{5}\int_2^4 \tfrac{15}{8}u\, du = \tfrac{9}{4}.$$

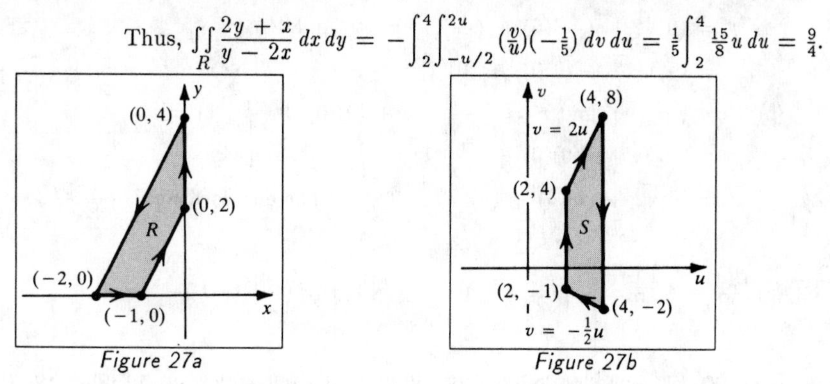

Figure 27a Figure 27b

[28] The vertices $(0, 0)$, $(4, 2)$, and $(4, 0)$ are transformed into the vertices $(0, 0)$, $(1, 0)$, and $(0, 4)$ in a negative direction. The three line segments in the xy-plane are transformed into three line segments in the uv-plane connecting the vertices as shown in the figures. Since $x = 4u + v$ and $y = 2u$, $J = -2$.

$$\text{Thus, } \iint_R \left(\sqrt{x - 2y} + \tfrac{1}{4}y^2\right) dx\, dy = -\int_0^1 \int_0^{4-4u} (v^{1/2} + u^2)(-2)\, dv\, du =$$

$$2\int_0^1 \left[\tfrac{2}{3}(4 - 4u)^{3/2} + 4u^2 - 4u^3\right] du = \tfrac{74}{15}.$$

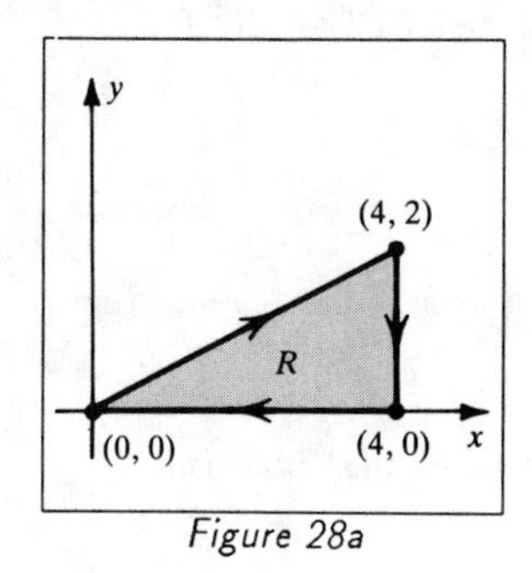

Figure 28a

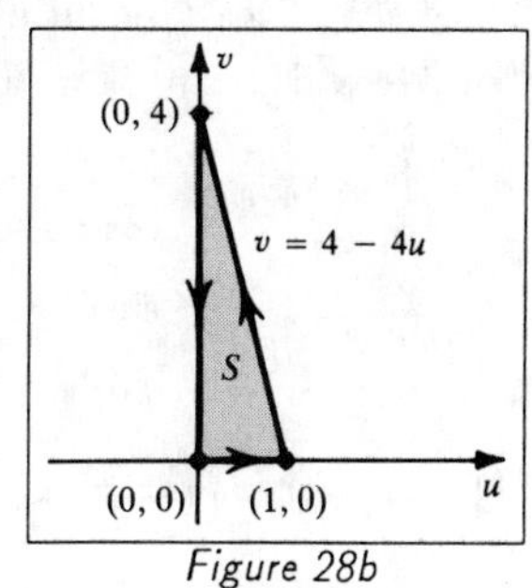

Figure 28b

[29] $\frac{x^2}{a^2} + \frac{y^2}{b^2} + \frac{z^2}{c^2} = 1$ $\{x = au,\ y = bv, \text{ and } z = cw\} \Rightarrow u^2 + v^2 + w^2 = 1$,

a sphere with radius 1. $J = \dfrac{\partial(x, y, z)}{\partial(u, v, w)} = abc$. $\iiint_R dx\, dy\, dz = \iiint_S (abc)\, du\, dv\, dw =$

$$abc\int_0^{2\pi} \int_0^{\pi} \int_0^1 (\rho^2 \sin\phi)\, d\rho\, d\phi\, d\theta = \tfrac{4}{3}\pi abc = \tfrac{4}{3}\pi(6378)^2(6356) \approx 1.08 \times 10^{12} \text{ km}^3.$$

[30] $W = P\Delta V = (1.01 \times 10^5)\left[\frac{4}{3}\pi(\sqrt{10}\sqrt{16}\sqrt{12} - 3^3)\right] \approx 7.1 \times 10^6$ joules

31 (a) We have $\phi = 45^\circ$, $\theta = 0^\circ$, $\psi = 0^\circ$, $x = \sqrt{2}$, $y = 0$, and $z = 1$. Thus,

$x' = (\frac{1}{\sqrt{2}})\sqrt{2} + (-\frac{1}{\sqrt{2}})(0) + (0)(1) = 1$, $y' = (\frac{1}{\sqrt{2}})\sqrt{2} + (\frac{1}{\sqrt{2}})(0) + (0)(1) = 1$,

and $z' = (0)\sqrt{2} + (0)(0) + (1)(1) = 1$. The new location is $(1, 1, 1)$.

(b) Now $\theta = 90^\circ$ and $x' = (0)\sqrt{2} + (-\frac{1}{\sqrt{2}})(0) + (\frac{1}{\sqrt{2}})(1) = \frac{1}{\sqrt{2}}$,

$y' = (0)\sqrt{2} + (\frac{1}{\sqrt{2}})(0) + (\frac{1}{\sqrt{2}})(1) = \frac{1}{\sqrt{2}}$, and

$z' = (-1)\sqrt{2} + (0)(0) + (0)(1) = -\sqrt{2}$. The new location is $(\frac{1}{\sqrt{2}}, \frac{1}{\sqrt{2}}, -\sqrt{2})$.

(c) Now $\psi = 90^\circ$ and $x' = (-\frac{1}{\sqrt{2}})\sqrt{2} + (0)(0) + (\frac{1}{\sqrt{2}})(1) = \frac{1}{\sqrt{2}} - 1$,

$y' = (\frac{1}{\sqrt{2}})\sqrt{2} + (0)(0) + (\frac{1}{\sqrt{2}})(1) = \frac{1}{\sqrt{2}} + 1$, and

$z' = (0)\sqrt{2} + (1)(0) + (0)(1) = 0$. The new location is $(\frac{1}{\sqrt{2}} - 1, \frac{1}{\sqrt{2}} + 1, 0)$.

32 (a) $$\frac{\partial(x', y', z')}{\partial(x, y, z)} = \begin{vmatrix} -1/\sqrt{2} & 0 & 1/\sqrt{2} \\ 1/\sqrt{2} & 0 & 1/\sqrt{2} \\ 0 & 1 & 0 \end{vmatrix} = (-1)(-\tfrac{1}{2} - \tfrac{1}{2}) = 1.$$

(b) Since the Jacobian is not zero, there is only one choice of motion. No.

33 $x = \rho \sin\phi \cos\theta$, $y = \rho \sin\phi \sin\theta$, $z = \rho \cos\phi \Rightarrow$

$$J = \frac{\partial(x, y, z)}{\partial(\rho, \phi, \theta)} = \begin{vmatrix} \sin\phi \cos\theta & \rho \cos\phi \cos\theta & -\rho \sin\phi \sin\theta \\ \sin\phi \sin\theta & \rho \cos\phi \sin\theta & \rho \sin\phi \cos\theta \\ \cos\phi & -\rho \sin\phi & 0 \end{vmatrix}$$

Evaluating J along the third row gives us

$$J = \cos\phi\,(\rho^2 \sin\phi \cos\phi \cos^2\theta + \rho^2 \sin\phi \cos\phi \sin^2\theta) + \rho \sin\phi\,(\rho \sin^2\phi \cos^2\theta + \rho \sin^2\phi \sin^2\theta)$$

$$= \rho^2 \sin\phi \cos^2\phi\,(\cos^2\theta + \sin^2\theta) + \rho^2 \sin\phi \sin^2\phi\,(\cos^2\theta + \sin^2\theta)$$

$$= \rho^2 \sin\phi\,(\cos^2\phi + \sin^2\phi) = \rho^2 \sin\phi.$$

34 $x = r\cos\theta$, $y = r\sin\theta$, $z = z \Rightarrow$

$$J = \frac{\partial(x, y, z)}{\partial(r, \theta, z)} = \begin{vmatrix} \cos\theta & -r\sin\theta & 0 \\ \sin\theta & r\cos\theta & 0 \\ 0 & 0 & 1 \end{vmatrix} = r\cos^2\theta + r\sin^2\theta = r.$$

Thus, $\iiint_R F(x, y, z)\,dx\,dy\,dz = \pm \iiint_S F(r\cos\theta, r\sin\theta, z)(r)\,dr\,d\theta\,dz$.

35 $$\frac{\partial(x, y)}{\partial(u, v)} \cdot \frac{\partial(u, v)}{\partial(x, y)} = \begin{vmatrix} f_u & f_v \\ g_u & g_v \end{vmatrix} \cdot \begin{vmatrix} u_x & u_y \\ v_x & v_y \end{vmatrix} \quad \{\text{since } x = f(u, v) \text{ and } y = g(u, v)\}$$

$$= \begin{vmatrix} f_u u_x + f_v v_x & f_u u_y + f_v v_y \\ g_u u_x + g_v v_x & g_u u_y + g_v v_y \end{vmatrix} \quad \{\text{using the hint}\}$$

$$= \begin{vmatrix} f_x & f_y \\ g_x & g_y \end{vmatrix} \quad \{\text{by a chain rule}\}$$

$$= \begin{vmatrix} 1 & 0 \\ 0 & 1 \end{vmatrix} \quad \{f(u, v) = x \Rightarrow f_x = 1,\ g(u, v) = y \Rightarrow g_y = 1\} = 1.$$

[36] $\dfrac{\partial(x, y)}{\partial(u, v)} \cdot \dfrac{\partial(u, v)}{\partial(r, s)} = \begin{vmatrix} x_u & x_v \\ y_u & y_v \end{vmatrix} \cdot \begin{vmatrix} u_r & u_s \\ v_r & v_s \end{vmatrix}$

{ since $x = f(u, v)$, $y = g(u, v)$, $u = h(r, s)$, $v = k(r, s)$ }

$= \begin{vmatrix} x_u u_r + x_v v_r & x_u u_s + x_v v_s \\ y_u u_r + y_v v_r & y_u u_s + y_v v_s \end{vmatrix}$ { using the hint in Exercise 35 }

$= \begin{vmatrix} x_r & x_s \\ y_r & y_s \end{vmatrix}$ { by a chain rule } $= \dfrac{\partial(x, y)}{\partial(r, s)}$.

[37] $x = \frac{1}{4}v^4 + u^2$, $y = u - \cos v \Rightarrow$

$$J = \frac{\partial(x, y)}{\partial(u, v)} = \begin{vmatrix} 2u & v^3 \\ 1 & \sin u \end{vmatrix} = 2u \sin u - v^3. \ J = 0 \Rightarrow v = \sqrt[3]{2u \sin u}.$$

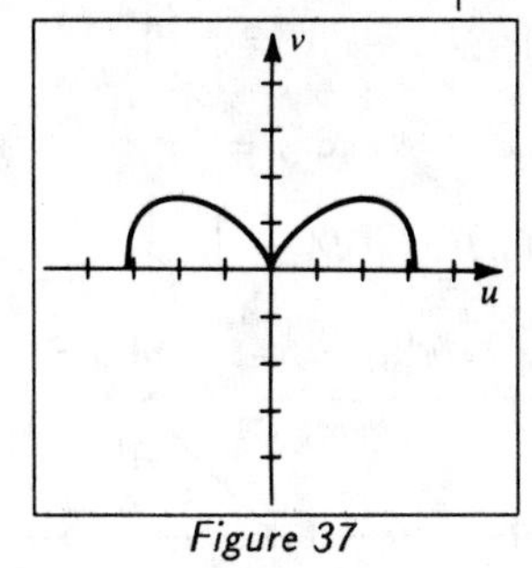

Figure 37

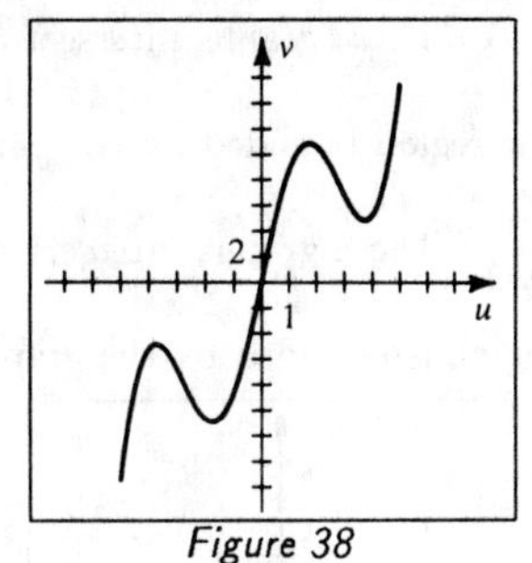

Figure 38

[38] $x = \frac{1}{20}u^4 - 10\cos u + \frac{1}{2}v^2$, $y = u + v \Rightarrow$

$$J = \frac{\partial(x, y)}{\partial(u, v)} = \begin{vmatrix} \frac{1}{5}u^3 + 10\sin u & v \\ 1 & 1 \end{vmatrix} = \tfrac{1}{5}u^3 + 10\sin u - v.$$

$$J = 0 \Rightarrow v = \tfrac{1}{5}u^3 + 10\sin u.$$

17.10 Review Exercises

[1] $\displaystyle\int_{-1}^{0}\int_{x+1}^{x^3} (x^2 - 2y)\,dy\,dx = \int_{-1}^{0}\Big[x^2(x^3 - x - 1) - x^6 + (x+1)^2\Big]\,dx = -\tfrac{5}{84}.$

[2] $\displaystyle\int_{1}^{2}\int_{1}^{y^2} \frac{1}{y}\,dx\,dy = \int_{1}^{2}\left(y - \frac{1}{y}\right)dy = \Big[\tfrac{1}{2}y^2 - \ln y\Big]_1^2 = \tfrac{3}{2} - \ln 2 \approx 0.81.$

[3] $\displaystyle\int_{0}^{3}\int_{r}^{r^2+1} r\,d\theta\,dr = \int_{0}^{3}(r^3 + r - r^2)\,dr = \tfrac{63}{4}.$

[4] $\displaystyle\int_{2}^{0}\int_{0}^{z^2}\int_{x}^{z} (x + z)\,dy\,dx\,dz = \int_{2}^{0}\int_{0}^{z^2} (x + z)(z - x)\,dx\,dz = \int_{2}^{0}(z^4 - \tfrac{1}{3}z^6)\,dz = -\tfrac{32}{105}.$

[5] $\displaystyle\int_{0}^{2}\int_{\sqrt{y}}^{1}\int_{z^2}^{y} xy^2z^3\,dx\,dz\,dy = \tfrac{1}{2}\int_{0}^{2}\int_{\sqrt{y}}^{1}(y^4z^3 - y^2z^7)\,dz\,dy =$

$$\tfrac{1}{16}\int_{0}^{2}(2y^4 - y^2 - y^6)\,dy = -\tfrac{107}{210}.$$

6 $\displaystyle\int_0^{\pi/2}\int_0^{\pi/4}\int_0^{a\cos\phi} \rho^2 \sin\phi \, d\rho \, d\phi \, d\theta = \frac{1}{3}\int_0^{\pi/2}\int_0^{\pi/4} a^3 \cos^3\phi \sin\phi \, d\phi \, d\theta =$

$$\frac{1}{3}a^3\int_0^{\pi/2}\Big[-\tfrac{1}{4}\cos^4\phi\Big]_0^{\pi/4} d\theta = -\tfrac{1}{12}a^3(-\tfrac{3}{4})\int_0^{\pi/2} d\theta = \tfrac{1}{32}a^3\pi.$$

7 R is the region bounded by the right branch of the hyperbola $(x^2/4) - (y^2/4) = 1$

and the vertical line $x = 4$. $\iint_R f(x, y)\, dA = \displaystyle\int_2^4\int_{-\sqrt{x^2-4}}^{\sqrt{x^2-4}} f(x, y)\, dy\, dx$

8 R is the region bounded by the right and left branches of the hyperbola $(x^2/4) - (y^2/4) = 1$ and the horizontal lines $y = 0$ and $y = 4$.

$$\iint_R f(x, y)\, dA = \int_0^4\int_{-\sqrt{y^2+4}}^{\sqrt{y^2+4}} f(x, y)\, dx\, dy$$

9 R is the region bounded by the parabolas $x = y^2 - 4$ and $x = -y^2 + 4$.

These graphs intersect at $(0, \pm 2)$. $\iint_R f(x, y)\, dA = \displaystyle\int_{-2}^2\int_{y^2-4}^{4-y^2} f(x, y)\, dx\, dy$

10 R is the region bounded by the parabolas $y = -x^2 + 4$ and $y = 3x^2$.

These graphs intersect at $(\pm 1, 3)$. $\iint_R f(x, y)\, dA = \displaystyle\int_{-1}^1\int_{3x^2}^{4-x^2} f(x, y)\, dy\, dx$

11 The region is bounded by the graphs of $x = e^y$, $x = y^3$, $y = -1$, and $y = 1$.

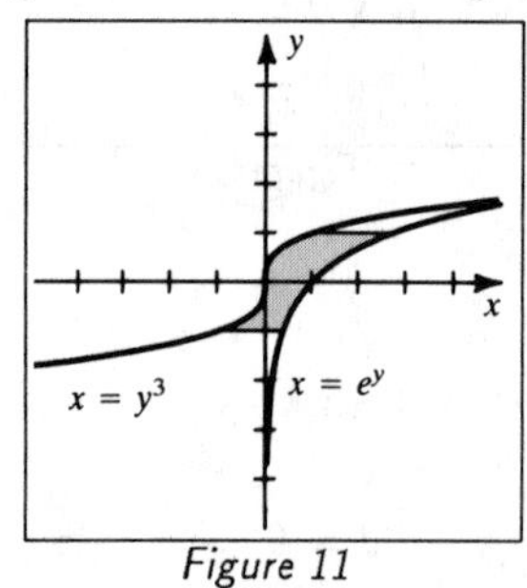

Figure 11

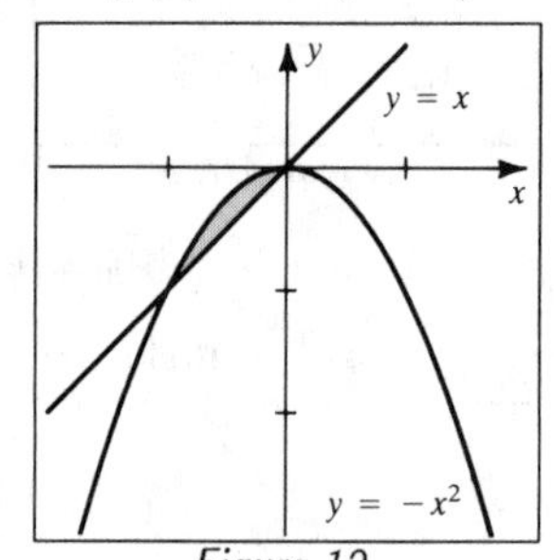

Figure 12

12 The region is bounded by the graphs of $y = x$, $y = -x^2$, $x = -1$, and $x = 0$.

13 $r = 2 + 4\cos\theta$ for $0 \le \theta \le \pi$ is the upper half of a limaçon with a loop.

It passes through the pole when $\theta = \frac{2\pi}{3}$.

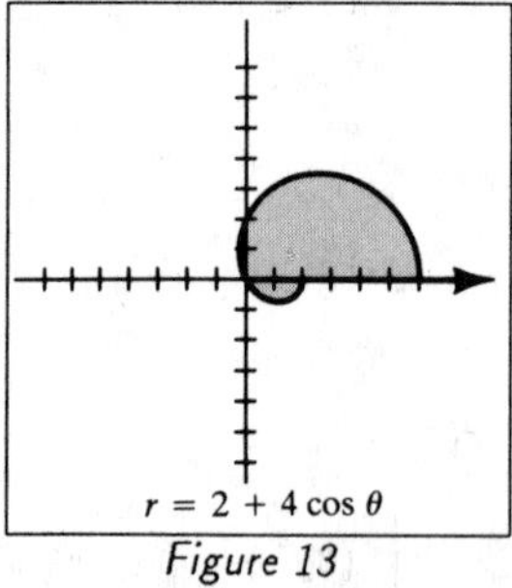

Figure 13

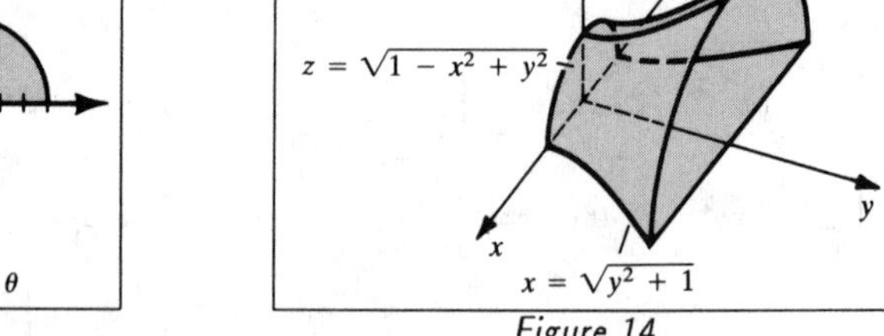

Figure 14

14 $z = \sqrt{1 - x^2 + y^2}$ is the upper half of a hyperboloid of one sheet with axis along the y-axis.

15 R is the region bounded by $x = y^2$ $\{y = \pm\sqrt{x}\}$ and $x = 9$ from $y = 0$ to 3.

$$\int_0^3\int_{y^2}^9 ye^{-x^2}\,dx\,dy = \int_0^9\int_0^{\sqrt{x}} ye^{-x^2}\,dy\,dx = \tfrac{1}{2}\int_0^9 xe^{-x^2}\,dx = \tfrac{1}{4}(1 - e^{-81}) \approx 0.25.$$

16 R is the region bounded by $y = x$ and $y = \sqrt{x}$ from $x = 0$ to 1.

$$\int_0^1\int_x^{\sqrt{x}} e^{x/y}\,dy\,dx = \int_0^1\int_{y^2}^{y} e^{x/y}\,dx\,dy = \int_0^1\Big[ye^{x/y}\Big]_{y^2}^{y}\,dy =$$

$$\int_0^1 (ye - ye^y)\,dy = \Big[\tfrac{1}{2}ey^2 - ye^y + e^y\Big]_0^1 = \tfrac{1}{2}e - 1 \approx 0.36.$$

17 $V = \displaystyle\int_1^2\int_1^3 xy^2\,dy\,dx = \tfrac{26}{3}\int_1^2 x\,dx = \tfrac{26}{3}\cdot\tfrac{3}{2} = 13.$

18 $V = \displaystyle\int_0^{2\pi}\int_0^2\int_0^{7-r^2} r\,dz\,dr\,d\theta = \int_0^{2\pi}\int_0^2 (7r - r^3)\,dr\,d\theta = 10\int_0^{2\pi} d\theta = 20\pi.$

19 The trace of the surface described in the xy-plane is the circle $(x - 2)^2 + y^2 = 4$.

$$z = f(x, y) = (x^2 + y^2)^{1/2} \Rightarrow f_x = \frac{x}{(x^2 + y^2)^{1/2}} \text{ and } f_y = \frac{y}{(x^2 + y^2)^{1/2}}.$$

$$S = \iint_R \sqrt{1 + 1}\,dA = \sqrt{2}\iint_R dA = 4\sqrt{2}\,\pi \approx 17.77$$

since the area of R (a circle with radius 2 in the xy-plane) is 4π.

20 For the hyperboloid, $x^2 + y^2 \geq 1$.

Hence, the surface area that lies over the circle $x^2 + y^2 = 1$ is 0.

21 $e^\theta = r = 2$ when $\theta = \ln 2$.

$$A = \int_0^{\ln 2}\int_{e^\theta}^2 r\,dr\,d\theta$$

$$= \tfrac{1}{2}\int_0^{\ln 2}(4 - e^{2\theta})\,d\theta = 2\ln 2 - \tfrac{3}{4} \approx 0.64.$$

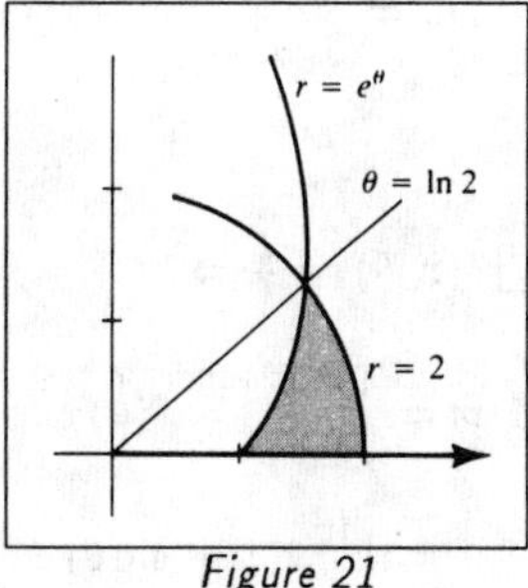

Figure 21

22 R is the quarter of the circle in quadrant III whose boundary is $x^2 + y^2 = a^2$.

$$\int_{-a}^0\int_{-\sqrt{a^2-x^2}}^0 \sqrt{x^2 + y^2}\,dy\,dx = \int_\pi^{3\pi/2}\int_0^a (r)\,r\,dr\,d\theta = \frac{a^3}{3}\int_\pi^{3\pi/2} d\theta = \tfrac{1}{6}\pi a^3.$$

23 R is the quarter of the circle in quadrant I whose boundary is $x^2 + y^2 = 16$. This integral represents the volume between the cone $z = \sqrt{x^2 + y^2}$ and the hemisphere $z = \sqrt{32 - x^2 - y^2}$. These intersect when $x^2 + y^2 = 16$, or $z = 4$.

$$\int_0^4\int_0^{\sqrt{16-x^2}}\int_{\sqrt{x^2+y^2}}^{\sqrt{32-x^2-y^2}} \sqrt{x^2 + y^2 + z^2}\,dz\,dy\,dx = \int_0^{\pi/2}\int_0^{\pi/4}\int_0^{4\sqrt{2}} \rho(\rho^2\sin\phi)\,d\rho\,d\phi\,d\theta =$$

$$256\int_0^{\pi/2}\int_0^{\pi/4}\sin\phi\,d\phi\,d\theta = 256\left(1 - \tfrac{\sqrt{2}}{2}\right)\int_0^{\pi/2} d\theta = 64(2 - \sqrt{2})\pi \approx 117.78.$$

24 (1) See *Figure 24* below. $\underline{z \text{ integrated first}}$ The trace in the xy-plane is $y = x^2$.

$$\int_{-2}^{2}\int_{x^2}^{4}\int_{-\sqrt{y-x^2}/2}^{\sqrt{y-x^2}/2} f(x, y, z)\, dz\, dy\, dx;\ \int_{0}^{4}\int_{-\sqrt{y}}^{\sqrt{y}}\int_{-\sqrt{y-x^2}/2}^{\sqrt{y-x^2}/2} f(x, y, z)\, dz\, dx\, dy$$

(2) $\underline{x \text{ integrated first}}$ The trace in the yz-plane is $y = 4z^2$.

$$\int_{0}^{4}\int_{-\sqrt{y}/2}^{\sqrt{y}/2}\int_{-\sqrt{y-4z^2}}^{\sqrt{y-4z^2}} f(x, y, z)\, dx\, dz\, dy;\ \int_{-1}^{1}\int_{4z^2}^{4}\int_{-\sqrt{y-4z^2}}^{\sqrt{y-4z^2}} f(x, y, z)\, dx\, dy\, dz$$

(3) $\underline{y \text{ integrated first}}$ The trace in the xz-plane is $x^2 + 4z^2 = 4$.

$$\int_{-2}^{2}\int_{-\sqrt{4-x^2}/2}^{\sqrt{4-x^2}/2}\int_{x^2+4z^2}^{4} f(x, y, z)\, dy\, dz\, dx;\ \int_{-1}^{1}\int_{-\sqrt{4-4z^2}}^{\sqrt{4-4z^2}}\int_{x^2+4z^2}^{4} f(x, y, z)\, dy\, dx\, dz$$

Note: Let k denote a constant of proportionality.

25 $\delta(x, y) = k|x| = kx$ since $x \geq 0$. $m = \int_0^3\int_x^{2x} (kx)\, dy\, dx = k\int_0^3 x^2\, dx = 9k$.

$$M_x = \int_0^3\int_x^{2x} (kx)\, y\, dy\, dx = \tfrac{3}{2}k\int_0^3 x^3\, dx = \tfrac{243}{8}k.$$

$$M_y = \int_0^3\int_x^{2x} (kx)\, x\, dy\, dx = k\int_0^3 x^3\, dx = \tfrac{81}{4}k. \quad \text{Thus, } \bar{x} = \frac{M_y}{m} = \tfrac{9}{4} \text{ and } \bar{y} = \frac{M_x}{m} = \tfrac{27}{8}.$$

26 $\delta(x, y) = k(x + 1)$ since $x \geq -1$. $y^2 = x = 4$ when $y = \pm 2$.

$$m = \int_{-2}^{2}\int_{y^2}^{4} k(x + 1)\, dx\, dy = k\int_{-2}^{2} (12 - \tfrac{1}{2}y^4 - y^2)\, dy = \tfrac{544}{15}k.$$

$$M_y = \int_{-2}^{2}\int_{y^2}^{4} k(x + 1)\, x\, dx\, dy = k\int_{-2}^{2} (\tfrac{64}{3} + 8 - \tfrac{1}{3}y^6 - \tfrac{1}{2}y^4)\, dy = \tfrac{3456}{35}k.$$

By symmetry, $\bar{y} = 0$. $\bar{x} = \frac{M_y}{m} = \frac{324}{119}$.

27 $\delta(r, \theta) = \frac{k}{r} \Rightarrow$

$$m = \int_0^{2\pi}\int_1^{2+\sin\theta} (\tfrac{k}{r})\, r\, dr\, d\theta$$

$$= k\int_0^{2\pi} (1 + \sin\theta)\, d\theta$$

$$= 2\pi k.$$

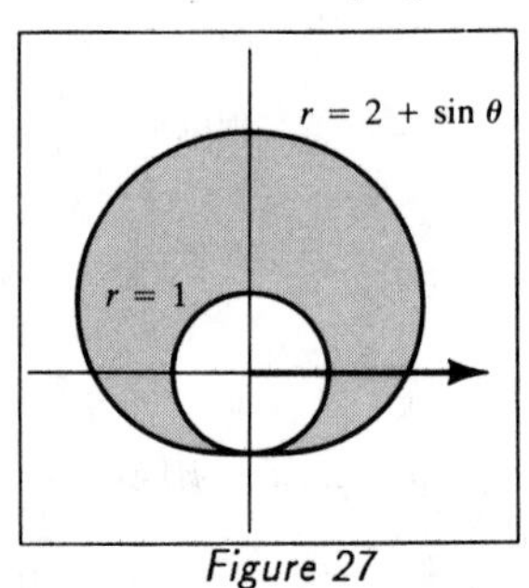

Figure 27

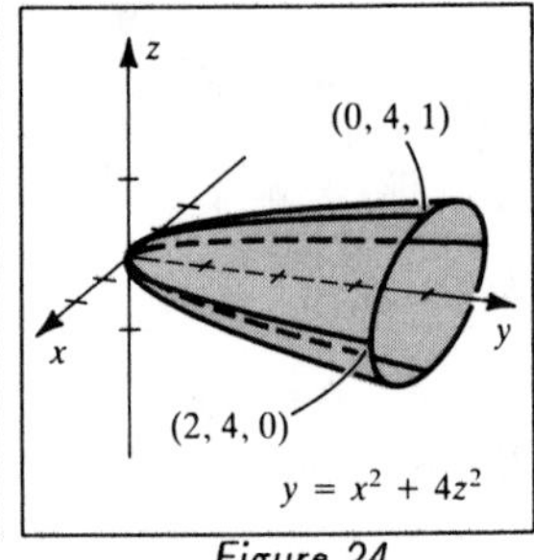

Figure 24

28 $\delta(x, y) = k|x| = kx$ since $x \geq 0$.

$$I_x = \int_0^1\int_{x^3}^{x^2} y^2(kx)\, dy\, dx = \tfrac{1}{3}k\int_0^1 (x^7 - x^{10})\, dx = (\tfrac{1}{3}k)(\tfrac{3}{88}) = \tfrac{1}{88}k.$$

$$I_y = \int_0^1\int_{x^3}^{x^2} x^2(kx)\, dy\, dx = k\int_0^1 (x^5 - x^6)\, dx = \tfrac{1}{42}k. \qquad I_O = I_x + I_y = \tfrac{65}{1848}k.$$

29 Let the triangle have vertices $(0, a)$, $(b, 0)$, and $(0, 0)$. An equation of the hypotenuse is $ax + by = ab$, or $y = -(ax/b) + a$. We need to find I_y.

$$I_y = \int_0^b\int_0^{-(ax/b)+a} x^2(kx)\, dy\, dx = k\int_0^b (-\tfrac{a}{b}x^4 + ax^3)\, dx = \tfrac{1}{20}ab^4k.$$

[30] Let the origin be the center of the circles and the x-axis the line through the center.

$\delta(r, \theta) = kr$. $I_x = \iint_R y^2\,\delta\,dA = \int_0^{2\pi}\int_a^b (r\sin\theta)^2(kr)\,r\,dr\,d\theta =$

$$\tfrac{1}{5}k(b^5 - a^5)\int_0^{2\pi}\sin^2\theta\,d\theta = \tfrac{1}{5}k(b^5 - a^5)\cdot\tfrac{1}{2}(2\pi) = \tfrac{1}{5}\pi(b^5 - a^5)k.$$

[31] Without loss of generality, let $\delta(x, y, z) = 1$.

$$m = V = \int_{-2}^{2}\int_{x^2}^{4}\int_0^{4-z} dy\,dz\,dx = \int_{-2}^{2}\int_{x^2}^{4}(4 - z)\,dz\,dx = \tfrac{1}{2}\int_{-2}^{2}(4 - x^2)^2\,dx = \tfrac{256}{15}.$$

$$M_{xz} = \int_{-2}^{2}\int_{x^2}^{4}\int_0^{4-z} y\,dy\,dz\,dx = \tfrac{1}{2}\int_{-2}^{2}\int_{x^2}^{4}(4 - z)^2\,dz\,dx$$

$$= \tfrac{1}{6}\int_{-2}^{2}(4 - x^2)^3\,dx = \tfrac{1}{6}\int_{-2}^{2}(64 - 48x^2 + 12x^4 - x^6)\,dx = \tfrac{2048}{105}.$$

$$M_{xy} = \int_{-2}^{2}\int_{x^2}^{4}\int_0^{4-z} z\,dy\,dz\,dx$$

$$= \int_{-2}^{2}\int_{x^2}^{4}(4z - z^2)\,dz\,dx = \int_{-2}^{2}\left(\tfrac{32}{3} - 2x^4 + \tfrac{1}{3}x^6\right)dx = \tfrac{1024}{35}.$$

By symmetry, $\bar{x} = 0$. $\bar{y} = \frac{M_{xz}}{m} = \frac{8}{7}$ and $\bar{z} = \frac{M_{xy}}{m} = \frac{12}{7}$.

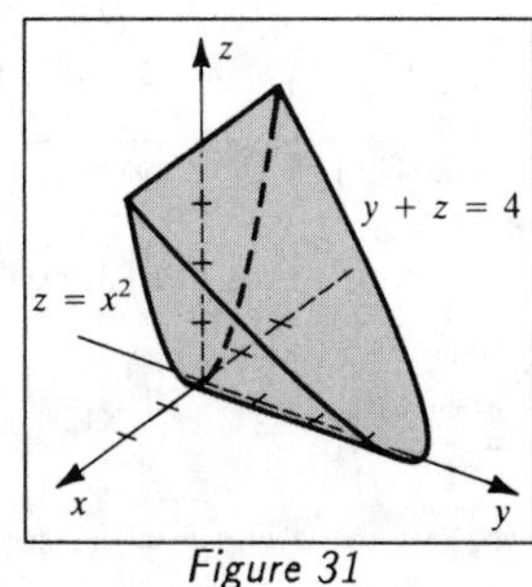

Figure 31

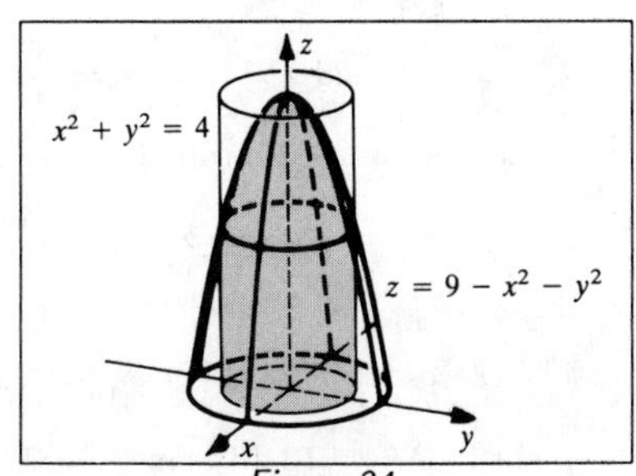

Figure 34

[32] $\delta(x, y, z) = \dfrac{k}{x^2 + y^2 + (z + 1)^2}$.

The trace in the xy-plane of the top of the paraboloid when $z = 9$ is $9x^2 + y^2 = 9$.

$$I_z = \int_{-1}^{1}\int_{-3\sqrt{1-x^2}}^{3\sqrt{1-x^2}}\int_{9x^2+y^2}^{9}\frac{k(x^2 + y^2)}{x^2 + y^2 + (z + 1)^2}\,dz\,dy\,dx.$$

[33] $x^2 - y^2 + z^2 = 1$ is a hyperboloid of one sheet with axis along the y-axis.

The trace of the solid in the xy-plane is $x^2 - y^2 = 1$. $\delta(x, y, z) = k\sqrt{x^2 + z^2}$.

$$I_y = \int_0^4\int_{-\sqrt{1+y^2}}^{\sqrt{1+y^2}}\int_{-\sqrt{1-x^2+y^2}}^{\sqrt{1-x^2+y^2}} k(x^2 + z^2)\sqrt{x^2 + z^2}\,dz\,dx\,dy.$$

[34] The solid lies between the cylinder $x^2 + y^2 = 4$ and the paraboloid $z = 9 - x^2 - y^2$. Let $\delta(r, \theta, z) = \delta$ be a constant density.

(a) $m = \int_2^3 \int_0^{2\pi} \int_0^{9-r^2} \delta\, r\, dz\, d\theta\, dr$

$= \delta \int_2^3 \int_0^{2\pi} (9r - r^3)\, d\theta\, dr = 2\pi\delta \int_2^3 (9r - r^3)\, dr = \frac{25}{2}\pi\delta.$

(b) $M_{xy} = \delta \int_2^3 \int_0^{2\pi} \int_0^{9-r^2} zr\, dz\, d\theta\, dr = \frac{1}{2}\delta \int_2^3 \int_0^{2\pi} (9 - r^2)^2\, r\, d\theta\, dr =$

$\pi\delta \int_2^3 (9 - r^2)^2\, r\, dr = \frac{125}{6}\pi\delta.$ By symmetry, $\bar{x} = \bar{y} = 0$. $\bar{z} = \frac{M_{xy}}{m} = \frac{5}{3}$.

(c) $I_z = \delta \int_2^3 \int_0^{2\pi} \int_0^{9-r^2} (r^2)\, r\, dz\, d\theta\, dr$

$= \delta \int_2^3 \int_0^{2\pi} (9r^3 - r^5)\, d\theta\, dr = 2\pi\delta \int_2^3 (9r^3 - r^5)\, dr = \frac{425}{6}\pi\delta.$

[35] Since $\delta(\rho, \theta, \phi) = k\rho$, $m = \int_0^{2\pi} \int_0^{\pi} \int_0^{a} (k\rho)(\rho^2 \sin\phi)\, d\rho\, d\phi\, d\theta =$

$\frac{1}{4}a^4 k \int_0^{2\pi} \int_0^{\pi} \sin\phi\, d\phi\, d\theta = \frac{1}{2}a^4 k \int_0^{2\pi} d\theta = \pi a^4 k.$

[36] $x = 2$, $y = -2$, and $z = 1 \Rightarrow r = \sqrt{2^2 + (-2)^2} = 2\sqrt{2}$; $\tan\theta = \frac{-2}{2} = -1 \Rightarrow$

$\theta = -\frac{\pi}{4}$. Cylindrical coordinates are $(2\sqrt{2}, -\frac{\pi}{4}, 1)$. $\rho = \sqrt{2^2 + (-2)^2 + 1^2} = 3$.

$\cos\phi = \frac{z}{\rho} = \frac{1}{3} \Rightarrow \phi = \cos^{-1}\frac{1}{3}$. Spherical coordinates are $(3, \cos^{-1}\frac{1}{3}, -\frac{\pi}{4})$.

[37] $\rho = 12$, $\phi = \frac{\pi}{6}$, and $\theta = \frac{3\pi}{4} \Rightarrow$

$x = \rho \sin\phi \cos\theta = 12 \sin\frac{\pi}{6} \cos\frac{3\pi}{4} = 12(\frac{1}{2})(-\frac{\sqrt{2}}{2}) = -3\sqrt{2}$;

$y = \rho \sin\phi \sin\theta = 12 \sin\frac{\pi}{6} \sin\frac{3\pi}{4} = 12(\frac{1}{2})(\frac{\sqrt{2}}{2}) = 3\sqrt{2}$; $z = \rho \cos\phi = 12 \cos\frac{\pi}{6} =$

$12(\frac{\sqrt{3}}{2}) = 6\sqrt{3}$. Rectangular coordinates are $(-3\sqrt{2}, 3\sqrt{2}, 6\sqrt{3})$.

$r = \sqrt{(-3\sqrt{2})^2 + (3\sqrt{2})^2} = 6$. Cylindrical coordinates are $(6, \frac{3\pi}{4}, 6\sqrt{3})$.

[38] $r = -\csc\theta \Rightarrow r\sin\theta = -1 \Rightarrow y = -1$,

the plane parallel to the xz-plane with y-intercept -1.

[39] $z + 3r^2 = 9 \Rightarrow z = 9 - 3r^2 \Rightarrow z = 9 - 3x^2 - 3y^2$,

a paraboloid with vertex $(0, 0, 9)$ and opening downward.

[40] $z = \frac{1}{4}r \Rightarrow r = 4z \Rightarrow r^2 = 16z^2 \Rightarrow x^2 + y^2 = 16z^2$, a cone.

[41] $\rho \sin\phi = 4 \Rightarrow \rho^2 \sin^2\phi = 16 \Rightarrow \rho^2 \sin^2\phi(\cos^2\theta + \sin^2\theta) = 16 \Rightarrow$

$(\rho \sin\phi \cos\theta)^2 + (\rho \sin\phi \sin\theta)^2 = 16 \Rightarrow x^2 + y^2 = 16$,

a right circular cylinder of radius 4 with axis along the z-axis.

[42] $\rho \sin\phi \cos\theta = 1 \Rightarrow x = 1$, the plane parallel to the yz-plane with x-intercept 1.

[43] $\rho^2 - 3\rho = 0 \Rightarrow \rho(\rho - 3) = 0 \Rightarrow \sqrt{x^2 + y^2 + z^2}(\sqrt{x^2 + y^2 + z^2} - 3) = 0$,

the sphere of radius 3 with center at the origin, together with its center.

[44] (a) $x^2 + y^2 = 1 \Rightarrow r^2 = 1 \Rightarrow r = 1$

(b) $x^2 + y^2 = 1 \Rightarrow \rho^2 \sin^2\phi \cos^2\theta + \rho^2 \sin^2\phi \sin^2\theta = 1 \Rightarrow \rho^2 \sin^2\phi = 1$

[45] (a) $z = x^2 - y^2 \Rightarrow z = r^2 \cos^2\theta - r^2 \sin^2\theta = r^2(\cos^2\theta - \sin^2\theta) = r^2 \cos 2\theta$

(b) $z = x^2 - y^2 \Rightarrow \rho \cos\phi = \rho^2 \sin^2\phi \cos^2\theta - \rho^2 \sin^2\phi \sin^2\theta \Rightarrow$

$$\cos\phi = \rho \sin^2\phi \cos 2\theta$$

[46] (a) $x^2 + y^2 + z^2 - 2z = 0 \Rightarrow r^2 + z^2 - 2z = 0$

(b) $x^2 + y^2 + z^2 - 2z = 0 \Rightarrow \rho^2 - 2\rho \cos\phi = 0 \Rightarrow \rho - 2\cos\phi = 0 \Rightarrow$

$$\rho = 2\cos\phi$$

[47] (a) $2x + y - 3z = 4 \Rightarrow 2r\cos\theta + r\sin\theta - 3z = 4$

(b) $2x + y - 3z = 4 \Rightarrow 2\rho \sin\phi \cos\theta + \rho \sin\phi \sin\theta - 3\rho \cos\phi = 4$

[48] (a) On $3 \le x \le 4$, $\sqrt{16 - x^2} \le y \le \sqrt{25 - x^2}$. On $4 \le x \le 5$, $0 \le y \le \sqrt{25 - x^2}$.

$$\int_3^4 \int_{\sqrt{16-x^2}}^{\sqrt{25-x^2}} dy\, dx + \int_4^5 \int_0^{\sqrt{25-x^2}} dy\, dx$$

(b) On $0 \le y \le \sqrt{7}$, $\sqrt{16 - y^2} \le x \le \sqrt{25 - y^2}$.

On $\sqrt{7} \le y \le 4$, $3 \le x \le \sqrt{25 - y^2}$. $\int_0^{\sqrt{7}} \int_{\sqrt{16-y^2}}^{\sqrt{25-y^2}} dx\, dy + \int_{\sqrt{7}}^4 \int_3^{\sqrt{25-y^2}} dx\, dy$

(c) For $0 \le \theta \le \arctan \frac{\sqrt{7}}{3}$, $4 \le r \le 5$. The vertical line $x = 3$ has equation $r = 3\sec\theta$ in polar coordinates. Hence, for $\arctan \frac{\sqrt{7}}{3} \le \theta \le \arctan \frac{4}{3}$,

$3\sec\theta \le r \le 5$. $\int_0^{\arctan(\sqrt{7}/3)} \int_4^5 r\, dr\, d\theta + \int_{\arctan(\sqrt{7}/3)}^{\arctan(4/3)} \int_{3\sec\theta}^5 r\, dr\, d\theta$

[49] (a) On $0 \le x \le 4$, $0 \le y \le \sqrt{25 - x^2}$. $\int_0^4 \int_0^{\sqrt{25-x^2}} dy\, dx$

(b) On $0 \le y \le 3$, $0 \le x \le 4$. On $3 \le y \le 5$, $0 \le x \le \sqrt{25 - y^2}$.

$$\int_0^3 \int_0^4 dx\, dy + \int_3^5 \int_0^{\sqrt{25-y^2}} dx\, dy$$

(c) For $0 \le \theta \le \arctan \frac{3}{4}$, $0 \le r \le 4\sec\theta$. For $\arctan \frac{3}{4} \le \theta \le \frac{\pi}{2}$, $0 \le r \le 5$.

$$\int_0^{\arctan(3/4)} \int_0^{4\sec\theta} r\, dr\, d\theta + \int_{\arctan(3/4)}^{\pi/2} \int_0^5 r\, dr\, d\theta$$

50 (a) We will subtract the volume bounded by the plane $z = 1$ and the cone $z = \sqrt{x^2 + y^2}$ (all in the first octant) from the volume bounded by the sphere $x^2 + y^2 + z^2 = 18$ and the cone. The sphere and the cone intersect $\Rightarrow$ $2x^2 + 2y^2 = 18$, or $x^2 + y^2 = 9$. The plane and the cone intersect $\Rightarrow$ $x^2 + y^2 = 1$.

$$\int_0^3 \int_0^{\sqrt{9-x^2}} \int_{\sqrt{x^2+y^2}}^{\sqrt{18-x^2-y^2}} dz\, dy\, dx - \int_0^1 \int_0^{\sqrt{1-x^2}} \int_{\sqrt{x^2+y^2}}^{1} dz\, dy\, dx$$

(b) We will add the volume inside the right circular cylinder $r = 1$ to the volume between the right circular cylinders $r = 1$ and $r = 3$.

$$\int_0^{\pi/2} \int_0^1 \int_1^{\sqrt{18-r^2}} r\, dz\, dr\, d\theta + \int_0^{\pi/2} \int_1^3 \int_r^{\sqrt{18-r^2}} r\, dz\, dr\, d\theta$$

(c) An equation of the plane $z = 1$ is $\rho \cos\phi = 1$, or $\rho = \sec\phi$.

$$\int_0^{\pi/2} \int_0^{\pi/4} \int_{\sec\phi}^{3\sqrt{2}} (\rho^2 \sin\phi)\, d\rho\, d\phi\, d\theta$$

51 (a) We will subtract the volume above the plane $z = 3$ from the volume of the sphere $x^2 + y^2 + z^2 = 25$ (all in the first octant). The plane and the sphere intersect $\Rightarrow$ $x^2 + y^2 = 16$.

$$\int_0^5 \int_0^{\sqrt{25-x^2}} \int_0^{\sqrt{25-x^2-y^2}} dz\, dy\, dx - \int_0^4 \int_0^{\sqrt{16-x^2}} \int_3^{\sqrt{25-x^2-y^2}} dz\, dy\, dx$$

(b) We will add the volume inside the right circular cylinder $r = 4$ between $z = 0$ and $z = 3$ to the volume outside the cylinder and inside the hemisphere $z = \sqrt{25 - r^2}$.

$$\int_0^{\pi/2} \int_0^4 \int_0^3 r\, dz\, dr\, d\theta + \int_0^{\pi/2} \int_4^5 \int_0^{\sqrt{25-r^2}} r\, dz\, dr\, d\theta$$

(c) In the yz-plane, the plane intersects the sphere at $(0, 4, 3)$. Thus, $\tan\phi = \frac{4}{3}$ (*not* $\frac{3}{4}$), and $\phi = \arctan\frac{4}{3}$. We will add the volume bounded by $\rho = 3\sec\phi$ $\{z = 3\}$ to that bounded by $\rho = 5$.

$$\int_0^{\pi/2} \int_0^{\arctan(4/3)} \int_0^{3\sec\phi} (\rho^2 \sin\phi)\, d\rho\, d\phi\, d\theta +$$

$$\int_0^{\pi/2} \int_{\arctan(4/3)}^{\pi/2} \int_0^5 (\rho^2 \sin\phi)\, d\rho\, d\phi\, d\theta$$

52 (a) $u = c \Rightarrow 2x + 5y = c \Rightarrow y = -\frac{2}{5}x + \frac{1}{5}c$, lines with slope $-\frac{2}{5}$.

(b) $v = d \Rightarrow 3x - 4y = d \Rightarrow y = \frac{3}{4}x - \frac{1}{4}d$, lines with slope $\frac{3}{4}$.

(c) $4(u = 2x + 5y)$ and $5(v = 3x - 4y) \Rightarrow x = \frac{4}{23}u + \frac{5}{23}v$.

$3(u = 2x + 5y)$ and $-2(v = 3x - 4y) \Rightarrow y = \frac{3}{23}u - \frac{2}{23}v$.

(d) $$\frac{\partial(x, y)}{\partial(u, v)} = \begin{vmatrix} \frac{4}{23} & \frac{5}{23} \\ \frac{3}{23} & -\frac{2}{23} \end{vmatrix} = -\frac{8}{23^2} - \frac{15}{23^2} = -\frac{23}{23^2} = -\frac{1}{23}$$

(e) $ax + by + c = 0 \Rightarrow a(\frac{4}{23}u + \frac{5}{23}v) + b(\frac{3}{23}u - \frac{2}{23}v) + c = 0 \Rightarrow$

$(4a + 3b)u + (5a - 2b)v + 23c = 0$, which are lines.

(f) $x^2 + y^2 = a^2 \Rightarrow (\frac{4}{23}u + \frac{5}{23}v)^2 + (\frac{3}{23}u - \frac{2}{23}v)^2 = a^2 \Rightarrow$

$25u^2 + 28uv + 29v^2 = 529a^2$, which are ellipses.

53 The region is bounded by $x + y = 2$, $x = y$, and $y = 0$, which forms a triangle with vertices $(0, 0)$, $(2, 0)$, and $(1, 1)$. They are transformed into a triangle in the uv-plane with vertices $(0, 0)$, $(2, 2)$, and $(0, 2)$ in a positive direction. $x = \frac{1}{2}(u + v)$ and $y = -\frac{1}{2}(u - v) \Rightarrow J = \frac{1}{2}$.

$$\int_0^1 \int_y^{2-y} e^{(x-y)/(x+y)}\, dx\, dy = \int_0^2 \int_0^v e^{u/v} (\tfrac{1}{2})\, du\, dv = \tfrac{1}{2}(e - 1)\int_0^2 v\, dv = e - 1 \approx 1.72.$$

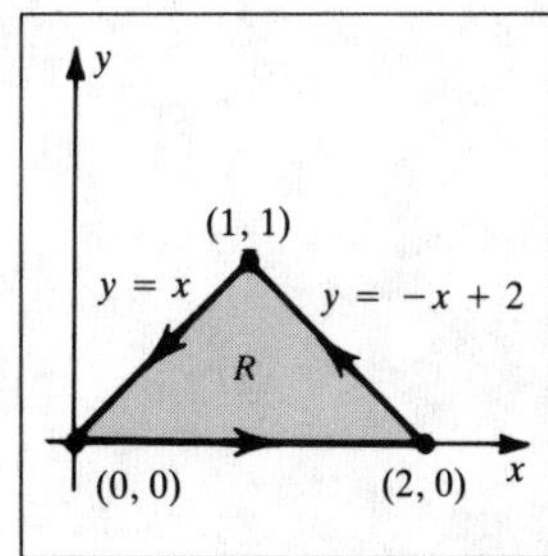

Figure 53a

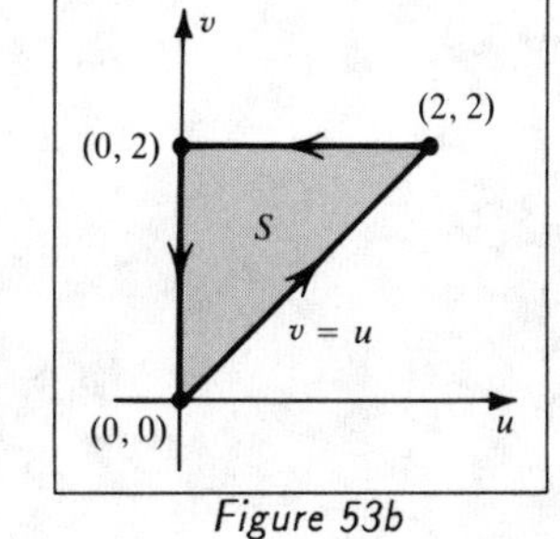

Figure 53b

Chapter 18: Vector Calculus

Exercises 18.1

[1] $\mathbf{F}(x, y) = x\mathbf{i} - y\mathbf{j}$

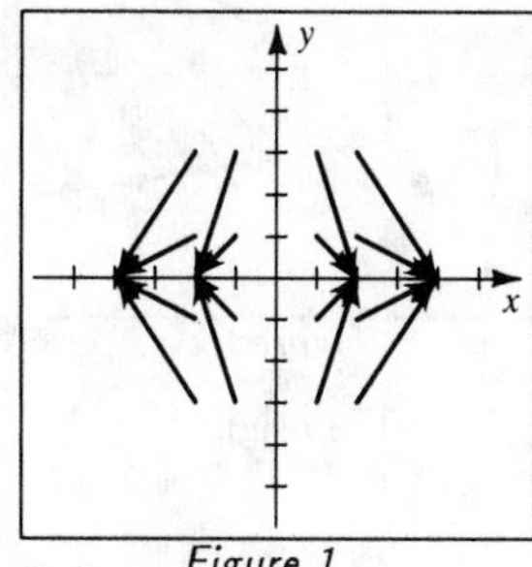

Figure 1

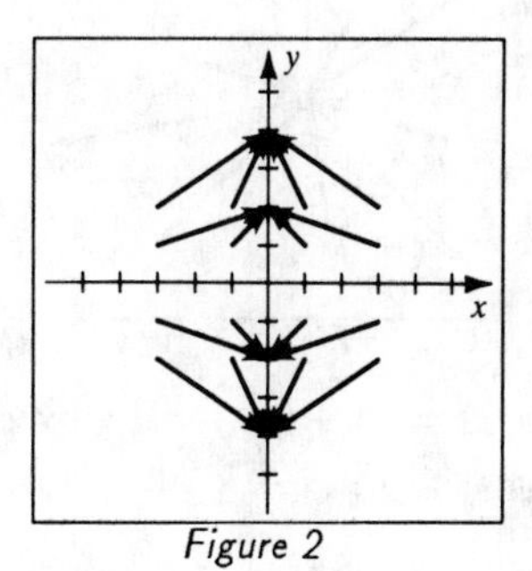

Figure 2

[2] $\mathbf{F}(x, y) = -x\mathbf{i} + y\mathbf{j}$

[3] $\mathbf{F}(x, y) = 2x\mathbf{i} + 3y\mathbf{j}$

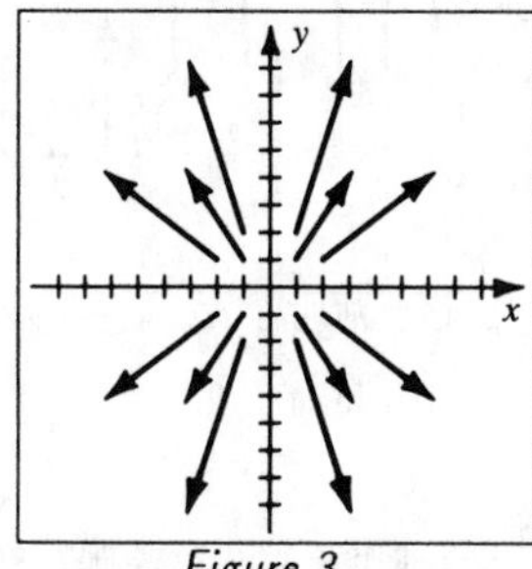

Figure 3

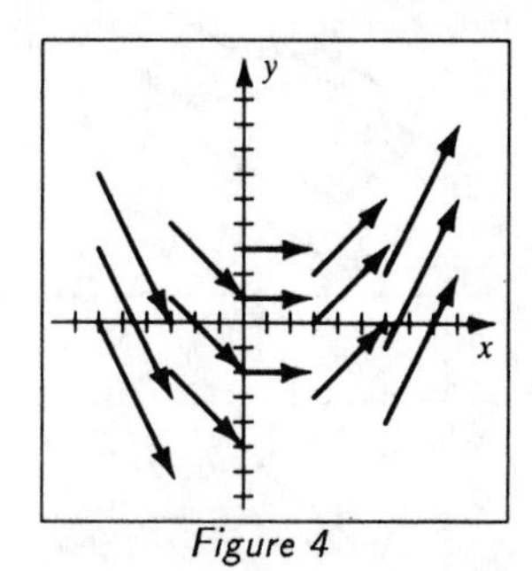

Figure 4

[4] $\mathbf{F}(x, y) = 3\mathbf{i} + x\mathbf{j}$

[5] $\mathbf{F}(x, y) = (x^2 + y^2)^{-1/2}(x\mathbf{i} + y\mathbf{j})$;

all vectors are unit vectors pointing away from the origin.

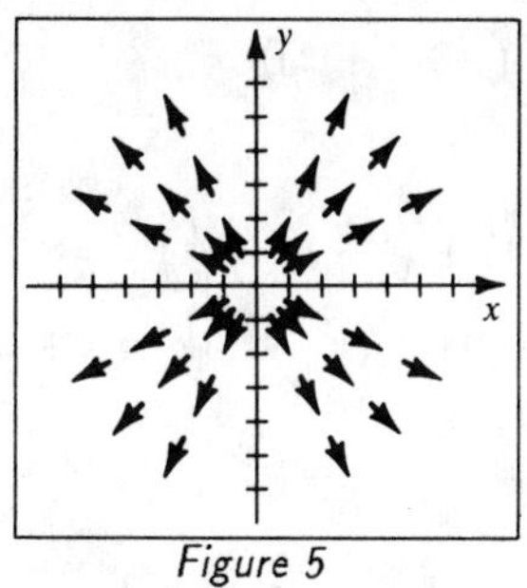

Figure 5

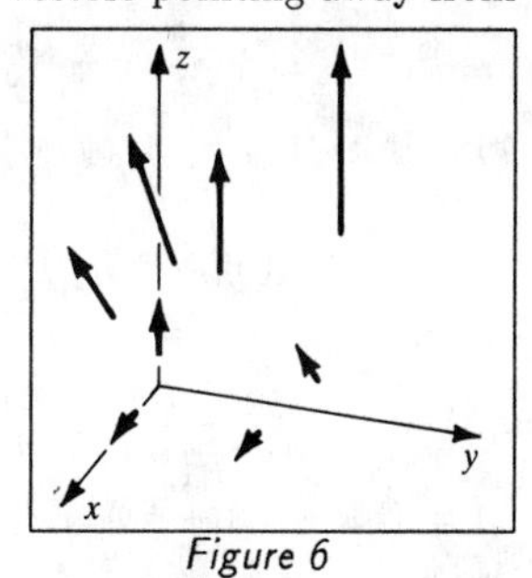

Figure 6

[6] $\mathbf{F}(x, y, z) = x\mathbf{i} + z\mathbf{k}$; all vectors are perpendicular to the y-axis.

[7] $\mathbf{F}(x, y, z) = -x\mathbf{i} - y\mathbf{j} - z\mathbf{k}$; all vectors point toward the origin.

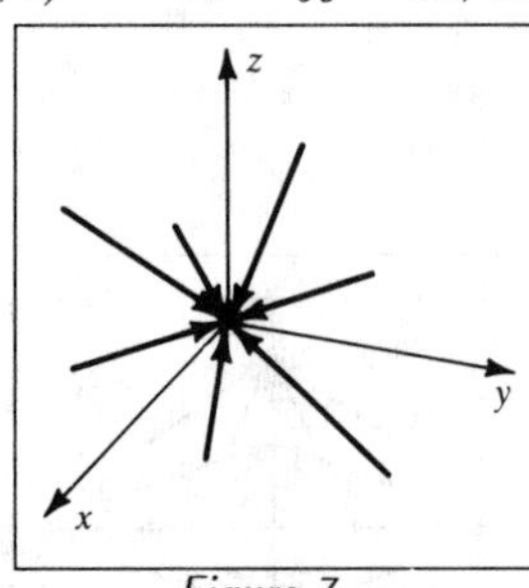

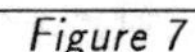
Figure 7

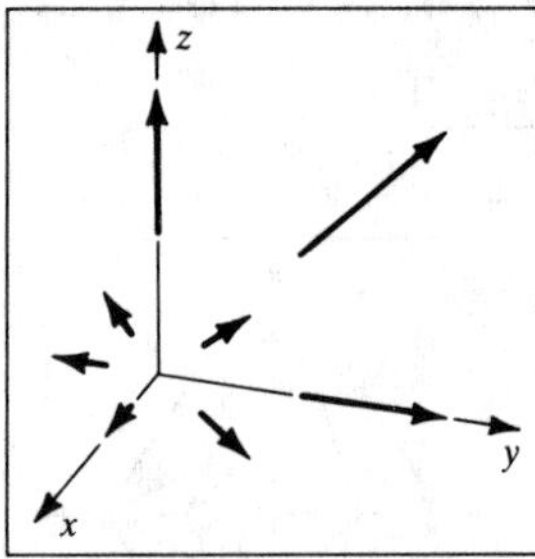

Figure 8

[8] $\mathbf{F}(x, y, z) = x\mathbf{i} + y\mathbf{j} + z\mathbf{k}$; all vectors point away from the origin.

[9] $\mathbf{F}(x, y, z) = \mathbf{i} + \mathbf{j} + \mathbf{k}$; all vectors are constant.

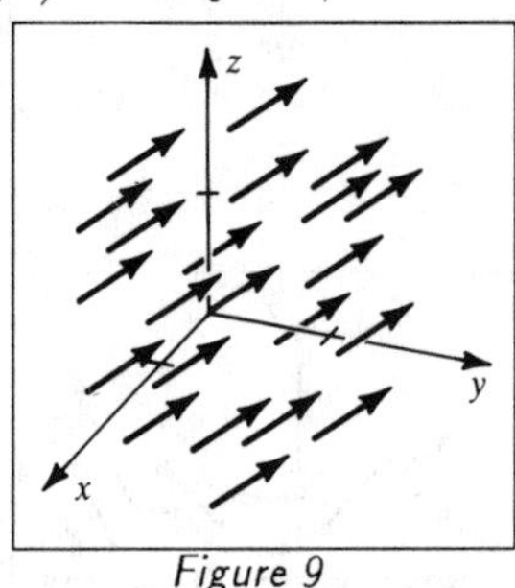

Figure 9

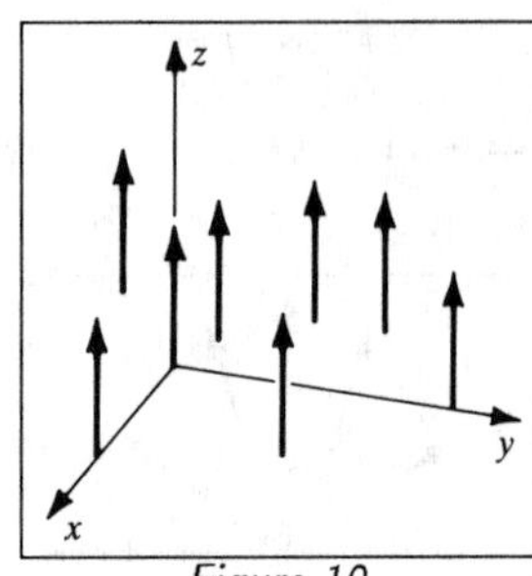

Figure 10

[10] $\mathbf{F}(x, y, z) = 2\mathbf{k}$; all vectors are parallel to the z-axis with length 2.

[11] Using (18.3) with $f(x, y, z) = x^2 - 3y^2 + 4z^2$,

$$\mathbf{F}(x, y, z) = \nabla f(x, y, z) = 2x\mathbf{i} - 6y\mathbf{j} + 8z\mathbf{k} \text{ is a conservative vector field.}$$

[12] $f(x, y, z) = \sin(x^2 + y^2 + z^2) \Rightarrow$

$$\mathbf{F}(x, y, z) = \nabla f(x, y, z) = 2\cos(x^2 + y^2 + z^2)(x\mathbf{i} + y\mathbf{j} + z\mathbf{k}).$$

[13] $f(x, y) = \arctan(xy) \Rightarrow \mathbf{F}(x, y) = \nabla f(x, y) = \dfrac{1}{1 + x^2y^2}(y\mathbf{i} + x\mathbf{j})$.

[14] $f(x, y) = y^2e^{-3x} \Rightarrow \mathbf{F}(x, y) = \nabla f(x, y) = ye^{-3x}(-3y\mathbf{i} + 2\mathbf{j})$.

[15] $\mathbf{F}(x, y, z) = x^2z\mathbf{i} + y^2x\mathbf{j} + (y + 2z)\mathbf{k} \Rightarrow$

$$\operatorname{curl}\mathbf{F} = \nabla \times \mathbf{F} = \left(\frac{\partial P}{\partial y} - \frac{\partial N}{\partial z}\right)\mathbf{i} + \left(\frac{\partial M}{\partial z} - \frac{\partial P}{\partial x}\right)\mathbf{j} + \left(\frac{\partial N}{\partial x} - \frac{\partial M}{\partial y}\right)\mathbf{k} =$$

$$(1 - 0)\mathbf{i} + (x^2 - 0)\mathbf{j} + (y^2 - 0)\mathbf{k} = \mathbf{i} + x^2\mathbf{j} + y^2\mathbf{k}.$$

$$\operatorname{div}\mathbf{F} = \nabla \cdot \mathbf{F} = \frac{\partial M}{\partial x} + \frac{\partial N}{\partial y} + \frac{\partial P}{\partial z} = 2xz + 2xy + 2.$$

[16] $\mathbf{F}(x, y, z) = (3x + y)\mathbf{i} + xy^2z\mathbf{j} + xz^2\mathbf{k} \Rightarrow$

$$\nabla \times \mathbf{F} = -xy^2\mathbf{i} - z^2\mathbf{j} + (y^2z - 1)\mathbf{k} \text{ and } \nabla \cdot \mathbf{F} = 3 + 2xyz + 2xz.$$

[17] $\mathbf{F}(x, y, z) = 3xyz^2\mathbf{i} + y^2\sin z\mathbf{j} + xe^{2z}\mathbf{k} \Rightarrow$

$$\nabla \times \mathbf{F} = -y^2\cos z\mathbf{i} + (6xyz - e^{2z})\mathbf{j} - 3xz^2\mathbf{k} \text{ and } \nabla \cdot \mathbf{F} = 3yz^2 + 2y\sin z + 2xe^{2z}.$$

[18] $\mathbf{F}(x, y, z) = x^3\ln z\mathbf{i} + xe^{-y}\mathbf{j} - (y^2 + 2z)\mathbf{k} \Rightarrow$

$$\nabla \times \mathbf{F} = -2y\mathbf{i} + \frac{x^3}{z}\mathbf{j} + e^{-y}\mathbf{k} \text{ and } \nabla \cdot \mathbf{F} = 3x^2\ln z - xe^{-y} - 2.$$

[19] (a) $\mathbf{r} = x\mathbf{i} + y\mathbf{j} + z\mathbf{k} \Rightarrow \nabla \cdot \mathbf{r} = 1 + 1 + 1 = 3.$

(b) $\nabla \times \mathbf{r} = (0 - 0)\mathbf{i} + (0 - 0)\mathbf{j} + (0 - 0)\mathbf{k} = \mathbf{0}.$

(c) $\|\mathbf{r}\| = (x^2 + y^2 + z^2)^{1/2} \Rightarrow$

$$\nabla\|\mathbf{r}\| = \frac{x}{(x^2 + y^2 + z^2)^{1/2}}\mathbf{i} + \frac{y}{(x^2 + y^2 + z^2)^{1/2}}\mathbf{j} + \frac{z}{(x^2 + y^2 + z^2)^{1/2}}\mathbf{k}$$

$$= \frac{1}{\|\mathbf{r}\|}(x\mathbf{i} + y\mathbf{j} + z\mathbf{k}) = \frac{\mathbf{r}}{\|\mathbf{r}\|}.$$

[20] (a) Let $\mathbf{a} = e\mathbf{i} + f\mathbf{j} + g\mathbf{k}$, where e, f, and g are constants.

$\mathbf{r} = x\mathbf{i} + y\mathbf{j} + z\mathbf{k} \Rightarrow \mathbf{a} \times \mathbf{r} = (fz - gy)\mathbf{i} - (ez - gx)\mathbf{j} + (ey - fx)\mathbf{k}$ and

$\text{curl}(\mathbf{a} \times \mathbf{r}) = \left[e - (-e)\right]\mathbf{i} + \left[f - (-f)\right]\mathbf{j} + \left[g - (-g)\right]\mathbf{k} =$

$$2e\mathbf{i} + 2f\mathbf{j} + 2g\mathbf{k} = 2(e\mathbf{i} + f\mathbf{j} + g\mathbf{k}) = 2\mathbf{a}.$$

(b) $\text{div}(\mathbf{a} \times \mathbf{r}) = 0 + 0 + 0 = 0.$

[21] (1) By (18.2), $\mathbf{F}(x, y, z) = \frac{c}{\|\mathbf{r}\|^3}\mathbf{r} = \frac{c}{(x^2 + y^2 + z^2)^{3/2}}(x\mathbf{i} + y\mathbf{j} + z\mathbf{k})$. Thus,

$$\text{curl}\,\mathbf{F} = \frac{c}{(x^2 + y^2 + z^2)^{5/2}}\left[(3yz - 3yz)\mathbf{i} + (3xz - 3xz)\mathbf{j} + (3xy - 3xy)\mathbf{k}\right] = \mathbf{0}.$$

(2) $$\text{div}\,\mathbf{F} = \frac{c}{(x^2 + y^2 + z^2)^{5/2}} \times$$

$$\left[(-2x^2 + y^2 + z^2) + (x^2 - 2y^2 + z^2) + (x^2 + y^2 - 2z^2)\right] = 0.$$

[22] If $\mathbf{F}$ is conservative, then by (18.3), $\mathbf{F}(x, y, z) = \nabla f(x, y, z)$ for some scalar function f. Thus, $\text{curl}\,\mathbf{F} = \nabla \times (\nabla f) = \nabla \times (f_x\mathbf{i} + f_y\mathbf{j} + f_z\mathbf{k}) =$

$$(f_{zy} - f_{yz})\mathbf{i} + (f_{xz} - f_{zx})\mathbf{j} + (f_{yx} - f_{xy})\mathbf{k} = 0\mathbf{i} + 0\mathbf{j} + 0\mathbf{k} = \mathbf{0}.$$

Note: In Exercises 23-26, let $\mathbf{F} = M\mathbf{i} + N\mathbf{j} + P\mathbf{k}$ and $\mathbf{G} = Q\mathbf{i} + R\mathbf{j} + S\mathbf{k}$.

[23]
$$\begin{aligned}\nabla \times (\mathbf{F} + \mathbf{G}) &= \nabla \times \left[(M + Q)\mathbf{i} + (N + R)\mathbf{j} + (P + S)\mathbf{k}\right] \\ &= \left[(P_y + S_y) - (N_z + R_z)\right]\mathbf{i} + \left[(M_z + Q_z) - (P_x + S_x)\right]\mathbf{j} + \\ &\qquad \left[(N_x + R_x) - (M_y + Q_y)\right]\mathbf{k} \\ &= \left[(P_y - N_z)\mathbf{i} + (M_z - P_x)\mathbf{j} + (N_x - M_y)\mathbf{k}\right] + \\ &\qquad \left[(S_y - R_z)\mathbf{i} + (Q_z - S_x)\mathbf{j} + (R_x - Q_y)\mathbf{k}\right] \\ &= \nabla \times \mathbf{F} + \nabla \times \mathbf{G}\end{aligned}$$

[24]
$$\begin{aligned}\nabla \cdot (\mathbf{F} + \mathbf{G}) &= \nabla \cdot \left[(M + Q)\mathbf{i} + (N + R)\mathbf{j} + (P + S)\mathbf{k}\right] \\ &= (M_x + Q_x) + (N_y + R_y) + (P_z + S_z) \\ &= (M_x + N_y + P_z) + (Q_x + R_y + S_z) = \nabla \cdot \mathbf{F} + \nabla \cdot \mathbf{G}\end{aligned}$$

25 $\nabla \times (f\mathbf{F}) = \nabla \times \left[(fM)\mathbf{i} + (fN)\mathbf{j} + (fP)\mathbf{k}\right]$

$$= \left[\frac{\partial}{\partial y}(fP) - \frac{\partial}{\partial z}(fN)\right]\mathbf{i} + \left[\frac{\partial}{\partial z}(fM) - \frac{\partial}{\partial x}(fP)\right]\mathbf{j} + \left[\frac{\partial}{\partial x}(fN) - \frac{\partial}{\partial y}(fM)\right]\mathbf{k}$$

$$= \left[(fP_y + f_y P) - (fN_z + f_z N)\right]\mathbf{i} + \left[(fM_z + f_z M) - (fP_x + f_x P)\right]\mathbf{j} + \left[(fN_x + f_x N) - (fM_y + f_y M)\right]\mathbf{k}$$

$$= f\left[(P_y - N_z)\mathbf{i} + (M_z - P_x)\mathbf{j} + (N_x - M_y)\mathbf{k}\right] + \left[(f_y P - f_z N)\mathbf{i} + (f_z M - f_x P)\mathbf{j} + (f_x N - f_y M)\mathbf{k}\right]$$

$$= f(\nabla \times \mathbf{F}) + (\nabla f) \times \mathbf{F}$$

26 $\nabla \cdot (\mathbf{F} \times \mathbf{G}) = \nabla \cdot \left[(NS - PR)\mathbf{i} + (PQ - MS)\mathbf{j} + (MR - NQ)\mathbf{k}\right]$

$$= \frac{\partial}{\partial x}(NS - PR) + \frac{\partial}{\partial y}(PQ - MS) + \frac{\partial}{\partial z}(MR - NQ)$$

$$= (NS_x + N_xS - PR_x - P_xR) + (PQ_y + P_yQ - MS_y - M_yS) + (MR_z + M_zR - NQ_z - N_zQ)$$

$$= \left[(P_y - N_z)\,Q + (M_z - P_x)\,R + (N_x - M_y)\,S\right] + \left[(R_z - S_y)\,M + (S_x - Q_z)\,N + (Q_y - R_x)\,P\right]$$

$$= (\nabla \times \mathbf{F}) \cdot \mathbf{G} - (\nabla \times \mathbf{G}) \cdot \mathbf{F}$$

27 $\operatorname{curl} \operatorname{grad} f = \nabla \times (\nabla f) = \nabla \times (f_x\mathbf{i} + f_y\mathbf{j} + f_z\mathbf{k})$

$$= (f_{zy} - f_{yz})\mathbf{i} + (f_{xz} - f_{zx})\mathbf{j} + (f_{yx} - f_{xy})\mathbf{k} = \mathbf{0}.$$

28 $\operatorname{div} \operatorname{curl} \mathbf{F} = \nabla \cdot (\nabla \times \mathbf{F}) = \nabla \cdot \left[(P_y - N_z)\mathbf{i} + (M_z - P_x)\mathbf{j} + (N_x - M_y)\mathbf{k}\right]$

$$= \frac{\partial}{\partial x}(P_y - N_z) + \frac{\partial}{\partial y}(M_z - P_x) + \frac{\partial}{\partial z}(N_x - M_y)$$

$$= P_{yx} - N_{zx} + M_{zy} - P_{xy} + N_{xz} - M_{yz}$$

$$= (P_{yx} - P_{xy}) + (N_{zx} - N_{xz}) + (M_{zy} - M_{yz}) = 0.$$

29 By Exercise 23, $\operatorname{curl}(\operatorname{grad} f + \operatorname{curl} \mathbf{F}) = \operatorname{curl} \operatorname{grad} f + \operatorname{curl} \operatorname{curl} \mathbf{F}$.

By Exercise 27, $\operatorname{curl} \operatorname{grad} f = \mathbf{0}$, and the result follows.

30 Since $\mathbf{a}$ is a constant vector, all partial derivatives are zero. Thus, $\operatorname{curl} \mathbf{a} = \mathbf{0}$.

31 By Exercise 25, $\operatorname{curl} \mathbf{F} = \nabla \times \mathbf{F} = \nabla \times \left(\frac{c}{r^k}\right)\mathbf{r} = \frac{c}{r^k}(\nabla \times \mathbf{r}) + \nabla\left(\frac{c}{r^k}\right) \times \mathbf{r}$.

The first term is $\mathbf{0}$ by Exercise 19. Now, $\nabla\left(\frac{c}{r^k}\right) = -\frac{kc}{r^{k+2}}(x\mathbf{i} + y\mathbf{j} + z\mathbf{k}) = -\frac{kc}{r^{k+2}}\mathbf{r}$. Since $\mathbf{r} \times \mathbf{r} = \mathbf{0}$, the second term is also $\mathbf{0}$.

32 (a) $\nabla \cdot (\nabla f) = \nabla \cdot (f_x\mathbf{i} + f_y\mathbf{j} + f_z\mathbf{k}) = f_{xx} + f_{yy} + f_{zz} = \nabla^2 f$.

(b) $\nabla^2(fg) = \nabla \cdot \nabla(fg) = \nabla \cdot \left[(fg)_x\mathbf{i} + (fg)_y\mathbf{j} + (fg)_z\mathbf{k}\right]$

$$= \nabla \cdot \left[(fg_x + f_xg)\mathbf{i} + (fg_y + f_yg)\mathbf{j} + (fg_z + f_zg)\mathbf{k}\right]$$

$$= \frac{\partial}{\partial x}(fg_x + f_xg) + \frac{\partial}{\partial y}(fg_y + f_yg) + \frac{\partial}{\partial z}(fg_z + f_zg)$$

$$= (fg_{xx} + f_xg_x + f_xg_x + f_{xx}g) + (fg_{yy} + f_yg_y + f_yg_y + f_{yy}g) + (fg_{zz} + f_zg_z + f_zg_z + f_{zz}g)$$

$$= f(g_{xx} + g_{yy} + g_{zz}) + g(f_{xx} + f_{yy} + f_{zz}) + 2(f_xg_x + f_yg_y + f_zg_z)$$

$$= f\nabla^2 g + g\nabla^2 f + 2\nabla f \cdot \nabla g$$

[33] $f_x = \dfrac{-x}{(x^2 + y^2 + z^2)^{3/2}} \Rightarrow f_{xx} = \dfrac{2x^2 - y^2 - z^2}{(x^2 + y^2 + z^2)^{5/2}}$.

Also, $f_{yy} = \dfrac{2y^2 - x^2 - z^2}{(x^2 + y^2 + z^2)^{5/2}}$ and $f_{zz} = \dfrac{2z^2 - x^2 - y^2}{(x^2 + y^2 + z^2)^{5/2}}$.

Thus, $f_{xx} + f_{yy} + f_{zz} = 0$, which is Laplace's equation, $\nabla^2 f = 0$.

[34] $f_{xx} + f_{yy} + f_{zz} = 2a + 2b + 2c = 2(a + b + c) = 0$.

[35] Define $\lim\limits_{(x,y,z)\to(x_0,\,y_0,\,z_0)} \mathbf{F}(x, y, z) = \mathbf{a}$ to be $\left[\lim\limits_{(x,y,z)\to(x_0,\,y_0,\,z_0)} M(x, y, z)\right]\mathbf{i}$ +

$\left[\lim\limits_{(x,y,z)\to(x_0,\,y_0,\,z_0)} N(x, y, z)\right]\mathbf{j} + \left[\lim\limits_{(x,y,z)\to(x_0,\,y_0,\,z_0)} P(x, y, z)\right]\mathbf{k} =$

$u\mathbf{i} + v\mathbf{j} + w\mathbf{k}$, where $\mathbf{a} = u\mathbf{i} + v\mathbf{j} + w\mathbf{k}$. Thus, $\lim\limits_{(x,y,z)\to(x_0,\,y_0,\,z_0)} \mathbf{F}(x, y, z) = \mathbf{a}$

means that for every $\epsilon > 0$, $\exists\delta > 0$, such that $\|\mathbf{F}(x, y, z) - \mathbf{a}\| < \epsilon$ whenever

$0 < \|\langle x, y, z\rangle - \langle x_0, y_0, z_0\rangle\| = \sqrt{(x - x_0)^2 + (y - y_0)^2 + (z - z_0)^2} < \delta$.

Geometrically this means that as the point (x, y, z) approaches (x_0, y_0, z_0),

$\mathbf{F}(x, y, z)$ has nearly the same magnitude and direction as $\mathbf{a}$.

[36] $\mathbf{F}$ is continuous at $P_0(x_0, y_0, z_0)$ iff $\lim\limits_{(x,y,z)\to(x_0,\,y_0,\,z_0)} \mathbf{F}(x, y, z) = \mathbf{F}(x_0, y_0, z_0)$.

This happens iff each component of $\mathbf{F}$ is continuous. Geometrically this means that the vectors $\mathbf{F}(x, y, z)$ for points near P_0 will have nearly the same direction and magnitude as $\mathbf{F}(x_0, y_0, z_0)$, i.e., there are no abrupt changes in the vector field near P_0.

[37] $f_{xx}(0.3, 0.5, 0.2) \approx \dfrac{f(0.35, 0.5, 0.2) - 2f(0.3, 0.5, 0.2) + f(0.25, 0.5, 0.2)}{(0.05)^2} \approx 0.0107$.

$f_{yy}(0.3, 0.5, 0.2) \approx \dfrac{f(0.3, 0.55, 0.2) - 2f(0.3, 0.5, 0.2) + f(0.3, 0.45, 0.2)}{(0.05)^2} \approx 0.0039$.

$f_{zz}(0.3, 0.5, 0.2) \approx \dfrac{f(0.3, 0.5, 0.25) - 2f(0.3, 0.5, 0.2) + f(0.3, 0.5, 0.15)}{(0.05)^2} \approx 0.1308$.

$\nabla^2 f(0.3, 0.5, 0.2) =$

$f_{xx}(0.3, 0.5, 0.2) + f_{yy}(0.3, 0.5, 0.2) + f_{zz}(0.3, 0.5, 0.2) \approx 0.1454$.

[38] $f_{xx}(0.3, 0.5, 0.2) \approx \dfrac{f(0.35, 0.5, 0.2) - 2f(0.3, 0.5, 0.2) + f(0.25, 0.5, 0.2)}{(0.05)^2} \approx -8.411$.

$f_{yy}(0.3, 0.5, 0.2) \approx \dfrac{f(0.3, 0.55, 0.2) - 2f(0.3, 0.5, 0.2) + f(0.3, 0.45, 0.2)}{(0.05)^2} \approx 0.5078$.

$f_{zz}(0.3, 0.5, 0.2) \approx \dfrac{f(0.3, 0.5, 0.25) - 2f(0.3, 0.5, 0.2) + f(0.3, 0.5, 0.15)}{(0.05)^2} \approx 7.6867$.

$\nabla^2 f(0.3, 0.5, 0.2) =$

$f_{xx}(0.3, 0.5, 0.2) + f_{yy}(0.3, 0.5, 0.2) + f_{zz}(0.3, 0.5, 0.2) \approx -0.2165$.

Exercises 18.2

1 (1) $ds = \sqrt{(dx)^2 + (dy)^2} = \sqrt{3^2 + (3t^2)^2}\,dt = 3\sqrt{1 + t^4}\,dt.$

$x^3 + y = (3t)^3 + t^3 = 28t^3.$ $\int_C f(x, y)\,ds =$

$$\int_0^1 (28t^3)(3\sqrt{1 + t^4})\,dt = 84\Big[\tfrac{1}{6}(1 + t^4)^{3/2}\Big]_0^1\,dt = 14(2^{3/2} - 1) \approx 25.60.$$

(2) $dx = 3\,dt.$ $\int_C f(x, y)\,dx = \int_0^1 (28t^3)\,3\,dt = 84\Big[\tfrac{1}{4}t^4\Big]_0^1 = 21.$

(3) $dy = 3t^2\,dt.$ $\int_C f(x, y)\,dy = \int_0^1 (28t^3)(3t^2)\,dt = 84\Big[\tfrac{1}{6}t^6\Big]_0^1 = 14.$

2 (1) $ds = \sqrt{(\frac{1}{2})^2 + (\frac{5}{2}t^{3/2})^2}\,dt = \frac{1}{2}\sqrt{1 + 25t^3}\,dt.$ $xy^{2/5} = (\frac{1}{2}t)(t^{5/2})^{2/5} = \frac{1}{2}t^2.$

$\int_C f(x, y)\,ds = \int_0^1 (\frac{1}{2}t^2)(\frac{1}{2}\sqrt{1 + 25t^3})\,dt = \frac{1}{4}\Big[\frac{2}{225}(1 + 25t^3)^{3/2}\Big]_0^1 =$

$$\tfrac{1}{450}(26^{3/2} - 1) \approx 0.29.$$

(2) $dx = \frac{1}{2}\,dt.$ $\int_C f(x, y)\,dx = \int_0^1 (\frac{1}{2}t^2)\,\frac{1}{2}\,dt = \frac{1}{12}.$

(3) $dy = \frac{5}{2}t^{3/2}\,dt.$ $\int_C f(x, y)\,dy = \int_0^1 (\frac{1}{2}t^2)(\frac{5}{2}t^{3/2})\,dt = \frac{5}{18}.$

3 $y = x^3 + 1 \Rightarrow dy = 3x^2\,dx.$

$\int_C 6x^2y\,dx + xy\,dy = \int_{-1}^1 \Big[6x^2(x^3 + 1) + x(x^3 + 1)(3x^2)\Big]dx =$

$$3\int_{-1}^1 (x^6 + 2x^5 + x^3 + 2x^2)\,dx = 6\int_0^1 (x^6 + 2x^2)\,dx = \tfrac{34}{7}.$$

4 $y = x^2 + 2x \Rightarrow dy = (2x + 2)\,dx.$ $\int_C y\,dx + (x + y)\,dy =$

$$\int_0^2 \Big[(x^2 + 2x) + (x + x^2 + 2x)(2x + 2)\Big]dx = \int_0^2 (2x^3 + 9x^2 + 8x)\,dx = 48.$$

5 $y^2 = x \Rightarrow dx = 2y\,dy.$ $\int_C (x - y)\,dx + x\,dy =$

$$\int_{-2}^2 \Big[(y^2 - y)(2y) + y^2\Big]dy = \int_{-2}^2 (2y^3 - y^2)\,dy = -2\int_0^2 y^2\,dy = -\tfrac{16}{3}.$$

6 $x = y^3 \Rightarrow dx = 3y^2\,dy.$ $\int_C xy\,dx + x^2y^3\,dy = \int_0^1 \Big[y^3(y)(3y^2) + (y^3)^2y^3\Big]dy = \frac{37}{70}.$

Note: Let I denote the required line integral.

7 (a) C_1: $x = t,\ y = 0,\ 0 \le t \le 1 \Rightarrow dx = dt$ and $dy = 0\,dt.$

C_2: $x = 1,\ y = t,\ 0 \le t \le 3 \Rightarrow dx = 0\,dt$ and $dy = dt.$

$\int_C xy\,dx + (x + y)\,dy = \int_{C_1} xy\,dx + (x + y)\,dy + \int_{C_2} xy\,dx + (x + y)\,dy =$

$$\int_0^1 (0 + t\cdot 0)\,dt + \int_0^3 \Big[t\cdot 0 + (1 + t)\Big]dt = 0 + \tfrac{15}{2} = \tfrac{15}{2}.$$

(b) C_1: $x = 0,\ y = t,\ 0 \le t \le 3 \Rightarrow dx = 0\,dt$ and $dy = dt.$

C_2: $x = t,\ y = 3,\ 0 \le t \le 1 \Rightarrow dx = dt$ and $dy = 0\,dt.$

$$\mathrm{I} = \int_0^3 \Big[0 + (0 + t)\Big]dt + \int_0^1 \Big[3t + (t + 3)\cdot 0\Big]dt = \tfrac{9}{2} + \tfrac{3}{2} = 6.$$

(c) $y = 3x \Rightarrow dy = 3\,dx$. $\mathrm{I} = \int_0^1 \left[x(3x) + (x + 3x)\cdot 3\right] dx = 3\int_0^1 (x^2 + 4x)\,dx = 7$.

(d) $y = 3x^2 \Rightarrow dy = 6x\,dx$.

$$\mathrm{I} = \int_0^1 \left[x(3x^2) + (x + 3x^2)(6x)\right] dx = 3\int_0^1 (7x^3 + 2x^2)\,dx = \tfrac{29}{4}.$$

[8] (a) C_1: $x = 1,\ y = t,\ 2 \le t \le 8 \Rightarrow dx = 0\,dt$ and $dy = dt$.

C_2: $x = t,\ y = 8,\ -2 \le t \le 1 \Rightarrow dx = dt$ and $dy = 0\,dt$.

$\int_C (x^2 + y^2)\,dx + 2x\,dy = \int_2^8 \left[(1 + t^2)\cdot 0 + 2\right] dt + \int_1^{-2} \left[(t^2 + 64) + 0\right] dt =$

$12 - 195 = -183$.

(b) C_1: $x = t,\ y = 2,\ -2 \le t \le 1 \Rightarrow dx = dt$ and $dy = 0\,dt$.

C_2: $x = -2,\ y = t,\ 2 \le t \le 8 \Rightarrow dx = 0\,dt$ and $dy = dt$.

$$\mathrm{I} = \int_1^{-2} \left[(t^2 + 4) + 0\right] dt + \int_2^8 \left[(4 + t^2)\cdot 0 + (-4)\right] dt = -15 - 24 = -39.$$

(c) $y = -2x + 4 \Rightarrow dy = -2\,dx$.

$$\mathrm{I} = \int_1^{-2} \left[x^2 + (-2x + 4)^2 + 2x(-2)\right] dx = \int_1^{-2} (5x^2 - 20x + 16)\,dx = -93.$$

(d) $y = 2x^2 \Rightarrow dy = 4x\,dx$.

$$\mathrm{I} = \int_1^{-2} \left[(x^2 + 4x^4) + (2x)(4x)\right] dx = \int_1^{-2} (4x^4 + 9x^2)\,dx = -\tfrac{267}{5}.$$

[9] $x = e^t,\ y = e^{-t},\ z = e^{2t} \Rightarrow dx = e^t\,dt,\ dy = -e^{-t}\,dt$, and $dz = 2e^{2t}\,dt \Rightarrow$

$\int_C xz\,dx + (y + z)\,dy + x\,dz = \int_0^1 \left[e^{3t}(e^t) + (e^{-t} + e^{2t})(-e^{-t}) + e^t(2e^{2t})\right] dt =$

$$\int_0^1 (e^{4t} - e^{-2t} - e^t + 2e^{3t})\,dt = \tfrac{1}{12}(3e^4 + 6e^{-2} - 12e + 8e^3 - 5) \approx 23.97.$$

[10] $x = \sin t,\ y = 2\sin t,\ z = \sin^2 t \Rightarrow dx = \cos t\,dt,\ dy = 2\cos t\,dt$, and

$dz = 2\sin t\cos t\,dt$. $\int_C y\,dx + z\,dy + x\,dz =$

$\int_0^{\pi/2} \left[2\sin t(\cos t) + \sin^2 t(2\cos t) + \sin t(2\sin t\cos t)\right] dt =$

$$\int_0^{\pi/2} \left[(2\sin t\cos t + 4\sin^2 t\cos t)\right] dt = \left[\sin^2 t + \tfrac{4}{3}\sin^3 t\right]_0^{\pi/2} = \tfrac{7}{3}.$$

[11] (a) C_1: $x = t,\ y = z = 0 \Rightarrow dx = dt,\ dy = dz = 0\,dt$. $0 \le t \le 2$.

C_2: $x = 2,\ y = t,\ z = 0 \Rightarrow dy = dt,\ dx = dz = 0\,dt$. $0 \le t \le 3$.

C_3: $x = 2,\ y = 3,\ z = t \Rightarrow dz = dt,\ dx = dy = 0\,dt$. $0 \le t \le 4$.

$\int_C (x + y + z)\,dx + (x - 2y + 3z)\,dy + (2x + y - z)\,dz =$

$$\int_0^2 t\,dt + \int_0^3 (2 - 2t)\,dt + \int_0^4 (7 - t)\,dt = 2 - 3 + 20 = 19.$$

(b) C_1: $x = 0,\ y = 0,\ z = t \Rightarrow dz = dt,\ dx = dy = 0\,dt$. $0 \le t \le 4$.

C_2: $x = t,\ y = 0,\ z = 4 \Rightarrow dx = dt,\ dy = dz = 0\,dt$. $0 \le t \le 2$.

C_3: $x = 2,\ y = t,\ z = 4 \Rightarrow dy = dt,\ dx = dz = 0\,dt$. $0 \le t \le 3$.

$$\mathrm{I} = \int_0^4 (-t)\,dt + \int_0^2 (t + 4)\,dt + \int_0^3 (14 - 2t)\,dt = -8 + 10 + 33 = 35.$$

(c) C: $x = 2t$, $y = 3t$, $z = 4t \Rightarrow dx = 2\,dt$, $dy = 3\,dt$, $dz = 4\,dt$. $0 \le t \le 1$.

$$\text{I} = \int_0^1 \left[(9t)(2) + (8t)(3) + (3t)(4)\right] dt = \int_0^1 54t\,dt = 27.$$

[12] (a) C_1: $x = t$, $y = -2$, $z = 3 \Rightarrow dx = dt$, $dy = dz = 0\,dt$. $-4 \le t \le 1$.

C_2: $x = -4$, $y = t$, $z = 3 \Rightarrow dy = dt$, $dx = dz = 0\,dt$. $-2 \le t \le 5$.

C_3: $x = -4$, $y = 5$, $z = t \Rightarrow dz = dt$, $dx = dy = 0\,dt$. $2 \le t \le 3$.

$\int_C (x - y)\,dx + (y - z)\,dy + x\,dz =$

$$\int_1^{-4} (t + 2)\,dt + \int_{-2}^{5} (t - 3)\,dt + \int_3^2 (-4)\,dt = -\tfrac{5}{2} + \tfrac{-21}{2} + 4 = -9.$$

(b) C_1: $x = 1$, $y = -2$, $z = t \Rightarrow dz = dt$, $dx = dy = 0\,dt$. $2 \le t \le 3$.

C_2: $x = t$, $y = -2$, $z = 2 \Rightarrow dx = dt$, $dy = dz = 0\,dt$. $-4 \le t \le 1$.

C_3: $x = -4$, $y = t$, $z = 2 \Rightarrow dy = dt$, $dx = dz = 0\,dt$. $-2 \le t \le 5$.

$$\text{I} = \int_3^2 (1)\,dt + \int_1^{-4} (t + 2)\,dt + \int_{-2}^{5} (t - 2)\,dt = -1 + \tfrac{-5}{2} + \tfrac{-7}{2} = -7.$$

(c) C: $x = 1 + \left[(-4) - 1\right]t = -5t + 1$, $y = -2 + \left[5 - (-2)\right]t = 7t - 2$,

$z = 3 + (2 - 3)t = -t + 3 \Rightarrow dx = -5\,dt$, $dy = 7\,dt$, $dz = -\,dt$. $0 \le t \le 1$.

$$\text{I} = \int_0^1 \left[(-12t + 3)(-5) + (8t - 5)(7) + (-5t + 1)(-1)\right] dt$$

$$= \int_0^1 (121t - 51)\,dt = \tfrac{19}{2}.$$

[13] C: $x = t$, $y = 2t$, $z = 3t \Rightarrow ds = \sqrt{1^2 + 2^2 + 3^2}\,dt = \sqrt{14}\,dt$. $0 \le t \le 1$.

$$\int_C xyz\,ds = \int_0^1 (6t^3)\sqrt{14}\,dt = \tfrac{3}{2}\sqrt{14}.$$

[14] $x = a\cos t$, $y = a\sin t$, $z = bt \Rightarrow ds = \sqrt{a^2\sin^2 t + a^2\cos^2 t + b^2}\,dt = \sqrt{a^2 + b^2}\,dt$.

$0 \le t \le 2\pi$. $\int_C (xy + z)\,ds = \int_0^{2\pi} (a^2 \sin t\cos t + bt)\sqrt{a^2 + b^2}\,dt =$

$$\sqrt{a^2 + b^2}\left[\tfrac{1}{2}a^2\sin^2 t + \tfrac{1}{2}bt^2\right]_0^{2\pi} = 2\pi^2 b\sqrt{a^2 + b^2}.$$

[15] Refer to Exercise 7 for the parametric equations.

For each part, $\int_C \mathbf{F}\cdot d\mathbf{r} = \int_C xy^2\,dx + x^2y\,dy$.

(a) $W = \int_0^1 (0 + 0)\,dt + \int_0^3 (t^2\cdot 0 + t)\,dt = 0 + \tfrac{9}{2} = \tfrac{9}{2}$.

(b) $W = \int_0^3 (0 + 0)\,dt + \int_0^1 (9t + 3t^2\cdot 0)\,dt = \tfrac{9}{2}$.

(c) $W = \int_0^1 \left[9x^3 + 3x^3(3)\right] dx = \tfrac{9}{2}$. (d) $W = \int_0^1 \left[9x^5 + (3x^4)(6x)\right] dx = \tfrac{9}{2}$.

[16] Refer to Exercise 8 for the parametric equations.

For each part, $\int_C \mathbf{F}\cdot d\mathbf{r} = \int_C (2x + y)\,dx + (x + 2y)\,dy$.

(a) $W = \int_2^8 (1 + 2t)\,dt + \int_1^{-2} (2t + 8)\,dt = 66 + (-21) = 45.$

(b) $W = \int_1^{-2} (2t + 2)\, dt + \int_2^8 (-2 + 2t)\, dt = -3 + 48 = 45.$

(c) $W = \int_1^{-2} \big[4 + (-3x + 8)(-2)\big]\, dx = \int_1^{-2} (6x - 12)\, dx = 45.$

(d) $W = \int_1^{-2} \big[(2x + 2x^2) + (x + 4x^2)(4x)\big]\, dx = \int_1^{-2} (16x^3 + 6x^2 + 2x)\, dx = 45.$

[17] $\mathbf{F}(x, y) = \frac{4}{\|\mathbf{r}\|^3}\mathbf{r} = \frac{4}{(x^2 + y^2)^{3/2}}(x\mathbf{i} + y\mathbf{j}) = \frac{4}{a^3}(x\mathbf{i} + y\mathbf{j}).$

Let $x = a\cos t$, $y = a\sin t$ as t varies from π to 0. $W = \int_C \mathbf{F}\cdot d\mathbf{r} =$

$$\frac{4}{a^3}\int_C x\, dx + y\, dy = \frac{4}{a^3}\int_\pi^0 \big[(a\cos t)(-a\sin t) + (a\sin t)(a\cos t)\big]\, dt = 0.$$

[18] $y = x^3 \Rightarrow dy = 3x^2\, dx.$

$$W = \int_C \mathbf{F}\cdot d\mathbf{r} = \int_C (x^2 + y^2)\, dx + xy\, dy = \int_0^2 \big[(x^2 + x^6) + (x^4)(3x^2)\big]\, dx = \tfrac{1592}{21}.$$

[19] $W = \int_C y\, dx + z\, dy + x\, dz = \int_0^2 \big[(t^2)(1) + (t^3)(2t) + (t)(3t^2)\big]\, dt = \frac{412}{15}.$

[20] $W = \int_C e^x\, dx + e^y\, dy + e^z\, dz = \int_0^2 \big[(e^t)(1) + (e^{t^2})(2t) + (e^{t^3})(3t^2)\big]\, dt =$

$$e^2 + e^4 + e^8 - 3 \approx 3039.95.$$

[21] Let the object's position be given by $\mathbf{r}(t)$. Thus, $\mathbf{r}'(t)$ is the object's velocity and $\mathbf{F}(x, y, z)\cdot\mathbf{r}'(t) = 0$. Since $d\mathbf{r} = \mathbf{r}'(t)\, dt$, $W = \int_C \mathbf{F}(x, y, z)\cdot d\mathbf{r} = 0.$

[22] Let $x = a\cos t$, $y = a\sin t$ for $0 \le t \le 2\pi$. Then, $dx = -a\sin t\, dt$ and $dy = a\cos t\, dt$. Thus, $W = \int_C \mathbf{F}\cdot d\mathbf{r} = c\int_0^{2\pi} (-a\sin t + a\cos t)\, dt = 0.$

[23] Let the wire be represented by a curve C as in Figure 18.10. Then the mass of the piece of wire between P_{k-1} and P_k can be considered concentrated at (u_k, v_k). Its mass is approximately $\delta(u_k, v_k)\Delta s_k$ and its moment with respect to the x-axis is approximately $v_k\,\delta(u_k, v_k)\Delta s_k$ and with respect to the y-axis is approximately $u_k\delta(u_k, v_k)\Delta s_k$. Using the limit of sums,

$$M_x = \lim_{\|P\|\to 0}\sum_k v_k\,\delta(u_k, v_k)\Delta s_k = \int_C y\,\delta(x, y)\, ds \text{ and}$$

$$M_y = \lim_{\|P\|\to 0}\sum_k u_k\,\delta(u_k, v_k)\Delta s_k = \int_C x\,\delta(x, y)\, ds.$$

In Example 3, $m = 2ka^2$. Now, $M_x = \int_0^\pi (a\sin t)(ka\sin t)\, a\, dt = \frac{1}{2}\pi ka^3$ and

$$M_y = \int_0^\pi (a\cos t)(ka\sin t)\, a\, dt = 0. \text{ Thus, } \bar{x} = \frac{M_y}{m} = 0 \text{ and } \bar{y} = \frac{M_x}{m} = \tfrac{1}{4}\pi a.$$

[24] Let $x = t$, $y = 4 - t^2$ for $-2 \le t \le 2$ and $\delta(x, y) = k|x|$. $ds = \sqrt{1 + 4t^2}\, dt.$

$$m = \int_{-2}^2 k|t|\sqrt{1 + 4t^2}\, dt = 2k\int_0^2 t\sqrt{1 + 4t^2}\, dt = \tfrac{1}{6}k(17^{3/2} - 1).$$

$$M_x = \int_{-2}^{2} (4 - t^2)\, k|t| \sqrt{1 + 4t^2}\, dt = 2k\int_0^2 (4 - t^2)\sqrt{1 + 4t^2}\,(t)\, dt.$$

Let $u = 1 + 4t^2$ and $du = 8t\, dt$. $t = 0, 2 \Rightarrow u = 1, 17$.

$$M_x = \tfrac{1}{4}k\int_1^{17} \Big[\tfrac{1}{4}(17 - u)\Big] u^{1/2}\, du = \tfrac{1}{16}k\Big[\tfrac{2}{3}(17)u^{3/2} - \tfrac{2}{5}u^{5/2}\Big]_1^{17} = \tfrac{1}{60}k(17^{5/2} - 41).$$

$$M_y = \int_{-2}^{2} t\,k|t|\sqrt{1 + 4t^2}\, dt = 0 \text{ since the integrand is odd.}$$

$$\text{Thus, } \bar{x} = \frac{M_y}{m} = 0 \text{ and } \bar{y} = \frac{M_x}{m} = \frac{17^{5/2} - 41}{10(17^{3/2} - 1)} \approx 1.67.$$

[25] Let the wire be represented by a curve C as in Figure 18.17.

Then in a manner similar to the one in Exercise 23, $m = \int_C \delta(x, y, z)\, ds$,

$M_{yz} = \int_C x\,\delta(x, y, z)\, ds$, $M_{xz} = \int_C y\,\delta(x, y, z)\, ds$, and $M_{xy} = \int_C z\,\delta(x, y, z)\, ds$.

Use $\bar{x} = \dfrac{M_{yz}}{m}$, $\bar{y} = \dfrac{M_{xz}}{m}$, and $\bar{z} = \dfrac{M_{xy}}{m}$ to find the center of mass.

[26] $x = a\cos t$, $y = a\sin t$, $z = bt \Rightarrow dx = -a\sin t\, dt$, $dy = a\cos t\, dt$, and $dz = b\, dt$.

$0 \le t \le 3\pi$. $ds = \sqrt{a^2\sin^2 t + a^2\cos^2 t + b^2}\, dt = \sqrt{a^2 + b^2}\, dt$.

Let $\delta(x, y, z) = \delta$, a constant. Then, $m = \displaystyle\int_0^{3\pi} \delta\sqrt{a^2 + b^2}\, dt = 3\pi\delta\sqrt{a^2 + b^2}$,

$$M_{yz} = \int_0^{3\pi} (a\cos t)\,\delta\sqrt{a^2 + b^2}\, dt = 0,$$

$$M_{xz} = \int_0^{3\pi} (a\sin t)\,\delta\sqrt{a^2 + b^2}\, dt = 2a\delta\sqrt{a^2 + b^2}, \text{ and}$$

$$M_{xy} = \int_0^{3\pi} (bt)\,\delta\sqrt{a^2 + b^2}\, dt = \tfrac{9}{2}\pi^2 b\,\delta\sqrt{a^2 + b^2}.$$

$$\text{Thus, } \bar{x} = \frac{M_{yz}}{m} = 0,\ \bar{y} = \frac{M_{xz}}{m} = \frac{2a}{3\pi}, \text{ and } \bar{z} = \frac{M_{xy}}{m} = \tfrac{3}{2}\pi b.$$

[27] Define $I_x = \int_C y^2\,\delta(x, y)\, ds$ and $I_y = \int_C x^2\,\delta(x, y)\, ds$. For Example 3,

$$I_x = \int_0^{\pi} (a\sin t)^2(ka\sin t)\, a\, dt = ka^4\int_0^{\pi} (1 - \cos^2 t)\sin t\, dt = \tfrac{4}{3}ka^4 \text{ and}$$

$$I_y = \int_0^{\pi} (a\cos t)^2(ka\sin t)\, a\, dt = ka^4\int_0^{\pi} \cos^2 t\sin t\, dt = \tfrac{2}{3}ka^4.$$

[28]
$$I_x = \int_{-2}^{2} (4 - t^2)^2\, k|t|\sqrt{1 + 4t^2}\, dt = 2k\int_0^2 (4 - t^2)^2\, t\sqrt{1 + 4t^2}\, dt$$

$$= \tfrac{1}{4}k\int_1^{17} \Big[\tfrac{1}{4}(17 - u)\Big]^2 u^{1/2}\, du = \tfrac{1}{64}k\int_1^{17} (289u^{1/2} - 34u^{3/2} + u^{5/2})\, du$$

$$= \tfrac{1}{64}k\Big[\tfrac{2}{3}(289)u^{3/2} - \tfrac{2}{5}(34)u^{5/2} + \tfrac{2}{7}u^{7/2}\Big]_1^{17} = \tfrac{1}{420}k(17^{7/2} - 1177) \approx 45.4k.$$

$$I_y = \int_{-2}^{2} t^2\, k|t|\sqrt{1 + 4t^2}\, dt = 2k\int_0^2 t^2\sqrt{1 + 4t^2}\,(t)\, dt = \tfrac{1}{4}k\int_1^{17} \Big[\tfrac{1}{4}(u - 1)\Big] u^{1/2}\, du$$

$$= \tfrac{1}{16}k\int_1^{17} (u^{3/2} - u^{1/2})\, du = \tfrac{1}{60}k\Big[23(17^{3/2}) + 1\Big] \approx 26.9k.$$

[29] If the density at (x, y, z) is $\delta(x, y, z)$, then $I_x = \int_C (y^2 + z^2)\,\delta(x, y, z)\,ds$,

$I_y = \int_C (x^2 + z^2)\,\delta(x, y, z)\,ds$, and $I_z = \int_C (x^2 + y^2)\,\delta(x, y, z)\,ds$.

[30] $I_z = \int_0^{3\pi} (a^2)\,\delta\,\sqrt{a^2 + b^2}\,dt = 3\pi a^2\,\delta\,\sqrt{a^2 + b^2}$.

[31] $u_k = \frac{1}{10}k - \frac{1}{20} \Rightarrow v_k = (\frac{1}{10}k - \frac{1}{20})^4 \Rightarrow u_k v_k = (\frac{1}{10}k - \frac{1}{20})^5$.

$$\sum_{k=1}^{10} \sin(u_k v_k)\,\Delta x_k = \sum_{k=1}^{10} \sin(\tfrac{1}{10}k - \tfrac{1}{20})^5\,(\tfrac{1}{10}) \approx 0.1554.$$

[32] $u_k = \frac{19}{20} + \frac{1}{10}k \Rightarrow v_k = 3(\frac{19}{20} + \frac{1}{10}k) + 4 \Rightarrow u_k v_k = (\frac{19}{20} + \frac{1}{10}k)(\frac{137}{20} + \frac{3}{10}k)$.

$$\sum_{k=1}^{10} \ln\sqrt{u_k v_k}\,\Delta y_k = \sum_{k=1}^{10} \ln\sqrt{\tfrac{3}{100}k^2 + \tfrac{194}{200}k + \tfrac{2603}{400}}\,(\tfrac{3}{10}) \approx 3.7821.$$

Exercises 18.3

[1] $f_x = 3x^2y + 2 \Rightarrow f = x^3y + 2x + g(y) \Rightarrow f_y = x^3 + g'(y) = x^3 + 4y^3$.

Thus, $g'(y) = 4y^3$ and $g(y) = y^4 + c$. So $f(x, y) = x^3y + 2x + y^4 + c$.

[2] $f_x = 6xy^2 + 2y \Rightarrow f = 3x^2y^2 + 2xy + g(y) \Rightarrow f_y = 6x^2y + 2x + g'(y) =$

$6x^2y + 2x$. Thus, $g'(y) = 0$ and $g(y) = c$. So $f(x, y) = 3x^2y^2 + 2xy + c$.

[3] $f_x = 2x\sin y + 4e^x \Rightarrow f = x^2\sin y + 4e^x + g(y) \Rightarrow f_y = x^2\cos y + g'(y) =$

$x^2\cos y$. Thus, $g'(y) = 0$ and $g(y) = c$. So $f(x, y) = x^2\sin y + 4e^x + c$.

[4] $f_x = 2xe^{2y} + 4y^3 \Rightarrow f = x^2e^{2y} + 4xy^3 + g(y) \Rightarrow f_y = 2x^2e^{2y} + 12xy^2 + g'(y) =$

$2x^2e^{2y} + 12xy^2$. Thus, $g'(y) = 0$ and $g(y) = c$. So $f(x, y) = x^2e^{2y} + 4xy^3 + c$.

[5] $f_x = -2y^3\sin x \Rightarrow f = 2y^3\cos x + g(y) \Rightarrow f_y = 6y^2\cos x + g'(y) = 6y^2\cos x + 5$.

Thus, $g'(y) = 5$ and $g(y) = 5y + c$. So $f(x, y) = 2y^3\cos x + 5y + c$.

[6] $f_x = 5y^3 + 4y^3\sec^2 x \Rightarrow f = 5xy^3 + 4y^3\tan x + g(y) \Rightarrow$

$f_y = 15xy^2 + 12y^2\tan x + g'(y) = 15xy^2 + 12y^2\tan x$.

Thus, $g'(y) = 0$ and $g(y) = c$. So $f(x, y) = 5xy^3 + 4y^3\tan x + c$.

[7] $f_x = 8xz \Rightarrow f = 4x^2z + g(y, z) \Rightarrow f_y = g_y(y, z) = 1 - 6yz^3$.

Thus, $g(y, z) = y - 3y^2z^3 + k(z)$. $f_z = 4x^2 - 9y^2z^2 + k'(z) = 4x^2 - 9y^2z^2 \Rightarrow$

$k'(z) = 0$ and $k(z) = c$. So $f(x, y, z) = 4x^2z + y - 3y^2z^3 + c$.

[8] $f_x = y + z \Rightarrow f = xy + xz + g(y, z) \Rightarrow f_y = x + g_y(y, z) = x + z$.

Thus, $g_y(y, z) = z$ and $g(y, z) = yz + k(z)$. $f_z = x + y + k'(z) = x + y \Rightarrow$

$k'(z) = 0$ and $k(z) = c$. So $f(x, y, z) = xy + xz + yz + c$.

[9] $f_x = y\sec^2 x - ze^x \Rightarrow f = y\tan x - ze^x + g(y, z) \Rightarrow f_y = \tan x + g_y(y, z) = \tan x$.

Thus, $g_y(y, z) = 0$ and $g(y, z) = k(z)$. $f_z = -e^x + k'(z) = -e^x \Rightarrow$

$k'(z) = 0$ and $k(z) = c$. So $f(x, y, z) = y\tan x - ze^x + c$.

[10] *Note*: The j term should be " $+ 2y\cos z\,\mathbf{j}$". $f_x = 2x\sin z \Rightarrow f = x^2\sin z + g(y, z) \Rightarrow$

$f_y = g_y(y, z) = 2y\cos z$. Thus, $g(y, z) = y^2\cos z + k(z)$.

$f_z = x^2\cos z - y^2\sin z + k'(z) = x^2\cos z - y^2\sin z \Rightarrow k'(z) = 0$ and $k(z) = c$.

So $f(x, y, z) = x^2\sin z + y^2\cos z + c$.

Note: In Exercises 11-14, we will show that the line integral is independent of path by finding a scalar function f such that $\mathbf{F} = \nabla f$, and then use (18.14) to evaluate the given integral I.

11 $f_x = y^2 + 2xy \Rightarrow f = xy^2 + x^2y + g(y) \Rightarrow f_y = 2xy + x^2 + g'(y) = x^2 + 2xy.$

So $g'(y) = 0$ and $g(y) = c$. Thus, let $f(x, y) = xy^2 + x^2y$ { we may omit " $+ c$" }.

$$\mathrm{I} = f(3, 1) - f(-1, 2) = 12 - (-2) = 14.$$

12 $f_x = e^x \sin y \Rightarrow f = e^x \sin y + g(y) \Rightarrow f_y = e^x \cos y + g'(y) = e^x \cos y.$

So $g'(y) = 0$ and $g(y) = c$. Thus, let $f(x, y) = e^x \sin y$.

$$\mathrm{I} = f(1, \tfrac{\pi}{2}) - f(0, 0) = e - 0 = e.$$

13 $f_x = 6xy^3 + 2z^2 \Rightarrow f = 3x^2y^3 + 2xz^2 + g(y, z) \Rightarrow$

$f_y = 9x^2y^2 + g_y(y, z) = 9x^2y^2$. So $g_y(y, z) = 0$ and $g(y, z) = k(z)$.

$f_z = 4xz + k'(z) = 4xz + 1 \Rightarrow k'(z) = 1$ and $k(z) = z + c$. Thus, let

$f(x, y, z) = 3x^2y^3 + 2xz^2 + z$. $\mathrm{I} = f(-2, 1, 3) - f(1, 0, 2) = -21 - 10 = -31.$

14 $f_x = yz + 1 \Rightarrow f = xyz + x + g(y, z) \Rightarrow f_y = xz + g_y(y, z) = xz + 1.$

So $g_y(y, z) = 1$ and $g(y, z) = y + k(z)$.

$f_z = xy + k'(z) = xy + 1 \Rightarrow k'(z) = 1$ and $k(z) = z + c$. Thus, let

$f(x, y, z) = xyz + x + y + z$. $\mathrm{I} = f(-1, 1, 2) - f(4, 0, 3) = 0 - 7 = -7.$

15 $M = 4xy^3$ and $N = 2xy^3$. $\dfrac{\partial M}{\partial y} = \dfrac{\partial}{\partial y}(4xy^3) = 12xy^2$. $\dfrac{\partial N}{\partial x} = \dfrac{\partial}{\partial x}(2xy^3) = 2y^3$.

Since $\dfrac{\partial M}{\partial y} \neq \dfrac{\partial N}{\partial x}$, $\int_C \mathbf{F} \cdot d\mathbf{r}$ is not independent of path.

16 $M = 6x^2 - 2xy^2$ and $N = 2x^2y + 5$.

$$\frac{\partial M}{\partial y} = \frac{\partial}{\partial y}(6x^2 - 2xy^2) = -4xy \neq 4xy = \frac{\partial}{\partial x}(2x^2y + 5) = \frac{\partial N}{\partial x}.$$

17 $M = e^x$ and $N = 3 - e^x \sin y$.

$$\frac{\partial M}{\partial y} = \frac{\partial}{\partial y}(e^x) = 0 \neq -e^x \sin y = \frac{\partial}{\partial x}(3 - e^x \sin y) = \frac{\partial N}{\partial x}.$$

18 $M = y^3 \cos x$ and $N = 3y^2 \sin x$.

$$\frac{\partial M}{\partial y} = \frac{\partial}{\partial y}(y^3 \cos x) = 3y^2 \cos x \neq -3y^2 \cos x = \frac{\partial}{\partial x}(-3y^2 \sin x) = \frac{\partial N}{\partial x}.$$

19 If the line integral is path independent,

then $\mathbf{F}(x, y, z) = \nabla f(x, y, z)$ and $M = f_x$, $N = f_y$, and $P = f_z$. Consequently,

$$\frac{\partial M}{\partial y} = f_{xy} = f_{yx} = \frac{\partial N}{\partial x}, \frac{\partial M}{\partial z} = f_{xz} = f_{zx} = \frac{\partial P}{\partial x}, \text{ and } \frac{\partial N}{\partial z} = f_{yz} = f_{zy} = \frac{\partial P}{\partial y}.$$

20 $M = 5y$, $N = 5x$, and $P = yz^2$. $\dfrac{\partial M}{\partial y} = 5 = \dfrac{\partial N}{\partial x}$, $\dfrac{\partial M}{\partial z} = 0 = \dfrac{\partial P}{\partial x}$,

but $\dfrac{\partial N}{\partial z} = 0 \neq z^2 = \dfrac{\partial P}{\partial y}$. Thus, I is not independent of path.

21 $M = 2xy$, $N = x^2 + z^2$, and $P = yz$. $\dfrac{\partial M}{\partial y} = 2x = \dfrac{\partial N}{\partial x}$, $\dfrac{\partial M}{\partial z} = 0 = \dfrac{\partial P}{\partial x}$,

but $\dfrac{\partial N}{\partial z} = 2z \neq z = \dfrac{\partial P}{\partial y}$.

[22] $M = e^y \cos z$, $N = xe^y \cos z$, and $P = xe^y \sin z$. $\frac{\partial M}{\partial y} = e^y \cos z = \frac{\partial N}{\partial x}$,

but $\frac{\partial M}{\partial z} = -e^y \sin z \neq e^y \sin z = \frac{\partial P}{\partial x}$, and $\frac{\partial N}{\partial z} = -xe^y \sin z \neq xe^y \sin z = \frac{\partial P}{\partial y}$.

[23] Let $\mathbf{r} = x\mathbf{i} + y\mathbf{j} + z\mathbf{k}$. Then, $-\frac{\mathbf{r}}{\|\mathbf{r}\|}$ is a unit vector in the direction of the origin.

The magnitude of $\mathbf{F}$ is inversely proportional to the distance from the origin.

Thus, $\|\mathbf{F}\| = K(x, y, z) = \frac{c}{\|\mathbf{r}\|} = \frac{c}{(x^2 + y^2 + z^2)^{1/2}}$ and $\mathbf{F}(x, y, z) = -K(x, y, z)\frac{\mathbf{r}}{\|\mathbf{r}\|}$,

with $c > 0$. Then, $f_x = \frac{-cx}{x^2 + y^2 + z^2}$, $f_y = \frac{-cy}{x^2 + y^2 + z^2}$, and $f_z = \frac{-cz}{x^2 + y^2 + z^2}$.

Now, $f = -\frac{1}{2}c\ln(x^2 + y^2 + z^2) + g(y, z) \Rightarrow f_y = \frac{-cy}{x^2 + y^2 + z^2} + g_y(y, z)$.

So $g_y(y, z) = 0$ and $g(y, z) = k(z)$. $f_z = \frac{-cz}{x^2 + y^2 + z^2} + k'(z) \Rightarrow k'(z) = 0$ and

$k(z) = d$. Thus, $f(x, y, z) = -\frac{1}{2}c\ln(x^2 + y^2 + z^2) + d$, and $\mathbf{F}$ is conservative.

[24] Let $\mathbf{F}(x, y, z) = K(x, y, z)\frac{\mathbf{r}}{\|\mathbf{r}\|}$, where $K(x, y, z) = c\|\mathbf{r}\|$ and $c > 0$. Then, $f_x = cx$,

$f_y = cy$, and $f_z = cz$. $f = \frac{1}{2}cx^2 + g(y, z) \Rightarrow f_y = g_y(y, z)$. So $g_y(y, z) = cy$ and

$g(y, z) = \frac{1}{2}cy^2 + k(z)$. $f_z = k'(z) \Rightarrow k'(z) = cz$ and $k(z) = \frac{1}{2}cz^2 + d$.

Thus, $f(x, y, z) = \frac{1}{2}c(x^2 + y^2 + z^2) + d$, and $\mathbf{F}$ is conservative.

[25] If $\mathbf{F}(x, y, z) = a\mathbf{i} + b\mathbf{j} + c\mathbf{k}$, then $f(x, y, z) = ax + by + cz + d$ and

$W = \int_C \mathbf{F} \cdot d\mathbf{r}$ is independent of path. Thus, if $P = (x_0, y_0, z_0)$ and

$Q = (x_1, y_1, z_1)$, then $W = f(x_1, y_1, z_1) - f(x_0, y_0, z_0) =$

$a(x_1 - x_0) + b(y_1 - y_0) + c(z_1 - z_0) = \mathbf{F} \cdot \overrightarrow{PQ}$.

[26] Let $G(x)$, $H(y)$, and $K(z)$ be the antiderivatives of $g(x)$, $h(y)$, and $k(z)$, respectively.

If we let $f(x, y, z) = G(x) + H(y) + K(z)$, then $\mathbf{F}(x, y, z) = \nabla f(x, y, z)$.

[27] $\frac{\partial M}{\partial y} = \frac{y^2 - x^2}{(x^2 + y^2)^2} = \frac{\partial N}{\partial x}$. Let C_1 be the upper semicircle with end points $(-1, 0)$

and $(1, 0)$, and C_2 the lower semicircle with the same end points. Then,

$\int_{C_1} \mathbf{F} \cdot d\mathbf{r} = \int_{C_1} \frac{-y\,dx + x\,dy}{x^2 + y^2} = \int_0^{\pi} d\theta = \pi$ and $\int_{C_2} \mathbf{F} \cdot d\mathbf{r} = \int_0^{-\pi} d\theta = -\pi$,

where $x = \cos\theta$ and $y = \sin\theta$. Since they are not equal, $\int_C \mathbf{F} \cdot d\mathbf{r}$ is not independent

of path. This does not violate (18.16) since D is not simply connected.

M and N are not continuous at $(0, 0)$.

[28] Since the satellite's orbit is circular, its velocity is orthogonal to the

gravitational force $\mathbf{F}$. By Exercise 21, §18.2, the work done is zero.

[29] From the proof of (18.4), $f(x, y, z) = \frac{-c}{\|\mathbf{r}\|}$, where $\mathbf{F}(x, y, z) = \nabla f(x, y, z)$.

Since $\mathbf{F}$ is conservative, $W = \frac{-c}{d_2} - \frac{-c}{d_1} = c\left(\frac{1}{d_1} - \frac{1}{d_2}\right) = c\,\frac{d_2 - d_1}{d_1 d_2}$.

[30] By (18.18), the kinetic energy must be *increasing* at a rate of k units per second.

Exercises 18.4

1 $M = x^2 + y$, $N = xy^2 \Rightarrow N_x = y^2$, $M_y = 1$. Thus, $\oint_C (x^2 + y)\,dx + (xy^2)\,dy =$

$$\int_{-1}^{0}\int_{y^2}^{-y} (y^2 - 1)\,dx\,dy = \int_{-1}^{0} (y^2 - 1)(-y - y^2)\,dy = -\tfrac{7}{60}.$$

2 $M = x + y^2$, $N = 1 + x^2 \Rightarrow N_x = 2x$, $M_y = 2y$. Thus,

$$\oint_C (x + y^2)\,dx + (1 + x^2)\,dy = \int_0^1\int_{x^3}^{x^2} (2x - 2y)\,dy\,dx = \int_0^1 (x^6 - 3x^4 + 2x^3)\,dx = \tfrac{3}{70}.$$

3 $M = x^2y^2$, $N = x^2 - y^2 \Rightarrow N_x = 2x$, $M_y = 2x^2y$.

Thus, $\oint_C x^2y^2\,dx + (x^2 - y^2)\,dy = \int_0^1\int_0^1 (2x - 2x^2y)\,dy\,dx = \int_0^1 (2x - x^2)\,dx = \tfrac{2}{3}$.

4 $M = \sqrt{y}$, $N = \sqrt{x} \Rightarrow N_x = \frac{1}{2\sqrt{x}}$, $M_y = \frac{1}{2\sqrt{y}}$. Thus, $\oint_C \sqrt{y}\,dx + \sqrt{x}\,dy$

$$= \int_1^2\int_y^{4-y} \left(\tfrac{1}{2}x^{-1/2} - \tfrac{1}{2}y^{-1/2}\right) dx\,dy$$

$$= \tfrac{1}{2}\int_1^2 \left[2(4 - y)^{1/2} - y^{-1/2}(4 - y) - 2y^{1/2} + y^{1/2}\right] dy$$

$$= \int_1^2 \left[(4 - y)^{1/2} - 2y^{-1/2}\right] dy = \left[-\tfrac{2}{3}(4 - y)^{3/2} - 4y^{1/2}\right]_1^2 =$$

$$-\tfrac{16}{3}\sqrt{2} + 2\sqrt{3} + 4 \approx -0.08.$$

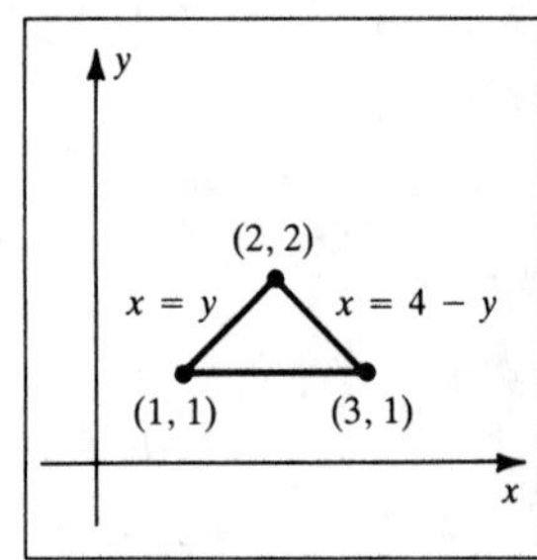

Figure 4

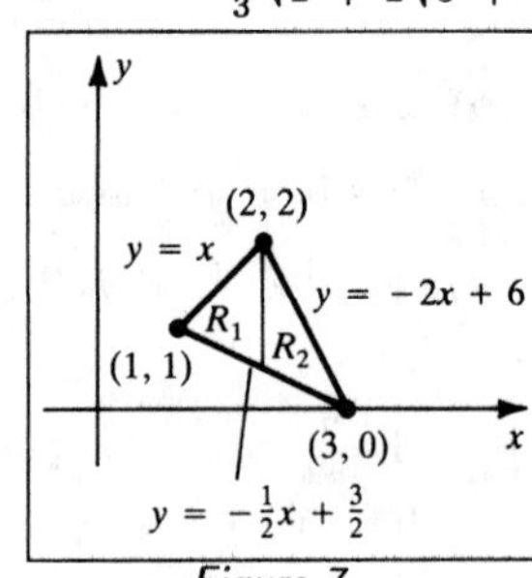

Figure 7

Note: In Exercises 5–14, let I denote the indicated line integral.

5 $M = xy$, $N = y + x \Rightarrow N_x = 1$, $M_y = x$.

Thus, $\mathrm{I} = \iint_R (1 - x)\,dA = \int_0^{2\pi}\int_0^1 (1 - r\cos\theta)\,r\,dr\,d\theta = \int_0^{2\pi} \left(\tfrac{1}{2} - \tfrac{1}{3}\cos\theta\right) d\theta = \pi$.

6 $M = y^2$, $N = x^2 \Rightarrow N_x = 2x$, $M_y = 2y$. Thus, I =

$$\iint_R (2x - 2y)\,dA = 2\int_0^{\pi}\int_0^2 (r\cos\theta - r\sin\theta)\,r\,dr\,d\theta = \tfrac{16}{3}\int_0^{\pi} (\cos\theta - \sin\theta)\,d\theta = -\tfrac{32}{3}.$$

7 $M = xy$, $N = \sin y \Rightarrow N_x = 0$, $M_y = x$.

Divide the triangle into the two regions R_1 and R_2 as shown in *Figure 7*. Thus,

$$\mathrm{I} = \iint_{R_1} (0 - x)\,dA + \iint_{R_2} (0 - x)\,dA$$

$$= \int_1^2\int_{-x/2+3/2}^{x} (-x)\,dy\,dx + \int_2^3\int_{-x/2+3/2}^{-2x+6} (-x)\,dy\,dx \text{ (cont.)}$$

$$= \int_1^2 (-x)(\tfrac{3}{2}x - \tfrac{3}{2})\,dx + \int_2^3 (-x)(-\tfrac{3}{2}x + \tfrac{9}{2})\,dx = -\tfrac{5}{4} + \tfrac{-7}{4} = -3.$$

[8] $M = \tan^{-1}x,\ N = 3x \Rightarrow N_x = 3,\ M_y = 0.$ Thus,

$$\mathrm{I} = \iint_R 3\,dA = 3\iint_R dA = 3(\sqrt{2})(\sqrt{8})\ \{\text{since the sides have length } \sqrt{2} \text{ and } \sqrt{8}\} = 12.$$

[9] $M = \dfrac{y^2}{1+x^2},\ N = 2y\tan^{-1}x \Rightarrow N_x = \dfrac{2y}{1+x^2} = M_y.$ Since $N_x - M_y = 0$, $\mathrm{I} = 0$.

[10] $M = x^2 + y^2,\ N = 2xy \Rightarrow N_x = 2y = M_y.$ Since $N_x - M_y = 0$, $\mathrm{I} = 0$.

[11] $M = x^4 + 4,\ N = xy \Rightarrow N_x = y,\ M_y = 0.$

Thus, $\mathrm{I} = \iint_R y\,dA = \int_0^{2\pi}\int_0^{1+\cos\theta} (r\sin\theta)\,r\,dr\,d\theta = \int_0^{2\pi} \tfrac{1}{3}(1+\cos\theta)^3\sin\theta\,d\theta =$

$$-\tfrac{1}{3}\int_2^2 u^3\,du\ \{u = 1 + \cos\theta\} = 0.$$

[12] $M = xy,\ N = x^2 + y^2 \Rightarrow N_x = 2x,\ M_y = x.$

Thus, $\mathrm{I} = \iint_R (2x - x)\,dA = \int_0^{\pi/2}\int_0^{\sin 2\theta} (r\cos\theta)\,r\,dr\,d\theta = \tfrac{1}{3}\int_0^{\pi/2} (\sin 2\theta)^3\cos\theta\,d\theta =$

$$\tfrac{1}{3}\int_0^{\pi/2} (2\sin\theta\cos\theta)^3\cos\theta\,d\theta = \tfrac{8}{3}\int_0^{\pi/2} \cos^4\theta(1-\cos^2\theta)\sin\theta\,d\theta = -\tfrac{8}{3}\int_1^0 u^4(1-u^2)\,du$$

$$= -\tfrac{8}{3}(-\tfrac{2}{35}) = \tfrac{16}{105}.$$

[13] $M = x + y,\ N = y + x^2 \Rightarrow N_x = 2x,\ M_y = 1.$ Thus,

$$\mathrm{I} = \iint_R (2x - 1)\,dA = \int_0^{2\pi}\int_1^2 (2r\cos\theta - 1)\,r\,dr\,d\theta = \int_0^{2\pi} (\tfrac{14}{3}\cos\theta - \tfrac{3}{2})\,d\theta = -3\pi.$$

[14] $M = 1 - x^2y,\ N = \sin y \Rightarrow N_x = 0,\ M_y = -x^2.$ Thus,

$$\mathrm{I} = \int_{-2}^2\int_{-2}^2 x^2\,dx\,dy - \int_{-1}^1\int_{-1}^1 x^2\,dx\,dy = \int_{-2}^2 \tfrac{16}{3}\,dy - \int_{-1}^1 \tfrac{2}{3}\,dy = \tfrac{64}{3} - \tfrac{4}{3} = 20.$$

[15] A positive direction for the curve is shown in *Figure 15*.

C_1: $y = 4x^2$ with $0 \le x \le 4$. C_2: $y = 16x$ with $0 \le x \le 4$.

$$A = \oint x\,dy = \int_0^4 x(8x)\,dx + \int_4^0 x(16)\,dx = \tfrac{512}{3} - 128 = \tfrac{128}{3}.$$

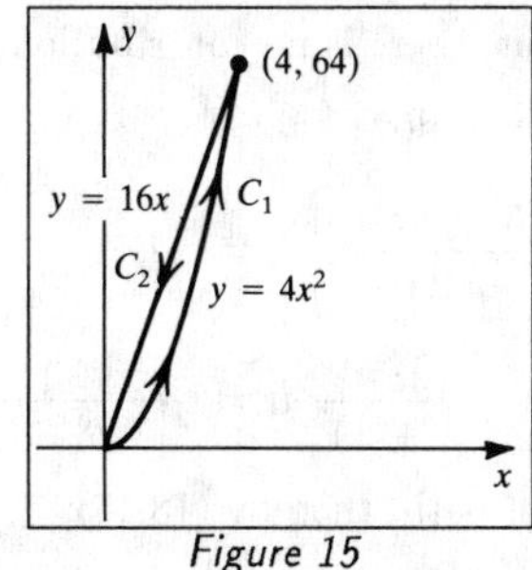

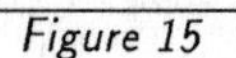
Figure 15

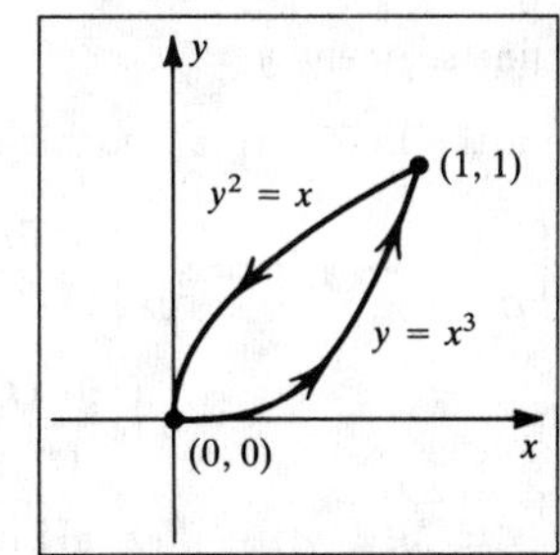

Figure 16

[16] A positive direction for the curve is shown in *Figure 16*.

C_1: $y = x^3$ with $0 \le x \le 1$. C_2: $x = y^2$ with $0 \le x \le 1$.

$$A = \oint x\,dy = \int_0^1 x(3x^2)\,dx + \int_1^0 x\cdot\frac{1}{2\sqrt{x}}\,dx = \tfrac{3}{4} - \tfrac{1}{3} = \tfrac{5}{12}.$$

[17] R is the region bounded by the upper branch of the hyperbola $y^2 - x^2 = 5$ and the line $y = 3$. C_1: $y = \sqrt{x^2 + 5}$ with $-2 \le x \le 2$. C_2: $y = 3$ with $-2 \le x \le 2$.

$$A = \oint_{C_1} x\,dy + \oint_{C_2} x\,dy = \int_{-2}^{2} x \cdot \frac{x}{\sqrt{x^2+5}}\,dx + \int_{2}^{-2} x \cdot 0\,dx$$

$$= 2\int_0^2 \frac{x^2}{\sqrt{x^2+5}}\,dx = 2\left[\tfrac{x}{2}\sqrt{x^2+5} - \tfrac{5}{2}\ln\left|x + \sqrt{x^2+5}\right|\right]_0^2 \quad \{\text{Formula } 26\}$$

$$= 2\left[(3 - \tfrac{5}{2}\ln 5) - (-\tfrac{5}{2}\ln\sqrt{5})\right] = 6 - \tfrac{5}{2}\ln 5 \approx 1.98.$$

[18] R is the region bounded by the first-quadrant branch of the hyperbola $y = 2/x$ and the line $y = 3 - x$. C_1: $y = 2/x$ with $1 \le x \le 2$. C_2: $y = 3 - x$ with $1 \le x \le 2$.

$$A = \oint_{C_1} x\,dy + \oint_{C_2} x\,dy = \int_1^2 x(-2/x^2)\,dx + \int_2^1 x(-1)\,dx =$$

$$-2\int_1^2 \tfrac{1}{x}\,dx - \int_2^1 x\,dx = -2\Big[\ln|x|\Big]_1^2 - \Big[\tfrac{1}{2}x^2\Big]_2^1 = \tfrac{3}{2} - 2\ln 2 \approx 0.11.$$

[19] $x = a\cos^3 t$, $y = a\sin^3 t \Rightarrow dx = -3a\cos^2 t\sin t\,dt$, $dy = 3a\sin^2 t\cos t\,dt$.

$$A = \tfrac{1}{2}\oint_C x\,dy - y\,dx$$

$$= \tfrac{1}{2}\int_0^{2\pi}\left[(a\cos^3 t)(3a\sin^2 t\cos t) - (a\sin^3 t)(-3a\cos^2 t\sin t)\right]dt$$

$$= \frac{3a^2}{2}\int_0^{2\pi} \sin^2 t\cos^2 t(\cos^2 t + \sin^2 t)\,dt$$

$$= \frac{3a^2}{2}\int_0^{2\pi}\left[\tfrac{1}{2}(1 - \cos 2t)\cdot\tfrac{1}{2}(1 + \cos 2t)\right]dt$$

$$= \frac{3a^2}{8}\int_0^{2\pi}(1 - \cos^2 2t)\,dt = \frac{3a^2}{8}\int_0^{2\pi}\left[1 - \tfrac{1}{2}(1 + \cos 4t)\right]dt = \tfrac{3}{8}\pi a^2.$$

[20] See the figure in the text for §3.7, Exercise 20 { *Late Trig. Ed.*—§3.6, #12 }.

R is the lower half of the loop in the first quadrant.

$$x = \frac{3t}{t^3+1} \Rightarrow dx = \frac{3 - 6t^3}{(t^3+1)^2}\,dt \text{ and } y = \frac{3t^2}{t^3+1} \Rightarrow dy = \frac{6t - 3t^4}{(t^3+1)^2}\,dt.$$

On the line segment $y = x$, $x\,dy - y\,dx = 0$. Thus, there is no contribution to the area from the line integral on this segment. It now follows that

$$A = \tfrac{1}{2}\oint_C x\,dy - y\,dx = \tfrac{1}{2}\int_0^1\left[\frac{3t(6t - 3t^4)}{(t^3+1)^3} - \frac{3t^2(3 - 6t^3)}{(t^3+1)^3}\right]dt =$$

$$\tfrac{1}{2}\int_0^1 \frac{9t^2(t^3+1)}{(t^3+1)^3}\,dt = \tfrac{3}{2}\int_0^1 \frac{3t^2}{(t^3+1)^2}\,dt = \tfrac{3}{2}\left[-\frac{1}{t^3+1}\right]_0^1 = \tfrac{3}{4}.$$

[21] If $\mathbf{F}(x, y) = M(x, y)\mathbf{i} + N(x, y)\mathbf{j}$ is independent of path, then by (18.16),

$$\frac{\partial M}{\partial y} = \frac{\partial N}{\partial x}, \text{ and hence, } \oint_C \mathbf{F}\cdot d\mathbf{r} = \oint_C M\,dx + N\,dy = \iint_R\left(\frac{\partial N}{\partial x} - \frac{\partial M}{\partial y}\right)dA = 0.$$

[22] Using Green's theorem,

$$\oint_C f(x)\,dx + g(y)\,dy = \iint_R\left(\frac{\partial g}{\partial x} - \frac{\partial f}{\partial y}\right)dA = \iint_R (0 - 0)\,dA = 0.$$

[23] $\iint_R \left(\frac{\partial N}{\partial x} - \frac{\partial M}{\partial y}\right) dA = \iint_R \left(\frac{x^2 - y^2}{(x^2 + y^2)^2} - \frac{x^2 - y^2}{(x^2 + y^2)^2}\right) dA = 0.$

However, $\oint_C M\,dx + N\,dy = \oint_C \frac{y\,dx - x\,dy}{x^2 + y^2} \{ x = \cos\theta,\ y = \sin\theta \} =$

$\int_0^{2\pi} -(\sin^2\theta + \cos^2\theta)\,d\theta = -2\pi$. Green's theorem does not apply since M and N are

undefined at (0, 0) and hence are not continuous everywhere inside the unit circle.

[24] If curl $\mathbf{F} = \nabla \times \mathbf{F} = \mathbf{0}$, then by (18.21),

$$\oint_C \mathbf{F} \cdot d\mathbf{r} = \oint_C \mathbf{F} \cdot \mathbf{T}\,ds = \iint_R (\nabla \times \mathbf{F}) \cdot \mathbf{k}\,dA = \iint_R 0\,dA = 0.$$

[25] Without loss of generality, let $\delta(x, y) = 1$. Also, let M_x and M_y denote moments.

Since $\delta(x, y) = 1$, the area A of the region is equal to the mass m. By definition,

$M_y = \iint_R x\,dA$. Let $\frac{\partial N}{\partial x} = x$ and $\frac{\partial M}{\partial y} = 0$. Then, $\left(\frac{\partial N}{\partial x} - \frac{\partial M}{\partial y}\right) = x$ and

$N = \frac{1}{2}x^2 + C_1$, $M = C_2$. Without loss of generality, let $C_1 = C_2 = 0$.

It now follows by Green's theorem, $\oint_C 0 \cdot dx + \frac{1}{2}x^2\,dy = \iint_R x\,dA = M_y$,

and similarly, $\oint_C -\frac{1}{2}y^2\,dx + 0 \cdot dy = \iint_R y\,dA = M_x$.

$$\text{So, } \bar{x} = \frac{M_y}{A} = \frac{1}{2A}\oint_C x^2\,dy \text{ and } \bar{y} = \frac{M_x}{A} = -\frac{1}{2A}\oint_C y^2\,dx.$$

[26] By Green's theorem, $-\frac{k}{3}\oint_C y^3\,dx + 0 \cdot dy = -\frac{1}{3}k\iint_R -3y^2\,dA = \iint_R ky^2\,dA = I_x$

$$\text{and } \frac{k}{3}\oint_C 0 \cdot dx + x^3\,dy = \frac{1}{3}k\iint_R 3x^2\,dA = \iint_R kx^2\,dA = I_y.$$

[27] Place the semicircle as shown in *Figure 27.*

On C_1: $x = a\cos t$, $y = a\sin t$, $0 \le t \le \pi \Rightarrow$

$dx = -a\sin t\,dt$, $dy = a\cos t\,dt$.

On C_2: $x = t$, $y = 0$, $-a \le t \le a \Rightarrow$

$dx = dt$, $dy = 0\,dt$.

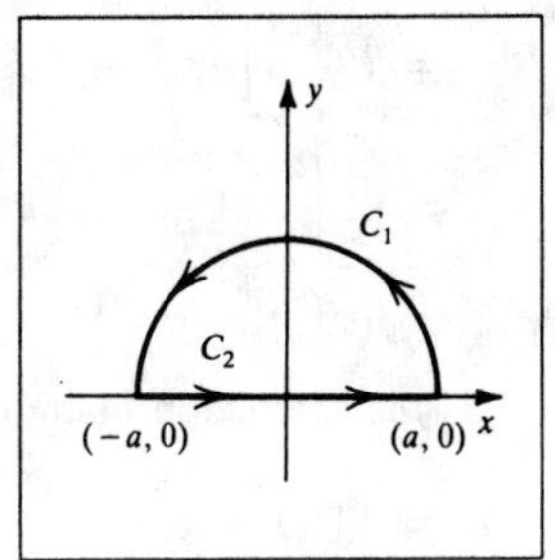

Figure 27

$$\bar{x} = \frac{1}{\pi a^2}\oint_C x^2\,dy = \frac{1}{\pi a^2}\left(\oint_{C_1} x^2\,dy + \oint_{C_2} x^2\,dy\right)$$

$$= \frac{1}{\pi a^2}\left(\int_0^\pi (a^2\cos^2 t)(a\cos t)\,dt + \int_{-a}^a t^2 \cdot 0\,dt\right)$$

$$= \frac{a}{\pi}\int_0^\pi \cos^3 t\,dt = 0. \text{ \{This can also be seen by symmetry.\}}$$

$$\bar{y} = -\frac{1}{\pi a^2}\left(\int_0^\pi (a^2\sin^2 t)(-a\sin t)\,dt + \int_{-a}^a 0\,dt\right) = \frac{a}{\pi}\int_0^\pi \sin^3 t\,dt =$$

$$\frac{2a}{\pi}\int_0^1 (1 - u^2)\,du\ \{u = \cos t\} = \frac{2a}{\pi} \cdot \frac{2}{3} = \frac{4a}{3\pi}.$$

[28] Let $x = a\cos t$, $y = a\sin t \Rightarrow dx = -a\sin t\,dt$, $dy = a\cos t\,dt$ for $0 \le t \le 2\pi$.

Also let the diameter lie on the x-axis. Then

$$I_x = -\frac{k}{3}\oint_C y^3\,dx = -\frac{k}{3}\int_0^{2\pi}(a^3\sin^3 t)(-a\sin t)\,dt = \tfrac{1}{3}a^4k\int_0^{2\pi}\sin^4 t\,dt$$

$$= \tfrac{1}{12}a^4k\int_0^{2\pi}\left[1 - 2\cos 2t + \tfrac{1}{2}(1 + \cos 4t)\right]dt = (\tfrac{1}{12}a^4k)(3\pi) = \tfrac{1}{4}\pi a^4 k.$$

Exercises 18.5

[1] $z = f(x, y) = (a^2 - x^2 - y^2)^{1/2} \Rightarrow f_x = \dfrac{-x}{(a^2 - x^2 - y^2)^{1/2}}$ and

$$f_y = \frac{-y}{(a^2 - x^2 - y^2)^{1/2}} \Rightarrow \sqrt{f_x^2 + f_y^2 + 1} = \frac{a}{(a^2 - x^2 - y^2)^{1/2}}.$$

R_{xy} is the region inside the circle $x^2 + y^2 = a^2$.

$$I = \iint_{R_{xy}} x^2\frac{a}{(a^2 - x^2 - y^2)^{1/2}}\,dA = \int_0^{2\pi}\int_0^a r^2\cos^2\theta\left(\frac{a}{(a^2 - r^2)^{1/2}}\right)r\,dr\,d\theta$$

$$= a\int_0^{2\pi}\cos^2\theta\left[\int_0^a\frac{r^3}{(a^2 - r^2)^{1/2}}\,dr\right]d\theta$$

$$= a\int_0^{2\pi}\cos^2\theta\left[\int_0^{\pi/2}a^3\sin^3 t\,dt\right]d\theta\ \{r = a\sin t\} = a\int_0^{2\pi}\cos^2\theta\,(\tfrac{2}{3}a^3)\,d\theta = \tfrac{2}{3}\pi a^4.$$

[2] $z = f(x, y) = y + 4 \Rightarrow f_x = 0$ and $f_y = 1 \Rightarrow \sqrt{f_x^2 + f_y^2 + 1} = \sqrt{2}$.

R_{xy} is the region inside the circle $x^2 + y^2 = 4$.

$$I = \iint_{R_{xy}}\left[x^2 + y^2 + (y + 4)^2\right]\sqrt{2}\,dA = \int_0^{2\pi}\int_0^2\left[r^2 + (r\sin\theta + 4)^2\right]\sqrt{2}\,r\,dr\,d\theta$$

$$= \sqrt{2}\int_0^{2\pi}\int_0^2(r^3 + r^3\sin^2\theta + 8r^2\sin\theta + 16r)\,dr\,d\theta$$

$$= \sqrt{2}\int_0^{2\pi}(4 + 4\sin^2\theta + \tfrac{64}{3}\sin\theta + 32)\,d\theta = 76\pi\sqrt{2}.$$

[3] $z = f(x, y) = 6 - 2x - 3y \Rightarrow f_x = -2$ and $f_y = -3 \Rightarrow \sqrt{f_x^2 + f_y^2 + 1} = \sqrt{14}$.

R_{xy} is the region bounded by $x = 0$, $y = 0$, and $2x + 3y = 6$.

$$I = \iint_{R_{xy}}(x + y)\sqrt{14}\,dA = \sqrt{14}\int_0^2\int_0^{(6-3y)/2}(x + y)\,dx\,dy$$

$$= \sqrt{14}\int_0^2\left[\tfrac{9}{8}(2 - y)^2 + \tfrac{3}{2}(2y - y^2)\right]dy = \sqrt{14}\int_0^2(\tfrac{9}{2} - \tfrac{3}{2}y - \tfrac{3}{8}y^2)\,dy = 5\sqrt{14}.$$

[4] $z = f(x, y) = \frac{1}{2}(x^2 + y^2) \Rightarrow f_x = x$ and $f_y = y \Rightarrow \sqrt{f_x^2 + f_y^2 + 1} = \sqrt{x^2 + y^2 + 1}$.

R_{xy} is the region inside the circle $x^2 + y^2 = 2y$, or $r = 2\sin\theta$. See *Figure 4*.

$$I = \iint_{R_{xy}}(x^2 + y^2 + 1)^{1/2}(x^2 + y^2 + 1)^{1/2}\,dA = \int_0^{\pi}\int_0^{2\sin\theta}(r^2 + 1)\,r\,dr\,d\theta$$

$$= \int_0^{\pi}(4\sin^4\theta + 2\sin^2\theta)\,d\theta = \int_0^{\pi}(\tfrac{5}{2} - 3\cos 2\theta + \tfrac{1}{2}\cos 4\theta)\,d\theta = \tfrac{5\pi}{2}.$$

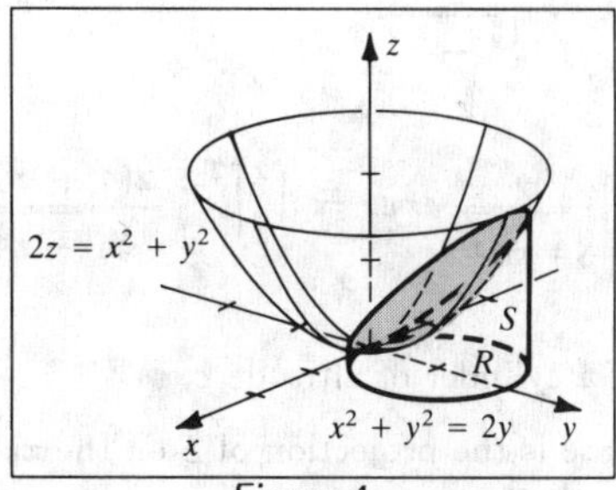

Figure 4

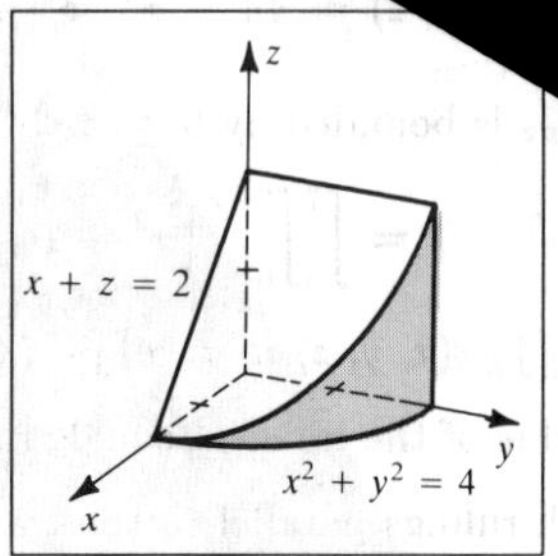

Figure 8

[5] (a) $x = k(y, z) = \frac{1}{2}(12 - 3y - 4z) \Rightarrow \sqrt{k_y^2 + k_z^2 + 1} = \sqrt{\frac{29}{4}}$.

R_{yz} is bounded by $y = 0$, $z = 0$, and $3y + 4z = 12$.

$$\mathrm{I} = \int_0^4 \int_0^{(12-3y)/4} \tfrac{1}{2}(12 - 3y - 4z)y^2 z^3 \left(\tfrac{1}{2}\sqrt{29}\right) dz\, dy.$$

(b) $y = h(x, z) = \frac{1}{3}(12 - 2x - 4z) \Rightarrow \sqrt{h_x^2 + h_z^2 + 1} = \sqrt{\frac{29}{9}}$.

R_{xz} is bounded by $x = 0$, $z = 0$, and $2x + 4z = 12$.

$$\mathrm{I} = \int_0^3 \int_0^{6-2z} x\left[\tfrac{1}{3}(12 - 2x - 4z)\right]^2 z^3 \left(\tfrac{1}{3}\sqrt{29}\right) dx\, dz.$$

[6] (a) $x = k(y, z) = y^{1/3} \Rightarrow \sqrt{k_y^2 + k_z^2 + 1} = \sqrt{\frac{1}{9}y^{-4/3} + 1}$.

R_{yz} is bounded by $0 \le y \le 8$ and $0 \le z \le 2$.

$$\mathrm{I} = \int_0^2 \int_0^8 \left(y^{1/3} z + 2y\right) \sqrt{\tfrac{1}{9}y^{-4/3} + 1}\, dy\, dz.$$

(b) $y = h(x, z) = x^3 \Rightarrow \sqrt{h_x^2 + h_z^2 + 1} = \sqrt{9x^4 + 1}$.

R_{xz} is bounded by $0 \le x \le 2$ and $0 \le z \le 2$.

$$\mathrm{I} = \int_0^2 \int_0^2 \left(xz + 2x^3\right) \sqrt{9x^4 + 1}\, dz\, dx.$$

[7] (a) $x = k(y, z) = \frac{1}{4}(8 - y) \Rightarrow \sqrt{k_y^2 + k_z^2 + 1} = \sqrt{\frac{17}{16}}$.

R_{yz} is bounded by $0 \le y \le 8$ and $0 \le z \le 6$. I =

$$\int_0^8 \int_0^6 \left[\left(2 - \tfrac{1}{4}y\right)^2 - 2y + z\right]\left(\tfrac{1}{4}\sqrt{17}\right) dz\, dy = \int_0^8 \int_0^6 \left(4 - 3y + \tfrac{1}{16}y^2 + z\right)\left(\tfrac{1}{4}\sqrt{17}\right) dz\, dy.$$

(b) $y = h(x, z) = 8 - 4x \Rightarrow \sqrt{h_x^2 + h_z^2 + 1} = \sqrt{17}$.

R_{xz} is bounded by $0 \le x \le 2$ and $0 \le z \le 6$.

$$\mathrm{I} = \int_0^2 \int_0^6 \left[x^2 - 2(8 - 4x) + z\right] \sqrt{17}\, dz\, dx.$$

[8] (a) $x = k(y, z) = \sqrt{4 - y^2} \Rightarrow \sqrt{k_y^2 + k_z^2 + 1} = \dfrac{2}{\sqrt{4 - y^2}}$.

R_{yz} is bounded by $0 \le y \le 2$ and $0 \le z \le 2 - \sqrt{4 - y^2}$. See *Figure 8*. I =

$$\int_0^2 \int_0^{2-\sqrt{4-y^2}} \left[(4 - y^2) + y^2 + z^2\right] \frac{2}{\sqrt{4 - y^2}}\, dz\, dy = \int_0^2 \int_0^{2-\sqrt{4-y^2}} \frac{2(4 + z^2)}{\sqrt{4 - y^2}}\, dz\, dy.$$

$\Rightarrow \sqrt{h_x^2 + h_z^2 + 1} = \dfrac{2}{\sqrt{4 - x^2}}$.

$\leq x \leq 2$ and $0 \leq z \leq 2 - x$.

$\left[x^2 + (4 - x^2) + z^2\right]\dfrac{2}{\sqrt{4 - x^2}}\,dz\,dx = \displaystyle\int_0^2\int_0^{2-x} \dfrac{2(4 + z^2)}{\sqrt{4 - x^2}}\,dz\,dx.$

$dS = c\iint_R dA$,

ıntegral equals the volume of a cylinder of altitude c,

ȝ parallel to the z-axis, whose base is the projection of S on the xy-plane.

the xy-plane, that is, let $z = 0$.

Then, $\displaystyle\iint_S g(x, y, z)\,dS = \iint_R g((x, y, 0)\,(1)\,dA = \iint_R f(x, y)\,dA.$

407

11 z - $f(x, y) = \sqrt{a^2 - x^2 - y^2} \Rightarrow f_x = \dfrac{-x}{(a^2 - x^2 - y^2)^{1/2}}$ and

$f_y = \dfrac{-y}{(a^2 - x^2 - y^2)^{1/2}}$. Also, $\mathbf{n} = \dfrac{-f_x\mathbf{i} - f_y\mathbf{j} + \mathbf{k}}{\sqrt{f_x^2 + f_y^2 + 1}}$.

$$\mathrm{I} = \iint_S \mathbf{F}\cdot\mathbf{n}\,dS = \iint_{R_{xy}} \frac{-xf_x - yf_y + z}{\sqrt{f_x^2 + f_y^2 + 1}}\sqrt{f_x^2 + f_y^2 + 1}\,dA$$

$$= \iint_{R_{xy}} \left[\frac{x^2 + y^2}{(a^2 - x^2 - y^2)^{1/2}} + (a^2 - x^2 - y^2)^{1/2}\right] dA$$

$= \displaystyle\iint_{R_{xy}} \frac{a^2}{(a^2 - x^2 - y^2)^{1/2}}\,dA$. Since R_{xy} is the interior of the circle $x^2 + y^2 = a^2$,

$$\mathrm{I} = \int_0^{2\pi}\int_0^a a^2(a^2 - r^2)^{-1/2}\,r\,dr\,d\theta = a^3\int_0^{2\pi} d\theta = 2\pi a^3.$$

12 As in Exercise 11,

$$\mathrm{I} = \iint_{R_{xy}} \frac{-xf_x + yf_y}{\sqrt{f_x^2 + f_y^2 + 1}}\sqrt{f_x^2 + f_y^2 + 1}\,dA$$

$$= \iint_{R_{xy}} \frac{x^2 - y^2}{(a^2 - x^2 - y^2)^{1/2}}\,dA = \int_0^a\int_0^{\pi/2} \frac{r^2\cos^2\theta - r^2\sin^2\theta}{(a^2 - r^2)^{1/2}}\,r\,d\theta\,dr$$

$$= \int_0^a\int_0^{\pi/2} \frac{r^3\cos 2\theta}{(a^2 - r^2)^{1/2}}\,d\theta\,dr = \int_0^a \frac{r^3}{2(a^2 - r^2)^{1/2}}\Big[\sin 2\theta\Big]_0^{\pi/2} dr = 0.$$

Note: Since the radical obtained in changing from dS to dA is the same as the radical in the denominator of $\mathbf{n}$, we will only compute $\mathbf{F}\cdot\langle -f_x, -f_y, 1\rangle$,

and denote this value by $\mathbb{A}$.

13 $z = f(x, y) = (x^2 + y^2)^{1/2} \Rightarrow f_x = \dfrac{x}{(x^2 + y^2)^{1/2}}$ and $f_y = \dfrac{y}{(x^2 + y^2)^{1/2}}$.

$$\mathbb{A} = \mathbf{F}\cdot\langle -f_x, -f_y, 1\rangle = \frac{-2x}{(x^2 + y^2)^{1/2}} + \frac{-5y}{(x^2 + y^2)^{1/2}} + 3.$$

$\mathrm{I} = \displaystyle\iint_{R_{xy}} \mathbb{A}\,dA$. R_{xy} is the interior of the circle $x^2 + y^2 = 1$.

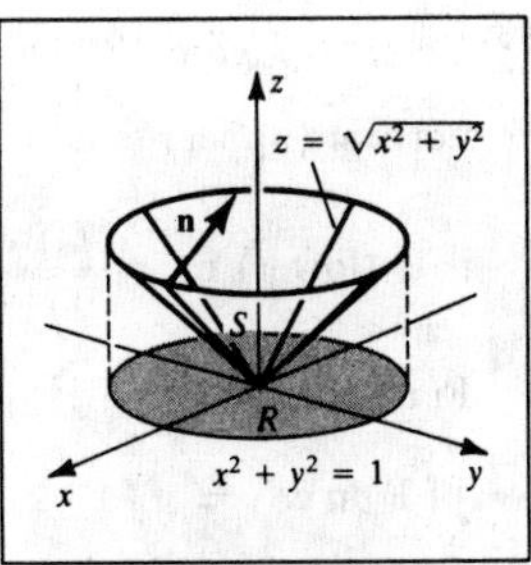

Figure 13

$$\text{I} = \int_0^1 \int_0^{2\pi} (-2\cos\theta - 5\sin\theta + 3)\, r\, d\theta\, dr$$

$$= 6\pi \int_0^1 r\, dr = 3\pi.$$

[14] $z = f(x, y) = 12 - 3x - 2y \Rightarrow f_x = -3$ and $f_y = -2$.

$\mathbb{A} = 3x + 2y + z = 12$.

$$\text{I} = \iint_{R_{xy}} \mathbb{A}\, dA = \int_0^1 \int_0^2 (12)\, dy\, dx = 24.$$

[15] $z = f(x, y) = 6 - 2x - 3y \Rightarrow f_x = -2$ and $f_y = -3$. $\mathbb{A} = 2x + 3y + z = 6$.

By (18.25), the flux of $\mathbf{F}$ through S is $\iint_S \mathbf{F} \cdot \mathbf{n}\, dS = \iint_{R_{xy}} \mathbb{A}\, dA =$

$$\int_0^3 \int_0^{(6-2x)/3} (6)\, dy\, dx = \int_0^3 (12 - 4x)\, dx = 18.$$

[16] $z = f(x, y) = x^2 + y^2 \Rightarrow f_x = 2x$ and $f_y = 2y$.

$\mathbb{A} = -2x(x^2 + z) - 2y(y^2 z) + (x^2 + y^2 + z)$.

$$\text{I} = \iint_{R_{xy}} \mathbb{A}\, dA = \iint_{R_{xy}} \left[-2x(2x^2 + y^2) - 2y^3(x^2 + y^2) + 2(x^2 + y^2)\right] dA.$$

R_{xy} is the quarter circle in the first quadrant determined by $x^2 + y^2 = 4$.

$$\text{I} = \int_0^{\pi/2} \int_0^2 (-4r^3 \cos^3\theta - 2r^3 \sin^2\theta \cos\theta - 2r^5 \sin^3\theta + 2r^2)\, r\, dr\, d\theta$$

$$= \int_0^{\pi/2} \left[-\tfrac{4}{5}r^5 \cos^3\theta - \tfrac{2}{5}r^5 \sin^2\theta \cos\theta - \tfrac{2}{7}r^7 \sin^3\theta + \tfrac{2}{4}r^4\right]_0^2 d\theta$$

$$= \int_0^{\pi/2} \left[-\tfrac{128}{5}(1 - \sin^2\theta)\cos\theta - \tfrac{64}{5}\sin^2\theta \cos\theta - \tfrac{256}{7}(1 - \cos^2\theta)\sin\theta + 8\right] d\theta$$

$$= \left[-\tfrac{128}{5}(\sin\theta - \tfrac{1}{3}\sin^3\theta) - \tfrac{64}{15}\sin^3\theta - \tfrac{256}{7}(-\cos\theta + \tfrac{1}{3}\cos^3\theta) + 8\theta\right]_0^{\pi/2}$$

$$= \left[-\tfrac{128}{5}(1 - \tfrac{1}{3}) - \tfrac{64}{15} - (\tfrac{256}{7} - \tfrac{256}{21}) + 4\pi\right] = 4\pi - \tfrac{320}{7}.$$

[17] The cube has six surfaces. The unit outer normal vectors are $\pm\mathbf{i}$, $\pm\mathbf{j}$, and $\pm\mathbf{k}$.

$$\iint_S \mathbf{F} \cdot \mathbf{n}\, dS = \iint_{S_1} \mathbf{F} \cdot \mathbf{i}\, dS + \iint_{S_2} \mathbf{F} \cdot (-\mathbf{i})\, dS + \iint_{S_3} \mathbf{F} \cdot \mathbf{j}\, dS + \iint_{S_4} \mathbf{F} \cdot (-\mathbf{j})\, dS +$$

$$\iint_{S_5} \mathbf{F} \cdot \mathbf{k}\, dS + \iint_{S_6} \mathbf{F} \cdot (-\mathbf{k})\, dS.$$

Let the six integrals be I_1 through I_6. $\mathbf{F}(x, y, z) = (x + y)\mathbf{i} + z\mathbf{j} + xz\mathbf{k}$.

(1) $x = 1 \Rightarrow$ $\text{I}_1 = \int_{-1}^1 \int_{-1}^1 (1 + y)(1)\, dy\, dz = 4$.

(2) $x = -1 \Rightarrow$ $\text{I}_2 = \int_{-1}^1 \int_{-1}^1 -(-1 + y)(1)\, dy\, dz = 4$.

(3) $y = 1 \Rightarrow$ $\text{I}_3 = \int_{-1}^1 \int_{-1}^1 z(1)\, dx\, dz = 0$.

(4) $y = -1 \Rightarrow$ $\text{I}_4 = \int_{-1}^1 \int_{-1}^1 -z(1)\, dx\, dz = 0$.

(5) $z = 1 \Rightarrow$ $\text{I}_5 = \int_{-1}^1 \int_{-1}^1 x(1)\, dx\, dy = 0$.

(6) $z = -1 \Rightarrow$ $\text{I}_6 = \int_{-1}^1 \int_{-1}^1 x(1)\, dx\, dy = 0$.

Thus, $\iint_S \mathbf{F} \cdot \mathbf{n}\, dS = 4 + 4 + 0 + 0 + 0 + 0 = 8$.

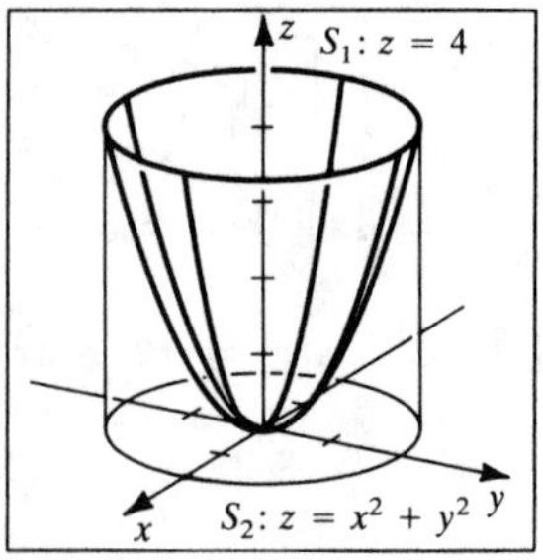

[18] Refer to *Figure 18*. On S_1 $\{ z = 4 \}$, the unit outer normal (upper) is $\mathbf{n}_1 = \mathbf{k}$. On S_2 $\{ z = x^2 + y^2 \}$,

it is (lower) $\mathbf{n}_2 = \dfrac{f_x\mathbf{i} + f_y\mathbf{j} - \mathbf{k}}{\sqrt{f_x^2 + f_y^2 + 1}} = \dfrac{2x\mathbf{i} + 2y\mathbf{j} - \mathbf{k}}{\sqrt{f_x^2 + f_y^2 + 1}}$.

$\mathbf{F}(x, y, z) = x\mathbf{i} - y\mathbf{j} + z\mathbf{k}$.

$$\iint_S \mathbf{F}\cdot\mathbf{n}\, dS = \iint_{S_1} \mathbf{F}\cdot\mathbf{k}\, dS + \iint_{S_2} \mathbf{F}\cdot\mathbf{n}_2\, dS$$
$$= \iint_{R_{xy}} 4\, dA + \iint_{R_{xy}} \left[2x^2 - 2y^2 - (x^2 + y^2)\right] dA.$$

S_1: $z = 4$

S_2: $z = x^2 + y^2$

Figure 18

R_{xy} is the interior of the circle $x^2 + y^2 = 4$.

$$\mathrm{I} = \int_0^{2\pi}\int_0^2 4\, r\, dr\, d\theta + \int_0^{2\pi}\int_0^2 (r^2\cos^2\theta - 3r^2\sin^2\theta)\, r\, dr\, d\theta$$
$$= \int_0^{2\pi} 8\, d\theta + 4\int_0^{2\pi} (\cos^2\theta - 3\sin^2\theta)\, d\theta = 16\pi + 4(-2\pi) = 8\pi.$$

[19] $x = k(y, z) \Rightarrow g(x, y, z) = x - k(y, z) \Rightarrow \mathbf{n} = \dfrac{\nabla g(x, y, z)}{\|\nabla g(x, y, z)\|} = \dfrac{\mathbf{i} - k_y\,\mathbf{j} - k_z\,\mathbf{k}}{\sqrt{k_y^2 + k_z^2 + 1}}$.

So, $\displaystyle\iint_S \mathbf{F}\cdot\mathbf{n}\, dS = \iint_{R_{yz}} \frac{M - Nk_y - Pk_z}{\sqrt{k_y^2 + k_z^2 + 1}}\sqrt{k_y^2 + k_z^2 + 1}\, dA =$

$$\iint_{R_{yz}} \left[M - Nk_y(y, z) - Pk_z(y, z)\right] dy\, dz.$$

[20] The unit outer normal is $\mathbf{n} = \frac{\mathbf{r}}{\|\mathbf{r}\|}$. $\mathbf{F}(x, y, z) = (cq/\|\mathbf{r}\|^3)\mathbf{r}$, with $\mathbf{r} = x\mathbf{i} + y\mathbf{j} + z\mathbf{k}$.

Thus, $\displaystyle\iint_S \mathbf{F}\cdot\mathbf{n}\, dS = \iint_S \frac{cq}{\|\mathbf{r}\|^4}\mathbf{r}\cdot\mathbf{r}\, dS = \iint_S \frac{cq}{\|\mathbf{r}\|^2}\, dS = \frac{cq}{\|\mathbf{r}\|^2}\iint_S dS.$

$\displaystyle\iint_S dS$ is a surface integral and equals the surface area of a sphere, $4\pi\|\mathbf{r}\|^2$.

$$\text{Hence, } \iint_S \mathbf{F}\cdot\mathbf{n}\, dS = \left(\frac{cq}{\|\mathbf{r}\|^2}\right)(4\pi\|\mathbf{r}\|^2) = 4\pi cq.$$

[21] Divide the surface into subregions $P = \{ S_k \}$ and let (x_k, y_k, z_k) be on S_k.

Let T_k be the corresponding tangent plane at (x_k, y_k, z_k). We approximate the mass m_k of S_k by $\delta(x_k, y_k, z_k)\Delta T_k$ and its moment with respect to the xy-plane by $z_k\,\delta(x_k, y_k, z_k)\Delta T_k$. In the limit, $\displaystyle M_{xy} = \lim_{\|P\|\to 0}\sum_k z_k\,\delta(x_k, y_k, z_k)\Delta T_k =$

$\displaystyle\iint_S z\delta(x_k, y_k, z_k)\, dS$. M_{xz} and M_{yz} are derived in a similar manner.

(a) For the metal funnel, $z = f(x, y) = (x^2 + y^2)^{1/2}$ and

$\delta(x, y, z) = z^2 = x^2 + y^2$. So, $\displaystyle m = \iint_S \delta(x, y, z)\, dS = \iint_R (x^2 + y^2)\sqrt{2}\, dx\, dy =$

$\displaystyle\sqrt{2}\int_1^4\int_0^{2\pi} (r^2)\, r\, d\theta\, dr = 2\pi\sqrt{2}\int_1^4 r^3\, dr = \frac{255}{2}\pi\sqrt{2}$. Since $\delta(x, y, z) = z^2$ is

independent of x and y, by symmetry $\bar{x} = \bar{y} = 0$. $\displaystyle M_{xy} = \iint_S z\delta(x, y, z)\, dS =$

$\displaystyle\sqrt{2}\iint_R (x^2 + y^2)^{3/2}\, dA = \sqrt{2}\int_1^4\int_0^{2\pi} r^4\, d\theta\, dr = 2\pi\sqrt{2}\int_1^4 r^4\, dr = \frac{2046}{5}\pi\sqrt{2}.$

$$\text{Thus, } \bar{z} = \frac{M_{xy}}{m} = \frac{1364}{425} \approx 3.21.$$

(b) $I_z = \iint_S (x^2 + y^2)\,\delta(x, y, z)\,dS = \sqrt{2}\iint_R (x^2 + y^2)^2\,dA$

$$= \sqrt{2}\int_1^4\int_0^{2\pi} r^5\,d\theta\,dr = 2\pi\sqrt{2}\int_1^4 r^5\,dr = 1365\pi\sqrt{2}.$$

[22] By symmetry, $\bar{x} = \bar{y} = 0$. $m = \iint_S \delta\,dS = k\iint_S dS = 2\pi a^2 k$ since $\iint_S dS$ gives the surface area of the hemisphere. $z = f(x, y) = (a^2 - x^2 - y^2)^{1/2} \Rightarrow$

$\sqrt{f_x^2 + f_y^2 + 1} = \dfrac{a}{(a^2 - x^2 - y^2)^{1/2}}$. $M_{xy} = \iint_S z\delta\,dS =$

$k\iint_R \dfrac{a\sqrt{a^2 - x^2 - y^2}}{\sqrt{a^2 - x^2 - y^2}}\,dA = ak\iint_R dA = ak(\pi a^2)$ {since $\iint_R dA$ gives the area of a circle with radius a} $= \pi a^3 k$. Thus, $\bar{z} = \dfrac{M_{xy}}{m} = \dfrac{a}{2}$, and its center of mass is $(0, 0, \frac{1}{2}a)$.

Exercises 18.6

[1] $\iint_S \mathbf{F}\cdot\mathbf{n}\,dS = \iiint_Q \operatorname{div}\mathbf{F}\,dV = \int_{-1}^1\int_{-1}^1\int_{-1}^1 (y\cos x + 2yz + 3)\,dz\,dy\,dx =$

$$\int_{-1}^1\int_{-1}^1 (2y\cos x + 6)\,dy\,dx = \int_{-1}^1 12\,dx = 24.$$

[2] $\iint_S \mathbf{F}\cdot\mathbf{n}\,dS = \iiint_Q \operatorname{div}\mathbf{F}\,dV = \int_0^1\int_0^{1-x}\int_0^{1-x-y} (-x)\,dz\,dy\,dx =$

$$\int_0^1\int_0^{1-x} (-x + x^2 + xy)\,dy\,dx = -\tfrac{1}{2}\int_0^1 (x^3 - 2x^2 + x)\,dx = -\tfrac{1}{24}.$$

[3] $\operatorname{div}\mathbf{F} = 2x + 1 + 2z$. Using cylindrical coordinates,

$\iint_S \mathbf{F}\cdot\mathbf{n}\,dS = \int_0^{2\pi}\int_0^2\int_0^{2-r\cos\theta} (2r\cos\theta + 1 + 2z)\,r\,dz\,dr\,d\theta =$

$$\int_0^{2\pi}\int_0^2 (6r - r^2\cos\theta - r^3\cos^2\theta)\,dr\,d\theta = \int_0^{2\pi} (12 - \tfrac{8}{3}\cos\theta - 4\cos^2\theta)\,d\theta = 20\pi.$$

[4] $\operatorname{div}\mathbf{F} = 2y + 2z$. Using cylindrical coordinates,

$\iint_S \mathbf{F}\cdot\mathbf{n}\,dS = \int_0^{2\pi}\int_0^1\int_{r^2}^1 (2r\sin\theta + 2z)\,r\,dz\,dr\,d\theta =$

$$\int_0^{2\pi}\int_0^1 (2r^2\sin\theta - 2r^4\sin\theta + r - r^5)\,dr\,d\theta = \int_0^{2\pi} (\tfrac{4}{15}\sin\theta + \tfrac{1}{3})\,d\theta = \tfrac{2\pi}{3}.$$

[5] $\mathbf{F}(x, y, z) = yz\mathbf{i} + xz\mathbf{j} + xy\mathbf{k} \Rightarrow \operatorname{div}\mathbf{F} = 0$. Thus, $\iint_S \mathbf{F}\cdot\mathbf{n}\,dS = 0$.

[6] $\operatorname{div}\mathbf{F} = (2x) + (2y - 2x) + (4 - 2y) = 4$.

$\iint_S \mathbf{F}\cdot\mathbf{n}\,dS = \iiint_Q 4\,dV = 4\left[\tfrac{1}{3}\pi(9)^2 9\right] = 972\pi$.

Note that $\frac{1}{3}\pi r^2 h$ has been used for the volume of the cone.

[7] Using cylindrical coordinates, $\iint_S \mathbf{F}\cdot\mathbf{n}\,dS = \iiint_Q \operatorname{div}\mathbf{F}\,dV =$

$$\int_0^{2\pi}\int_0^2\int_0^{4-r^2} (3 + 2z)\,r\,dz\,dr\,d\theta = \int_0^{2\pi}\int_0^2 (28r - 11r^3 + r^5)\,dr\,d\theta = \int_0^{2\pi} \tfrac{68}{3}\,d\theta = \tfrac{136\pi}{3}.$$

[8] $\operatorname{div}\mathbf{F} = y^2 + z^2 + x^2$. Using cylindrical coordinates, $\iint_S \mathbf{F}\cdot\mathbf{n}\,dS =$

$$\int_0^{2\pi}\int_2^3\int_{-1}^2 (z^2 + r^2)\,r\,dz\,dr\,d\theta = \int_0^{2\pi}\int_2^3 (3r + 3r^3)\,dr\,d\theta = \int_0^{2\pi} \tfrac{225}{4}\,d\theta = \tfrac{225\pi}{2}.$$

[9] $\iint_S \mathbf{F}\cdot\mathbf{n}\,dS = \iiint_Q \operatorname{div}\mathbf{F}\,dV$

$$= \int_0^2\int_0^4\int_0^{(4-x)/2} (2z + xz + y)\,dz\,dx\,dy$$

$$= \tfrac{1}{8}\int_0^2\int_0^4 (32 - 6x^2 + x^3 + 16y - 4xy)\,dx\,dy$$

$$= 4\int_0^2 (2 + y)\,dy = 24.$$

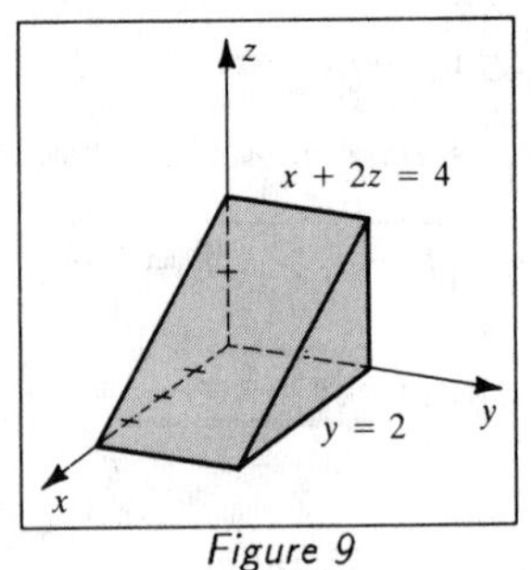

Figure 9

[10] $\operatorname{div}\mathbf{F} = 3(x^2 + y^2 + z^2)$. Using spherical coordinates,

$$\iint_S \mathbf{F}\cdot\mathbf{n}\,dS = \int_0^{2\pi}\int_0^{\pi/4}\int_0^5 (3\rho^2)(\rho^2\sin\phi)\,d\rho\,d\phi\,d\theta = \int_0^{2\pi}\int_0^{\pi/4} 1875\sin\phi\,d\phi\,d\theta$$

$$= -1875\int_0^{2\pi} (\tfrac{\sqrt{2}}{2} - 1)\,d\theta = 1875\pi(2 - \sqrt{2}) \approx 3450.57.$$

[11] (1) The unit outer normal for a sphere is $\mathbf{n} = \frac{\mathbf{r}}{\|\mathbf{r}\|} = \frac{x\mathbf{i} + y\mathbf{j} + z\mathbf{k}}{\sqrt{x^2 + y^2 + z^2}}$ and

$\mathbf{F}\cdot\mathbf{n} = \sqrt{x^2 + y^2 + z^2} = a$. The upper half of the sphere is given by

$z = \sqrt{a^2 - x^2 - y^2}$ and the lower half is given by $z = -\sqrt{a^2 - x^2 - y^2}$.

On both the upper and lower halves, $\sqrt{z_x^2 + z_y^2 + 1} = \dfrac{a}{(a^2 - x^2 - y^2)^{1/2}}$.

$\iint_S \mathbf{F}\cdot\mathbf{n}\,dS = \iint_{S(\text{upper})} \mathbf{F}\cdot\mathbf{n}\,dS + \iint_{S(\text{lower})} \mathbf{F}\cdot\mathbf{n}\,dS = 2\iint_{R_{xy}} \dfrac{a^2}{(a^2 - x^2 - y^2)^{1/2}}\,dA =$

$2(2\pi a^3)$ {Exercise 11, §18.5} $= 4\pi a^3$.

(2) $\operatorname{div}\mathbf{F} = 3$. $\iiint_Q 3\,dV = 3(\frac{4}{3}\pi a^3)$ {since the volume of a sphere is $\frac{4}{3}\pi r^3$} $= 4\pi a^3$.

[12] (1) The six surfaces of the cube have unit outer normals of $\pm\mathbf{i}$, $\pm\mathbf{j}$, $\pm\mathbf{k}$.

For the top $(z = a)$ and bottom $(z = 0)$, $\mathbf{F}\cdot\mathbf{n} = a^2$ and $\mathbf{F}\cdot\mathbf{n} = 0$, respectively.

For the top, $\iint_{S(\text{top})} \mathbf{F}\cdot\mathbf{n}\,dS = a^2 \iint_{S(\text{top})} dS = a^2\cdot a^2 = a^4$, since $\iint_S dS$ is simply the surface area of a face. By symmetry, $\iint_S \mathbf{F}\cdot\mathbf{n}\,dS = 3a^4$.

(2) $\iiint_Q \operatorname{div}\mathbf{F}\,dV = \int_0^a\int_0^a\int_0^a (2x + 2y + 2z)\,dx\,dy\,dz =$

$$\int_0^a\int_0^a (a^2 + 2ay + 2az)\,dy\,dz = \int_0^a (2a^3 + 2a^2 z)\,dz = 3a^4.$$

[13] (1) The region Q has 3 surfaces as shown in *Figure 13a*.

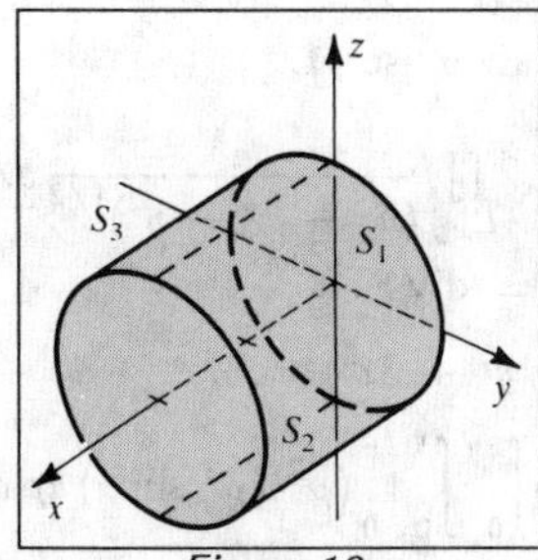

Figure 13a

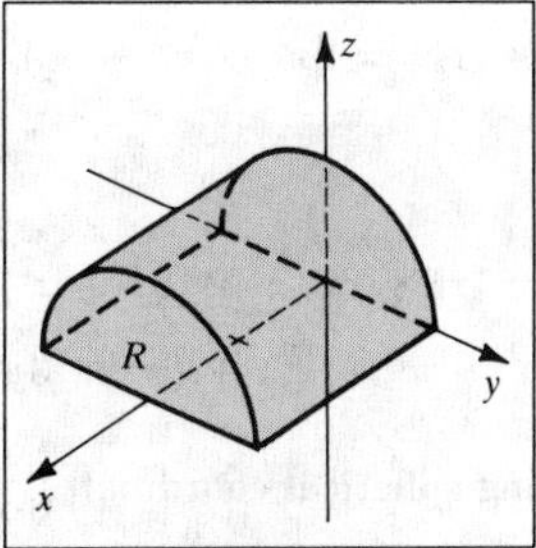

Figure 13b

On S_1, $\mathbf{n} = -\mathbf{i}$ and $\mathbf{F}\cdot\mathbf{n} = -(x + z) = -z$ since $x = 0$ on S_1.

Now S_1 is R_{yz} and so $\iint_{S_1} \mathbf{F}\cdot\mathbf{n}\,dS = -\int_0^1\int_0^{2\pi} (r\sin\theta)\,r\,d\theta\,dr = 0$.

(Note that we have used polar coordinates where $y = r\cos\theta$ and $z = r\sin\theta$.)

On S_3, $\mathbf{n} = \mathbf{i}$ and $\mathbf{F}\cdot\mathbf{n} = x + z = 2 + z$ since $x = 2$ on S_3.

$$\iint_{S_3} \mathbf{F}\cdot\mathbf{n}\,dS = \iint_{S_3} (2 + z)\,dS = \iint_{S_3} 2\,dS + \iint_{S_3} z\,dS = 2\left[\pi(1)^2\right] + 0 = 2\pi,$$

where the first integral is twice the area of S_3,

and the value of the second integral was already shown to be zero.

On S_2, $\mathbf{n} = \dfrac{\nabla g(x, y, z)}{\|\nabla g(x, y, z)\|} = \dfrac{2y\mathbf{j} + 2z\mathbf{k}}{\sqrt{4y^2 + 4z^2}} = y\mathbf{j} + z\mathbf{k}$ since $y^2 + z^2 = 1$ on S_2.

$\mathbf{F}\cdot\mathbf{n} = y(y + z) + z(x + y) = y^2 + 2yz + xz$. The region R_{xy}, which the upper and lower halves of S_2 project onto, is the rectangle $0 \le x \le 2$, $-1 \le y \le 1$ as shown in *Figure 13b*. For the upper half $z = \sqrt{1 - y^2}$ and the lower half $z = -\sqrt{1 - y^2}$. In both cases, $\sqrt{z_x^2 + z_y^2 + 1} = \dfrac{1}{\sqrt{1 - y^2}}$. Now,

$$\iint_{S_2} \mathbf{F}\cdot\mathbf{n}\,dS = \iint_{S_2\text{(upper)}} \mathbf{F}\cdot\mathbf{n}\,dS + \iint_{S_2\text{(lower)}} \mathbf{F}\cdot\mathbf{n}\,dS$$

$$= \int_{-1}^1\int_0^2 \left(y^2 + 2y\sqrt{1 - y^2} + x\sqrt{1 - y^2}\right)\cdot\frac{1}{\sqrt{1 - y^2}}\,dx\,dy +$$

$$\int_{-1}^1\int_0^2 \left(y^2 - 2y\sqrt{1 - y^2} - x\sqrt{1 - y^2}\right)\cdot\frac{1}{\sqrt{1 - y^2}}\,dx\,dy$$

$$= 2\int_{-1}^1\int_0^2 \frac{y^2}{\sqrt{1 - y^2}}\,dx\,dy = 8\int_0^1 \frac{y^2}{\sqrt{1 - y^2}}\,dy = 8\int_0^{\pi/2} \sin^2\theta\,d\theta = 2\pi,$$

where the substitution $y = \sin\theta$ has been used.

Thus, $\iint_S \mathbf{F}\cdot\mathbf{n}\,dS = 0 + 2\pi + 2\pi = 4\pi$.

(2) $\operatorname{div}\mathbf{F} = 1 + 1 + 0 = 2$.

$\iiint_Q 2\,dV = 2\left[\pi(1)^2(2)\right] = 4\pi$ since the volume of a cylinder is $\pi r^2 h$.

14 (1) On the sphere, $\mathbf{F}\cdot\mathbf{n} = \|\mathbf{r}\|^2\,\mathbf{r}\cdot\frac{\mathbf{r}}{\|\mathbf{r}\|} = \|\mathbf{r}\|^3$ since $\mathbf{r}\cdot\mathbf{r} = \|\mathbf{r}\|^2$.

Now $\|\mathbf{r}\|^3 = (x^2 + y^2 + z^2)^{3/2} = a^3$. Using Exercise 11,

$$\iint_S \mathbf{F}\cdot\mathbf{n}\,dS = 2\iint_{R_{xy}} \frac{a^4}{(a^2 - x^2 - y^2)^{1/2}}\,dA = 4\pi a^5.$$

(2) $\mathbf{F} = \|\mathbf{r}\|^2\,\mathbf{r} = (x^2 + y^2 + z^2)(x\mathbf{i} + y\mathbf{j} + z\mathbf{k}) \Rightarrow \operatorname{div}\mathbf{F} =$

$(3x^2 + y^2 + z^2) + (x^2 + 3y^2 + z^2) + (x^2 + y^2 + 3z^2) = 5(x^2 + y^2 + z^2)$.

Using spherical coordinates, $\iiint_Q \operatorname{div}\mathbf{F}\,dV = 5\int_0^{2\pi}\int_0^{\pi}\int_0^{a} (\rho^2)(\rho^2\sin\phi)\,d\rho\,d\phi\,d\theta =$

$$a^5\int_0^{2\pi}\int_0^{\pi}\sin\phi\,d\phi\,d\theta = 2a^5\int_0^{2\pi} d\theta = 4\pi a^5.$$

15 Let $P(x, y, z)$ be arbitrary. If $\iint_S \mathbf{F}\cdot\mathbf{n}\,dS = 0$ for every closed surface, then it equals zero for every sphere S_k with radius k and center P. By (18.27), $[\operatorname{div}\mathbf{F}]_P =$

$\lim_{k\to 0}\frac{1}{V_k}\iint_{S_k}\mathbf{F}\cdot\mathbf{n}\,dS$, and both sides equal 0. Since P was arbitrary, $\operatorname{div}\mathbf{F} = 0$.

16 Since $\mathbf{n}$ is a unit vector, by (16.32), $D_{\mathbf{n}}f = \nabla f\cdot\mathbf{n}$.

Then, $\iint_S D_{\mathbf{n}}f\,dS = \iint_S \nabla f\cdot\mathbf{n}\,dS = \iiint_Q \nabla\cdot\nabla f\,dV = \iiint_Q \nabla^2 f\,dV$.

17 RHS $= \iint_S (f\nabla g)\cdot\mathbf{n}\,dS = \iint_S \mathbf{F}\cdot\mathbf{n}\,dS$ {hint} $= \iiint_Q \operatorname{div}\mathbf{F}\,dV$ {(18.26)} $=$

$\iiint_Q \nabla\cdot(f\nabla g)\,dV = \iiint_Q \left[f[\nabla\cdot(\nabla g)] + (\nabla f)\cdot(\nabla g)\right]dV$ {Example 5, §18.1} $=$

$$\iiint_Q (f\nabla^2 g + \nabla f\cdot\nabla g)\,dV = \text{LHS}.$$

18 Using Exercise 17, $\iiint_Q (f\nabla^2 g + \nabla f\cdot\nabla g)\,dV = \iint_S (f\nabla g)\cdot\mathbf{n}\,dS$.

Interchanging f and g gives $\iiint_Q (g\nabla^2 f + \nabla g\cdot\nabla f)\,dV = \iint_S (g\nabla f)\cdot\mathbf{n}\,dS$.

By subtracting the second equation from the first, we obtain the desired result.

19 RHS $= \iint_S f\mathbf{F}\cdot\mathbf{n}\,dS = \iiint_Q \operatorname{div} f\mathbf{F}\,dV$ {(18.26)}

$= \iiint_Q (f\operatorname{div}\mathbf{F} + \nabla f\cdot\mathbf{F})\,dV$ {Example 5, §18.1}

$= \iiint_Q \mathbf{F}\cdot\mathbf{F}\,dV$ {$\operatorname{div}\mathbf{F} = 0$ and $\mathbf{F} = \nabla f$} $=$ LHS.

20 $\frac{1}{3}\iint_S \mathbf{r}\cdot\mathbf{n}\,dS = \frac{1}{3}\iiint_Q \operatorname{div}\mathbf{r}\,dV = \frac{1}{3}\iiint_Q (1 + 1 + 1)\,dV = \iiint_Q dV = V$.

21 $\iint_S \operatorname{curl}\mathbf{F}\cdot\mathbf{n}\,dS = \iiint_Q \operatorname{div}\operatorname{curl}\mathbf{F}\,dV = 0$ since $\operatorname{div}\operatorname{curl}\mathbf{F} = 0$ by Exercise 28, §18.1.

22 $\iint_S \mathbf{a}\cdot\mathbf{n}\,dS = \iiint_Q \operatorname{div}\mathbf{a}\,dV = 0$ since $\mathbf{a}$ is constant and $\operatorname{div}\mathbf{a} = 0$.

Note: In Exercises 23–24, Exercise 50 in §15.2 is used.

[23] By (14.33(v)), $(\mathbf{c} \times \mathbf{F}) \cdot \mathbf{n} = \mathbf{c} \cdot (\mathbf{F} \times \mathbf{n})$ and by Exercise 26, §18.1, $\nabla \cdot (\mathbf{c} \times \mathbf{F}) = (\nabla \times \mathbf{c}) \cdot \mathbf{F} - (\nabla \times \mathbf{F}) \cdot \mathbf{c} = -\mathbf{c} \cdot (\nabla \times \mathbf{F})$ since $\mathbf{c}$ is constant and $\nabla \times \mathbf{c} = \mathbf{0}$. Now by substituting $\mathbf{c} \times \mathbf{F}$ into the divergence theorem we have the following equalities. $\mathbf{c} \cdot \iint_S \mathbf{F} \times \mathbf{n}\, dS = \iint_S \mathbf{c} \cdot (\mathbf{F} \times \mathbf{n})\, dS = \iint_S (\mathbf{c} \times \mathbf{F}) \cdot \mathbf{n}\, dS =$ $\iiint_Q \nabla \cdot (\mathbf{c} \times \mathbf{F})\, dV = \iiint_Q -\mathbf{c} \cdot (\nabla \times \mathbf{F})\, dV = -\mathbf{c} \cdot \iiint_Q \nabla \times \mathbf{F}\, dV$. Since $\mathbf{c}$ is an arbitrary vector, $\iint_S \mathbf{F} \times \mathbf{n}\, dS = -\iiint_Q \nabla \times \mathbf{F}\, dV$.

(To see this, let $\mathbf{c} = \langle 1, 0, 0\rangle$, $\langle 0, 1, 0\rangle$, and $\langle 0, 0, 1\rangle$ and compare components.)

[24] From Example 5, §18.1, $\nabla \cdot (\mathbf{c}f) = f(\nabla \cdot \mathbf{c}) + \mathbf{c} \cdot \nabla f = \mathbf{c} \cdot \nabla f$. $\{\nabla \cdot \mathbf{c} = 0$ since $\mathbf{c}$ is constant.$\}$ Substituting $f\mathbf{c}$ into the divergence theorem we have the following equalities. $\mathbf{c} \cdot \iint_S f\mathbf{n}\, dS = \iint_S \mathbf{c}f \cdot \mathbf{n}\, dS = \iiint_Q \nabla \cdot (\mathbf{c}f)\, dV = \iiint_Q \mathbf{c} \cdot \nabla f\, dV =$ $\mathbf{c} \cdot \iiint_Q \nabla f\, dV$. Since $\mathbf{c}$ is arbitrary, $\iint_S f\mathbf{n}\, dS = \iiint_Q \nabla f\, dV$.

(To see this, let $\mathbf{c} = \langle 1, 0, 0\rangle$, $\langle 0, 1, 0\rangle$, and $\langle 0, 0, 1\rangle$ and compare components.)

[25] If $\mathbf{F}$ is orthogonal to S at each point, then $\mathbf{F} \times \mathbf{n} = \mathbf{0}$ for all normals $\mathbf{n}$ to S.

Using Exercise 23, $\iiint_Q \operatorname{curl} \mathbf{F}\, dV = -\iint_S \mathbf{F} \times \mathbf{n}\, dS = -\iint_S \mathbf{0}\, dS = \mathbf{0}$.

[26] RHS $= \frac{1}{2}\iint_S r^2 \mathbf{n}\, dS = \frac{1}{2}\iiint_Q \nabla r^2\, dV$ {Exercise 24} $=$

$\frac{1}{2}\iiint_Q 2\mathbf{r}\, dV$ $\{r^2 = x^2 + y^2 + z^2\}$ $= \iiint_Q \mathbf{r}\, dV =$ LHS.

[27] $-\iint_S p\mathbf{n}\, dS = -\iiint_Q \nabla p\, dV = -\iiint_Q (-62.5\mathbf{k})\, dV = \left(62.5 \iiint_Q dV\right)\mathbf{k} =$ $62.5\left[\pi(1)^2(10)\right]\mathbf{k}$ {volume of the cylindrical tank} $= (625\pi \text{ lb})\mathbf{k}$.

The force is directed upward since it is in the direction of the unit vector $\mathbf{k}$.

Exercises 18.7

[1] (1) $\mathbf{F} = y^2\mathbf{i} + z^2\mathbf{j} + x^2\mathbf{k} \Rightarrow \operatorname{curl} \mathbf{F} = -2z\mathbf{i} - 2x\mathbf{j} - 2y\mathbf{k}$.

A unit normal to S $(g(x, y, z) = x + y + z - 1)$, pointing into the first octant, is $\dfrac{\nabla g(x, y, z)}{\|\nabla g(x, y, z)\|} = \frac{1}{\sqrt{3}}(\mathbf{i} + \mathbf{j} + \mathbf{k})$. $\operatorname{curl} \mathbf{F} \cdot \mathbf{n} = -\frac{2}{\sqrt{3}}(z + x + y) = -\frac{2}{\sqrt{3}}$ since $x + y + z = 1$ on S. If we let $z = 1 - x - y$, then $\sqrt{z_x^2 + z_y^2 + 1} = \sqrt{3}$. Thus, $\iint_S \operatorname{curl} \mathbf{F} \cdot \mathbf{n}\, dS = \iint_S -\frac{2}{\sqrt{3}}\, dS = \iint_{R_{xy}} -\frac{2}{\sqrt{3}} \cdot \sqrt{3}\, dA = -2 \iint_{R_{xy}} dA = -2(\frac{1}{2}) = -1$

since R_{xy} is a triangular region with area $\frac{1}{2}$.

(2) To calculate $\oint_C \mathbf{F}\cdot\mathbf{T}\,ds = \oint_C \mathbf{F}\cdot d\mathbf{r} = \oint_C y^2\,dx + z^2\,dy + x^2\,dz$, we express $C = C_1 \cup C_2 \cup C_3$ as shown in *Figure 1.*

C_1: $x = 1 - t,\ y = t,\ z = 0,\ 0 \le t \le 1$, and $\int_{C_1} \mathbf{F}\cdot d\mathbf{r} = \int_0^1 -t^2\,dt = -\frac{1}{3}$.

C_2: $x = 0,\ y = 1 - t,\ z = t,\ 0 \le t \le 1$, and $\int_{C_2} \mathbf{F}\cdot d\mathbf{r} = \int_0^1 -t^2\,dt = -\frac{1}{3}$.

C_3: $x = t,\ y = 0,\ z = 1 - t,\ 0 \le t \le 1$, and $\int_{C_3} \mathbf{F}\cdot d\mathbf{r} = \int_0^1 -t^2\,dt = -\frac{1}{3}$.

Their sum is -1.

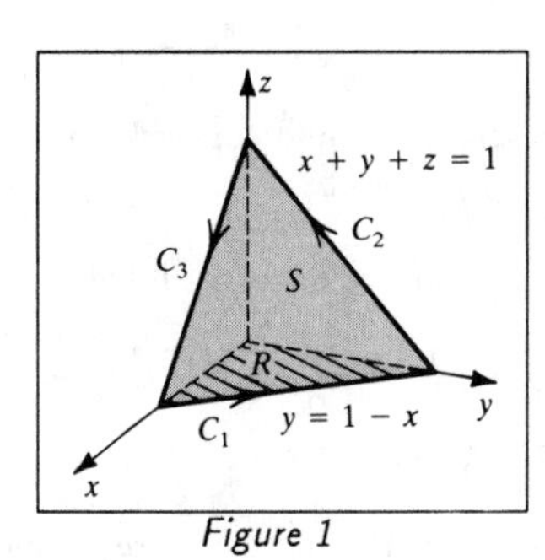

Figure 1

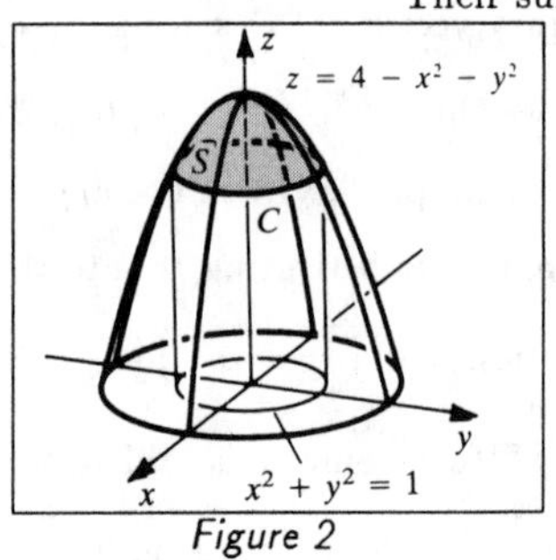

Figure 2

[2] (1) $\mathbf{F} = 2y\mathbf{i} - z\mathbf{j} + 3\mathbf{k} \Rightarrow \operatorname{curl}\mathbf{F} = \mathbf{i} - 2\mathbf{k}$. $z = 4 - x^2 - y^2 \Rightarrow$ $z_x = -2x,\ z_y = -2y$, and $\mathbf{n} = \dfrac{2x\mathbf{i} + 2y\mathbf{j} + \mathbf{k}}{\sqrt{z_x^2 + z_y^2 + 1}}$. $\iint_S \operatorname{curl}\mathbf{F}\cdot\mathbf{n}\,dS =$

$$\iint_{R_{xy}} (2x - 2)\,dA = 2\int_0^{2\pi}\int_0^1 (r\cos\theta - 1)\,r\,dr\,d\theta = 2\int_0^{2\pi} \left(\tfrac{1}{3}\cos\theta - \tfrac{1}{2}\right)d\theta = -2\pi.$$

(2) To calculate $\oint_C \mathbf{F}\cdot d\mathbf{r} = \oint_C 2y\,dx - z\,dy + 3\,dz$, we express C (the intersection of the paraboloid and the cylinder) as $x = \cos t,\ y = \sin t,\ z = 3,\ 0 \le t \le 2\pi$.

Then, $\oint_C \mathbf{F}\cdot d\mathbf{r} = \displaystyle\int_0^{2\pi} \Big[2\sin t(-\sin t) - 3\cos t + 3(0)\Big]\,dt = -2\pi$.

[3] (1) $\mathbf{F} = z\mathbf{i} + x\mathbf{j} + y\mathbf{k} \Rightarrow \operatorname{curl}\mathbf{F} = \mathbf{i} + \mathbf{j} + \mathbf{k}$. $z = f(x, y) = \sqrt{a^2 - x^2 - y^2} \Rightarrow$

$f_x = \dfrac{-x}{\sqrt{a^2 - x^2 - y^2}},\ f_y = \dfrac{-y}{\sqrt{a^2 - x^2 - y^2}}$, and $\mathbf{n} = \dfrac{-f_x\mathbf{i} - f_y\mathbf{j} + \mathbf{k}}{\sqrt{f_x^2 + f_y^2 + 1}}$. Then,

$$\begin{aligned}
\iint_S \operatorname{curl}\mathbf{F}\cdot\mathbf{n}\,dS &= \iint_{R_{xy}} \frac{x + y + \sqrt{a^2 - x^2 - y^2}}{\sqrt{a^2 - x^2 - y^2}}\,dA \\
&= \int_0^{2\pi}\int_0^a \frac{r\cos\theta + r\sin\theta + \sqrt{a^2 - r^2}}{\sqrt{a^2 - r^2}}\,r\,dr\,d\theta \\
&= \int_0^a\int_0^{2\pi} \frac{r^2(\cos\theta + \sin\theta)}{\sqrt{a^2 - r^2}}\,d\theta\,dr + \int_0^{2\pi}\int_0^a r\,dr\,d\theta \\
&= \int_0^a 0\,dr + \int_0^{2\pi} \tfrac{1}{2}a^2\,d\theta = \pi a^2.
\end{aligned}$$

(2) To calculate $\oint_C \mathbf{F}\cdot d\mathbf{r} = \oint_C z\,dx + x\,dy + y\,dz$, we express C as $x = a\cos t$, $y = a\sin t,\ z = 0,\ 0 \le t \le 2\pi$. $\{C$ is the circle $x^2 + y^2 = a^2.\}$ Then, $\oint_C \mathbf{F}\cdot d\mathbf{r}$ $= \displaystyle\int_0^{2\pi} x\,dy = \pi a^2$ since by (18.20), the integral equals the area of the circle.

4 (1) $\mathbf{F} = x^2\,\mathbf{i} + y^2\,\mathbf{j} + z^2\,\mathbf{k} \Rightarrow \operatorname{curl}\mathbf{F} = \mathbf{0} \Rightarrow \iint_S \operatorname{curl}\mathbf{F}\cdot\mathbf{n}\,dS = 0.$

(2) To calculate $\oint_C \mathbf{F}\cdot d\mathbf{r} = \oint_C x^2\,dx + y^2\,dy + z^2\,dz$,

we express C as $x = \cos t$, $y = \sin t$, $z = 1$, $0 \le t \le 2\pi$.

Then, $\oint_C \mathbf{F}\cdot d\mathbf{r} = \int_0^{2\pi}\left[\cos^2 t(-\sin t) + \sin^2 t\cos t\right]dt = 0.$

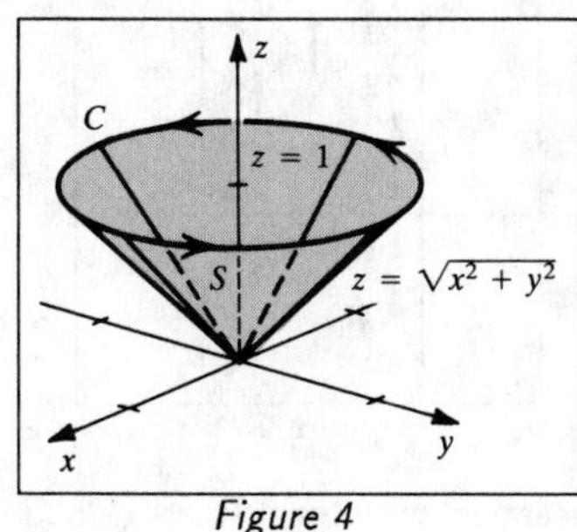

Figure 4

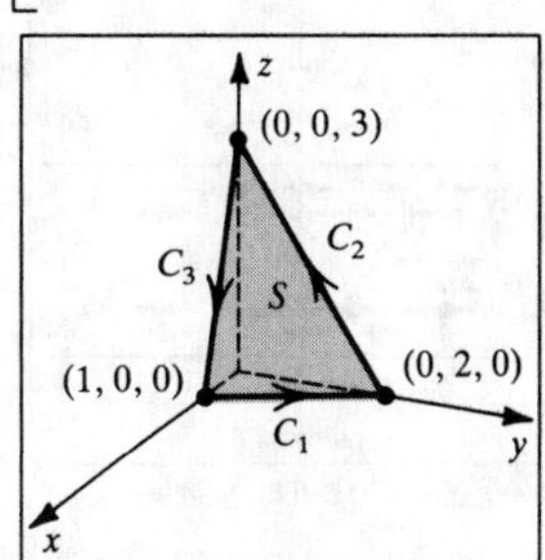

Figure 8

5 Let the surface S be given by $z = 1$ bounded by C with $\mathbf{n} = \mathbf{k}$.

Then, $\oint_C \mathbf{F}\cdot d\mathbf{r} = \iint_S \operatorname{curl}\mathbf{F}\cdot\mathbf{n}\,dS = \iint_S (3y^2\,\mathbf{i} + 3\,\mathbf{j} + 2x\,\mathbf{k})\cdot\mathbf{k}\,dS = \iint_S 2x\,dS =$

$$\iint_{R_{xy}} 2x(1)\,dA = \int_0^1\int_0^{2\pi}(2r\cos\theta)\,r\,d\theta\,dr = \int_0^1 0\,dr = 0.$$

6 Let the surface S be given by $z = 2$ bounded by C with $\mathbf{n} = \mathbf{k}$. Then, $\oint_C \mathbf{F}\cdot d\mathbf{r} =$

$\iint_S \operatorname{curl}\mathbf{F}\cdot\mathbf{n}\,dS = \iint_S \left[(y - z)\,\mathbf{j} + (y - z)\,\mathbf{k}\right]\cdot\mathbf{k}\,dS = \iint_S (y - z)\,dS =$

$$\iint_{R_{xy}} (y - 2)(1)\,dA \;\{\text{since } z = 2 \text{ on } S\} = \int_0^1\int_0^1 (y - 2)\,dx\,dy = \int_0^1 (y - 2)\,dy = -\tfrac{3}{2}.$$

7 C is the circle $x^2 + y^2 = 4$ in the xy-plane. To obtain a positive direction for C, let $x = 2\cos t$, $y = 2\sin t$, $z = 0$, $0 \le t \le 2\pi$. Then,

$\iint_S \operatorname{curl}\mathbf{F}\cdot\mathbf{n}\,dS = \oint_C \mathbf{F}\cdot d\mathbf{r} = \oint_C 2y\,dx + e^z\,dy - \arctan x\,dz = \oint_C 2y\,dx + dy$

$$\{\text{since } z = 0\} = \int_0^{2\pi}\left[4\sin t(-2\sin t) + 2\cos t\right] = -8\pi.$$

8 The curve C can be split up into 3 curves as shown in *Figure 8*.

C_1 is given by $y = -2x + 2$, $z = 0$, $0 \le x \le 1$.

C_2 is given by $z = -\frac{3}{2}y + 3$, $x = 0$, $0 \le y \le 2$.

C_3 is given by $x = -\frac{1}{3}z + 1$, $y = 0$, $0 \le z \le 3$.

Thus, $\iint_S \operatorname{curl}\mathbf{F}\cdot\mathbf{n}\,dS = \oint_C \mathbf{F}\cdot d\mathbf{r} = \oint_C xy^2\,dx + yz^2\,dy + zx^2\,dz$

$$= \int_1^0 x(-2x + 2)^2\,dx + \int_2^0 y(-\tfrac{3}{2}y + 3)^2\,dy + \int_3^0 z(-\tfrac{1}{3}z + 1)^2\,dz$$

$$= 4\int_1^0 (x^3 - 2x^2 + x)\,dx + 9\int_2^0 (\tfrac{1}{4}y^3 - y^2 + y)\,dy + \int_3^0 (\tfrac{1}{9}z^3 - \tfrac{2}{3}z^2 + z)\,dz$$

$$= -\tfrac{1}{3} - 3 - \tfrac{3}{4} = -\tfrac{49}{12}.$$

[9] Refer to *Figure 9*. The curl meter rotates counterclockwise for $0 < y < 1$ and clockwise for $1 < y < 2$. There is no rotation if $y = 1$. $\operatorname{curl} \mathbf{F} = 2(1 - y)\,\mathbf{k}$; $|(\operatorname{curl} \mathbf{F}) \cdot \mathbf{k}| = |2(1 - y)|$ has a maximum value 2 at $y = 0$ and $y = 2$ and a minimum value 0 at $y = 1$.

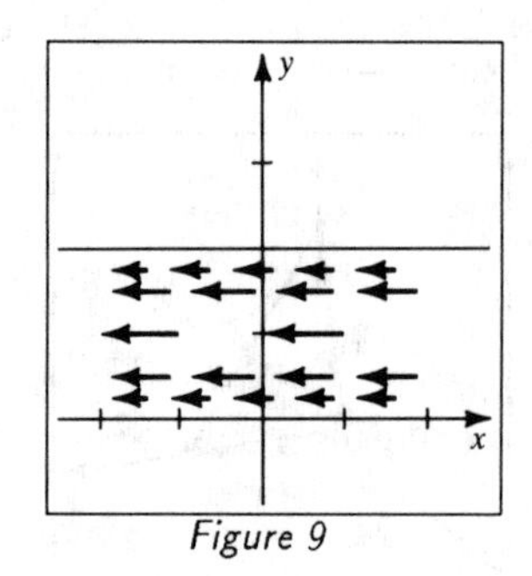

Figure 9

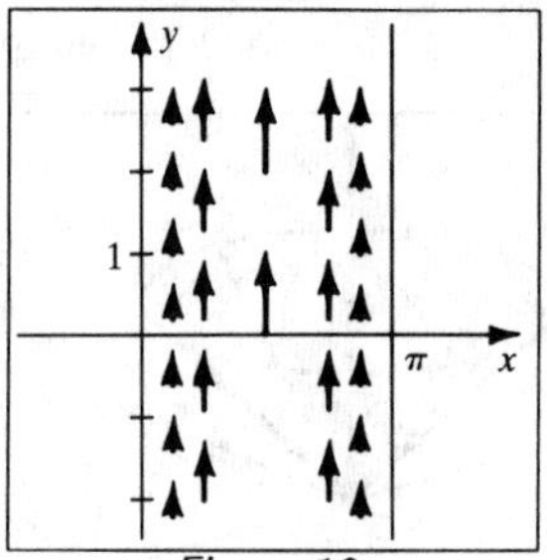

Figure 10

[10] Refer to *Figure 10*. The curl meter rotates counterclockwise for $0 < x < \frac{\pi}{2}$ and clockwise for $\frac{\pi}{2} < x < \pi$. There is no rotation at $x = \frac{\pi}{2}$. $\operatorname{curl} \mathbf{F} = \cos x\,\mathbf{k}$; $|(\operatorname{curl} \mathbf{F}) \cdot \mathbf{k}| = |\cos x|$ has a maximum value 1 at $x = 0$ and $x = \pi$ and a minimum value 0 at $x = \frac{\pi}{2}$.

[11] Typical field vectors are shown in Figure 18.5. A curl meter rotates counterclockwise for every $(x, y) \neq (0, 0)$. $\operatorname{curl} \mathbf{F} = 2\mathbf{k}$; $|(\operatorname{curl} \mathbf{F}) \cdot \mathbf{k}| = 2$ for every (x, y).

[12] Typical field vectors are shown in Figure 18.7. By Exercise 21 in §18.1, $\operatorname{curl} \mathbf{F} = \mathbf{0}$, and hence, by (18.31)(iv), the field is irrotational.

[13] $\operatorname{curl} \mathbf{F} = (e^z - e^z)\mathbf{i} - (0 - 0)\mathbf{j} + (1 - 1)\mathbf{k} = \mathbf{0} \Rightarrow \mathbf{F}$ is irrotational.

[14] $\mathbf{F} = f(r)\,x\mathbf{i} + f(r)\,y\mathbf{j} + f(r)\,z\mathbf{k}$ and $r = \sqrt{x^2 + y^2 + z^2}$. The $\mathbf{i}$ component of $\operatorname{curl} \mathbf{F}$ is $\frac{\partial}{\partial y}[f(r)\,z] - \frac{\partial}{\partial z}[f(r)\,y]$. Using the product and chain rules we obtain

$$\frac{\partial}{\partial y}[f(r)\,z] = \left[f(r) \cdot 0 + z \cdot f'(r) \cdot \tfrac{1}{2}(x^2 + y^2 + z^2)^{-1/2} \cdot 2y\right] = zy\frac{f'(r)}{r} \text{ and}$$

$$\frac{\partial}{\partial z}[f(r)\,y] = \left[f(r) \cdot 0 + y \cdot f'(r) \cdot \tfrac{1}{2}(x^2 + y^2 + z^2)^{-1/2} \cdot 2z\right] = zy\frac{f'(r)}{r}.$$

Hence, the $\mathbf{i}$ component is zero. Similarly, the $\mathbf{j}$ and $\mathbf{k}$ components are zero. Thus, $\operatorname{curl} \mathbf{F} = \mathbf{0}$ and $\mathbf{F}$ is irrotational.

[15] (a) $\iint_S \operatorname{curl} \mathbf{F} \cdot \mathbf{n}\, dS = \iiint_Q \operatorname{div} \operatorname{curl} \mathbf{F}\, dV = 0$ since $\operatorname{div} \operatorname{curl} \mathbf{F} = 0$ by Exercise 28, §18.1.

(b) We may assume that S is a sphere of radius a with center at the origin. Let S_1 be the upper hemisphere ($z \geq 0$) and S_2 be the lower hemisphere ($z \leq 0$). Let C be the circle $x^2 + y^2 = a^2$ in the xy-plane. To apply Stokes' theorem, we let $\mathbf{N}$ denote the unit upper normal. Then, $\mathbf{N} = \mathbf{n}$ on S_1, and $\mathbf{N} = -\mathbf{n}$ on S_2.

Thus, $\iint_S \operatorname{curl} \mathbf{F} \cdot \mathbf{n}\, dS = \iint_{S_1} \operatorname{curl} \mathbf{F} \cdot \mathbf{N}\, dS + \iint_{S_2} \operatorname{curl} \mathbf{F} \cdot \mathbf{N}\, dS =$

$\iint_{S_1} \operatorname{curl} \mathbf{F} \cdot \mathbf{n}\, dS + \iint_{S_2} \operatorname{curl} \mathbf{F} \cdot (-\mathbf{n})\, dS = \oint_C \mathbf{F} \cdot d\mathbf{r} - \oint_C \mathbf{F} \cdot d\mathbf{r} = 0.$

Note that on S_1, we traverse C in a counterclockwise direction, and on S_2, we traverse C in the opposite direction.

[16] F constant $\Rightarrow$ curl $\mathbf{F} = \mathbf{0}$. Thus, $\oint_C \mathbf{F} \cdot \mathbf{T}\, ds = \iint_S \text{curl}\,\mathbf{F} \cdot \mathbf{n}\, dS = \iint_S 0\, dS = 0$.

[17] By Exercise 25, §18.1, we have $\nabla \times (f\nabla g) = f(\nabla \times \nabla g) + (\nabla f \times \nabla g)$.

By Exercise 27, §18.1, $\nabla \times \nabla g = \mathbf{0}$, so $\nabla \times (f\nabla g) = \nabla f \times \nabla g$.

Now by Stokes' theorem, $\oint_C f\nabla g \cdot d\mathbf{r} = \iint_S \nabla \times (f\nabla g) \cdot \mathbf{n}\, dS = \iint_S (\nabla f \times \nabla g) \cdot \mathbf{n}\, dS$.

[18] $\oint_C \mathbf{a} \times \mathbf{r} \cdot d\mathbf{r} = \iint_S \nabla \times (\mathbf{a} \times \mathbf{r}) \cdot \mathbf{n}\, dS = \iint_S 2\mathbf{a} \cdot \mathbf{n}\, dS$ { Exercise 20(a), §18.1 } $=$

$2\mathbf{a} \cdot \iint_S \mathbf{n}\, dS$ { Exercise 50, §15.2 }. See the explanation in the text of a surface integral of a vector function before Exercises 23–24 of §18.6.

[19] Let $\mathbf{c}$ be an arbitrary constant vector.

$$\begin{aligned}
\mathbf{c} \cdot \oint_C f\, d\mathbf{r} &= \oint_C f\mathbf{c} \cdot d\mathbf{r} \ \{\text{Exercise 50, §15.2}\} = \iint_S \nabla \times (f\mathbf{c}) \cdot \mathbf{n}\, dS \\
&= \iint_S \left[f(\nabla \times \mathbf{c}) + \nabla f \times \mathbf{c}\right] \cdot \mathbf{n}\, dS \ \{\text{by Exercise 25, §18.1}\} \\
&= \iint_S (\nabla f \times \mathbf{c}) \cdot \mathbf{n}\, dS \ \{\text{since } \mathbf{c} \text{ is constant, } \nabla \times \mathbf{c} = \mathbf{0}\} \\
&= \iint_S \mathbf{c} \cdot (\mathbf{n} \times \nabla f)\, dS \ \{\text{by (14.33)(i) and (14.33)(v)}\} = \mathbf{c} \cdot \iint_S \mathbf{n} \times \nabla f\, dS.
\end{aligned}$$

Since $\mathbf{c}$ is arbitrary, $\oint_C f\, d\mathbf{r} = \iint_S \mathbf{n} \times \nabla f\, dS$.

(To see this, let $\mathbf{c} = \langle 1, 0, 0 \rangle$, $\langle 0, 1, 0 \rangle$, and $\langle 0, 0, 1 \rangle$ and compare components.)

[20] The implication in one direction (independence of path $\Rightarrow$ the given set of equalities) was established in Exercise 19, §18.3. Conversely, given the equality of the partials and using the definition of curl F (18.5), we have curl $\mathbf{F} = \mathbf{0}$.

Then by (18.31), the integral is independent of path.

18.8 Review Exercises

[1] $\mathbf{F}(x, y) = 2x\mathbf{i} + y\mathbf{j}$

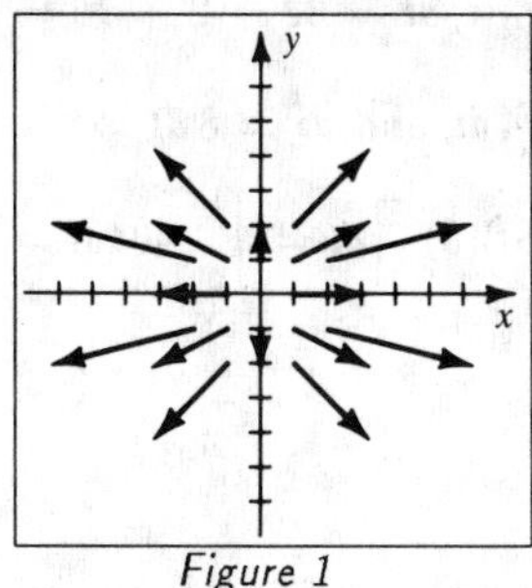

Figure 1

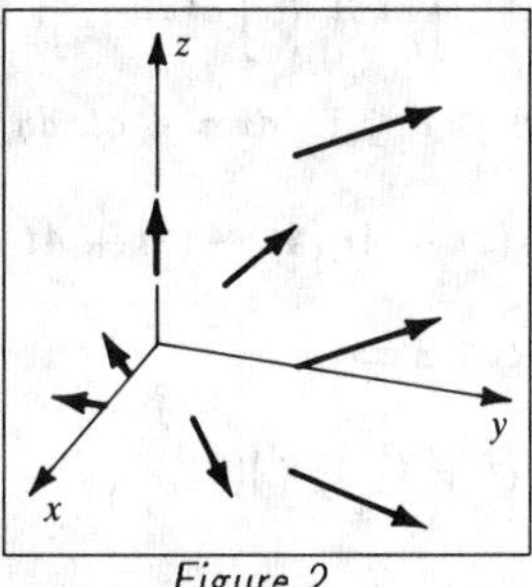

Figure 2

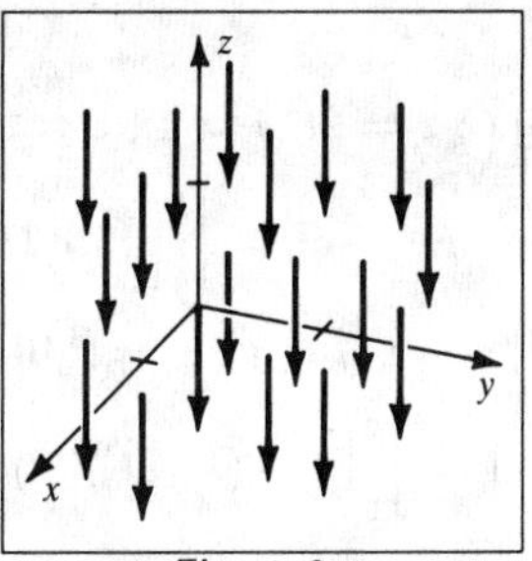

Figure 3

[2] $\mathbf{F}(x, y, z) = x\mathbf{i} + y\mathbf{j} + \mathbf{k}$

[3] $\mathbf{F}(x, y, z) = -\mathbf{k}$

[4] $\mathbf{F}(x, y, z) = \nabla(x^2 + y^2 + z^2)^{-1/2} = -\frac{\mathbf{r}}{\|\mathbf{r}\|^3}$, where $\mathbf{r} = x\mathbf{i} + y\mathbf{j} + z\mathbf{k}$. By (18.2), F is an inverse square field with $c < 0$. It's vector field is similar to Figure 18.7(i).

[5] $f(x, y) = y^2 \tan x \Rightarrow \mathbf{F}(x, y) = \nabla f(x, y) = (y^2 \sec^2 x)\mathbf{i} + (2y \tan x)\mathbf{j}$.

[6] $f(x, y, z) = \ln(x + y + z) \Rightarrow \mathbf{F}(x, y, z) = \nabla f(x, y, z) = \dfrac{1}{x + y + z}(\mathbf{i} + \mathbf{j} + \mathbf{k})$.

[7] Let C be divided into two parts. On C_1: $x = t$, $y = 0$, $-1 \leq t \leq 1$. On C_2:

$$x = -1,\ y = t,\ 0 \leq t \leq 4. \text{ Thus, } \int_C y^2\,dx + xy\,dy = \int_1^{-1} 0\,dt + \int_0^4 (-t)\,dt = -8.$$

[8] The line segment is given by $y = -2x + 2$ for $-1 \leq x \leq 1$. $\{dy = -2\,dx\}$

$$\mathrm{I} = \int_1^{-1} (-2x + 2)^2\,dx + x(-2x + 2)(-2)\,dx = \int_1^{-1} (8x^2 - 12x + 4)\,dx =$$

$$-8\int_0^1 (2x^2 + 1)\,dx = -\tfrac{40}{3}.$$

[9] $x = 1 - t$, $y = t^2$, $0 \leq t \leq 2 \Rightarrow$

$$\mathrm{I} = \int_0^2 \Big[t^4(-1) + (1 - t)(t^2)(2t)\Big]\,dt = \int_0^2 (-3t^4 + 2t^3)\,dt = -\tfrac{56}{5}.$$

[10] $y = 2 - 2x^3$, $dy = -6x^2$, $-1 \leq x \leq 1 \Rightarrow$

$$\mathrm{I} = \int_1^{-1} \Big[(2 - 2x^3)^2 + x(2 - 2x^3)(-6x^2)\Big]\,dx = \int_1^{-1} (4 - 20x^3 + 16x^6)\,dx =$$

$$-8\int_0^1 (4x^6 + 1)\,dx = -\tfrac{88}{7}.$$

[11] $y = x^4 \Rightarrow ds = \sqrt{(dx)^2 + (dy)^2} = \sqrt{1 + (4x^3)^2}\,dx \Rightarrow$

$$\int_C xy\,ds = \int_{-1}^2 x(x^4)\sqrt{1 + 16x^6}\,dx = \Big[\tfrac{1}{96}\cdot\tfrac{2}{3}(1 + 16x^6)^{3/2}\Big]_{-1}^2 =$$

$$\tfrac{1}{144}(1025^{3/2} - 17^{3/2}) \approx 227.40.$$

[12] C_1: $x = 0$, $y = 0$, $z = t$, $0 \leq t \leq 8$. C_2: $x = t$, $y = 0$, $z = 8$, $0 \leq t \leq 2$.

C_3: $x = 2$, $y = t$, $z = 8$, $0 \leq t \leq 4$. $\int_C x\,dx + (x + y)\,dy + (x + y + z)\,dz =$

$$\int_0^8 t\,dt + \int_0^2 t\,dt + \int_0^4 (2 + t)\,dt = 32 + 2 + 16 = 50.$$

[13] C: $x = 2t$, $y = 4t$, $z = 8t$, $0 \leq t \leq 1$. $dx = 2\,dt$, $dy = 4\,dt$, and $dz = 8\,dt \Rightarrow$

$$\mathrm{I} = \int_0^1 \Big[(2t)(2) + (2t + 4t)(4) + (2t + 4t + 8t)(8)\Big]\,dt = \int_0^1 140t\,dt = 70.$$

[14] $x = t$, $y = t^2$, $z = t^3$, $0 \leq t \leq 2 \Rightarrow$

$$\mathrm{I} = \int_0^2 \Big[t + (t + t^2)(2t) + (t + t^2 + t^3)(3t^2)\Big]\,dt =$$

$$\int_0^2 (t + 2t^2 + 5t^3 + 3t^4 + 3t^5)\,dt = \tfrac{1178}{15}.$$

[15] $$\int_C (x + y)\,dx + (x - y)\,dy = \int_{-\pi}^0 \Big[(\cos t + \sin t)(-\sin t) + (\cos t - \sin t)(\cos t)\Big]\,dt$$

$$= \int_{-\pi}^0 \Big[\cos 2t - \sin 2t\Big]\,dt = 0.$$

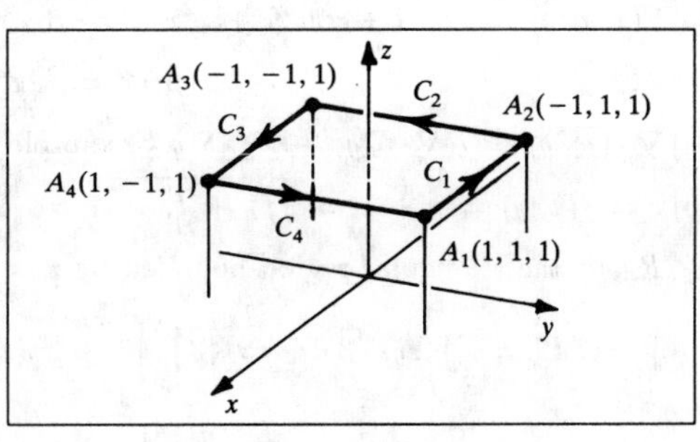

Figure 16

[16] $W = \oint \mathbf{F} \cdot d\mathbf{r}$

$= \oint_C xy\,dx + y^2 z\,dy + xz^2\,dz,$

where C is as shown in *Figure 16.*

Each C_i has $z = 1$, and $0 \le t \le 2$.

C_1: $x = 1 - t$, $\quad y = 1$.

C_2: $x = -1$, $\quad y = 1 - t$.

C_3: $x = t - 1$, $\quad y = -1$.

C_4: $x = 1$, $\quad y = t - 1$.

Thus, $W = \int_0^2 (1 - t)(1)(-1)\,dt + \int_0^2 (1 - t)^2(1)(-1)\,dt +$

$\int_0^2 (t - 1)(-1)(1)\,dt + \int_0^2 (t - 1)^2(1)(1)\,dt = 0 + (-\frac{2}{3}) + 0 + \frac{2}{3} = 0.$

[17] $f_x = x + y \Rightarrow f = \frac{1}{2}x^2 + xy + g(y) \Rightarrow$

$f_y = x + g'(y) = x + y \Rightarrow g'(y) = y$ and $g(y) = \frac{1}{2}y^2 + c$.

Thus, let $f(x, y) = \frac{1}{2}x^2 + xy + \frac{1}{2}y^2$. Then, $f(2, 3) - f(1, -1) = \frac{25}{2} - 0 = \frac{25}{2}$.

[18] *Note*: This exercise uses the extensions of (18.13) and (18.14) to 3 dimensions.

$f_x = 8x^3 + z^2 \Rightarrow f = 2x^4 + xz^2 + g(y, z) \Rightarrow$

$f_y = g_y(y, z) = -3z \Rightarrow g(y, z) = -3yz + k(z)$.

$f_z = 2xz - 3y + k'(z) = 2xz - 3y \Rightarrow k'(z) = 0$ and $k(z) = c$. Thus,

let $f(x, y, z) = 2x^4 + xz^2 - 3yz$. Then, $f(2, 1, 3) - f(0, 0, 0) = 41 - 0 = 41$.

[19] $f_x = 2xe^{2y} \Rightarrow f = x^2 e^{2y} + g(y, z) \Rightarrow f_y = 2x^2 e^{2y} + g_y(y, z) = 2x^2 e^{2y} + 2y \cot z$

$\Rightarrow g_y(y, z) = 2y \cot z$ and $g(y, z) = y^2 \cot z + k(z)$.

$f_z = g_z(y, z) = -y^2 \csc^2 z + k'(z) = -y^2 \csc^2 z \Rightarrow k'(z) = 0$ and $k(z) = c$. Thus,

$f(x, y, z) = x^2 e^{2y} + y^2 \cot z + c$, and by (18.31), $\int_C \mathbf{F} \cdot d\mathbf{r}$ is independent of path.

Note: In Exercises 20–22, $M = xy$, $N = x^2 + y^2 \Rightarrow N_x = 2x$, $M_y = x$.

Thus, $\mathrm{I} = \oint_C xy\,dx + (x^2 + y^2)\,dy = \iint_R (2x - x)\,dA = \iint_R x\,dA$.

[20] $\mathrm{I} = \int_{-1}^{2}\int_{x^2}^{x+2} x\,dy\,dx = \int_{-1}^{2} (x^2 + 2x - x^3)\,dx = \left[\frac{1}{3}x^3 + x^2 - \frac{1}{4}x^4\right]_{-1}^{2} = \frac{9}{4}$.

[21] $\mathrm{I} = \int_0^1\int_0^{1-x} x\,dy\,dx = \int_0^1 (x - x^2)\,dx = \left[\frac{1}{2}x^2 - \frac{1}{3}x^3\right]_0^1 = \frac{1}{6}$.

[22] In polar coordinates, $x^2 + y^2 - 2x = 0$ has equation $r = 2\cos\theta$.

$\mathrm{I} = \int_{-\pi/2}^{\pi/2}\int_0^{2\cos\theta} (r\cos\theta)\,r\,dr\,d\theta = \frac{8}{3}\int_{-\pi/2}^{\pi/2} \cos^4\theta\,d\theta$

$= \frac{2}{3}\int_{-\pi/2}^{\pi/2} (\frac{3}{2} + 2\cos 2\theta + \frac{1}{2}\cos 4\theta)\,d\theta = \frac{2}{3}\int_0^{\pi/2} (3 + 4\cos 2\theta + \cos 4\theta)\,d\theta =$

$\frac{2}{3}(\frac{3\pi}{2}) = \pi$.

[23] $\mathbf{F}(x,\, y,\, z) = x^3z^4\,\mathbf{i} + xyz^2\,\mathbf{j} + x^2y^2\,\mathbf{k} \Rightarrow \operatorname{div}\mathbf{F} = 3x^2z^4 + xz^2$ and

$$\operatorname{curl}\mathbf{F} = (2x^2y - 2xyz)\,\mathbf{i} + (4x^3z^3 - 2xy^2)\,\mathbf{j} + (yz^2)\,\mathbf{k}.$$

[24] $\nabla \cdot (f\nabla g) = f(\nabla \cdot \nabla g) + \nabla f \cdot \nabla g$ {Example 5, §18.1} $= f\nabla^2 g + \nabla f \cdot \nabla g$.

[25] $z = f(x,\, y) = x + y \Rightarrow \sqrt{f_x^2 + f_y^2 + 1} = \sqrt{3}$.

R_{xy} is the triangular region bounded by $x = 0$, $y = 0$, and $y = 2 - 2x$. Thus,

$$\iint\limits_S xyz\,dS = \iint\limits_{R_{xy}} xyz\sqrt{3}\,dA = \sqrt{3}\int_0^1\int_0^{2-2x} xy(x + y)\,dy\,dx$$

$$= \sqrt{3}\int_0^1 \left[\tfrac{1}{2}x^2(2 - 2x)^2 + \tfrac{1}{3}x(2 - 2x)^3\right]dx$$

$$= \tfrac{1}{3}\sqrt{3}\int_0^1 (-2x^4 + 12x^3 - 18x^2 + 8x)\,dx = \tfrac{1}{5}\sqrt{3}.$$

[26] $z = f(x,\, y) = \sqrt{4 - y^2} \Rightarrow f_x = 0,\ f_y = \dfrac{-y}{(4 - y^2)^{1/2}}$, and

$\sqrt{f_x^2 + f_y^2 + 1} = \dfrac{2}{(4 - y^2)^{1/2}}$. R_{xy} is the rectangle $0 \le x \le 1$ and $-2 \le y \le 2$.

Thus, $$\iint\limits_S x^2z^2\,dS = \int_{-2}^{2}\int_0^1 x^2(4 - y^2)\left[\frac{2}{(4 - y^2)^{1/2}}\right]dx\,dy$$

$$= \tfrac{2}{3}\int_{-2}^{2}\sqrt{4 - y^2}\,dy = \tfrac{4}{3}\int_0^2\sqrt{4 - y^2}\,dy = \tfrac{4\pi}{3} \text{ {Formula 30}}.$$

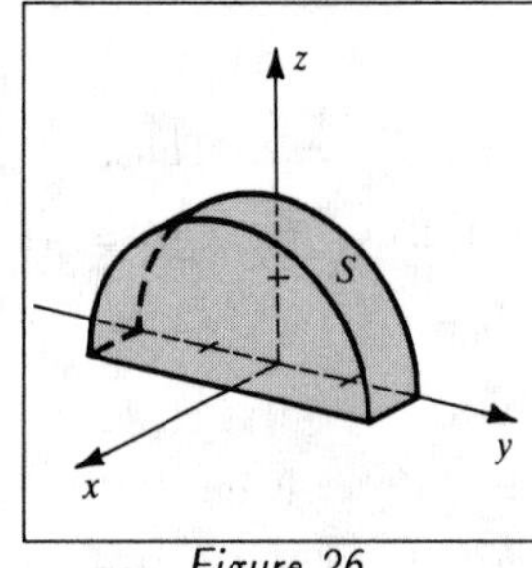

Figure 26

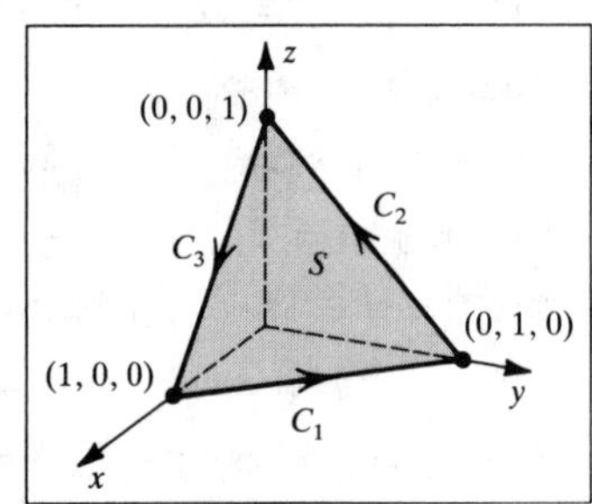

Figure 30

[27] $\mathbf{F} = x^3\,\mathbf{i} + y^3\,\mathbf{j} + z^3\,\mathbf{k} \Rightarrow \operatorname{div}\mathbf{F} = 3(x^2 + y^2 + z^2)$.

Using cylindrical coordinates, $\iint\limits_S \mathbf{F}\cdot\mathbf{n}\,dS = \iiint\limits_Q \operatorname{div}\mathbf{F}\,dV =$

$$3\int_0^{2\pi}\int_0^1\int_0^1 (r^2 + z^2)\,r\,dz\,dr\,d\theta = 3\int_0^{2\pi}\int_0^1 (r^3 + \tfrac{1}{3}r)\,dr\,d\theta = 3\cdot\tfrac{5}{12}\int_0^{2\pi} d\theta = \tfrac{5\pi}{2}.$$

[28] (1) The six surfaces of the box have unit outer normals of $\pm\,\mathbf{i}$, $\pm\,\mathbf{j}$, $\pm\,\mathbf{k}$.

For the top ($z = 3$) and bottom ($z = -3$), $\mathbf{F}\cdot\mathbf{n} = -3$. Thus,

$\iint\limits_{\text{top}} \mathbf{F}\cdot\mathbf{n}\,dS = \iint\limits_{\text{bottom}} \mathbf{F}\cdot\mathbf{n}\,dS = -3\iint\limits_{\text{top}} dS = -3(2\cdot 4) = -24$ since the area of the top is $2\cdot 4 = 8$. For the two sides ($y = \pm\,2$), $\mathbf{F}\cdot\mathbf{n} = 2$ and their contribution is $2(2\cdot 6) = 24$ each. For the front ($x = 1$) and back ($x = -1$), $\mathbf{F}\cdot\mathbf{n} = 2$ and their contribution is $2(4\cdot 6) = 48$.

Thus, $\iint\limits_S \mathbf{F}\cdot\mathbf{n}\,dS = 2(-24) + 2(24) + 2(48) = 96$.

(2) $\iiint\limits_Q \operatorname{div}\mathbf{F}\,dV = \iiint\limits_Q (2 + 1 - 1)\,dV = 2\iiint\limits_Q dV = 2(2\cdot 4\cdot 6) = 96.$

29 (1) $\operatorname{curl}\mathbf{F} = 5\mathbf{i} + (2 - 2y)\,\mathbf{k}$. $z = f(x, y) = (4 - x^2 - y^2)^{1/2} \Rightarrow$

$$f_x = \frac{-x}{(4 - x^2 - y^2)^{1/2}},\ f_y = \frac{-y}{(4 - x^2 - y^2)^{1/2}},\ \text{and } \mathbf{n} = \frac{-f_x\mathbf{i} - f_y\mathbf{j} + \mathbf{k}}{\sqrt{f_x^2 + f_y^2 + 1}}.$$

Then, $\iint\limits_S \operatorname{curl}\mathbf{F}\cdot\mathbf{n}\,dS = \iint\limits_{R_{xy}} \left[\dfrac{5x}{(4 - x^2 - y^2)^{1/2}} + (2 - 2y)\right] dA$

{where R_{xy} is the region bounded by $x^2 + y^2 = 4$}

$$= \int_0^2\int_0^{2\pi}\left[\frac{5r\cos\theta}{(4 - r^2)^{1/2}} + (2 - 2r\sin\theta)\right] r\,d\theta\,dr = \int_0^2 4\pi r\,dr = 8\pi.$$

(2) To calculate $\oint_C \mathbf{F}\cdot d\mathbf{r} = \oint_C y^2\,dx + 2x\,dy + 5y\,dz$, we express C as $x = 2\cos t$, $y = 2\sin t$, $z = 0$ for $0 \le t \le 2\pi$. Then, $\oint_C \mathbf{F}\cdot d\mathbf{r} =$

$$\int_0^{2\pi}(-8\sin^3 t + 8\cos^2 t)\,dt = 8\int_0^{2\pi}\left[-(1 - \cos^2 t)\sin t + \tfrac{1}{2}(1 + \cos 2t)\right] dt =$$

$$8\int_1^1 (1 - u^2)\,du\ \{u = \cos t\} + 4\Big[t + \tfrac{1}{2}\sin 2t\Big]_0^{2\pi} = 0 + 4(2\pi) = 8\pi.$$

30 (1) $\operatorname{curl}\mathbf{F} = -(\mathbf{i} + \mathbf{j} + \mathbf{k})$. An equation of S is $z = f(x, y) = 1 - x - y$,

and hence $\sqrt{f_x^2 + f_y^2 + 1} = \sqrt{3}$. A unit normal is $\mathbf{n} = \frac{1}{\sqrt{3}}(\mathbf{i} + \mathbf{j} + \mathbf{k})$. Thus,

$\iint\limits_S \operatorname{curl}\mathbf{F}\cdot\mathbf{n}\,dS = \iint\limits_{R_{xy}} -\frac{3}{\sqrt{3}}(\sqrt{3})\,dA = -3\iint\limits_{R_{xy}} dA = -3(\frac{1}{2}) = -\frac{3}{2}$,

since the area of the triangular region R_{xy} is $\frac{1}{2}$.

(2) To calculate $\oint_C \mathbf{F}\cdot d\mathbf{r} = \oint_C (x + y)\,dx + (y + z)\,dy + (x + z)\,dz$,

we express $C = C_1 \cup C_2 \cup C_3$ as shown in *Figure 30.*

C_1: $z = 0$, $y = 1 - x$, $0 \le x \le 1$. C_2: $x = 0$, $z = 1 - y$, $0 \le y \le 1$.

C_3: $x = 1 - z$, $y = 0$, $0 \le z \le 1$.

$$\begin{aligned}\oint_C \mathbf{F}\cdot d\mathbf{r} &= \oint_{C_1}\mathbf{F}\cdot d\mathbf{r} + \oint_{C_2}\mathbf{F}\cdot d\mathbf{r} + \oint_{C_3}\mathbf{F}\cdot d\mathbf{r}\\ &= \int_1^0 (1)\,dx + (1 - x)(-dx) + \int_1^0 (1)\,dy + (1 - y)(-dy) +\\ &\qquad\int_1^0 (1 - z)(-dz) + (1)\,dz\\ &= \int_1^0 x\,dx + \int_1^0 y\,dy + \int_1^0 z\,dz = \tfrac{-1}{2} + \tfrac{-1}{2} + \tfrac{-1}{2} = -\tfrac{3}{2}.\end{aligned}$$

Chapter 19: Differential Equations

Exercises 19.1

1 (a) $y' = 3x^2 \Rightarrow y = x^3 + C.$

The solutions are vertical translations of $y = x^3$ with y-intercept C.

(b) $y = 2$ when $x = 0 \Rightarrow 2 = 0^3 + C \Rightarrow C = 2$, and hence, $y = x^3 + 2$.

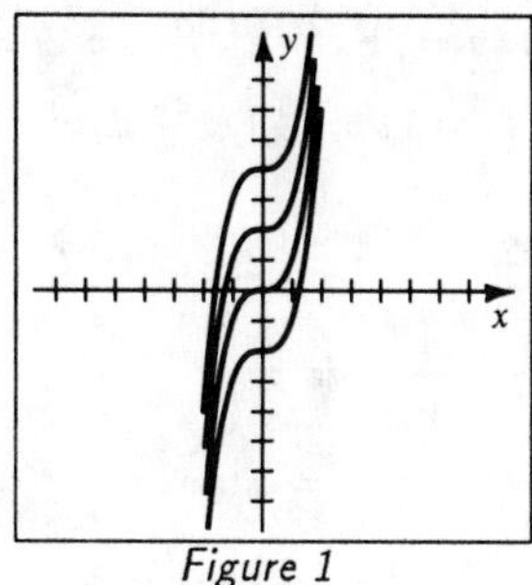

Figure 1

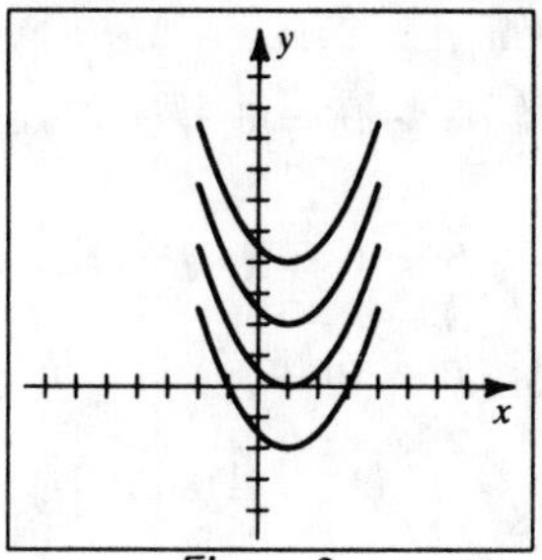

Figure 2

2 (a) $y' = x - 1 \Rightarrow y = \frac{1}{2}x^2 - x + C = \frac{1}{2}(x - 1)^2 + D$. The solutions are vertical translations of $y = \frac{1}{2}(x - 1)^2$ with vertices on the line $x = 1$.

(b) $y = 2$ when $x = 0 \Rightarrow C = 2$, and hence, $y = \frac{1}{2}x^2 - x + 2$.

3 (a) $y' = \dfrac{-x}{\sqrt{4 - x^2}} \Rightarrow y = \sqrt{4 - x^2} + C$. The solutions are vertical translations of the semicircle $y = \sqrt{4 - x^2}$ with centers on the y-axis.

(b) $y = 2$ when $x = 0 \Rightarrow C = 0$, and hence $y = \sqrt{4 - x^2}$.

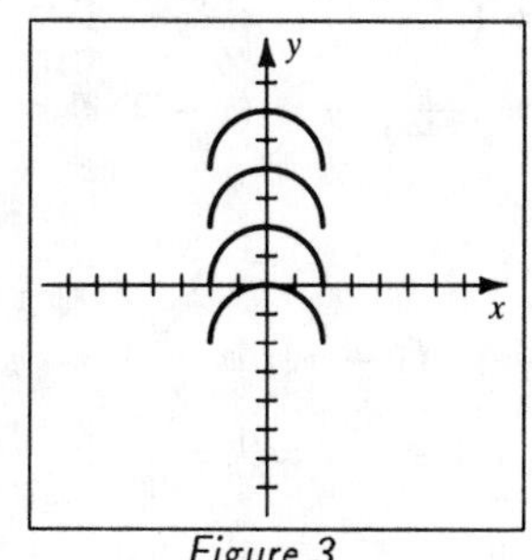

Figure 3

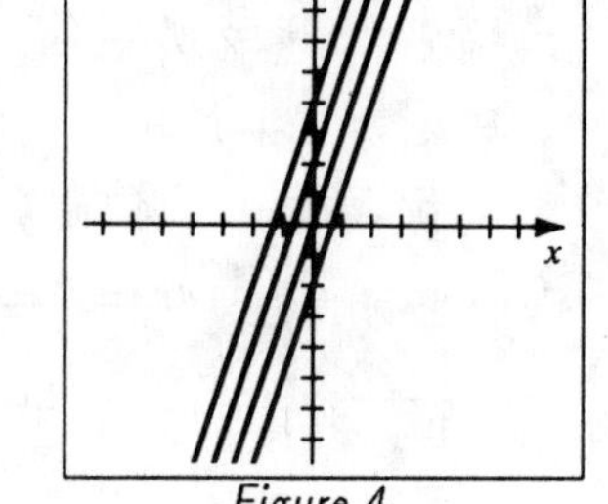

Figure 4

4 (a) $y' = 3 \Rightarrow y = 3x + C$. The solutions are lines with slope 3.

(b) $y = 2$ when $x = 0 \Rightarrow C = 2$, and hence, $y = 3x + 2$.

5 $y = C_1e^x + C_2e^{2x} \Rightarrow y' = C_1e^x + 2C_2e^{2x}$ and $y'' = C_1e^x + 4C_2e^{2x}$.

$$y'' - 3y' + 2y = C_1e^x + 4C_2e^{2x} - 3C_1e^x - 6C_2e^{2x} + 2C_1e^x + 2C_2e^{2x} = 0.$$

6 $y = Ce^{-3x} \Rightarrow y' = -3Ce^{-3x} = -3y$. $y' + 3y = -3y + 3y = 0$.

7 $y = Cx^{-2/3} \Rightarrow y' = -\frac{2}{3}Cx^{-5/3}$.

$$2xy^3 + 3x^2y^2y' = 2xC^3x^{-2} + 3x^2C^2x^{-4/3}(-\tfrac{2}{3}Cx^{-5/3}) = 2C^3x^{-1} - 2C^3x^{-1} = 0.$$

8 $y = Cx^3 \Rightarrow y' = 3Cx^2 \Rightarrow y'' = 6Cx \Rightarrow y''' = 6C$. $x^3y''' + x^2y'' - 3xy' - 3y =$

$$x^3(6C) + x^2(6Cx) - 3x(3Cx^2) - 3(Cx^3) = 6Cx^3 + 6Cx^3 - 9Cx^3 - 3Cx^3 = 0.$$

[9] Differentiating $(y^2 - x^2 - xy = c)$ implicitly yields $2yy' - 2x - y - xy' = 0 \Rightarrow$

$$(2y - x)y' - (2x + y) = 0 \Rightarrow (x - 2y)y' + 2x + y = 0.$$

[10] Differentiating $(x^2 - y^2 = c)$ implicitly yields $2x - 2yy' = 0 \Rightarrow yy' = x$.

Note: Let k and C denote constants in the following exercises.

[11] $\sec x\,dy - 2y\,dx = 0 \Rightarrow \frac{1}{2y}\,dy = \cos x\,dx \Rightarrow \frac{1}{2}\ln|y| = \sin x + k \Rightarrow$

$$\ln|y| = 2\sin x + 2k \Rightarrow |y| = e^{2\sin x + 2k} \Rightarrow y = \pm\, e^{2k}\, e^{2\sin x} \Rightarrow y = Ce^{2\sin x}.$$

[12] $x^2\,dy - \csc 2y\,dx = 0 \Rightarrow \sin 2y\,dy = x^{-2}\,dx \Rightarrow -\frac{1}{2}\cos 2y = -1/x + k \Rightarrow$

$$\cos 2y = 2/x + C.$$

[13] $x\,dy - y\,dx = 0 \Rightarrow \frac{1}{y}\,dy = \frac{1}{x}\,dx \Rightarrow \ln|y| = \ln|x| + k \Rightarrow y = \pm\, e^k\, x = Cx.$

[14] $(4 + y^2)\,dx + (9 + x^2)\,dy = 0 \Rightarrow \dfrac{1}{4 + y^2}\,dy = -\dfrac{1}{9 + x^2}\,dx \Rightarrow$

$$\tfrac{1}{2}\tan^{-1}(\tfrac{1}{2}y) = -\tfrac{1}{3}\tan^{-1}(\tfrac{1}{3}x) + k \Rightarrow y = 2\tan\left[C - \tfrac{2}{3}\tan^{-1}(\tfrac{1}{3}x)\right].$$

[15] $3y\,dx + (xy + 5x)\,dy = 0 \Rightarrow \frac{3}{x}\,dx + \left(1 + \frac{5}{y}\right)dy = 0$ {divide by xy} $\Rightarrow$

$3\ln|x| + y + 5\ln|y| = k \Rightarrow \ln|x^3| + y + \ln|y^5| = k \Rightarrow$

$|x^3| \cdot e^y \cdot |y^5| = e^k \Rightarrow x^3 e^y y^5 = C.$

Assume $x \neq 0$, but note that $y = 0$ is a solution to the original equation.

[16] $(xy - 4x)\,dx + (x^2y + y)\,dy = 0 \Rightarrow x(y - 4)\,dx + y(x^2 + 1)\,dy = 0 \Rightarrow$

$\dfrac{x}{x^2 + 1}\,dx + \dfrac{y}{y - 4}\,dy = 0 \Rightarrow \dfrac{x}{x^2 + 1}\,dx + \left(1 + \dfrac{4}{y - 4}\right)dy = 0 \Rightarrow$

$$\tfrac{1}{2}\ln(x^2 + 1) + y + 4\ln|y - 4| = k \Rightarrow (x^2 + 1)^{1/2}\, e^y\, (y - 4)^4 = e^k = C.$$

[17] $y' = x - 1 + xy - y \Rightarrow y' = (x - 1)(1 + y) \Rightarrow \dfrac{1}{1 + y}\,dy = (x - 1)\,dx \Rightarrow$

$\ln|1 + y| = \frac{1}{2}x^2 - x + k \Rightarrow$

$$|y + 1| = e^{(x^2/2) - x + k} \Rightarrow y = -1 \pm e^k e^{(x^2/2) - x} = -1 + Ce^{(x^2/2) - x}.$$

[18] $(y + yx^2)\,dy + (x + xy^2)\,dx = 0 \Rightarrow y(1 + x^2)\,dy + x(1 + y^2)\,dx = 0 \Rightarrow$

$\dfrac{y}{1 + y^2}\,dy + \dfrac{x}{1 + x^2}\,dx = 0 \Rightarrow \frac{1}{2}\ln(1 + y^2) + \frac{1}{2}\ln(1 + x^2) = k \Rightarrow$

$$(1 + y^2)(1 + x^2) = e^{2k} = C, \text{ where } C \geq 1.$$

[19] $e^{x+2y}\,dx - e^{2x-y}\,dy = 0 \Rightarrow e^x e^{2y}\,dx - e^{2x} e^{-y}\,dy = 0 \Rightarrow e^{-x}\,dx = e^{-3y}\,dy \Rightarrow$

$-e^{-x} = -\frac{1}{3}e^{-3y} + k \Rightarrow e^{-3y} = 3e^{-x} - 3k \Rightarrow -3y = \ln(C + 3e^{-x}) \Rightarrow$

$$y = -\tfrac{1}{3}\ln(C + 3e^{-x}).$$

[20] $\cos x\,dy - y\,dx = 0 \Rightarrow \frac{1}{y}\,dy - \sec x\,dx = 0 \Rightarrow \ln|y| - \ln|\sec x + \tan x| = k \Rightarrow$

$$\ln\left|\frac{y}{\sec x + \tan x}\right| = k \Rightarrow \frac{y}{\sec x + \tan x} = \pm\, e^k \Rightarrow y = C(\sec x + \tan x).$$

[21] $y(1 + x^3)\,y' + x^2(1 + y^2) = 0 \Rightarrow \dfrac{y}{1 + y^2}\,dy + \dfrac{x^2}{1 + x^3}\,dx = 0 \Rightarrow$

$\frac{1}{2}\ln(1 + y^2) + \frac{1}{3}\ln|1 + x^3| = k \Rightarrow (1 + y^2)^{1/2}|1 + x^3|^{1/3} = e^k \Rightarrow$

$$(1 + y^2)^{1/2} = \pm\, e^k\,(1 + x^3)^{-1/3} \Rightarrow y^2 = C(1 + x^3)^{-2/3} - 1.$$

[22] $x^2 y' - yx^2 = y \Rightarrow x^2\,dy - y(x^2 + 1)\,dx = 0 \Rightarrow \frac{1}{y}\,dy - \left(1 + \frac{1}{x^2}\right)dx = 0 \Rightarrow$

$$\ln|y| - \left(x - \frac{1}{x}\right) = k \Rightarrow |y| = e^{x-(1/x)+k} \Rightarrow y = \pm\, e^k e^{x-(1/x)} = Ce^{x-(1/x)}.$$

[23] $x\tan y - y'\sec x = 0 \Rightarrow x\tan y\,dx - \sec x\,dy = 0 \Rightarrow x\cos x\,dx - \cot y\,dy = 0 \Rightarrow$

$$x\sin x + \cos x - \ln|\sin y| = C.$$

[24] $xy + y'\,e^{(x^2)}\ln y = 0 \Rightarrow xy\,dx + e^{(x^2)}\ln y\,dy = 0 \Rightarrow$

$$xe^{-x^2}\,dx + \frac{\ln y}{y}\,dy = 0 \Rightarrow -\tfrac{1}{2}e^{-x^2} + \tfrac{1}{2}(\ln y)^2 = k \Rightarrow (\ln y)^2 - e^{-x^2} = C.$$

[25] $e^y\sin x\,dx - \cos^2 x\,dy = 0 \Rightarrow \sec x\tan x\,dx - e^{-y}\,dy = 0 \Rightarrow \sec x + e^{-y} = C.$

[26] $\sin y\cos x\,dx + (1 + \sin^2 x)\,dy = 0 \Rightarrow \frac{\cos x}{1 + \sin^2 x}\,dx + \csc y\,dy = 0 \Rightarrow$

$$\tan^{-1}(\sin x) + \ln|\csc y - \cot y| = C.$$

[27] $2y^2 y' = 3y - y' \Rightarrow (2y^2 + 1)\,dy - 3y\,dx = 0 \Rightarrow \left(2y + \frac{1}{y}\right)dy - 3\,dx = 0 \Rightarrow$

$y^2 + \ln|y| - 3x = C$. Letting $y = 1$ and $x = 3 \Rightarrow C = -8$. Thus,

$y^2 + \ln y = 3x - 8$ for $y > 0$. (We have $y > 0$ since the condition has $y = 1 > 0$.)

[28] $x^{1/2}y' - y^{1/2} = xy^{1/2} \Rightarrow x^{1/2}\,dy = y^{1/2}(x + 1)\,dx \Rightarrow$

$y^{-1/2}\,dy = (x^{1/2} + x^{-1/2})\,dx \Rightarrow 2y^{1/2} = \frac{2}{3}x^{3/2} + 2x^{1/2} + C.$

Letting $y = 4$ and $x = 9 \Rightarrow C = -20$. Thus, $y = (\frac{1}{3}x^{3/2} + x^{1/2} - 10)^2$.

[29] $x\,dy - (2x + 1)e^{-y}\,dx = 0 \Rightarrow e^y\,dy - \left(2 + \frac{1}{x}\right)dx = 0 \Rightarrow e^y - 2x - \ln|x| = C.$

Letting $y = 2$ and $x = 1 \Rightarrow C = e^2 - 2$. Thus, $y = \ln(2x + \ln x + e^2 - 2)$

for $x > 0$. (We have $x > 0$ since the condition has $x = 1 > 0$.)

[30] $\sec 2y\,dx - \cos^2 x\,dy = 0 \Rightarrow \sec^2 x\,dx - \cos 2y\,dy = 0 \Rightarrow \tan x - \frac{1}{2}\sin 2y = C.$

Letting $y = \frac{\pi}{6}$ and $x = \frac{\pi}{4} \Rightarrow C = \frac{1}{4}(4 - \sqrt{3})$. Thus, $4\tan x = 2\sin 2y + 4 - \sqrt{3}$.

[31] $(xy + x)\,dx + \sqrt{4 + x^2}\,dy = 0 \Rightarrow \frac{x}{(4 + x^2)^{1/2}}\,dx + \frac{1}{y + 1}\,dy = 0 \Rightarrow$

$\sqrt{4 + x^2} + \ln|y + 1| = C$. Letting $y = 1$ and $x = 0 \Rightarrow C = 2 + \ln 2$. Thus,

$\ln|y + 1| = 2 + \ln 2 - \sqrt{4 + x^2} \Rightarrow |y + 1| = 2e^{2-\sqrt{4+x^2}} \Rightarrow y = 2e^{2-\sqrt{4+x^2}} - 1$

for $y > -1$. (We pick the positive solution since the condition has $y = 1$, and hence, $y + 1 > 0$.)

[32] $x\,dy - \sqrt{1 - y^2}\,dx = 0 \Rightarrow \frac{dy}{\sqrt{1 - y^2}} - \frac{dx}{x} = 0 \Rightarrow \sin^{-1}y - \ln|x| = C.$

Letting $y = \frac{1}{2}$ and $x = 1 \Rightarrow C = \frac{\pi}{6}$. Thus, $\sin^{-1}y = \frac{\pi}{6} + \ln x$.

[33] $\cot x\,dy - (1 + y^2)\,dx = 0 \Rightarrow \frac{1}{1 + y^2}\,dy - \tan x\,dx = 0 \Rightarrow \tan^{-1}y - \ln|\sec x| = C.$

Letting $y = 1$ and $x = 0 \Rightarrow C = \frac{\pi}{4}$. Thus, $\tan^{-1}y - \ln\sec x = \frac{\pi}{4}$.

[34] $\csc y\,dx - e^x\,dy = 0 \Rightarrow e^{-x}\,dx - \sin y\,dy = 0 \Rightarrow -e^{-x} + \cos y = C$. Letting

$y = 0$ and $x = 0 \Rightarrow C = 0$. Thus, $\cos y = e^{-x}$, where $x \geq 0$ since $\cos y \leq 1$.

[35] Differentiating ($x^2 - y^2 = c$) implicitly yields $2x - 2yy' = 0 \Rightarrow y' = \frac{x}{y}$.

Orthogonal trajectories will have slopes $y' = -\frac{y}{x} \Rightarrow \frac{1}{y}\,dy + \frac{1}{x}\,dx = 0 \Rightarrow$

$\ln|y| + \ln|x| = C \Rightarrow |xy| = e^C \Rightarrow xy = k$; hyperbolas.

[36] From Exercise 35,

we see that the orthogonal trajectories will be $x^2 - y^2 = k$; hyperbolas.

Note: In Exercises 37–40, it is easiest to solve for c first,

thereby eliminating c upon differentiation.

[37] Differentiating implicitly, $\frac{y^2}{x} = c \Rightarrow \frac{(x)(2yy') - (y^2)(1)}{x^2} = 0 \Rightarrow$

$2xy\,y' - y^2 = 0 \Rightarrow y' = \frac{y}{2x}$. Orthogonal trajectories will have slopes

$y' = -\frac{2x}{y} \Rightarrow y\,dy + 2x\,dx = 0 \Rightarrow y^2 + 2x^2 = k$; ellipses.

[38] Differentiating implicitly, $\frac{y}{x^2} = c \Rightarrow \frac{(x^2)(y') - (y)(2x)}{x^4} = 0 \Rightarrow x^2y' - 2xy = 0 \Rightarrow$

$y' = \frac{2y}{x}$. Orthogonal trajectories will have slopes $y' = -\frac{x}{2y} \Rightarrow 2y\,dy + x\,dx = 0 \Rightarrow$

$2y^2 + x^2 = k$; ellipses.

[39] Differentiating implicitly, $\frac{y^2}{x^3} = c \Rightarrow \frac{(x^3)(2y\,y') - (y^2)(3x^2)}{x^6} = 0 \Rightarrow$

$2yx^3y' = 3x^2y^2 \Rightarrow y' = \frac{3y}{2x}$. Orthogonal trajectories will have slopes $y' = -\frac{2x}{3y} \Rightarrow$

$3y\,dy + 2x\,dx = 0 \Rightarrow 3y^2 + 2x^2 = k$; ellipses.

[40] Differentiating implicitly, $ye^x = c \Rightarrow ye^x + y'e^x = 0 \Rightarrow y' = -y$. Orthogonal

trajectories will have slopes $y' = \frac{1}{y} \Rightarrow y\,dy = 1\,dx \Rightarrow y^2 = 2x + k$; parabolas.

[41] (a) Step 1: Let $f(x, y) = xy$, $k = 0$, $a = x_0 = 0$, $b = 1$, and $n = 8$.

Hence, $h = \frac{b-a}{n} = \frac{1-0}{8} = \frac{1}{8}$. Step 2: Let $y_0 = 1$.

Step 3: $y_1 = y_0 + hf(x_0, y_0) = 1 + \frac{1}{8}(0)(1) = 1$. Step 4: $x_1 \not\approx b$, so go back to Step 3 with $x_1 = \frac{1}{8}$. $y_2 = y_1 + hf(x_1, y_1) = 1 + \frac{1}{8}(\frac{1}{8})(1) = 1.015625$.

In a similar manner, $y_3 \approx 1.0474$, $y_4 = 1.0965$, $y_5 = 1.1650$,

$y_6 = 1.2560$, $y_7 \approx 1.3738$, and $y_8 \approx 1.5240$.

(b) $y' = xy \Rightarrow \frac{1}{y}\,dy = x\,dx \Rightarrow \ln y = \frac{1}{2}x^2 + C$ {no absolute value since $y > 0$} $\Rightarrow$

$y = e^{x^2/2+C}$. $y = 1$ at $x = 0 \Rightarrow C = 0$.

Thus, $y = e^{x^2/2}$ and $y = e^{1/2} \approx 1.648721$ at $x = 1$.

[42] (a) $y_1 = y_0 + hf(x_0, y_0) = y_0 + \frac{b-a}{n}(1/y_0) = 1 + \frac{1}{8}(1/1) = 1.125$.

$y_2 = y_1 + hf(x_1, y_1) = 1.125 + \frac{1}{8}(1/1.125) = 1.2361$.

In a similar manner, $y_3 \approx 1.3372$, $y_4 = 1.4307$, $y_5 = 1.5181$,

$y_6 = 1.6004$, $y_7 \approx 1.6785$, and $y_8 \approx 1.7530$.

(b) $y' = 1/y \Rightarrow y\,dy = dx \Rightarrow \frac{1}{2}y^2 = x + C \Rightarrow y = \pm\sqrt{2x + 2C}$. $y = 1$ at $x = 0 \Rightarrow C = \frac{1}{2}$. Thus, $y = \sqrt{2x + 1}$ and $y = \sqrt{3} \approx 1.732051$ at $x = 1$.

[43] $y_1 = y_0 + \frac{1}{2}h\big[f(x_0, y_0) + f(x_0 + h, y_0 + hf(x_0, y_0))\big] =$

$1 + \frac{1}{16}\big[f(0, 1) + f(0 + \frac{1}{8}, 1 + \frac{1}{8}f(0, 1))\big] = 1 + \frac{1}{16}\big[0 + f(\frac{1}{8}, 1)\big] = 1 + \frac{1}{16}\cdot\frac{1}{8} =$

1.0078125. In a similar manner, $y_2 \approx 1.031679$, $y_3 \approx 1.072735$, $y_4 \approx 1.132971$, $y_5 \approx 1.215399$, $y_6 \approx 1.324299$, $y_7 \approx 1.465587$, and $y_8 \approx 1.647355$.

[44] $y_1 = y_0 + \frac{1}{2}h\big[f(x_0, y_0) + f(x_0 + h, y_0 + hf(x_0, y_0))\big] =$

$1 + \frac{1}{16}\big[f(0, 1) + f(0 + \frac{1}{8}, 1 + \frac{1}{8}f(0, 1))\big] = 1 + \frac{1}{16}\big[1 + f(\frac{1}{8}, \frac{9}{8})\big] = 1 + \frac{1}{16}(1 + \frac{8}{9})$

≈ 1.118056. In a similar manner, $y_2 \approx 1.224775$, $y_3 \approx 1.322909$, $y_4 \approx 1.414249$, $y_5 \approx 1.500035$, $y_6 \approx 1.581174$, $y_7 \approx 1.658347$, and $y_8 \approx 1.732085$.

Exercises 19.2

Note: Let *IF* denote the integrating factor.

[1] The form in (19.1) is $y' + P(x)y = Q(x)$. $y' + 2y = e^{2x} \Rightarrow P(x) = 2$ and $IF = e^{\int 2\,dx} = e^{2x}$. Multiplying both sides by *IF* gives us $y'e^{2x} + 2e^{2x}y = e^{4x}$. Since the left side of the last equation equals $D_x(ye^{2x})$, we have

$$D_x(ye^{2x}) = e^{4x} \Rightarrow ye^{2x} = \tfrac{1}{4}e^{4x} + C \Rightarrow y = \tfrac{1}{4}e^{2x} + Ce^{-2x}.$$

[2] $IF = e^{\int -3\,dx} = e^{-3x}$. $y'e^{-3x} - 3ye^{-3x} = 2e^{-3x} \Rightarrow D_x(ye^{-3x}) = 2e^{-3x} \Rightarrow$

$$ye^{-3x} = -\tfrac{2}{3}e^{-3x} + C \Rightarrow y = Ce^{3x} - \tfrac{2}{3}.$$

[3] $y' - \frac{3}{x}y = x^4 \Rightarrow IF = e^{\int(-3/x)\,dx} = |x|^{-3}$. $x^{-3}y' - 3x^{-4}y = x \Rightarrow$

$$D_x(yx^{-3}) = x \Rightarrow x^{-3}y = \tfrac{1}{2}x^2 + C \Rightarrow y = \tfrac{1}{2}x^5 + Cx^3.$$

[4] $IF = e^{\int \cot x\,dx} = e^{\ln|\sin x|} = |\sin x|$.

$$y'\sin x + y\cos x = 1 \Rightarrow D_x(y\sin x) = 1 \Rightarrow y\sin x = x + C \Rightarrow y = (x + C)\csc x.$$

[5] $y' + \frac{y}{x} = \frac{e^x}{x} - 1 \Rightarrow IF = e^{\int(1/x)\,dx} = |x|$, and the original equation is already in proper form. $xy' + y = e^x - x \Rightarrow xy = e^x - \frac{1}{2}x^2 + C \Rightarrow y = \frac{e^x}{x} - \frac{1}{2}x + \frac{C}{x}$.

[6] $y' + \left(\frac{1}{x} + 1\right)y = \frac{5}{x} \Rightarrow IF = e^{\int(1/x + 1)\,dx} = |x|e^x$.

$$xe^x y' + (e^x + xe^x)y = 5e^x \Rightarrow xe^x y = 5e^x + C \Rightarrow y = \tfrac{5}{x} + \tfrac{Ce^{-x}}{x}.$$

[7] $x^2y' + (2xy - e^x) = 0 \Rightarrow y' + \frac{2}{x}y = \frac{1}{x^2}e^x \Rightarrow IF = e^{\int(2/x)\,dx} = x^2$.

$$x^2y' + 2xy = e^x \Rightarrow x^2y = e^x + C \Rightarrow y = (e^x + C)/x^2.$$

[8] $x^2y' + (x - 3xy + 1) = 0 \Rightarrow y' - \frac{3}{x}y = -\frac{1 + x}{x^2} \Rightarrow IF = e^{\int(-3/x)\,dx} = |x|^{-3}$.

$x^{-3}y' - 3x^{-4}y = -x^{-5} - x^{-4} \Rightarrow x^{-3}y = \frac{1}{4}x^{-4} + \frac{1}{3}x^{-3} + C \Rightarrow$

$$y = \tfrac{1}{4}x^{-1} + \tfrac{1}{3} + Cx^3.$$

[9] $IF = e^{\int \cot x\,dx} = |\sin x|$. $y' \sin x + y \cos x = 4x^2 \Rightarrow y \sin x = \frac{4}{3}x^3 + C \Rightarrow$
$y = (\frac{4}{3}x^3 + C)\csc x$.

[10] $IF = e^{\int \tan x\,dx} = |\sec x|$. $y' \sec x + y \tan x \sec x = \tan x \Rightarrow$
$y \sec x = \ln|\sec x| + C \Rightarrow y = (\ln|\sec x| + C)\cos x$.

[11] $y' \cos x + y \sin x = 2 \Rightarrow y' + y \tan x = 2\sec x \Rightarrow IF = e^{\int \tan x\,dx} = |\sec x|$.
$y' \sec x + y \sec x \tan x = 2\sec^2 x \Rightarrow y \sec x = 2\tan x + C \Rightarrow y = 2\sin x + C\cos x$.

[12] $x^3 y' + (x^2 y - 1) = 0 \Rightarrow y' + \frac{1}{x}y = x^{-3} \Rightarrow IF = e^{\int (1/x)\,dx} = |x|$.
$xy' + y = x^{-2} \Rightarrow xy = -\frac{1}{x} + C \Rightarrow y = -\frac{1}{x^2} + \frac{C}{x} = (Cx - 1)/x^2$.

[13] $-xy' + (x^2 \cos x + y) = 0 \Rightarrow y' - \frac{1}{x}y = x \cos x \Rightarrow IF = e^{\int (-1/x)\,dx} = |x|^{-1}$.
$x^{-1}y' - x^{-2}y = \cos x \Rightarrow x^{-1}y = \sin x + C \Rightarrow y = x \sin x + Cx$.

[14] $IF = e^{\int 1\,dx} = e^x$. $e^x y' + e^x y = e^x \sin x \Rightarrow$
$e^x y = \frac{1}{2}(e^x \sin x - e^x \cos x) + C$ {Formula 98} $\Rightarrow y = \frac{1}{2}(\sin x - \cos x) + Ce^{-x}$.

[15] $y' + \left(\frac{2}{x} + 3\right)y = e^{-3x} \Rightarrow IF = e^{\int [(2/x) + 3]\,dx} = x^2 e^{3x}$.
$x^2 e^{3x} y' + (2xe^{3x} + 3x^2 e^{3x})\,y = x^2 \Rightarrow x^2 e^{3x} y = \frac{1}{3}x^3 + C \Rightarrow$
$y = \frac{1}{3}xe^{-3x} + Cx^{-2}e^{-3x} = \left(\frac{1}{3}x + \frac{C}{x^2}\right)e^{-3x}$.

[16] $y' + \frac{5}{x+4}y = x + 4 \Rightarrow IF = e^{\int [5/(x+4)]\,dx} = |x + 4|^5$.
$(x + 4)^5 y' + 5(x + 4)^4 y = (x + 4)^6 \Rightarrow (x + 4)^5 y = \frac{1}{7}(x + 4)^7 + C \Rightarrow$
$y = \frac{1}{7}(x + 4)^2 + C(x + 4)^{-5}$.

[17] $y' + 2xy = 3x \Rightarrow IF = e^{\int 2x\,dx} = e^{(x^2)}$.
$e^{(x^2)}y' + 2xe^{(x^2)}y = 3xe^{(x^2)} \Rightarrow ye^{(x^2)} = \frac{3}{2}e^{(x^2)} + C \Rightarrow y = \frac{3}{2} + Ce^{-x^2}$.

[18] $IF = e^{\int -5\,dx} = e^{-5x}$.
$y'e^{-5x} - 5e^{-5x}y = 1 \Rightarrow ye^{-5x} = x + C \Rightarrow y = (x + C)e^{5x}$.

[19] $y' \tan x + y = \sin x \Rightarrow y' + y \cot x = \cos x \Rightarrow IF = e^{\int \cot x\,dx} = |\sin x|$.
$y' \sin x + y \cos x = \sin x \cos x \Rightarrow y \sin x = \frac{1}{2}\sin^2 x + C \Rightarrow y = \frac{1}{2}\sin x + C\csc x$.

[20] $y' \cos x - y \sin x = -e^{-x}$ {this is in integrable form} $\Rightarrow$
$y \cos x = e^{-x} + C \Rightarrow y = \sec x(C + e^{-x})$.

[21] $IF = e^{\int 3x^2\,dx} = e^{(x^3)}$. $e^{(x^3)}y' + 3x^2 e^{(x^3)}y = x^2 e^{(x^3)} + 1 \Rightarrow$
$ye^{(x^3)} = \frac{1}{3}e^{(x^3)} + x + C \Rightarrow y = \frac{1}{3} + (x + C)e^{-x^3}$.

[22] $IF = e^{\int \tan x\,dx} = |\sec x|$. $y' \sec x + y \sec x \tan x = \cos^2 x \Rightarrow$
$y \sec x = \frac{1}{2}x + \frac{1}{4}\sin 2x + C \Rightarrow y = (\frac{1}{2}x + \frac{1}{4}\sin 2x + C)\cos x$.

[23] $y' - \frac{1}{x}y = x + 1 \Rightarrow IF = e^{\int (-1/x)\,dx} = |x|^{-1}$.
$x^{-1}y' - x^{-2}y = 1 + x^{-1} \Rightarrow x^{-1}y = x + \ln|x| + C \Rightarrow y = x(x + \ln|x| + C)$.
Letting $y = 2$ and $x = 1 \Rightarrow C = 1$. Thus, $y = x(x + \ln x + 1)$, for $x > 0$.

[24] $IF = e^{\int 2\,dx} = e^{2x}$.

$e^{2x}y' + 2e^{2x}y = e^{-x} \Rightarrow ye^{2x} = -e^{-x} + C \Rightarrow y = -e^{-3x} + Ce^{-2x}$.

Letting $y = 2$ and $x = 0 \Rightarrow C = 3$. Thus, $y = -e^{-3x} + 3e^{-2x}$.

[25] $y' + \left(\frac{1}{x} + 1\right)y = \frac{1}{x}e^{-x} \Rightarrow IF = e^{\int[(1/x)+1]\,dx} = xe^{x}$.

$xe^{x}y' + (e^{x} + xe^{x})y = 1 \Rightarrow xe^{x}y = x + C \Rightarrow y = e^{-x} + Cx^{-1}e^{-x}$.

Letting $y = 0$ and $x = 1 \Rightarrow C = -1$. Thus, $y = e^{-x}(1 - x^{-1})$.

[26] $IF = e^{\int 2x\,dx} = e^{(x^2)}$. $e^{(x^2)}y' + 2xe^{(x^2)}y = xe^{(x^2)} + 1 \Rightarrow$

$ye^{(x^2)} = \frac{1}{2}e^{(x^2)} + x + C \Rightarrow y = \frac{1}{2} + xe^{-x^2} + Ce^{-x^2}$.

Letting $y = 1$ and $x = 0 \Rightarrow C = \frac{1}{2}$. Thus, $y = \frac{1}{2} + xe^{-x^2} + \frac{1}{2}e^{-x^2}$.

[27] (a) $\frac{dQ}{dt} + \frac{1}{RC}Q = \frac{V}{R} \Rightarrow IF = e^{\int(1/(RC))\,dt} = e^{t/RC}$.

$e^{t/RC}\frac{dQ}{dt} + \frac{1}{RC}e^{t/RC}Q = e^{t/RC}\left(\frac{V}{R}\right) \Rightarrow Qe^{t/RC} = CVe^{t/RC} + k \Rightarrow$

$Q = CV + ke^{-t/RC}$. $Q(0) = 0 \Rightarrow k = -CV$. Thus, $Q = CV(1 - e^{-t/RC})$.

(b) $R\,dQ + \left(\frac{Q}{C} - V\right)dt = 0 \Rightarrow \frac{C}{Q - CV}dQ + \frac{1}{R}dt = 0 \Rightarrow$

$C\ln(Q - CV) + \frac{t}{R} = k \Rightarrow \ln(Q - CV) = \frac{k}{C} - \frac{t}{RC} \Rightarrow$

$Q - CV = e^{k/C}e^{-t/RC} \Rightarrow Q = CV + me^{-t/RC}$ $\{m = e^{k/C}\}$.

$Q(0) = 0 \Rightarrow m = -CV$. Thus, $Q = CV(1 - e^{-t/RC})$.

[28] (a) V constant $\Rightarrow \frac{dV}{dt} = 0$. $\frac{dI}{dt} + \frac{1}{RC}I = 0 \Rightarrow IF = e^{\int(1/(RC))\,dt} = e^{t/RC}$.

$e^{t/RC}\frac{dI}{dt} + \frac{1}{RC}e^{t/RC}I = 0 \Rightarrow Ie^{t/RC} = k \Rightarrow I = ke^{-t/RC}$.

$I(0) = I_0 \Rightarrow I = I_0e^{-t/RC}$.

(b) $\frac{1}{I}\,dI + \frac{1}{RC}dt = 0 \Rightarrow \ln I + \frac{t}{RC} = k \Rightarrow \ln I = k - \frac{t}{RC} \Rightarrow I = e^{k}e^{-t/RC}$.

$I(0) = I_0 \Rightarrow I = I_0e^{-t/RC}$.

[29] The rate at which salt comes into the tank is 6 gal/min × $\frac{1}{3}$ lb/gal = 2 lb/min.

Since $f(t)$ denotes the total amount of salt in the tank, the rate at which salt leaves the tank is 6 gal/min × $\frac{1}{80}f(t)$ lb/gal = $\frac{3}{40}f(t)$ lb/gal.

The net rate of change is $f'(t) = 2 - \frac{3}{40}f(t) \Rightarrow f'(t) + 0.075f(t) = 2 \Rightarrow$

$IF = e^{0.075t}$, $f(t)e^{0.075t} = \frac{2}{0.075}e^{0.075t} + C$, and $f(t) = \frac{80}{3} + Ce^{-0.075t}$.

$f(0) = K \Rightarrow C = K - \frac{80}{3}$. Thus, $f(t) = \frac{80}{3}(1 - e^{-0.075t}) + Ke^{-0.075t}$ lb.

[30] The force on the object is given by $F = e^{-t} - 2v$. Since $F = ma = m\frac{dv}{dt}$,

we have $m\frac{dv}{dt} + 2v = e^{-t} \Rightarrow \frac{dv}{dt} + \frac{2}{m}v = \frac{1}{m}e^{-t} \Rightarrow IF = e^{2t/m}$.

$e^{2t/m}\frac{dv}{dt} + \frac{2}{m}e^{2t/m}v = \frac{1}{m}e^{t[(2-m)/m]} \Rightarrow e^{2t/m}v = \frac{1}{2-m}e^{t[(2-m)/m]} + k \Rightarrow$

$v = \frac{1}{2-m}e^{-t} + ke^{-2t/m}$. $v(0) = 0 \Rightarrow k = -\frac{1}{2-m} \Rightarrow$

$$v = \frac{1}{2-m}(e^{-t} - e^{-2t/m}).$$

[31] (a) $f'(t) = k[M - f(t)] \Rightarrow f'(t) + kf(t) = kM$.

$IF = e^{kt}$, $f(t)\,e^{kt} = Me^{kt} + C$, and $f(t) = M + Ce^{-kt}$. $f(1) = A \Rightarrow$

$C = (A - M)\,e^{k}$. Thus, $f(t) = M + (A - M)\,e^{k(1-t)}$ (k a constant).

(b) $M = 30$, $A = 5 \Rightarrow f(t) = 30 - 25e^{k(1-t)}$. $f(2) = 8 \Rightarrow 8 = 30 - 25e^{-k} \Rightarrow$

$e^{k} = \frac{25}{22}$. So $f(t) = 30 - 25(\frac{25}{22})^{1-t}$ and $f(20) \approx 27.8 \approx 28$ items.

[32] (a) The rate at which CO enters the room is $(0.05)(0.12) = 0.006$ ft^3/min. The rate at which CO leaves the room is $\frac{f(t)}{10 \times 15 \times 8} \cdot (0.12) = 0.0001\,f(t)$ ft^3/min.

Thus, the net rate of change is $f'(t) = 0.006 - 0.0001\,f(t) \Rightarrow$

$f'(t) + 0.0001\,f(t) = 0.006 \Rightarrow IF = e^{0.0001t}$ and $f(t) = 60 + ke^{-0.0001t}$.

At $t = 0$, 0.001% of the 1200 ft^3 is CO.

Thus, $f(0) = 0.012$ ft^3 $\Rightarrow k = -59.988$ and $f(t) = 60 - 59.988e^{-0.0001t}$.

(b) $60 - 59.988e^{-0.0001t} = (0.00015)(1200) = 0.18 \Rightarrow$

$$t = -10{,}000\ln\left(\tfrac{59.82}{59.988}\right) \approx 28 \text{ min.}$$

[33] $y' = k(y_L - y)$, $k > 0 \Rightarrow y' + ky = ky_L \Rightarrow IF = e^{kt}$, $ye^{kt} = e^{kt}y_L + C$, and $y = y_L + Ce^{-kt}$, where $C < 0$. If we let $C = -cy_L$, where $0 < c < 1$, then $y = y_L(1 - ce^{-kt})$. Note that c gives the initial percentage of y_L that the animal must grow to reach its maximum length.

[34] $c' = k(c_0 - c)$, $k > 0 \Rightarrow c' + kc = kc_0 \Rightarrow IF = e^{kt}$, $ce^{kt} = c_0e^{kt} + C$,

and $c = c_0 + Ce^{-kt}$. $c(0) = 0 \Rightarrow C = -c_0$. Thus, $c(t) = c_0(1 - e^{-kt})$.

[35] (a) $y' = -ky$, $k > 0 \Rightarrow y' + ky = 0 \Rightarrow IF = e^{kt}$, $ye^{kt} = C$, and $y = Ce^{-kt}$.

$$y(0) = y_0 \Rightarrow y = y_0e^{-kt}.$$

(b) $y' + ky = I \Rightarrow ye^{kt} = \frac{I}{k}e^{kt} + C \Rightarrow y = \frac{I}{k} + Ce^{-kt}$.

$y(0) = 0 \Rightarrow C = -\frac{I}{k}$ so $y = \frac{I}{k}(1 - e^{-kt})$. As $t \to \infty$, $y = \frac{I}{k}$.

(c) $y(0) = \frac{I}{k}$ and a half-life of 2 hr $\Rightarrow y(2) = \frac{1}{2} \cdot \frac{I}{k} = \frac{I}{k}(1 - e^{-2k}) \Rightarrow$

$\frac{1}{2} = (1 - e^{-2k}) \Rightarrow k = \ln\sqrt{2}$. Since the long-term amount is

$\frac{I}{k}$, $y = 100 = \frac{I}{\ln\sqrt{2}} \Rightarrow I = 100\ln\sqrt{2} \approx 34.7$ mg/hr ≈ 0.58 mg/min.

[36] (a) At any instant in time, $\frac{5}{50}$, or $\frac{1}{10}$, of the dye present is leaving Tank 1 each minute. Thus, $x' = -0.1x \Rightarrow x' + 0.1x = 0 \Rightarrow IF = e^{0.1t}$, $xe^{0.1t} = C$,

and $x = Ce^{-0.1t}$. $x(0) = 1 \Rightarrow C = 1$. Thus, $x(t) = e^{-0.1t}$.

(b) As in part (a), the dye is leaving Tank 2 at a rate of $-0.1y(t)$.

Since the dye is entering Tank 2 at a rate of $0.1e^{-0.1t}$, the net rate of change is

$y'(t) = -0.1y(t) + 0.1e^{-0.1t}$. $y' + 0.1y = 0.1e^{-0.1t} \Rightarrow$

$IF = e^{0.1t}$, $ye^{0.1t} = 0.1t + C$, and $y = 0.1te^{-0.1t} + Ce^{-0.1t}$.

$y(0) = 0 \Rightarrow C = 0$ and $y(t) = 0.1te^{-0.1t}$.

(c) $y'(t) = -0.01te^{-0.1t} + 0.1e^{-0.1t} = 0 \Rightarrow 0.1e^{-0.1t}(1 - 0.1t) = 0 \Rightarrow t = 10$.

Since $y' > 0$ for $0 < t < 10$ and $y' < 0$ for $t > 10$,

this will give a *MAX* of $y(10) = e^{-1} \approx 0.37$ lb.

(d) As in part (b), the dye is still entering Tank 2 at a rate of $0.1e^{-0.1t}$. However, it is now leaving Tank 2 at a rate of $-\frac{5}{40}y(t) = -0.125\,y(t)$. The net rate of change is $y' = -0.125y + 0.1e^{-0.1t}$. $y' + 0.125y = 0.1e^{-0.1t} \Rightarrow IF = e^{0.125t}$,

$ye^{0.125t} = 4e^{0.025t} + C$, and $y = 4e^{-0.1t} + Ce^{-0.125t}$.

$y(0) = 0 \Rightarrow C = -4$ and $y(t) = 4e^{-0.1t} - 4e^{-0.125t}$.

[37] $y = e^{-x} \Rightarrow y' = -e^{-x}$. $y' = 10y - 11e^{-x} = 10(e^{-x}) - 11e^{-x} = -e^{-x}$.

At $x = 0$, $y = e^{-0} = 1$. Thus, $y = e^{-x}$ is the solution for the differential equation.

With $h = \frac{1}{8}$, $x_0 = 0$, $y_0 = 1$, and $f(x, y) = 10y - 11e^{-x}$, Euler's method gives:

$y_0 = 1.000000$, $y_1 = 0.875000$, $y_2 \approx 0.755317$, $y_3 \approx 0.628612$, $y_4 \approx 0.469353$,

$y_5 \approx 0.222065$, $y_6 \approx -0.236338$, $y_7 \approx -1.181264$, and $y_8 \approx -3.231030$.

[38] With $h = \frac{1}{8}$, $x_0 = 0$, $y_0 = 1$, and $f(x, y) = 10y - 11e^{-x}$,

the improved Euler's method gives: $y_0 = 1.000000$, $y_1 = 0.877658$, $y_2 \approx 0.759864$,

$y_3 \approx 0.626119$, $y_4 \approx 0.417782$, $y_5 \approx -0.039817$, $y_6 \approx -1.273429$, $y_7 \approx -4.877367$,

and $y_8 \approx -15.68227$.

[39] $h = \frac{1}{8}$, $x_0 = 0$, and $y_0 = 0.1 \Rightarrow K_1 = 0.01$, $K_2 \approx 0.014032$, $K_3 \approx 0.014082$,

$K_4 \approx 0.025980 \Rightarrow y_1 = y_0 + \frac{1}{48}(K_1 + 2K_2 + 2K_3 + K_4) \approx 0.101921$.

Similarly, $y_2 \approx 0.107841$, $y_3 \approx 0.121837$, and $y_4 \approx 0.148170$.

[40] $h = \frac{1}{8}$, $x_0 = 0$, and $y_0 = 0.1 \Rightarrow K_1 \approx 0.01$, $K_2 \approx 0.072562$, $K_3 \approx 0.073616$,

$K_4 \approx 0.136491 \Rightarrow y_1 = y_0 + \frac{1}{48}(K_1 + 2K_2 + 2K_3 + K_4) \approx 0.109132$.

Similarly, $y_2 \approx 0.134197$, $y_3 \approx 0.175401$, and $y_4 \approx 0.232949$.

Exercises 19.3

[1] By (19.5), the auxiliary equation of $y'' - 5y' + 6y = 0$ is $m^2 - 5m + 6$.

$m^2 - 5m + 6 = (m - 2)(m - 3) = 0 \Rightarrow m = 2, 3$.

By (19.6), the general solution is $y = C_1e^{2x} + C_2e^{3x}$.

[2] $m^2 - m - 2 = (m + 1)(m - 2) = 0 \Rightarrow m = -1, 2;$ $\quad y = C_1 e^{-x} + C_2 e^{2x}.$

[3] $m^2 - 3m = m(m - 3) = 0 \Rightarrow m = 0, 3;$ $\quad y = C_1 + C_2 e^{3x}.$

[4] $m^2 + 6m + 8 = (m + 2)(m + 4) = 0 \Rightarrow m = -2, -4;$ $\quad y = C_1 e^{-2x} + C_2 e^{-4x}.$

[5] $m^2 + 4m + 4 = (m + 2)^2 = 0 \Rightarrow m = -2$ is a double root.

By (19.7), the general solution is $y = C_1 e^{-2x} + C_2 x e^{-2x} = e^{-2x}(C_1 + C_2 x).$

[6] $m^2 - 4m + 4 = (m - 2)^2 = 0 \Rightarrow m = 2$ is a double root.

$y = C_1 e^{2x} + C_2 x e^{2x} = e^{2x}(C_1 + C_2 x).$

[7] $m^2 - 4m + 1 = 0 \Rightarrow m = \dfrac{4 \pm \sqrt{12}}{2} = 2 \pm \sqrt{3};$ $\quad y = C_1 e^{(2+\sqrt{3})x} + C_2 e^{(2-\sqrt{3})x}.$

[8] $6m^2 - 7m - 3 = (2m - 3)(3m + 1) = 0 \Rightarrow m = \frac{3}{2}, -\frac{1}{3};$ $y = C_1 e^{3x/2} + C_2 e^{-x/3}.$

[9] $m^2 + 2\sqrt{2}\, m + 2 = (m + \sqrt{2})^2 \Rightarrow m = -\sqrt{2}$ is a double root.

$y = C_1 e^{-\sqrt{2}\, x} + C_2 x e^{-\sqrt{2}\, x} = e^{-\sqrt{2}\, x}(C_1 + C_2 x).$

[10] $4m^2 + 20m + 25 = (2m + 5)^2 = 0 \Rightarrow m = -\frac{5}{2}$ is a double root.

$y = C_1 e^{-5x/2} + C_2 x e^{-5x/2} = e^{-5x/2}(C_1 + C_2 x).$

[11] $8m^2 + 2m - 15 = (2m + 3)(4m - 5) = 0 \Rightarrow m = -\frac{3}{2}, \frac{5}{4};$

$y = C_1 e^{-3x/2} + C_2 e^{5x/4}.$

[12] $m^2 + 4m + 1 = 0 \Rightarrow m = \dfrac{-4 \pm \sqrt{12}}{2} = -2 \pm \sqrt{3};$

$y = C_1 e^{(-2+\sqrt{3})\,x} + C_2 e^{(-2-\sqrt{3})\,x}.$

[13] $9m^2 - 24m + 16 = (3m - 4)^2 = 0 \Rightarrow m = \frac{4}{3}$ is a double root.

$y = C_1 e^{4x/3} + C_2 x e^{4x/3} = e^{4x/3}(C_1 + C_2 x).$

[14] $4m^2 - 8m + 7 = 0 \Rightarrow m = \dfrac{8 \pm \sqrt{-48}}{8} = 1 \pm \frac{1}{2}\sqrt{3}\, i.$

By (19.10), $y = e^x\left[C_1 \cos\left(\frac{1}{2}\sqrt{3}\, x\right) + C_2 \sin\left(\frac{1}{2}\sqrt{3}\, x\right)\right].$

[15] $2m^2 - 4m + 1 = 0 \Rightarrow m = (2 \pm \sqrt{2})/2;$ $\quad y = C_1 e^{(2+\sqrt{2})x/2} + C_2 e^{(2-\sqrt{2})x/2}.$

[16] $2m^2 + 7m = m(2m + 7) = 0 \Rightarrow m = 0, -\frac{7}{2};$ $\quad y = C_1 + C_2 e^{-7x/2}.$

[17] $m^2 - 2m + 2 = 0 \Rightarrow m = \dfrac{2 \pm \sqrt{-4}}{2} = 1 \pm i.$

By (19.10), with $s = t = 1$, $y = e^x(C_1 \cos x + C_2 \sin x).$

[18] $m^2 - 2m + 5 = 0 \Rightarrow m = \dfrac{2 \pm \sqrt{-16}}{2} = 1 \pm 2i;$ $\quad y = e^x(C_1 \cos 2x + C_2 \sin 2x).$

[19] $m^2 - 4m + 13 = 0 \Rightarrow m = \dfrac{4 \pm \sqrt{-36}}{2} = 2 \pm 3i;$ $\quad y = e^{2x}(C_1 \cos 3x + C_2 \sin 3x).$

[20] $m^2 + 4 = 0 \Rightarrow m = \pm 2i.$

By (19.10), with $s = 0$ and $t = 2$, $y = C_1 \cos 2x + C_2 \sin 2x.$

[21] $m^2 + 6m + 2 = 0 \Rightarrow m = \dfrac{-6 \pm \sqrt{28}}{2} = -3 \pm \sqrt{7}.$

$y = C_1 e^{(-3+\sqrt{7})\,x} + C_2 e^{(-3-\sqrt{7})\,x}.$

[22] $m^2 + 2m + 6 = 0 \Rightarrow m = \dfrac{-2 \pm \sqrt{-20}}{2} = -1 \pm \sqrt{5}\, i.$

$y = e^{-x}(C_1 \cos \sqrt{5}\, x + C_2 \sin \sqrt{5}\, x).$

[23] $m^2 - 3m + 2 = 0 \Rightarrow m = 1, 2$, so $y = C_1 e^x + C_2 e^{2x}$.

$x = 0, y = 0 \Rightarrow C_1 + C_2 = 0$. $y' = C_1 e^x + 2C_2 e^{2x}$ and $x = 0, y' = 2 \Rightarrow$

$C_1 + 2C_2 = 2$. Thus, $C_2 = 2$, $C_1 = -2$, and $y = -2e^x + 2e^{2x}$.

[24] $m^2 - 2m + 1 = (m-1)^2 = 0 \Rightarrow m = 1$ is a double root so $y = C_1 e^x + C_2 x e^x$.

$x = 0, y = 1 \Rightarrow C_1 = 1$. $y' = e^x + C_2(xe^x + e^x)$ and $x = 0, y' = 2 \Rightarrow$

$1 + C_2 = 2 \Rightarrow C_2 = 1$. Thus, $y = e^x + xe^x$.

[25] $m^2 + 1 = 0 \Rightarrow m = \pm i$, so $y = C_1 \cos x + C_2 \sin x$. $x = 0, y = 1 \Rightarrow C_1 = 1$.

$y' = C_2 \cos x - \sin x$ and $x = 0, y' = 2 \Rightarrow C_2 = 2$. Thus, $y = \cos x + 2\sin x$.

[26] $m^2 - m - 6 = 0 \Rightarrow m = 3, -2$, so $y = C_1 e^{3x} + C_2 e^{-2x}$. $x = 0, y = 0 \Rightarrow$

$C_1 + C_2 = 0$. $y' = 3C_1 e^{3x} - 2C_2 e^{-2x}$ and $x = 0, y' = 1 \Rightarrow 3C_1 - 2C_2 = 1$.

Thus, $C_1 = \frac{1}{5}$, $C_2 = -\frac{1}{5}$, and $y = \frac{1}{5}e^{3x} - \frac{1}{5}e^{-2x}$.

[27] $m^2 + 8m + 16 = 0 \Rightarrow m = -4$ is a double root, so $y = C_1 e^{-4x} + C_2 x e^{-4x}$.

$x = 0, y = 2 \Rightarrow C_1 = 2$. $y' = -8e^{-4x} + C_2(e^{-4x} - 4xe^{-4x})$ and $x = 0$,

$y' = 1 \Rightarrow -8 + C_2 = 1 \Rightarrow C_2 = 9$. Thus, $y = 2e^{-4x} + 9xe^{-4x} = e^{-4x}(2 + 9x)$.

[28] $m^2 + 5 = 0 \Rightarrow m = \pm\sqrt{5}\,i$, so $y = C_1 \cos\sqrt{5}\,x + C_2 \sin\sqrt{5}\,x$.

$x = 0, y = 4 \Rightarrow C_1 = 4$. $y' = \sqrt{5}\,C_2 \cos\sqrt{5}\,x - 4\sqrt{5}\sin\sqrt{5}\,x$ and

$x = 0, y' = 2 \Rightarrow C_2 = \frac{2}{\sqrt{5}}$. Thus, $y = 4\cos\sqrt{5}\,x + \frac{2}{\sqrt{5}}\sin\sqrt{5}\,x$.

[29] $m^2 - 2m + 5 = 0 \Rightarrow m = 1 \pm 2i$, so $y = e^x(C_1 \cos 2x + C_2 \sin 2x)$.

$x = 0, y = 0 \Rightarrow C_1 = 0$. $y' = C_2(e^x \sin 2x + 2e^x \cos 2x)$ and

$x = 0, y' = 1 \Rightarrow C_2 = \frac{1}{2}$. Thus, $y = \frac{1}{2}e^x \sin 2x$.

[30] $m^2 - 6m + 13 = 0 \Rightarrow m = 3 \pm 2i$, so $y = e^{3x}(C_1 \cos 2x + C_2 \sin 2x)$.

$x = 0, y = 2 \Rightarrow C_1 = 2$.

$y' = 3e^{3x}(2\cos 2x + C_2 \sin 2x) + e^{3x}(-4\sin 2x + 2C_2 \cos 2x)$ and

$x = 0, y' = 3 \Rightarrow C_2 = -\frac{3}{2}$. Thus, $y = e^{3x}(2\cos 2x - \frac{3}{2}\sin 2x)$.

Exercises 19.4

[1] $m^2 + 1 = 0 \Rightarrow m = \pm i$. The complementary solution is $y_c = C_1 \cos x + C_2 \sin x$.

To find a particular solution y_p, we use (19.13) with $y_1 = \cos x$ and $y_2 = \sin x$.

$$\begin{cases} u' \cos x + v' \sin x = 0 \\ -u' \sin x + v' \cos x = \tan x \end{cases}$$

Solving this as a system of equations in the unknowns u' and v',

we multiply the first equation by $\sin x$ and the second by $\cos x$, yielding the system

$$\begin{cases} u' \cos x \sin x + v' \sin^2 x = 0 \\ -u' \cos x \sin x + v' \cos^2 x = \sin x \end{cases}$$

Adding yields $v' = \sin x$ and then $u' = -\dfrac{\sin^2 x}{\cos x} = -\dfrac{1 - \cos^2 x}{\cos x} = -\sec x + \cos x$.

Integrating gives us $u = -\ln|\sec x + \tan x| + \sin x$ and $v = -\cos x$.

Thus, $y_p = u\,y_1 + v\,y_2 = -\cos x \ln|\sec x + \tan x| + \sin x \cos x - \cos x \sin x$.

The general solution is $y = y_c + y_p = C_1 \cos x + C_2 \sin x - \cos x \ln|\sec x + \tan x|$.

[2] $m^2 + 1 = 0 \Rightarrow m = \pm i$. $y_c = C_1 \cos x + C_2 \sin x$. Using (19.13),

$$\begin{cases} u' \cos x + v' \sin x = 0 \\ -u' \sin x + v' \cos x = \sec x \end{cases} \Rightarrow \begin{cases} v' = 1 \\ u' = -\tan x \end{cases}$$

Integrating gives us $u = \ln|\cos x|$ and $v = x$.

The general solution is $y = (C_2 + x)\sin x + (C_1 + \ln|\cos x|)\cos x$.

[3] $m^2 - 6m + 9 = 0 \Rightarrow m = 3$. $y_c = C_1 e^{3x} + C_2 x e^{3x}$. Using (19.13),

$$\begin{cases} u'(e^{3x}) + v'(xe^{3x}) = 0 \\ u'(3e^{3x}) + v'(1 + 3x)e^{3x} = x^2 e^{3x} \end{cases} \Rightarrow \begin{cases} v' = x^2 \\ u' = -x^3 \end{cases}$$

Integrating gives us $u = -\frac{1}{4}x^4$ and $v = \frac{1}{3}x^3$. The general solution is

$$y = (C_1 - \tfrac{1}{4}x^4)e^{3x} + (C_2 + \tfrac{1}{3}x^3)xe^{3x} = (C_1 + C_2 x + \tfrac{1}{12}x^4)e^{3x}.$$

[4] $m^2 + 3m = 0 \Rightarrow m = 0, -3$. $y_c = C_1(1) + C_2 e^{-3x}$. Using (19.13),

$$\begin{cases} u'(1) + v' e^{-3x} = 0 \\ u'(0) + v'(-3e^{-3x}) = e^{-3x} \end{cases} \Rightarrow \begin{cases} u' = \frac{1}{3}e^{-3x} \\ v' = -\frac{1}{3} \end{cases}$$

Integrating gives us $u = -\frac{1}{9}e^{-3x}$ and $v = -\frac{1}{3}x$.

The general solution is $y = C_1 + C_3 e^{-3x} - \frac{1}{3}xe^{-3x}$, where $C_3 = C_2 - \frac{1}{9}$.

[5] $m^2 - 1 = 0 \Rightarrow m = \pm 1$. $y_c = C_1 e^{x} + C_2 e^{-x}$. Using (19.13),

$$\begin{cases} u' e^{x} + v' e^{-x} = 0 \\ u' e^{x} - v' e^{-x} = e^{x} \cos x \end{cases} \Rightarrow \begin{cases} u' = \frac{1}{2}\cos x \\ v' = -\frac{1}{2} e^{2x} \cos x \end{cases}$$

Integrating gives us $u = \frac{1}{2}\sin x$ and $v = -\frac{1}{5} e^{2x} \cos x - \frac{1}{10} e^{2x} \sin x$.

The general solution is $y = C_1 e^{x} + C_2 e^{-x} + \frac{2}{5} e^{x} \sin x - \frac{1}{5} e^{x} \cos x$.

[6] $m^2 - 4m + 4 = 0 \Rightarrow m = 2$. $y_c = C_1 e^{2x} + C_2 x e^{2x}$. Using (19.13),

$$\begin{cases} u' e^{2x} + v' xe^{2x} = 0 \\ u'(2e^{2x}) + v'(1 + 2x)e^{2x} = x^{-2} e^{2x} \end{cases} \Rightarrow \begin{cases} v' = x^{-2} \\ u' = -x^{-1} \end{cases}$$

Integrating gives us $u = -\ln|x|$ and $v = -x^{-1}$.

The general solution is $y = (C_3 + C_2 x - \ln|x|)e^{2x}$, where $C_3 = C_1 - 1$.

[7] $m^2 - 9 = 0 \Rightarrow m = \pm 3$. $y_c = C_1 e^{3x} + C_2 e^{-3x}$. Using (19.13),

$$\begin{cases} u' e^{3x} + v' e^{-3x} = 0 \\ u'(3e^{3x}) + v'(-3e^{-3x}) = e^{3x} \end{cases} \Rightarrow \begin{cases} u' = \frac{1}{6} \\ v' = -\frac{1}{6}e^{6x} \end{cases}$$

Integrating gives us $u = \frac{1}{6}x$ and $v = -\frac{1}{36}e^{6x}$.

The general solution is $y = (C_3 + \frac{1}{6}x)\, e^{3x} + C_2 e^{-3x}$, where $C_3 = C_1 - \frac{1}{36}$.

[8] $m^2 + 1 = 0 \Rightarrow m = \pm i$. $y_c = C_1 \cos x + C_2 \sin x$. Using (19.13),

$$\begin{cases} u' \cos x + v' \sin x = 0 \\ -u' \sin x + v' \cos x = \sin x \end{cases} \Rightarrow \begin{cases} v' = \sin x \cos x \\ u' = -\sin^2 x \end{cases}$$

Integrating gives us $u = -\frac{1}{2}x + \frac{1}{4}\sin 2x$ and $v = -\frac{1}{2}\cos^2 x$. The general solution is

$y = C_1 \cos x + C_2 \sin x + (-\frac{1}{2}x + \frac{1}{2}\sin x \cos x)\cos x - \frac{1}{2}\cos^2 x \sin x \Rightarrow$

$$y = C_1 \cos x + C_2 \sin x - \tfrac{1}{2}x \cos x.$$

[9] $m^2 - 3m - 4 = 0 \Rightarrow m = -1, 4.$ $y_c = C_1 e^{-x} + C_2 e^{4x}$. Using (19.13),

$$\begin{cases} u' e^{-x} + v' e^{4x} = 0 \\ -u' e^{-x} + v'(4e^{4x}) = 2 \end{cases} \Rightarrow \begin{cases} v' = \frac{2}{5}e^{-4x} \\ u' = -\frac{2}{5}e^{x} \end{cases}$$

Integrating gives us $u = -\frac{2}{5}e^{x}$ and $v = -\frac{1}{10}e^{-4x}$.

The general solution is $y = C_1 e^{-x} + C_2 e^{4x} - \frac{1}{2}$.

[10] $m^2 - m = 0 \Rightarrow m = 0, 1.$ $y_c = C_1(1) + C_2 e^{x}$. Using (19.13),

$$\begin{cases} u'(1) + v' e^{x} = 0 \\ u'(0) + v' e^{x} = x + 1 \end{cases} \Rightarrow \begin{cases} v' = (x+1)e^{-x} \\ u' = -(x+1) \end{cases}$$

Integrating gives us $u = -\frac{1}{2}x^2 - x$ and $v = -xe^{-x} - 2e^{-x}$.

The general solution is $y = C_3 + C_2 e^{x} - \frac{1}{2}x^2 - 2x$, where $C_3 = C_1 - 2$.

[11] $m^2 - 3m + 2 = 0 \Rightarrow m = 1, 2.$ $y_c = C_1 e^{x} + C_2 e^{2x}$. Let $y_p = Ae^{-x}$.

Then, $y_p'' - 3y_p' + 2y_p = (A + 3A + 2A)e^{-x} = 4e^{-x} \Rightarrow A = \frac{2}{3}$.

The general solution is $y = C_1 e^{x} + C_2 e^{2x} + \frac{2}{3}e^{-x}$.

[12] $m^2 + 6m + 9 = 0 \Rightarrow m = -3.$ $y_c = C_1 e^{-3x} + C_2 xe^{-3x}$. Let $y_p = Ae^{2x}$.

Then, $y_p'' + 6y_p' + 9y_p = (4A + 12A + 9A)e^{2x} = e^{2x} \Rightarrow A = \frac{1}{25}$. The general solution is $y = C_1 e^{-3x} + C_2 xe^{-3x} + \frac{1}{25}e^{2x} = (C_1 + C_2 x)e^{-3x} + \frac{1}{25}e^{2x}$.

[13] $m^2 + 2m = 0 \Rightarrow m = 0, -2.$ $y_c = C_1(1) + C_2 e^{-2x}$.

Let $y_p = A\cos 2x + B\sin 2x$.

Then, $y_p'' + 2y_p' = \left[-4A\cos 2x - 4B\sin 2x + 2(-2A\sin 2x + 2B\cos 2x)\right] =$
$(-4A + 4B)\cos 2x + (-4A - 4B)\sin 2x = \cos 2x \Rightarrow A = -\frac{1}{8}$ and $B = \frac{1}{8}$.

The general solution is $y = C_1 + C_2 e^{-2x} - \frac{1}{8}\cos 2x + \frac{1}{8}\sin 2x$.

[14] $m^2 + 1 = 0 \Rightarrow m = \pm i.$ $y_c = C_1\cos x + C_2\sin x$. Let $y_p = A\cos 5x + B\sin 5x$.

Then, $y_p'' + y_p = \left[(-25A\cos 5x - 25B\sin 5x) + (A\cos 5x + B\sin 5x)\right]$
$= -24A\cos 5x - 24B\sin 5x = \sin 5x \Rightarrow A = 0$ and $B = -\frac{1}{24}$.

The general solution is $y = C_1\cos x + C_2\sin x - \frac{1}{24}\sin 5x$.

[15] $m^2 - 1 = 0 \Rightarrow m = \pm 1.$ $y_c = C_1 e^{x} + C_2 e^{-x}$. Let $y_p = (A + Bx)e^{2x}$.

Then, $y_p'' - y_p = (4B + 4A)e^{2x} + 4Bxe^{2x} - (A + Bx)e^{2x} =$
$(4B + 3A)e^{2x} + 3Bxe^{2x} = xe^{2x} \Rightarrow B = \frac{1}{3}$ and $A = -\frac{4}{9}$.

The general solution is $y = C_1 e^{x} + C_2 e^{-x} + \frac{1}{9}(-4 + 3x)e^{2x}$.

[16] $m^2 + 3m - 4 = 0 \Rightarrow m = 1, -4.$ $y_c = C_1 e^{x} + C_2 e^{-4x}$. Let $y_p = (A + Bx)e^{-x}$.

Then, $y_p'' + 3y_p' - 4y_p = \{[(A - 2B) + 3(B - A) - 4A] + [B - 3B - 4B]x\}e^{-x}$
$= [(-6A + B) - 6Bx]e^{-x} = xe^{-x} \Rightarrow B = -\frac{1}{6}$ and $A = -\frac{1}{36}$.

The general solution is $y = C_1 e^{x} + C_2 e^{-4x} - (\frac{1}{36} + \frac{1}{6}x)e^{-x}$.

[17] $m^2 - 6m + 13 = 0 \Rightarrow m = 3 \pm 2i$. $y_c = e^{3x}(C_1 \cos 2x + C_2 \sin 2x)$.

Let $y_p = Ae^x \cos x + Be^x \sin x$. Then, $y_p'' - 6y_p' + 13y_p$

$= \left[2B - 6(A + B) + 13A\right] e^x \cos x + \left[-2A - 6(B - A) + 13B\right] e^x \sin x$

$= (7A - 4B)\, e^x \cos x + (4A + 7B)\, e^x \sin x = e^x \cos x \Rightarrow A = \frac{7}{65}$ and $B = -\frac{4}{65}$.

The general solution is $y = e^{3x}(C_1 \cos 2x + C_2 \sin 2x) + \frac{1}{65} e^x (7 \cos x - 4 \sin x)$.

[18] $m^2 - 2m + 2 = 0 \Rightarrow m = 1 \pm i$. $y_c = e^x(C_1 \cos x + C_2 \sin x)$.

Let $y_p = Ae^{-x} \cos 2x + Be^{-x} \sin 2x$. Then,

$y_p'' - 2y_p' + 2y_p = \left[(-3A - 4B) - 2(2B - A) + 2A\right] e^{-x} \cos 2x +$

$\left[(4A - 3B) - 2(-2A - B) + 2B\right] e^{-x} \sin 2x$

$= (A - 8B)e^{-x} \cos 2x + (8A + B)e^{-x} \sin 2x = e^{-x} \sin 2x \Rightarrow A = \frac{8}{65}$ and $B = \frac{1}{65}$.

The general solution is $y = e^x(C_1 \cos x + C_2 \sin x) + \frac{1}{65} e^{-x} (8 \cos 2x + \sin 2x)$.

[19] $L(Cy) = (D^2 + bD + c)(Cy)$

$= D^2(Cy) + bD(Cy) + c(Cy)$

$= CD^2 y + CbDy + Ccy$

$= C(D^2 y + bDy + cy) = CL(y)$.

[20] $L(y_1 \pm y_2) = (D^2 + bD + c)(y_1 \pm y_2)$

$= D^2(y_1 \pm y_2) + bD(y_1 \pm y_2) + c(y_1 \pm y_2)$

$= D^2 y_1 \pm D^2 y_2 + bDy_1 \pm bDy_2 + cy_1 \pm cy_2$

$= (D^2 + bD + c)y_1 \pm (D^2 + bD + c)y_2 = L(y_1) \pm L(y_2)$.

[21] Since we are approximating y at $x = \frac{1}{2}$, let $b = \frac{1}{2}$.

$x_0 = 0$, $y_0 = 1$, and $h = (b - a)/n = (\frac{1}{2} - 0)/4 = \frac{1}{8} \Rightarrow$

$y_1 = 2y_0 - y_{-1} + h^2 f(x_0, y_0) = 2(1) - (0.984496) + (\frac{1}{8})^2(-2)(1) = 0.984254$.

In a similar manner, $y_2 \approx 0.938711$, $y_3 \approx 0.867501$, and $y_4 \approx 0.776805$.

[22] $y_1 = 2y_0 - y_{-1} + h^2 f(x_0, y_0) = 2(1) - (0.882823) + (\frac{1}{8})^2(1 - 0) = 1.132802$.

In a similar manner, $y_2 \approx 1.281351$, $y_3 \approx 1.446015$, and $y_4 \approx 1.627413$.

Exercises 19.5

[1] $F = ky \Rightarrow 5 = k(6 \text{ in.} = \frac{1}{2} \text{ ft}) \Rightarrow k = 10$. $W = mg \Rightarrow m = \frac{W}{g} = \frac{5}{32}$.

Also, $\omega^2 = \frac{k}{m} = 64 \Rightarrow \omega = 8$. By Example 1, $y = C_1 \cos 8t + C_2 \sin 8t$. $y = -\frac{1}{3}$ when $t = 0 \Rightarrow C_1 = -\frac{1}{3}$ and $y' = 0$ when $t = 0 \Rightarrow C_2 = 0$. Thus, $y = -\frac{1}{3} \cos 8t$.

[2] $k(\frac{2}{3}) = 10 \Rightarrow k = 15$ and $m = \frac{W}{g} = \frac{10}{32}$. Also, $\omega^2 = \frac{k}{m} = 48 \Rightarrow \omega = \sqrt{48}$.

By Example 1, $y = C_1 \cos \sqrt{48}\, t + C_2 \sin \sqrt{48}\, t$. $y = \frac{1}{4}$ when $t = 0 \Rightarrow$ $C_1 = \frac{1}{4}$ and $y' = \frac{1}{2}$ when $t = 0 \Rightarrow \frac{1}{2} = \sqrt{48}\, C_2 \Rightarrow C_2 = \frac{1}{24}\sqrt{3}$.

Thus, $y = \frac{1}{4} \cos (4\sqrt{3}\, t) + \frac{1}{24}\sqrt{3} \sin (4\sqrt{3}\, t)$.

[3] $k(1) = 10 \Rightarrow k = 10$ and $m = \frac{W}{g} = \frac{10}{32}$. Also, $\omega^2 = \frac{k}{m} = 32$. Using (19.16) with $c = 5$ and $2p = \frac{c}{m} = 16$ or $p = 8 \Rightarrow p^2 - \omega^2 = 32 > 0$, and the motion is overdamped. By Case 1, $y = e^{-8t}(C_1 e^{\sqrt{32}\,t} + C_2 e^{-\sqrt{32}\,t})$. $y = 0$ when $t = 0 \Rightarrow C_1 + C_2 = 0$ and $t = 0$, $y' = 2 \Rightarrow (\sqrt{32} - 8)\,C_1 + (-\sqrt{32} - 8)\,C_2 = 2$.

So $C_1 = \frac{1}{8}\sqrt{2}$ and $C_2 = -\frac{1}{8}\sqrt{2}$. Thus, $y = \frac{1}{8}\sqrt{2}\,e^{-8t}(e^{4\sqrt{2}\,t} - e^{-4\sqrt{2}\,t})$.

[4] $k = 48$ and $m = \frac{W}{g} = \frac{6}{32}$. Also, $\omega^2 = \frac{k}{m} = 256 \Rightarrow \omega = 16$.

By Example 1, $y = C_1 \cos 16t + C_2 \sin 16t$. $y = \frac{5}{12}$ when $t = 0 \Rightarrow C_1 = \frac{5}{12}$ and $y' = 4$ when $t = 0 \Rightarrow C_2 = \frac{1}{4}$. Thus, $y = \frac{5}{12}\cos 16t + \frac{1}{4}\sin 16t$.

[5] $k(\frac{1}{4}) = 4 \Rightarrow k = 16$ and $m = \frac{W}{g} = \frac{4}{32}$. Also, $\omega^2 = \frac{k}{m} = 128$.

Using (19.16) with $c = 2$ and $2p = \frac{c}{m} = 16$ or $p = 8 \Rightarrow p^2 - \omega^2 = -64 < 0$, and the motion is underdamped. The roots of the auxiliary equation $m^2 + 2pm + \omega^2 = 0$ are $a \pm bi = -8 \pm 8i$.

By Case 3, $y = e^{-8t}(C_1 \cos 8t + C_2 \sin 8t)$. $y = \frac{1}{3}$ when $t = 0 \Rightarrow C_1 = \frac{1}{3}$ and $y' = 0$ when $t = 0 \Rightarrow 8C_2 = \frac{8}{3}$, or $C_2 = \frac{1}{3}$. Thus, $y = \frac{1}{3}e^{-8t}(\cos 8t + \sin 8t)$.

[6] $k(\frac{1}{2}) = 8 \Rightarrow k = 16$ and $m = \frac{W}{g} = \frac{8}{32}$. Also, $\omega^2 = \frac{k}{m} = 64$.

Using (19.16) with $c = 4$ and $2p = \frac{c}{m} = 16$ or $p = 8 \Rightarrow p^2 - \omega^2 = 0$, and the motion is critically damped. By Case 2, $y = e^{-8t}(C_1 + C_2 t)$. $y = 0$ when $t = 0 \Rightarrow C_1 = 0$ and $y' = -1$ when $t = 0 \Rightarrow C_2 = -1$. Thus, $y = -te^{-8t}$.

[7] Rewriting as $\frac{d^2y}{dt^2} + 4\frac{dy}{dt} + 24y = 0$ and comparing with (19.16), we see that $\frac{c}{m} = 2p = 4$, or $p = 2$, and $\frac{k}{m} = \omega^2 = 24$. If m is the mass of the weight, then the spring constant k is $24m$ and the damping force, $-c\frac{dy}{dt}$, is $-4m\frac{dy}{dt}$. The motion is begun by releasing the weight from 2 ft above the equilibrium position with an initial velocity of 1 ft/sec in the upward direction.

[8] $k(1) = 4 \Rightarrow k = 4$ and $m = \frac{W}{g} = \frac{4}{32}$. Also, $\omega^2 = \frac{k}{m} = 32$ and $2p = \frac{c}{m} = 8c$.

Thus, $p^2 - \omega^2 = 16c^2 - 32$. (a) $16c^2 - 32 > 0 \Rightarrow c > \sqrt{2}$.

(b) $16c^2 - 32 = 0 \Rightarrow c = \sqrt{2}$. (c) $16c^2 - 32 < 0 \Rightarrow 0 < c < \sqrt{2}$.

[9] Let the damping force be $-c\frac{dy}{dt}$. $2p = \frac{4}{3}c$ and $\omega^2 = 32$.

$p^2 - \omega^2 = \frac{4}{9}c^2 - 32 = 0$ when $c = 6\sqrt{2}$.

[10] (a) The total force on the spring is found by combining the spring force $-k(l_1 + y)$, the weight $mg = kl_1$, and the external force $F \sin \alpha t$. Therefore,

$$ma = m\frac{d^2y}{dt^2} = -k(l_1 + y) + kl_1 + F \sin \alpha t \Rightarrow \frac{d^2y}{dt^2} + \frac{k}{m}y = \frac{F}{m}\sin \alpha t,$$

which is the required form if $\omega^2 = k/m$.

(b) From part (a), $m^2 + \frac{k}{m} = m^2 + \omega^2 = 0 \Rightarrow m = \pm\,\omega i \Rightarrow$

$y_c = C_1 \cos \omega t + C_2 \sin \omega t$. Let $y_p = A \cos \alpha t + C \sin \alpha t$. Then,

$y_p'' + \omega^2 y_p = (-\alpha^2 + \omega^2)\, y_p = (\omega^2 - \alpha^2)(A \cos \alpha t + C \sin \alpha t) = \frac{F}{m} \sin \alpha t \Rightarrow$

$A = 0$ and $C = \dfrac{F}{m(\omega^2 - \alpha^2)}$. Thus, $y = C_1 \cos \omega t + C_2 \sin \omega t + C \sin \alpha t$.

Exercises 19.6

[1] $y'' + y = \sum_{n=2}^{\infty} n(n-1) a_n x^{n-2} + \sum_{n=0}^{\infty} a_n x^n = 0$. We replace n with $n + 2$ in the first series so that both series have the same powers of x. Hence

$$\sum_{n=0}^{\infty} (n+2)(n+1) a_{n+2} x^n + \sum_{n=0}^{\infty} a_n x^n = \sum_{n=0}^{\infty} \left[(n+2)(n+1) a_{n+2} + a_n\right] x^n = 0.$$

Since the coefficients of x are equal to 0, $a_{n+2} = -\dfrac{1}{(n+2)(n+1)} a_n$.

The even-numbered terms corresponding to $n = 0, 2, 4, \ldots$, are $a_2 = -\frac{1}{2 \cdot 1} a_0$,

$$a_4 = -\frac{1}{4 \cdot 3} a_2 = \frac{1}{4!} a_0,\ a_6 = -\frac{1}{6 \cdot 5} a_4 = -\frac{1}{6!} a_0,\ \ldots, \text{ and } a_{2k} = (-1)^k \frac{1}{(2k)!} a_0.$$

The odd-numbered terms corresponding to $n = 1, 3, 5, \ldots$, are $a_3 = -\frac{1}{3 \cdot 2} a_1$,

$$a_5 = -\frac{1}{5 \cdot 4} a_3 = \frac{1}{5!} a_1,\ a_7 = -\frac{1}{7 \cdot 6} a_5 = -\frac{1}{7!} a_1,\ \ldots, \text{ and } a_{2k+1} = (-1)^k \frac{1}{(2k+1)!} a_1.$$

$y = \sum_{n=0}^{\infty} a_n x^n$ can be represented as the sum of two series.

$$\text{Thus, } y = a_0 \sum_{n=0}^{\infty} \frac{(-1)^n}{(2n)!} x^{2n} + a_1 \sum_{n=0}^{\infty} \frac{(-1)^n}{(2n+1)!} x^{2n+1} = a_0 \cos x + a_1 \sin x.$$

[2] Similar to Exercise 1, we have $a_{n+2} = \dfrac{4}{(n+2)(n+1)} a_n$, and each a_i will be 4 times a_{i-2} rather than -1 times it. Hence, $a_{2k} = \dfrac{4^k}{(2k)!} a_0$, and $a_{2k+1} = \dfrac{4^k}{(2k+1)!} a_1$.

$$\text{Thus, } y = a_0 \sum_{n=0}^{\infty} \frac{4^n}{(2n)!} x^{2n} + a_1 \sum_{n=0}^{\infty} \frac{4^n}{(2n+1)!} x^{2n+1}.$$

[3] $$\begin{aligned} y'' - 2xy &= \sum_{n=2}^{\infty} n(n-1) a_n x^{n-2} - 2x \sum_{n=0}^{\infty} a_n x^n \ \{\text{replace } n \text{ with } n+3\} \\ &= \sum_{n=-1}^{\infty} (n+3)(n+2) a_{n+3} x^{n+1} - 2 \sum_{n=0}^{\infty} a_n x^{n+1} \\ &= 2a_2 + \sum_{n=0}^{\infty} \left[(n+3)(n+2) a_{n+3} - 2a_n\right] x^{n+1} \\ &= 0 \Rightarrow a_2 = 0 \text{ and } a_{n+3} = \frac{2}{(n+3)(n+2)} a_n. \end{aligned}$$

Since $a_2 = 0$, we also have $a_5 = a_8 = a_{11} = \cdots = 0$.

For $n = 0, 3, 6, \ldots$: $a_3 = \frac{2}{3 \cdot 2} a_0$, $a_6 = \frac{2}{6 \cdot 5} a_3 = \frac{2^2 \cdot 4}{6!} a_0$, $a_9 = \frac{2}{9 \cdot 8} a_6 = \frac{2^3 \cdot 7 \cdot 4}{9!} a_0$,

$\ldots$. We multiplied a_6 by $\frac{4}{4}$, a_9 by $\frac{7}{7}$, and a_{3k} by $\frac{3k-2}{3k-2}$, to get $(3k)!$ in the

denominator. Therefore, $a_{3k} = \dfrac{2^k (3k-2)(3k-5)\cdots 7 \cdot 4 \cdot 1}{(3k)!} a_0$.

For $n = 1, 4, 7, \ldots$: $a_4 = \frac{2}{4 \cdot 3} a_1 = \frac{2 \cdot 2}{4!} a_1$, $a_7 = \frac{2}{7 \cdot 6} a_4 = \frac{2^2 \cdot 5 \cdot 2}{7!} a_1$,

$a_{10} = \frac{2}{10 \cdot 9} a_7 = \frac{2^3 \cdot 8 \cdot 5 \cdot 2}{10!} a_1$. Therefore, $a_{3k+1} = \dfrac{2^k (3k-1)(3k-4)\cdots 8 \cdot 5 \cdot 2}{(3k+1)!} a_1$.

$y = \sum_{n=0}^{\infty} a_n x^n = a_0 + a_1 x + \sum_{n=3}^{\infty} a_n x^n$ can be represented as the sum of $(a_0 + a_1 x)$

and the sum of two series.

Thus, $y = a_0 \left[1 + \sum_{n=1}^{\infty} \frac{2^n (3n-2)(3n-5)\cdots 7 \cdot 4 \cdot 1}{(3n)!} x^{3n} \right] +$

$$a_1 \left[x + \sum_{n=1}^{\infty} \frac{2^n (3n-1)(3n-4)\cdots 8 \cdot 5 \cdot 2}{(3n+1)!} x^{3n+1} \right].$$

4

$$\begin{aligned}
y'' + 2xy' + y &= \sum_{n=2}^{\infty} n(n-1) a_n x^{n-2} + 2x \sum_{n=1}^{\infty} n a_n x^{n-1} + \sum_{n=0}^{\infty} a_n x^n \\
&= \sum_{n=0}^{\infty} (n+2)(n+1) a_{n+2} x^n + 2 \sum_{n=1}^{\infty} n a_n x^n + \sum_{n=0}^{\infty} a_n x^n \\
&= (2a_2 + a_0) + \sum_{n=1}^{\infty} \left[(n+2)(n+1) a_{n+2} + 2n a_n + a_n \right] x^n \\
&= 0 \Rightarrow a_2 = -\frac{1}{2 \cdot 1} a_0 \text{ and } a_{n+2} = -\frac{2n+1}{(n+2)(n+1)} a_n.
\end{aligned}$$

The even-numbered terms are $a_4 = -\frac{5}{4 \cdot 3} a_2 = \frac{5 \cdot 1}{4!} a_0$,

$a_6 = -\frac{9}{6 \cdot 5} a_4 = -\frac{9 \cdot 5 \cdot 1}{6!} a_0$, $\ldots$, and $a_{2k} = (-1)^k \dfrac{(4k-3)(4k-7)\cdots 9 \cdot 5 \cdot 1}{(2k)!} a_0$.

The odd-numbered terms are $a_3 = -\frac{3}{3 \cdot 2} a_1$, $a_5 = -\frac{7}{5 \cdot 4} a_3 = \frac{7 \cdot 3}{5!} a_1$,

$a_7 = -\frac{11}{7 \cdot 6} a_5 = -\frac{11 \cdot 7 \cdot 3}{7!} a_1$, $\ldots$, and $a_{2k+1} = (-1)^k \dfrac{(4k-1)(4k-5)\cdots 7 \cdot 3}{(2k+1)!} a_1$.

$y = \sum_{n=0}^{\infty} a_n x^n = a_0 + a_1 x + \sum_{n=2}^{\infty} a_n x^n$.

Thus, $y = a_0 \left[1 + \sum_{n=1}^{\infty} \frac{(-1)^n (4n-3)(4n-7)\cdots 9 \cdot 5 \cdot 1}{(2n)!} x^{2n} \right] +$

$$a_1 \left[x + \sum_{n=1}^{\infty} \frac{(-1)^n (4n-1)(4n-5)\cdots 7 \cdot 3}{(2n+1)!} x^{2n+1} \right].$$

5 $y'' - xy' + 2y = \sum_{n=2}^{\infty} n(n-1)a_n x^{n-2} - x\sum_{n=1}^{\infty} na_n x^{n-1} + 2\sum_{n=0}^{\infty} a_n x^n$

$= \sum_{n=0}^{\infty} (n+2)(n+1)a_{n+2} x^n - \sum_{n=1}^{\infty} na_n x^n + 2\sum_{n=0}^{\infty} a_n x^n$

$= (2a_2 + 2a_0) + \sum_{n=1}^{\infty} \left[(n+2)(n+1)a_{n+2} - na_n + 2a_n\right]x^n$

$= 0 \Rightarrow a_2 = -a_0$ and $a_{n+2} = \dfrac{n-2}{(n+2)(n+1)}a_n.$

When $n = 2$, $a_4 = 0$, and hence $a_6 = a_8 = a_{10} = \cdots = 0$.

The odd-numbered terms are $a_3 = \dfrac{-1}{3\cdot 2}a_1 = -\dfrac{1}{3!}a_1$, $a_5 = \dfrac{1}{5\cdot 4}a_3 = -\dfrac{1}{5!}a_1$,

$a_7 = \dfrac{3}{7\cdot 6}a_5 = -\dfrac{3\cdot 1}{7!}a_1$, ..., and $a_{2k+1} = -\dfrac{(2k-3)(2k-5)\cdots 5\cdot 3\cdot 1}{(2k+1)!}a_1$, for $k \geq 2$.

$y = \sum_{n=0}^{\infty} a_n x^n = a_0 + a_1 x + a_2 x^2 + a_3 x^3 + \sum_{n=5}^{\infty} a_n x^n.$

Since $a_2 = -a_0$ and $a_3 = -\frac{1}{6}a_1$, we have

$$y = a_0(1 - x^2) + a_1\left[x - \tfrac{1}{6}x^3 - \sum_{n=2}^{\infty} \frac{(2n-3)(2n-5)\cdots 5\cdot 3\cdot 1}{(2n+1)!}x^{2n+1}\right].$$

6 $y'' + x^2 y = \sum_{n=2}^{\infty} n(n-1)a_n x^{n-2} + x^2\sum_{n=0}^{\infty} a_n x^n$ {replace n with $n+4$}

$= \sum_{n=-2}^{\infty} (n+4)(n+3)a_{n+4} x^{n+2} + \sum_{n=0}^{\infty} a_n x^{n+2}$

$= 2a_2 + 6a_3 x + \sum_{n=0}^{\infty} \left[(n+4)(n+3)a_{n+4} + a_n\right]x^{n+2}$

$= 0 \Rightarrow a_2 = 0$, $a_3 = 0$, and $a_{n+4} = -\dfrac{1}{(n+4)(n+3)}a_n.$

$a_2 = 0 \Rightarrow a_6 = a_{10} = a_{14} = \cdots = 0$. $a_3 = 0 \Rightarrow a_7 = a_{11} = a_{15} = \cdots = 0$.

For $n = 0, 4, 8, \ldots$: $a_4 = -\dfrac{1}{4\cdot 3}a_0$, $a_8 = -\dfrac{1}{8\cdot 7}a_4 = \dfrac{1}{8\cdot 7\cdot 4\cdot 3}a_0$,

$a_{12} = -\dfrac{1}{12\cdot 11}a_8 = -\dfrac{1}{12\cdot 11\cdot 8\cdot 7\cdot 4\cdot 3}a_0, \ldots,$

and $a_{4k} = (-1)^k \dfrac{1}{(4k)(4k-1)(4k-4)(4k-5)\cdots 4\cdot 3}a_0.$

For $n = 1, 5, 9, \ldots$: $a_5 = -\dfrac{1}{5\cdot 4}a_1$, $a_9 = -\dfrac{1}{9\cdot 8}a_5 = \dfrac{1}{9\cdot 8\cdot 5\cdot 4}a_1$,

$a_{13} = -\dfrac{1}{13\cdot 12}a_9 = -\dfrac{1}{13\cdot 12\cdot 9\cdot 8\cdot 5\cdot 4}a_1, \ldots,$

and $a_{4k+1} = (-1)^k \dfrac{1}{(4k+1)(4k)(4k-3)(4k-4)\cdots 5\cdot 4}a_1.$

$y = \sum_{n=0}^{\infty} a_n x^n = a_0 + a_1 x + \sum_{n=4}^{\infty} a_n x^n.$

Thus, $y = a_0\left[1 + \sum_{n=1}^{\infty} \dfrac{(-1)^n}{4n(4n-1)(4n-4)(4n-5)\cdots 4\cdot 3}x^{4n}\right] +$

$$a_1\left[x + \sum_{n=1}^{\infty} \frac{(-1)^n}{(4n+1)(4n)(4n-3)(4n-4)\cdots 5\cdot 4}x^{4n+1}\right].$$

7 $xy' + y' - 3y = x\sum_{n=1}^{\infty} na_n x^{n-1} + \sum_{n=1}^{\infty} na_n x^{n-1} - 3\sum_{n=0}^{\infty} a_n x^n$

$$= \sum_{n=1}^{\infty} na_n x^n + \sum_{n=0}^{\infty} (n+1)a_{n+1} x^n - 3\sum_{n=0}^{\infty} a_n x^n$$

$$= \sum_{n=1}^{\infty} na_n x^n + a_1 + \sum_{n=1}^{\infty} (n+1)a_{n+1} x^n - 3a_0 - 3\sum_{n=1}^{\infty} a_n x^n$$

$$= -3a_0 + a_1 + \sum_{n=1}^{\infty} \left[na_n + (n+1)a_{n+1} - 3a_n\right] x^n$$

$$= 0 \Rightarrow a_1 = 3a_0 \text{ and } a_{n+1} = \frac{3-n}{n+1} a_n.$$

Hence, $a_2 = \frac{2}{2}a_1 = 3a_0$, $a_3 = \frac{1}{3}a_2 = a_0$, $a_4 = 0$,

and since every other a_i is $\frac{3-n}{n+1}$ times its predecessor, the rest of the terms are 0.

Thus, $y = \sum_{n=0}^{3} a_n x^n = a_0 + 3a_0 x + 3a_0 x^2 + a_0 x^3 = a_0(x+1)^3$.

8 $y' - 4x^3 y = \sum_{n=1}^{\infty} na_n x^{n-1} - 4x^3 \sum_{n=0}^{\infty} a_n x^n$ {replace n with $n + 4$}

$$= \sum_{n=-3}^{\infty} (n+4)a_{n+4} x^{n+3} - 4\sum_{n=0}^{\infty} a_n x^{n+3}$$

$$= a_1 + 2a_2 x + 3a_3 x^2 + \sum_{n=0}^{\infty} \left[(n+4)a_{n+4} - 4a_n\right] x^{n+3}$$

$$= 0 \Rightarrow a_1 = a_2 = a_3 = 0 \text{ and } a_{n+4} = \frac{4}{n+4} a_n.$$

For $n = 0, 4, 8, \ldots$: $a_4 = \frac{4}{4}a_0 = a_0$, $a_8 = \frac{4}{8}a_4 = \frac{1}{2}a_0$, $a_{12} = \frac{4}{12}a_8 = \frac{1}{3\cdot 2}a_0$,

$a_{16} = \frac{4}{16}a_{12} = \frac{1}{4\cdot 3\cdot 2}a_0$, ..., and $a_{4k} = \frac{1}{k!}a_0$. All other terms are 0.

Thus, $y = a_0 \sum_{n=0}^{\infty} \frac{x^{4n}}{n!} = a_0 \sum_{n=0}^{\infty} \frac{(x^4)^n}{n!} = a_0 e^{(x^4)}$.

9 $y'' - y - 5x = \sum_{n=2}^{\infty} n(n-1)a_n x^{n-2} - \sum_{n=0}^{\infty} a_n x^n - 5x$

$$= \sum_{n=0}^{\infty} (n+2)(n+1)a_{n+2} x^n - \sum_{n=0}^{\infty} a_n x^n - 5x$$

$$= -5x + \sum_{n=0}^{\infty} \left[(n+2)(n+1)a_{n+2} - a_n\right] x^n$$

{let $n = 0, 1$ to determine the coefficient of x}

$$= (2a_2 - a_0) + (-5 + 6a_3 - a_1)x + \sum_{n=2}^{\infty} \left[(n+2)(n+1)a_{n+2} - a_n\right] x^n$$

$$= 0 \Rightarrow a_2 = \frac{1}{2\cdot 1}a_0,\ a_3 = \frac{a_1+5}{3\cdot 2}, \text{ and } a_{n+2} = \frac{1}{(n+2)(n+1)} a_n.$$

For $n = 2, 4, 6, \ldots$: $a_4 = \frac{1}{4\cdot 3}a_2 = \frac{1}{4!}a_0$, $a_6 = \frac{1}{6\cdot 5}a_4 = \frac{1}{6!}a_0$, ...,

and $a_{2k} = \frac{1}{(2k)!}a_0$.

For $n = 3, 5, 7, \ldots$: $a_5 = \frac{1}{5\cdot 4}a_3 = \frac{a_1+5}{5!}$, $a_7 = \frac{1}{7\cdot 6}a_5 = \frac{a_1+5}{7!}$,

and $a_{2k+1} = \frac{a_1+5}{(2k+1)!}$.

$$\text{Thus, } y = -5x + a_0 \sum_{n=0}^{\infty} \frac{1}{(2n)!} x^{2n} + (a_1 + 5) \sum_{n=0}^{\infty} \frac{1}{(2n+1)!} x^{2n+1}$$
$$= -5x + a_0 \cosh x + (a_1 + 5) \sinh x$$
$$= -5x + a_0 \left(\frac{e^x + e^{-x}}{2} \right) + (a_1 + 5) \left(\frac{e^x - e^{-x}}{2} \right)$$
$$= -5x + (\tfrac{1}{2}a_0 + \tfrac{1}{2}a_1 + \tfrac{5}{2})e^x + (\tfrac{1}{2}a_0 - \tfrac{1}{2}a_1 - \tfrac{5}{2})e^{-x}$$

10 $y'' - xy - x^4 = \sum_{n=2}^{\infty} n(n-1)a_n x^{n-2} - x \sum_{n=0}^{\infty} a_n x^n - x^4$

$$= \sum_{n=-1}^{\infty} (n+3)(n+2)a_{n+3} x^{n+1} - \sum_{n=0}^{\infty} a_n x^{n+1} - x^4$$
$$= 2a_2 - x^4 + \sum_{n=0}^{\infty} \left[(n+3)(n+2)a_{n+3} - a_n \right] x^{n+1} = 0 \Rightarrow$$

$a_2 = 0$, $6 \cdot 5 a_6 - a_3 - 1 = 0$ {x^4 terms}, and $a_{n+3} = \dfrac{1}{(n+3)(n+2)} a_n$, $n \neq 3$.

Since $a_2 = 0$, $a_5 = a_8 = a_{11} = \cdots = 0$. For $n = 0, 3, 6, \ldots$:

$a_3 = \dfrac{1}{3 \cdot 2} a_0$, $a_6 = \dfrac{1}{6 \cdot 5} + \dfrac{1}{6 \cdot 5} a_3$ {from above} $= \dfrac{4 \cdot 3!}{6 \cdot 5 \cdot 4!} + \dfrac{4}{6!} a_0 = \dfrac{4(6 + a_0)}{6!}$,

$a_9 = \dfrac{1}{9 \cdot 8} a_6 = \dfrac{7 \cdot 4(6 + a_0)}{9!}$, and $a_{3k} = \dfrac{(3k-2)(3k-5) \cdots 7 \cdot 4(6 + a_0)}{(3k)!}$, for $k \geq 2$.

For $n = 1, 4, 7, \ldots$: $a_4 = \dfrac{1}{4 \cdot 3} a_1 = \dfrac{2}{4!} a_1$, $a_7 = \dfrac{1}{7 \cdot 6} a_4 = \dfrac{5 \cdot 2}{7!} a_1$,

and $a_{3k+1} = \dfrac{(3k-1)(3k-4) \cdots 5 \cdot 2}{(3k+1)!} a_1$.

$$y = \sum_{n=0}^{\infty} a_n x^n = a_0 + a_1 x + a_3 x^3 + \sum_{n=4}^{\infty} a_n x^n.$$

$$\text{Thus, } y = a_0 + a_1 x + \tfrac{1}{6} a_0 x^3 + (a_0 + 6) \sum_{n=2}^{\infty} \frac{(3n-2)(3n-5) \cdots 7 \cdot 4}{(3n)!} x^{3n} +$$
$$a_1 \sum_{n=1}^{\infty} \frac{(3n-1)(3n-4) \cdots 5 \cdot 2}{(3n+1)!} x^{3n+1}.$$

11 $x^2 y'' - y'' + 6xy' + 4y$ {we will leave -4 on the right side}

$$= x^2 \sum_{n=2}^{\infty} n(n-1)a_n x^{n-2} - \sum_{n=2}^{\infty} n(n-1)a_n x^{n-2} + 6x \sum_{n=1}^{\infty} n a_n x^{n-1} + 4 \sum_{n=0}^{\infty} a_n x^n$$
$$= \sum_{n=2}^{\infty} n(n-1)a_n x^n - \sum_{n=0}^{\infty} (n+2)(n+1)a_{n+2} x^n + 6 \sum_{n=1}^{\infty} n a_n x^n + 4 \sum_{n=0}^{\infty} a_n x^n$$
$$= (-2a_2 + 4a_0) + (-3 \cdot 2a_3 + 6a_1 + 4a_1)x +$$
$$\sum_{n=2}^{\infty} \left[n(n-1)a_n - (n+2)(n+1)a_{n+2} + 6na_n + 4a_n \right] x^n = -4 \Rightarrow$$

$a_2 = 2a_0 + 2 = 2(a_0 + 1)$, $a_3 = \tfrac{5}{3} a_1$, and $a_{n+2} = \dfrac{n^2 + 5n + 4}{(n+2)(n+1)} a_n = \dfrac{n+4}{n+2} a_n$.

For $n = 2, 4, 6, \ldots$: $a_4 = \tfrac{6}{4} a_2 = 3(a_0 + 1)$, $a_6 = \tfrac{8}{6} a_4 = 4(a_0 + 1)$, $\ldots$,

and $a_{2k} = (k+1)(a_0 + 1) = (k+1)a_0 + (k+1)$.

For $n = 3, 5, 7, \ldots$: $a_5 = \tfrac{7}{5} a_3 = \tfrac{7}{3} a_1$, $a_7 = \tfrac{9}{7} a_5 = \tfrac{9}{3} a_1$, $\ldots$, and $a_{2k+1} = \dfrac{2k+3}{3} a_1$.

$$\text{Thus, } y = a_0 \sum_{n=0}^{\infty} (n+1) x^{2n} + \sum_{n=1}^{\infty} (n+1) x^{2n} + a_1 \sum_{n=0}^{\infty} \left(\frac{2n+3}{3} \right) x^{2n+1}.$$

[12] $y'' + y - e^x = \sum_{n=0}^{\infty}\left[a_n + (n+2)(n+1)a_{n+2}\right]x^n - \sum_{n=0}^{\infty}\frac{1}{n!}x^n$ {Exercise 1 and (11.48)(c)} $= 0 \Rightarrow a_{n+2} = \dfrac{(1/n!) - a_n}{(n+2)(n+1)} = \dfrac{1}{(n+2)!} - \dfrac{1}{(n+2)(n+1)}a_n$.

For $n = 2, 4, 6, \ldots$: $a_2 = \frac{1}{2!} - \frac{1}{2\cdot 1}a_0$, $a_4 = \frac{1}{4!} - \frac{1}{4\cdot 3}a_2 = \frac{1}{4!} - \frac{1}{4!} + \frac{1}{4!}a_0 = \frac{1}{4!}a_0$,

$a_6 = \frac{1}{6!} - \frac{1}{6\cdot 5}a_4 = \frac{1}{6!} - \frac{1}{6!}a_0$, $a_8 = \frac{1}{8!} - \frac{1}{8\cdot 7}a_6 = \frac{1}{8!}a_0, \ldots,$

and $a_{4k+2} = \dfrac{1}{(4k+2)!} - \dfrac{1}{(4k+2)!}a_0$, $a_{4k+4} = \dfrac{1}{(4k+4)!}a_0$.

For $n = 1, 3, 5, \ldots$: $a_3 = \frac{1}{3!} - \frac{1}{3\cdot 2}a_1$, $a_5 = \frac{1}{5!} - \frac{1}{5\cdot 4}a_3 = \frac{1}{5!} - \frac{1}{5!} + \frac{1}{5!}a_1 = \frac{1}{5!}a_1$,

$a_7 = \frac{1}{7!} - \frac{1}{7\cdot 6}a_5 = \frac{1}{7!} - \frac{1}{7!}a_1$, $a_9 = \frac{1}{9!} - \frac{1}{9\cdot 8}a_7 = \frac{1}{9!}a_1, \ldots,$

and $a_{4k+3} = \dfrac{1}{(4k+3)!} - \dfrac{1}{(4k+3)!}a_1$, $a_{4k+5} = \dfrac{1}{(4k+5)!}a_1$.

Thus, $y = a_0 + a_1 x$

$$+ \sum_{n=0}^{\infty}\frac{1}{(4n+2)!}x^{4n+2} - a_0\sum_{n=0}^{\infty}\frac{1}{(4n+2)!}x^{4n+2} + a_0\sum_{n=0}^{\infty}\frac{1}{(4n+4)!}x^{4n+4}$$

$$+ \sum_{n=0}^{\infty}\frac{1}{(4n+3)!}x^{4n+3} - a_1\sum_{n=0}^{\infty}\frac{1}{(4n+3)!}x^{4n+3} + a_1\sum_{n=0}^{\infty}\frac{1}{(4n+5)!}x^{4n+5}.$$

This can be simplified to

$$y = \tfrac{1}{2}\sum_{n=0}^{\infty}\frac{x^n}{n!} + \left(a_0 - \tfrac{1}{2}\right)\sum_{n=0}^{\infty}\frac{(-1)^n}{(2n)!}x^{2n} + \left(a_1 - \tfrac{1}{2}\right)\sum_{n=0}^{\infty}\frac{(-1)^n}{(2n+1)!}x^{2n+1}.$$

19.7 Review Exercises

[1] $xe^y\,dx - \csc x\,dy = 0 \Rightarrow x\sin x\,dx - e^{-y}\,dy = 0 \Rightarrow \sin x - x\cos x + e^{-y} = C.$

[2] $x\,dy - (x+1)y\,dx = 0 \Rightarrow \frac{1}{y}\,dy = \left(1 + \frac{1}{x}\right)dx \Rightarrow \ln|y| = x + \ln|x| + k \Rightarrow$

$|y| = e^k\,|x|\,e^x \Rightarrow y = Cxe^x.$

[3] $x(1+y^2)\,dx + \sqrt{1-x^2}\,dy = 0 \Rightarrow \dfrac{x}{\sqrt{1-x^2}}\,dx + \dfrac{1}{1+y^2}\,dy = 0 \Rightarrow$

$-\sqrt{1-x^2} + \tan^{-1}y = C \Rightarrow y = \tan\left(\sqrt{1-x^2} + C\right).$

[4] $IF = e^{\int 4\,dx} = e^{4x}$. $y'e^{4x} + 4ye^{4x} = e^{3x} \Rightarrow ye^{4x} = \frac{1}{3}e^{3x} + C \Rightarrow$

$y = \frac{1}{3}e^{-x} + Ce^{-4x}.$

[5] $IF = e^{\int \sec x\,dx} = |\sec x + \tan x|.$

$y'(\sec x + \tan x) + y\sec x(\sec x + \tan x) = 2\cos x(\sec x + \tan x) = 2 + 2\sin x \Rightarrow$

$y(\sec x + \tan x) = 2x - 2\cos x + C \Rightarrow y = \dfrac{2x - 2\cos x + C}{\sec x + \tan x}.$

[6] $x^2(y+1)\,dy + y\,dx = 0 \Rightarrow \left(1 + \frac{1}{y}\right)dy + \frac{1}{x^2}\,dx = 0 \Rightarrow y + \ln|y| - \frac{1}{x} = C.$

[7] $IF = e^{\int \tan x\,dx} = |\sec x|$. $y'\sec x + y\sec x\tan x = 2\sec^2 x \Rightarrow$

$y\sec x = 2\tan x + C \Rightarrow y = 2\sin x + C\cos x.$

[8] $IF = e^{\int \tan x\,dx} = |\sec x|$. $y'\sec x + y\sec x\tan x = 3\sec x \Rightarrow$

$y\sec x = 3\ln|\sec x + \tan x| + C \Rightarrow y = (3\ln|\sec x + \tan x| + C)\cos x.$

[9] $\dfrac{y}{\sqrt{1-y^2}}\,dy = \dfrac{1}{\sqrt{1-x^2}}\,dx \Rightarrow -\sqrt{1-y^2} = \sin^{-1}x + C \Rightarrow \sqrt{1-y^2} + \sin^{-1}x = C.$

[10] $(2y + x^3)\,dx - x\,dy = 0 \Rightarrow y' - \frac{2}{x}y = x^2.$ $IF = e^{\int(-2/x)\,dx} = |x|^{-2}.$

$$x^{-2}y' - \frac{2}{x^3}y = 1 \Rightarrow \frac{y}{x^2} = x + C \Rightarrow y = x^3 + Cx^2.$$

[11] $e^x \cos y\,dy - \sin^2 y\,dx = 0 \Rightarrow \dfrac{\cos y}{\sin^2 y}\,dy - e^{-x}\,dx = 0 \Rightarrow$

$$-\cot y \csc y\,dy = -e^{-x}\,dx \Rightarrow \csc y = e^{-x} + C.$$

[12] $y' + y\tan x = 3\cos x \Rightarrow IF = |\sec x|.$

$$y'\sec x + y\sec x\tan x = 3 \Rightarrow y\sec x = 3x + C \Rightarrow y = (3x + C)\cos x.$$

[13] $IF = e^{\int 2\cos x\,dx} = e^{2\sin x}.$ $y'e^{2\sin x} + 2y\cos x\,e^{2\sin x} = \cos x\,e^{2\sin x} \Rightarrow$

$$y\,e^{2\sin x} = \tfrac{1}{2}e^{2\sin x} + C \Rightarrow y = \tfrac{1}{2} + Ce^{-2\sin x}.$$

[14] $m^2 + m - 6 = (m+3)(m-2) = 0 \Rightarrow m = -3, 2.$ Thus, $y = C_1e^{-3x} + C_2e^{2x}.$

[15] $m^2 - 8m + 16 = (m-4)^2 = 0 \Rightarrow m = 4$ is a double root.

Thus, $y = C_1e^{4x} + C_2xe^{4x} = e^{4x}(C_1 + C_2x).$

[16] $m^2 - 6m + 25 = 0 \Rightarrow m = \dfrac{6 \pm \sqrt{-64}}{2} = 3 \pm 4i.$

Thus, $y = e^{3x}(C_1\cos 4x + C_2\sin 4x).$

[17] $m^2 - 2m = m(m-2) = 0 \Rightarrow m = 0, 2.$ Thus, $y = C_1 + C_2e^{2x}.$

[18] $m^2 - 1 = 0 \Rightarrow m = \pm 1.$ $y_c = C_1e^{-x} + C_2e^x.$ Let $y_p = A\cos x + B\sin x.$

Then, $y_p'' - y_p = \big[(-A\cos x - B\sin x) - (A\cos x + B\sin x)\big] =$

$-2A\cos x - 2B\sin x = \sin x \Rightarrow A = 0$ and $B = -\frac{1}{2}.$

The general solution is $y = C_1e^{-x} + C_2e^x - \frac{1}{2}\sin x.$

[19] $m^2 - 1 = 0 \Rightarrow m = \pm 1.$ $y_c = C_1e^{-x} + C_2e^x.$ Let $y_p = Ae^x\cos x + Be^x\sin x.$

Then, $y_p'' - y_p = (2B - A)e^x\cos x + (-2A - B)e^x\sin x = e^x\sin x \Rightarrow A = -\frac{2}{5}$

and $B = -\frac{1}{5}.$ The general solution is $y = C_1e^{-x} + C_2e^x - \frac{1}{5}e^x(2\cos x + \sin x).$

[20] $m^2 - m - 6 = 0 \Rightarrow m = -2, 3.$ $y_c = C_1e^{-2x} + C_2e^{3x}.$ Let $y_p = Ae^{2x}.$

Then, $y_p'' - y_p' - 6y_p = (4A - 2A - 6A)e^{2x} = e^{2x} \Rightarrow A = -\frac{1}{4}.$

The general solution is $y = C_1e^{-2x} + C_2e^{3x} - \frac{1}{4}e^{2x}.$

[21] $IF = e^{\int 1\,dx} = e^x.$ $y'e^x + e^xy = e^{5x} \Rightarrow ye^x = \frac{1}{5}e^{5x} + C \Rightarrow y = \frac{1}{5}e^{4x} + Ce^{-x}.$

[22] $m^2 + 2m = m(m+2) = 0 \Rightarrow m = 0, -2.$ Thus, $y = C_1 + C_2e^{-2x}.$

[23] $m^2 - 3m + 2 = 0 \Rightarrow m = 1, 2.$ $y_c = C_1e^x + C_2e^{2x}.$ Let $y_p = Ae^{5x}.$

Then, $y_p'' - 3y_p' + 2y_p = (25A - 15A + 2A)e^{5x} = e^{5x} \Rightarrow A = \frac{1}{12}.$

The general solution is $y = C_1e^x + C_2e^{2x} + \frac{1}{12}e^{5x}.$

[24] $xe^y\,dx - (x+1)y\,dy = 0 \Rightarrow \left(1 - \dfrac{1}{x+1}\right)dx - ye^{-y}\,dy = 0 \Rightarrow$

$$x - \ln|x+1| + ye^{-y} + e^{-y} = C.$$

[25] $xy' + y = (x-2)^2 \Rightarrow xy = \frac{1}{3}(x-2)^3 + C \Rightarrow y = \dfrac{(x-2)^3}{3x} + \dfrac{C}{x}.$

[26] $\sec^2 y\,dx + x\sec^2 y\,dx = \sqrt{1-x^2}\,dy \Rightarrow \dfrac{1}{\sqrt{1-x^2}}\,dx + \dfrac{x}{\sqrt{1-x^2}}\,dx = \cos^2 y\,dy \Rightarrow$

$$\sin^{-1}x - \sqrt{1-x^2} - \tfrac{1}{2}y - \tfrac{1}{4}\sin 2y = C.$$

[27] $m^2 + 5m + 7 = 0 \Rightarrow m = \dfrac{-5 \pm \sqrt{-3}}{2} = -\frac{5}{2} \pm \frac{1}{2}\sqrt{3}\,i.$

$$\text{Thus, } y = e^{-5x/2}\left[C_1\cos\left(\tfrac{1}{2}\sqrt{3}\,x\right) + C_2\sin\left(\tfrac{1}{2}\sqrt{3}\,x\right)\right].$$

[28] $m^2 + 1 = 0 \Rightarrow m = \pm i.$ $y_c = C_1\cos x + C_2\sin x.$ Using (19.13),

$$\begin{cases} u'\cos x + v'\sin x = 0 \\ -u'\sin x + v'\cos x = \csc x \end{cases} \Rightarrow \begin{cases} v' = \cot x \\ u' = -1 \end{cases}$$

Integrating gives us $u = -x$ and $v = \ln|\sin x|$.

$$y = (C_1 - x)\cos x + (C_2 + \ln|\sin x|)\sin x.$$

[29] $e^x e^y\,dx - \csc x\,dy = 0 \Rightarrow e^x\sin x\,dx - e^{-y}\,dy = 0 \Rightarrow \frac{1}{2}e^x(\sin x - \cos x) + e^{-y} = C.$

[30] $m^2 + 10m + 25 = (m+5)^2 = 0 \Rightarrow m = -5$ is a double root.

$$\text{Thus, } y = C_1e^{-5x} + C_2xe^{-5x} = e^{-5x}(C_1 + C_2x).$$

[31] $\cot x\,dy = (y - \cos x)\,dx \Rightarrow y' - y\tan x = -\sin x \Rightarrow$

$IF = e^{\int -\tan x\,dx} = |\cos x|.$ $y'\cos x - y\sin x = -\sin x\cos x \Rightarrow$

$$y\cos x = \tfrac{1}{2}\cos^2 x + C \Rightarrow y = \tfrac{1}{2}\cos x + C\sec x.$$

[32] $m^2 + m + 1 = 0 \Rightarrow m = -\frac{1}{2} \pm \frac{1}{2}\sqrt{3}\,i.$

$y_c = e^{-x/2}\left[C_1\cos\left(\frac{1}{2}\sqrt{3}\,x\right) + C_2\sin\left(\frac{1}{2}\sqrt{3}\,x\right)\right].$ Let $y_p = Ae^x\cos x + Be^x\sin x.$ Then,

$$\begin{aligned} y_p'' + y_p' + y_p &= \left[2B + (A+B) + A\right]e^x\cos x + \left[-2A + (B-A) + B\right]e^x\sin x \\ &= (2A + 3B)\,e^x\cos x + (2B - 3A)\,e^x\sin x = e^x\cos x \Rightarrow \end{aligned}$$

$A = \frac{2}{13}$ and $B = \frac{3}{13}$. The general solution is

$$y = e^{-x/2}\left[C_1\cos\left(\tfrac{1}{2}\sqrt{3}\,x\right) + C_2\sin\left(\tfrac{1}{2}\sqrt{3}\,x\right)\right] + \tfrac{1}{13}e^x(2\cos x + 3\sin x).$$

[33] $IF = |\csc x - \cot x|.$ $y'(\csc x - \cot x) + y(\csc^2 x - \csc x\cot x) = \sec x - 1 \Rightarrow$

$$y(\csc x - \cot x) = \ln|\sec x + \tan x| - x + C \Rightarrow y = \frac{\ln|\sec x + \tan x| - x + C}{\csc x - \cot x}.$$

[34] $m^2 - m - 20 = 0 \Rightarrow m = -4, 5.$ $y_c = C_1e^{-4x} + C_2e^{5x}.$

Let $y_p = (A + Bx)e^{-x}$. Then, $y_p'' - y_p' - 20y_p$

$= \{[(A - 2B) - (B - A) - 20A] + [B - (-B) - 20B]\,x\}\,e^{-x}$

$= [(-18A - 3B) - 18Bx]\,e^{-x} = xe^{-x} \Rightarrow B = -\frac{1}{18}$ and $A = \frac{1}{108}.$

$$\text{The general solution is } y = C_1e^{-4x} + C_2e^{5x} + \left(\tfrac{1}{108} - \tfrac{1}{18}x\right)e^{-x}.$$

35 $y' - (3\sin 2\pi t)y = 2000\sin 2\pi t \Rightarrow IF = e^{\int -3\sin 2\pi t\,dt} = e^{(3/(2\pi))\cos 2\pi t}$.

$y'\,e^{(3/(2\pi))\cos 2\pi t} - (3\sin 2\pi t)\,e^{(3/(2\pi))\cos 2\pi t} = (2000\sin 2\pi t)\,e^{(3/(2\pi))\cos 2\pi t} \Rightarrow$

$ye^{(3/(2\pi))\cos 2\pi t} = -\frac{2000}{3}\,e^{(3/(2\pi))\cos 2\pi t} + C \Rightarrow y = -\frac{2000}{3} + Ce^{-(3/(2\pi))\cos 2\pi t}$.

$y = 500$ when $t = 0 \Rightarrow C = \frac{3500}{3}\,e^{3/(2\pi)}$. $y = \frac{3500}{3}\,e^{(3/(2\pi))(1-\cos 2\pi t)} - \frac{2000}{3}$.

The maximum value of y will occur when the exponent is maximum, that is,

when $\cos 2\pi t = -1$. Thus, the maximum is $\frac{3500}{3}\,e^{3/\pi} - \frac{2000}{3} \approx 2365$.

36 $y' = k(10 - y)^2 \Rightarrow \dfrac{dy}{(10 - y)^2} = k\,dt \Rightarrow \dfrac{1}{10 - y} = kt + C$. $y = 0$ when $t = 0 \Rightarrow$

$C = \frac{1}{10}$. $y = 2$ when $t = 30 \Rightarrow \frac{1}{8} = 30k + \frac{1}{10} \Rightarrow k = \frac{1}{1200}$. Thus, $y = \dfrac{10t}{t + 120}$.

37 $\dfrac{dy}{dt} = k(a - y)(b - y) \Rightarrow \dfrac{dy}{(a - y)(b - y)} = k\,dt$. Using partial fractions and

integrating, we have $\dfrac{1}{b - a}\ln\left(\dfrac{b - y}{a - y}\right) = kt + C$. $y = 0$ when $t = 0 \Rightarrow$

$C = \dfrac{\ln(b/a)}{b - a}$. Thus, $\dfrac{1}{b - a}\ln\left[\frac{a}{b}\left(\dfrac{b - y}{a - y}\right)\right] = kt \Rightarrow \frac{a}{b}\left(\dfrac{b - y}{a - y}\right) = e^{k(b-a)t} \Rightarrow$

$bye^{k(b-a)t} - ay = abe^{k(b-a)t} - ab \Rightarrow y = f(t) = \dfrac{ab\left[e^{k(b-a)t} - 1\right]}{be^{k(b-a)t} - a}$.

38 (a) $\frac{dy}{dt} = y(c - by) \Rightarrow \frac{dy}{y(c - by)} = dt$. Using partial fractions, we have

$\frac{1}{c}\ln\left|\frac{y}{c - by}\right| = t + D$. $f(0) = k \Rightarrow D = \frac{1}{c}\ln\left|\frac{k}{c - bk}\right|$. Substituting and

solving gives $t = \frac{1}{c}\ln\left[\frac{y(c - bk)}{k(c - by)}\right] \Rightarrow y = f(t) = \frac{ck}{bk + (c - bk)e^{-ct}}$.

(b) Since $\lim_{t\to\infty} e^{-ct} = 0$, $\lim_{t\to\infty} f(t) = \frac{c}{b}$.

(c) $f'(t)$ will be increasing when $f''(t) > 0$ and decreasing when $f''(t) < 0$.

$\frac{dy}{dt} = f'(t) = cy - by^2 \Rightarrow f''(t) = (c - 2by)\,y' = (c - 2by)(y)(c - by)$.

Since b and c are positive constants, $f''(t)$ is positive when $y < \frac{c}{2b}$ and

$f''(t)$ is negative when $y > \frac{c}{2b}$. From part (b), we know that $0 \le y < \frac{c}{b}$,

so $f''(t) > 0$ for $y > \frac{c}{b}$ doesn't apply to this problem.

(d) In general, there is a *PI* when $y = \frac{c}{2b}$.

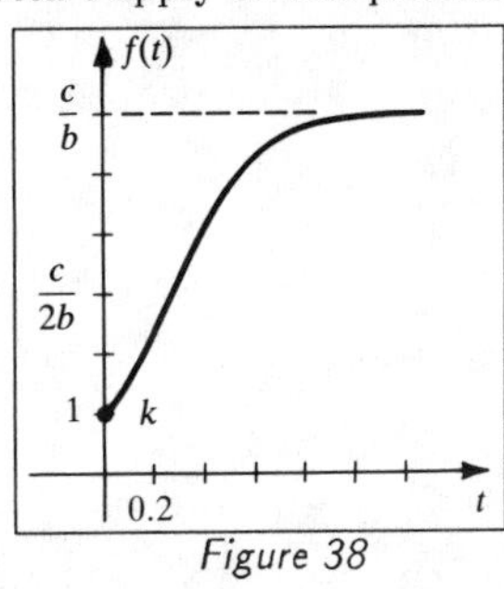

Figure 38

The figure shown has $k = b = 1$, and $c = 6$.

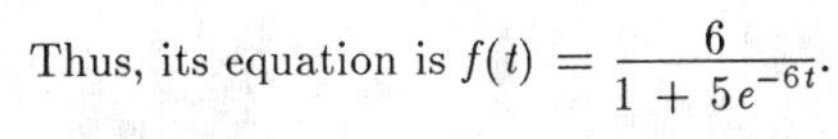

Thus, its equation is $f(t) = \frac{6}{1 + 5e^{-6t}}$.

39 The slope of the line connecting the origin with the point $P(x, y)$ is $m = \frac{y}{x}$.

If $y = f(x)$ has a tangent line at $P(x, y)$ perpendicular to this line, then $\frac{dy}{dx} = -\frac{x}{y}$.

Thus, $y\,dy + x\,dx = 0 \Rightarrow y^2 + x^2 = C$, which is a circle with center at the origin.

40 $\frac{d}{d\theta}\left(\sin\theta\,\frac{dV}{d\theta}\right) = 0 \Rightarrow \sin\theta\,\frac{d^2V}{d\theta^2} + \cos\theta\,\frac{dV}{d\theta} = 0$. If we let $z = \frac{dV}{d\theta}$,

then $\frac{dz}{d\theta} + z\cot\theta = 0$. *IF* $= |\sin\theta|$, so $\frac{dz}{d\theta}\sin\theta + z\cos\theta = 0$, and $z\sin\theta = C$.

Moreover, $\frac{dV}{d\theta}\sin\theta = C \Rightarrow dV = C\csc\theta\,d\theta \Rightarrow V = C\ln|\csc\theta - \cot\theta| + k$.

$V(\frac{\pi}{2}) = 0 \Rightarrow k = 0$ and $V(\frac{\pi}{4}) = V_0 \Rightarrow C = \frac{V_0}{\ln(\sqrt{2} - 1)}$.

Thus, $V = \frac{V_0}{\ln(\sqrt{2} - 1)}\ln|\csc\theta - \cot\theta|$.

Appendix I: Mathematical Induction

Note: P_n is the statement in the text for Exercises 1–22.

[1] (i) P_1 is true, since $2(1) = 1(1 + 1) = 2$.

(ii) Assume P_k is true:

$2 + 4 + 6 + \cdots + 2k = k(k + 1)$. Hence

$$\begin{aligned} 2 + 4 + 6 + \cdots + 2k + 2(k + 1) &= k(k + 1) + 2(k + 1) \\ &= (k + 1)(k + 2) = (k + 1)(k + 1 + 1). \end{aligned}$$

Thus, P_{k+1} is true and the proof is complete.

[2] (i) P_1 is true, since $3(1) - 2 = \dfrac{1[3(1) - 1]}{2} = 1$.

(ii) Assume P_k is true:

$1 + 4 + 7 + \cdots + (3k - 2) = \dfrac{k(3k - 1)}{2}$. Hence

$$\begin{aligned} 1 + 4 + 7 + \cdots + (3k - 2) + 3(k + 1) - 2 &= \frac{k(3k - 1)}{2} + 3(k + 1) - 2 \\ &= \frac{3k^2 + 5k + 2}{2} \\ &= \frac{(k + 1)(3k + 2)}{2} \\ &= \frac{(k + 1)[3(k + 1) - 1]}{2}. \end{aligned}$$

Thus, P_{k+1} is true and the proof is complete.

[3] (i) P_1 is true, since $2(1) - 1 = (1)^2 = 1$.

(ii) Assume P_k is true:

$1 + 3 + 5 + \cdots + (2k - 1) = k^2$. Hence

$$\begin{aligned} 1 + 3 + 5 + \cdots + (2k - 1) + 2(k + 1) - 1 &= k^2 + 2(k + 1) - 1 \\ &= k^2 + 2k + 1 \\ &= (k + 1)^2. \end{aligned}$$

Thus, P_{k+1} is true and the proof is complete.

[4] (i) P_1 is true, since $6(1) - 3 = 3(1)^2 = 3$.

(ii) Assume P_k is true:

$3 + 9 + 15 + \cdots + (6k - 3) = 3k^2$. Hence

$$\begin{aligned} 3 + 9 + 15 + \cdots + (6k - 3) + 6(k + 1) - 3 &= 3k^2 + 6(k + 1) - 3 \\ &= 3k^2 + 6k + 3 \\ &= 3(k^2 + 2k + 1) \\ &= 3(k + 1)^2. \end{aligned}$$

Thus, P_{k+1} is true and the proof is complete.

5 (i) P_1 is true, since $5(1) - 3 = \frac{1}{2}(1)[5(1) - 1] = 2$.

(ii) Assume P_k is true:

$2 + 7 + 12 + \cdots + (5k - 3) = \frac{1}{2}k(5k - 1)$. Hence

$$\begin{aligned} 2 + 7 + 12 + \cdots + (5k - 3) + 5(k + 1) - 3 \\ &= \tfrac{1}{2}k(5k - 1) + 5(k + 1) - 3 \\ &= \tfrac{5}{2}k^2 + \tfrac{9}{2}k + 2 \\ &= \tfrac{1}{2}(5k^2 + 9k + 4) \\ &= \tfrac{1}{2}(k + 1)(5k + 4) \\ &= \tfrac{1}{2}(k + 1)[5(k + 1) - 1]. \end{aligned}$$

Thus, P_{k+1} is true and the proof is complete.

6 (i) P_1 is true, since $2 \cdot 3^{1-1} = 3^1 - 1 = 2$.

(ii) Assume P_k is true:

$2 + 6 + 18 + \cdots + 2 \cdot 3^{k-1} = 3^k - 1$. Hence

$$\begin{aligned} 2 + 6 + 18 + \cdots + 2 \cdot 3^{k-1} + 2 \cdot 3^k &= 3^k - 1 + 2 \cdot 3^k \\ &= 1 \cdot 3^k + 2 \cdot 3^k - 1 \\ &= 3^1 \cdot 3^k - 1 = 3^{k+1} - 1. \end{aligned}$$

Thus, P_{k+1} is true and the proof is complete.

7 (i) P_1 is true, since $1 \cdot 2^{1-1} = 1 + (1 - 1) \cdot 2^1 = 1$.

(ii) Assume P_k is true:

$1 + 2 \cdot 2 + 3 \cdot 2^2 + \cdots + k \cdot 2^{k-1} = 1 + (k - 1) \cdot 2^k$. Hence

$$\begin{aligned} 1 + 2 \cdot 2 + 3 \cdot 2^2 + \cdots + k \cdot 2^{k-1} + (k + 1) \cdot 2^k \\ &= 1 + (k - 1) \cdot 2^k + (k + 1) \cdot 2^k \\ &= 1 + k \cdot 2^k - 2^k + k \cdot 2^k + 2^k \\ &= 1 + k \cdot 2^1 \cdot 2^k \\ &= 1 + [(k + 1) - 1] \cdot 2^{k+1}. \end{aligned}$$

Thus, P_{k+1} is true and the proof is complete.

8 (i) P_1 is true, since $(-1)^1 = \dfrac{(-1)^1 - 1}{2} = -1$.

(ii) Assume P_k is true:

$(-1)^1 + (-1)^2 + (-1)^3 + \cdots + (-1)^k = \dfrac{(-1)^k - 1}{2}$. Hence

$$\begin{aligned} (-1)^1 + (-1)^2 + (-1)^3 + \cdots + (-1)^k + (-1)^{k+1} \\ &= \frac{(-1)^k - 1}{2} + (-1)^{k+1} \\ &= \frac{1(-1)^k}{2} - \frac{1}{2} - \frac{2(-1)^k}{2} \\ &= \frac{(-1)^k \cdot (-1) - 1}{2} \\ &= \frac{(-1)^{k+1} - 1}{2}. \end{aligned}$$

Thus, P_{k+1} is true and the proof is complete.

9 (i) P_1 is true, since $(1)^1 = \dfrac{1(1+1)[2(1)+1]}{6} = 1$.

(ii) Assume P_k is true:

$$1^2 + 2^2 + 3^2 + \cdots + k^2 = \frac{k(k+1)(2k+1)}{6}. \text{ Hence}$$

$$\begin{aligned} 1^2 + 2^2 + 3^2 + \cdots + k^2 + (k+1)^2 &= \frac{k(k+1)(2k+1)}{6} + (k+1)^2 \\ &= (k+1)\left[\frac{k(2k+1)}{6} + \frac{6(k+1)}{6}\right] \\ &= \frac{(k+1)(2k^2+7k+6)}{6} \\ &= \frac{(k+1)(k+2)(2k+3)}{6}. \end{aligned}$$

Thus, P_{k+1} is true and the proof is complete.

10 (i) P_1 is true, since $(1)^3 = \left[\dfrac{1(1+1)}{2}\right]^2 = 1$.

(ii) Assume P_k is true:

$$1^3 + 2^3 + 3^3 + \cdots + k^3 = \left[\frac{k(k+1)}{2}\right]^2. \text{ Hence}$$

$$\begin{aligned} 1^3 + 2^3 + 3^3 + \cdots + k^3 + (k+1)^3 &= \left[\frac{k(k+1)}{2}\right]^2 + (k+1)^3 \\ &= \frac{(k+1)^2}{2^2}[k^2 + 4(k+1)] \\ &= \frac{(k+1)^2}{2^2}(k+2)^2 \\ &= \left[\frac{(k+1)[(k+1)+1]}{2}\right]^2. \end{aligned}$$

Thus, P_{k+1} is true and the proof is complete.

11 (i) P_1 is true, since $\dfrac{1}{1(1+1)} = \dfrac{1}{1+1} = \dfrac{1}{2}$.

(ii) Assume P_k is true:

$$\frac{1}{1\cdot 2} + \frac{1}{2\cdot 3} + \frac{1}{3\cdot 4} + \cdots + \frac{1}{k(k+1)} = \frac{k}{k+1}. \text{ Hence}$$

$$\begin{aligned} &\frac{1}{1\cdot 2} + \frac{1}{2\cdot 3} + \frac{1}{3\cdot 4} + \cdots + \frac{1}{k(k+1)} + \frac{1}{(k+1)(k+2)} \\ &\quad = \frac{k}{k+1} + \frac{1}{(k+1)(k+2)} \\ &\quad = \frac{k(k+2)+1}{(k+1)(k+2)} \\ &\quad = \frac{k^2+2k+1}{(k+1)(k+2)} = \frac{k+1}{(k+1)+1}. \end{aligned}$$

Thus, P_{k+1} is true and the proof is complete.

[12] (i) P_1 is true, since $\dfrac{1}{1(1+1)(1+2)} = \dfrac{1(1+3)}{4(1+1)(1+2)} = \dfrac{1}{6}$.

(ii) Assume P_k is true:

$\dfrac{1}{1\cdot2\cdot3} + \dfrac{1}{2\cdot3\cdot4} + \dfrac{1}{3\cdot4\cdot5} + \cdots + \dfrac{1}{k(k+1)(k+2)} = \dfrac{k(k+3)}{4(k+1)(k+2)}$. Hence

$$\frac{1}{1\cdot2\cdot3} + \frac{1}{2\cdot3\cdot4} + \frac{1}{3\cdot4\cdot5} + \cdots + \frac{1}{k(k+1)(k+2)} + \frac{1}{(k+1)(k+2)(k+3)}$$
$$= \frac{k(k+3)}{4(k+1)(k+2)} + \frac{1}{(k+1)(k+2)(k+3)}$$
$$= \frac{k(k+3)^2+4}{4(k+1)(k+2)(k+3)} = \frac{k(k^2+6k+9)+4}{4(k+1)(k+2)(k+3)}$$
$$= \frac{k^3+6k^2+9k+4}{4(k+1)(k+2)(k+3)} = \frac{(k+1)(k^2+5k+4)}{4(k+1)(k+2)(k+3)}$$
$$= \frac{(k+1)(k+4)}{4(k+2)(k+3)}.$$

Thus, P_{k+1} is true and the proof is complete.

[13] (i) P_1 is true, since $3^1 = \frac{3}{2}(3^1 - 1) = 3$.

(ii) Assume P_k is true:

$3 + 3^2 + 3^3 + \cdots + 3^k = \frac{3}{2}(3^k - 1)$. Hence

$$3 + 3^2 + 3^3 + \cdots + 3^k + 3^{k+1} = \tfrac{3}{2}(3^k - 1) + 3^{k+1}$$
$$= \tfrac{3}{2}\cdot 3^k - \tfrac{3}{2} + 3\cdot 3^k = \tfrac{9}{2}\cdot 3^k - \tfrac{3}{2}$$
$$= \tfrac{3}{2}(3\cdot 3^k - 1) = \tfrac{3}{2}(3^{k+1} - 1).$$

Thus, P_{k+1} is true and the proof is complete.

[14] (i) P_1 is true, since $[2(1) - 1]^3 = (1)^2(2\cdot 1^2 - 1) = 1$.

(ii) Assume P_k is true:

$1^3 + 3^3 + 5^3 + \cdots + (2k-1)^3 = k^2(2k^2 - 1)$. Hence

$$1^3 + 3^3 + 5^3 + \cdots + (2k-1)^3 + [2(k+1) - 1]^3$$
$$= k^2(2k^2 - 1) + [2(k+1) - 1]^3$$
$$= 2k^4 - k^2 + (2k+1)^3 = 2k^4 + 8k^3 + 11k^2 + 6k + 1$$
$$= (k+1)^2(2k^2 + 4k + 1) = (k+1)^2[2(k+1)^2 - 1].$$

Thus, P_{k+1} is true and the proof is complete.

[15] (i) P_1 is true, since $1 < 2^1$.

(ii) Assume P_k is true: $k < 2^k$. Now $k + 1 < k + k = 2(k)$ for $k > 1$.

From P_k, we see that $2(k) < 2(2^k) = 2^{k+1}$ and conclude that $k + 1 < 2^{k+1}$.

Thus, P_{k+1} is true and the proof is complete.

[16] (i) P_1 is true, since $1 + 2(1) \le 3^1$.

(ii) Assume P_k is true: $1 + 2k \le 3^k$.

$1 + 2(k+1) = 2k + 3 < 6k + 3$ which is $3(1 + 2k)$; From P_k, we see that $3(1 + 2k) < 3(3^k) = 3^{k+1}$ and conclude that $1 + 2(k+1) \le 3^{k+1}$.

Thus, P_{k+1} is true and the proof is complete.

17 (i) P_1 is true, since $1 < \frac{1}{8}[2(1) + 1]^2 = \frac{9}{8}$.

(ii) Assume P_k is true: $1 + 2 + 3 + \cdots + k < \frac{1}{8}(2k + 1)^2$. Hence

$$\begin{aligned} 1 + 2 + 3 + \cdots + k + (k + 1) &< \tfrac{1}{8}(2k + 1)^2 + (k + 1) \\ &= \tfrac{1}{2}k^2 + \tfrac{3}{2}k + \tfrac{9}{8} = \tfrac{1}{8}(4k^2 + 12k + 9) \\ &= \tfrac{1}{8}(2k + 3)^2 = \tfrac{1}{8}[2(k + 1) + 1]^2. \end{aligned}$$

Thus, P_{k+1} is true and the proof is complete.

18 (i) If $0 < a < b$, then $a^2 b < ab^2$ {multiply by ab} and $\frac{a^2}{b^2} < \frac{a}{b}$ {divide by b^3}.

This is P_1: $(\frac{a}{b})^2 < (\frac{a}{b})^1$.

(ii) Assume P_k is true: $(\frac{a}{b})^{k+1} < (\frac{a}{b})^k$. Hence, $a^{k+1}b^k < a^k b^{k+1} \Rightarrow$

$a^{k+2}b^{k+1} < a^{k+1}b^{k+2}$ {multiply by ab} $\Rightarrow \frac{a^{k+2}}{b^{k+2}} < \frac{a^{k+1}}{b^{k+1}}$ {divide by b^{2k+3}}.

This is P_{k+1}: $(\frac{a}{b})^{k+2} < (\frac{a}{b})^{k+1}$.

Thus, P_{k+1} is true and the proof is complete.

19 (i) For $n = 1$, $n^3 - n + 3 = 3$ and 3 is a factor of 3.

(ii) Assume 3 is a factor of $k^3 - k + 3$. The $(k + 1)$st term is

$$\begin{aligned} (k + 1)^3 - (k + 1) + 3 &= k^3 + 3k^2 + 2k + 3 \\ &= (k^3 - k + 3) + 3k^2 + 3k \\ &= (k^3 - k + 3) + 3(k^2 + k). \end{aligned}$$

By the induction hypothesis, 3 is a factor of $k^3 - k + 3$ and

3 is a factor of $3(k^2 + k)$, so 3 is a factor of the $(k + 1)$st term.

Thus, P_{k+1} is true and the proof is complete.

20 (i) For $n = 1$, $n^2 + n = 2$ and 2 is a factor of 2.

(ii) Assume 2 is a factor of $k^2 + k$. The $(k + 1)$st term is

$$\begin{aligned} (k + 1)^2 + (k + 1) &= k^2 + 3k + 2 \\ &= (k^2 + k) + 2k + 2 = (k^2 + k) + 2(k + 1). \end{aligned}$$

By the induction hypothesis, 2 is a factor of $k^2 + k$ and

2 is a factor of $2(k + 1)$, so 2 is a factor of the $(k + 1)$st term.

Thus, P_{k+1} is true and the proof is complete.

21 (i) For $n = 1$, $5^n - 1 = 4$ and 4 is a factor of 4.

(ii) Assume 4 is a factor of $5^k - 1$. The $(k + 1)$st term is

$5^{k+1} - 1 = 5 \cdot 5^k - 1 = 5 \cdot 5^k - 5 + 4 = 5(5^k - 1) + 4$.

By the induction hypothesis, 4 is a factor of 5^{k-1} and

4 is a factor of 4, so 4 is a factor of the $(k + 1)$st term.

Thus, P_{k+1} is true and the proof is complete.

22 (i) For $n = 1$, $10^{n+1} + 3 \cdot 10^n + 5 = 135$ and 9 is a factor of 135.

(ii) Assume 9 is a factor of $10^{k+1} + 3 \cdot 10^k + 5$. The $(k + 1)$st term is

$$\begin{aligned} 10^{k+2} + 3 \cdot 10^{k+1} + 5 &= 10 \cdot 10^{k+1} + 10 \cdot 3 \cdot 10^k + 5 \\ &= 10^{k+1} + 9 \cdot 10^{k+1} + 3 \cdot 10^k + 9 \cdot 3 \cdot 10^k + 5 \\ &= (10^{k+1} + 3 \cdot 10^k + 5) + 9(10^{k+1} + 3 \cdot 10^k). \end{aligned}$$

By the induction hypothesis, 9 is a factor of $10^{k+1} + 3 \cdot 10^k + 5$ and

9 is a factor of $9(10^{k+1} + 3 \cdot 10^k)$, so 9 is a factor of the $(k + 1)$st term.

Thus, P_{k+1} is true and the proof is complete.

Note: For Exercises 23-30, there are several ways to find j. Possibilities include solving the inequality, sketching the graphs of the functions that represent each side, and trial and error. Trial and error may be the easiest to use.

23 $n + 12 \leq n^2$ •

For j: $n^2 \geq n + 12 \Rightarrow n^2 - n - 12 \geq 0 \Rightarrow (n - 4)(n + 3) \geq 0 \Rightarrow n \geq 4 \{n>0\}$

(i) P_4 is true, since $4 + 12 \leq 4^2$.

(ii) Assume P_k is true: $k + 12 \leq k^2$. Hence

$$(k + 1) + 12 = (k + 12) + 1 \leq (k^2) + 1 < k^2 + 2k + 1 = (k + 1)^2.$$

Thus, P_{k+1} is true and the proof is complete.

24 $n^2 + 18 \leq n^3$ • For j: By trial and error, $j = 3$.

(i) P_3 is true, since $3^2 + 18 \leq 3^3$.

(ii) Assume P_k is true: $k^2 + 18 \leq k^3$. Hence

$$(k + 1)^2 + 18 = (k^2 + 18) + 2k + 1$$
$$\leq (k^3) + 2k + 1 < k^3 + 3k^2 + 3k + 1 = (k + 1)^3.$$

Thus, P_{k+1} is true and the proof is complete.

25 $5 + \log_2 n \leq n$ •

For j: By sketching $y = 5 + \log_2 x$ and $y = x$, we see that the solution for $x > 1$ must be larger than 5. See *Figure 25*. By trial and error, $j = 8$.

(i) P_8 is true, since $5 + \log_2 8 \leq 8$.

(ii) Assume P_k is true: $5 + \log_2 k \leq k$. Hence

$$5 + \log_2 (k + 1) < 5 + \log_2 (k + k) = 5 + \log_2 2k$$
$$= 5 + \log_2 2 + \log_2 k$$
$$= (5 + \log_2 k) + 1 \leq k + 1.$$

Thus, P_{k+1} is true and the proof is complete.

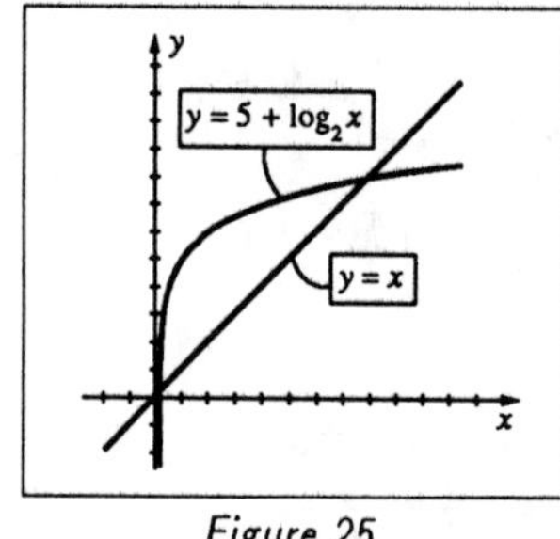

Figure 25

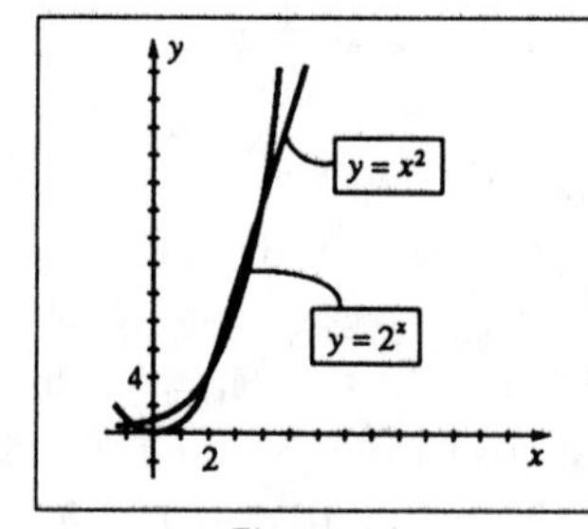

Figure 26

26 $n^2 \leq 2^n$ • For j: By sketching $y = x^2$ and $y = 2^x$, we see that there are 3 intersection points, the largest being 4.

(i) P_4 is true, since $4^2 \leq 2^4$.

(ii) Assume P_k is true: $k^2 \leq 2^k$. Hence

$$(k + 1)^2 = k^2 + 2k + 1 = k(k + 2 + \tfrac{1}{k}) < k(k + k) = 2k^2 \leq 2 \cdot 2^k = 2^{k+1}.$$

Thus, P_{k+1} is true and the proof is complete.

[27] $2^n \le n!$ • For j: Examining the pattern formed by letting $n = 1, 2, 3, 4$ leads us to the conclusion that $j = 4$.

(i) P_4 is true, since $2^4 \le 4!$.

(ii) Assume P_k is true: $2^k \le k!$. Hence

$2^{k+1} = 2 \cdot 2^k \le 2 \cdot k! < (k + 1) \cdot k! = (k + 1)!$.

Thus, P_{k+1} is true and the proof is complete.

[28] $10^n \le n^n$ •

For j: $10^n \le n^n \Rightarrow \left(\frac{n}{10}\right)^n \ge 1$. This is true if $\frac{n}{10} \ge 1$ or $n \ge 10$. Thus, $j = 10$.

(i) P_{10} is true, since $10^{10} \le 10^{10}$.

(ii) Assume P_k is true: $10^k \le k^k$. Hence

$10^{k+1} = 10 \cdot 10^k \le 10 \cdot k^k < (k + 1) \cdot k^k < (k + 1) \cdot (k + 1)^k = (k + 1)^{k+1}$.

Thus, P_{k+1} is true and the proof is complete.

[29] $2n + 2 \le 2^n$ • For j: By sketching $y = 2x + 2$ and $y = 2^x$, we see there is one positive solution. By trial and error, $j = 3$.

(i) P_3 is true, since $2(3) + 2 \le 2^3$.

(ii) Assume P_k is true: $2k + 2 \le 2^k$. Hence

$2(k + 1) + 2 = (2k + 2) + 2 \le 2^k + 2^k = 2 \cdot 2^k = 2^{k+1}$.

Thus, P_{k+1} is true and the proof is complete.

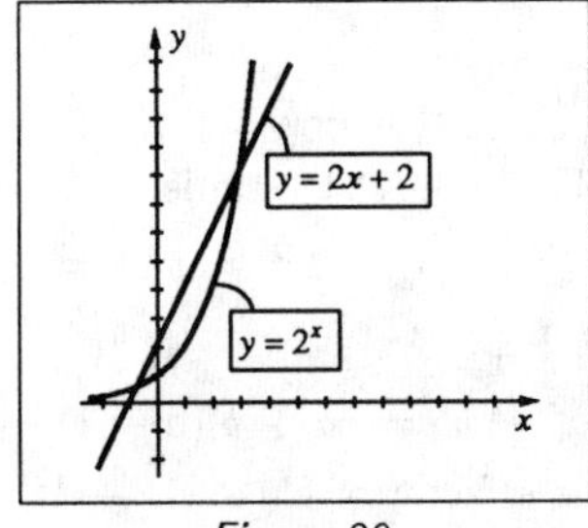

Figure 29

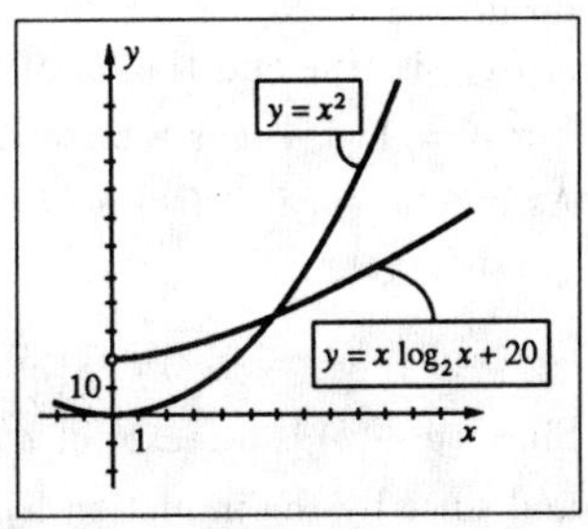

Figure 30

[30] $n \log_2 n + 20 \le n^2$ • For j: Since $n^2 < 20$ if $n = 4$, a reasonable first guess would be $j = 5$. See *Figure 30*. By trial and error, $j = 6$.

(i) P_6 is true, since $6 \log_2 6 + 20 \le 6^2$.

(ii) Assume P_k is true: $k \log_2 k + 20 \le k^2$.

$$\begin{aligned}(k + 1)\log_2(k + 1) + 20 &= k\log_2(k + 1) + \log_2(k + 1) + 20 \\ &< k\log_2 2k + \log_2 2k + 20 \\ &= k\log_2 k + k + 1 + \log_2 k + 20 \\ &\le k^2 + k + 1 + \log_2 k \\ &< k^2 + 2k + 1 = (k + 1)^2.\end{aligned}$$

Thus, P_{k+1} is true and the proof is complete.

[31] (i) If $a > 1$ then $a^1 = a > 1$, so P_1 is true.

(ii) Assume P_k is true: $a^k > 1$.

Multiply both sides by a to obtain $a^{k+1} > a$, but since $a > 1$, we have $a^{k+1} > 1$.

Thus, P_{k+1} is true and the proof is complete.

[32] (i) For $n = 1$, $ar^{1-1} = a$ and $\dfrac{a(1 - r^1)}{1 - r} = a$, so P_1 is true.

(ii) Assume P_k is true:

$a + ar + ar^2 + \cdots + ar^{k-1} = \dfrac{a(1 - r^k)}{1 - r}$. Hence

$$\begin{aligned} a + ar + ar^2 + \cdots + ar^{k-1} + ar^k &= \frac{a(1 - r^k)}{1 - r} + ar^k \\ &= a\left(\frac{1 - r^k}{1 - r} + \frac{r^k(1 - r)}{1 - r}\right) \\ &= a\left(\frac{1 - r^k + r^k - r^{k+1}}{1 - r}\right) \\ &= \frac{a(1 - r^{k+1})}{1 - r}. \end{aligned}$$

Thus, P_{k+1} is true and the proof is complete.

[33] (i) For $n = 1$, $a - b$ is a factor of $a^1 - b^1$.

(ii) Assume $a - b$ is a factor of $a^k - b^k$. Following the hint for the $(k + 1)$st term,
$a^{k+1} - b^{k+1} = a^k \cdot a - b \cdot a^k + b \cdot a^k - b^k \cdot b = a^k(a - b) + (a^k - b^k)b$.
Since $(a - b)$ is a factor of $a^k(a - b)$ and since by the induction hypothesis, $a - b$ is a factor of $(a^k - b^k)$, it follows that $a - b$ is a factor of the $(k + 1)$st term.

Thus, P_{k+1} is true and the proof is complete.

[34] (i) For $n = 1$, $a + b$ is a factor of $a^{2(1)-1} + b^{2(1)-1} = a + b$.

(ii) Assume $a + b$ is a factor of $a^{2k-1} + b^{2k-1}$. The $(k + 1)$st term is

$$\begin{aligned} a^{2k+1} + b^{2k+1} &= a^{2k-1} \cdot a^2 - a^{2k-1} \cdot b^2 + a^{2k-1} \cdot b^2 + b^{2k-1} \cdot b^2 \\ &= a^{2k-1}(a^2 - b^2) + b^2(a^{2k-1} + b^{2k-1}). \end{aligned}$$

Since $(a + b)$ is a factor of $a^{2k-1}(a^2 - b^2)$ $\{a^2 - b^2 = (a + b)(a - b)\}$ and since by the induction hypothesis, $a + b$ is a factor of $b^2(a^{2k-1} + b^{2k-1})$, it follows that $a + b$ is a factor of the $(k + 1)$st term.

Thus, P_{k+1} is true and the proof is complete.

bollywood dreams

AN EXPLORATION OF THE MOTION PICTURE INDUSTRY
AND ITS CULTURE IN INDIA

bollywood dreams

BY JONATHAN TORGOVNIK

LIBRARY
FRANKLIN PIERCE UNIVERSITY
RINDGE, NH 03461

for Tali

PN
1993.5
.I8
T67
2003

A Way of Life: An Introduction to Indian Cinema
by Nasreen Munni Kabir

Indian movies have a curiously infectious quality and have held a special place in Indian life ever since the birth of the Indian movie industry in 1913. Nothing in today's Indian popular culture is as pervasive as Bollywood movies – the Hindi and Urdu commercial films made in Mumbai (formerly Bombay) – with their distinctive approach to storytelling. Usually woven together by six songs and at least two lavish dance numbers, the movies are about unconditional love, the conflict between fathers and sons, revenge, redemption, survival against the odds, the importance of honour and self-respect, and the mission to uphold religious and moral values – grand themes that Hollywood generally leaves to the now rarely produced epic. Not so in India, where film directors routinely tackle the big questions head-on, even when making a formulaic run-of-the-mill entertainer. It is this particular kind of storytelling that has offered people of Indian origin their most beloved form of popular entertainment.

Bollywood movies are characterized by a selective number of cinematic ingredients that are reworked in each film. Indeed, repetition is part of the often predictable plots. To satisfy an audience, however, the right buttons must be pressed. These include great performances by glamorous stars; melodious, rhythmic music; exquisite sets and exotic locations. How the audience responds to the weaving together of these ingredients determines whether or not a film will be a blockbuster hit. Other key ingredients include elaborate, loud action scenes and a sense that the social or moral order will not be challenged. A happy ending is a mandatory requirement to conclude the two-and-a-half to three-hour movies.

The majority of the moviegoing audience in India consists of young men from a variety of regional, linguistic, religious and social backgrounds. Today there are around 500 million Indians under the age of twenty-five, out of a total population of over one billion, and films are made primarily to appeal to this age group. But of course, for a film to be genuinely popular it must also entertain the whole family, from grandmother to grandson, all of whom may also be avid moviegoers.

Watching a movie in an Indian cinema hall is a lively experience. The audience makes itself seen and heard at every turn of the plot – whistling at a sexy 'wet saree' number, egging on the hero as he takes on ten bad guys, and applauding melodramatic dialogue about lost values. Once it becomes clear that there will be a happy ending, the audience often does not bother to wait for the last scene but starts making its way out of the cinema before the film actually finishes. However, it would be wrong to assume that Indian audiences are passive consumers of whatever Bollywood offers. In fact, fewer than eight out of the more than 800 films made each year will make serious money (India is by far the world leader in terms of the sheer number of films produced).

Presented in a seamless mix of Hindi and Urdu (the two north Indian sister languages understood by over 400 million people, about half the population of India), the classic Bollywood movie may appear simplistic. Even the familiar boy-meets-girl saga, however, can contain many layers of Indian culture, manifested in some form or other relating to class, religion and tradition. Popular cinema in India may borrow plots from Hollywood, but these are so transformed by the must-have ingredients of the Bollywood film that only the bare outlines of the originals can be discerned.

This sense that every film must address the theme of what it means to be Indian or reflect Indian thinking can be traced to the beginnings of Indian cinema. The early silent films were based on well-known Hindu epic tales from the *Mahabharata* and the *Ramayan*. The first cinema audiences loved seeing familiar mythological stories involving gods combating demons brought to life on the screen. The new Western invention was perfectly suited to the Indian context of storytelling which relied on oral tradition. The fact that cinema techniques, such as special effects or low-angle shots, could enhance the mythical was seen as a great asset in the telling of heroic tales. This remains a major reason why Bollywood films continue to capture the popular imagination in India.

Theatrical forms such as *Ram Leela*, an enactment of the exploits and adventures of Ram) and *Ras Leela* (based on the exploits of Krishna and episodes from his life) have had a great impact on the evolution of Indian cinema. This is still apparent both in the way music and drama work together and in the portrayal of the stock characters of Indian cinema. The villain, for example, is still given a curling moustache and a sinister laugh, an instantly recognizable version of the stage demons associated with *Ram Leela*. Early film screenings from 1913 onwards took place in tents beside temples in villages and small towns, where, after prayers, devotees made their way to see Lord Ram or Lord Krishna come alive on the screen.

Such devotion can still be seen in the hero-worship accorded to Bollywood stars. People want to act, talk and look like their idols. Barbers down the decades have been asked to give their customers an Ashok Kumar, Dilip Kumar or Shahrukh Khan cut, and tailors have been told to copy the clothes of the beautiful Madhubala or Aishwarya Rai. Until the early 1990s, star gossip was almost exclusively reported in India's dozens of film magazines, but interest in the world of cinema is now so extensive that virtually every daily newspaper devotes substantial space to who is doing what in Bollywood.

The style, content and pace of Indian movies have changed greatly over the years, as has the way in which the film industry operates. The studio era ended in the late 1940s, and freelancing became the norm in cities where the bulk of Indian films continue to be produced, including Mumbai, Calcutta, Chennai (formerly Madras) and Hyderabad. Erratic start-stop shooting schedules and complex financing have meant that, from the stars, music directors and choreographers at the top down to the lowest-paid technicians, people work on several productions at the same time. The leading music director A.R. Rahman commented, 'How do I know that the film I'm working on will ever get released? Or even how long it will take to be completed? So I have to compose several soundtracks at the same time in order to make a living.' This juggling of many projects has become commonplace since the 1960s, and no one is surprised when a star travels from one set to another, playing a cop in the morning and a psychopath in the afternoon.

Other key players are the action scene directors and the set and costume designers. There is a huge demand for exciting action scenes, as this has great appeal for young male audiences. Yet there are only a handful of action directors (known as 'stunt masters' in India) working in the film industry. Stunt masters involved in any one film are usually members of the same family and, like the stars, work on several films at the same time. This is also how the relatively few set and costume designers work. In a Bollywood movie, set design can range from the rickety and makeshift to the elaborate and lavish. Costume design has always been important, but never as much as it is in today's culture of glamour and beauty. Bollywood designers have become so trendy that many create exclusive wedding clothes for the ultra rich as a sideline.

Budgets today are higher than they have ever been, with star fees tripling costs. This has put a lot of pressure on filmmakers to succeed at the box office. It is no coincidence that Indian cinema had its golden age from the 1950s to the mid-1960s, at a time when budgets were generally lower and directors were encouraged to be inventive rather than play safe. Even minor films of this period have a special quality, whether it is a stunning romantic scene, an atmospheric song sequence, or a fabulous performance by Dilip Kumar or the comedian Johnny Walker. The era produced immensely popular stars and fine directors, of whom the most influential on the aesthetic of Indian cinema were Mehboob Khan, Bimal Roy, Raj Kapoor and Guru Dutt. These extraordinary filmmakers worked within the conventions of Indian cinema while making deeply personal classics. They set the standard, not only in their choice of theme and subject, but also in their approach to black-and-white photography, set design and editing. They avoided the usual stereotypes and stock figures, and the layered psychology and sophistication of their heroes and heroines have given Indian cinema its most enduring characters. These directors mastered the use of film music and choreography. Their song sequences rival the best in world cinema and in many cases surpass the Hollywood musical in their subtle linking of dialogue and lyrics. These directors transformed the film song into an art form and confirmed that music was Indian cinema's greatest strength. Even today, Indian filmmakers are aware that their moment of cinematic glory may well come from the songs. During every decade since the 1950s, a large number of films that would otherwise have been forgotten have been saved by a marvellous musical sequence in which melody, lyrics, cinematography, choreography and performance combine to magical effect.

After the golden age of the 1950–60s, the form of popular films started to change. By the 1970s, Hindi films began to combine all genres into a single movie, with song and dance firmly at the heart of the narrative. This 'mixed' approach is still the way the stories unfold today. In a Bollywood movie, such mixing and matching can translate into the hero fighting a sinister politician in one scene and serenading his heroine, with forty dancers moving in unison behind him, in the next.

Despite the popularity of television in India since the early 1990s, there is still a demand in the remoter areas of the country for touring cinemas, which involve a projectionist travelling in a truck with an assistant, eighteen reels of film and a tent that he will set up in the village. In nearly every city street, too, there are signs of Bollywood's influence. Postcards of the current movie heart-throbs are proudly displayed for sale on pavement stalls next to images of the most revered gods and national icons such as Mahatma Gandhi and Mother Teresa. At nearly every major roundabout and road junction – anywhere where there is space for a large billboard – gigantic hand-painted images of stars stare down at the passing traffic. Outdoing Mumbai in this respect are the cut-outs and billboards that line the streets in south India, where the popularity of stars is such that in recent years the leading names of Tamil and Telugu cinema have successfully transferred their appeal from screen to voting booth and have become chief ministers in Tamil Nadu and Andhra Pradesh.

Nearly every Indian, whether living in a village or a city, feels connected to the movies in some way either because they love a star (Amitabh Bachchan has broken all records with his fan following) or because of their love of film songs. The Bollywood movie is also an active link to homeland culture for those who have made Europe, the United States or Canada their home. When a movie with an A-list cast, such as *Lagaan* (Ashutosh Gowarikar, 2001) or *Devdas* (Sanjay Leela Bhansali, 2002), is released, people of Indian origin, whether they live in Lucknow or Leicester, head to the cinema at virtually the same time. Impassioned fans can also be found in the Middle East, Russia, China and many parts of Africa. Indian cinema is unique to Indian culture and history. At the same time, its energetic style, the emotional appeal of its themes, the glamorous lifestyles it portrays, its enduring melodies and lush settings all contribute to its increasing popularity world-wide. Jonathan Torgovnik's lyrical photographs show us the human face of Indian moviegoers, as well as those working behind the scenes, who together have made Indian cinema as alive as it is today.

the touring cinema

Amar Touring Cinema on the road between the villages of Pussegaon and Palli in the state of Maharashtra. It is one of the few remaining mobile cinemas in existence in the remote rural areas of India.

Right: Kisan the projectionist sits on the cinema truck in Palli.

मध्ये
देवकी
गुडस-कॅरीअर
अमर टूरिंग
MRS
5198
अमर

The touring cinema arrives at the village with three films to show. Crew members advertise the next show with posters in front of the cinema's tent.

One of the projectionists fixes a torn reel of film before the show starts. The touring cinema's two projectors can be seen in the foreground.

In the village of Pussegaon, a young man examines a film strip left by the touring cinema, hoping to find an image of his favourite actor.

Right: A crew member of the Amar Touring Cinema rewinds film reels between shows by the riverbank where the tent is set up.

Metal ticket booths are set up near the tent of the Amar Touring Cinema. A young boy purchases a 10 rupee ($0.25) ticket for his first film ever.

Right: Eager to enter the cinema, a group of young men wait for the previous screening to end. For some it will be the first time they have ever seen a film.

Family-oriented films attract mostly women with their children, while action films are normally geared to a male audience.

Right: The interior of Amar Touring Cinema's tent. At many shows, including this 12 noon screening, audiences can number more than 1,000.

Villagers sit in clusters on the ground for the duration of the three-hour film. Many have travelled from neighbouring villages for the screening.

The end of a screening at around 3 a.m.

The films are shown late into the night and the projectionists take turns sleeping between shows. *Right:* The film is projected through a hole cut out of the back of the touring cinema's truck.

STANLEY
32 AMPS
415 VOLTS
ON
OFF
ON
इंजिन
STANLEY
32 AMPS
240 VOLTS
ON
OFF
मेन
११०
मेन बोर्ड
रेक्टीफायर-२

Left: One of the ticket sellers from the Amar Touring Cinema counts the money from the matinee show. The Indian government regulates ticket prices for touring cinemas.

After a long night of showing back-to-back films to villagers, crew members sleep either in the truck or in the tent. They keep warm by wrapping themselves with the tent's detachable canvas sides.

on the set

Stars Govinda and Sonali Bendre during a song-and-dance sequence on the set of *Jis Desh Mein Ganga Rehta Hai*. Locations are an important element of Hindi musical cinema. This was shot in Mahabaleshwar, a hill station in the state of Maharashtra. Songs are typically performed by the hero and heroine with dozens of dancers moving in unison behind them.

Following pages: These action scenes on the set of *Ab Ke Baras* take place at Film City in Mumbai, a large tract of land owned by the state that provides indoor and outdoor locations and post-production facilities. Action scenes are an important part of popular Indian cinema, and Indian films often end with long and elaborate fight sequences between hero and villain.

Stuntman Srikan Shetty dangles upside down during an action shot on the set of *Dil Ke Aas Paas*. This film was shot at Filmalaya Studios in north Mumbai where the Bollywood industry is centred.

Right: Leading actors Manoj Bajpai and Tabu during a love scene on a fantasy set at Mehboob Studios, Mumbai. Bollywood love scenes are often discreet, with little overt sexuality. Physicality and intimacy are usually woven into the film's song-and-dance numbers.

Above and right: Madhuri Dixit, the most popular star of the 1990s, at Filmistan Studios, Mumbai. Female stars have shorter film careers in Indian films than their male counterparts. By the time they are in their thirties, they are often regarded by the industry as too old to play the virgin heroines.

Left: On the set of *Mere Sapnon Ki Rani*, Rajahmundry. *Above:* Megastar Shahrukh Khan and his make-up artist prepare for the next scene at Film City in Mumbai. Essential ingredients for a successful film are its stars, music, score, choreography and singers. A film featuring Khan will most likely be a success, which explains his high fee (up to the equivalent of US $1 million). His film *Devdas*, released in 2002 and shown worldwide, was the most expensive Hindi film ever produced.

Actresses Madhoo and Urmila Matondkar embrace while a member of the film crew blocks the light from the camera with a black cloth. On the set of *Mere Sapnon Ki Rani*, Rajahmundry.

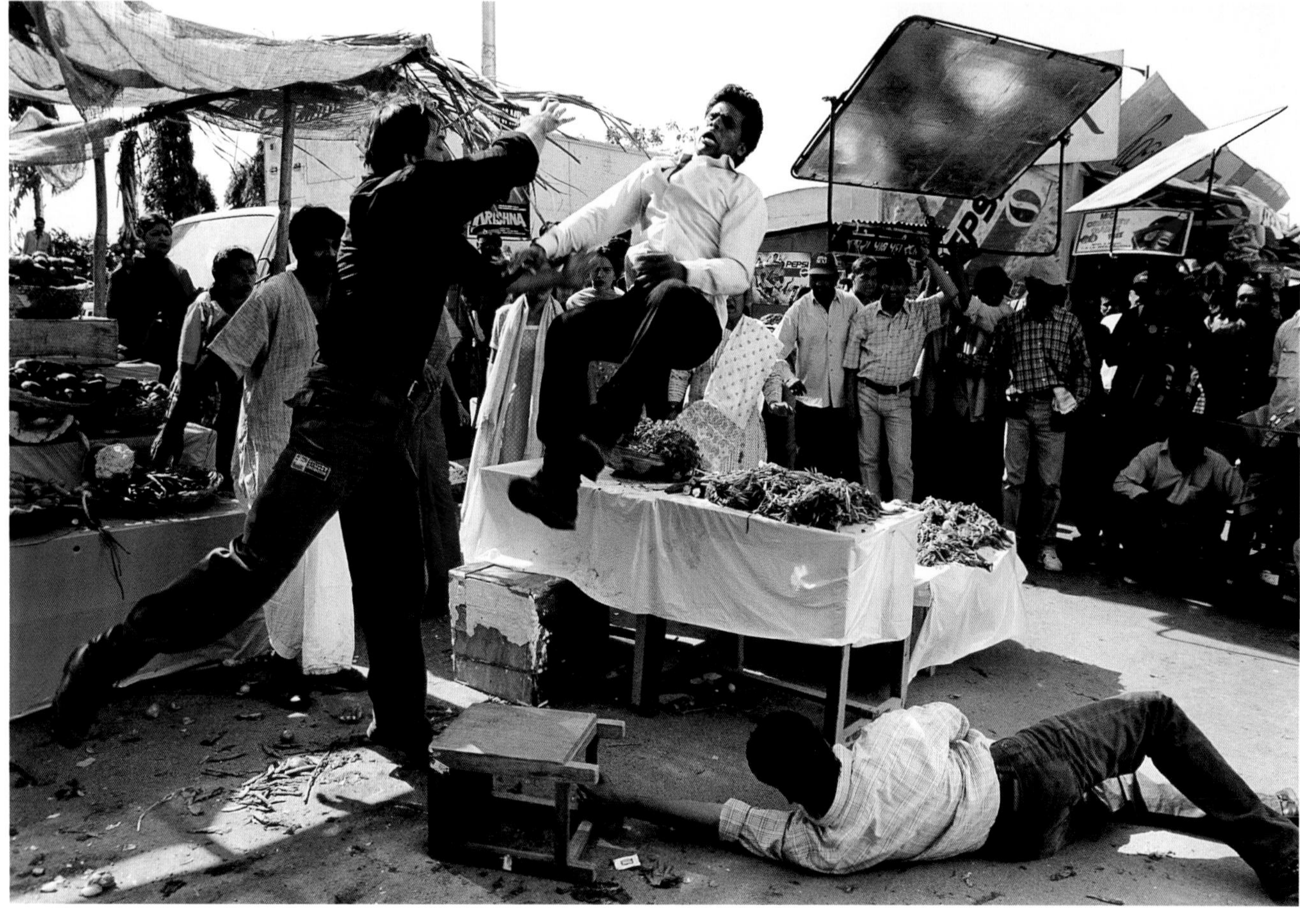

Previous pages, clockwise from top left: On many of the sets, the day begins with a prayer ('puja') and an offering where a coconut is cracked and drops of its water are sprayed onto the camera to pray for a successful day of filming. Improvisation is common practice in building sets, as seen here (top and bottom centre); a cameraman prepares for the shoot to begin; an editing suite in use; a stunt scene caught in mid-action.

Above: Dancers prepare for a song-and-dance sequence on the set of *Hum To Mohabbat Karega* at Raj Kamal Studios in Mumbai.

Children from nearby villages gather to watch the filming of *Mere Sapnon Ki Rani* in a small village near Rajahmundry. The children climb trees to get the best possible view of the film stars.

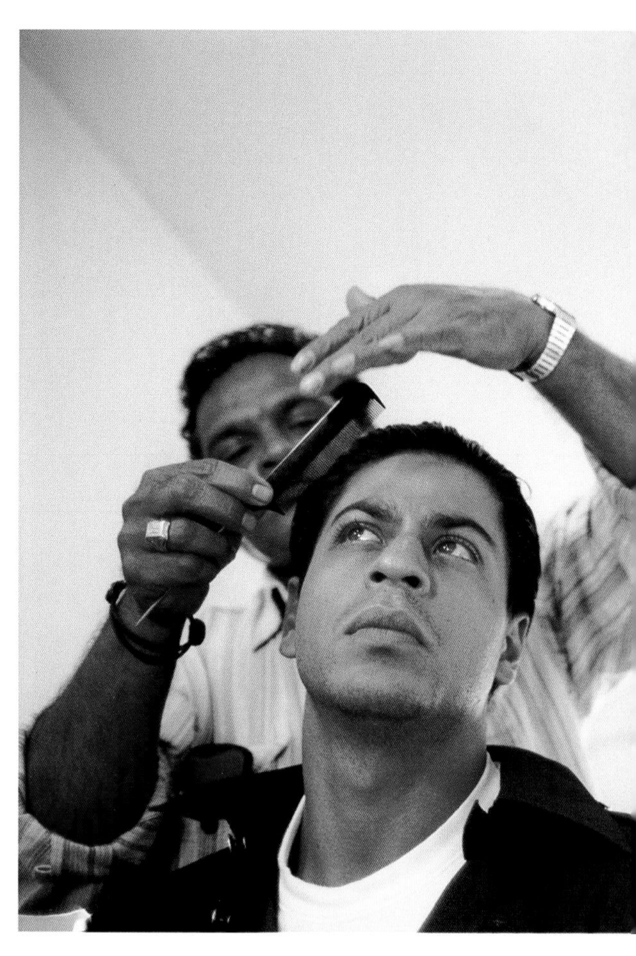

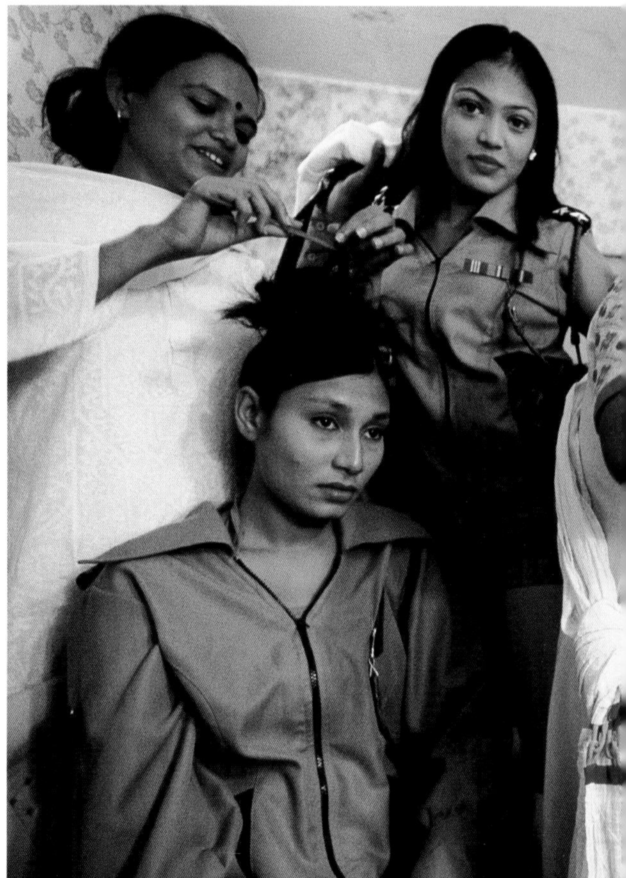

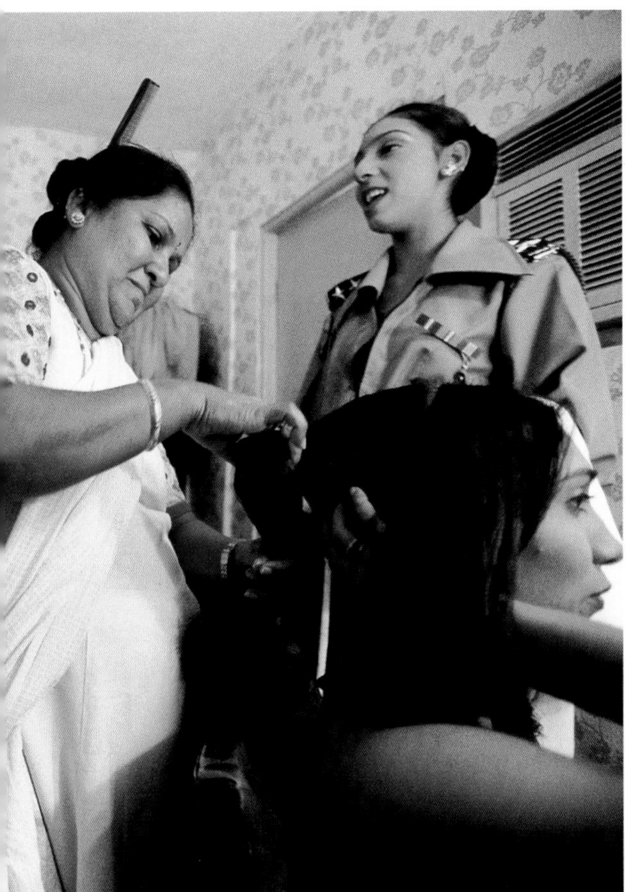

Previous pages, clockwise from top left: Character actors and stars (Shahrukh Khan, centre top, and Manisha Koirala, bottom left) receive last-minute touch-ups from their make-up artists before filming begins; dancers from *Chor Machai Shor* (bottom centre) prepare for a song-and-dance sequence.

Dancers prepare for a song-and-dance sequence on the set of *Jis Desh Mein Ganga Rehta Hai* in Panjgani, Maharashtra. Song-and-dance routines are one of the most important components in Indian films.

Actors Bobby Deol and Karishma Kapoor work through elaborate dance steps. Both stars come from families of actors. Bobby Deol's father, Dharmendra, enjoyed enormous fame in the 1970s as an actor and director. Karishma's grandfather, Raj Kapoor, had a huge following in Russia, China and the Middle East and was one of India's first international stars.

Right: A lovesong-and-dance sequence starring Bobby Deol and Bipasha Basu on the set of *Chor Machai Shor* at Mumbai's Filmalaya Studios. Bollywood films tend to be spectacular melodramas about love and romance. Kissing scenes are allowed in the movies but explicit eroticism is strictly regulated by the country's censorship laws.

RAJ
LITE

INDERJIT FILMS COMBINE
'AB KE BARAS'
Cinemascope
SCENE
SHOT
TAKE
95
61
3
DATE
DAY/NIGHT

Left: On the set of *Ab Ke Baras* at Film City, Mumbai.
Above: Villagers gather to watch the filming of *Mere Sapnon Ki Rani* in a small village near Rajahmundry.

A lighting technician at Filmistan Studios, Mumbai.

the characters

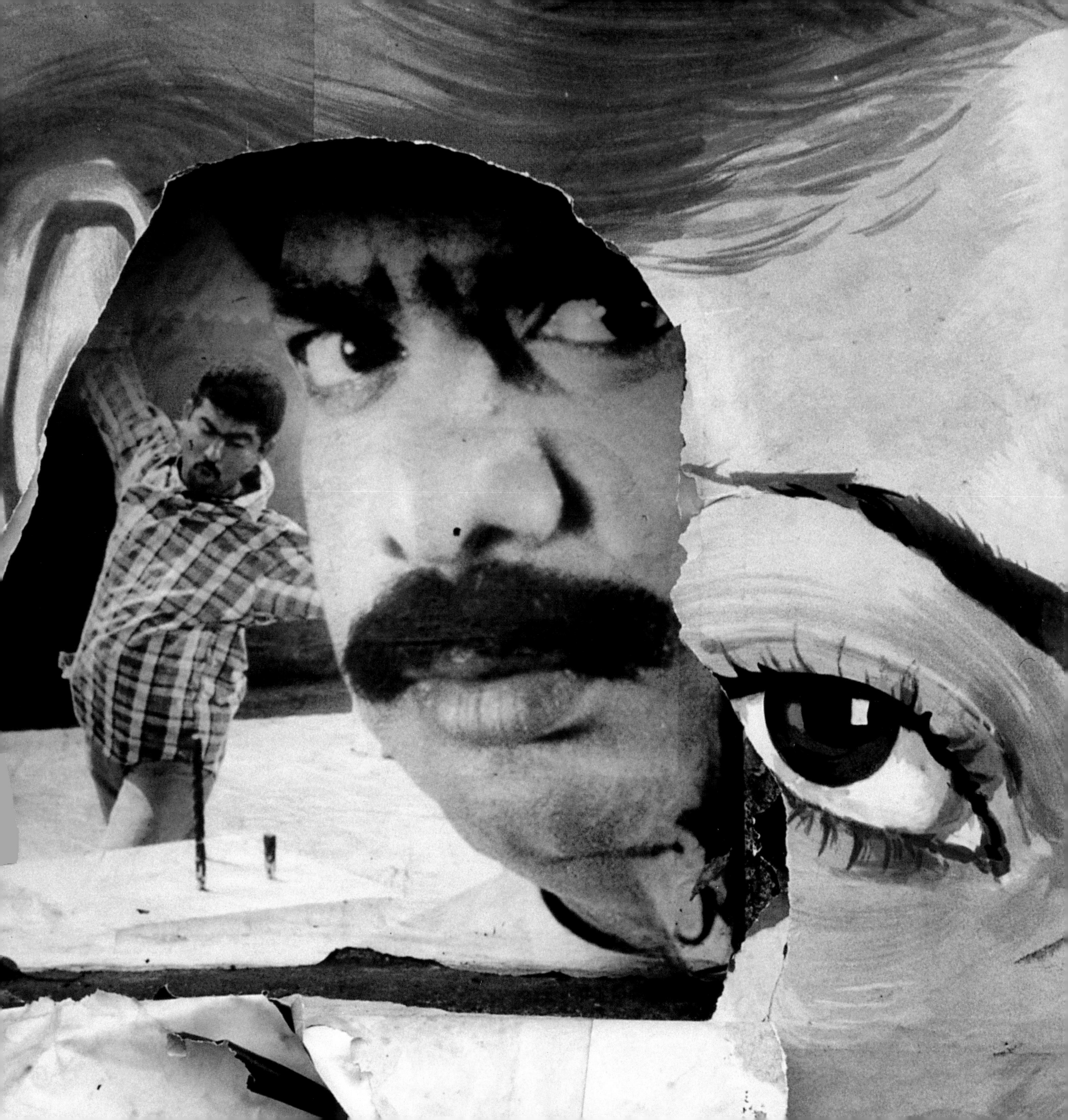

Actress Madhoo on the set of *Mere Sapnon Ki Rani* on location in Rajahmundry.

In the courtyard of Roopam Cinema in Chennai, a boy reaches up to touch the lips of a painted image of his favourite film actress.

Character actor Razak Khan, best known for his gangster sidekick roles, on the set of *Dil Ke Aas Paas* in Filmalaya Studios, Mumbai.

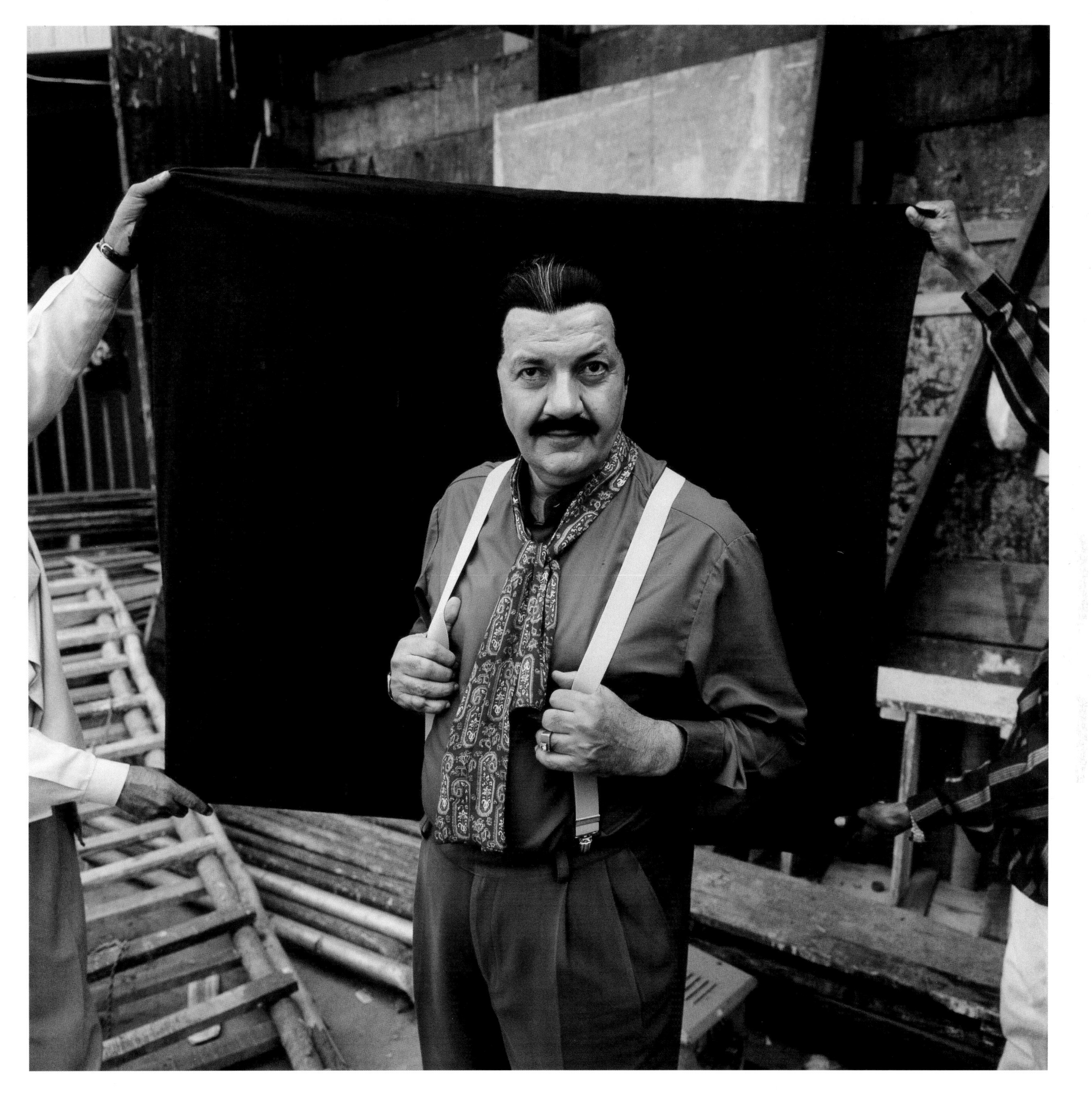

Actor Prem Chopra on the set of *Dil Ke Aas Paas* at Filmalaya Studios, Mumbai. Chopra has played a villain for most of his thirty-seven years in film and is one of the most famous villains in Hindi cinema. 'When I was a kid, just mentioning Mr Chopra's name made me scared,' says action-scene director Mahendra Verma.

Left: On the set of *Dil Ke Aas Paas* at Filmalaya Studios in Mumbai with action-scene director and stunt-master Mahendra Verma. Verma is one of five brothers who, together with their father, are all action-scene directors and stuntmen. *Above:* A character actor playing a villain in Studio No. 12 in Film City, Mumbai.

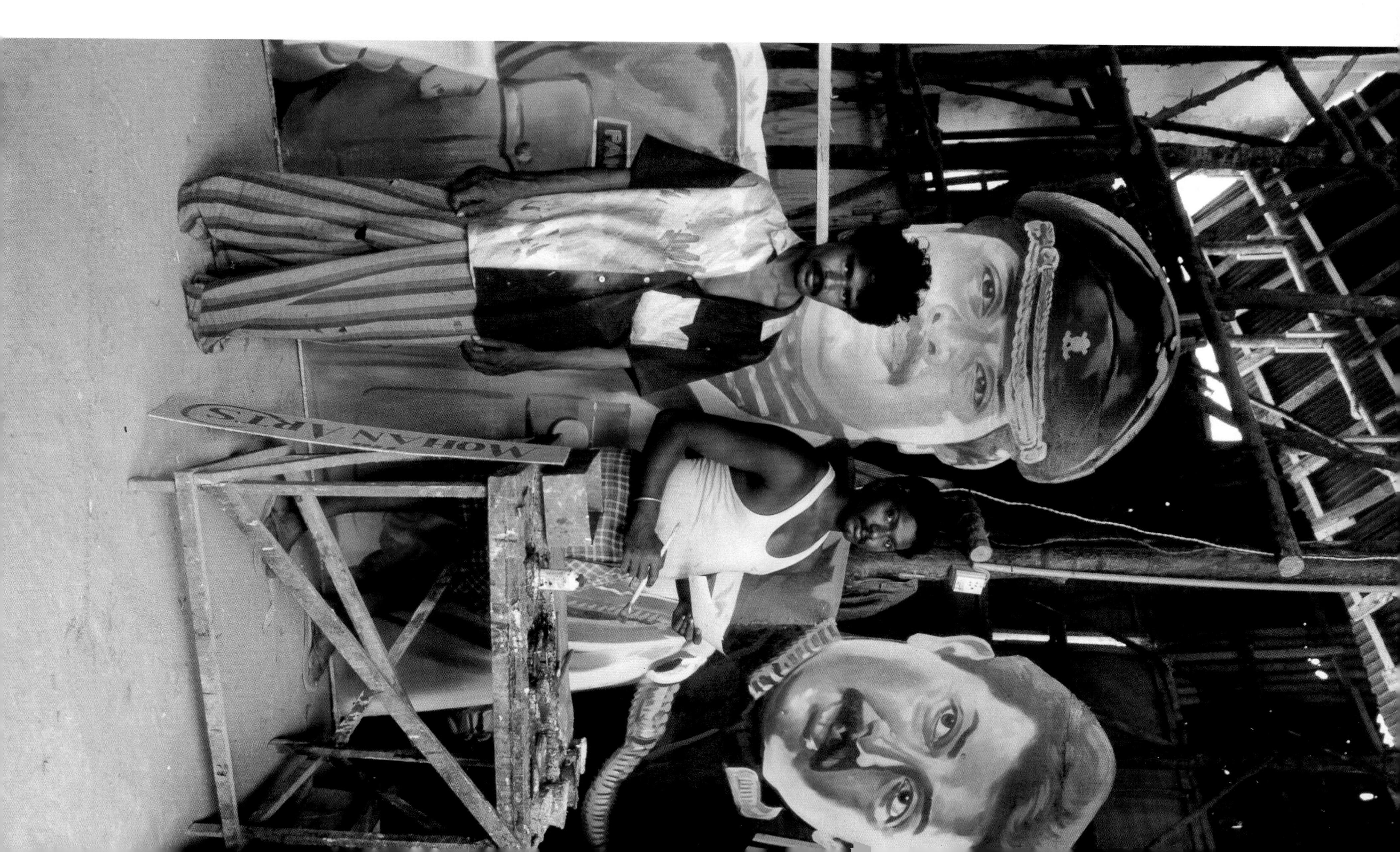

Artists at Mohan Arts Studio in Chennai take a break in front of the hand-painted cut-outs they are working on. The cut-outs, which can be as high as 18 metres, will be placed outside cinemas in Chennai. These larger-than-life advertisements typically cost less to hand-paint than to print. In southern India, particularly in Chennai and Hyderabad, hand-painted film advertisements like these still outnumber printed ones.

Following pages: Giant film cut-outs and advertising billboards dominate the streets of large cities in India. In Chennai, hand-painted cut-outs are placed near the cinemas, and on the main commercial roads (left and centre). In Mumbai, hand-painted banners have been almost entirely replaced by printed ones (right).

OSCAR FILMS(P)LTD
V.ரவிச்சந்திரன்
A THEATRE

தர்மசக்கரம்
சிம்பி.

skyliners
CO-PRODUCED BY BHARAT SHAH LYRICS
A RAJKUMAR SANTOSHI
SKIN
Weight
Loss

Selvam, an artist at Mohan Arts Studio in Chennai, stands in front of a freshly painted banner that will hang outside a cinema.

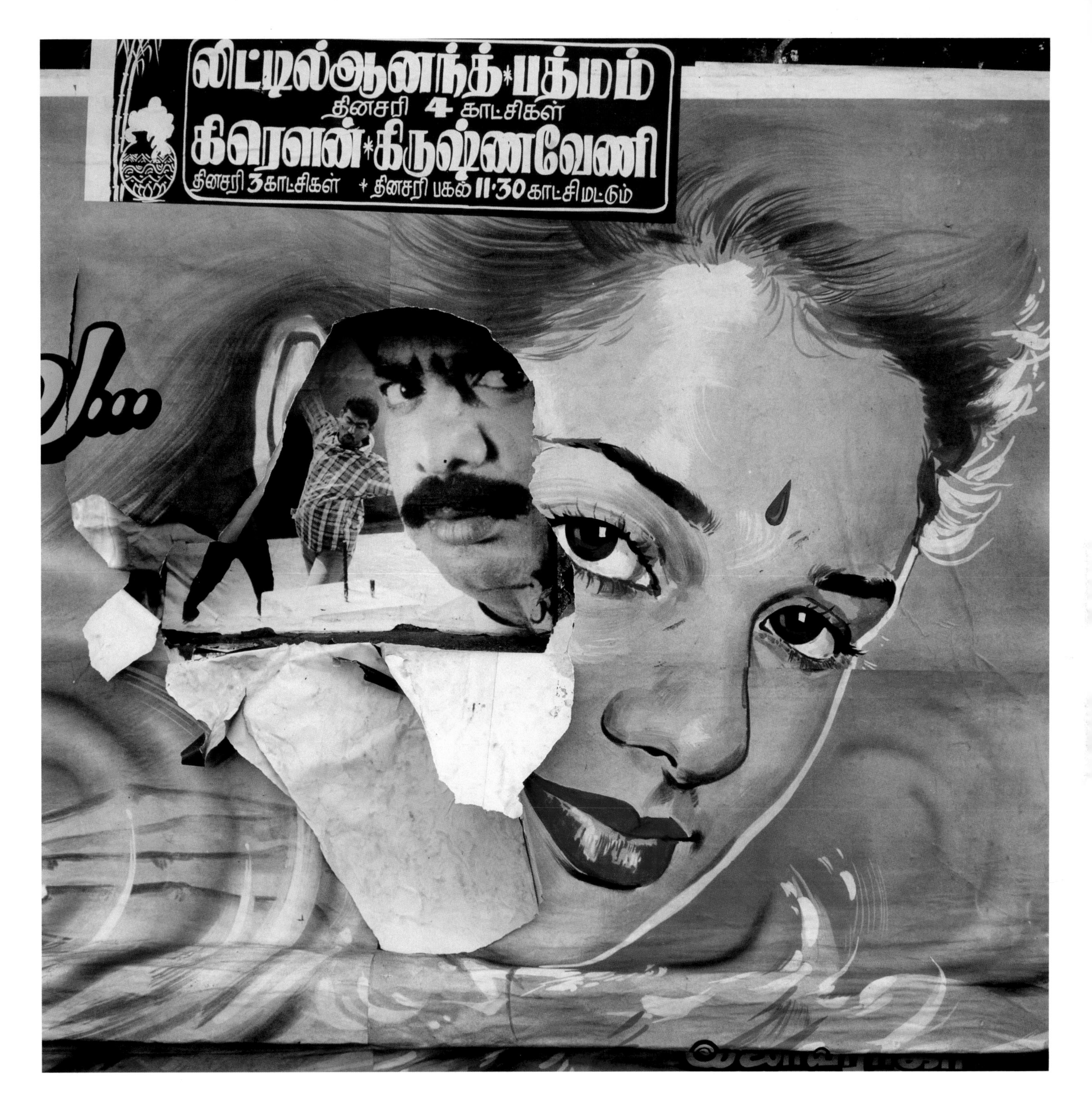

Torn cinema posters on a wall in Chennai. Cinema banners and printed posters are placed on almost every bare wall, turning the streets of Chennai and Mumbai into gigantic collages of India's film stars.

Left: Character actor Noshad Abbas on the set of *Ab Ke Baras* and, *above,* actress Lata Haya at Stage No. 9 in Film City, Mumbai.

Actress Manisha Koirala takes a break during the filming of *Champion* at Film City in Hyderabad. Koirala is originally from Nepal and has acted in a number of key Indian films by directors from both Mumbai and Chennai, including in the hugely successful films *Bombay* and *Dil Se* directed by Mani Ratnam.

Actor Jackie Shroff during the making of the political melodrama *Mission Kashmir* at Film City in Mumbai.

Heroes in Indian films are almost always opposed by equally dynamic villains. These usually have huge moustaches and teary red eyes, and are modelled on demonic characters from Hindu epics such as the *Mahabharata* and the *Ramayana*.

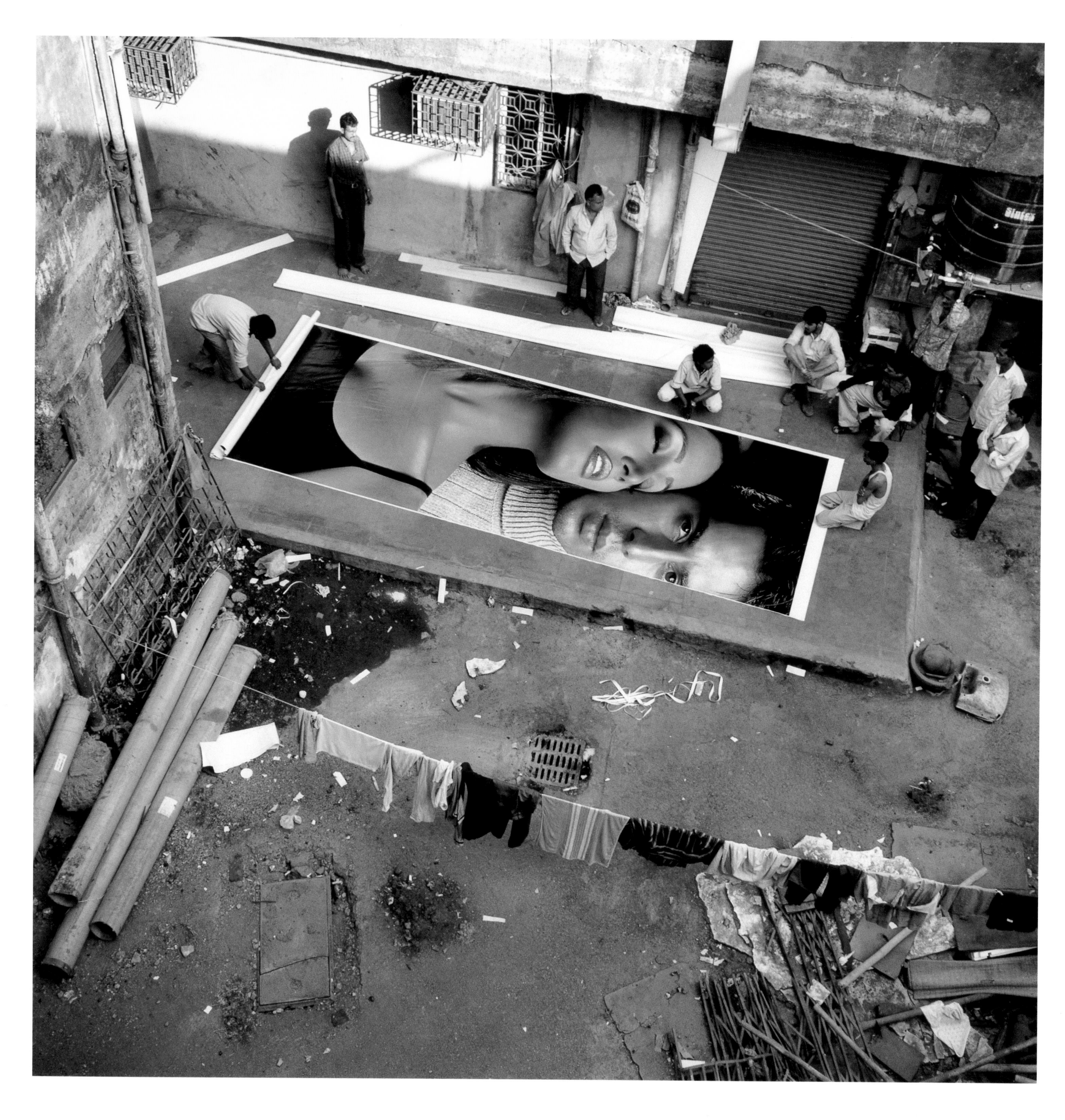

Left: The courtyard of a cinema billboard printing house in Mumbai. The billboard is printed in several sections and then reassembled as one enormous image. Here, the hero and heroine are far more Westernized in appearance than was the case in the past. *Above:* Actors Bobby Deol and Karishma Kapoor in the makeshift studio of celebrity photographer Avi Gowariker at Mumbai's Raj Kamal Studios during a publicity photo shoot for their forthcoming film *Hum To Mohabbat Karega.*

Following pages: On the streets of Mumbai and Chennai cinema banners and posters cover almost every bare street wall, and postcards and memorabilia of film stars are sold on the streets. The idea of 'Darshan', the sighting of a god, is a vital part of prayer in India. By virtue of their superstardom and godlike status in Indian society, images of film stars take on powerful meaning and can be revered as much as images of icons, such as Mahatma Gandhi, and Hindi gods and goddesses.

MOTHER
TERESA

AAROHI presents
15026·89·C

Filmfare
VERVE
MADHUR JAFFREY
NASEERUDDIN SHAH'S Secret Desire
A MOUTHFUL OF SOAP
For Alisha Chinai love is just a four-letter word
Kamal Sidhu's out-of-sync fashion fusion
CHASTITY
STARDUST

OVE BIRD
Bhattacharya's

THE SAFIRE CINE

Torn cinema posters on an outdoor wall in Chennai.

Actor Indravadan Purohit and his wife Rekha. Indravadan Purohit has acted in over 250 films in six languages.

S P PAGARE

Left: Character actors on Bollywood sets at Film City and Filmistan Studios in Mumbai.

The most popular star of Indian cinema is Amitabh Bachchan, now in his sixties. When he had a near fatal accident in 1982, one of his millions of fans walked backwards for more than 300 miles as an offering to God to pray for his recovery.

A character actor in devil costume at Kamalistan Studios, Mumbai.

Dancers practise a dance sequence on the set of *Chor Mach Shor* at Filmalaya Studios, Mumbai.

L
HE
play-dialogue-directio
VINA
nusic RA

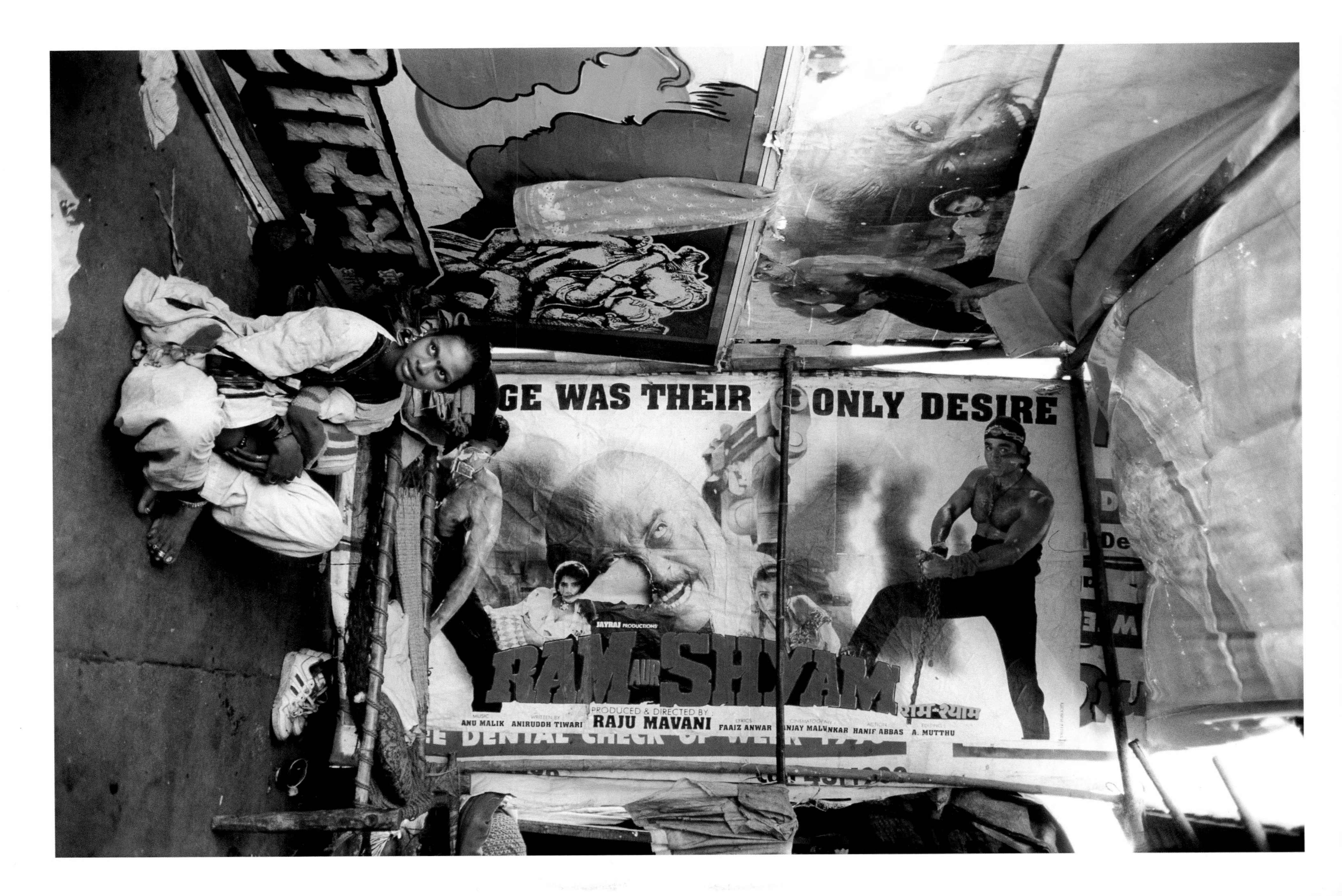

Left: A food stall near the Alfred Talkies Cinema in Mumbai is covered wall-to-wall with cinema posters.

Near Khar Road Railway Station in Mumbai, a woman sits with her newborn baby in a makeshift home that has been constructed from old cinema banners.

The memorial stone of M.G. Ramachandran, a great actor and politician from the state of Tamil Nadu. His devotees believe that if you are lucky you might hear his voice at his memorial site. Every day hundreds of fans flock to the site near the beach in Chennai.

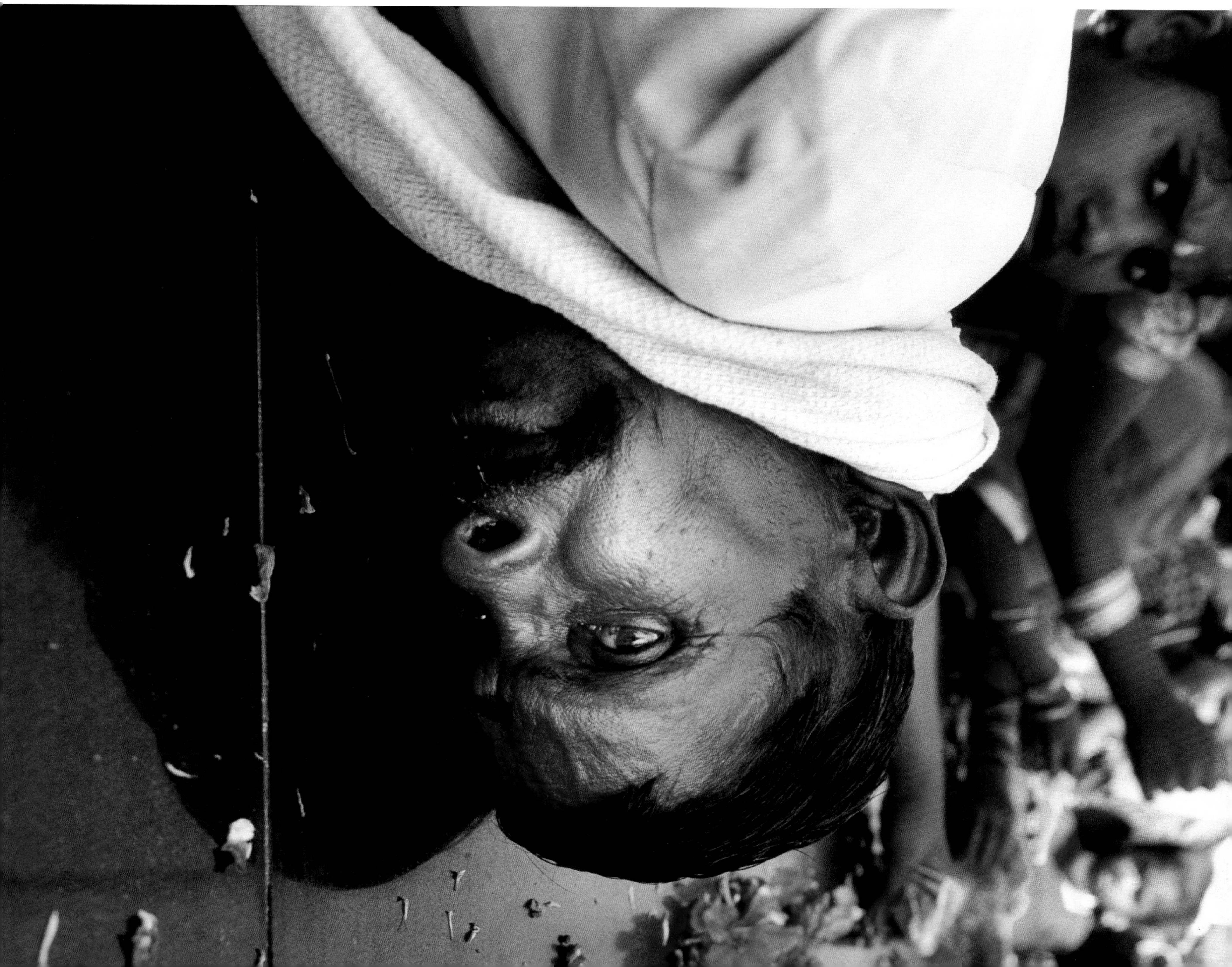

at the cinema

A typical scene outside the Melody Cinema, Chennai. This film, *Kaho Naa Pyaar Hai*, launched the career of heart-throb Hritik Roshan in 2000.

akesh Roshan's
Music
RAJESH ROSHAN
Kaho Naa... Pyaar Hai
Coca-Cola
Always Coca-Cola
MELODY
(AIR-CONDITIONED)
THEATRE
dts
Dress Circle

Clockwise from top left: Women in line reach into the ticket booth at the Abirami Cinema in Chennai; young men queue outside Chennai's Roopam Cinema, and one of them displays a photo of film megastar Rajnikanth that he carries in his wallet; at the box office of the Imperial Cinema in Mumbai, competition to get a good position in the queue to buy tickets is so fierce that the cinema owners hire guards to keep order.

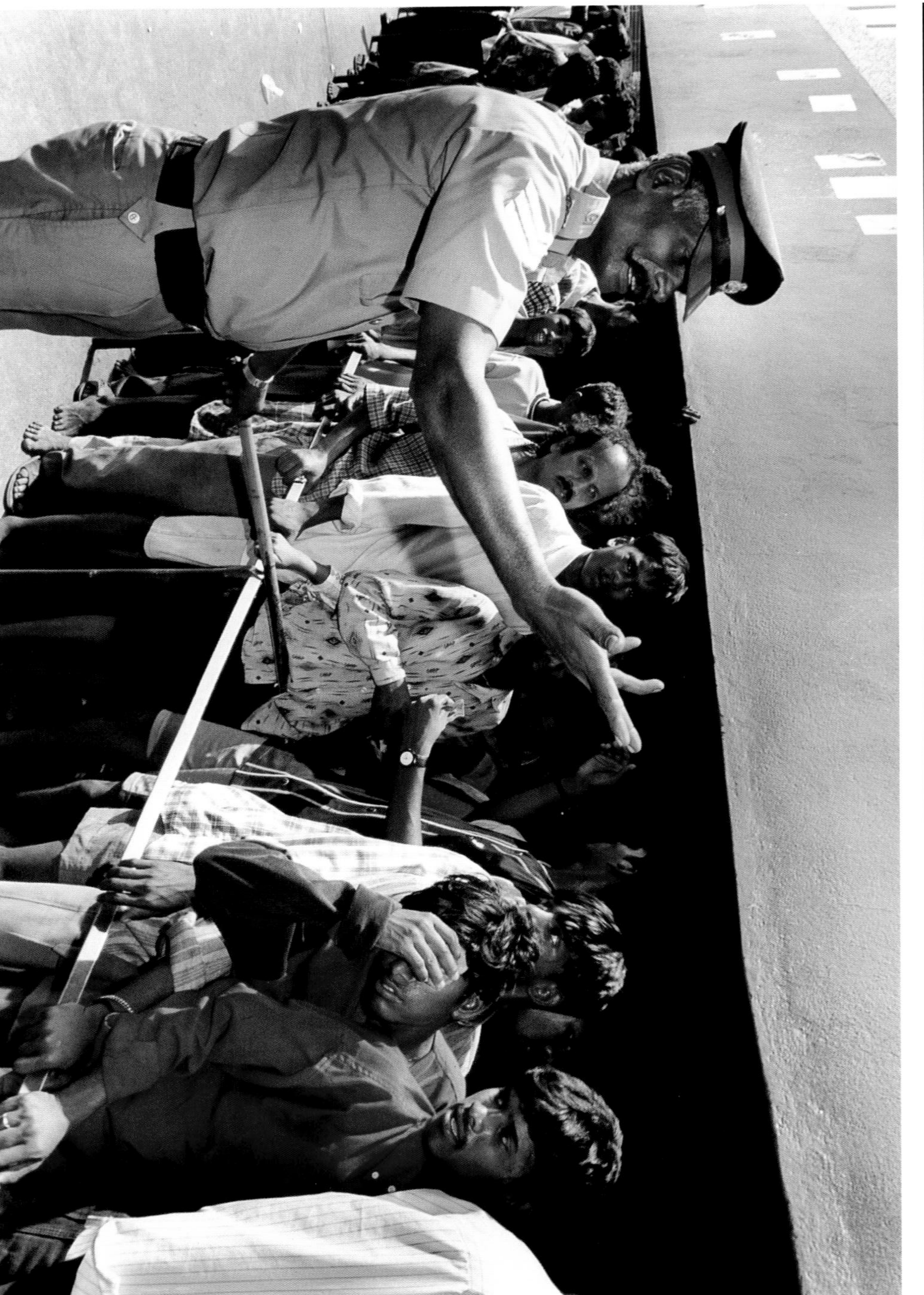

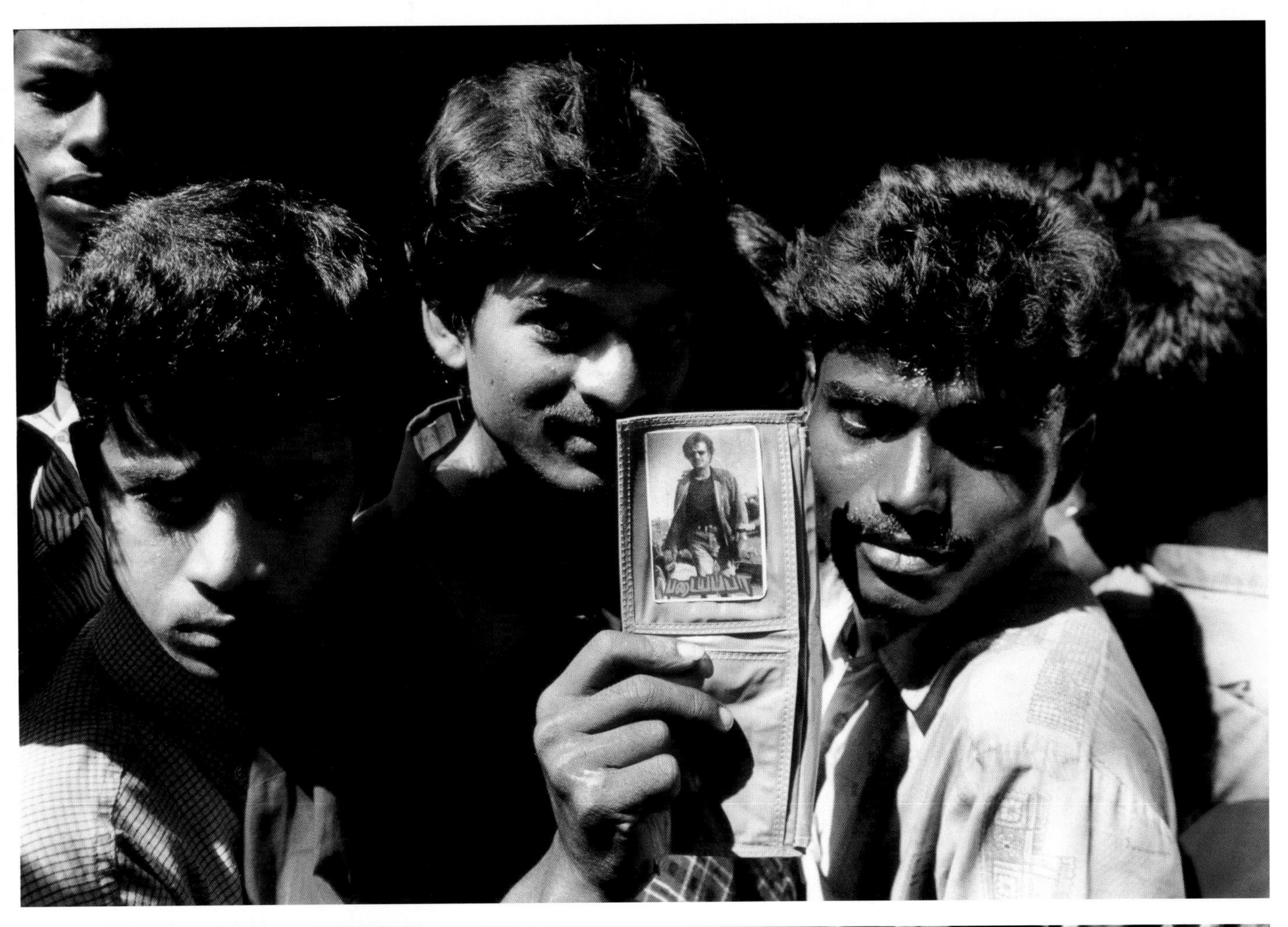

Approximately 14 million Indians queue each day to go to the movies. This queue has formed outside the Abirami Cinema in Chennai for the matinee show.

An usher displays tickets collected at the eighty-year-old Imperial Cinema, one of the oldest cinemas in Mumbai. Tickets cost 12 rupees (about US $0.25) for a three-hour show.

At many Indian cinemas there are separate ticket booths
for women and men. This stems from the social convention
within much of Indian society that women prefer not to
be in close proximity to men they do not know.

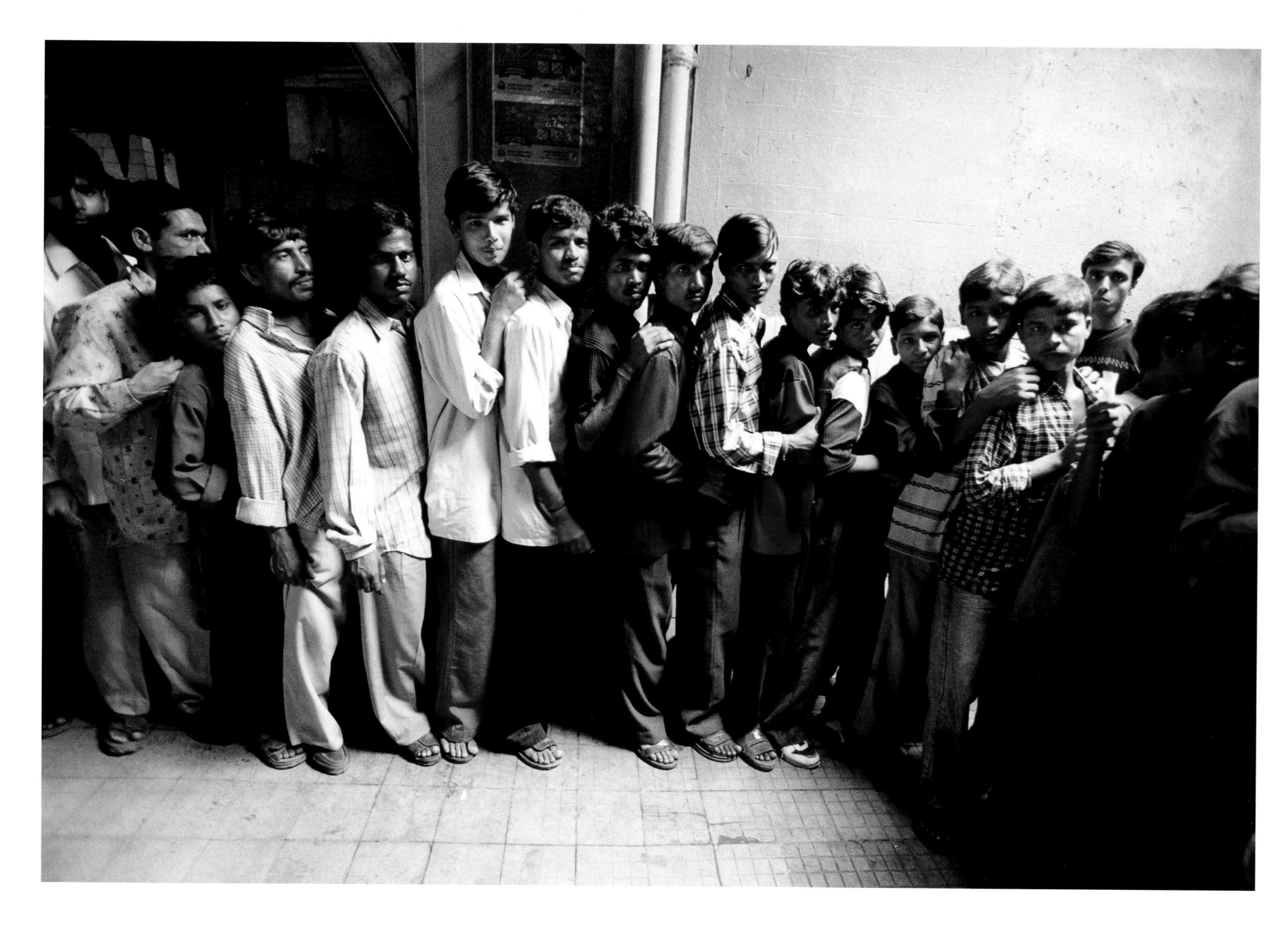

It is mostly men who gather for this action movie showing at the New Roshan Cinema near Grant Road Station in Mumbai, and there is fierce competition for a good position in the ticket queue.

New Shirin Talkies Cinema in Mumbai. Cinemas are usually divided into sections according to ticket price. The biggest fans sit in the front row.

A view from the balcony of the Capital Cinema, Mumbai. This single-screen cinema shows three to four films each day.

A child peeps through a hole in a wall covered with film posters in the courtyard of the Abirami Cinema in Chennai.

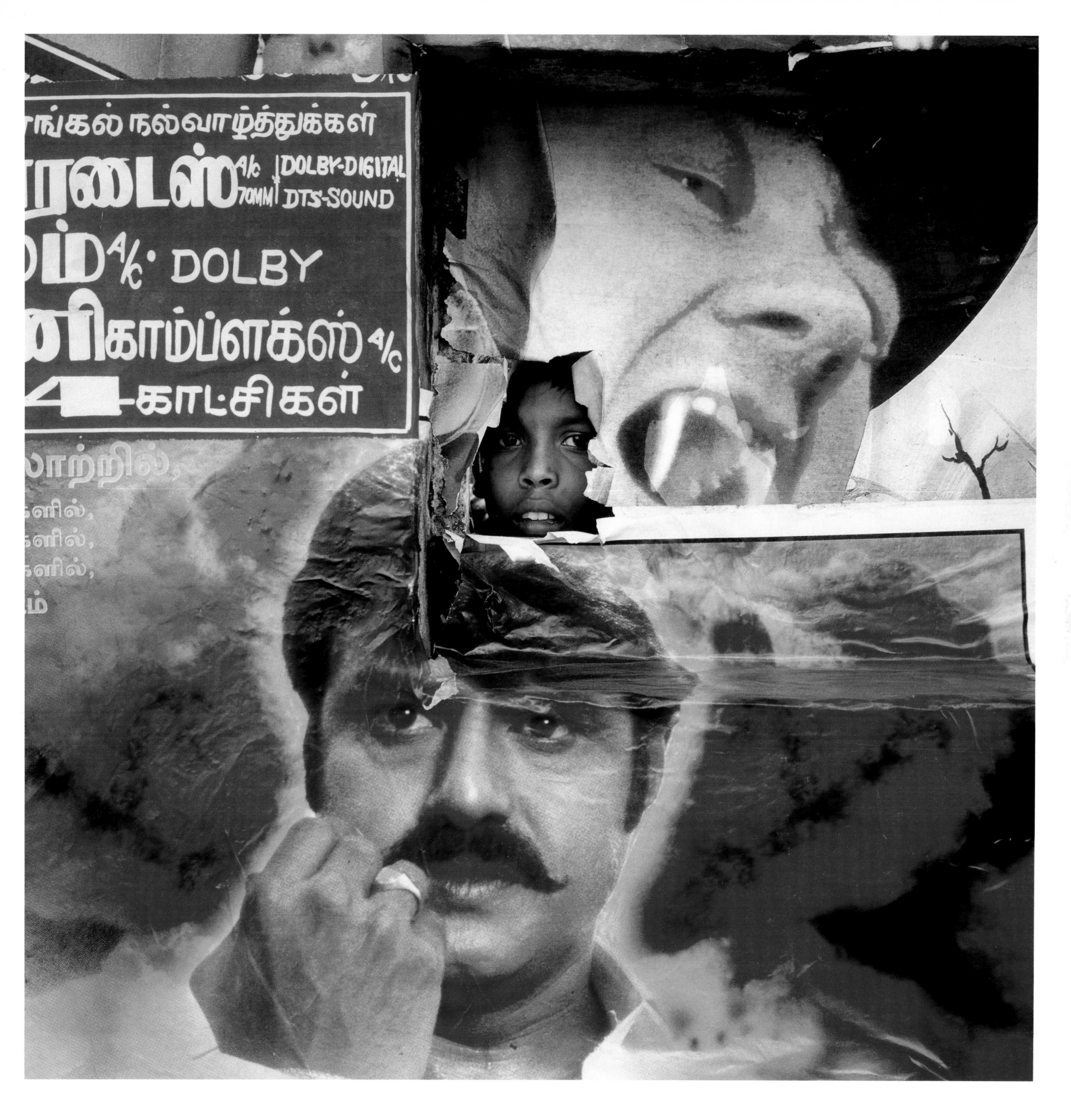
நல்வாழ்த்துக்கள்
A/C 70MM
DOLBY-DIGITAL
DTS-SOUND
A/C DOLBY
காம்ப்ளக்ஸ் A/C
4-காட்சிகள்

A view from the balcony of the New Shirin Talkies Cinema in Mumbai during a film.

With the absence of air conditioning in many Indian cinemas, electric fans are used to keep the audience cool. The noise of the fans often competes with the movie's sound, forcing the film to be played at a very high volume. Padmam Cinema, Chennai.

A sold-out matinee at the Alfred Talkies Cinema, Mumbai. Working-class men make up most of the audience at the many cinemas near Grant Road station.

ALFRED TALKIES

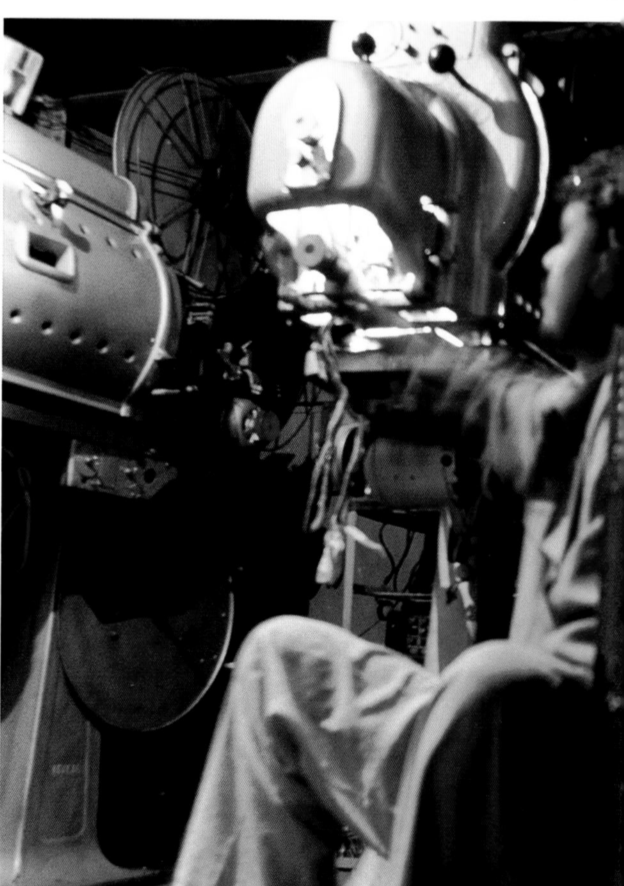

30X30

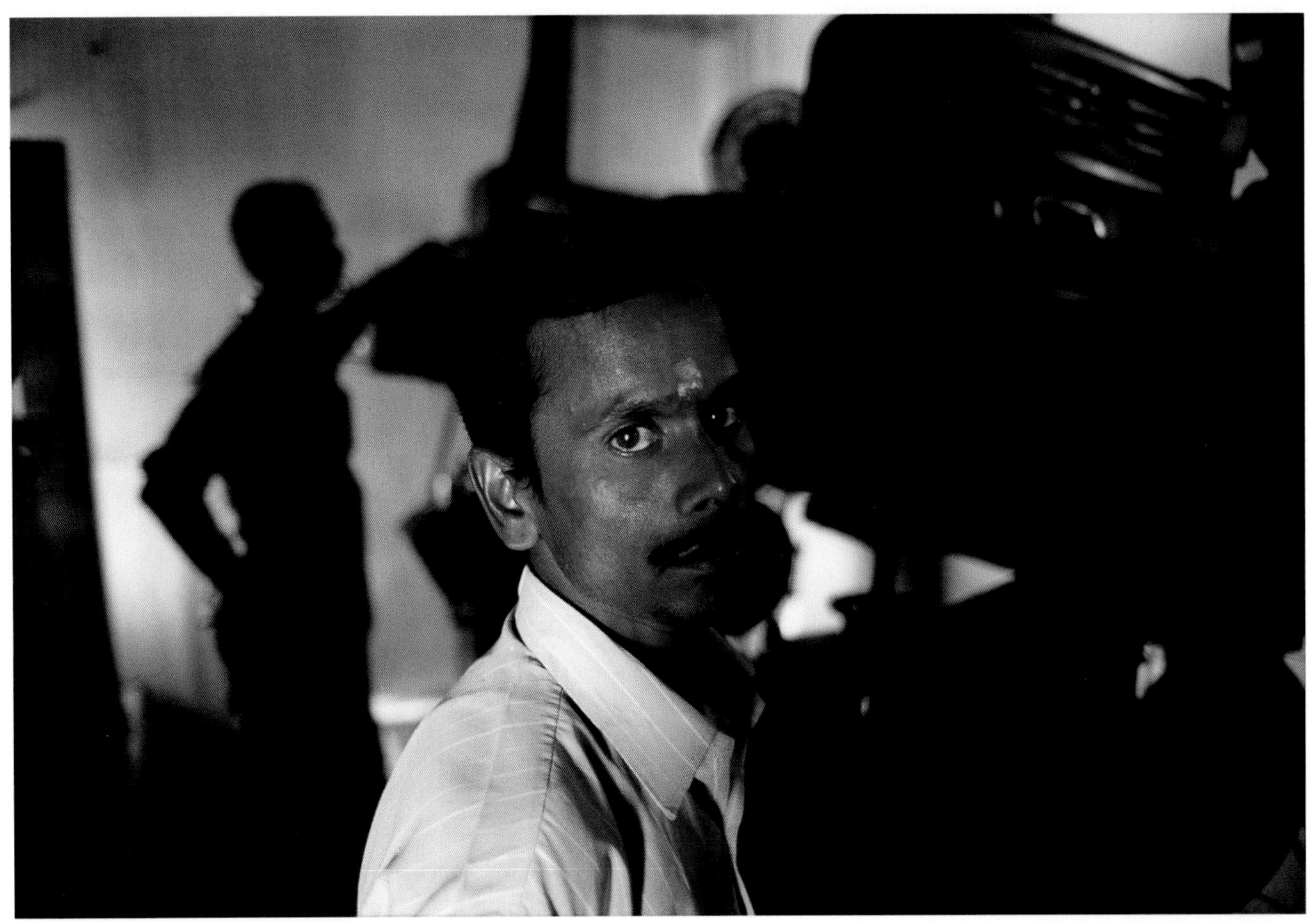

Previous pages, clockwise from top left: The Alfred Talkies Cinema in Mumbai during intermission; behind the scenes with the projectionists; at the Prarthana Drive-In Beach Cinema in Chennai.

Above left: An usher directs movie-goers to their seats at the Alfred Talkies Cinema, Mumbai. *Above:* A film's reflection on the floor and wall of the projection room in the Sangam Cinema, Chennai.

Mr K. Venugopal, manager of the Shanti Cinema in Chennai.

Projectionists at the Padmam Cinema in Chennai. Older projectors such as this one often produce dim images with poor focus and sound.

An old 'Enarc' arc lamp projector at the New Shirin Talkies Cinema in Mumbai, still in daily use.

Right: A projectionist at the Shanti Cinema in Chennai. Projectionists tend to work long hours and each typically shows a minimum of four shows each day. At times, the heat given off by the projectors makes the room unbearably hot.

Chennai's Prarthana Drive-In Beach Theatre is the only drive-in cinema in southern India. Many viewers prefer to sit outside their cars in chairs that they bring along with them. Chennai has its own thriving film industry, with films made in the Tamil language.

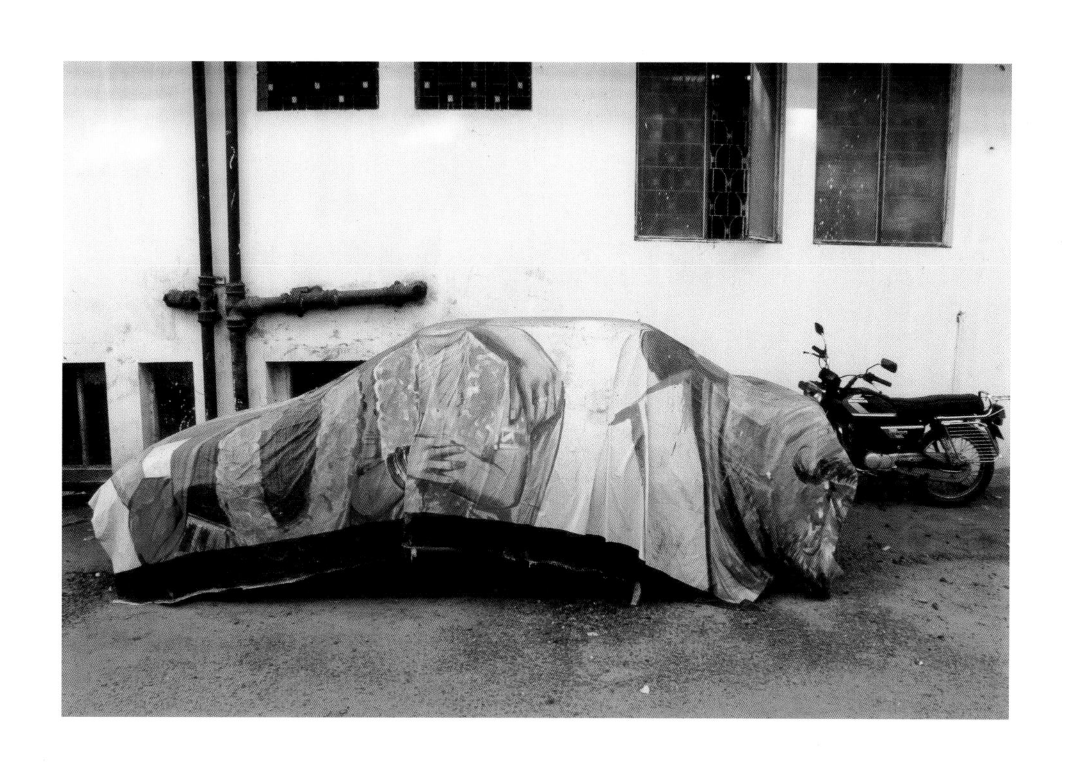

Acknowledgements

For the following people and organizations in India who went beyond expectations to make this photographic journey successful: Manisha Koirala, Jackie Shroff, Anupam Kher, Mahendra Verma, Indu Mirani, Leo Mirani, Ralphy Jhirad,N.D. Rangan, Dr N. Devanathan, the Government of India Tourist Office, Air India, and M. Rafiullah at Maharashtra Film, Stage & Cultural Development in Film City, Mumbai.

My deepest appreciation to all the actors, technicians, directors, producers, cinema owners, and movie-goers who allowed me to photograph them and who appear in this book.

A very special thanks to Audrey Jonckheer from Kodak for supporting this project from its very beginning stages, and for becoming a true friend. Thank you for believing in this project.

This project would not have been possible without continuous support from KODAK PROFESSIONAL.

I would like to extend my appreciation to Stephen Cohen at the Stephen Cohen Gallery, and Alan Klotz and Janet Sirmon at the Klotz/Sirman Gallery for their support and belief in this work.

I would like to thank the following people for their friendship, support and faith in my work: David Friend, Nick Hall, Tom Reynolds, Robert Peacock, Howard Greenberg, Celina Lunsford, Enrica Vigano, Alison Morley, Stephanie Heiman, James Wellford, Sarah Harbutt, Michelle Molloy, Simon Barnett, Ed Rich, Cheryl Newman, Elinor Carucci, Bill Hunt, Katie Webb, Susan Miklas, Jean-Francois Leroy, Meredith Kennedy, Preminda Jacob, Mira Nair, Bert Sun, Netta Navot and Geula Goldberg.

My gratitude goes to Richard Schlagman, Amanda Renshaw, Noel Daniel, Fran Johnson, Gudrun Hughes and Julia Joern at Phaidon Press for making this book possible, and to Nasreen Munni Kabir for her wonderful text and advice on India's cinema culture.

My thanks to Frederic Brenner for always stimulating and challenging me, and for encouraging me not to give up. My warmest thanks to Elaine Matczak, my long time friend, teacher and source of inspiration, for her valuable advice and encouragement and her active presence at every stage of this project. To my dearest friends Gadi Dotan, Jonathan Saacks, Assaf Lerman, Ziv Koren, Yoni Koenig and Tsadok Yecheskeli, thank you for your unconditional friendship and for being there for me.

I am most grateful to Karl, Liz and Jonathan Katz for their love, their generosity and for being family to me. Thank you Karl for all your help, advice and creative direction.

To my parents Virginia and Efraim and my sister Dana for their unconditional love, support and encouragement throughout my life.

To my wife and best friend, Tali, for always being supportive and patient, and for being my best creative advisor, editor and listener. Your sensitivity and creativity gave me comfort during the most difficult hours. My true friend and love. Thank you.

Phaidon Press Limited
Regent's Wharf
All Saints Street
London N1 9PA

Phaidon Press Inc.
180 Varick Street
New York, NY 10014

www.phaidon.com

First published 2003
© 2003 Phaidon Press Limited

ISBN 0 7148 4307 5

A CIP catalogue record for this book is available from the British Library.

All rights reserved. No part of this publication may be reproduced, stored in a retrieval system or transmitted, in any form or by any means, electronic, mechanical, photocopying, recording or otherwise, without the written permission of Phaidon Press Limited.

Designed by SEA Design
Printed in Italy

Franklin Pierce University
00179295